SCHAUM'S SOLVED PROBLEMS SERIES

2500 SOLVED PROBLEMS IN

FLUID MECHANICS AND HYDRAULICS

by

Jack B. Evett, Ph.D

Cheng Liu, Ph.D

University of North Carolina at Charlotte

McGRAW-HILL BOOK COMPANY

New York St. Louis San Francisco Auckland Bogotá Caracas
Colorado Springs Hamburg Lisbon London Madrid Mexico
Milan Montreal New Delhi Oklahoma City Panama Paris
San Juan São Paulo Singapore Sydney Tokyo Toronto

▌ Jack B. Evett, Ph.D., *Professor of Civil Engineering*, and Cheng Liu, M.S., *Associate Professor of Civil Engineering Technology*, both at the *University of North Carolina at Charlotte*.

Both authors have extensive teaching experience in the domain of fluid mechanics and hydraulics. They are coauthors of a textbook in fluid mechanics for the McGraw-Hill College Division.

Other Contributors to This Volume

▌ Robert L. Daugherty, Ph.D., *California Institute of Technology*

▌ E. John Finnemore, Ph.D., *University of Santa Clara*

▌ Joseph B. Franzini, Ph.D., *Stanford University*

▌ Ranald V. Giles, Ph.D., *Drexel Institute of Technology*

▌ Max A. Kohler, Ph.D., *U.S. National Weather Service*

▌ Ray K. Linsley, Ph.D., *Stanford University*

▌ Alan Mironer, Ph.D., *University of Lowell*

▌ Irving H. Shames, Ph.D., *State University of New York at Buffalo*

▌ Victor L. Streeter, Ph.D., *University of Michigan*

▌ Frank M. White, Ph.D., *University of Rhode Island*

▌ E. Benjamin Wylie, Ph.D., *University of Michigan*

Project supervision by The Total Book.

Library of Congress Cataloging-in-Publication Data

Evett, Jack B., 1942–
   2500 solved problems in fluid mechanics and hydraulics / by Jack B. Evett, Cheng Liu.
     p.   cm. — (Schaum's solved problems series)
   ISBN 0-07-019783-0
   1. Fluid mechanics—Problems, exercises, etc. 2. Hydraulics—Problems, exercises, etc. I. Liu, Cheng, 1937– . II. Title. III. Title: Twenty-five hundred solved problems in fluid mechanics and hydraulics. IV. Series.
TA357.3.E84 1988
620.1′06′076—dc 19

                           88-13373
                            CIP

1 2 3 4 5 6 7 8 9 0 SHP/SHP 8 9 3 2 1 0 9 8

ISBN 0-07-019783-0

CONTENTS

To the Student   v

List of Abbreviations   vii

List of Conversion Factors   xi

Chapter 1 PROPERTIES OF FLUIDS   1

Chapter 2 FLUID STATICS   25

Chapter 3 FORCES ON SUBMERGED PLANE AREAS   53

Chapter 4 DAMS   77

Chapter 5 FORCES ON SUBMERGED CURVED AREAS   85

Chapter 6 BUOYANCY AND FLOTATION   108

Chapter 7 KINEMATICS OF FLUID MOTION   132

Chapter 8 FUNDAMENTALS OF FLUID FLOW   157

Chapter 9 FLOW IN CLOSED CONDUITS   197

Chapter 10 SERIES PIPELINE SYSTEMS   269

Chapter 11 PARALLEL PIPELINE SYSTEMS   278

Chapter 12 BRANCHING PIPELINE SYSTEMS   302

Chapter 13 PIPE NETWORKS   315

Chapter 14 FLOW IN OPEN CHANNELS   356

Chapter 15 FLOOD ROUTING   459

Chapter 16 FLOW OF COMPRESSIBLE FLUIDS   469

Chapter 17 FLOW MEASUREMENT   520

Chapter 18 DIMENSIONAL ANALYSIS AND SIMILITUDE   574

Chapter 19 UNSTEADY FLOW   589

Chapter 20 PUMPS AND FANS   610

Chapter 21 TURBINES   638

Chapter 22 HYDRAULIC AND ENERGY GRADE LINES   657

Chapter 23 FORCES DEVELOPED BY FLUIDS IN MOTION **664**

Chapter 24 DYNAMIC DRAG AND LIFT **684**

Chapter 25 BASIC HYDRODYNAMICS **703**

Appendix **711**

To the Student

This book contains precisely 2500 completely solved problems in the areas of fluid mechanics and hydraulics. Virtually all types of problems ordinarily encountered in study and practice in these areas are covered. Not only you, but teachers, practitioners, and graduates reviewing for engineering licensing examinations should find these problems valuable.

To acquaint you with our "approach," particular steps taken in presenting the problems and their solutions are itemized below.

• First and most important of all, each problem and its solution are essentially independent and self-contained. That is to say, each contains all the data, equations, and computations necessary to find the answers. Thus, you should be able to pick a problem anywhere and follow its solution without having to review whatever precedes it. The exception to this is the occasional problem that specifically refers to, and carries over information from, a previous problem.

• In the solutions, our objective has been to present any needed equation first and then clearly to evaluate each term in the equation in order to find the answer. The terms may be evaluated separately or within the equation itself. For example, when solving an equation that has the parameter "area" as one of its terms, the area term (A) may be evaluated separately and its value substituted into the equation [as in Prob. 14.209], or it may be evaluated within the equation itself [as in Prob. 14.94].

• Virtually every number appearing in a solution is either "given" information (appearing as data in the statement of the problem or on an accompanying illustration), a previously computed value within the problem, a conversion factor (obtainable from the List of Conversion Factors), or a physical property (obtainable from a table or illustration in the Appendix). For example, in Prob. 1.77, the number 1.49, which does not appear elsewhere in the problem, is the dynamic viscosity (μ) of glycerin; it was obtained from Fig. A-3 in the Appendix.

• We have tried to include all but the most familiar items in the List of Abbreviations and Symbols. Hence, when an unknown sign is encountered in a problem or its solution, a scan of that list should prove helpful. Thus, the infrequently used symbol ψ is encountered in Prob. 25.6. According to the list, ψ represents the stream function, and you are quickly on your way to a solution.

Every problem solution in this book has been checked, but, with 2500 in all, it is inevitable that some mistakes will slip through. We would appreciate it if you would take the time to communicate any mistakes you find to us, so that they may be corrected in future printings. We wish to thank Bill Langley, of The University of North Carolina at Charlotte, who assisted us with some of the problem selection and preparation.

Acknowledgments

Some illustrations in this volume come from the following McGraw-Hill books:

Fundamentals of Fluid Mechanics, Jack B. Evett and Cheng Liu, 1987.
Fluid Mechanics, 2d ed., Frank M. White, 1986.
Fluid Mechanics with Engineering Applications, 8th ed., Robert L. Daugherty, Joseph B. Franzini, and E. John Finnemore, 1985.
Fluid Mechanics, 8th ed., Victor L. Streeter and E. Benjamin Wylie, 1985.
Mechanics of Fluids, 2d ed., Irving H. Shames, 1982.

Schaum's Outline Series: *Theory and Problems of Fluid Mechanics and Hydraulics,* 2d ed., Ranald V. Giles, 1962.

Water Resources Engineering, 3d ed., Ray K. Linsley and Joseph B. Franzini, 1979.

Engineering Fluid Mechanics, Alan Mironer, 1979.

Hydrology for Engineers, 3d ed., Ray K. Linsley, Max A. Kohler, and Joseph L. H. Paulhus, 1982.

Abbreviations and Symbols

a	acceleration or area
A	area
abs	absolute
α (alpha)	angle between absolute velocity of fluid in hydraulic machine and linear velocity of a point on a rotating body or coefficient of thermal expansion or dimensionless ratio of similitude
atm	atmosphere
atmos	atmospheric
β (beta)	angle between relative velocity in hydraulic machines and linear velocity of a point on a rotating body or coefficient of compressibility or ratio of obstruction diameter to duct diameter
b	surface width or other width
B	surface width or other width
bhp	brake horsepower
bp	brake power
Btu	British thermal unit
c	speed of sound or wave speed (celerity)
C	Celsius or discharge coefficient or speed of propagation
cal	calorie
c.b. or CB	center of buoyancy
C_c	coefficient of contraction
C_d	coefficient of discharge
C_D	drag coefficient
C_f	friction-drag coefficient
C_F	force coefficient
cfs	cubic foot per second
c.g. or CG	center of gravity
C_I	Pitot tube coefficient
C_L	lift coefficient
cm	centimeter $(10^{-2}\,\text{m})$
cP	centipoise
c.p.	center of pressure
c_P	specific heat at constant pressure
c_v	specific heat at constant volume
C_v	coefficient of velocity
C_w	weir coefficient
d	depth or diameter
D	depth or diameter or drag force
δ (delta)	thickness of boundary layer
δ_1 (delta)	thickness of the viscous sublayer
Δ (Delta)	change in (or difference between)
d_c	critical depth
D_{eff}	effective diameter
D_h	hydraulic diameter
d_m	mean depth
d_n	normal depth
d_N	normal depth
E	modulus of elasticity or specific energy or velocity approach factor
e_h	hydraulic efficiency
el	elevation
η (eta)	pump or turbine efficiency
ϵ (epsilon)	height or surface roughness
E_p	pump energy
E_t	turbine energy
exp	exponential
f	frequency of oscillation (cycles per second) or friction factor

F	Fahrenheit or force
F_b	buoyant force
F_D	drag force
F_H	horizontal force
F_L	lift force
fps	foot per second
F.S.	factor of safety
ft	foot
F_U	uplift force on a dam
F_V	vertical force
g	acceleration due to gravity or gage height or gram
G	weight flow rate
gal	gallon
γ (gamma)	specific (or unit) weight
Γ (Gamma)	circulation
GN	giganewton (10^9 N)
GPa	gigapascal (10^9 Pa)
gpm	gallons per minute
h	enthalpy per unit mass or height or depth or pressure head or hour
$\bar{h}$	average height or depth or head
$\hat{h}$	enthalpy per unit weight
H	energy head or total energy head
h_1	unit head loss
h_{cg}	vertical depth to center of gravity
h_{cp}	vertical depth to center of pressure
h_f	head loss due to friction
Hg	mercury
HGL	hydraulic grade line
h_L	total head loss
h_m	head loss due to minor losses
hp	horsepower
Hz	hertz (cycles per second)
I	inflow or moment of inertia
ID	inside diameter
in	inch
∞ (infinity)	sometimes used as a subscript to indicate upstream
J	joule
K	bulk modulus of elasticity or Kelvin or minor loss coefficient
k	specific heat ratio
kcal	kilocalorie (10^3 cal)
kg	kilogram (10^3 g)
kJ	kilojoule (10^3 J)
km	kilometer (10^3 m)
kN	kilonewton (10^3 N)
kPa	kilopascal (10^3 Pa)
kW	kilowatt (10^3 W)
L	length or lift force or liter
λ (lambda)	model ratio or wave length
lb	pound
lb_m	pound mass
L_e	equivalent length
L_m	linear dimension in model
L_p	linear dimension in prototype
m	mass or meter
$\dot{m}$	mass flow rate
$\underline{M}$	mass flow rate or molecular weight or moment or torque
$\overline{MB}$	distance from center of buoyancy to metacenter
mbar	millibar (10^{-3} bar)
mc	metacenter
mgd	million gallons per day

ml	milliliter (10^{-3} L)
min	minute
mm	millimeter (10^{-3} meter)
MN	meganewton (10^6 N)
MPa	megapascal (10^6 Pa)
mph	mile per hour
MR	manometer reading
μ (mu)	absolute or dynamic viscosity
MW	megawatt (10^6 W)
n	Manning roughness coefficient or number of moles
N	newton or rotational speed
N_B	Brinkman number
N_F	Froude number
N_M	Mach number
NPSH	net positive suction head
N_R	Reynolds number
N_s	specific speed of pump or turbine
ν (nu)	kinematic viscosity
N_W	Weber number
O	outflow
OD	outside diameter
Ω (ohm)	rotational rate
ω (omega)	angular velocity
p	pressure or poise
P	force (usually resulting from an applied pressure) or power
Pa	pascal
ϕ (phi)	peripheral-velocity factor
π (pi)	constant $= 3.14159265$
Π (pi)	dimensionless parameter
P_r	power ratio
p_s	stagnation pressure
psi	pound per square inch
ψ (psi)	stream function
psia	pound per square inch absolute
psig	pound per square inch gage
$p^{*'}$	pressure for condition at $N_M = 1/\sqrt{k}$
p_v	vapor pressure
p_w	wetted perimeter
q	flow rate per unit width or heat per unit mass
Q	discharge or heat or volume flow rate
Q_H	heat transferred per unit weight of fluid
Q/w	volume flow rate per unit width of channel
qt	quart
r	radius
R	gas constant or Rankine or resultant force or hydraulic radius
R'	manometer reading
rad	radian
R_c	critical hydraulic radius
R_h	hydraulic radius
ρ (rho)	mass density
r_i	inside radius
r_o	outside radius
rpm	revolutions per minute
R_u	universal gas constant
s	entropy of a substance or second or slope
S	slope or storage
s_c	critical slope
s.g.	specific gravity
$s.g._M$	specific gravity of manometer fluid
$s.g._F$	specific gravity of flowing fluid

σ (sigma)	pump cavitation parameter or stress or surface tension
σ'	cavitation index
Σ (sigma)	summation
S	specific gravity of flowing fluid
S_0	specific gravity of manometer fluid
t	thickness or time
T	surface width or temperature or torque or tension
τ (tau)	shear stress
τ_0 (tau)	shear stress at the wall
T_s	stagnation temperature
u	velocity
$u_{\mathbb{C}}$	centerline velocity
U	velocity
v	velocity
v_c	critical velocity
V	velocity or volume
v_{av}	average velocity
$V_{\mathbb{C}}$	centerline velocity
V_d	volume of fluid displaced
V_m	velocity in model
V_p	velocity in prototype
V_s	specific volume
v_*	shear velocity
v_t	tangential velocity
v_T	terminal velocity
w	width
W	watt or weight or weight flow rate or work
x_{cp}	distance from center of gravity to center of pressure in x direction
ξ (xi)	vorticity
y	depth
y_c	critical depth
y_{cp}	distance from center of gravity to center of pressure in y direction
y_n	normal depth
y_N	normal depth
z_{cg}	inclined distance from liquid surface to center of gravity
z_{cp}	inclined distance from liquid surface to center of pressure

Conversion Factors

$0.00001667 \text{ m}^3/\text{s} = 1 \text{ L/min}$

$0.002228 \text{ ft}^3/\text{s} = 1 \text{ gal/min}$

$0.0145 \text{ lb/in}^2 = 1 \text{ mbar}$

$0.100 \text{ kN/m}^2 = 1 \text{ mbar}$

$0.3048 \text{ m} = 1 \text{ ft}$

$2.54 \text{ cm} = 1 \text{ in}$

$3.281 \text{ ft} = 1 \text{ m}$

$4 \text{ qt} = 1 \text{ gal}$

$4.187 \text{ kJ} = 1 \text{ kcal}$

$4.448 \text{ N} = 1 \text{ lb}$

$6.894 \text{ kN/m}^2 = 1 \text{ lb/in}^2$

$7.48 \text{ gal} = 1 \text{ ft}^3$

$12 \text{ in} = 1 \text{ ft}$

$14.59 \text{ kg} = 1 \text{ slug}$

$25.4 \text{ mm} = 1 \text{ in}$

$60 \text{ min} = 1 \text{ h}$

$60 \text{ s} = 1 \text{ min}$

$100 \text{ cm} = 1 \text{ m}$

$100 \text{ N/m}^2 = 1 \text{ bar}$

$101.3 \text{ kPa} = 1 \text{ atm}$

$144 \text{ in}^2 = 1 \text{ ft}^2$

$550 \text{ ft-lb/s} = 1 \text{ hp}$

$778 \text{ ft-lb} = 1 \text{ Btu}$

$1000 \text{ N} = 1 \text{ kN}$

$1000 \text{ L} = 1 \text{ m}^3$

$1000 \text{ mm} = 1 \text{ m}$

$1000 \text{ Pa} = 1 \text{ kPa}$

$1728 \text{ in}^3 = 1 \text{ ft}^3$

$2000 \text{ lb} = 1 \text{ ton}$

$3600 \text{ s} = 1 \text{ h}$

$4187 \text{ J} = 1 \text{ kcal}$

$5280 \text{ ft} = 1 \text{ mile}$

$86\,400 \text{ s} = 1 \text{ day}$

$1\,000\,000 \text{ N} = 1 \text{ MN}$

$1\,000\,000 \text{ Pa} = 1 \text{ MPa}$

$1\,000\,000\,000 \text{ N} = 1 \text{ GN}$

$1\,000\,000\,000 \text{ Pa} = 1 \text{ GPa}$

CHAPTER 1

Properties of Fluids

Note: For many problems in this chapter, values of various physical properties of fluids are obtained from Tables A-1 through A-8 in the Appendix.

1.1 A reservoir of glycerin (glyc) has a mass of 1200 kg and a volume of 0.952 m³. Find the glycerin's weight (W), mass density (ρ), specific weight (γ), and specific gravity (s.g.).

$$F = W = ma = (1200)(9.81) = 11\,770\,\text{N} \quad \text{or} \quad 11.77\,\text{kN}$$
$$\rho = m/V = 1200/0.952 = 1261\,\text{kg/m}^3$$
$$\gamma = W/V = 11.77/0.952 = 12.36\,\text{kN/m}^3$$
$$\text{s.g.} = \gamma_{\text{glyc}}/\gamma_{\text{H}_2\text{O at 4 °C}} = 12.36/9.81 = 1.26$$

1.2 A body requires a force of 100 N to accelerate it at a rate of 0.20 m/s². Determine the mass of the body in kilograms and in slugs.

$$F = ma$$
$$100 = (m)(0.20)$$
$$m = 500\,\text{kg} = 500/14.59 = 34.3\,\text{slugs}$$

1.3 A reservoir of carbon tetrachloride (CCl_4) has a mass of 500 kg and a volume of 0.315 m³. Find the carbon tetrachloride's weight, mass density, specific weight, and specific gravity.

$$F = W = ma = (500)(9.81) = 4905\,\text{N} \quad \text{or} \quad 4.905\,\text{kN}$$
$$\rho = m/V = 500/0.315 = 1587\,\text{kg/m}^3$$
$$\gamma = W/V = 4.905/0.315 = 15.57\,\text{kN/m}^3$$
$$\text{s.g.} = \gamma_{CCl_4}/\gamma_{\text{H}_2\text{O at 4 °C}} = 15.57/9.81 = 1.59$$

1.4 The weight of a body is 100 lb. Determine (a) its weight in newtons, (b) its mass in kilograms, and (c) the rate of acceleration [in both feet per second per second (ft/s²) and meters per second per second (m/s²)] if a net force of 50 lb is applied to the body.

(a)
$$W = (100)(4.448) = 444.8\,\text{N}$$

(b)
$$F = W = ma \qquad 444.8 = (m)(9.81) \qquad m = 45.34\,\text{kg}$$

(c)
$$m = 45.34/14.59 = 3.108\,\text{slugs}$$
$$F = ma \qquad 50 = 3.108a \qquad a = 16.09\,\text{ft/s}^2 = (16.09)(0.3048) = 4.904\,\text{m/s}^2$$

1.5 The specific gravity of ethyl alcohol is 0.79. Calculate its specific weight (in both pounds per cubic foot and kilonewtons per cubic meter) and mass density (in both slugs per cubic foot and kilograms per cubic meter).

$$\gamma = (0.79)(62.4) = 49.3\,\text{lb/ft}^3 \qquad \gamma = (0.79)(9.79) = 7.73\,\text{kN/m}^3$$
$$\rho = (0.79)(1.94) = 1.53\,\text{slugs/ft}^3 \qquad \rho = (0.79)(1000) = 790\,\text{kg/m}^3$$

1.6 A quart of water weights about 2.08 lb. Compute its mass in slugs and in kilograms.

$$F = W = ma \qquad 2.08 = (m)(32.2)$$
$$m = 0.0646\,\text{slug} \qquad m = (0.0646)(14.59) = 0.943\,\text{kg}$$

1.7 One cubic foot of glycerin has a mass of 2.44 slugs. Find its specific weight in both pounds per cubic foot and kilonewtons per cubic meter.

$F = W = ma = (2.44)(32.2) = 78.6\,\text{lb}$. Since the glycerin's volume is 1 ft³, $\gamma = 78.6\,\text{lb/ft}^3 = (78.6)(4.448)/(0.3048)^3 = 12\,350\,\text{N/m}^3$, or 12.35 kN/m³.

1.8 A quart of SAE 30 oil at 68 °F weighs about 1.85 lb. Calculate the oil's specific weight, mass density, and specific gravity.

$$V = 1/[(4)(7.48)] = 0.03342 \text{ ft}^3$$

$$\gamma = W/V = 1.85/0.03342 = 55.4 \text{ lb/ft}^3$$

$$\rho = \gamma/g = 55.4/32.2 = 1.72 \text{ slugs/ft}^3$$

$$\text{s.g.} = \gamma_{\text{oil}}/\gamma_{\text{H}_2\text{O at 4 °C}} = 55.4/62.4 = 0.888$$

1.9 The volume of a rock is found to be 0.00015 m³. If the rock's specific gravity is 2.60, what is its weight?

$$\gamma_{\text{rock}} = (2.60)(9.79) = 25.5 \text{ kN/m}^3 \qquad W_{\text{rock}} = (25.5)(0.00015) = 0.00382 \text{ kN} \quad \text{or} \quad 3.82 \text{ N}$$

1.10 A certain gasoline weighs 46.0 lb/ft³. What are its mass density, specific volume, and specific gravity?

$$\rho = \gamma/g = 46.0/32.2 = 1.43 \text{ slugs/ft}^3 \qquad V_s = 1/\rho = 1/1.43 = 0.699 \text{ ft}^3/\text{slug}$$

$$\text{s.g.} = 1.43/1.94 = 0.737$$

1.11 If the specific weight of a liquid is 8000 N/m³, what is its mass density?

$$\rho = \gamma/g = 8000/9.81 = 815 \text{ kg/m}^3$$

1.12 An object at a certain location has a mass of 2.0 kg and weighs 19.0 N on a spring balance. What is the acceleration due to gravity at this location?

$$F = W = ma \qquad 19.0 = 2.0a \qquad a = 9.50 \text{ m/s}^2$$

1.13 If an object has a mass of 2.0 slugs at sea level, what would its mass be at a location where the acceleration due to gravity is 30.00 ft/s²?

Since the mass of an object does not change, its mass will be 2.0 slugs at that location.

1.14 What would be the weight of a 3-kg mass on a planet where the acceleration due to gravity is 10.00 m/s²?

$$F = W = ma = (3)(10.00) = 30.00 \text{ N}$$

1.15 Determine the weight of a mass of 3 slugs at a place where the acceleration due to gravity is 31.7 ft/s².

$$F = W = ma = (3)(31.7) = 95.1 \text{ lb}$$

1.16 If 200 ft³ of oil weighs 10 520 lb, calculate its specific weight, density, and specific gravity.

$$\gamma = W/V = 10\,520/200 = 52.6 \text{ lb/ft}^3 \qquad \rho = \gamma/g = 52.6/32.2 = 1.63 \text{ slugs/ft}^3$$

$$\text{s.g.} = \gamma_{\text{oil}}/\gamma_{\text{H}_2\text{O at 4 °C}} = 52.6/62.4 = 0.843$$

1.17 How high will the free surface be if 1 ft³ of water is poured into a container that is a right circular cone 18 in high with a base radius of 10 in? How much additional water is required to fill the container?

$$V_{\text{cone}} = \pi r^2 h/3 = \pi(10)^2(18)/3 = 1885 \text{ in}^3 \qquad V_{\text{H}_2\text{O}} = 1 \text{ ft}^3 = 1728 \text{ in}^3$$

Additional water to fill container = 1885 − 1728 = 157 in³. From Fig. 1-1, $r_o/10 = h_o/18$, or $r_o = h_o/1.8$; $V_{\text{empty (top) cone}} = \pi(h_o/1.8)^2 h_o/3 = 157$; $h_o = 7.86$ in. Free surface will be 18 − 7.86, or 10.14 in above bottom of container.

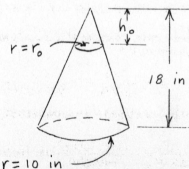

Fig. 1-1

1.18   If the conical container in Prob. 1.17 can be filled with 27.0 kg of a certain oil, what is the density of the oil?

$$V_{cone} = 1885 \text{ in}^3 \quad \text{(from Prob. 1.17)}$$
$$= \tfrac{1885}{1728}(0.3048)^3 = 0.03089 \text{ m}^3$$
$$\rho = m/V = 27.0/0.03089 = 874 \text{ kg/m}^3$$

1.19   A certain gas weights 0.10 lb/ft^3 at a certain temperature and pressure. What are the values of its density, specific volume, and specific gravity relative to air weighing 0.075 lb/ft^3?

$$\rho = \gamma/g = 0.10/32.2 = 0.00311 \text{ slug/ft}^3 \qquad V_s = 1/\rho = 1/0.00311 = 322 \text{ ft}^3/\text{slug}$$
$$\text{s.g.} = 0.10/0.075 = 1.333$$

1.20   If the specific volume of a gas is $350 \text{ ft}^3/\text{slug}$, what is its specific weight?

$$\rho = 1/V_s = \tfrac{1}{350} = 0.002857 \text{ slug/ft}^3 \qquad \gamma = \rho g = (0.002857)(32.2) = 0.0920 \text{ lb/ft}^3$$

1.21   Initially, when 1000.00 ml of water at $10\,°C$ is poured into a glass cylinder, the depth of the water column is 100.00 cm. The water and its container are heated to $80\,°C$. Assuming no evaporation, what will be the depth of the water column if the cofficient of thermal expansion (α) for the glass is 3.6×10^{-6} per $°C$?

$$\text{Mass of water} = \rho V = \rho_{10}V_{10} = \rho_{80}V_{80} \qquad (1000)(1000.00/1000) = 971V_{80} \qquad V_{80} = 1.030 \text{ m}^3 \quad \text{or} \quad 1030 \text{ cm}^3$$
$$A_{10} = V_{10}/h_{10} = 1000.00/100.00 = 10.000 \text{ cm}^2$$
$$A_{10} = \pi r_{10}^2 \qquad 10.000 = \pi r_{10}^2 \qquad r_{10} = 1.7841 \text{ cm}$$
$$r_{80} = r_{10}[1 + (\Delta T)(\alpha)] = (1.7841)[1 + (80 - 10)(3.6 \times 10^{-6})] = 1.7845 \text{ cm}$$
$$A_{80} = \pi r_{80}^2 = \pi(1.7845)^2 = 10.004 \text{ cm}^2 \qquad h_{80} = V_{80}/A_{80} = 1030/10.004 = 102.96 \text{ cm}$$

1.22   A vessel contains 3.000 ft^3 of water at $50\,°F$ and atmospheric pressure. If it is heated to $160\,°F$, what will be the percentage change in its volume? What weight of water must be removed to maintain the volume at the original value?

$$\text{Weight of water} = \gamma V = \gamma_{50}V_{50} = \gamma_{160}V_{160} \qquad (62.4)(3.000) = 61.0V_{160} \qquad V_{160} = 3.0689 \text{ ft}^3$$

Change in volume $= (3.0689 - 3.000)/3.000 = 0.023$, or 2.3% (increase). Must remove $(3.0689 - 3.000)(61.0)$, or 4.20 lb.

1.23   A vertical, cylindrical tank with a diameter of 10.00 m and a depth of 5.00 m contains water at $20\,°C$ and is filled to the brim. If the water is heated to $50\,°C$, how must water will spill over the edge of the tank?

$$V_{tank} = (V_{H_2O})_{20} = \pi(10.00/2)^2(5.00) = 392.7 \text{ m}^3$$
$$W_{H_2O} = (9.79)(392.7) = 3845 \text{ kN} \qquad (V_{H_2O})_{50} = 3845/9.69 = 396.8 \text{ m}^3$$
$$\text{Amount of water spilled} = 396.8 - 392.7 = 4.1 \text{ m}^3$$

1.24   A closed heavy steel chamber is filled with water at $50\,°F$ and atmospheric pressure. If the temperature of water and chamber is raised to $90\,°F$, what will be the new pressure of the water? Assume the chamber is unaffected by the water pressure. The coefficient of thermal expansion of steel (α) is 6.5×10^{-6} per $°F$.

The volume of water would attempt to expand proportional to the cube of the linear thermal expansion. Hence, $V_{90} = V_{50}[1 + (90 - 50)(6.5 \times 10^{-6})]^3 = 1.000780V_{50}$; weight of water $= \gamma V = \gamma_{50}V_{50} = \gamma_{90}V_{90}$, $62.4V_{50} = \gamma_{90}(1.000780V_{50})$, $\gamma_{90} = 62.35 \text{ lb/ft}^3$. From Fig. A-3, $p_{90} = 1300 \text{ psia}$ (approximately).

1.25   A liquid compressed in a cylinder has a volume of 1000 cm^3 at 1 MN/m^2 and a volume of 995 cm^3 at 2 MN/m^2. What is its bulk modulus of elasticity (K)?

$$K = -\frac{\Delta p}{\Delta V/V} = -\frac{2 - 1}{(995 - 1000)/1000} = 200 \text{ MPa}$$

1.26   Find the bulk modulus of elasticity of a liquid if a pressure of 150 psi applied to 10 ft^3 of the liquid causes a volume reduction of 0.02 ft^3.

$$K = -\frac{\Delta p}{\Delta V/V} = -\frac{(150 - 0)(144)}{-0.02/10} = 10\,800\,000 \text{ lb/ft}^2 \quad \text{or} \quad 75\,000 \text{ psi}$$

1.27 For $K = 2.2$ GPa for the bulk modulus of elasticity for water, what pressure is required to reduce its volume by 0.5 percent?

∎ $\qquad K = -\dfrac{\Delta p}{\Delta V/V} \qquad 2.2 = -\dfrac{p_2 - 0}{-0.005} \qquad p_2 = 0.0110$ GPa or 11.0 MPa

1.28 Find the change in volume of 1.00000 ft³ of water at 80 °F when subjected to a pressure increase of 300 psi. Water's bulk modulus of elasticity at this temperature is 325 000 psi.

∎ $\qquad K = -\dfrac{\Delta p}{\Delta V/V} \qquad 325\,000 = -\dfrac{300 - 0}{\Delta V/1.00000} \qquad \Delta V = -0.00092$ ft³

1.29 From the following test data, determine the bulk modulus of elasticity of water: at 500 psi the volume was 1.000 ft³, and at 3500 psi the volume was 0.990 ft³.

∎ $\qquad K = -\dfrac{\Delta p}{\Delta V/V} = -\dfrac{500 - 3500}{(1.000 - 0.990)/1.000} = 300\,000$ psi

1.30 A high-pressure steel container is partially full of a liquid at a pressure of 10 atm. The volume of the liquid is 1.23200 L. At a pressure of 25 atm, the volume of the liquid equals 1.23100 L. What is the average bulk modulus of elasticity of the liquid over the given range of pressure if the temperature after compression is allowed to return to the original temperature? What is the coefficient of compressibility (β)?

∎ $\qquad K = -\dfrac{\Delta p}{\Delta V/V} = -\dfrac{(25 - 10)(101.3)}{(1.23100 - 1.23200)/1.23200} = 1.872 \times 10^6$ kN/m² or 1872 MN/m²

$$\beta = 1/K = \tfrac{1}{1872} = 0.000534 \text{ m}^2/\text{MN}$$

1.31 A heavy tank contains oil (A) and water (B) over which air pressure is varied. The dimensions shown in Fig. 1-2 correspond to the atmospheric pressure of the air. If air is slowly added from a pump to bring pressure p up to 1 MPa gage, what will be the total downward movement of the free surface of oil and air? Take average values of bulk moduli of elasticity of the liquids to be, for the pressure range, 2050 MN/m² for oil and 2075 MN/m² for water. Assume the container does not change volume. Neglect hydrostatic pressures.

∎ $\qquad K = -\dfrac{\Delta p}{\Delta V/V} \qquad 2050 = -\dfrac{1 - 0}{\Delta V_{\text{oil}}/[\frac{500}{1000}\pi(\frac{300}{1000})^2/4]} \qquad \Delta V_{\text{oil}} = -0.00001724 \text{ m}^3$

$$2075 = -\dfrac{1 - 0}{\Delta V_{\text{H}_2\text{O}}/[\frac{800}{1000}\pi(\frac{300}{1000})^2/4]} \qquad \Delta V_{\text{H}_2\text{O}} = -0.00002725 \text{ m}^3$$

$$\Delta V_{\text{total}} = -0.00001724 + (-0.00002725) = -0.00004449 \text{ m}^3$$

Let x = the distance the upper free surface moves. $-0.00004449 = -[\pi(\frac{300}{1000})^2/4]x$, $x = 0.000629$ m or 0.629 mm.

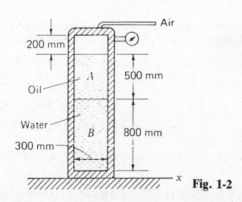

Fig. 1-2

1.32 Water at a pressure of 4442 psig is forced into a thin-walled spherical tank. If the water is then released from the tank, how much water will be collected at atmospheric pressure? The deformed inside volume in the tank is

800.4069 in^3 when the pressure is 4442 psig. Use a value of 305 000 psi for an average value of the bulk modulus of elasticity.

$$K = -\frac{\Delta p}{\Delta V/V} \qquad 305\,000 = -\frac{0-4442}{(V_2 - 800.4069)/800.4069} \qquad V_2 = 812.06 \text{ in}^3$$

$$W = (62.4)(812.06/1728) = 29.3 \text{ lb}$$

1.33 Water in a hydraulic press is subjected to a pressure of 15 000 psia at 68 °F. If the initial pressure is 15 psia, what will be the percentage decrease in specific volume? Use an average bulk modulus of elasticity of 365 000 psi for this pressure range.

$$K = -\frac{\Delta p}{\Delta V/V} \qquad 365\,000 = -\frac{15\,000 - 15}{\Delta V/V_1} \qquad \frac{\Delta V}{V_1} = -0.0411 \quad \text{or} \quad 4.11\% \text{ decrease}$$

1.34 At a depth of 8 km in the ocean, the pressure is 81.8 MPa. Assume specific weight at the surface is 10 050 N/m^3 and the average bulk modulus of elasticity is 2.34×10^9 N/m^2 for that pressure range. (*a*) What will be the change in specific volume between that at the surface and at that depth? (*b*) What will be the specific volume at that depth? (*c*) What will be the specific weight at that depth?

(a)
$$(V_s)_1 = 1/\rho_1 = g/\gamma_1 = 9.81/10\,050 = 0.0009761 \text{ m}^3/\text{kg}$$

$$K = -\frac{\Delta p}{\Delta V_s/V_s} \qquad 2.34 \times 10^9 = -\frac{81.8 \times 10^6 - 0}{\Delta V_s/0.0009761} \qquad \Delta V_s = -0.0000341 \text{ m}^3/\text{kg}$$

(b) $\qquad (V_s)_2 = (V_s)_1 + \Delta V_s = 0.0009761 - 0.0000341 = 0.000942 \text{ m}^3/\text{kg}$
(c) $\qquad \gamma_2 = g/V_2 = 9.81/0.000942 = 10\,414 \text{ N/m}^3$

1.35 Approximately what pressure must be applied to water at 60 °F to reduce its volume 2 percent?

$$K = -\frac{\Delta p}{\Delta V/V} \qquad 311\,000 = -\frac{p_2 - 0}{0.02} \qquad p_2 = 6220 \text{ psi}$$

1.36 A gas at 20 °C and 0.2×10^6 Pa abs has a volume of 40 L and a gas constant (R) of 210 m · N/(kg · K). Determine the density and mass of the gas.

$$\rho = p/RT = 0.2 \times 10^6/[(210)(20 + 273)] = 3.25 \text{ kg/m}^3 \qquad m = \rho V = (3.25)(\tfrac{40}{1000}) = 0.130 \text{ kg}$$

1.37 What is the specific weight of air at 60 psia and 90 °F?

$\gamma = p/RT$. From Table A-6, $R = 53.3$ ft/°R; $\gamma = (60)(144)/[(53.3)(90 + 460)] = 0.295 \text{ lb/ft}^3$.

Note: p/RT gives ρ (Prob. 1.36) or γ (Prob. 1.37), depending on the value of R used. Corresponding values of R in Table A-6 differ by a factor of g.

1.38 What is the density of water vapor at 400 000 Pa abs and 15 °C? Its gas constant (R) is 462 m · N/(kg · K).

$$\rho = p/RT = 400\,000/[(462)(15 + 273)] = 3.01 \text{ kg/m}^3$$

1.39 A gas with molecular weight 28 has a volume of 4.0 ft^3 and a pressure and temperature of 2000 lb/ft^2 abs and 600 °R, respectively. What are its specific volume and specific weight?

$$R = R_u/M = 49\,709/28 = 1775 \text{ ft} \cdot \text{lb}/(\text{slug} \cdot \text{°R})$$

[where R_u, the universal gas constant, $= 49\,709$ ft · lb/(slug · °R)]

$$\rho = 1/V_s = p/RT = 2000/[(1775)(600)] \qquad V_s = 532.5 \text{ ft}^3/\text{slug}$$

$$\gamma = \rho g = (1/V_s)(g) = (1/532.5)(32.2) = 0.0605 \text{ lb/ft}^3$$

1.40 One kilogram of hydrogen is confined in a volume of 150 L at −40 °C. What is the pressure if R is 4115 m · N/(kg · K)?

$$p = \rho RT = (m/V)RT = (1/\tfrac{150}{1000})(4115)(-40 + 273) = 6.392 \times 10^6 \text{ N/m}^2 \quad \text{or} \quad 6392 \text{ kPa abs}$$

1.41 What is the specific weight of air at a temperature of 30 °C and a pressure of 470 kPa abs?

$$\gamma = p/RT = 470/[(29.3)(30 + 273)] = 0.0529 \text{ kN/m}^3$$

1.42 Find the mass density of helium at a temperature of 39 °F and a pressure of 26.9 psig, if atmospheric pressure is 14.9 psia.

▮
$$\rho = p/RT = (14.9 + 26.9)(144)/[(12\,420)(39 + 460)]$$
$$= 0.000971\ \text{lb} \cdot \text{s}^2/\text{ft}^4 \quad \text{or} \quad 0.000971\ \text{slug/ft}^3$$

1.43 The temperature and pressure of nitrogen in a tank are 28 °C and 600 kPa abs, respectively. Determine the specific weight of the nitrogen.

▮
$$\gamma = p/RT = 600/[(30.3)(28 + 273)] = 0.0658\ \text{kN/m}^3$$

1.44 The temperature and pressure of oxygen in a container are 60 °F and 20.0 psig, respectively. Determine the oxygen's mass density if atmospheric pressure is 14.7 psia.

▮
$$\rho = p/RT = (20.0 + 14.7)(144)/[(1552)(60 + 460)] = 0.00619\ \text{slug/ft}^3$$

1.45 Calculate the specific weight and density of methane at 100 °F and 120 psia.

▮
$$\gamma = p/RT = (120)(144)/[(96.2)(100 + 460)] = 0.321\ \text{lb/ft}^3$$
$$\rho = \gamma/g = 0.321/32.2 = 0.00997\ \text{slug/ft}^3$$

1.46 At 90 °F and 30.0 psia, the specific weight of a certain gas was 0.0877 lb/ft³. Determine the gas constant and density of this gas.

▮
$$\gamma = p/RT \qquad 0.0877 = (30.0)(144)/[(R)(90 + 460)] \qquad R = 89.6\ \text{ft/°R}$$
$$\rho = \gamma/g = 0.0877/32.2 = 0.00272\ \text{slug/ft}^3$$

1.47 A cylinder contains 12.5 ft³ of air at 120 °F and 40 psia. The air is then compressed to 2.50 ft³. (**a**) Assuming isothermal conditions, what are the pressure at the new volume and the bulk modulus of elasticity? (**b**) Assuming adiabatic conditions, what are the final pressure and temperature and the bulk modulus of elasticity?

▮ (**a**)
$$p_1 V_1 = p_2 V_2 \qquad \text{(for isothermal conditions)}$$
$$(40)(12.5) = (p_2')(2.50)$$
$$p_2' = 200\ \text{psia}$$
$$K = -\frac{\Delta p}{\Delta V/V} = -\frac{40 - 200}{(12.5 - 2.5)/12.5} = 200\ \text{psi}$$

(**b**) $p_1 V_1^k = p_2 V_2^k$ (for adiabatic conditions). From Table A-6, $k = 1.40$. $(40)(12.5)^{1.40} = (p_2')(2.50)^{1.40}$, $p_2' = 381$ psia; $T_2/T_1 = (p_2/p_1)^{(k-1)/k}$, $T_2/(120 + 460) = (\frac{381}{40})^{(1.40-1)/1.40}$, $T_2 = 1104$ °R, or 644 °F; $K = kp' = (1.40)(381) = 533$ psi.

1.48 Air is kept at a pressure of 200 kPa and a temperature of 30 °C in a 500-L container. What is the mass of the air?

▮
$$\rho = p/RT = [(200)(1000)]/[(287)(30 + 273)] = 2.300\ \text{kg/m}^3 \qquad m = (2.300)(\tfrac{500}{1000}) = 1.15\ \text{kg}$$

1.49 A perfect gas undergoes a process whereby its pressure is doubled and its specific volume is decreased by two-thirds. If the initial temperature is 100 °F, what is the final temperature?

▮
$$\rho = 1/V_s = p/RT \qquad pV_s = RT \qquad p_1(V_s)_1 = RT_1 \qquad p_2(V_s)_2 = RT_2$$
$$(p_2/p_1)[(V_s)_2/(V_s)_1] = (R/R)(T_2/T_1) \qquad (2)(\tfrac{1}{3}) = T_2/(100 + 460) \qquad T_2 = 373\ \text{°R} \quad \text{or} \quad -87\ \text{°F}$$

1.50 In order to reduce gasoline consumption in city driving, the Department of Energy of the federal government is studying the so-called "inertial transmission" system. In this system, when drivers want to slow up, the wheels are made to drive pumps which pump oil into the compressor tank so as to increase the pressure of the trapped air in the tank. The pumps thus act as brakes. As long as the pressure in the tank stays above a certain minimum value, the tank can supply energy to the pumps, which then act as motors to drive the wheels when a driver wishes to accelerate. If insufficient braking takes place to keep the air pressure up, a conventional gas engine cuts in to build up the pressure in the tank. It is expected that a doubling of mileage per gallon can take place in city driving by this system.

Suppose the volume of air initially in the tank is 80 L and the temperature is 30 °C with a pressure of 200 kPa. As a result of braking on going down a long hill, the volume decreases to 40 L and the air reaches a pressure of 500 kPa. What is the final temperature of the air if there is a loss of air due to a leak of 0.003 kg?

$$\rho_1 = p_1/RT_1 = (200)(1000)/[(287)(30 + 273)] = 2.300 \text{ kg/m}^3 \qquad m = (2.300)(\tfrac{80}{1000}) = 0.1840 \text{ kg}$$

$$\rho_2 = p_2/RT_2 \qquad (0.1840 - 0.003)/\tfrac{40}{1000} = (500)(1000)/(287T_2) \qquad T_2 = 385 \text{ K} \quad \text{or} \quad 112 \text{ °C}$$

1.51 For the same conditions given in Prob. 1.50, suppose the initial volume of air in the tank is 80 L at a pressure of 120 kPa and temperature of 20 °C. The gasoline engine cuts in to double the pressure in the tank while the volume is decreased to 50 L. What are the final temperature and density of the air?

$$\rho_1 = p_1/RT_1 = (120)(1000)/[(287)(20 + 273)] = 1.427 \text{ kg/m}^3 \qquad m = (1.427)(\tfrac{80}{1000}) = 0.1142 \text{ kg}$$

$$\rho_2 = p_2/RT_2 \qquad 0.1142/\tfrac{50}{1000} = (2)(120)(1000)/(287T_2) \qquad T_2 = 366 \text{ K} \quad \text{or} \quad 93 \text{ °C}$$

$$\rho = 0.1142/(\tfrac{50}{1000}) = 2.28 \text{ kg/m}^3$$

1.52 For 2 lb mol of air with a molecular weight of 29, a temperature of 100 °F, and a pressure of 2 atm, what is the volume?

$$pV/nM = RT \qquad [(2)(14.7)(144)]\{V/[(2)(29)]\} = (53.3)(100 + 460) \qquad V = 409 \text{ ft}^3$$

1.53 If nitrogen has a molecular weight of 28, what is its density according to the perfect gas law when $p = 200\,000$ Pa and $T = 50$ °C?

$$R = R_u/M = 8312/28 = 297 \text{ N} \cdot \text{m/(kg} \cdot \text{K)} \qquad [\text{where } R_u = 8312 \text{ N} \cdot \text{m/(kg} \cdot \text{K)}]$$

$$\rho = p/RT = 200\,000/[(297)(50 + 273)] = 2.08 \text{ kg/m}^3$$

1.54 If a gas occupies 1 m³ at 1 atm pressure, what pressure is required to reduce the volume of the gas by 1 percent under isothermal conditions if the fluid is (a) air, (b) argon, and (c) hydrogen?

$pV = nRT =$ constant for isothermal conditions. Therefore, if V drops to $0.99V_o$, p must rise to $(1/0.99)p_o$, or $1.010p_o$. This is true for any perfect gas.

1.55 (a) Calculate the density, specific weight, and specific volume of oxygen at 100 °F and 15 psia. (b) What would be the temperature and pressure of this gas if it were compressed isentropically to 40 percent of its original volume? (c) If the process described in (b) had been isothermal, what would the temperature and pressure have been?

(a)
$$\rho = p/RT = (15)(144)/[(1552)(100 + 460)] = 0.00248 \text{ slug/ft}^3$$

$$\gamma = \rho g = (0.00248)(32.2) = 0.0799 \text{ lb/ft}^3 \qquad V_s = 1/\rho = 1/0.00248 = 403 \text{ ft}^3/\text{slug}$$

(b)
$$p_1(V_s)_1^k = p_2(V_s)_2^k \qquad [(15)(144)](403)^{1.40} = [(p_2)(144)][(0.40)(403)]^{1.40} \qquad p_2 = 54.1 \text{ psia}$$

$$p_2 = \rho_2 RT_2 \qquad (54.1)(144) = (0.00248/0.40)(1552)(T_2 + 460) \qquad T_2 = 350 \text{ °F}$$

(c) If isothermal, $T_2 = T_1 = 100$ °F and $pV =$ constant.

$$[(15)(144)](403) = [(p_2)(144)][(0.40)(403)] \qquad p_2 = 37.5 \text{ psia}$$

1.56 Calculate the density, specific weight, and volume of chloride gas at 25 °C and pressure of 600 000 N/m² abs.

$$\rho = p/RT = 600\,000/[(118)(25 + 273)] = 17.1 \text{ kg/m}^3$$

$$\gamma = \rho g = (17.1)(9.81) = 168 \text{ N/m}^3 \qquad V_s = 1/\rho = 1/17.1 = 0.0585 \text{ m}^3/\text{kg}$$

1.57 If natural gas has a specific gravity of 0.60 relative to air at 14.7 psia and 60 °F, what are its specific weight and specific volume at that same pressure and temperature? What is the value of R for the gas?

$$\gamma_{\text{air}} = p/RT = (14.7)(144)/[(53.3)(60 + 460)] = 0.07637 \text{ lb/ft}^3$$

$$\gamma_{\text{gas}} = (0.60)(0.07637) = 0.0458 \text{ lb/ft}^3$$

$$V_s = 1/\rho = g/\gamma \qquad (V_s)_{\text{gas}} = 32.3/0.0458 = 703 \text{ ft}^3/\text{slug}$$

Since R varies inversely with density for the same pressure and temperature, $R_{\text{gas}} = 53.3/0.60 = 88.8 \text{ ft/°R}$.

1.58 A gas at 40 °C under a pressure of 20 000 mbar abs has a unit weight of 332 N/m³. What is the value of R for this gas? What gas might this be?

$$\gamma = p/RT \qquad 332 = (20\,000)(100)/[(R)(40 + 273)] \qquad R = 19.2 \text{ m/K}$$

This gas might be carbon dioxide, since its gas constant is 19.3 m/K (from Table A-6).

1.59 If water vapor in the atmosphere has a partial pressure of 0.50 psia and the temperature is 90 °F, what is its specific weight? Use R for water vapor = 85.7 ft/°R.

$$\gamma = p/RT = (0.50)(144)/[(85.7)(90 + 460)] = 0.00153 \text{ lb/ft}^3$$

1.60 If the barometer reads 14.50 psia in Prob. 1.59, what is the partial pressure of the air and what is its specific weight? What is the specific weight of the atmosphere (i.e., air plus water vapor present)?

$$p_{air} = 14.50 - 0.50 = 14.00 \text{ psia} \qquad \gamma = p/RT$$

$$\gamma_{air} = (14.00)(144)/[(53.3)(90 + 460)] = 0.0688 \text{ lb/ft}^3 \qquad \gamma_{atm} = \gamma_{air} + \gamma_{H_2O(vap)}$$

$$\gamma_{H_2O(vap)} = 0.00153 \text{ lb/ft}^3 \quad \text{(from Prob. 1.59)} \qquad \gamma_{atm} = 0.0688 + 0.00153 = 0.0703 \text{ lb/ft}^3$$

1.61 (a) Calculate the density, specific weight, and specific volume of oxygen at 10 °C and 30 kN/m² abs. (b) If the oxygen is enclosed in a rigid container of constant volume, what will be the pressure if the temperature is reduced to −120 °C?

(a)
$$\rho = p/RT = (30)(1000)/[(260)(10 + 273)] = 0.408 \text{ kg/m}^3$$

$$\gamma = \rho g = (0.408)(9.81) = 4.00 \text{ N/m}^3 \qquad V_s = 1/\rho = 1/0.408 = 2.45 \text{ m}^3/\text{kg}$$

(b) $\rho = 1/V_s = p/RT$. Since V_s and R are constants, $V_s/R = T/p = \text{constant}$, $(10 + 273)/30 = (-120 + 273)/p_2$, $p_2 = 16.2 \text{ kN/m}^2$.

1.62 Helium at 140 kN/m² abs and 11 °C is isentropically compressed to one-fifth of its original volume. What is its final pressure?

$$p_1 V_1^k = p_2 V_2^k \qquad 140 V_1^{1.66} = (p_2)(V_1/5)^{1.66} \qquad p_2 = 2025 \text{ kN/m}^2 \text{ abs}$$

1.63 (a) If 10 ft³ of carbon dioxide at 80 °F and 20 psia is compressed isothermally to 2 ft³, what is the resulting pressure? (b) What would the pressure and temperature have been if the process had been isentropic?

(a)
$$p_1 V_1 = p_2 V_2 \qquad (20)(10) = (p_2)(2) \qquad p_2 = 100 \text{ psia}$$

(b)
$$p_1 V_1^k = p_2 V_2^k \qquad (20)(10)^{1.30} = (p_2)(2)^{1.30} \qquad p_2 = 162 \text{ psia}$$

$$T_2/T_1 = (p_2/p_1)^{(k-1)/k} \qquad T_2/(80 + 460) = (\tfrac{162}{20})^{(1.30-1)/1.30} \qquad T_2 = 875 \text{ °R} \quad \text{or} \quad 415 \text{ °F}$$

1.64 (a) If 10 m³ of nitrogen at 30 °C and 150 kPa abs are permitted to expand isothermally to 25 m³, what is the resulting pressure? (b) What would the pressure and temperature have been if the process had been isentropic?

(a)
$$p_1 V_1 = p_2 V_2 \qquad (150)(10) = (p_2)(25) \qquad p_2 = 60.0 \text{ kPa abs}$$

(b)
$$p_1 V_1^k = p_2 V_2^k \qquad (150)(10)^{1.40} = (p_2)(25)^{1.40} \qquad p_2 = 41.6 \text{ kPa abs}$$

$$T_2/T_1 = (p_2/p_1)^{(k-1)/k} \qquad T_2/(30 + 273) = (41.6/150)^{(1.40-1)/1.40} \qquad T_2 = 210 \text{ K} \quad \text{or} \quad -63 \text{ °C}$$

1.65 If the viscosity of water at 68 °F is 0.01008 poise, compute its absolute viscosity (μ) in pound-seconds per square foot. If the specific gravity at 68 °F is 0.998, compute its kinematic viscosity (ν) in square feet per second.

The poise is measured in dyne-seconds per square centimeter. Since 1 lb = 444 800 dynes and 1 ft = 30.48 cm, $1 \text{ lb} \cdot \text{s/ft}^2 = 444\,800 \text{ dyne} \cdot \text{s}/(30.48 \text{ cm})^2 = 478.8 \text{ poises}$

$$\mu = \frac{0.01008}{478.8} = 2.11 \times 10^{-5} \text{ lb} \cdot \text{s/ft}^2 \qquad \nu = \frac{\mu}{\rho} = \frac{\mu}{\gamma/g} = \frac{\mu g}{\gamma} = \frac{(2.11 \times 10^{-5})(32.2)}{(0.998)(62.4)} = 1.09 \times 10^{-5} \text{ ft}^2/\text{s}$$

1.66 Convert 15.14 poises to kinematic viscosity in square feet per second if the liquid has a specific gravity of 0.964.

$$1 \text{ lb} \cdot \text{s/ft}^2 = 478.8 \text{ poises} \qquad \text{(from Prob. 1.65)}$$

$$\mu = 15.14/478.8 = 0.03162 \text{ lb} \cdot \text{s/ft}^2 \qquad \nu = \mu g/\gamma = (0.03162)(32.2)/[(0.964)(62.4)] = 0.0169 \text{ ft}^2/\text{s}$$

1.67 The fluid flowing in Fig. 1-3 has an absolute viscosity (μ) of 0.0010 lb · s/ft² and specific gravity of 0.913. Calculate the velocity gradient and intensity of shear stress at the boundary and at points 1 in, 2 in, and 3 in from the boundary, assuming (a) a straight-line velocity distribution and (b) a parabolic velocity distribution. The parabola in the sketch has its vertex at A and origin at B.

❚ (a) For the straight-line assumption, the relation between velocity v and distance y is $v = 15y$, $dv = 15dy$. The velocity gradient $= dv/dy = 15$. Since $\mu = \tau/(dv/dy)$, $\tau = \mu\,(dv/dy)$. For $y = 0$ (i.e., at the boundary), $v = 0$ and $dv/dy = 15\,\text{s}^{-1}$; $\tau = (0.0010)(15) = 0.015\,\text{lb/ft}^2$. For $y = 1$ in, 2 in, and 3 in, dv/dy and τ are also $15\,\text{s}^{-1}$ and $0.015\,\text{lb/ft}^2$, respectively. **(b)** For the parabolic assumption, the parabola passes through the points $v = 0$ when $y = 0$ and $v = 45$ when $y = 3$. The equation of this parabola is $v = 45 - 5(3 - y)^2$, $dv/dy = 10(3 - y)$, $\tau = 0.0010\,(dv/dy)$. For $y = 0$ in, $v = 0$ in/s, $dv/dy = 30\,\text{s}^{-1}$, and $\tau = 0.030\,\text{lb/ft}^2$. For $y = 1$ in, $v = 25$ in/s, $dv/dy = 20\,\text{s}^{-1}$, and $\tau = 0.020\,\text{lb/ft}^2$. For $y = 2$ in, $v = 40$ in/s, $dv/dy = 10\,\text{s}^{-1}$, and $\tau = 0.010\,\text{lb/ft}^2$. For $y = 3$ in, $v = 45$ in/s, $dv/dy = 0\,\text{s}^{-1}$, and $\tau = 0\,\text{lb/ft}^2$.

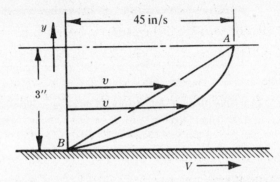

Fig. 1-3

1.68 A cylinder of 0.40-ft radius rotates concentrically inside a fixed cylinder of 0.42-ft radius. Both cylinders are 1.00 ft long. Determine the viscosity of the liquid that fills the space between the cylinders if a torque of 0.650 lb · ft is required to maintain an angular velocity of 60 rpm.

❚ The torque is transmitted through the field layers to the outer cylinder. Since the gap between the cylinders is small, the calculations may be made without integration. The tangential velocity v_t of the inner cylinder $= r\omega$, where $r = 0.40$ ft and $\omega = 2\pi$ rad/s. Hence, $v_t = (0.40)(2\pi) = 2.51$ ft/s. For the small space between cylinders, the velocity gradient may be assumed to be a straight line and the mean radius can be used. Then, $dv/dy = (2.51 - 0)/(0.42 - 0.40) = 125.5\,\text{s}^{-1}$. Since applied torque equals resisting torque, applied torque $= (\tau)(\text{area})(\text{arm})$, $0.650 = \tau[(1.00)(2\pi)(0.40 + 0.42)/2][(0.40 + 0.42)/2]$, $\tau = 0.615\,\text{lb/ft}^2 = \mu\,(dv/dy)$, $0.615 = (\mu)(125.5)$, $\mu = 0.00490\,\text{lb} \cdot \text{s/ft}^2$.

1.69 Water is moving through a pipe. The velocity profile at some section is shown in Fig. 1-4 and is given mathematically as $v = (\beta/4\mu)(d^2/4 - r^2)$, where $v =$ velocity of water at any position r, $\beta =$ a constant, $\mu =$ viscosity of water, $d =$ pipe diameter, and $r =$ radial distance from centerline. What is the shear stress at the wall of the pipe due to the water? What is the shear stress at a position $r = d/4$? If the given profile persists a distance L along the pipe, what drag is induced on the pipe by the water in the direction of flow over this distance?

❚
$$v = (\beta/4\mu)(d^2/4 - r^2) \qquad dv/dr = (\beta/4\mu)(-2r) = -2\beta r/4\mu$$
$$\tau = \mu\,(dv/dr) = \mu(-2\beta r/4\mu) = -2\beta r/4$$

At the wall, $r = d/2$. Hence,

$$\tau_{\text{wall}} = \frac{-2\beta(d/2)}{4} = -\frac{\beta d}{4} \qquad \tau_{r=d/4} = \frac{-2\beta(d/4)}{4} = -\frac{\beta d}{8}$$

$$\text{Drag} = (\tau_{\text{wall}})(\text{area}) = (\tau_{\text{wall}})(\pi dL) = (\beta d/4)(\pi dL) = \beta d^2 \pi L/4$$

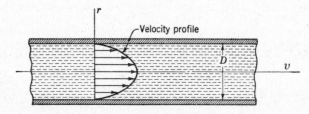

Velocity profile

Fig. 1-4

1.70   A large plate moves with speed v_0 over a stationary plate on a layer of oil (see Fig. 1-5). If the velocity profile is that of a parabola, with the oil at the plates having the same velocity as the plates, what is the shear stress on the moving plate from the oil? If a linear profile is assumed, what is the shear stress on the upper plate?

┃   For a parabolic profile, $v^2 = ay$. When $y = d$, $v = v_0$. Hence, $v_0^2 = ad$, $a = v_0^2/d$. Therefore,

$$v^2 = (v_0^2/d)(y) = (v_0^2)(y/d) \qquad v = v_0\sqrt{y/d} \qquad dv/dy = [(v_0)(1/\sqrt{d})(\tfrac{1}{2})(y^{-1/2})]$$

$$\tau = \mu\,(dv/dy) = \mu[(v_0)(1/\sqrt{d})(\tfrac{1}{2})(y^{-1/2})]$$

For $y = d$, $\tau = \mu[(v_0)(1/\sqrt{d})(\tfrac{1}{2})(d^{-1/2})] = \mu v_0/(2d)$. For a linear profile, $dv/dy = v_0/d$, $\tau = \mu(v_0/d)$.

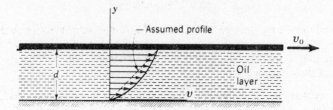

Fig. 1-5

1.71   A block weighing 1 kN and having dimensions 200 mm on an edge is allowed to slide down an incline on a film of oil having a thickness of 0.0050 mm (see Fig. 1-6). If we use a linear velocity profile in the oil, what is the terminal speed of the block? The viscosity of the oil is $7 \times 10^{-3}\,\text{N} \cdot \text{s/m}^2$.

┃ $\tau = \mu\,(dv/dy) = (7 \times 10^{-3})[v_T/(0.005/1000)] = 1400v_T \qquad F_f = \tau A = (1400v_T)(\tfrac{200}{1000})^2 = 56.0v_T$

At the terminal condition, equilibrium occurs. Hence, $1000 \sin 20° = 56.0v_T$, $v_T = 6.11$ m/s.

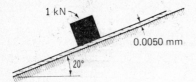

Fig. 1-6(a)

Fig. 1-6(b)

1.72   A cylinder of weight 20 lb slides in a lubricated pipe, as shown in Fig. 1-7. The clearance between cylinder and pipe is 0.001 in. If the cylinder is observed to decelerate at a rate of 2 ft/s² when the speed is 20 ft/s, what is the viscosity of the oil?

┃

$$\tau = \mu\,(dv/dy) = \mu[v/(0.001/12)] = 12\,000\mu v$$

$$F_f = \tau A = 12\,000\mu v[(\pi)(\tfrac{6}{12})(\tfrac{5}{12})] = 7854\mu v$$

$$\Sigma F = ma \qquad 20 - (7854)(\mu)(20) = (20/32.2)(-2) \qquad \mu = 1.35 \times 10^{-4}\,\text{lb} \cdot \text{s/ft}^2$$

Fig. 1-7

1.73   A plunger is moving through a cylinder at a speed of 20 ft/s, as shown in Fig. 1-8. The film of oil separating the plunger from the cylinder has a viscosity of $0.020\,\text{lb} \cdot \text{s/ft}^2$. What is the force required to maintain this motion?

┃   Assume the thickness of the film is uniform over the entire peripheral surface of the plunger. Because the film is thin, assume a linear velocity profile for the flow of oil in the film. To find the frictional resistance, compute the shear stress at the plunger surface.

$$\tau = \mu\frac{dv}{dr} = 0.020\left[\frac{20}{(5.000 - 4.990)/2}\right](12) = 960\,\text{lb/ft}^2 \qquad F_f = \tau A = 960\left[\pi\left(\frac{4.990}{12}\right)\left(\frac{3}{12}\right)\right] = 314\,\text{lb}$$

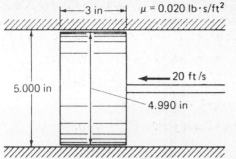

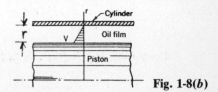

Fig. 1-8(a)

Fig. 1-8(b)

1.74 In some electrical measuring devices, the motion of a pointer mechanism is dampened by having a circular disc turn (with the pointer) in a container of oil (see Fig. 1-9). In this way, extraneous rotations are damped out. What is the damping torque for $\omega = 0.2$ rad/s if the oil has a viscosity of 8×10^{-3} N · s/m²? Neglect effects on the outer edge of the rotating plate.

❙ Assume at any point that the velocity profile of the oil is linear $dv/dn = r\omega/(0.5/1000) = (r)(0.2)/(0.5/1000) = 400r$; $\tau = \mu\,(dv/dn) = \mu(400r) = (8 \times 10^{-3})(400r) = 3.20r$. The force dF_f on dA on the upper face of the disc is then $dF_f = \tau\,dA = (3.20r)(r\,d\theta\,dr) = 3.20r^2\,d\theta\,dr$. The torque dT for dA on the upper face is then $dT = r\,dF_f = r(3.20r^2\,d\theta\,dr) = 3.20r^3\,d\theta\,dr$. The total resisting torque on both faces is

$$T = 2\left[\int_0^{0.075/2}\int_0^{2\pi} 3.20r^3\,d\theta\,dr\right] = (6.40)(2\pi)\left[\frac{r^4}{4}\right]_0^{0.075/2} = 1.99 \times 10^{-5}\,\text{N} \cdot \text{m}$$

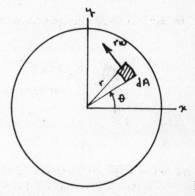

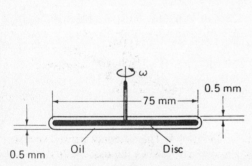

Fig. 1-9(a)

Fig. 1-9(b)

1.75 For the apparatus in Prob. 1.74, develop an expression giving the damping torque as a function of x, the distance that the midplane of the rotating plate is from its center position (see Fig. 1-10). Do this for an angular rotation of 0.2 rad/s.

❙ Assume at any point that the velocity profile of the oil is linear; $\tau = \mu\,(dv/dn)$. For the upper face, $dv/dn = r\omega/[(0.5 - x)/1000] = (r)(0.2)/[(0.5 - x)/1000]$; $\tau = (8 \times 10^{-3})\{(r)(0.2)/[(0.5 - x)/1000]\} = 1.60r/(0.5 - x)$. The force dF_f on dA on the upper face of the disc is then $dF_f = \tau\,dA = [1.60r/(0.5 - x)](r\,d\theta\,dr) = [1.60r^2/(0.5 - x)]\,(d\theta\,dr)$. The torque dT for dA on the upper face is then $dT = r\,dF_f = r[1.60r^2/(0.5 - x)]\,(d\theta\,dr) = [1.60r^3/(0.5 - x)]\,(d\theta\,dr)$. For the lower face, $dv/dn = r\omega/[(0.5 + x)/1000] = r(0.2)/[(0.5 + x)/1000]$; $\tau = (8 \times 10^{-3})\{r(0.2)/[(0.5 + x)/1000]\} = 1.60r/(0.5 + x)$. The force dF_f on dA on the lower face of the disc is then $dF_f = \tau\,dA = [1.60r/(0.5 + x)](r\,d\theta\,dr) = [1.60r^2/(0.5 + x)]\,(d\theta\,dr)$. The torque dT for dA on the lower face is then $dT = r\,dF_f = r[1.60r^2/(0.5 + x)]\,(d\theta\,dr) = [1.60r^3/(0.5 + x)]\,(d\theta\,dr)$. The total resisting torque on both faces is

$$T = \int_0^{0.075/2}\int_0^{2\pi} \frac{1.60r^3}{0.5 - x}\,d\theta\,dr + \int_0^{0.075/2}\int_0^{2\pi} \frac{1.60r^3}{0.5 + x}\,d\theta\,dr$$

$$= \left(\frac{1}{0.5 - x} + \frac{1}{0.5 + x}\right)(1.60)(2\pi)\left[\frac{r^4}{4}\right]_0^{0.075/2} = \left(\frac{0.5 + x + 0.5 - x}{0.25 - x^2}\right)(4.97 \times 10^{-6})$$

$$= \frac{4.97 \times 10^{-6}}{0.25 - x^2}$$

Note: x in millimeters gives torque in newton-meters.

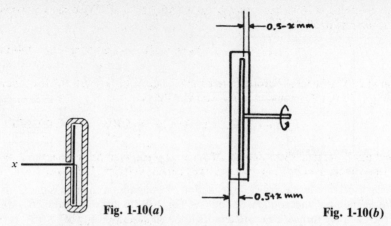

Fig. 1-10(a)   Fig. 1-10(b)

1.76 A conical body is made to rotate in a container, as shown in Fig. 1-11, at a constant speed of 10 rad/s. A film of oil having a viscosity of 4.5×10^{-5} lb · s/ft² (3.125×10^{-7} lb · s/in²) separates the cone from the container. The film thickness is 0.01 in. What torque is required to maintain this motion? The cone has a 2-in radius at its base and is 4 in tall.

▌ Consider the conical surface first ($r/2 = z/4$, $r = z/2$). The stress on this element is $\tau = \mu (dv/dx) = \mu(r\omega/0.01) = (3.125 \times 10^{-7})[(z/2)(10)/0.01] = 1.562 \times 10^{-4}z$. The area of the strip shown is $dA = 2\pi r \, ds = (2\pi z/2)[dz/(4/\sqrt{20})] = 3.512z \, dz$. The torque on the strip is $dT = \tau (dA)(r) = (1.562 \times 10^{-4}z)(3.512z \, dz)(z/2) = 2.743 \times 10^{-4}z^3 \, dz$.

$$T_1 = \int_0^4 2.743 \times 10^{-4}z^3 \, dz = 2.743 \times 10^{-4}\left[\frac{z^4}{4}\right]_0^4 = 0.01756 \text{ in} \cdot \text{lb}$$

Next consider the base: $dF_f = \tau \, dA$, $\tau = \mu(r\omega/0.01) = (3.125 \times 10^{-7})[(r)(10)/0.01] = 3.125 \times 10^{-4}r$, $dF_f = (3.125 \times 10^{-4}r)(r \, d\theta \, dr) = 3.125 \times 10^{-4}r^2 \, d\theta \, dr$, $dT_2 = (3.125 \times 10^{-4}r^2 \, d\theta \, dr)(r) = 3.125 \times 10^{-4}r^3 \, d\theta \, dr$.

$$T_2 = \int_0^2 \int_0^{2\pi} 3.125 \times 10^{-4}r^3 \, d\theta dr = (3.125 \times 10^{-4})(2\pi)\left[\frac{r^4}{4}\right]_0^2 = 0.00785 \text{ in} \cdot \text{lb}$$

$$T_{\text{tot}} = 0.01756 + 0.00785 = 0.0254 \text{ in} \cdot \text{lb}$$

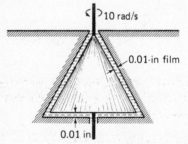

Fig. 1-11(a)

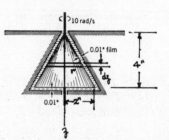

Fig. 1-11(b)

1.77 In Fig. 1-12, if the fluid is glycerin at 20 °C and the width between plates is 6 mm, what shear stress is required to move the upper plate at 2.5 m/s? What is the Reynolds number if D is taken to be the distance between plates?

▌

$$\tau = \mu (dv/dh) = (1.49)[2.5/(\tfrac{6}{1000})] = 621 \text{ N/m}^2 \quad \text{or} \quad 621 \text{ Pa}$$

$$N_R = \rho D v/\mu = (1258)(\tfrac{6}{1000})(2.5)/1.49 = 12.7$$

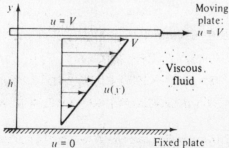

Fig. 1-12

1.78 Carbon tetrachloride at 20 °C has a viscosity of 0.000967 N · s/m². What shear stress is required to deform this fluid at a strain rate of 5000 s⁻¹?

$$\tau = \mu\,(dv/dx) = (0.000967)(5000) = 4.84\ \text{N/m}^2 \quad \text{or} \quad 4.84\ \text{Pa}$$

1.79 SAE 10 oil at 20 °C is sheared between two parallel plates 0.01 in apart with the lower plate fixed and the upper plate moving at 15 ft/s. Compute the shear stress in the oil.

$$\tau = \mu\,(dv/dh) = (1.70 \times 10^{-3})[15/(0.01/12)] = 30.6\ \text{lb/ft}^2$$

1.80 An 8-kg flat block of metal slides down a 20° inclined plane while lubricated by a 2-mm-thick film of SAE 30 oil at 20 °C. The contact area is 0.2 m². What is the terminal velocity of the block?

∎ See Fig. 1-13.

$$\Sigma F_x = 0 \qquad W \sin\theta - \tau A_{\text{bottom}} = 0$$

$$\tau = \mu\,(dv/dy) = (4.40 \times 10^{-1})(v_T/\tfrac{2}{1000}) = 220 v_T$$

$$[(8)(9.81)](\sin 20°) - (220 v_T)(0.2) = 0 \qquad v_T = 0.610\ \text{m/s}$$

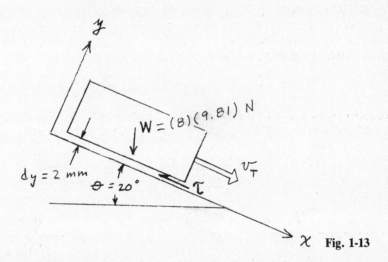

Fig. 1-13

1.81 A shaft 8.00 cm in diameter is being pushed through a bearing sleeve 8.02 cm in diameter and 30 cm long. The clearance, assumed uniform, is filled with oil at 20 °C with $v = 0.005\ \text{m}^2/\text{s}$ and s.g. = 0.9. If the shaft moves axially at 0.5 m/s, estimate the resistance force exerted by the oil on the shaft.

$$F = \tau A \qquad \tau = \mu\,(dv/dy) \qquad \mu = \rho v = [(0.9)(998)](0.005) = 4.49\ \text{kg/(m·s)}$$

$$dy = (8.02 - 8.00)/2 = 0.01\ \text{cm} = 0.0001\ \text{m} \qquad \tau = (4.49)(0.5/0.0001) = 22\,450\ \text{N/m}^2$$

$$A = (\pi)(8.00/100)(30/100) = 0.07540\ \text{m}^2 \qquad F = (22\,450)(0.07540) = 1693\ \text{N}$$

1.82 If the shaft in Prob. 1.81 is now fixed axially and rotated inside the sleeve at 1800 rpm, estimate the resisting torque exerted by the oil and the power required to rotate the shaft.

$$T = \tau A r \qquad \tau = \mu\,(dv/dr)$$

$$v = r\omega = [(8.00/2)/100][(1800)(2\pi/60)] = 7.540\ \text{m/s} \qquad dy = (8.02 - 8.00)/2 = 0.01\ \text{cm} = 0.0001\ \text{m}$$

$$\tau = (4.49)(7.540/0.0001) = 338.6 \times 10^3\ \text{N/m}^2 \qquad A = (\pi)(8.00/100)(\tfrac{30}{100}) = 0.07540\ \text{m}^2$$

$$T = (338.6 \times 10^3)(0.07540)[(8.00/2)/100] = 1021\ \text{N·m}$$

$$P = \omega T = [(1800)(2\pi/60)](1021) = 192.5 \times 10^3\ \text{W} \quad \text{or} \quad 192.5\ \text{kW}$$

1.83 A steel (7850-kg/m³) shaft 3.00 cm in diameter and 40 cm long falls of its own weight inside a vertical open tube

3.02 cm in diameter. The clearance, assumed uniform, is a film of glycerin at 20 °C. How fast will the cylinder fall at terminal conditions?

$$W_{shaft} = \tau A = [(7850)(9.81)][(\tfrac{40}{100})(\pi)(3.00/100)^2/4] = 21.77 \text{ N}$$

$$dr = [(3.02 - 3.00)/2]/100 = 0.0001 \text{ m}$$

$$\tau = \mu \, (dy/dr) = (1.49)(v_T/0.0001) = 14\,900 v_T$$

$$A = (\pi)(3.00/100)(\tfrac{40}{100}) = 0.03770 \text{ m}^2 \qquad 21.77 = (14\,900 v_T)(0.03770) \qquad v_T = 0.0388 \text{ m/s}$$

1.84 Air at 20 °C forms a boundary layer near a solid wall of sine-wave-shaped velocity profile (see Fig. 1-14). The boundary-layer thickness is 6 mm and the peak velocity is 10 m/s. Compute the shear stress in the boundary layer at y equal to (*a*) 0, (*b*) 3 mm, and (*c*) 6 mm.

$$\tau = \mu \, (dv/dy) \qquad v = v_{max} \sin [\pi y/(2\delta)]$$

$$dv/dy = [\pi v_{max}/(2\delta)] \cos [\pi y/(2\delta)] = \{(\pi)(10)/[(2)(\tfrac{6}{1000})]\} \cos \{\pi y/[(2)(\tfrac{6}{1000})]\} = 2618 \cos (261.8y)$$

Note: "261.8y" in the above equation is in radians.

$$\tau = (1.81 \times 10^{-5})[2618 \cos (261.8y)] = 0.04739 \cos (261.8y)$$

(*a*) At $y = 0$, $\tau = 0.04739 \cos [(261.8)(0)] = 0.0474 \text{ N/m}^2$. (*b*) At $y = 3$ mm, or 0.003 m, $\tau = 0.04739 \cos [(261.8)(0.003)] = 0.0335 \text{ N/m}^2$. (*c*) At $y = 6$ mm, or 0.006 m, $\tau = 0.04739 \cos [(261.8)(0.006)] = 0$.

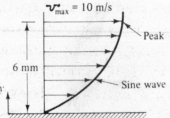

Fig. 1-14

1.85 A disk of radius r_0 rotates at angular velocity ω inside an oil bath of viscosity μ, as shown in Fig. 1-15. Assuming a linear velocity profile and neglecting shear on the outer disk edges, derive an expression for the viscous torque on the disk.

$$\tau = \mu \, (dv/dy) = \mu(r\omega/h) \qquad \text{(on both sides)}$$

$$dT = (2)(r\tau \, dA) = (2)\{(r)[\mu(r\omega/h)](2\pi r \, dr)\} = (4\mu\omega\pi/h)(r^3 \, dr)$$

$$T = \int_0^{r_0} \frac{4\mu\omega\pi}{h} (r^3 \, dr) = \frac{4\mu\omega\pi}{h} \left[\frac{r^4}{4}\right]_0^{r_0} = \frac{\pi\mu\omega r_0^4}{h}$$

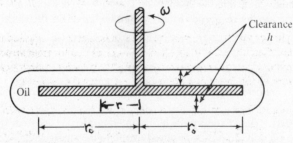

Fig. 1-15

1.86 A flat plate 30 cm by 50 cm slides on oil ($\mu = 0.8 \text{ N} \cdot \text{s/m}^2$) over a large plane surface. What force is required to drag the plate at 2 m/s, if the separating oil film is 0.5 mm thick?

$$\tau = \mu \, (dv/dx) = (0.8)[2/(0.5/1000)] = 3200 \text{ N/m}^2 \qquad F = \tau A = (3200)[(\tfrac{30}{100})(\tfrac{50}{100})] = 480 \text{ N}$$

1.87 A space of 1-in (0.0833-ft) width between two large plane surfaces is filled with SAE 30 western lubricating oil at 80 °F ($\mu = 0.0063 \text{ lb} \cdot \text{s/ft}^2$). What force is required to drag a very thin plate of 4-ft^2 area between the surfaces at a speed of 0.333 ft/s if this plate is equally spaced between the two surfaces?

$$\tau = \mu \, (dv/dx) = (0.0063)[0.333/(0.0833/2)] = 0.0504 \text{ lb/ft}^2 \qquad F = \tau A = (0.0504)(4) = 0.2016 \text{ lb}$$

Since there are two sides, $F_{required} = (2)(0.2016)$, or 0.403 lb.

1.88   Rework Prob. 1.87 if the plate is at a distance of 0.33 in (0.0275 ft) from one surface.

∎ $\tau = \mu\,(dv/dx)$ $\tau_1 = (0.0063)(0.333/0.0275) = 0.0763\ \text{lb/ft}^2$

$F = \tau A$ $F_1 = (0.0763)(4) = 0.3052\ \text{lb}$ $\tau_2 = (0.0063)[0.333/(0.0833 - 0.0275)] = 0.0376\ \text{lb/ft}^2$

$F_2 = (0.0376)(4) = 0.1504\ \text{lb}$ $F_{\text{required}} = F_1 + F_2 = 0.3052 + 0.1504 = 0.456\ \text{lb}$

1.89   A hydraulic lift of the type commonly used for greasing automobiles consists of a 10.000-in-diameter ram which slides in a 10.006-in-diameter cylinder, the annular space being filled with oil having a kinematic viscosity of $0.004\ \text{ft}^2/\text{s}$ and specific gravity of 0.85. If the rate of travel of the ram is 0.5 ft/s, find the frictional resistance when 10 ft of the ram are engaged in the cylinder.

∎ $\tau = \mu\,(dv/dx)$ $\rho = \gamma/g = [(0.85)(62.4)]/32.2 = 1.647\ \text{slugs/ft}^3$

$\mu = \rho v = (1.647)(0.004) = 0.006588\ \text{lb}\cdot\text{s/ft}^2$ $dx = [(10.006 - 10.000)/2]/12 = 0.000250\ \text{ft}$

$\tau = (0.006588)(0.5/0.000250) = 13.18\ \text{lb/ft}^2$ $F_f = \tau A = (13.18)[(10)(\pi)(\frac{10}{12})] = 345\ \text{lb}$

1.90   A journal bearing consists of a 6.00-in shaft in a 6.01-in sleeve 8 in long, the clearance space (assumed to be uniform) being filled with SAE 30 eastern lubricating oil at 100 °F ($\mu = 0.0018\ \text{lb}\cdot\text{s/ft}^2$). Calculate the rate at which heat is generated at the bearing when the shaft turns at 100 rpm.

∎ $dv = \omega(\text{circumference}) = \frac{100}{60}[\pi(6.00/12)] = 2.618\ \text{ft/s}$

$dx = [(6.01 - 6.00)/2]/12 = 0.0004167\ \text{ft}$

$\tau = \mu\,(dv/dx) = (0.0018)(2.618/0.0004167) = 11.31\ \text{lb/ft}^2$

$F_f = \tau A = 11.31[\pi(8.00/12)(\frac{6}{12})] = 11.84\ \text{lb}$

Rate of energy loss $= F_f v = (11.84)(2.618) = 31.00\ \text{ft}\cdot\text{lb/s}$

Rate of heat generation $= (31.00)(3600)/778 = 143\ \text{Btu/h}$

1.91   A journal bearing consists of an 8.00-cm shaft in an 8.03-cm sleeve 10 cm long, the clearance space (assumed to be uniform) being filled with SAE 30 western lubricating oil at 40 °C ($\mu = 0.11\ \text{N}\cdot\text{s/m}^2$). Calculate the rate at which heat is generated at the bearing when the shaft turns at 120 rpm.

∎ $dv = \omega(\text{circumference}) = \frac{120}{60}[\pi(8.00/100)] = 0.5027\ \text{m/s}$ $dx = [(8.03 - 8.00)/2]/100 = 0.000150\ \text{m}$

$\tau = \mu\,(dv/dx) = (0.11)(0.5027)/0.000150 = 368.6\ \text{N/m}^2$

$F_f = \tau A = 368.6[\pi(10.00/100)(8.00/100)] = 9.264\ \text{N}$

Rate of energy loss $= F_f v = (9.264)(0.5027) = 4.657\ \text{N}\cdot\text{m/s}$ Rate of heat generation $= 4.657\ \text{J/s}$

1.92   In using a rotating-cylinder viscometer, a bottom correction must be applied to account for the drag on the flat bottom of the cylinder. Calculate the theoretical amount of this torque correction, neglecting centrifugal effects, for a cylinder of diameter d, rotated at a constant angular velocity ω, in a liquid of viscosity μ, with a clearance Δh between the bottom of the inner cylinder and the floor of the outer one.

∎   Let r = variable radius. $T = \int r\tau\,dA$, $\tau = \mu\,(dv/dx) = \mu(r\omega/\Delta h)$, $dA = 2\pi r\,dr$.

$$T = \int_0^{d/2} r\left[\mu\left(\frac{r\omega}{\Delta h}\right)\right](2\pi r\,dr) = \frac{2\pi\mu\omega}{\Delta h}\int_0^{d/2} r^3\,dr = \frac{2\pi\mu\omega}{\Delta h}\left[\frac{r^4}{4}\right]_0^{d/2} = \frac{\pi\mu\omega d^4}{32\,\Delta h}$$

1.93   Assuming a velocity distribution as shown in Fig. 1-16, which is a parabola having its vertex 4 in from the boundary, calculate the shear stresses for $y = 0$, 1 in, 2 in, 3 in, and 4 in. Use $\mu = 0.00835\ \text{lb}\cdot\text{s/ft}^2$.

∎ $\tau = \mu\,(dv/dy)$. At $y = 0$, $v = 0$ and at $y = 4$ in, $v = 8$ ft/s, or 96 in/s. The equation of the parabola is $v = 96 - (6)(4 - y)^2$ (y in inches gives v in inches per second); $dv/dy = (12)(4 - y)$; $\tau = (0.00835)[(12)(4 - y)] = 0.4008 - 0.1002y$. At $y = 0$, $\tau = 0.4008 - (0.1002)(0) = 0.401\ \text{lb/ft}^2$. At $y = 1$ in, $\tau = 0.4008 - (0.1002)(1) = 0.301\ \text{lb/ft}^2$. At $y = 2$ in, $\tau = 0.4008 - (0.1002)(2) = 0.200\ \text{lb/ft}^2$. At $y = 3$ in, $\tau = 0.4008 - (0.1002)(3) = 0.100\ \text{lb/ft}^2$. At $y = 4$ in, $\tau = 0.4008 - (0.1002)(4) = 0$.

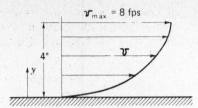

Fig. 1-16

1.94 In Fig. 1-17a, oil of viscosity μ fills the small gap of thickness Y. Determine an expression for the torque T required to rotate the truncated cone at constant speed ω. Neglect fluid stress exerted on the circular bottom.

❚ See Fig. 1-17b. $\tau = \mu(dv/dy)$, $v = r\omega = (y\tan\alpha)(\omega)$, $dv/dy = (y\tan\alpha)(\omega)/Y$.

$$\tau = \mu\left[\frac{(y\tan\alpha)(\omega)}{Y}\right] = \frac{\mu y\omega\tan\alpha}{Y}$$

$$dA = 2\pi r\,ds = 2\pi(y\tan\alpha)(dy/\cos\alpha) = 2\pi y(\tan\alpha/\cos\alpha)(dy)$$

$$dF = \tau\,dA = \left(\frac{\mu y\omega\tan\alpha}{Y}\right)\left[2\pi y\left(\frac{\tan\alpha}{\cos\alpha}\right)(dy)\right] = \left(\frac{2\pi\mu\omega\tan^2\alpha}{Y\cos\alpha}\right)y^2\,dy$$

$$dT = r\,dF = (y\tan\alpha)\left(\frac{2\pi\mu\omega\tan^2\alpha}{Y\cos\alpha}\right)y^2\,dy = \left(\frac{2\pi\mu\omega\tan^3\alpha}{Y\cos\alpha}\right)y^3\,dy$$

$$T = \int_a^{a+b}\left(\frac{2\pi\mu\omega\tan^3\alpha}{Y\cos\alpha}\right)y^3\,dy = \left(\frac{2\pi\mu\omega\tan^3\alpha}{Y\cos\alpha}\right)\left[\frac{y^4}{4}\right]_a^{a+b} = \left(\frac{2\pi\mu\omega\tan^3\alpha}{Y\cos\alpha}\right)\left[\frac{(a+b)^4}{4} - \frac{a^4}{4}\right]$$

$$= \left(\frac{\pi\mu\omega\tan^3\alpha}{2Y\cos\alpha}\right)[(a+b)^4 - a^4]$$

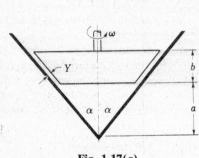

Fig. 1-17(a)

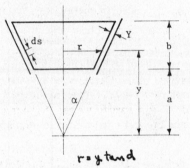

$r = y\tan\alpha$

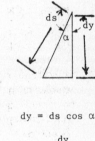

$dy = ds\cos\alpha$

$ds = \dfrac{dy}{\cos\alpha}$

Fig. 1-17(b)

1.95 A Newtonian fluid is in the clearance between a shaft and a concentric sleeve. When a force of 600 N is applied to the sleeve parallel to the shaft, the sleeve attains a speed of 1 m/s. If a 1500-N force is applied, what speed will the sleeve attain? The temperature of the sleeve remains constant.

❚ $\tau = F/A = \mu(dv/dx)$; $F/dv = \mu A/dx = $ constant. Therefore, $F_1/dv_1 = F_2/dv_2$, $\frac{600}{1} = 1500/dv_2$, $dv_2 = 2.50$ m/s.

1.96 A plate 0.5 mm distant from a fixed plate moves at 0.25 m/s and requires a force per unit area of 2.0 Pa to maintain this speed. Determine the viscosity of the fluid between the plates.

❚ $\tau = \mu(dv/dx)$ $2.0 = \mu[0.25/(0.5/1000)]$ $\mu = 0.00400$ N · s/m^2

1.97 Determine the viscosity of fluid between shaft and sleeve in Fig. 1-18.

❚ $\tau = F/A = \mu(dv/dx)$ $20/[(\pi)(\frac{3}{12})(\frac{8}{12})] = \mu[0.4/(0.003/12)]$ $\mu = 0.0239$ lb · s/ft^2

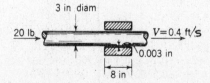

3 in diam

20 lb $V = 0.4$ ft/s

0.003 in

8 in

Fig. 1-18

1.98 A 1-in-diameter steel cylinder 12 in long falls, because of its own weight, at a uniform rate of 0.5 ft/s inside a tube of slightly larger diameter. A castor-oil film of constant thickness is between the cylinder and the tube. Determine the clearance between the cylinder and the tube. The temperature is 100 °F. Use s.g. = 7.85 for steel and $\mu = 6 \times 10^{-3}$ lb · s/ft² for castor oil.

$$\tau = F/A = \mu \, (dv/dx)$$

$$F = W = \gamma V = [(7.85)(62.4)][(\tfrac{12}{12})(\pi)(\tfrac{1}{12})^2/4] = 2.672 \text{ lb} \quad 2.672/[(\tfrac{12}{12})(\pi)(\tfrac{1}{12})] = (6 \times 10^{-3})(0.5/dx)$$

$$dx = 0.0002939 \text{ ft} \quad \text{or} \quad 0.00353 \text{ in}$$

1.99 A piston of diameter 60.00 mm moves inside a cylinder of 60.10-mm diameter. Determine the percent decrease in force necessary to move the piston when the lubricant warms up from 0 to 120 °C. Values of μ for the lubricant are 0.01820 N · s/m² at 0 °C and 0.00206 N · s/m² at 120 °C.

▮ $\tau = F/A = \mu \, (dv/dx)$; $F/\mu = A \, (dv/dx) =$ constant. Therefore, $\Delta F/F_{0\,°C} = \Delta \mu / \mu_{0\,°C} = (0.01820 - 0.00206)/0.01820 = 0.887$, or 88.7%.

1.100 A body weighing 120 lb with a flat surface area of 2 ft² slides down a lubricated inclined plane making a 30° angle with the horizontal. For viscosity of 0.002089 lb · s/ft² and a body speed of 3.0 ft/s, determine the lubricant film thickness.

▮ $F =$ weight of body along inclined plane $= 120 \sin 30° = 60.0$ lb

$$\tau = F/A = \mu \, (dv/dx) \qquad 60.0/2 = (0.002089)(3.0/dx) \qquad dx = 0.0002089 \text{ ft} \quad \text{or} \quad 0.00251 \text{ in}$$

1.101 A small drop of water at 80 °F is in contact with the air and has a diameter of 0.0200 in. If the pressure within the droplet is 0.082 psi greater than the atmosphere, what is the value of the surface tension?

▮ $p(\pi d^2/4) = (\pi d)(\sigma) \qquad \sigma = pd/4 = [(0.082)(144)](0.0200/12)/4 = 0.00492$ lb/ft

1.102 Estimate the height to which water at 70 °F will rise in a capillary tube of diameter 0.120 in.

▮ $h = 4\sigma \cos \theta /(\gamma d)$. From Table A-1, $\sigma = 0.00500$ lb/ft and $\gamma = 62.3$ lb/ft³ at 70 °F. Assume $\theta = 0°$ for a clean tube. $h = (4)(0.00500)(\cos 0°)/[(62.3)(0.120/12)] = 0.0321$ ft, or 0.385 in.

1.103 The shape of a hanging drop of liquid is expressible by the following formulation developed from photographic studies of the drop: $\sigma = (\gamma - \gamma_0)(d_e)^2/H$, where $\sigma =$ surface tension, i.e., force per unit length, $\gamma =$ specific weight of liquid drop, $\gamma_0 =$ specific weight of vapor around it, $d_e =$ diameter of drop at its equator, and $H = $ a function determined by experiment. For this equation to be dimensionally homogeneous, what dimensions must H possess?

▮ Dimensionally, $(F/L) = (F/L^3)(L^2)/\{H\}$, $\{H\} = (1)$. Therefore, H is dimensionless.

1.104 Two parallel, wide, clean, glass plates separated by a distance d of 1 mm are placed in water, as shown in Fig. 1-19a and b. How far does the water rise due to capillary action away from the ends of the plate? Use $\sigma = 0.0730$ N/m.

▮ Because the plates are clean, the angle of contact between water and glass is taken as zero. Consider the free-body diagram of a unit width of the raised water away from the ends (Fig. 1-19c). Summing forces in the vertical direction gives $(2)[(\sigma)(\tfrac{1}{1000})] - (\tfrac{1}{1000})^2(h)(\gamma) = 0$, $(2)[(0.0730)(\tfrac{1}{1000})] - (\tfrac{1}{1000})^2(h)(9790) = 0$, $h = 0.0149$ m, or 14.9 mm.

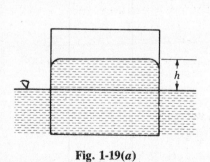

Fig. 1-19(a)

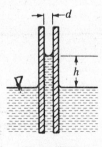

Fig. 1-19(b)

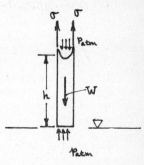

Fig. 1-19(c)

1.105 A glass tube is inserted in mercury. What is the upward force on the glass as a result of surface effects? Note in Fig. 1-20 that the contact angle is 50° inside and outside. The temperature is 20 °C.

∎ $F = (\sigma)(\pi d_o)(\cos 50°) + (\alpha)(\pi d_i)(\cos 50°) = (0.514)[(\pi)(\frac{45}{1000})](\cos 50°) + (0.514)[(\pi)(\frac{35}{1000})](\cos 50°) = 0.0830$ N

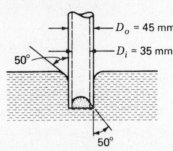

D_o = 45 mm

D_i = 35 mm

50°

50°

Fig. 1-20

1.106 In Fig. 1-21a, compute an approximate distance h for mercury in the glass capillary tube. Angle θ is 40°. *Hint*: Pressure p_{gage} below the main free surface is the specific weight times the depth below the free surface. Do your assumptions render the actual h larger or smaller than the computed h?

∎ Consider the meniscus of the mercury as a free body (see Fig. 1-21b). Neglect the weight of the free body. Summing forces in the vertical direction gives $-(\sigma)(\pi d)(\cos \theta) + (p)(\pi d^2/4) = 0$, $-(0.514)[(\pi)(\frac{1}{1000})](\cos 40°) + [(13.6)(9790)(h)][(\pi)(\frac{1}{1000})^2/4] = 0$, $h = 0.01183$ m, or 11.83 mm. Actual h must be larger because the weight of the meniscus was neglected.

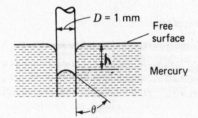

D = 1 mm

Free surface

Mercury

θ

Fig. 1-21(a)

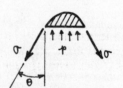

σ p σ

θ

Fig. 1-21(b)

1.107 A narrow tank with one end open (Fig. 1-22) is filled with water at 20 °C carefully and slowly to get the maximum amount of water in without spilling any water. If the pressure gage measures a gage pressure of 2943.70 Pa, what is the radius of curvature of the water surface at the top of the surface away from the ends?

∎
$$p = \sigma/r = p_{gage} - \gamma d = 2943.70 - (9790)(\tfrac{300}{1000}) = 6.70 \text{ Pa gage}$$

$$6.70 = 0.0728/r \qquad r = 0.01087 \text{ m} \quad \text{or} \quad 10.87 \text{ mm}$$

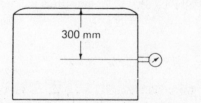

300 mm

D

Fig. 1-22

1.108 Water at 10 °C is poured into a region between concentric cylinders until water appears above the top of the open end (see Fig. 1-23). If the pressure measured by the gage is 3970.80 Pa gage, what is the curvature of the water at the top?

∎
$$p = \sigma/r = p_{gage} - \gamma d = 3970.80 - (9810)(\tfrac{402}{1000}) = 27.18 \text{ Pa gage}$$

$$27.18 = 0.0742/r \qquad r = 0.00273 \text{ m} \quad \text{or} \quad 2.73 \text{ mm}$$

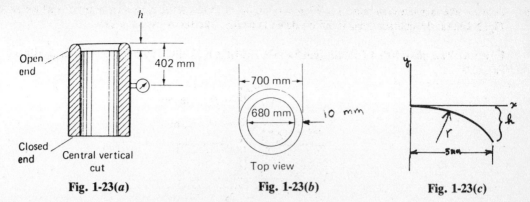

Fig. 1-23(a)   Fig. 1-23(b)   Fig. 1-23(c)

1.109 In structural mechanics, we can determine the rate of twist α of a shaft of any shape by using *Prandtl's soap-film* analogy. A soap film is attached to a sharp edge having the shape of the outside boundary of the shaft cross section (a rectangle here, as shown in Fig. 1-24). Air pressure (Δp) gage is increased under the film so that it forms an elevated curved surface above the boundary. The rate of twist α is then given as

$$\alpha = \frac{M_x \Delta p}{4\sigma GV} \qquad \text{(radians per unit length)}$$

where p = air pressure gage, M_x = torque transmitted by actual shaft, G = shear modulus of actual shaft, and V = volume of air under the soap film and above the cross section formed by the sharp edge. For the case at hand, the Δp used is 0.4 lb/ft^2 gage. Volume V equals the volume of air forced in to raise the membrane and is measured during the experiment to be 0.5 in^3. The angle θ along the long edge of the cross section is measured optically to be 30°. For a torque of 500 ft · lb on a shaft having $G = 10 \times 10^6$ lb/in^2, what angle of twist does this analogy predict?

❚
$$\alpha = \frac{M_x \Delta p}{4\sigma GV}$$

To get σ, consider a unit length of the long side of the shaft cross section away from the ends (see Fig. 1-24c). Considering the membrane for equilibrium in the vertical direction (remembering there are two surfaces on each side) $(-4)[(\sigma)(L)(\cos\theta)] + pA = 0$, $(-4)[(\sigma)(\frac{1}{12})(\cos 30°)] + (0.4)[(0.5)(1)/144] = 0$, $\sigma = 0.00481$ lb/ft;

$$\alpha = \frac{(500)(0.4)}{(4)(0.00481)[(10 \times 10^6)(144)](0.5/1728)} = 0.0249 \text{ rad/ft}$$

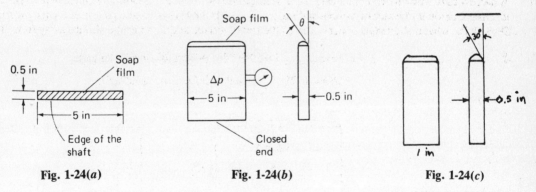

Fig. 1-24(a)   Fig. 1-24(b)   Fig. 1-24(c)

1.110 In using Prandtl's soap-film analogy (see Prob. 1.109), we wish to check the mechanism for measuring the pressure Δp under the soap film. Accordingly, we use a circular cross section (Fig. 1-25) for which we have an accurate theory for determining the rate of twist α. The surface tension for the soap film is 0.1460 N/m and volume V under the film is measured to be 0.001120 m^3. Compute Δp from consideration of the soap film and from solid mechanics using the equation given in Prob. 1.109 and the well-known formula from strength of materials

$$\alpha = \frac{M_x}{GJ}$$

where J, the polar moment of inertia, is $\pi r^4/2$. Compare the results.

▮ From consideration of the film (see Fig. 1-25), $-2\sigma\pi d \cos 45° + (\Delta p)(\pi d^2)/4 = 0$,
$-(2)(0.1460)(\pi)(\frac{200}{1000})(\cos 45°) + (\Delta p)[(\pi)(\frac{200}{1000})^2/4] = 0$, $\Delta p = 4.13$ Pa gage. From strength of materials, equate α's for the equations given in this problem and in Prob. 1.109.

$$\frac{M_x \Delta p}{4\sigma GV} = \frac{M_x}{GJ} \qquad J = \frac{\pi[(\frac{200}{2})/1000]^4}{2} = 0.0001571 \text{ m}^4 \qquad \frac{\Delta p}{(4)(0.1460)(0.001120)} = \frac{1}{0.0001571} \qquad \Delta p = 4.16 \text{ Pa gage}$$

The pressure measurement is quite close to what is expected from theory.

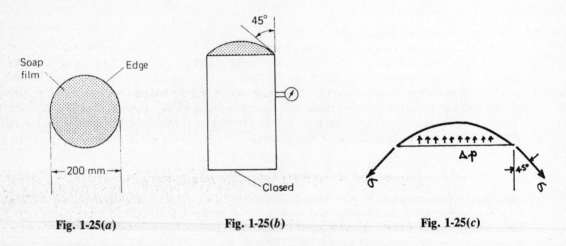

Fig. 1-25(a)   Fig. 1-25(b)   Fig. 1-25(c)

1.111 Find the capillary rise in the tube shown in Fig. 1-26 for a water–air–glass interface ($\theta = 0°$) if the tube radius is 1 mm and the temperature is 20 °C.

▮
$$h = \frac{2\sigma \cos \theta}{\rho g r} = \frac{(2)(0.0728)(\cos 0°)}{(1000)(9.81)(\frac{1}{1000})} = 0.0148 \text{ m} \quad \text{or} \quad 14.8 \text{ mm}$$

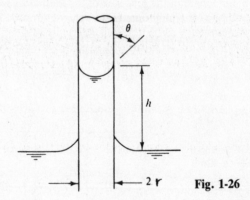

Fig. 1-26

1.112 Find the capillary rise in the tube shown in Fig. 1-26 for a mercury–air–glass interface with $\theta = 130°$ if the tube radius is 1 mm and the temperature is 20 °C.

▮
$$h = \frac{2\sigma \cos \theta}{\rho g r} = \frac{(2)(0.514)(\cos 130°)}{(13\,570)(9.81)(\frac{1}{1000})} = -0.0050 \text{ m} \quad \text{or} \quad -5.0 \text{ mm}$$

1.113 Assuming that a soda-water bubble is equivalent to an air–water interface with $\sigma = 0.005$ lb/ft, what is the pressure difference between the inside and outside of a bubble whose diameter is 0.004 in?

▮
$$p = 2\sigma/r = (2)(0.005)/[(0.004/2)/12] = 60.0 \text{ lb/ft}^2$$

1.114 A small circular jet of mercury 0.1 mm in diameter issues from an opening. What is the pressure difference between the inside and outside of the jet at 20 °C?

▮ See Fig. 1-27. Equating the force due to surface tension ($2\sigma L$) and the force due to pressure (pdL), $2\sigma L = pdL$, $p = 2\sigma/d = (2)(0.514)/(0.1/1000) = 10\,280$ N/m^2.

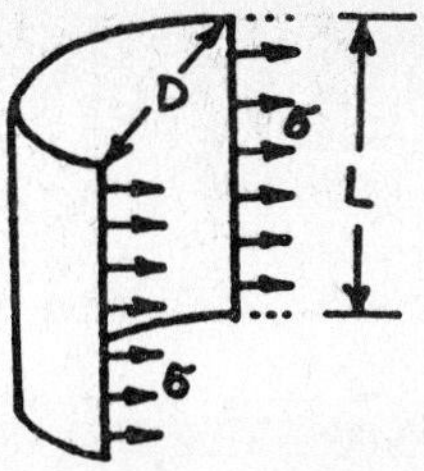

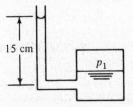

Fig. 1-27

1.115 At 60 °C the surface tension of mercury and water is 0.47 N/m and 0.0662 N/m, respectively. What capillary height changes will occur in these two fluids when they are in contact with air in a glass tube of radius 0.25 mm? Use $\theta = 130°$ for mercury and $0°$ for water and $\gamma = \rho g = 132\,300$ N/m³ for mercury and 9650 N/m³ for water.

▮
$$h = \frac{2\sigma \cos \theta}{\rho g r}$$

For mercury:

$$h = \frac{(2)(0.47)(\cos 130°)}{(132\,300)(0.25/1000)} = -0.0183 \text{ m} \quad \text{or} \quad -18.3 \text{ mm}$$

For water:

$$h = \frac{(2)(0.0662)(\cos 0°)}{(9650)(0.25/1000)} = 0.0549 \text{ m} \quad \text{or} \quad 54.9 \text{ mm}$$

1.116 At 30 °C what diameter glass tube is necessary to keep the capillary-height change of water less than 1 mm?

▮
$$h = \frac{2\sigma \cos \theta}{\rho g r} \qquad \frac{1}{1000} = \frac{(2)(0.0712)(\cos 0°)}{(996)(9.81)(r)}$$
$$r = 0.0146 \text{ m} \quad \text{or} \quad 14.6 \text{ mm} \qquad d = (2)(14.6) = 29.2 \text{ mm (or greater)}$$

1.117 A 1-in-diameter soap bubble has an internal pressure 0.004 lb/in² greater than that of the outside atmosphere. Compute the surface tension of the soap–air interface. Note that a soap bubble has two interfaces with air, an inner and outer surface of nearly the same radius.

▮
$$p = 4\sigma/r \qquad (0.004)(144) = (4)(\sigma)/[(\tfrac{1}{2})/12] \qquad \sigma = 0.0060 \text{ lb/ft}$$

1.118 Neglecting the weight of the wire, what force is required to lift a thin wire ring 4 cm in diameter from a water surface at 20 °C?

▮ $F = \sigma L$. Since there is resistance on the inside and outside of the ring, $F = (2)(\sigma)(\pi d) = (2)(0.0728)[(\pi)(\tfrac{4}{100})] = 0.0183$ N.

1.119 The glass tube in Fig. 1-28 is used to measure pressure p_1 in the water tank. The tube diameter is 1 mm and the water is at 30 °C. After correcting for surface tension, what is the true water height in the tube? What percent error is made if no correction is computed?

▮
$$h = \frac{2\sigma \cos \theta}{\rho g r} = \frac{(2)(0.0712)(\cos 0°)}{(996)(9.81)[(\tfrac{1}{2})/1000]} = 0.029 \text{ m} \quad \text{or} \quad 2.9 \text{ cm}$$

True water height in the tube $= 15 - 2.9 = 12.1$ cm. Neglecting capillary correction causes $2.9/15 = 0.19$, or 19% error.

Fig. 1-28

1.120 An atomizer forms water droplets with a diameter of 5×10^{-5} m. What excess pressure exists in the interior of these droplets for water at 30 °C?

▮ $$p = 2\sigma/r = (2)(0.0712)/[(5 \times 10^{-5})/2] = 5696 \text{ Pa}$$

1.121 What is the pressure within a droplet of water of 0.002 in diameter at 68 °F if the pressure outside the droplet is standard atmospheric pressure of 14.70 psi?

▮ $$p = 2\sigma/r = (2)(0.005)/[(0.002/2)/12] = 120 \text{ lb/ft}^2 \quad \text{or} \quad 0.83 \text{ lb/in}^2$$
$$p_{\text{inside}} = 0.83 + 14.70 = 15.53 \text{ psia}$$

1.122 At low speeds the jet issuing from a faucet approximates a liquid cylinder with an air–water interface. What is the pressure difference between the inside and outside of the jet when the diameter is 2 mm and the temperature is 10 °C? (See Fig. 1-27.)

▮ $$p = \sigma/r = 0.0742/[(\tfrac{2}{2})/1000] = 74.2 \text{ Pa}$$

1.123 Find the angle the surface tension film leaves the glass for a vertical tube immersed in water if the diameter is 0.2 in and the capillary rise is 0.09 in. Use $\sigma = 0.005$ lb/ft.

▮ $$h = \frac{2\sigma \cos \theta}{\rho g r} \qquad \frac{0.09}{12} = \frac{(2)(0.005)(\cos \theta)}{(1.94)(32.2)[(0.2/2)/12]} \qquad \cos \theta = 0.390425 \qquad \theta = 67.0°$$

1.124 Develop a formula for capillary rise between two concentric glass tubes of radii r_o and r_i and contact angle θ.

▮ See Fig. 1-29. Equating the force due to pressure and the force due to surface tension,

$$(h)(\gamma)(\pi r_o^2 - \pi r_i^2) = \sigma(2\pi r_i + 2\pi r_o)(\cos \theta)$$
$$h = \frac{(2)(\sigma)(r_i + r_o)(\cos \theta)}{\gamma(r_o^2 - r_i^2)} = \frac{2\sigma \cos \theta}{\gamma(r_o - r_i)}$$

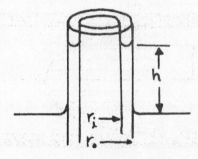

Fig. 1-29

1.125 Distilled water at 10 °C stands in a glass tube of 8.0-mm diameter at a height of 25.0 mm. What is the true static height?

▮ $$h = \frac{2\sigma \cos \theta}{\rho g r} = \frac{(2)(0.0742)(\cos 0°)}{(1000)(9.81)[(8.0/2)/1000]} = 0.0038 \text{ m} \quad \text{or} \quad 3.8 \text{ mm}$$

True static height $= 25.0 - 3.8 = 21.2$ mm.

1.126 Compute the capillary depression of mercury ($\theta = 140°$) to be expected in a 0.10-in-diameter tube at a temperature of 68 °F.

▮ $$h = \frac{2\sigma \cos \theta}{\rho g r} = \frac{(2)(0.0352)(\cos 140°)}{(26.34)(32.2)[(0.10/2)/12]} = -0.01526 \text{ ft} \quad \text{or} \quad -0.18 \text{ in}$$

1.127 Early mountaineers used boiling water to estimate their height. If they reach the top and find that water boils at 80 °C, approximately how high is the mountain?

▮ From Table A-2, water boiling at 80 °C corresponds to a vapor pressure of 47.4 kPa. From Table A-8, this corresponds to a standard atmosphere elevation of approximately 6000 m.

1.128 At approximately what temperature will water boil at an elevation of 10 000 ft?

▮ From Table A-7, the pressure of the standard atmosphere at 10 000-ft elevation is 10.11 psia, or 1456 lb/ft² abs. From Table A-1, the saturation pressure of water is 1456 lb/ft² abs at about 193 °F. Hence, the water will boil at 193 °F; this explains why it takes longer to cook at high altitudes.

1.129 At approximately what temperature will water boil in Mexico City (elevation 7400 ft)?

▮ From Table A-7, the pressure of the standard atmosphere at 7400-ft elevation is 11.23 psia, or 1617 lb/ft² abs. From Table A-1, the saturation pressure of water is 1617 lb/ft² abs at about 198 °F. Hence, the water will boil at 198 °F.

1.130 Water at 100 °F is placed in a beaker within an airtight container. Air is gradually pumped out of the container. What reduction below standard atmospheric pressure of 14.7 psia must be achieved before the water boils?

▮ From Table A-1, $p_v = 135$ lb/ft² abs, or 0.94 psia at 100 °F. Hence, pressure must be reduced by $14.7 - 0.94$, or 13.76 psi.

1.131 At what pressure in millibars will 40 °C water boil?

▮ From Table A-2, $p_v = 7.38$ kN/m² at 40 °C. Hence, water will boil at 7.38 kN/m², or 7380 N/m². Since 1 mbar = 100 N/m², 7380 N/m² = 73.8 mbar.

1.132 At what pressure can cavitation be expected at the inlet of a pump that is handling water at 20 °C?

▮ Cavitation occurs when the internal pressure drops to the vapor pressure. From Table A-2, the vapor pressure of water at 20 °C is 2.34 kPa; hence, cavitation can be expected at that pressure.

1.133 For low-speed (laminar) flow through a circular pipe, as shown in Fig. 1-30, the velocity distribution takes the form $v = (B/\mu)(r_0^2 - r^2)$, where μ is the fluid viscosity. What are the units of the constant B?

▮ Dimensionally, $(L/T) = [\{B\}/(M/LT)](L^2)$, $\{B\} = ML^{-2}T^{-2}$. In SI units, B could be kg/(m² · s²), or Pa/m.

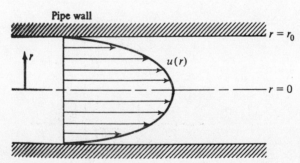

Pipe wall

$r = r_0$

$u(r)$

$r = 0$

Fig. 1-30

1.134 The mean free path L of a gas is defined as the mean distance traveled by molecules between collisions. According to kinetic theory, the mean free path of an ideal gas is given by $L = 1.26(\mu/\rho)(RT)^{-1/2}$, where R is the gas constant and T is the absolute temperature. What are the units of the constant 1.26?

▮ Dimensionally, $L = \{1.26\}[(M/LT)/(M/L^3)][(L^2/T^2D)(D)]^{-1/2}$, $L = \{1.26\}(L)$, $\{1.26\} = 1$. Therefore, the constant 1.26 is dimensionless.

1.135 The Stokes–Oseen formula for the drag force F on a sphere of diameter d in a fluid stream of low velocity v is $F = 3\pi\mu dv + (9\pi/16)(\rho v^2 d^2)$. Is this formula dimensionally consistent?

▮ Dimensionally, $(F) = (1)(M/LT)(L)(L/T) + (1)(M/L^3)(L/T)^2(L)^2 = (ML/T^2) + (ML/T^2) = (F) + (F)$. Therefore, the formula is dimensionally consistent.

1.136 The speed of propagation C of waves traveling at the interface between two fluids is given by $C = (\pi\sigma/\rho_a\lambda)^{1/2}$, where λ is the wavelength and ρ_a is the average density of the two fluids. If the formula is dimensionally consistent, what are the units of σ? What might it represent?

▮ Dimensionally, $(L/T) = [(1)\{\sigma\}/(M/L^3)(L)]^{1/2} = [\{\sigma\}(L^2/M)]^{1/2}$, $\{\sigma\} = M/T^2 = F/L$. In SI units, σ could be N/m. (In this formula, σ is actually the surface tension.)

1.137 Is the following equation dimensionally homogeneous? $a = 2d/t^2 - 2v_0/t$, where a = acceleration, d = distance, v_0 = velocity, and t = time.

∎ $L/T^2 = (L)/(T^2) - (L/T)/(T) = (L/T^2) - (L/T^2)$. Therefore, the equation is homogeneous.

1.138 A popular formula in the hydraulics literature is the Hazen-Williams formula for volume flow rate Q in a pipe of diameter D and pressure gradient dp/dx: $Q = 61.9 D^{2.63}(dp/dx)^{0.54}$. What are the dimensions of the constant 61.9?

∎
$$\frac{L^3}{T} = \{61.9\}(L)^{2.63}\left(\frac{M}{L^2 T^2}\right)^{0.54} \qquad \{61.9\} = L^{1.45} T^{0.08} M^{-0.54}$$

2.1 For the dam shown in Fig. 2-1, find the horizontal pressure acting at the face of the dam at 20-ft depth.

$$p = \gamma h = (62.4)(20) = 1248 \text{ lb/ft}^2$$

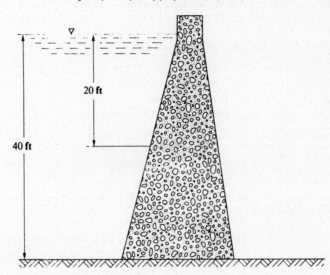

Fig. 2-1. Dam.

2.2 For the vessel containing glycerin under pressure as shown in Fig. 2-2, find the pressure at the bottom of the tank.

$$p = 50 + \gamma h = 50 + (12.34)(2.0) = 74.68 \text{ kN/m}^2 \quad \text{or} \quad 74.68 \text{ kPa}$$

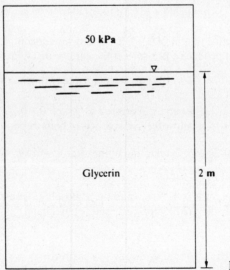

50 kPa

Glycerin   2 m

Fig. 2-2

2.3 If the pressure in a tank is 50 psi, find the equivalent pressure head of (*a*) water, (*b*) mercury, and (*c*) heavy fuel oil with a specific gravity of 0.92.

$$h = p/\gamma$$

(*a*) $h = [(50)(144)]/62.4 = 115.38 \text{ ft}$

(*b*) $h = [(50)(144)]/847.3 = 8.50 \text{ ft}$

(*c*) $h = [(50)(144)]/[(0.92)(62.4)] = 125.42 \text{ ft}$

2.4 A weather report indicates the barometric pressure is 29.75 in of mercury. What is the atmospheric pressure in pounds per square inch?

▌ $$p = \gamma h = [(13.6)(62.4)][(29.75/12)]/144 = 14.61 \text{ lb/in}^2 \quad \text{or} \quad 14.61 \text{ psi}$$

2.5 Find the atmospheric pressure in kilopascals if a mercury barometer reads 742 mm.

▌ $$p = \gamma h = (133.1)(\tfrac{742}{1000}) = 98.8 \text{ kN/m}^2 \quad \text{or} \quad 98.8 \text{ kPa}$$

2.6 A pressure gage at elevation 8.0 m on the side of a tank containing a liquid reads 57.4 kN/m². Another gage at elevation 5.0 m reads 80.0 kN/m². Compute the specific weight and mass density of the fluid.

▌ $$\gamma = \Delta p/\Delta h = (80.0 - 57.4)/(8.0 - 5.0) = 7.53 \text{ kN/m}^3 \quad \text{or} \quad 7530 \text{ N/m}^3$$
$$\rho = \gamma/g = 7530/9.81 = 786 \text{ kg/m}^3$$

2.7 A pressure gage at elevation 20.0 ft on the side of a tank containing a liquid reads 12.8 psi. Another gage at elevation 13.0 ft reads 15.5 psi. Compute the specific weight, mass density, and specific gravity of the liquid.

▌ $$\Delta p = \gamma(\Delta h) \quad (15.5 - 12.8)(144) = (\gamma)(20.0 - 13.0) \quad \gamma = 55.5 \text{ lb/ft}^3$$
$$\rho = \gamma/g = 55.5/32.2 = 1.724 \text{ slugs/ft}^3 \qquad \text{s.g.} = 55.5/62.4 = 0.889$$

2.8 An open tank contains 5.0 m of water covered with 2.0 m of oil ($\gamma = 8.0 \text{ kN/m}^3$). Find the pressure at the interface and at the bottom of the tank.

▌ $$p = \gamma h = (8.0)(2.0) = 16.0 \text{ kN/m}^2 \text{ at the interface}$$
$$p = 16.0 + (9.79)(5.0) = 65.0 \text{ kN/m}^2 \text{ at the tank bottom}$$

2.9 An open tank contains 10.0 ft of water covered with 2.0 ft of oil (s.g. = 0.86). Find the pressure at the interface and at the bottom of the tank.

▌ $$p = \gamma h = [(0.86)(62.4)](2.0)/144 = 0.745 \text{ psi at the interface}$$
$$p = 0.745 + (62.4)(10.0)/144 = 5.08 \text{ psi at the tank bottom}$$

2.10 If air had a constant specific weight of 0.076 lb/ft³ and were incompressible, what would be the height of air surrounding the earth to produce a pressure at the surface of 14.7 psia?

▌ $$h = p/\gamma = (14.7)(144)/0.076 = 27\,900 \text{ ft}$$

2.11 If the specific weight of a sludge can be expressed as $\gamma = 65.0 + 0.2h$, determine the pressure in pounds per square inch at a depth of 15 ft below the surface. γ is in pounds per cubic foot and h in feet below the surface.

▌ $dp = \gamma \, dh = (65.0 + 0.2h) \, dh$. Integrating both sides: $p = 65.0h + 0.1h^2$. For $h = 15$ ft:
$p = (65.0)(15)/144 + (0.1)(15)^2/144 = 6.93$ psi.

2.12 The absolute pressure on a gas is 42.5 psia and the atmospheric pressure is 840 mbar abs. Find the gage pressure in (a) pounds per square inch, (b) kilopascals, and (c) millibars.

▌ (a) $\qquad p_{\text{atm}} = (840)(0.0145) = 12.2 \text{ lb/in}^2 \qquad p_{\text{gage}} = 42.5 - 12.2 = 30.3 \text{ lb/in}^2$

(b) $p_{\text{abs}} = (42.5)(6.894) = 293 \text{ kN/m}^2 \qquad p_{\text{atm}} = (840)(0.100) = 84 \text{ kN/m}^2 \qquad p_{\text{gage}} = 293 - 84 = 209 \text{ kN/m}^2$

(c) $\qquad p_{\text{abs}} = 42.5/0.0145 = 2931 \text{ mbar} \qquad p_{\text{gage}} = 2931 - 840 = 2091 \text{ mbar}$

2.13 If the atmospheric pressure is 920 mbar abs and a gage attached to a tank reads 400 mmHg vacuum, what is the absolute pressure within the tank?

▌ $$p = \gamma h \qquad p_{\text{atm}} = (920)(0.100) = 92.0 \text{ kN/m}^2$$
$$p_{\text{gage}} = [(13.6)(9.79)](\tfrac{400}{1000}) = 53.3 \text{ kN/m}^2 \text{ vacuum} \quad \text{or} \quad -53.3 \text{ kN/m}^2$$
$$p_{\text{abs}} = 92.0 + (-53.3) = 38.7 \text{ kN/m}^2$$

2.14 If the atmospheric pressure is 13.70 psia and a gage attached to a tank reads 8.0 inHg vacuum, what is the absolute pressure within the tank?

$$p = \gamma h \qquad p_{gage} = [(13.6)(62.4)][(8.0/12)/144] = 3.93 \text{ lb/in}^2 \text{ vacuum} \quad \text{or} \quad -3.93 \text{ lb/in}^2$$

$$p_{abs} = 13.70 + (-3.93) = 9.77 \text{ psia}$$

2.15 The closed tank in Fig. 2-3 is at 20 °C. If the pressure at point A is 90 000 Pa abs, what is the absolute pressure at point B? What percent error do you make by neglecting the specific weight of the air?

$p_A + \gamma_{air}h_{AC} - \gamma_{H_2O}h_{DC} - \gamma_{air}h_{DB} = p_B$, $90\,000 + (11.8)(4) - (9790)(4-2) - (11.8)(2) = p_B = 70\,444$ Pa. Neglecting air, $p_B = 90\,000 - (9790)(4-2) = 70\,420$ N; error $= (70\,444 - 70\,420)/70\,444 = 0.00034$, or 0.034%.

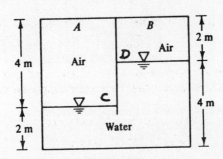

Fig. 2-3

2.16 The system in Fig. 2-4 is at 70 °F. If the pressure at point A is 2000 lb/ft², determine the pressures at points B, C, and D.

$$p_B = 2000 - (62.4)(3-2) = 1938 \text{ lb/ft}^2 \qquad p_D = 2000 + (62.4)(5) = 2312 \text{ lb/ft}^2$$

$$p_C = 2000 + (62.4)(5-2) - (0.075)(4+2) = 2187 \text{ lb/ft}^2$$

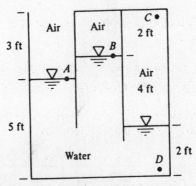

Fig. 2-4

2.17 The system in Fig. 2-5 is at 20 °C. If atmospheric pressure is 101.33 kPa and the absolute pressure at the bottom of the tank is 237 kPa, what is the specific gravity of fluid X?

$$101.33 + (0.89)(9.79)(1) + (9.79)(2) + (\text{s.g.}_X)(9.79)(3) + (13.6)(9.79)(0.5) = 237 \qquad \text{s.g.}_X = 1.39$$

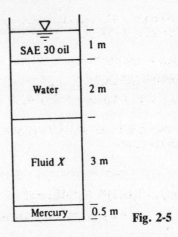

Fig. 2-5

2.18 The container shown in Fig. 2-6 holds water and air as shown. What are the pressures at A, B, C, and D?

▌ $p_A = (62.4)(3+1) = 250 \text{ lb/ft}^2$, $p_B = -(62.4)(1) = -62.4 \text{ lb/ft}^2$. Neglecting air, $p_C = p_B = -62.4 \text{ lb/ft}^2$; $p_D = -62.4 - (62.4)(3+1+1) = -374 \text{ lb/ft}^2$.

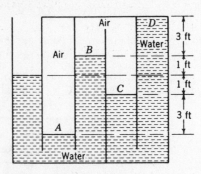

Fig. 2-6

2.19 The tube shown in Fig. 2-7 is filled with oil. Determine the pressure head at A and at B in meters of water.

▌ $(h_{H_2O})(\gamma_{H_2O}) = (h_{oil})(\gamma_{oil}) = (h_{oil})[(s.g._{oil})(\gamma_{H_2O})]$. Therefore, $h_{H_2O} = (h_{oil})(s.g._{oil})$, $h_A = -(2+0.5)(0.85) = -2.125$ m of water, $h_B = (-0.5)(0.85) = -0.425$ m of water.

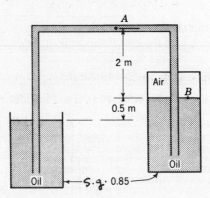

Fig. 2-7

2.20 Calculate the pressures at A, B, C, and D in Fig. 2-8.

▌ $p_A = -(0.3+0.3)(9790) = -5874$ Pa; $p_B = (0.6)(9790) = 5874$ Pa. Neglecting air, $p_C = p_B = 5874$ Pa; $p_D = 5874 + (0.9)(9790)(1+0.6+0.3) = 22\,615$ Pa.

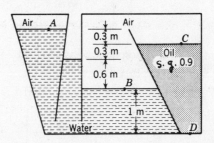

Fig. 2-8

2.21 Express a pressure of 8 psi in (a) inches of mercury, (b) feet of water, (c) feet of acetylene tetrabromide (s.g. = 2.94), and (d) pascals.

▌ (a) $h = p/\gamma = [(8)(144)]/[(13.6)(62.4)] = 1.357$ ft, or 16.28 inHg
(b) $h = [(8)(144)]/62.4 = 18.46$ ft of water
(c) $h = [(8)(144)]/[(2.94)(62.4)] = 6.28$ ft of acetylene tetrabromide
(d) $(8)(4.448)(144)/0.3048^2 = 55\,155$ Pa

2.22 Express an absolute pressure of 4 atm in meters of water gage when the barometer reads 750 mmHg.

▌ $\qquad p_{abs} = (4)(101.3)/9.79 = 41.39$ m of water $\qquad p_{atm} = (\frac{750}{1000})(13.6) = 10.20$ m of water

$$p_{gage} = 41.39 - 10.20 = 31.19 \text{ m of water}$$

2.23 Bourdan gage A inside a pressure tank (Fig. 2-9) reads 12.0 psi. Another Bourdan gage B outside the pressure tank and connected with it reads 20.0 psi, and an aneroid barometer reads 29.00 inHg. What is the absolute pressure at A in inches of mercury?

▌
$$h = p/\gamma \qquad h_A = 12.0/[(13.6)(62.4)/(12)^3] = 24.43 \text{ inHg}$$
$$h_B = 20.0/[(13.6)(62.4)/(12)^3] = 40.72 \text{ inHg} \qquad (h_A)_{\text{abs}} = 29.00 + 24.43 + 40.72 = 94.15 \text{ inHg}$$

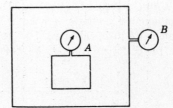

Fig. 2-9

2.24 Determine the heights of columns of water, kerosene (ker), and acetylene tetrabromide ($CHBr_2CHBr_2$) (s.g. $= 2.94$) equivalent to 300 mmHg.

▌
$$(h_{\text{Hg}})(\gamma_{\text{Hg}}) = (H_{\text{H}_2\text{O}})(\gamma_{\text{H}_2\text{O}}) = (h_{\text{ker}})(\gamma_{\text{ker}}) = (h_{\text{CHBr}_2\text{CHBr}_2})(\gamma_{\text{CHBr}_2\text{CHBr}_2})$$
$$\tfrac{300}{1000}[(13.6)(9.79)] = (h_{\text{H}_2\text{O}})(9.79) \qquad h_{\text{H}_2\text{O}} = 4.08 \text{ m}$$
$$\tfrac{300}{1000}[(13.6)(9.79)] = (h_{\text{ker}})[(0.82)(9.79)] \qquad h_{\text{ker}} = 4.98 \text{ m}$$
$$\tfrac{300}{1000}[(13.6)(9.79)] = (h_{\text{CHBr}_2\text{CHBr}_2})[(2.94)(9.79)] \qquad h_{\text{CHBr}_2\text{CHBr}_2} = 1.39 \text{ m}$$

2.25 In Fig. 2-10, for a reading of h of 20.0 in, determine the pressure at A. The liquid has a specific gravity of 1.90.

▌
$$p = \gamma h = [(1.90)(62.4)][20.0/12] = 197.6 \text{ lb/ft}^2 \quad \text{or} \quad 1.37 \text{ psi}$$

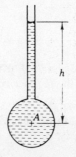

Fig. 2-10

2.26 For the pressure vessel containing glycerin, with piezometer attached, as shown in Fig. 2-11, what is the pressure at point A?

▌
$$p = \gamma h = [(1.26)(62.4)](40.8/12) = 267 \text{ lb/ft}^2$$

Open to atmosphere

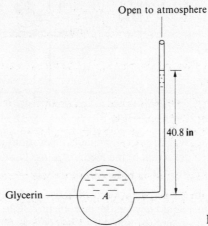

40.8 in

Glycerin —————— A

Fig. 2-11

2.27 For the open tank, with piezometers attached on the side, containing two different immiscible liquids, as shown in Fig. 2-12, find the **(a)** elevation of the liquid surface in piezometer A, **(b)** elevation of the liquid surface in piezometer B, and **(c)** total pressure at the bottom of the tank.

▮ **(a)** Liquid A will simply rise in piezometer A to the same elevation as liquid A in the tank (i.e., to elevation 2 m). **(b)** Liquid B will rise in piezometer B to elevation 0.3 m (as a result of the pressure exerted by liquid B) plus an additional amount as a result of the overlying pressure of liquid A. The overlying pressure can be determined by $p = \gamma h = [(0.72)(9.79)](2 - 0.3) = 11.98$ kN/m^2. The height liquid B will rise in piezometer B as a result of the overlying pressure of liquid A can be determined by $h = p/\gamma = 11.98/[(2.36)(9.79)] = 0.519$ m. Hence, liquid B will rise in piezometer B to an elevation of 0.3 m + 0.519 m, or 0.819 m.
(c) $p_{\text{bottom}} = [(0.72)(9.79)](2 - 0.3) + [(2.36)(9.79)](0.3) = 18.9$ kPa.

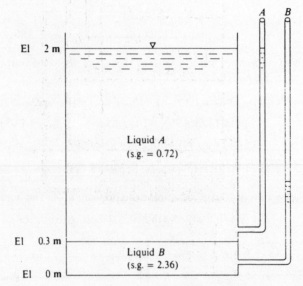

Fig. 2-12

2.28 The air–oil–water system shown in Fig. 2-13 is at 70 °F. Knowing that gage A reads 15.0 lb/in^2 abs and gage B reads 1.25 lb/in^2 less than gage C, compute the **(a)** specific weight of the oil and **(b)** actual reading of gage C.

▮ **(a)** $(15.0)(144) + (0.0750)(2) + (\gamma_{\text{oil}})(1) = p_B$, $p_B + (\gamma_{\text{oil}})(1) + (62.4)(2) = p_C$. Since $p_C - p_B = 1.25$, $(\gamma_{\text{oil}})(1) + (62.4)(2) = (1.25)(144)$, $\gamma_{\text{oil}} = 55.2$ lb/ft^3. **(b)** $(15.0)(144) + (0.0750)(2) + (55.2)(1) = p_B$, $p_B = 2215$ lb/ft^2; $p_C = 2215 + (1.25)(144) = 2395$ lb/ft^2, or 16.63 lb/in^2.

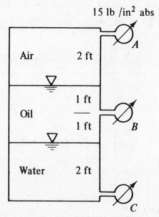

Fig. 2-13

2.29 For a gage reading at A of −2.50 psi, determine the **(a)** elevations of the liquids in the open piezometer columns E, F, and G and **(b)** deflection of the mercury in the U-tube gage in Fig. 2-14. Neglect the weight of the air.

▮ **(a)** The liquid between the air and the water would rise to elevation 49.00 ft in piezometer column E as a result of its weight. The actual liquid level in the piezometer will be lower, however, because of the vacuum in the air above the liquid. The amount the liquid level will be lowered (h in Fig. 2-14) can be determined by

$(-2.50)(144) + [(0.700)(62.4)](h) = 0$, $h = 8.24$ ft. Elevation at $L = 49.00 - 8.24 = 40.76$ ft; $(-2.50)(144) + [(0.700)(62.4)][49.00 - 38.00] = p_M$, $p_M = 120.5$ lb/ft². Hence, pressure head at $M = 120.5/62.4 = 1.93$ ft of water. Elevation at $N = 38.00 + 1.93 = 39.93$ ft; $120.5 + (62.4)(38.00 - 26.00) = p_O$, $p_O = 869.3$ lb/ft². Hence, pressure head at $O = 869.3/[(1.600)(62.4)] = 8.71$ ft (of the liquid with s.g. = 1.600). Elevation at $Q = 26.00 + 8.71 = 34.71$ ft. (b) $869.3 + (62.4)(26.00 - 14.00) - [(13.6)(62.4)](h_1) = 0$, $h_1 = 1.91$ ft.

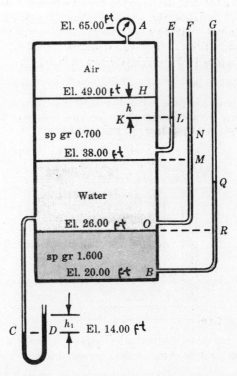

Fig. 2-14

2.30 A vessel containing oil under pressure is shown in Fig. 2-15. Find the elevation of the oil surface in the attached piezometer.

⦚ Elevation of oil surface in piezometer $= 2 + 35/[(0.83)(9.79)] = 6.31$ m

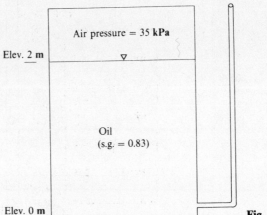

Fig. 2-15

2.31 The fuel gage for a gasoline tank in a car reads proportional to the bottom gage pressure as in Fig. 2-16. If the tank is 30 cm deep and accidentally contains 2 cm of water in addition to the gasoline, how many centimeters of air remain at the top when the gage erroneously reads "full?" Use $\gamma_{gasoline} = 6670$ N/m³ and $\gamma_{air} = 11.8$ N/m³.

⦚ When full of gasoline, $p_{gage} = (6670)(\frac{30}{100}) = 2001$ N/m². With water added, $2001 = (9790)(\frac{2}{100}) + (6670)[(30 - 2)/100 - h] + (11.8)(h)$, $h = 0.0094$ m, or 0.94 cm.

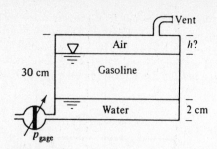

Fig. 2-16

2.32 The hydraulic jack shown in Fig. 2-17 is filled with oil at 56 lb/ft³. Neglecting the weight of the two pistons, what force F on the handle is required to support the 2000-lb weight for this design?

▌ The pressure against the large and the small piston is the same. $p = W/A_{\text{large}} = 2000/[\pi(\frac{3}{12})^2/4] = 40\,740$ lb/ft². Let P be the force from the small piston onto the handle. $P = pA_{\text{small}} = (40\,740)[\pi(\frac{1}{12})^2/4] = 222$ lb. For the handle, $\Sigma M_A = 0 = (15+1)(F) - (1)(222)$, $F = 13.9$ lb.

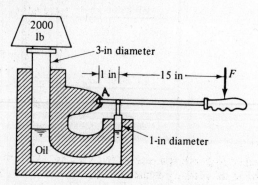

Fig. 2-17

2.33 Figure 2-18 shows a setup with a vessel containing a plunger and a cylinder. What force F is required to balance the weight of the cylinder if the weight of the plunger is negligible?

▌ $$10\,000/500 - [(0.78)(62.4)](15)/144 = F/5 \qquad F = 74.6 \text{ lb}$$

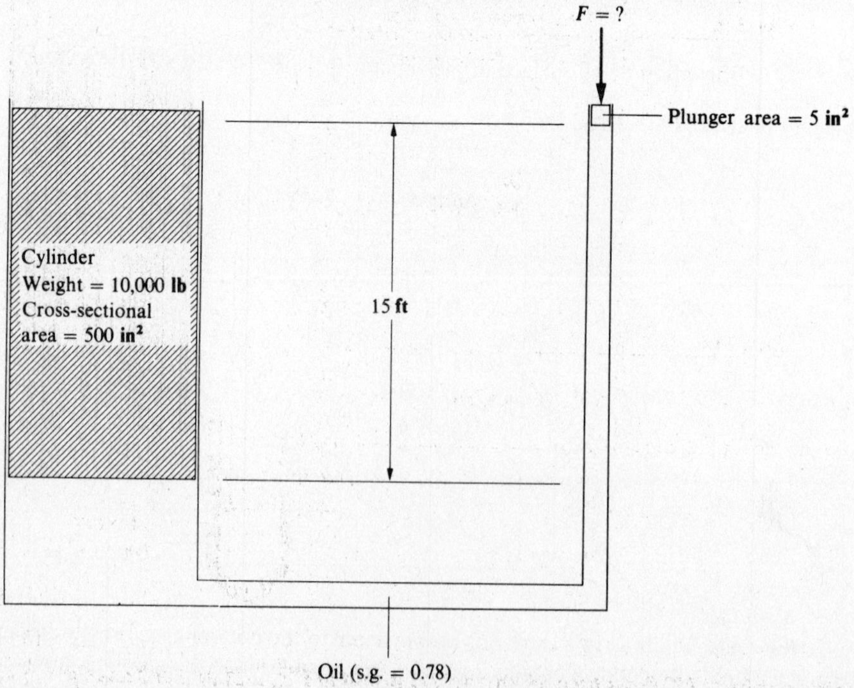

$F = ?$

Plunger area = 5 in²

15 ft

Cylinder
Weight = 10,000 lb
Cross-sectional
area = 500 in²

Oil (s.g. = 0.78)

Fig. 2-18

2.34 For the vertical pipe with manometer attached, as shown in Fig. 2-19, find the pressure in the oil at point A.

▌ $$p_A + [(0.91)(62.4)](7.22) - [(13.6)(62.4)](1.00) = 0 \qquad p_A = 438.7 \text{ lb/ft}^2 \quad \text{or} \quad 3.05 \text{ lb/in}^2$$

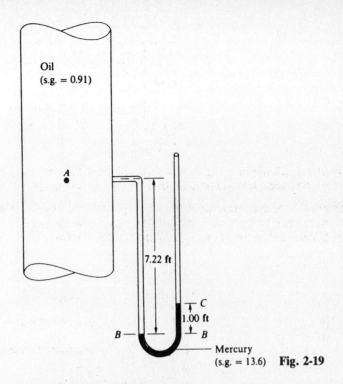

Fig. 2-19

2.35 A monometer is attached to a tank containing three different fluids, as shown in Fig. 2-20. What will be the difference in elevation of the mercury column in the manometer (i.e., y in Fig. 2-20)?

$$30 + [(0.82)(9.79)](5-2) + (9.79)(2-0) + (9.79)(1.00) - [(13.6)(9.79)]y = 0 \qquad y = 0.627 \text{ m}$$

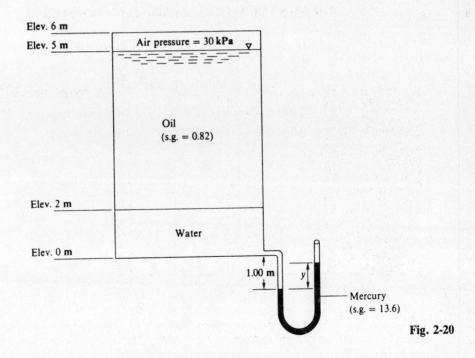

Fig. 2-20

2.36 Oil of specific gravity 0.750 flows through the nozzle shown in Fig. 2-21 and deflects the mercury in the U-tube gage. Determine the value of h if the pressure at A is 20.0 psi.

$$20.0 + [(0.750)(62.4)](2.75 + h)/144 - [(13.6)(62.4)](h)/144 = 0 \qquad h = 3.75 \text{ ft}$$

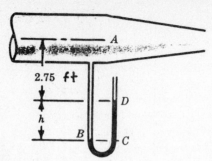

Fig. 2-21

2.37 Determine the reading h in Fig. 2-22 for $p_A = 30$ kPa suction if the liquid is kerosene (s.g. = 0.83).

❚ $$-30 + [(0.83)(9.97)]h = 0 \qquad h = 3.692 \text{ m}$$

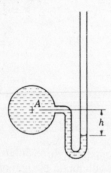

Fig. 2-22

2.38 In Fig. 2-22, for $h = 8$ in and barometer reading 29.0 inHg, with water the liquid, find p_A in feet of water absolute.

❚ $$p_A + \tfrac{8}{12} = (13.6)(29.0/12) \qquad p_A = 32.2 \text{ ft of water absolute}$$

2.39 In Fig. 2-23, s.g.$_1$ = 0.86, s.g.$_2$ = 1.0, $h_2 = 90$ mm, and $h_1 = 150$ mm. Find p_A in millimeters of mercury gage. If the barometer reading is 720 mmHg, what is p_A in meters of water absolute?

❚ $$p_A + (0.86)(90) - (1.0)(150) = 0$$

$$p_A = 72.6 \text{ mm of water gage} = 72.6/13.6 = 5.34 \text{ mmHg gage}$$

$$= 72.6/1000 + (13.6)(\tfrac{720}{1000}) = 9.865 \text{ m of water absolute}$$

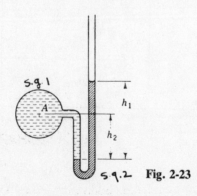

Fig. 2-23

2.40 At 20 °C, gage A in Fig. 2-24 reads 300 kPa abs. What is the height h of water? What should gage B read?

❚ $$300 - [(13.6)(9.79)](\tfrac{80}{100}) - 9.79h = 180 \qquad h = 1.377 \text{ m}$$

$$p_B - (9.79)(\tfrac{80}{100} + 1.377) = 180 \qquad p_B = 201 \text{ kPa}$$

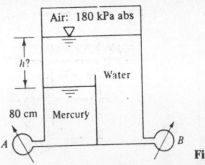

Fig. 2-24

2.41 The U-tube shown in Fig. 2-25a has a 1-cm diameter and contains mercury as shown. If 10.0 cm³ of water is poured into the right-hand leg, what will the free-surface height in each leg be after the sloshing has died down?

❚ After the water is poured, the orientation of the liquids will be as shown in Fig. 2-25b; $h = 10.0/[\pi(1)^2/4] = 12.73$ cm, $(13.6)(20 - L) = 13.6L + 12.73$, $L = 9.53$ cm. Left leg height above bottom of U-tube $= 20 - 9.53 = 10.47$ cm; right leg height above bottom of U-tube $= 9.53 + 12.73 = 22.26$ cm.

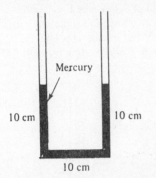

Fig. 2-25(a)

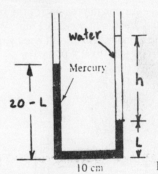

Fig. 2-25(b)

2.42 The deepest known point in the ocean is 11 034 m in the Mariana Trench in the Pacific. Assuming sea water to have a constant specific weight of 10.05 kN/m³, what is the absolute pressure at this point, in atmospheres?

❚ $$p = 1 + (10.05)(11\,034)/101.3 = 1096 \text{ atm}$$

2.43 In Fig. 2-26, fluid 2 is carbon tetrachloride and fluid 1 is glycerin. If p_{atm} is 101.0 kPa, determine the absolute pressure at point A.

❚ $$101.0 + (15.57)(\tfrac{32}{100}) - (12.34)(\tfrac{10}{100}) = p_A \qquad p_A = 104.7 \text{ kPa}$$

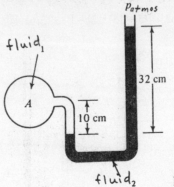

Fig. 2-26

2.44 In Fig. 2-27a, originally there is a 4-in manometer reading. Atmospheric pressure is 14.7 psia. If the absolute pressure at A is doubled, what then would be the manometer reading?

❚ $p_A + (62.4)(3) - [(13.6)(62.4)](\tfrac{4}{12}) = (14.7)(144)$, $p_A = 2212$ lb/ft². If p_A is doubled to 4424 kPa, the mercury level will fall some amount, x, on the left side of the manometer and will rise by that amount on the right side of the manometer (see Fig. 2-27b). Hence, $4424 + (62.4)(3 + x/12) - [(13.6)(62.4)][(4 + 2x)/12] = (14.7)(144)$, $x = 16.2$ in. New manometer reading $= 4 + (2)(16.2) = 36.4$ in.

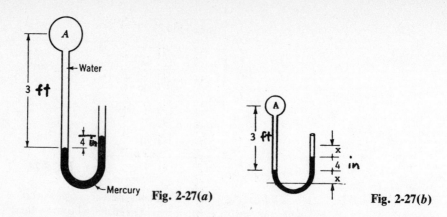

Fig. 2-27(a)

Fig. 2-27(b)

2.45 In Fig. 2-28a, A contains water, and the manometer fluid has a specific gravity of 2.94. When the left meniscus is at zero on the scale, $p_A = 100$ mm of water. Find the reading of the right meniscus for $p_A = 8$ kPa with no adjustment of the U-tube or scale.

▮ First, determine the reading of the right meniscus for $p_A = 100$ mm of water (see Fig. 2-28b):
$100 + 600 - 2.94h = 0$, $h = 238.1$ mm. When $p_A = 8$ kPa, the mercury level will fall some amount, d, on the left side of the manometer and will rise by that amount on the right side of the manometer (see Fig. 2-28b). Hence, $8/9.79 + (600 + d)/1000 - [(238.1 + 2d)/1000](2.94) = 0$, $d = 147.0$ mm. Scale reading for $p_A = 8$ kPa is $238.1 + 147.0$, or 385.1 mm.

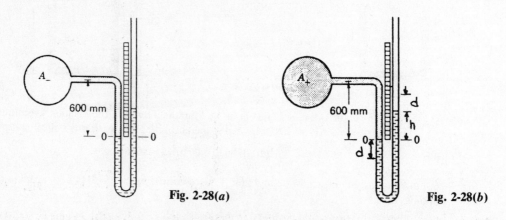

Fig. 2-28(a)　　　　Fig. 2-28(b)

2.46 A manometer is attached to a conduit, as shown in Fig. 2-29. Calculate the pressure at point A.

▮ $p_A + (62.4)[(5 + 15)/12] - [(13.6)(62.4)](\frac{15}{12}) = 0$　　$p_A = 957$ lb/ft²

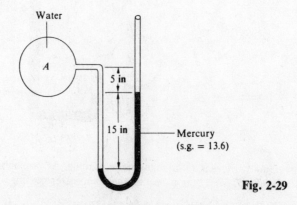

Fig. 2-29

2.47 A manometer is attached to a pipe containing oil, as shown in Fig. 2-30. Calculate the pressure at point A.

▮ $p_A + [(0.85)(9.79)](0.2) - (9.79)(1.5) = 0$　　$p_A = 13.02$ kN/m²

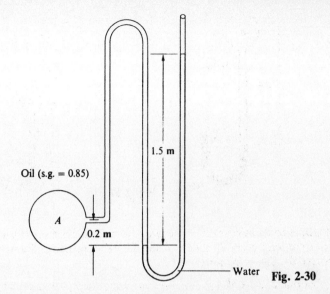

1.5 m

Oil (s.g. = 0.85)

A

0.2 m

Water **Fig. 2-30**

2.48 A monometer is attached to a pipe to measure pressure, as shown in Fig. 2-31. Calculate the pressure at point A.

$$p_A + (62.4)(\tfrac{18}{12}) - [(13.6)(62.4)](\tfrac{6}{12}) = 0 \qquad p_A = 331\ \text{lb/ft}^2$$

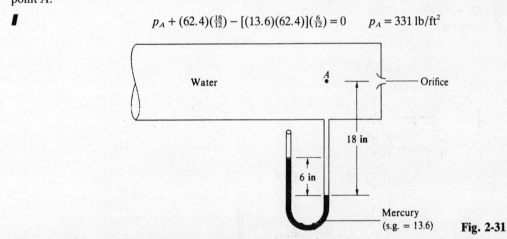

Water

A

Orifice

18 in

6 in

Mercury
(s.g. = 13.6) **Fig. 2-31**

2.49 A glass U-tube open to the atmosphere at both ends is shown in Fig. 2-32. if the U-tube contains oil and water as shown, determine the specific gravity of the oil.

$$[(\text{s.g.}_{\text{oil}})(9.79)](0.35) - (9.79)(0.30) = 0 \qquad \text{s.g.}_{\text{oil}} = 0.86$$

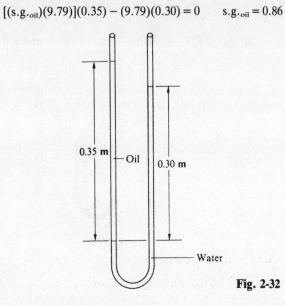

0.35 m

Oil

0.30 m

Water

Fig. 2-32

2.50 A differential manometer is shown in Fig. 2-33. Calculate the pressure difference between points A and B.

$\blacksquare$ $\qquad p_A + [(0.92)(62.4)][(x + 12)/12] - [(13.6)(62.4)](\frac{12}{12}) - [(0.92)(62.4)][(x + 24)/12] = p_B$

$$p_A - p_B = 906 \text{ lb/ft}^2$$

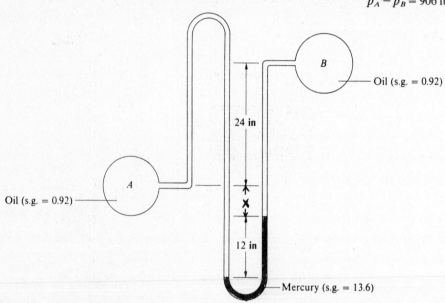

Fig. 2-33

2.51 A differential manometer is attached to a pipe, as shown in Fig. 2.34. Calculate the pressure difference between points A and B.

$\blacksquare$ $\qquad p_A + [(0.91)(62.4)](y/12) - [(13.6)(62.4)](\frac{4}{12}) - [(0.91)(62.4)][(y - 4)/12] = p_B$

$$p_A - p_B = 264 \text{ lb/ft}^2$$

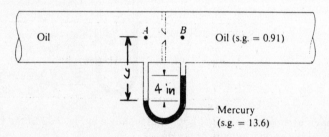

Fig. 2-34

2.52 A differential manometer is attached to a pipe, as shown in Fig. 2-35. Calculate the pressure difference between points A and B.

$\blacksquare$ $\qquad p_A - [(0.91)(62.4)](y/12) - [(13.6)(62.4)](\frac{4}{12}) + [(0.91)(62.4)][(y + 4)/12] = p_B$

$$p_A - p_B = 264 \text{ lb/ft}^2$$

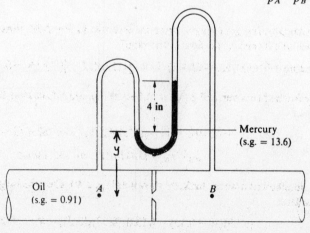

Fig. 2-35

2.53 For the configuration shown in Fig. 2-36, calculate the weight of the piston if the gage pressure reading is 70.0 kPa.

▌ Let W = weight of the piston. $W/[(\pi)(1)^2/4] - [(0.86)(9.79)](1) = 70.0$, $W = 61.6$ kN.

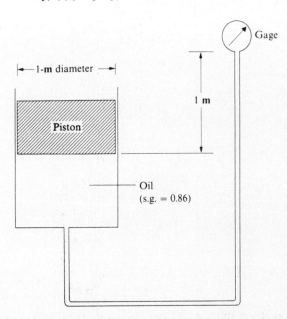

Fig. 2-36

2.54 A manometer is attached to a horizontal oil pipe, as shown in Fig. 2-37. If the pressure at point A is 10 psi, find the distance between the two mercury surfaces in the manometer (i.e., determine the distance y in Fig. 2-37).

▌ $(10)(144) + [(0.90)(62.4)](3 + y) - [(13.6)(62.4)]y = 0$ $y = 2.03$ ft   or   24.4 in

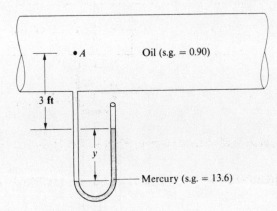

Fig. 2-37

2.55 A vertical pipe with attached gage and manometer is shown in Fig. 2-38. What will be the gage reading in pounds per square inch if there is no flow in the pipe?

▌ Gage reading + $[(0.85)(62.4)](2 + 8)/144 - [(13.6)(62.4)](\frac{18}{12})/144 = 0$ Gage reading = 5.16 psi

2.56 A monometer is attached to a vertical pipe, as shown in Fig. 2-39. Calculate the pressure difference between points A and B.

▌
$$p_A - (62.4)(5 + 1) - [(13.6)(62.4)](2) + (62.4)(2 + 1) = p_B$$
$$p_A - p_B = 1884 \text{ lb/ft}^2 \quad \text{or} \quad 13.1 \text{ lb/in}^2$$

2.57 A manometer is attached to a water tank, as shown in Fig. 2-40. Find the height of the free water surface above the bottom of the tank.

▌ $(9.79)(H - 0.15) - [(13.6)(9.79)](0.20) = 0$ $H = 2.87$ m

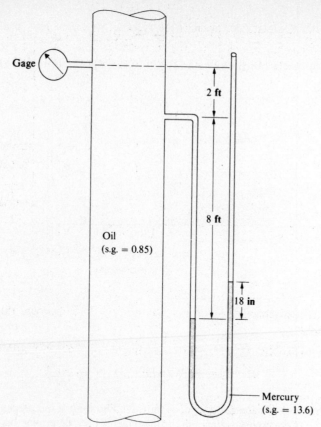

Gage

2 ft

8 ft

Oil
(s.g. = 0.85)

18 in

Mercury
(s.g. = 13.6)

Fig. 2-38

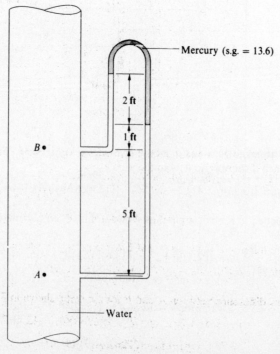

Mercury (s.g. = 13.6)

2 ft

1 ft

B •

5 ft

A •

Water

Fig. 2-39

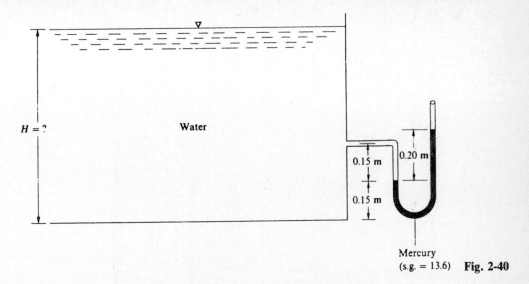

Water

$H = ?$

0.15 m

0.20 m

0.15 m

Mercury
(s.g. = 13.6)   **Fig. 2-40**

2.58 A differential manometer is attached to two tanks, as shown in Fig. 2-41. Calculate the pressure difference between chambers A and B.

▌ $$p_A + [(0.89)(9.79)](1.1) + [(13.6)(9.79)](0.3) - [(1.59)(9.79)](0.8) = p_B$$

$$p_A - p_B = -37.1 \text{ kN/m}^2 \quad \text{(i.e., } p_B > p_A)$$

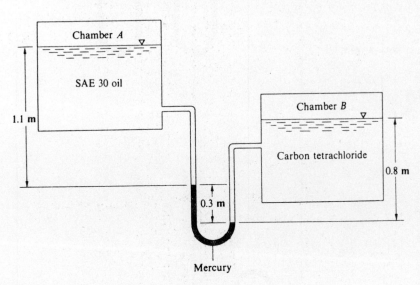

Chamber A

SAE 30 oil

Chamber B

1.1 m

Carbon tetrachloride

0.8 m

0.3 m

Mercury   **Fig. 2-41**

2.59 Calculate the pressure difference between A and B for the setup shown in Fig. 2-42.

▌ $$p_A + (62.4)(66.6/12) - [(13.6)(62.4)](40.3/12) + (62.4)(22.2/12)$$
$$- [(13.6)(62.4)](30.0/12) - (62.4)(10.0/12) = p_B$$

$$p_A - p_B = 4562 \text{ lb/ft}^2 \quad \text{or} \quad 31.7 \text{ lb/in}^2$$

2.60 Calculate the pressure difference between A and B for the setup shown in Fig. 2-43.

▌ $$p_A - (9.79)x - [(0.8)(9.79)](0.70) + (9.79)(x - 0.80) = p_B \qquad p_A - p_B = 13.3 \text{ kN/m}^2$$

2.61 Calculate the pressure difference between A and B for the setup shown in Fig. 2-44.

▌ $$p_A + (62.4)(x + 4) - [(13.6)(62.4)](4) + (62.4)(7 - x) = p_B$$

$$p_A - p_B = 2708 \text{ lb/ft}^2 \quad \text{or} \quad 18.8 \text{ lb/in}^2$$

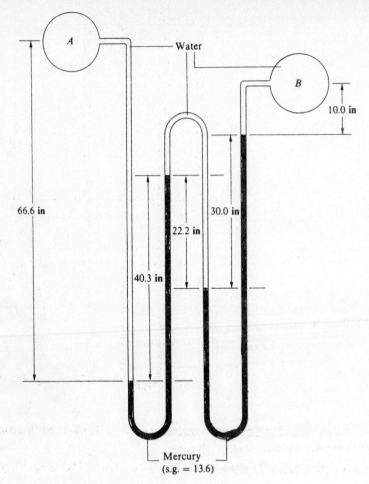

Fig. 2-42

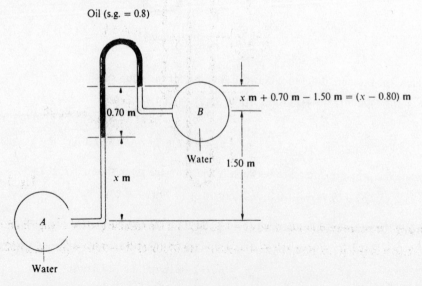

Fig. 2-43

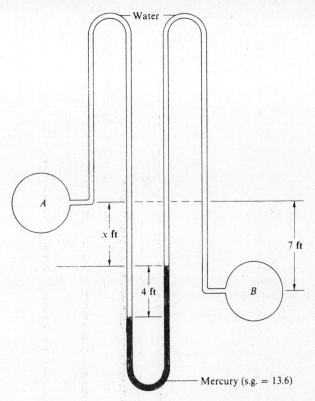

Water

Mercury (s.g. = 13.6)

Fig. 2-44

A

x ft

7 ft

4 ft

B

2.62 Vessels A and B in Fig. 2-45 contain water under pressures of 40.0 psi and 20.0 psi, respectively. What is the deflection of the mercury in the differential gage?

▌ $(40.0)(144) + (62.4)(x + h) - [(13.6)(62.4)]h + 62.4y = (20.0)(144)$. Since $x + y = 16.00 - 10.00$, or 6.00 ft, $h = 4.14$ ft.

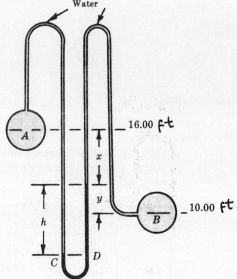

Water

A

16.00 ft

x

y

10.00 ft

B

h

C   D

Fig. 2-45

2.63 For a gage pressure at A in Fig. 2-46 of -1.58 psi, find the specific gravity of gage liquid B.

▌ $(-1.58)(144) + [(1.60)(62.4)](10.50 - 9.00) - (0.0750)(11.25 - 9.00) + [(\text{s.g.}_{\text{liq.} B})(62.4)](11.25 - 10.00) = 0$

$\text{s.g.}_{\text{liq.} B} = 1.00$

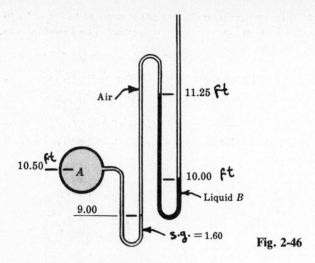

Fig. 2-46

2.64 In Fig. 2-47, liquid A weights 53.5 lb/ft³ and liquid B weighs 78.8 lb/ft³. Manometer liquid M is mercury. If the pressure at B is 30 psi, find the pressure at A.

$$p_A - (53.5)(6.5 + 1.3) + [(13.6)(62.4)](1.3) + (78.8)(6.5 + 10.0) = (30)(144)$$

$$p_A = 2334 \text{ lb/ft}^2 \quad \text{or} \quad 16.2 \text{ lb/in}^2$$

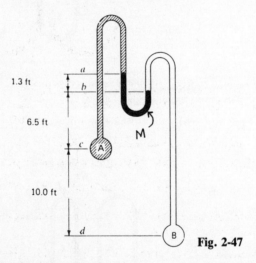

Fig. 2-47

2.65 What would be the manometer reading in Fig. 2-47 if $p_B - p_A$ is 150 kN/m²? (Use the unit weights given in Prob. 2.64.)

■ 150 kN/m² = (150)(1000/4.448)(0.3048)², or 3133 lb/ft². With $p_B - p_A$ = 150 kN/m², or 3133 lb/ft², the mercury level will rise some amount, x, on the left side of the manometer and will fall by that amount on the right side of the manometer (see Fig. 2.48). Hence, $p_A - (53.5)(6.5 + 1.3 + x) + [(13.6)(62.4)](1.3 + 2x) + (78.8)(6.5 + 10.0 - x) = p_B$, $1644x + 1986 = p_B - p_A = 3133$, $x = 0.70$ ft; manometer reading = $1.3 + (2)(0.70) = 2.70$ ft.

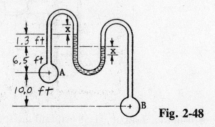

Fig. 2-48

2.66 In Fig. 2-49, A and B are at the same elevation. Water is contained in A and rises in the tube to a level 76 in above A. Glycerin is contained in B. The inverted U-tube is filled with air at 20 psi and 70 °F. Atmospheric pressure is 14.7 psia. Determine the difference in pressure between A and B if y is 14 in. Express the answer in pounds per square inch. What is the absolute pressure in B in inches of mercury and in feet of glycerin?

$$p_A - (62.4)(\tfrac{76}{12}) = (20)(144) \qquad p_A = 3275.2 \text{ lb/ft}^2$$

$$p_B - [(1.26)(62.4)][(76-14)/12] = (20)(144) \qquad p_B = 3286.2 \text{ lb/ft}^2$$

$$p_A - p_B = 3275.2 - 3286.4 = -11.2 \text{ lb/ft}^2 \quad \text{or} \quad -0.078 \text{ lb/in}^2$$

$$(p_{abs})_B = (3286.2/144 + 14.7)/[(13.6)(62.4)/(12)^3] = 76.4 \text{ inHg}$$

$$(p_{abs})_B = (3286.2/144 + 14.7)/[(1.26)(62.4)/(12)^3] = 824.6 \text{ in} \quad \text{or} \quad 68.7 \text{ ft of glycerin}$$

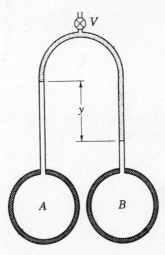

Fig. 2-49

2.67 Gas confined in a rigid container exerts a pressure of 140 kN/m² when its temperature is 5 °C. What pressure would the gas exert if the temperature were raised to 60 °C? Barometric pressure remains constant at 711 mmHg.

$$p_{atm} = [(13.6)(9.79)](\tfrac{711}{1000}) = 94.7 \text{ kN/m}^2 \qquad p_{abs} = 94.7 + 140 = 234.7 \text{ kN/m}^2$$

$$p_1 V_1/T_1 = p_2 V_2/T_2 \qquad (234.7)(V)/(273+5) = (p_2)(V)/(273+60) \qquad [V \text{ (volume) is constant}]$$

$$p_2 = 281.1 \text{ kN/m}^2 \text{ (absolute)} = 281.1 - 94.7 = 186.4 \text{ kN/m}^2 \text{ (gage)}$$

2.68 In Fig. 2-50, atmospheric pressure is 14.7 psia, the gage reading at A is 5.0 psi, and the vapor pressure of the alcohol is 1.7 psia. Compute x and y.

❚ Working in terms of absolute pressure heads, $[(5.0 + 14.7)(144)]/[(0.90)(62.4)] - x = (1.7)(144)/[(0.90)(62.4)]$, $x = 46.15$ ft; $[(5.0 + 14.7)(144)]/[(0.90)(62.4)] + (y + 4) - (4)(13.6/0.90) = 0$, $y = 5.93$ ft.

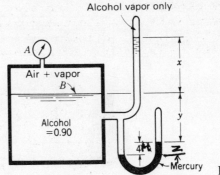

Alcohol vapor only

Air + vapor
B

Alcohol
$= 0.90$

Mercury   **Fig. 2-50**

2.69 In Fig. 2-50, assume the following: atmospheric pressure = 850 mbar abs, vapor pressure of the alcohol = 160 mbar abs, $x = 2.80$ m, $y = 2.00$ m. Compute the reading on the pressure gage (p_A) and on the manometer (z).

▮ Working in terms of absolute pressure heads, $[(p_A)_{\text{gage}} + 850](0.100)/[(0.90)(9.79)] - 2.80 = (160)(0.100)/[(0.90)(9.79)]$, $(p_A)_{\text{gage}} = -443$ mbar, or -44.3 kN/m^2; $[(-443 + 850)(0.100)]/[(0.90)(9.79)] + (2.00 + z) - (z)(13.6/0.90) = 0$, $z = 0.469$ m.

2.70 At a certain point, A, the pressure in a pipeline containing gas ($\gamma = 0.05$ lb/ft^3) is 4.50 in of water. The gas is not flowing. What is the pressure in inches of water at another point, B, in the line where the elevation is 500 ft greater than the first point?

▮ Assume $\gamma_{\text{gas}} = 0.05$ lb/ft^3 is constant. Note that change in pressure in the atmosphere must be considered. Assume $\gamma_{\text{air}} = 0.076$ lb/ft^3 is constant.

$$(p_A/\gamma)_{\text{abs}} = (p_A/\gamma)_{\text{atm}} + 4.50/12 \text{ ft of water} \tag{1}$$

$$(p_B/\gamma)_{\text{abs}} = (p_B/\gamma)_{\text{atm}} + x/12 \text{ ft of water} \tag{2}$$

Subtracting Eq. (2) from Eq. (1),

$$(p_A/\gamma)_{\text{abs}} - (p_B/\gamma)_{\text{abs}} = (p_A/\gamma)_{\text{atm}} - (p_B/\gamma)_{\text{atm}} + 4.50/12 - x/12 \tag{3}$$

$$(p_A/\gamma)_{\text{atm}} - (p_B/\gamma)_{\text{atm}} = 500 \text{ ft of air} = (500)(0.076/62.4) = 0.609 \text{ ft of water}$$

$$(p_A/\gamma)_{\text{abs}} - (p_B/\gamma)_{\text{abs}} = 500 \text{ ft of gas} = (500)(0.05/62.4) = 0.401 \text{ ft of water}$$

Substituting these relationships into Eq. (3), $0.401 = 0.609 + 4.50/12 - x/12$, $x = 7.00$ in of water.

2.71 Determine the pressure difference between points A and B in Fig. 2-51.

▮ $p_A + [(0.88)(9.79)](\frac{20}{100}) - [(13.6)(9.79)](\frac{8}{100}) - [(0.82)(9.79)][(40-8)/100]$
$$+ (9.79)[(40-14)/100)] - (0.0118)(\tfrac{9}{100}) = p_B$$

$$p_A - p_B = 8.95 \text{ kN/m}^2$$

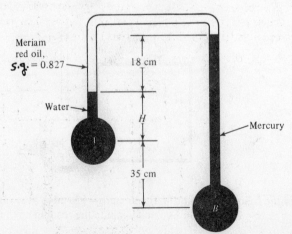

Fig. 2-51

2.72 In Fig. 2-52, if $p_B - p_A = 99.0$ kPa, what must the height H be?

▮
$$p_A - (9.79)(H/100) - [(0.827)(9.79)](\tfrac{18}{100}) + [(13.6)(9.79)][(35 + H + 18)/100] = p_B$$

$$1.234H + 69.11 = p_B - p_A = 99.0 \qquad H = 24.2 \text{ cm}$$

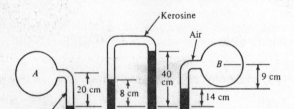

Fig. 2-52

2.73 For Fig. 2-53, if fluid 1 is water and fluid 2 is mercury, and $z_A = 0$ and $z_1 = -10$ cm, what is level z_2 for which $p_A = p_{atm}$?

▌ $\quad\quad 0 + (9.79)[0-(-10)]/100 - [(13.6)(9.79)][z_2-(-10)]/100 = 0 \quad\quad z_2 = -9.26$ cm

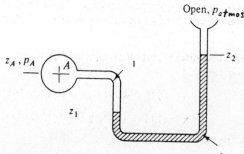

Fig. 2-53

2.74 The inclined manometer in Fig. 2-54a contains Meriam red manometer oil (s.g. = 0.827). Assume the reservoir is very large. If the inclined arm is fitted with graduations 1 in apart, what should the angle θ be if each graduation corresponds to 1 lb/ft^2 gage pressure for p_A?

▌ $\Delta L = 1$ in corresponds to 1 lb/ft^2 scale (see Fig. 2-54b). $\Delta z = \Delta L \sin\theta$, $\Delta p = \gamma\Delta z = [(0.827)(62.4)](\frac{1}{12}\sin\theta) = 1$ lb/ft^2, $\theta = 13.45°$.

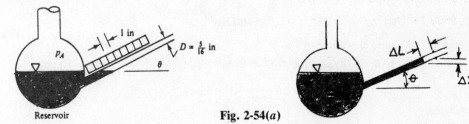

Fig. 2-54(a)   **Fig. 2-54(b)**

2.75 The system in Fig. 2-55 is at 20 °C. Compute the absolute pressure at point A.

▌ $\quad p_A + [(0.85)(62.4)](\frac{6}{12}) - [(13.6)(62.4)](\frac{10}{12}) + (62.4)(\frac{5}{12}) = (14.7)(144) \quad\quad p_A = 2771$ lb/ft^2 abs

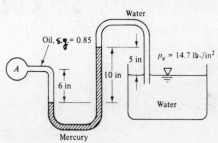

Fig. 2-55

2.76 Very small pressure differences $p_A - p_B$ can be measured accurately by the two-fluid differential manometer shown in Fig. 2-56. Density ρ_2 is only slightly larger than the upper fluid ρ_1. Derive an expression for the proportionality between h and $p_A - p_B$ if the reservoirs are very large.

▌ $\quad p_A + \rho_1 g h_1 - \rho_2 g h - \rho_1 g(h_1 - h) = p_B$, $\quad p_A - p_B = (\rho_2 - \rho_1)gh$. If $(\rho_2 - \rho_1)$ is small, h will be large (sensitive).

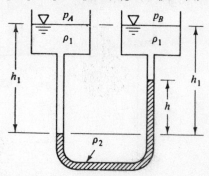

Fig. 2-56

2.77 Water flows downward in a pipe at 45°, as shown in Fig. 2-57. The pressure drop $p_1 - p_2$ is partly due to gravity and partly due to friction. The mercury manometer reads a 6-in height difference. What is the total pressure drop $p_1 - p_2$? What is the pressure drop due to friction only between 1 and 2? Does the manometer reading correspond only to friction drop?

$$p_1 + (62.4)(5 \cos 45° + x/12 + \tfrac{6}{12}) - [(13.6)(62.4)](\tfrac{6}{12}) - (62.4)(x/12) = p_2$$

$$p_1 - p_2 = 172.5 \text{ lb/ft}^2 \quad \text{(total pressure drop)}$$

Pressure drop due to friction only $= [(13.6)(62.4) - 62.4](\tfrac{6}{12}) = 393.1 \text{ lb/ft}^2$

Manometer reads only the friction loss.

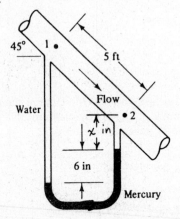

Fig. 2-57

2.78 Determine the gage pressure at point A in Fig. 2-58.

$$p_A - (9.79)(\tfrac{45}{100}) + (0.0118)(\tfrac{30}{100}) + [(13.6)(9.79)](\tfrac{15}{100}) - [(0.85)(9.79)](\tfrac{40}{100}) = 0 \qquad p_A = -12.24 \text{ kPa}$$

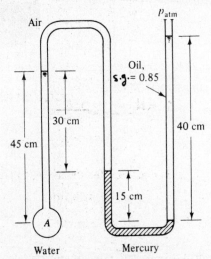

Fig. 2-58

2.79 In Fig. 2-59, what will be level h of the oil in the right-hand tube? Both tubes are open to the atmosphere.

$$0 + (9.79)[(10 + 20)/100] - [(0.82)(9.79)][(20 + h)/100] = 0 \qquad h = 16.59 \text{ cm}$$

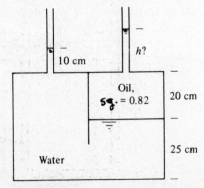

Fig. 2-59

2.80 The inclined manometer of Fig. 2-60a reads zero when A and B are the same pressure. The reservoir diameter is 2.0 in and that of the inclined tube is $\frac{1}{4}$ in. For $\theta = 30°$ and gage fluid with s.g. = 0.832, find $p_A - p_B$ in pounds per square inch as a function of gage reading R in feet.

$$p_A = \gamma(\Delta h + \Delta y) + p_B \quad \text{(see Fig. 2-60b)} \quad p_A - p_B = \gamma(\Delta h + \Delta y)$$

From Fig. 2-60b, $(A_A)(\Delta y) = (A_B)(R)$ or $\Delta y = A_B R / A_A$, $\Delta h = R \sin \theta$, $p_A - p_B = \gamma(R \sin \theta + A_B R / A_A) = \gamma R(\sin \theta + A_B / A_A)$, $A_B / A_A = [\pi(\frac{1}{4})^2/4]/[\pi(2)^2/4] = \frac{1}{64}$; $p_A - p_B = [(0.832)(62.4)](R)(\sin 30° + \frac{1}{64})/144 = 0.1859R$ [R in feet gives $(p_A - p_B)$ in pounds per square inch].

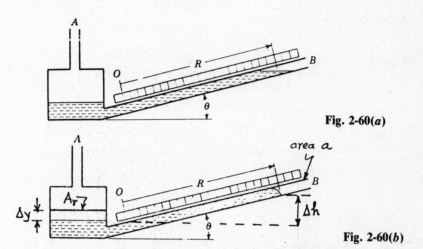

Fig. 2-60(a)

Fig. 2-60(b)

2.81 Determine the weight W that can be sustained by the force acting on the piston of Fig. 2-61.

$$p_1 = p_2 = F_1/A_1 = F_2/A_2 \quad 1/[\pi(40)^2/4] = W/[\pi(240)^2/4] \quad W = 36.0 \text{ kN}$$

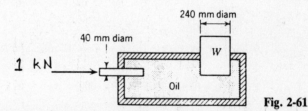

Fig. 2-61

2.82 Neglecting the container's weight in Fig. 2-62, find the force tending to lift the circular top CD.

$$p_{CD} - [(0.8)(62.4)](3) = 0 \quad p_{CD} = 149.8 \text{ lb/ft}^2 \quad F = pA = (149.8)[\pi(3)^2/4] = 1059 \text{ lb}$$

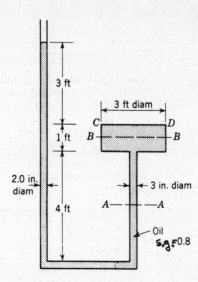

Fig. 2-62

2.83 Find the force of oil on the top surface CD of Fig. 2-62 if the liquid level in the open pipe is reduced by 1 m.

▌ $p_{CD} - [(0.8)(62.4)](3 - 3.281) = 0$ $p_{CD} = -14.03$ lb/ft^2

$$F = pA = (-14.03)[\pi(3)^2/4] = -99.2 \text{ lb} \quad \text{(i.e., negative force acting inward)}$$

2.84 A barrel 2 ft in diameter filled with water has a vertical pipe of 0.60-in diameter attached to the top. Neglecting compressibility, how many pounds of water must be added to the pipe to exert a force of 1000 lb on the top of the barrel?

▌ $\quad p = F/A = 1000/[\pi(2)^2/4] = 318.3 \text{ lb/ft}^2$ $h = p/\gamma = 318.3/62.4 = 5.10 \text{ ft}$

$$W_{H_2O} = (5.10)[\pi(0.60/12)^2/4](62.4) = 0.625 \text{ lb}$$

2.85 In Fig. 2-63, the liquid at A and B is water and the manometer liquid is oil with s.g. = 0.80, $h_1 = 300$ mm, $h_2 = 200$ mm, and $h_3 = 600$ mm. (a) Determine $p_A - p_B$. (b) If $p_B = 50$ kPa and the barometer reading is 730 mmHg, find the absolute pressure at A in meters of water.

▌ (a) $p_A - (9.79)(\frac{300}{1000}) - [(0.80)(9.79)](\frac{200}{1000}) + (9.79)(\frac{600}{1000}) = p_B$ $p_A - p_B = -1.37$ kPa

(b) $p_A - (9.79)(\frac{300}{1000}) - [(0.80)(9.79)](\frac{200}{1000}) + (9.79)(\frac{600}{1000}) = 50$

$$p_A = 48.63 \text{ kPa} \quad \text{(gage)} = 48.63/9.79 + \tfrac{730}{1000}(13.6) = 14.90 \text{ m water} \quad \text{(absolute)}$$

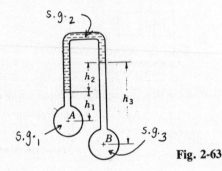

Fig. 2-63

2.86 In Fig. 2-63, s.g.$_1$ = 1.0, s.g.$_2$ = 0.95, s.g.$_3$ = 1.0, $h_1 = h_2 = 280$ mm, and $h_3 = 1$ m. Compute $p_A - p_B$ in millimeters of water.

▌ $p_A - (1.0)(280) - (0.95)(280) + (1.0)(1000) = p_B$ $p_A - p_B = -454$ mm of water

2.87 In Fig. 2-63, s.g.$_1$ = 1.0, s.g.$_2$ = 0.95, s.g.$_3$ = 1.0, $h_1 = 280$ mm, $h_3 = 1$ m, and $p_A - p_B = -350$ mm of water. Find the gage difference (h_2).

▌ $p_A - (1.0)(280) - (0.95)(h_2) + (1.0)(1000) = p_B$ $p_A - p_B = -350 = -720 + (0.95)(h_2)$ $h_2 = 389$ mm

2.88 The Empire State Building is 1250 ft high. What is the pressure difference in pounds per square inch of a water column of the same height?

▌ $p = \gamma h = (62.4)(1250)/144 = 542$ psi

2.89 What is the pressure at a point 10 m below the free surface in a fluid that has a variable density in kilograms per cubic meter given by $\rho = 450 + ah$, in which $a = 12$ kg/m^4 and h is the distance in meters measured from the free surface.

▌ $dp = \gamma \, dh = \rho g \, dh = (g)(450 + ah) \, dh$. Integrating both sides: $p = (g)(450h + ah^2/2)$. For $h = 10$ m:
$p = (9.81)[(450)(10) + (12)(10)^2/2] = 50\,030$ kN/m^2.

2.90 If atmospheric pressure is 29.92 inHg, what will be the height of water in a water barometer if the temperature of the water is (a) 40 °F, (b) 90 °F, and (c) 140 °F?

▌ $p = \gamma h = [(13.6)(62.4)](29.92/12) = 2116 \text{ lb/ft}^2 \quad \text{or} \quad 14.69 \text{ lb/in}^2$

(a) At 40 °F, $\gamma = 62.4$ lb/ft^3 and $p_{vapor} = 18.5/144$, or 0.128 lb/in^2, $h_{H_2O} = (14.69 - 0.128)(144)/62.4 = 33.60$ ft.
(b) At 90 °F, $\gamma = 62.1$ lb/ft^3 and $p_{vapor} = \frac{101}{144}$, or 0.70 lb/in^2, $h_{H_2O} = (14.69 - 0.70)(144)/62.1 = 32.44$ ft.
(c) At 140 °F, $\gamma = 61.4$ lb/ft^3 and $p_{vapor} = \frac{416}{144}$, or 2.89 lb/in^2, $h_{H_2O} = (14.69 - 2.89)(144)/61.4 = 27.67$ ft.

2.91 The tire of an airplane is inflated at sea level to 60.0 psi. Here, atmospheric pressure is 14.7 psia and the temperature is 59 °F. Assuming the tire does not expand, what is the pressure within the tire at elevation 30 000 ft, where atmospheric pressure is 4.4 psia and the temperature is −48 °F? Express the answer in psi and psia.

❙ Let subscript 1 indicate sea level and subscript 2 indicate elevation 30 000 ft.

$$(p_1)_{abs} = 14.7 + 60.0 = 74.7 \text{ psia} \qquad p_1 V_1 / T_1 = p_2 V_2 / T_2$$

$$(74.7)(V)/(460 + 59) = (p_2)(V)/[460 + (-48)] \qquad (V \text{ is constant})$$

$$p_2 = 59.3 \text{ psia} \qquad (p_2)_{gage} = 59.3 - 4.4 = 54.9 \text{ psi}$$

2.92 Find the difference in pressure between tanks A and B in Fig. 2-64 if $d_1 = 300$ mm, $d_2 = 150$ mm, $d_3 = 460$ mm, and $d_4 = 200$ mm.

❙
$$p_A + (9.79)(\tfrac{300}{1000}) - [(13.6)(9.79)](\tfrac{460}{1000} + \tfrac{200}{1000}\sin 45°) = p_B \qquad p_A - p_B = 77.14 \text{ kN/m}^2$$

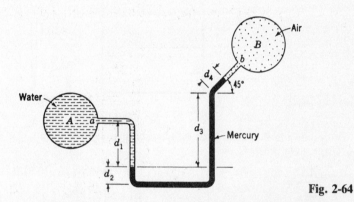

Fig. 2-64

2.93 A cylindrical tank contains water at a height of 50 mm, as shown in Fig. 2-65. Inside is a smaller open cylindrical tank containing kerosene (s.g. = 0.8) at height h. The following pressures are known from the indicated gages: $p_B = 13.80$ kPa gage and $p_C = 13.82$ kPa gage. What are gage pressure p_A and height h of kerosene? Assume that the kerosene is prevented from moving to the top of the tank.

❙
$$p_A + (9.79)(\tfrac{50}{1000}) = 13.82 \qquad p_A = 13.33 \text{ kPa}$$

$$13.33 + (9.79)[(50 - h)/1000] + [(0.8)(9.79)](h/1000) = 13.80 \qquad h = 10.0 \text{ mm}$$

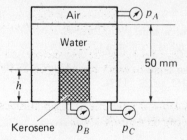

Fig. 2-65

2.94 An open tube is attached to a tank, as shown in Fig. 2-66. If the water rises to a height of 900 mm in the tube, what are the pressures p_A and p_B of the air above the water? Neglect capillary effects in the tube.

❙
$$p_A - (9.79)[(900 - 400 - 200)/1000] = 0 \qquad p_A = 2.94 \text{ kPa}$$

$$p_B - (9.79)[(900 - 400)/1000] = 0 \qquad p_B = 4.90 \text{ kPa}$$

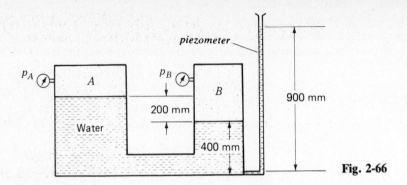

Fig. 2-66

2.95 For the setup shown in Fig. 2-67, what is the pressure p_A if the specific gravity of the oil is 0.8?

$$p_A + [(0.8)(9.79)](3) + (9.79)(4.6-3) - [(13.6)(9.79)](\tfrac{300}{1000}) = 0 \qquad p_A = 0.783 \text{ kN/m}^2$$

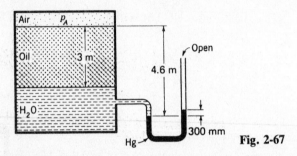

Fig. 2-67

2.96 For the setup shown in Fig. 2-68, what is the absolute pressure in drum A at position a? Assume an atmospheric pressure of 101.3 kPa.

$$101.3 + (9.79)[(600-300)/1000] - [(13.6)(9.79)](\tfrac{150}{1000}) + [(0.8)(9.79)][(150+100)/1000] = p_A$$

$$p_A = 86.22 \text{ kPa}$$

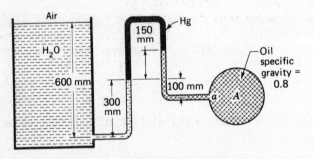

Fig. 2-68

2.97 A force of 445 N is exerted on lever AB, as shown in Fig. 2-69. End B is connected to a piston which fits into a cylinder having a diameter of 50 mm. What force F_D must be exerted on the larger piston to prevent it from moving in its cylinder, which has a 250-mm diameter?

▮ Let F_C = force exerted on smaller piston at C: $F_C = (445)(\tfrac{200}{100}) = 890$ N. $F_C/A_C = F_D/A_D$, $(890)/[\pi(\tfrac{50}{1000})^2/4] = F_D/[\pi(\tfrac{250}{1000})^2/4]$, $F_D = 22\,250$ N, or 22.25 kN.

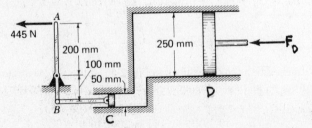

Fig. 2-69

CHAPTER 3

Forces on Submerged Plane Areas

3.1 If a triangle of height d and base b is vertical and submerged in liquid with its vertex at the liquid surface (see Fig. 3-1), derive an expression for the depth to its center of pressure.

$$h_{cp} = h_{cg} + \frac{I_{cg}}{h_{cg}A} = \frac{2d}{3} + \frac{bd^3/36}{(2d/3)(bd/2)} = \frac{3d}{4}$$

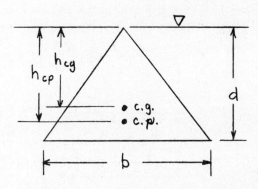

Fig. 3-1

3.2 If a triangle of height d and base b is vertical and submerged in liquid with its vertex a distance a below the liquid surface (see Fig. 3-2), derive an expression for the depth to its center of pressure.

$$h_{cp} = h_{cg} + \frac{I_{cg}}{h_{cg}A} = \left(a + \frac{2d}{3}\right) + \frac{bd^3/36}{(a + 2d/3)(bd/2)} = \left(a + \frac{2d}{3}\right) + \frac{d^2}{18(a + 2d/3)}$$

$$= \frac{18(a^2 + 4ad/3 + 4d^2/9) + d^2}{18(a + 2d/3)} = \frac{6a^2 + 8ad + 3d^2}{6(a + 2d/3)}$$

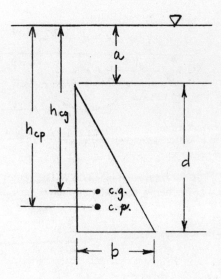

Fig. 3-2

3.3 If a triangle of height d and base b is vertical and submerged in liquid with its base at the liquid surface (see Fig. 3-3), derive an expression for the depth to its center of pressure.

$$h_{cp} = h_{cg} + \frac{I_{cg}}{h_{cg}A} = \frac{d}{3} + \frac{bd^3/36}{(d/3)(bd/2)} = \frac{d}{3} + \frac{d}{6} = \frac{d}{2}$$

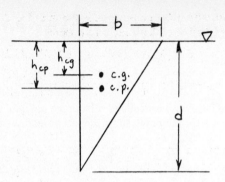

Fig. 3-3

3.4 A circular area of diameter d is vertical and submerged in a liquid. Its upper edge is coincident with the liquid surface (see Fig. 3-4). Derive an expression for the depth to its center of pressure.

▌
$$h_{cp} = h_{cg} + \frac{I_{cg}}{h_{cg}A} = \frac{d}{2} + \frac{\pi d^4/64}{(d/2)(\pi d^2/4)} = \frac{d}{2} + \frac{d}{8} = \frac{5d}{8}$$

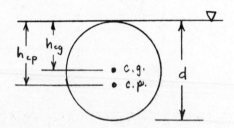

Fig. 3-4

3.5 A vertical semicircular area of diameter d and radius r is submerged and has its diameter in a liquid surface (see Fig. 3-5). Derive an expression for the depth to its center of pressure.

▌
$$h_{cp} = h_{cg} + \frac{I_{cg}}{h_{cg}A} \qquad h_{cg} = \frac{4r}{3\pi} \qquad I_x = \frac{1}{2}\left(\frac{\pi d^4}{64}\right) = \frac{1}{2}\left[\frac{\pi(2r)^4}{64}\right] = \frac{\pi r^4}{8}$$

$$I_{cg} = \frac{\pi r^4}{8} - \left(\frac{\pi r^2}{2}\right)\left(\frac{4r}{3\pi}\right)^2 = \left(\frac{\pi}{8} - \frac{8}{9\pi}\right)(r^4) \qquad h_{cp} = \frac{4r}{3\pi} + \frac{[\pi/8 - 8/(9\pi)](r^4)}{[4r/(3\pi)][(\pi r^2/2)]} = 0.589r$$

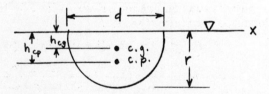

Fig. 3-5

3.6 A dam 20 m long retains 7 m of water, as shown in Fig. 3-6. Find the total resultant force acting on the dam and the location of the center of pressure.

▌ $F = \gamma hA = (9.79)[(0 + 7)/2][(20)(7/\sin 60°)] = 5339$ kN. The center of pressure is located at two-thirds the total water depth of 7 m, or 4.667 m below the water surface (i.e., $h_{cp} = 4.667$ m in Fig. 3-6).

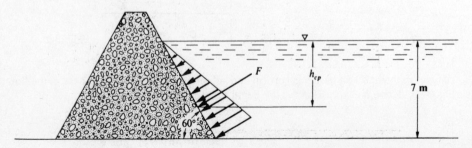

Fig. 3-6

3.7 A vertical, rectangular gate with water on one side is shown in Fig. 3-7. Determine the total resultant force acting on the gate and the location of the center of pressure.

$$F = \gamma h_{cg} A = (9.79)(3 + 1.2/2)[(2)(1.2)] = 84.59 \text{ kN}$$

$$h_{cp} = h_{cg} + \frac{I_{cg}}{h_{cg}A} = \left(3 + \frac{1.2}{2}\right) + \frac{(2)(1.2)^3/12}{(3 + 1.2/2)[(2)(1.2)]} = 3.633 \text{ m}$$

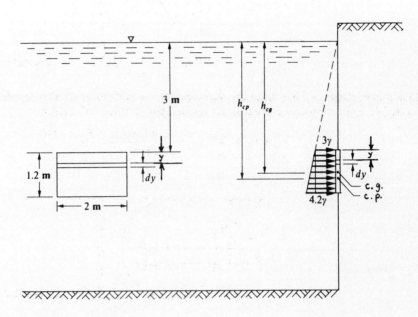

Fig. 3-7

3.8 Solve Prob. 3.7 by the integration method.

$$F = \int \gamma h \, dA = \int_0^{1.2} (9.79)(3 + y)(2 \, dy) = (19.58)\left[3y + \frac{y^2}{2}\right]_0^{1.2} = 84.59 \text{ kN}$$

$$h_{cp} = \frac{\int \gamma h^2 \, dA}{F} = \frac{\int_0^{1.2} (9.79)(3 + y)^2(2 \, dy)}{84.59} = \frac{(19.58)[9y + 3y^2 + y^3/3]_0^{1.2}}{84.59} = 3.633 \text{ m}$$

3.9 A vertical, triangular gate with water on one side is shown in Fig. 3-8. Determine the total resultant force acting on the gate and the location of the center of pressure.

$$F = \gamma h_{cg} A = (62.4)(6 + 3/3)[(2)(3)/2] = 1310 \text{ lb}$$

$$h_{cp} = h_{cg} + \frac{I_{cg}}{h_{cg}A} = \left(6 + \frac{3}{3}\right) + \frac{(2)(3)^3/36}{(6 + 3/3)[(2)(3)/2]} = 7.07 \text{ ft}$$

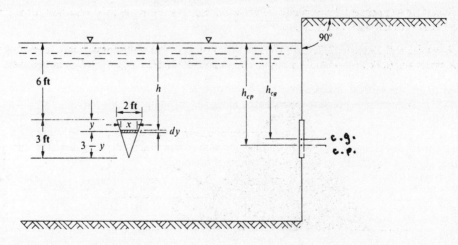

Fig. 3-8

3.10 Solve Prob. 3.9 by the integration method.

$$F = \gamma h_{cg}A = [(0.82)(9.79)][4 + (1 + 1.2/2)(\sin 40°)][(0.8)(1.2)] = 38.75 \text{ kN}$$

$$F = \int_0^3 (62.4)(6 + y)[(2 - 2y/3) \, dy] = \int_0^3 (62.4)(12 - 2y - 2y^2/3) \, dy = (62.4)[12y - y^2 - 2y^3/9]_0^3 = 1310 \text{ lb}$$

$$h_{cp} = \frac{\int \gamma h^2 \, dA}{F} = \frac{\int_0^3 (62.4)(6 + y)^2(2 - 2y/3) \, dy}{1310} = \frac{\int_0^3 (62.4)(72 - 6y^2 - 2y^3/3) \, dy}{1310}$$

$$= \frac{(62.4)[72y - 2y^3 - y^4/6]_0^3}{1310} = 7.07 \text{ ft}$$

3.11 An inclined, rectangular gate with water on one side is shown in Fig. 3-9. Determine the total resultant force acting on the gate and the location of the center of pressure.

$$F = \gamma h_{cg}A = (62.4)[8 + \tfrac{1}{2}(4 \cos 60°)][(4)(5)] = 11\,230 \text{ lb}$$

$$z_{cp} = z_{cg} + \frac{I_{cg}}{z_{cg}A} = \left(\frac{8}{\cos 60°} + \frac{4}{2}\right) + \frac{(5)(4)^3/12}{(8/\cos 60° + \tfrac{4}{2})[(4)(5)]} = 18.07 \text{ ft}$$

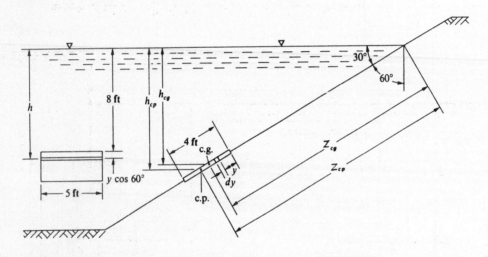

Fig. 3-9

3.12 Solve Prob. 3.11 by the integration method.

$$F = \int \gamma h \, dA = \int_0^4 (62.4)(8 + y \cos 60°)(5 \, dy) = (312)\left[8y + \frac{y^2}{4}\right]_0^4 = 11\,230 \text{ lb}$$

$$h_{cp} = \frac{\int \gamma h^2 \, dA}{F} = \frac{\int_0^4 (62.4)(8 + y \cos 60°)^2(5 \, dy)}{11\,230} = \frac{\int_0^4 (312)(64 + 8y + y^2/4) \, dy}{11\,230}$$

$$= \frac{(312)[64y + 4y^2 + y^3/12]_0^4}{11\,230} = 9.04 \text{ ft}$$

Note: h_{cp} is the vertical distance from the water surface to the center of pressure. The distance from the water surface to the center of pressure as measured along the inclination of the gate (z_{cp}) would be 9.04/cos 60°, or 18.08 ft.

3.13 An inclined, circular gate with water on one side is shown in Fig. 3-10. Determine the total resultant force acting on the gate and the location of the center of pressure.

$$F = \gamma h_{cg}A = (9.79)[1.5 + \tfrac{1}{2}(1.0 \sin 60°)][\pi(1.0)^2/4] = 14.86 \text{ kN}$$

$$z_{cp} = z_{cg} + \frac{I_{cg}}{z_{cg}A} = \left[\frac{1.5}{\sin 60°} + \frac{1}{2}(1.0)\right] + \frac{\pi(1.0)^4/64}{[1.5/\sin 60° + \tfrac{1}{2}(1.0)][\pi(1.0)^2/4]} = 2.260 \text{ m}$$

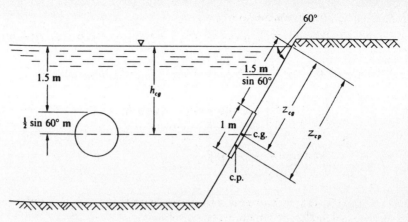

Fig. 3-10

3.14 A vertical, triangular gate with water on one side is shown in Fig. 3-11. Determine the total resultant force acting on the gate and the location of the center of pressure.

$$F = \gamma h_{cg}A = (9.79)[3 + \tfrac{2}{3}(1)][(1.2)(1)/2] = 21.54 \text{ kN}$$

$$h_{cp} = h_{cg} + \frac{I_{cg}}{h_{cg}A} = [3 + (\tfrac{2}{3})(1)] + \frac{(1.2)(1)^3/36}{[3 + \tfrac{2}{3}(1)][(1.2)(1)/2]} = 3.68 \text{ m}$$

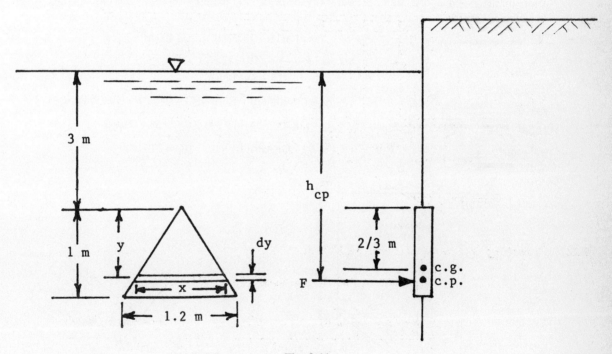

Fig. 3-11

3.15 Solve Prob. 3.14 by the integration method.

$F = \int \gamma h \, dA$. From Fig. 3-11, $y/x = 1/1.2$. Therefore, $x = 1.2y$.

$$F = \int_0^1 (9.79)(3 + y)(1.2y \, dy) = \int_0^1 (11.75)(3y + y^2) \, dy = (11.75)\left[\frac{3y^2}{2} + \frac{y^3}{3}\right]_0^1 = 21.54 \text{ kN}$$

$$h_{cp} = \frac{\int \gamma h^2 \, dA}{F} = \frac{\int_0^1 (9.79)(3 + y)^2(1.2y \, dy)}{21.54} = \frac{\int_0^1 (11.75)(9y + 6y^2 + y^3) \, dy}{21.54}$$

$$= \frac{(11.75)[9y^2/2 + 2y^3 + y^4/4]_0^3}{21.54} = 3.68 \text{ m}$$

3.16 A tank containing water is shown in Fig. 3-12. Calculate the total resultant force acting on side $ABCD$ of the container and the location of the center of pressure.

$$F = \gamma hA = (62.4)[(0 + 6)/2][(20)(6)] = 22\,500 \text{ lb}$$

$$h_{cp} = (\tfrac{2}{3})(6) = 4.00 \text{ ft} \quad \text{(vertically below the water surface)}$$

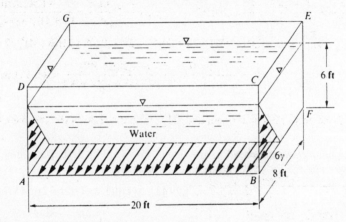

Fig. 3-12

3.17 The gate in Fig. 3-13 is 5 ft wide, is hinged at point B, and rests against a smooth wall at A. Compute (a) the force on the gate due to seawater pressure, (b) the horizontal force P exerted by the wall at point A, and (c) the reaction at hinge B.

(a) $$F = \gamma h_{cg}A = (64)(15 - \tfrac{6}{2})[(5)(10)] = 38\,400 \text{ lb}$$

(b) $$y_{cp} = \frac{-I_{xx}\sin\theta}{h_{cg}A} = \frac{-[(5)(10)^3/12](\tfrac{6}{10})}{(15 - \tfrac{6}{2})[(5)(10)]} = -0.417 \text{ ft}$$

$$\sum M_B = 0 \quad (P)(6) - (38\,400)(10 - 5 - 0.417) = 0 \quad P = 29\,330 \text{ lb}$$

(c) $$\sum F_x = 0 \quad B_x + (38\,400)(\tfrac{6}{10}) - 29\,300 = 0 \quad B_x = 6260 \text{ lb}$$

$$\sum F_y = 0 \quad B_y - (38\,400)(\tfrac{8}{10}) = 0 \quad B_y = 30\,720 \text{ lb}$$

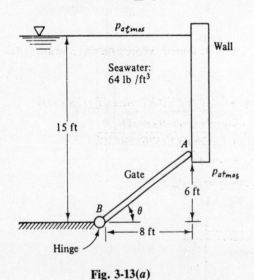

Fig. 3-13(a)　　　　　　　**Fig. 3-13(b)**

3.18 Repeat Prob. 3.17, but instead let the hinge be at point A and let point B rest against a smooth bottom.

(a) From Prob. 3.17, $F = 38\,400$ lb. (b) From Prob. 3.17, $y_{cp} = -0.417$ ft; $\sum M_A = 0$;
$(B_y)(8) - (38\,400)(5 + 0.417) = 0$, $B_y = 26\,000$ lb.

(c) $$\sum F_x = 0 \quad (38\,400)(\tfrac{6}{10}) - A_x = 0 \quad A_x = 23\,040 \text{ lb}$$

$$\sum F_y = 0 \quad A_y - (38\,400)(\tfrac{8}{10}) + 26\,000 = 0 \quad A_y = 4720 \text{ lb}$$

3.19 A tank of oil has a right-triangular panel near the bottom as shown in Fig. 3-14. Calculate the total resultant force on the triangular panel and the location of the center of pressure on the panel.

▌ $F = \gamma h_{cg} A = \rho g h_{cg} A = (800)(9.81)(5+4)[\frac{1}{2}(8+4)(2+4)] = 2\,543\,000\text{ N}$ or 2543 kN

$$I_{xx} = \frac{bh^3}{36} = \frac{(4+2)(8+4)^3}{36} = 288\text{ m}^4 \qquad y_{cp} = \frac{-I_{xx}\sin\theta}{h_{cg}A} = \frac{-(288)(\sin 30°)}{(5+4)[\frac{1}{2}(8+4)(2+4)]} = -0.444\text{ m}$$

$$I_{xy} = b(b-2s)(h)^2/72 = (4+2)[(4+2)-(2)(4+2)](8+4)^2/72 = -72.0\text{ m}^4$$

$$x_{cp} = \frac{-I_{xy}\sin\theta}{h_{cg}A} = \frac{-(-72)(\sin 30°)}{(5+4)[\frac{1}{2}(8+4)(2+4)]} = +0.111\text{ m}$$

(The resultant force acts at 0.444 m down and 0.111 m to the right of the centroid.)

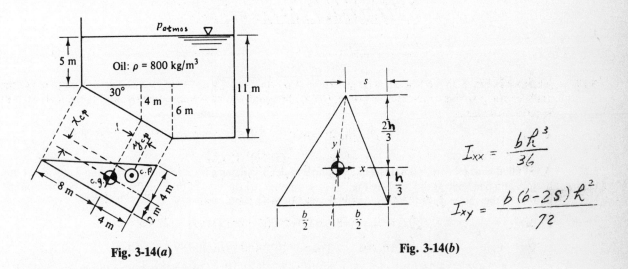

Fig. 3-14(a)　　　　　　　　　　　**Fig. 3-14(b)**

3.20 Gate AB in Fig. 3-15 is 1.2 m long and 0.8 m wide. Calculate force F on the gate and its center of pressure position X.

▌ $F = \gamma h_{cg} A = [(0.82)(9.79)][4 + (1 + 1.2/2)(\sin 40°)][(0.8)(1.2)] = 38.75\text{ kN}$

$$y_{cp} = \frac{-I_{xx}\sin\theta}{h_{cg}A} = \frac{-[(0.8)(1.2)^3/12](\sin 40°)}{[4 + (1 + 1.2/2)(\sin 40°)][(0.8)(1.2)]}$$

$$= -0.015\text{ m from the centroid}$$

$$X = 1.2/2 + 0.015 = 0.615\text{ m from point } A$$

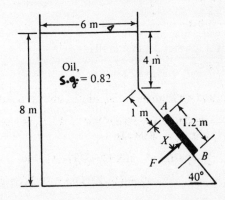

Fig. 3-15

3.21 A vertical submerged gate 8 ft wide and 10 ft high is hinged at the top and held closed by water, as shown in Fig. 3-16. What horizontal force applied at the bottom of the gate is required to open the gate?

$$F = \gamma h_{cg} A = (62.4)(5+5)[(8)(10)] = 49\ 920\ \text{lb}$$

$$h_{cp} = h_{cg} + \frac{I_{cg}}{h_{cg}A} = (5+5) + \frac{(8)(10)^3/12}{(5+5)[(8)(10)]} = 10.83\ \text{ft}$$

$$\sum M_A = 0 \qquad (P)(10) - (49\ 920)(10.83 - 5) = 0 \qquad P = 29\ 100\ \text{lb}$$

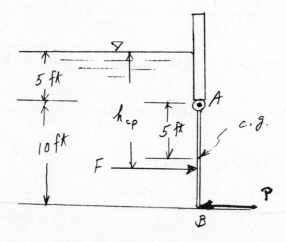

Fig. 3-16

3.22 A vat filled with oil (s.g. = 0.82) is 7 m long and 3 m deep and has a trapezoidal cross section 4 m wide at the bottom and 6 m wide at the top (see Fig. 3-17). Compute (a) the weight of oil in the vat, (b) the force on the bottom of the vat, and (c) the force on the trapezoidal end panel.

(a)
$$W = \gamma V = [(0.82)(9.79)][(7)(3)(6+4)/2] = 843\ \text{kN}$$

(b)
$$F = \gamma h_{cg} A \qquad F_{\text{bottom}} = [(0.82)(9.79)](3)[(4)(7)] = 674\ \text{kN}$$

(c)
$$F_{\text{end}} = F_{\text{square}} + 2F_{\text{triangle}} = [(0.82)(9.79)][(0+3)/2][(4)(3)] + (2)[(0.82)(9.79)](\tfrac{3}{3})[(3)(1)/2] = 169\ \text{kN}$$

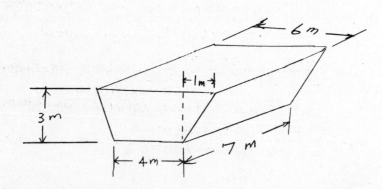

Fig. 3-17

3.23 Gate AB in Fig. 3-18 is 6 ft wide, hinged at point A, and restrained by a stop at point B. Compute the force on the stop and the reactions at A if water depth h is 10 ft.

$$F = \gamma h_{cg} A = (62.4)(10 - \tfrac{4}{2})[(4)(6)] = 11\ 980\ \text{lb}$$

$$y_{cp} = \frac{-I_{xx} \sin\theta}{h_{cg}A} = \frac{-[(6)(4)^3/12](\sin 90°)}{(10 - 4/2)[(4)(6)]} = -0.167\ \text{ft}$$

$$\sum M_A = 0 \qquad (B_x)(4) - (11\ 980)(2 + 0.167) = 0 \qquad B_x = 6490\ \text{lb}$$

$$\sum F_x = 0 \qquad 11\ 980 - 6490 - A_x = 0 \qquad A_x = 5490\ \text{lb}$$

If gate weight is neglected, $A_y = 0$.

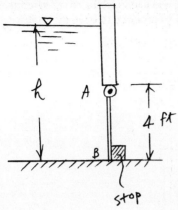 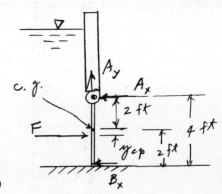

Fig. 3-18(a)   **Fig. 3-18(b)**

3.24 In Fig. 3-18, stop B will break if the force on it equals or exceeds 10 000 lb. For what water depth h is this condition reached?

❚ $\quad F = \gamma h_{cg} A = (62.4)(h_{cg})[(4)(6)] = 1498 h_{cg} \qquad y_{cp} = \dfrac{-I_{xx} \sin \theta}{h_{cg} A} = \dfrac{-[(6)(4)^3/12](\sin 90°)}{(h_{cg})[(4)(6)]} = -\dfrac{1.333}{h_{cg}}$

$$\sum M_A = 0 \qquad (10\,000)(4) - (1498 h_{cg})(2 + 1.333/h_{cg}) = 0$$

$$h_{cg} = 12.68 \text{ ft} \qquad h = 12.68 + 2 = 14.68 \text{ ft}$$

3.25 In Fig. 3-18, hinge A will break if its horizontal reaction equals or exceeds 9000 lb. For what water depth h is this condition reached?

❚ From Prob. 3.24, $F = 1498 h_{cg}$ and $y_{cp} = -1.333/h_{cg}$; $\sum M_B = 0$; $(1498 h_{cg})(2 - 1.333/h_{cg}) - (9000)(4) = 0$, $h_{cg} = 12.68$ ft; $h = 12.68 + 2 = 14.68$ ft.

3.26 Calculate the total resultant force on triangular gate ABC in Fig. 3-19 and locate the position of its center of pressure.

❚ $$F = \gamma h_{cg} A = (8.64)[\tfrac{15}{100} + (\tfrac{2}{3})(\tfrac{45}{100})][(\tfrac{30}{100})(\tfrac{45}{100})/2] = 0.262 \text{ kN}$$

$$I_{xx} = bh^3/36 = (\tfrac{30}{100})(\tfrac{45}{100})^3/36 = 0.000759 \text{ m}^4$$

$$y_{cp} = \dfrac{-I_{xx} \sin \theta}{h_{cg} A} = \dfrac{-(0.000759)(\sin 90°)}{[\tfrac{15}{100} + (\tfrac{2}{3})(\tfrac{45}{100})][(\tfrac{30}{100})(\tfrac{45}{100})/2]} = -0.025 \text{ m} \qquad \text{(i.e., below the centroid)}$$

$$I_{xy} = b(b - 2s)(h)^2/72 = \tfrac{30}{100}[\tfrac{30}{100} - (2)(\tfrac{30}{100})](\tfrac{45}{100})^2/72 = -0.000253 \text{ m}^4$$

$$x_{cp} = \dfrac{-I_{xy} \sin \theta}{h_{cg} A} = \dfrac{-(-0.000253)(\sin 90°)}{[\tfrac{15}{100} + (\tfrac{2}{3})(\tfrac{45}{100})][(\tfrac{30}{100})(\tfrac{45}{100})/2]} = +0.008 \text{ m} \qquad \text{(i.e., right of the centroid)}$$

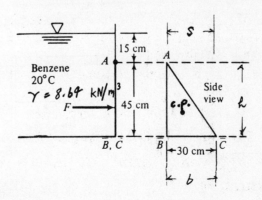

Fig. 3-19

3.27 When freshly poured between wooden forms, concrete approximates a fluid with s.g. = 2.40. Figure 3-20 shows a slab poured between wooden forms which are connected by four corner bolts A, B, C, and D. Neglecting end effects, compute the forces in the four bolts.

❚
$$F = \gamma h_{cg} A = [(2.40)(62.4)](\tfrac{10}{2})[(8)(10)] = 59\,900 \text{ lb}$$

$$y_{cp} = \frac{-I_{xx} \sin \theta}{h_{cg} A} = \frac{-[(8)(10)^3/12](\sin 90°)}{[(\tfrac{10}{2})][(8)(10)]} = -1.67 \text{ ft}$$

$$\sum M_A = 0 \qquad (2)(F_C)(10) - (59\,900)(5 + 1.67) = 0 \qquad F_C = F_D = 19\,980 \text{ lb}$$

$$\sum M_C = 0 \qquad (59\,900)(5 - 1.67) - (2)(F_A)(10) = 0 \qquad F_A = F_B = 9970 \text{ lb}$$

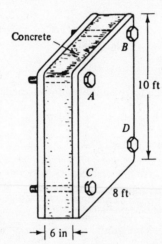

Fig. 3-20

3.28 Find the net hydrostatic force per unit width on rectangular gate AB in Fig. 3-21 and its line of action.

❚
$$F_{H_2O} = (9.79)(1.8 + 1.2 + \tfrac{2}{2})[(2)(1)] = 78.32 \text{ kN} \qquad F_{glyc} = (12.36)(1.2 + \tfrac{2}{2})[(2)(1)] = 54.38 \text{ kN}$$

$$F_{net} = F_{H_2O} - F_{glyc} = 78.32 - 54.38 = 23.94 \text{ kN}$$

$$y_{cp} = \frac{-I_{xx} \sin \theta}{h_{cg} A}$$

$$(y_{cp})_{H_2O} = \frac{-[(1)(2)^3/12](\sin 90°)}{(1.8 + 1.2 + \tfrac{2}{2})[(2)(1)]} = -0.0833 \text{ m}$$

$$(y_{cp})_{glyc} = \frac{-[(1)(2)^3/12](\sin 90°)}{[(1.2 + \tfrac{2}{2})][(2)(1)]} = -0.1515 \text{ m}$$

$$\sum M_B = 0 \qquad (78.32)(1 - 0.0833) - (54.38)(1 - 0.1515) = 23.94D$$

$$D = 1.072 \text{ m} \qquad \text{(above point } B, \text{ as shown in Fig. 3-21}c)$$

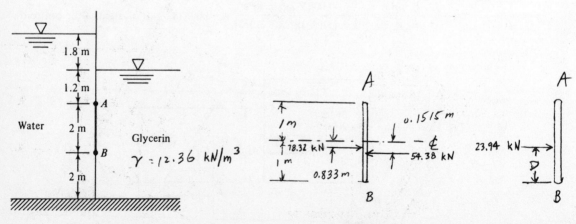

Fig. 3-21(a)   Fig. 3-21(b)   Fig. 3-21(c)

3.29 A cylindrical, wooden-stave barrel is 4 ft in diameter and 6 ft high, as shown in Fig. 3-22. It is held together by steel hoops at the top and bottom, each with a cross section of 0.35 in². If the barrel is filled with apple juice (s.g. = 1.02), compute the tension stress in each hoop.

$$F = \gamma h_{cg} A = [(1.02)(62.4)][(\tfrac{6}{2})][(4)(6)] = 4583 \text{ lb}$$

$$y_{cp} = \frac{-I_{xx} \sin \theta}{h_{cg} A} = \frac{-[(4)(6)^3/12](\sin 90°)}{\tfrac{6}{2}[(4)(6)]} = -1.00 \text{ ft}$$

$$\sum M_B = 0 \qquad 4583(\tfrac{6}{2} - 1.00) - 2(F_{upper})(6) = 0 \qquad F_{upper} = 764 \text{ lb}$$

$$\sum M_A = 0 \qquad 2(F_{lower})(6) - 4583(\tfrac{6}{2} + 1.00) = 0 \qquad F_{lower} = 1528 \text{ lb}$$

$$\sigma_{upper} = 764/0.35 = 2183 \text{ psi} \qquad \sigma_{lower} = 1528/0.35 = 4366 \text{ psi}$$

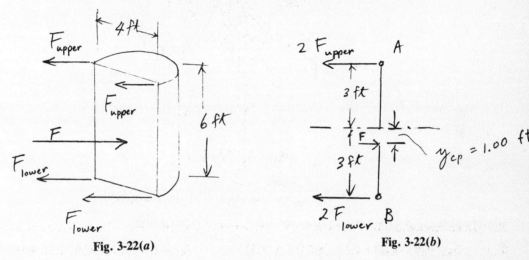

Fig. 3-22(a)   **Fig. 3-22(b)**

3.30 Gate AB in Fig. 3-23 is 15 ft long and 10 ft wide and hinged at B with a stop at A. Neglecting the weight of the gate, compute the water level h for which the gate will start to fall.

$$F = \gamma h_{cg} A = (62.4)(h/2)[(10)(h/\sin 60°)] = 360.2h^2$$

$$y_{cp} = \frac{-I_{xx} \sin \theta}{h_{cg} A} = \frac{-[10(h/\sin 60°)^3/12](\sin 60°)}{(h/2)[10(h/\sin 60°)]} = -0.1925h$$

$$\sum M_B = 0 \qquad (10\,000)(15) - (360.3h^2)[(h/\sin 60°)/2 - 0.1925h] = 0 \qquad h = 10.3 \text{ ft}$$

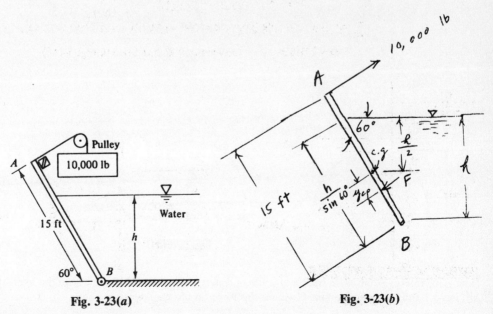

Fig. 3-23(a)   **Fig. 3-23(b)**

3.31 Repeat Prob. 3.30 by including the weight of the gate, which is 1-in-thick steel (s.g. = 7.85). (See Fig. 3-24.)

▮ $W_{gate} = [(7.85)(62.4)][(15)(10)(\frac{1}{12})] = 6123$ lb.   From Prob. 3.30, $F = 360.3h^2$; $\sum M_B = 0$, $(10\,000)(15) - (360.3h^2)[(h/\sin 60°)/2 - 0.1925h] - 6123(\frac{15}{2}\cos 60°) = 0$, $h = 9.71$ ft.

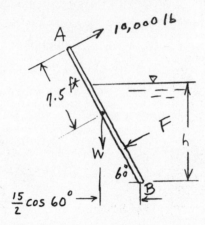

Fig. 3-24

3.32 The turbine inlet duct coming from a large dam is 3 m in diameter. It is closed by a vertical, circular gate whose center point (i.e., centroid) is 50 m below the dam's water level. Compute the force on the gate and its center of pressure.

▮ $$F = \gamma h_{cg}A = (9.79)(50)[\pi(3)^2/4] = 3460 \text{ kN} \qquad I_{xx} = \pi r^4/4 = \pi(\tfrac{3}{2})^4/4 = 3.976 \text{ m}^4$$

$$y_{cp} = \frac{-I_{xx}\sin\theta}{h_{cg}A} = \frac{-(3.976)(\sin 90°)}{(50)[\pi(3)^2/4]} = -0.0112 \text{ m}$$

Line of action of F is 0.0112 m below the centroid of the turbine inlet duct.

3.33 Gate AB in Fig. 3-25 is semicircular, hinged at B, and held by horizontal force P at A. What force P is required for equilibrium?

▮ $$4r/(3\pi) = (4)(3)/(3\pi) = 1.273 \text{ m} \qquad F = \gamma h_{cg}A = (9.79)(5 + 3 - 1.273)[\pi(3)^2/2] = 931 \text{ kN}$$

$$y_{cp} = \frac{-I_{xx}\sin\theta}{h_{cg}A} = \frac{-[(0.10976)(3)^4](\sin 90°)}{(5 + 3 - 1.273)[\pi(3)^2/2]} = -0.0935 \text{ m}$$

$$\sum M_B = 0 \qquad (931)(1.273 - 0.0935) - 3P = 0 \qquad P = 366 \text{ kN}$$

$$I_{xx} = 0.10976 r^4$$

$$I_{xy} = 0$$

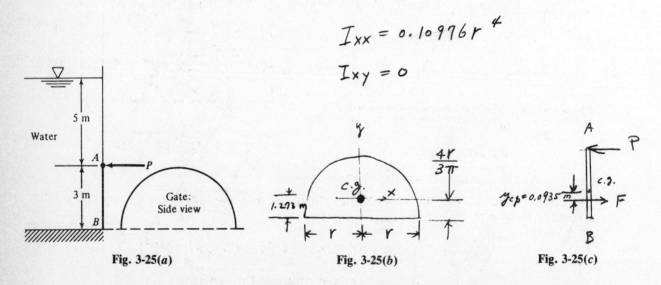

Fig. 3-25(a)   **Fig. 3-25(b)**   **Fig. 3-25(c)**

3.34 Dam ABC in Fig. 3-26 is 40 m wide and made of concrete weighing 22 kN/m³. Find the hydrostatic force on surface AB and its moment about C. Could this force tip the dam over?

▌ $F = \gamma h_{cg} A = (9.79)(\frac{80}{2})[(40)(100)] = 1\,566\,000$ kN. F acts at $(\frac{2}{3})(100)$, or 66.67 ft from A along surface AB (see Fig. 3-26b). For the given triangular shape, the altitude from C to AB intersects AB 64.00 m from A (see Fig. 3-26b). Hence, $M_C = (1\,566\,000)(66.67 - 64.00) = 4\,181\,000$ kN. Since the moment of F about point C is counterclockwise, there is no danger of tipping.

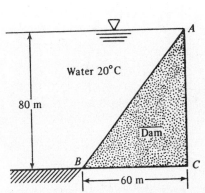

Fig. 3-26(a)

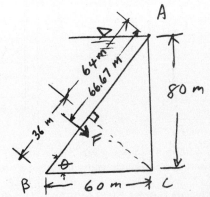

Fig. 3-26(b)

3.35 Isosceles traingular gate AB in Fig. 3-27 is hinged at A and taken to be weightless. What horizontal force P is required at point B for equilibrium?

▌ $\quad AB = 2/\sin 50° = 2.611$ m $\qquad F = \gamma h_{cg} A = [(0.83)(9.79)](3 + 0.667)[(1)(2.611)/2] = 38.90$ kN

$$y_{cp} = \frac{-I_{xx}\sin\theta}{h_{cg}A} = \frac{-[(1)(2.611)^3/36](\sin 50°)}{(3 + 0.667)[(1)(2.611)/2]} = -0.0791 \text{ m}$$

$$\sum M_A = 0 \qquad 2P - (38.90)(2.611/3 + 0.0791) = 0 \qquad P = 18.47 \text{ kN}$$

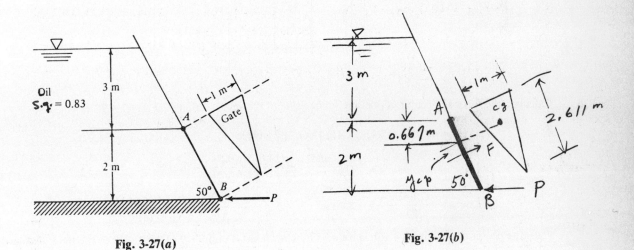

Fig. 3-27(a)

Fig. 3-27(b)

3.36 The tank in Fig. 3-28 is 50 cm wide. Compute the hydrostatic force on lower panel BC and on upper panel AD. Neglect atmospheric pressure.

▌ $\qquad p = \gamma h \qquad p_{BC} = [(0.82)(9.79)](\frac{30}{100} + \frac{40}{100}) + (9.79)(\frac{20}{100}) = 7.577$ kPa

$$F = pA \qquad F_{BC} = (7.577)[(\frac{100}{100})(\frac{50}{100})] = 3.788 \text{ kN}$$

$$p_{AD} = [(0.82)(9.79)](\frac{30}{100}) = 2.408 \text{ kPa} \qquad F_{AD} = (2.408)[(\frac{45}{100})(\frac{50}{100})] = 0.542 \text{ kN}$$

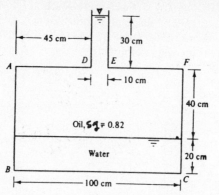

Fig. 3-28

3.37 The water tank shown in Fig. 3-29 is pressurized as shown by the manometer. Determine the hydrostatic pressure force per unit width on gate AB (i.e., use gate width of 1 m perpendicular to the paper).

▮ On gate AB, $p_{cg} = [(13.6)(9.79)](\frac{80}{100}) + (9.79)(5 + \frac{2}{2}) = 165.3$ kPa, $F_{AB} = (165.3)[(2)(1)] = 330.6$ kN.

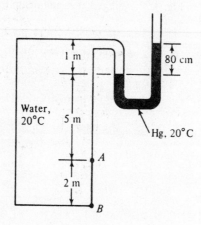

Fig. 3-29

3.38 Calculate the force and center of pressure on one side of the vertical triangular panel ABC in Fig. 3-30.

▮
$$F = \gamma h_{cg}A = (62.4)(2 + 4)[(4)(6)/2] = 4493 \text{ lb} \qquad I_{xx} = (4)(6)^3/36 = 24.0 \text{ ft}^4$$

$$y_{cp} = \frac{-I_{xx}\sin\theta}{h_{cg}A} = \frac{-(24.0)(\sin 90°)}{(2+4)[(4)(6)/2]} = -0.33 \text{ ft}$$

$$I_{xy} = \frac{4[4-(2)(4)](6)^2}{72} = -8.00 \text{ ft}^4 \qquad x_{cp} = \frac{-I_{xy}\sin\theta}{h_{cg}A} = \frac{-(-8.00)(\sin 90°)}{(2+4)[(4)(6)/2]} = 0.11 \text{ ft}$$

Thus, the center of pressure is $4 + 0.33$, or 4.33 ft below point A and $\frac{4}{3} + 0.11$, or 1.44 ft to the right of point B.

3.39 In Fig. 3-31, gate AB is 3 m wide and is connected by a rod and pulley to a concrete sphere. What diameter of sphere is just sufficient to keep the gate closed?

▮
$$F = \gamma h_{cg}A = (9.79)(8 + \frac{4}{2})[(4)(3)] = 1175 \text{ kN}$$

$$y_{cp} = \frac{-I_{xx}\sin\theta}{h_{cg}A} = \frac{-[(3)(4)^3/12](\sin 90°)}{(8+\frac{4}{2})[(4)(3)]} = -0.133 \text{ m}$$

$$\sum M_B = 0 \qquad (W_{sphere})(6+8+4) - (1175)(4-2-0.133) = 0 \qquad W_{sphere} = 121.9 \text{ kN}$$

$$W_{sphere} = \gamma(\pi d^3/6) \qquad 121.9 = [(2.40)(9.79)](\pi d^3/6) \qquad d = 2.148 \text{ m}$$

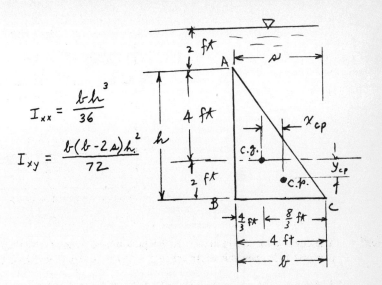

Fig. 3-30(a)

Fig. 3-30(b)

$$I_{xx} = \frac{bh^3}{36}$$

$$I_{xy} = \frac{b(b-2a)h_i^2}{72}$$

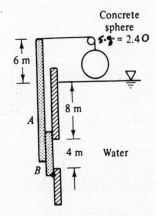

Fig. 3-31

3.40 The V-shaped container in Fig. 3-32 is hinged at A and held together by cable BC at the top. If cable spacing is 1 m into the paper, what is the cable tension?

I

$$F = \gamma h_{cg}A = (9.79)(\tfrac{3}{2})[(5.230)(1)] = 76.80 \text{ kN}$$

$$y_{cp} = \frac{-I_{xx}\sin\theta}{h_{cg}A} = \frac{-[(1)(5.230)^3/12](\sin 35°)}{\tfrac{3}{2}[(5.230)(1)]} = -0.872 \text{ m}$$

$$\sum M_A = 0 \qquad (T)(1+3) - (76.80)(2.615 - 0.872) = 0 \qquad T = 33.5 \text{ kN}$$

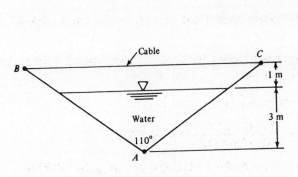

Fig. 3-32(a)

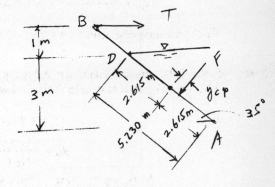

Fig. 3-32(b)

3.41 In Fig. 3-33, gate AB is 5 ft wide and opens to let fresh water out when the ocean tide is dropping. The hinge at A is 2 ft above the fresh water level. At what ocean level h will the gate open? Neglect the gate's weight.

$$F = \gamma h_{cg}A \qquad F_1 = (62.4)(\tfrac{10}{2})[(10)(5)] = 15\,600 \text{ lb} \qquad F_2 = [(1.025)(62.4)](h/2)[(5)(h)] = 159.9h^2$$

$$\sum M_A = 0 \qquad (159.9h^2)(10 + 2 - h/3) - (15\,600)(2 + 6.67) = 0 \qquad h = 9.85 \text{ ft}$$

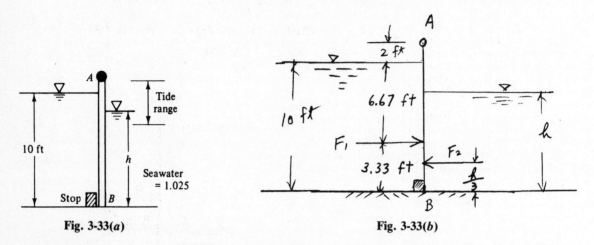

Fig. 3-33(a)   Fig. 3-33(b)

3.42 For the conditions given in Prob. 3.41, investigate analytically whether the required height h is independent of gate width b (perpendicular to the paper).

$$F = \gamma h_{cg}A \qquad F_1 = (62.4)(\tfrac{10}{2})[(10)(b)] = 3120b \qquad F_2 = [(1.025)(62.4)](h/2)(bh) = 31.98bh^2$$

$$\sum M_A = 0 \qquad (31.98bh^2)(10 + 2 - h/3) - (3120b)(2 + 6.67) = 0 \qquad h = 9.85 \text{ ft}$$

Hence, required h is independent of gate width b.

3.43 Compute the force on one side of parabolic panel ABC in Fig. 3-34 and the vertical distance down to the center of pressure.

$$F = \gamma h_{cg}A = (9.79)(2 + 3)[(\tfrac{2}{3})(5)(3)] = 489.5 \text{ kN}$$

$$I_{xx} = I_{x'} - A(\Delta h)^2 = \tfrac{2}{7}(bh^3) - [\tfrac{2}{3}(bh)][\tfrac{3}{5}(h)]^2 = (\tfrac{2}{7})(3)(5)^3 - [(\tfrac{2}{3})(3)(5)][(\tfrac{3}{5})(5)]^2 = 17.14 \text{ m}^4$$

$$y_{cp} = \frac{-I_{xx}\sin\theta}{h_{cg}A} = \frac{-(17.14)(\sin 90°)}{(2 + 3)[(\tfrac{2}{3})(5)(3)]} = -0.343 \text{ m}$$

Hence, the center of pressure is $3 + 0.343$, or 3.343 m below point A.

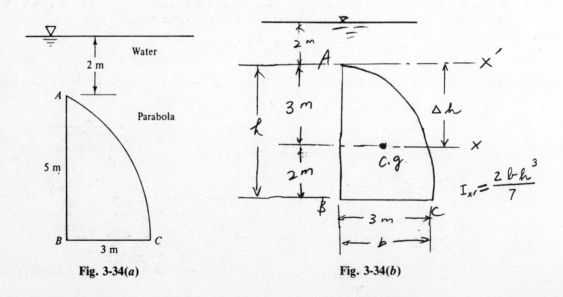

Fig. 3-34(a)   Fig. 3-34(b)

3.44 Circular gate ABC in Fig. 3-35 has a 1-m radius and is hinged at B. Compute the force P just sufficient to keep the gate from opening when h is 10 m.

$$F = \gamma h_{cg} A = (9.79)(10)[\pi)(1)^2] = 307.6 \text{ kN} \qquad I_{xx} = (\pi)(r)^4/4 = (\pi)(1)^4/4 = 0.7854 \text{ m}^4$$

$$y_{cp} = \frac{-I_{xx} \sin \theta}{h_{cg} A} = \frac{-(0.7854)(\sin 90°)}{(10)[(\pi)(1)^2]} = -0.025 \text{ m}$$

$$\sum M_B = 0 \qquad (P)(1) - (307.6)(0.025) = 0 \qquad P = 7.69 \text{ kN}$$

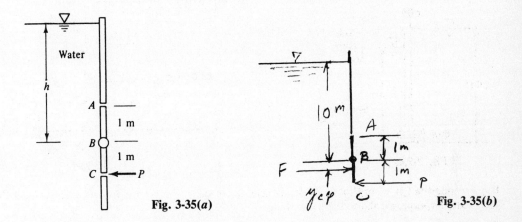

Fig. 3-35(a)   Fig. 3-35(b)

3.45 For the conditions given in Prob. 3.44, derive an analytical expression for P as a function of h.

$$F = \gamma h_{cg} A = \gamma h_{cg}[\pi(r)^2] \qquad I_{xx} = \pi(r)^4/4$$

$$y_{cp} = \frac{-I_{xx} \sin \theta}{h_{cg} A} = \frac{-[(\pi)(r)^4/4](\sin 90°)}{h[(\pi)(r)^2]} = \frac{-r^2}{4h}$$

$$\sum M_B = 0 \qquad Pr - [\gamma h_{cg}(\pi)(r)^2][(r)^2/(4r)] = 0 \qquad P = \gamma \pi r^3/4$$

(Note that force P is independent of depth h.)

3.46 Gate ABC in Fig. 3-36 is 1 m square and hinged at B. It will open automatically when water depth h becomes high enough. Determine the lowest height (i.e., minimum value of h) at which the gate will open.

▌ The gate will open when resultant force F is above point B—i.e., when $|y_{cp}| < 0.5 - 0.4$, or 0.1 m. (Note in Fig. 3-36b that y_{cp} is the distance between F and the centroid of gate ABC.)

$$y_{cp} = \frac{-I_{xx} \sin \theta}{h_{cg} A} = \frac{-[(1)(1)^3/12](\sin 90°)}{(h + 0.5)[(1)(1)]} = \frac{-0.0833}{h + 0.5}$$

For $|y_{cp}| < 0.1$, $0.0833/(h + 0.5) < 0.1$, $h > 0.333$ m. (Note that this result is independent of fluid weight.)

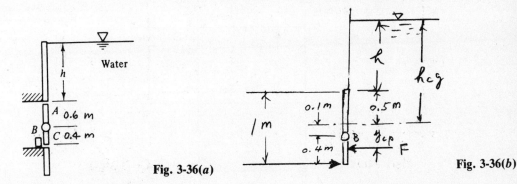

Fig. 3-36(a)   Fig. 3-36(b)

3.47 Gate *AB* in Fig. 3-37 is 7 ft wide and weighs 3000 lb when submerged. It is hinged at *B* and rests against a smooth wall at *A*. Determine the water level *h* which will just cause the gate to open.

$$F = \gamma h_{cg} A \qquad F_1 = 62.4(h + \tfrac{8}{2})[(10)(7)] = 4368h + 17\,472 \qquad F_2 = 62.4(4 + \tfrac{8}{2})[(10)(7)] = 34\,944 \text{ lb}$$

$$y_{cp} = \frac{-I_{xx} \sin \theta}{h_{cg} A} \qquad (y_{cp})_1 = \frac{-[(7)(10)^3/12](\tfrac{8}{10})}{(h + \tfrac{8}{2})[(10)(7)]} = \frac{-6.67}{h + 4}$$

$$(y_{cp})_2 = \frac{-[(7)(10)^3/12](\tfrac{8}{10})}{(4 + \tfrac{8}{2})[(10)(7)]} = -0.833 \text{ ft}$$

$$\sum M_B = 0 \qquad (4368h + 17\,472)[5 - 6.67/(h + 4)] - (34\,944)(5 - 0.833) - (3000)(\tfrac{6}{2}) = 0 \qquad h = 4.41 \text{ ft}$$

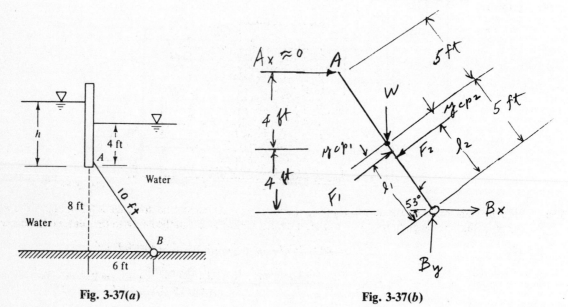

Fig. 3-37(*a*) **Fig. 3-37(*b*)**

3.48 The tank in Fig. 3-38 contains oil and water as shown. Find the resultant force on side *ABC*, which is 4 ft wide.

$$F = \gamma h_{cg} A \qquad F_{AB} = [(0.80)(62.4)][(\tfrac{10}{2})][(10)(4)] = 9980 \text{ lb}$$

F_{AB} acts at a point $(\tfrac{2}{3})(10)$, or 6.67 ft below point *A*. Water is acting on area *BC*, and any superimposed liquid can be converted to an equivalent depth of water. Employ an imaginary water surface (IWS) for this calculation, locating IWS by changing 10 ft of oil to (0.80)(10), or 8 ft of water. Thus, $F_{BC} = (62.4)(8 + \tfrac{6}{2})[(6)(4)] = 16\,470$ lb.

$$y_{cp} = \frac{-I_{xx} \sin \theta}{h_{cg} A} = \frac{-[(4)(6)^3/12](\sin 90°)}{(8 + \tfrac{6}{2})[(6)(4)]} = -0.27 \text{ ft} \qquad \text{(i.e., below the centroid of } BC)$$

F_{BC} acts at a point $(2 + 8 + \tfrac{6}{2} + 0.27)$, or 13.27 ft below *A*. $\sum M_A = 0$; $(9980 + 16\,470)(h_{cp}) - (9980)(6.67) - (16\,470)(13.27) = 0$, $h_{cp} = 10.78$ ft from *A*. Thus, the total resultant force on side *ABC* is 9980 + 16 470, or 26 450 lb acting 10.78 ft below *A*.

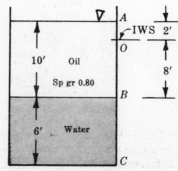

Fig. 3-38

3.49 Gate AB in Fig. 3-39 is 4 ft wide and hinged at A. Gage G reads -2.17 psi, while oil (s.g. = 0.75) is in the right tank. What horizontal force must be applied at B for equilibrium of gate AB?

$$F = \gamma h_{cg}A \qquad F_{oil} = [(0.75)(62.4)](\tfrac{6}{2})[(6)(4)] = 3370 \text{ lb}$$

F_{oil} acts $(\tfrac{2}{3})(6)$, or 4.0 ft from A. For the left side, the negative pressure due to the air can be converted to its equivalent head in feet of water. $h = p/\gamma = (-2.17)(144)/62.4 = -5.01$ ft. This negative pressure head is equivalent to having 5.01 ft less water above A. Hence, $F_{H_2O} = (62.4)(6.99 + \tfrac{6}{2})[(6)(4)] = 14\,960$ lb.

$$y_{cp} = \frac{-I_{xx}\sin\theta}{h_{cg}A} = \frac{-[(4)(6)^3/12](\sin 90°)}{(6.99 + \tfrac{6}{2})[(6)(4)]} = -0.30 \text{ ft}$$

F_{H_2O} acts at $(0.30 + \tfrac{6}{2})$, or 3.30 ft below A. $\sum M_A = 0$; $(3370)(4.0) + 6F - (14\,960)(3.30) = 0$, $F = 5980$ lb (acting leftward).

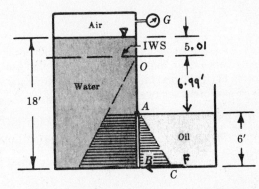

Fig. 3-39

3.50 A plane surface is circular and 1.2 m in diameter. If it is vertical and the top edge is 0.3 m below the water surface, find the magnitude of the hydrostatic force on one side and the depth to the center of pressure.

$$F = \gamma h_{cg}A = (9.79)(0.3 + 1.2/2)[(\pi)(1.2)^2/4] = 9.97 \text{ kN}$$

$$h_{cp} = h_{cg} + \frac{I_{cg}}{h_{cg}A} = \left(0.3 + \frac{1.2}{2}\right) + \frac{(\pi)(1.2)^4/64}{(0.3 + 1.2/2)[(\pi)(1.2)^2/4]} = 1.00 \text{ m}$$

3.51 The Utah-shaped plate shown in Fig. 3-40 is submerged in oil (s.g. = 0.82) and lies in a vertical plane. Find the magnitude of the hydrostatic force on one side and the depth to the center of pressure.

$$F = \gamma h_{cg}A \qquad F_1 = [(0.82)(9.79)](1 + \tfrac{6}{2})[(3)(6)] = 578 \text{ kN} \qquad h_{cp} = h_{cg} + \frac{I_{cg}}{h_{cg}A}$$

$$(h_{cp})_1 = 1 + \tfrac{6}{2} + \frac{(3)(6)^3/12}{(1 + \tfrac{6}{2})[(3)(6)]} = 4.75 \text{ m} \qquad F_2 = [(0.82)(9.79)][1 + 2.5 + 3.5/2][(2)(3.5)] = 295 \text{ kN}$$

$$(h_{cp})_2 = [1 + 2.5 + 3.5/2] + \frac{(2)(3.5)^3/12}{(1 + 2.5 + 3.5/2)[(2)(3.5)]} = 5.44 \text{ m}$$

$$F = 578 + 295 = 873 \text{ kN} \qquad 873 h_{cp} = (578)(4.75) + (295)(5.44) \qquad h_{cp} = 4.98 \text{ m}$$

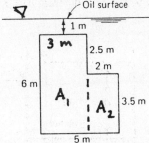

Fig. 3-40

3.52 The irrigation head gate shown in Fig. 3-41a is a plate which slides over the opening to a culvert. The coefficient of friction between the gate and its sliding ways is 0.6. Find the force required to slide open this 900-lb gate if it is set (**a**) vertically and (**b**) on a 2:1 slope ($n = 2$ in Fig. 3-41a), as is common.

▌ (a) $F = \gamma h_{cg}A = (62.4)[15 + (\frac{60}{12})/2][(\frac{60}{12})(\frac{60}{12})] = 27\,300$ lb. Let T = force parallel to gate required to open it. $\sum F_y = 0$; $T - 900 - (0.6)(27\,300) = 0$, $T = 17\,280$ lb. (b) See Fig. 3-41b. $F = (62.4)[15 + \frac{60}{12}(1/\sqrt{5})/2][(\frac{60}{12})(\frac{60}{12})] = 25\,140$ lb. Let N = total force normal to gate; $N = 25\,140 + (900)(2/\sqrt{5}) = 25\,940$ lb. $\sum F_y = 0$; $T - (900)(1/\sqrt{5}) - (0.6)(25\,940) = 0$, $T = 15\,970$ lb.

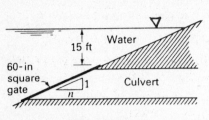

Fig. 3-41(a)

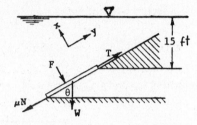

Fig. 3-41(b)

3.53 In the drainage of irrigated lands, it is frequently desirable to install automatic flap gates to prevent a flood from backing up the lateral drains from a river. Suppose a 60-in-square flap gate, weighing 2555 lb, is hinged 40 in above the center, as shown in Fig. 3-42, and the face is sloped 4° from the vertical. Find the depth to which water will rise behind the gate before it will open.

▌ Closing moment of gate about hinge = $(2555)[(\frac{40}{12})(\sin 4°)] = 594$ lb · ft

$$F = \gamma h_{cg}A = (62.4)(h/2)[(\frac{60}{12})(h)/\cos 4°] = 156.4h^2$$

$$\sum M_{hinge} = 0 \qquad (156.4h^2)[(60 + 10)/12 - (h/\cos 4°)/3] - 594 = 0 \qquad h = 0.827 \text{ ft}$$

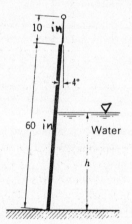

Fig. 3-42

3.54 Gate MN in Fig. 3-43 rotates about an axis through N. If the width of the gate is 4 ft, what torque applied to the shaft through N is required to hold the gate closed?

▌ $F = \gamma h_{cg}A$ $F_1 = 62.4[5 + (3 + 2)/2][(3 + 2)(4)] = 9360$ lb $F_2 = (62.4)(\frac{3}{2})[(3)(4)] = 1123$ lb

$$y_{cp} = \frac{-I_{xx}\sin\theta}{h_{cg}A} \qquad (y_{cp})_1 = \frac{-[(4)(3 + 2)^3/12](\sin 90°)}{[5 + (3 + 2)/2][(3 + 2)(4)]} = 0.278 \text{ ft}$$

F_2 acts at $(\frac{1}{3})(3)$, or 1 ft from N. $\sum M_N = 0$; $(9360)[(3 + 2)/2 - 0.278] - (1123)(1) - \text{torque}_N = 0$, $\text{torque}_N = 19\,670$ lb · ft.

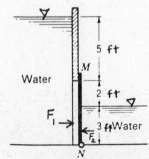

Fig. 3-43

3.55 Find the minimum value of z for which the gate in Fig. 3-44 will rotate counterclockwise if the gate is (*a*) rectangular and (*b*) triangular, 4-ft base as axis and 4-ft height.

▌ **(a)**
$$F = \gamma h_{cg} A \qquad F_{H_2O} = (62.4)(z - \tfrac{4}{2})[(4)(4)] = (998.4)(z - 2)$$

$$y_{cp} = \frac{-I_{xx} \sin \theta}{h_{cg} A} \qquad (y_{cp})_{H_2O} = \frac{-[(4)(4)^3/12](\sin 90°)}{(z - \tfrac{4}{2})[(4)(4)]} = \frac{-1.333}{z - 2}$$

Moment due to water $= [(998.4)(z - 2)][\tfrac{4}{2} + 1.333/(z - 2)] = (998.4)(2z - 2.667)$

$F_{gas} = pA = [(4)(144)][(4)(4)] = 9216$ lb. F_{gas} acts at $\tfrac{4}{2}$, or 2 ft below hinge. Moment due to gas $= (9216)(2) = 18\,432$ lb · ft. Equating moments gives $(998.4)(2z - 2.667) = 18\,432$, $z = 10.56$ ft.

(b)
$$F_{H_2O} = (62.4)[z - (\tfrac{2}{3})(4)][(4)(4)/2] = (499.2)(z - 2.667)$$

$$(y_{cp})_{H_2O} = \frac{-[(4)(4)^3/36](\sin 90°)}{[z - (\tfrac{2}{3})(4)][(4)(4)/2]} = \frac{0.8889}{z - 2.667}$$

Moment due to water $= [(499.2)(z - 2.667)][\tfrac{4}{3} + 0.8889/(z - 2.667)] = 665.6z - 1331$

$F_{gas} = [(4)(144)][(4)(4)/2] = 4608$ lb. F_{gas} acts at $\tfrac{4}{3}$, or 1.333 ft below hinge. Moment due to gas $= (4608)(1.333) = 6142$ lb · ft. Equating moments gives $(665.6z - 1331) = 6142$, $z = 11.23$ ft.

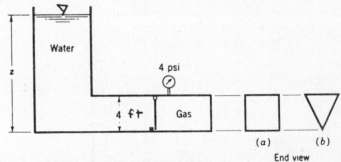

Water

4 psi

z

4 ft   Gas

(a)   (b)

End view   **Fig. 3-44**

3.56 The triangular gate *CDE* in Fig. 3-45 is hinged along *CD* and is opened by a normal force *P* applied at *E*. It holds oil (s.g. = 0.80) above it and is open to the atmosphere on its lower side. Neglecting the weight of the gate, find (*a*) the magnitude of force exerted on the gate, by integration and by the moment of inertia method; (*b*) the location of the center of pressure; and (*c*) the force *P* needed to open the gate.

▌ **(a)** By integration, $F = \int \gamma h \, dA = \int (\gamma)(y \sin \theta)(x \, dy)$. When $y = 8$, $x = 0$, and when $y = 8 + \tfrac{10}{2}$, or 13, $x = 6$, with x varying linearly with y. Hence, $x = \tfrac{6}{5}(y - 8)$. When $y = 13$, $x = 6$, and when $y = 8 + 10$, or 18, $x = 0$, with x varying linearly with y. Hence, $x = \tfrac{6}{5}(18 - y)$.

$$F = \int_8^{13} [(0.80)(62.4)](y \sin 30°)[\tfrac{6}{5}(y - 8) \, dy] + \int_{13}^{18} [(0.80)(62.4)](y \sin 30°)[\tfrac{6}{5}(18 - y) \, dy]$$

$$= [(0.80)(62.4)](\sin 30°)(\tfrac{6}{5}) \left[\frac{y^3}{3} - 4y^2 \right]_8^{13} + \left[9y^2 - \frac{y^3}{3} \right]_{13}^{18} = 9734 \text{ lb}$$

By the moment of inertia method: $F = \gamma h_{cg} A = [(0.80)(62.4)][(8 + \tfrac{10}{2})(\sin 30°)][(10)(6)/2] = 9734$ lb.

(b)
$$x_{cp} = \frac{-I_{xy} \sin \theta}{h_{cg} A}$$

Since $I_{xy} = 0$, $x_{cp} = 0$,

$$y_{cp} = \frac{-I_{xx} \sin \theta}{h_{cg} A} = \frac{-[(2)(6)(\tfrac{10}{2})^3/12](\sin 30°)}{[(8 + \tfrac{10}{2})(\sin 30°)][(10)(6)/2]} = -0.32 \text{ ft}$$

(i.e., the pressure center is 0.32 ft below the centroid, measured in the plane of the area).

(c)
$$\sum M_{CD} = 0 \qquad 6P = (9734)(\tfrac{6}{3}) \qquad P = 3245 \text{ lb}$$

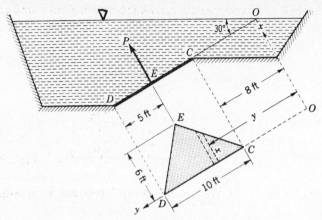

Fig. 3-45

3.57 Determine the force acting on one side of vertical surface $OACO$ in Fig. 3-46 and the location of the center of pressure if $\gamma = 9.0 \text{ kN/m}^3$.

$$F = \int \gamma y \, dA = \int_0^1 (9.0)(y)(2x \, dy)$$

From the given equation of the curve, $x = \sqrt{8y}$. Hence,

$$F = \int_0^1 (9.0)(y)(2\sqrt{8y}) \, dy = \int_0^1 50.91 y^{3/2} \, dy = [20.36 y^{5/2}]_0^1 = 20.36 \text{ kN}$$

$$y_{cp} = \frac{\int \gamma y^2 \, dA}{F} = \frac{\int_0^1 (9.0)(y)^2(2x \, dy)}{20.36} = \frac{\int_0^1 (9.0)(y)^2(2\sqrt{8y}) \, dy}{20.36} = \frac{\int_0^1 50.91 y^{5/2} \, dy}{20.36} = \frac{[14.55 y^{7/2}]_0^1}{20.36} = 0.714 \text{ m}$$

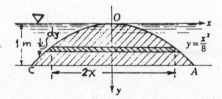

Fig. 3-46

3.58 Find the force exerted by water on one side of the vertical annular area shown in Fig. 3-47 and the location of the center of pressure.

$$F = \gamma h_{cg} A = (9.79)(2)[(\pi)(1)^2 - (\pi)(\tfrac{500}{1000})^2] = 46.13 \text{ kN}$$

$$I_{cg} = (\pi)(1)^4/4 - (\pi)(\tfrac{500}{1000})^4/4 = 0.7363 \text{ m}^4$$

$$h_{cp} = h_{cg} + \frac{I_{cg}}{h_{cg}A} = 2 + \frac{0.7363}{2[(\pi)(1)^2 - (\pi)(\tfrac{500}{1000})^2]} = 2.156 \text{ m}$$

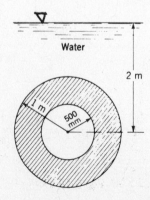

Water

2 m

1 m

500 mm

Fig. 3-47

3.59 Determine y in Fig. 3-48 so that the flashboards will tumble when water reaches their top.

   ❚ The flashboards will tumble when y is at the center of pressure. Hence, $y = \frac{1}{3}$, or 0.333 m.

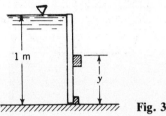

Fig. 3-48

3.60 Determine the pivot location y of the square gate in Fig. 3-49 so that it will open when the liquid surface is as shown.

   ❚ The gate will open when the pivot location is at the center of pressure.

$$h_{cp} = h_{cg} + \frac{I_{cg}}{h_{cg}A} = (2 - \tfrac{1}{2}) + \frac{(1)(1)^3/12}{(2 - \tfrac{1}{2})[(1)(1)]} = 1.556 \text{ m} \qquad y = 2 - 1.556 = 0.444 \text{ m}$$

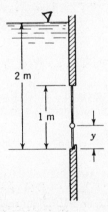

Fig. 3-49

3.61 The gate in Fig. 3-50a weighs 300 lb/ft normal to the paper. Its center of gravity is 1.5 ft from the left face and 2.0 ft above the lower face. It is hinged at O. Determine the water-surface position for the gate just to start to come up when the water surface is below the hinge at O.

   ❚ Refer to Fig. 3-50b and consider 1 ft of length. $F = \gamma hA = (62.4)[(h_O/2)][(h_O)(1)] = 31.2h_O^2;\ \Sigma M_O = 0;$
$(2)(300) - (5 - h_O/3)(31.2h_O^2) = 0,\ h_O = 2.12$ ft.

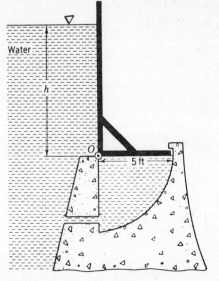

Fig. 3-50(a)

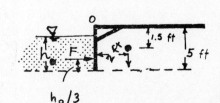

$h_o/3$

Fig. 3-50(b)

3.62 For the gate described in Prob. 3.61 and Fig. 3-50a, find h for the gate just to come up to the vertical position shown in Fig. 3-50a.

▌ See Fig. 3-51. $F_1 = \gamma h A = (62.4)(h)[(5)(1)] = 312h$, $F_2 = (62.4)(h/2)[(h)(1)] = 31.2h^2$; $\sum M_O = 0$; $(1.5)(300) + (h/3)(31.2h^2) - (2.5)(312h) = 0$, $h = 0.58$ ft.

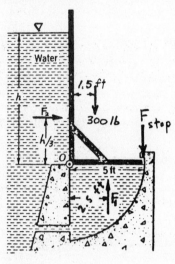

Fig. 3-51

3.63 For the gate described in Prob. 3.61 and Fig. 3-50a, find h and the force against the stop when this force is a maximum for the gate.

▌ See Fig. 3-51. $F_1 = \gamma h A = (62.4)(h)[(5)(1)] = 312h$, $F_2 = (62.4)(h/2)[(h)(1)] = 31.2h^2$; $\sum M_O = 0$; $(1.5)(300) + (h/3)(31.2h^2) - (2.5)(312h) + (5)(F_{\text{stop}}) = 0$, $F_{\text{stop}} = 156h - 2.08h^3 - 90$.

$$\frac{dF_{\text{stop}}}{dh} = 156 - 6.24h^2 = 0 \qquad h = 5.00 \text{ ft}$$

$$F_{\text{stop}} = (156)(5.00) - (2.08)(5.00)^3 - 90.0 = 430 \text{ lb}$$

3.64 Compute the air pressure required to keep the 700-mm-diameter gate of Fig. 3-52 closed. The gate is a circular plate that weighs 1800 N.

▌ $\qquad F = \gamma h A \qquad F_{\text{liq}} = [(2)(9.79)][1.5 + (\frac{1}{2})(\frac{700}{1000})(\sin 45°)][\pi(\frac{700}{1000})^2/4] = 13.17 \text{ kN}$

$$z_{\text{cp}} = z_{\text{cg}} + \frac{I_{\text{cg}}}{z_{\text{cg}}A} = \left[\frac{1.5}{\cos 45°} + \left(\frac{1}{2}\right)\left(\frac{700}{1000}\right)\right] + \frac{\pi[(\frac{1}{2})(\frac{700}{1000})]^4/4}{[1.5/\cos 45° + (\frac{1}{2})(\frac{700}{1000})][\pi(\frac{700}{1000})^2/4]} = 2.484 \text{ m}$$

$\sum M_{\text{hinge}} = 0 \qquad (13.17)(2.484 - 1.5/\cos 45°) + \frac{1800}{1000}[(\frac{1}{2})(\frac{700}{1000})(\cos 45°)] - [\pi(\frac{700}{1000})^2/4](p_{\text{air}})[(\frac{1}{2})(\frac{700}{1000})] = 0$

$$p_{\text{air}} = 38.77 \text{ kPa}$$

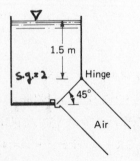

Fig. 3-52

4.1 In Fig. 4-1, calculate the width of concrete dam that is necessary to prevent the dam from sliding. The specific weight of the concrete is 150 lb/ft^3, and the coefficient of friction between the base of the dam and the foundation is 0.42. Use 1.5 as the factor of safety (F.S.) against sliding. Will it also be safe against overturning?

▮ Working with a 1-ft "slice" (i.e., dimension perpendicular to the paper) of the dam, $W_{dam} = (20)(w)(1)(150) = 3000w$, $F = \gamma hA$, $F_H = (62.4)[(0 + 15)/2][(15)(1)] = 7020$ lb.

$$\text{F.S.}_{\text{sliding}} = \frac{\text{sliding resistance}}{\text{sliding force}} \qquad 1.5 = \frac{(0.42)(3000w)}{7020} \qquad w = 8.36 \text{ ft}$$

$$\text{F.S.}_{\text{overturning}} = \frac{\text{total righting moment}}{\text{overturning moment}} = \frac{[(3000)(8.36)](8.36/2)}{(7020)(\frac{15}{3})} = 2.99$$

Therefore, it should be safe against overturning.

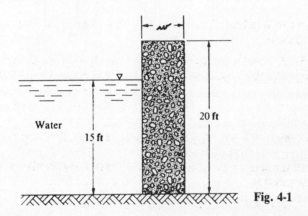

Water

15 ft

20 ft

Fig. 4-1

4.2 Figure 4-2 shows a cross section through a concrete dam where the specific weight of concrete is 2.5γ and γ is the specific weight of water. Assuming that hydrostatic uplift varies linearly from one-half the hydrostatic head at the upstream edge of the dam to zero at the downstream edge, find the maximum and minimum pressure intensity in the base of the dam.

▮ $\qquad F = \gamma hA \qquad F_H = (62.4)[(0 + 100)/2][(100)(1)] = 312\,000$ lb

For equilibrium, $R_x = 312\,000$ lb.

$$W_1 = [(2.5)(62.4)][(1)(10)(90 + 15)] = 163\,800 \text{ lb} \qquad W_2 = [(2.5)(62.4)][(1)(60)(90)/2] = 421\,200 \text{ lb}$$

$$F_U = [(62.4)(50 + 0)/2][(60 + 10)(1)] = 109\,200 \text{ lb} \qquad R_y = 163\,800 + 421\,200 - 109\,200 = 475\,800 \text{ lb}$$

$$\sum M_0 = 0 \qquad (312\,000)(33.33) + (163\,800)(5) + (421\,200)(30) - (109\,200)[(60 + 10)/3] - 475\,800x = 0$$

$$x = 44.78 \text{ ft} \qquad \text{Eccentricity} = 44.78 - (60 + 10)/2 = 9.78 \text{ ft}$$

Since the eccentricity is less than one-sixth the base of the dam, the resultant acts within the middle third of the base.

$$p = \frac{F}{A} \pm \frac{M_y x}{I_y} \pm \frac{M_x y}{I_x} = \frac{475\,800}{(60 + 10)(1)} \pm \frac{[(475\,800)(9.78)](60 + 10)/2}{(1)(60 + 10)^3/12} \pm 0 = 6797 \pm 5698$$

$$p_{\max} = 6797 + 5698 = 12\,495 \text{ lb/ft}^2 \qquad p_{\min} = 6797 - 5698 = 1099 \text{ lb/ft}^2$$

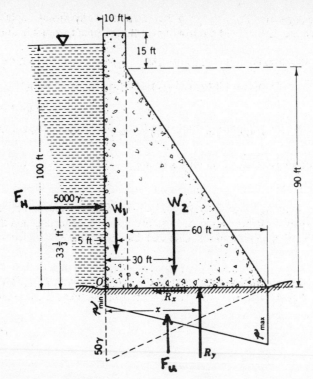

Fig. 4-2

4.3 For linear stress variation over the base of the dam of Fig. 4-3a, locate where the resultant crosses the base and compute the maximum and minimum pressure intensity at the base. Neglect hydrostatic uplift.

▌ Figure 4-3b shows the forces acting on the dam. $F_1 = \gamma[(20+7)/2][(20+7)(1)] = 364\gamma$, $F_2 = \gamma[(7)(3)(1)] = 21\gamma$, $F_3 = \gamma[(1)(20)(3)/2] = 30\gamma$, $F_4 = [(2.5)(\gamma)][(4)(20+7)(1)] = 270\gamma$, $F_5 = [(2.5)(\gamma)][(1)(20)(3)/2] = 75\gamma$, $F_6 = [(2.5)(\gamma)][(1)(20)(11)/2] = 275\gamma$; $R_y = 21\gamma + 30\gamma + 270\gamma + 75\gamma + 275\gamma = 671\gamma$. $\sum M_A = 0$; $(671\gamma)(x) - (364\gamma)[(20+7)/3] - (21\gamma)(1.5) - (30\gamma)(1) - (270\gamma)(3+2) - (75\gamma)(3-1) - (275\gamma)(4+3+\frac{11}{3}) = 0$, $x = 11.58$ m. Eccentricity $= 11.58 - (11+4+3)/2 = 2.58$ ft. Since the eccentricity is less than one-sixth the base of the dam, the resultant acts within the middle third of the base.

$$p = \frac{F}{A} \pm \frac{M_y x}{I_y} \pm \frac{M_x y}{I_x} = \frac{(671)(9.79)}{(11+4+3)(1)} \pm \frac{[(671)(9.79)(2.58)][(11+4+3)/2]}{(1)(11+4+3)^3/12} \pm 0 = 365 \pm 314$$

$$p_{max} = 365 + 314 = 679 \text{ kN/m}^2 \qquad p_{min} = 365 - 314 = 51 \text{ kN/m}^2$$

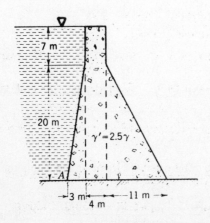

Fig. 4-3(a)

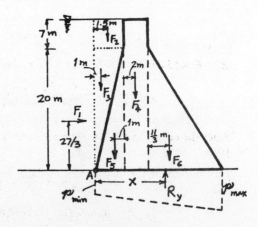

Fig. 4-3(b)

4.4 For the conditions given in Prob. 4.3 with the addition that hydrostatic uplift varies linearly from 20 m at A to zero at the toe of the dam, would the resultant still act within the middle third of the base?

▋ $F_U = \gamma[(20 + 0)/2][(4 + 3 + 11)(1)] = 180\gamma$ $\quad R_y = 21\gamma + 30\gamma + 270\gamma + 75\gamma + 275\gamma - 180\gamma = 491\gamma$

$\sum M_A = 0$ $\quad (491\gamma)(x) - (364\gamma)[(20 + 7)/3] - (21\gamma)(1.5) - (30\gamma)(1) - (270\gamma)(3 + 2) - (75\gamma)(3 - 1)$

$$- (275\gamma)(4 + 3 + \tfrac{11}{3}) + (180\gamma)[(4 + 3 + 11)/3] = 0$$

$x = 13.63$ m $\quad$ Eccentricity $= 13.63 - (11 + 4 + 3)/2 = 4.63$ ft

Since the eccentricity is greater than one-sixth the base of the dam, the resultant acts outside the middle third of the base.

4.5 A concrete dam retaining water is shown in Fig. 4-4a. If the specific weight of the concrete is 150 lb/ft³, find the factor of safety against sliding, the factor of safety against overturning, and the pressure intensity on the base. Assume the foundation soil is impermeable and that the coefficient of friction between dam and foundation soil is 0.45.

▋ The forces acting on the dam are shown in Fig. 4.4b. $F = \gamma h A$, $F_x = (62.4)[(0 + 42)/2][(42)(1)] = 55\,040$ lb. From Fig. 4-4b, $CD/42 = \tfrac{10}{50}$, $CD = 8.40$ ft; $F_y = (62.4)[(8.40)(42)/2](1) = 11\,010$ lb.

component	weight of component (kips)	moment arm from toe, B (ft)	righting moment about toe, B (kip · ft)
1	$(\tfrac{1}{2})(10 \times 50)(0.15)(1) = 37.50$	$20 + \tfrac{10}{3} = 23.33$	875
2	$(10 \times 50)(0.15)(1) = 75.00$	$10 + \tfrac{10}{2} = 15.00$	1125
3	$(\tfrac{1}{2})(10 \times 50)(0.15)(1) = 37.50$	$(\tfrac{2}{3})(10) = 6.67$	250
F_y	11.01	$30 - (\tfrac{1}{3})(8.40) = 27.20$	299
	$\sum V = 161.01$ **kips**		$\sum M_r = 2549$ **kip · ft**

$M_{\text{overturning}} = (55.04)(\tfrac{42}{3}) = 771$ kip-ft $\quad$ F.S.$_{\text{·sliding}} = \dfrac{\text{sliding resistance}}{\text{sliding force}} = \dfrac{(0.45)(161.01)}{55.04} = 1.32$

$$\text{F.S.}_{\text{·overturning}} = \frac{\text{total righting moment}}{\text{overturning moment}} = \frac{2549}{771} = 3.31$$

$R_x = F_x = 55.04$ kips and $R_y = \sum V = 161.01$ kips; hence, $R = \sqrt{55.04^2 + 161.01^2} = 170.16$ kips.

$$x = \frac{\sum M_B}{R_y} = \frac{\sum M_r - M_0}{\sum V} = \frac{2549 - 771}{161.01} = 11.04 \text{ ft} \quad \text{Eccentricity} = \tfrac{30}{2} - 11.04 = 3.96 \text{ ft}$$

Since the eccentricity is less than one-sixth the base of the dam, the resultant acts within the middle third of the base.

$$p = \frac{F}{A} \pm \frac{M_y x}{I_y} \pm \frac{M_x y}{I_x} = \frac{161.01}{(30)(1)} \pm \frac{[(161.01)(3.96)](15)}{(1)(30)^3/12} \pm 0 = 5.37 \pm 4.25$$

$$p_B = 5.37 + 4.25 = 9.62 \text{ kips/ft}^2 \quad p_A = 5.37 - 4.25 = 1.12 \text{ kips/ft}^2$$

The complete pressure distribution on the base of the dam is given in Fig. 4-4c.

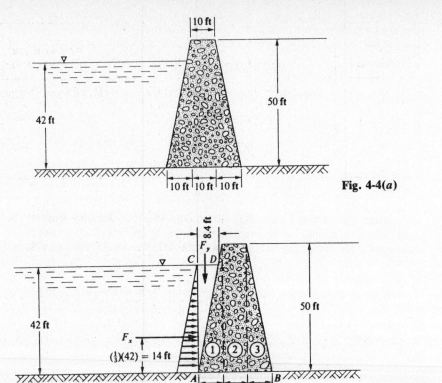

Fig. 4-4(a)

Fig. 4-4(b)

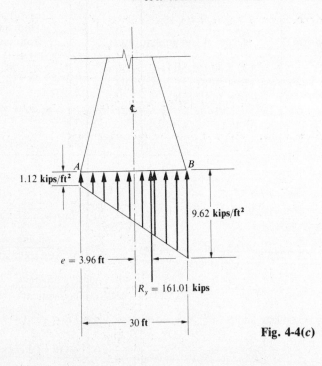

Fig. 4-4(c)

4.6   A concrete dam retaining water is shown in Fig. 4-5a. If the specific weight of the concrete is 23.5 kN/m³, find the factor of safety against sliding, the factor of safety against overturning, and the pressure intensity on the base. Assume there is a hydrostatic uplift that varies uniformly from full hydrostatic head at the heel of the dam to zero at the toe and that the coefficient of friction between dam and foundation soil is 0.45.

▐   The forces acting on the dam are shown in Fig. 4-5b. $F = \gamma h A$, $F_x = (9.79)[(0 + 14)/2][(14)(1)] = 959.4$ kN, $F_y = (9.79)[(3)(14 - 3)(1)] = 323.1$ kN. Hydrostatic uplift varies from $(14)(9.79)$, or 137.1 kN/m² at the heel to zero at the toe, as shown in Fig. 4-5b. $F_U = (137.1/2)(15)(1) = 1028$ kN. It acts at $(\frac{1}{3})(15)$, or 5.0 m from point A, as shown in Fig. 4-5b.

component	weight of component (kN)	moment arm from toe, B (m)	righting moment about toe, B (kN · m)
1	$(\frac{1}{2})(15-3-4)(12)(23.5)(1) = 1128$	$(\frac{2}{3})(15-3-4) = 5.333$	6 016
2	$(4)(12+3)(23.5)(1) = 1410$	$(15-3-\frac{4}{2}) = 10.000$	14 100
3	$(15)(3)(23.5)(1) = 1058$	$\frac{15}{2} = 7.500$	7 935
F_y	$= 323$	$(15-\frac{3}{2}) = 13.500$	4 360
	$\sum V = 3919 \text{ kN}$		$\sum M_r = 32\,411 \text{ kN} \cdot \text{m}$

$$M_{\text{overturning}} = (959.4)(\tfrac{14}{3}) + (1028)(10) = 14\,760 \text{ kN}$$

$$\text{F.S.}_{\text{sliding}} = \frac{\text{sliding resistance}}{\text{sliding force}} = \frac{(0.45)(3919-1028)}{959.4} = 1.36$$

$$\text{F.S.}_{\text{overturning}} = \frac{\text{total righting moment}}{\text{overturning moment}} = \frac{32\,411}{14\,760} = 2.20$$

$R_x = F_x = 959.4 \text{ kN}$ and $R_y = \sum V - F_U = 3919 - 1028 = 2891 \text{ kN}$; hence, $R = \sqrt{959.4^2 + 2891^2} = 3046 \text{ kN}$.

$$x = \frac{\sum M_B}{R_y} = \frac{\sum M_r - M_0}{\sum V} = \frac{32\,411 - 14\,760}{2891} = 6.105 \text{ m} \qquad \text{Eccentricity} = \tfrac{15}{2} - 6.105 = 1.395 \text{ m}$$

Since the eccentricity is less than one-sixth the base of the dam, the resultant acts within the middle third of the base.

$$p = \frac{F}{A} \pm \frac{M_y x}{I_y} \pm \frac{M_x y}{I_x} = \frac{2891}{(15)(1)} \pm \frac{[(2891)(1.395)](\frac{15}{2})}{(1)(15)^3/12} \pm 0 = 192.7 \pm 107.5$$

$$p_B = 192.7 + 107.5 = 300.2 \text{ kN/m}^2 \qquad p_A = 192.7 - 107.5 = 85.2 \text{ kN/m}^2$$

The complete pressure distribution on the base of the dam is given in Fig. 4-5c.

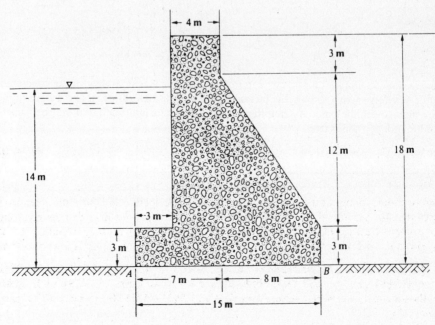

Fig. 4-5(a)

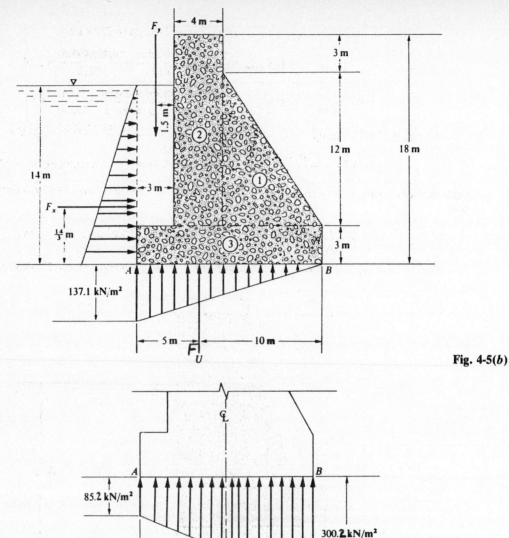

Fig. 4-5(b)

Fig. 4-5(c)

4.7 A concrete dam retaining water is shown in Fig. 4-6a. If the specific weight of the concrete is 23.5 kN/m³, find the factor of safety against sliding, the factor of safety against overturning, and the maximum and minimum pressure intensity on the base. Assume there is no hydrostatic uplift and that the coefficient of friction between dam and foundation soil is 0.48.

▌ The forces acting on the dam are shown in Fig. 4-6b. $F = \gamma h A$, $F_H = (9.79)[(0 + 6)/2][(6)(1)] = 176.2$ kN.

component	weight of component (kN)	moment arm from toe, A (m)	righting moment about toe, A (kN · m)
1	$(\frac{1}{2})(2)(7)(23.5) = 164.5$	$(\frac{2}{3})(2) = 1.333$	219
2	$(2)(7)(23.5) = 329.0$	$2 + \frac{2}{2} = 3.000$	987
	$\sum V = 493.5$ **kN**		$\sum M_r = 1206$ **kN · m**

$$M_{\text{overturning}} = (176.2)(\tfrac{6}{3}) = 352.4 \text{ kN}$$

$$\text{F.S.}_{\cdot\text{sliding}} = \frac{\text{slinding resistance}}{\text{sliding force}} = \frac{(0.48)(493.5)}{176.2} = 1.34$$

$$\text{F.S.}_{\cdot\text{overturning}} = \frac{\text{total righting moment}}{\text{overturning moment}} = \frac{1206}{352.4} = 3.42$$

$R_x = F_H = 176.2 \text{ kN}$ and $R_y = \sum V = 493.5 \text{ kN}$; hence, $R = \sqrt{176.2^2 + 493.5^2} = 524 \text{ kN}$.

$$x = \frac{\sum M_A}{R_y} = \frac{\sum M_r - M_0}{\sum V} = \frac{1206 - 352.4}{493.5} = 1.730 \text{ m} \qquad \text{Eccentricity} = \tfrac{4}{2} - 1.730 = 0.270 \text{ m}$$

Since the eccentricity is less than one-sixth the base of the dam, the resultant acts within the middle third of the base.

$$p = \frac{F}{A} \pm \frac{M_y x}{I_y} \pm \frac{M_x y}{I_x} = \frac{493.5}{(4)(1)} \pm \frac{[(493.5)(0.270)](\tfrac{4}{2})}{(1)(4)^3/12} \pm 0 = 123.4 \pm 50.0$$

$$p_{\text{max}} = 123.4 + 50.0 = 173.4 \text{ kN/m}^2 \qquad p_{\text{min}} = 123.4 - 50.0 = 73.4 \text{ kN/m}^2$$

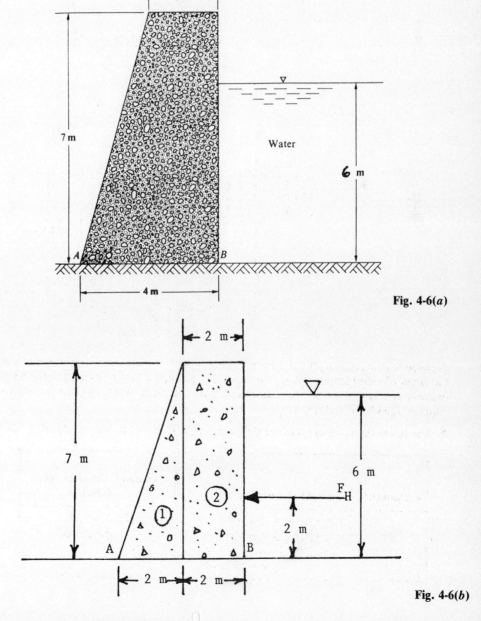

Fig. 4-6(a)

Fig. 4-6(b)

4.8   For the dam shown in Fig. 4-7, what is the minimum width b for the base of a dam 100 ft high if hydrostatic uplift is assumed to vary uniformly from full hydrostatic head at the heel to zero at the toe, and also assuming an ice thrust P_i of 12 480 lb per linear foot of dam at the top? For this study, make the resultant of the reacting forces cut the base at the downstream edge of the middle third of the base (i.e., at O in Fig. 4-7) and take the weight of the masonry as 2.50γ.

$$F = \gamma h A \qquad F_H = (62.4)[(100 + 0)/2][(100)(1)] = 312\,000\,\text{lb} \qquad F_U = [(100)(62.4)/2][(1)(b)] = 3120b$$

$$W_1 = [(2.50)(62.4)][(20)(100)(1)] = 312\,000\,\text{lb} \qquad W_2 = [(2.50)(62.4)][(b - 20)(100)(1)/2] = 7800b - 156\,000$$

$$\sum M_O = 0$$

$$(312\,000)(\tfrac{100}{3}) + (3120b)(b/3) - (312\,000)[(\tfrac{2}{3})(b) - \tfrac{20}{2}] - (7800b - 156\,000)[(\tfrac{2}{3})(b - 20) - b/3] + (12\,480)(100) = 0$$

$$3b^2 + 100b - 24\,400 = 0 \qquad b = 75.0\,\text{ft}$$

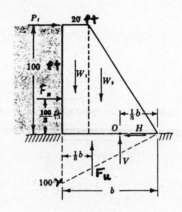

Fig. 4-7

CHAPTER 5

Forces on Submerged Curved Areas

5.1 The submerged, curved surface AB in Fig. 5-1a is one-quarter of a circle of radius 4 ft. The tank's length (distance perpendicular to the plane of the figure) is 6 ft. Find the horizontal and vertical components of the total resultant force acting on the curved surface and their locations.

   I The horizontal component of the total resultant force acting on the curved surface is equal to the total resultant force, F_H, acting on the vertical projection of curved surface AB (i.e., BF in Fig. 5-1b). This projection is a rectangle 6 ft long and 4 ft high. For the portion of F_H resulting from horizontal pressure of $BHEF$ in Fig. 5-1b, $p_1 = (8)(62.4) = 499$ lb/ft^2, $A = (6)(4) = 24$ ft^2, $F_1 = (499)(24) = 11\,980$ lb. For the portion of F_H resulting from horizontal pressure of HGE in Fig. 5-1b, $p_2 = (62.4)[(0 + 4)/2] = 125$ lb/ft^2, $F_2 = (125)(24) = 3000$ lb; $F_H = F_1 + F_2 = 11\,980 + 3000 = 14\,980$ lb. The vertical component of the total resultant force acting on the curved surface is equal to the weight of the volume of water vertically above curved surface AB. This volume consists of a rectangular area ($AFCD$ in Fig. 5-1c) 4 ft by 8 ft and a quarter-circular area (ABF in Fig. 5-1c) of radius 4 ft, both areas being 6 ft long. This volume (V) is $V = [(4)(8) + (\pi)(4)^2/4](6) = 267.4$ ft^3, $F_V =$ weight of water in $V = (267.4)(62.4) = 16\,690$ lb. The location of the horizontal component (F_H) is along a (horizontal) line through the center of pressure for the vertical projection (i.e., the center of gravity of $EFBG$ in Fig. 5-1b). This can be determined by equating the sum of the moments of F_1 and F_2 about point C to the moment of F_H about the same point. $(11\,980)(8 + \frac{4}{2}) + (3000)[8 + (\frac{2}{3})(4)] = 14\,980h_{cp}$, $h_{cp} = 10.13$ ft. (This is the depth from the water surface to the location of the horizontal component. Stated another way, the horizontal component acts at a distance of $12 - 10.13$, or 1.87 ft above point B in Fig. 5-1b.) The location of the vertical component (F_V) is

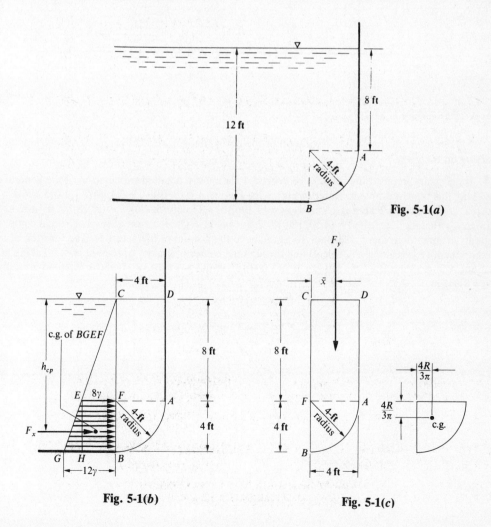

Fig. 5-1(a)

Fig. 5-1(b)

Fig. 5-1(c)

along a (vertical) line through the center of gravity of the liquid volume vertically above surface AB (i.e., the center of gravity of $ABCD$ in Fig. 5-1c). This can be determined by referring to Fig. 5-1c and equating the sum of the moments of the rectangular area ($AFCD$ in Fig. 5-1c) and of the quarter-circular area (ABF in Fig. 5-1c) about a vertical line through point B to the moment of the total area about the same line. $(x)[(8)(4) + (\pi)(4)^2/4] = [(8)(4)](\frac{4}{2}) + [(\pi)(4)^2/4][(4)(4)/(3\pi)]$, $x = 1.91$ ft. (This is the distance from point B to the line of action of the vertical component.)

5.2   Solve Prob. 5.1 for the same given conditions except that water is on the other side of curved surface AB, as shown in Fig. 5-2.

❚ If necessary, refer to the solution of Prob. 5.1 for a more detailed explanation of the general procedure for solving this type of problem. $p = p_{avg} = (\gamma)[(h_1 + h_2)/2] = (62.4)[(8 + 12)/2] = 624$ lb/ft^2, $A = (6)(4) = 24$ ft^2, $F_H = pA = (624)(24) = 14\,980$ lb. The vertical component (F_V) is equal to the weight of the imaginary volume of water vertically above surface AB. Hence, $F_V = [(4)(8) + (\pi)(4)^2/4](6)(62.4) = 16\,690$ lb. The location of the horizontal component is 10.13 ft below the water surface (same as in Prob. 5.1 except that F_H acts toward the left). The location of the vertical component is 1.91 ft from point B (same as in Prob. 5.1 except that F_V acts upward).

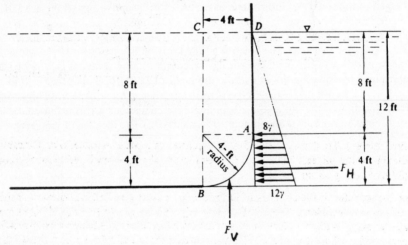

Fig. 5-2

5.3   The submerged sector gate AB shown in Fig. 5-3a is one-sixth of a circle of radius 6 m. The length of the gate is 10 m. Determine the amount and location of the horizontal and vertical components of the total resultant force acting on the gate.

❚ If necessary, refer to the solution of Prob. 5.1 for a more detailed explanation of the general procedure for solving this type of problem. Refer to Fig. 5-3b. $F_H = \gamma \bar{h} A = (9.79)[(0 + 5.196)/2][(10)(5.196)] = 1322$ kN, Area$_{ABC}$ = area$_{ACBD}$ + area$_{BDO}$ − area$_{ABO}$ = $(5.196)(3) + (3.000)(5.196)/2 - (\pi)(6)^2/6 = 4.532$ m^2, F_V = (area$_{ABC}$)(length of gate)(γ) = $(4.532)(10)(9.79) = 444$ kN. The location of the horizontal component (F_H) is along a (horizontal) line 5.196/3, or 1.732 m above the bottom of the gate (A). The location of the vertical component (F_V) is along a (vertical) line through the center of gravity of section ABC. Taking area moments about AC, $4.532x = [(5.196)(3)](\frac{3}{2}) + [(\frac{1}{2})(3.000)(5.196)](3 + 3.000/3) - [(\pi)(6)^2/6]\{6 - [\cos(60°/2)](2)(6)/\pi\}$, $x = 0.842$ m.

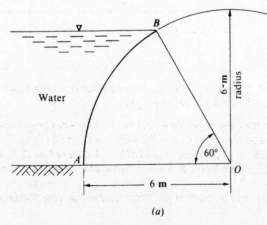

(a)

Fig. 5-3(a)

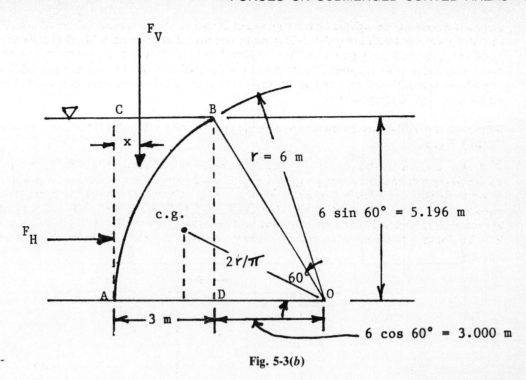

Fig. 5-3(b)

5.4 The curved surface AB shown in Fig. 5-4a is a quarter of a circle of radius 5 ft. Determine, for an 8-ft length perpendicular to the paper, the amount and location of the horizontal and vertical components of the total resultant force acting on surface AB.

▮ If necessary, refer to the solution of Prob. 5.1 for a more detailed explanation of the general procedure for solving this type of problem. Refer to Fig. 5-4b. $F_H = \gamma \bar{h} A = (62.4)[(0+5)/2][(5)(8)] = 6240$ lb, area$_{ABD} =$ area$_{ACBD} -$ area$_{ABC} = (5)(5) - (\pi)(5)^2/4 = 5.365$ ft^2, $F_V = ($area$_{ABD})($length$)(\gamma) = (5.365)(8)(62.4) = 2678$ lb. F_H is located at $\frac{5}{3}$, or 1.67 ft above C. F_V is located at x from line AD. $5.365x = [(5)(5)](\frac{5}{2}) - [(\pi)(5)^2/4][5 - (4)(5)/(3\pi)]$, $x = 1.12$ ft.

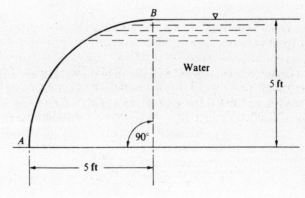

Fig. 5-4(a)

5.5 Determine the value and location of the horizontal and vertical components of the force due to water acting on curved surface AB in Fig. 5-5, per foot of its length.

▮ If necessary, refer to the solution of Prob. 5.1 for a more detailed explanation of the general procedure for solving this type of problem. $F_H = \gamma \bar{h} A = (62.4)[(0+6)/2][(6)(1)] = 1123$ lb, $F_V = ($area$)($length$)(\gamma) = [(\pi)(6)^2/4](1)(62.4) = 1764$ lb. F_H is located at $(\frac{2}{3})(6)$, or 4.00 ft below C. F_V is located at the center of gravity of area ABC, or distance x from line CB. $x = 4r/(3\pi) = (4)(6)/(3\pi) = 2.55$ ft.

5.6 The 6-ft-diameter cylinder in Fig. 5-6 weighs 5000 lb and is 5 ft long. Determine the reactions at A and B, neglecting friction.

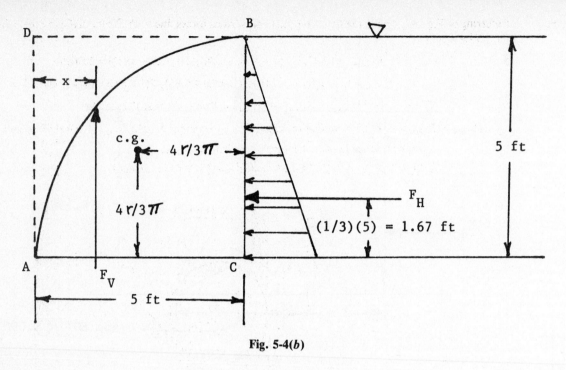

Fig. 5-4(b)

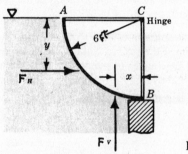

Fig. 5-5

▌ The reaction at A is due to the horizontal component of the liquid force acting on the cylinder (F_H).
$F_H = \gamma \bar{h} A = [(0.800)(62.4)][(0+6)/2][(3+3)(5)] = 4493$ lb. F_H acts to the right; hence, the reaction at A is
4493 lb to the left. The reaction at B is the algebraic sum of the weight of the cylinder and the net vertical
component of the force due to the liquid. $(F_V)_{up} = (\text{area}_{ECOBDE})(\text{length})(\gamma)$, $(F_V)_{down} = (\text{area}_{ECDE})(\text{length})(\gamma)$,
$(F_V)_{net} = (F_V)_{up} - (F_V)_{down} = (\text{area}_{COBDC})(\text{length})(\gamma) = [(\pi)(3)^2/2](5)[(0.800)(62.4)] = 3529$ lb (upward). The
reaction at B is $5000 - 3529$, or 1471 lb upward.

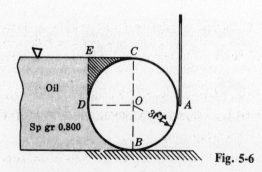

Fig. 5-6

5.7 Referring to Fig. 5-7, determine the horizontal and vertical forces due to the water acting on the cylinder per foot of its length.

$$(F_H)_{CDA} = 62.4\{[4 + (4 + 4.24 + 0.88)]/2\}[(2.12 + 3)(1)] = 2096 \text{ lb}$$

$$(F_H)_{AB} = (62.4)\{[(4 + 4.24) + (4 + 4.24 + 0.88)]/2\}[(0.88)(1)] = 477 \text{ lb}$$

$$(F_H)_{net} = (F_H)_{CDA} - (F_H)_{AB} = 2096 - 477 = 1619 \text{ lb (right)}$$

$$(F_V)_{net} = (F_V)_{DAB} - (F_V)_{DC} = \text{weight of volume}_{DABFED} - \text{weight of volume}_{DCGED} = \text{weight of volume}_{DABFGCD}$$

$$= \text{weight of (rectangle}_{GFJC} + \text{triangle}_{CJB} + \text{semicircle}_{CDAB})$$

$$= 62.4[(4)(4.24) + (4.24)(4.24)/2 + (\pi)(3)^2/2](1) = 2501 \text{ lb (upward)}$$

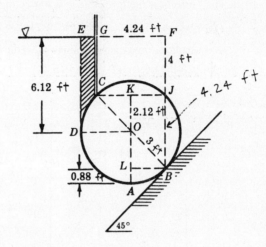

Fig. 5-7

5.8 In Fig. 5-8, an 8-ft-diameter cylinder plugs a rectangular hole in a tank that is 3 ft long. With what force is the cylinder pressed against the bottom of the tank due to the 9-ft depth of water?

$$(F_V)_{net} = (F_V)_{CDE} - (F_V)_{CA} - (F_V)_{BE} = 62.4[(4 + 4)(7) - (\pi)(4)^2/2](3)$$

$$- 62.4[(7)(0.54) + (\tfrac{30}{360})(\pi)(4)^2 - (2)(3.46)/2](3)$$

$$- 62.4[(7)(0.54) + (\tfrac{30}{360})(\pi)(4)^2 - (2)(3.46)/2](3) = 4090 \text{ lb (downward)}$$

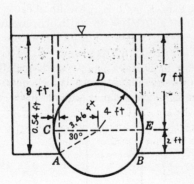

Fig. 5-8

5.9 In Fig. 5-9, the 8-ft-diameter cylinder weighs 500 lb and rests on the bottom of a tank that is 3 ft long. Water and oil are poured into the left- and right-hand portions of the tank to depths of 2 ft and 4 ft, respectively. Find the magnitudes of the horizontal and vertical components of the force that will keep the cylinder touching the tank at B.

$$(F_H)_{net} = (F_H)_{AB} - (F_H)_{CB} = [(0.750)(62.4)][(0 + 4)/2][(4)(3)] - (62.4)[(0 + 2)/2][(2)(3)] = 749 \text{ lb (left)}$$

$$(F_V)_{net} = (F_V)_{AB} + (F_V)_{CB} = [(0.750)(62.4)][(\pi)(4)^2/4](3) + (62.4)[(\tfrac{60}{360})(\pi)(4)^2 - (2)(\sqrt{12})/2](3)$$

$$= 2684 \text{ lb (upward)}$$

The components to hold the cylinder in place are 749 lb to the right and 2684 − 500, or 2184 lb down.

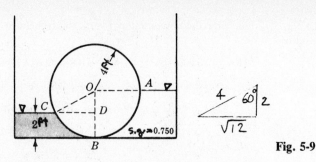

Fig. 5-9

5.10   The half-conical buttress ABE shown in Fig. 5-10 is used to support a half-cylindrical tower $ABCD$. Calculate
the horizontal and vertical components of the force due to water acting on the buttress.

$$F_H = \gamma h_{cg} A = (62.4)(3 + \tfrac{6}{3})[(6)(2+2)/2] = 3744 \text{ lb (right)}$$

$$F_V = \text{weight of (imaginary) volume of water above curved surface}$$

$$= (62.4)[(\tfrac{1}{3})(6)(\pi)(2)^2/3 + (\tfrac{1}{2})(\pi)(2)^2(3)] = 1960 \text{ lb (up)}$$

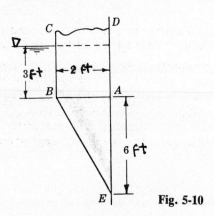

Fig. 5-10

5.11   A dam has a parabolic shape $z = z_0(x/x_0)^2$, as shown in Fig. 5-11a. The fluid is water and atmospheric pressure
may be neglected. If $x_0 = 10$ ft and $z_0 = 24$ ft, compute forces F_H and F_V on the dam and the position c.p. where
they act. The width of the dam is 50 ft.

$F_H = \gamma \bar{h} A = 62.4[(24+0)/2][(24)(50)] = 898\,600$ lb. The location of F_H is along a (horizontal) line $\tfrac{24}{3}$, or
8.00 ft above the bottom of the dam. $F_V = (\text{area}_{0AB})(\text{width of dam})(\gamma)$. (See Fig. 5-11b.) $\text{Area}_{0AB} = 2x_0 Z_0/3 =$
$(2)(10)(24)/3 = 160$ ft^2, $F_V = (160)(50)(62.4) = 499\,200$ lb. The location of F_V is along a (vertical) line through the
center of gravity of area_{0AB}. From Fig. 5-11b, $x = 3x_0/8 = (3)(10)/8 = 3.75$ ft, $z = 3z_0/5 = (3)(24)/5 = 14.4$ ft,
$F_{\text{resultant}} = \sqrt{499\,200^2 + 898\,600^2} = 1\,028\,000$ lb. As seen in Fig. 5-11c, $F_{\text{resultant}}$ acts down and to the right at an
angle of arctan $(499\,200/898\,600)$, or 29.1°. $F_{\text{resultant}}$ passes through the point $(x, z) = (3.75$ ft, 8 ft). If we move
down alone the 29.1° line until we strike the dam, we find an equivalent center of pressure on the dam at
$x = 5.43$ ft and $z = 7.07$ ft. This definition of c.p. is rather artificial, but this is an unavoidable complication of
dealing with a curved surface.

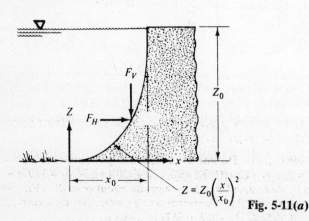

Fig. 5-11(a)

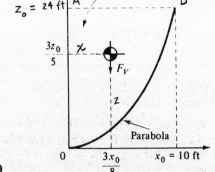

Fig. 5-11(b)

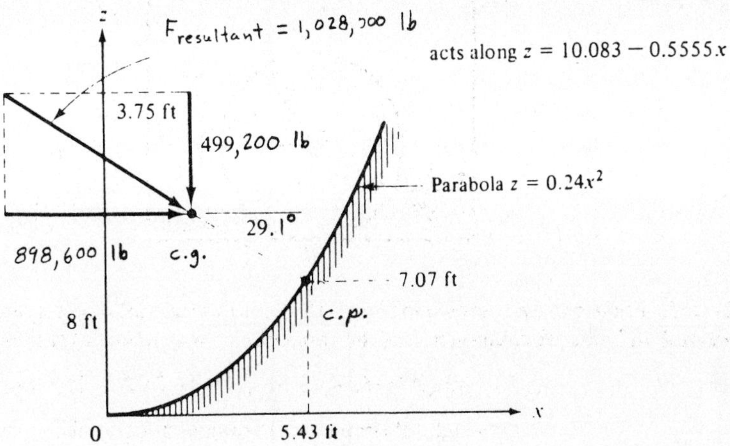

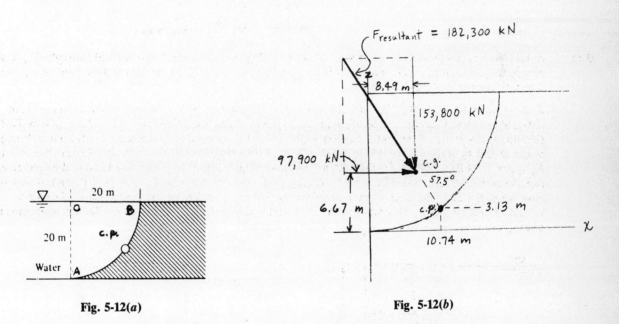

Fig. 5-11(c)

5.12 The dam in Fig. 5-12a is a quarter circle 50 m wide into the paper. Determine the horizontal and vertical components of the hydrostatic force against the dam and the point c.p. where the resultant strikes the dam.

┃ $F_H = \gamma \bar{h} A = 9.79[(20 + 0)/2][(20)(50)] = 97\,900$ kN. The location of F_H is along a (horizontal) line $\frac{20}{3}$, or 6.67 m above the bottom of the dam. $F_V = 9.79[(50)(\pi)(20)^2/4] = 153\,800$ kN. The location of F_V is along a (vertical) line through the center of gravity of area$_{OAB}$. $x = 4r/(3\pi) = (4)(20)/(3\pi) = 8.49$ m, $F_{resultant} = \sqrt{97\,900^2 + 153\,800^2} = 182\,300$ kN. As seen in Fig. 5-12b, $F_{resultant}$ acts down and to the right at an angle of arctan $(153\,800/97\,900)$, or 57.5°. $F_{resultant}$ passes through the point $(x, z) = (8.49$ m, 6.67 m). If we move down along the 57.5° line until we strike the dam, we find an equivalent center of pressure on the dam at $x = 10.74$ m and $z = 3.13$ m.

Fig. 5-12(a)   **Fig. 5-12(b)**

5.13 Gate AB in Fig. 5-13a is a quarter circle 10 ft wide into the paper and hinged at B. Find the force F just sufficient to keep the gate from opening. Neglect the weight of the gate.

┃ $F_H = \gamma \bar{h} A = 62.4[(8 + 0)/2][(8)(10)] = 19\,968$ lb (left). The location of F_H is along a (horizontal) line $\frac{8}{3}$, or 2.667 ft above point B. (See Fig. 5-13b.) $F_V = F_1 - F_2 = 62.4[(10)(8)(8)] - 62.4[(10)(\pi)(8)^2/4] = 39\,936 - 31\,366 = 8570$ lb (up). The location of F_V can be determined by taking moments about point B in Fig. 5-13b. $8570x = (39\,936)(\frac{8}{2}) - (31\,366)[8 - (4)(8)/(3\pi)]$, $x = 1.787$ ft. The forces acting on the gate are shown in Fig. 5-13c. $\sum M_B = 0$; $8F - (2.667)(19\,968) - (1.787)(8570) = 0$, $F = 8571$ lb (down).

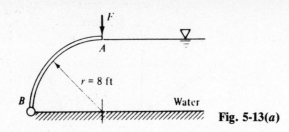

Fig. 5-13(*a*)

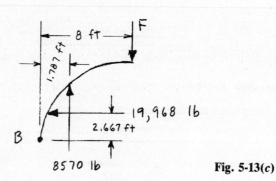

Fig. 5-13(*b*)

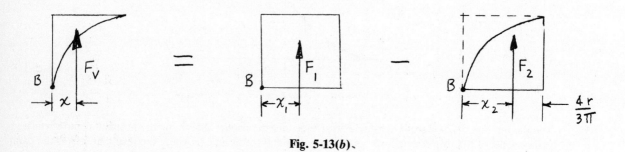

Fig. 5-13(*c*)

5.14 Repeat Prob. 5.13 by including the 4000-lb weight of the gate, which is a steel plate of uniform thickness.

▌ The weight of the gate acts at the center of gravity of the gate shown in Fig. 5-14. $2r/\pi =$ $(2)(8)/\pi = 5.093$ ft; $\sum M_B = 0$. From Prob. 5.14, $8F - (2.667)(19\,968) - (1.787)(8570) +$ $(4000)(8 - 5.093) = 0$, $F = 7118$ lb.

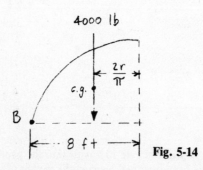

Fig. 5-14

5.15 Compute the horizontal and vertical components of the hydrostatic force on the quarter-circle panel of the water tank shown in Fig. 5-15*a*.

▌
$$F_H = \gamma h_{cg} A = 9.79[5 + \tfrac{2}{2}][(2)(6)] = 705 \text{ kN}$$

$$F_V = F_1 - F_2 = (9.79)[(6)(2)(7)] - (9.79)[(6)(\pi)(2)^2/4] = 638 \text{ kN} \qquad \text{(See Fig. 5-15}b.)$$

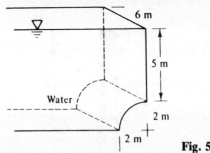

Fig. 5-15(a)

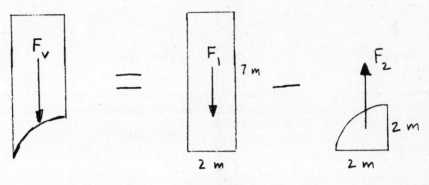

Fig. 5-15(b)

5.16 Compute the horizontal and vertical components of the hydrostatic force on the hemispherical bulge at the bottom of the tank shown in Fig. 5-16a.

∎ From symmetry, $F_H = 0$, $F_V = F_1 - F_2$ (see Fig. 5-16b). $F_V = 62.4[(\pi)(2)^2(10)] - (62.4)[(\frac{1}{2})(\frac{4}{3})(\pi)(2)^3] = 6796$ lb.

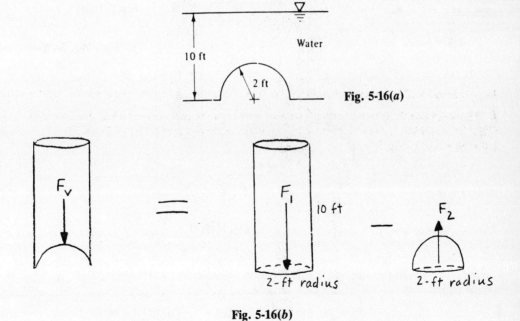

Fig. 5-16(a)

Fig. 5-16(b)

5.17 The bottle of champagne (s.g. = 0.96) in Fig. 5-17 is under pressure, as shown by the manometer reading. Compute the net force on the 2-in-radius hemispherical end cap at the bottom of the bottle.

∎ From symmetry, $F_H = 0$, $p_{AA} + [(0.96)(62.4)](\frac{2}{12}) - [(13.6)(62.4)](\frac{4}{12}) = p_{atm} = 0$, $p_{AA} = 273$ lb/ft^2 (gage); $F_V = p_{AA}A_{bottom} +$ weight of champagne below $AA = 273[(\pi)(\frac{4}{12})^2/4] + [(0.96)(62.4)][(\frac{6}{12})(\pi)(\frac{4}{12})^2/4] - [(0.96)(62.4)][(\frac{1}{2})(\frac{4}{3})(\pi)(\frac{2}{12})^3] = 25.9$ lb.

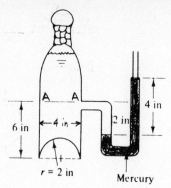

Fig. 5-17

5.18 Half-cylinder ABC in Fig. 5-18a is 10 ft wide into the paper. Calculate the net moment of the hydrostatic oil forces on the cylinder about point C.

▮ From symmetry, the horizontal forces balance and produce no net moment about point C. (See Fig. 5-18b.)
$F_V = F_1 - F_2 = F_{\text{buoyancy of body } ABC} = [(0.88)(62.4)][(10)(\pi)(\frac{8}{2})^2/2] = 13\,800$ lb, $x = 4r/(3\pi) = (4)(\frac{8}{2})/(3\pi) = 1.698$ ft, $M_C = (13\,800)(1.698) = 23\,430$ lb $\cdot$ ft (clockwise).

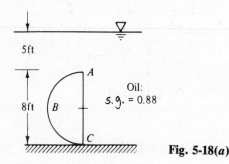

Fig. 5-18(a)

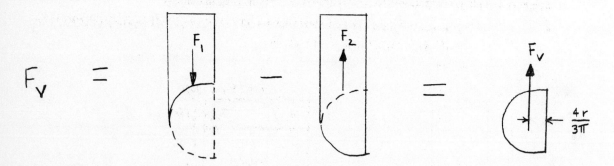

Fig. 5-18(b)

5.19 Compute the hydrostatic force and its line of action on semicylindrical bulge ABC in Fig. 5-19a for a 1-m width into the paper.

▮
$$F_H = \gamma h_{cg} A = [(0.88)(9.79)][(1 + 3 + \tfrac{2}{2})][(2)(1)] = 86.15 \text{ kN}$$

$$y_{cp} = \frac{-I_{xx} \sin\theta}{h_{cg}A} = \frac{-[(1)(2)^3/12](\sin 90°)}{(1 + 3 + \tfrac{2}{2})[(2)(1)]} = -0.067 \text{ m}$$

As demonstrated in Prob. 5.18, $F_V = F_{\text{buoyancy of body } ABC}$ and it acts at $4r/(3\pi)$ from point C. $F_V = [(0.88)(9.79)][(1)(\pi)(\frac{2}{2})^2/2] = 13.53$ kN, $x = 4r/(3\pi) = (4)(\frac{2}{2})/(3\pi) = 0.424$ m. The forces acting on the bulge are shown in Fig. 5-19b. $F_{\text{resultant}} = \sqrt{13.53^2 + 86.15^2} = 87.21$ kN. As shown in Fig. 5-19b, $F_{\text{resultant}}$ passes through point O and acts up and to the right at an angle of arctan $(13.53/86.15)$, or $8.93°$.

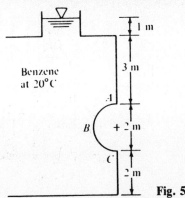

Fig. 5-19(a)

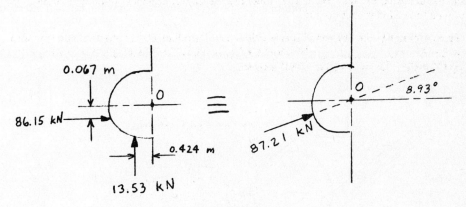

Fig. 5-19(b)

5.20   A 1-ft-diameter hole in the bottom of the tank in Fig. 5-20 is plugged by a conical 45° plug. Neglecting the weight of the plug, compute the force F required to keep the plug in the hole.

$F = pA_{\text{bottom}} + \text{weight of water above cone} = [(3)(144)][(\pi)(1)^2/4] + (62.4)[(3)(\pi)(1)^2/4]$
$- (62.4)[(\frac{1}{3})(1.207)(\pi)(1)^2/4] = 467 \text{ lb}$

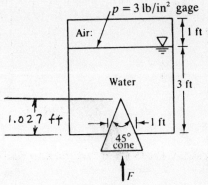

Fig. 5-20

5.21   The hemispherical dome in Fig. 5-21 weighs 30 kN, is filled with water, and is attached to the floor by six equally spaced bolts. What is the force in each bolt required to hold the dome down?

$F_V = \text{weight of (imaginary) water above the container}$
$= 9.79[(4 + 2)(\pi)(2)^2] - 9.79[(4)(\pi)(\frac{3}{100})^2/4] - 9.79[(\frac{1}{2})(\frac{4}{3})(\pi)(2)^3] = 574.1 \text{ kN (up)}$
$F_{\text{each bolt}} = 574.1/6 = 95.7 \text{ kN}$

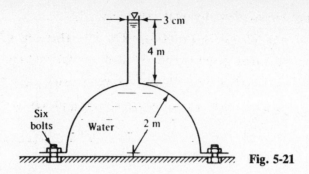

Fig. 5-21

5.22 A 4-m-diameter water tank consists of two half-cylinders, each weighing 4.5 kN/m, bolted together as shown in Fig. 5-22a. If support of the end caps is neglected, determine the force induced in each bolt.

▌ See Fig. 5-22b. Assuming the bottom half is properly supported, only the top half affects the bolt force. $p_1 = (9.79)(2 + 2) = 39.16 \text{ kN/m}^2$; $\sum F_y = p_1 A_1 - 2F_{\text{bolt}} - W_{\text{H}_2\text{O}} - W_{\text{tank half}} = 0$, $39.16[(4)(\frac{25}{100})] - 2F_{\text{bolt}} - 9.79[(\frac{25}{100})(\pi)(2)^2/2] - 4.5/4 = 0$, $F_{\text{bolt}} = 11.3 \text{ kN}$.

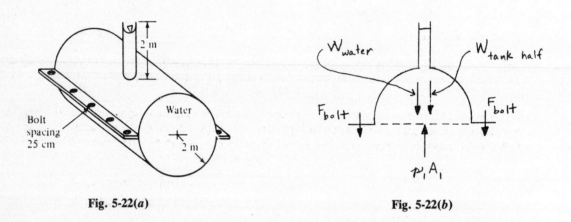

Fig. 5-22(a)   Fig. 5-22(b)

5.23 The 2-ft-radius cylinder in Fig. 5-23a is 8 ft wide into the paper. Compute the horizontal and vertical components of the water against the cylinder.

▌ See Fig. 5-23b. Note that the net horizontal force is based on the projected vertical area with depth AB. $F_H = \gamma h_{\text{cg}} A = 62.4[(2 + 1.414)/2][(2 + 1.414)(8)] = 2909 \text{ lb}$; F_V = equivalent weight of fluid in regions 1, 2, 3, and 4 = $(62.4)(8)[(\pi)(2)^2/2 + (1.414)(2) + (1.414)(1.414)/2 + (\pi)(2)^2/8] = 5831 \text{ lb}$.

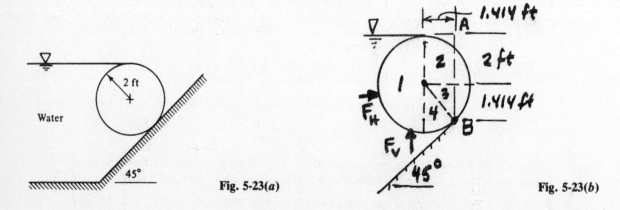

Fig. 5-23(a)   Fig. 5-23(b)

5.24 The 4-ft-diameter log (s.g. = 0.80) in Fig. 5-24a is 10 ft long into the paper and dams water as shown. Compute the net vertical and horizontal reactions at point C.

▌ $F = \gamma hA$. Figure 5-24b shows the forces acting on the log.

$$(F_H)_1 = 62.4[(0+4)/2][(2+2)(10)] = 4992 \text{ lb} \qquad (F_H)_2 = 62.4[(0+2)/2][(2)(10)] = 1248 \text{ lb}$$

$$(F_V)_1 = 62.4[(10)(\pi)(2)^2/2] = 3921 \text{ lb} \qquad (F_V)_2 = 62.4[(10)(\pi)(2)^2/4] = 1960 \text{ lb}$$

$$\sum F_x = 0 \qquad 4992 - 1248 - C_x = 0 \qquad C_x = 3744 \text{ lb (left)}$$

$$\sum F_y = 0 \qquad 3921 + 1960 - [(0.80)(62.4)][(10)(\pi)(2)^2] + C_y = 0 \qquad C_y = 392 \text{ lb (up)}$$

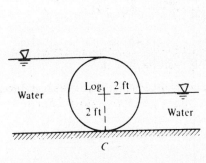

Fig. 5-24(a)

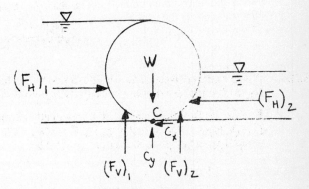

Fig. 5-24(b)

5.25 The 2-m-diameter cylinder in Fig. 5-25a is 5 m long into the paper and rests in static equilibrium against the smooth wall at point B. Compute the weight and specific gravity of the cylinder. Assume zero wall friction at point B.

▌ See Fig. 5-25b. If B is "smooth" it has no vertical force. Then the log weight must exactly balance the vertical hydrostatic force, which equals the equivalent weight of water in the shaded area. $W_{\text{log}} = F_V = (9.79)(5)[(\frac{3}{4})(\pi)(1)^2 + (1)(1)] = 164.3 \text{ kN}$, $\gamma_{\text{log}} = 164.3/[(5)(\pi)(1)^2] = 10.46 \text{ kN/m}^3$, s.g. $= 10.46/9.79 = 1.07$.

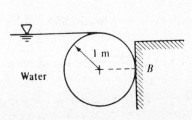

Fig. 5-25(a)

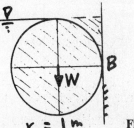

Fig. 5-25(b)

5.26 The tank in Fig. 5-26a is 2 m wide into the paper. Neglecting atmospheric pressure, compute the hydrostatic horizontal, vertical, and resultant force on quarter-circle panel BC.

▌ $F_H = \gamma h_{cg}A = (9.79)(6 + \frac{4}{2})[(4)(2)] = 626.6 \text{ kN}$, $F_V = $ weight of water above panel $BC = (9.79)[(2)(6)(4)] + (9.79)[(2)(\pi)(4)^2/4] = 716.0 \text{ kN}$, $F_{\text{resultant}} = \sqrt{626.6^2 + 716.0^2} = 951.5 \text{ kN}$. As seen in Fig. 5-26b, $F_{\text{resultant}}$ passes through point O and acts down and to the right at an angle of arctan (716.0/626.6), or 48.8°.

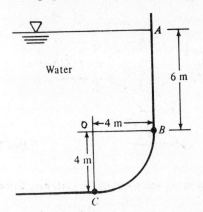

Fig. 5-26(a)

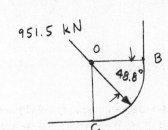

Fig. 5-26(b)

5.27 Gate AB in Fig. 5-27a is a quarter circle 8 ft wide, hinged at B and resting against a smooth wall at A. Compute the reaction forces at A and B.

▮ $F_H = \gamma h_{cg} A = (64)(12 - \frac{5}{2})[(8)(5)] = 24\,320$ lb $\qquad y_{cp} = \dfrac{-I_{xx} \sin \theta}{h_{cg} A} = \dfrac{-[(8)(5)^3/12](\sin 90°)}{(12 - \frac{5}{2})[(8)(5)]} = -0.219$ ft

Thus, F_H acts at $\frac{5}{2} - 0.219$, or 2.281 ft above point B. F_V = weight of seawater above gate AB = $(64)(8)[(12)(5)] - (64)(8)[(\pi)(5)^2/4] = 30\,720 - 10\,053 = 20\,667$ lb. The location of F_V can be determined by taking moments about point A in Fig. 5-27b. $(30\,720)(\frac{5}{2}) - (10\,053)[(4)(5)/(3\pi)] = 20\,667x$, $x = 2.684$ ft. The forces acting on the gate are shown in Fig. 5-27c.

$$\sum M_B = 0 \qquad (24\,320)(2.281) + (20\,667)(5 - 2.684) - 5A_x = 0 \qquad A_x = 20\,668 \text{ lb}$$

$$\sum F_x = 0 \qquad 24\,320 - B_x - 20\,668 = 0 \qquad B_x = 3652 \text{ lb}$$

$$\sum F_y = 0 \qquad B_y - 20\,667 = 0 \qquad B_y = 20\,667 \text{ lb}$$

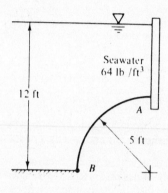

Fig. 5-27(a)

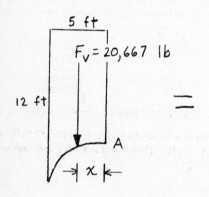 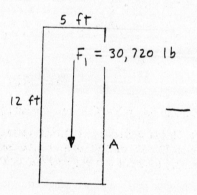 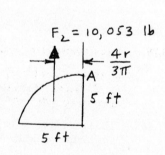

Fig. 5-27(b)

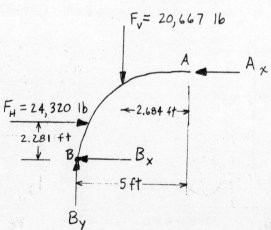

Fig. 5-27(c)

5.28 Gate *ABC* in Fig. 5-28*a* is a quarter circle 10 ft wide into the paper. Compute the horizontal and vertical hydrostatic forces on the gate and the line of action of the resultant force.

❚ See Fig. 5-28*b*. $F_H = \gamma h_{cg} A = (62.4)(2.828)[(5.656)(10)] = 9981$ lb, $F_V =$ weight of (imaginary) water in crosshatched area in Fig. 5-28*b* $= (62.4)(10)[(\pi)(4)^2/4 - (2)(4 \sin 45°)(4 \cos 45°)/2] = 2849$ lb; $F_{resultant} = \sqrt{2849^2 + 9981^2} = 10\,380$ lb. $F_{resultant}$ passes through point *O* and acts at an angle of arctan $\frac{2849}{9981}$, or 15.9°, as shown in Fig. 5-28*c*.

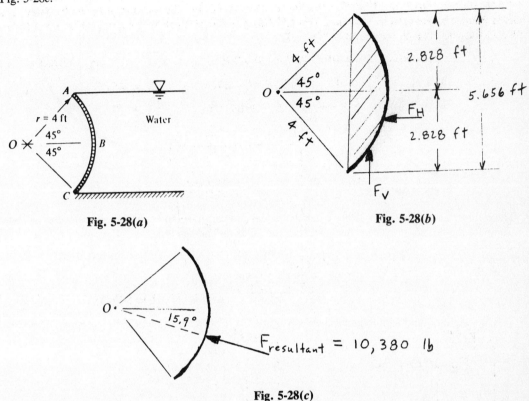

Fig. 5-28(*a*) Fig. 5-28(*b*)

Fig. 5-28(*c*)

5.29 Pressurized water fills the tank in Fig. 5-29*a*. Compute the net hydrostatic force on conical surface *ABC*.

❚ From symmetry, $F_H = 0$. The gage pressure of 150 kPa corresponds to a fictitious water level at 150/9.79, or 15.322 m above the gage or 15.322 − 7, or 8.322 m above *AC* (see Fig. 5-29*b*). $F_V =$ weight of fictitious water above cone *ABC* $= 9.79[(8.322)(\pi)(2)^2/4 + (\frac{1}{3})(4)(\pi)(2)^2/4] = 297$ kN (up).

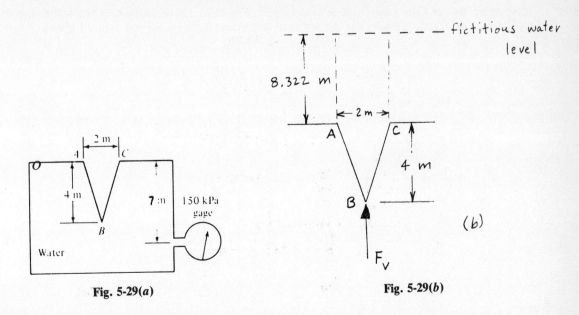

Fig. 5-29(*a*) Fig. 5-29(*b*)

5.30   Gate AB in Fig. 5-30a is 10 m wide into the paper, parabolic in shape, and hinged at B. Compute the force F required to hold the gate in equilibrium. Neglect atmospheric pressure.

▌ $F_H = \gamma \bar{h} A = 9.79[(8+0)/2][(8)(10)] = 3133$ kN. F_H acts at $\frac{8}{3}$, or 2.667 m above B (see Fig. 5-30b). F_V = weight of water above the gate = $9.79[(\frac{2}{3})(5)(8)(10)] = 2611$ kN. F_V acts at $\frac{15}{8}$, or 1.875 m right of B (see Fig. 5-30b). $\Sigma M_B = 0$; $(2.667)(3133) + (1.875)(2611) - 8F = 0$, $F = 1656$ kN.

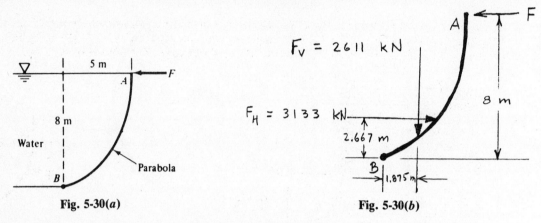

Fig. 5-30(a)   Fig. 5-30(b)

5.31   The cylindrical tank in Fig. 5-31 has a hemispherical end cap ABC and contains oil and water as shown. Compute the horizontal and vertical forces of the fluids on end cap ABC.

▌ $$F = \gamma h_{cg} A \qquad (F_H)_1 = [(0.9)(9.79)](2 + \tfrac{3}{2})[(\pi)(3)^2/2] = 436 \text{ kN} \quad \text{(left)}$$

$$(F_H)_2 = \{[(0.9)(9.79)](3+2) + (9.79)(\tfrac{3}{2})\}[(\pi)(3)^2]/2 = 830 \text{ kN} \quad \text{(left)}$$

$$(F_H)_{\text{total}} = 436 + 830 = 1266 \text{ kN} \quad \text{(left)}$$

$$F_V = \text{weight of fluid within hemisphere} = [(0.9)(9.79)][(\tfrac{1}{4})(\tfrac{4}{3})(\pi)(3)^3] + (9.79)[(\tfrac{1}{4})(\tfrac{4}{3})(\pi)(3)^3] = 526 \text{ kN} \quad \text{(down)}$$

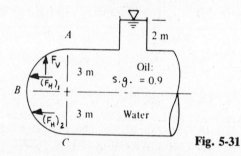

Fig. 5-31

5.32   A cylindrical barrier holds water, as shown in Fig. 5-32. The contact between cylinder and wall is smooth. Consider a 1-m length of cylinder and determine its weight and the force exerted against the wall.

▌ $$(F_V)_{BCD} = (9.79)(1)[(\pi)(2)^2/2 + (2)(2) + (2)(2)] = 139.8 \text{ kN} \quad \text{(up)}$$

$$(F_V)_{AB} = (9.79)(1)[(2)(2) - (\pi)(2)^2/4] = 8.4 \text{ kN} \quad \text{(down)}$$

$$\Sigma F_y = 0 \qquad 139.8 - W_{\text{cylinder}} - 8.4 = 0 \qquad W_{\text{cylinder}} = 131.4 \text{ kN}$$

$$F_H = \gamma h_{cg} A \qquad (F_H)_{ABC} = (9.79)(2)[(2+2)(1)] = 78.3 \text{ kN} \quad \text{(right)}$$

$$(F_H)_{DC} = (9.79)(2 + \tfrac{2}{2})[(2)(1)] = 58.7 \text{ kN} \quad \text{(left)} \qquad (F_H)_{\text{against wall}} = 78.3 - 58.7 = 19.6 \text{ kN} \quad \text{(right)}$$

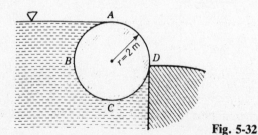

Fig. 5-32

5.33 Determine the horizontal and vertical components of force acting on radial gate ABC in Fig. 5-33a and their lines of action. What force F is required to open the gate (neglecting the weight of the gate)?

$$F_H = \gamma h_{cg} A = (9.79)(3 + \tfrac{2}{2})[(2)(2)] = 156.6 \text{ kN}$$

$$y_{cp} = \frac{-I_{xx} \sin \theta}{h_{cg} A} = \frac{-[(2)(2)^3/12](\sin 90°)}{(3 + \tfrac{2}{2})[(2)(2)]} = -0.083 \text{ m}$$

Hence, $b = \tfrac{2}{2} + 0.083 = 1.083$ m in Fig. 5-33b. F_V = weight of imaginary water above gate = $9.79[(2)(2)(3)] + 9.79[(2)(\pi)(2)^2/4] = 117.5 + 61.5 = 179.0$ kN. Equating moments about OC, $(117.5)(\tfrac{2}{2}) + 61.5[(4)(2)/(3\pi)] = 179.0a$, $a = 0.948$ m (see Fig. 5-33b). $\sum M_O = 0$; $2F - (156.6)(1.083) + (179.0)(0.948) = 0$, $F = 0$.

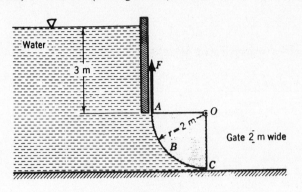

Fig. 5-33(a)

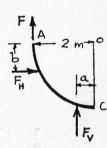

Fig. 5-33(b)

5.34 Find the vertical component of force on the curved gate of Fig. 5-34a and its line of action.

$$F_V = \text{weight of imaginary liquid above gate} = \gamma L \int (H - y)\, dx \quad \text{(see Fig. 5-34b)}$$

$$= (9.00)(2) \int_0^{0.8} (2 - \sqrt{5}x)\, dx = (9.00)(2)\left[2x - \frac{\sqrt{5}x^{3/2}}{\frac{3}{2}}\right]_0^{0.8} = 9.60 \text{ kN}$$

$$x_{cp} = \frac{\gamma L \int (H - y)x\, dx}{F_V} \quad \text{(see Fig. 5-34b)}$$

$$= \frac{(9.00)(2) \int_0^{0.8} (2 - \sqrt{5}x)x\, dx}{9.60} = \frac{(9.00)(2) \int_0^{0.8} (2x - \sqrt{5}x^{3/2})\, dx}{9.60}$$

$$= (9.00)(2)[x^2 - (\sqrt{5}x^{5/2})/\tfrac{5}{2}]_0^{0.8}/9.60 = 0.240 \text{ m}$$

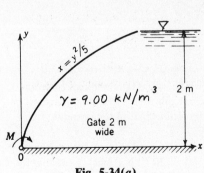

Fig. 5-34(a)

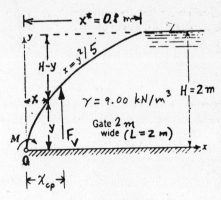

Fig. 5-34(b)

5.35 Determine the moment M to hold the gate of Fig. 5-34a. Neglect its weight.

$F_H = \gamma \bar{h} A = 9.00[(0 + 2)/2][(2)(2)] = 36.0$ kN (left). F_H acts at $\tfrac{2}{3}$, or 0.667 m above point 0. $F_V = 9.60$ kN (up) and $x_{cp} = 0.240$ m (from Prob. 5.34 and Fig. 5-34b). $\sum M_0 = 0$; $M - (9.60)(0.240) - (36.0)(0.667) = 0$, $M = 26.3$ kN · m.

5.36 What is the force on the surface whose trace is OA of Fig. 5-35? The length normal to the paper (L) is 3 m, and γ is 9.00 kN/m³.

$$F_H = \gamma \bar{h} A = (9.00)(\tfrac{1}{2})[(1)(3)] = 13.50 \text{ kN}$$

$$F_V = \text{weight of liquid above } OA = \int \gamma L y \, dx = \int_0^{\sqrt{8}} (9.00)(3)\left(\frac{x^2}{8}\right) dx = (9.00)(3)\left[\frac{x^3}{24}\right]_0^{\sqrt{8}} = 25.46 \text{ kN}$$

$$F_{\text{resultant}} = \sqrt{25.46^2 + 13.50^2} = 28.82 \text{ kN}$$

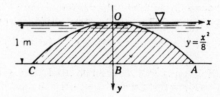

Fig. 5-35

5.37 Find the resultant force, including its line of action, acting on the outer surface of the first quadrant of a spherical shell of radius 600 mm with center at the origin. Its center is 1.2 m below the water surface.

■ See Fig. 5-36. $F_H = \gamma \bar{h} A = \gamma[H - 4r/(3\pi)](\pi r^2/4)$, $F_x = F_z = F_H = 9.79[1.2 - (4)(0.6)/(3\pi)][(\pi)(0.6)^2/4] = 2.617$ kN (both F_x and F_z act toward 0); $F_y = F_V = $ weight of water above curved surface $= \gamma[(H)(\pi)(r)^2/4 - (\tfrac{4}{3})(\pi)(r)^3/8] = 9.79[(1.2)(\pi)(0.6)^2/4 - (\tfrac{4}{3})(\pi)(0.6)^3/8] = 2.214$ kN. $F_{\text{resultant}}$ acts on a line through 0 making a 45° angle with the x and z axes because of symmetry; $F_{\text{resultant}} = \sqrt{2.214^2 + 2.617^2 + 2.617^2} = 4.313$ kN. It acts at an angle $\theta = \arccos(2.214/4.313) = 59.1°$.

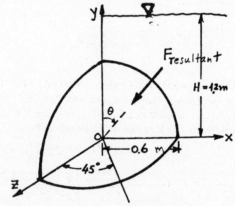

Fig. 5-36

5.38 Find the horizontal and vertical components of the force per unit width exerted by fluids on the horizontal cylinder in Fig. 5-37a if the fluid to the left of the cylinder is (a) a gas confined in a closed tank at a pressure of 35.0 kN/m² and (b) water with a free surface at an elevation coincident with the uppermost part of the cylinder. Assume in both instances that atmospheric pressure occurs to the right of the cylinder.

■ (a) The "net vertical projection" (see Fig. 5-37a) of the portion of the cylinder surface under consideration is $4 - (2 - 2\cos 30°)$, or 3.732 m. $F_H = pA = 35.0[(1)(3.732)] = 130.6$ kN (right). Note that the vertical force of the gas on surface ab is equal and opposite to that on surface bc. Hence, the "net horizontal projection" with regard to the gas is ae (see Fig. 5-38b), which is $2\sin 30°$, or 1.000 m. $F_V = 35.0[(1)(1.000)] = 35.0$ kN (up).

(b) $$F_H = \gamma \bar{h} A = (9.79)(3.732/2)[(1)(3.732)] = 68.2 \text{ kN} \quad \text{(right)}$$

$$F_V = \text{weight of crosshatched volume of water} \quad \text{(Fig. 5-37b)}$$

$$= (9.79)(1)[(\tfrac{210}{360})(\pi)(4)^2/4 + (\tfrac{1}{2})(1.000)(3.732 - \tfrac{4}{2}) + (1)(\tfrac{4}{2})] = 99.8 \text{ kN} \quad \text{(up)}$$

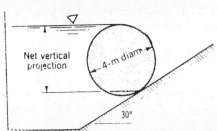

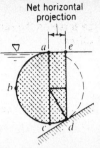

Fig. 5-37(a)   Fig. 5-37(b)

5.39 A vertical-thrust bearing for a large hydraulic gate is composed of an 11-in-radius bronze hemisphere mating into a steel hemispherical shell in the gate bottom. At what pressure must lubricant be supplied to the bearing so that a complete oil film is present if the vertical thrust on the bearing is 800 000 lb?

$$\text{Projected area} = \pi r^2 = (\pi)(11)^2 = 380.1 \text{ in}^2 \qquad p = F/A = 800\,000/380.1 = 2105 \text{ lb/in}^2$$

5.40 Find horizontal and vertical forces per foot of width on the Tainter gate shown in Fig. 5-38. Locate the horizontal force and indicate the line of action of the vertical force without actually computing its location.

$F_H = \gamma \bar{h} A = (62.4)[(0+20)/2][(20)(1)] = 12\,480$ lb. F_H acts at $(\frac{2}{3})(20)$, or 13.33 ft below the water surface. F_V = weight of imaginary water in $ACBA$ = $(62.4)(1)[(\pi)(20)^2/6 - (2)(20\cos 30°)(20\sin 30°)/2] = 2261$ lb. F_V acts through the centroid of segment $ABCA$.

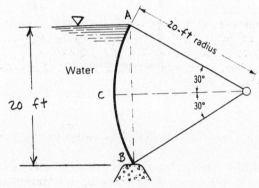

Fig. 5-38

5.41 A tank with vertical ends contains water and is 8 m long normal to the plane in Fig. 5-39. The sketch shows a portion of its cross section where MN is one-quarter of an ellipse with semiaxes b and d. If $b = 4$ m, $d = 6$ m, and $a = 1.5$ m, find, for the surface represented by MN, the magnitude and position of the line of action of the horizontal and vertical components of force and of the resultant force.

$$F_H = \gamma h_{cg} A = 9.79(1.5 + \tfrac{6}{2})[(6)(8)] = 2115 \text{ kN}$$

$$h_{cp} = h_{cg} + \frac{I_{cg}}{h_{cg}A} = (1.5 + \tfrac{6}{2}) + \frac{(8)(6)^3/12}{(1.5 + \tfrac{6}{2})[(6)(8)]} = 5.167 \text{ m below water surface}$$

$$F_V = \text{weight of water above surface } MN = (9.79)(8)[(\pi)(4)(6)/4 + (1.5)(4)] = 1946 \text{ kN}$$

$$x_{cp} = 4b/(3\pi) = (4)(4)/(3\pi) = 1.698 \text{ m to the right of } N \qquad F_{\text{resultant}} = \sqrt{1946^2 + 2115^2} = 2874 \text{ kN}$$

$F_{\text{resultant}}$ acts through the intersection of F_H and F_V at an angle of arctan (1946/2115), or 42.6°.

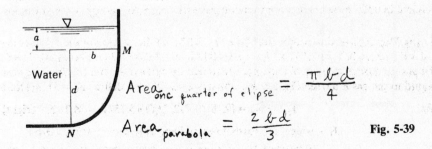

$$\text{Area}_{\text{one quarter of elipse}} = \frac{\pi\,b\,d}{4}$$

$$\text{Area}_{\text{parabola}} = \frac{2\,b\,d}{3}$$

Fig. 5-39

5.42 Solve Prob. 5.41 if $a = 1.5$ ft, $b = 4$ ft, $d = 6$ ft, and MN represents a parabola with vertex at N.

❚
$$F_H = \gamma h_{cg} A = (62.4)(1.5 + \tfrac{6}{2})[(6)(8)] = 13\,480 \text{ lb}$$

$$h_{cp} = h_{cg} + \frac{I_{cg}}{h_{cg}A} = (1.5 + \tfrac{6}{2}) + \frac{(8)(6)^3/12}{(1.5 + \tfrac{6}{2})[(6)(8)]} = 5.167 \text{ ft below water surface}$$

$$F_V = \text{weight of water above surface } MN = (62.4)(8)[(\tfrac{2}{3})(6)(4) + (1.5)(4)] = 10\,980 \text{ lb}$$

$$x_{cp} = (\tfrac{3}{8})(b) = (\tfrac{3}{8})(4) = 1.50 \text{ ft to the right of } N \qquad F_{\text{resultant}} = \sqrt{10\,980^2 + 13\,480^2} = 17\,390 \text{ lb}$$

$F_{\text{resultant}}$ acts through the intersection of F_H and F_V at an angle of arctan $(10\,980/13\,480)$, or $39.2°$.

5.43 For the cross section of the tank shown in Fig. 5-40, BC is a cylindrical surface. If the tank contains water to a depth of 7 ft, determine the magnitude and location of the horizontal and vertical components on wall ABC per 1 ft width.

❚
$$F_H = \gamma \bar{h} A = (62.4)[(0 + 7)/2][(1)(7)] = 1529 \text{ lb} \qquad h_{cp} = (\tfrac{2}{3})(7) = 4.67 \text{ ft}$$

$$F_V = \text{weight of water above surface } BC = (62.4)(1)[(7)(4)] - (62.4)(1)[(\pi)(4)^2/4] = 1747 - 784 = 963 \text{ lb}$$

The location of F_V can be determined by taking moments about point B. $(1747)(\tfrac{4}{2}) - (784)[(4)(4)/(3\pi)] = 963 x_{cp}$, $x_{cp} = 2.25$ ft.

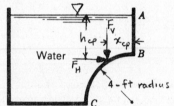

Fig. 5-40

5.44 Rework Prob. 5.43 where the tank is closed and contains gas at a pressure of 8.0 psi.

❚
$$F_H = pA = [(8)(144)][(1)(7)] = 8064 \text{ lb} \qquad h_{cp} = \tfrac{7}{2} = 3.50 \text{ ft}$$

$$F_V = [(8)(144)][(1)(4)] = 4608 \text{ lb} \qquad x_{cp} = \tfrac{4}{2} = 2.00 \text{ ft}$$

5.45 A spherical steel tank of 20 m diameter contains gas under a pressure of 350 kN/m². The tank consists of two half-spheres joined together with a weld. What will be the tensile force across the weld? If the steel is 20.0 mm thick, what is the tensile stress in the steel? Neglect the effects of cross bracing and stiffeners.

❚
$$F = pA = 350[(\pi)(20)^2/4] = 110\,000 \text{ kN} \qquad \sigma = \frac{\text{force/length}}{\text{thickness}} = \frac{110\,000/(20\pi)}{20.0/1000} = 87\,500 \text{ kN/m}^2$$

5.46 Determine the force required to hold the cone shown in Fig. 5-41a in position.

❚ Figure 5-41b shows the vertical projection above the opening. $p_{\text{gas}} = 0.5 - [(0.8)(62.4)](5)/144 = -1.23$ psi, $F_{\text{gas}} = [(1.23)(144)][(\pi)(0.804)^2] = 360$ lb, $F_{\text{cylinder}} = (62.4)(0.8)[(\pi)(0.804)^2(5 + 3)] = 811$ lb, $F_{\text{cone}} = (62.4)(0.8)[(3)(\pi)(0.804)^2/3] = 101$ lb; $\Sigma F_y = 0$, $360 - 811 + 101 + F = 0$, $F = 350$ lb.

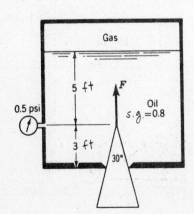

Fig. 5-41(a)

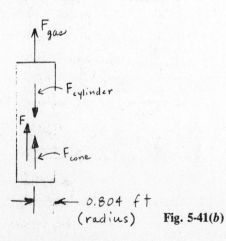

Fig. 5-41(b)

5.47 The cross section of a gate is shown in Fig. 5-42. Its dimension normal to the plane of the paper is 10 m, and its shape is such that $x = 0.2y^2$. The gate is pivoted about O. Find the horizontal and vertical forces and the clockwise moment acting on the gate if the water depth is 2.0 m.

I

$$F_H = \gamma \bar{h} A = 9.79[(0 + 2)/2][(10)(2)] = 195.8 \text{ kN}$$

$$F_v = \text{weight of water above the gate} = \int_0^2 (9.79)(10)(x\, dy) = (9.79)(10) \int_0^2 0.2y^2\, dy = (9.79)(10) \left[\frac{0.2y^3}{3}\right]_0^2 = 52.2 \text{ kN}$$

$$M_O = (195.8)(\tfrac{2}{3}) + \int_0^2 (9.79)(10)\left(\frac{x}{2}\right)(x\, dy) = 130.5 + (9.79)(10) \int_0^2 \frac{(0.2y^2)^2}{2}\, dy$$

$$= 130.5 + (9.79)(10)\left[\frac{0.04y^5}{10}\right]_0^2 = 143.0 \text{ kN} \cdot \text{m}$$

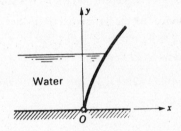

Water

Fig. 5-42

5.48 Find the wall thickness of steel pipe needed to resist the static pressure in a 36-in-diameter steel pipe carrying water under a head of 750 ft of water. Use an allowable working stress for steel pipe of 16 000 psi.

I

$$p = \gamma h = (62.4)(750) = 46\,800 \text{ lb/ft}^2 \quad \text{or} \quad 325 \text{ lb/in}^2$$

$$T = pd/2 = (325)(36)/2 = 5850 \text{ lb/in of pipe length} \qquad t = 5850/16\,000 = 0.366 \text{ in}$$

5.49 A vertical cylindrical tank is 6 ft in diameter and 10 ft high. Its sides are held in position by means of two steel hoops, one at the top and one at the bottom. The tank is filled with water up to 9 ft high. Determine the tensile stress in each hoop.

I See Fig. 5-43. $F = \gamma \bar{h} A = 62.4[(0 + 9)/2][(9)(6)] = 15\,163$ lb, $T = F/2 = 15\,163/2 = 7582$ lb; stress in top hoop = $(7582)(\tfrac{3}{10}) = 2275$ lb, stress in bottom hoop = $(7582)[(10 - 3)/10] = 5307$ lb.

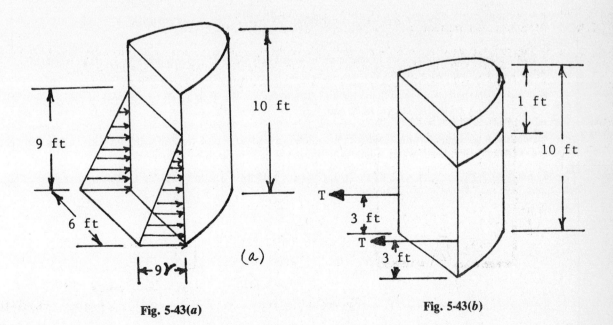

Fig. 5-43(a)   **Fig. 5-43(b)**

5.50 A 48-in-diameter steel pipe, $\frac{1}{4}$ in thick, carries oil of s.g. = 0.822 under a head of 400 ft of oil. Compute the **(a)** stress in the steel and **(b)** thickness of steel required to carry a pressure of 250 psi with an allowable stress of 18 000 psi.

∎
$$p = \gamma h = [(0.822)(62.4)](400) = 20\,517 \text{ lb/ft}^2 \quad \text{or} \quad 142.5 \text{ lb/in}^2 \qquad \sigma = \frac{pr}{t}$$

(a)
$$\sigma = \frac{(142.5)(48/2)}{\frac{1}{4}} = 13\,680 \text{ psi}$$

(b)
$$18\,000 = \frac{(250)(48/2)}{t} \qquad t = 0.333 \text{ in}$$

5.51 A wooden storage vat, 20 ft in outside diameter, is filled with 24 ft of brine, s.g. = 1.06. The wood staves are bound by flat steel bands, 2 in wide by $\frac{1}{4}$ in thick, whose allowable stress is 16 000 psi. What is the spacing of the bands near the bottom of the vat, neglecting any initial stress? Refer to Fig. 5-44.

∎ Force P represents the sum of all the horizontal components of small forces dP acting on length y of the vat, and forces T represent the total tension carried in a band loaded by the same length y.

$$\sum F_x = 0 \qquad 2T - P = 0 \qquad T = A_{\text{steel}}\sigma_{\text{steel}} = [(2)(\tfrac{1}{4})](16\,000) = 8000 \text{ lb}$$

$$p = \gamma h A = [(1.06)(62.4)](24)(20y) = 31\,749y \qquad (2)(8000) - 31\,749y = 0 \qquad y = 0.504 \text{ ft} \quad \text{or} \quad 6.05 \text{ in}$$

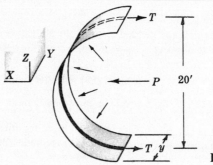

Fig. 5-44

5.52 A 4.0-in-ID steel pipe has a $\frac{1}{4}$-in wall thickness. For an allowable tensile stress of 10 000 psi, what is the maximum pressure?

∎
$$\sigma = \frac{pr}{t} \qquad 10\,000 = \frac{(p)(4.0/2)}{\frac{1}{4}} \qquad p = 1250 \text{ lb/in}^2$$

5.53 A thin-walled hollow sphere 3 m in diameter holds gas at 1500 kPa. For an allowable stress of 60 000 kPa, determine the minimum wall thickness.

∎ Considering half a sphere of diameter d (3 m) and thickness t, $(\pi dt)(\sigma) = (p)(\pi d^2/4)$, $[(\pi)(3)(t)](60\,000) = 1500[(\pi)(3)^2/4]$, $t = 0.01875$ m, or 18.75 mm.

5.54 A cylindrical container 7 ft high and 4 ft in diameter provides for pipe tension with two hoops a foot from each end. When it is filled with water, what is the tension in each hoop due to the water?

∎ See Fig. 5-45. $F = \gamma \bar{h} A = 62.4[(0 + 7)/2][(7)(4)] = 6115$ lb. F acts at $(\frac{2}{3})(7)$, or 4.667 ft from the top of the container.

$$\sum F_x = 0$$

$$2T_1 + 2T_2 - 6115 = 0 \tag{1}$$

$$\sum M_A = 0 \qquad (2T_2)(1.333) - (2T_1)(3.667) = 0$$

$$T_2 = 2.75T_1 \tag{2}$$

Solve simultaneous equations (1) and (2). $2T_1 + (2)(2.75T_1) - 6115 = 0$, $T_1 = 815$ lb, $T_2 = (2.75)(815) = 2241$ lb.

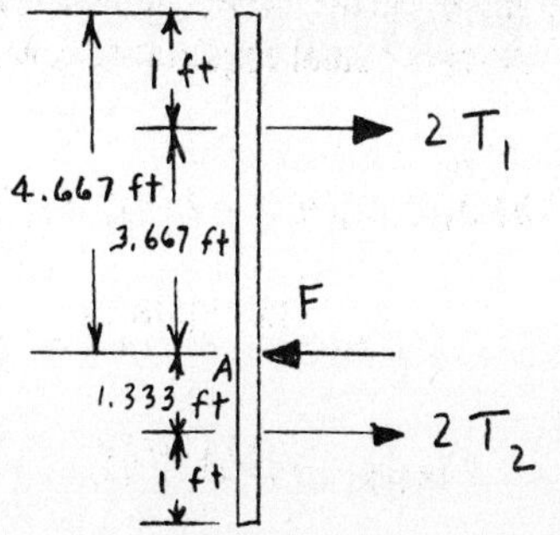

Fig. 5-45

5.55 A 20-mm-diameter steel ($\gamma = 77.0$ kN/m³) ball covers a 10-mm-diameter hole in a pressure chamber where the pressure is 30 000 kPa. What force is required to lift the ball from the opening?

$$F = pA + \text{weight of ball} = 30\,000[(\pi)(\tfrac{10}{1000})^2/4] + [(\tfrac{4}{3})(\pi)(\tfrac{10}{1000})^3](77.0) = 2.357 \text{ kN}$$

CHAPTER 6
Buoyancy and Flotation

6.1 A stone weighs 105 lb in air. When submerged in water, it weighs 67.0 lb. Find the volume and specific gravity of the stone.

Buoyant force (F_b) = weight of water displaced by stone $(W) = 105 - 67.0 = 38.0$ lb

$$W = \gamma V = 62.4V \qquad 38.0 = 62.4V \qquad V = 0.609 \text{ ft}^3$$

$$\text{s.g.} = \frac{\text{weight of stone in air}}{\text{weight of equal volume of water}} = \frac{105}{(0.609)(62.4)} = 2.76$$

6.2 A piece of irregularly shaped metal weighs 300.0 N in air. When the metal is completely submerged in water, it weighs 232.5 N. Find the volume of the metal.

$$F_b = W \qquad 300.0 - 232.5 = [(9.79)(1000)](V) \qquad V = 0.00689 \text{ m}^3$$

6.3 A cube of timber 1.25 ft on each side floats in water as shown in Fig. 6-1. The specific gravity of the timber is 0.60. Find the submerged depth of the cube.

$$F_b = W \qquad 62.4[(1.25)(1.25)(D)] = [(0.60)(62.4)][(1.25)(1.25)(1.25)] \qquad D = 0.750 \text{ ft}$$

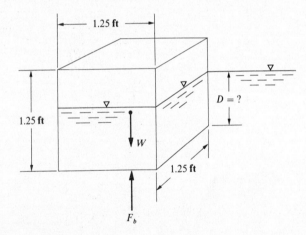

Fig. 6-1

6.4 Determine the magnitude and direction of the force necessary to hold a concrete cube, 0.300 m on each side, in equilibrium and completely submerged (*a*) in mercury (Hg) and (*b*) in water. Use s.g.$_\text{concrete} = 2.40$.

(*a*) Since s.g.$_\text{Hg} = 13.6$ and s.g.$_\text{concrete} = 2.40$, it is evident that the concrete will float in mercury. Therefore, a force F acting downward will be required to hold the concrete in equilibrium and completely submerged in mercury. The forces acting on the concrete are shown in Fig. 6-2a, where F is the force required to hold the concrete cube in equilibrium and completely submerged, W is the weight of the concrete cube in air, and F_b is the buoyant force. $\sum F_y = 0$, $F + W - F_b = 0$, $F + [(2.40)(9.79)][(0.300)(0.300)(0.300)] - [(13.6)(9.79)][(0.300)(0.300)(0.300)] = 0$, $F = 2.96$ kN (downward). (*b*) Since s.g.$_\text{concrete} = 2.40$, it will sink in water. Therefore, a force F acting upward will be required to hold the concrete in equilibrium and completely submerged in water. The forces acting on the concrete in this case are shown in Fig. 6-2b. $\sum F_y = 0$, $W - F - F_b = 0$, $[(2.40)(9.79)][(0.300)(0.300)(0.300)] - F - 9.79[(0.300)(0.300)(0.300)] = 0$, $F = 0.370$ kN (upward).

6.5 A concrete cube 10.0 in on each side is to be held in equilibrium under water by attaching a lightweight foam buoy to it, as shown in Fig. 6-3. (In theory, the attached foam buoy and concrete cube, when placed under water, will neither rise nor sink.) If the specific weight of concrete and foam are 150 lb/ft^3 and 5.0 lb/ft^3, respectively, what minimum volume of foam is required?

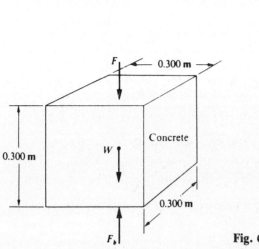

Fig. 6-2(a)

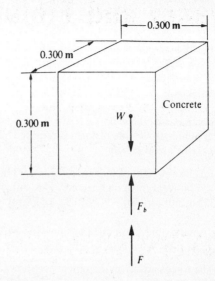

Fig. 6-2(b)

▮ The forces acting in this problem are shown in Fig. 6-3, where W_f and W_c are the respective weights of the foam and the concrete, and F_{bf} and F_{bc} are the respective buoyant forces on the foam and the concrete. $\sum F_y = 0$, $W_f - F_{bf} + W_c - F_{bc} = 0$, $5.0 V_{\text{foam}} - 62.4 V_{\text{foam}} + 150[(\frac{10}{12})(\frac{10}{12})(\frac{10}{12})] - 62.4[(\frac{10}{12})(\frac{10}{12})(\frac{10}{12})] = 0$, $V_{\text{foam}} = 0.883$ ft³.

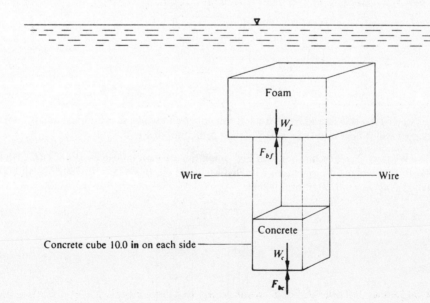

Fig. 6-3

6.6 A barge is loaded with 150 tons of coal. The weight of the empty barge in air is 35 tons. If the barge is 18 ft wide, 52 ft long, and 9 ft high, what is its draft (i.e., its depth below the water surface)?

▮ $\qquad F_b = W \qquad 62.4[(18)(52)(D)] = (150 + 35)(2000) \qquad D = 6.33$ ft

6.7 Determine the submerged depth of a cube of steel 0.30 m on each side floating in mercury. The specific gravities of steel and mercury are 7.8 and 13.6, respectively.

▮ $\qquad F_b = W \qquad [(13.6)(9.79)][(0.3)(0.3)(D)] = [(7.8)(9.79)][(0.3)(0.3)(0.3)] \qquad D = 0.172$ m

6.8 A cube of wood (s.g. = 0.60) has 9-in sides. Compute the magnitude and direction of the force F required to hold the wood completely submerged in water.

▮ Since s.g.$_{\text{wood}} = 0.60$, it is evident that the wood will float in water. Therefore, a force F acting downward will be required to hold the wood in equilibrium and completely submerged. The forces acting on the wood are essentially the same as those shown acting on the concrete cube in Fig. 6-2a: $\sum F_y = 0$, $F + W - F_b = 0$, $F + [(0.60)(62.4)][(\frac{9}{12})(\frac{9}{12})(\frac{9}{12})] - 62.4[(\frac{9}{12})(\frac{9}{12})(\frac{9}{12})] = 0$, $F = 10.5$ lb (downward).

6.9 A hollow cube 1.0 m on each side weighs 2.4 kN. The cube is tied to a solid concrete block weighing 10.0 kN. Will these two objects tied together float or sink in water? The specific gravity of the concrete is 2.40.

▐ Let W = weight of hollow cube plus solid concrete block, $(F_b)_1$ = buoyant force on hollow cube, and $(F_b)_2$ = buoyant force on solid concrete block. $W = 2.4 + 10.0 = 12.4$ kN, $(F_b)_1 = 9.79[(1)(1)(1)] = 9.79$ kN, $V_{block} = 10/[(2.40)(9.79)] = 0.4256$ m^3, $(F_b)_2 = (9.79)(0.4256) = 4.17$ kN, $(F_b)_1 + (F_b)_2 = 9.79 + 4.17 = 13.96$ kN. Since $[W = 12.4] < [(F_b)_1 + (F_b)_2 = 13.96$ kN$]$, the two objects tied together will float in water.

6.10 A concrete cube 0.5 m on each side is to be held in equilibrium under water by attaching a light foam buoy to it. What minimum volume of foam is required? The specific weights of concrete and foam are 23.58 kN/m^3 and 0.79 kN/m^3, respectively.

▐ Let W_f = weight of foam in air, $(F_b)_f$ = buoyant force on foam, W_c = weight of concrete in air, and $(F_b)_c$ = buoyant force on concrete. $\sum F_y = 0$, $W_f - (F_b)_f + W_c - (F_b)_c = 0$, $0.79V_{foam} - 9.79V_{foam} + 23.58[(0.5)(0.5)(0.5)] - 9.79[(0.5)(0.5)(0.5)] = 0$, $V_{foam} = 0.192$ m^3.

6.11 A prismatic object 8 in thick by 8 in wide by 16 in long is weighed in water at a depth of 20 in and found to weigh 11.0 lb. What is its weight in air and its specific gravity?

▐ The forces acting on the object are shown in Fig. 6-4. $\sum F_y = 0$, $T + F_b - W = 0$, F_b = weight of displaced water = $62.4[(8)(8)(16)/1728] = 37.0$ lb, $11.0 + 37.0 - W = 0$, $W = 48.0$ lb, s.g. = $48.0/37.0 = 1.30$.

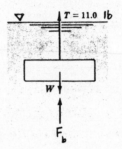

Fig. 6-4

6.12 A hydrometer weighs 0.00485 lb and has a stem at the upper end which is cylindrical and 0.1100 in in diameter. How much deeper will it float in oil of s.g. 0.780 than in alcohol of s.g. 0.821?

▐ $W_{hydrometer} = W_{displaced \ liquid}$. For position 1 in Fig. 6-5 in the alcohol, $0.00485 = [(0.821)(62.4)](V_1)$, $V_1 = 0.0000947$ ft^3 (in alcohol). For position 2 in Fig. 6-5 in the oil, $0.00485 = [(0.780)(62.4)][0.0000947 + (h)(\pi)(0.1100/12)^2/4]$, $h = 0.0750$ ft, or 0.900 in.

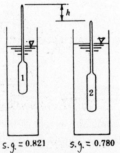

s.g. = 0.821   s.g. = 0.780   **Fig. 6-5**

6.13 A piece of wood of s.g. 0.651 is 3 in square and 5 ft long. How many pounds of lead weighing 700 lb/ft^3 must be fastened at one end of the stick so that it will float upright with 1 ft out of water?

▐
$$W_{wood \ and \ lead} = W_{displaced \ water}$$
$$[(0.651)(62.4)][(5)(\tfrac{3}{12})(\tfrac{3}{12})] + 700V = 62.4[(5-1)(\tfrac{3}{12})(\tfrac{3}{12}) + V]$$
$$V = 0.00456 \ \text{ft}^3 \qquad W_{lead} = (0.00456)(700) = 3.19 \ \text{lb}$$

6.14 What fraction of the volume of a solid piece of metal of s.g. 7.25 floats above the surface of a container of mercury?

▐ Let V = volume of the metal and V' = volume of mercury displaced. $F_b = W$, $[(13.6)(62.4)](V') = [(7.25)(62.4)](V)$, $V'/V = 0.533$. Fraction of volume above mercury = $1 - 0.533 = 0.467$.

6.15 A rectangular open box 25 ft by 10 ft in plan and 12 ft deep weighs 40 tons and is launched in fresh water. (*a*) How deep will it sink? (*b*) If the water is 12 ft deep, what weight of stone placed in the box will cause it to rest on the bottom?

$$F_b = W$$

(*a*) $$62.4[(25)(10)(D)] = (40)(2000) \qquad D = 5.13 \text{ ft}$$

(*b*) $$62.4[(25)(10)(12)] = (40 + W_{\text{stone}})(2000) \qquad W_{\text{stone}} = 53.6 \text{ tons}$$

6.16 A block of wood floats in water with 2.0 in projecting above the water surface. When placed in glycerin of s.g. 1.35, the block projects 3.0 in above the liquid surface. Determine the specific gravity of the wood.

▮ Let A = area of block and h = height of block. $W_{\text{block}} = [(\text{s.g.})(62.4)](Ah/12)$, $W_{\text{displaced water}} = 62.4[(A)(h-2)/12]$, $W_{\text{displaced glycerin}} = [(1.35)(62.4)][(A)(h-3)/12]$. Since the weight of each displaced liquid equals the weight of the block, $W_{\text{displaced water}} = W_{\text{displaced glycerin}}$: $62.4[(A)(h-2)/12] = [(1.35)(62.4)][(A)(h-3)/12]$, $h = 5.86$ in. Also, $W_{\text{block}} = W_{\text{displaced water}}$: $[(\text{s.g.})(62.4)][(A)(5.86/12)] = 62.4[(A)(5.86-2)/12]$, s.g. $= 0.659$.

6.17 To what depth will an 8-ft-diameter log 15 ft long and of s.g. 0.425 sink in fresh water?

▮ The log is sketched in Fig. 6-6 with center O of the log above the water surface because the specific gravity is less than 0.5. (Had the specific gravity been equal to 0.5, the log would be half submerged.) $F_b = W$, F_b = weight of displaced liquid = $62.4\{(15)[(2\theta/360)(\pi 4^2) - (2)(\frac{1}{2})(4 \sin \theta)(4 \cos \theta)]\} = 261.4\theta - (14\,976)(\sin \theta)(\cos \theta)$, $W = [(0.425)(62.4)][(15)(\pi 4^2) = 19\,996$.

$$261.4\theta - (14\,976)(\sin \theta)(\cos \theta) = 19\,996$$

This equation can be solved by successive trials.

Try $\theta = 85°$: $\qquad (261.4)(85) - (14\,976)(\sin 85°)(\cos 85°) = 20\,919 \qquad (\neq 19\,996)$

Try $\theta = 83°$: $\qquad (261.4)(83) - (14\,976)(\sin 83°)(\cos 83°) = 19\,885 \qquad (\neq 19\,996)$

Try $\theta = 83.2°$: $\qquad (261.4)(83.2) - (14\,976)(\sin 83.2°)(\cos 83.2°) = 19\,988 \qquad (\neq 19\,996)$

Try $\theta = 83.22°$: $\quad (261.4)(83.22) - (14\,976)(\sin 83.22°)(\cos 83.22°) = 19\,998 \qquad$ (close enough)

$$\text{Depth of flotation} = DC = OC - OD = 4.00 - 4.00 \cos 83.22° = 3.53 \text{ ft}$$

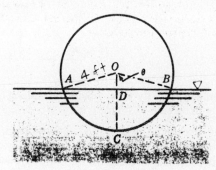

Fig. 6-6

6.18 (*a*) Neglecting the thickness of the tank walls in Fig. 6-7*a*, if the tank floats in the position shown what is its weight? (*b*) If the tank is held so that the top is 10 ft below the water surface, as shown in Fig. 6-7*b*, what is the force on the inside top of the tank? Use an atmospheric pressure equivalent to a 34.0-ft head of water.

▮ (*a*) $$W_{\text{tank}} = W_{\text{displaced liquid}} = 62.4[(1)(\pi 4^2/4)] = 784 \text{ lb}$$

(*b*) The space occupied by the air will be less at the new depth shown in Fig. 6-7*b*. Assuming that the temperature of the air is constant, then for positions a and b, $p_A V_A = p_D V_D$, $[62.4(34.0+1)][(4)(\pi 4^2/4)] = [(62.4)(34.0+10+y)][(y)(\pi 4^2/4)]$, $y^2 + 44.0y - 140 = 0$, $y = 2.98$ ft. The pressure at D is $10 + 2.98$, or 12.98 ft of water (gage), which is essentially the same as the pressure on the inside top of the cylinder. Hence, the force on the inside top of the cylinder is given by $F = \gamma hA = (62.4)(12.98)(\pi 4^2/4) = 10\,180$ lb.

6.19 A ship, with vertical sides near the water line, weighs 4000 tons and draws 22 ft in salt water ($\gamma = 64.0$ lb/ft³)(see Fig. 6-8). Discharge of 200 tons of water ballast decreases the draft to 21 ft. What would be the draft d of the ship in fresh water?

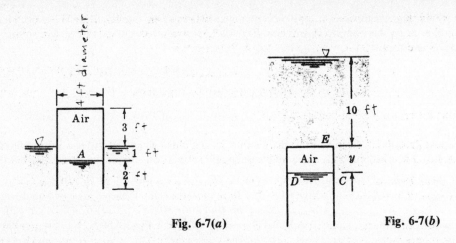

Fig. 6-7(a)   **Fig. 6-7(b)**

▌ Because the shape of the underwater section of the ship is not known, it is best to solve this problem on the basis of volumes displaced. A 1-ft decrease in draft was caused by a reduction in weight of 200 tons, or $\gamma V_d = 64.0[(1)(A)] = (200)(2000)$ where V_d represents the volume between drafts 22 ft and 21 ft, and $[(1)(A)]$ represents the water-line area times 1 ft, or the same volume V_d. From the equation above, $V_d = (200)(2000)/64.0 = 6250$ ft³ (this is per foot depth), F_b = weight of displaced liquid = γV_d, $V_d = F_b/\gamma$. In Fig. 6-8, the vertically crosshatched volume is the difference in displaced fresh water and salt water. This difference in volume can be expressed as $W/\gamma_{\text{fresh H}_2\text{O}} - W/\gamma_{\text{salt H}_2\text{O}}$, or $(4000 - 200)(2000)/62.4 - (4000 - 200)(2000)/64.0$. Since $V_d = 6250$ ft³/ft depth, the vertically crosshatched volume can also be expressed as $6250y$. Hence, $6250y = (4000 - 200)(2000)/62.4 - (4000 - 200)(2000)/64.0$, $y = 0.49$ ft; $d = 21 + 0.49 = 21.49$ ft.

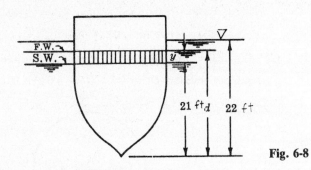

Fig. 6-8

6.20   A barrel containing water weighs 283.5 lb. What will be the reading on the scales if a 2 in by 2 in piece of wood is held vertically in the water to a depth of 2.0 ft?

▌ For every acting force, there must be an equal and opposite reacting force. The buoyant force exerted by the water upward against the bottom of the piece of wood is opposed by the 2 in by 2 in area of wood acting downward on the water with equal magnitude. This force will measure the increase in scale reading. $F_b = 62.4[(2)(\frac{2}{12})(\frac{2}{12})] = 3.5$ lb, new scale reading = $283.5 + 3.5 = 287.0$ lb.

6.21   The can in Fig. 6-9 floats in the position shown. What is its weight?

▌ $\qquad\qquad F_b = W \qquad 9.79[(\frac{8}{100})(\pi)(\frac{9}{100})^2/4] = W \qquad W = 0.00498$ kN   or   4.98 N

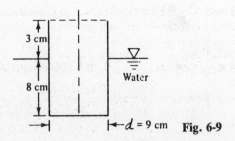

$\leftarrow d = 9$ cm   **Fig. 6-9**

6.22 It is said that Archimedes discovered the buoyancy laws when asked by King Hiero of Syracuse to determine whether or not his new crown was gold (s.g. = 19.3). Archimedes found the weight of the crown in air to be 13.0 N and its weight in water to be 11.8 N. Was it gold?

▌ $F_b = 13.0 - 11.8 = 1.2\,\text{N}$ $\qquad V_{\text{displaced H}_2\text{O}} = V_{\text{crown}} = 1.2/[(9.79)(1000)] = 0.0001226\,\text{m}^3$

$\gamma_{\text{crown}} = 13.0/0.0001226 = 106\,000\,\text{N/m}^3$ or $106.0\,\text{kN/m}^3$ $\qquad$ s.g.$_{\text{crown}} = 106.0/9.79 = 10.83$

Thus the crown was not pure gold.

6.23 Repeat Prob. 6.22 assuming the crown is an alloy of gold (s.g. = 19.3) and silver (s.g. = 10.5). For the same measured weights, compute the percentage silver in the crown.

▌ From Prob. 6.22, s.g.$_{\text{crown}} = 10.83$. Let α = percentage of silver in crown. $(\alpha)(10.5) + (1 - \alpha)(19.3) = 10.83$, $10.5\alpha + 19.3 - 19.3\alpha = 10.83$, $\alpha = 0.962$ (that is, the crown is 96.2 percent silver).

6.24 A sphere of buoyant, solid, molded foam is immersed in sea water ($\gamma = 64.0\,\text{lb/ft}^3$) and moored at the bottom. The sphere radius is 14 in. The mooring line has a tension of 150 lb. What is the specific weight of the sphere?

▌ The mooring line tension (T) and sphere weight (W) act downward on the sphere, while the buoyant force (F_b) acts upward. $\Sigma F_y = 0$; $F_b - T - W = 0$, $64.0[(\frac{4}{3})(\pi)(\frac{14}{12})^3] - 150 - (\gamma_{\text{sphere}})[(\frac{4}{3})(\pi)(\frac{14}{12})^3] = 0$, $\gamma_{\text{sphere}} = 41.4\,\text{lb/ft}^3$.

6.25 If the total weight of the hydrometer in Fig. 6-10 is 0.030 lb and the stem diameter is 0.30 in, compute the height h where it will float when the liquid has a specific gravity of 1.3.

▌ Let ΔV = submerged volume between s.g. = 1 and s.g. = 1.3, V_0 = submerged total volume when s.g. = 1.0, and W = weight of hydrometer. $W = \gamma V_0 = (\text{s.g.})(\gamma)(V_0 - \Delta V) = (\text{s.g.})(\gamma)(V_0) - (\text{s.g.})(\gamma)(\Delta V)$. Since $(\gamma)(V_0) = W$ and $\Delta V = hA = h(\pi d^2/4)$, $W = (\text{s.g.})(W) - (\text{s.g.})(\gamma)[(h)(\pi d^2/4)]$, $0.030 = (1.3)(0.030) - (1.3)(62.4)[(h)(\pi)(0.30/12)^2/4]$, $h = 0.226\,\text{ft}$, or 2.71 in.

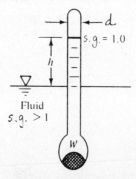

Fig. 6-10

6.26 For the hydrometer of Fig. 6-10, derive a formula for float position h as a function of s.g., W, d, and the specific weight γ of pure water. Are the scale markings linear or nonlinear as a function of s.g.?

▌ From Prob. 6.25, $W = (\text{s.g.})(W) - (\text{s.g.})(\gamma)[(h)(\pi d^2/4)]$.

$$h = \frac{(\text{s.g.})(W) - W}{(\text{s.g.})(\gamma)(\pi d^2/4)} = \frac{(W)(\text{s.g.} - 1)}{(\text{s.g.})(\gamma)(\pi d^2/4)}$$

When plotted in Fig. 6-11 (in arbitrary units), it is slightly nonlinear.

6.27 A hydrometer has a weight of 0.16 N and a stem diameter of 1.0 cm. What is the difference between scale markings for s.g. = 1.0 and s.g. = 1.1? Between 1.1 and 1.2?

▌ Let h_1 = difference between markings for s.g. = 1.0 and s.g. = 1.1 and h_2 = difference between scale markings for s.g. = 1.1 and s.g. = 1.2. From Prob. 6.26,

$$h = \frac{(W)(\text{s.g.} - 1)}{(\text{s.g.})(\gamma)(\pi d^2/4)} \qquad h_1 = \frac{(0.16)(1.1 - 1)}{1.1[(9.79)(1000)][(\pi)(1.0/100)^2/4]} = 0.0189\,\text{m} \quad \text{or} \quad 1.89\,\text{cm}$$

$$h_1 + h_2 = \frac{0.16(1.2 - 1)}{1.2[(9.79)(1000)][(\pi)(1.0/100)^2/4]} = 0.0347\,\text{m} \quad \text{or} \quad 3.47\,\text{cm}$$

$$h_2 = 3.47 - 1.89 = 1.58\,\text{cm}$$

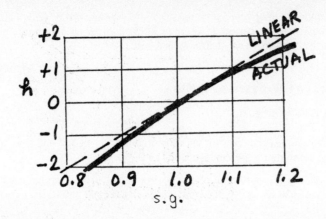

Fig. 6-11

6.28 A wooden pole (s.g. = 0.65), 9 cm by 9 cm by 5 m long, hangs vertically from a string in such a way that 3 m are submerged in water and 2 m are above the surface. What is the tension in the string?

❚ String tension (T) and buoyant force (F_b) act upward on the pole, while pole weight (W) acts downward. $\Sigma F_y = 0$; $T + F_b - W = 0$, $T + 9.79[(\frac{9}{100})(\frac{9}{100})(3)] - [(0.65)(9.79)][(\frac{9}{100})(\frac{9}{100})(5)] = 0$, $T = 0.0198$ kN, or 19.8 N.

6.29 A spar buoy is a buoyant rod weighted at the bottom so that it floats upright and can be used for measurements or markers. The spar in Fig. 6-12 is wood (s.g. = 0.60), 2 in by 2 in by 12 ft, and floats in sea water (s.g. = 1.025). How many pounds of steel (s.g. = 7.85) should be added to the bottom so that exactly $h = 2$ ft of the spar is exposed?

❚
$$V_{spar} = (\tfrac{2}{12})(\tfrac{2}{12})(12) = 0.3333 \text{ ft}^3 \qquad V_{submerged} = (\tfrac{2}{12})(\tfrac{2}{12})(10) = 0.2778 \text{ ft}^3$$

$$V_{steel} = W_{steel}/[(7.85)(62.4)] = 0.002041 W_{steel} \qquad F_b = W_{wood} + W_{steel}$$

$$[(1.025)(62.4)](0.2778 + 0.002041 W_{steel}) = [(0.60)(62.4)](0.3333) + W_{steel} \qquad W_{steel} = 6.08 \text{ lb}$$

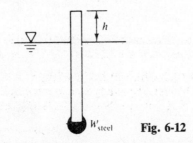

Fig. 6-12

6.30 A right circular cone is 4 cm in radius and 15 cm high and weighs 1.0 N in air. How much force is required to push this cone vertex-downward into benzene so that its base is exactly at the surface? How much additional force will push the base 10 cm below the surface?

❚ Downward force (F) and cone weight (W) act downward on the cone, while buoyant force (F_b) acts upward. $\Sigma F_y = 0$; $F_b - F - W = 0$, $[(0.88)(9.79)(1000)][(\pi)(\frac{4}{100})^2(\frac{15}{100})/3] - F - 1.0 = 0$, $F = 1.17$ N. Once the cone is fully submerged, F is constant at 1.17 N.

6.31 The 2-in by 2-in by 12-ft spar buoy of Fig. 6-12 has 6 lb of steel weight attached and has gone aground on a rock 8 ft deep, as depicted in Fig. 6-13. Compute the angle θ at which the buoy will lean, assuming the rock exerts no moment on the buoy.

❚ From Prob. 6.29, $V_{spar} = 0.3333$ ft³. $W_{wood} = [(0.60)(62.4)](0.3333) = 12.48$ lb and $F_b = 62.4[(\frac{2}{12})(\frac{2}{12})(L)] = 1.733L$. W_{wood} acts downward at a distance of $6 \sin \theta$ to the right of A, and F_b acts upward at a distance of $(L/2)(\sin \theta)$ to the right of A; while the steel force passes through point A. Hence, $\Sigma M_A = 0$, $12.48(6 \sin \theta) - (1.733L)[(L/2)(\sin \theta)] = 0$, $L = 9.296$ ft; $\cos \theta = 8/L = 8/9.296 = 0.86059$, $\theta = 30.6°$.

6.32 The solid cube 12 cm on a side in Fig. 6-14 is balanced by a 2-kg mass on the beam scale when the cube is immersed in water. What is the specific weight of the cube material?

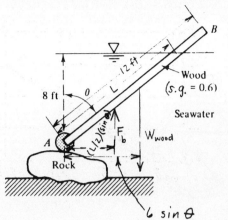

Fig. 6-13

▮ $F = ma = (2)(9.81) = 19.62$ N. Upward force (F) and buoyant force (F_b) act upward on the solid cube, while cube weight (W) acts downward. $\sum F_y = 0$; $F_b + F - W = 0$, $[(9.79)(1000)][(\frac{12}{100})(\frac{12}{100})(\frac{12}{100})] + 19.62 - W = 0$, $W = 36.54$ N; $\gamma = 36.54/[(\frac{12}{100})(\frac{12}{100})(\frac{12}{100})] = 21\,150$ N/m³, or 21.15 kN/m³.

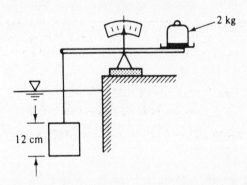

Fig. 6-14

6.33 The balloon in Fig. 6-15 is filled with helium (He) (molecular weight = 4.00) and pressurized to 110 kPa. The atmospheric conditions are as shown. Compute the tension in the mooring line.

▮ $\gamma = p/RT$; $\gamma_{air} = [(100)(1000)]/[(29.3)(273 + 20)] = 11.65$ N/m³, $\gamma_{He} = [(110)(1000)]/[(212.0)(273 + 20)] = 1.771$ N/m³. Weight of helium (W) and tension in mooring line (T) act downward on the balloon, while buoyant force (F_b) acts upward. $\sum F_y = 0$; $F_b - W - T = 0$, $11.65[(\frac{4}{3})(\pi)(\frac{10}{2})^3] - 1.771[(\frac{4}{3})(\pi)(\frac{10}{2})^3] - T = 0$, $T = 5173$ N.

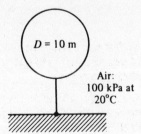

Fig. 6-15

6.34 A 1-ft-diameter hollow sphere is made of steel (s.g. = 7.85) with 0.16-in (0.01333-ft) wall thickness. How high will the sphere float in water (i.e., find h in Fig. 6-16)? How much weight must be added inside to make the sphere neutrally buoyant?

▮
$$F_b = W = \text{weight of displaced water} = \gamma[(\pi/3)(h)^2(3r - h)]$$
$$= 62.4\{(\pi/3)(h^2)[(3)(\tfrac{1}{2}) - h)]\} = 98.02h^2 - 65.35h^3$$
$$W = (\gamma_{steel})(V_{steel}) \qquad \gamma_{steel} = (7.85)(62.4) = 489.8 \text{ lb/ft}^3$$
$$V_{steel} = (\tfrac{4}{3})(\pi)(\tfrac{1}{2})^3 - (\tfrac{4}{3})(\pi)\{[1 - (2)(0.01333)]/2\}^3 = 0.04077 \text{ ft}^3$$
$$W = (489.8)(0.04077) = 19.97 \text{ lb} \qquad 98.02h^2 - 65.35h^3 = 19.97$$

Two roots of this equation are complex. The other, obtained by trial and error, is $h = 0.575$ ft. For neutral buoyancy, the total weight of the sphere plus added weight must equal the weight of water displaced by the entire sphere. Hence, $19.97 + W_{added} = 62.4[(\frac{4}{3})(\pi)(\frac{1}{2})^3]$, $W_{added} = 12.70$ lb.

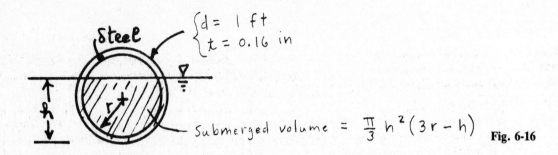

steel $\begin{cases} d = 1 \text{ ft} \\ t = 0.16 \text{ in} \end{cases}$

Submerged volume $= \frac{\pi}{3} h^2 (3r - h)$

Fig. 6-16

6.35 When a 5-lb weight is placed on the end of a floating 4-in by 4-in by 9-ft wooden beam, the beam tilts at $1.5°$ with its right upper corner at the surface, as shown in Fig. 6-17. What is the specific weight of the wood?

$$\tan 1.5° = h/9 \qquad h = 0.2357 \text{ ft} \qquad V_{wood} = (\frac{4}{12})(\frac{4}{12})(9) = 1.000 \text{ ft}^3$$

$$F_b = W = 62.4[1.000 - (\frac{1}{2})(0.2357)(\frac{4}{12})(9)] = 40.34 \text{ lb}$$

$$W = (\gamma_{wood})(1.000) + 5 \qquad 40.34 = (\gamma_{wood})(1.000) + 5 \qquad \gamma_{wood} = 35.3 \text{ lb/ft}^3$$

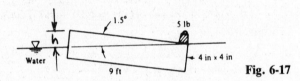

1.5°   5 lb
h
Water
9 ft   4 in x 4 in

Fig. 6-17

6.36 A wooden beam (s.g. $= 0.60$) is 15 cm by 15 cm by 4 m and is hinged at A, as shown in Fig. 6-18. At what angle θ will the beam float in water?

| The forces acting on the beam are shown in Fig. 6-18. $W_{beam} = [(0.60)(9.79)][(\frac{15}{100})(\frac{15}{100})(4)] = 0.5287$ kN and $F_b = 9.79[(\frac{15}{100})(\frac{15}{100})(L)] = 0.2203L$. $\Sigma M_A = 0$; $(0.2203L)[(4 - L/2)(\cos \theta)] - (0.5287)[(\frac{4}{2})(\cos \theta)] = 0$, $-0.1102L^2 + 0.8812L - 1.057 = 0$, $L = 1.470$ m; $\sin \theta = 1/(4 - 1.470) = 0.39526$, $\theta = 23.3°$.

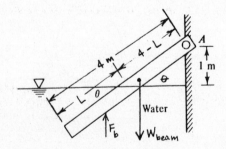

4 m   4 - L   A
L   θ   1 m
Water
F_b W_{beam}

Fig. 6-18

6.37 A barge weighs 50 tons empty and is 20 ft wide, 50 ft long, and 8 ft high. What will be its draft when loaded with 130 tons of gravel and floating in sea water (s.g. $= 1.025$)?

| $F_b = W$ $[(1.025)(62.4)][(20)(50)(h)] = (50 + 130)(2000)$ $h = 5.63$ ft

6.38 A block of steel (s.g. $= 7.85$) will "float" at a mercury-water interface as in Fig. 6-19. What will be the ratio of distances a and b for this condition?

| Let $w =$ width of block and $L =$ length of block. $F_b = W$, $(\gamma_{H_2O})(aLw) + (13.6)(\gamma_{H_2O})(bLw) = (7.85)(\gamma_{H_2O})(a + b)(Lw)$, $a + 13.6b = 7.85a + 7.85b$, $a/b = 0.839$.

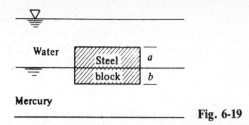

Fig. 6-19

6.39 A balloon weighing 3.0 lb is 5 ft in diameter. It is filled with hydrogen (H) (molecular weight = 2.02) at 15 psia and 60° F and released. At what altitude in the standard atmosphere will this balloon be neutrally buoyant?

$$F_b = W \qquad \gamma = p/RT \qquad \gamma_H = (15)(144)/[(765.5)(460 + 60)] = 0.005426 \text{ lb/ft}^3$$

$$(\gamma_{air})[(\tfrac{4}{3})(\pi)(\tfrac{5}{2})^3] = 3 + (0.005426)[(\tfrac{4}{3})(\pi)(\tfrac{5}{2})^3] \qquad \gamma_{air} = 0.05126 \text{ lb/ft}^3$$

From Table A-7, altitude = approximately 13 100 ft.

6.40 A rectangular barge 20 ft wide by 50 ft long by 8 ft deep floats empty with a draft of 3 ft in a canal lock 30 ft wide by 60 ft long and water depth 6 ft when the empty barge is present. If 200 000 lb of steel are loaded onto the barge, what are the new draft of the barge (h) and water depth in the lock (H)?

▌ The weight of the barge (W_b) is equal to the buoyant force when the draft is 3 ft. $W_b = 62.4[(20)(50)(3)] = 187\,200$ lb; $F_b = W$, $62.4[(20)(50)(h)] = 187\,200 + 200\,000$, $h = 6.205$ ft. Volume of water in lock $= (6)(30)(60) - (3)(20)(50) = 7800$ ft^3. After steel is added, $(H)(30)(60) - (6.205)(20)(50) = 7800$, $H = 7.78$ ft.

6.41 A 4-in-diameter solid cylinder of height 3.75 in weighing 0.85 lb is immersed in liquid ($\gamma = 52.0$ lb/ft^3) contained in a tall, upright metal cylinder having a diameter of 5 in. Before immersion the liquid was 3.0 in deep. At what level will the solid cylinder float? See Fig. 6-20.

▌ Let x = distance solid cylinder falls below original liquid surface, y = distance liquid rises above original liquid surface, and $x + y$ = depth of submergence. $V_A = V_B$, $x[(\pi)(4)^2/4] = y[(\pi)(5)^2/4] - y[(\pi)(4)^2/4]$, $x = 0.5625y$. $F_b = W$, $52.0[(\pi)(\tfrac{4}{12})^2/4][(x + y)/12] = 0.85$, $x + y = 2.248$, $0.5625y + y = 2.248$, $y = 1.44$ in, $x = (0.5625)(1.44) = 0.81$ in. The bottom of the solid cylinder will be $3.0 - 0.81$, or 2.19 in above the bottom of the hollow cylinder.

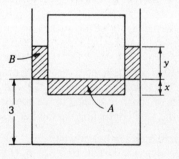

Fig. 6-20

6.42 An iceberg in the ocean floats with one-seventh of its volume above the surface. What is its specific gravity relative to ocean water? What portion of its volume would be above the surface if ice were floating in pure water? $\gamma_{ocean\,H_2O} = 64.0$ lb/ft^3.

$$(V_{iceberg})_{submerged} = (1 - \tfrac{1}{7})V_{iceberg} = 0.857 V_{iceberg}$$

$$F_b = W \qquad (\gamma_{ocean\,H_2O})(V_{iceberg})_{submerged} = (\gamma_{iceberg})(V_{iceberg})$$

$$s.g._{iceberg} = \gamma_{iceberg}/\gamma_{ocean\,H_2O} = (V_{iceberg})_{submerged}/V_{iceberg}$$

$$= 00.857 V_{iceberg}/V_{iceberg} = 0.857 \qquad \text{(relative to ocean water)}$$

$$\gamma_{iceberg} = (0.857)(64.0) = 54.85 \text{ lb/ft}^3 \qquad s.g._{iceberg} = 54.85/62.4 = 0.879 \qquad \text{(relative to pure water)}$$

Therefore, $1 - 0.879 = 0.121$, or 12.1 percent of its volume would be above the water surface in pure water.

6.43 A hydrometer consists of an 8-mm-diameter cylinder of length 20 cm attached to a 25-mm-diameter weighted sphere. The cylinder has a mass of 1.2 g, and the mass of the sphere is 12.8 g. At what level will this device float in liquids having specific gravities of 0.8, 1.0, and 1.2? Is the scale spacing on the cylindrical stem uniform? Why or why not?

▌ Let y = submerged length of cylinder in millimeters. $V_{sphere} = (\frac{4}{3})(\pi)(\frac{25}{2})^3 = 8181$ mm³, $V_{submerged\ cylinder} = (y)(\pi)(\frac{8}{2})^2 = 50.27y$, $F_b = W = (s.g.)[(9.79)(1000)/1000^3](8181 + 50.27y) = (0.08009)(s.g.) + (0.0004921)(s.g.)(y)$, $W = mg = (1.2 + 12.8)(9.81) = 137.3$ g · m/s², or 0.1373 N, $(0.08009)(s.g.) + (0.0004921)(s.g.)(y) = 0.1373$, $y = [0.1373 - (0.08009)(s.g.)]/[(0.0004921)(s.g.)]$.

For s.g. = 0.8 $y = [0.1373 - (0.08009)(0.8)]/[(0.0004921)(0.8)] = 186.0$ mm

For s.g. = 1.0 $y = [0.1373 - (0.08009)(1.0)]/[(0.0004921)(1.0)] = 116.3$ mm

For s.g. = 1.2 $y = [0.1373 - (0.08009)(1.2)]/[(0.0004921)(1.2)] = 69.8$ mm

Scale spacing is not uniform because buoyant force is not directly proportional to submergence.

6.44   Determine the volume of an object that weighs 5 lb in water and 7 lb in oil of specific gravity 0.82. What is the specific weight of the object?

▌ $F_b = W$, $62.4V_{object} = W_{object} - 5$, $[(0.82)(62.4)](V_{object}) = W_{object} - 7$. Subtracting the second equation from the first gives $62.4V_{object} - [(0.82)(62.4)](V_{object}) = -5 - (-7)$, $V_{object} = 0.178$ ft³; $(62.4)(0.178) = W_{object} - 5$, $W_{object} = 16.1$ lb. $\gamma = 16.1/0.178 = 90.4$ lb/ft³.

6.45   A balloon weighs 250 lb and has a volume of 14 000 ft³. It is filled with helium, which weighs 0.0112 lb/ft³ at the temperature and pressure of the air, which weighs 0.0807 lb/ft³. What load will the balloon support?

$$F_b = W \qquad (0.0807)(14\,000) = 250 + (0.0112)(14\,000) + \text{load} \qquad \text{Load} = 723\text{ lb}$$

6.46   A cylindrical bucket 30 cm in diameter and 50 cm high weighing 25.0 N contains oil (s.g. = 0.80) to a depth of 20 cm. (*a*) When placed in water, what will be the depth y to the bottom of the bucket? (*b*) What is the maximum volume of oil the bucket can hold and still float?

▌ (*a*) $F_b = W$ $9.79[(y)(\pi)(\frac{30}{100})^2/4] = 25.0/1000 + [(0.80)(9.79)][(\frac{20}{100})(\pi)(\frac{30}{100})^2/4]$

$$y = 0.196\text{ m} \quad \text{or} \quad 19.6\text{ cm}$$

(*b*) $9.79[(\frac{50}{100})(\pi)(\frac{30}{100})^2/4] = 25.0/1000 + [(0.80)(9.79)][(h)(\pi)(\frac{30}{100})^2/4]$ $h = 0.580$ m   or   58.0 cm

Since $h = 58.0$ cm is greater than the height of the bucket (50 cm), the bucket will float when full of oil. Therefore, $V_{max} = V_{bucket} = (\frac{50}{100})(\pi)(\frac{30}{100})^2/4 = 0.0353$ m³, or 35.3 L.

6.47   A metal block 1 ft square and 10 in deep is floated on a body of liquid which consists of an 8-in layer of water above a layer of mercury. The block weighs 120 lb/ft³. (*a*) What is the position of the bottom of the block? (*b*) If a downward vertical force of 250 lb is applied to the center of this block, what is the new position of the bottom of the block? Assume the tank containing the fluid is of infinite dimensions.

▌ (*a*) $F_b = W$. Let x = depth into mercury below water–mercury interface. $[(13.6)(62.4)][(1)(1)(x)] + 62.4[(1)(1)(\frac{8}{12})] = (120)[(1)(1)(\frac{10}{12})]$, $x = 0.0688$ ft, or 0.826 in. (*b*) In this case the top of the block will be below the water surface. Hence, $[(13.6)(62.4)][(1)(1)(x)] + 62.4[(1)(1)(\frac{10}{12} - x)] = 120[(1)(1)(\frac{10}{12})] + 250$, $x = 0.379$ ft, or 4.55 in.

6.48   Two spheres, each 1.2 m in diameter, weigh 4 kN and 12 kN, respectively. They are connected with a short rope and placed in water. What is the tension (T) in the rope and what portion of the lighter sphere protrudes from the water?

▌ For the lower (heavier) sphere, the buoyant force and T act upward and its weight acts downward. Hence, $\Sigma F_y = 0$, $F_b = 9.79[(\frac{4}{3})(\pi)(1.2/2)^3] = 8.86$ kN, $8.86 + T - 12 = 0$, $T = 3.14$ kN. For the upper (lighter) sphere, the buoyant force acts upward and its weight and T act downward. Hence, $F_b - 4 - 3.14 = 0$, $F_b = 7.14$ kN. Portion above water = $(8.86 - 7.14)/8.86 = 0.194$, or 19.4 percent of volume.

6.49   A wooden pole weighing 2 lb/ft has a cross-sectional area of 7 in² and is supported as shown in Fig. 6-21. The hinge is frictionless. Find θ.

▌ The forces acting on the beam are shown in Fig. 6-21. $W_{pole} = (2)(10) = 20.0$ lb; $F_b = (52)[(\frac{7}{144})(x)] = 2.528x$, $\Sigma M_{hinge} = 0$, $(20.0)[(\frac{10}{2})(\sin\theta)] - (2.528x)[(10 - x/2)(\sin\theta)] = 0$, $1.264x^2 - 25.28x + 100 = 0$; $x_1 = 14.57$ ft and

$x_2 = 5.43$ ft. Using $x = 5.43$ ft (since $x = 14.57$ ft is impossible for this situation), $\cos \theta = 4/(10 - 5.43) = 0.87527$, $\theta = 28.9°$.

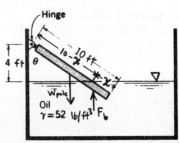

Fig. 6-21

6.50 A cube 2 ft on an edge has its lower half of s.g. = 1.4 and upper half of s.g. = 0.6. It is submerged into a two-layer fluid, the lower s.g. = 1.2 and the upper s.g. = 0.9. Determine the height of the top of the cube above the interface (i.e., h in Fig. 6-22).

$$F_b = W$$

$$[(1.2)(62.4)][(2)(2)(2-h)] + [(0.9)(62.4)][(2)(2)(h)] = [(1.4)(62.4)][(2)(2)(\tfrac{2}{2})] + [(0.6)(62.4)][(2)(2)(\tfrac{2}{2})]$$

$$h = 1.33 \text{ ft}$$

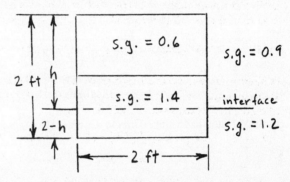

Fig. 6-22

6.51 Determine the volume, density, and specific volume of an object that weighs 3 N in water and 4 N in oil of s.g. 0.83.

$$F_b = W \qquad [(9.79)(1000)](V_{\text{object}}) = W_{\text{object}} - 3 \qquad [(0.83)(9.79)(1000)](V_{\text{object}}) = W_{\text{object}} - 4$$

Subtracting the second equation from the first gives $9790V_{\text{object}} - 8126V_{\text{object}} = 1$, $V_{\text{object}} = 0.0006010 \text{ m}^3$. $[(9.79)(1000)](0.0006010) = W_{\text{object}} - 3$, $W_{\text{object}} = 8.884 \text{ N}$; $\gamma = 8.884/0.0006010 = 14\,780 \text{ N/m}^3$.

$$\rho = \frac{\gamma}{g} = \frac{14\,780}{9.81} = 1507 \frac{\text{N/m}^3}{\text{m/s}^2} \quad \text{or} \quad 1507 \text{ kg/m}^3 \qquad V_s = \frac{1}{\rho} = \frac{1}{1507} = 0.000664 \text{ m}^3/\text{kg}$$

6.52 How many pounds of concrete, $\gamma = 25 \text{ kN/m}^3$, must be attached to a beam having a volume of 0.1 m^3 and s.g. = 0.65 to cause both to sink in water?

$$F_b = W \qquad (9.79)(0.1) + 9.79V_{\text{concrete}} = [(0.65)(9.79)](0.1) + 25V_{\text{concrete}} \qquad V_{\text{concrete}} = 0.02253 \text{ m}^3$$

$$W_{\text{concrete}} = (0.02253)(25) = 0.563 \text{ kN} \quad \text{or} \quad 563 \text{ N} \quad \text{or} \quad 563/4.448 = 127 \text{ lb}$$

6.53 The gate of Fig. 6-23 weighs 150 lb/ft normal to the page. It is in equilibrium as shown. Neglecting the weight of the arm and brace supporting the counterweight, find W (weight in air). The weight is made of concrete, s.g. = 2.50.

$$F_H = \gamma h A = (62.4)(\tfrac{5}{2})[(5)(1)] = 780 \text{ lb} \qquad \sum M_{\text{hinge}} = 0 \qquad (780)(\tfrac{5}{3}) - (W)(4 \sin 30°) = 0$$

$$W = 650 \text{ lb}$$

This is the submerged weight.

$$F_b = W \qquad 62.4V_{\text{concrete}} = [(2.50)(62.4)](V_{\text{concrete}}) - 650 \qquad V_{\text{concrete}} = 6.944 \text{ ft}^3$$

$$W_{\text{concrete}} = [(2.50)(62.4)](6.944) = 1083 \text{ lb}$$

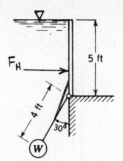

Fig. 6-23

6.54 A wooden cylinder 600 mm in diameter, s.g. = 0.50, has a concrete cylinder 600 mm long of the same diameter, s.g. = 2.50, attached to one end. Determine the minimum length of wooden cylinder for the system to float in static equilibrium with axis vertical.

▮ The system will float at minimum length of wooden cylinder as shown in Fig. 6-24. $F_b = W$, $(\gamma)(A)(L + \frac{600}{1000}) = [(0.50)(\gamma)](L) + [(2.50)(\gamma)](\frac{600}{1000})$, $L = 1.800$ m, or 1800 mm.

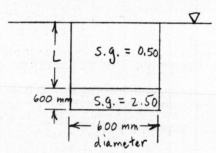

Fig. 6-24

6.55 A hydrometer weighs 0.035 N and has a stem 6 mm in diameter. Compute the distance between specific gravity markings 1.0 and 1.1.

▮ From Prob. 6.26,

$$h = \frac{(W)(\text{s.g.} - 1)}{(\text{s.g.})(\gamma)(\pi d^2/4)} = \frac{(0.035)(1.1 - 1)}{(1.1)[(9.79)(1000)][(\pi)(\frac{6}{1000})^2/4]} = 0.0115 \text{ m} \quad \text{or} \quad 11.5 \text{ mm}$$

6.56 What is the total weight of barge and load in Fig. 6-25? The barge is 6 m in width.

▮ $\qquad F_b = W \qquad 9.79\{(6)[(12)(2.4) + (2)(2.4)(2.4)/2]\} = W \qquad W = 2030$ kN

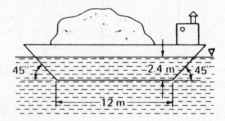

Fig. 6-25

6.57 In Fig. 6-26, a wedge of wood having a specific gravity of 0.60 is forced into water by a 150-lb force. The wedge is 2 ft in width. What is depth d?

▮ The 150-lb force and the weight of the wood (W) act downward on the wedge, while the buoyant force (F_b) acts upward. $\sum F_y = 0$, $F_b - 150 - W = 0$, $62.4[(2)(2)(d)(d \tan 30°)/2] - 150 - [(0.60)(62.4)]\{(2)(2)(\frac{3}{2})[(\frac{3}{2})/\tan 30°]/2\} = 0$, $d = 2.48$ ft.

6.58 The tank in Fig. 6-27 is filled to the edge with water. If a cube 600 mm on an edge and weighing 445 N is lowered slowly into the water until it floats, how much water flows over the edge of the tank if no appreciable waves are formed during the action? Neglect effects of adhesion at the tank's edge.

▮ $F_b = W$. Let h = the depth to which the cube will sink in the water. $[(9.79)(1000)][(\frac{600}{1000})(\frac{600}{1000})(h)] = 445$, $h = 0.126$ m, $V_{\text{displaced}} = [(\frac{600}{1000})(\frac{600}{1000})(0.126)] = 0.0454$ m³. This must be the amount of water that will overflow.

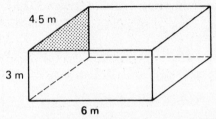

Fig. 6-26

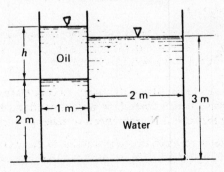

Fig. 6-27

6.59 A cube of material 300 mm on a side and weighing 445 N is lowered into a tank containing a layer of water over a layer of mercury. Determine the position of the block when it has reached equilibrium.

▌ $\gamma_{\text{cube}} = 445/[(\frac{300}{1000})(\frac{300}{1000})(\frac{300}{1000})] = 16\,481$ N/m³. Since the cube is heavier than water but lighter than mercury, it will sink beneath the water surface and come to rest at the water–mercury interface, as shown in Fig. 6-28. $F_b = W$, $9.79[(\frac{300}{1000})(\frac{300}{1000})(0.3-x)] + [(13.6)(9.79)][(\frac{300}{1000})(\frac{300}{1000})(x)] = \frac{445}{1000}$, $x = 0.0163$ m, or 16.3 mm. Thus, the bottom of the cube will come to rest 16.3 mm below the water–mercury interface.

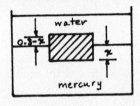

Fig. 6-28

6.60 An iceberg has a specific weight of 9000 N/m³ in ocean water, which has a specific weight of 10 000 N/m³. If we observe a volume of 2800 m³ of the iceberg protruding above the free surface, what is the volume of the iceberg below the free surface of the ocean?

▌ $\qquad F_b = W \qquad 10\,000 V_{\text{below}} = 9000(V_{\text{below}} + 2800) \qquad V_{\text{below}} = 25\,000$ m³

6.61 A rectangular tank of internal width 6 m is partitioned as shown in Fig. 6-29 and contains oil and water. If the oil's specific gravity is 0.82, what must h be? Next, if a 1000-N block of wood is placed in flotation in the oil, what is the rise in free surface of the water in contact with air?

▌ $P_{\text{atm}} + [(0.82)(9.79)](h) + (9.79)(2) - (9.79)(3) = p_{\text{atm}}$, $h = 1.220$ m. Let $h' =$ the new value of h with the 1000-N block in flotation. Since the volume of oil does not change, $(1.220)(1)(6) = (h')(1)(6) - 1000/[(0.82)(9.79)(1000)]$, $h' = 1.241$ m. If the oil–water interface drops by a distance δ, the free surface of water with air will rise by $\delta/2$. $p_{\text{atm}} + [(0.82)(9.79)](1.241) + 9.79(2-\delta) - 9.79(3+\delta/2) = p_{\text{atm}}$, $\delta = 0.01175$ m, or 11.75 mm. The free surface of the water will rise by 11.75/2, or 5.88 mm.

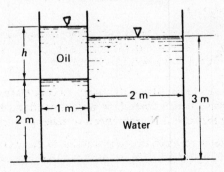

Fig. 6-29

6.62 A balloon of 2800 m³ is filled with hydrogen having a specific weight of 1.1 N/m³. (*a*) What lift is the balloon capable of at the earth's surface if the balloon weighs 1335 N and the temperature is 15 °C? (*b*) What lift is the balloon capable of at 9150 m standard atmosphere, assuming that the volume has increased 5 percent?

▌ From Table A-8, $\gamma_{air} = 12.01$ N/m³ at elevation 0 and 4.51 N/m³ at elevation 9150 m. $\sum F_y = 0$, $F_b - W_{balloon} - W_H = 0$.

(*a*) $\qquad (12.01)(2800) - 1335 - (1.1)(2800) - \text{lift} = 0 \qquad \text{Lift} = 29\,200$ N or 29.2 kN

(*b*) $\qquad 4.51[(1.05)(2800)] - 1335 - (1.1)(2800) - \text{lift} = 0 \qquad \text{Lift} = 8840$ N or 8.84 kN

6.63 A wooden rod weighing 5 lb is mounted on a hinge below the free surface, as shown in Fig. 6-30. The rod is 10 ft long and uniform in cross section, and the support is 5 ft below the free surface. At what angle α will it come to rest when allowed to drop from a vertical position? The cross section of the stick is 1.5 in² in area.

▌ The forces acting on the beam are shown in Fig. 6-30.

$$F_b = 62.4[(10 - e)(1.5/144)] = 6.500 - 0.6500e \qquad \sum M_A = 0$$

$$5(5 \cos \alpha) - (6.500 - 0.6500e)[(10 - e)/2](\cos \alpha) = 0 \qquad -0.325e^2 + 6.50e - 7.5 = 0 \qquad e = 1.229 \text{ ft}$$

$$\sin \alpha = 5/(10 - e) = 5/(10 - 1.229) = 0.57006 \qquad \alpha = 34.8°$$

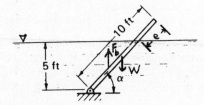

Fig. 6-30

6.64 A block of wood having a volume of 0.028 m³ and weighing 290 N is allowed to sink in water as shown in Fig. 6-31. A wooden rod of length 3.3 m and cross section of 1935 mm² is attached to the weight and also to the wall. If the rod weighs 13 N, what will angle θ be for equilibrium?

▌ $(F_b)_{block} = [(9.79)(1000)](0.028) = 274$ N $\qquad (F_b)_{rod} = [(9.79)(1000)][(AC)(1935/10^6)] = 18.94AC$ N

$$\sum M_B = 0$$

$$274(3.3 \cos \theta) + (18.94AC)[(AC/2) + (\tfrac{300}{1000})/\sin \theta](\cos \theta) - 290(3.3 \cos \theta) - (13)(3.3/2)(\cos \theta) = 0$$

$$AC = 3.3 - (\tfrac{300}{1000})/\sin \theta$$

$$274(3.3 \cos \theta) + 18.94[3.3 - (\tfrac{300}{1000})/\sin \theta]$$

$$\times \{[3.3 - (\tfrac{300}{1000})/\sin \theta]/2 + (\tfrac{300}{1000})/\sin \theta\}(\cos \theta) - 290(3.3 \cos \theta) - (13)(3.3/2)(\cos \theta) = 0$$

$$3.920 = [3.3 - (\tfrac{300}{1000})/\sin \theta][1.650 + (\tfrac{300}{1000})/(2 \sin \theta)] \qquad 3.920 = 5.445 - 0.045/\sin^2 \theta$$

$$\sin^2 \theta = 0.029508 \qquad \sin \theta = 0.17178 \qquad \theta = 9.9°$$

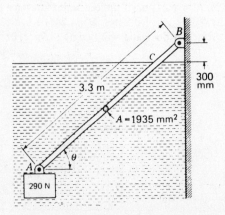

Fig. 6-31

6.65 A barge with a flat bottom and square ends has a draft of 6.0 ft when fully loaded and floating in an upright position, as shown in Fig. 6-32a. The center of gravity (*CG*) of the barge when fully loaded is on the axis of symmetry and 1.0 ft above the water surface. Is the barge stable? If it is stable, what is the righting moment when the angle of heel is 12°?

∎ $\overline{MB} = I/V_d = [(42)(25)^3/12]/[(25)(42)(6)] = 8.68$ ft. Therefore, the metacenter (mc) is located 8.68 ft above the center of buoyancy (*CB*), as shown in Fig. 6-32b. Hence, it (the metacenter) is located $8.68 - 3 - 1$, or 4.68 ft above the barge's center of gravity and the barge is stable. The end view of the barge when the angle of heel is 12° is shown in Fig. 6-32c. Righting moment $= (F_b)(x)$, $F_b = 62.4[(25)(42)(6)] = 393\,120$ lb, $x = (\sin 12°)(\text{distance from mc to } CG) = (\sin 12°)(4.68) = 0.973$ ft, righting moment $= (393\,120)(0.973) = 382\,500$ lb · ft.

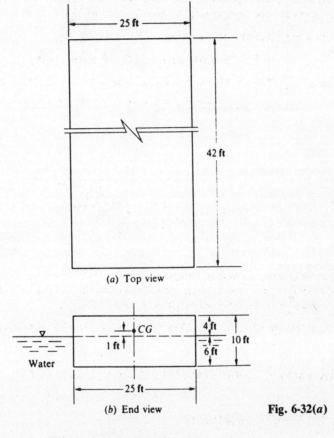

(a) Top view

(b) End view   Fig. 6-32(a)

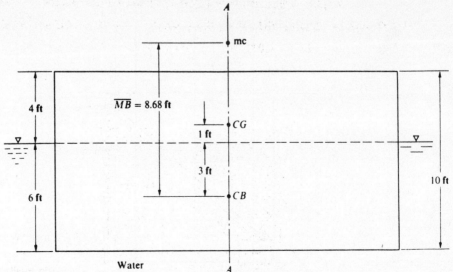

Water   Fig. 6-32(b)

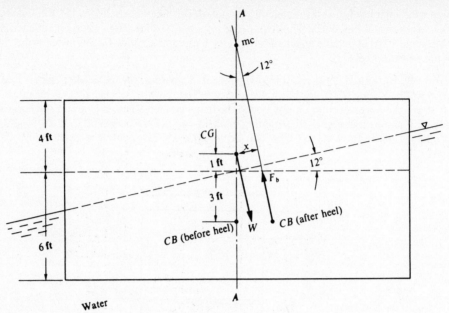

Fig. 6-32(c)

Water

6.66 Would the wooden cylinder (s.g. = 0.61) in Fig. 6-33a be stable if placed vertically in oil as shown in the figure?

▌ The first step is to determine the submerged depth of the cylinder when placed in the oil. $F_b = W$,
$[(0.85)(9.79)][(D)(\pi)(0.666)^2/4] = [(0.61)(9.79)][(1.300)(\pi)(0.666)^2/4]$, $D = 0.9333$ m. The center of buoyancy
is located at a distance of 0.933/2, or 0.466 m from the bottom of the cylinder (see Fig. 6-33b).
$\overline{MB} = I/V_d = [(\pi)(0.666)^4/64]/[(0.933)(\pi)(0.666)^2/4] = 0.030$ m. The metacenter is located 0.030 m above the
center of buoyancy, as shown in Fig. 6-33b. This places the metacenter $1.300/2 - 0.466 - 0.030$, or 0.154 m
below the center of gravity. Therefore, the cylinder is not stable.

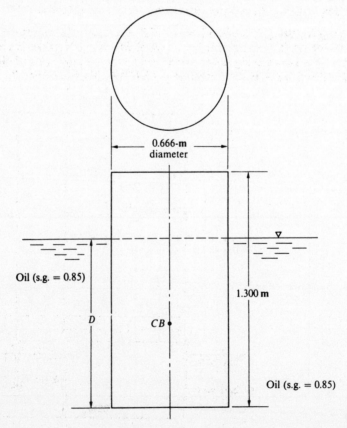

0.666-m
diameter

Oil (s.g. = 0.85)

1.300 m

D

CB ●

Oil (s.g. = 0.85)

Fig. 6-33(a)

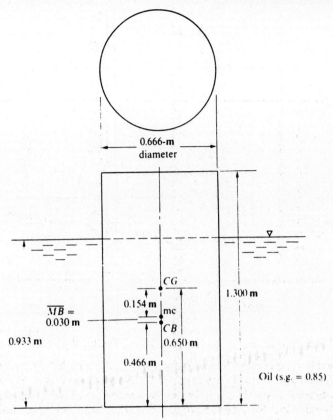

0.666-m
diameter

1.300 m

CG

$\overline{MB} =$
0.030 m

0.154 m

mc

CB

0.933 m

0.650 m

0.466 m

Oil (s.g. = 0.85)

Fig. 6-33(b)

6.67 Figure 6-34a shows the cross section of a boat. The hull of the boat is solid. Show if the boat is stable or not. If the boat is stable, compute the righting moment when the angle of heel is 10°?

▌ $\overline{MB} = I/V_d = [(20)(10)^3/12]/[(10)(5)(20)] = 1.67$ ft. Therefore, the metacenter is located $1.67 - 0.5$, or 1.17 ft above the center of gravity, as shown in Fig. 6-34b, and the barge is stable. The end view of the barge when the angle of heel is 10° is shown in Fig. 6-34c. Righting moment $= (F_b)(x)$, $F_b = 62.4[(10)(5)(20)] = 62\ 400$ lb, $x = (\sin 10°)(1.17) = 0.203$ ft, righting moment $= (62\ 400)(0.203) = 12\ 670$ lb · ft.

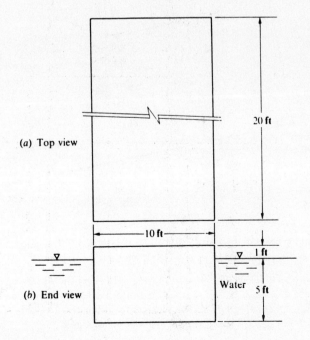

20 ft

(a) Top view

10 ft

1 ft

Water   5 ft

(b) End view

Fig. 6-34(a)

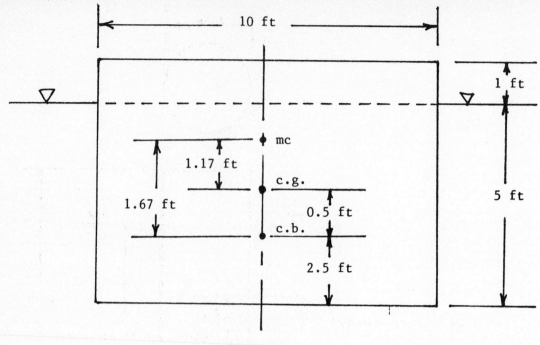

Fig. 6-34(b)

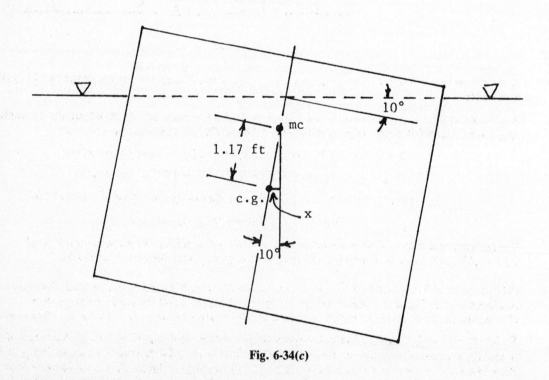

Fig. 6-34(c)

6.68 A solid wood cylinder has a diameter of 2.0 ft and a height of 4.0 ft. The specific gravity of the wood is 0.60. If the cylinder is placed vertically in oil (s.g. = 0.85), would it be stable?

▌ $F_b = W$, $[(0.85)(62.4)][(D)(\pi)(2)^2/4] = [(0.60)(62.4)][(4)(\pi)(2)^2/4]$, $D = 2.82$ ft. The center of buoyancy is located at a distance of 2.82/2, or 1.41 ft from the bottom of the cylinder (see Fig. 6-35). $\overline{MB} = I/V_d = [(\pi)(2)^4/64]/[(2.82)(\pi)(2)^2/4] = 0.09$ ft. The metacenter is located $2 - 1.41 - 0.09$, or 0.50 ft below the center of gravity, as shown in Fig. 6-35. Therefore, the cylinder is not stable.

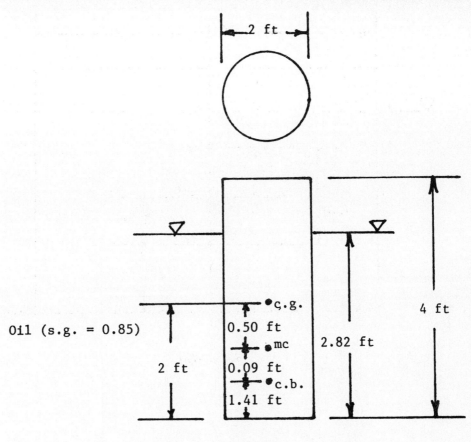

Fig. 6-35

6.69 A wood cone floats in water in the position shown in Fig. 6-36a. The specific gravity of the wood is 0.60. Would it be stable?

▮ The center of gravity is located $\frac{10}{4}$, or 2.50 in from the base of the cone or 7.50 in from the tip, as shown in Fig. 6-36b. $W_{\text{cone}} = [(0.60)(62.4)][(10)(\pi)(7)^2/12]/1728 = 2.779$ lb. Let x = submerged depth.

$$D_x = 0.700x \qquad V_d = (x)(\pi)(D_x)^2/12 = (x)(\pi)(0.700x)^2/12 = 0.1283x^3$$

$$F_b = W \qquad 62.4(0.1283x^3) = 2.779 \qquad x = 0.703 \text{ ft} \quad \text{or} \quad 8.44 \text{ in}$$

$$D_x = (0.700)(8.44) = 5.91 \text{ in} \qquad V_d = (0.1283)(0.703)^3 = 0.0446 \text{ ft}^3 \quad \text{or} \quad 77.1 \text{ in}^3$$

$$\overline{MB} = I/V_d = [(\pi)(5.91)^4/64]/77.1 = 0.78 \text{ in}$$

The metacenter is located 0.78 in above the center of buoyancy. Hence, the metacenter is located $7.50 - 6.33 - 0.78$, or 0.39 in below the cone's center of gravity, and the cone is not stable.

6.70 A block of wood 6 ft by 8 ft floats on oil of specific gravity 0.751. A clockwise couple holds the block in the position shown in Fig. 6-37. Determine the **(a)** buoyant force acting on the block and its position, **(b)** magnitude of the couple acting on the block, and **(c)** location of the metacenter for the tilted position.

▮ **(a)** $F_b = W = [(0.751)(62.4)][(10)(4+4)(4.618)/2] = 8656$ lb. F_b acts upward through the center of gravity O' of the displaced oil. The center of gravity lies 5.333 ft from A and 1.540 ft from D, as shown in Fig. 6-37. $AC = AR + RC = AR + LO' = (5.333)(\cos 30°) + (1.540)(\sin 30°) = 5.388$ ft. Hence, the buoyant force of 8650 lb acts upward through the center of gravity of the displaced oil, which is 5.388 ft to the right of A.

(b) One method of obtaining the magnitude of the righting couple (which must equal the magnitude of the external couple for equilibrium) is to find the eccentricity e. This dimension is the distance between the two parallel, equal forces W and F_b, which form the righting couple. $e = FC = AC - AF$, $AF = AR + RF = (5.333)(\cos 30°) + GR \sin 30° = 4.619 + (0.691)(\sin 30°) = 4.964$ ft, $e = 5.388 - 4.964 = 0.424$ ft; couple $= (8656)(0.424) = 3670$ ft · lb.

(c) Metacentric distance $MG = MR - GR = RC/\sin 30° - GR = 0.770/\sin 30° - 0.691 = 0.85$ ft

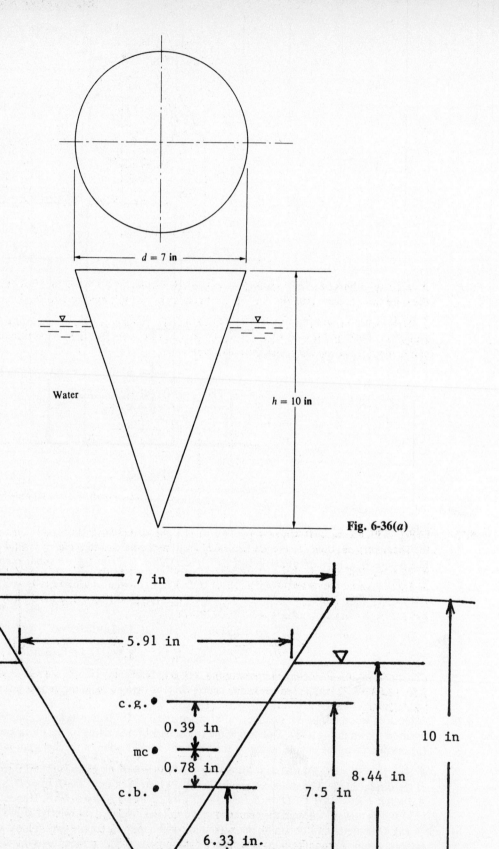

Water

$d = 7$ in

$h = 10$ in

Fig. 6-36(a)

7 in

5.91 in

c.g.

0.39 in

mc

0.78 in

c.b.

6.33 in.

7.5 in

8.44 in

10 in

Fig. 6-36(b)

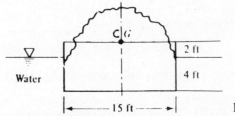

All lengths in feet.   **Fig. 6-37**

6.71 A rectangular barge 6 ft by 15 ft by 30 ft long is piled so high with gravel that its center of gravity is 2 ft above the waterline, as shown in Fig. 6-38. Is the barge stable for this configuration?

▮ $\overline{MB} = I/V_d = [(30)(15)^3/12]/[(4)(15)(30)] = 4.69$ ft. The metacenter is located 4.69 ft above the center of buoyancy, which is 2 ft above the bottom of the barge. Hence, the metacenter is located at $4.69 - 4$, or 0.69 ft above the center of gravity, and the barge is stable.

Fig. 6-38

6.72 In Fig. 6-39, a scow 20 ft wide and 60 ft long has a gross weight of 225 tons. Its center of gravity is 1.0 ft above the water surface. Find the metacentric height and restoring couple when $\Delta y = 1.0$ ft.

▮ $F_b = W$, $62.4[(60)(20)(h)] = (225)(2000)$, $h = 6.01$ ft. To locate CB', the center of buoyancy in the tipped position, take moments about AB and BC. $(6.01)(20)(x) = (6.01 - 1.0)(20)(10) + [(1.0 + 1.0)(20)/2](\frac{20}{3})$, $x = 9.45$ ft; $(6.01)(20)(y) = (6.01 - 1.0)(20)[(6.01 - 1.0)/2] + [(1.0 + 1.0)(20)/2][(6.01 - 1.0) + (1.0 + 1.0)/3]$, $y = 3.03$ ft. By similar triangles AEO and $\overline{CB'}PM$,

$$\frac{\Delta y}{b/2} = \frac{\overline{CB'}P}{MP} \qquad \frac{1.0}{20/2} = \frac{10 - 9.45}{MP} \qquad MP = 5.50 \text{ ft}$$

CG is $6.01 + 1.0$, or 7.01 ft from the bottom. Hence, $\overline{CGP} = 7.01 - 3.03 = 3.98$ ft, $\overline{MCG} = \overline{MP} - \overline{CGP} = 5.50 - 3.98 = 1.52$ ft. The scow is stable, since $\overline{MCG}$ is positive. Righting moment $= W(\overline{MCG})\sin\theta = [(225)(2000)](1.52)[1/\sqrt{(\frac{20}{2})^2 + 1.0^2}] = 68\,060$ ft · lb.

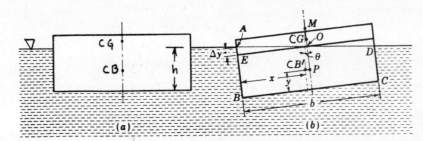

Fig. 6-39

6.73 What are the proportions of radius to height (r_0/h) of a right-circular cylinder of specific gravity s.g. so that it will float in water with end faces horizontal in stable equilibrium?

∎ See Fig. 6-40.

$$h_G = h/2 \qquad h_B = h_1/2 \qquad F_b = W \qquad \gamma(h_1 \pi r_0^2) = [(s.g.)(\gamma)](h \pi r_0^2) \qquad h_1 = (s.g.)(h)$$

$$h_B = (s.g.)(h)/2 \qquad MG = MB - GB$$

$$\overline{MB} = I/V_d = (\pi r_0^4/4)/(h_1 \pi r_0^2) = r_0^2/(4h_1) = r_0^2/[(4)(s.g.)(h)]$$

$$GB = h_G - h_B = h/2 - (s.g.)(h)/2 = (h)(1 - s.g.)/2 \qquad MG = r_0^2[(4)(s.g.)(h)] - (h)(1 - s.g.)/2$$

For stable equilibrium, $MG \geq 0$, in which case $r_0^2/[(4)(s.g.)(h)] \geq (h)(1 - s.g.)/2$, $r_0/h \geq \sqrt{(2)(s.g.)(1 - s.g.)}$.

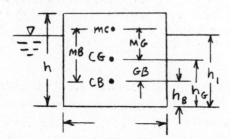

Fig. 6-40

6.74 The plane gate in Fig. 6-41a weighs 2000 N/m normal to the paper, and its center of gravity is 2 m from the hinge at O. **(a)** Find h as a function of θ for equilibrium of the gate. **(b)** Is the gate in stable equilibrium for any values of θ?

∎ Consider a unit width of gate, as shown in Fig. 6-41b.

(a) $\qquad F = \gamma \bar{h} A \qquad F_x = [(9.79)(1000)](h/2)(h/\sin \theta) = 4895h^2/\sin \theta \qquad \sum M_O = 0$

$$(4895h^2/\sin \theta)[(h/\sin \theta)/3] - 2000[(\tfrac{4}{2})(\cos \theta)] = 0 \qquad h^3 = 2.451 \sin^2 \theta \cos \theta \qquad h = 1.348(\sin^2 \theta \cos \theta)^{1/3}$$

(b) From part **(a)** $\sum M_O = (1632h^3)/\sin^2 \theta - 4000 \cos \theta$, $dM/d\theta = -3264h^3 \sin^{-3} \theta \cos \theta + 4000 \sin \theta$. Substituting $h = 1.348(\sin^2 \theta \cos \theta)^{1/3}$ [from part **(a)**], $dM/d\theta = -(3264)(1.348)^3(\cos^2 \theta/\sin \theta) + 4000 \sin \theta = -(7995)(\cos^2 \theta/\sin \theta) + 4000 \sin \theta$. For stability, $dM/d\theta < 0$, in which case $4000 \sin \theta < 7995(\cos^2 \theta/\sin \theta)$, $0.500 \sin \theta < \cos^2 \theta/\sin \theta$, $\tan^2 \theta < (1/0.500 = 2.00)$. This occurs for $\theta \leq 54.7°$ (upper limit). For the lower limit (when water spills over the top of the gate), $h = 4 \sin \theta$, $\sum M_O = (1632h^3)/\sin^2 \theta - 4000 \cos \theta$. Substituting $h = 4 \sin \theta$, $\sum M_O = 1632(4 \sin \theta)^3/\sin^2 \theta - 4000 \cos \theta = 104448 \sin \theta - 4000 \cos \theta$. In this case, $\sum M_O = 0$, $\tan \theta = 4000/104448 = 0.038297$, $\theta = 2.2°$. Thus for stable equilibrium, θ must be between 2.2° and 54.7°.

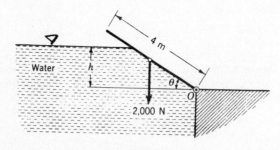

Fig. 6-41(a)

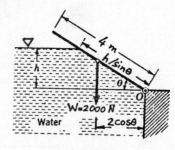

Fig. 6-41(b)

6.75 The barge shown in Fig. 6-42 has the form of a parallelopiped having dimensions 10 m by 26.7 m by 3 m. The barge weighs 4450 kN when loaded and has a center of gravity 4 m from the bottom. Find the metacentric height for a rotation about its longest centerline, and determine whether or not the barge is stable.

∎ First, find the center of buoyancy of the barge. $F_b = W$, $9.79[(10)(26.7)(D)] = 4450$, $D = 1.702$ m. Hence, the center of buoyancy (CB) is at a distance 1.702/2, or 0.851 m above the bottom of the barge. $\overline{MB} = I/V_d = [(26.7)(10)^3/12]/[(10)(26.7)(1.702)] = 4.896$. The distance from CB to CG is $4 - 0.851$, or 3.149 m. Therefore, the metacenter is located $4.896 - 3.149$, or 1.747 m above the CG, and the barge is stable.

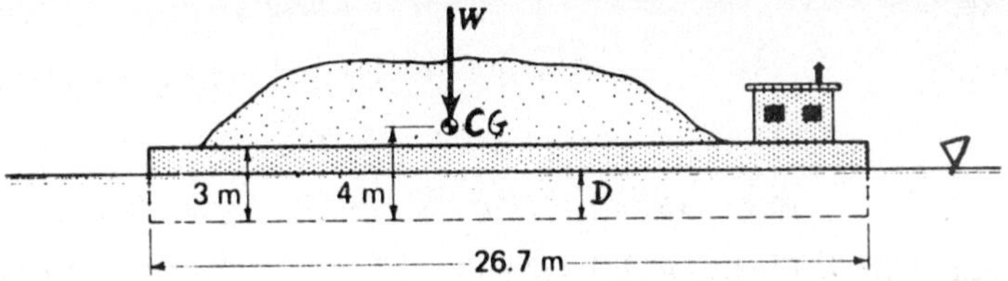

Fig. 6-42

6.76 A cube of side length L and specific gravity 0.8 floats in water. Is the cube stable?

❚ The cube's center of gravity is at $0.5L$ above its bottom. If the cube has a s.g. $= 0.8$, it will float at a submerged depth of $0.8L$, and its center of buoyancy will be at $0.4L$ above its bottom. $\overline{MB} = I/V_d = [(L)(L)^3/12]/[(L)(L)(0.8L)] = 0.1042L$. Therefore, the metacenter is located $0.1042L$ above the center of buoyancy and $0.1042L + 0.4L - 0.5L$, or $0.0042L$ above the center of gravity, and the cube is stable (although just barely).

6.77 For the cube specified in Prob. 6.76, determine the range of values of specific gravity between 0 and 1.0 for which the cube is stable.

❚ The cube's center of gravity is at $0.5L$ above its bottom. For any specific gravity s.g., the cube will float at a submerged depth of $(\text{s.g.})(L)$, and its center of buoyancy will be at $(\text{s.g.})(L)/2$ above its bottom. $\overline{MB} = I/V_d = [(L)(L)^3/12]/\{(L)(L)[(\text{s.g.})(L)]\} = 0.08333L/(\text{s.g.})$. Therefore, if the cube is stable, the metacenter must be located $0.08333L/(\text{s.g.})$ above the center of buoyancy and $0.08333L/(\text{s.g.}) + (\text{s.g.})(L)/2 - 0.5L$ above the center of gravity. For this to occur, $0.08333L/(\text{s.g.}) + (\text{s.g.})(L)/2 - 0.5L > 0$, $(\text{s.g.})^2/2 - (0.5)(\text{s.g.}) + 0.08333 > 0$. This condition is true (i.e., the cube is stable) for s.g. > 0.789 and s.g. < 0.211.

CHAPTER 7
Kinematics of Fluid Motion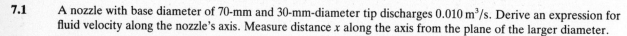

7.1 A nozzle with base diameter of 70-mm and 30-mm-diameter tip discharges $0.010 \text{ m}^3/\text{s}$. Derive an expression for fluid velocity along the nozzle's axis. Measure distance x along the axis from the plane of the larger diameter.

▍ Let L = length of nozzle and D = diameter of nozzle at any point. $D = \frac{70}{1000} - (\frac{70}{1000} - \frac{30}{1000})(x/L) = 0.070 - 0.040x/L$, $v = Q/A = 0.010/[\pi(0.070 - 0.040x/L)^2/4] = 1.273/(0.70 - 0.40x/L)^2$. *Note*: x and L in millimeters gives v in m/s.

7.2 What angle α of jet is required to reach the roof of the building in Fig. 7-1 with minimum jet velocity v_0 at the nozzle? What is the value of v_0?

▍ $d^2y/dt^2 = -g$, $dy/dt = -gt + c_1$. At $t = 0$, $dy/dt = v_0 \sin \alpha$. Therefore, $c_1 = v_0 \sin \alpha$, and $dy/dt = -gt + v_0 \sin \alpha$, $y = -gt^2/2 + tv_0 \sin \alpha + c_2$. At $t = 0$, $y = 0$. Therefore $c_2 = 0$, and $y = -gt^2/2 + tv_0 \sin \alpha$, $L = tv_0 \cos \alpha$, $t = L/(v_0 \cos \alpha)$.

$$H = -g[L/(v_0 \cos \alpha)]^2/2 + [L/(v_0 \cos \alpha)](v_0 \sin \alpha) \tag{1}$$

Let $F = gL^2/(2v_0^2)$. Then, from Eq. (1), $F = (\cos \alpha)(L \sin \alpha - H \cos \alpha) = L \cos \alpha \sin \alpha - H \cos^2 \alpha$. Find maximum F for minimum v_0.

$$dF/d\alpha = L(\cos^2 \alpha - \sin^2 \alpha) + 2H \sin \alpha \cos \alpha = 0 \qquad 2H/L = -(\cos^2 \alpha - \sin^2 \alpha)/(\sin \alpha \cos \alpha) = -2 \cot 2\alpha$$

$$(2)(3)/25 = -2 \cot 2\alpha \qquad \alpha = 70.1°$$

Substituting into Eq. (1), $30 = -(9.807)[25/(v_0 \cos 70.1°)]^2/2 + [25/(v_0 \cos 70.1°)](v_0 \sin 70.1°)$, $v_0 = 26.0 \text{ m/s}$.

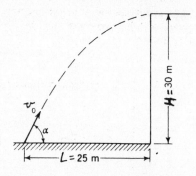

Fig. 7-1

7.3 Given the velocity field, $\mathbf{V}(x, y, z, t) = (6xy^2 + t)\mathbf{i} + (3z + 10)\mathbf{j} + 20\mathbf{k} \text{ m/s}$, with x, y, z in meters and t in seconds, what is the velocity vector at position $x = 10 \text{ m}$, $y = -1 \text{ m}$, and $z = 2 \text{ m}$ when $t = 5 \text{ s}$? What is the magnitude of this velocity?

▍
$$\mathbf{V} = [(6)(10)(-1)^2 + 5]\mathbf{i} + [(3)(2) + 10]\mathbf{j} + 20\mathbf{k} = 65\mathbf{i} + 16\mathbf{j} + 20\mathbf{k} \quad \text{m/s}$$
$$|\mathbf{V}| = \sqrt{65^2 + 16^2 + 20^2} = 69.9 \text{ m/s}$$

Note: Boldface letters are used herein to denote vectors.

7.4 The velocity components in a flow of fluid are known to be $v_x = 6xt + y^2z + 15 \text{ m/s}$, $v_y = 3xy^2 + t^2 + y \text{ m/s}$, and $v_z = 2 + 3ty \text{ m/s}$, where x, y, and z are given in meters and t in seconds. What is the velocity vector at $(3, 2, 4) \text{ m}$ at time $t = 3 \text{ s}$? What is the magnitude of this vector at this point and time?

▍
$$\mathbf{V} = [(6)(3)(3) + (2)^2(4) + 15]\mathbf{i} + [(3)(3)(2)^2 + 3^2 + 2]\mathbf{j} + [2 + (3)(3)(2)]\mathbf{k} = 85\mathbf{i} + 47\mathbf{j} + 20\mathbf{k} \text{ m/s}$$
$$|\mathbf{V}| = \sqrt{85^2 + 47^2 + 20^2} = 99.2 \text{ m/s}$$

7.5 A *path line* is the curve traversed by any one particle in the flow and corresponds to the *trajectory*. Given the velocity field $\mathbf{V} = (6x)\mathbf{i} + (16y + 10)\mathbf{j} + (20t^2)\mathbf{k} \text{ m/s}$, what is the path line of a particle which is at $(2, 4, 6) \text{ m}$ at time $t = 2 \text{ s}$?

$$v_x = dx/dt = 6x \tag{1}$$

$$v_y = dy/dt = 16y + 10 \tag{2}$$

$$v_z = dz/dt = 20t^2 \tag{3}$$

From (1), $dx/x = 6\,dt$, $\ln x = 6t + c_1$. At $t = 2$, $x = 2$. Hence, $\ln 2 = (6)(2) + c_1$, $c_1 = -11.3$.

$$\ln x = 6t - 11.3 \tag{4}$$

From (2), $dy/(16y + 10) = dt$, $\ln(16y + 10) = 16t + c_2$. At $t = 2$, $y = 4$. Hence, $\ln[(16)(4) + 10] = (16)(2) + c_2$, $c_2 = -27.7$.

$$\ln(16y + 10) = 16t - 27.7 \tag{5}$$

From (3), $dz = 20t^2\,dt$, $z = 20t^3/3 + c_3$. At $t = 2$, $z = 6$. Hence, $6 = (20)(2)^3/3 + c_3$, $c_3 = -47.3$.

$$z = 20t^3/3 - 47.3 \tag{6}$$

Add Eqs. (4) and (5) to get

$$\ln x + \ln(16y + 10) = 22t - 39.0 \tag{7}$$

Solve for t in Eq. (6): $t = [(z + 47.3)(\frac{3}{20})]^{1/3}$. Substitute this value of t into Eq. (7): $\ln x + \ln(16y + 10) = 22[(z + 47.3)(\frac{3}{20})]^{1/3} - 39.0$, $\ln[(x)(16y + 10)] = 11.69(z + 47.3)^{1/3} - 39.0$.

7.6 An incompressible ideal fluid flows at 0.5 cfs through a circular pipe into a conically converging nozzle, as shown in Fig. 7-2. Determine the average velocity of flow at sections A and B.

▌ As a first step, an approximate flow net is sketched to provide a general picture of the flow. Since this is an axially symmetric flow, the net is not a true two-dimensional flow net. At section A, the streamlines are parallel; hence, the area at right angles to the velocity vectors is a circle. Thus, $v_A = Q/A_A = 0.5/[(\pi)(\frac{8}{12})^2/4] = 1.43$ ft/s. At section B, however, the area at right angles to the streamlines is not clearly defined; it is a curved, dish-shaped section. As a rough approximation, it might be assumed to be the portion of the surface of a sphere of radius 2.0 in that is intersected by a circle of diameter 2.82 in. $v_B = Q/A_B = Q/(2\pi rh) = 0.5/[(2)(\pi)(\frac{2}{12})(0.59/12)] = 9.71$ ft/s.

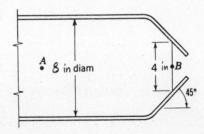

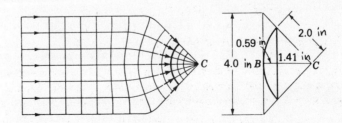

Fig. 7-2

7.7 Water flows at 5 gal/min through a small circular hole in the bottom of a large tank. Assuming the water in the tank approaches the hole radially, find the velocity in the tank at 2, 4, and 8 in from the hole.

▌ The area through which flow occurs is a hemispherical surface, with $A = 2\pi r^2$. $Q = 5/[(7.48)(60)] = 0.01114$ ft³/s, $v = Q/A$. At 2 in from the hole, $v = 0.01114/[(2)(\pi)(\frac{2}{12})^2] = 0.0638$ ft/s. At 4 in from the hole, $v = 0.01114/[(2)(\pi)(\frac{4}{12})^2] = 0.0160$ ft/s. At 8 in from the hole, $v = 0.01114/[(2)(\pi)(\frac{8}{12})^2] = 0.00399$ ft/s.

7.8 Given the eulerian velocity-vector field $\mathbf{V}(x, y, z, t) = 3t\mathbf{i} + xz\mathbf{j} + ty^2\mathbf{k}$, find the acceleration of a particle.

$$\frac{d\mathbf{V}}{dt} = \frac{\partial \mathbf{V}}{\partial t} + \left(u\frac{\partial \mathbf{V}}{\partial x} + v\frac{\partial \mathbf{V}}{\partial y} + w\frac{\partial \mathbf{V}}{\partial z}\right) \qquad u = 3t \qquad v = xz \qquad w = ty^2$$

$$\frac{\partial \mathbf{V}}{\partial t} = \mathbf{i}\frac{\partial u}{\partial t} + \mathbf{j}\frac{\partial v}{\partial t} + \mathbf{k}\frac{\partial w}{\partial t} = 3\mathbf{i} + y^2\mathbf{k} \qquad \frac{\partial \mathbf{V}}{\partial x} = z\mathbf{j} \qquad \frac{\partial \mathbf{V}}{\partial y} = 2ty\mathbf{k} \qquad \frac{\partial \mathbf{V}}{\partial z} = x\mathbf{j}$$

$$\frac{d\mathbf{V}}{dt} = (3\mathbf{i} + y^2\mathbf{k}) + (3t)(z\mathbf{j}) + (xz)(2ty\mathbf{k}) + (ty^2)(x\mathbf{j}) = 3\mathbf{i} + (3tz + txy^2)\mathbf{j} + (y^2 + 2xyzt)\mathbf{k}$$

If V is valid everywhere as given, this acceleration applies to all positions and times within the flow field.

7.9 Flow through a converging nozzle can be approximated by a one-dimensional velocity distribution $u = u(x)$. For the nozzle shown in Fig. 7-3, assume the velocity varies linearly from $u = v_0$ at the entrance to $u = 3v_0$ at the exit: $u(x) = v_0(1 + 2x/L)$; $\partial u/\partial x = 2v_0/L$. (**a**) Compute the acceleration du/dt as a general function of x, and (**b**) evaluate du/dt at the entrance and exit if $v_0 = 10$ ft/s and $L = 1$ ft.

(**a**)
$$\frac{du}{dt} = \frac{\partial u}{\partial t} + u\frac{\partial u}{\partial x} + v\frac{\partial u}{\partial y} + w\frac{\partial u}{\partial z} \qquad \frac{\partial u}{\partial t} = v\frac{\partial u}{\partial y} = w\frac{\partial u}{\partial z} = 0 \qquad u = v_0\left(1 + \frac{2x}{L}\right) \qquad \frac{\partial u}{\partial x} = \frac{2v_0}{L}$$

$$\frac{du}{dt} = 0 + \left[v_0\left(1 + \frac{2x}{L}\right)\right]\left(\frac{2v_0}{L}\right) + 0 + 0 = \left(\frac{2v_0^2}{L}\right)\left(1 + \frac{2x}{L}\right)$$

(**b**) At the entrance, where $x = 0$, $du/dt = [(2)(10)^2(1)][1 + (2)(0)/(1)] = 200$ ft/s². At the exit, where $x = 1$ ft, $du/dt = [(2)(10)^2(1)][1 + (2)(1)/(1)] = 600$ ft/s².

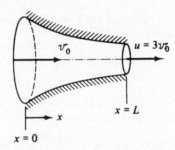

$$v_0 \qquad u = 3v_0$$
$$x = L$$
$$x = 0$$

Fig. 7-3

7.10 A velocity field is given by $u = 3y^2$, $v = 2x$, $w = 0$, in arbitrary units. Is this flow steady or unsteady? Is it two or three dimensional? At $(x, y, z) = (2, 1, 0)$, compute the (**a**) velocity, (**b**) local acceleration, and (**c**) convective acceleration.

Flow is steady because time t does not appear. Flow is two-dimensional because $w = 0$ everywhere.

$$\frac{du}{dt} = \frac{\partial u}{\partial t} + u\frac{\partial u}{\partial x} + v\frac{\partial u}{\partial y} + w\frac{\partial u}{\partial z} = 0 + (3y^2)(0) + (2x)(6y) + (0)(0) = 12xy$$

$$\frac{dv}{dt} = \frac{\partial v}{\partial t} + u\frac{\partial v}{\partial x} + v\frac{\partial v}{\partial y} + w\frac{\partial v}{\partial z} = 0 + (3y^2)(2) + (2x)(0) + (0)(0) = 6y^2$$

$$\mathbf{a} = \frac{d\mathbf{V}}{dt} = 12xy\mathbf{i} + 6y^2\mathbf{j} \qquad \text{(convective only)}$$

(**a**) $$\mathbf{v} = \mathbf{i}[(3)(1)^2] + \mathbf{j}[(2)(2)] = 3\mathbf{i} + 4\mathbf{j}$$

(**b**) $$\frac{\partial \mathbf{V}}{\partial t} = 0$$

(**c**) $$\mathbf{a} = (12)(2)(1)\mathbf{i} + (6)(1)^2\mathbf{j} = 24\mathbf{i} + 6\mathbf{j}$$

7.11 For the velocity field described in Prob. 7.10, at $(2, 1, 0)$ compute the (**a**) acceleration component parallel to the velocity vector and (**b**) component normal to the velocity vector.

From Prob. 7.10, $\mathbf{V} = 3\mathbf{i} + 4\mathbf{j}$ and $\mathbf{a} = 24\mathbf{i} + 6\mathbf{j}$ at $(2, 1, 0)$.
(**a**) Tangential acceleration:

$$\mathbf{n}_v = \mathbf{V}/|\mathbf{V}| = \tfrac{3}{5}\mathbf{i} + \tfrac{4}{5}\mathbf{j} \qquad a_t = \mathbf{a} \cdot \mathbf{n}_v = (24\mathbf{i} + 6\mathbf{j}) \cdot (\tfrac{3}{5}\mathbf{i} + \tfrac{4}{5}\mathbf{j}) = 14.4 + 4.8 = 19.2 \text{ units parallel to } \mathbf{V}$$

(b) From Fig. 7-4, the angle θ between **V** and **a** is indicated by $\cos\theta = a_t/|a| = 19.2/(24^2 + 6^2)^{1/2} = 0.77611$, $\theta = 39.1°$, $a_n = |a|\sin\theta = (24^2 + 6^2)^{1/2}(\sin 39.1°) = 15.6$ units normal to **V**.

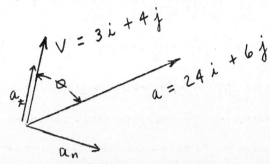

Fig. 7-4

7.12 An idealized velocity field is given by the formula $\mathbf{V} = 3tx\mathbf{i} - t^2y\mathbf{j} + 2xz\mathbf{k}$. Is this flow steady or unsteady? Is it two- or three-dimensional? At the point $(x, y, z) = (1, -1, 0)$, compute the **(a)** total acceleration vector and **(b)** unit vector normal to the acceleration.

▌ Flow is unsteady because time t appears explicitly. Flow is three-dimensional because u, v, $w \neq 0$.

$$\frac{du}{dt} = \frac{\partial u}{\partial t} + u\frac{\partial u}{\partial x} + v\frac{\partial u}{\partial y} + w\frac{\partial u}{\partial z} = 3x + (3tx)(3t) + (-t^2y)(0) + (2xy)(0) = 3x + 9t^2x$$

$$\frac{dv}{dt} = \frac{\partial v}{\partial t} + u\frac{\partial v}{\partial x} + v\frac{\partial v}{\partial y} + w\frac{\partial v}{\partial z} = -2ty + (3tx)(0) - (-t^2y)(-t^2) + (2xz)(0) = -2ty + t^4y$$

$$\frac{dw}{dt} = \frac{\partial w}{\partial t} + u\frac{\partial w}{\partial x} + v\frac{\partial w}{\partial y} + w\frac{\partial w}{\partial z} = 0 + (3tx)(2z) + (-t^2y)(0) + (2xz)(2x) = 6txz + 4x^2z$$

(a)
$$\mathbf{a} = \mathbf{i}\frac{du}{dt} + \mathbf{j}\frac{dv}{dt} + \mathbf{k}\frac{dw}{dt}$$

At point $(1, -1, 0)$, $du/dt = (3)(1) + (9)(t^2)(1) = 3 + 9t^2$, $dv/dt = -(2)(t)(-1) + (t^4)(-1) = 2t - t^4$, $dw/dt = (6)(t)(1)(0) + (4)(1)^2(0) = 0$. Hence, $\mathbf{a} = (3 + 9t^2)\mathbf{i} + (2t - t^4)\mathbf{j}$. **(b)** The unit vector normal to **a** must satisfy $\mathbf{a} \cdot \mathbf{n} = 0 = n_x(3 + 9t^2) + n_y(2t - t^4) + n_z(0)$ plus $n_x^2 + n_y^2 + n_z^2 = 1$. A special case solution is $n = \pm\mathbf{k}$.

7.13 For steady flow through a conical nozzle, the axial velocity is approximately $u = U_0(1 - x/L)^{-2}$, where U_0 is the entrance velocity and L is the distance to the apparent vertex of the cone. Compute **(a)** a general expression for the axial acceleration du/dt and **(b)** the acceleration at the entrance and at $x = 1$ m if $U_0 = 5$ m/s and $L = 2$ m.

▌ **(a)**
$$\frac{du}{dt} = \frac{\partial u}{\partial t} + u\frac{\partial u}{\partial x} + v\frac{\partial u}{\partial y} + w\frac{\partial u}{\partial z} = 0 + \left[U_0\left(1 - \frac{x}{L}\right)^{-2}\right]\left[\left(1 - \frac{x}{L}\right)^{-3}(-2U_0)\left(-\frac{1}{L}\right)\right] + 0 + 0 = \left(1 - \frac{x}{L}\right)^{-5}\left(\frac{2U_0^2}{L}\right)$$

(b) At entrance $(x = 0)$:
$$\frac{du}{dt} = (1 - \tfrac{0}{2})^{-5}[(2)(5)^2/2] = 25.0 \text{ m/s}^2$$

At $x = 1$ m:
$$\frac{du}{dt} = (1 - \tfrac{1}{2})^{-5}[(2)(5)^2/2] = 800 \text{ m/s}^2$$

7.14 A two-dimensional velocity field is given by $\mathbf{V} = (x^2 - y^2 + x)\mathbf{i} - (2xy + y)\mathbf{j}$ in arbitrary units. At $x = 2$ and $y = 1$, compute the **(a)** accelerations a_x and a_y, **(b)** velocity component in the direction $\theta = 30°$, and **(c)** directions of maximum acceleration and maximum velocity.

▌
$$\frac{du}{dt} = a_x = \frac{\partial u}{\partial t} + u\frac{\partial u}{\partial x} + v\frac{\partial u}{\partial y} + w\frac{\partial u}{\partial z} = 0 + (x^2 - y^2 + x)(2x + 1) + (-2xy - y)(-2y) + 0$$

$$\frac{dv}{dt} = a_y = \frac{\partial v}{\partial t} + u\frac{\partial v}{\partial x} + v\frac{\partial v}{\partial y} + w\frac{\partial v}{\partial z} = 0 + (x^2 - y^2 + x)(-2y) + (-2xy - y)(-2x - 1) + 0$$

(a)
$$a_x = (2^2 - 1^2 + 2)[(2)(2) + 1] + [(-2)(2)(1) - 1][(-2)(1)] = 35$$

$$a_y = (2^2 - 1^2 + 2)[(-2)(1)] + [(-2)(2)(1) - 1][(-2)(2) - 1] = 15$$

(b)
$$v_{30°} = \mathbf{V} \cdot \mathbf{n}_{30°} \qquad \mathbf{V} = (2^2 - 1^2 + 2)\mathbf{i} - [(2)(2)(1) + 1]\mathbf{j} = 5\mathbf{i} - 5\mathbf{j}$$

$$\mathbf{n}_{30°} = \frac{\sqrt{3}}{2}\mathbf{i} + \frac{1}{2}\mathbf{j} \qquad v_{30°} = (5\mathbf{i} - 5\mathbf{j})\left(\frac{\sqrt{3}}{2}\mathbf{i} + \frac{1}{2}\mathbf{j}\right) = 4.33 - 2.50 = 1.83$$

(c) For maximum acceleration, $\alpha = \arctan \frac{15}{35} = \arctan 0.42857 = 23.2°$. Hence, the direction of maximum acceleration is

For maximum velocity, $\beta = \arctan \frac{-5}{5} = \arctan(-1.0000) = -45.0°$. Hence, the direction of maximum velocity is

7.15   The velocity field in the neighborhood of a stagnation point is given by $u = U_0 x/L$, $v = -U_0 y/L$, $w = 0$.
(a) Show that the acceleration vector is purely radial. (b) If $L = 2$ ft, what is the magnitude of U_0 if the total acceleration at $(x, y) = (L, L)$ is 30 ft/s²?

∎
$$\frac{du}{dt} = a_x = \frac{\partial u}{\partial t} + u\frac{\partial u}{\partial x} + v\frac{\partial u}{\partial y} + w\frac{\partial u}{\partial z} = 0 + \left(\frac{U_0 x}{L}\right)\left(\frac{U_0}{L}\right) + 0 + 0 = \frac{U_0^2 x}{L^2}$$

$$\frac{dv}{dt} = a_y = \frac{\partial v}{\partial t} + u\frac{\partial v}{\partial x} + v\frac{\partial v}{\partial y} + w\frac{\partial v}{\partial z} = 0 + 0 + \left(\frac{-U_0 y}{L}\right)\left(\frac{-U_0}{L}\right) + 0 = \frac{U_0^2 y}{L^2}$$

(a) $\mathbf{a} = a_x\mathbf{i} + a_y\mathbf{j} = (U_0^2/L^2)(x\mathbf{i} + y\mathbf{j}) = (U_0^2/L^2)(\mathbf{r})$. (Hence, purely radial.)
(b) $|\mathbf{a}| = a(L, L) = (U_0^2/L^2)|L\mathbf{i} + L\mathbf{j}| = U_0^2\sqrt{2}/L$. If $L = 2$ ft and $|\mathbf{a}| = 30$ ft/s², $30 = U_0^2\sqrt{2}/2$, $U_0 = 6.51$ ft/s.

7.16   Suppose that a particle moves around the circular path $x^2 + y^2 = 4$ m² at a uniform tangential velocity of 3 m/s. Express the motion in terms of u and v components. Compute the tangential and radial acceleration at the point $(x, y) = (2, 0)$. See Fig. 7-5.

∎
$$u = u_r \cos\theta - u_\theta \sin\theta = -3\sin\theta \quad \text{m/s} \qquad v = v_r\sin\theta + v_\theta\cos\theta = +3\cos\theta \quad \text{m/s}$$

$$d\theta/dt = 2\pi/(4\pi/3) = 1.5 \text{ rad/s} \qquad \theta = 1.5t$$

Tangential acceleration: $\qquad\qquad\qquad a_\theta = dv_\theta/dt = 0$

Radial acceleration: $\qquad a_r = -v_\theta^2/r = -(3)^2/2 = -4.50$ m/s² (i.e., toward the center)

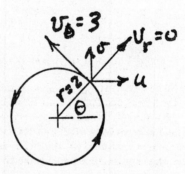

Fig. 7-5

7.17   A nearly frictionless liquid flows from the bottom of a large tank through a 1-cm-diameter hole at a rate of 800 cm³/s. If the fluid flows radially toward the hole with the same volume flow across every section, compute the convective acceleration at points 20 cm and 40 cm from the hole.

∎   Consider the radial velocity (v_r): $v_r = -Q/A_r = -Q/(2\pi r^2)$ (A_r is the area of a hemisphere).

$$a_r = v_r\frac{\partial v_r}{\partial r} = \left(\frac{-Q}{2\pi r^2}\right)\left(\frac{Q}{\pi r^3}\right) = -\frac{Q^2}{2\pi^2 r^5}$$

At $r = 20$ cm, $a_r = -(800)^2/[(2)(\pi)^2(20)^5] = -0.0101$ cm/s². At $r = 40$ cm, $a_r = -(800)^2/[(2)(\pi)^2(40)^5] = -0.000317$ cm/s².

7.18 Given the velocity field $\mathbf{V}(x, y, z, t) = 10x^2\mathbf{i} - 20yx\mathbf{j} + 100t\mathbf{k}$, determine the velocity and acceleration of a particle at position $x = 1$ m, $y = 2$ m, $z = 5$ m, and $t = 0.1$ s.

▮ $$\mathbf{V} = (10)(1)^2\mathbf{i} - (20)(2)(1)\mathbf{j} + (100)(0.1)\mathbf{k} = 10\mathbf{i} - 40\mathbf{j} + 10\mathbf{k} \text{ m/s}$$

$$\mathbf{a}(x, y, z, t) = \frac{\partial V}{\partial t} + \left(v_x \frac{\partial V}{\partial x} + v_y \frac{\partial V}{\partial y} + v_z \frac{\partial V}{\partial z}\right)$$

$$\mathbf{a} = 100\mathbf{k} + [(10x^2)(20x\mathbf{i} - 20y\mathbf{j}) + (-20yx)(-20x\mathbf{j}) + (100t)(0)] = 200x^3\mathbf{i} + (-200x^2y + 400yx^2)\mathbf{j} + 100\mathbf{k}$$

At position $x = 1$ m, $y = 2$ m, $z = 5$ m, and $t = 0.1$ s, $\mathbf{a} = (200)(1)^3\mathbf{i} + [(-200)(1)^2(2) + (400)(2)(1)^2]\mathbf{j} + 100\mathbf{k} = 200\mathbf{i} + 400\mathbf{j} + 100\mathbf{k} \text{ m/s}^2$.

7.19 If the flow in Fig. 7-2 is steady at 0.50 cfs. find the acceleration in the flow at sections A and B.

▮ Since the flow at section A is uniform and also steady, $\mathbf{a}_A = 0$.

$$\mathbf{a} = \frac{\partial \mathbf{V}}{\partial t} + u\frac{\partial \mathbf{V}}{\partial x} + v\frac{\partial \mathbf{V}}{\partial y} + w\frac{\partial \mathbf{V}}{\partial z} \qquad \mathbf{a}_B = 0 + u\frac{\partial \mathbf{V}}{\partial x} + v\frac{\partial \mathbf{V}}{\partial y} + 0 = u\frac{\partial \mathbf{V}}{\partial x} + v\frac{\partial \mathbf{V}}{\partial y}$$

For point B on the axis of the pipe at section B, $v = 0$; hence,

$$\mathbf{a}_B = u\frac{\partial \mathbf{V}}{\partial x}$$

The effective area through which the flow is occurring in the converging section of the nozzle may be expressed approximately as $A = 2\pi hr$, where $h = r(1 - \cos 45°) = 0.293r$ and r is the distance from point C. Thus $A = (2\pi)(0.293r^2) = 1.84r^2$, and the velocity in the converging nozzle (assuming the streamlines flow radially toward C) may be expressed approximately as $v = Q/A = 0.50/(1.84r^2)$. At section B, $r = 2$ in $= 0.167$ ft; hence, $v = 0.50/[(1.84)(0.167)^2] = 9.744$ fps.

$$\frac{\partial \mathbf{V}}{\partial x} = -\frac{\partial \mathbf{V}}{\partial r} = -\left[\frac{(-2)(0.50)}{1.84r^3}\right] = \frac{(2)(0.50)}{(1.84)(0.167^3)} = 116.7 \text{ fps/ft} \qquad a_B = u\frac{\partial \mathbf{V}}{\partial x} = (9.744)(116.7) = 1137 \text{ ft/s}^2$$

7.20 A two-dimensional flow field is given by $u = 2y$, $v = x$. Sketch the flow field. Derive a general expression for the velocity and acceleration (x and y are in units of length L; u and v are in units of L/T). Find the acceleration in the flow field at point A ($x = 3.5$, $y = 1.2$).

▮ The flow field is sketched in Fig. 7-6a. Velocity components u and v are plotted to scale, and streamlines are sketched tangentially to the resultant velocity vectors. This gives a general picture of the flow field.

$$V = (u^2 + v^2)^{1/2} = (4y^2 + x^2)^{1/2} \qquad a_x = u\frac{\partial u}{\partial x} + v\frac{\partial u}{\partial y} = 2y(0) + x(2) = 2x \qquad a_y = u\frac{\partial v}{\partial x} + v\frac{\partial v}{\partial y} = 2y(1) + x(0) = 2y$$

$$a = (a_x^2 + a_y^2)^{1/2} = (4x^2 + 4y^2)^{1/2} \qquad (a_A)_x = 2x = 7.0 L/T^2 \qquad (a_A)_y = 2y = 2.4 L/T^2$$

$$a_A = [(a_A)_x^2 + (a_A)_y^2]^{1/2} = [(7.0)^2 + (2.4)^2]^{1/2} = 7.4 L/T^2$$

To get a rough check on the acceleration imagine a velocity vector at point A. This vector would have a magnitude approximately midway between that of the adjoining vectors, or $V_A \approx 4L/T$. The radius of curvature of the sketched streamline at A is roughly $3L$. Thus $(a_A)_n \approx 4^2/3 \approx 5.3 L/T^2$. The tangential acceleration of the particle at A may be approximated by noting that the velocity along the streamline increases from about $3.2L/T$, where it crosses the x axis, to about $8L/T$ at B. The distance along the streamline between these two points is roughly $4L$. Hence a very approximate value of the tangential acceleration at A is

$$(a_A)_t = V\frac{\partial V}{\partial s} \approx 4\left(\frac{8 - 3.2}{4}\right) \approx 4.8 L/T^2$$

Vector diagrams of these roughly computed normal and tangential acceleration components are plotted (Fig. 7-6b) for comparison with the true acceleration as given by the analytic expressions (Fig. 7-6c).

7.21 The velocity along a streamline lying on the x axis is given by $u = 10 + x^{1/2}$. What is the convective acceleration at $x = 3$? Specify units in terms of L and T. Assuming the fluid is incompressible, is the flow converging or diverging?

▮ $$a_x = \frac{\partial u}{\partial t} + u\frac{\partial u}{\partial x} + v\frac{\partial u}{\partial y} + w\frac{\partial u}{\partial z}$$

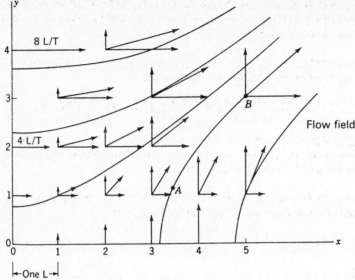

Fig. 7-6(a)

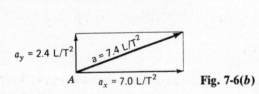

Fig. 7-6(b)

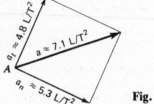

Fig. 7-6(c)

Since

$$\frac{\partial u}{\partial t} = v\frac{\partial u}{\partial y} = w\frac{\partial u}{\partial z} = 0 \qquad a_x = u\frac{\partial u}{\partial x} = (10 + x^{1/2})\left(\frac{x^{-1/2}}{2}\right) = 5x^{-1/2} + \tfrac{1}{2}$$

At $x = 3$, $a_x = (5)(3)^{-1/2} + \tfrac{1}{2} = 3.39 L/T^2$. For incompressible flow, the flow is converging.

7.22 A large tank contains an ideal liquid which flows out of the bottom of the tank through a 6-in-diameter hole. The flow is unsteady and $Q = 10 - 0.5t$, where Q is in cubic feet per second and t is in seconds. Find the local acceleration at a point 2 ft from the center of the hole at $t = 10$ s. What is the local acceleration at this point at $t = 15$ s? Find the total acceleration at a point 3 ft from the center of the hole at $t = 15$ s. Assume that liquid approaches the center of the hole radially.

▮ $v = Q/A$. The area through which flow occurs is a hemispherical surface. $v = (10 - 0.5t)/(2\pi r^2)$.

$$a_{\text{local}} = \frac{\partial v}{\partial t} = -\frac{0.5}{2\pi r^2}$$

At $r = 2$ ft and $t = 10$ s, $a_{\text{local}} = -0.5/[(2)(\pi)(2)^2] = -0.0199$ fps/s. Since a_{local} is independent of t, $a_{\text{local}} = -0.0199$ fps/s at $t = 15$ s.

$$a_{\text{total}} = v\frac{\partial v}{\partial r} + \frac{\partial v}{\partial t} = \left[\frac{(10 - 0.5t)}{2\pi r^2}\right]\left[\frac{-(10 - 0.5t)}{\pi r^3}\right] - \frac{0.5}{2\pi r^2}$$

At $r = 3$ ft and $t = 15$ s, $a_{\text{total}} = \{[10 - (0.5)(15)]/[(2)(\pi)(3)^2]\}\{-[10 - (0.5)(15)]/[(\pi)(3)^3]\} - 0.5/[(2)(\pi)(3)^2] = -0.0101$ ft/s².

7.23 Under what conditions does the velocity field $\mathbf{V} = (a_1x + b_1y + c_1z)\mathbf{i} + (a_2x + b_2y + c_2z)\mathbf{j} + (a_3x + b_3y + c_3z)\mathbf{k}$, where a_1, a_2, etc. = constant, represent an incompressible flow which conserves mass?

▮ $\frac{\partial u}{\partial x} + \frac{\partial v}{\partial y} + \frac{\partial w}{\partial z} = 0 \qquad \frac{\partial}{\partial x}(a_1x + b_1y + c_1z) + \frac{\partial}{\partial y}(a_2x + b_2y + c_2z) + \frac{\partial}{\partial z}(a_3x + b_3y + c_3z) = 0$

or $a_1 + b_2 + c_3 = 0$. At least two of the constants a_1, b_2, and c_3 must have opposite signs. Continuity imposes no restrictions whatever on the constants b_1, c_1, a_2, c_2, a_3, and b_3, which do not contribute to a mass increase or decrease of a differential element.

7.24 An incompressible velocity field is given by $u = a(x^2 - y^2)$, v unknown, $w = b$, where a and b are constants. What must the form of the velocity component v be?

$$\frac{\partial u}{\partial x} + \frac{\partial v}{\partial y} + \frac{\partial w}{\partial z} = 0 \qquad \frac{\partial}{\partial x}(ax^2 - ay^2) + \frac{\partial v}{\partial y} + \frac{\partial b}{\partial z} = 0 \qquad \frac{\partial v}{\partial y} = -2ax$$

This is integrated partially with respect to y: $v(x, y, z, t) = -2axy + f(x, z, t)$. This is the only possible form for v that satisfies the incompressible continuity equation. The function of integration f is entirely arbitrary since it vanishes when v is differentiated with respect to y.

7.25 An incompressible flow field is given by $u = Kxz^2$ and $w = Cy$, where K and C are constants. What does continuity tell us about the form of the velocity component v?

$$\frac{\partial u}{\partial x} + \frac{\partial v}{\partial y} + \frac{\partial w}{\partial z} = 0 \qquad Kz^2 + \frac{\partial v}{\partial y} + 0 = 0 \qquad \frac{\partial v}{\partial y} = -Kz^2 \qquad v = -Kz^2 y + f(x, z)$$

7.26 A two-dimensional incompressible velocity field has $u = K(1 - e^{-ay})$, for $x \leq L$ and $0 \leq L \leq \infty$. What is the most general form of $v(x, y)$ for which continuity is satisfied and $v = v_0$ at $y = 0$? What are the proper dimensions for the constants K and a?

Dimensions of constants: $\{K\} = \{L/T\}$, $\{a\} = \{1/L\}$.

$$\frac{\partial u}{\partial x} + \frac{\partial v}{\partial y} + \frac{\partial w}{\partial z} = 0 \qquad 0 + \frac{\partial v}{\partial y} + 0 = 0 \qquad \frac{\partial v}{\partial y} = 0 \qquad v = f(x) \quad \text{only}$$

If $v = v_0$ at $y = 0$ for all x, then $v = v_0$ everywhere.

7.27 Which of the following velocity fields satisfies conservation of mass for incompressible plane flow?
(a) $u = x$, $v = y$ (b) $u = y$, $v = x$ (c) $u = 2x$, $v = -2y$
(d) $u = 3xt$, $v = -3yt$ (e) $u = xy + y^2 t$, $v = xy + x^2 t$ (f) $u = 3x^2 y^2$, $v = -2xy^3$

Ignore dimensional inconsistencies.

In order to satisfy continuity,

$$\frac{\partial u}{\partial x} + \frac{\partial v}{\partial t} = 0 \qquad \text{or} \qquad \frac{\partial u}{\partial x} = -\frac{\partial v}{\partial y}$$

(a) $\dfrac{\partial u}{\partial x} = 1$ and $\dfrac{\partial v}{\partial y} = 1$ therefore, it does not satisfy continuity.

(b) $\dfrac{\partial u}{\partial x} = 0$ and $\dfrac{\partial v}{\partial y} = 0$ therefore, it does satisfy continuity.

(c) $\dfrac{\partial u}{\partial x} = 2$ and $\dfrac{\partial v}{\partial y} = -2$ therefore, it does satisfy continutiy.

(d) $\dfrac{\partial u}{\partial x} = 3t$ and $\dfrac{\partial v}{\partial y} = -3t$ therefore, it does satisfy continuity.

(e) $\dfrac{\partial u}{\partial x} = y$ and $\dfrac{\partial v}{\partial y} = x$ therefore, it does not satisfy continuity.

(f) $\dfrac{\partial u}{\partial x} = 6xy^2$ and $\dfrac{\partial v}{\partial y} = -6xy^2$ therefore, it does satisfy continuity.

7.28 If the radial velocity for incompressible flow is given by $v_r = b \cos \theta / r^2$, $b = $ constant, what is the most general form of $v_\theta(r, \theta)$ that satisfies continuity?

$$\frac{1}{r}\frac{\partial}{\partial r}(r v_r) + \frac{1}{r}\frac{\partial}{\partial \theta}(v_\theta) + \frac{\partial v_z}{\partial z} = 0 \qquad \frac{1}{r}\frac{\partial}{\partial r}\left[(r)\left(\frac{b \cos \theta}{r^2}\right)\right] + \frac{1}{r}\frac{\partial v_\theta}{\partial \theta} + 0 = 0$$

$$\frac{\partial v_\theta}{\partial \theta} = \frac{b \cos \theta}{r^2} \qquad v_\theta = \frac{b \sin \theta}{r^2} + f(r)$$

7.29 A two-dimensional velocity field is given by

$$u = -\frac{Ky}{x^2+y^2} \qquad v = \frac{Kx}{x^2+y^2}$$

where K is constant. Does this field satisfy incompressible continuity? Transform these velocities into polar components v_r and v_θ. What might the flow represent?

∎
$$\frac{\partial u}{\partial x} + \frac{\partial v}{\partial y} + \frac{\partial w}{\partial z} = 0 \qquad \frac{2xKy}{(x^2+y^2)^2} - \frac{2yKx}{(x^2+y^2)^2} + 0 = 0$$

Therefore, continuity is satisfied. $x^2 + y^2 = r^2$, $\cos\theta = x/r$, $\sin\theta = y/r$.

$$v_r = u\cos\theta + v\sin\theta = -\left(\frac{Ky}{x^2+y^2}\right)\left(\frac{x}{r}\right) + \left(\frac{Kx}{x^2+y^2}\right)\left(\frac{y}{r}\right) = -\frac{Kyx}{r^3} + \frac{Kxy}{r^3} = 0$$

$$v = -u\sin\theta + v\cos\theta = -\left[-\left(\frac{Ky}{x^2+y^2}\right)\right]\left(\frac{y}{r}\right) + \left(\frac{Kx}{x^2+y^2}\right)\left(\frac{x}{r}\right) = \frac{Ky^2}{r^3} + \frac{Kx^2}{r^3} = (y^2+x^2)\left(\frac{K}{r^3}\right) = \frac{K}{r}$$

Hence, in polar coordinates $v_r(r, \theta)$ and $v_\theta(r, \theta)$, $v_r = 0$, $v_\theta = K/r$. (This represents a potential vortex.)

7.30 For incompressible polar coordinate flow, what is the most general form of a purely circulatory motion, $v_\theta = v(r, \theta, t)$ and $v_r = 0$, which satisfies continuity?

∎
$$\frac{1}{r}\frac{\partial}{\partial r}(rv_r) + \frac{1}{r}\frac{\partial}{\partial\theta}(v_\theta) + \frac{\partial v_z}{\partial z} = 0 \qquad \frac{1}{r}\frac{\partial}{\partial r}[(r)(0)] + \frac{1}{r}\frac{\partial}{\partial\theta}(v_\theta) + 0 = 0 \qquad \frac{1}{r}\frac{\partial}{\partial\theta}(v_\theta) = 0 \qquad v_\theta = f(r)$$

7.31 What is the most general form of a purely radial polar coordinate incompressible flow pattern, $v_r = v_r(r, \theta, t)$ and $v_\theta = 0$ that satisfies continuity?

∎
$$\frac{1}{r}\frac{\partial}{\partial r}(rv_r) + \frac{1}{r}\frac{\partial}{\partial\theta}(v_\theta) + \frac{\partial v_z}{\partial z} = 0 \qquad \frac{1}{r}\frac{\partial}{\partial r}(rv_r) + \left(\frac{1}{r}\right)(0) + 0 = 0 \qquad \frac{1}{r}\frac{\partial}{\partial r}(rv_r) = 0 \qquad v_r = \left(\frac{1}{r}\right)f(\theta)$$

7.32 An incompressible steady flow pattern is given by $u = x^3 + 2z^2$ and $w = y^3 - 2yz$. What is the most general form of the third component, $v(x, y, z)$, that satisfies continuity?

∎
$$\frac{\partial u}{\partial x} + \frac{\partial v}{\partial y} + \frac{\partial w}{\partial z} = 0 \qquad 3x^2 + \frac{\partial v}{\partial y} - 2y = 0 \qquad \frac{\partial v}{\partial y} = 2y - 3x^2 \qquad v = y^2 - 3x^2y + f(x, z)$$

7.33 A certain two-dimensional shear flow near a wall, as in Fig. 7-7, has the velocity component

$$u = U\left(\frac{2y}{ax} - \frac{y^2}{a^2x^2}\right)$$

where a is constant. Derive from continuity the velocity component $v(x, y)$ assuming that $v = 0$ at the wall, $y = 0$.

∎
$$\frac{\partial u}{\partial x} + \frac{\partial v}{\partial y} + \frac{\partial w}{\partial z} = 0 \qquad \frac{\partial u}{\partial x} + \frac{\partial v}{\partial y} + 0 = 0 \qquad \frac{\partial v}{\partial y} = -\frac{\partial u}{\partial x} = -U\left(\frac{-2y}{ax^2} + \frac{2y^2}{a^2x^3}\right) \qquad v = U\left(\frac{y^2}{a^2x^2} - \frac{2y^3}{3a^2x^3}\right) + f(x)$$

Enforce no-slip condition: $v(x, 0) = U(0 - 0) + f(x) = 0$, $f(x) = 0$.

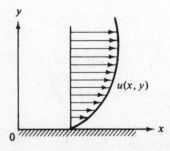

Fig. 7-7

7.34 Consider the flat-plate boundary-layer flow in Fig. 7-8. From the no-slip condition $v = 0$ all along the wall $y = 0$, and $u = U =$ constant outside the layer. If the layer thickness δ increases with x as shown, prove with incompressible two-dimensional continuity that (**a**) the component $v(x, y)$ is everywhere positive within the layer; (**b**) v increases parabolically with y very near the wall; and (**c**) v reaches a positive maximum at $y = \delta$.

∎ (**a**) If δ increases with x, the streamlines in the shear layer must everywhere move upward to satisfy continuity. Therefore, $\partial u/\partial x < 0$ everywhere inside the shear layer. Since continuity requires

$$\frac{\partial u}{\partial x} + \frac{\partial v}{\partial y} = 0$$

everywhere, it follows that $\partial v/\partial y > 0$ everywhere in the shear layer.
(**b**) Near the wall, $u = y f(x)$, $\partial v/\partial y = -\partial u/\partial x = -f'(x)$; therefore,

$$v = -\frac{y^2}{2} f'(x) \qquad \text{(parabolic)}$$

(**c**) At $y \geq \delta(x)$, $\partial u/\partial x \approx 0$; therefore $\partial v/\partial y = 0$, and $v = \text{maximum}$.

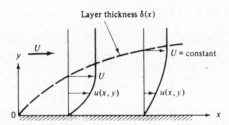

Fig. 7-8

7.35 The axial velocity field for fully developed laminar flow in a pipe is $v_z = u_{\max}(1 - r^2/R^2)$ and there is no swirl, $v_\theta = 0$. Determine the radial velocity field $v_r(r, z)$ from the incompressible relation if $u_{\max}$ is constant and $v_r = 0$ at $r = R$. (r denotes radial distance from the pipe's center; R denotes the pipe's radius.)

∎ $\quad \dfrac{1}{r}\dfrac{\partial}{\partial r}(rv_r) + \dfrac{1}{r}\dfrac{\partial}{\partial \theta}(v_\theta) + \dfrac{\partial v_z}{\partial z} = 0 \qquad \dfrac{1}{r}\dfrac{\partial}{\partial r}(rv_r) + 0 + 0 = 0 \qquad \dfrac{\partial}{\partial r}(rv_r) = 0 \qquad v_r = \dfrac{f(\theta, z)}{r}$

if $v_r(R) = 0$ for all θ, z, $v_r = 0$.

7.36 An incompressible flow field has the cylindrical components $v_\theta = Cr$, $v_z = K(R^2 - r^2)$, $v_r = 0$, where C and K are constants and $r \leq R$, $z \leq L$. Does this flow satisfy continuity? What might it represent physically?

∎ $\quad \dfrac{1}{r}\dfrac{\partial}{\partial r}(rv_r) + \dfrac{1}{r}\dfrac{\partial}{\partial \theta}(v_\theta) + \dfrac{\partial v_z}{\partial z} = 0 \qquad \dfrac{1}{r}\dfrac{\partial}{\partial r}[(r)(0)] + \dfrac{1}{r}\dfrac{\partial}{\partial \theta}(Cr) + \dfrac{\partial}{\partial z}[K(R^2 - r^2)] = 0$

$$0 + 0 + 0 = 0 \qquad \text{(satisfies continuity)}$$

This flow represents pressure-driven, laminar, steady flow in a rotating tube (fully developed).

7.37 An incompressible flow in polar coordinates is given by $v_r = K \cos\theta(1 - b/r^2)$, $v_\theta = -K \sin\theta(1 + b/r^2)$. Does this field satisfy continuity? For consistency, what should the dimensions of the constants K and b be?

∎ $\quad \dfrac{1}{r}\dfrac{\partial}{\partial r}(rv_r) + \dfrac{1}{r}\dfrac{\partial}{\partial \theta}(v_\theta) + \dfrac{\partial v_z}{\partial z} = 0 \qquad \dfrac{1}{r}\dfrac{\partial}{\partial r}\left[rK\cos\theta\left(1 - \dfrac{b}{r^2}\right)\right] + \dfrac{1}{r}\dfrac{\partial}{\partial \theta}\left[-K\sin\theta\left(1 + \dfrac{b}{r^2}\right)\right] + 0 = 0$

$$\frac{1}{r}K\cos\theta\left(1 + \frac{b}{r^2}\right) - \frac{1}{r}K\cos\theta\left(1 + \frac{b}{r^2}\right) = 0 \qquad 0 = 0 \qquad \text{(satisfies continuity)}$$

Dimensions of constants: $\{K\} = \{L/T\}$, $\{b\} = \{L^2\}$.

7.38 The x component of velocity is $u = x^2 + z^2 + 5$, and the y component is $v = y^2 + z^2$. Find the simplest z component of velocity that satisfies continuity.

∎ $\quad \dfrac{\partial u}{\partial x} + \dfrac{\partial v}{\partial y} + \dfrac{\partial w}{\partial z} = 0 \qquad 2x + 2y + \dfrac{\partial w}{\partial z} = 0 \qquad \dfrac{\partial w}{\partial z} = -2(x + y) \qquad w = -2z(x + y)$

7.39 Is the continuity equation for steady, **incompressible** flow satisfied if the following velocity components are involved?

$$u = 2x^2 - xy + z^2 \qquad v = x^2 - 4xy + y^2 \qquad w = -2xy - yz + y^2$$

▮ $\dfrac{\partial u}{\partial x} + \dfrac{\partial v}{\partial y} + \dfrac{\partial w}{\partial z} = 0 \qquad (4x - y) + (-4x + 2y) + (-y) = 0 \qquad$ (satisfies continuity)

7.40 For steady, incompressible flow, are the following values of u and v possible?
   (*a*) $u = 4xy + y^2$, $v = 6xy + 3x$ (*b*) $u = 2x^2 + y^2$, $v = -4xy$

▮ $$\dfrac{\partial u}{\partial x} + \dfrac{\partial v}{\partial y} + \dfrac{\partial w}{\partial z} = 0$$

(*a*) $4y + 6x + 0 \neq 0$ (Flow is not possible.)
(*b*) $4x - 4x + 0 = 0$ (Flow is possible.)

7.41 Determine whether the velocity field $\mathbf{V} = 3t\mathbf{i} + xz\mathbf{j} + ty^2\mathbf{k}$ is incompressible, irrotational, both, or neither.

▮ The divergence of this velocity field is

$$\nabla \cdot V = \frac{\partial}{\partial x}(3t) + \frac{\partial}{\partial y}(xz) + \frac{\partial}{\partial z}(ty^2) = 0$$

Therefore, this velocity field is incompressible. The curl of this velocity field is

$$\nabla \times V = \begin{vmatrix} \mathbf{i} & \mathbf{j} & \mathbf{k} \\ \dfrac{\partial}{\partial x} & \dfrac{\partial}{\partial y} & \dfrac{\partial}{\partial z} \\ 3t & xz & ty^2 \end{vmatrix} = (2ty - x)\mathbf{i} + z\mathbf{k}$$

This is not zero; hence, the flow field is rotational, not irrotational.

7.42 If a velocity potential exists for the velocity field $u = a(x^2 - y^2)$, $v = -2axy$, $w = 0$, find it and plot it.

▮ Since $w = 0$, the curl of $\mathbf{V}$ has only one (z) component, and we must show that it is zero.

$$(\nabla \times V)_z = 2\omega_z = \frac{\partial v}{\partial x} - \frac{\partial u}{\partial y} = \frac{\partial}{\partial x}(-2axy) - \frac{\partial}{\partial y}(ax^2 - ay^2) = -2ay + 2ay = 0 \qquad \text{checks}$$

The flow is indeed irrotational. A potential exists. To find $\phi(x, y)$, set

$$u = \frac{\partial \phi}{\partial x} = ax^2 - ay^2 \tag{1}$$

$$v = \frac{\partial \phi}{\partial y} = -2axy \tag{2}$$

Integrate (*1*)

$$\phi = \frac{ax^3}{3} - axy^2 + f(y) \tag{3}$$

Differentiate (*3*) and compare with (*2*)

$$\frac{\partial \phi}{\partial y} = -2axy + f'(y) = -2axy \tag{4}$$

Therefore $f' = 0$, or $f = $ constant. The velocity potential is $\phi = ax^3/3 - axy^2 + C$. Letting $C = 0$, we can plot the ϕ lines as shown in Fig. 7-9.

7.43 Given the velocity field $\mathbf{V} = 10x^2y\mathbf{i} + 20(yz + x)\mathbf{j} + 13\mathbf{k}$, what is the total angular velocity of a fluid particle at $(1, 4, 3)$ m?

▮ $\omega_x = \dfrac{1}{2}\left(\dfrac{\partial V_z}{\partial y} - \dfrac{\partial V_y}{\partial z}\right) = \frac{1}{2}(0 - 20y) = -10y \qquad \omega_y = \dfrac{1}{2}\left(\dfrac{\partial V_x}{\partial z} - \dfrac{\partial V_z}{\partial x}\right) = \frac{1}{2}(0 - 0) = 0$

$\omega_z = \dfrac{1}{2}\left(\dfrac{\partial V_y}{\partial x} - \dfrac{\partial V_x}{\partial y}\right) = \frac{1}{2}(20 - 10x^2) = 10 - 5x^2 \qquad \omega = -10y\mathbf{i} + 0\mathbf{j} + (10 - 5x^2)\mathbf{k}$

At point $(1, 4, 3)$ m, $\omega = (-10)(4)\mathbf{i} + [10 - (5)(1)^2]\mathbf{k} = -40\mathbf{i} + 5\mathbf{k}$ rad/s; $|\omega| = \sqrt{40^2 + 5^2} = 40.3$ rad/s.

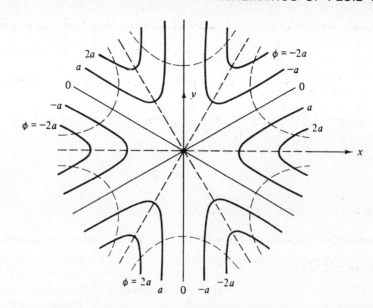

Fig. 7-9

7.44 Given the velocity field $\mathbf{V} = 5x^2y\mathbf{i} - (3x - 3z)\mathbf{j} + 10z^2\mathbf{k}$ m/s, compute the angular velocity field $\omega(x, y, z)$.

▌
$$\omega_x = \frac{1}{2}\left(\frac{\partial V_z}{\partial y} - \frac{\partial V_y}{\partial z}\right) = \tfrac{1}{2}(0 - 3) = -1.5 \qquad \omega_y = \frac{1}{2}\left(\frac{\partial V_x}{\partial z} - \frac{\partial V_z}{\partial x}\right) = \tfrac{1}{2}(0 - 0) = 0$$

$$\omega_z = \frac{1}{2}\left(\frac{\partial V_y}{\partial x} - \frac{\partial V_x}{\partial y}\right) = \tfrac{1}{2}(-3 - 5x^2) = -(1.5 + 5x^2) \qquad \omega = -1.5\mathbf{i} + 0\mathbf{j} - (1.5 + 5x^2)\mathbf{k} \text{ rad/s}$$

7.45 Show that any velocity field $\mathbf{V}$ expressible as the gradient of a scaler ϕ must be an irrotational field.

▌ Show curl (grad ϕ) = **0**

$$\text{curl}\left(\frac{\partial\phi}{\partial x}\mathbf{i} + \frac{\partial\phi}{\partial y}\mathbf{j} + \frac{\partial\phi}{\partial z}\mathbf{k}\right) = \mathbf{0}$$

$$\left[\frac{\partial}{\partial y}\left(\frac{\partial\phi}{\partial z}\right) - \frac{\partial}{\partial z}\left(\frac{\partial\phi}{\partial y}\right)\right]\mathbf{i} + \left[\frac{\partial}{\partial z}\left(\frac{\partial\phi}{\partial x}\right) - \frac{\partial}{\partial x}\left(\frac{\partial\phi}{\partial z}\right)\right]\mathbf{j} + \left[\frac{\partial}{\partial x}\left(\frac{\partial\phi}{\partial y}\right) - \frac{\partial}{\partial y}\left(\frac{\partial\phi}{\partial x}\right)\right]\mathbf{k} = 0$$

Since $\partial^2\phi/\partial y\,\partial z = \partial^2\phi/\partial z\,\partial y$, etc., we see that we have proven our point provided the partial derivatives of ϕ are continuous.

7.46 Is the following flow field irrotational or not? $\mathbf{V} = 6x^2y\mathbf{i} + 2x^3\mathbf{j} + 10\mathbf{k}$ ft/s.

▌
$$\frac{\partial V_x}{\partial y} - \frac{\partial V_y}{\partial x} = 6x^2 - 6x^2 = 0 \qquad \frac{\partial V_y}{\partial z} - \frac{\partial V_z}{\partial y} = 0 \qquad \frac{\partial V_z}{\partial x} - \frac{\partial V_x}{\partial z} = 0$$

Therefore, the flow is irrotational.

7.47 For the velocity vector $\mathbf{V} = 3t\mathbf{i} + xz\mathbf{j} + ty^2\mathbf{k}$ evaluate the volume flow and the average velocity through the square surface whose vertices are at $(0, 1, 0)$, $(0, 1, 2)$, $(2, 1, 2)$, and $(2, 1, 0)$. See Fig. 7-10.

▌ The surface S is shown in Fig. 7-10 and is such that $\mathbf{n} = \mathbf{j}$ and $dA = dx\,dz$ everywhere. The velocity field is $\mathbf{V} = 3t\mathbf{i} + xz\mathbf{j} + ty^2\mathbf{k}$. The normal component to S is $\mathbf{V} \cdot \mathbf{n} = \mathbf{V} \cdot \mathbf{j} = v$, the y component, which equals xz. The limits on the integral for Q are 0 to 2 for both dx and dz. The volume flow is thus

$$Q = \int_S V_n\,dA = \int_0^2\int_0^2 xz\,dx\,dz = 4.0 \text{ units}$$

The area of the surface is $(2)(2) = 4$ units. Then the average velocity is $V_{av} = Q/A = 4.0/4.0 = 1.0$ unit.

7.48 At low velocities, the flow through a long circular tube has a paraboloid velocity distribution $u = u_{max}(1 - r^2/R^2)$, where R is the tube radius and u_{max} is the maximum velocity, which occurs at the tube centerline. (**a**) Find a general expression for volume flow and average velocity through the tube; (**b**) compute the volume flow if $R = 3$ cm and $u_{max} = 8$ m/s; and (**c**) compute the mass flow if $\rho = 1000$ kg/m^3.

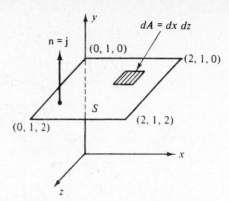

Fig. 7-10

$\blacksquare$ (*a*) The area S is the cross section of the tube, and $\mathbf{n} = \mathbf{i}$. The normal component $\mathbf{V} \cdot \mathbf{n} = \mathbf{V} \cdot \mathbf{i} = u$. Since u varies only with r, the element dA can be taken to be the annular strip $dA = 2\pi r\,dr$. The volume flow becomes

$$Q = \int_S u\,dA = \int_0^R u_{max}\left(1 - \frac{r^2}{R^2}\right) 2\pi r\,dr$$

Carrying out the integration over r, we obtain $Q = \frac{1}{2}u_{max}\pi R^2$. The average velocity is $u_{av} = Q/A = \frac{1}{2}u_{max}\pi R^2/\pi R^2 = \frac{1}{2}u_{max}$. The average velocity is half the maximum, which is an accepted result for low-speed, or *laminar*, flow through a long tube. (*b*) For the given numerical values $Q = \frac{1}{2}(8)\pi(0.03)^2 = 0.0113$ m³/s. (*c*) For the given density, assumed constant, $\dot{m} = \rho Q = (1000)(0.0113) = 11.3$ kg/s.

7.49 For low-speed (laminar) flow through a circular pipe, as shown in Fig. 7-11, the velocity distribution takes the form $u = (B/\mu)(r_0^2 - r^2)$, where μ is the fluid viscosity. Determine (*a*) the maximum velocity in terms of B, μ, and r_0 and (*b*) the mass flow rate in terms of B, μ, and r_0.

$\blacksquare$ (*a*) u_{max} occurs when $du/dr = 0$. $du/dr = -2Br/\mu = 0$, $r = 0$, $u_{max} = Br_0^2/\mu$.

(*b*)
$$\dot{m} = \int \rho v_n\,dA = \int_0^{r_0} \rho \frac{B}{\mu}(r_0^2 - r^2)(2\pi r\,dr) = 2\pi\rho \frac{B}{\mu}\left[\frac{r_0^2 r^2}{2} - \frac{r^4}{4}\right]_0^{r_0}$$
$$= \frac{\rho}{2}\left(\frac{Br_0^2}{\mu}\right)\pi r_0^2 = \left(\frac{\rho}{2}\right)u_{max}(\pi r_0^2)$$

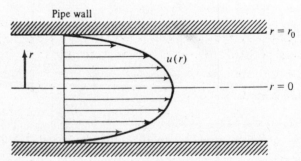

Pipe wall

Fig. 7-11

7.50 If the fluid in Fig. 7-11 is water at 20 °C and 1 atm, what is the centerline velocity U_0 if the tube radius is 1 cm and the mass flow through the tube is 1.2 kg/s?

$\blacksquare$ From Prob. 7.49, $\dot{m} = (\rho/2)u_{max}(\pi r_0^2)$, $1.2 = (\frac{998}{2})(u_{max})[(\pi)(\frac{1}{100})^2]$, $u_{max} = U_0 = 7.65$ m/s. (Actually, this is unrealistic. At this μ, $N_R > 2000$, so the flow is probably turbulent.)

7.51 A velocity field in arbitrary units is given by $\mathbf{V} = 2x^2\mathbf{i} - xy\mathbf{j} - 3xz\mathbf{k}$. Find the volume flow Q passing through the unit square bounded by $(x, y, z) = (1, 0, 0)$, $(1, 1, 0)$, $(1, 1, 1)$, and $(1, 0, 1)$. See Fig. 7-12.

$\blacksquare$ $Q = \iint (u)_{x=1}\,dy\,dz$. Since $\mathbf{n} = \mathbf{i}$, $\mathbf{V} \cdot \mathbf{n} = u = 2x^2$. $Q = \int_0^1 \int_0^1 (2)\,dy\,dz = 2$ units.

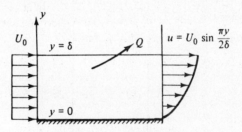

Fig. 7-12

7.52 If the volume flow Q across the upper surface in Fig. 7-13 equals the difference between inlet and outlet flows past the plate, what is this flow Q in terms of the inlet velocity U_0 and the height δ of the fluid region shown in the figure?

$$Q = \int_0^\delta U_0 b\,dy - \int_0^\delta U_0 \sin\left(\frac{\pi y}{2\delta}\right) b\,dy = U_0 b [y]_0^\delta - U_0 b\left(\frac{2\delta}{\pi}\right)\left[-\cos\left(\frac{\pi y}{2\delta}\right)\right]_0^\delta$$

$$= U_0 b\delta - U_0 b\left(\frac{2\delta}{\pi}\right)[-0+1] = U_0 b\delta - U_0 b\left(\frac{2\delta}{\pi}\right)$$

$$= U_0 b\delta(1 - 2/\pi) = 0.363 U_0 b\delta$$

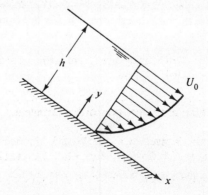

Plate (width b into paper)   Fig. 7-13

7.53 The velocity profile in water flow down a spillway is given approximately by $u = (U_0)(y/h)^{1/7}$ where $y = 0$ denotes the bottom and the depth is h (see Fig. 7-14). If $U_0 = 1.5$ m/s, $h = 2$ m, and the width is 20 m, how many hours will it take 10^6 m^3 of water to pass this section of the spillway?

$$Q = \int u\,dA = \int_0^h (U_0)\left(\frac{y}{h}\right)^{1/7} (b\,dy) = U_0 b h^{-1/7}\left[\frac{y^{8/7}}{\frac{8}{7}}\right]_0^h = \tfrac{7}{8}U_0 bh = (\tfrac{7}{8})(1.5)(20)(2) = 52.5 \text{ m}^3/\text{s}$$

$$t = V/Q = 10^6/52.5 = 19\,050 \text{ s} \quad \text{or} \quad 5.29 \text{ h}$$

Fig. 7-14

7.54   The two-dimensional pressure field, in arbitrary units, $p = 4x^3 - 2y^2$ is associated with the velocity field $\mathbf{V} = (x^2 - y^2 + x)\mathbf{i} - (2xy + y)\mathbf{j}$. Compute the rate of change dp/dt at $x = 2$ and $y = 1$.

I
$$\frac{dp}{dt} = \frac{\partial p}{\partial t} + u\frac{\partial p}{\partial x} + v\frac{\partial p}{\partial y} + w\frac{\partial p}{\partial z} = 0 + (x^2 - y^2 + x)(12x^2) - (2xy + y)(-4y) + 0$$

At $x = 2$ and $y = 1$, $dp/dt = (2^2 - 1^2 + 2)[(12)(2)^2] - [(2)(2)(1) + 1][(-4)(1)] = 260$ units.

7.55   A frictionless, incompressible steady flow field is given by $\mathbf{V} = 2xy\mathbf{i} - y^2\mathbf{j}$ in arbitrary units. Let the density be $\rho_0 = $ constant and neglect gravity. Find an expression for the pressure gradient in the x direction and evaluate this gradient at $(1, 2, 0)$.

I $\rho\dfrac{d\mathbf{V}}{dt} = \rho_0\left(u\dfrac{\partial \mathbf{V}}{\partial x} + v\dfrac{\partial \mathbf{V}}{\partial y} + w\dfrac{\partial \mathbf{V}}{\partial z}\right) = \rho_0[(2xy)(2y\mathbf{i}) + (-y^2)(2x\mathbf{i} - 2y\mathbf{j}) + 0] = -\nabla p \qquad \nabla p = -2\rho_0(xy^2\mathbf{i} + y^3\mathbf{j})$

At $(1, 2, 0)$, $\nabla p = -2\rho_0[(1)(2)^2\mathbf{i} + (2)^3\mathbf{j}] = -\rho_0(8\mathbf{i} + 16\mathbf{j})$.

7.56   A velocity field given by $u = 3y^2$, $v = 2x$, $w = 0$, in arbitrary units, is associated with the temperature field $T = 4xy^2$, in arbitrary units. Compute the rate of change dT/dt at the point $(x, y) = (2, 1)$.

I
$$\frac{\partial T}{\partial t} = u\frac{\partial T}{\partial x} + v\frac{\partial T}{\partial y} + w\frac{\partial T}{\partial z} = (3y^2)(4y^2) + (2x)(8xy) + 0 = 12y^4 + 16x^2y$$

At $(2, 1)$,
$$\frac{dT}{dt} = (12)(1)^4 + (16)(2)^2(1) = 76 \text{ units}$$

7.57   Take the velocity field $u = a(x^2 - y^2)$, $v = -2axy$, $w = 0$ and determine under what conditions it is a solution to the Navier–Stokes momentum equation. Assuming that these conditions are met, determine the resulting pressure distribution when z is "up" ($g_x = 0$, $g_y = 0$, $g_z = -g$).

I $\rho g_x - \dfrac{\partial p}{\partial x} + \mu\left(\dfrac{\partial^2 u}{\partial x^2} + \dfrac{\partial^2 u}{\partial y^2} + \dfrac{\partial^2 u}{\partial z^2}\right) = \rho\dfrac{du}{dt} \qquad \rho g_y - \dfrac{\partial p}{\partial y} + \mu\left(\dfrac{\partial^2 v}{\partial x^2} + \dfrac{\partial^2 v}{\partial y^2} + \dfrac{\partial^2 v}{\partial z^2}\right) = \rho\dfrac{dv}{dt}$

$$\rho g_z - \frac{\partial p}{\partial z} + \mu\left(\frac{\partial^2 w}{\partial x^2} + \frac{\partial^2 w}{\partial y^2} + \frac{\partial^2 w}{\partial z^2}\right) = \rho\frac{dw}{dt}$$

Make a direct substitution of u, v, w.

$$\rho(0) - \frac{\partial p}{\partial x} + \mu(2a - 2a) = 2a^2\rho(x^3 + xy^2) \tag{1}$$

$$\rho(0) - \frac{\partial p}{\partial y} + \mu(0) = 2a^2\rho(x^2y + y^3) \tag{2}$$

$$\rho(-g) - \frac{\partial p}{\partial z} + \mu(0) = 0 \tag{3}$$

The viscous terms vanish identically (although μ is *not* zero). Equation (3) can be integrated partially to obtain

$$p = -\rho gz + f_1(x, y) \tag{4}$$

i.e., the pressure is hydrostatic in the z direction, which follows anyway from the fact that the flow is two-dimensional ($w = 0$). Now the question is: Do Eqs. (1) and (2) show that the given velocity field *is* a solution? One way to find out is to form the mixed derivative $\partial^2 p/(\partial x\,\partial y)$ from (1) and (2) separately and then compare them.
    Differentiate Eq. (1) with respect to y

$$\frac{\partial^2 p}{\partial x\,\partial y} = -4a^2\rho xy \tag{5}$$

Now differentiate Eq. (2) with respect to x

$$-\frac{\partial^2 p}{\partial x\,\partial y} = \frac{\partial}{\partial x}[2a^2\rho(x^2y + y^3)] = -4a^2\rho xy \tag{6}$$

Since these are identical, the given velocity field is an *exact* solution to the Navier–Stokes equation.

To find the pressure distribution, substitute Eq. (4) into Eqs. (1) and (2), which will enable us to find $f_1(x, y)$

$$\frac{\partial f_1}{\partial x} = -2a^2\rho(x^3 + xy^2) \tag{7}$$

$$\frac{\partial f_1}{\partial y} = -2a^2\rho(x^2y + y^3) \tag{8}$$

Integrate Eq. (7) partially with respect to x

$$f_1 = -\tfrac{1}{2}a^2\rho(x^4 + 2x^2y^2) + f_2(y) \tag{9}$$

Differentiate this with respect to y and compare with Eq. (8)

$$\frac{\partial f_1}{\partial y} = -2a^2\rho x^2 y + f_2'(y) \tag{10}$$

Comparing (8) and (10), we see they are equivalent if

$$f_2'(y) = -2a^2\rho y^3$$

or

$$f_2(y) = \tfrac{1}{2}a^2\rho y^4 + C \tag{11}$$

where C is a constant. Combine Eqs. (4), (9), and (11) to give the complete expression for pressure distribution

$$p(x, y, z) = -\rho g z - \tfrac{1}{2}a^2\rho(x^4 + y^4 + 2x^2y^2) + C \tag{12}$$

This is the desired solution. Do you recognize it? Not unless you go back to the beginning and square the velocity components:

$$u^2 + v^2 + w^2 = V^2 = a^2(x^4 + y^4 + 2x^2y^2) \tag{13}$$

Comparing with Eq. (12), we can rewrite the pressure distribution as

$$p + \tfrac{1}{2}\rho V^2 + \rho g z = C \tag{14}$$

7.58 The sprinkler shown in Fig. 7-15 on p. 148 discharges water upward and outward from the horizontal plane so that it makes an angle of $\theta°$ with the t axis when the sprinkler arm is at rest. It has a constant cross-sectional flow area of A_0 and discharges q cfs starting with $\omega = 0$ and $t = 0$. The resisting torque due to bearings and seals is the constant T_0, and the moment of inertia of the rotating empty sprinkler head is I_s. Determine the equation for ω as a function of time.

▌ The control volume is the cylindrical area enclosing the rotating sprinkler head. The inflow is along the axis, so that it has no moment of momentum; hence, the torque $-T_0$ due to friction is equal to the time rate of change of moment of momentum of sprinkler head and fluid within the sprinkler head plus the net efflux of moment of momentum from the control volume. Let $V_r = q/2A_0$.

$$-T_0 = 2\frac{d}{dt}\int_0^{r_0} A_o \rho \omega r^2 \, dr + I_s \frac{d\omega}{dt} - \left[\frac{2\rho q r_0}{2}(V_r \cos\theta - \omega r_0)\right]$$

The total derivative can be used. Simplifying gives

$$\frac{d\omega}{dt}(I_s + \tfrac{2}{3}\rho A_0 r_0^3) = \rho q r_0(V_r \cos\theta - \omega r_0) - T_0$$

For rotation to start, $\rho q r_0 V_r \cos\theta$ must be greater than T_0. The equation is easily integrated to find ω as a function of t. The final value of ω is obtained by setting $d\omega/dt = 0$ in the equation.

7.59 A turbine discharging 10 m³/s is to be so designed that a torque of 10 000 N · m is to be exerted on an impeller turning at 200 rpm that takes all the moment of momentum out of the fluid. At the outer periphery of the impeller, $r = 1$ m. What must the tangential component of velocity be at this location?

▌ $\qquad T = \rho Q[(rv_t)_2 - (rv_t)_1] \qquad 10\,000 = (1000)(10)[(1)(v_t)_{in} - 0] \qquad (v_t)_{in} = 1.00$ m/s

7.60 The sprinkler of Fig. 7-16 discharges 0.01 cfs through each nozzle. Neglecting friction, find its speed of rotation. The area of each nozzle opening is 0.001 ft².

▌ The fluid entering the sprinkler has no moment of momentum, and no torque is exerted on the system externally; hence the moment of momentum of fluid leaving must be zero. Let ω be the speed of rotation; then

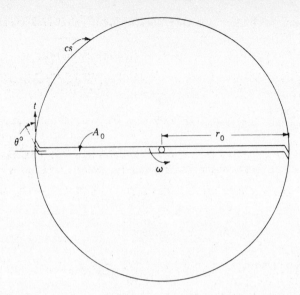

Fig. 7-15

the moment of momentum leaving is $\rho Q_1 r_1 v_{t1} + \rho Q_2 r_2 v_{t2}$ in which v_{t1} and v_{t2} are absolute velocities. Then $v_{t1} = v_{r1} - \omega r_1 = Q_1/0.001 - \omega r_1 = 10 - \omega$ and $v_{t2} = v_{r2} - \omega r_2 = 10 - \frac{2}{3}\omega$. For the moment of momentum to be zero, $\rho Q(r_1 v_{t1} + r_2 v_{t2}) = 0$ or $(1)(10 - \omega) + (\frac{2}{3})(10 - \frac{2}{3}\omega) = 0$ and $\omega = 11.54$ rad/s, or 110.2 rpm.

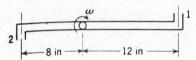

Fig. 7-16

7.61 The velocity profile for laminar flow between two plates, as in Fig. 7-17, is

$$u = \frac{4u_{max} y(h - y)}{h^2} \qquad v = w = 0$$

If the wall temperature is T_w at both walls, use the incompressible-flow energy equation to solve for the temperature distribution $T(y)$ between the walls for steady flow.

$$\rho c_v \frac{dT}{dt} = K\frac{d^2 T}{dy^2} + \mu\left(\frac{du}{dy}\right)^2 \qquad 0 = K\frac{d^2 T}{dy^2} + \mu\left(\frac{du}{dy}\right)^2$$

$$\frac{d^2 T}{dy^2} = -\frac{\mu}{K}\left(\frac{du}{dy}\right)^2 = -\frac{\mu}{K}\left[\frac{4u_{max}}{h^2}(h - 2y)\right]^2 = \left(\frac{-16\mu u_{max}^2}{Kh^4}\right)(h^2 - 4hy + 4y^2)$$

$$\frac{dT}{dy} = \left(\frac{-16\mu u_{max}^2}{Kh^4}\right)[h^2 y - 2hy^2 + (\tfrac{4}{3})(y)^3 + C_1]$$

Since $dT/dy = 0$ at $y = h/2$, $C_1 = -h^3/6$.

$$\frac{dT}{dy} = \left(\frac{-16\mu u_{max}^2}{Kh^4}\right)\left[h^2 y - 2hy^2 + (\tfrac{4}{3})(y)^3 - \frac{h^3}{6}\right] \qquad T = \left(\frac{-16\mu u_{max}^2}{Kh^4}\right)\left[\frac{h^2 y^2}{2} - (\tfrac{2}{3})(hy^3) + \frac{y^4}{3} - \frac{yh^3}{6} + C_2\right]$$

If $T = T_w$ at $y = 0$ and $y = h$, then $C_2 = T_w$.

$$T = T_w + \left(\frac{8\mu u_{max}^2}{K}\right)\left[(\tfrac{1}{3})\left(\frac{y}{h}\right) - \frac{y^2}{h^2} + (\tfrac{4}{3})\left(\frac{y^3}{h^3}\right) - (\tfrac{2}{3})\left(\frac{y^4}{h^4}\right)\right]$$

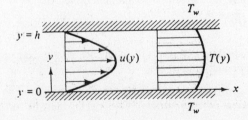

Fig. 7-17

7.62 Consider a viscous, steady flow through a pipe (Fig. 7-18a). The velocity profile forms a paraboloid about the pipe centerline, given as

$$V = -C(r^2 - D^2/4) \qquad \text{m/s} \tag{1}$$

where C is a constant. (**a**) What is the flow of mass through the left end of the control surface shown dashed? (**b**) What is the flow of kinetic energy through the left end of the control surface? Assume that the velocity profile does not change along the pipe.

❚ In Fig. 7-18b, we have shown a cross section of the pipe. For an infinitesimal strip, we can say noting that $\mathbf{V}$ and $d\mathbf{A}$ are collinear but of opposite sense: $\rho\mathbf{V} \cdot d\mathbf{A} = \rho[C(r^2 - D^2/4)]2\pi r\, dr$. For the whole cross section, we have

$$\iint \rho\mathbf{V} \cdot d\mathbf{A} = \rho \int_0^{D/2} C\!\left(r^2 - \frac{D^2}{4}\right)2\pi r\, dr = 2\pi\rho C\!\left[\frac{r^4}{4} - \frac{D^2}{4}\frac{r^2}{2}\right]_0^{D/2} = -\frac{\rho C\pi D^4}{32} \qquad \text{kg/s} \tag{2}$$

We now turn to the flow of kinetic energy through the left end of the control surface. The kinetic energy for an element of fluid is $\frac{1}{2}dm\,V^2$. This corresponds to an infinitesimal amount of an extensive property N. To get η, the corresponding intensive property, we divide by dm to get

$$\eta = \tfrac{1}{2}V^2 \tag{3}$$

We accordingly wish to compute $\iint \eta\rho\mathbf{V} \cdot d\mathbf{A} = \iint (\frac{1}{2}V^2)\{\rho\mathbf{V} \cdot d\mathbf{A}\}$. Employing Eq. (1) for V, and noting again that $\mathbf{V}$ and $d\mathbf{A}$ are collinear but of opposite sense, we get

$$\iint \eta\rho\mathbf{V} \cdot d\mathbf{A} = \int_0^{D/2} \frac{1}{2}C^2\!\left(r^2 - \frac{D^2}{4}\right)^2\!\left\{\rho\!\left[C\!\left(r^2 - \frac{D^2}{4}\right)2\pi r\, dr\right]\right\}$$

$$= \rho C^3\pi \int_0^{D/2}\!\left(r^2 - \frac{D^2}{4}\right)^3 r\, dr = \frac{\rho C^3\pi D^8}{2048}\,\text{N}\cdot\text{m/s} \tag{4}$$

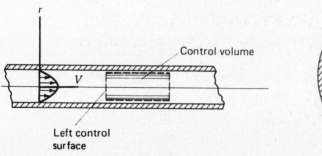

Control volume

V

Left control surface

Fig. 7-18(a)

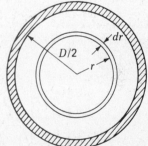

$D/2$

dr

r

Fig. 7-18(b)

7.63 In Prob. 7.62, assume a *one-dimensional* model with the same mass flow. Compute the kinetic energy flow through a section of the pipe for this model. That is, compute kinetic energy flow with an average constant velocity. What is the ratio of the actual kinetic energy to the kinetic energy flow for the one-dimensional model flow?

❚ We first compute the constant velocity at a section for the one-dimensional model. Hence, using Eq. (2) of Prob. 7.62,

$$-(V_{\text{av}})\!\left(\frac{\rho\pi D^2}{4}\right) = -\frac{\rho C D^4\pi}{32}$$

$$V_{\text{av}} = \frac{CD^2}{8} \qquad \text{m/s} \tag{1}$$

The kinetic energy flow for the one-dimensional model is then

$$\iint \frac{V^2}{2}(\rho\mathbf{V} \cdot d\mathbf{A}) = -\frac{\rho}{2}\!\left(\frac{CD^2}{8}\right)^3\!\left(\frac{\pi D^2}{4}\right) = -\frac{\rho C^3 D^8\pi}{4096} \qquad \text{N}\cdot\text{m/s} \tag{2}$$

We now define the *kinetic-energy correction* factor α as the ratio of the actual flow of kinetic energy through a cross section to the flow of kinetic energy for one-dimensional model for the same mass flow. That is

$$\alpha = \frac{\text{KE flow for section}}{\text{KE flow for 1-}D \text{ model}}$$

(3)

For the case at hand, we have from Eq. (2) of this problem and Eq. (4) of Prob. 7.62

$$\alpha = \frac{-\rho C^3 \pi D^8 / 2048}{-\rho C^3 \pi D^8 / 4096} = 2$$

(4)

The factor α exceeds unity, so there is an underestimation of kinetic energy flow for a one-dimensional model.

7.64 The velocity field in a diffuser is $u = U_0 e^{-2x/L}$, and the density field is $\rho = \rho_0 e^{-x/L}$. Find the rate of change of density at $x = L$.

▌ $$\frac{d\rho}{dt} = \frac{\partial \rho}{\partial t} + u \frac{\partial \rho}{\partial x} + v \frac{\partial \rho}{\partial y} + w \frac{\partial \rho}{\partial z} = 0 + (U_0 e^{-2x/L}) \left(\frac{-\rho_0 e^{-x/L}}{L} \right) + 0 + 0 = -\frac{\rho_0 U_0}{L} e^{-3x/L}$$

At $x = L$,

$$\frac{d\rho}{dt} = -\frac{\rho_0 U_0}{L} e^{-3L/L} = -\frac{0.0498 \rho_0 U_0}{L}$$

7.65 Gas is flowing in a long 6-in-diameter pipe from A to B. At section A the flow is 0.35 lb/s while at the same instant at section B the flow is 0.38 lb/s. The distance between A and B is 800 ft. Find the mean value of the time rate of change of the specific weight of the gas between sections A and B at that instant.

▌ $\gamma_1 A_1 v_1 - \gamma_2 A_2 v_2 = \left(\frac{\partial \gamma}{\partial t} \right)(v_{\text{pipe}})$. Since $G = \gamma A v$,

$$0.35 - 0.38 = \left(\frac{\partial \gamma}{\partial t} \right)[(800)(\pi)(\tfrac{6}{12})^2 / 4] \qquad \frac{\partial \gamma}{\partial t} = -0.000191 \quad \text{lb/ft}^3/\text{s}$$

7.66 An incompressible flow field is given by $\mathbf{V} = x^2 \mathbf{i} - z^2 \mathbf{j} - 2xz \mathbf{k}$ with V in meters per second and (x, y, z) in meters. If the fluid viscosity is 0.05 kg/(m · s), evaluate the entire viscous stress tensor at the point $(x, y, z) = (1, 2, 3)$.

▌ $$\tau_{ij} = \begin{vmatrix} \tau_{xx} & \tau_{yx} & \tau_{zx} \\ \tau_{xy} & \tau_{yy} & \tau_{zy} \\ \tau_{xz} & \tau_{yz} & \tau_{zz} \end{vmatrix} \qquad \tau_{xx} = 2\mu \frac{\partial u}{\partial x} = 4\mu x \qquad \tau_{yy} = 2\mu \frac{\partial v}{\partial y} = 0 \qquad \tau_{zz} = 2\mu \frac{\partial w}{\partial z} = -4\mu x$$

$$\tau_{xy} = \tau_{yx} = (\mu) \left(\frac{\partial u}{\partial y} + \frac{\partial v}{\partial x} \right) = 0 \qquad \tau_{yz} = \tau_{zy} = (\mu) \left(\frac{\partial v}{\partial z} + \frac{\partial w}{\partial y} \right) = -2\mu z \qquad \tau_{xz} = \tau_{zx} = (\mu) \left(\frac{\partial y}{\partial z} + \frac{\partial w}{\partial x} \right) = -2\mu z$$

At $(x, y, z) = (1, 2, 3)$ for $\mu = 0.05$ kg/(m · s):

$$\tau_{ij} = \begin{vmatrix} 0.2 & 0 & -0.3 \\ 0 & 0 & -0.3 \\ -0.3 & -0.3 & -0.2 \end{vmatrix} \qquad \text{Pa}$$

7.67 Given the velocity distribution

$$u = Kx \qquad v = -Ky \qquad w = 0$$

(1)

where k is constant, compute and plot the streamlines of flow, including directions, and give some possible interpretations of the pattern.

▌ Since time does not appear explicitly in Eqs. (1), the motion is steady, so that streamlines, path lines, and streaklines will coincide. Since $w = 0$ everywhere, the motion is two-dimensional, in the xy plane. The streamlines can be computed by substituting the expressions for u and v into

$$\frac{dx}{u} = \frac{dy}{v} = \frac{dz}{w} = \frac{dr}{V} \qquad \frac{dx}{Kx} = -\frac{dy}{Ky} \qquad \text{or} \qquad \int \frac{dx}{x} = -\int \frac{dy}{y}$$

Integrating, we obtain $\ln x = -\ln y + \ln C$, or

$$xy = C$$

(2)

This is the general expression for the streamlines, which are hyperbolas. The complete pattern is plotted in Fig. 7-19 by assigning various values to the constant C. The arrowheads can be determined only by returning to Eqs. (1) to ascertain the velocity component directions, assuming K is positive. For example, in the upper right quadrant ($x > 0$, $y > 0$), u is positive and v is negative; hence the flow moves down and to the right, establishing the arrowheads as shown.

Note that the streamline pattern is entirely independent of the constant K. It could represent the impingement of two opposing streams, or the upper half could simulate the flow of a single downward stream against a flat wall. Taken in isolation, the upper right quadrant is similar to the flow in a 90° corner.

Finally note the peculiarity that the two streamlines ($C = 0$) have opposite directions and intersect each other. This is possible only at a point where $u = v = w = 0$, which occurs at the origin in this case. Such a point of zero velocity is called a *stagnation point*.

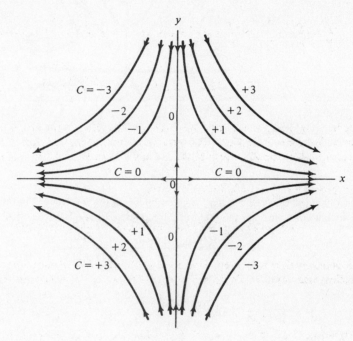

Fig. 7-19

7.68 A velocity field is given by $u = V \cos \theta$, $v = V \sin \theta$, and $w = 0$, where V and θ are constants. Find an expression for the streamlines of this flow.

$$\frac{dx}{u} = \frac{dy}{v} = \frac{dz}{w} = \frac{dr}{V} \qquad \frac{dx}{V \cos \theta} = \frac{dy}{V \sin \theta} = \frac{dz}{0}$$

(*Note*: $dz/0$ indicates that the streamlines do not vary with z.)

$$\frac{dy}{dx} = \frac{V \sin \theta}{V \cos \theta} = \tan \theta \qquad y = x \tan \theta + C$$

Hence, the streamlines are straight and inclined at angle θ, as illustrated in Fig. 7-20.

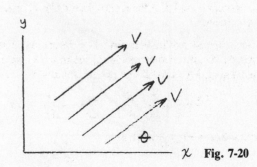

Fig. 7-20

7.69 A two-dimensional steady velocity field is given by $u = x^2 - y^2$, $v = -2xy$. Derive the streamline pattern and sketch a few streamlines in the upper half-plane.

$$\frac{dx}{u} = \frac{dy}{v} = \frac{dz}{w} = \frac{dr}{V} \qquad \frac{dx}{x^2 - y^2} = \frac{dy}{-2xy}$$

$$-2xy\,dx = (x^2 - y^2)\,dy \qquad df = 2xy\,dx + (x^2 - y^2)\,dy \qquad f(x, y) = x^2y - y^3/3 = \text{constant}$$

Hence, the streamlines represent inviscid flow in six 60° corners, as illustrated in Fig. 7-21.

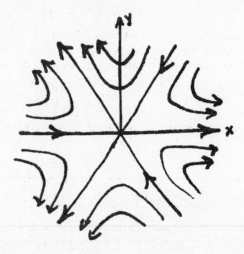

Fig. 7-21

7.70 A two-dimensional unsteady velocity field is given by $u = x(1 + 2t)$, $v = y$. Find the equation of the time-varying streamlines which all pass through the point (x_0, y_0) at some time t. Sketch some.

$$\frac{dx}{u} = \frac{dy}{v} = \frac{dz}{w} = \frac{dr}{V} \qquad \frac{dx}{x(1 + 2t)} = \frac{dy}{y}$$

Integrate, holding t constant.

$$\frac{\ln x}{1 + 2t} = \ln y + C \qquad y = Cx^{1/(1+2t)}$$

If $y = y_0$ at $x = x_0$, $y_0 = Cx_0^{1/(1+2t)}$.

$$C = \frac{y_0}{x_0^{1/(1+2t)}} \qquad y = \left[\frac{y_0}{x_0^{1/(1+2t)}}\right][x^{1/(1+2t)}] = (y_0)\left(\frac{x}{x_0}\right)^{1/(1+2t)}$$

Some streamlines are sketched in Fig. 7-22.

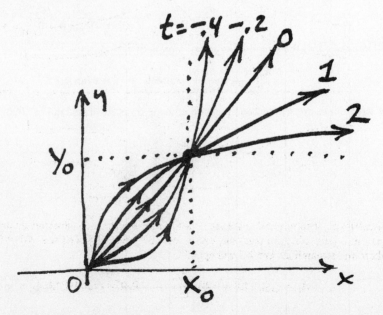

Fig. 7-22

7.71   Investigate the stream function in polar coordinates

$$\psi = U \sin \theta (r - R^2/r) \qquad (1)$$

where U and R are constants, a velocity and a length, respectively. Plot the streamlines. What does the flow represent? Is it a realistic solution to the basic equations?

❚   The streamlines are lines of constant ψ, which has units of square meters per second. Note that ψ/UR is dimensionless. Rewrite Eq. (1) in dimensionless form

$$\psi/UR = \sin \theta (\eta - 1/\eta) \qquad \eta = r/R \qquad (2)$$

Of particular interest is the special line $\psi = 0$. From Eq. (1) or (2) this occurs when (a) $\theta = 0°$ or $180°$ and (b) $r = R$. Case (a) is the x axis and case (b) is a circle of radius R, both of which are plotted in Fig. 7-23.
   For any other nonzero value of ψ it is easiest to pick a value of r and solve for θ:

$$\sin \theta = \frac{\psi/UR}{r/R - R/r} \qquad (3)$$

In general, there will be two solutions for θ because of the symmetry about the y axis. For example take $\psi/UR = +1.0$:

Guess r/R	3.0	2.5	2.0	1.8	1.7	1.618
Compute θ	22°	28°	42°	54°	64°	90°
	158°	152°	138°	126°	116°	

This line is plotted in Fig. 7-23 and passes over the circle $r = R$. You have to watch it, though, because there is a second curve for $\psi/UR = +1.0$ for small $r < R$ below the x axis:

Guess r/R	0.618	0.6	0.5	0.4	0.3	0.2	0.1
Compute θ	−90°	−70°	−42°	−28°	−19°	−12°	−6°
		−110°	−138°	−152°	−161°	−168°	−174°

This second curve plots as a closed curve inside the circle $r = R$. There is a singularity of infinite velocity and indeterminate flow direction at the origin. Figure 7-23 shows the full pattern.
   The given stream function, Eq. (1), is an exact and classic solution to the momentum equation for frictionless flow. Outside the circle $r = R$ it represents two-dimensional inviscid flow of a uniform stream past a circular cylinder. Inside the circle it represents a rather unrealistic trapped circulating motion of what is called a *line doublet.*

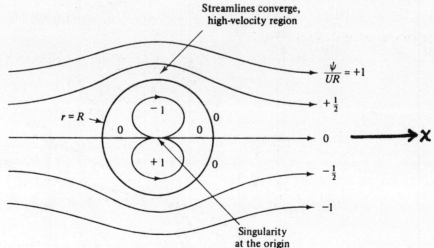

Fig. 7-23

7.72   In two-dimensional, incompressible steady flow around an airfoil, the streamlines are drawn so that they are 10 mm apart at a great distance from the airfoil, where the velocity is 40 m/s. What is the velocity near the airfoil, where the streamlines are 7.5 mm apart?

❚ $$Q = Av = (\tfrac{10}{1000})(40) = 0.40 \, \frac{m^3/s}{m} \qquad 0.40 = (7.5/1000)(v) \qquad v = 53.3 \, m/s$$

7.73 A three-dimensional velocity distribution is given by $u = -x$, $v = 2y$, $w = 5 - z$. Find the equation of the streamline through $(2, 1, 1)$.

▮ $$\frac{dx}{u} = \frac{dy}{v} = \frac{dz}{w} \qquad \frac{dx}{-x} = \frac{dy}{2y} \qquad -\ln x = \ln \sqrt{y} + \ln C_1 \qquad x\sqrt{y} = -C_1$$

At $x = 2$, $y = 1$, $2\sqrt{1} = -C_1$, $C_1 = -2$; $x\sqrt{y} = 2$.

$$\frac{dx}{-x} = \frac{dz}{5-z} \qquad -\ln x = -\ln(5-z) + \ln C_2 \qquad \frac{5-z}{x} = C_2$$

At $x = 2$, $z = 1$, $(5-1)/2 = C_2$, $C_2 = 2$; $(5-z)/x = 2$. Therefore, $x\sqrt{y} = 2$ and $(5-z)/x = 2$ is the equation of the streamline.

7.74 A two-dimensional flow can be described by $u = -y/b^2$, $v = x/a^2$. Verify that this is the flow of an incompressible fluid and that the ellipse $x^2/a^2 + y^2/b^2 = 1$ is a streamline.

▮ $$\frac{\partial u}{\partial x} + \frac{\partial v}{\partial y} = 0 + 0 = 0 \qquad \text{(Therefore, continuity is satisfied.)}$$

$$\frac{dx}{u} = \frac{dy}{v} = \frac{dz}{w} \qquad \frac{dx}{-y/b^2} = \frac{dy}{x/a^2} \qquad \frac{x\,dx}{a^2} = \frac{-y\,dy}{b^2}$$

$$x^2/a^2 + y^2/b^2 = \text{constant} \qquad \text{(Therefore, ellipse } x^2/a^2 + y^2/b^2 = 1 \text{ is a streamline.)}$$

7.75 A velocity potential in two-dimensional flow is $\phi = y + x^2 - y^2$. Find the stream function for this flow.

▮ $$\frac{\partial \phi}{\partial x} = \frac{\partial \psi}{\partial y} \qquad \frac{\partial \phi}{\partial x} = 2x \qquad \frac{\partial \psi}{\partial y} = 2x \qquad \psi = 2xy + f(x)$$

$$\frac{\partial \phi}{\partial y} = -\frac{\partial \psi}{\partial x} \qquad \frac{\partial \phi}{\partial y} = -2y + 1 \qquad -\frac{\partial \psi}{\partial x} = -2y + 1 \qquad \psi = 2xy - x + f(y)$$

Therefore, $f(x) = -x + C$ and $\psi = 2xy - x + C$.

7.76 Examine the two-dimensional flow shown in Fig. 7-24 with the upper boundary that of a rectangular hyperbola, given by the equation $xy = K$. The scaler components of the velocity field are $V_x = -Ax$, $V_y = Ay$, $V_z = 0$, where A is a constant. Note that the flow is steady. Find the equations of the streamlines and the components of acceleration.

▮ $(dy/dx)_{\text{stream}} = V_y/V_x = -y/x$, $dy/y = -dx/x$, $\ln y = -\ln x + \ln C$. Hence, $xy = C$. Note that the streamlines form a family of rectangular hyperbolas. The wetted boundaries are part of the family, as is to be expected.

$$a_x = \frac{\partial V_x}{\partial t} + \left(V_x \frac{\partial V_x}{\partial x} + V_y \frac{\partial V_x}{\partial y} + V_z \frac{\partial V_x}{\partial z} \right) = 0 + (-Ax)(-A) + (Ay)(0) + 0 = A^2 x$$

$$a_y = \frac{\partial V_y}{\partial t} + \left(V_x \frac{\partial V_y}{\partial x} + V_y \frac{\partial V_y}{\partial y} + V_z \frac{\partial V_y}{\partial z} \right) = 0 + (-Ax)(0) + (Ay)(A) + 0 = A^2 y$$

$$a_z = 0 \qquad \mathbf{a} = A^2 x \mathbf{i} + A^2 y \mathbf{j}$$

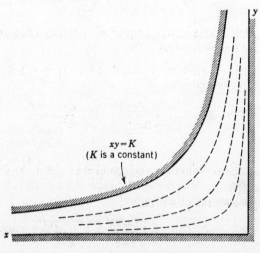

xy = K
(K is a constant)

Fig. 7-24

7.77 The equation for streamlines corresponding to a two-dimensional doublet is given in meters as

$$x^2 + y^2 - \frac{\chi}{C} y = 0$$

where χ is a constant for the flow and C is a constant for a streamline. What is the direction of the velocity of a particle at position $x = 5$ m and $y = 10$ m? If $V_x = 5$ m/s, what is V_y at the point of interest?

▌ At position $x = 5$ m and $y = 10$ m, $5^2 + 10^2 - \frac{\chi}{C}(10) = 0$, $C = \chi/12.5$ for the streamline of interest.

Differentiating the equation for streamlines, $2x\,dx + 2y\,dy - \chi/C\,dy = 0$

$$\left(\frac{dy}{dx}\right)_{\text{stream}} = \frac{2x}{-2y + \chi/C}$$

At position $x = 5$ m and $y = 10$ m,

$$\left(\frac{dy}{dx}\right)_{\text{stream}} = \frac{(2)(5)}{(-2)(10) + \chi/(\chi/12.5)} = -1.333$$

Therefore, the slope is arctan (-1.333), or $-53.1°$. $(dy/dx)_{\text{stream}} = V_y/V_x = -1.333$, $V_y/5 = -1.333$, $V_y = -6.67$ m/s.

7.78 In Prob. 7.77, it should be apparent from analytic geometry that the streamlines represent circles. For a given value of χ and for different values of C, along what axes do the centers of the aforementioned circles lie? Show that all circles go through the origin. Sketch a system of streamlines.

▌ For any value of y, there are two values of x having the same magnitude but different signs. Hence, the centers must lie along the y axis. If $x = y = 0$, clearly the equation for streamlines given in Prob. 7.77 is satisfied; and all circles therefore go through the origin. A sketch of the system is shown in Fig. 7-25.

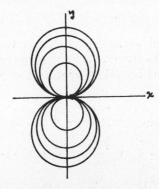

Fig. 7-25

7.79 In Prob. 7.76, what is the equation of the streamline passing through position $x = 2$, $y = 4$? Remembering that the radius of curvature of a curve is

$$R = \frac{[1 + (dy/dx)^2]^{3/2}}{|d^2y/dx^2|}$$

determine the acceleration of a particle in a direction normal to the streamline and toward the center of curvature at the aforementioned position.

▌ The family of streamlines is given by $xy = K$. At $x = 2$, $y = 4$, $(2)(4) = K$, $K = 8$. The equation of the streamline through this point is $xy = 8$.

$$R = \frac{[1 + (dy/dx)^2]^{3/2}}{|d^2y/dx^2|} \qquad \frac{dy}{dx} = -\frac{8}{x^2} \qquad \frac{d^2y}{dx^2} = \frac{16}{x^3}$$

At $x = 2$, $y = 4$,

$$\frac{dy}{dx} = -\frac{8}{2^2} = -2 \qquad \frac{d^2y}{dx^2} = \frac{16}{2^3} = 2 \qquad R = \frac{[1 + (-2)^2]^{3/2}}{2} = 5.59$$

The acceleration component desired is V^2/R.

$$a_N = \frac{V_x^2 + V_y^2}{5.59} = \frac{4A^2 + 16A^2}{5.59} = 3.58A^2$$

7.80 Repeat Prob. 7.70 to find the equation of the path line that passes through (x_0, y_0) at time $t = 0$. Sketch it.

▌ $u = dx/dt = x(1 + 2t)$ $dx/x = (1 + 2t)\, dt$ $\ln x = t + t^2 + C_1$ $x = \exp(t + t^2 + C_1) = [\exp(t + t^2)](C_2)$

At $x = x_0$ and $t = 0$, $x_0 = e^0 C_2$, $C_2 = x_0$; $x = x_0 \exp(t + t^2)$, $v = dy/dt = y$, $dy/y = dt$, $\ln y = t + C_3$, $y = \exp(t + C_3) = e^t C_4$.

At $y = y_0$ and $t = 0$, $y_0 = e^0 C_4$, $C_4 = y_0$; $y = y_0 e^t$, $t = \ln(y/y_0)$, $x = x_0 \exp(t + t^2) = x_0 \exp[\ln(y/y_0) + \ln^2(y/y_0)]$. This pathline is sketched in Fig. 7-26.

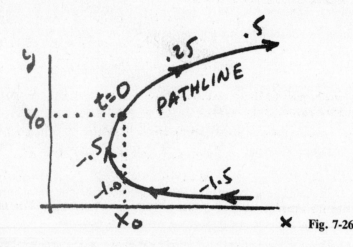

Fig. 7-26

CHAPTER 8

Fundamentals of Fluid Flow

8.1 Water flows through a 3-in-diameter pipe at a velocity of 10 ft/s. Find the **(a)** volume flow rate in cfs and gpm, **(b)** weight flow rate, and **(c)** mass flow rate.

▮ **(a)** $$Q = Av = [(\pi)(\tfrac{3}{12})^2/4](10) = 0.4909 \text{ cfs} = 0.4909/0.002228 = 220 \text{ gpm}$$

(b) $$W = \gamma Av = 62.4[(\pi)(\tfrac{3}{12})^2/4](10) = 30.6 \text{ lb/s}$$

(c) $$M = \rho Av = 1.94[(\pi)(\tfrac{3}{12})^2/4](10) = 0.952 \text{ slug/s}$$

8.2 Benzene flows through a 100-mm-diameter pipe at a mean velocity of 3.00 m/s. Find the **(a)** volume flow rate in m^3/s and L/min, **(b)** weight flow rate, and **(c)** mass flow rate.

▮ **(a)** $$Q = Av = [(\pi)(\tfrac{100}{1000})^2/4](3.00) = 0.0236 \text{ m}^3/\text{s} = 0.0236/0.00001667 = 1416 \text{ L/min}$$

(b) $$W = \gamma Av = 8.62[(\pi)(\tfrac{100}{1000})^2/4](3.00) = 0.203 \text{ kN/s}$$

(c) $$M = \rho Av = 879[(\pi)(\tfrac{100}{1000})^2/4](3.00) = 20.7 \text{ kg/s}$$

8.3 The flow rate of air moving through a square 0.50-m by 0.50-m duct is 160 m^3/min. What is the mean velocity of the air?

▮ $$v = Q/A = 160/[(0.50)(0.50)] = 640 \text{ m/min} \quad \text{or} \quad 10.7 \text{ m/s}$$

8.4 Assume the conduit shown in Fig. 8-1 has (inside) diameters of 12 in and 18 in at sections 1 and 2, respectively. If water is flowing in the conduit at a velocity of 16.6 ft/s at section 2, find the **(a)** velocity at section 1, **(b)** volume flow rate at section 1, **(c)** volume flow rate at section 2, **(d)** weight flow rate, and **(e)** mass flow rate.

▮ **(a)** $$A_1v_1 = A_2v_2 \quad [(\pi)(\tfrac{12}{12})^2/4](v_1) = [(\pi)(\tfrac{18}{12})^2/4](16.6) \quad v_1 = 37.3 \text{ ft/s}$$

(b) $$Q_1 = A_1v_1 = [(\pi)(\tfrac{12}{12})^2/4](37.3) = 29.3 \text{ ft}^3/\text{s}$$

(c) $Q_2 = A_2v_2 = [(\pi)(\tfrac{18}{12})^2/4](16.6) = 29.3 \text{ ft}^3/\text{s}$. (Since the flow in incompressible, the flow rate is the same at sections 1 and 2.)

(d) $$W = \gamma A_1v_1 = 62.4[(\pi)(\tfrac{12}{12})^2/4](37.3) = 1828 \text{ lb/s}$$

(e) $$M = \rho A_1v_1 = 1.94[(\pi)(\tfrac{12}{12})^2/4](37.3) = 56.8 \text{ slugs/s}$$

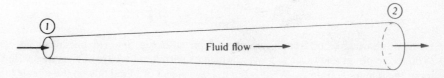

Fig. 8-1

8.5 A gas flows through a square conduit. At one point along the conduit, the conduit sides are 0.100 m, the velocity is 7.55 m/s, and the gas's mass density is (for its particular pressure and temperature) 1.09 kg/m^3. At a second point, the conduit sides are 0.250 m, and the velocity is 2.02 m/s. Find the mass flow rate of the gas and its mass density at the second point.

▮ $$M = \rho_1 A_1 v_1 = 1.09[(0.100)(0.100)](7.55) = 0.0823 \text{ kg/s} \qquad \rho_1 A_1 v_1 = \rho_2 A_2 v_2$$

$$1.09[(0.100)(0.100)](7.55) = (\rho_2)[(0.250)(0.250)](2.02) \qquad \rho_2 = 0.652 \text{ kg/m}^3$$

8.6 Water is forced into the device shown in Fig. 8-2 at the rate of 0.1 m^3/s through pipe A, while oil with specific gravity 0.8 is forced in at the rate of 0.03 m^3/s through pipe B. If the liquids are incompressible and form a homogeneous mixture of oil globules in water, what are the average velocity and density of the mixture leaving through pipe C having a 0.3-m diameter?

▮ $$M = \rho Av = \rho Q \qquad \sum (\text{mass flow in unit time})_{\text{in}} = \sum (\text{mass flow in unit time})_{\text{out}}$$

$$(1000)(0.1) + [(0.8)(1000)](0.03) = (\rho)[(\pi)(0.3)^2/4](v) \qquad \rho v = 1754$$

We can assume no chemical reaction between oil and water and its mixture is incompressible; it is clear that volume is conserved. Hence, $Q = 0.1 + 0.03 = 0.13$ m³/s; $Q = Av$, $0.13 = [(\pi)(0.3)^2/4](v)$, $v_C = 1.84$ m/s; $\rho_C = 1754/1.84 = 953$ kg/m³.

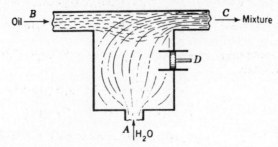

Fig. 8-2

8.7 Water flows into a cylindrical tank (Fig. 8-3) through pipe 1 at the rate of 20 ft/s and leaves through pipes 2 and 3 at the rates of 8 ft/s and 10 ft/s, respectively. At 4, we have an open air vent. The following are the inside pipe diameters: $D_1 = 3$ in, $D_2 = 2$ in, $D_3 = 2.5$ in, $D_4 = 2$ in. Using the entire inside volume of the tank as a control volume, what is dh/dt? What is the average velocity of airflow through vent 4, assuming that the flow is incompressible?

❚ $\qquad M = \rho Av = \rho Q \qquad \sum (\text{mass flow in unit time})_{in} = \sum (\text{mass flow in unit time})_{out}$

$$(\rho)[(\pi)(\tfrac{3}{12})^2/4](20) = (\rho)[(\pi)(\tfrac{2}{12})^2/4](8) + (\rho)[(\pi)(2.5/12)^2/4](10) + (\rho)[(\pi)(2)^2/4](dh/dt)$$

$$dh/dt = 0.1484 \text{ ft/s}$$

Consider only air in the control volume. It must be conserved. Hence, $(\rho_{air})[(\pi)(\tfrac{2}{12})^2/4](v) = (\rho_{air})[(\pi)(2)^2/4](0.1484)$, $v = 21.4$ ft/s.

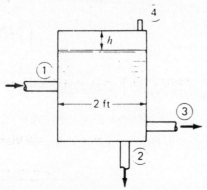

Fig. 8-3

8.8 A nurse is withdrawing blood from a patient, as depicted in Fig. 8-4. The piston is being withdrawn at a speed of 0.25 in/s. The piston allows air to move through its peripheral region of clearance with the glass at the cylinder at the rate of 0.001 in³/s. What is the average speed of blood flow in the needle? Choose as a control volume the region just to the right of the piston to the tip of the needle.

❚ $\qquad M = \rho Av = \rho Q \qquad \sum (\text{mass flow in unit time})_{in} = \sum (\text{mass flow in unit time})_{out}$

$$(\rho_{blood})[(\pi)(0.02/12)^2/4](v) + (\rho_{blood})(0.001/1728) = (\rho_{blood})[(\pi)(0.2/12)^2/4](0.25/12) \qquad v = 1.82 \text{ ft/s}$$

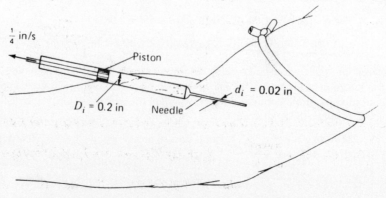

Fig. 8-4

8.9   Air at 30 °C and 100 kPa flows through a 15-cm by 30-cm rectangular duct at 15 N/s. Compute the average velocity and volume flux.

$\blacksquare$

$$W = \gamma A v \qquad \gamma = p/RT = (100)(1000)/[(29.3)(30 + 273)] = 11.26 \text{ N/m}^3$$

$$15 = 11.26[(\tfrac{15}{100})(\tfrac{30}{100})](v) \qquad v = 29.6 \text{ m/s} \qquad Q = Av = [(\tfrac{15}{100})(\tfrac{30}{100})](29.6) = 1.33 \text{ m}^3/\text{s}$$

8.10   Oil (s.g. = 0.86) flows through a 30-in-diameter pipeline at 8000 gpm. Compute the **(a)** volume flux, **(b)** average velocity, and **(c)** mass flux.

$\blacksquare$ **(a)**

$$Q = 8000/[(7.48)(60)] = 17.8 \text{ ft}^3/\text{s}$$

(b)

$$Q = Av \qquad 17.8 = [(\pi)(\tfrac{30}{12})^2/4](v) \qquad v = 3.63 \text{ ft/s}$$

(c)

$$M = \rho A v = [(0.86)(1.94)][(\pi)(\tfrac{30}{12})^2/4](3.63) = 29.7 \text{ slugs/s}$$

8.11   Water flows steadily through a box at three sections, as shown in Fig. 8-5. Section 1 has a diameter of 3 in and the flow in is 1 cfs. Section 2 has a diameter of 2 in and the flow out is 30 fps average velocity. Compute the average velocity and volume flux at section 3 if $D_3 = 1$ in. Is the flow at 3 in or out?

$\blacksquare$

$$Q_1 = Q_2 + Q_3 \qquad \text{(assuming } Q_3 \text{ is out)}$$

$$1 = [(\pi)(\tfrac{2}{12})^2/4](30) + Q_3 \qquad Q_3 = 0.346 \text{ cfs} \qquad \text{(out)} \qquad v = Q/A = 0.346/[(\pi)(\tfrac{1}{12})^2/4] = 63.4 \text{ fps}$$

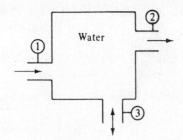

Water

Fig. 8-5

8.12   The water tank in Fig. 8-6 is being filled through section 1 at $v_1 = 4$ m/s and through section 3 at $Q_3 = 0.01$ m^3/s. If water level h is constant, determine exit velocity v_2.

$\blacksquare$

$$Q_1 + Q_3 = Q_2 \qquad [(\pi)(\tfrac{5}{100})^2/4](4) + 0.01 = Q_2$$

$$Q_2 = 0.01785 \text{ m}^3/\text{s} \qquad v_2 = Q_2/A_2 = 0.01785/[(\pi)(\tfrac{7}{100})^2/4] = 4.64 \text{ m/s}$$

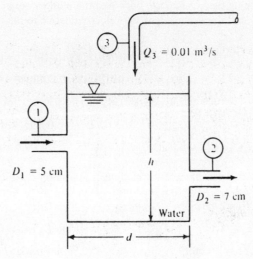

$Q_3 = 0.01 \text{ m}^3/\text{s}$

$D_1 = 5$ cm

h

$D_2 = 7$ cm

Water

d

Fig. 8-6

8.13   If the water level varies in Prob. 8.12 and $v_2 = 6$ m/s, find rate of change dh/dt if the tank diameter $d = 95$ cm.

$\blacksquare$

$$Q_1 + Q_3 = Q_2 + \frac{d}{dt}\frac{h\pi d^2}{4} \qquad [(\pi)(\tfrac{5}{100})^2/4](4) + 0.01 = [(\pi)(\tfrac{7}{100})^2/4](6) + \frac{dh}{dt}[(\pi)(\tfrac{95}{100})^2/4]$$

$$dh/dt = -0.00739 \text{ m/s} \qquad \text{(i.e., falling)}$$

8.14 For the general case of the flow depicted in Fig. 8-6, derive an expression for dh/dt in terms of tank size and volume flows Q_1, Q_2, and Q_3 at the three ports.

$$Q_1 + Q_3 = Q_2 + \frac{d}{dt}\frac{h\pi d^2}{4} \qquad \frac{dh}{dt} = \frac{4(Q_1 - Q_2 + Q_3)}{\pi d^2}$$

8.15 Water at 20 °C flows steadily through the nozzle in Fig. 8-7 at 50 kg/s. The diameters are $D_1 = 20$ cm and $D_2 = 6$ cm. Compute the average velocities at sections 1 and 2.

$$Q = M/\rho = \tfrac{50}{998} = 0.0501 \text{ m}^3/\text{s}$$

$$v_1 = Q/A_1 = 0.0501/[(\pi)(\tfrac{20}{100})^2/4] = 1.59 \text{ m/s} \qquad v_2 = Q/A_2 = 0.0501/[(\pi)(\tfrac{6}{100})^2/4] = 17.7 \text{ m/s}$$

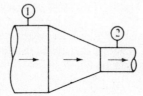

Fig. 8-7

8.16 The hypodermic needle in Fig. 8-8 contains some sort of serum (s.g. = 1.02). If the plunger is pushed in steadily at 0.8 in/s, what is exit velocity V_2? Assume no leakage past the plunger.

$$\gamma_1 A_1 V_1 = \gamma_2 A_2 V_2$$

$$[(1.02)(62.4)][(\pi)(0.75/12)^2/4](0.8/12) = [(1.02)(62.4)][(\pi)(0.030/12)^2/4](V_2)$$

$$V_2 = 41.7 \text{ ft/s}$$

(Note that the answer is independent of the serum's specific gravity.)

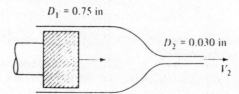

Fig. 8-8

8.17 Repeat Prob. 8.16 assuming there is leakage back past the plunger equal to one-third of the volume flux out of the needle. Compute V_2 and the average leakage velocity relative to the needle walls if the plunger diameter is 0.746 in.

$$Q = Q_1 = A_1 V_1 = [(\pi)(0.746/12)^2/4](0.8/12) = 0.0002024 \text{ ft}^3/\text{s}$$

$$Q_2 = (\tfrac{2}{3})(0.0002024) = 0.0001349 \text{ ft}^3/\text{s}$$

$$V_2 = Q_2/A_2 = 0.0001349/[(\pi)(0.030/12)^2/4] = 27.5 \text{ ft/s}$$

$$Q_{\text{leak}} = (\tfrac{1}{3})(0.0002024) = 0.00006747 \text{ ft}^3/\text{s}$$

$$V_{\text{leak}} = Q_{\text{leak}}/A_{\text{leak}} = 0.00006747/[(\pi)(0.75/12)^2/4 - (\pi)(0.746/12)^2/4] = 2.07 \text{ ft/s}$$

8.18 A full gasoline tank has an 8-cm-diameter steel plunger at section 1, which is being pushed into the tank at 5 cm/s. Assuming the fluid (s.g. = 0.68) is incompressible, how many pounds of gasoline per second are being forced out at section 2, $D_2 = 2$ cm?

$$A_1 v_1 = A_2 v_2 \qquad [(\pi)(\tfrac{8}{100})^2/4](\tfrac{5}{100}) = [(\pi)(\tfrac{2}{100})^2/4](v_2) \qquad v_2 = 0.800 \text{ m/s}$$

$$W = \gamma A v = [(0.68)(9.79)][(\pi)(\tfrac{2}{100})^2/4](0.800) = 0.001673 \text{ kN/s} \quad \text{or} \quad 1.673 \text{ N/s}$$

$$= 1.673/4.448 = 0.376 \text{ lb/s}$$

8.19 A gasoline pump fills a 75-L tank in 1 min 10 s. If the pump exit diameter is 3 cm, what is the average pump-flow exit velocity?

$$Q = V/t = (\tfrac{75}{1000})/(60 + 10) = 0.001071 \text{ m}^3/\text{s} \qquad v = Q/A = 0.001071/[(\pi)(\tfrac{3}{100})^2/4] = 1.52 \text{ m/s}$$

8.20 The tank in Fig. 8-9 is admitting water at 90 N/s and ejecting gasoline (s.g. = 0.68) at 50 N/s. If all fluids are incompressible, how much air is passing through the vent? In which direction?

$$Q_1 = Q_2 + Q_3 \qquad \text{(assuming airflow is out)}$$

$$Q_1 = W_1/\gamma_{\text{H}_2\text{O}} = 90/[(9.79)(1000)] = 0.009193 \text{ m}^3/\text{s}$$

$$Q_2 = W_2/\gamma_{\text{gas}} = 50/[(0.68)(9.79)(1000)] = 0.007511 \text{ m}^3/\text{s}$$

$$0.009193 = 0.007511 + Q_3 \qquad Q_3 = 0.001682 \text{ m}^3/\text{s} \qquad \text{(out)}$$

$$\gamma_{\text{air}} = p/RT = (1)(101.3)/[(29.3)(20 + 273)] = 0.01180 \text{ kN/m}^3$$

$$W_3 = (\gamma_{\text{air}})(Q_3) = (0.01180)(0.001682) = 0.0000198 \text{ kN/s} \quad \text{or} \quad 0.0198 \text{ N/s}$$

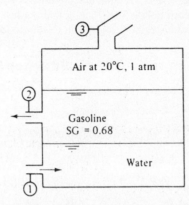

Air at 20°C, 1 atm

Gasoline
SG = 0.68

Water

Fig. 8-9

8.21 Air at 70 °F and 15 psia enters a chamber at section 1 at velocity 200 fps and leaves section 2 at 1200 °F and 200 psia. What is the exit velocity if $D_1 = 6$ in and $D_2 = 2$ in? Assume the flow is steady.

$$\rho_1 A_1 v_1 = \rho_2 A_2 v_2 \qquad \rho = p/RT \qquad \rho_1 = (15)(144)/[(1716)(460 + 70)] = 0.002375 \text{ slug/ft}^3$$

$$\rho_2 = (200)(144)/[(1716)(460 + 1200)] = 0.01011 \text{ slug/ft}^3$$

$$0.002375[(\pi)(\tfrac{6}{12})^2/4](200) = 0.01011[(\pi)(\tfrac{2}{12})^2/4](v_2) \qquad v_2 = 423 \text{ fps}$$

8.22 Oil (s.g. = 0.86) enters at section 1 in Fig. 8-10 at 0.06 N/s to lubricate a thrust bearing. The 10-cm-diameter bearing plates are 2 mm apart. Assuming steady flow, compute the inlet average velocity v_1, outlet average velocity v_2 assuming radial flow, and outlet volume flux.

$$W_1 = \gamma_{\text{oil}} A_1 v_1 \qquad 0.06 = [(0.86)(9.79)(1000)][(\pi)(\tfrac{3}{1000})^2/4](v_1) \qquad v_1 = 1.01 \text{ m/s}$$

$$Q_1 = A_1 v_1 = [(\pi)(\tfrac{3}{1000})^2/4](1.01) = 0.00000714 \text{ m}^3/\text{s} \qquad Q_2 = Q_1 = 0.00000714 \text{ m}^3/\text{s} \quad \text{or} \quad 7.14 \text{ mL}$$

$$v_2 = Q_2/A_2 = 0.00000714/[(\pi)(\tfrac{10}{100})(\tfrac{2}{1000})] = 0.0114 \text{ m/s} \quad \text{or} \quad 1.14 \text{ cm/s}$$

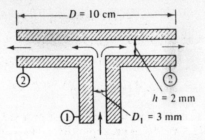

Fig. 8-10

8.23 In Fig. 8-11, pipes 1 and 2 are of diameter 2 cm; $D_3 = 3$ cm. Alcohol (s.g. = 0.79) enters section 1 at 8 m/s while water enters section 2 at 12 m/s. Assuming ideal mixing of incompressible fluids, compute the exit velocity and density of the mixture at section 3. The temperature is 20 °C.

$$Q_1 + Q_2 = Q_3 \qquad [(\pi)(\tfrac{2}{100})^2/4](8) + [(\pi)(\tfrac{2}{100})^2/4](12) = Q_3 \qquad Q_3 = 0.006283 \text{ m}^3/\text{s}$$

$$v_3 = Q_3/A_3 = 0.006283/[(\pi)(\tfrac{3}{100})^2/4] = 8.89 \text{ m/s} \qquad M_1 + M_2 = M_3$$

$$\rho_{\text{alcohol}} A_1 v_1 + \rho_{\text{H}_2\text{O}} A_2 v_2 = \rho_{\text{mixture}} A_3 v_3$$

$$[(0.79)(998)][(\pi)(\tfrac{2}{100})^2/4](8) + 998[(\pi)(\tfrac{2}{100})^2/4](12) = (\rho_{\text{mixture}})[(\pi)(\tfrac{3}{100})^2/4](8.99) \qquad \rho_{\text{mixture}} = 914 \text{ kg/m}^3$$

Fig. 8-11

8.24 In some wind tunnels, the test-section wall is porous or perforated; fluid is sucked out to provide a thin viscous boundary layer. The wall in Fig. 8-12 is 4 m long and contains 800 holes of 6-mm diameter per square meter of area. The suction velocity out each hole is $V_s = 10$ m/s, and the test section entrance velocity is $V_1 = 45$ m/s. Assuming incompressible flow of air at 20 °C and 1 atm, compute *(a)* V_0, *(b)* the total wall suction volume flow, *(c)* V_2, and *(d)* V_f.

▌ *(a)* $A_0 V_0 = A_1 V_1$ $[(\pi)(2.5)^2/4](V_0) = [(\pi)(0.8)^2/4](45)$ $V_0 = 4.6$ m/s

(b) $Q_{\text{suction}} = N_{\text{holes}} Q_{\text{hole}}$ $N_{\text{holes}} = 800[(\pi)(0.8)(4)] = 8042$

$$Q_{\text{hole}} = A_{\text{hole}} V_{\text{hole}} = [(\pi)(\tfrac{6}{1000})^2/4](10) = 0.0002827 \text{ m}^3/\text{s}$$

$$Q_{\text{suction}} = (8042)(0.0002827) = 2.27 \text{ m}^3/\text{s}$$

(c) $Q_1 = Q_2 + Q_{\text{suction}}$ $[(\pi)(0.8)^2/4](45) = Q_2 + 2.27$ $Q_2 = 20.35 \text{ m}^3/\text{s}$

$$v_2 = Q_2/A_2 = 20.35/[(\pi)(0.8)^2/4] = 40.5 \text{ m/s}$$

(d) $A_f V_f = A_2 V_2$ $[(\pi)(2.2)^2/4](V_f) = [(\pi)(0.8)^2/4](40.5)$ $V_f = 5.36$ m/s

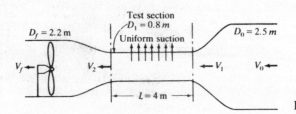

Fig. 8-12

8.25 A rocket motor is operating steadily as shown in Fig. 8-13. The products of combustion flowing out the exhaust nozzle approximate a perfect gas with a molecular weight of 26. For the given conditions, calculate v_2.

▌ $M_2 = M_1 + M_3 = 0.6 + 0.2 = 0.6 \text{ slug/s} = \rho_2 A_2 v_2$ $R = 49\,709/26 = 1912 \text{ lb} \cdot \text{ft}/(\text{slug} \cdot {}^\circ\text{R})$

$$\rho_2 = p/RT = (15)(144)/[(1912)(1100 + 460)] = 0.0007242 \text{ slug/ft}^3$$

$$0.8 = 0.0007242[(\pi)(5.5/12)^2/4](v_2)   v_2 = 6695 \text{ ft/s}$$

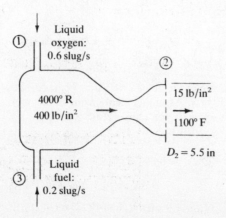

Fig. 8-13

8.26 In contrast to the liquid rocket in Fig. 8-13, the solid-propellant rocket in Fig. 8-14 is self-contained and has no entrance ducts. Using a control-volume analysis for the conditions shown in Fig. 8-14, compute the rate of mass loss of the propellant, assuming the exit gas has a molecular weight of 31.

∎

$$M_{in} = M_{out} + \frac{d}{dt}(m_{propellant}) \qquad 0 = M_{out} + \frac{d}{dt}(m_{propellant})$$

$$\frac{d}{dt}(m_{propellant}) = -M_{out} = -\rho_e A_e V_e$$

$$R = \tfrac{8312}{31} = 268 \text{ N} \cdot \text{m}/(\text{kg} \cdot \text{K}) \qquad \rho_e = p_e/RT_e = (100)(1000)/[(268)(800)] = 0.4664 \text{ kg/m}^3$$

$$\frac{d}{dt}(m_{propellant}) = -(0.4664)\left[\frac{(\pi)(\tfrac{18}{100})^2}{4}\right](1200) = -14.2 \text{ kg/s}$$

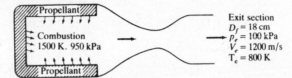

Combustion
1500 K. 950 kPa

Propellant
Propellant

Exit section
$D_f = 18$ cm
$p_e = 100$ kPa
$V_e = 1200$ m/s
$T_c = 800$ K

Fig. 8-14

8.27 The water-jet pump in Fig. 8-15 injects water at $U_1 = 100$ ft/s through a 3-in pipe and entrains a secondary flow of water $U_2 = 10$ ft/s in the annular region around the small pipe. The two flows become fully mixed downstream, where U_3 is approximately constant. If the flow is steady and incompressible, compute U_3.

∎ $Q_1 + Q_2 = Q_3 \qquad [(\pi)(\tfrac{3}{12})^2/4](100) + \{(\pi)[(\tfrac{10}{12})^2 - (\tfrac{3}{12})^2]/4\}(10) = [(\pi)(\tfrac{10}{12})^2/4](U_3) \qquad U_3 = 18.1$ ft/s

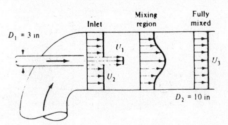

Inlet

Mixing region

Fully mixed

$D_1 = 3$ in

U_1

U_2

U_3

$D_2 = 10$ in

Fig. 8-15

8.28 The flow in the inlet between parallel plates in Fig. 8-16 is uniform $u = U_0 = 4$ cm/s, while downstream the flow develops into the parabolic laminar profile $u = az(z_0 - z)$, where a is a constant. If $z_0 = 1$ cm and the fluid is glycerin at 20 °C, for steady flow what is the value of u_{max}?

∎ Let b = width of plates (into paper).

$$Q_{in} = Q_{out} \qquad z_0 b U_0 = \int u \, dA$$

$$z_0 b U_0 = \int_0^{z_0} az(z_0 - z)b \, dz = ab\left[\frac{z_0 z^2}{2} - \frac{z^3}{3}\right]_0^{z_0} = ab\left(\frac{z_0 z_0^2}{2} - \frac{z_0^3}{3}\right) = \frac{abz_0^3}{6}$$

$$a = 6U_0/z_0^2 \qquad u = az(z_0 - z) = (6U_0/z_0^2)(z)(z_0 - z)$$

u_{max} occurs at $z = z_0/2 = \tfrac{1}{100}/2 = 0.005$ m: $u_{max} = [(6)(\tfrac{4}{100})/(\tfrac{1}{100})^2](0.005)(\tfrac{1}{100} - 0.005) = 0.0600$ m/s or 6.00 cm/s.

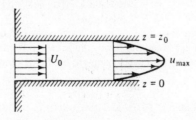

$z = z_0$

U_0

u_{max}

$z = 0$

Fig. 8-16

8.29 Assuming the container in Fig. 8-17a is large and losses are negligible, derive an expression for the distance X where the free jet leaving horizontally will strike the floor, as a function of h and H. Sketch the three trajectories for $h/H = 0.25$, 0.50, and 0.75.

∎ $v_0 = \sqrt{2g(H-h)}$ $h = gt^2/2$ $t = \sqrt{2h/g}$ $X = v_0 t = \sqrt{2g(H-h)}\sqrt{2h/g} = 2\sqrt{h(H-h)}$

For $h/H = 0.25$, or $h = 0.25H$, $X = 2\sqrt{(0.25H)(H - 0.25H)} = 0.866H$. For $h/H = 0.50$, or $h = 0.50H$, $X = 2\sqrt{(0.50H)(H - 0.50H)} = H$. For $h/H = 0.75$, or $h = 0.75H$, $X = 2\sqrt{(0.75H)(H - 0.75)} = 0.866H$. These three trajectories are sketched in Fig. 8-17b.

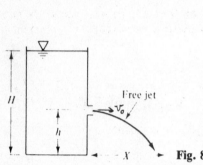
Fig. 8-17(a)

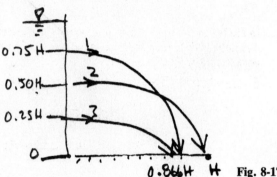
Fig. 8-17(b)

8.30 In Fig. 8-18 what should the water level h be for the free jet to just clear the wall?

∎ $v_0 = \sqrt{2gh}$ Fall distance $= gt^2/2 = \frac{30}{100}$ $t = 0.7746/\sqrt{g}$

Horizontal distance $= v_0 t = (\sqrt{2gh})(0.7746/\sqrt{g}) = \frac{40}{100}$ $h = 0.133$ m

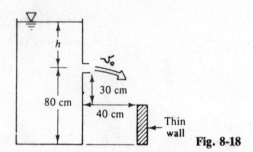
Fig. 8-18

8.31 When 500 gpm flows through a 12-in pipe which later reduces to a 6-in pipe, calculate the average velocities in the two pipes.

∎ $Q = 500/[(7.48)(60)] = 1.114 \text{ ft}^3/\text{s}$

$v_{12} = Q/A_{12} = 1.114/[(\pi)(\frac{12}{12})^2/4] = 1.42 \text{ ft/s}$ $v_6 = Q/A_6 = 1.114/[(\pi)(\frac{6}{12})^2/4] = 5.67 \text{ ft/s}$

8.32 If the velocity in a 12-in pipe is 1.65 ft/s, what is the velocity in a 3-in-diameter jet issuing from a nozzle attached to the pipe?

∎ $A_1 v_1 = A_2 v_2$ $[(\pi)(\frac{12}{12})^2/4](1.65) = [(\pi)(\frac{3}{12})^2/4](v_2)$ $v_2 = 26.4 \text{ ft/s}$

8.33 Air flows in a 6-in pipe at a pressure of 30.0 psig and a temperature of 100 °F. If barometric pressure is 14.7 psia and velocity is 10.5 ft/s, how many pounds of air per second are flowing?

∎ $\gamma = p/RT = (30.0 + 14.7)(144)/[(53.3)(100 + 460)] = 0.2157 \text{ lb/ft}^3$

$W = \gamma A v = 0.2157[(\pi)(\frac{6}{12})^2/4](10.5) = 0.445 \text{ lb/s}$

8.34 Carbon dioxide passes point A in a 3-in pipe at a velocity of 15.0 ft/s. The pressure at A is 30 psig and the temperature is 70 °F. At point B downstream, the pressure is 20 psig and the temperature is 90 °F. For a barometric pressure reading of 14.7 psia, calculate the velocity at B and compare the flows at A and B.

∎ $\gamma = p/RT$ $\gamma_A = (30 + 14.7)(144)/[(35.1)(70 + 460)] = 0.3460 \text{ lb/ft}^3$

$\gamma_B = (20 + 14.7)(144)/[(35.1)(90 + 460)] = 0.2588 \text{ lb/ft}^3$ $\gamma_A A_A v_A = \gamma_B A_B v_B$

Since $A_A = A_B$, $(0.3460)(15.0) = (0.2588)(v_B)$, $v_B = 20.1$ ft/s. The number of pounds per second flowing is constant, but the flow in cubic feet per second will differ because the specific weight is not constant. $Q_A = A_A v_A = [(\pi)(\frac{3}{12})^2/4](15.0) = 0.736$ ft³/s; $Q_B = A_B v_B = [(\pi)(\frac{3}{12})^2/4](20.1) = 0.987$ ft³/s.

8.35 What minimum diameter of pipe is necessary to carry 0.500 lb/s of air with a maximum velocity of 18.5 ft/s? The air is at 80 °F and under an absolute pressure of 34.0 psi.

▎ $W = \gamma A v$ $\qquad \gamma_{\text{air}} = p/RT = (34.0)(144)/[(53.3)(80 + 460)] = 0.170$ lb/ft³

$\qquad 0.500 = (0.170)[(\pi)(d)^2/4](18.5) \qquad d = 0.450$ ft or 5.40 in

8.36 In the laminar flow of a fluid in a circular pipe, the velocity profile is exactly a true parabola. The rate of discharge is then represented by the volume of a paraboloid. Prove that for this case the ratio of the mean velocity to the maximum velocity is 0.5.

▎ See Fig. 8-19. For a paraboloid, $u = u_{\max}[1 - (r/r_0)^2]$.

$$Q = \int u \, dA = \int_0^{r_0} u_{\max}\left[1 - \left(\frac{r}{r_0}\right)^2\right](2\pi r \, dr) = 2\pi u_{\max}\left[\frac{r^2}{2} - \frac{r^4}{4r_0^2}\right]_0^{r_0} = 2\pi u_{\max}\left[\frac{r_0^2}{2} - \frac{r_0^4}{4r_0^2}\right] = u_{\max}\left(\frac{\pi r_0^2}{2}\right)$$

$$V_{\text{mean}} = Q/A = u_{\max}(\pi r_0^2/2)/(\pi r_0^2) = u_{\max}/2. \text{ Thus } V_{\text{mean}}/u_{\max} = 0.5.$$

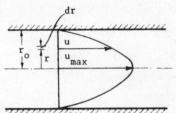

Fig. 8-19

8.37 A gas ($\gamma = 0.04$ lb/ft³) flows at the rate of 1.0 lb/s past section A through a long rectangular duct of uniform cross section 2 ft by 2 ft. At section B some distance along the duct, the gas weighs 0.065 lb/ft³. What is the average velocity of flow at sections A and B?

▎ $W = \gamma A v$ $\quad 1.0 = (0.04)[(2)(2)](v_A) \quad v_A = 6.25$ ft/s $\quad 1.0 = (0.065)[(2)(2)](v_B) \quad v_B = 3.85$ ft/s

8.38 The velocity of a liquid (s.g. = 1.26) in a 4-in pipeline is 1.5 fps. Calculate the rate of flow in cubic feet per second, gallons per minute, pounds per second, and slugs per second.

▎ $\qquad Q = Av = [(\pi)(\frac{4}{12})^2/4](1.5) = 0.131$ ft³/s $= (0.131)(7.48)(60) = 58.8$ gal/min

$\qquad W = \gamma A v = [(1.26)(62.4)][(\pi)(\frac{4}{12})^2/4](1.5) = 10.3$ lb/s

$\qquad M = \rho A v = [(1.26)(1.94)][(\pi)(\frac{4}{12})^2/4](1.5) = 0.320$ slug/s

8.39 Oxygen flows in a 2-in by 2-in duct at a pressure of 40 psi and a temperature of 100 °F. If atmospheric pressure is 13.4 psia and the velocity of flow is 18 fps, calculate the weight flow rate.

▎ $\qquad \gamma = p/RT = (40 + 13.4)(144)/[(48.2)(460 + 100)] = 0.2849$ lb/ft³

$\qquad W = \gamma A v = (0.2849)[(\frac{2}{12})(\frac{2}{12})](18) = 0.142$ lb/s

8.40 Air at 40 °C and under a pressure of 3000 mbar abs flows in a 250-mm-diameter conduit at a mean velocity of 10 m/s. Find the mass flow rate.

▎ $\rho = p/RT = 3000(100)/[(287)(273 + 40)] = 3.340$ kg/m³ $\quad M = \rho A v = 3.340[(\pi)(\frac{250}{1000})^2/4](10) = 1.64$ kg/s

8.41 Gas flows at a steady rate in a 10-cm-diameter pipe that enlarges to a 15-cm-diameter pipe. At a certain section of the 10-cm pipe, the density of the gas is 200 kg/m³ and the velocity is 20 m/s. At a certain section of the 15-cm pipe the velocity is 14 m/s. What must be the density of the gas at that section? If these same data were given for the case of unsteady flow at a certain instant, could the problem have been solved? Discuss.

▎ $\qquad \rho_1 A_1 v_1 = \rho_2 A_2 v_2 \quad 200[(\pi)(\frac{10}{100})^2/4](20) = (\rho_2)[(\pi)(\frac{15}{100})^2/4](14) \quad \rho_2 = 127$ kg/m³

If the flow were unsteady, the problem could not be solved, for no information is given on $\partial \rho/\partial t$. Also, the volume between the two sections is unknown.

8.42   Water flows in a river. At 9 a.m. the flow past bridge 1 is 55 m³/s. At the same instant the flow past bridge 2 is 45 m³/s. At what rate is water being stored between the two bridges at this instant? Assume zero seepage and negligible evaporation.

▮ $$Q_1 - Q_2 = dS/dt \qquad 55 - 45 = dS/dt \qquad dS/dt = 10 \text{ m}^3/\text{s}$$

8.43   Some steam locomotives had scoops installed that took water from a tank between the tracks and lifted it into the water reservoir in the tender. To lift the water 10 ft with a scoop, neglecting all losses, what speed is required? Consider the locomotive stationary and the water moving toward it to reduce to a steady-flow situation.

▮ $$v = \sqrt{2gh} = \sqrt{(2)(32.2)(10)} = 25.4 \text{ ft/s}$$

8.44   At section 1 of a pipe system carrying water the velocity is 3.0 fps and the diameter is 2.0 ft. At section 2 the diameter is 3.0 ft. Find the discharge and velocity at section 2.

▮ $$Q_1 = Q_2 = Av = [(\pi)(2.0)^2/4](3.0) = 9.42 \text{ cfs} \qquad v_2 = Q_2/A_2 = 9.42/[(\pi)(3.0)^2/4] = 1.33 \text{ fps}$$

8.45   In two-dimensional flow around a circular cylinder (Fig. 8-20), the discharge between streamlines is 0.01 cfs per foot of depth. At a great distance the streamlines are 0.25 in apart, and at a point near the cylinder they are 0.12 in apart. Calculate the magnitude of the velocity at these two points.

▮ $v = Q/A$. At great distance, $v = 0.01/(0.25/12) = 0.480$ fps. Near the cylinder, $v = 0.01/(0.12/12) = 1.00$ fps.

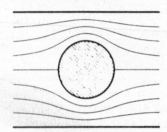

Fig. 8-20

8.46   A pipeline carries oil (s.g. = 0.86) at $v = 2$ m/s through a 200-mm-ID pipe. At another section the diameter is 70 mm. Find the velocity at this section and the mass flow rate.

▮ $$Q = A_1 v_1 = [(\pi)(\tfrac{200}{1000})^2/4](2) = 0.06283 \text{ m}^3/\text{s} \qquad v_2 = Q/A_2 = 0.06283/[(\pi)(\tfrac{70}{1000})^2/4] = 16.3 \text{ m/s}$$
$$M = \rho Av = [(0.86)(1000)][(\pi)(\tfrac{200}{1000})^2/4](2) = 54.0 \text{ kg/s}$$

8.47   Hydrogen is flowing in a 2.0-in-diameter pipe at the mass flow rate of 0.03 lbm/s. At section 1 the pressure is 30 psia and the temperature is 80 °F. What is the average velocity?

▮ $$M = \rho Av \qquad \rho = p/RT = (30)(144)/[(765.5)(460 + 80)] = 0.01045 \text{ lbm/ft}^3$$
$$0.03 = (0.01045)[(\pi)(2.0/12)^2/4](v) \qquad v = 132 \text{ ft/s}$$

8.48   If a jet is inclined upward 30° from the horizontal, what must be its velocity to reach over a 10-ft wall at a horizontal distance of 60 ft, neglecting friction?

▮ $(v_x)_0 = v_0 \cos 30° = 0.8660 v_0$, $(v_z)_0 = v_0 \sin 30° = 0.5000 v_0$. From Newton's laws, $x = (0.8660 v_0)(t) = 60$, $z = 0.5000 v_0 t - 32.2 t^2/2 = 10$. From the first equation, $t = 69.28/v_0$. Substituting this into the second equation, $(0.5000)(v_0)(69.28/v_0) - (32.2)(69.28/v_0)^2/2 = 10$, $v_0 = 56.0$ fps.

8.49   Water flows at 10 m³/s in a 150-cm-diameter pipe; the head loss in a 1000-m length of this pipe is 20 m. Find the rate of energy loss due to pipe friction.

▮ $$\text{Rate of energy loss} = \gamma QH = (9.79)(10)(20) = 1958 \text{ kW}$$

8.50   Oil with specific gravity 0.750 is flowing through a 6-in pipe under a pressure of 15.0 psi. If the total energy relative to a datum plane 8.00 ft below the center of the pipe is 58.6 ft · lb/lb, determine the flow rate of the oil.

▮ $$H = z + v^2/2g + p/\gamma \qquad 58.6 = 8.00 + v^2/[(2)(32.2)] + (15)(144)/[(0.750)(62.4)]$$
$$v = 16.92 \text{ ft/s} \qquad Q = Av = [(\pi)(\tfrac{6}{12})^2/4](16.92) = 3.32 \text{ ft}^3/\text{s}$$

8.51 In Fig. 8-21, water flows from A, where the diameter is 12 in, to B, where the diameter is 24 in, at the rate of 13.2 cfs. The pressure head at A is 22.1 ft. Considering no loss of energy from A to B, find the pressure head at B.

▮ $p_A/\gamma + v_A^2/2g + z_A = p_B/\gamma + v_B^2/2g + z_B$ $v_A = Q/A_A = 13.2/[(\pi)(\frac{12}{12})^2/4] = 16.81$ ft/s

$v_B = Q/A_B = 13.2/[(\pi)(\frac{24}{12})^2/4] = 4.202$ ft/s

$22.1 + 16.81^2/[(2)(32.2)] + 0 = p_B/\gamma + 4.202^2/[(2)(32.2)] + (25.0 - 10.0)$ $p_B/\gamma = 11.2$ ft of water

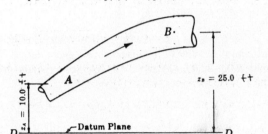

D ———— Datum Plane ———— D **Fig. 8-21**

8.52 A pipe carrying oil with specific gravity 0.877 changes in size from 6 in at section E to 18 in at section R. Section E is 12 ft lower than R, and the pressures are 13.2 psi and 8.75 psi, respectively. If the discharge is 5.17 cfs, determine the lost head and the direction of flow.

▮ $H = z + v^2/2g + p/\gamma$. Use the lower section (E) as the datum plane. $v_E = Q/A_E = 5.17/[(\pi)(\frac{6}{12})^2/4] = 26.33$ ft/s, $v_R = Q/A_R = 5.17/[(\pi)(\frac{18}{12})^2/4] = 2.926$ ft/s; $H_E = 0 + 26.33^2/[(2)(32.2)] + (13.2)(144)/[(0.877)(62.4)] = 45.50$ ft, $H_R = 12 + 2.926^2/[(2)(32.2)] + (8.75)(144)/[(0.877)(62.4)] = 35.16$ ft. Since the energy at E exceeds that at R, flow occurs from E to R. The lost head is $45.50 - 35.16$, or 10.34 ft, E to R.

8.53 A horizontal air duct is reduced in cross-sectional area from 0.75 ft² to 0.20 ft². Assuming no losses, what pressure change will occur when 1.50 lb/s of air flows? Use $\gamma = 0.200$ lb/ft³ for the pressure and temperature conditions involved.

▮ $Q = 1.50/0.200 = 7.500$ ft³/s $p_A/\gamma + v_A^2/2g + z_A = p_B/\gamma + v_B^2/2g + z_B$

$p_A/\gamma + (7.500/0.75)^2/[(2)(32.2)] + 0 = p_B/\gamma + (7.500/0.20)^2/[(2)(32.2)] + 0$

$p_A/\gamma - p_B/\gamma = 20.28$ ft of air $p_A - p_B = (20.28)(0.200)/144 = 0.0282$ psi

8.54 A turbine is rated at 600 hp when the flow of water through it is 21.5 cfs. Assuming an efficiency of 87 percent, what head is acting on the turbine?

▮ Rated horsepower = (extracted horsepower)(efficiency) = $(\gamma QH/550)$(efficiency)

$600 = [(62.4)(21.5)(H)/550](0.87)$ $H = 283$ ft

8.55 A standpipe 30 ft in diameter and 40 ft high is filled with water. How much potential energy (PE) is in this water if the elevation datum is taken 10 ft below the base of the standpipe?

▮ PE = $Wz = [(62.4)(40)(\pi)(30)^2/4](10 + 40/2) = 5.29 \times 10^7$ ft · lb

8.56 How much work could be obtained from the water in Prob. 8.55 if run through a 100 percent efficient turbine that discharged into a reservoir with elevation 30 ft below the base of the standpipe?

▮ Work = PE = $Wz = [(62.4)(40)(\pi)(30)^2/4](30 + \frac{40}{2}) = 8.82 \times 10^7$ ft · lb

8.57 What is the kinetic energy (KE) flux of 0.01 m³/s of oil (s.g. = 0.80) discharging through a 30-mm-diameter nozzle?

▮ $v = Q/A = 0.01/[(\pi)(\frac{30}{1000})^2/4] = 14.5$ m/s

KE = $mv^2/2 = \rho Q v^2/2 = [(0.80)(1000)](0.01)(14.15)^2/2 = 801$ N · m

8.58 By neglecting air resistance determine the height a vertical jet of water will rise with velocity 60 ft/s.

▮ PE = KE $Wz = mv^2/2 = (W/32.2)(60)^2/2$ $z = 55.9$ ft

8.59   If the water jet of Prob. 8.58 is directed upward 45° with the horizontal and air resistance is neglected, how high will it rise and what is its velocity at its high point?

▌ At 45°, $v_H = v_V = (60)(0.7071) = 42.43$ ft/s; $Wz = mv^2/2 = (W/32.2)(42.43)^2/2$, $z = 28.0$ ft. At its high point, $v_V = 0$ and $v_H = 42.43$ ft/s.

8.60   Show that the work a liquid can do by virtue of its pressure is $\int p\, dV$, in which V is the volume of liquid displaced.

▌ Work $= \int F\, ds$. Since $F = pA$, work $= \int pA\, ds$. Since $A\, ds = dV$, work $= \int p\, dV$.

8.61   A fluid is flowing in a 6-in-diameter pipe at a pressure of 4.00 lb/in² with a velocity of 8.00 ft/s. As shown in Fig. 8-22, the elevation of the center of the pipe above a given datum is 10.0 ft. Find the total energy head above the given datum if the fluid is (a) water, (b) oil with a specific gravity of 0.82, and (c) gas with a specific weight of 0.042 lb/ft³.

▌ $$H = z + v^2/2g + p/\gamma$$

(a) $$H = 10.0 + 8.00^2/[(2)(32.2)] + (4.00)(144)/62.4 = 20.22 \text{ ft}$$

(b) $$H = 10.0 + 8.00^2/[(2)(32.2)] + (4.00)(144)/[(0.82)(62.4)] = 22.25 \text{ ft}$$

(c) $$H = 10.0 + 8.00^2/[(2)(32.2)] + (4.00)(144)/(0.042) = 13\,725 \text{ ft}$$

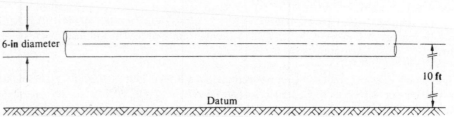

6-in diameter

10 ft

Datum

Fig. 8-22

8.62   A 100-mm-diameter suction pipe leading to a pump, as shown in Fig. 8-23, carries a discharge of 0.0300 m³/s of oil (s.g. = 0.85). If the pressure at point A in the suction pipe is a vacuum of 180 mmHg, find the total energy head at point A with respect to a datum at the pump.

▌ $v = Q/A = 0.0300/[(\pi)(\frac{100}{1000})^2/4] = 3.820$ m/s $p = \gamma h = [(13.6)(9.79)](-\frac{180}{1000}) = -23.97$ kN/m²

$H = z + v^2/2g + p/\gamma = -1.200 + 3.820^2/[(2)(9.807)] + (-23.97)/[(0.85)(9.79)] = -3.337$ m

8.63   Figure 8-24 shows a pump drawing water from a reservoir and discharging it into the air at point B. The pressure at point A in the suction pipe is a vacuum of 10 in mercury, and the discharge is 3.00 ft³/s. Determine the total head at point A and at point B with respect to a datum at the base of the reservoir.

▌ $$H = z + v^2/2g + p/\gamma \qquad v_A = Q/A_A = 3.00/[(\pi)(\tfrac{10}{12})^2/4] = 5.50 \text{ ft/s}$$

$$H_A = 25 + 5.50^2/[(2)(32.2)] + [(13.6)(62.4)](-\tfrac{10}{12})/62.4 = 14.14 \text{ ft}$$

$v_B = Q/A_B = 3.00/[(\pi)(\tfrac{8}{12})^2/4] = 8.59 \text{ ft/s}$ $H_B = (25 + 15 + 40) + 8.59^2/[(2)(32.2)] + 0 = 81.15 \text{ ft}$

8.64   If the total available head of a stream flowing at a rate of 300 ft³/s is 25.0 ft, what is the theoretical horsepower available?

▌ $$P = Q\gamma H = (300)(62.4)(25.0) = 468\,000 \text{ ft} \cdot \text{lb/s} = 468\,000/550 = 851 \text{ hp}$$

8.65   A 150-mm-diameter jet of water is discharging from a nozzle into the air at a velocity of 36.0 m/s. Find the power in the jet with respect to a datum at the jet.

▌ $$Q = Av = [(\pi)(\tfrac{150}{1000})^2/4](36.0) = 0.6262 \text{ m}^3/\text{s}$$

$$H = z + v^2/2g + p/\gamma = 0 + 36.0^2/[(2)(9.807)] + 0 = 66.08 \text{ m}$$

$$P = Q\gamma H = (0.6362)(9.79)(66.08) = 412 \text{ kN} \cdot \text{m/s} \quad \text{or} \quad 412 \text{ kW}$$

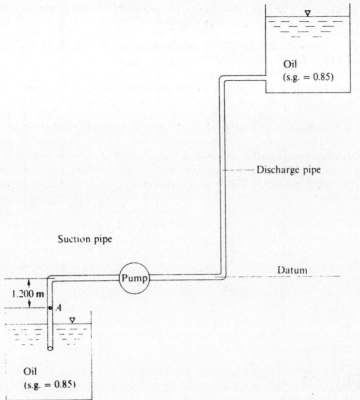

Fig. 8-23

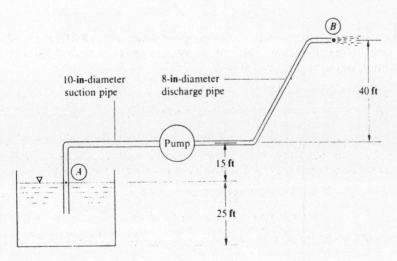

Fig. 8-24

8.66 Oil (s.g. = 0.84) is flowing in a pipe under the conditions shown in Fig. 8-25. If the total head loss (h_L) from point 1 to point 2 is 3.0 ft, find the pressure at point 2.

$$p_1/\gamma + v_1^2/2g + z_1 = p_2/\gamma + v_2^2/2g + z_2 + h_L$$

$$v_1 = Q/A_1 = 2.08/[(\pi)(\tfrac{6}{12})^2/4] = 10.59 \text{ ft/s} \qquad v_2 = Q/A_2 = 2.08/[(\pi)(\tfrac{9}{12})^2/4] = 4.71 \text{ ft/s}$$

$$(65)(144)/[(0.84)(62.4)] + 10.59^2/[(2)(32.2)] + 10.70 = p_2/\gamma + 4.71^2/[(2)(32.2)] + 4.00 + 3.00$$

$$p_2/\gamma = 183.67 \text{ ft} \qquad p_2 = [(0.84)(62.4)](183.67) = 9627 \text{ lb/ft}^2 \quad \text{or} \quad 66.9 \text{ lb/in}^2$$

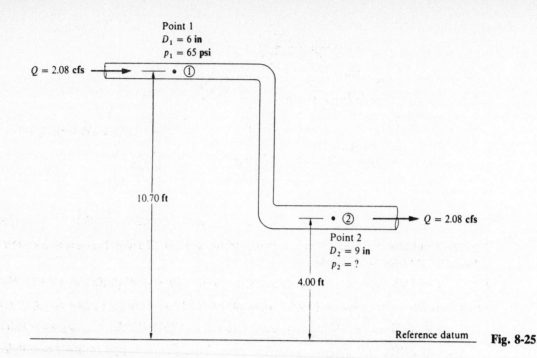

Fig. 8-25

8.67 An 8-in-diameter horizontal pipe is attached to a reservoir, as shown in Fig. 8-26. If the total head loss between the water surface in the reservoir and the water jet at the end of the pipe is 6.0 ft, what are the velocity and flow rate of the water being discharged from the pipe?

❚ $p_1/\gamma + v_1^2/2g + z_1 = p_2/\gamma + v_2^2/2g + z_2 + h_L$ $0 + 0 + 15 = 0 + v_2^2/[(2)(32.2)] + 0 + 6.0$

$v_2 = 24.1 \text{ ft/s}$ $Q = A_2 v_2 = [(\pi)(\tfrac{8}{12})^2/4](24.1) = 8.41 \text{ ft}^3/\text{s}$

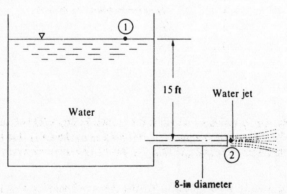

Fig. 8-26

8.68 A 50-mm-diameter siphon is drawing oil (s.g. = 0.82) from an oil reservoir, as shown in Fig. 8-27. If the head loss from point 1 to point 2 is 1.50 m and from point 2 to point 3 is 2.40 m, find the discharge of oil from the siphon and the oil pressure at point 2.

❚ $p_1/\gamma + v_1^2/2g + z_1 = p_3/\gamma + v_3^2/2g + z_3 + h_L$ $0 + 0 + 5.00 = 0 + v_3^2/[(2)(9.807)] + 0 + 3.90$

$v_3 = 4.645 \text{ m/s}$ $Q = A_3 v_3 = [(\pi)(\tfrac{50}{1000})^2/4](4.645) = 0.00912 \text{ m}^3/\text{s}$

$p_1/\gamma + v_1^2/2g + z_1 = p_2/\gamma + v_2^2/2g + z_2 + h_L$ $0 + 0 + 5.00 = p_2/\gamma + 4.645^2/[(2)(9.807)] + 7.00 + 1.50$

$p_2/\gamma = -4.60 \text{ m}$ $p_2 = [(0.82)(9.79)](-4.60) = -36.9 \text{ kN/m}^2$ or -36.9 kPa

8.69 Figure 8-28 shows a siphon discharging oil (s.g. = 0.84) from a reservoir into open air. If the velocity of flow in the pipe is v, the head loss from point 1 to point 2 is $2.0v^2/2g$, and the head loss from point 2 to point 3 is

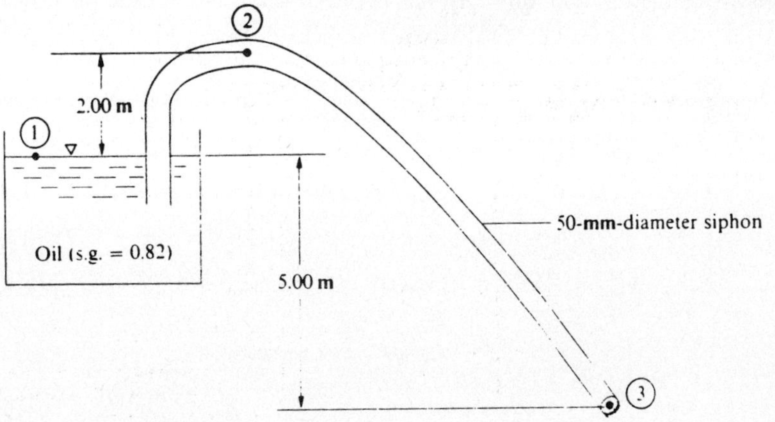

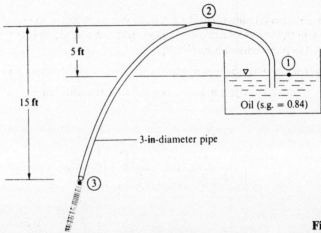

Fig. 8-27

$3.0v^2/2g$, determine the volume flow rate in the siphon pipe and the absolute pressure at point 2. Assume an atmospheric pressure of 14.70 psia.

$\blacksquare$ $\quad p_1/\gamma + v_1^2/2g + z_1 = p_3/\gamma + v_3^2/2g + z_3 + h_L \qquad 0 + 0 + 10 = 0 + v_3^2/[(2)(32.2)] + 0 + 5\{v_3^2/[(2)(32.2)]\}$

$v_3 = 10.36 \text{ ft/s} \qquad Q = A_3 v_3 = [(\pi)(\tfrac{3}{12})^2/4](10.36) = 0.509 \text{ ft}^3/\text{s} \qquad p_1/\gamma + v_1^2/2g + z_1 = p_2/\gamma + v_2^2/2g + z_2 + h_L$

$\qquad 0 + 0 + 10 = p_2/\gamma + 10.36^2/[(2)(32.2)] + 15 + 2\{10.36^2/[(2)(32.2)]\} \qquad p_2/\gamma = -10.0 \text{ ft of oil}$

$\qquad p_2 = [(0.84)(62.4)](-10.0) = -524 \text{ lb/ft}^2 \quad \text{or} \quad -3.64 \text{ lb/in}^2 \qquad p_2 = 14.70 - 3.64 = 11.06 \text{ lb/in}^2 \text{ abs}$

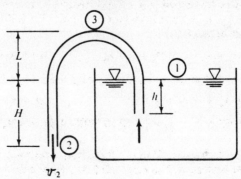

Fig. 8-28

8.70 Once it has been started by sufficient suction, the siphon in Fig. 8-29 will run continuously as long as reservoir fluid is available. Using Bernoulli's equation with no losses, show (**a**) that the exit velocity v_2 depends only upon gravity and the distance H and (**b**) that the lowest (vacuum) pressure occurs at point 3 and depends on the distance $L + H$.

$\blacksquare$ $\quad p_1/\gamma + v_1^2/2g + z_1 = p_2/\gamma + v_2^2/2g + z_2 + h_L, \; 0 + 0 + z_1 = 0 + v_2^2/2g + z_2 + 0, \; v_2 = \sqrt{2g(z_1 - z_2)} = \sqrt{2gH}.$
For any point B in the tube, $p_B/\gamma + v_B^2/2g + z_B = p_2/\gamma + v_2^2/2g + z_2 + h_L$. Since $v_B = v_2$ and $p_2 = p_{\text{atm}}$, $p_B = p_{\text{atm}} - \gamma(z_B - z_2)$. The lowest pressure occurs at the highest z_B, or $p_{\min} = p_3 = p_{\text{atm}} - \gamma(L + H)$.

Fig. 8-29

8.71 The siphon of Fig. 8-30 is filled with water and discharging at 150 L/s. Find the losses from point 1 to point 3 in terms of velocity head $v^2/2g$. Find the pressure at point 2 if two-thirds of the losses occur between points 1 and 2.

$$p_1/\gamma + v_1^2/2g + z_1 = p_3/\gamma + v_3^2/2g + z_3 + h_L \qquad 0 + 0 + 1.5 = 0 + v_3^2/2g + 0 + (K)(v_3^2/2g)$$

$$v_3 = Q/A_3 = (\tfrac{150}{1000})/[(\pi)(\tfrac{200}{1000})^2/4] = 4.775 \text{ m/s} \qquad 1.5 = 4.775^2/[(2)(9.807)] + K\{4.775^2/[(2)(9.807)]\}$$

$$K = 0.2904 \qquad h_L = (0.2904)\{4.775^2/[(2)(9.807)]\} = 0.338 \text{ m}$$

$$p_1/\gamma + v_1^2/2g + z_1 = p_2/\gamma + v_2^2/2g + z_2 + h_L \qquad 0 + 0 + 0 = p_2/\gamma + 4.775^2/[(2)(9.807)] + (\tfrac{2}{3})(0.338)$$

$$p_2/\gamma = -3.388 \text{ m of water} \qquad p_2 = (-3.388)(9.79) = -33.2 \text{ kN/m}^2$$

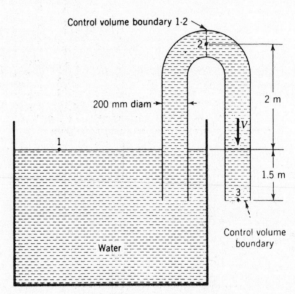

Fig. 8-30

8.72 For the water shooting out of the pipe and nozzle under the conditions shown in Fig. 8-31, find the height above the nozzle to which the water jet will "shoot" (i.e., distance h in Fig. 8-31). Assume negligible head loss.

$$p_A/\gamma + v_A^2/2g + z_A = p_{top}/\gamma + v_{top}^2/2g + z_{top} + h_L$$

$$55.0/9.79 + v_A^2/2g + 0 = 0 + 0 + (1.00 + h) + 0 \qquad h = 4.518 + v_A^2/2g$$

$$p_A/\gamma + v_A^2/2g + z_A = p_{nozzle}/\gamma + v_{nozzle}^2/2g + z_{nozzle} + h_L \qquad 55.0/9.79 + v_A^2/2g + 0 = 0 + v_{nozzle}^2/2g + 1.100 + 0$$

$$A_A v_A = A_{nozzle} v_{nozzle} \qquad [(\pi)(\tfrac{200}{1000})^2/4]v_A = [(\pi)(\tfrac{100}{1000})^2/4]v_{nozzle} \qquad v_{nozzle} = 4.00 v_A$$

$$55.0/9.79 + v_A^2/[(2)(9.807)] + 0 = 0 + (4.00 v_A)^2/[(2)(9.807)] + 1.100 + 0$$

$$v_A = 2.431 \text{ m/s} \qquad h = 4.518 + 2.431^2/[(2)(9.807)] = 4.82 \text{ m}$$

8.73 Water flows from section 1 to section 2 in the pipe shown in Fig. 8-32. Determine the velocity of flow and the fluid pressure at section 2. Assume the total head loss from section 1 to section 2 is 3.00 m.

$$Q = A_1 v_1 = A_2 v_2 \qquad [(\pi)(\tfrac{100}{1000})^2/4](2.0) = [(\pi)(\tfrac{50}{1000})^2/4](v_2) \qquad v_2 = 8.00 \text{ m/s}$$

$$p_1/\gamma + v_1^2/2g + z_1 = p_2/\gamma + v_2^2/2g + z_2 + h_L$$

$$300/9.79 + 2.0^2/[(2)(9.807)] + 2 = p_2/9.79 + 8.00^2/[(2)(9.807)] + 0 + 3.00 \qquad p_2 = 260 \text{ kPa}$$

8.74 A nozzle is attached to a pipe as shown in Fig. 8-33. The inside diameter of the pipe is 100 mm, while the water jet exiting from the nozzle has a diameter of 50 mm. If the pressure at section 1 is 500 kPa, determine the water jet's velocity. Assume head loss in the jet is negligible.

$$Q = A_1 v_1 = A_2 v_2 \qquad [(\pi)(\tfrac{100}{1000})^2/4](v_1) = [(\pi)(\tfrac{50}{1000})^2/4](v_2) \qquad v_1 = 0.250 v_2$$

$$p_1/\gamma + v_1^2/2g + z_1 = p_2/\gamma + v_2^2/2g + z_2 + h_L$$

$$500/9.79 + (0.250 v_2)^2/[(2)(9.807)] + 0 = 0 + v_2^2/[(2)(9.807)] + 0 + 0 \qquad v_2 = 32.7 \text{ m/s}$$

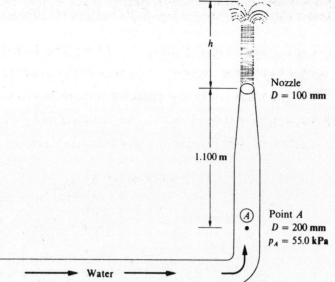

Fig. 8-31

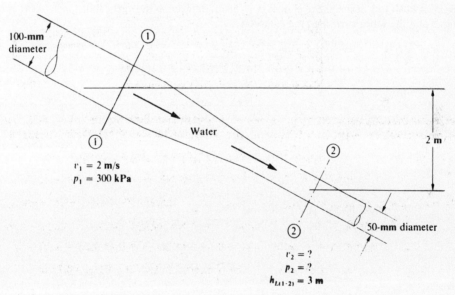

Fig. 8-32

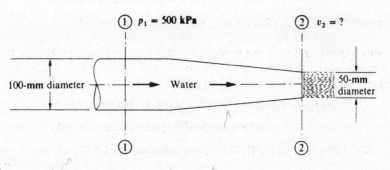

Fig. 8-33

8.75 Oil flows from a tank through 500 ft of 6-in-diameter pipe and then discharges into the air, as shown in Fig. 8-34. If the head loss from point 1 to point 2 is 1.95 ft of oil, determine the pressure needed at point 1 to cause 0.60 ft^3 of oil to flow.

$$v_2 = Q/A = 0.60/[(\pi)(\tfrac{6}{12})^2/4] = 3.06 \text{ ft/s} \qquad p_1/\gamma + v_1^2/2g + z_1 = p_2/\gamma + v_2^2/2g + z_2 + h_L$$

$$p_1/\gamma + 0 + 80 = 0 + 3.06^2/[(2)(32.2)] + 100 + 1.95 \qquad p_1/\gamma = 22.10 \text{ ft of oil}$$

$$p_1 = [(0.84)(62.4)](22.10)/144 = 8.04 \text{ lb/in}^2$$

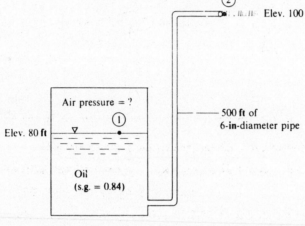

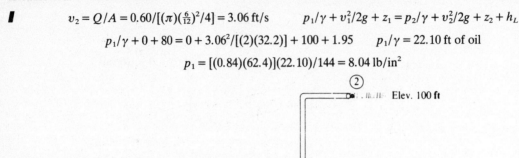

Fig. 8-34

8.76 Water is to be delivered from a reservoir through a pipe to a lower level and discharged into the air, as shown in Fig. 8-35. If head loss in the entire system is 11.58 m, determine the vertical distance between the point of water discharge and the water surface in the reservoir.

$$v_2 = Q/A_2 = 0.00631/[(\pi)(\tfrac{50}{1000})^2/4] = 3.214 \text{ m/s} \qquad p_1/\gamma + v_1^2/2g + z_1 = p_2/\gamma + v_2^2/2g + z_2 + h_L$$

$$0 + 0 + z_1 = 0 + 3.214^2/[(2)(9.807)] + 0 + 11.58 \qquad z_1 = 12.11 \text{ m}$$

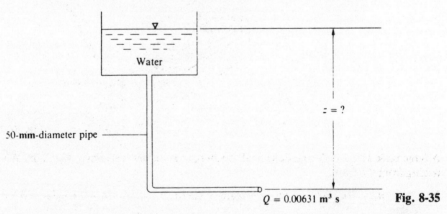

Fig. 8-35

8.77 Determine the velocity and pressure at section 2 and section 3 if water flows steadily through the pipe system shown in Fig. 8-36. Assume a head loss of 6.0 ft from section 1 to section 2 and of 15.0 ft from section 2 to section 3.

$$A_1v_1 = A_2v_2 \qquad [(\pi)(\tfrac{16}{12})^2/4](5.0) = [(\pi)(\tfrac{8}{12})^2/4](v_2) \qquad v_2 = 20.0 \text{ ft/s}$$

$$p_1/\gamma + v_1^2/2g + z_1 = p_2/\gamma + v_2^2/2g + z_2 + h_L$$

$$(25)(144)/62.4 + 5.0^2/[(2)(32.2)] + 20 = (p_2)(144)/62.4 + 20.0^2/[(2)(32.2)] + 15 + 6.0$$

$$p_2 = 22.0 \text{ lb/in}^2 \qquad A_1v_1 = A_3v_3$$

$$[(\pi)(\tfrac{16}{12})^2/4](5.0) = [(\pi)(\tfrac{12}{12})^2/4](v_3) \qquad v_3 = 8.99 \text{ ft/s}$$

$$p_1/\gamma + v_1^2/2g + z_1 = p_3/\gamma + v_3^2/2g + z_3 + h_L$$

$$(25)(144)/62.4 + 5.0^2/[(2)(32.2)] + 20 = (p_3)(144)/62.4 + 8.89^2/[(2)(32.2)] + 10 + (15.0 + 6.0)$$

$$p_3 = 19.9 \text{ lb/in}^2$$

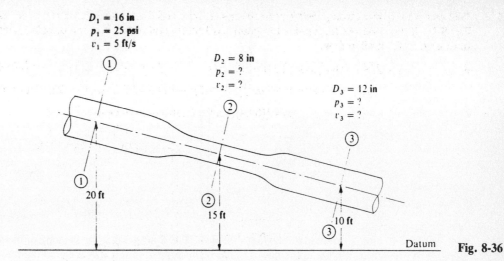

$D_1 = 16$ in
$p_1 = 25$ psi
$v_1 = 5$ ft/s

$D_2 = 8$ in
$p_2 = ?$
$v_2 = ?$

$D_3 = 12$ in
$p_3 = ?$
$v_3 = ?$

20 ft

15 ft

10 ft

Datum **Fig. 8-36**

8.78 Compute the ideal flow rate through the pipe system shown in Fig. 8-37.

$p_1/\gamma + v_1^2/2g + z_1 = p_2/\gamma + v_2^2/2g + z_2 + h_L$ $p_1/\gamma + v_1^2/[(2)(9.807)] + 0.6\sin 30° = p_2/\gamma + 0 + 0 + 0$

$v_1^2/[(2)(9.807)] = p_2/\gamma - p_1/\gamma - 0.300$

From the manometer reading, $p_1 - 9.79(1.2\sin 60°) = p_2$, $p_1 - p_2 = 10.17$ kN/m²; $v_1^2/[(2)(9.807)] = 10.17/9.79 - 0.300$, $v_1 = 3.807$ m/s; $Q = A_1 v_1 = [(\pi)(\frac{200}{1000})^2/4](3.807) = 0.120$ m³/s.

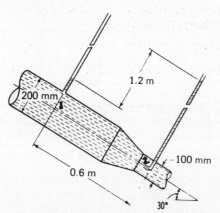

1.2 m

200 mm

100 mm

0.6 m

30° **Fig. 8-37**

8.79 A large tank with a well-rounded, small opening as an outlet is shown in Fig. 8-38. What is the velocity of a jet issuing from the tank?

$p_1/\gamma + v_1^2/2g + z_1 = p_2/\gamma + v_2^2/2g + z_2 + h_L$ $0 + 0 + h = 0 + v_2^2/2g + 0 + 0$ $v_2 = \sqrt{2gh}$

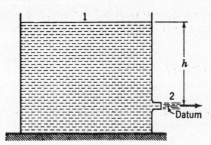

1

h

2

Datum

Fig. 8-38

8.80 If friction is neglected, what is the velocity of water issuing from the tank as a free jet in Fig. 8-39? What is the discharge rate?

$p_1/\gamma + v_1^2/2g + z_1 = p_2/\gamma + v_2^2/2g + z_2 + h_L$ $0 + 0 + (2.5 + 0.6 + 1.5) = 0 + v_2^2/[(2)(9.807)] + 0 + 0$

$v_2 = 9.50$ m/s $Q = Av = [(\pi)(\frac{150}{1000})^2/4](9.50) = 0.168$ m³/s

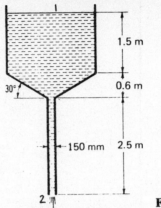

Fig. 8-39

8.81 One end of a U-tube is oriented directly into the flow (Fig. 8-40) so that the velocity of the stream is zero at this point. The pressure at a point in the flow that has been stopped in this way is called the *stagnation pressure*. The other end of the U-tube measures the "undisturbed" pressure at a section in the flow. Neglecting friction, determine the volume flow of water in the pipe.

$\blacksquare$ $p_1/\gamma + v_1^2/2g + z_1 = p_2/\gamma + v_2^2/2g + z_2 + h_L$, $p_1/\gamma + 0 + 0 = p_2/\gamma + v_2^2/2g + 0 + 0$, $v_2^2/2g = p_1/\gamma - p_2/\gamma$. From the manometer reading, $p_1 + (62.4)(\frac{2}{12}) - [(13.6)(62.4)](\frac{2}{12}) = p_2$, $p_1 - p_2 = 131.0$ lb/ft²; $v_2^2/[(2)(32.2)] = 131.0/62.4$, $v_2 = 11.63$ ft/s; $Q = Av = [(\pi)(\frac{8}{12})^2/4](11.63) = 4.06$ ft³/s.

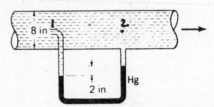

Fig. 8-40

8.82 A cylindrical tank contains air, oil, and water, as shown in Fig. 8-41. On the oil a pressure p of 5 lb/in² gage is maintained. What is the velocity of the water leaving if we neglect both friction everywhere and the kinetic energy of the fluid above elevation A? The jet of water leaving has a diameter of 1 ft.

$\blacksquare$ $\qquad p_1/\gamma + v_1^2/2g + z_1 = p_2/\gamma + v_2^2/2g + z_2 + h_L \qquad p_1 = (5)(144) + [(0.8)(62.4)](3) = 869.8$ lb/ft²

$\qquad 869.8/62.4 + 0 + 10 = 0 + v_2^2/[(2)(32.2)] + 0 + 0 \qquad v_2 = 39.3$ ft/s

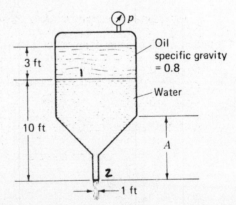

Fig. 8-41

8.83 A large tank contains compressed air, gasoline at specific gravity 0.68, light oil at specific gravity 0.80, and water, as shown in Fig. 8-42. The pressure p of the air is 150 kPa gage. If we neglect friction, what is the mass flow of oil from a 20-mm-diameter jet?

$\blacksquare$ $\qquad p_1/\gamma + v_1^2/2g + z_1 = p_2/\gamma + v_2^2/2g + z_2 + h_L \qquad p_1 = 150 + [(0.68)(9.79)](2) = 163.3$ kN/m²

$\qquad 163.3/[(0.80)(9.79)] + 0 + 0 = 0 + v_2^2/[(2)(9.807)] + 3 + 0 \qquad v_2 = 18.71$ m/s

$\qquad M = \rho Av = [(0.80)(1000)][(\pi)(\frac{20}{1000})^2/4](18.71) = 4.70$ kg/s

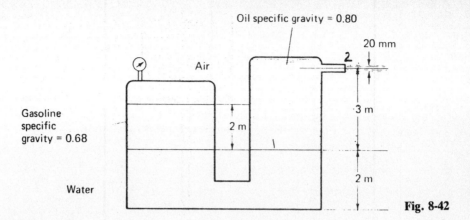

Fig. 8-42

8.84 A flow nozzle is a device inserted into a pipe as shown in Fig. 8-43. If A_2 is the exit area of the flow nozzle, show that for incompressible flow we get for Q,

$$Q = C_d \left[\frac{A_2}{\sqrt{1 - (A_2/A_1)^2}} \sqrt{2g\left(\frac{p_1 - p_2}{\gamma}\right)} \right]$$

where C_d is the *coefficient of discharge*, which takes into account frictional effects and is determined experimentally.

∎ $\dfrac{p_A}{\gamma} + \dfrac{v_A^2}{2g} + z_A = \dfrac{p_B}{\gamma} + \dfrac{v_B^2}{2g} + z_B + h_L$ $\quad \dfrac{p_A}{\gamma} + \dfrac{v_1^2}{2g} + 0 = \dfrac{p_B}{\gamma} + \dfrac{v_2^2}{2g} + 0 + 0$ $\quad v_1^2 = v_2^2 + 2g\left(\dfrac{p_B - p_A}{\gamma}\right)$

But $p_B - p_A = p_2 - p_1$ and $v_2 = (v_1)(A_1/A_2)$; hence,

$$v_1^2 = \left[v_1\left(\frac{A_1}{A_2}\right) \right]^2 + 2g\left(\frac{p_2 - p_1}{\gamma}\right) \qquad v_1 = \sqrt{\frac{1}{1 - (A_1/A_2)^2}} \sqrt{2g\left(\frac{p_2 - p_1}{\gamma}\right)}$$

$$Q = Av = C_d A_1 \sqrt{\frac{1}{1 - (A_1/A_2)^2}} \sqrt{2g\left(\frac{p_2 - p_1}{\gamma}\right)} = C_d \sqrt{\frac{A_1^2}{1 - (A_1/A_2)^2}} \sqrt{2g\left(\frac{p_2 - p_1}{\gamma}\right)}$$

Multiplying by A_2^2/A_1^2 in the numerator and denominator of the radical gives

$$Q = C_d \left[\frac{A_2}{\sqrt{1 - (A_2/A_1)^2}} \sqrt{2g\left(\frac{p_1 - p_2}{\gamma}\right)} \right]$$

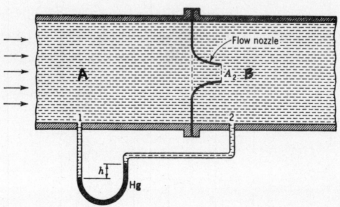

Fig. 8-43

8.85 In Prob. 8.84, express Q in terms of h, the height of the mercury column (Fig. 8-43), and the diameters of the pipe and flow nozzle.

∎ From Prob. 8.84,

$$Q = C_d \left[\frac{A_2}{\sqrt{1 - (A_2/A_1)^2}} \sqrt{2g\left(\frac{p_1 - p_2}{\gamma}\right)} \right]$$

From the manometer, $p_1 - p_2 = (\gamma_{Hg} - \gamma_{H_2O})(h)$.

$$Q = C_d \left[\frac{\pi d_2^2/4}{\sqrt{1 - (d_2^2/d_1^2)^2}} \sqrt{\frac{2g(\gamma_{Hg} - \gamma_{H_2O})(h)}{\gamma_{H_2O}}} \right]$$

8.86   A hump of height δ is placed on the channel bed in a rectangular channel of uniform width over its entire width (see Fig. 8-44). The free surface has a dip d as shown. If we neglect friction, we can consider that we have one-dimensional flow. Compute the flow q for the channel per unit width. This system is called a *venturi flume*.

▌

$$p_1/\gamma + v_1^2/2g + z_1 = p_2/\gamma + v_2^2/2g + z_2 + h_L \qquad 0 + v_1^2/2g + h = 0 + v_2^2/2g + (h-d) + 0$$

$$v_1^2 = v_2^2 - 2gd \qquad A_1 v_1 = A_2 v_2 \qquad [(1)(h)](v_1) = [(1)(h - d - \delta)](v_2)$$

$$v_2 = \left(\frac{h}{h-d-\delta} \right)(v_1) \qquad v_1^2 = (v_1)^2 \left(\frac{h}{h-d-\delta} \right)^2 - 2gd \qquad v_1^2 \left[1 - \left(\frac{h}{h-d-\delta} \right)^2 \right] = -2gd$$

$$v_1 = \sqrt{\frac{-2gd}{1 - [h/(h-d-\delta)]^2}} \qquad q = hv_1 = \sqrt{\frac{-2gd}{1/h^2 - [1/(h-d-\delta)]^2}}$$

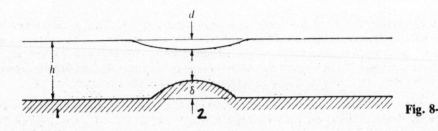

Fig. 8-44

8.87   Water flows steadily up the vertical pipe in Fig. 8-45 and enters the annular region between the circular plates as shown. It then moves out radially, issuing out as a free sheet of water. If we neglect friction entirely, what is the flow of water through the pipe if the pressure at A is 69 kPa gage?

▌

$$p_A/\gamma + v_A^2/2g + z_A = p_E/\gamma + v_E^2/2g + z_E + h_L$$

$$69/9.79 + v_A^2/[(2)(9.807)] + 0 = 0 + v_E^2/[(2)(9.807)] + 1.5 + 0$$

$$A_A v_A = A_E v_E \qquad [(\pi)(\tfrac{200}{1000})^2/4](v_A) = [(\tfrac{13}{1000})(\pi)(0.3 + 0.3)](v_E) \qquad v_A = 0.780 v_E$$

$$69/9.79 + (0.780 v_E)^2/[(2)(9.807)] + 0 = 0 + v_E^2/[(2)(9.807)] + 1.5 + 0 \qquad v_E = 16.67 \text{ m/s}$$

$$Q = A_E v_E = [(\tfrac{13}{1000})(\pi)(0.3 + 0.3)](16.67) = 0.408 \text{ m}^3/\text{s}$$

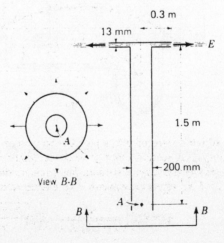

Fig. 8-45

8.88 If the velocity at point A in Fig. 8-46 is 18 m/s, what is the pressure at point B if we neglect friction?

$$p_B/\gamma + v_B^2/2g + z_B = p_A/\gamma + v_A^2/2g + z_A + h_L$$

$$p_B/9.79 + v_B^2/[(2)(9.807)] + 0 = 0 + 18^2/[(2)(9.807)] + (0.5 + 21) + 0$$

$$p_B = -0.4991v_B^2 + 372.2 \qquad p_C/\gamma + v_C^2/2g + z_C = p_A/\gamma + v_A^2/2g + z_A + h_L$$

$$0 + v_C^2/[(2)(9.807)] + 0 = 0 + 18^2/[(2)(9.807)] + 21 + 0 \qquad v_C = 27.13 \text{ m/s} \qquad A_B v_B = A_C v_C$$

$$[(\pi)(\tfrac{200}{1000})^2/4](v_B) = [(\pi)(\tfrac{75}{1000})^2/4](27.13) \qquad v_B = 3.815 \text{ m/s} \qquad p_B = (-0.4991)(3.815)^2 + 372.2 = 365 \text{ kN/m}^2$$

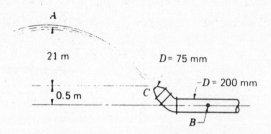

21 m

D = 75 mm

0.5 m

C

$-D$ = 200 mm

B

Fig. 8-46

8.89 A rocket-powered test sled slides over rails, as shown in Fig. 8-47. The test sled is used for experimentation on the ability of human beings to undergo large persistent accelerations. To brake the sled from high speeds, small scoops are lowered to deflect water from a stationary tank of water placed near the end of the run. If the sled is moving at a speed of 100 km/h at the instant of interest, compute h of the deflected stream of water as seen from the sled. Assume no loss in speed of the water relative to the scoop.

▮ Use Bernoulli's equation between the scoop (point 1) and the highest point in the trajectory (point 2); take the datum at the scoop. $p_1/\gamma + v_1^2/2g + z_1 = p_2/\gamma + v_2^2/2g + z_2 + h_L$, $0 + v_1^2/[(2)(9.807)] + 0 = 0 + (v_1 \cos 20°)^2/[(2)(9.807)] + (h - \tfrac{200}{1000}) + 0$. From the given information, $v_1 = (100)(1000)/3600 = 27.78$ m/s, $27.78^2/[(2)(9.807)] = (27.78 \cos 20°)^2/[(2)(9.807)] + (h - \tfrac{200}{1000})$, $h = 4.80$ m.

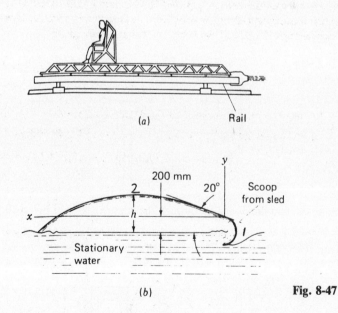

(a)

Rail

200 mm

y

20°

Scoop from sled

2

x

h

Stationary water

1

(b)

Fig. 8-47

8.90 A *venturi meter* is a device which is inserted into a pipe line to measure incompressible flow rates. As shown in Fig. 8-48, it consists of a convergent section which reduces the diameter to between one-half and one-fourth the pipe diameter. This is followed by a divergent section. The pressure difference between the position just before the venturi and at the throat of the venturi is measured by a differential manometer as shown. Show that

$$Q = C_d \left[\frac{A_2}{\sqrt{1 - (A_2/A_1)^2}} \sqrt{2g\left(\frac{p_1 - p_2}{\gamma}\right)} \right]$$

where C_d is the *coefficient of discharge*, which takes into account frictional effects and is determined experimentally.

$$\frac{p_1}{\gamma} + \frac{v_1^2}{2g} + z_1 = \frac{p_2}{\gamma} + \frac{v_2^2}{2g} + z_2 + h_L \qquad \frac{p_1}{\gamma} + \frac{v_1^2}{2g} + 0 = \frac{p_2}{\gamma} + \frac{v_2^2}{2g} + 0 + 0 \qquad v_1^2 - v_2^2 = 2g\left(\frac{p_2 - p_1}{\gamma}\right)$$

$$A_1 v_1 = A_2 v_2 \qquad v_1 = (v_2)\left(\frac{A_2}{A_1}\right) \qquad \left[(v_2)\left(\frac{A_2}{A_1}\right)\right]^2 - v_2^2 = 2g\left(\frac{p_2 - p_1}{\gamma}\right) \qquad \left[\left(\frac{A_2}{A_1}\right)^2 - 1\right](v_2^2) = 2g\left(\frac{p_2 - p_1}{\gamma}\right)$$

$$v_2 = \sqrt{\frac{1}{1 - (A_2/A_1)^2}} \sqrt{2g\left(\frac{p_1 - p_2}{\gamma}\right)}$$

$$Q = Av = C_d A_2 \sqrt{\frac{1}{1 - (A_2/A_1)^2}} \sqrt{2g\left(\frac{p_1 - p_2}{\gamma}\right)} = C_d\left[\frac{A_2}{\sqrt{1 - (A_2/A_1)^2}} \sqrt{2g\left(\frac{p_1 - p_2}{\gamma}\right)}\right]$$

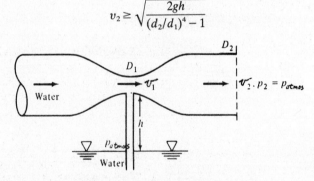

Venturi meter   **Fig. 8-48**

8.91 A necked-down, or venturi, section of a pipe flow develops a low pressure which can be used to aspirate fluid upward from a reservoir, as shown in Fig. 8-49. Using Bernoulli's equation with no losses, derive an expression for the exit velocity v_2 that is just sufficient to cause the reservoir fluid to rise in the tube up to section 1.

$$p_1/\gamma + v_1^2/2g + z_1 = p_2/\gamma + v_2^2/2g + z_2 + h_L \qquad p_1/\gamma + v_1^2/2g + 0 = p_{atm}/\gamma + v_2^2/2g + 0 + 0$$

$$A_1 v_1 = A_2 v_2 \qquad (\pi d_1^2/4)(v_1) = (\pi d_2^2/4)(v_2) \qquad v_1 = (v_2)(d_2/d_1)^2$$

$$p_1/\gamma + [(v_2)(d_2/d_1)^2]^2/2g = p_{atm}/\gamma + v_2^2/2g \qquad p_{atm} - p_1 = (\gamma/2g)(v_2^2)[(d_2/d_1)^4 - 1]$$

For fluid to rise in the tube, $p_{atm} - p_1 \geq \gamma h$; hence, $(\gamma/2g)(v_2^2)[(d_2/d_1)^4 - 1] \geq \gamma h$,

$$v_2 \geq \sqrt{\frac{2gh}{(d_2/d_1)^4 - 1}}$$

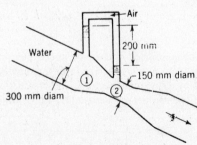

Fig. 8-49

8.92 Neglecting losses, find the discharge through the venturi meter of Fig. 8-50.

$p_1/\gamma + v_1^2/2g + z_1 = p_2/\gamma + v_2^2/2g + z_2 + h_L$. From the manometer, $p_1/\gamma - (k + \frac{200}{1000}) + (z_1 - z_2 + k) = p_2/\gamma$; $A_1 v_1 = A_2 v_2$, $[(\pi)(\frac{300}{1000})^2/4](v_1) = [(\pi)(\frac{150}{1000})^2/4](v_2)$, $v_1 = 0.250 v_2$; $p_1/\gamma + (0.250 v_2)^2/[(2)(9.807)] + z_1 = [p_1/\gamma - (k + \frac{200}{1000}) + (z_1 - z_2 + k)] + v_2^2/[(2)(9.807)] + z_2 + 0$, $v_2 = 2.046$ m/s; $Q = A_2 v_2 = [(\pi)(\frac{150}{1000})^2/4](2.046) = 0.0362$ m³/s.

Fig. 8-50

8.93 With losses of $0.2v_1^2/2g$ between sections 1 and 2 of Fig. 8-50, calculate the flow in gallons per minute.

▌ From Prob. 8.92, $v_1 = 0.250v_2 = (0.250)(2.046) = 0.5112$ m/s; $p_1/\gamma + (0.250v_2)^2/[(2)(9.807)] + z_1 = [p_1/\gamma - (k + \frac{200}{1000}) + (z_1 - z_2 + k)] + v_2^2/[(2)(9.807)] + z_2 + 0$. For Prob. 8.93, add a term $0.2v_1^2/2g$ to the previous equation, giving $p_1/\gamma + (0.250v_2)^2/[(2)(9.807)] + z_1 = [p_1/\gamma - (k + \frac{200}{1000}) + (z_1 - z_2 + k)] + v_2^2/[(2)(9.807)] + z_2 + (0.2)\{0.5112^2/[(2)(9.807)]\}$, $v_2 = 2.032$ m/s; $Q = A_2v_2 = [(\pi)(\frac{150}{1000})^2/4](2.032) = 0.0359$ m³/s $= [0.0359/(0.3048)^3](7.48)(60) = 569$ gpm.

8.94 The device shown in Fig. 8-51 is used to determine the velocity of liquid at point 1. It is a tube with its lower end directed upstream and its other leg vertical and open to the atmosphere. The impact of liquid against opening 2 forces liquid to rise in the vertical leg to the height z above the free surface. Determine the velocity at 1.

▌ $p_1/\gamma + v_1^2/2g + z_1 = p_2/\gamma + v_2^2/2g + z_2 + h_L$ $k + v_1^2/2g + 0 = 0 + (k + \Delta z) + 0 + 0$ $v_1 = \sqrt{2g\,\Delta z}$

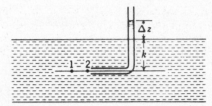

Fig. 8-51

8.95 The liquid in Fig. 8-52 is gasoline (s.g. = 0.85). The losses in the exit pipe equal $Kv^2/2g$, where $K = 6.0$. The tank reservoir is large. Compute the flow rate in cubic feet per minute.

▌
$$p_1/\gamma + v_1^2/2g + z_1 = p_2/\gamma + v_2^2/2g + z_2 + h_L$$

$$(20)(144)/[(0.85)(62.4)] + 0 + 5 = (14.7)(144)/[(0.85)(62.4)] + v_2^2/[(2)(32.2)] + 0 + (6.0)\{v_2^2/[(2)(32.2)]\}$$

$$v_2 = 13.36 \text{ ft/s} \qquad Q = Av = [(\pi)(\tfrac{1}{12})^2/4](13.36) = 0.07287 \text{ ft}^3/\text{s} \quad \text{or} \quad 4.37 \text{ ft}^3/\text{min}$$

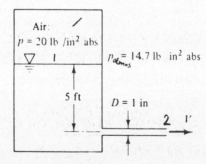

Fig. 8-52

8.96 The manometer fluid in Fig. 8-53 is mercury. Neglecting losses, calculate the flow rate in the tube if the flowing fluid is **(a)** water, **(b)** air. Use 60 °F as the fluid temperature.

▌ $p_1/\gamma + v_1^2/2g + z_1 = p_2/\gamma + v_2^2/2g + z_2 + h_L$, $p_1/\gamma + v_1^2/2g + 0 = p_2/\gamma + 0 + 0 + 0$, $p_2 - p_1 = (\gamma)(v_1^2/2g)$. From the manometer, $p_1 + (62.4)(y/12) + [(13.6)(62.4)](\frac{1}{12}) - (62.4)(\frac{1}{12}) - (62.4)(y/12) = p_2$, $p_2 - p_1 = 65.52$ lb/ft²; $(\gamma)(v_1^2/2g) = 65.52$.

(a) $(62.4)\{v_1^2/[(2)(32.2)]\} = 65.52$ $v_1 = 8.223$ ft/s $Q = A_1v_1 = [(\pi)(\frac{3}{12})^2/4](8.223) = 0.404$ ft³/s

(b) $(0.0763)\{v_1^2/[(2)(32.2)]\} = 65.52$ $v_1 = 235.2$ ft/s $Q = A_1v_1 = [(\pi)(\frac{3}{12})^2/4](235.2) = 11.5$ ft³/s

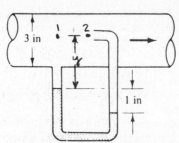

Fig. 8-53

8.97 In Fig. 8-54 the fluid flowing is air ($\gamma = 12\,\text{N/m}^3$), and the manometer fluid is Meriam red oil (s.g. = 0.827). Assuming no losses, compute the flow rate.

▌ $p_1/\gamma + v_1^2/2g + z_1 = p_2/\gamma + v_2^2/2g + z_2 + h_L$, $p_1/\gamma + 0 + 0 = p_2/\gamma + v_2^2/2g + 0 + 0$, $p_1 - p_2 = (\gamma)(v_2^2/2g)$. From the manometer, $p_1 + (\frac{12}{1000})(y+8)/100 - [(0.827)(9.79)](\frac{8}{100}) - (\frac{12}{1000})(y/100) = p_2$, $p_1 - p_2 = 0.6467\,\text{kN/m}^2$; $(\gamma)(v_2^2/2g) = 0.6467$, $(\frac{12}{1000})\{v_2^2/[(2)(9.807)]\} = 0.6467$, $v_2 = 32.51\,\text{m/s}$; $Q = A_2 v_2 = [(\pi)(\frac{6}{100})^2/4](32.51) = 0.0919\,\text{m}^3/\text{s}$.

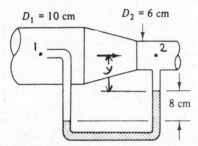

$D_1 = 10$ cm $\qquad D_2 = 6$ cm

Fig. 8-54

8.98 The flow from two reservoirs mixes together and flows through a common pipe. The elevations and pipe diameters are indicated in Fig. 8-55. Both reservoirs contain the same liquid and are open to the atmosphere. The common pipe empties to the atmosphere. Neglecting any frictional effects, find the flow rate through the common pipe.

▌
$$p_1/\gamma + v_1^2/2g + z_1 = p_2/\gamma + v_2^2/2g + z_2 + h_L \qquad p_{\text{atm}}/\gamma + 0 + h_1 = p_2/\gamma + v_2^2/2g + 0 + 0$$

$$P_{\text{atm}} + \gamma h_1 = p_2 + \gamma v_2^2/2g \tag{1}$$

$$p_3/\gamma + v_3^2/2g + z_3 = p_4/\gamma + v_4^2/2g + z_4 + h_L \qquad p_{\text{atm}}/\gamma + 0 + h_2 = p_4/\gamma + v_4^2/2g + 0 + 0$$

$$p_{\text{atm}} + \gamma h_2 = p_4 + \gamma v_4^2/2g \tag{2}$$

$$p_5/\gamma + v_5^2/2g + z_5 = p_6/\gamma + v_6^2/2g + z_6 + h_L \qquad p_5/\gamma + v_5^2/2g + 0 = p_{\text{atm}}/\gamma + v_6^2/2g + (-h_3) + 0$$

Since $v_5 = v_6$,

$$p_5 = p_{\text{atm}} - \gamma h_3 \tag{3}$$

Assume $p_5 = p_2 = p_4$. Substituting this common value of pressure back into Eqs. (1) and (2) and solving for the velocity in each branch, we get $v_2 = \sqrt{2g(h_1 + h_3)}$, $v_4 = \sqrt{2g(h_2 + h_3)}$ $Q = Av = (\pi d_1^2/4)\sqrt{2g(h_1 + h_3)} + (\pi d_2^2/4)\sqrt{2g(h_2 + h_3)} = (\pi/4)[d_1^2\sqrt{2g(h_1 + h_3)} + d_2^2\sqrt{2g(h_2 + h_3)}]$.

8.99 A steady jet of water comes from a hydrant and hits the ground some distance away, as shown in Fig. 8-56. If the water outlet is 1 m above the ground and the hydrant water pressure is 862 kPa, what distance from the hydrant does the jet hit the ground? Atmospheric pressure is 101 kPa.

▌ The magnitude of v_x can be obtained by noting that at the hydrant outlet the flow is entirely in the x direction, $v_x = V_2$. Applying the Bernoulli equation between the interior of the hydrant and the outlet gives $p_1 + \rho g y_1 + \frac{1}{2}\rho V_1^2 = p_2 + \rho g y_2 + \frac{1}{2}\rho V_2^2$. The pressure in the hydrant p_1 is given, and the outlet is open to the atmosphere, $p_2 = p_{\text{atm}}$. The elevation of points 1 and 2 is the same, $y_1 = y_2$. We assume that the outlet area is small enough compared with the hydrant cross-sectional area for the hydrant to be essentially a reservoir, $V_1^2 \ll V_2^2$. Neglecting V_1, we get $v_x = \sqrt{(2/\rho)(p_1 - p_{\text{atm}})}$. Since v_x is constant, it can be brought outside the integral for l, giving $l = v_x T$.

To find the time T required for a fluid particle to hit the ground, we apply the Bernoulli equation between point 1 and some arbitrary point on the jet having elevation y and velocity V: $p_1 + \rho g h = p_{\text{atm}} + \rho g y + \frac{1}{2}\rho V^2$. Now $V^2 = v_x^2 + v_y^2$. When we use the previously determined value of v_x and note that $v_y = dy/dt$, the Bernoulli equation becomes $(dy/dt)^2 = 2g(h - y)$. We take the square root (the negative root is the appropriate one since dy/dt must be negative): $dy/dt = -\sqrt{2g(h - y)}$. Then we separate variables and integrate between the limits of $y = h$, $t = 0$ and $y = 0$, $t = T$:

$$\int_h^0 \frac{dy}{\sqrt{h - y}} = -\sqrt{2g} \int_0^T dt$$

Integrating and solving for T gives $T = \sqrt{2h/g}$. The y component of the fluid motion is that of a body freely falling under the influence of gravity.

Finally, we substitute numerical values to get $v_x = \sqrt{\frac{2}{1000}[(862 - 101)(1000)]} = 39.0\,(\text{m} \cdot \text{N/kg})^{1/2}$, or 39.0 m/s; $T = \sqrt{(2)(1.0)/9.807} = 0.452\,\text{s}$; $L = vt = (39.0)(0.452) = 17.6\,\text{m}$.

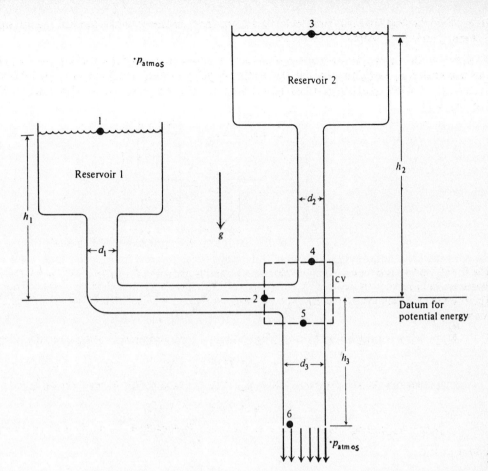

Fig. 8-55

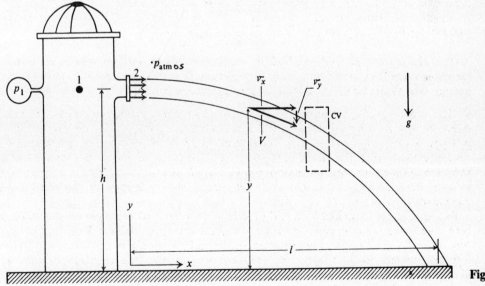

Fig. 8-56

8.100 For the water flow between two reservoirs in Fig. 8-57, the flow rate is found to be 0.016 m³/s. What is the head loss in the pipe? If atmospheric pressure is 100 kPa and the vapor pressure is 8 kPa, for what constriction diameter d will cavitation occur? Assume no additional losses due to changes in the constriction.

▌ $p_1/\gamma + v_1^2/2g + z_1 = p_2/\gamma + v_2^2/2g + z_2 + h_L$, $0 + 0 + 25 = 0 + 0 + 10 + h_L$, $h_L = 15$ m; $v_{\text{throat}} = Q/A_{\text{throat}} = 0.016/(\pi d^2/4)$. Assume a central constriction, with $\frac{15}{2}$, or 7.5-m head loss on each side. Apply Bernoulli's equation between point 1 and the constriction, with $p_1 = p_{\text{atm}} = 100$ kPa and $p_v = 8$ kPa at the constriction.

$100/9.79 + 0 + 25 = 8/9.79 + v_{max}^2/[(2)(9.807)] + 0 + 7.5$, $v_{max} = 23.0$ m/s; $23.0 = 0.016/(\pi d^2/4)$, $d = 0.0298$ m, or 2.98 cm.

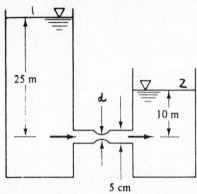

Fig. 8-57

8.101 The horizontal wye fitting in Fig. 8-58 splits Q_1 into two equal flow rates. At section 1, $Q_1 = 4$ ft³/s and $p_1 = 20$ psig. Neglecting losses, compute pressures p_2 and p_3.

$$p_1/\gamma + v_1^2/2g + z_1 = p_2/\gamma + v_2^2/2g + z_2 + h_L \qquad v_1 = Q_1/A_1 = 4/[(\pi)(\tfrac{6}{12})^2/4] = 20.37 \text{ ft/s}$$

$$v_2 = Q_2/A_2 = \tfrac{4}{2}/[(\pi)(\tfrac{3}{12})^2/4] = 40.74 \text{ ft/s}$$

$$(20)(144)/62.4 + 20.37^2/[(2)(32.2)] + 0 = (p_2)(144)/62.4 + 40.74^2/[(2)(32.2)] + 0 + 0$$

$p_2 = 11.6$ psig $\qquad p_1/\gamma + v_1^2/2g + z_1 = p_3/\gamma + v_3^2/2g + z_3 + h_L \qquad v_3 = Q_3/A_3 = \tfrac{4}{2}/[(\pi)(\tfrac{4}{12})^2/4] = 22.92$ ft/s

$$(20)(144)/62.4 + 20.37^2/[(2)(32.2)] + 0 = (p_3)(144)/62.4 + 22.92^2/[(2)(32.2)] + 0 + 0 \qquad p_3 = 19.3 \text{ psig}$$

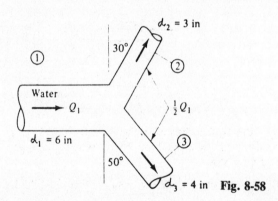

Fig. 8-58

8.102 A cylindrical tank of diameter d_0 contains liquid to an initial height h_0. At time $t = 0$ a small stopper of diameter d is removed from the bottom. Using Bernoulli's equation with no losses, derive a differential equation for the free-surface height h during draining and an expression for the time t_0 to drain the entire tank.

Letting point 1 be the liquid surface and point 2 at the exit, $p_1/\gamma + v_1^2/2g + z_1 = p_2/\gamma + v_2^2/2g + z_2 + h_L$,
$0 + v_1^2/2g + h = 0 + v_2^2/2g + 0 + 0$, $v_1^2 = v_2^2 - 2gh$; $A_1v_1 = A_2v_2$, $(\pi d_0^2/4)(v_1) = (\pi d^2/4)(v_2)$, $v_2 = (d_0/d)^2(v_1)$;
$v_1^2 = [(d_0/d)^2(v_1)]^2 - 2gh$, $v_1 = \sqrt{2gh/(d_0^4/d^4 - 1)}$. Or $v_1 = \sqrt{Kh}$ where $K = 2g/[(d_0/d)^4 - 1]$. But also,
$v_1 = -dh/dt$, $dh/dt = -\sqrt{Kh}$

$$\int_{h_0}^{h} \frac{dh}{\sqrt{h}} = \int_0^t -\sqrt{K}\, dt \qquad (2)[h^{1/2}]_{h_0}^{h} = -\sqrt{K}\,[t]_0^t \qquad (2)(h^{1/2} - h_0^{1/2}) = -\sqrt{K}\,t \qquad h = \left(h_0^{1/2} - \frac{\sqrt{K}\,t}{2}\right)^2$$

Or, $h = \{h_0^{1/2} - [g/(2)(d_0^4/d^4 - 1)]^{1/2}t\}^2$.

8.103 In the water flow over the spillway in Fig. 8-59, the velocity is uniform at sections 1 and 2 and the pressure approximately hydrostatic. Neglecting losses, compute v_1 and v_2. Assume unit width.

$$p_1/\gamma + v_1^2/2g + z_1 = p_2/\gamma + v_2^2/2g + z_2 + h_L \qquad 0 + v_1^2/2g + 5 = 0 + v_2^2/2g + 0.7 + 0$$

$$A_1v_1 = A_2v_2 \qquad [(5)(1)](v_1) = [(0.7)(1)](v_2) \qquad v_2 = 7.143v_1$$

$$v_1^2/[(2)(9.807)] + 5 = (7.143v_1)^2/[(2)(9.807)] + 0.7 \qquad v_1 = 1.30 \text{ m/s} \qquad v_2 = (7.143)(1.30) = 9.29 \text{ m/s}$$

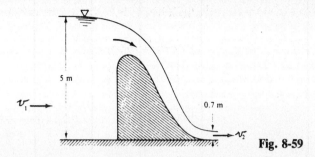

Fig. 8-59

8.104 For the water channel flow down the sloping ramp of Fig. 8-60, $h_1 = 1$ m, $H = 3$ m, and $v_1 = 4$ m/s. The flow is uniform at 1 and 2. Neglecting losses, find the downstream depth h_2 and show that three solutions are possible, of which only two are realistic. Neglect friction.

▌ $p_1/\gamma + v_1^2/2g + z_1 = p_2/\gamma + v_2^2/2g + z_2 + h_L$ $0 + 4^2/[(2)(9.807)] + (3 + 1) = 0 + v_2^2/[(2)(9.807)] + h_2 + 0$

$A_1v_1 = A_2v_2$ $[(1)(1)](4) = [(h_2)(1)](v_2)$ $v_2 = 4/h_2$

$4^2/[(2)(9.807)] + 4 = (4/h_2)^2/[(2)(9.807)] + h_2$ $h_2^3 - 4.816h_2^2 + 0.8157 = 0$

There are three mathematical solutions to this equation:

$$h_2 = 4.78 \text{ m} \quad \text{(subcritical)}$$
$$h_2 = 0.432 \text{ m} \quad \text{(supercritical)}$$
$$h_2 = -0.396 \text{ m} \quad \text{(impossible)}$$

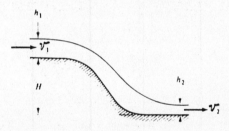

Fig. 8-60

8.105 For water flow up the sloping channel in Fig. 8-61, $h_1 = 0.5$ ft, $v_1 = 15$ ft/s, and $H = 2$ ft. Neglect losses and assume uniform flow at 1 and 2. Find the downstream depth h_2 and show that three solutions are possible, of which only two are realistic.

▌ $p_1/\gamma + v_1^2/2g + z_1 = p_2/\gamma + v_2^2/2g + z_2 + h_L$ $0 + 15^2/[(2)(32.2)] + 0.5 = 0 + v_2^2/[(2)(32.2)] + (2 + h_2) + 0$

$A_1v_1 = A_2v_2$ $[(0.5)(1)](15) = [(h_2)(1)](v_2)$ $v_2 = 7.5/h_2$

$15^2/[(2)(32.2)] + 0.5 = (7.5/h_2)^2/[(2)(32.2)] + (2 + h_2)$ $h_2^3 - 1.994h_2^2 + 0.8734 = 0$

There are three mathematical solutions to this equation:

$$h_2 = 1.69 \text{ ft} \quad \text{(subcritical)}$$
$$h_2 = 0.887 \text{ ft} \quad \text{(supercritical)}$$
$$h_2 = -0.582 \text{ ft} \quad \text{(impossible)}$$

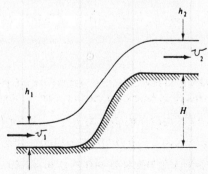

Fig. 8-61

8.106 If a 10-lb force is applied steadily to the piston in Fig. 8-62 and flow losses are negligible, what will the water-jet exit velocity v_2 be?

▮ $p_1/\gamma + v_1^2/2g + z_1 = p_{atm}/\gamma + v_2^2/2g + z_2 + h_L$. Considering the force acting on the piston, $10 = (p_1 - p_{atm})[(\pi)(\frac{8}{12})^2/4]$, $p_1 - p_{atm} = 28.65$ lb/ft^2; $A_1v_1 = A_2v_2$, $[(\pi)(\frac{8}{12})^2/4](v_1) = [(\pi)(\frac{4}{12})^2/4](v_2)$, $v_1 = 0.250v_2$; $(p_1 - p_{atm})/\gamma + (0.250v_2)^2/[(2)(32.2)] + 0 = v_2^2/[(2)(32.2)] + 0 + 0$, $28.65/62.4 + (0.250v_2)^2/[(2)(32.2)] + 0 = v_2^2/[(2)(32.2)] + 0 + 0$, $v_2 = 5.62$ ft/s.

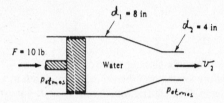

Fig. 8-62

8.107 The three-arm lawn sprinkler of Fig. 8-63 receives water through the center at 1200 mL/s. If collar friction is negligible, what is the steady rotation rate for (**a**) $\theta = 0°$ and (**b**) $\theta = 30°$?

▮ $$v_0 = \frac{Q}{A} = \frac{(1200/1\,000\,000)/3}{(\pi)(\frac{7}{1000})^2/4} = 10.39 \text{ m/s}$$

If collar friction is negligible, the steady rotation rate (ω) is $v_0 \cos \theta/r$, independent of flow rate.

(**a**) $\omega = (10.39)(\cos 0°)/(\frac{15}{100}) = 69.27$ rad/s   or   661 rpm

(**b**) $\omega = (10.39)(\cos 30°)/(\frac{15}{100}) = 59.99$ rad/s   or   573 rpm

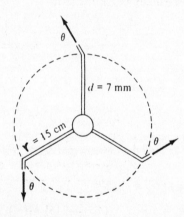

Fig. 8-63

8.108 Water flows at 10 ft/s through a pipe 1000 ft long with diameter 1 in. The inlet pressure $p_1 = 200$ psig, and the exit section is 100 ft higher than the inlet. What is the exit pressure p_2 if the friction head loss is 350 ft?

▮ $$p_1/\gamma + v_1^2/2g + z_1 = p_2/\gamma + v_2^2/2g + z_2 + h_L$$

$(200)(144)/62.4 + v_1^2/2g + 0 = (p_2)(144)/62.4 + v_2^2/2g + 100 + 350$ $v_1^2/2g = v_2^2/2g$ $p_2 = 5.00$ psig

8.109 A 30-in-diameter pipeline carries oil (s.g. $= 0.86$) at 500 000 barrels per day. The friction head loss is 8.4 ft per 1000 ft of pipe. It is planned to place pumping stations every 12 statute miles along the pipe. Compute the pressure drop between pumping stations.

▮ $$p_1/\gamma + v_1^2/2g + z_1 = p_2/\gamma + v_2^2/2g + z_2 + h_L$$

$p_1/[(0.86)(62.4)] + v_1^2/2g + 0 = p_2/[(0.86)(62.4)] + v_2^2/2g + 0 + (8.4/1000)[(12)(5280)]$

$v_1^2/2g = v_2^2/2g$ $p_1 - p_2 = 28\,560$ lb/ft^2 or   198 lb/in^2

8.110 The long pipe in Fig. 8-64 is filled with water. When valve A is closed, $p_2 - p_1 = 10$ psi. When the valve is open and water flows at 10 ft^3/s, $p_1 - p_2 = 34$ psi. What is the friction head loss between 1 and 2 for the flowing condition?

▮ $$p_1/\gamma + v_1^2/2g + z_1 = p_2/\gamma + v_2^2/2g + z_2 + h_L \qquad v_1^2/2g = v_2^2/2g$$

Valve closed:
$$(p_2 - p_1)(144)/62.4 = z_1 - z_2 \qquad z_1 - z_2 = (10)(144)/62.4 = 23.08 \text{ ft}$$
Valve open:
$$(p_1 - p_2)(144)/62.4 + (z_1 - z_2) = h_L \qquad (34)(144)/62.4 + 23.08 = h_L \qquad h_L = 101.5 \text{ ft}$$

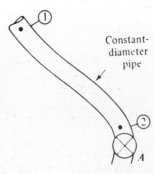

Constant-diameter pipe

Fig. 8-64

8.111 If losses are neglected for the water-flow system in Fig. 8-65, what should the mercury manometer reading be? The left manometer leg is open to the atmosphere and $v_1 = 2$ ft/s.

$$p_1/\gamma + v_1^2/2g + z_1 = p_2/\gamma + v_2^2/2g + z_2 + h_L$$

$$A_1 v_1 = A_2 v_2 \qquad [(\pi)(\tfrac{3}{12})^2/4](2) = [(\pi)(\tfrac{1}{12})^2/4](v_2) \qquad v_2 = 18.0 \text{ ft/s}$$

$$p_1/\gamma + 2^2/[(2)(32.2)] + 0 = 0 + 18.0^2/[(2)(32.2)] + 10 + 0 \qquad p_1/\gamma = 14.97 \text{ ft}$$

For the manometer, $14.97 + 2 - 13.6h = 0$, $h = 1.25$ ft.

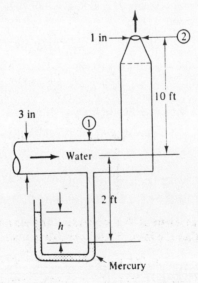

1 in

2

10 ft

3 in

1

Water

2 ft

h

Mercury   **Fig. 8-65**

8.112 In Fig. 8-66 on p. 188 the pipe exit losses are $Kv^2/2g$, where v is the exit velocity and $K = 1.5$. What is the exit weight flux of water?

$$p_1/\gamma + v_1^2/2g + z_1 = p_2/\gamma + v_2^2/2g + z_2 + h_L \qquad p_1 = (20)(144) + [(0.68)(62.4)](4) = 3050 \text{ lb/ft}^2$$

$$3050/62.4 + 0 + 5 = 0 + v_2^2/[(2)(32.2)] + 0 + (1.5)\{v_2^2/[(2)(32.2)]\} \qquad v_2 = 37.25 \text{ ft/s}$$

$$W = \gamma A v = (62.4)[(\pi)(\tfrac{1}{12})^2/4](37.25) = 12.7 \text{ lb/s}$$

8.113 In Fig. 8-67 the fluid is water, and the pressure gage reads $p_1 = 170$ kPa gage. If the mass flux is 15 kg/s, what is the head loss between 1 and 2?

$$p_1/\gamma + v_1^2/2g + z_1 = p_2/\gamma + v_2^2/2g + z_2 + h_L$$

$$M = \rho A v \qquad 15 = 1000[(\pi)(\tfrac{8}{100})^2/4](v_1) \qquad v_1 = 2.984 \text{ m/s} \qquad 15 = 1000[(\pi)(\tfrac{5}{1000})^2/4](v_2) \qquad v_2 = 7.639 \text{ m/s}$$

$$170/9.79 + 2.984^2/[(2)(9.807)] + 0 = 0 + 7.639^2/[(2)(9.807)] + 12 + h_L \qquad h_L = 2.84 \text{ m}$$

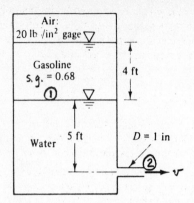

Fig. 8-66

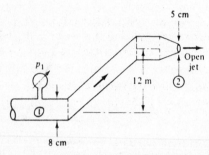

Fig. 8-67

8.114 Oil at specific gravity 0.761 flows from tank A to tank E, as shown in Fig. 8-68. Lost head items may be assumed to be as follows: A to $B = 0.60v_{12}^2/2g$; B to $C = 9.0v_{12}^2/2g$; C to $D = 0.40v_6^2/2g$; D to $E = 9.0v_6^2/2g$. Find the flow rate and the pressure at C.

$$p_A/\gamma + v_A^2/2g + z_A = p_E/\gamma + v_E^2/2g + z_E + h_L$$

$$0 + 0 + 40.0 = 0 + 0 + 0 + (0.60v_{12}^2 + 9.0v_{12}^2 + 0.40v_6^2 + 9.0v_6^2)/[(2)(32.2)]$$

$$9.60v_{12}^2 + 9.40v_6^2 = 2576 \qquad A_1v_1 = A_2v_2 \qquad [(\pi)(\tfrac{12}{12})^2/4](v_{12}) = [(\pi)(\tfrac{6}{12})^2/4](v_6) \qquad v_6 = 4.00v_{12}$$

$$9.60v_{12}^2 + (9.40)(4.00v_{12})^2 = 2576 \qquad v_{12} = 4.012 \text{ ft/s}$$

$$Q = Av = [(\pi)(\tfrac{12}{12})^2/4](4.012) = 3.15 \text{ ft}^3/\text{s}$$

$$p_A/\gamma + v_A^2/2g + z_A = p_C/\gamma + v_C^2/2g + z_C + h_L$$

$$0 + 0 + 40.0 = (p_C)(144)/[(0.761)(62.4)] + 4.012^2/[(2)(32.2)]$$
$$+ (40.0 + 2) + [(0.60)(4.012)^2 + (9.0)(4.012)^2]/[(2)(32.2)]$$

$$p_C = -1.53 \text{ lb/in}^2$$

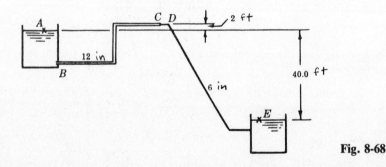

Fig. 8-68

8.115 (a) What is the pressure on the nose of a torpedo moving in salt water at 100 ft/s at a depth of 30.0 ft? (b) If the pressure at point C on the side of the torpedo at the same elevation as the nose is 10.0 psig, what is the relative velocity at that point?

▮ (a) In this case, greater clarity in the application of the Bernoulli equation may be attained by considering the relative motion of a stream of water past the stationary torpedo. The velocity of the nose of the torpedo will then be zero. Assume no lost head in the streamtube from a point A in the undistrubed water just ahead of the torpedo to a point B on the nose of the torpedo: $p_A/\gamma + v_A^2/2g + z_A = p_B/\gamma + v_B^2/2g + z_B + h_L$, $30.0 + 100^2/[(2)(32.2)] + 0 = p_B/\gamma + 0 + 0 + 0$, $p_B/\gamma = 185.3$ ft, $p_B = (64.2)(185.3)/144 = 82.6$ psi. This pressure is called the stagnation pressure and may be expressed as $p_s = p_0 + \rho v_0^2/2$.

(b)
$$p_A/\gamma + v_A^2/2g + z_A = p_C/\gamma + v_C^2/2g + z_C + h_L$$

$$30.0 + 100^2/[(2)(32.2)] + 0 = (10.0)(144)/64.2 + v_C^2/[(2)(32.2)] + 0 + 0 \qquad v_C = 102.4 \text{ ft/s}$$

8.116 A sphere is placed in an air stream which is at atmospheric pressure and is moving at 100.0 ft/s. Using the density of air constant at 0.00238 slug/ft³, calculate the stagnation pressure and the pressure on the surface of the sphere at a point B, 75° from the stagnation point, if the velocity there is 220.0 ft/s.

❚ From Prob. 8.115, $p_s = p_0 + \rho v_0^2/2 = (14.7)(144) + (0.00238)(100.0)^2/2 = 2129$ lb/ft², or 14.8 lb/in². $p_s/\gamma + v_s^2/2g + z_s = p_B/\gamma + v_B^2/2g + z_B + h_L$, $2129/[(0.00238)(32.2)] + 0 + 0 = p_B/\gamma + 220.0^2/[(2)(32.2)] + 0 + 0$, $p_B/\gamma = 27\,029$ ft of air, $p_B = [(0.00238)(32.2)](27\,029) = 2071$ lb/ft², or 14.4 lb/in².

8.117 A large closed tank is filled with ammonia (NH_3) under a pressure of 5.30 psig and at 65 °F. The ammonia discharges into the atmosphere through a small opening in the side of the tank. Neglecting friction losses, calculate the velocity of the ammonia leaving the tank assuming constant density. The gas constant for ammonia is 89.5 ft/°R.

❚ Apply Bernoulli's equation between the tank (1) and the atmosphere (2). $p_1/\gamma + v_1^2/2g + z_1 = p_2/\gamma + v_2^2/2g + z_2 + h_L$, $\gamma = p/RT$, $\gamma_{NH_3} = (5.30 + 14.7)(144)/[(89.5)(460 + 65)] = 0.06129$ lb/ft³, $(5.30)(144)/0.06129 + 0 + 0 = 0 + v_2^2/[(2)(32.2)] + 0 + 0$, $v_2 = 896$ ft/s.

8.118 Water at 90 °F is to be lifted from a sump at a velocity of 6.50 ft/s through the suction pipe of a pump. Calculate the theoretical maximum height of the pump setting under the following conditions: $p_{atm} = 14.25$ psia, $p_v = 0.70$ psia, and h_L in the suction pipe = 3 velocity heads.

❚ The minimum pressure at the entrance to the pump cannot be less than the vapor pressure of the liquid. Apply Bernoulli's equation between the water surface outside the suction pipe and the entrance to the pump: $p_1/\gamma + v_1^2/2g + z_1 = p_2/\gamma + v_2^2/2g + z_2 + h_L$, $(14.25)(144)/62.1 + 0 + 0 = (0.70)(144)/62.1 + 6.50^2/[(2)(32.2)] + z_2 + 3\{6.50^2/[(2)(32.2)]\}$, $z_2 = 28.8$ ft. (Under these conditions, serious damage due to cavitation will probably occur.)

8.119 For the Venturi meter shown in Fig. 8-69, the deflection of mercury in the differential gage is 14.3 in. Determine the flow of water through the meter if no energy is lost between A and B.

❚
$$p_A/\gamma + v_A^2/2g + z_A = p_B/\gamma + v_B^2/2g + z_B + h_L$$

$p_A/\gamma + v_A^2/[(2)(32.2)] + 0 = p_B/\gamma + v_B^2/[(2)(32.2)] + 30.0/12 + 0 \qquad p_A/\gamma - p_B/\gamma = 0.01553(v_B^2 - v_A^2) + 2.500$

$$A_A v_A = A_B v_B \qquad [(\pi)(\tfrac{12}{12})^2/4](v_A) = [(\pi)(\tfrac{6}{12})^2/4](v_B) \qquad v_A = 0.250 v_B$$

From the manometer, $p_A/\gamma + z + 14.3/12 - (13.6)(14.3/12) - z - 30.0/12 = p_B/\gamma$, $p_A/\gamma - p_B/\gamma = 17.52$ ft, $17.52 = 0.01553[v_B^2 - (0.250 v_B)^2] + 2.500$, $v_B = 32.12$ ft/s; $Q = Av = [(\pi)(\tfrac{6}{12})^2/4](32.12) = 6.31$ ft³/s.

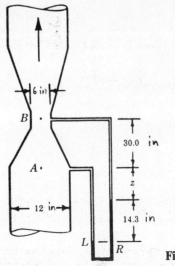

Fig. 8-69

8.120 For the meter in Fig. 8-69, consider air at 80 °F with the pressure at $A = 37.5$ psig. Consider a deflection of the gage of 14.3 in of water. Assuming that the specific weight of the air does not change between A and B and that the energy loss is negligible, determine the amount of air flowing in pounds per second.

$$p_A/\gamma + v_A^2/2g + z_A = p_B/\gamma + v_B^2/2g + z_B + h_L$$

$$p_A/\gamma + v_A^2/[(2)(32.2)] + 0 = p_B/\gamma + v_B^2/[(2)(32.2)] + 30.0/12 + 0 \qquad p_A/\gamma - p_B/\gamma = 0.01553(v_B^2 - v_A^2) + 2.500$$

$$A_A v_A = A_B v_B \qquad [(\pi)(\tfrac{12}{12})^2/4](v_A) = [(\pi)(\tfrac{6}{12})^2/4](v_B) \qquad v_A = 0.250 v_B$$

$$\gamma = p/RT \qquad \gamma_{air} = (37.5 + 14.7)(144)/[(53.3)(460 + 80)] = 0.2612 \text{ lb/ft}^3$$

From the manometer, $p_A/\gamma + z + 14.3/12 - (62.4/0.2612)(14.3/12) - z - 30.0/12 = p_B/\gamma$, $p_A/\gamma - p_B/\gamma = 286.0$ ft of air; $286.0 = 0.01553[v_B^2 - (0.250 v_B)^2] + 2.500$, $v_B = 139.5$ ft/s; $W = \gamma A v = 0.2612[(\pi)(\tfrac{6}{12})^2/4](139.5) = 7.15$ lb/s.

8.121 Assume frictionless flow in a long, horizontal, conical pipe, the diameter of which is 2 ft at one end and 4 ft at the other. The pressure head at the smaller end is 16 ft of water. If water flows through the cone at a rate of 125.6 cfs, find the velocities at the two ends and the pressure head at the larger end.

$$v_1 = Q/A_1 = 125.6/[(\pi)(2)^2/4] = 39.98 \text{ ft/s} \qquad v_2 = Q/A_2 = 125.6/[(\pi)(4)^2/4] = 9.99 \text{ ft/s}$$

$$p_1/\gamma + v_1^2/2g + z_1 = p_2/\gamma + v_2^2/2g + z_2 + h_L$$

$$16 + 39.98^2/[(2)(32.2)] + 0 = p_2/\gamma + 9.99^2/[(2)(32.2)] + 0 + 0 \qquad p_2/\gamma = 39.3 \text{ ft of water}$$

8.122 Water flows through a long, horizontal, conical diffuser at the rate of 4.0 m³/s. The diameter of the diffuser changes from 1.0 m to 1.5 m. The pressure at the smaller end is 7.5 kPa. Find the pressure at the downstream end of the diffuser, assuming frictionless flow. Assume also that the angle of the cone is so small that separation of the flow from the walls of the diffuser does not occur.

$$p_1/\gamma + v_1^2/2g + z_1 = p_2/\gamma + v_2^2/2g + z_2 + h_L$$

$$v_1 = Q/A_1 = 4.0/[(\pi)(1.0)^2/4] = 5.093 \text{ m/s} \qquad v_2 = Q/A_2 = 4.0/[(\pi)(1.5)^2/4] = 2.264 \text{ m/s}$$

$$7.5/9.79 + 5.093^2/[(2)(9.807)] + 0 = p_2/9.79 + 2.264^2/[(2)(9.807)] + 0 + 0 \qquad p_2 = 17.9 \text{ kPa}$$

8.123 A vertical pipe 6 ft in diameter and 60 ft long has a pressure head at the upper end of 18 ft of water. When the flow of water through it is such that the mean velocity is 15 fps, the friction loss is 4 ft. Find the pressure head at the lower end of the pipe when the flow is (a) downward and (b) upward.

(a)
$$p_1/\gamma + v_1^2/2g + z_1 = p_2/\gamma + v_2^2/2g + z_2 + h_L$$

$$18 + 15^2/[(2)(32.2)] + 60 = p_2/\gamma + 15^2/(2)(32.2)] + 0 + 4 \qquad p_2/\gamma = 74.0 \text{ ft}$$

(b)
$$p_2/\gamma + v_2^2/2g + z_2 = p_1/\gamma + v_1^2/2g + z_1 + h_L$$

$$p_2/\gamma + 15^2/[(2)(32.2)] + 0 = 18 + 15^2/[(2)(32.2)] + 60 + 4 \qquad p_2/\gamma = 82.0 \text{ ft}$$

8.124 A conical pipe has diameters at the two ends of 1.5 ft and 4.5 ft and is 50 ft long. It is vertical, and the friction loss is 8 ft for flow in either direction when the velocity at the smaller section is 30 fps. If the smaller section is at the top and the pressure head there is 6.5 ft of water, find the pressure head at the lower end when the flow is (a) downward and (b) upward.

$$Q = A_1 v_1 = [(\pi)(1.5)^2/4](30) = 53.01 \text{ ft}^3/\text{s} \qquad v_2 = Q/A_2 = 53.01/[(\pi)(4.5)^2/4] = 3.333 \text{ ft/s}$$

(a)
$$p_1/\gamma + v_1^2/2g + z_1 = p_2/\gamma + v_2^2/2g + z_2 + h_L$$

$$6.5 + 30^2/[(2)(32.2)] + 50 = p_2/\gamma + 3.333^2/[(2)(32.2)] + 0 + 8 \qquad p_2/\gamma = 62.3 \text{ ft}$$

(b)
$$p_2/\gamma + v_2^2/2g + z_2 = p_1/\gamma + v_1^2/2g + z_1 + h_L$$

$$p_2/\gamma + 3.333^2/[(2)(32.2)] + 0 = 6.5 + 30^2/[(2)(32.2)] + 50 + 8 \qquad p_2/\gamma = 78.3 \text{ ft}$$

8.125 In Fig. 8-70, pipe AB is of uniform diameter. The pressure at A is 20 psi and at B, 30 psi. In which direction is the flow, and what is the friction loss of the fluid if the liquid has a specific weight of (a) 30 lb/ft³ and (b) 100 lb/ft³?

$$p_A/\gamma + v_A^2/2g + z_A = p_B/\gamma + v_B^2/2g + z_B + h_L. \text{ Assume flow is from } A \text{ to } B.$$

(a) $(20)(144)/30 + v_A^2/2g + 30 = (30)(144)/30 + v_B^2/2g + 0 + h_L$, $v_A^2/2g = v_B^2/2g$, $h_L = -18.0$ ft. Since h_L is negative, flow is actually from B to A.

(b) $(20)(144)/100 + v_A^2/2g + 30 = (30)(144)/100 + v_B^2/2g + 0 + h_L$, $v_A^2/2g = v_B^2/2g$, $h_L = 15.6$ ft. Since h_L is positive, flow is from A to B, as assumed.

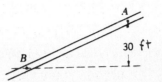

Fig. 8-70

8.126   In Fig. 8-70, if the difference in elevation between A and B is 8 m and the pressures at A and B are 150 kPa and 250 kPa, respectively, find the direction of flow and the head loss. Assume the liquid has a specific gravity of 0.85.

▋ $p_A/\gamma + v_A^2/2g + z_A = p_B/\gamma + v_B^2/2g + z_B + h_L$. Assume flow is from A to B. $150/[(0.85)(9.79)] + v_A^2/2g + 8 = 250/[(0.85)(9.79)] + v_B^2/2g + 0 + h_L$, $v_A^2/2g = v_B^2/2g$, $h_L = -4.02$ m. Since h_L is negative, flow is actually from B to A.

8.127   A pipe line conducts water from a reservoir to a powerhouse, the elevation of which is 800 ft lower than that of the surface of the reservoir. The water is discharged through a nozzle with a jet velocity of 220 fps and the diameter of the jet is 8 in. Find the power of the jet and the power lost in friction between reservoir and jet.

▋
$$p_A/\gamma + v_A^2/2g + z_A = p_B/\gamma + v_B^2/2g + z_B + h_L \qquad 0 + 0 + 800 = 0 + 220^2/[(2)(32.2)] + 0 + h_L$$

$$h_L = 48.45 \text{ ft} \qquad Q = Av = [(\pi)(\tfrac{8}{12})^2/4](220) = 76.79 \text{ ft}^3/\text{s}$$

$$P_{\text{jet}} = Q\gamma v_2^2/2g = (76.79)(62.4)\{220^2/[(2)(32.2)]\} = 3\,601\,000 \text{ ft} \cdot \text{lb/s} = 3\,601\,000/550 = 6547 \text{ hp}$$

$$P_{\text{lost}} = Q\gamma h_L = (76.79)(62.4)(48.45) = 232\,200 \text{ ft} \cdot \text{lb/s} = 232\,200/550 = 422 \text{ hp}$$

8.128   Water is flowing in a channel, as shown in Fig. 8-71. Neglecting all losses, determine the two possible depths of flow y_1 and y_2.

▋
$$p_A/\gamma + v_A^2/2g + z_A = p_B/\gamma + v_B^2/2g + z_B + h_L$$

$$Q = A_A v_A = [(4)(10)](16.1) = 644 \text{ ft}^3/\text{s} \qquad v_B = Q/A_B = 644/(10y) = 64.4/y$$

$$0 + 16.1^2/[(2)(32.2)] + (8 + 4) = 0 + (64.4/y)^2/[(2)(32.2)] + y + 0 \qquad 64.4/y^2 + y - 16.02 = 0$$

$$y^3 - 16.02y^2 + 64.4 = 0 \qquad y_1 = 2.16 \text{ ft} \qquad y_2 = 15.8 \text{ ft}$$

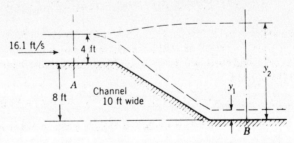

Fig. 8-71

8.129   Neglecting all losses, in Fig. 8-71 the channel narrows in the drop to 6 ft wide at section B. For uniform flow across section B, determine the two possible depths of flow.

▋
$$p_A/\gamma + v_A^2/2g + z_A = p_B/\gamma + v_B^2/2g + z_B + h_L$$

$$Q = A_A v_A = [(4)(10)](16.1) = 644 \text{ ft}^3/\text{s} \qquad v_B = Q/A_B = 644/(6y) = 107.3/y$$

$$0 + 16.1^2/[(2)(32.2)] + (8 + 4) = 0 + (107.3/y)^2/[(2)(32.2)] + y + 0 \qquad 178.8/y^2 + y - 16.02 = 0$$

$$y^3 - 16.02y^2 + 178.8 = 0 \qquad y_1 = 3.83 \text{ ft} \qquad y_2 = 15.3 \text{ ft}$$

8.130 If the losses from section A to section B of Fig. 8-71 are 1.9 ft, determine the two possible depths at section B.

$$p_A/\gamma + v_A^2/2g + z_A = p_B/\gamma + v_B^2/2g + z_B + h_L$$

$$Q = A_A v_A = [(4)(10)](16.1) = 644 \text{ ft}^3/\text{s} \qquad v_B = Q/A_B = 644/(10y) = 64.4/y$$

$$0 + 16.1^2/[(2)(32.2)] + (8+4) = 0 + (64.4/y)^2/[(2)(32.2)] + y + 1.9 \qquad 64.4/y^2 + y - 14.12 = 0$$

$$y^3 - 14.12y^2 + 64.4 = 0 \qquad y_1 = 2.34 \text{ ft} \qquad y_2 = 13.8 \text{ ft}$$

8.131 High-velocity water flows up an inclined plane, as shown in Fig. 8-72. Neglecting all losses, calculate the two possible depths of flow at section B.

$$p_A/\gamma + v_A^2/2g + z_A = p_B/\gamma + v_B^2/2g + z_B + h_L$$

$$Q = A_A v_A = [(\tfrac{500}{1000})(2)](9.806) = 9.806 \text{ m}^3/\text{s} \qquad v_B = Q/A_B = 9.806/(2y) = 4.903/y$$

$$0 + 9.806^2/[(2)(9.807)] + \tfrac{500}{1000} = 0 + (4.903/y)^2/[(2)(9.807)] + (2.5+y) + 0 \qquad 1.226/y^2 + y - 2.903 = 0$$

$$y^3 - 2.903y^2 + 1.226 = 0 \qquad y_1 = 0.775 \text{ m} \qquad y_2 = 2.74 \text{ m}$$

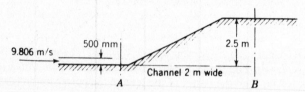

Fig. 8-72

8.132 In Fig. 8-72, the channel changes in width from 2 m at section A to 3 m at section B. For losses of 0.3 m between sections A and B, find the two possible depths at section B.

$$p_A/\gamma + v_A^2/2g + z_A = p_B/\gamma + v_B^2/2g + z_B + h_L \qquad Q = A_A v_A = [(\tfrac{500}{1000})(2)](9.806) = 9.806 \text{ m}^3/\text{s}$$

$$v_B = Q/A_B = 9.806/(3y) = 3.269/y$$

$$0 + 9.806^2/[(2)(9.807)] + \tfrac{500}{1000} = 0 + (3.269/y)^2/[(2)(9.807)] + (2.5+y) + 0.3$$

$$0.5448/y^2 + y - 2.603 = 0 \qquad y^3 - 2.603y^2 + 0.5448 = 0 \qquad y_1 = 0.510 \text{ m} \qquad y_2 = 2.52 \text{ m}$$

8.133 For losses of $0.1H$ through the nozzle of Fig. 8-73, what is the gage difference R in terms of H?

$$1.2H + 1.2y + 1.2R - 3.0R - 1.2y = (0.9)(1.2)(H) \qquad R = 0.06667H$$

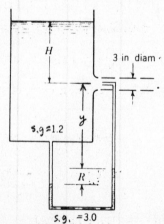

Fig. 8-73

8.134 Neglecting losses, calculate H in terms of R for Fig. 8-73.

$1.2H + 1.2y + 1.2R - 3.0R - 1.2y = 1.2H$. Therefore, $R = 0$ for all H.

8.135 At point A in a pipeline carrying water, the diameter is 1 m, the pressure 98 kPa, and the velocity 1 m/s. At point B, 2 m higher than A, the diameter is 0.5 m and the pressure is 20 kPa. Determine the head loss and the direction of flow.

❚ Assume the direction of flow is from A to B. $p_A/\gamma + v_A^2/2g + z_A = p_B/\gamma + v_B^2/2g + z_B + h_L$, $Q = A_1v_1 = [(\pi)(1)^2/4](1) = 0.7854 \text{ m}^3/\text{s}$, $v_2 = Q/A_2 = 0.7854/[(\pi)(0.5)^2/4] = 4.00 \text{ m/s}$, $98/9.79 + 1^2/[(2)(9.807)] + 0 = 20/9.79 + 4.00^2/[(2)(9.807)] + 2 + h_L$, $h_L = 5.20 \text{ m}$. Since h_L is positive, flow is from A to B as assumed.

8.136 Water is flowing in an open channel at a depth of 2 m and velocity of 3 m/s, as shown in Fig. 8-74. It then flows down a contracting chute into another channel where the depth is 1 m and the velocity is 10 m/s. Assuming frictionless flow, determine the difference in elevation of the channel floors.

❚ The velocities are assumed to be uniform over the cross sections, and the pressures hydrostatic.
$p_A/\gamma + v_A^2/2g + z_A = p_B/\gamma + v_B^2/2g + z_B + h_L$, $0 + 3^2/[(2)(9.807)] + (y + 2) = 0 + 10^2/[(2)(9.807)] + 1 + 0$, $y = 3.64 \text{ m}$.

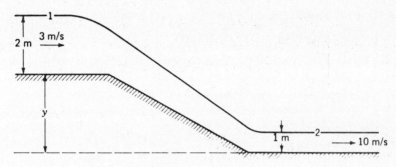

Fig. 8-74

8.137 For losses of 0.1 m, find the velocity at A in Fig. 8-75. The barometer reading is 750 mmHg.

❚
$$p_B/\gamma + v_B^2/2g + z_B = p_A/\gamma + v_A^2/2g + z_A + h_L$$
$$70/9.79 + 0 + 4 = [(13.6)(9.79)](\tfrac{750}{1000})/9.79 + v_A^2/[(2)(9.807)] + 0 + 0.1 \qquad v_A = 4.08 \text{ m/s}$$

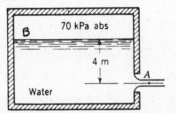

Fig. 8-75

8.138 For flow of 750 gpm in Fig. 8-76, determine H for losses of $10v^2/2g$.

❚
$$\frac{p_B}{\gamma} + \frac{v_B^2}{2g} + z_B = \frac{p_A}{\gamma} + \frac{v_A^2}{2g} + z_A + h_L \qquad v = \frac{Q}{A} = \frac{(750/7.48)/60}{(\pi)(\tfrac{6}{12})^2/4} = 8.511 \text{ ft/s}$$
$$0 + 0 + H = 0 + 8.511^2/[(2)(32.2)] + 0 + 10\{8.511^2/[(2)(32.2)]\} \qquad H = 12.37 \text{ ft}$$

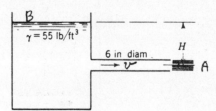

Fig. 8-76

8.139 For 1500-gpm flow and $H = 32$ ft in Fig 8-76, calculate the losses through the system in velocity heads, $Kv^2/2g$.

❚
$$\frac{p_B}{\gamma} + \frac{v_B^2}{2g} + z_B = \frac{p_A}{\gamma} + \frac{v_A^2}{2g} + z_A + h_L \qquad v = \frac{Q}{A} = \frac{(1500/7.48)/60}{(\pi)(\tfrac{6}{12})^2/4} = 17.02 \text{ ft/s}$$
$$0 + 0 + 32 = 0 + 17.02^2/[(2)(32.2)] + 0 + K\{17.02^2/[(2)(32.2)]\} \qquad K = 6.11 \qquad \text{(i.e., 6.11 velocity heads)}$$

8.140 The losses in Fig. 8-76 for $H = 25$ ft are $3v^2/2g$. What is the discharge?

▮ $p_B/\gamma + v_B^2/2g + z_B = p_A/\gamma + v_A^2/2g + z_A + h_L$ $0 + 0 + 25 = 0 + v_A^2/[(2)(32.2)] + 0 + 3\{v_A^2/[(2)(32.2)]\}$

$v_A = 20.06$ ft/s $Q = Av = [(\pi)(\frac{6}{12})^2/4](20.06) = 3.94$ ft³/s

8.141 In Fig. 8-77, the losses up to section A are $5v_1^2/2g$ and the nozzle losses are $0.05v_2^2/2g$. Determine the discharge and pressure at A, if $H = 8$ m.

▮
$$p_B/\gamma + v_B^2/2g + z_B = p_C/\gamma + v_C^2/2g + z_C + h_L$$

$$0 + 0 + 8 = 0 + v_2^2/[(2)(9.807)] + 0 + 5\{v_1^2/[(2)(9.807)]\} + 0.05\{v_2^2/[(2)(9.807)]\}$$

$0.05353v_2^2 + 0.2549v_1^2 - 8 = 0$ $A_1v_1 = A_2v_2$ $[(\pi)(\frac{150}{1000})^2/4](v_1) = [(\pi)(\frac{50}{1000})^2/4](v_2)$ $v_1 = 0.1111v_2$

$0.05353v_2^2 + (0.2549)(0.1111v_2)^2 - 8 = 0$ $v_2 = 11.88$ m/s $v_1 = (0.1111)(11.88) = 1.320$ m/s

$Q = A_2v_2 = [(\pi)(\frac{50}{1000})^2/4](11.88) = 0.0233$ m³/s $p_B/\gamma + v_B^2/2g + z_B = p_A/\gamma + v_A^2/2g + z_A + h_L$

$0 + 0 + 8 = p_A/9.79 + 1.320^2/[(2)(9.807)] + 0 + 5\{1.320^2/[(2)(9.807)]\}$ $p_A = 73.1$ kN/m²

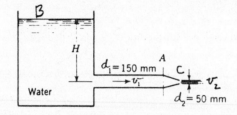

Fig. 8-77

8.142 For pressure at A of 25 kPa in Fig. 8-77 with the losses given in Prob. 8.141, determine the discharge and the head H.

▮
$$p_B/\gamma + v_B^2/2g + z_B = p_A/\gamma + v_A^2/2g + z_A + h_L$$

$$0 + 0 + H = 25/9.79 + v_1^2/[(2)(9.807)] + 0 + 5\{v_1^2/[(2)(9.807)]\}$$

$$H = 0.3059v_1^2 + 2.554 \qquad (1)$$

$$p_B/\gamma + v_B^2/2g + z_B = p_C/\gamma + v_C^2/2g + z_C + h_L$$

$$0 + 0 + H = 0 + v_2^2/[(2)(9.807)] + 0 + 5\{v_1^2/[(2)(9.807)]\} + 0.05\{v_2^2/[(2)(9.807)]\}$$

$0.05353v_2^2 + 0.2549v_1^2 - H = 0$ $A_1v_1 = A_2v_2$ $[(\pi)(\frac{150}{1000})^2/4](v_1) = [(\pi)(\frac{50}{1000})^2/4](v_2)$ $v_2 = 9.000v_1$

$$(0.05353)(9.000v_1)^2 + 0.2549v_1^2 = H \qquad (2)$$

Solving Eqs. (1) and (2) simultaneously, $(0.05353)(9.000v_1)^2 + 0.2549v_1^2 = 0.3059v_1^2 + 2.554$, $v_1 = 0.7720$ m/s;
$Q = A_1v_1 = [(\pi)(\frac{150}{1000})^2/4](0.7720) = 0.0136$ m³/s, $H = (0.3059)(0.7720)^2 + 2.554 = 2.736$ m.

8.143 The internal diameter of the pipe system shown in Fig. 8-78 is 6 in. The exit nozzle diameter is 3 in. What is the velocity v_e of flow leaving the nozzle? Neglect losses. (Do not consider the flow inside the pipe proper to be inviscid.)

▮
$$p_1/\gamma + v_1^2/2g + z_1 = p_A/\gamma + v_A^2/2g + z_A + h_L$$

$$(14.7)(144)/62.4 + 0 + 50 = 4500/62.4 + v_A^2/[(2)(32.2)] + 0 + 0$$

$v_A = 27.58$ ft/s $A_Av_A = A_Dv_D$ $[(\pi)(\frac{6}{12})^2](27.58) = [(\pi)(\frac{3}{12})^2](v_e)$ $v_e = 110.3$ ft/s

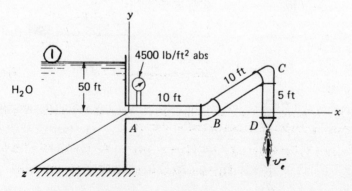

Fig. 8-78

8.144 In Fig. 8-79, $H = 6$ m and $h = 5.75$ m. Calculate the discharge and the losses in meters and in watts.

$v = \sqrt{2gh} = \sqrt{(2)(9.807)(5.75)} = 10.62$ m/s $\quad Q = Av = [(\pi)(\frac{80}{1000})^2/4](10.62) = 0.0534$ m³/s

$p_1/\gamma + v_1^2/2g + z_1 = p_2/\gamma + v_2^2/2g + z_2 + h_L \quad 0 + 0 + 6 = 0 + 10.62^2/[(2)(9.807)] + 0 + h_L$

$h_L = 0.250$ m $\quad$ Losses $= Q\gamma h_L = 0.0534[(1.05)(9.79)](0.250) = 0.137$ kW $\quad$ or $\quad$ 137 W

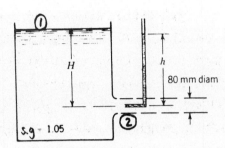

Fig. 8-79

8.145 In Fig. 8-80, 100 L/s of water flows from section 1 to section 2 with losses of $0.4(v_1 - v_2)^2/2g$. If $p_1 = 80$ kPa, find p_2.

$$p_1/\gamma + v_1^2/2g + z_1 = p_2/\gamma + v_2^2/2g + z_2 + h_L$$

$v_1 = Q/A_1 = (\frac{100}{1000})/[(\pi)(\frac{300}{1000})^2/4] = 1.415$ m/s $\quad v_2 = Q/A_2 = (\frac{100}{1000})/[(\pi)(\frac{450}{1000})^2/4] = 0.629$ m/s

$80/9.79 + 1.415^2/[(2)(9.807)] + 0 = p_2/9.79 + 0.629^2/[(2)(9.807)] + 0 + 0 \quad p_2 = 80.8$ kPa

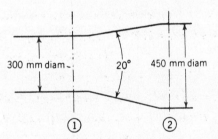

Fig. 8-80

8.146 Neglecting losses, determine the discharge in Fig. 8-81.

$p_1/\gamma + v_1^2/2g + z_1 = p_2/\gamma + v_2^2/2g + z_2 + h_L \quad [(0.75)(62.4)](3)/62.4 + 0 + 4 = 0 + v_2^2/[(2)(32.2)] + 0 + 0$

$v_2 = 20.06$ ft/s $\quad Q = Av = [(\pi)(\frac{4}{12})^2/4](20.06) = 1.75$ ft³/s

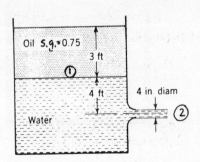

Fig. 8-81

8.147 A pipeline leads from one reservoir to another which has its water surface 12 m lower. For a discharge of 0.6 m³/s, determine the losses in meters and in kilowatts.

$p_1/\gamma + v_1^2/2g + z_1 = p_2/\gamma + v_2^2/2g + z_2 + h_L \quad 0 + 0 + 12 = 0 + 0 + 0 + h_L \quad h_L = 12$ m

Losses $= Q\gamma h_L = (0.6)(9.79)(12) = 70.5$ kW

8.148 In the siphon of Fig. 8-82, $h_1 = 1$ m, $h_2 = 3$ m, $d_1 = 3$ m, and $d_2 = 5$ m, and the losses to section 2 are $2.6v_2^2/2g$, with 10 percent of the losses occurring before section 1. Find the discharge and the pressure at section 1.

$$p_B/\gamma + v_B^2/2g + z_B = p_2/\gamma + v_2^2/2g + z_2 + h_L$$

$$0 + 0 + (1 + 3) = 0 + v_2^2/[(2)(9.807)] + 0 + 2.6v_2^2/[(2)(9.807)] \qquad v_2 = 4.668 \text{ m/s}$$

$$Q = A_2 v_2 = [(\pi)(5)^2/4](4.668) = 91.7 \text{ m}^3/\text{s} \qquad v_1 = Q/A_1 = 91.7/[(\pi)(3)^2/4] = 12.97 \text{ m/s}$$

$$p_B/\gamma + v_B^2/2g + z_B = p_1/\gamma + v_1^2/2g + z_1 + h_L$$

$$0 + 0 + (1 + 3) = p_1/9.79 + 12.97^2/[(2)(9.807)] + 3 + 0.10\{(2.6)(4.668)^2/[(2)(9.807)]\}$$

$$p_1 = -77.0 \text{ kPa}$$

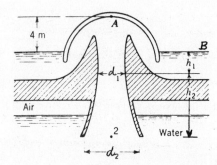

Fig. 8-82

8.149 Find the pressure at A of Prob. 8.148 if it is a stagnation point (velocity zero).

$$p_B/\gamma + v_B^2/2g + z_B = p_A/\gamma + v_A^2/2g + z_A + h_L \qquad 0 + 0 + 0 = p_A/9.79 + 0 + 4 + 0 \qquad p_A = -39.2 \text{ kPa}$$

8.150 If we neglect friction entirely for the siphon shown in Fig. 8-83, what is the velocity of the water leaving at C as a free jet? What are the pressures of the water in the tube at B and at A?

$$p_D/\gamma + v_D^2/2g + z_D = p_C/\gamma + v_C^2/2g + z_C + h_L$$

$$0 + 0 + 2.4 = 0 + v_C^2/[(2)(9.807)] + 0 + 0 \qquad v_C = 6.86 \text{ m/s}$$

$$p_D/\gamma + v_D^2/2g + z_D = p_B/\gamma + v_B^2/2g + z_B + h_L$$

$$0 + 0 + 2.4 = p_B/9.79 + 6.86^2/[(2)(9.807)] + (2.4 + 1.2) + 0 \qquad p_B = -35.2 \text{ kPa}$$

$$p_D/\gamma + v_D^2/2g + z_D = p_A/\gamma + v_A^2/2g + z_A + h_L$$

$$0 + 0 + 2.4 = p_A/9.79 + 6.86^2/[(2)(9.807)] + 2.4 + 0 \qquad p_A = -23.5 \text{ kPa}$$

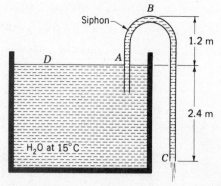

Fig. 8-83

8.151 If the vapor pressure of water is 0.1799 m of water, how high (h) above the free surface can point B be in Prob. 8.150 before the siphon action breaks down? Assume atmospheric pressure is 101 kPa.

$p_D/\gamma + v_D^2/2g + z_D = p_B/\gamma + v_B^2/2g + z_B + h_L$. Using absolute pressures and considering that $v_B = 0$ at maximum h when the siphon action breaks down, $101/9.79 + 0 + 2.4 = 0.1799 + 0 + (2.4 + h) + 0$, $h = 10.14$ m.

CHAPTER 9
Flow in Closed Conduits

9.1 Water at 10 °C flows in a 150-mm-diameter pipe at a velocity of 5.5 m/s. Is this flow laminar or turbulent?

   $N_R = dv/v = (\frac{150}{1000})(5.5)/(1.30 \times 10^{-6}) = 634\,615$. Since $634\,615 > 4000$, the flow is turbulent.

9.2 SAE10 oil at 68 °F flows in a 9-in-diameter pipe. Find the maximum velocity for which the flow will be laminar.

   $N_R = \rho dv/\mu$. For laminar flow, assume $N_R \le 2000$. $2000 = (1.68)(\frac{9}{12})(v)/(1.70 \times 10^{-3})$, $v = 2.70$ ft/s.

9.3 The accepted transition Reynolds number for flow past a smooth sphere is 250 000. At what velocity will this occur for airflow at 20 °C past a 12-cm-diameter sphere?

$$N_R = dv/v \qquad 250\,000 = (\tfrac{12}{100})(v)/(1.51 \times 10^{-5}) \qquad v = 31.5 \text{ m/s}$$

9.4 Repeat Prob. 9.3 if the fluid is (a) water at 20° and (b) hydrogen at 20 °C ($v = 1.08 \times 10^{-4}$ m^2/s).

   (a) $N_R = dv/v$ $250\,000 = (\frac{12}{100})(v)/(1.02 \times 10^{-6})$ $v = 2.12$ m/s
   (b) $N_R = dv/v$ $250\,000 = (\frac{12}{100})(v)/(1.08 \times 10^{-4})$ $v = 225$ m/s

9.5 A $\frac{1}{2}$-in-diameter water pipe is 60 ft long and delivers water at 5 gpm at 20 °C. What fraction of this pipe is taken up by the entrance region?

   $Q = (5)(0.002228) = 0.01114$ ft^3/s $V = Q/A = 0.01114/[(\pi)(0.5/12)^2/4] = 8.170$ ft/s $N_R = dv/v$

   From Table A-2, $v = 1.02 \times 10^{-6}$ m^2/s at 20 °C, which equals 1.10×10^{-5} ft^2/s; hence, $N_R = (0.5/12)(8.170)/(1.10 \times 10^{-5}) = 30\,947$. Since $30\,947 > 4000$, the flow is turbulent and for entrance length, $L_e/d = 4.4N_R^{1/6} = (4.4)(30\,947)^{1/6} = 25$. The actual pipe has $L/d = 60/[(\frac{1}{2})/12] = 1440$; hence,

$$\frac{L_e/d}{L/d} = \frac{L_e}{L} = \tfrac{25}{1440} = 0.017 \quad \text{or} \quad 1.7 \text{ percent}$$

9.6 An oil with $\rho = 900$ kg/m^3 and $v = 0.0002$ m^2/s flows upward through an inclined pipe as shown in Fig. 9-1. Assuming steady laminar flow, (a) verify that the flow is up and find the (b) head loss between section 1 and section 2, (c) flow rate, (d) velocity, and (e) Reynolds number. Is the flow really laminar?

   (a) $HGL = z + p/\rho g$ $HGL_1 = 0 + 350\,000/[(900)(9.807)] = 39.65$ m

   $$HGL_2 = (10)(\sin 40°) + 250\,000/[(900)(9.807)] = 34.75 \text{ m}$$

   Since $HGL_1 > HGL_2$, the flow is upward.

   (b) $h_f = HGL_1 - HGL_2 = 39.65 - 34.75 = 4.90$ m
   (c) $\mu = \rho v = (900)(0.0002) = 0.180$ kg/(m $\cdot$ s)

   $$Q = \frac{\pi \rho g d^4 h_f}{128 \mu L} = \frac{(\pi)(900)(9.807)(\frac{6}{100})^4(4.90)}{(128)(0.180)(10)} = 0.00764 \text{ m}^3\text{/s}$$

   (d) $v = Q/A = 0.00764/[(\pi)(\frac{6}{100})^2/4] = 2.70$ m/s
   (e) $N_R = dv/v = (\frac{6}{100})(2.70)/0.0002 = 810$

   This value of N_R is well within the laminar range; hence, the flow is most likely laminar.

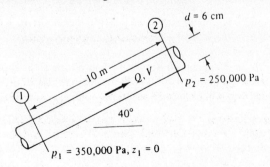

Fig. 9-1

9.7 For flow of SAE10 oil through a 5-cm-diameter pipe, for what flow rate in cubic meters per hour would we expect transition to turbulence at (a) 20 °C $[\mu = 0.104 \text{ kg}/(\text{m} \cdot \text{s})]$, and (b) 100 °C $[\mu = 0.0056 \text{ kg}/(\text{m} \cdot \text{s})]$?

⬛ Assume transition to turbulence occurs at $N_R = 2300$. $N_R = \rho d v / \mu$.

(a) $2300 = (869)(\frac{5}{100})(v)/0.104$ $v = 5.505 \text{ m/s}$

$$Q = Av = [(\pi)(\tfrac{5}{100})^2/4](5.505) = 0.0108 \text{ m}^3/\text{s} \quad \text{or} \quad 38.9 \text{ m}^3/\text{h}$$

(b) $2300 = (869)(\frac{5}{100})(v)/0.0056$ $v = 0.2964 \text{ m/s}$

$$Q = Av = [(\pi)(\tfrac{5}{100})^2/4](0.2964) = 0.000582 \text{ m}^3/\text{s} \quad \text{or} \quad 2.10 \text{ m}^3/\text{h}$$

9.8 A fluid at 20 °C flows at 700 cm³/s through an 8-cm-diameter pipe. Determine whether the flow is laminar or turbulent if the fluid is (a) hydrogen ($v = 1.08 \times 10^{-4} \text{ m}^2/\text{s}$), (b) air, (c) gasoline ($v = 4.06 \times 10^{-7} \text{ m}^2/\text{s}$), (d) water, (e) mercury ($v = 1.15 \times 10^{-7} \text{ m}^2/\text{s}$), or (f) glycerin.

⬛ $N_R = dv/v$ $v = Q/A = (700/1\,000\,000)/[(\pi)(\tfrac{8}{100})^2/4] = 0.1393 \text{ m/s}$

(a) $N_R = (\frac{8}{100})(0.1393)/(1.08 \times 10^{-4}) = 103$ (laminar)

(b) $N_R = (\frac{8}{100})(0.1393)/(1.51 \times 10^{-5}) = 738$ (laminar)

(c) $N_R = (\frac{8}{100})(0.1393)/(4.06 \times 10^{-7}) = 27\,448$ (turbulent)

(d) $N_R = (\frac{8}{100})(0.1393)/(1.02 \times 10^{-6}) = 10\,925$ (turbulent)

(e) $N_R = (\frac{8}{100})(0.1393)/(1.15 \times 10^{-7}) = 96\,904$ (turbulent)

(f) $N_R = (\frac{8}{100})(0.1393)/(1.18 \times 10^{-3}) = 9$ (laminar)

9.9 Oil (s.g. = 0.9, $v = 0.0003 \text{ m}^2/\text{s}$) enters a 4-cm-diameter tube. Estimate the entrance length if the flow rate is (a) 0.001 m³/s and (b) 1 m³/s.

⬛ $N_R = dv/v$

(a) $v = Q/A = 0.001/[(\pi)(\tfrac{4}{100})^2/4] = 0.7958 \text{ m/s}$ $N_R = (\frac{4}{100})(0.7958)/0.0003 = 106$ (laminar)

$$L_e/d = 0.06N_R \qquad L_e = (\tfrac{4}{100})(0.06)(106) = 0.254 \text{ m}$$

(b) $v = Q/A = 1/[(\pi)(\tfrac{4}{100})^2/4] = 795.8 \text{ m/s}$ $N_R = (\frac{4}{100})(795.8)/0.0003 = 106\,107$ (turbulent)

$$L_e/d = 4.4N_R^{1/6} \qquad L_e = (\tfrac{4}{100})(4.4)(106\,107)^{1/6} = 1.21 \text{ m}$$

9.10 What is the Reynolds number for a flow of oil (s.g. = 0.8, $\mu = 0.00200 \text{ lb} \cdot \text{s/ft}^2$) in a 6-in-diameter pipe at a flow rate of 20 ft³/s. Is the flow laminar or turbulent?

⬛
$$v = Q/A = 20/[(\pi)(\tfrac{6}{12})^2/4] = 101.9 \text{ ft/s}$$

$$N_R = \rho d v/\mu = [(0.8)(1.94)](\tfrac{6}{12})(101.9)/0.00200 = 39\,537 \quad \text{(turbulent)}$$

9.11 Gasoline at a temperature of 20 °C flows through a flexible pipe from the gas pump to the gas tank of a car. If 3 L/s is flowing and the pipe has an inside diameter of 60 mm, what is the Reynolds number?

⬛
$$v = Q/A = (\tfrac{3}{1000})/[(\pi)(\tfrac{60}{1000})^2/4] = 1.061 \text{ m/s}$$

$$N_R = \rho d v/\mu = (719)(\tfrac{60}{1000})(1.061)/(2.92 \times 10^{-4}) = 156\,752$$

9.12 The Reynolds number for fluid in a pipe of 10 in diameter is 1800. What will be the Reynolds number in a 6-in-diameter pipe forming an extension of the 10-in pipe? Take the flow as incompressible.

⬛ $N_R = dv/v$. Since v is constant, $[N_R/(dv)]_1 = [N_R/(dv)]_2$, $A_1v_1 = A_2v_2$, $[(\pi)(\tfrac{10}{12})^2/4](v_1) = [(\pi)(\tfrac{6}{12})^2/4](v_2)$, $v_1 = 0.360v_2$, $1800/[(\tfrac{10}{12})(0.360v_2)] = (N_R)_2/[(\tfrac{6}{12})(v_2)]$, $(N_R)_2 = 3000$.

9.13 Water is flowing through capillary tubes A and B into tube C, as shown in Fig. 9-2. If $Q_A = 2 \times 10^{-3}$ L/s in tube A, what is the largest Q_B allowable in tube B for laminar flow in tube C? The water is at a temperature of 40 °C. With the calculated Q_B, what kind of flow exists in tubes A and B?

⬛ For laminar flow, assume $N_R \le 2300$. $N_R = dv/v$. In tube C, $2300 = (\frac{6}{1000})(v_C)/(6.56 \times 10^{-7})$, $v_C = 0.2515$ m/s; $Q_C = A_C v_C = [(\pi)(\tfrac{6}{1000})^2/4](0.2515) = 7.11 \times 10^{-6}$ m³/s, or 7.11×10^{-3} L/s, $Q_B = (7.11 - 2) \times 10^{-3} = 5.11 \times 10^{-3}$ L/s. In tube A, $v_A = Q_A/A_A = [(2 \times 10^{-3})/1000]/[(\pi)(\tfrac{5}{1000})^2/4] = 0.1019$ m/s,

$$N_R = (\tfrac{5}{1000})(0.1019)/(6.56 \times 10^{-7}) = 777 \quad \text{(laminar)}$$

In tube B, $v_B = Q_B/A_B = [(5.11 \times 10^{-3})/1000]/[(\pi)(\frac{4}{1000})^2/4] = 0.4066$ m/s,

$$N_R = (\tfrac{4}{1000})(0.4066)/(6.56 \times 10^{-7}) = 2479 \quad \text{(turbulent)}$$

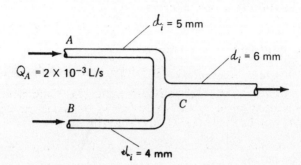

Fig. 9-2

9.14 An incompressible steady flow of water is present in a tube of constant cross section, as shown in Fig. 9-3. What is the head loss between sections A and B along the tube?

$$p_A/\gamma + v_A^2/2g + z_A = p_B/\gamma + v_B^2/2g + z_B + h_L$$

$$(100)(144)/62.4 + v_A^2/2g + 0 = (30)(144)/62.4 + v_B^2/2g + 100 + h_L$$

$$v_A^2/2g = v_B^2/2g \qquad h_L = 61.5 \text{ ft}$$

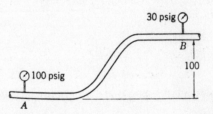

Fig. 9-3

9.15 Water is flowing through a pipe at the rate of 5 L/s, as shown in Fig. 9-4. If gage pressures of 12.0 kPa, 11.5 kPa, and 10.3 kPa are measured for p_1, p_2 and p_3, respectively, what are the head losses between 1 and 2 and 1 and 3?

$$p_1/\gamma + v_1^2/2g + z_1 = p_2/\gamma + v_2^2/2g + z_2 + (h_L)_{1-2} \qquad 12.0/9.79 + v_1^2/2g + 10 = 11.5/9.79 + v_2^2/2g + 10 + (h_L)_{1-2}$$

$$v_1^2/2g = v_2^2/2g \qquad (h_L)_{1-2} = 0.0511 \text{ m} \qquad p_1/\gamma + v_1^2/2g + z_1 = p_3/\gamma + v_3^2/2g + z_3 + (h_L)_{1-3}$$

$$v_1 = Q/A_1 = (\tfrac{5}{1000})/[(\pi)(\tfrac{50}{1000})^2/4] = 2.546 \text{ m/s} \qquad v_3 = Q/A_3 = (\tfrac{5}{1000})/[(\pi)(\tfrac{30}{1000})^2/4] = 7.074 \text{ m/s}$$

$$12.0/9.79 + 2.546^2/[(2)(9.807)] + 10 = 10.3/9.79 + 7.074^2/[(2)(9.807)] + 0 + (h_L)_{1-3} \qquad (h_L)_{1-3} = 7.95 \text{ m}$$

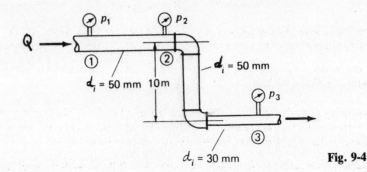

Fig. 9-4

9.16 A large oil reservoir has a pipe of 3 in diameter and 7000-ft length connected to it, as shown in Fig. 9-5. Assuming laminar flow through the pipe, compute the amount of oil issuing out of the pipe as a free jet.

Compute the velocity and Reynolds number to see if the flow is laminar. $\nu_{oil} = 1 \times 10^{-4}$ ft²/s. Neglect entrance losses to the pipe.

❚ $p_2/\gamma + v_2^2/2g + z_2 = p_3/\gamma + v_3^2/2g + z_3 + h_L$ $p_2/62.4 + v_2^2/2g + 0 = 0 + v_3^2/2g + 0 + h_f$ $v_2^2/2g = v_3^2/2g$

$$p_2/62.4 = h_f \tag{1}$$

$$Q = \frac{\pi \rho g d^4 h_f}{128 \mu L}$$

$$h_f = \frac{128 Q \mu L}{\pi \rho g d^4} = \frac{128[(\pi d^2/4)(v_2)]\mu L}{\pi \rho g d^4} = \frac{32 v_2 \nu L}{g d^2} = \frac{(32)(v_2)(1 \times 10^{-4})(7000)}{(32.2)(3/12)^2} = 11.13 v_2 \tag{2}$$

$$p_1/\gamma + v_1^2/2g + z_1 = p_2/\gamma + v_2^2/2g + z_2 + h_L \qquad 0 + 0 + 10 = p_2/62.4 + v_2^2/[(2)(32.2)] + 0 + 0$$

$$p_2/62.4 = 10 - v_2^2/[(2)(32.2)] \tag{3}$$

Equating h_f from Eqs. (1) and (2),

$$p_2/62.4 = 11.13 v_2 \tag{4}$$

Equating $p_2/62.4$ from Eqs. (3) and (4), $11.13 v_2 = 10 - v_2^2/[(2)(32.2)]$, $v_2^2 + 716.8 v_2 - 644 = 0$, $v_2 = 0.8973$ ft/s; $Q = Av = [(\pi)(\frac{3}{12})^2/4](0.8973) = 0.0440$ ft³/s; $N_R = dv/\nu = (\frac{3}{12})(0.8973)/(1 \times 10^{-4}) = 2243$ (barely laminar).

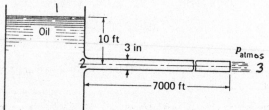

Fig. 9-5

9.17 If 140 L/s of water flows through the pipe system shown in Fig. 9-6, what total head loss is being developed over the length of the pipe?

❚
$$p_2/\gamma + v_2^2/2g + z_2 = p_3/\gamma + v_3^2/2g + z_3 + h_L$$

$$v_2 = Q/A_2 = (\tfrac{140}{1000})/[(\pi)(\tfrac{300}{1000})^2/4] = 1.981 \text{ m/s} \qquad v_3 = Q/A_3 = (\tfrac{140}{1000})/[(\pi)(\tfrac{150}{1000})^2/4] = 7.922 \text{ m/s}$$

$$p_2/9.97 + 1.981^2/[(2)(9.807)] + 0 = 0 + 7.922^2/[(2)(9.807)] + 16 + h_L \qquad h_L = p_2/9.79 - 19.00$$

$$p_1/\gamma + v_1^2/2g + z_1 = p_2/\gamma + v_2^2/2g + z_2 + h_L \qquad 0 + 0 + 30 = p_2/9.79 + 1.981^2/[(2)(9.807)] + 0 + 0$$

$$p_2/9.79 = 29.80 \text{ m} \qquad h_L = 29.80 - 19.00 = 10.80 \text{ m}$$

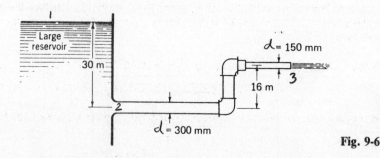

Fig. 9-6

9.18 Determine the maximum velocity for laminar flow for (a) medium fuel oil at 60 °F ($\nu = 4.75 \times 10^{-5}$ ft²/s) flowing through a 6-in pipe and (b) water at 60 °F flowing in the 6-in pipe.

❚ For laminar flow, assume $N_R \le 2000$. $N_R = dv/\nu$.
(a) $2000 = (\tfrac{6}{12})(v)/(4.75 \times 10^{-5})$ $v = 0.190$ ft/s
(b) $2000 = (\tfrac{6}{12})(v)(1.21 \times 10^{-5})$ $v = 0.0484$ ft/s

9.19   Determine the type of flow occurring in a 12-in pipe when (*a*) water at 60 °F flows at a velocity of 3.50 ft/s and (*b*) heavy fuel oil at 60 °F ($v = 221 \times 10^{-5}$ ft²/s) flows at the same velocity.

> $N_R = dv/v$
>
> (*a*) $\qquad\qquad\qquad\qquad N_R = (\tfrac{12}{12})(3.50)/(1.21 \times 10^{-5}) = 289\,256 \qquad$ (turbulent)
>
> (*b*) $\qquad\qquad\qquad\qquad N_R = (\tfrac{12}{12})(3.50)/(221 \times 10^{-5}) = 1584 \qquad$ (laminar)

9.20   For laminar flow conditions, what size pipe will deliver 90 gpm of medium fuel oil at 40 °F ($v = 6.55 \times 10^{-5}$ ft²/s)?

> $Q = (90)(0.002228) = 0.2005$ ft³/s. For laminar flow, assume $N_R \le 2000$. $N_R = dv/v$, $v = Q/A = 0.2005/(\pi\, d^2/4) = 0.2553 d^{-2}$, $2000 = (d)(0.2553 d^{-2})/(6.55 \times 10^{-5})$, $d = 1.95$ ft.

9.21   What is the Reynolds number of flow of 0.3 m³/s of oil (s.g. = 0.86, $\mu = 0.025$ N-s/m²) through a 450-mm-diameter pipe?

> $$v = Q/A = 0.3/[(\pi)(\tfrac{450}{1000})^2/4] = 1.886 \text{ m/s}$$
> $$N_R = \rho\, dv/\mu = [(0.86)(1000)](\tfrac{450}{1000})(1.886)/0.025 = 29\,195$$

9.22   An oil with s.g. = 0.85 and $v = 1.8 \times 10^{-5}$ m²/s flows in a 10-cm-diameter pipe at 0.50 L/s. Is the flow laminar or turbulent?

> $$v = Q/A = (0.50)(1000)/[(\pi)(\tfrac{10}{100})^2/4] = 0.06366 \text{ m/s}$$
> $$N_R = dv/v = (\tfrac{10}{100})(0.06366)/(1.8 \times 10^{-5}) = 354 \qquad \text{(laminar)}$$

9.23   An oil with a kinematic viscosity of 0.00015 ft²/s flows through a pipe of diameter 6 in. Below what velocity will the flow be laminar?

> For laminar flow, assume $N_R \le 2000$. $N_R = dv/v$, $2000 = (\tfrac{6}{12})(v)/0.00015$, $v = 0.600$ ft/s.

9.24   An oil with a kinematic viscosity of 0.005 ft²/s flows through a 3-in-diameter pipe with a velocity of 10 fps. Is the flow laminar or turbulent?

> $$N_R = dv/v = (\tfrac{3}{12})(10)/0.005 = 500 \qquad \text{(laminar)}$$

9.25   Hydrogen at atmospheric pressure and a temperature of 50 °F has a kinematic viscosity of 0.0011 ft²/s. Determine the maximum laminar flow rate in pounds per second in a 2-in-diameter pipe. $\gamma = 0.00540$ lb/ft³.

> For laminar flow, assume $N_R \le 2000$. $N_R = dv/v$, $2000 = (\tfrac{2}{12})(v)/0.0011$, $v = 13.20$ ft/s; $W = \gamma Av = (0.00540)[(\pi)(\tfrac{2}{12})^2/4](13.20) = 0.00156$ lb/s.

9.26   Air at a pressure of approximately 1500 kPa abs and a temperature of 100 °C flows in a 1.5-cm-diameter tube. What is the maximum laminar flow rate?

> For laminar flow, assume $N_R \le 2000$. $N_R = \rho dv/\mu$, $\rho = p/RT = (1500)(1000)/[(287)(273 + 100)] = 14.01$ kg/m³, $2000 = (14.01)(1.5/100)(v)/(2.17 \times 10^{-5})$, $v = 0.2065$ m/s; $Q = Av = [(\pi)(1.5/100)^2/4](0.2065) = 0.0000365$ m³/s, or 0.0365 L/s.

9.27   What is the hydraulic radius of a rectangular air duct 6 in by 14 in?

> $$R_h = A/p_w = [(6)(14)]/(6 + 6 + 14 + 14) = 2.10 \text{ in} \quad \text{or} \quad 0.175 \text{ ft}$$

9.28   What is the percentage difference between the hydraulic radii of a 20-cm-diameter circular and a 20-cm square duct?

> $$R_h = A/p_w$$
>
> $(R_h)_{\text{circular}} = [(\pi)(20)^2/4]/[(\pi)(20)] = 5.00$ cm $\qquad (R_h)_{\text{square}} = (20)(20)/(20 + 20 + 20 + 20) = 5.00$ cm

Since they are equal, the percentage difference is zero. Note that the hydraulic radius of a circular section is one-fourth its diameter.

9.29   Two pipes, one circular and one square, have the same cross-sectional area. Which has the larger hydraulic radius, and by what percentage?

▮ Let d = diameter of the circular pipe and a = the side of the square one. Since they have the same cross-sectional area, $\pi d^2/4 = a^2$, $a = \sqrt{\pi}d/2$; $(R_h)_{circular} = d/4 = 0.2500d$, $(R_h)_{square} = A/p_w = [(a)(a)]/(a + a + a + a) = a/4$. Since $a = \sqrt{\pi}d/2$, $(R_h)_{square} = (\sqrt{\pi}d/2)/4 = 0.2216d$, hence, the circular pipe has the larger hydraulic radius by $(0.2500 - 0.2216)/0.2216 = 0.128$, or 12.8 percent.

9.30   Steam with a specific weight of 0.25 lb/ft³ flows with a velocity of 100 fps through a circular pipe. The friction factor f was found to have a value of 0.016. What is the shearing stress at the wall?

▮ $\tau_0 = (f/4)(\gamma)(v^2/2g) = (0.016/4)(0.25)\{100^2/[(2)(32.2)]\} = 0.155$ lb/ft²

9.31   Glycerin at 68 °F flows 120 ft through a 6-in-diameter new wrought iron pipe at a velocity of 10.0 ft/s. Determine the head loss due to friction.

▮ $h_f = (f)(L/d)(v^2/2g)$ $N_R = \rho dv/\mu = (2.44)(\frac{6}{12})(10.0)/(3.11 \times 10^{-2}) = 392$

Since $N_R < 2000$, the flow is laminar and $f = 64/N_R = \frac{64}{392} = 0.1633$, $h_f = 0.1633[120/(\frac{6}{12})]\{10.0^2/[(2)(32.2)]\} = 60.9$ ft.

9.32   SAE10 oil flows through a cast iron pipe at a velocity of 1.0 m/s. The pipe is 45.0 m long and has a diameter of 150 mm. Find the head loss due to friction.

▮ $h_f = (f)(L/d)(v^2/2g)$ $N_R = \rho dv/\mu = (869)(\frac{150}{1000})(1.0)/0.0814 = 1601$

Since $N_R < 2000$, the flow is laminar and $f = 64/N_R = \frac{64}{1601} = 0.0400$, $h_f = 0.0400[45.0/(\frac{150}{1000})]\{1.0^2/[(2)(9.807)]\} = 0.612$ m.

9.33   The 6-cm-diameter pipe in Fig. 9-7 contains glycerin at 20 °C flowing at a rate of 8 m³/h. Verify that the flow is laminar. For the pressure measurements shown, is the flow up or down? What is the indicated head loss for these pressures?

▮ $v = Q/A = (\frac{8}{3600})/[(\pi)(\frac{6}{100})^2/4] = 0.7860$ m/s $N_R = \rho dv/\mu = (1258)(\frac{6}{100})(0.7860)/1.49 = 40$ (laminar)

           $HGL = z + p/\rho g$ $HGL_A = 0 + (2.1)(101\,400)/[(1258)(9.807)] = 17.26$ m

                  $HGL_B = 12 + (3.7)(101\,400)/[(1258)(9.807)] = 42.41$ m

Hence, the flow is from B to A (i.e., downward). Head loss $= 42.41 - 17.26 = 25.15$ m.

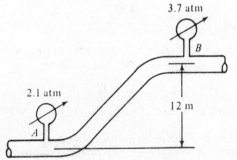

3.7 atm

B

2.1 atm

12 m

A

Fig. 9-7

9.34   In Prob. 9.33, compute the theoretical head loss if the pipe length is 30 m between A and B. Compare with the indicated pressure drop $p_B - p_A$.

▮ $Q = \dfrac{\pi \rho g d^4 h_f}{128 \mu L}$ $\dfrac{8}{3600} = \dfrac{(\pi)(1258)(9.807)(\frac{6}{100})^4(h_f)}{(128)(1.49)(30)}$

       $h_f = 25.13$ m $p_B/\rho g + v_B^2/2g + z_B = p_A/\rho g + v_A^2/2g + z_A + h_f$

       $(3.7)(101\,400)/[(1258)(9.807)] + v_B^2/2g + 12 = p_A/\rho g + v_A^2/2g + 0 + 25.31$

              $v_B^2/2g = v_A^2/2g$ $p_A/\rho g = 17.10$ m

   $p_A = (1258)(9.807)(17.10)/101\,400 = 2.08$ atm   (very close to original value of 2.1 atm)

9.35   Two infinite plates a distance h apart are parallel to the xz plane with the upper plate moving at speed V, as in Fig. 9-8. There is a fluid of viscosity μ and constant pressure between the plates. If $V = 4$ m/s and $h = 1.8$ cm, compute the shear stress at the upper and lower walls if the fluid is SAE30 oil at 20 °C.

▌ $N_R = \rho h v / \mu = (888)(1.8/100)(4)/0.440 = 145$. Since the flow is laminar,

$$\tau = \mu \frac{\partial u}{\partial y}$$

where $u = (V/h)(y)$, $\tau = (\mu)(V/h) = 0.440[4/(1.8/100)] = 97.8$ Pa.

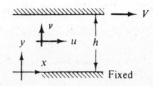

Fig. 9-8

9.36 Find the head loss per unit length when oil (s.g. = 0.9) of kinematic viscosity 0.007 ft²/s flows in a 3-in-diameter pipe at a rate of 5 gpm.

▌ $Q = (5)(0.002228) = 0.01114 \text{ ft}^3/\text{s}$ $h_f = (f)(L/d)(v^2/2g)$ $v = Q/A = 0.01114/[(\pi)(\frac{3}{12})^2/4] = 0.2269 \text{ ft/s}$

$N_R = dv/v = (\frac{3}{12})(0.2269)/0.007 = 8.10$

Since $N_R < 2000$, the flow is laminar and $f = 64/N_R = 64/8.10 = 7.90$, $h_f = 7.90[1/(\frac{3}{12})]\{0.2269^2/[(2)(32.2)]\} = 0.0253$ ft per unit length.

9.37 Tests made on a certain 12-in-diameter pipe showed that, when $V = 10$ fps, $f = 0.015$. The fluid used was water at 60 °F. Find the unit shear at the wall and at radii of 0, 0.2, 0.3, 0.5, and 0.75 times the pipe radius.

▌ $\tau_0 = (f/4)(\gamma)(V^2/2g) = (0.015/4)(62.4)\{10^2/[(2)(32.2)]\} = 0.3634 \text{ lb/ft}^2$

The stress distribution is linear; hence,

r/r_o	τ, lb/ft²
0	0
0.2	0.0727
0.3	0.1090
0.5	0.1817
0.75	0.2726

9.38 If oil with a kinematic viscosity of 0.005 ft²/s weighs 58 lb/ft³, what will be the flow rate and head loss in a 3000-ft length of 4-in-diameter pipe when the Reynolds number is 800?

▌ $N_R = dv/v$ $800 = (\frac{4}{12})(v)/0.005$ $v = 12.00 \text{ ft/s}$ $Q = Av = [(\pi)(\frac{4}{12})^2/4](12.00) = 1.047 \text{ ft}^3/\text{s}$

$f = 64/N_R = \frac{64}{800} = 0.0800$ $h_f = (f)(L/d)(v^2/2g) = 0.0800[3000/(\frac{4}{12})]\{12.00^2/[(2)(32.2)]\} = 1610 \text{ ft}$

9.39 How much power is lost per meter of pipe length when oil with a viscosity of 0.20 N · s/m² flows in a 20-cm-diameter pipe at 0.50 L/s? The oil has a density of 840 kg/m³.

▌ $v = Q/A = (0.50/1000)/[(\pi)(\frac{20}{100})^2/4] = 0.01592 \text{ m/s}$ $N_R = \rho dv / \mu = (840)(\frac{20}{100})(0.01592)/0.20 = 13.37$

$$h_f = (f)(L/d)(v^2/2g)$$

Since $N_R < 2000$, the flow is laminar and $f = 64/N_R = 64/13.37 = 4.787$, $h_f = 4.787[1/(\frac{20}{100})]\{0.01592^2/[(2)(9.807)]\} = 0.0003093 \text{ m}$, $P = Q\gamma h_f = Q\rho g h_f = (0.50/1000)(840)(9.807)(0.0003093) = 0.00127$ W per meter of pipe.

9.40 Calculate the discharge of the system in Fig. 9-9, neglecting all losses except through the pipe.

▌ Assuming laminar flow, $\mu = 0.1$ poise $= 0.01$ N · s/m² $= 0.0002089$ lb · s/ft²,

$$v = \frac{\gamma \, \Delta h \, d^2}{32 \mu L} = \frac{(55)(20)[(\frac{1}{4})/12]^2}{(32)(0.0002089)(16)} = 4.464 \text{ ft/s}$$

$N_R = \rho dv / \mu = (\gamma/g)(d)(v)/\mu = (55/32.2)[(\frac{1}{4})/12](4.464)/0.0002089 = 760$ (laminar)

$$Q = Av = \{(\pi)[(\frac{1}{4})/12]^2/4\}(4.464) = 0.00152 \text{ ft}^3/\text{s}$$

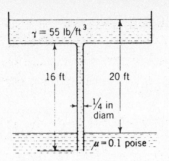

Fig. 9-9

9.41 In Fig. 9-10, $H = 24$ m, $L = 40$ m, $\theta = 30°$, $d = 8$ mm, $\gamma = 10$ kN/m³, and $\mu = 0.08$ kg/(m · s). Find the head loss per unit length of pipe and the discharge in liters per minute.

▌ Assuming laminar flow,

$$v = \frac{\gamma \, \Delta h \, d^2}{32 \mu L} = \frac{[(10)(1000)](24)(\frac{8}{1000})^2}{(32)(0.08)(40)} = 0.1500 \text{ m/s}$$

$$N_R = \rho dv/\mu = (\gamma/g)(d)(v)/\mu = [(10)(1000)/9.807](\tfrac{8}{1000})(0.1500)/0.08 = 15 \quad \text{(laminar)}$$

$$Q = Av = [(\pi)(\tfrac{8}{1000})^2/4](0.1500) = 0.000007540 \text{ m}^3/\text{s} = (0.000007540)(1000)(60) = 0.452 \text{ L/min}$$

$$\Delta h/L = \tfrac{24}{40} = 0.600 \text{ m/m}$$

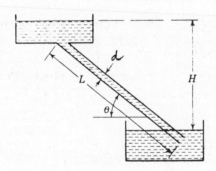

Fig. 9-10

9.42 For Fig. 9-10 and the data given in Prob. 9.41, find H if the velocity is 0.1 m/s.

▌ $$v = \frac{\gamma \, \Delta h \, d^2}{32 \mu L} \qquad 0.1 = \frac{[(10)(1000)](H)(\frac{8}{1000})^2}{(32)(0.08)(40)} \qquad H = 16.0 \text{ m}$$

9.43 A test on a 300-mm-diameter pipe with water showed a gage difference of 280 mm on a mercury–water manometer connected to two piezometer rings 120 m apart. The flow was 0.23 m³/s. What is the friction factor?

▌ $h_f = (f)(L/d)(v^2/2g)$. From the manometer, $h_f = (13.6/1 - 1)(\frac{280}{1000}) = 3.528$ m; $v = Q/A = 0.23/[(\pi)(\frac{300}{1000})^2/4] = 3.254$ m/s, $3.528 = (f)[120/(\frac{300}{1000})]\{3.254^2/[(2)(9.807)]\}$, $f = 0.0163$.

9.44 Use the Blasius equation for determination of friction factor to find the horsepower per mile required to pump 3.0 cfs liquid, $v = 3.3 \times 10^{-4}$ ft²/s, $\gamma = 55$ lb/ft³, through an 18-in pipeline.

▌ $$h_f = (f)(L/d)(v^2/2g) \qquad f = 0.316/N_R^{1/4} \qquad v = Q/A = 3.0/[(\pi)(\tfrac{18}{12})^2/4] = 1.698 \text{ ft/s}$$

$$N_R = dv/v = (\tfrac{18}{12})(1.698)/(3.3 \times 10^{-4}) = 7718 \qquad f = 0.316/7718^{1/4} = 0.03371$$

$$h_f = 0.03371[5280/(\tfrac{18}{12})]\{1.698^2/[(2)(32.2)]\} = 5.312 \text{ ft}$$

$$P = Q\gamma h_f = (3.0)(55)(5.312) = 876.5 \text{ ft} \cdot \text{lb/s per mile} \qquad 876.5/550 = 1.59 \text{ hp per mile}$$

9.45 Determine the head loss per kilometer required to maintain a velocity of 3 m/s in a 10-mm-diameter pipe. $v = 4 \times 10^{-5}$ m²/s.

▌ $$h_f = (f)(L/d)(v^2/2g) \qquad N_R = dv/v = (\tfrac{10}{1000})(3)/(4 \times 10^{-5}) = 750 \quad \text{(laminar)}$$

$$f = 64/N_R = \tfrac{64}{750} = 0.08533 \qquad h_f = 0.08533[1000/(\tfrac{10}{1000})]\{3^2/[(2)(9.807)]\} = 3915 \text{ m per km}$$

9.46 Fluid flows through a 10-mm-diameter tube at a Reynolds number of 1800. The head loss is 30 m in a 100-m length of tubing. Calculate the discharge in liters per minute.

▌ $h_f = (f)(L/d)(v^2/2g)$. Since $N_R < 2000$, flow is laminar and $f = 64/N_R = \frac{64}{1800} = 0.03556$, $30 = 0.03556[100/(\frac{10}{1000})]\{v^2/[(2)(9.807)]\}$, $v = 1.286$ m/s; $Q = Av = [(\pi)(\frac{10}{1000})^2/4](1.286) = 0.0001010$ m^3/s $= (0.0001010)(1000)(60) = 6.06$ L/min.

9.47 Oil of absolute viscosity 0.00210 lb · s/ft^2 and specific gravity 0.850 flows through 10 000 ft of 12-in-diameter cast iron pipe at the rate of 1.57 cfs. What is the lost head in the pipe?

▌
$$h_f = (f)(L/d)(v^2/2g) \qquad v = Q/A = 1.57/[(\pi)(\tfrac{12}{12})^2/4] = 1.999 \text{ ft/s}$$

$$N_R = \rho dv/\mu = [(0.850)(1.94)](\tfrac{12}{12})(1.999)/0.00210 = 1570 \qquad \text{(laminar)}$$

$$f = 64/N_R = \tfrac{64}{1570} = 0.04076 \qquad h_f = 0.04076[10\,000/(\tfrac{12}{12})]\{1.999^2/[(2)(32.2)]\} = 25.3 \text{ ft}$$

9.48 When first installed between two reservoirs, a 4-in-diameter metal pipe of length 6000 ft conveyed 0.20 cfs of water. (**a**) If after 15 years a chemical deposit had reduced the effective diameter of the pipe to 3.0 in, what then would be the flow rate? Assume f remains constant. Assume no change in reservoir levels. (**b**) What would be the flow rate if in addition to the diamater change, f had doubled in value?

▌ (**a**) $(f_1)(L_1/d_1)(v_1^2/2g) = (f_2)(L_2/d_2)(v_2^2/2g)$. Since f, L, and g are constant and $v = Q/A = Q/(\pi d^2/4)$, $Q_1^2/d_1^5 = Q_2^2/d_2^5$, $0.20^2/4^5 = Q_2^2/3.0^5$, $Q_2 = 0.0974$ cfs.
 (**b**) $(f_1)(Q_1^2/d_1^5) = (f_2)(Q_2^2/d_2^5)$. Since $f_2 = 2f_1$, $Q_1^2/d_1^5 = (2)(Q_2^2/d_2^5)$, $0.20^2/4^5 = (2)(Q_2^2/3.0^5)$, $Q_2 = 0.0689$ cfs.

9.49 A liquid with $\gamma = 58$ lb/ft^3 flows by gravity through a 1-ft tank and a 1-ft capillary tube at a rate of 0.15 ft^3/h, as shown in Fig. 9-11. Sections 1 and 2 are at atmospheric pressure. Neglecting entrance effects, compute the viscosity of the liquid in slugs per foot-second.

▌ $p_1/\gamma + v_1^2/2g + z_1 = p_2/\gamma + v_2^2/2g + z_2 + h_L \qquad v_2 = Q/A_2 = (0.15/3600)/[(\pi)(0.004)^2/4] = 3.316$ ft/s

$$0 + 0 + (1 + 1) = 0 + 3.316^2/[(2)(32.2)] + 0 + h_f \qquad h_f = 1.829 \text{ ft}$$

Assuming laminar flow,

$$h_f = \frac{32\mu L v}{\gamma d^2} \qquad 1.829 = \frac{(32)(\mu)(1)(3.316)}{(58)(0.004)^2} \qquad \mu = 1.600 \times 10^{-5} \text{ slug/(ft} \cdot \text{s)}$$

$$N_R = \rho dv/\mu = (\gamma/g)(d)(v)/\mu = (58/32.2)(0.004)(3.316)/(1.600 \times 10^{-5}) = 1493 \qquad \text{(laminar)}$$

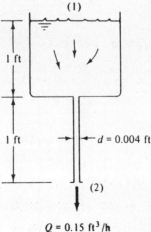

$Q = 0.15$ ft^3/h **Fig. 9-11**

9.50 In Prob. 9.49, suppose the flow rate is unknown but the liquid viscosity is 2×10^{-5} slug/(ft · s). What will be the flow rate in cubic feet per hour? Is the flow still laminar?

▌
$$p_1/\gamma + v_1^2/2g + z_1 = p_2/\gamma + v_2^2/2g + z_2 + h_L$$

$$0 + 0 + (1 + 1) = 0 + v^2/[(2)(32.2)] + 0 + h_f \qquad h_f = 2 - 0.01553v^2$$

Assuming laminar flow,

$$h_f = 32\mu Lv/\gamma d^2 \qquad 2 - 0.01553v^2 = (32)(2 \times 10^{-5})(1)(v)/[(58)(0.004)^2]$$

$$v^2 + 44.41v - 128.8 = 0 \qquad v = 2.732 \text{ ft/s}$$

$$Q = Av = [(\pi)(0.004)^2/4](2.732) = 0.00003433 \text{ ft}^3/\text{s} = (0.00003433)(3600) = 0.124 \text{ ft}^3/\text{h}$$

$$N_R = \rho dv/\mu = (\gamma/g)(d)(v)/\mu = (58/32.2)(0.004)(2.732)/(2 \times 10^{-5}) = 984 \qquad \text{(laminar)}$$

9.51 A steady push on the piston in Fig. 9-12 causes a flow rate of 0.4 cm³/s through the needle. The fluid has $\rho = 900 \text{ kg/m}^3$ and $\mu = 0.002 \text{ kg/(m} \cdot \text{s)}$. What force F is required to maintain the flow? Neglect head loss in the larger cylinder.

▮ $\quad p_A/\rho g + v_A^2/2g + z_A = p_B/\rho g + v_B^2/2g + z_B + h_L \qquad v_B = Q/A_B = 0.4/[(\pi)(0.25/10)^2/4] = 814.9 \text{ cm/s}$

$$N_R = \rho dv/\mu = (900)(0.25/1000)(814.9/100)/0.002 = 917 \qquad \text{(laminar)}$$

$$h_f = \frac{32\mu Lv}{\rho g d^2} = \frac{(32)(0.002)(1.5/100)(814.9/100)}{(900)(9.807)(0.25/1000)^2} = 14.18 \text{ m}$$

$p_A/[(900)(9.807)] + 0 + 0 = p_B/[(900)(9.807)] + (814.9/100)^2/[(2)(9.807)] + 0 + 14.18 \qquad p_A - p_B = $ 155 040 N/m²

$$F = (p_A - p_B)(A_{\text{piston}}) = 155\,040[(\pi)(\tfrac{1}{100})^2/4] = 12.2 \text{ N}$$

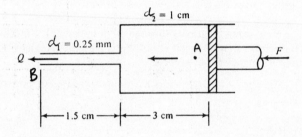

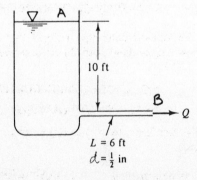

Fig. 9-12

9.52 An oil (s.g. = 0.9) issues from the pipe in Fig. 9-13 at $Q = 45 \text{ ft}^3/\text{h}$. What is the kinematic viscosity of the oil? Is the flow laminar?

▮ $\quad p_A/\rho g + v_A^2/2g + z_A = p_B/\rho g + v_B^2/2g + z_B + h_L \qquad v_B = Q/A_B = (\tfrac{45}{3600})/[(\pi)(0.5/12)^2/4] = 9.167 \text{ ft/s}$

$$0 + 0 + 10 = 0 + 9.167^2/[(2)(32.2)] + 0 + h_f \qquad h_f = 8.695 \text{ ft}$$

Assuming laminar flow,

$$h_f = \frac{128\nu LQ}{\pi g d^4} \qquad 8.695 = \frac{(128)(\nu)(6)(\tfrac{45}{3600})}{(\pi)(32.2)(0.5/12)^4} \qquad \nu = 0.000276 \text{ ft}^2/\text{s}$$

$$N_R = dv/\nu = (0.5/12)(9.167)/0.000276 = 1384 \qquad \text{(laminar)}$$

9.53 In Prob. 9.52, what will the flow rate be if the fluid is SAE10 oil at 20 °C? Use $\rho = 1.78 \text{ slugs/ft}^3$ and $\mu = 0.00217 \text{ lb} \cdot \text{s/ft}^2$.

▮ $\quad p_A/\rho g + v_A^2/2g + z_A = p_B/\rho g + v_B^2/2g + z_B + h_L \qquad 0 + 0 + 10 = 0 + v^2/[(2)(32.2)] + 0 + h_f$

Assuming laminar flow,

$$h_f = \frac{32\mu Lv}{\rho g d^2} = \frac{(32)(0.00217)(6)(v)}{(1.78)(32.2)(0.5/12)^2} = 4.187v$$

$$0 + 0 + 10 = 0 + v^2/[(2)(32.2)] + 0 + 4.187v \qquad v^2 = 269.6v - 644.0 = 0 \qquad v = 2.368 \text{ ft/s}$$

$$Q = Av = [(\pi)(0.5/12)^2/4](2.368) = 0.003229 \text{ ft}^3/\text{s} \quad \text{or} \quad 11.6 \text{ ft}^3/\text{h}$$

$$N_R = \rho dv/\mu = (1.78)(0.5/12)(2.368)/0.00217 = 81 \qquad \text{(laminar)}$$

9.54 For the configuration shown in Fig. 9-14, the fluid is ethanol at 20 °C and the tanks are very wide. Find the flow rate. Is the flow laminar?

I

$$p_A/\gamma + v_A^2/2g + z_A = p_B/\gamma + v_B^2/2g + z_B + h_L \qquad 0 + 0 + (40 + 50)/100 = 0 + 0 + 0 + h_f$$

$$h_f = 0.9000 \text{ m} = \frac{128\mu LQ}{\pi\rho g d^4} \qquad 0.9000 = \frac{(128)(1.20 \times 10^{-3})[(80 + 40)/100](Q)}{(\pi)(788)(9.807)(\frac{2}{1000})^4}$$

$$Q = 1.897 \times 10^{-6} \text{ m}^3/\text{s} \quad \text{or} \quad 0.00683 \text{ m}^3/\text{h}$$

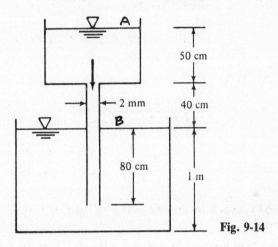

Fig. 9-14

9.55 For the system in Fig. 9-14, if the fluid has density of 920 kg/m³ and the flow rate is unknown, for what value of viscosity will the capillary Reynolds number exactly equal the critical value of 2300?

I

$$h_f = 0.9000 \text{ m} = \frac{32\mu Lv}{\rho g d^2} \qquad \text{(from Prob. 9.54)}$$

$$0.9000 = \frac{(32)(\mu)[(80 + 40)/100](v)}{(920)(9.807)(\frac{2}{1000})^2} \qquad v = \frac{0.0008459}{\mu}$$

$$N_R = \rho dv/\mu \qquad 2300 = (920)(\tfrac{2}{1000})(0.0008459/\mu)/\mu \qquad \mu = 0.000823 \text{ kg/(m} \cdot \text{s)}$$

9.56 SAE30 oil at 20 °C flows in the 3-cm-diameter pipe in Fig. 9-15. For the pressure measurements shown, determine (*a*) whether the flow is up or down, and (*b*) the flow rate. Use $\rho = 917$ kg/m³ and $\mu = 0.290$ kg/(m · s).

I HGL $= z + p/\rho g$

(*a*)
$$\text{HGL}_B = 15 + (180)(1000)/[(917)(9.807)] = 35.02 \text{ m}$$
$$\text{HGL}_A = 0 + (500)(1000)/[(917)(9.807)] = 55.60 \text{ m}$$

Since HGL$_A$ > HGL$_B$, the flow is from A to B (i.e., up).
(*b*) Assume flow is laminar.

$$h_f = \frac{128\mu LQ}{\pi\rho g d^4} = 55.60 - 35.02 = 20.58 \text{ m} \qquad L = \sqrt{15^2 + 20^2} = 25.00 \text{ m} \qquad 20.58 = \frac{(128)(0.290)(25.00)(Q)}{(\pi)(917)(9.807)(\frac{3}{100})^4}$$

$$Q = 0.000508 \text{ m}^3/\text{s} \quad \text{or} \quad 1.83 \text{ m}^3/\text{h} \qquad v = Q/A = 0.000508/[(\pi)(\tfrac{3}{100})^2/4] = 0.7187 \text{ m/s}$$

$$N_R = \rho dv/\mu = (917)(\tfrac{3}{100})(0.7187)/0.290 = 68 \qquad \text{(laminar)}$$

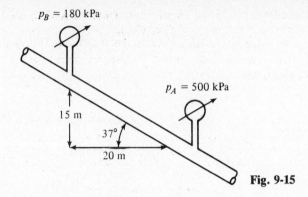

$p_B = 180$ kPa

$p_A = 500$ kPa

15 m

37°

20 m

Fig. 9-15

9.57 Modify and repeat Prob. 9.56 if the pressures are the same but there is a pump between A and B which adds a 10-m head rise in the flow direction. Verify that the flow is laminar.

▌ $h_f = \text{HGL}_A - \text{HGL}_B + h_{\text{pump}}$. Using values of HGL_A and HGL_B from Prob. 9.56,

$$h_f = 55.60 - 35.02 + 10 = 30.58 \text{ m} = \frac{128\mu LQ}{\pi\rho g d^4} \qquad L = 25.00 \text{ m} \qquad \text{(from Prob. 9.56)}$$

$$30.58 = \frac{(128)(0.290)(25.00)(Q)}{(\pi)(917)(9.807)(\frac{3}{100})^4} \qquad Q = 0.000754 \text{ m}^3/\text{s} \quad \text{or} \quad 2.71 \text{ m}^3/\text{h}$$

$$v = Q/A = 0.000754/[(\pi)(\tfrac{3}{100})^2/4] = 1.067 \text{ m/s} \qquad N_R = \rho dv/\mu = (917)(\tfrac{3}{100})(1.067)/0.290 = 101 \qquad \text{(laminar)}$$

9.58 Water at 40 °C flows from tank A to tank B as shown in Fig. 9-16. What is the volumetric flow at the configuration shown? Neglect entrance losses to the capillary tube as well as exit losses.

▌ $\qquad p_A/\gamma + v_A^2/2g + z_A = p_B/\gamma + v_B^2/2g + z_B + h_L \qquad 0 + 0 + (0.3 + 0.1) = 0 + 0 + 0 + h_f \qquad h_f = 0.4$

Assume laminar flow.

$$h_f = \frac{128\mu LQ}{\pi\rho g d^4} \qquad 0.4 = \frac{(128)(6.51 \times 10^{-4})(0.3 + 0.08)(Q)}{(\pi)(992)(9.807)(\frac{1}{1000})^4} \qquad Q = 3.86 \times 10^{-7} \text{ m}^3/\text{s} \quad \text{or} \quad 3.86 \times 10^{-4} \text{ L/s}$$

$$v = Q/A = 3.86 \times 10^{-7}/[(\pi)(\tfrac{1}{1000})^2/4] = 0.4915 \text{ m/s}$$

$$N_R = \rho dv/\mu = (992)(\tfrac{1}{1000})(0.4915)/(6.51 \times 10^{-4}) = 749 \qquad \text{(laminar)}$$

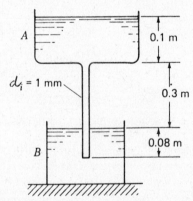

A

0.1 m

$d_i = 1$ mm

0.3 m

0.08 m

B

Fig. 9-16

9.59 In Prob. 9.58, what should the internal diameter of the tube be to permit a flow of 6×10^{-4} L/s?

▌ $\qquad\qquad\qquad\qquad h_f = 0.4 \qquad \text{(from Prob. 9.58)}$

Assume laminar flow.

$$h_f = \frac{128\mu LQ}{\pi\rho g d^4} \qquad 0.4 = \frac{(128)(6.51 \times 10^{-4})(0.3 + 0.08)[(6 \times 10^{-4})/1000]}{(\pi)(992)(9.807)(d/1000)^4} \qquad d = 1.117 \text{ mm}$$

$$v = Q/A = (6 \times 10^{-4}/1000)/[(\pi)(1.117/1000)^2/4] = 0.6123 \text{ m/s}$$

$$N_R = \rho dv/\mu = (992)(1.117/1000)(0.6123)/(6.51 \times 10^{-4}) = 1042 \qquad \text{(laminar)}$$

9.60 A hypodermic needle has an inside diameter of 0.3 mm and is 60 mm in length, as shown in Fig. 9-17. If the piston moves to the right at a speed of 18 mm/s and there is no leakage, what force F is needed on the piston? The medicine in the hypodermic has a viscosity μ of $0.980 \times 10^{-3}\,\text{N} \cdot \text{s/m}^2$ and its density ρ is 800 kg/m³. Consider flows in both needle and cylinder. Neglect exit losses from the needle as well as losses at the juncture of the needle and cylinder.

▌ For cylinder:

$$Q = Av = [(\pi)(\tfrac{5}{1000})^2/4](\tfrac{18}{1000}) = 3.534 \times 10^{-7}\,\text{m}^3/\text{s}$$

$$N_R = \rho dv/\mu = (800)(\tfrac{5}{1000})(\tfrac{18}{1000})/(0.980 \times 10^{-3}) = 73 \quad \text{(laminar)}$$

$$p = \frac{128\mu LQ}{\pi d^4} \qquad p_1 = \frac{(128)(0.980 \times 10^{-3})(\tfrac{50}{1000})(3.534 \times 10^{-7})}{(\pi)(\tfrac{5}{1000})^4} = 1.129\,\text{Pa}$$

For needle:

$$v = Q/A = 3.534 \times 10^{-7}/[(\pi)(0.3/1000)^2/4] = 5.000\,\text{m/s}$$

$$N_R = (800)(0.3/1000)(5.000)/(0.980 \times 10^{-3}) = 1224 \quad \text{(laminar)}$$

$$p_2 = \frac{(128)(0.980 \times 10^{-3})(\tfrac{60}{1000})(3.534 \times 10^{-7})}{(\pi)(0.3/1000)^4} = 104\,525\,\text{Pa}$$

$$F = (\Delta p)(A_{\text{cylinder}}) = (104\,525 - 1.129)[(\pi)(\tfrac{5}{1000})^2/4] = 2.05\,\text{N}$$

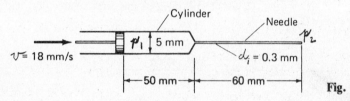

Fig. 9-17

9.61 In Prob. 9.60, suppose that you are drawing the medicine from a container at atmospheric pressure. What is the largest flow of fluid if the fluid has a vapor pressure of 4700 Pa abs? Neglect losses in the cylinder. What is the speed of the cylinder for the maximum flow of medicine if there is a 10 percent leakage around the piston for the pressure of 4700 Pa abs in the cylinder?

▌
$$\Delta p = 101\,400 - 4700 = 96\,700\,Pa = \frac{128\mu LQ}{\pi d^4} \qquad 96\,700 = \frac{(128)(0.980 \times 10^{-3})(\tfrac{60}{1000})(Q_{\text{max}})}{(\pi)(0.3/1000)^4}$$

$$Q_{\text{max}} = 3.27 \times 10^{-7}\,\text{m}^3/\text{s}$$

$$v_{\text{max}} = (Q_{\text{max}}/A)/0.90 = \{3.27 \times 10^{-7}/[(\pi)(\tfrac{5}{1000})^2/4]\}/0.90 = 0.0185\,\text{m/s} \quad \text{or} \quad 18.5\,\text{mm/s}$$

9.62 In Prob. 9.60, it took a force of 2.05 N to move the piston to the right at a speed of 18 mm/s. What should the inside diameter be for the cylinder if the force needed is only 1 N for the same piston speed? Neglect losses in cylinder.

▌
$$p_1 = \frac{F}{A} = \frac{1}{(\pi)(\tfrac{5}{1000})^2/4} = 50\,930\,\text{Pa} \qquad p = \frac{128\mu LQ}{\pi d^4} \qquad 50\,930 = \frac{(128)(0.980 \times 10^{-3})(\tfrac{60}{1000})(3.534 \times 10^{-7})}{(\pi)(d/1000)^4}$$

$$d = 0.359\,\text{mm} \qquad N_R = \rho dv/\mu = (800)(0.350/1000)(\tfrac{18}{1000})/(0.980 \times 10^{-3}) = 5 \quad \text{(laminar)}$$

9.63 Water at 70 °F flows through a new cast iron pipe at a velocity of 9.7 ft/s. The pipe is 1200 ft long and has a diameter of 6 in. Find the head loss due to friction.

▌
$$h_f = (f)(L/d)(v^2/2g) \qquad N_R = dv/v = (\tfrac{6}{12})(9.7)/(1.05 \times 10^{-5}) = 461\,905$$

From Table A-9, $\epsilon = 0.00085$ ft for new cast iron pipe; $\epsilon/d = 0.00085/(\tfrac{6}{12}) = 0.0017$. From Fig. A-5, $f = 0.0230$; $h_f = 0.0230[1200/(\tfrac{6}{12})]\{9.7^2/[(2)(32.2)]\} = 80.6$ ft.

9.64 A 96-in-diameter new cast iron pipe carries water at 60 °F. The head loss due to friction is 1.5 ft per 1000 ft of pipe. What is the discharge capacity of the pipe?

▌
$$h_f = (f)(L/d)(v^2/2g) \qquad 1.5 = (f)[1000/(\tfrac{96}{12})]\{v^2/[(2)(32.2)]\} \qquad fv^2 = 0.7728$$

Assume $f = 0.0150$; $(0.0150)(v^2) = 0.7728$, $v = 7.178$ ft/s; $N_R = dv/v = (\tfrac{96}{12})(7.178)/(1.21 \times 10^{-5}) = 4.75 \times 10^6$.

From Table A-9, $\epsilon = 0.00085$ ft for new cast iron pipe, $\epsilon/d = 0.00085/(\frac{96}{12}) = 0.000106$. From Fig. A-5, $f = 0.0124$. Evidently, the assumed value of f of 0.0150 was not the correct one. Try a value of f of 0.0124. $(0.0124)(v^2) = 0.7728$, $v = 7.894$ ft/s; $N_R = (\frac{96}{12})(7.894)/(1.21 \times 10^{-5}) = 5.22 \times 10^6$. From Fig. A-5, $f = 0.0124$. Hence, 0.0124 must be the correct value of f, and $v = 7.894$ ft/s. $Q = Av = [(\pi)(\frac{96}{12})^2/4](7.894) = 397$ ft³/s.

9.65 Water at 70 °F is being drained from an open tank through a 24-in-diameter, 130-ft-long new cast iron pipe, as shown in Fig. 9-18. Find the flow rate at which water is being discharged from the pipe. Neglect minor losses.

┃

$$p_1/\gamma + v_1^2/2g + z_1 = p_2/\gamma + v_2^2/2g + z_2 + h_L$$

$$h_L = h_f = (f)(L/d)(v^2/2g) = (f)[130/(\tfrac{24}{12})]\{v_2^2/[(2)(32.2)]\} = 1.009fv_2^2$$

$$0 + 0 + 150.5 = 0 + v_2^2/[(2)(32.2)] + 98.4 + 1.009fv_2^2$$

Assume $f = 0.0240$.

$$150.5 = v_2^2/[(2)(32.2)] + 98.4 + (1.009)(0.0240)(v_2^2) \qquad v_2 = 36.21 \text{ ft/s}$$

$$N_R = dv/\nu = (\tfrac{24}{12})(36.21)/(1.05 \times 10^{-5}) = 6.90 \times 10^6$$

From Table A-9, $\epsilon = 0.00085$, $\epsilon/d = 0.00085/(\tfrac{24}{12}) = 0.000425$. From Fig. A-5, $f = 0.0162$. Evidently, the assumed value of f of 0.0240 was not the correct one. Try a value of f of 0.0162.

$$150.5 = v_2^2/[(2)(32.2)] + 98.4 + (1.009)(0.0162)(v_2^2) \qquad v_2 = 40.43 \text{ ft/s}$$

$$N_R = (\tfrac{24}{12})(40.43)/(1.05 \times 10^{-5}) = 7.70 \times 10^6$$

From Fig. A-5, $f = 0.0162$. Hence, 0.0162 must be the correct value of f, and $v = 40.43$ ft/s. $Q = Av = [(\pi)(\tfrac{24}{12})^2/4](40.43) = 127$ ft³/s.

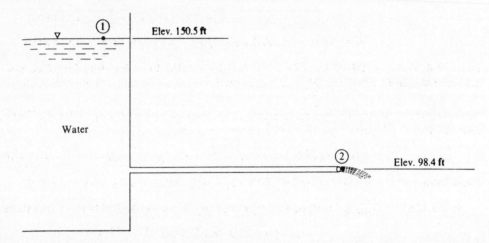

Water

① Elev. 150.5 ft

② Elev. 98.4 ft

Fig. 9-18

9.66 Gasoline is being discharged from a pipe, as shown in Fig. 9-19. The pipe roughness (ϵ) is 0.500 mm, and the pressure at point 1 is 2500 kPa. Find the pipe diameter needed to discharge gasoline at a rate of 0.10 m³/s. Neglect any minor losses.

┃ $\quad p_1/\gamma + v_1^2/2g + z_1 = p_2/\gamma + v_2^2/2g + z_2 + h_L$

$$h_L = h_f = (f)(L/d)(v^2/2g) = (f)(965.5/d)\{v_2^2/[(2)(9.807)]\} = 49.23fv_2^2/d$$

$$2.500/7.05 + v_1^2/2g + 82.65 = 0 + v_2^2/2g + 66.66 + 49.23fv_2^2/d \qquad v_1^2/2g = v_2^2/2g$$

$$fv_2^2/d = 0.3320 \qquad v_2 = Q/A_2 = 0.10/(\pi d^2/4) = 0.1273/d^2 \qquad (f)(0.1273/d^2)^2/d = 0.3320 \qquad d = (0.04881f)^{1/5}$$

Assume $f = 0.0200$. $d = [(0.04881)(0.0200)]^{1/5} = 0.2500$ m, $v_2 = 0.1273/0.2500^2 = 2.037$ m/s; $N_R = \rho dv/\mu = (719)(0.2500)(2.037)/(2.92 \times 10^{-4}) = 1.25 \times 10^6$. From Table A-9, $\epsilon = 0.00050$ m. $\epsilon/d = 0.00050/0.2500 = 0.0020$. From Fig. A-5, $f = 0.0235$. Evidently, the assumed value of f of 0.0200 was not the correct one. Try a value of f of 0.0235.

$$d = [(0.04881)(0.0235)]^{1/5} = 0.2582 \text{ m} \qquad v = 0.1273/0.2582^2 = 1.909 \text{ m/s}$$

$$N_R = (719)(0.2582)(1.909)/(2.92 \times 10^{-4}) = 1.21 \times 10^6$$

$$\epsilon/d = 0.00050/0.2582 = 0.00194 \qquad f = 0.0235$$

Hence, 0.0235 must be the correct value of f, and $d = 0.2582$ m.

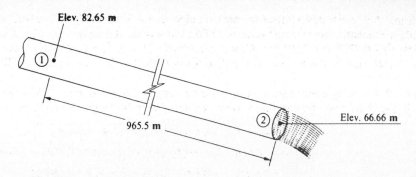

Fig. 9-19

9.67 Water at 20 °C flows through a new cast iron pipe at a velocity of 4.2 m/s. The pipe is 400 m long and has a diameter of 150 mm. Determine the head loss due to friction.

$$h_f = (f)(L/d)(v^2/2g) \qquad N_R = dv/v = (\tfrac{150}{1000})(4.2)/(1.02 \times 10^{-6}) = 6.18 \times 10^5$$

From Table A-9, $\epsilon = 0.00026$ m. $\epsilon/d = 0.00026/0.150 = 0.00173$. From Fig. A-5, $f = 0.0226$. $h_f = 0.0226[400/(\tfrac{150}{1000})]\{4.2^2/[(2)(9.807)]\} = 54.20$ m.

9.68 SAE10 oil at 68 °F is to be pumped at a flow rate of 2.0 ft³/s through a level 6-in-diameter new wrought iron pipe. Determine the pressure loss in pounds per square inch per mile of pipe and compute the horsepower lost to friction.

$$h_f = (f)(L/d)(v^2/2g) \qquad v = Q/A = 2.0/[(\pi)(\tfrac{6}{12})^2/4] = 10.19 \text{ ft/s}$$

$$N_R = \rho dv/\mu = (1.68)(\tfrac{6}{12})(10.19)/(1.70 \times 10^{-3}) = 5035 \qquad \text{(turbulent)}$$

From Table A-9, $\epsilon = 0.00015$ ft. $\epsilon/d = 0.00015/(\tfrac{6}{12}) = 0.00030$. From Fig. A-5, $f = 0.038$. $h_f = 0.038[5280/(\tfrac{6}{12})]\{10.19^2/[(2)(32.2)]\} = 647$ ft of oil; $p = \gamma h = (54.2)(647)/144 = 244$ psi/mile.

9.69 Water at 20 °C flows in a 100-mm-diameter new cast iron pipe with a velocity of 5.0 m/s. Determine the pressure drop in kilopascals per 100 m of pipe and the power lost to friction.

$$h_f = (f)(L/d)(v^2/2g) \qquad N_R = dv/v = (\tfrac{100}{1000})(5.0)/(1.02 \times 10^{-6}) = 4.90 \times 10^5$$

From Table A-9, $\epsilon = 0.00026$ m. $\epsilon/d = 0.00026/(\tfrac{100}{1000}) = 0.0026$. From Fig. A-5, $f = 0.0252$.

$$h_f = 0.0252[100/(\tfrac{100}{1000})]\{5.0^2/[(2)(9.807)]\} = 32.12 \text{ m} \qquad p = (9.79)(32.12) = 314 \text{ kN/m}^2 \text{ per 100 m of pipe}$$

$$Q = Av = [(\pi)(\tfrac{100}{1000})^2/4](5.0) = 0.03927 \text{ m}^3/\text{s}$$

$$\text{Power lost} = Q\gamma h_f = (0.03927)(9.79)(32.12) = 12.35 \text{ kW per 100 m of pipe}$$

9.70 Determine the discharge capacity of a 150-mm-diameter new wrought iron pipe to carry water at 20 °C if the pressure loss due to friction may not exceed 35 kPa per 100 m of level pipe.

$$N_R = dv/v = (\tfrac{150}{1000})(v)/(1.02 \times 10^{-6}) = 1.47 \times 10^5 v$$

Trial No. 1
Assume $v = 3.0$ m/s: $N_R = (1.47 \times 10^5)(3.0) = 4.41 \times 10^5$, $\epsilon/d = 0.000046/(\tfrac{150}{1000}) = 0.000307$. From Fig. A-5, $f = 0.0164$, $h_f = (f)(L/d)(v^2/2g) = p/\gamma = 35/9.79 = 3.575$ m, $3.575 = 0.0164[100/(\tfrac{150}{1000})]\{v^2/[(2)(9.807)]\}$, $v = 2.53$ m/s.

Trial No. 2
Assume $v = 2.53$ m/s: $N_R = (1.47 \times 10^5)(2.53) = 3.72 \times 10^5$, $f = 0.0166$, $3.575 = 0.0166[100/(\tfrac{150}{1000})]\{v^2/[(2)(9.807)]\}$, $v = 2.52$ m/s; $Q = Av = [(\pi)(\tfrac{150}{1000})^2/4](2.52) = 0.0445$ m³/s.

9.71 SAE30 oil at 68 °F is to be pumped at a flow rate of 3.0 ft³/s through a level new cast iron pipe. Allowable pipe friction loss is 10 psi per 1000 ft of pipe. What size commercial pipe should be used?

$$p = \gamma h \qquad (10)(144) = 55.4 h_f \qquad h_f = 26.0 \text{ ft of oil per 1000 ft of pipe}$$

Trial No. 1
Assume $v = 5.0$ ft/s:

$$Q = A/v \qquad 3.0 = (\pi d^2/4)(5.0) \qquad d = 0.874 \text{ ft}$$
$$N_R = \rho dv/\mu = (1.72)(0.874)(5.0)/(9.2 \times 10^{-3}) = 817 \qquad \text{(laminar)}$$
$$h_f = (f)(L/d)(v^2/2g) \qquad f = 64/N_R = \tfrac{64}{817} = 0.0783$$
$$26.0 = (0.0783)(1000/0.874)\{v^2/[(2)(32.2)]\} \qquad v = 4.32 \text{ ft/s}$$

Trial No. 2
Assume $v = 4.32$ ft/s:

$$3.0 = (\pi d^2/4)(4.32) \qquad d = 0.940 \text{ ft} \qquad N_R = (1.72)(0.940)(4.32)/(9.2 \times 10^{-3}) = 759 \qquad \text{(laminar)}$$
$$f = \tfrac{64}{759} = 0.0843 \qquad 26.0 = (0.0843)(1000/0.940)\{v^2/[(2)(32.2)]\} \qquad v = 4.32 \text{ ft/s}$$

Hence, a pipe diameter of 0.940 ft, or 11.28 in, would be required. A 12-in-diameter commercial pipe should be used, which would result in a pipe friction loss somewhat less than the allowable 10 psi per 1000 ft of pipe.

9.72 SAE10 oil at 20 °C is to flow through a 300-m level concrete pipe. What size pipe will carry 0.0142 m³/s with a pressure drop due to friction of 23.94 kPa?

▮ $\qquad\qquad p = \gamma h \qquad 23.94 = 8.52 h_f \qquad h_f = 2.81 \text{ m}$

Trial No. 1
Assume $v = 1.5$ m/s:

$$Q = A/v \qquad 0.0142 = (\pi d^2/4)(1.5) \qquad d = 0.110 \text{ m}$$
$$N_R = \rho dv/\mu = (869)(0.110)(1.5)/(8.14 \times 10^{-2}) = 1761 \qquad \text{(laminar)}$$
$$h_f = (f)(L/d)(v^2/2g) \qquad f = 64/N_R = \tfrac{64}{1761} = 0.0363$$
$$2.81 = (0.0363)(300/0.110)\{v^2/[(2)(9.807)]\} \qquad v = 0.746 \text{ m/s}$$

Trial No. 2
Assume $v = 0.746$ m/s:

$$0.0142 = (\pi d^2/4)(0.746) \qquad d = 0.156 \text{ m} \qquad N_R = (869)(0.156)(0.746)/(8.14/10^{-2}) = 1242 \qquad \text{(laminar)}$$
$$f = \tfrac{64}{1242} = 0.0515 \qquad 2.81 = (0.0515)(300/0.156)\{v^2/[(2)(9.807)]\} \qquad v = 0.746 \text{ m/s}$$

Hence, a pipe diameter of 0.156 m, or 156 mm, would be required.

9.73 Compute the friction factor for flow having a Reynolds number of 5×10^3 and relative roughness (ϵ/d) of 0.015 (transition zone). Use the Colebrook formula, the Swamee–Jain formula, and the Moody diagram.

▮ **Colebrook formula:**
$$1/\sqrt{f} = 1.14 - 2.0 \log [\epsilon/d + 9.35/(N_F\sqrt{f})] = 1.14 - 2.0 \log [0.015 + 9.35/(5 \times 10^3\sqrt{f})]$$
$$f = 0.0515 \qquad \text{(by trial and error)}$$

Swamee–Jain formula:
$$f = 0.25/[\log (\epsilon/3.7d) + (5.47/N_R^{0.9})]^2 = 0.25/\{\log (0.015/3.7) + [5.47/(5 \times 10^3)^{0.9}]\}^2 = 0.0438$$

Moody diagram (Fig. A-5): $\qquad\qquad\qquad f = 0.0512$

9.74 Repeat Prob. 9.73 for flow having a Reynolds number of 4×10^6 and relative roughness (ϵ/d) of 0.0001 (rough-pipe zone).

▮ **Colebrook formula:**
$$f = 1/[1.14 - 2.0 \log (\epsilon/d)]^2 = 1/[1.14 - 2.0 \log (0.0001)]^2 = 0.0120$$

Swamee–Jain formula:
$$f = 0.25/[\log (\epsilon/3.7d) + (5.74/N_R^{0.9})]^2 = 0.25/\{\log (0.0001/3.7) + [5.74/(4 \times 10^6)^{0.9}]\}^2 = 0.0120$$

Moody diagram (Fig. A-5):
$$f = 0.0125$$

9.75 We have oil of kinematic viscosity 8×10^{-5} ft²/s going through an 80-ft horizontal pipe. If the initial pressure is 5.0 psig and the final pressure is 3.5 psig, compute the mass flow if the pipe has a diameter of 3 in. At a point 10 ft from the end of the pipe a vertical tube is attached to be flush with the inside radius of the pipe. How high will the oil rise in the tube? $\rho = 50$ lbm/ft³. Pipe is commercial steel ($\epsilon = 0.000145$ ft).

$$p_1/\gamma + v_1^2/2g + z_1 = p_2/\gamma + v_2^2/2g + z_2 + h_L$$

$$(5.0)(144)/(50/32.2) + v_1^2/2g + 0 = (3.5)(144)/(50/32.2) + v_2^2/2g + 0 + h_f$$

$$v_1^2/2g = v_2^2/2g \qquad h_f = 139.1 \text{ ft} = (f)(L/d)(v^2/2) \qquad 139.1 = (f)[80/(\tfrac{3}{12})](v^2/2) \qquad fv^2 = 0.8694$$

Try $f = 0.020$: $0.020v^2 = 0.8694$, $v = 6.593$ ft/s; $N_R = dv/v = (\tfrac{3}{12})(6.593)/(8 \times 10^{-5}) = 2.06 \times 10^4$; $\epsilon/d = 0.000145/(\tfrac{3}{12}) = 0.000580$. From Fig. A-5, $f = 0.0265$. Try $f = 0.0265$: $0.0265v^2 = 0.8694$, $v = 5.728$ ft/s; $N_R = (\tfrac{3}{12})(5.728)/(8 \times 10^{-5}) = 1.79 \times 10^4$; $f = 0.0267$, $0.0267v^2 = 0.8694$, $v = 5.706$ ft/s; $M = \rho A v = 50[(\pi)(\tfrac{3}{12})^2/4](5.706) = 14.0$ lbm/s. To find the pressure at the point 10 ft from the end of the pipe (call it point A), apply the Bernoulli equation between point 1 and point A:

$$(5.0)(144)/(50/32.2) + v_1^2/2g + 0 = p_A/(50/32.2) + v_2^2/2g + 0 + h_f$$

$$h_f = 463.7 - 0.6440p_A = 0.0267[70/(\tfrac{3}{12})](5.706^2/2) = 121.7 \text{ ft}$$

$$121.7 = 463.7 - 0.6440p_A \qquad p_A = 531.1 \text{ lbm/ft}^2 \qquad h = p/\rho = 531.1/50 = 10.62 \text{ ft}$$

9.76 How much water is flowing through the 150-mm commercial steel pipe shown in Fig. 9-20? Take $v = 0.113 \times 10^{-5}$ m²/s and $\epsilon = 0.0000442$ m.

$$p_1/\gamma + v_1^2/2g + z_1 = p_2/\gamma + v_2^2/2g + z_2 + h_L \qquad 1.6 + v_1^2/2g + 0 = 0.3 + v_2^2/2g + 0 + h_f$$

$$v_1^2/2g = v_2^2/2g \qquad h_f = 1.3 \text{ m} = (f)(L/d)(v^2/2g)$$

Try $f = 0.015$:

$$1.3 = 0.015[10/(\tfrac{150}{1000})]\{v^2/[(2)(9.807)]\} \qquad v = 5.050 \text{ m/s}$$

$$N_R = dv/v = (\tfrac{150}{1000})(5.050)/(0.113 \times 10^{-5}) = 6.70 \times 10^5 \qquad \epsilon/d = 0.0000442/(\tfrac{150}{1000}) = 0.000295$$

From Fig. A-5, $f = 0.016$.

$$1.3 = 0.016[10/(\tfrac{150}{1000})]\{v^2/[(2)(9.807)]\} \qquad v = 4.889 \text{ m/s} \qquad M = \rho A v = 1000[(\pi)(\tfrac{150}{1000})^2/4](4.889) = 86.4 \text{ kg/s}$$

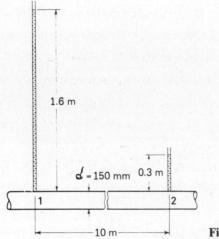

1.6 m

$d = 150$ mm   0.3 m

1   2

10 m   **Fig. 9-20**

9.77 Gasoline at 20 °C (use $v = 4.8 \times 10^{-7}$ m²/s) is being siphoned from a tank through a rubber hose having an inside diameter of 25 mm (see Fig. 9-21). The relative roughness for the hose is 0.0004. What is the flow of gasoline? What is the minimum pressure in the hose? The total length of hose is 9 m and the length of hose to point A is 3.25 m. Neglect minor losses at head entrance.

$$p_1/\gamma + v_1^2/2g + z_1 = p_2/\gamma + v_2^2/2g + z_2 + h_L \qquad 0 + 0 + (5 - 1.5) = 0 + v_2^2/[(2)(9.807)] + 0 + h_f$$

$$h_f = 3.5 - 0.05098v_2^2 = (f)(L/d)(v^2/2g)$$

Try $f = 0.016$:

$$h_f = 0.016[9/(\tfrac{25}{1000})]\{v_2^2/[(2)(9.807)]\} = 0.2937v_2^2 \qquad 0.2937v_2^2 = 3.5 - 0.05098v_2^2 \qquad v = 3.187 \text{ m/s}$$

$$N_R = dv/v = (\tfrac{25}{1000})(3.187)/(4.8 \times 10^{-7}) = 1.66 \times 10^5$$

From Fig. A-5 with $\epsilon/d = 0.0004$, $f = 0.019$. Try $f = 0.019$:

$$h_f = 0.019[9/(\tfrac{25}{1000})]\{v_2^2/[(2)(9.807)]\} = 0.3487v_2^2 \qquad 0.3487v_2^2 = 3.5 - 0.05098v_2^2 \qquad v = 2.959 \text{ m/s}$$

$$N_R = dv/v = (\tfrac{25}{1000})(2.959)/(4.8 \times 10^{-7}) = 1.54 \times 10^5 \qquad f = 0.019$$

$$Q = Av = [(\pi)(\tfrac{25}{1000})^2/4](2.959) = 0.00145 \text{ m}^3/\text{s} \qquad p_1/\gamma + v_1^2/2g + z_1 = p_A/\gamma + v_A^2/2g + z_A + h_L$$

$$0 + 0 + (5 - 1.5) = p_A/[(0.6)(9.79)] + 2.959^2/[(2)(9.807)] + 5 + h_f \qquad p_A = -11.43 - 5.874h_f$$

$$h_f = 0.019[3.25/(\tfrac{25}{1000})]\{2.959^2/[(2)(9.807)]\} = 1.103 \text{ m} \qquad p_A = -11.43 - (5.874)(1.103) = -17.91 \text{ kPa}$$

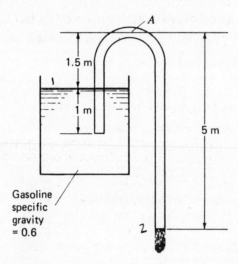

Fig. 9-21

9.78 In using the Darcy–Weisbach equation for flow in a pressure conduit, what percentage error is introduced in Q when f is misjudged by 20 percent?

∎ $$h_f = (f)(L/d)(v^2/2g) = K_1fQ^2 \qquad \text{(where } K_1 \text{ is a constant)}$$

Assume h_f is constant. $Q = K_2/\sqrt{f} = K_2f^{-1/2}$, $dQ = -\tfrac{1}{2}(K_2)(f^{-3/2})(df)$

$$\frac{dQ}{Q} = \frac{-\tfrac{1}{2}(K_2)(f^{-3/2})(df)}{K_2f^{-1/2}} = -\frac{1}{2}\frac{df}{f} = -(\tfrac{1}{2})(0.20) = -0.10 \quad \text{or} \quad -10 \text{ percent}$$

9.79 For the system in Fig. 9-13, find the flow rate if the liquid is water at 20 °C.

∎ Assume smooth-wall turbulent flow. $p_A/\gamma + v_A^2/2g + z_A = p_B/\gamma + v_B^2/2g + z_B + h_L$, $0 + 0 + 10 = 0 +$ $v_B^2/[(2)(32.2)] + 0 + h_f$, $h_f = 10 - 0.01553v_B^2 = (f)(L/d)(v^2/2g)$. Try $f = 0.02$:

$$h_f = 0.02[6/(0.5/12)]\{v_B^2/[(2)(32.2)]\} = 0.04472v_B^2 \qquad 10 - 0.01553v_B^2 = 0.04472v_B^2 \qquad v_B = 12.88 \text{ ft/s}$$

$$N_R = \rho dv/\mu = (1.93)(0.5/12)(12.88)/(2.04 \times 10^{-5}) = 5.08 \times 10^4 \qquad \text{(turbulent)}$$

From Fig. A-5, $f = 0.0208$. Try $f = 0.0208$:

$$h_f = 0.0208[6/(0.5/12)]\{v_B^2/[(2)(32.2)]\} = 0.04651v_B^2 \qquad 10 - 0.01553v_B^2 = 0.04651v_B^2 \qquad v_B = 12.70 \text{ ft/s}$$

$$N_R = \rho dv/\mu = (1.93)(0.5/12)(12.70)/(2.04 \times 10^{-5}) = 5.01 \times 10^4 \qquad \text{(turbulent)}$$

$$f = 0.0208 \qquad Q = Av = [(\pi)(0.5/12)^2/4](12.70) = 0.0173 \text{ ft}^3/\text{s}$$

9.80 If 1 mile of 3-in-diameter wrought iron pipe carries water at 20 °C and $v = 7$ m/s, compute the head loss and the pressure drop.

∎ $$h_f = (f)(L/d)(v^2/2g) \qquad N_R = \rho dv/\mu = (1.93)(\tfrac{3}{12})(7/0.3048)/(2.04 \times 10^{-5}) = 5.43 \times 10^5$$

$$\epsilon/d = 0.00015/(\tfrac{3}{12}) = 0.000600$$

From Fig. A-5, $f = 0.0182$.

$$h_f = 0.0182[5280/(\tfrac{3}{12})]\{(7/0.3048)^2/[(2)(32.2)]\} = 3148 \text{ ft} \qquad p = \gamma h_f = (62.4)(3148)/144 = 1364 \text{ lb/in}^2$$

9.81 Mercury at 20 °C flows through 3 m of 6-mm-diameter glass tubing at an average velocity of 2.5 m/s. Compute the head loss and the pressure drop.

▌ $\qquad h_f = (f)(L/d)(v^2/2g) \qquad N_R = \rho dv/\mu = (13\,570)(\tfrac{6}{1000})(2.5)/(1.56 \times 10^{-3}) = 1.30 \times 10^5$

From Fig. A-5, $f = 0.0170$ (assuming glass to be "smooth").

$$h_f = 0.0170[3/(\tfrac{6}{1000})]\{2.5^2/[(2)(9.807)]\} = 2.709 \text{ m} \qquad p = \gamma h_f = [(13.6)(9.79)](2.709) = 361 \text{ kN/m}^2$$

9.82 Gasoline at 20 °C is pumped at 0.2 m³/s through 15 km of 18-cm-diameter cast iron pipe. Compute the power required if the pumps are 80 percent efficient.

▌ $\qquad h_f = (f)(L/d)(v^2/2g) \qquad v = Q/A = 0.2/[(\pi)(\tfrac{18}{100})^2/4] = 7.860 \text{ m/s}$

$$N_R = \rho dv/\mu = (719)(\tfrac{18}{100})(7.860)/(2.92 \times 10^{-4}) = 3.48 \times 10^6 \qquad \epsilon/d = 0.00026/(\tfrac{18}{100}) = 0.00144$$

From Fig. A-5, $f = 0.0216$.

$$h_f = 0.0216[(15)(1000)/(\tfrac{18}{100})]\{7.860^2/[(2)(9.807)]\} = 5670 \text{ m}$$
$$P = \rho g Q h_f/\eta = (719)(9.807)(0.2)(5670)/0.80 = 10.0 \times 10^6 \text{ W} \quad \text{or} \quad 10\,000 \text{ kW}$$

9.83 Oil (s.g. = 0.9, $v = 0.00003$ ft²/s) flows at 1 cfs through a 6-in asphalted cast iron pipe. The pipe is 2000 ft long and slopes upward at 5° in the flow direction. Compute the head loss and the pressure change.

▌ $\qquad h_f = (f)(L/d)(v^2/2g) \qquad v = Q/A = 1/[(\pi)(\tfrac{6}{12})^2/4] = 5.093 \text{ ft/s}$

$$N_R = dv/v = (\tfrac{6}{12})(5.093)/0.00003 = 8.49 \times 10^4 \qquad \epsilon/d = 0.0004/(\tfrac{6}{12}) = 0.000800$$

From Fig. A-5, $f = 0.0219$.

$$h_f = 0.0219[2000/(\tfrac{6}{12})]\{5.093^2/[(2)(32.2)]\} = 35.28 \text{ ft} \qquad h_L = 35.28 + 2000 \sin 5° = 209.6 \text{ ft}$$
$$p = \gamma h_f = [(0.9)(62.4)](209.6)/144 = 81.7 \text{ lb/in}^2$$

9.84 The pipe flow in Fig. 9-22 is driven by pressurized air in the tank. What gage pressure p_1 is needed to provide a flow rate of 50 m³/h of water? Assume a "smooth" pipe.

▌ $\qquad p_1/\gamma + v_1^2/2g + z_1 = p_2/\gamma + v_2^2/2g + z_2 + h_L \qquad v_2 = Q/A_2 = (\tfrac{50}{3600})/[(\pi)(\tfrac{5}{100})^2/4] = 7.074 \text{ m/s}$

$$h_L = h_f = (f)(L/d)(v^2/2g) \qquad N_R = \rho dv/\mu = (998)(\tfrac{5}{100})(7.074)/(1.02 \times 10^{-3}) = 3.46 \times 10^5$$

From Fig. A-5, $f = 0.0140$.

$$h_L = 0.0140[(60 + 80 + 30)/(\tfrac{5}{100})]\{7.074^2/[(2)(9.807)]\} = 121.4 \text{ m}$$
$$p_1/9.79 + 0 + 10 = 0 + 7.074^4/[(2)(9.807)] + 80 + 121.4 \qquad p_1 = 1899 \text{ kPa gage}$$

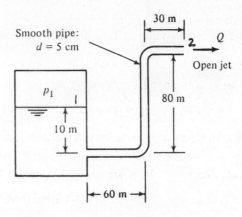

Fig. 9-22

9.85   In Fig. 9-22 suppose the fluid is methanol at 20 °C and $p_1 = 900$ kPa gage. What flow rate Q results?

▮ $p_1/\gamma + v_1^2/2g + z_1 = p_2/\gamma + v_2^2/2g + z_2 + h_L$ $900/7.73 + 0 + 10 = 0 + v_2^2/[(2)(9.807)] + 80 + h_L$

$$h_L = 46.43 - 0.05098v_2^2 = h_f = (f)(L/d)(v^2/2g)$$

Try $f = 0.02$:

$$h_L = 0.02[(60 + 80 + 30)/(\tfrac{5}{100})]\{v_2^2/[(2)(9.807)]\} = 3.467v_2^2 \qquad 3.467v_2^2 = 46.43 - 0.05098v_2^2 \qquad v_2 = 3.633 \text{ m/s}$$

$$N_R = \rho dv/\mu = (788)(\tfrac{5}{100})(3.633)/(5.98 \times 10^{-4}) = 2.39 \times 10^5$$

From Fig. A-5, $f = 0.0150$. Try $f = 0.0150$:

$$h_L = 0.0150[(60 + 80 + 30)/(\tfrac{5}{100})]\{v_2^2/[(2)(9.807)]\} = 2.600v_2^2 \qquad 2.600v_2^2 = 46.43 - 0.05098v_2^2 \qquad v_2 = 4.185 \text{ m/s}$$

$$N_R = (788)(\tfrac{5}{100})(4.185)/(5.98 \times 10^{-4}) = 2.76 \times 10^5 \qquad f = 0.0146$$

Try $f = 0.0146$:

$$h_L = 0.0146[(60 + 80 + 30)/(\tfrac{5}{100})]\{v_2^2/[(2)(9.807)]\} = 2.531v_2^2 \qquad 2.531v_2^2 = 46.43 - 0.05098v_2^2 \qquad v_2 = 4.241 \text{ m/s}$$

$$N_R = (788)(\tfrac{5}{100})(4.241)/(5.98 \times 10^{-4}) = 2.79 \times 10^5 \qquad f = 0.0146$$

$$Q = Av = [(\pi)(\tfrac{5}{100})^2/4](4.241) = 0.00833 \text{ m}^3/\text{s} \quad \text{or} \quad 30.0 \text{ m}^3/\text{h}$$

9.86   In Fig. 9-22 suppose the fluid is carbon tetrachloride at 20 °C and $p_1 = 1300$ kPa. What pipe diameter is needed to provide a flow rate of 20 m³/h?

▮ $p_1/\gamma + v_1^2/2g + z_1 = p_2/\gamma + v_2^2/2g + z_2 + h_L$ $p_1/\gamma + 0 + z_1 = 0 + v_2^2/2g + z_2 + (f)(L/d)(v_2^2/2g)$

$$v_2^2 = \frac{(2g)(p_1/\gamma + z_1 - z_2)}{1 + fL/d} = \frac{(2)(9.807)(1300/15.57 + 10 - 80)}{1 + (f)(60 + 80 + 30)/d} = \frac{264.7}{1 + 170f/d}$$

$$= (Q/A_2)^2 = [(\tfrac{20}{3600})/(\pi d^2/4)]^2 = 0.00005004/d^4$$

$$v_2 = 0.007074/d^2$$

$$\frac{0.00005004}{d^4} = \frac{264.7}{1 + 170f/d} \qquad d = \frac{(1 + 170f/d)^{1/4}}{47.96}$$

Try $d = 5$ cm, or 0.05 m: $N_R = \rho dv/\mu = (1588)(0.05)(0.007074/0.05^2)/(9.67 \times 10^{-4}) = 2.32 \times 10^5$. From Fig. A-5, $f = 0.0151$. $d = [1 + (170)(0.0151/0.05)]^{1/4}/47.96 = 0.0561$ m. Try $d = 0.0561$ m:

$$N_R = (1588)(0.0561)(0.007074/0.0561^2)/(9.67 \times 10^{-4}) = 2.07 \times 10^5 \qquad f = 0.0155$$

$$d = [1 + (170)(0.0155/0.0561)]^{1/4}/47.96 = 0.0549 \text{ m}$$

Try $d = 0.0549$ m:

$$N_R = (1588)(0.0549)(0.007074/0.0549^2)/(9.67 \times 10^{-4}) = 2.12 \times 10^5 \qquad f = 0.0155$$

$$d = [1 + (170)(0.0155/0.0549)]^{1/4}/47.96 = 0.0552 \text{ m}$$

Try $d = 0.0552$ m:

$$N_R = (1588)(0.0552)(0.007074/0.0552^2)/(9.67 \times 10^{-4}) = 2.10 \times 10^5 \qquad f = 0.0155$$

$$d = [1 + (170)(0.0155/0.0552)]^{1/4}/47.96 = 0.0551 \text{ m}$$

Hence, use $d = 0.055$ m, or 5.5 cm.

9.87   The reservoirs in Fig. 9-23 contain water at 20 °C. If the pipe is smooth with $L = 7000$ m and $d = 5$ cm, what will the flow rate be for $\Delta z = 100$ m?

▮ $p_1/\gamma + v_1^2/2g + z_1 = p_2/\gamma + v_2^2/2g + z_2 + h_L$ $0 + 0 + 100 = 0 + 0 + 0 + h_f$ $h_f = 100 \text{ m} = (f)(L/d)(v^2/2g)$

$$100 = (f)[7000/(\tfrac{5}{100})]\{v^2/[(2)(9.807)]\} \qquad v = 0.1184/\sqrt{f}$$

Try $f = 0.02$: $v = 0.1184/\sqrt{0.02} = 0.8372$ m/s, $N_R = \rho dv/\mu = (998)(\tfrac{5}{100})(0.8372)/(1.02 \times 10^{-3}) = 4.10 \times 10^4$. From Fig. A-5, $f = 0.022$. Try $f = 0.022$:

$$v = 0.1184/\sqrt{0.022} = 0.7983 \text{ m/s} \qquad N_R = (998)(\tfrac{5}{100})(0.7983)/(1.02 \times 10^{-3}) = 3.91 \times 10^4$$

$$f = 0.022 \qquad \text{(O.K.)}$$

$$Q = Av = [(\pi)(\tfrac{5}{100})^2/4](0.7983) = 0.00157 \text{ m}^3/\text{s} \quad \text{or} \quad 5.64 \text{ m}^3/\text{h}$$

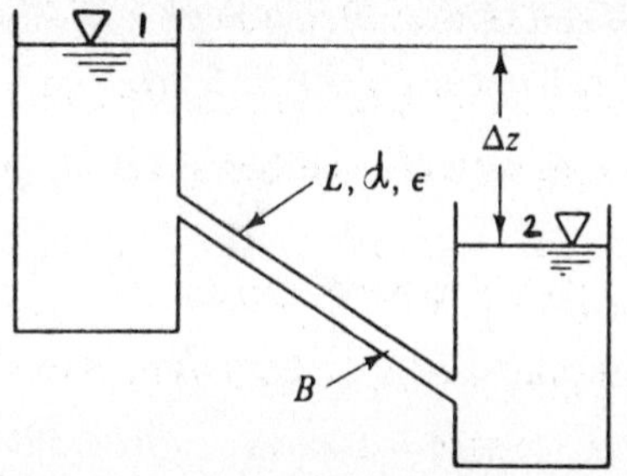

Fig. 9-23

9.88 Repeat Prob. 9.87 to find Q if $L = 2500$ ft, $d = 3$ in, and $\Delta z = 80$ ft.

$$p_1/\gamma + v_1^2/2g + z_1 = p_2/\gamma + v_2^2/2g + z_2 + h_L \qquad 0 + 0 + 80 = 0 + 0 + 0 + h_f \qquad h_f = 80 \text{ ft} = (f)(L/d)(v^2/2g)$$

$$80 = (f)[2500/(\tfrac{3}{12})]\{v^2/[(2)(32.2)]\} \qquad v = 0.7178/\sqrt{f}$$

Try $f = 0.02$:

$$v = 0.7178/\sqrt{0.02} = 5.076 \text{ ft/s} \qquad N_R = \rho dv/\mu = (1.93)(\tfrac{3}{12})(5.076)/(2.04 \times 10^{-5}) = 1.20 \times 10^5$$

From Fig. A-5, $f = 0.0175$. Try $f = 0.0175$:

$$v = 0.7178/\sqrt{0.0175} = 5.426 \text{ ft/s} \qquad N_R = (1.93)(\tfrac{3}{12})(5.426)/(2.04 \times 10^{-5}) = 1.28 \times 10^5 \qquad f = 0.0170$$

Try $f = 0.0170$:

$$v = 0.7178/\sqrt{0.0170} = 5.505 \text{ ft/s} \qquad N_R = (1.93)(\tfrac{3}{12})(5.505)/(2.04 \times 10^{-5}) = 1.30 \times 10^5$$

$$f = 0.0170 \qquad \text{(O.K.)} \qquad Q = Av = [(\pi)(\tfrac{3}{12})^2/4](5.505) = 0.2702 \text{ ft}^3/\text{s} \quad \text{or} \quad 973 \text{ ft}^3/\text{h}$$

9.89 Repeat Prob. 9.87 to find Q if $L = 2500$ ft, $d = 3$ in, $\Delta z = 80$ ft, and the pipe has a roughness of 0.2 mm. Compare the answer with that of Prob. 9.88 where the pipe is "smooth."

$$p_1/\gamma + v_1^2/2g + z_1 = p_2/\gamma + v_2^2/2g + z_2 + h_L \qquad 0 + 0 + 80 = 0 + 0 + 0 + h_f$$

$$h_f = 80 \text{ ft} = (f)(L/d)(v^2/2g) \qquad 80 = (f)[2500/(\tfrac{3}{12})]\{v^2/[(2)(32.2)]\} \qquad v = 0.7178/\sqrt{f}$$

Try $f = 0.02$:

$$v = 0.7178/\sqrt{0.02} = 5.076 \text{ ft/s} \qquad N_R = \rho dv/\mu = (1.93)(\tfrac{3}{12})(5.076)/(2.04 \times 10^{-5}) = 1.20 \times 10^5$$

$$\epsilon/d = 0.2/[(3)(25.4)] = 0.00262$$

From Fig. A-5, $f = 0.0265$. Try $f = 0.0265$:

$$v = 0.7178/\sqrt{0.0265} = 4.409 \text{ ft/s} \qquad N_R = (1.93)(\tfrac{3}{12})(4.409)/(2.04 \times 10^{-5}) = 1.04 \times 10^5$$

$$f = 0.0265 \qquad \text{(O.K.)} \qquad Q = Av = [(\pi)(\tfrac{3}{12})^2/4](4.409) = 0.2164 \text{ ft}^3/\text{s} \quad \text{or} \quad 779 \text{ ft}^3/\text{h}$$

This is $(973 - 779)/973 = 0.199$, or 19.9 percent less than when the pipe is smooth.

9.90 Water at 20 °C flows through a 600-m pipe 15 cm in diameter at a flow rate of 0.06 m³/s. If the head loss is 50 m, estimate the pipe roughness in millimeters. If this roughness is doubled, what will the percentage increase in head loss be?

$$h_L = (f)(L/d)(v^2/2g) \qquad v = Q/A = 0.06/[(\pi)(\tfrac{15}{100})^2/4] = 3.395 \text{ m/s}$$

$$50 = (f)[600/(\tfrac{15}{100})]\{3.395^2/[(2)(9.807)]\} \qquad f = 0.0213$$

$$N_R = \rho dv/\mu = (998)(\tfrac{15}{100})(3.395)/(1.02 \times 10^{-3}) = 4.98 \times 10^5$$

From Fig. A-5 with $f = 0.0213$ and $N_R = 4.98 \times 10^5$, $\epsilon/d = 0.0013$; $\epsilon = (15)(0.0013) = 0.0195$ cm, or 0.195 mm. If roughness is doubled, $\epsilon/d = 0.0026$, $f = 0.0254$ (from Fig. A-5): $h_L = 0.0254[600/(\tfrac{15}{100})]\{3.395^2/[(2)(9.807)]\} = 59.70$ m. Increase in head loss = $(59.70 - 50)/50 = 0.194$, or 19.4 percent.

9.91 A long 4-in commercial steel pipe is to be laid on a slope so that 200 gpm of water at 20 °C flows through it due to gravity only. What should the angle of slope of the pipe be?

$$Q = (200)(0.002228) = 0.4456 \text{ ft}^3/\text{s} \qquad v = Q/A = 0.4456/[(\pi)(\tfrac{4}{12})^2/4] = 5.106 \text{ ft/s}$$

$$\epsilon/d = 0.00015/(\tfrac{4}{12}) = 0.00045 \qquad N_R = dv/\nu = (\tfrac{4}{12})(5.106)/(1.05 \times 10^{-5}) = 1.62 \times 10^5$$

From Fig. A-5, $f = 0.0190$. For gravity flow through a constant-diameter pipe, $h_f = \Delta z$, $(f)(L/d)(v^2/2g) = L \sin \theta$, $0.0190[L/(\frac{4}{12})]\{5.106^2/[(2)(32.2)]\} = L \sin \theta$, $\theta = 1.32°$.

9.92 In Prob. 9.91 how many gallons per minute of water will flow by gravity if the pipe is laid at a slope angle of 4°?

▌ From Prob. 9.91, $(f)(L/d)(v^2/2g) = L \sin \theta$, $(f)[L/(\frac{4}{12})]\{v^2/[(2)(32.2)]\} = L \sin 4°$, $v = 1.224/\sqrt{f}$. Try $f = 0.02$:

$$v = 1.224/\sqrt{0.02} = 8.655 \text{ ft/s} \qquad N_R = dv/v = (\tfrac{4}{12})(8.655)/(1.05 \times 10^{-5}) = 2.75 \times 10^5$$

$$\epsilon/d = 0.00045 \qquad \text{(from Prob. 9.91)}$$

From Fig. A-5, $f = 0.018$. Try $f = 0.018$:

$$v = 1.224/\sqrt{0.018} = 9.123 \text{ ft/s} \qquad N_R = (\tfrac{4}{12})(9.123)/(1.05/10^{-5}) = 2.90 \times 10^5$$

$$f = 0.018 \quad \text{(O.K.)} \qquad Q = Av = [(\pi)(\tfrac{4}{12})^2/4](9.123) = 0.7961 \text{ ft}^3/\text{s} = 0.7961/0.002228 = 357 \text{ gpm}$$

9.93 A tank contains 1 m³ of water at 20 °C and has a drawn-capillary outlet tube at the bottom, as shown in Fig. 9-24. Find the outlet volume flux Q at this instant. $\epsilon = 0.0000015$ m.

▌
$$p_1/\gamma + v_1^2/2g + z_1 = p_2/\gamma + v_2^2/2g + z_2 + h_L$$

$$h_L = h_f = (f)(L/d)(v^2/2g) = (f)[(\tfrac{80}{100})/(\tfrac{4}{100})]\{v_2^2/[(2)(9.807)]\} = 1.020fv_2^2$$

$$0 + 0 + (1 + \tfrac{80}{100}) = 0 + v_2^2/[(2)(9.807)] + 0 + 1.020fv_2^2 \qquad v_2 = 1.342/\sqrt{0.05098 + 1.020f}$$

Try $f = 0.02$:

$$v_2 = 1.342/\sqrt{0.05098 + (1.020)(0.02)} = 5.023 \text{ m/s}$$

$$N_R = dv/v = (\tfrac{4}{100})(5.023)/(1.02 \times 10^{-6}) = 1.97 \times 10^5 \qquad \epsilon/d = 0.0000015/(\tfrac{4}{100}) = 0.0000375$$

From Fig. A-5, $f = 0.016$. Try $f = 0.016$:

$$v_2 = 1.342/\sqrt{0.05098 + (1.020)(0.016)} = 5.173 \text{ m/s} \qquad N_R = (\tfrac{4}{100})(5.173)/(1.02 \times 10^{-6}) = 2.03 \times 10^5$$

$$f = 0.016 \quad \text{(O.K.)} \qquad Q = Av = [(\pi)(\tfrac{4}{100})^2/4](5.173) = 0.006501 \text{ m}^3/\text{s} \quad \text{or} \quad 23.4 \text{ m}^3/\text{h}$$

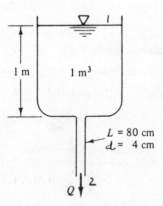

1 m   1 m³

$L = 80$ cm
$d = 4$ cm

Q 2

Fig. 9-24

9.94 For the system in Fig. 9-24, solve for the flow rate if the fluid is SAE30 oil at 20 °C. Is the flow laminar or turbulent? Use $\rho = 917 \text{ kg/m}^3$ and $\mu = 0.29 \text{ N} \cdot \text{s/m}^2$.

▌ $p_1/\gamma + v_1^2/2g + z_1 = p_2/\gamma + v_2^2/2g + z_2 + h_L$. Assume laminar flow.

$$h_L = h_f = \frac{32\mu Lv}{\rho g d^2} = \frac{(32)(0.29)(\tfrac{80}{100})(v)}{(917)(9.807)(\tfrac{4}{100})^2} = 0.5160v_2$$

$$0 + 0 + (1 + \tfrac{80}{100}) = 0 + v_2^2/[(2)(9.807)] + 0 + 0.5160v_2 \qquad v_2^2 + 10.12v_2 - 35.31 = 0 \qquad v = 2.745 \text{ m/s}$$

$$N_R = \rho dv/\mu = (917)(\tfrac{4}{100})(2.745)/0.29 = 347 \qquad \text{(laminar)}$$

$$Q = Av = [(\pi)(\tfrac{4}{100})^2/4](2.745) = 0.003449 \text{ m}^3/\text{s} \quad \text{or} \quad 12.4 \text{ m}^3/\text{h}$$

9.95 What level h must be maintained in the tank in Fig. 9-25 to deliver a flow rate of 0.02 ft³/s through the 0.5-in commercial steel exit pipe?

I
$$p_1/\gamma + v_1^2/2g + z_1 = p_2/\gamma + v_2^2/2g + z_2 + h_L \qquad v_2 = Q/A = 0.02/[(\pi)(0.5/12)^2/4] = 14.67 \text{ ft/s}$$
$$h_L = h_f = (f)(L/d)(v^2/2g) \qquad N_R = dv/v = (0.5/12)(14.67)/(1.05 \times 10^{-5}) = 5.82 \times 10^4$$
$$\epsilon/d = 0.00015/(0.5/12) = 0.00360$$

From Fig. A-5, $f = 0.0295$: $h_L = 0.0295[80/(0.5/12)]\{14.67^2/[(2)(32.2)]\} = 189.3$ ft, $0 + 0 + h = 0 + 14.67^2/[(2)(32.2)] + 0 + 189.3$, $h = 192.6$ ft.

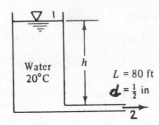

Fig. 9-25

9.96 In Fig. 9-25, suppose the fluid is benzene at 20 °C and $h = 100$ ft. What pipe diameter is required for the flow rate to be 0.02 ft³/s?

I
$$p_1/\gamma + v_1^2/2g + z_1 = p_2/\gamma + v_2^2/2g + z_2 + h_L$$
$$h_L = h_f = (f)(L/d)(v^2/2g) = (f)(80/d)\{v_2^2/[(2)(32.2)]\} = 1.242 f v_2^2/d$$
$$0 + 0 + 100 = 0 + v_2^2/[(2)(32.2)] + 0 + 1.242 f v_2^2/d \qquad v_2 = Q/A_2 = 0.02/(\pi d^2/4) = 0.02546/d^2$$
$$100 = (0.02546/d^2)^2/[(2)(32.2)] + (1.242)(f)(0.02546/d^2)^2/d \qquad d = [(1.007 \times 10^{-7}) + (8.051 \times 10^{-6} f/d)]^{1/4}$$

Try $d = 0.05$ ft: $N_R = \rho dv/\mu = (1.70)(0.05)(0.02546/0.05^2)/(1.36 \times 10^{-5}) = 6.36 \times 10^4$, $\epsilon/d = 0.00015/0.05 = 0.00300$. From Fig. A-5, $f = 0.028$. $d = [(1.007 \times 10^{-7}) + (8.051 \times 10^{-6})(0.028)/0.05]^{1/4} = 0.0463$ ft. Try $d = 0.0463$ ft:

$$N_R = (1.70)(0.0463)(0.02546/0.0463^2)/(1.36 \times 10^{-5}) = 6.87 \times 10^4 \qquad f = 0.028 \qquad \text{(O.K.)}$$
$$d = 0.0463 \text{ ft} \quad \text{or} \quad 0.56 \text{ in}$$

9.97 You wish to water your garden with 100 ft of 0.5-in-diameter hose whose roughness if 0.01 in. What will the delivery be if the pressure at the faucet is 60 psig? Use a water temperature of 70 °F.

I
$$p_1/\gamma + v_1^2/2g + z_1 = p_2/\gamma + v_2^2/2g + z_2 + h_L$$
$$h_L = h_f = (f)(L/d)(v^2/2g) = (f)[100/(0.5/12)]\{v^2/[(2)(32.2)]\} = 37.27 f v^2$$
$$(60)(144)/62.3 + v_1^2/2g + 0 = C + v_2^2/2g + 0 + 37.27 f v^2 \qquad v_1^2/2g = v_2^2/2g \qquad v_2 = 1.929/\sqrt{f}$$

Try $f = 0.05$:

$$v_2 = 1.929/\sqrt{0.05} = 8.627 \text{ ft/s} \qquad N_R = dv/v = (0.5/12)(8.627)/(1.05 \times 10^{-5}) = 3.42 \times 10^4$$
$$\epsilon/d = 0.01/0.5 = 0.0200$$

From Fig. A-5, $f = 0.050$. Try $f = 0.050$:

$$v_2 = 1.929/\sqrt{0.050} = 8.627 \text{ ft/s} \qquad N_R = (0.5/12)(8.627)/(1.05 \times 10^{-5}) = 3.42 \times 10^4$$
$$f = 0.050 \quad \text{(O.K.)} \qquad Q = Av = [(\pi)(0.5/12)^2/4](8.627) = 0.0118 \text{ ft}^3/\text{s}$$

9.98 Ethanol at 20 °C flows at 300 cm/s through 10-cm-diameter drawn tubing ($\epsilon = 0.0000015$ m). Compute (a) the head loss per 100 m of tube and (b) the wall shear stress.

I *(a)*
$$h_f = (f)(L/d)(v^2/2g) \qquad N_R = \rho dv/\mu = (788)(\tfrac{10}{100})(\tfrac{300}{100})/(1.20 \times 10^{-3}) = 1.97 \times 10^5$$
$$\epsilon/d = 0.0000015/(\tfrac{10}{100}) = 0.0000150$$

From Fig. A-5, $f = 0.0158$. $h_f = 0.0158[100/(\tfrac{10}{100})]\{(\tfrac{300}{100})^2/[(2)(9.807)]\} = 7.250$ m per 100 m.

(b) $\qquad u^* = v\sqrt{f/8} = (\tfrac{300}{100})\sqrt{0.0158/8} = 0.1333$ m/s $\qquad \tau_{\text{wall}} = (\rho)(u^*)^2 = (788)(0.1333)^2 = 14.0$ Pa

9.99 In Fig. 9-26, the connecting pipe is commercial steel 6 cm in diameter. Compute the flow rate if the fluid is SAE30 oil at 20 °C. Which way is the flow? Use $\rho = 917$ kg/m³ and $\mu = 0.29$ N · s/m².

▮ Assume laminar flow from 1 to 2.

$$p_1/\rho g + v_1^2/2g + z_1 = p_2/\rho g + v_2^2/2g + z_2 + h_L$$

$$h_L = h_f = \frac{128\mu LQ}{\pi\rho g d^4} = \frac{(128)(0.29)(50)(Q)}{(\pi)(917)(9.807)(\frac{6}{100})^4} = 5069Q$$

$$0 + 0 + 15 = (200)(1000)/[(917)(9.807)] + 0 + 0 + 5069Q \qquad Q = -0.001428 \text{ m}^3/\text{s} \quad \text{or} \quad -5.14 \text{ m}^3/\text{h}$$

Since Q is negative, flow is from 2 to 1.

$$v = Q/A = 0.001428/[(\pi)(\tfrac{6}{100})^2/4] = 0.5051 \text{ m/s} \qquad N_R = \rho dv/\mu = (917)(\tfrac{6}{100})(0.5051)/0.29 = 96 \qquad \text{(laminar)}$$

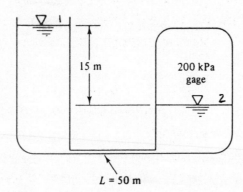

$L = 50$ m   **Fig. 9-26**

9.100 In Fig. 9-26 if the fluid is SAE30 oil at 20 °C, what should the pipe diameter be to maintain a flow rate of 25 m³/h?

▮

$$\frac{p_2}{\rho g} + \frac{v_2^2}{2g} + z_2 = \frac{p_1}{\rho g} + \frac{v_1^2}{2g} + z_1 + h_L \qquad h_L = h_f = \frac{128\mu LQ}{\pi\rho g d^4} = \frac{(128)(0.29)(50)(\frac{25}{3600})}{(\pi)(917)(9.807)(d/100)^4} = \frac{45\,621}{d^4}$$

$$(200)(1000)/[(917)(9.807)] + 0 + 0 = 0 + 0 + 15 + 45\,621/d^4 \qquad d = 8.91 \text{ cm}$$

$$v = Q/A = (\tfrac{25}{3600})/[(\pi)(8.91/100)^2/4] = 1.114 \text{ m/s}$$

$$N_R = \rho dv/\mu = (917)(8.91/100)(1.114)/0.29 = 314 \qquad \text{(laminar)}$$

9.101 A commercial steel annulus 40 ft long, with $a = 1$ in and $b = 0.5$ in, connects two reservoirs which differ in surface height by 20 ft. Compute the flow rate through the annulus if the fluid is water at 20 °C.

▮ Hydraulic diameter $= D_h = (2)(a - b) = (2)(1 - 0.5) = 1.00$ in $p_1/\gamma + v_1^2/2g + z_1 = p_2/\gamma + v_2^2/2g + z_2 + h_L$

$$0 + 0 + 20 = 0 + 0 + 0 + h_L \qquad h_L = 20 \text{ ft} = h_f = (f)(L/d)(v^2/2g)$$

$$20 = (f)[40/(1.00/12)]\{v^2/[(2)(32.2)]\} \qquad v = 1.638/\sqrt{f}$$

Try $f = 0.02$:

$$v = 1.638/\sqrt{0.02} = 11.58 \text{ ft/s} \qquad N_R = dv/\nu = (1.00/12)(11.58)/(1.05 \times 10^{-5}) = 9.19 \times 10^4$$

$$\epsilon/d = 0.00015/(1.00/12) = 0.00180$$

From Fig. A-5, $f = 0.0246$. Try $f = 0.0246$:

$$v = 1.638/\sqrt{0.0246} = 10.44 \text{ ft/s} \qquad N_R = (1.00/12)(10.44)/(1.05 \times 10^{-5}) = 8.29 \times 10^4 \qquad f = 0.0250$$

Try $f = 0.0250$:

$$v = 1.638/\sqrt{0.0250} = 10.36 \text{ ft/s} \qquad N_R = (1.00/12)(10.36)/(1.05 \times 10^{-5}) = 8.22 \times 10^4$$

$$f = 0.0250 \quad \text{(O.K.)} \qquad Q = Av = \{(\pi)[(1.00/12)^2 - (0.5/12)^2]\}(10.36) = 0.170 \text{ ft}^3/\text{s}$$

9.102 An annulus of narrow clearance causes a very large pressure drop and is useful as an accurate measurement of viscosity. If a smooth annulus 1 m long with $a = 0.050$ m and $b = 0.049$ m carries an oil flow at 0.001 m³/s, what is the oil viscosity if the pressure drop is 250 000 Pa?

$$Q = (\pi/8\mu)(\Delta p/L)[a^4 - b^4 - (a^2 - b^2)^2/\ln(a/b)]$$

$$0.001 = (\pi/8\mu)(250\,000/1)[0.050^4 - 0.049^4 - (0.050^2 - 0.049^2)^2/\ln(0.050/0.049)] \qquad \mu = 0.00648 \text{ N} \cdot \text{s/m}^2$$

9.103 A sheet-metal ventilation duct carries air at 20 °C and 1 atm (approximately). The duct section is an equilateral triangle 12 in on a side, and its length is 120 ft. If a blower can deliver 1 hp to the air, what flow rate can occur? Use $\epsilon = 0.00015$ ft.

▌ From Fig. 9-27,

$$A = (12)(12 \sin 60°)/2 = 62.35 \text{ in}^2 \quad \text{or} \quad 0.4330 \text{ ft}^2$$

$$P = Q\gamma h_f \qquad Q = Av = 0.4330v \qquad h_f = (f)(L/d)(v^2/2g) \qquad d = D_h = 4A/p_w$$

$$D_h = (4)(0.4330)/[(12 + 12 + 12)/12] = 0.5773 \text{ ft} \qquad h_f = (f)(120/0.5773)\{v^2/[(2)(32.2)]\} = 3.228fv^2$$

$$(1)(550) = (0.4330v)(0.0750)(3.228fv^2) \qquad v = 17.38/f^{1/3}$$

Try $f = 0.02$:

$$v = 17.38/0.02^{1/3} = 64.03 \text{ ft/s} \qquad N_R = dv/\nu = (0.5773)(64.03)/(1.64 \times 10^{-4}) = 2.25 \times 10^5$$

$$\epsilon/D_h = 0.00015/0.5773 = 0.000260$$

From Fig. A-5, $f = 0.0172$. Try $f = 0.0172$:

$$v = 17.38/0.0172^{1/3} = 67.33 \text{ ft/s} \qquad N_R = (0.5773)(67.33)/(1.64 \times 10^{-4}) = 2.37 \times 10^5$$

$$f = 0.0172 \qquad \text{(O.K.)} \qquad Q = (0.4330)(67.33) = 29.2 \text{ ft}^3/\text{s}$$

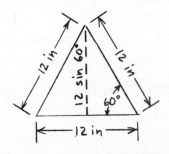

Fig. 9-27

9.104 SAE30 oil at 20 °C flows between two smooth parallel plates 4 cm apart ($h = 4$ cm) at an average velocity of 3 m/s. Compute the pressure drop, centerline velocity, and head loss for each 100 m of length. Use $\rho = 917$ kg/m³ and $\mu = 0.29$ N · s/m².

▌ $\quad d = D_h = 2h = (2)(\frac{4}{100}) = 0.0800 \text{ m} \qquad N_R = \rho dv/\mu = (917)(0.0800)(3)/0.29 = 759 \qquad \text{(laminar)}$

$$v = (h^2/12\mu)(\Delta p/L) \qquad 3 = \{(\tfrac{4}{100})^2/[(12)(0.29)]\}(\Delta p/100) \qquad \Delta p = 652\,500 \text{ Pa}$$

$$h_f = \Delta p/\rho g = 652\,500/[(917)(9.807)] = 72.56 \text{ m} \qquad u_\mathcal{C} = 1.5v = (1.5)(3) = 4.50 \text{ m/s}$$

9.105 Air at sea-level standard conditions is blown through a 30-cm-square steel duct 150 m long at 25 m/s. Compute the head loss, pressure drop, and power required if the blower efficiency is 60 percent. Use $\epsilon = 0.000046$ m.

▌ $\quad d = D_h = 4A/p_w = (4)[(\tfrac{30}{100})(\tfrac{30}{100})]/[(4)(\tfrac{30}{100})] = 0.3000 \text{ m}$

$$h_f = (f)(L/d)(v^2/2g) = (f)(150/0.3000)\{25^2/[(2)(9.807)]\} = 15\,932f$$

$$N_R = \rho dv/\mu = (1.20)(0.3000)(25)/(1.81 \times 10^{-5}) = 4.97 \times 10^5 \qquad \epsilon/d = 0.000046/0.3000 = 0.000153$$

From Fig. A-5, $f = 0.0150$.

$$h_f = (15\,923)(0.0150) = 238.8 \text{ m} \qquad \Delta p = \rho g h_f = (1.20)(9.807)(238.8) = 2810 \text{ Pa}$$

$$Q = Av = [(\tfrac{30}{100})(\tfrac{30}{100})](25) = 2.25 \text{ m}^3/\text{s}$$

$$P = \rho g Q h_f/\eta = (1.20)(9.807)(2.25)(238.8)/0.60 = 10\,500 \text{ W} \quad \text{or} \quad 10.5 \text{ kW}$$

9.106 A wind tunnel has a wooden ($\epsilon = 0.0001$ m) rectangular section 40 cm by 1 m by 50 m long. The average flow velocity is 45 m/s for air at sea-level standard conditions. Compute the pressure drop, assuming fully developed conditions, and the power required if the fan has 65 percent efficiency.

▮

$$d = D_h = 4A/p_w = (4)[(\tfrac{40}{100})(1)]/(\tfrac{40}{100} + \tfrac{40}{100} + 1 + 1) = 0.5714 \text{ m}$$

$$h_f = (f)(L/d)(v^2/2g) = (f)(50/0.5714)\{45^2/[(2)(9.807)]\} = 9034f$$

$$N_R = \rho dv/\mu = (1.20)(0.5714)(45)/(1.81 \times 10^{-5}) = 1.70 \times 10^6 \qquad \epsilon/d = 0.0001/0.5714 = 0.000175$$

From Fig. A-5, $f = 0.0140$.

$$h_f = (9034)(0.0140) = 126.5 \text{ m} \qquad \Delta p = \rho g h_f = (1.20)(9.807)(126.5) = 1489 \text{ Pa}$$

$$Q = Av = [(\tfrac{40}{100})(1)](45) = 18.00 \text{ m}^3/\text{s}$$

$$P = \rho g Q h_f/\eta = (1.20)(9.807)(18.00)(126.5)/0.65 = 41\,200 \text{ W} \quad \text{or} \quad 41.2 \text{ kW}$$

9.107 It is desired to pump hydrogen at 20 °C and 1 atm through a smooth rectangular duct 80 m long of aspect ratio 6:1. If the flow rate is 0.6 m³/s and the pressure drop 75 Pa, what should the width and height of the duct cross section be? For hydrogen, $\rho = 0.0838$ kg/m³ and $\mu = 9.05 \times 10^{-6}$ N · s/m².

▮ Let h = height of duct; then duct width = $6h$.

$$d = D_h = 4A/p_w = (4)[(6h)(h)]/(6h + 6h + h + h) = 1.714h \qquad v = Q/A = 0.6/[(6h)(h)] = 0.1000/h^2$$

$$h_f = (f)(L/d)(v^2/2g) = (f)(80/1.714h)\{(0.1000/h^2)^2/[(2)(9.807)]\} = 0.02380f/h^5$$

$$\Delta p = \rho g h_f \qquad 75 = (0.0838)(9.807)(0.02380f/h^5) \qquad h = 0.1920f^{1/5}$$

Try $f = 0.02$: $h = (0.1920)(0.02)^{1/5} = 0.08780$ m, $N_R = \rho dv/\mu =$ $0.0838[(1.714)(0.08780)](0.1000/0.08780^2)/(9.05 \times 10^{-6}) = 1.81 \times 10^4$. From Fig. A-5, $f = 0.0265$. Try $f = 0.0265$: $h = (0.1920)(0.0265)^{1/5} = 0.09289$ m, $N_R = 0.0838[(1.714)(0.09289)](0.1000/0.09289^2)/(9.05 \times 10^{-6}) = 1.71 \times 10^4$, $f = 0.0267$. Try $f = 0.0267$: $h = (0.1920)(0.0267)^{1/5} = 0.09303$ m, $N_R = 0.0838[(1.714)(0.09303)](0.1000/0.09303^2)/(9.05 \times 10^{-6}) = 1.71 \times 10^4$.

$$f = 0.0267 \qquad \text{(O.K.)}$$

Hence, $h = 0.0930$ m and width = $6h = (6)(0.0930) = 0.558$ m.

9.108 Kerosene at 10 °C flows steadily at 15 L/min through a 150-m-long horizontal length of 5.5-cm-diameter cast iron pipe. Compare the pressure drop of the kerosene flow with that of the same flow rate of benzene at 10 °C through the same pipe. At 10 °C, $\rho = 820$ kg/m³ and $\mu = 0.0025$ N · s/m² for kerosene, and $\rho = 899$ kg/m³ and $\mu = 0.0008$ N · s/m² for benzene.

▮ For kerosene:

$$v = Q/A = [(\tfrac{15}{1000})/60]/[(\pi)(5.5/100)^2/4] = 0.1052 \text{ m/s}$$

$$N_R = \rho dv/\mu = (820)(5.5/100)(0.1052)/0.0025 = 1898 \qquad \text{(laminar)}$$

$$\Delta p = 8\mu v L/r_0^2 = (8)(0.0025)(0.1052)(150)/[(5.5/100)/2]^2 = 417 \text{ Pa}$$

For benzene:

$$N_R = (899)(5.5/100)(0.1052)/0.0008 = 6502 \qquad \text{(turbulent)}$$

$$\Delta p = (f)(L/d)(\rho)(v^2/2) \qquad \epsilon/d = 0.00026/(5.5/100) = 0.00473$$

From Fig. A-5, $f = 0.040$: $\Delta p = (0.040)[150/(5.5/100)](899)(0.1052^2/2) = 543$ Pa. The pressure drop is greater for the benzene than the kerosene, although the benzene has a lower viscosity. If both flows had been laminar or both turbulent, the lower-viscosity fluid, benzene, would have had the lower pressure drop. However, the viscosity of the kerosene is high enough to give laminar flow, while the lower viscosity of the benzene causes a high enough Reynolds number for turbulent flow.

9.109 Water at 60 °F flows through a 250-ft length of horizontal 2-in-diameter drawn tubing. If the pressure drop across the tubing is 10 psi, what is the flow rate?

▮ Assume turbulent flow. $h_f = (f)(L/d)(v^2/2g)$, $h_f = (10)(144)/62.4 = 23.08$ ft, $23.08 =$ $(f)[250/(\tfrac{2}{12})]\{v^2/[(2)(32.2)]\}$, $v = 0.9954/\sqrt{f}$. Try $v = 5$ ft/s: $N_R = \rho dv/\mu = (1.94)(\tfrac{2}{12})(5)/(2.35 \times 10^{-5}) = 6.88 \times 10^4$ (turbulent). $\epsilon/d = 0.000005/(\tfrac{2}{12}) = 0.000030$. From Fig. A-5, $f = 0.0195$. $v = 0.9954/\sqrt{0.0195} = 7.13$ ft/s.

Try $v = 7.13$ ft/s: $N_R = (1.94)(\frac{2}{12})(7.13)/(2.35 \times 10^{-5}) = 9.81 \times 10^4$, $f = 0.0180$, $v = 0.9954/\sqrt{0.0180} = 7.42$ ft/s.
Try $v = 7.42$ ft/s: $N_R = (1.94)(\frac{2}{12})(7.42)/(2.35 \times 10^{-5}) = 1.02 \times 10^5$, $f = 0.0180$ (O.K.); $Q = Av = [(\pi)(\frac{2}{12})^2/4](7.42) = 0.162$ ft^2/s.

9.110 Air at 200 °F and 15 psig is to be passed through a 150-ft length of new galvanized iron pipe at a rate of 15 ft^3/s. If the maximum allowable pressure drop is 5 psi, estimate the minimum pipe diameter.

■ $\quad h_f = (f)(L/d)(v^2/2g) = p/\gamma \qquad \gamma = p/RT = (15 + 14.7)(144)/[(53.3)(200 + 460)] = 0.1216$ lb/ft^3

$$h_f = (5)(144)/0.1216 = 5921 \text{ ft} \qquad v = Q/A = 15/(\pi d^2/4) = 19.10/d^2$$

$$5921 = (f)(150/d)\{(19.10/d^2)^2/[(2)(32.2)]\} \qquad d = 0.6782 f^{1/5}$$

Try $d = 0.5$ ft:

$$N_R = \rho dv/\mu = (\gamma/g)(dv)/\mu = (0.1216/32.2)(0.5)(19.10/0.5^2)/(4.49 \times 10^{-7}) = 3.21 \times 10^5$$

$$\epsilon/d = 0.0005/0.5 = 0.0010$$

From Fig. A-5, $f = 0.0205$; $d = (0.6782)(0.0205)^{1/5} = 0.3117$ ft.

Try $d = 0.3117$ ft:

$$N_R = (0.1216/32.2)(0.3117)(19.10/0.3117^2)/(4.49 \times 10^{-7}) = 5.15 \times 10^5 \qquad \epsilon/d = 0.0005/0.3117 = 0.0016$$

$$f = 0.0225 \qquad d = (0.6782)(0.0225)^{1/5} = 0.3175 \text{ ft}$$

Try $d = 0.3175$ ft:

$$N_R = (0.1216/32.2)(0.3175)(19.10/0.3175^2)/(4.49 \times 10^{-7}) = 5.06 \times 10^5 \qquad \epsilon/d = 0.0005/0.3175 = 0.0016$$

$$f = 0.0225 \qquad \text{(O.K.)}$$

Hence, $d = 0.3175$ ft, or 3.81 in.

9.111 Compute the loss of head and pressure drop in 200 ft of horizontal 6-in-diameter asphalted cast iron pipe carrying water with a mean velocity of 6 ft/s.

■ $\quad h_f = (f)(L/d)(v^2/2g) \qquad N_R = dv/v = (\frac{6}{12})(6)/(1.05 \times 10^{-5}) = 2.86 \times 10^5 \qquad \epsilon/d = 0.0004/(\frac{6}{12}) = 0.00080$

From Fig. A-5, $f = 0.020$:

$$h_f = (0.020)[200/(\tfrac{6}{12})]\{6^2/[(2)(32.2)]\} = 4.47 \text{ ft} \qquad \Delta p = \gamma h_f = (62.4)(4.47) = 279 \text{ lb/ft}^2$$

9.112 Oil with $\rho = 900$ kg/m^3 and $v = 0.00001$ m^2/s flows at 0.2 m^3/s through 500 m of 200-mm-diameter cast iron pipe. Determine (a) the head loss and (b) the pressure drop if the pipe slopes down at 10° in the flow direction.

■ (a) $\qquad\qquad h_f = (f)(L/d)(v^2/2g) \qquad v = Q/A = 0.2/[(\pi)(\tfrac{200}{1000})^2/4] = 6.366$ m/s

$$N_R = dv/v = (\tfrac{200}{1000})(6.366)/0.00001 = 1.27 \times 10^5 \qquad \epsilon/d = 0.00026/(\tfrac{200}{1000}) = 0.00130$$

From Fig. A-5, $f = 0.0225$, $h_f = 0.0225[500/(\tfrac{200}{1000})]\{6.366^2/[(2)(9.807)]\} = 116.2$ m.

(b) $\qquad\qquad h_f = \Delta p/(\rho g) + L \sin 10° \qquad 116.2 = \Delta p/[(900)(9.807)] + (500)(\sin 10°)$

$$\Delta p = 259\,300 \text{ Pa} \quad \text{or} \quad 259.3 \text{ kPa}$$

9.113 Oil with $\rho = 950$ kg/m^3 and $v = 0.00002$ m^2/s flows through a 30-cm-diameter pipe 100 m long with a head loss of 8 m. The roughness ratio ϵ/d is 0.0002. Find the average velocity and flow rate.

■ $\qquad\qquad h_f = (f)(L/d)(v^2/2g) \qquad 8 = (f)[100/(\tfrac{30}{100})]\{v^2/[(2)(9.807)]\} \qquad v = 0.6861/\sqrt{f}$

Try $v = 5$ m/s: $N_R = dv/v = (\tfrac{30}{100})(5)/0.00002 = 7.50 \times 10^4$. From Fig. A-5, $f = 0.020$, $v = 0.6861/\sqrt{0.020} = 4.851$ m/s. Try $v = 4.851$ m/s: $N_R = (\tfrac{30}{100})(4.851)/0.00002 = 7.28 \times 10^4$, $f = 0.020$ (O.K.). $Q = Av = [(\pi)(\tfrac{30}{100})^2/4](4.851) = 0.343$ m^3/s.

9.114 Fluid flows at an average velocity of 6 ft/s between horizontal parallel plates a distance of 2.4 in apart ($h = 2.4$ in). Estimate the head loss and pressure drop for each 100 ft of length for $\rho = 1.9$ slugs/ft^3 and $v = 0.00002$ ft^2/s. Assume smooth walls.

■ $\qquad\qquad h_f = (f)(L/d)(v^2/2g) \qquad d = D_h = 2h \qquad D_h = (2)(2.4/12) = 0.400$ ft

$$N_R = dv/v = (0.400)(6)/0.00002 = 1.20 \times 10^5 \qquad \text{(turbulent)}$$

From Fig. A-5, $f = 0.0173$.

$$h_f = (0.0173)(100/0.400)\{6^2/[(2)(32.2)]\} = 2.42 \text{ ft} \qquad \Delta p = \rho g h_f = (1.9)(32.2)(2.42) = 148 \text{ lb/ft}^2$$

9.115 Repeat Prob. 9.114 if $v = 0.002 \text{ ft}^2/\text{s}$.

∎ $$N_R = dv/v = (0.400)(6)/0.002 = 1200 \qquad \text{(laminar)} \qquad f = 96/N_R = \tfrac{96}{1200} = 0.0800$$

$$h_f = (f)(L/d)(v^2/2g) = (0.0800)(100/0.400)\{6^2/[(2)(32.2)]\} = 11.18 \text{ ft}$$

$$\Delta p = \rho g h_f = (1.9)(32.2)(11.18) = 684 \text{ lb/ft}^2$$

9.116 Estimate the reservoir level h needed to maintain a flow of $0.01 \text{ m}^3/\text{s}$ through the commercial steel annulus 30 m long shown in Fig. 9-28. Neglect entrance effects and take $\rho = 1000 \text{ kg/m}^3$ and $v = 1.02 \times 10^{-6} \text{ m}^2/\text{s}$ for water.

∎ $$p_1/\gamma + v_1^2/2g + z_1 = p_2/\gamma + v_2^2/2g + z_2 + h_f \qquad v = Q/A = 0.01/\{(\pi)[(\tfrac{5}{100})^2 - (\tfrac{3}{100})^2]\} = 1.989 \text{ m/s}$$

$$h_f = (f)(L/d)(v^2/2g) \qquad d = D_h = (2)(a-b) \qquad D_h = (2)(\tfrac{5}{100} - \tfrac{3}{100}) = 0.0400 \text{ m}$$

$$h_f = (f)(30/0.0400)\{1.989^2/[(2)(9.807)]\} = 151.3f \qquad 0 + 0 + h = 0 + 1.989^2/[(2)(9.807)] + 0 + 151.3f$$

$$h = 0.2017 + 151.3f \qquad N_R = dv/v = (0.0400)(1.989)/(1.02 \times 10^{-6}) = 7.80 \times 10^4$$

$$\epsilon/d = 0.000046/0.0400 = 0.00115$$

From Fig. A-5, $f = 0.0232$, $h = 0.2017 + (151.3)(0.0232) = 3.71 \text{ m}$.

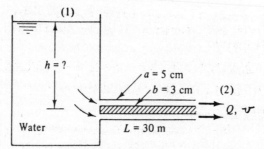

Fig. 9-28

9.117 Air with $\rho = 0.00237 \text{ slug/ft}^3$ and $v = 0.000157 \text{ ft}^2/\text{s}$ is forced through a horizontal square 9-in by 9-in duct 100 ft long at $25 \text{ ft}^3/\text{s}$. Find the presssure drop if $\epsilon = 0.0003 \text{ ft}$.

∎ $$h_f = (f)(L/d)(v^2/2g) \qquad N_R = dv/v \qquad D_h = 4A/p_w = 4[(\tfrac{9}{12})(\tfrac{9}{12})]/[(4)(\tfrac{9}{12})] = 0.7500 \text{ ft}$$

From Table A-10, for $b/a = \tfrac{9}{9}$, or 1.0, the effective diameter is $d = D_{\text{eff}} = (64/56.91)(D_h) = (64/56.91)(0.7500) = 0.8434 \text{ ft}$, $v = Q/A = 25/[(\tfrac{9}{12})(\tfrac{9}{12})] = 44.44 \text{ ft/s}$, $N_R = (0.8434)(44.44)/0.000157 = 2.39 \times 10^5$, $\epsilon/d = 0.0003/0.8434 = 0.000354$. From Fig. A-5, $f = 0.0177$; $h_f = (0.0177)(100/0.7500)\{44.44^2/[(2)(32.2)]\} = 72.37 \text{ ft}$, $\Delta p = \rho g h_f = (0.00237)(32.2)(72.37) = 5.52 \text{ lb/ft}^2$.

9.118 Find the head loss per foot in a 6-in-diameter pipe having $\epsilon = 0.042$ in when oil (s.g. = 0.90) having a viscosity of $0.0008 \text{ lb} \cdot \text{s/ft}^2$ flows at a rate of 15 cfs. Determine the shear stress at the wall of the pipe and the velocity 2.0 in from the centerline.

∎ $$h_f = (f)(L/d)(v^2/2g) \qquad v = Q/A = 15/[(\pi)(\tfrac{6}{12})^2/4] = 76.39 \text{ ft/s}$$

$$N_R = \rho dv/\mu = [(0.90)(1.94)](\tfrac{6}{12})(76.39)/0.0008 = 83.4 \times 10^4 \qquad \epsilon/d = 0.042/6 = 0.00700$$

From Fig. A-5, $f = 0.034$.

$$h_f = 0.034[1/(\tfrac{6}{12})]\{76.39^2/[(2)(32.2)]\} = 6.16 \text{ ft per foot}$$

$$\tau_0 = f\rho v^2/8 = 0.034[(0.90)(1.94)](76.39)^2/8 = 43.3 \text{ lb/ft}^2$$

$$u_{\max} = (v)(1 + 1.33f^{1/2}) = (76.39)[1 + (1.33)(0.034)^{1/2}] = 95.12 \text{ ft/s}$$

$$u = u_{\max} - (5.75)(\tau_0/\rho)^{1/2} \log [r_0/(r_0 - r)] = 95.12 - (5.75)\{43.3/[(0.90)(1.94)]\}^{1/2} \log [3/(3 - 2)] = 81.5 \text{ ft/s}$$

9.119 When water at 150 °F flows in a 0.75-in-diameter copper tube at 1.0 gpm, find the head loss per 100 ft. What is the centerline velocity, and what is the value of the nominal thickness of the viscous sublayer?

$\blacksquare$ $\quad Q = (1.0)(0.002228) = 0.002228 \text{ ft}^3/\text{s} \qquad v = Q/A = 0.002228/[(\pi)(0.75/12)^2/4] = 0.7262 \text{ ft/s}$

$$h_f = (f)(L/d)(v^2/2g) \qquad N_R = dv/v = (0.75/12)(0.7262)/(4.68 \times 10^{-6}) = 9.70 \times 10^3$$

$$\epsilon/d = 0.000005/(0.75/12) = 0.0000800$$

From Fig. A-5, $f = 0.030$.

$$h_f = (0.030)[100/(0.75/12)]\{0.7262^2/[(2)(32.2)]\} = 0.393 \text{ ft}$$

$$u_{max} = (v)(1 + 1.33 f^{1/2}) = 0.7262[1 + (1.33)(0.030)^{1/2}] = 0.893 \text{ ft/s}$$

$$\delta_1 = 32.8v/(vf^{1/2}) = (32.8)(4.68 \times 10^{-6})/[(0.7262)(0.030)^{1/2}] = 0.00122 \text{ ft}$$

9.120 Repeat Prob. 9.119 for a flowrate of 20 gpm.

$\blacksquare$ $\quad Q = (20)(0.002228) = 0.04456 \text{ ft}^3/\text{s} \qquad v = Q/A = 0.04456/[(\pi)(0.75/12)^2/4] = 14.52 \text{ ft/s}$

$$h_f = (f)(L/d)(v^2/2g)$$

$$N_R = dv/v = (0.75/12)(14.52)/(4.68 \times 10^{-6}) = 1.94 \times 10^5 \qquad \epsilon/d = 0.000005/(0.75/12) = 0.0000800$$

From Fig. A-5, $f = 0.016$.

$$h_f = 0.016[100/(0.75/12)]\{14.52^2/[(2)(32.2)]\} = 83.81 \text{ ft}$$

$$u_{max} = (v)(1 + 1.33 f^{1/2}) = 14.52[1 + (1.33)(0.016)^{1/2}] = 16.96 \text{ ft/s}$$

$$\delta_1 = 32.8v/(vf^{1/2}) = (32.8)(4.68 \times 10^{-6})/[(14.52)(0.016)^{1/2}] = 8.36 \times 10^{-5} \text{ ft}$$

9.121 Find the head loss per meter in a 10-cm-diameter pipe having $\epsilon = 0.00085$ m when oil (s.g. $= 0.82$) having a viscosity of $0.0052 \text{ N} \cdot \text{s/m}^2$ flows at a rate of 40 L/s. Determine the shear stress at the wall of the pipe. Find the velocity 2 cm from the centerline.

$\blacksquare$ $\quad h_f = (f)(L/d)(v^2/2g) \qquad v = Q/A = (\frac{40}{1000})/[(\pi)(\frac{10}{100})^2/4] = 5.093 \text{ m/s}$

$$N_R = \rho dv/\mu = [(0.82)(1000)](\tfrac{10}{100})(5.093)/0.0052 = 8.03 \times 10^4 \qquad \epsilon/d = 0.00085/(\tfrac{10}{100}) = 0.0085$$

From Fig. A-5, $f = 0.037$.

$$h_f = 0.037[1/(\tfrac{10}{100})]\{5.093^2/[(2)(9.807)]\} = 0.489 \text{ m per meter}$$

$$\tau_0 = f\rho v^2/8 = 0.037[(0.82)(1000)](5.093)^2/8 = 98.4 \text{ N/m}^2$$

$$u_{max} = (v)(1 + 1.33 f^{1/2}) = 5.093[1 + (1.33)(0.037)^{1/2}] = 6.396 \text{ m/s}$$

$$u = u_{max} - (5.75)(\tau_0/\rho)^{1/2} \log [r_0/(r_0 - r)] = 6.396 - 5.75\{98.4/[(0.82)(1000)]\}^{1/2} \log [5/(5 - 2)] = 5.95 \text{ m/s}$$

$$\delta_1 = (32.8)(\mu/\rho)/(vf^{1/2}) = 32.8\{0.0052/[(0.82)(1000)]\}/[(5.093)(0.037)^{1/2}] = 0.000212 \text{ m} \quad \text{or} \quad 0.212 \text{ mm}$$

9.122 The head loss in 200 ft of 6-in-diameter pipe is known to be 25 ft when oil (s.g. $= 0.90$) flows at 2.0 cfs. Determine the centerline velocity, the shear stress at the wall of the pipe, and the velocity at 2 in from the centerline.

$\blacksquare$ $\quad v = Q/A = 2.0/[(\pi)(\frac{6}{12})^2/4] = 10.19 \text{ ft/s}$

$$N_R = \rho dv/\mu = [(0.90)(1.94)](\tfrac{6}{12})(10.19)/0.0008 = 1.11 \times 10^4 \qquad \text{(turbulent)}$$

$$h_f = (f)(L/d)(v^2/2g) \qquad 25 = (f)[200/(\tfrac{6}{12})]\{10.19^2/(2)(32.2)]\} \qquad f = 0.0388$$

$$\tau_0 = f\rho v^2/8 = 0.0388[(0.90)(1.94)](10.19)^2/8 = 0.879 \text{ lb/ft}^2$$

$$u_{max} = (v)(1 + 1.33 f^{1/2}) = 10.19[1 + (1.33)(0.0388)^{1/2}] = 12.86 \text{ ft/s}$$

$$u = u_{max} - (5.75)(\tau_0/\rho)^{1/2} \log [r_0/(r_0 - r)] = 12.86 - (5.75)\{0.879/[(0.90)(1.94)]\}^{1/2} \log [3/(3 - 2)] = 10.9 \text{ ft/s}$$

9.123 The velocities in a 90-cm-diameter pipe are measured as 5.00 m/s on the centerline and 4.82 m/s at $r = 10$ cm. Approximately what is the flow rate?

$\blacksquare$ $\quad u = u_{max} - (5.75)(\tau_0/\rho)^{1/2} \log [r_0/(r_0 - r)]$

$$4.82 = 5.00 - (5.75)(\tau_0/\rho)^{1/2} \log [45/(45 - 10)] \qquad (\tau_0/\rho)^{1/2} = 0.2868$$

$$v = u_{max} - (\tfrac{3}{2})(2.5\sqrt{\tau_0/\rho}) = 5.00 - (\tfrac{3}{2})(2.5)(0.2868) = 3.924 \text{ m/s} \qquad Q = Av = [(\pi)(\tfrac{90}{100})^2/4](3.924) = 2.50 \text{ m}^3/\text{s}$$

9.124 The velocities in a 36-in-diameter pipe are measured as 15.0 fps at $r = 0$ and 14.5 fps at $r = 4$ in. Approximately what is the flow rate?

$$u = u_{max} - (5.75)(\tau_0/\rho)^{1/2} \log [r_0/(r_0 - r)] \qquad 14.5 = 15.0 - (5.75)(\tau_0/\rho)^{1/2} \log [18/(18 - 4)]$$

$$(\tau_0/\rho)^{1/2} = 0.7967 \qquad \tau_0/\rho = fv^2/8 \qquad (\tau_0/\rho)^{1/2} = (fv^2/8)^{1/2} = 0.7967 \qquad f = 5.078/v^2$$

$$u_{max} = (v)(1 + 1.33f^{1/2}) \qquad 15.0 = (v)(1 + 1.33f^{1/2}) \qquad 15.0 = (v)[1 + (1.33)(5.078/v^2)^{1/2}]$$

$$15.0 = v + 2.997 \qquad v = 12.00 \text{ ft/s} \qquad Q = Av = [(\pi)(\tfrac{36}{12})^2/4](12.00) = 84.8 \text{ ft}^3/\text{s}$$

9.125 With turbulent flow in a circular pipe, prove that the mean velocity occurs at a distance of approximately $0.78r_0$ from the centerline of the pipe.

$$u = (1 + 1.33f^{1/2})(v) - (2.04)(f^{1/2})(v) \log [r_0/(r_0 - r)] \qquad 0 = 1.33f^{1/2}v - (2.04f^{1/2}v)\{\log [r_0/(r_0 - r)]\}$$

$$\log [r_0/(r_0 - r)] = 0.65196 \qquad r_0/(r_0 - r) = \text{antilog } 0.65196 = 4.487 \qquad r = 0.776r_0$$

9.126 The flow rate in a 12-in-diameter pipe is 8 cfs. The flow is known to be turbulent, and the centerline velocity is 12.0 fps. Determine the velocity profile, and determine the head loss per foot of pipe.

$$v = Q/A = 8/[(\pi)(\tfrac{12}{12})^2/4] = 10.19 \text{ ft/s} \qquad u_{max} = (v)(1 + 1.33f^{1/2}) \qquad 12.0 = (10.19)(1 + 1.33f^{1/2})$$

$$f = 0.01784 \qquad h_f = (f)(L/d)(v^2/2g) = 0.01784[1/(\tfrac{12}{12})]\{10.19^2/[(2)(32.2)]\} = 0.0288 \text{ ft per foot}$$

$$\tau_0 = (f/4)(\rho)(v^2/2) = (0.01784/4)(\rho)(10.19^2/2) \qquad (\tau_0/\rho)^{1/2} = 0.4812$$

$$u = u_{max} - 5.75(\tau_0/\rho)^{1/2} \log [r_0/(r_0 - r)]$$

For $r = 0$, $u = 12.0 - (5.75)(0.4812) \log [6/(6 - 0)] = 12.0 \text{ ft/s}$. For $r = 2$, $u = 12.0 - (5.75)(0.4812) \log [6/(6 - 2)] = 11.5 \text{ ft/s}$. For $r = 4$, $u = 12.0 - (5.75)(0.4812) \log [6/(6 - 4)] = 10.7 \text{ ft/s}$. For $r = 5$, $u = 12.0 - (5.75)(0.4812) \log [6/(6 - 5)] = 9.85 \text{ ft/s}$. For $r = 5.5$, $u = 12.0 - (5.75)(0.4812) \log [6/(6 - 5.5)] = 9.01 \text{ ft/s}$. For $r = 5.9$, $u = 12.0 - (5.75)(0.4812) \log [6/(6 - 5.9)] = 7.08 \text{ ft/s}$. For $r = 5.99$, $u = 12.0 - (5.75)(0.4812) \log [6/(6 - 5.99)] = 4.31 \text{ ft/s}$.

9.127 Kerosene (s.g. = 0.81) flows at a temperature of 80 °F ($v = 2.21 \times 10^{-5} \text{ ft}^2/\text{s}$) in a 2-in-diameter smooth brass pipeline at a rate of 10 gpm. (*a*) Find the head loss per foot. (*b*) For the same head loss, what would be the rate of flow if the temperature of the kerosene were raised to 120 °F, when $v = 1.60 \times 10^{-5} \text{ ft}^2/\text{s}$?

$$Q = (10)(0.002228) = 0.02228 \text{ ft}^3/\text{s} \qquad v = Q/A = 0.02228/[(\pi)(\tfrac{2}{12})^2/4] = 1.021 \text{ ft/s}$$

(*a*) $$h_f = (f)(L/d)(v^2/2g) \qquad N_R = dv/v = (\tfrac{2}{12})(1.021)/(2.21 \times 10^{-5}) = 7.70 \times 10^3$$

From Fig. A-5 for smooth pipe, $f = 0.0333$. $h_f = 0.0333[1/(\tfrac{2}{12})]\{1.021^2/[(2)(32.2)]\} = 0.00323 \text{ ft per foot}$.

(*b*) $$0.00323 = (f)[1/(\tfrac{2}{12})]\{v^2/[(2)(32.2)]\} \qquad v = 0.1862/\sqrt{f}$$

Try $v = 1.05$ ft/s: $N_R = (\tfrac{2}{12})(1.05)/(1.60 \times 10^{-5}) = 1.09 \times 10^4$, $f = 0.0297$, $v = 0.1862/\sqrt{0.0297} = 1.080$ ft/s. Try $v = 1.080$ ft/s: $N_R = (\tfrac{2}{12})(1.080)/(1.60 \times 10^{-5}) = 1.12 \times 10^4$, $f = 0.0297$ (O.K.); $Q = Av = [(\pi)(\tfrac{2}{12})^2/4](1.080) = 0.02356 \text{ ft}^3/\text{s}$, or 10.6 gpm.

9.128 Water at 40 °C flows in a 20-cm-diameter pipe with $v = 5$ m/s. Head loss measurements indicate that $f = 0.022$. Determine the value of ϵ. Find the shear stress at the wall of the pipe and at $r = 4$ cm.

$N_R = dv/v = (\tfrac{20}{100})(5)/(6.56 \times 10^{-7})$. From Fig. A-5, $\epsilon/d = 0.0015$, $\epsilon = (20)(0.0015) = 0.0300$ cm; $\tau_0 = f\rho v^2/8 = (0.022)(992)(5)^2/8 = 68.2 \text{ N/m}^2$; $\tau = (\tau_0)(r/r_0) = (68.2)(\tfrac{4}{10}) = 27.3 \text{ N/m}^2$.

9.129 Water at 15 °C flows through a 20-cm-diameter pipe with $\epsilon = 0.01$ mm. (*a*) If the mean velocity is 3.5 m/s, what is the nominal thickness δ_1 of the viscous sublayer? (*b*) What will be the thickness of the viscous sublayer if the velocity is increased to 5.8 m/s?

(*a*) $$\delta_1 = 32.8v/(vf^{1/2}) \qquad N_R = dv/v = (\tfrac{20}{100})(3.5)/(1.16 \times 10^{-6}) = 6.03 \times 10^5$$

$$\epsilon/d = (0.01/10)/20 = 0.0000500$$

From Fig. A-5, $f = 0.0133$. $\delta_1 = (32.8)(1.16 \times 10^{-6})/[(3.5)(0.0133)^{1/2}] = 9.43 \times 10^{-5}$ m, or 0.0943 mm.

(*b*) $$N_R = (\tfrac{20}{100})(5.8)/(1.16 \times 10^{-6}) = 1.00 \times 10^6 \qquad f = 0.0126$$

$$\delta_1 = (32.8)(1.16 \times 10^{-6})/[(5.8)(0.0126)^{1/2}] = 5.84 \times 10^{-5} \text{ m} \quad \text{or} \quad 0.0584 \text{ mm}$$

9.130 When water at 50 °F flows at 3.0 cfs in a 24-in pipeline, the head loss is 0.0003 ft per foot. What will be the head loss when glycerin at 68 °F flows through the same pipe at the same rate?

$$h_f = (f)(L/d)(v^2/2g) \qquad v = Q/A = 3.0/[(\pi)(\tfrac{24}{12})^2/4] = 0.9549 \text{ ft/s}$$

For water:

$$0.0003 = (f)[1/(\tfrac{24}{12})]\{0.9549^2/[(2)(32.2)]\} \qquad f = 0.04238$$

$$N_R = \rho dv/\mu = (1.94)(\tfrac{24}{12})(0.9549)/(2.72 \times 10^{-5}) = 1.36 \times 10^5 \qquad \text{(turbulent)}$$

From Fig. A-5, $\epsilon/d = 0.014$.

For glycerin:

$$N_R = (2.44)(\tfrac{24}{12})(0.9549)/(3.11 \times 10^{-2}) = 150 \qquad \text{(laminar)}$$

$$f = 64/N_R = \tfrac{64}{150} = 0.4267 \qquad h_f = 0.4267[1/(\tfrac{24}{12})]\{0.9549^2/[(2)(32.2)]\} = 0.00302 \text{ ft per ft}$$

9.131 Air flows at 50 lb/min in a 4-in-diameter welded steel pipe ($\epsilon = 0.0018$ in) at 100 psia and 60 °F. Determine the head loss and pressure drop in 150 ft of this pipe. Assume the air to be of constant density.

$$h_f = (f)(L/d)(v^2/2g) \qquad N_R = \rho dv/\mu \qquad \rho = p/RT = (100)(144)/[(1716)(460 + 60)] = 0.01614 \text{ slug/ft}^3$$

$$W = \gamma Av = \rho g Av \qquad \tfrac{50}{60} = (0.01614)(32.2)[(\pi)(\tfrac{4}{12})^2/4](v) \qquad v = 18.37 \text{ ft/s}$$

$$N_R = (0.01614)(\tfrac{4}{12})(18.37)/(3.74 \times 10^{-7}) = 2.64 \times 10^5 \qquad \epsilon/d = 0.0018/4 = 0.00045$$

From Fig. A-5, $f = 0.018$. $h_f = 0.018[150/(\tfrac{4}{12})]\{18.37^2/[(2)(32.2)]\} = 42.4$ ft of air, $\Delta p = \rho g h_f = (0.01614)(32.2)(42.4) = 22.0 \text{ lb/ft}^2$, or 0.153 lb/in^2.

9.132 Air flows at an average velocity of 0.5 m/s through a long 3.2-m-diameter circular tunnel. Find the head-loss gradient at a point where the air temperature and pressure are 15 °C and 108 kN/m² abs, respectively. Assume $\epsilon = 2$ mm. Find also the shear stress at the wall and the thickness of the viscous sublayer.

$$h_f/L = (f/d)(v^2/2g) \qquad \rho = p/RT = (108)(1000)/[(287)(273 + 15)] = 1.307 \text{ kg/m}^3$$

$$N_R = \rho dv/\mu = (1.307)(3.2)(0.5)/(1.79 \times 10^{-5}) = 1.17 \times 10^5 \qquad \epsilon/d = (\tfrac{2}{1000})/3.2 = 0.000625$$

From Fig. A-5, $f = 0.021$.

$$h_f/L = (0.021/3.2)\{0.5^2/[(2)(9.807)]\} = 8.36 \times 10^{-5} \text{ m/m}$$

$$\tau_0 = f\rho v^2/8 = (0.021)(1.307)(0.5)^2/8 = 8.58 \times 10^{-4} \text{ N/m}^2$$

$$\delta_1 = 32.8v/(vf^{1/2}) = (32.8)(\mu/\rho)/(vf^{1/2}) = (32.8)(1.79 \times 10^{-5}/1.307)/[(0.5)(0.021)^{1/2}] = 0.00620 \text{ m} \quad \text{or} \quad 6.20 \text{ mm}$$

9.133 Repeat Prob. 9.132 for the case where the average velocity is 5.0 m/s.

$$h_f/L = (f/d)(v^2/2g) \qquad \rho = 1.307 \text{ kg/m}^3 \qquad \text{(from Prob. 9.132)}$$

$$N_R = \rho dv/\mu = (1.307)(3.2)(5.0)/(1.79 \times 10^{-5}) = 1.17 \times 10^6$$

$$\epsilon/d = 0.000625 \qquad \text{(from Prob. 9.132)}$$

From Fig. A-5, $f = 0.018$.

$$h_f/L = (0.018/3.2)\{5.0^2/[(2)(9.807)]\} = 7.17 \times 10^{-3} \text{ m/m}$$

$$\tau_0 = f\rho v^2/8 = (0.018)(1.307)(5.0)^2/8 = 0.0735 \text{ N/m}^2$$

$$\delta_1 = 32.8v/(vf^{1/2}) = (32.8)(\mu/\rho)/(vf^{1/2})$$

$$= (32.8)(1.79 \times 10^{-5}/1.307)/[(5.0)(0.018)^{1/2}] = 6.70 \times 10^{-4} \text{ m} \quad \text{or} \quad 0.670 \text{ mm}$$

9.134 Air at 20 °C and atmospheric pressure flows with a velocity of 5 m/s through a 50-mm-diameter pipe. Find the head loss in 20 m of pipe having $\epsilon = 0.0025$ mm.

$$h_f = (f)(L/d)(v^2/2g) \qquad N_R = dv/v = (\tfrac{50}{1000})(5)/(1.51 \times 10^{-5}) = 1.66 \times 10^4 \qquad \epsilon/d = 0.0025/50 = 0.000050$$

From Fig. A-5, $f = 0.027$. $h_f = 0.027[20/(\tfrac{50}{1000})]\{5^2/[(2)(9.807)]\} = 13.77$ m.

9.135 What is the head loss per foot of pipe when oil (s.g. = 0.90) having a viscosity of 2×10^{-4} lb · s/ft^2 flows in a 2-in-diameter welded steel pipe at 0.15 cfs?

$$h_f/L = (f/d)(v^2/2g) \qquad v = Q/A = 0.15/[(\pi)(\tfrac{2}{12})^2/4] = 6.875 \text{ ft/s}$$

$$N_R = \rho dv/\mu = [(0.90)(1.94)](\tfrac{2}{12})(6.875)/(2 \times 10^{-4}) = 1.00 \times 10^4 \qquad \epsilon/d = 0.00015/(\tfrac{2}{12}) = 0.00090$$

From Fig. A-5, $f = 0.033$. $h_f/L = [0.033/(\tfrac{2}{12})]\{6.875^2/[(2)(32.2)]\} = 0.145$ ft/ft.

9.136 Consider water at 50 °F flowing in a 36-in-diameter concrete pipe ($\epsilon = 0.02$ in). Determine N_R, τ_0, u_{max}/v, δ_1, δ_1/ϵ, and the flow regime (hydraulic smoothness) for a flow rate of 200 cfs.

$$v = Q/A = 200/[(\pi)(\tfrac{36}{12})^2/4] = 28.29 \text{ ft/s} \qquad N_R = dv/v = (\tfrac{36}{12})(28.29)/(1.40 \times 10^{-5}) = 6.06 \times 10^6$$

$$\tau_0 = f\rho v^2/8 \qquad \epsilon/d = 0.02/36 = 0.000556$$

From Fig. A-5, $f = 0.0175$.

$$\tau_0 = (0.0175)(1.94)(28.29)^2/8 = 3.40 \text{ lb/ft}^2 \qquad u_{max}/v = 1 + 1.33f^{1/2} = 1 + (1.33)(0.0175)^{1/2} = 1.18$$

$$\delta_1 = 32.8v/(vf^{1/2}) = (32.8)(1.40 \times 10^{-5})/[(28.29)(0.0175)^{1/2}] = 0.000123 \text{ ft} \qquad \delta_1/\epsilon = 0.000123/(0.02/12) = 0.0738$$

Since $[\delta_1 = 0.000123] < [0.3\epsilon = (0.3)(0.02/12) = 0.000500]$, regime is "rough."

9.137 Repeat Prob. 9.136 for a flow rate of 0.2 cfs.

$$v = Q/A = 0.2/[(\pi)(\tfrac{36}{12})^2/4] = 0.02829 \text{ ft/s} \qquad N_R = dv/v = (\tfrac{36}{12})(0.02829)/(1.40 \times 10^{-5}) = 6062$$

$$\tau_0 = f\rho v^2/8 \qquad \epsilon/d = 0.02/36 = 0.000556$$

From Fig. A-5, $f = 0.036$.

$$\tau_0 = (0.036)(1.94)(0.02829)^2/8 = 6.99 \times 10^{-6} \text{ lb/ft}^2 \qquad u_{max}/v = 1 + 1.33f^{1/2} = 1 + (1.33)(0.036)^{1/2} = 1.25$$

$$\delta_1 = 32.8v/(vf^{1/2}) = (32.8)(1.40 \times 10^{-5})/[(0.02829)(0.036)^{1/2}] = 0.0856 \text{ ft} \qquad \delta_1/\varepsilon = 0.0856/(0.02/12) = 51.4$$

Since $[\delta_1 = 0.0856] > [6\epsilon = (6)(0.02/12) = 0.0100]$, regime is "smooth."

9.138 Find the flow rate when water at 60 °F causes a head loss of 0.25 ft in 100 ft of average cast iron pipe. Diameter of pipe is 6 in.

$$h_f = (f)(L/d)(v^2/2g) \qquad 0.25 = (f)[100/(\tfrac{6}{12})]\{v^2/[(2)(32.2)]\} \qquad v = 0.2837/\sqrt{f}$$

Try $v = 2$ ft/s: $N_R = dv/v = (\tfrac{6}{12})(2)/(1.21 \times 10^{-5}) = 8.26 \times 10^4$, $\epsilon/d = 0.00085/(\tfrac{6}{12}) = 0.00170$. From Fig. A-5, $f = 0.0245$. $v = 0.2837/\sqrt{0.0245} = 1.81$ ft/s. Try $v = 1.81$ ft/s: $N_R = (\tfrac{6}{12})(1.81)/(1.21 \times 10^{-5}) = 7.48 \times 10^4$, $f = 0.025$, $v = 0.2837/\sqrt{0.025} = 1.79$ ft/s. Try $v = 1.79$ ft/s: $N_R = (\tfrac{6}{12})(1.79)/(1.21 \times 10^{-5}) = 7.40 \times 10^4$, $f = 0.025$ (O.K.); $Q = Av = [(\pi)(\tfrac{6}{12})^2/4](1.79) = 0.351$ ft^3/s.

9.139 Gasoline with a kinematic viscosity of 5×10^{-7} m^2/s flows in a 30-cm-diameter smooth pipe. Find the flow rate when the head loss is 0.4 m per 100 m.

$$h_f = (f)(L/d)(v^2/2g) \qquad 0.4 = (f)[100/(\tfrac{30}{100})]\{v^2/[(2)(9.807)]\} \qquad v = 0.1534/\sqrt{f}$$

Try $v = 1$ m/s: $N_R = dv/v = (\tfrac{30}{100})(1)/(5 \times 10^{-7}) = 6.00 \times 10^5$. From Fig. A-5, $f = 0.0128$. $v = 0.1534/\sqrt{0.0128} = 1.36$ m/s. Try $v = 1.36$ m/s: $N_R = (\tfrac{30}{100})(1.36)/(5 \times 10^{-7}) = 8.16 \times 10^5$, $f = 0.0122$, $v = 0.1534/\sqrt{0.0122} = 1.39$ m/s. Try $v = 1.39$ m/s: $N_R = (\tfrac{30}{100})(1.39)/(5 \times 10^{-7}) = 8.34 \times 10^5$, $f = 0.0122$ (O.K.); $Q = Av = [(\pi)(\tfrac{30}{100})^2/4](1.39) = 0.0983$ m^3/s.

9.140 What size pipe is required to carry oil having a kinematic viscosity of 0.0002 ft^2/s at a rate of 8.0 cfs if the head loss is to be 0.4 ft per 100 ft of pipe length? Assume $\epsilon = 0.00015$ ft.

$$h_f = (f)(L/d)(v^2/2g) \qquad v = Q/A = 8.0/(\pi d^2/4) = 10.19/d^2 \qquad 0.4 = (f)(100/d)\{(10.19/d^2)^2/[(2)(32.2)]\}$$

$$d = 3.320f^{1/5} \qquad N_R = dv/v = (d)(10.19/d^2)/0.0002 = 50\,950/d$$

Try $d = 1$ ft: $N_R = 50\,950/1 = 5.09 \times 10^4$, $\epsilon/d = 0.00015/1 = 0.00015$. From Fig. A-5, $f = 0.0215$. $d = (3.320)(0.0215)^{1/5} = 1.54$ ft. Try $d = 1.54$ ft: $N_R = 50\,950/1.54 = 3.31 \times 10^4$, $\epsilon/d = 0.00015/1.54 = 0.0000974$, $f = 0.0235$, $d = (3.320)(0.0235)^{1/5} = 1.57$ ft. Try $d = 1.57$ ft: $N_R = 50\,950/1.57 = 3.25 \times 10^4$, $\epsilon/d = 0.00015/1.57 = 0.0000955$, $f = 0.0235$ (O.K.). Hence, $d = 1.57$ ft, or 18.8 in.

9.141 A straight, new, asphalted cast iron pipe is 42 in in diameter and 1000 ft long. (**a**) Find the shear force on the pipe if the fluid is water at 72 °F and the average velocity is 10 fps. (**b**) What will be the shear force if the average velocity is reduced to 5 fps?

$$\tau_0 = f\rho v^2/8 \qquad N_R = dv/\nu$$

(**a**)
$$N_R = (\tfrac{42}{12})(10)/(1.02 \times 10^{-5}) = 3.43 \times 10^6 \qquad \epsilon/d = 0.0004/(\tfrac{42}{12}) = 0.000114$$

From Fig. A-5, $f = 0.013$.

$$\tau_0 = (0.013)(1.93)(10)^2/8 = 0.314 \,\text{lb/ft}^2 \qquad F_{\text{shear}} = \tau_0 A = 0.314[(\pi)(\tfrac{42}{12})(1000)] = 3452 \,\text{lb}$$

(**b**)
$$N_R = (\tfrac{42}{12})(5)/(1.02 \times 10^{-5}) = 1.72 \times 10^6 \qquad f = 0.0132$$

$$\tau_0 = (0.0132)(1.93)(5)^2/8 = 0.0796 \,\text{lb/ft}^2 \qquad F_{\text{shear}} = \tau_0 A = 0.0796[(\pi)(\tfrac{42}{12})(100)] = 875 \,\text{lb}$$

9.142 A steel pipe ($\epsilon = 0.0002$ ft) of length 15 000 ft is to convey oil ($\nu = 0.0006$ ft^2/s) at a rate of 10 cfs from a reservoir of surface elevation 625 ft to one of surface elevation 400 ft. What pipe size would be required?

$$h_f = (f)(L/d)(v^2/2g) \qquad v = Q/A = 10/(\pi d^2/4) = 12.73/d^2$$

$$625 - 400 = (f)(15\,000/d)\{(12.73/d^2)^2/[(2)(32.2)]\}$$

$$d = 2.786 f^{1/5} \qquad N_R = dv/\nu = (d)(12.73/d^2)/0.0006 = 21\,217/d$$

Try $d = 1$ ft: $N_R = 21\,217/1 = 2.12 \times 10^4$, $\epsilon/d = 0.00015/1 = 0.00015$. From Fig. A-5, $f = 0.0265$. $d = (2.786)(0.0265)^{1/5} = 1.35$ ft. Try $d = 1.35$ ft: $N_R = 21\,217/1.35 = 1.57 \times 10^4$, $\epsilon/d = 0.00015/1.35 = 0.00011$, $f = 0.0284$, $d = (2.786)(0.0284)^{1/5} = 1.37$ ft. Try $d = 1.37$ ft: $N_R = 21\,217/1.37 = 1.55 \times 10^4$, $\epsilon/d = 0.00015/1.37 = 0.00011$, $f = 0.0284$ (O.K.). Hence, $d = 1.37$ ft, or 16.4 in.

9.143 Water at 140 °F flows in a 0.824-in-diameter iron pipe ($\epsilon = 0.00015$ ft) of length 400 ft between points A and B. At point A the elevation of the pipe is 104.0 ft and the pressure is 8.20 psi. At point B the elevation of the pipe is 99.5 ft and the pressure is 9.05 psi. Compute the flow rate.

$$p_A/\gamma + v_A^2/2g + z_A = p_B/\gamma + v_B^2/2g + z_B + h_L$$

$$(8.20)(144)/61.4 + v_A^2/2g + 104.0 = (9.05)(144)/61.4 + v_B^2/2g + 99.5 + h_L \qquad v_A^2/2g = v_B^2/2g$$

$$h_L = 2.51 \,\text{ft} \qquad h_f = h_L = (f)(L/d)(v^2/2g) \qquad 2.51 = (f)[400/(0.824/12)]\{v^2/[(2)(32.2)]\} \qquad v = 0.1666/\sqrt{f}$$

Try $v = 1$ ft/s: $N_R = dv/\nu = (0.824/12)(1)/(5.03 \times 10^{-6}) = 1.37 \times 10^4$, $\epsilon/d = 0.00015/(0.824/12) = 0.0022$. From Fig. A-5, $f = 0.0325$. $v = 0.1666/\sqrt{0.0325} = \underline{0.924}$ ft/s. Try $v = 0.924$ ft/s: $N_R = (0.824/12)(0.924)/(5.03 \times 10^{-6}) = 1.26 \times 10^4$, $f = 0.0330$, $v = 0.1666/\sqrt{0.0330} = 0.917$ ft/s. Try $v = 0.917$ ft/s: $N_R = (0.824/12)(0.917)/(5.03 \times 10^{-6}) = 1.25 \times 10^4$, $f = 0.0330$ (O.K.); $Q = Av = [(\pi)(0.824/12)^2/4](0.917) = 0.00339$ ft^3/s.

9.144 Air at 50 psia and temperature of 150 °F flows in a 12-in by 18-in rectangular air duct at the rate of 1 lb/min. Find the head loss per 100 ft of duct. Express answer in feet of air flowing and in pounds per square inch. Assume $\epsilon = 0.0005$ in.

$$h_f = (f)(L/d)(v^2/2g) \qquad d = 4R_h \qquad R_h = A/p_w = (\tfrac{12}{12})(\tfrac{18}{12})/(\tfrac{12}{12} + \tfrac{12}{12} + \tfrac{18}{12} + \tfrac{18}{12}) = 0.300 \,\text{ft}$$

$$d = (4)(0.300) = 1.20 \,\text{ft} \qquad W = \gamma A v \qquad \gamma = p/RT = (50)(144)/[(53.3)(460 + 150)] = 0.2214 \,\text{lb/ft}^3$$

$$\tfrac{1}{60} = 0.2214[(\tfrac{12}{12})(\tfrac{18}{12})](v) \qquad v = 0.05019 \,\text{ft/s} \qquad h_f = (f)(100/1.20)\{0.05019^2/[(2)(32.2)]\} = 0.003260 f$$

$$N_R = dv/\nu = (1.20)(0.05019)/(2.06 \times 10^{-4}) = 292 \qquad \text{(laminar)}$$

$$f = 64/N_R = \tfrac{64}{292} = 0.2192 \qquad h_f = (0.003260)(0.2192) = 0.000715 \,\text{ft}$$

$$\Delta p = \gamma h_f = (0.2214)(0.000715)/144 = 1.10 \times 10^{-6} \,\text{lb/in}^2$$

9.145 Find the approximate rate at which 60 °F water will flow in a conduit shaped in the form of an equilateral triangle if the head loss is 2 ft per 100 ft. The cross-sectional area of the duct is 120 in^2, and $\epsilon = 0.0018$ in.

First, find the length of each side (s) of the cross section (see Fig. 9-29):

$$A = bh/2 \qquad 120 = (x + x)(\sqrt{3}x)/2 \qquad x = 8.324 \,\text{in} \qquad s = 2x = (2)(8.324) = 16.65 \,\text{in}$$

$$h_f = (f)(L/d)(v^2/2g) \qquad R_h = A/p_w = 120/[(3)(16.65)] = 2.402 \,\text{in} \quad \text{or} \quad 0.2002 \,\text{ft}$$

$$d = 4R_h = (4)(0.2002) = 0.8008 \,\text{ft} \qquad 2 = (f)(100/0.8008)\{v^2/[(2)(32.2)]\} \qquad v = 1.016/\sqrt{f}$$

Try $v = 10$ ft/s: $N_R = dv/v = (0.8008)(10)/(1.21 \times 10^{-5}) = 6.62 \times 10^5$, $\epsilon/d = (0.0018/12)/0.8008 = 0.000187$. From Fig. A-5, $f = 0.0150$. $v = 1.016/\sqrt{0.0150} = 8.30$ ft/s. Try $v = 8.30$ ft/s: $N_R = (0.8008)(8.30)/(1.21 \times 10^{-5}) = 5.49 \times 10^5$, $f = 0.0155$, $v = 1.016/\sqrt{0.0155} = 8.16$ ft/s. Try $v = 8.16$ ft/s: $N_R = (0.8008)(8.16)/(1.21 \times 10^{-5}) = 5.40 \times 10^5$, $f = 0.0155$ (O.K.); $Q = Av = (\frac{120}{144})(8.16) = 6.80$ ft^3/s.

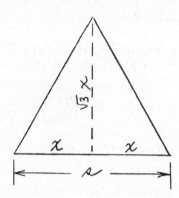

Fig. 9-29

9.146 When fluid of specific weight 50 lb/ft^3 flows in a 6-in-diameter pipe, the frictional stress between the fluid and the pipe wall is 0.5 lb/ft^2. Calculate the head loss per foot of pipe. If the flow rate is 2.0 cfs, how must power is lost per foot of pipe?

$$R_h = d/4 = (\tfrac{6}{12})/4 = 0.1250 \text{ ft} \qquad h_f = (\tau_0)(L/R_h\gamma) = 0.5\{1/[0.1250)(50)]\} = 0.0800 \text{ ft per foot}$$

$$P = Q\gamma h_f = (2.0)(50)(0.0800) = 8.000 \text{ ft} \cdot \text{lb/s per foot} = 8.000/550 = 0.0145 \text{ hp per foot}$$

9.147 Prove that for a constant rate of discharge and a constant value of f the friction head loss in a pipe varies inversely as the fifth power of the diameter.

$$v = Q/A = 4Q/\pi d^2 \qquad h_f = (f)(L/d)(v^2/2g) = (f)(L/d)[(4Q/\pi d^2)^2/2g] = (f)(L/d^5)(8Q^2/\pi^2 g)$$

Thus for constant f and constant Q, $h_f \propto 1/d^5$.

9.148 Two long pipes are used to convey water between two reservoirs whose water surfaces are at different elevations. One pipe has a diameter twice that of the other. If both pipes have the same value of f and if minor losses are neglected, what is the ratio of the flow rates through the two pipes?

$$h_f = (f)(L/d)(v^2/2g) = \Delta \text{ elevation} \qquad h_f \propto Q^2/d^5 \qquad \text{(from Prob. 9.147)}$$

$$(h_f)_1 = (h_f)_2$$

Therefore, $Q_1^2/d_1^5 = Q_2^2/d_2^5$, $Q_2/Q_1 = (d_2/d_1)^{5/2} = 2^{5/2} = 5.66$. Thus the flow in the larger pipe will be 5.66 times that in the smaller one.

9.149 Points C and D, at the same elevation, are 500 ft apart in an 8-in pipe and are connected to a differential gage by means of small tubing. When the flow of water is 6.31 cfs, the deflection of mercury in the gage is 6.43 ft. Determine the friction factor f.

$p_C/\gamma + v_C^2/2g + z_C = p_D/\gamma + v_D^2/2g + z_D + h_f$. Since $v_C^2/2g = v_D^2/2g$ and $z_C = z_D$, $P_C/\gamma - p_D/\gamma = h_f = (6.43)(13.6 - 1) = 81.02$ ft, $v = Q/A = 6.31/[(\pi)(\tfrac{8}{12})^2/4] = 18.08$ ft/s, $h_f = (f)(L/d)(v^2/2g) = (f)[500/(\tfrac{8}{12})]\{18.08^2/[(2)(32.2)]\} = 3807f$, $81.02 = 3807f$, $f = 0.0213$.

9.150 Oil flows from tank A through 500 ft of 6-in new asphalt-dipped cast iron pipe to point B, as shown in Fig. 9-30. What pressure in pounds per square inch will be needed at A to cause 0.450 cfs of oil to flow? (s.g. = 0.840; $v = 2.27 \times 10^{-5}$ ft^2/s; $\epsilon = 0.0004$ ft.)

$$p_A/\gamma + v_A^2/2g + z_A = p_B/\gamma + v_B^2/2g + z_B + h_L \qquad v_B = Q/A_B = 0.450/[(\pi)(\tfrac{6}{12})^2/4] = 2.292 \text{ ft/s}$$

$$h_L = h_f = (f)(L/d)(v^2/2g) = (f)[500/(\tfrac{6}{12})]\{2.292^2/[(2)(32.2)]\} = 81.57f$$

$$p_A/[(0.840)(62.4)] + 0 + 80.0 = 0 + 2.292^2/[(2)(32.2)] + 100.0 + 81.57f \qquad p_A = 1053 + 4276f$$

$$N_R = dv/v = (\tfrac{6}{12})(2.292)/(2.27 \times 10^{-5}) = 5.05 \times 10^4 \qquad \epsilon/d = 0.0004/(\tfrac{6}{12}) = 0.000800$$

From Fig. A-5, $f = 0.0235$. $p_A = 1053 + (4276)(0.0235) = 1153$ lb/ft^2, or 8.01 lb/in^2.

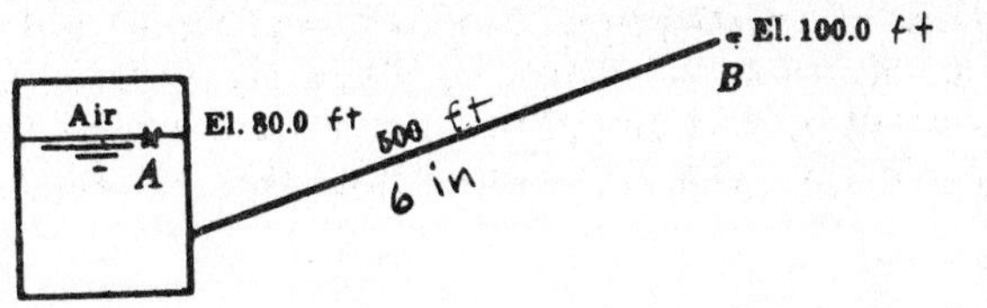

Fig. 9-30

9.151 An old 12-in by 18-in rectangular duct carries air at 15.2 psia and 68 °F through 1500 ft with an average velocity of 9.75 ft/s. Determine the loss of head and the pressure drop, assuming the duct to be horizontal and the size of the surface imperfections is 0.0018 ft.

$$R_h = A/p_w = (\tfrac{12}{12})(\tfrac{18}{12})/(\tfrac{12}{12} + \tfrac{12}{12} + \tfrac{18}{12} + \tfrac{18}{12}) = 0.300 \text{ ft} \qquad d = 4R_h = (4)(0.300) = 1.20 \text{ ft}$$

$$h_f = (f)(L/d)(v^2/2g) = (f)(1500/1.20)\{9.75^2/[(2)(32.2)]\} = 1845f$$

$$N_R = dv/v = (1.20)(9.75)/[(14.7/15.2)(1.64 \times 10^{-4})] = 7.38 \times 10^4 \quad \text{(turbulent)} \qquad \epsilon/d = 0.0018/1.20 = 0.00150$$

From Fig. A-5, $f = 0.024$.

$$h_f = (1845)(0.024) = 44.28 \text{ ft of air} \qquad \Delta p = \gamma h_f = [(15.2/14.7)(0.0750)](44.28)/144 = 0.0238 \text{ lb/in}^2$$

9.152 What size of new cast iron pipe, 8000 ft long, will deliver 37.5 cfs of water at 70 °F with a drop in the hydraulic grade line of 215 ft?

$$p_A/\gamma + v_A^2/2g + z_A = p_B/\gamma + v_B^2/2g + z_B + h_L \qquad [(p_A/\gamma + z_A) - (p_B/\gamma + z_B)] = h_L$$

$$[(p_A/\gamma + z_A) - (p_B/\gamma + z_B)] = \text{HGL} = 215 \text{ ft} \qquad v = Q/A = 37.5/(\pi d^2/4) = 47.75/d^2$$

$$h_L = h_f = (f)(L/d)(v^2/2g) = (f)(8000/d)\{(47.75/d^2)^2/[(2)(32.2)]\} = 283\,238f/d^5$$

$$215 = 283\,238f/d^5 \qquad d = 4.207f^{1/5} \qquad N_R = dv/v = (d)(47.75/d^2)/(1.05 \times 10^{-5}) = 4.55 \times 10^6/d$$

Try $d = 2$ ft: $N_R = (4.55 \times 10^6)/2 = 2.28 \times 10^6$, $\epsilon/d = 0.00085/2 = 0.000425$. From Fig. A-5, $f = 0.0164$. $d = (4.207)(0.0164)^{1/5} = 1.85$ ft. Try $d = 1.85$ ft: $N_R = (4.55 \times 10^6)/1.85 = 2.46 \times 10^6$, $\epsilon/d = 0.00085/1.85 = 0.000459$, $f = 0.0164$ (O.K.). Hence, $d = 1.85$ ft, or 22.2 in.

9.153 What rate of flow of air at 68 °F will be carried by a new horizontal 2-in-diameter steel pipe at an absolute pressure of 3 atm and with a drop of 0.150 psi in 100 ft of pipe? Use $\epsilon = 0.00025$ ft.

At 68 °F and standard atmospheric pressure, $\gamma = 0.0752 \text{ lb/ft}^3$ and $v = 1.60 \times 10^{-4} \text{ ft}^2/\text{s}$. At a pressure of 3 atm, $\gamma = (0.0752)(3) = 0.2256 \text{ lb/ft}^3$ and $v = (1.60 \times 10^{-4})/3 = 5.333 \times 10^{-5} \text{ ft}^2/\text{s}$. $h_f = (f)(L/d)(v^2/2g)$, $(0.150)(144)/0.2256 = (f)[100/(\tfrac{2}{12})]\{v^2/[(2)(32.2)]\}$, $v = 3.206/\sqrt{f}$. Try $v = 10$ ft/s: $N_R = dv/v = (\tfrac{2}{12})(10)/(5.333 \times 10^{-5}) = 3.13 \times 10^4$, $\epsilon/d = 0.00025/(\tfrac{2}{12}) = 0.00150$. From Fig. A-5, $f = 0.027$. $v = 3.206/\sqrt{0.027} = 19.51$ ft/s. Try $v = 19.51$ ft/s: $N_R = (\tfrac{2}{12})(19.51)/(5.333 \times 10^{-5}) = 6.10 \times 10^4$, $f = 0.0248$, $v = 3.206/\sqrt{0.0248} = 20.36$ ft/s. Try $v = 20.36$ ft/s: $N_R = (\tfrac{2}{12})(20.36)/(5.333 \times 10^{-5}) = 6.36 \times 10^4$, $f = 0.0248$ (O.K.); $Q = Av = [(\pi)(\tfrac{2}{12})^2/4](20.36) = 0.444 \text{ ft}^3/\text{s}$.

9.154 Determine the nature of the distribution of shear stress at a cross section in a horizontal, circular pipe under steady flow conditions.

For the free body in Fig. 9-31a, since the flow is steady, each particle moves to the right without acceleration. Hence, the summation of the forces in the x direction must equal zero. $(p_1)(\pi r^2) - (p_2)(\pi r^2) - (\tau)(2\pi rL) = 0$ or

$$\tau = (p_1 - p_2)(r)/(2L) \qquad (1)$$

When $r = 0$, the shear stress τ is zero; and when $r = r_0$, the stress τ_0 at the wall is a maximum. The variation is linear and is indicated in Fig. 9-31b. Equation (1) holds for laminar and turbulent flows as no limitations concerning flow were imposed in the derivation. Since $(p_1 - p_2)/\gamma$ represents the drop in the energy line, or the lost head h_L, multiplying Eq. (1) by γ/γ yields $\tau = (\gamma r/2L)[(p_1 - p_2)/\gamma]$ or

$$\tau = (\gamma h_L/2L)(r) \qquad (2)$$

9.155 Develop the expression for shear stress at a pipe wall.

$h_L = (f)(L/d)(v^2/2g)$. From Prob. 9.154, $h_L = 2\tau_0 L/\gamma r_0 = 4\tau_0 L/\gamma d$, $4\tau_0 L/\gamma d = (f)(L/d)(v^2/2g)$, $\tau_0 = f\gamma v^2/8g = f\rho v^2/8$.

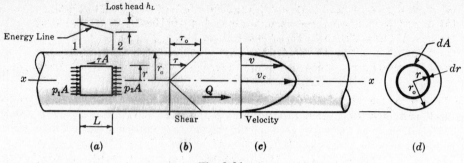

Fig. 9-31

9.156 For steady, laminar flow (*a*) what is the relationship between the velocity at a point in the cross section and the velocity at the center of the pipe, and (*b*) what is the equation for velocity distribution?

▌ (*a*)
$$\tau = -(\mu)(dv/dr) = (p_1 - p_2)(r)/(2L) \qquad \text{(from Prob. 9.154)}$$

$$-(\mu)(dv/dr) = (p_1 - p_2)(r)/(2L)$$

Since $(p_1 - p_2)/L$ is not a function of r,

$$-\int_{v_{\mathbb{C}}}^{v} dv = \frac{p_1 - p_2}{2\mu L}\int_0^r r\, dr \qquad -[v]_{v_{\mathbb{C}}}^v = \frac{p_1 - p_2}{4\mu L}[r^2]_0^r \qquad -(v - v_{\mathbb{C}}) = \frac{(p_1 - p_2)r^2}{4\mu L}$$

$$v = v_{\mathbb{C}} - \frac{(p_1 - p_2)r^2}{4\mu L} \tag{1}$$

But the lost head in L feet is $h_L = (p_1 - p_2)/\gamma$; hence,

$$v = v_{\mathbb{C}} - \frac{\gamma h_L r^2}{4\mu L} \tag{2}$$

(*b*) Since the velocity at the boundary is zero, when $r = r_o$, $v = 0$ in Eq. (*1*), we have

$$v_c = \frac{(p_1 - p_2)r_o^2}{4\mu L} \qquad \text{(at centerline)} \tag{3}$$

Thus, in general,

$$v = \left(\frac{p_1 - p_2}{4\mu L}\right)(r_o^2 - r^2) \tag{4}$$

9.157 Develop the expression for the loss of head in a pipe for steady, laminar flow of an incompressible fluid. Refer to Fig. 9-31*d*.

▌
$$v_{av} = \frac{Q}{A} = \frac{\displaystyle\int v\, dA}{\displaystyle\int dA} = \frac{\displaystyle\int_0^{r_o}(v)(2\pi r\, dr)}{\pi r_o^2} = \frac{(2\pi)(p_1 - p_2)}{(\pi r_o^2)(4\mu L)}\int_0^{r_o}(r_o^2 - r^2)(r\, dr)$$

$$v_{av} = \frac{(p_1 - p_2)(r_o^2)}{8\mu L} \tag{1}$$

Thus for laminar flow, the average velocity is half the maximum velocity v_c in Eq. (*3*) of Prob. 9.156. Rearranging Eq. (*1*), we obtain

$$h_L = \frac{p_1 - p_2}{\gamma} = \frac{8\mu L v_{av}}{\gamma r_o^2} = \frac{32\mu L v_{av}}{\gamma d^2} \tag{2}$$

These expressions apply for laminar flow of all fluids in all pipes and conduits.

9.158 Determine (*a*) the shear stress at the walls of a 12-in-diameter pipe when water flowing causes a measured head loss of 15 ft in 300 ft of pipe length, (*b*) the shear stress 2 in from the centerline of the pipe, (*c*) the shear velocity, (*d*) the average velocity for an f value of 0.50, and (*e*) the ratio v/v_*.

(a) $\tau_o = \gamma h_L r_o/2L = (62.4)(15)[(\frac{12}{12})/2]/[(2)(300)] = 0.780 \text{ lb/ft}^2$, or 0.00542 lb/in^2.
(b) Since τ varies linearly from centerline to wall, $\tau = (0.00542)(\frac{2}{6}) = 0.00181 \text{ lb/in}^2$.
(c) $v_* = \sqrt{\tau_o/\rho} = \sqrt{0.780/1.94} = 0.634 \text{ ft/s}$.
(d) $\tau_o = f\rho v^2/8$, $0.780 = (0.050)(1.94)(v^2)/8$, $v = 8.02 \text{ ft/s}$.
(e) $\tau_o = (\mu)(v/y)$, $v = \mu/\rho$, $\tau_0 = \rho v(v/y)$, $\tau_o/\rho = (v)(v/y) = v_*^2 = (v)(v/y)$, $v/v_*^2 = y/v$, $v/v_* = v_* y/v$.

9.159 If in Prob. 9.158 the water is flowing through a 3-ft by 4-ft rectangular conduit of the same length with the same lost head, what is the shear stress between the water and the pipe wall?

$$R_h = A/p_w = (3)(4)/(3 + 3 + 4 + 4) = 0.8571 \text{ ft}$$

$$\tau = (\gamma h_L/L)(R_h) = [(62.4)(15)/300](0.8571) = 2.67 \text{ lb/ft}^2 \quad \text{or} \quad 0.0186 \text{ lb/in}^2$$

9.160 Medium lubricating oil (s.g. = 0.860) is pumped through 1000 ft of horizontal 2-in pipe at the rate of 0.0436 cfs. If the drop in pressure is 30.0 psi, what is the absolute viscosity of the oil?

Assuming laminar flow,

$$\frac{p_1 - p_2}{\gamma} = \frac{32\mu L v_{av}}{\gamma d^2} \quad \text{(from Prob. 9.157)}$$

$$v_{av} = Q/A = 0.0436/[(\pi)(\tfrac{2}{12})^2/4] = 1.998 \text{ ft/s}$$

$$(30.0)(144)/[(0.860)(62.4)] = (32)(\mu)(1000)(1.998)/\{[(0.860)(62.4)](\tfrac{2}{12})^2\} \quad \mu = 0.00188 \text{ lb} \cdot \text{s/ft}^2$$

$$N_R = \rho dv/\mu = [(0.860)(1.94)](\tfrac{2}{12})(1.998)/0.00188 = 296 \quad \text{(laminar)}$$

9.161 A horizontal wrought iron pipe, 6-in inside diameter and somewhat corroded, is transporting 4.50 lb of air per second from A to B. At A the pressure is 70 psia and at B the pressure must be 65 psia. Flow is isothermal at 68 °F. What is the length of pipe from A to B? Use $\epsilon = 0.0013 \text{ ft}$.

$$\frac{p_1 - p_2}{\gamma_1} = \frac{2[2 \ln (v_2/v_1) + (f)(L/d)](v_1^2/2g)}{1 + p_2/p_1}$$

$$\gamma_1 = (0.0752)(70/14.7) = 0.3581 \text{ lb/ft}^3 \qquad \gamma_2 = (0.0752)(65/14.7) = 0.3325 \text{ lb/ft}^3$$

$$W = \gamma A v \quad 4.50 = (0.3581)[(\pi)(\tfrac{6}{12})^2/4](v_1) \quad v_2 = 64.00 \text{ ft/s}$$

$$4.50 = 0.3325[(\pi)(\tfrac{6}{12})^2/4](v_2) \quad v_1 = 68.93 \text{ ft/s}$$

$$N_R = dv/v = (\tfrac{6}{12})(64.00)/[(14.7/70.0)(1.60 \times 10^{-4})] = 9.52 \times 10^5 \qquad \epsilon/d = 0.0013/(\tfrac{6}{12}) = 0.0026$$

From Fig. A-5, $f = 0.025$.

$$\frac{(70 - 65)(144)}{0.3581} = \frac{2\{2 \ln (68.93/64.00) + 0.025[L/(\tfrac{6}{12})]\}\{64.00^2/[(2)(32.2)]\}}{1 + \tfrac{65}{70}} \qquad L = 607 \text{ ft}$$

9.162 Heavy fuel oil flows from A to B through 3000 ft of horizontal 6-in steel pipe. The pressure at A is 155 psi and at B is 5.0 psi. The kinematic viscosity is 0.00444 ft²/s and the specific gravity is 0.918. What is the flow rate?

Assuming laminar flow, from Eq. (2) of Prob. 9.157,

$$\frac{p_1 - p_2}{\gamma} = \frac{32\mu L v_{av}}{\gamma d^2} = \frac{(32)(v\rho)(L v_{av})}{\gamma d^2} \qquad \frac{(155 - 5.0)(144)}{(0.918)(62.4)} = \frac{32\{(0.00444)[(0.918)(1.94)]\}(3000)(v_{av})}{[(0.918)(62.4)](\tfrac{6}{12})^2}$$

$$v_{av} = 7.11 \text{ ft/s} \qquad N_R = dv/v = (\tfrac{6}{12})(7.11)/0.00444 = 808 \quad \text{(laminar)}$$

$$Q = Av = [(\pi)(\tfrac{6}{12})^2/4](7.11) = 1.40 \text{ ft}^3/\text{s}$$

9.163 What size pipe should be installed to carry 0.785 cfs of heavy fuel oil ($v = 0.00221$ ft²/s, s.g. = 0.912) at 60 °F if the available lost head in the 1000-ft length of horizontal pipe is 22.0 ft?

Assuming laminar flow,

$$h_f = \frac{32\mu L v}{\gamma d^2} \qquad \mu = \rho v = [(0.912)(1.94)](0.00221) = 0.003910 \text{ lb} \cdot \text{s/ft}^2$$

$$v = Q/A = 0.785/(\pi d^2/4) = 0.9995/d^2 \qquad 22.0 = (32)(0.003910)(1000)(0.9995/d^2)/\{[(0.912)(62.4)](d^2)\}$$

$$d = 0.562 \text{ ft} \quad \text{or} \quad 6.75 \text{ in} \qquad N_R = dv/v = (0.562)(0.9995/0.562^2)/0.00221 = 805 \quad \text{(laminar)}$$

9.164 Determine the head loss in 1000 ft of new, uncoated 12-in-ID cast iron pipe when water at 60 °F flows at 5.00 ft/s. Use $\epsilon/d = 0.0008$.

$$h_f = (f)(L/d)(v^2/2g) \qquad N_R = dv/v = (\tfrac{12}{12})(5.00)/(1.21 \times 10^{-5}) = 4.13 \times 10^5$$

From Fig. A-5, $f = 0.0194$. $h_f = 0.0194[1000/(\tfrac{12}{12})]\{5.00^2/[(2)(32.2)]\} = 7.53$ ft.

9.165 Rework Prob. 9.164 if the liquid is medium fuel oil at 60 °F ($v = 4.75 \times 10^{-5}$ ft^2/s) flowing at the same velocity.

$$h_f = (f)(L/d)(v^2 2g) \qquad N_R = dv/v = (\tfrac{12}{12})(5.00)/(4.75 \times 10^{-5}) = 1.05 \times 10^5$$

From Fig. A-5, $f = 0.0213$. $h_f = 0.0213[1000/(\tfrac{12}{12})]\{5.00^2/[(2)(32.2)]\} = 8.27$ ft.

9.166 Points A and B are 4000 ft apart along a new 6-in-ID steel pipe. Point B is 50.5 ft higher than A and the pressures at A and B are 123 psi and 48.6 psi, respectively. How much medium fuel oil at 70 °F will flow from A to B? Use s.g. = 0.854, $v = 4.12 \times 10^{-5}$ ft^2/s, $\epsilon = 0.0002$ ft.

$$p_A/\gamma + v_A^2/2g + z_A = p_B/\gamma + v_B^2/2g + z_B + h_L$$

$$(123)(144)/[(0.854)(62.4)] + v_A^2/2g + 0 = (48.6)(144)/[(0.854)(62.4)] + v_B^2/2g + 50.5 + h_L$$

$$v_A^2/2g = v_B^2/2g \qquad h_L = 150.5 \text{ ft} = h_f = (f)(L/d)(v^2/2g)$$

$$150.5 = (f)[4000/(\tfrac{6}{12})]\{v^2/[(2)(32.2)]\} \qquad v = 1.101/\sqrt{f}$$

Try $v = 10$ ft/s: $N_R = dv/v = (\tfrac{6}{12})(10)/(4.12 \times 10^{-5}) = 1.21 \times 10^5$, $\epsilon/d = 0.0002/(\tfrac{6}{12}) = 0.000400$. From Fig. A-5, $f = 0.0195$. $v = 1.101/\sqrt{0.0195} = 7.884$ ft/s. Try $v = 7.884$ ft/s: $N_R = (\tfrac{6}{12})(7.884)/(4.12 \times 10^{-5}) = 9.57 \times 10^4$, $f = 0.0195$ (O.K.); $Q = Av = [(\pi)(\tfrac{6}{12})^2/4](7.884) = 1.55$ ft^3/s.

9.167 How much water (60 °F) would flow under the conditions of Prob. 9.166.

$$p_A/\gamma + v_A^2/2g + z_A = p_B/\gamma + v_B^2/2g + z_B + h_L$$

$$(123)(144)/62.4 + v_A^2/2g + 0 = (48.6)(144)/62.4 + v_B^2/2g + 50.5 + h_L$$

$$v_A^2/2g = v_B^2/2g \qquad h_L = 121.2 \text{ ft} = h_f = (f)(L/d)(v^2/2g)$$

$$121.2 = (f)[4000/(\tfrac{6}{12})]\{v^2/[(2)(32.2)]\} \qquad v = 0.9878/\sqrt{f}$$

Try $v = 10$ ft/s: $N_R = dv/v = (\tfrac{6}{12})(10)/(1.21 \times 10^{-5}) = 4.13 \times 10^5$, $\epsilon/d = 0.0002/(\tfrac{6}{12}) = 0.000400$. From Fig. A-5, $f = 0.0172$. $v = 0.9878/\sqrt{0.0172} = 7.532$ ft/s. Try $v = 7.532$ ft/s: $N_R = (\tfrac{6}{12})(7.532)/(1.21 \times 10^{-5}) = 3.11 \times 10^5$, $f = 0.0176$, $v = 0.9878/\sqrt{0.0176} = 7.446$ ft/s. Try $v = 7.446$ ft/s: $N_R = (\tfrac{6}{12})(7.446)/(1.21 \times 10^{-5}) = 3.08 \times 10^5$, $f = 0.0176$ (O.K.); $Q = Av = [(\pi)(\tfrac{6}{12})^2/4](7.446) = 1.46$ ft^3/s.

9.168 If 300 cfs of air, $p = 16$ psia, $T = 70$ °F, is to be delivered to a mine with a head loss of 3 in of water per 1000 ft, what size galvanized pipe is needed? ($\epsilon = 0.0005$ ft.)

$$d = 0.66[(\epsilon^{1.25})(LQ^2/gh_f)^{4.75} + (v)(Q)^{9.4}(L/gh_f)^{5.2}]^{0.04}$$

$$\gamma = p/RT = (16)(144)/[(53.3)(460 + 70)] = 0.08156 \text{ lb/ft}^3 \qquad h_f = (\tfrac{3}{12})(62.4/0.08156) = 191.3 \text{ ft of air}$$

$$d = 0.66\left\{(0.0005^{1.25})\left[\frac{(1000)(300)^2}{(32.2)(191.3)}\right]^{4.75} + (1.64 \times 10^{-4})(300)^{9.4}\left[\frac{1000}{(32.2)(191.3)}\right]^{5.2}\right\}^{0.04} = 2.84 \text{ ft}$$

9.169 Two oil reservoirs with a difference in elevation of 5 m are connected by 300 m of commercial steel pipe. What size must the pipe be to convey 50 L/s? ($\mu = 0.05$ kg/m · s, $\gamma = 8$ kN/m^3).

$$d = 0.66[(\epsilon^{1.25})(LQ^2/gh_f)^{4.75} + (v)(Q)^{9.4}(L/gh_f)^{5.2}]^{0.04}$$

$$v = \mu/\rho = \mu g/\gamma = (0.05)(9.807)/[(8)(1000)] = 6.129 \times 10^{-5} \text{ m}^2/\text{s}$$

$$d = 0.66\left\{(0.000046^{1.25})\left[\frac{(300)(\tfrac{50}{1000})^2}{(9.807)(5)}\right]^{4.75} + (6.129 \times 10^{-5})(\tfrac{50}{1000})^{9.4}\left[\frac{300}{(9.807)(5)}\right]^{5.2}\right\}^{0.04} = 0.212 \text{ m}$$

9.170 Calculate the diameter of new wood-stave pipe in excellent condition needed to convey 300 cfs of water at 60 °F with a head loss of 1 ft per 1000 ft of pipe.

$$d = 0.66[(\epsilon^{1.25})(LQ^2/gh_f)^{4.75} + (v)(Q)^{9.4}(L/gh_f)^{5.2}]^{0.04}$$

$$d = 0.66\left\{(0.0006^{1.25})\left[\frac{(1000)(300)^2}{(32.2)(1)}\right]^{4.75} + (1.21 \times 10^{-5})(300)^{9.4}\left[\frac{1000}{(32.2)(1)}\right]^{5.2}\right\}^{0.04} = 7.73 \text{ ft}$$

9.171 An old pipe 2 m in diameter has a roughness of $\epsilon = 30$ mm. A 12-mm-thick lining would reduce the roughness to $\epsilon = 1$ mm. How much in actual pumping costs would be saved per kilometer of pipe for water at 20 °C with discharge of 6 m^3/s? The pumps and motor are 80 percent efficient, and power costs 4 cents per kilowatthour.

$$v_1 = Q/A_1 = 6/[(\pi)(2)^2/4] = 1.910 \text{ m/s} \qquad N_R = dv/\nu$$

$$(N_R)_1 = (2)(1.910)/(1.02 \times 10^{-6}) = 3.75 \times 10^6 \qquad \epsilon_1/d_1 = (\tfrac{30}{1000})/2 = 0.0150$$

From Fig. A-5, $f_1 = 0.044$.

$$d_2 = [2 - (2)(\tfrac{12}{1000})] = 1.976 \text{ m} \qquad v_2 = Q/A_2 = 6/[(\pi)(1.976)^2/4] = 1.957 \text{ m/s}$$

$$(N_R)_2 = (1.976)(1.957)/(1.02 \times 10^{-6}) = 3.79 \times 10^6 \qquad \epsilon_2/d_2 = (\tfrac{1}{1000})/1.976 = 0.000506$$

$$f_2 = 0.017 \qquad h_f = (f)(L/d)(v^2/2g)$$

$$(h_f)_1 = 0.044[(1)(1000)/2]\{1.910^2/[(2)(9.807)]\} = 4.902 \text{ m}$$

$$(h_f)_2 = 0.017[(1)(1000)/1.976]\{1.957^2/[(2)(9.807)]\} = 1.680 \text{ m}$$

$$\text{Saving in head} = 4.092 - 1.680 = 2.412 \text{ m} \qquad P = Q\gamma h_f/\eta = (6)(9.79)(2.412)/0.80 = 177.1 \text{ kW}$$

$$\text{Savings per year} = 177.1[(365)(24)](0.04) = \$62\,056$$

9.172 What size of new cast iron pipe is needed to transport 400 L/s of water at 25 °C for 1000 m with head loss of 2 m?

$$d = 0.66[(\epsilon^{1.25})(LQ^2/gh_f)^{4.75} + (\nu)(Q)^{9.4}(L/gh_f)^{5.2}]^{0.04}$$

$$d = 0.66\left\{(0.00026^{1.25})\left[\frac{(1000)(\tfrac{400}{1000})^2}{(9.807)(2)}\right]^{4.75} + (9.02 \times 10^{-7})(\tfrac{400}{1000})^{9.4}\left[\frac{1000}{(9.807)(2)}\right]^{5.2}\right\}^{0.04} = 0.655 \text{ m}$$

9.173 In a process, 10 000 lb/h of distilled water at 70 °F is conducted through a smooth tube between two reservoirs 30 ft apart and having a difference in elevation of 4 ft. What size tubing is needed?

$$Q = 10\,000/[(62.3)(3600)] = 0.04459 \text{ ft}^3/\text{s} \qquad h_f = (f)(L/d)(v^2/2g)$$

$$v = Q/A = 0.04459/(\pi d^2/4) = 0.05677/d^2 \qquad 4 = (f)(30/d)\{(0.05677/d^2)^2/[(2)(32.2)]\} \qquad d = 0.2065f^{1/5}$$

$$N_R = dv/\nu = (d)(0.05677/d^2)/(1.05 \times 10^{-5}) = 5.41 \times 10^3/d$$

Try $f = 0.020$: $d = (0.2065)(0.020)^{1/5} = 0.09443$ ft, $N_R = 5.41 \times 10^3/0.09443 = 5.73 \times 10^4$. From Fig. A-5, $f = 0.0205$. Try $f = 0.0205$: $d = (0.2065)(0.0205)^{1/5} = 0.09490$ ft, $N_R = 5.41 \times 10^3/0.09490 = 5.70 \times 10^4$, $f = 0.0205$ (O.K.). Hence, $d = 0.09490$ ft, or 1.14 in.

9.174 In Fig. 9-10, $H = 20$ m, $L = 150$ m, $d = 50$ mm, s.g. $= 0.85$, $\mu = 0.400$ N $\cdot$ s/m^2, and $\epsilon = 1$ mm. Find the newtons per second flowing. Neglect minor losses.

$$h_f = (f)(L/d)(v^2/2g) \qquad 20 = (f)[150/(\tfrac{50}{1000})]\{v^2/[(2)(9.807)]\} \qquad v = 0.03616/\sqrt{f}$$

$$N_R = \rho dv/\mu = [(0.85)(1000)](\tfrac{50}{1000})(v)/0.004 = 10\,625v$$

Try $f = 0.050$: $v = 0.3616/\sqrt{0.050} = 1.617$ m/s, $N_R = (10\,625)(1.617) = 1.72 \times 10^4$, $\epsilon/d = (\tfrac{1}{1000})/(\tfrac{50}{1000}) = 0.0200$. From Fig. A-5, $f = 0.051$.

$$v = 0.3616/\sqrt{0.051} = 1.601 \text{ m/s} \qquad N_R = (10\,625)(1.601) = 1.70 \times 10^4 \qquad f = 0.051 \qquad \text{(O.K.)}$$

$$W = \gamma Av = [(0.85)(9.79)][(\pi)(\tfrac{50}{1000})^2/4](1.601) = 0.0262 \text{ kN/s} \quad \text{or} \quad 26.2 \text{ N/s}$$

9.175 Determine the head loss for flow of 140 L/s of oil, $\nu = 0.00001$ m^2/s, through 400 m of 200-mm-diameter cast iron pipe.

$$h_f = (f)(L/d)(v^2/2g) \qquad v = Q/A = (\tfrac{140}{1000})/[(\pi)(\tfrac{200}{1000})^2/4] = 4.456 \text{ m/s}$$

$$N_R = dv/\nu = (\tfrac{200}{1000})(4.456)/0.00001 = 8.91 \times 10^4 \qquad \epsilon/d = 0.00026/(\tfrac{200}{1000}) = 0.00130$$

From Fig. A-5, $f = 0.023$. $h_f = 0.023[400/(\tfrac{200}{1000})]\{4.456^2/[(2)(9.807)]\} = 46.6$ m.

9.176 Water at 15 °C flows through a 300-mm-diameter riveted steel pipe, $\epsilon = 3$ mm, with a head loss of 6 m in 300 m. Determine the flow.

▌ $h_f = (f)(L/d)(v^2/2g)$. Try $f = 0.040$: $6 = 0.040[300/(\frac{300}{1000})]\{v^2/[(2)(9.807)]\}$, $v = 1.715$ m/s; $N_R = dv/v = (\frac{300}{1000})(1.715)/(1.16 \times 10^{-6}) = 4.44 \times 10^5$; $\epsilon/d = \frac{3}{300} = 0.0100$. From Fig. A-5, $f = 0.038$. Try $f = 0.038$: $6 = (0.038)[300/(\frac{300}{1000})]\{v^2/[(2)(9.807)]\}$, $v = 1.760$ m/s; $N_R = (\frac{300}{1000})(1.760)/(1.16 \times 10^{-6}) = 4.55 \times 10^5$; $f = 0.038$ (O.K.); $Q = Av = [(\pi)(\frac{300}{1000})^2/4](1.760) = 0.124$ m³/s.

9.177 Determine the size of clean wrought iron pipe required to convey 4000 gpm of oil, $v = 0.0001$ ft²/s, 10 000 ft with a head loss of 75 ft.

▌
$$Q = (4000)(0.002228) = 8.912 \text{ ft}^3/\text{s} \qquad h_f = (f)(L/d)(v^2/2g)$$

$$v = Q/A = 8.912/(\pi d^2/4) = 11.35/d^2 \qquad 75 = (f)(10\,000/d)\{(11.35/d^2)^2/[(2)(32.2)]\} \qquad d = 3.056 f^{1/5}$$

Try $f = 0.020$: $d = (3.056)(0.020)^{1/5} = 1.398$ ft, $N_R = dv/v = (1.398)(11.35/1.398^2)/0.0001 = 8.12 \times 10^4$, $\epsilon/d = 0.00015/1.398 = 0.000107$. From Fig. A-5, $f = 0.019$. Try $f = 0.019$: $d = (3.056)(0.019)^{1/5} = 1.383$ ft, $N_R = (1.383)(11.35/1.383^2)/0.0001 = 8.21 \times 10^4$, $f = 0.019$ (O.K.). Hence, $d = 1.383$ ft, or 16.6 in.

9.178 In Prob. 9.177, for $d = 16.6$ in, if the specific gravity is 0.85, $p_1 = 40$ psi, $z_1 = 200$ ft, and $z_2 = 50$ ft, determine the pressure at point 2.

▌
$$p_1/\gamma + v_1^2/2g + z_1 = p_2/\gamma + v_2^2/2g + z_2 + h_L$$

$$(40)(144)/[(0.85)(62.4)] + v_1^2/2g + 200 = (p_2)(144)/[(0.85)(62.4)] + v_2^2/2g + 50 + 75$$

$$v_1^2/2g = v_2^2/2g \qquad p_2 = 67.6 \text{ lb/in}^2$$

9.179 What size galvanized iron pipe is needed to be "hydraulically smooth" at $N_R = 3.5 \times 10^5$? (A pipe is said to be hydraulically smooth when it has the same losses as a smoother pipe under the same conditions.)

▌ From Fig. A-5, $\epsilon/d = 0.00001$ is equivalent to smooth pipe. $0.0005/d = 0.00001$, $d = 50.0$ ft.

9.180 Above what Reynolds number is the flow through a 3-m-diameter riveted steel pipe, $\epsilon = 3$ mm, independent of the viscosity of the fluid?

▌ $\epsilon/d = (\frac{3}{1000})/3 = 0.00100$. From Fig. A-5, $N_R = 1.3 \times 10^6$ at complete turbulence.

9.181 Determine the absolute roughness of a 1-ft-diameter pipe that has a friction factor of 0.03 for a Reynolds number of 10^6.

▌ From Fig. A-5, $\epsilon/d = 0.005$. $\epsilon/1 = 0.005$, $\epsilon = 0.005$ ft.

9.182 What diameter clean galvanized iron pipe has the same friction factor for $N_R = 100\,000$ as a 300-mm-diameter cast iron pipe?

▌ **For cast iron:** $\epsilon/d = 0.00026/(\frac{300}{1000}) = 0.000867$.

For galvanized iron: $\epsilon/d = 0.00015/d = 0.000867$, $d = 0.173$ m, or 173 mm.

9.183 Calculate the friction factor for atmospheric air at 80 °F, $v = 50$ ft/s in a 3-ft-diameter galvanized pipe.

▌
$$f = 1.325/[\ln(\epsilon/3.7d + 5.74/N_R^{0.9})]^2 \qquad N_R = dv/v = (3)(50)/(1.69 \times 10^{-4}) = 887\,574$$

$$f = 1.325/\{\ln[0.0005/(3.7)(3) + 5.74/887\,574^{0.9}]\}^2 = 0.0145$$

9.184 If 16 000 ft³/min of atmospheric air at 90 °F is conveyed 1000 ft through a 4-ft-diameter wrought iron pipe, what is the head loss in inches of water?

▌
$$h_f = (f)(L/d)(v^2/2g) \qquad \rho = p/RT = (14.7)(144)/[(1716)(460 + 90)] = 0.002243 \text{ slug/ft}^3$$

$$v = Q/A = (16\,000/60)/[(\pi)(4)^2/4] = 21.22 \text{ ft/s}$$

$$N_R = \rho dv/\mu = (0.002243)(4)(21.22)/(3.90 \times 10^{-7}) = 4.88 \times 10^5 \qquad \epsilon/d = 0.00015/4 = 0.0000375$$

From Fig. A-5, $f = 0.013$. $h_f = (0.013)(\frac{1000}{4})\{21.22^2/[(2)(32.2)]\} = 22.72$ ft of air, or

$$(h_f)_{\text{H}_2\text{O}}(\rho_{\text{H}_2\text{O}}) = (h_f)_{\text{air}}(\rho_{\text{air}}) \qquad (h_f)_{\text{H}_2\text{O}}(1.94) = (22.72)(0.002243) \qquad (h_f)_{\text{H}_2\text{O}} = 0.02627 \text{ ft} \quad \text{or} \quad 0.315 \text{ in}$$

9.185 What power motor for a fan must be purchased to circulate standard air in a wind tunnel at 500 km/h at a temperature of 15 °C? The tunnel is a closed loop 60 m long, and it can be assumed to have a constant cross section with a 2 m diameter. Assume smooth pipe.

∎
$$h_f = (f)(L/d)(v^2/2g) \qquad v = (500)(1000)/3600 = 138.9 \text{ m/s}$$
$$N_R = dv/\nu = (2)(138.9)/(1.46 \times 10^{-5}) = 1.90 \times 10^7$$

From Fig. A-5, $f = 0.0072$ (extrapolated).

$$h_f = (0.0072)(\tfrac{60}{2})\{138.9^2/[(2)(9.807)]\} = 212.5 \text{ m} \qquad Q = Av = [(\pi)(2^2/4)](138.9) = 436.4 \text{ m}^3/\text{s}$$
$$P = Q\gamma h_f = (436.4)(12.0)(212.5) = 1.11 \times 10^6 \text{ W} \quad \text{or} \quad 1.11 \text{ MW}$$

9.186 Assume that 2.0 cfs of oil, $\mu = 0.16$ P, $\gamma = 54 \text{ lb/ft}^3$, is pumped through a 12-in pipeline of cast iron. If each pump produces 80 psi, how far apart can they be placed?

∎
$$h_f = (f)(L/d)(v^2/2g) = p/\gamma = (80)(144)/54 = 213.3 \text{ ft} \qquad v = Q/A = 2.0/[(\pi)(\tfrac{12}{12})^2/4] = 2.546 \text{ ft/s}$$
$$\rho = \gamma/g = 54/32.2 = 1.677 \text{ lb} \cdot \text{s/ft}^2 \qquad \mu = 0.16 \text{ P} = 0.016 \text{ N} \cdot \text{s/m}^2 = (0.016)(0.3048)^2/4.448 = 0.0003342 \text{ lb} \cdot \text{s/ft}^2$$
$$N_R = \rho dv/\mu = (1.667)(\tfrac{12}{12})(2.546)/0.0003342 = 1.27 \times 10^4 \qquad \epsilon/d = 0.00085/(\tfrac{12}{12}) = 0.00085$$

From Fig. A-5, $f = 0.031$. $213.3 = 0.031[L/(\tfrac{12}{12})]\{2.546^2/[(2)(32.2)]\}$, $L = 68\,359$ ft, or 12.9 miles.

9.187 A 60-mm-diameter smooth pipe 150 m long conveys 10 L/s of water at 25 °C from a water main, $p = 1.6 \text{ MN/m}^2$, to the top of a building 25 m above the main. What pressure can be maintained at the top of the building?

∎
$$p_1/\gamma + v_1^2/2g + z_1 = p_2/\gamma + v_2^2/2g + z_2 + h_L \qquad h_L = h_f = (f)(L/d)(v^2/2g)$$
$$v = Q/A = (\tfrac{10}{1000})/[(\pi)(60/1000)^2/4] = 3.537 \text{ m/s} \qquad N_R = dv/\nu = (\tfrac{60}{1000})(3.537)/(9.10 \times 10^{-7}) = 2.33 \times 10^5$$

From Fig. A-5, $f = 0.016$.

$$h_L = 0.016[150/(\tfrac{60}{1000})]\{3.537^2/[(2)(9.807)]\} = 25.51 \text{ m}$$
$$(1.6)(1000)/9.79 + v_1^2/2g + 0 = (p_2)(1000)/9.79 + v_2^2/2g + 25 + 25.51 \qquad v_1^2/2g = v_2^2/2g \qquad p_2 = 1.11 \text{ MN/m}^2$$

9.188 For water at 150 °F, calculate the discharge for the pipe of Fig. 9-32. The pipe discharges to a reservoir at the bottom of the pipe.

∎
$$h_L = h_f = (f)(L/d)(v^2/2g) \qquad 260 = (f)[240/(\tfrac{2}{12})]\{v^2/[(2)(32.2)]\} \qquad v = 3.410/\sqrt{f}$$

Try $f = 0.019$: $v = 3.410/\sqrt{0.019} = 24.74 \text{ ft/s}$, $\epsilon/d = 0.00015/(\tfrac{2}{12}) = 0.000900$, $N_R = dv/\nu = (\tfrac{2}{12})(24.74)/(4.68 \times 10^{-6}) = 8.81 \times 10^5$. From Fig. A-5, $f = 0.019$ (O.K.); $Q = Av = [(\pi)(\tfrac{2}{12})^2/4](24.74) = 0.540 \text{ ft}^3/\text{s}$.

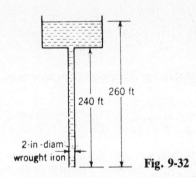

260 ft
240 ft
2-in-diam
wrought iron

Fig. 9-32

9.189 In Fig. 9-32 how much power would be required to pump 160 gpm of water at 60 °F from a reservoir at the bottom of the pipe to the reservoir shown?

∎
$$Q = (160)(0.002228) = 0.3565 \text{ ft}^3/\text{s} \qquad p_1/\gamma + v_1^2/2g + z_1 = p_2/\gamma + v_2^2/2g + z_2 + h_L$$
$$v_1 = Q/A_1 = 0.3565/[(\pi)(\tfrac{2}{12})^2/4] = 16.34 \text{ ft/s} \qquad h_L = h_f = (f)(L/d)(v^2/2g)$$
$$N_R = dv/\nu = (\tfrac{2}{12})(16.34)/(1.21 \times 10^{-5}) = 2.25 \times 10^5 \qquad \epsilon/d = 0.00015/(\tfrac{2}{12}) = 0.000900$$

From Fig. A-5, $f = 0.0205$.

$$h_L = 0.0205[240/(\tfrac{2}{12})]\{16.34^2/[(2)(32.2)]\} = 122.4 \text{ ft} \qquad p_1/\gamma + 16.34^2/[(2)(32.2)] + 0 = 0 + 0 + 260 + 122.4$$

$$p_1/\gamma = 378.3 \text{ ft} \qquad P = Q\gamma(\Delta p/\gamma) = (0.3565)(62.4)(378.3) = 8416 \text{ ft} \cdot \text{lb/s} = \tfrac{8416}{550} = 15.3 \text{ hp}$$

9.190 A 12-mm-diameter commercial steel pipe 15 m long is used to drain an oil tank. Determine the discharge when the oil level in the tank is 2 m above the exit end of the pipe. ($\mu = 0.10 \text{ P}$, $\gamma = 8 \text{ kN/m}^3$.)

$\blacksquare$ Assuming laminar flow,

$$v = \frac{h_L \gamma d^2}{32 \mu L} = \frac{2[(8)(1000)](\tfrac{12}{1000})^2}{(32)[(0.1)(0.10)](15)} = 0.4800 \text{ m/s}$$

$$N_R = \rho d v/\mu = (\gamma/g)(dv)/\mu = [(8)(1000)/9.807](\tfrac{12}{1000})(0.4800)/[(0.1)(0.10)] = 470 \qquad \text{(laminar)}$$

$$Q = Av = [(\pi)(\tfrac{12}{1000})^2/4](0.4800) = 0.0000543 \text{ m}^3/\text{s} \quad \text{or} \quad 0.0543 \text{ L/s}$$

9.191 Two liquid reservoirs are connected by 200 ft of 2-in-diameter smooth tubing. What is the flow rate when the difference in elevation is 50 ft? ($\nu = 0.001 \text{ ft}^2/\text{s}$).

$\blacksquare$
$$p_1/\gamma + v_1^2/2g + z_1 = p_2/\gamma + v_2^2/2g + z_2 + h_L \qquad h_L = h_f = (f)(L/d)(v^2/2g)$$

Assuming laminar flow,

$$f = 64/N_R \qquad N_R = dv/\nu = (\tfrac{2}{12})(v)/0.001 = 166.7v \qquad h_L = (64/166.7v)[200/(\tfrac{2}{12})]\{v^2/[(2)(32.2)]\} = 7.154v$$

$$0 + 0 + 50 = 0 + 0 + 0 + 7.154v \qquad v = 6.989 \text{ ft/s} \qquad N_R = (\tfrac{2}{12})(6.989)/0.001 = 1165 \qquad \text{(laminar)}$$

$$Q = Av = [(\pi)(\tfrac{2}{12})^2/4](6.989) = 0.152 \text{ ft}^3/\text{s}$$

6.192 For a head loss of 80 mm of water in a length of 200 m for flow of atmospheric air at 15 °C through a 1.25-m-diameter duct, $\epsilon = 1$ mm, calculate the flow in cubic meters per minute.

$\blacksquare$ $h_f = (f)(L/d)(v^2/2g) \qquad (\gamma h_f)_{\text{air}} = (\gamma h_f)_{\text{H}_2\text{O}} \qquad \gamma = p/RT \qquad \gamma_{\text{air}} = 101.4/[(29.3)(273 + 15)] = 0.01202 \text{ kN/m}^3$

$$(0.01202)(h_f)_{\text{air}} = (9.79)(\tfrac{80}{1000}) \qquad (h_f)_{\text{air}} = 65.16 \text{ m} \qquad 65.16 = (f)(200/1.25)\{v^2/[(2)(9.807)]\} \qquad v = 2.826/\sqrt{f}$$

Try $f = 0.020$: $v = 2.826/\sqrt{0.020} = 19.98 \text{ m/s}$, $N_R = dv/\nu = (1.25)(19.98)/(1.46 \times 10^{-5}) = 1.71 \times 10^6$, $\epsilon/d = (\tfrac{1}{1000})/1.25 = 0.000800$. From Fig. A-5, $f = 0.0205$. Try $f = 0.0205$: $v = 2.826/\sqrt{0.0205} = 19.74 \text{ m/s}$, $N_R = (1.25)(19.74)/(1.46 \times 10^{-5}) = 1.69 \times 10^6$, $f = 0.0205$ (O.K.); $Q = Av = [(\pi)(1.25^2/4)](19.74) = 24.2 \text{ m}^3/\text{s}$.

9.193 Water at 20 °C is to be pumped through 1 km of 200-mm-diameter wrought iron pipe at the rate of 60 L/s. Compute the head loss and power required.

$\blacksquare$
$$h_f = (f)(L/d)(v^2/2g) \qquad v = Q/A = (\tfrac{60}{1000})/[(\pi)(\tfrac{200}{1000})^2/4] = 1.910 \text{ m/s}$$

$$N_R = dv/\nu = (\tfrac{200}{1000})(1.910)/(1.02 \times 10^{-6}) = 3.75 \times 10^5 \qquad \epsilon/d = 0.000046/(\tfrac{200}{1000}) = 0.000230$$

From Fig. A-5, $f = 0.016$.

$$h_f = 0.016[(1)(1000)/(\tfrac{200}{1000})]\{1.910^2/[(2)(9.807)]\} = 14.88 \text{ m} \qquad P = Q\gamma h_f = (\tfrac{60}{1000})(9.79)(14.88) = 8.74 \text{ kW}$$

9.194 The 100 lbm/min of air required to ventilate a mine is admitted through 3000 ft of 12-in-diameter galvanized pipe. Neglecting minor losses, what head, in inches of water, does a blower have to produce to furnish the flow? ($p = 14$ psia; $T = 90$ °F.)

$\blacksquare$ $h_f = (f)(L/d)(v^2/2g) \qquad N_R = \rho dv/\mu \qquad \rho = p/RT = (14)(144)/[(1716)(460 + 90)] = 0.002136 \text{ slug/ft}^3$

$$M = \rho Av \qquad \tfrac{100}{60} = [(0.002136)(32.2)][(\pi)(\tfrac{12}{12})^2/4](v) \qquad v = 30.85 \text{ ft/s}$$

$$N_R = (0.002136)(\tfrac{12}{12})(30.85)/(3.90 \times 10^{-7}) = 1.69 \times 10^5 \qquad \epsilon/d = 0.0005/(\tfrac{12}{12}) = 0.000500$$

From Fig. A-5, $f = 0.019$.

$$h_f = 0.019[3000/(\tfrac{12}{12})]\{30.85^2/[(2)(32.2)]\} = 842.4 \text{ ft of air}$$

$$h = 842.4[(0.002136)(32.2)](12)/62.4 = 11.1 \text{ in of water}$$

9.195 A 2.0-m-diameter pipe of length 1560 m for which $\epsilon = 1.5$ mm conveys water at 12 °C between two reservoirs at a rate of 8.0 m³/s. What must be the difference in water-surface elevations between the two reservoirs? Neglect minor losses.

$$p_1/\gamma + v_1^2/2g + z_1 = p_2/\gamma + v_2^2/2g + z_2 + h_L \qquad 0+0+z_1 = 0+0+z_2+h_f$$
$$z_1 - z_2 = h_f = (f)(L/d)(v^2 2g)$$
$$v = Q/A = 8.0/[(\pi)(2.0)^2/4] = 2.546 \text{ m/s} \qquad N_R = dv/v = (2.0)(2.546)/(1.24 \times 10^{-6}) = 4.11 \times 10^6$$
$$\epsilon/d = (1.5/1000)/2.0 = 0.000750$$

From Fig. A-5, $f = 0.018$. $h_f = (0.018)(1560/2.0)\{2.546^2/[(2)(9.807)]\} = 4.64$ m. Hence, the difference in water-surface elevations between the two reservoirs is 4.64 m.

9.196 Water flows from reservoir 1 to reservoir 2 through a 4-in-diameter, 500-ft-length pipe, as shown in Fig. 9-33. Assume an initial friction factor (f) of 0.037 and a roughness (ϵ) of 0.003 ft for the pipe. Find the flow rate.

$$p_1/\gamma + v_1^2/2g + z_1 = p_2/\gamma + v_2^2/2g + z_2 + h_L \qquad 0+0+700.6 = 0+0+655.5+h_L \qquad h_L = 45.2 \text{ ft} = h_f + h_m$$

(I) **Friction loss:** $h_f = (f)(L/d)(v^2/2g) = 0.037[500/(\frac{4}{12})](v^2/2g) = 55.50v^2/2g$.

(II) **Minor losses:**
   (a) **Due to entrance:** From Fig. A-7, take $K_1 = 0.45$.
   (b) **Due to globe valve:** From Table A-11, $K_{open} = 5.7$. From Table A-12, take $K_2/K_{open} = 1.75$. Hence, $K_2 = (5.7)(1.75) = 9.98$.
   (c) **Due to bend:** $R/D = \frac{12}{4} = 3.0$, $\epsilon/D = 0.003/(\frac{4}{12}) = 0.00900$. From Fig. A-12, $K_3 = 0.45$.
   (d) **Due to elbow:** From Table A-11, $K_4 = 0.23$.
   (e) **Due to exit:** From Fig. A-7, $K_5 = 1.0$.
Thus,
$$h_f + h_m = \{v^2/[(2)(32.2)]\}(55.50 + 0.45 + 9.98 + 0.45 + 0.23 + 1.0) = 1.050v^2$$
$$1.050v^2 = 45.2 \qquad v = 6.561 \text{ ft/s}$$
$$N_R = Dv/v = (\tfrac{4}{12})(6.561)/(1.90 \times 10^{-5}) = 1.15 \times 10^5 \qquad \epsilon/D = 0.003/(\tfrac{4}{12}) = 0.00900$$

From Fig. A-5, $f = 0.037$. (Assumed value of f O.K.) $Q = Av = [(\pi)(\tfrac{4}{12})^2/4](6.561) = 0.573$ ft³/s.

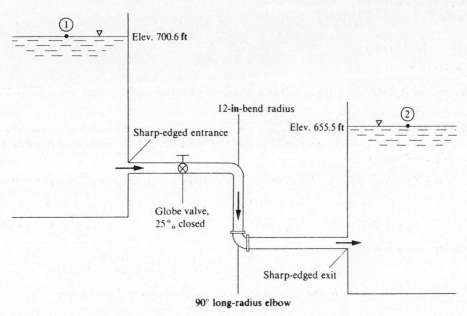

Fig. 9-33

9.197 Determine the head loss in the system shown in Fig. 9-34 and the discharge in the pipe.

$$\blacksquare \qquad p_1/\gamma + v_1^2/2g + z_1 = p_2/\gamma + v_2^2/2g + z_2 + h_L \qquad 0 + 0 + 100 = 0 + v_2^2/[(2)(32.2)] + 0 + h_L$$

$$h_L = h_f + h_m \qquad h_f = (f)(L/D)(v^2/2g) = f[500/(\tfrac{6}{12})]\{v_2^2/[(2)(32.2)]\} = 15.53fv_2^2 \qquad h_m = (K)(v^2/2g) \qquad K = 0.5$$

$$h_m = 0.5\{v_2^2/[(2)(32.2)]\} = 0.007764v_2^2 \qquad h_L = 15.53fv_2^2 + 0.007764v_2^2$$

$$100 = 0.01553v_2^2 + 15.53fv_2^2 + 0.007764v_2^2 \qquad v_2 = \sqrt{100/(0.02329 + 15.53f)}$$

Try $v_2 = 10$ ft/s: $N_R = Dv/v = (\tfrac{6}{12})(10)/(1.21 \times 10^{-5}) = 4.13 \times 10^5$, $\epsilon/D = 0.00015/(\tfrac{6}{12}) = 0.00030$. From Fig. A-5, $f = 0.0165$. $v_2 = \sqrt{100/[(0.02329 + (15.53)(0.0165)]} = 18.91$ ft/s. Try $v_2 = 18.91$ ft/s: $N_R = (\tfrac{6}{12})(18.91)/(1.21 \times 10^{-5}) = 7.81 \times 10^5$, $f = 0.0157$, $v_2 = \sqrt{100/[0.02329 + (15.53)(0.0157)]} = 19.35$ ft/s. Try $v_2 = 19.35$ ft/s: $N_R = (\tfrac{6}{12})(19.35)/(1.21 \times 10^{-5}) = 8.00 \times 10^5$, $f = 0.0157$ (O.K.); $h_L = (15.53)(0.0157)(19.35)^2 + (0.007764)(19.35)^2 = 94.2$ ft of water, $Q = Av = [(\pi)(\tfrac{6}{12})^2/4](19.35) = 3.80$ ft^3/s.

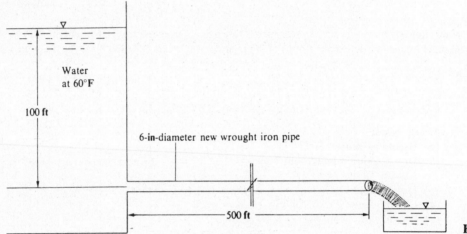

Water at 60°F

100 ft

6-in-diameter new wrought iron pipe

500 ft

Fig. 9-34

9.198 Figure 9-35 shows two reservoirs containing water at 60 °F. The water in the upper reservoir is to be drained to the other reservoir at a lower level as shown. The total length of commercial steel pipe is 100 ft, and the diameter of the pipe is 4 in. What will be the flow rate of water through the pipe when the water surface elevations are as shown in Fig. 9-35?

$$\blacksquare \qquad p_1/\gamma + v_1^2/2g + z_1 = p_2/\gamma + v_2^2/2g + z_2 + h_L \qquad 0 + 0 + 335 = 0 + 0 + 300 + h_L$$

$$h_L = h_f + h_m \qquad h_f = (f)(L/D)(v^2/2g) = f[100/(\tfrac{4}{12})]\{v_2^2/[(2)(32.2)]\} = 4.658fv_2^2 \qquad h_m = (K)(v^2/2g)$$

Due to entrance, take $K_1 = 0.45$. Due to elbow, $K_2 = 0.64$. Due to gate valve, $K_{\text{open}} = 0.11$.
Take $K_3/K_{\text{open}} = 17$. Hence, $K_3 = (0.11)(17) = 1.87$. Due to exit, $K_4 = 1.0$.

$$h_m = (0.45 + 0.64 + 1.87 + 1.0)\{v^2/[(2)(32.2)]\} = 0.06149v^2 \qquad h_L = 4.658fv^2 + 0.06149v^2$$

$$335 = 300 + 4.658fv^2 + 0.06149v^2 \qquad v = \sqrt{35/(4.658f + 0.06149)}$$

Try $f = 0.019$: $v = \sqrt{35/[(4.658)(0.019) + 0.06149]} = 15.28$ ft/s, $N_R = Dv/v = (\tfrac{4}{12})(15.28)/(1.21 \times 10^{-5}) = 4.21 \times 10^5$. From Fig. A-5, $f = 0.0175$. Try $f = 0.0175$: $v = \sqrt{35/[(4.658)(0.0175) + 0.06149]} = 15.64$ ft/s, $N_R = (\tfrac{4}{12})(15.64)/(1.21 \times 10^{-5}) = 4.31 \times 10^5$, $f = 0.0175$ (O.K.); $Q = Av = [(\pi)(\tfrac{4}{12})^2/4](15.64) = 1.36$ ft^3/s.

9.199 A 15-in-diameter new cast iron pipe connecting two reservoirs as shown in Fig. 9-36 carries water at 60 °F. The pipe is 120 ft long, and the discharge is 20 ft/s. Determine the difference in elevation between water surfaces in the two reservoirs.

$$\blacksquare \qquad p_1/\gamma + v_1^2/2g + z_1 = p_2/\gamma + v_2^2/2g + z_2 + h_L \qquad 0 + 0 + z_1 = 0 + 0 + z_2 + h_L$$

Since $z_1 - z_2 = H$ and $h_L = h_f + h_m$, $H = h_f + h_m$, $h_f = (f)(L/D)(v^2/2g)$.

$$v = Q/A = 20/[(\pi)(\tfrac{15}{12})^2/4] = 16.30 \text{ ft/s}$$

$$N_R = Dv/v = (\tfrac{15}{12})(16.30)/(1.21 \times 10^{-5}) = 1.68 \times 10^6 \qquad \epsilon/D = 0.00085/(\tfrac{15}{12}) = 0.00068$$

From Fig. A-5, $f = 0.018$. $h_f = 0.018[120/(\tfrac{15}{12})]\{16.30^2/[(2)(32.2)]\} = 7.13$ ft, $h_m = (K)(v^2/2g)$. For entrance, take $K_1 = 0.45$. For exit, $K_2 = 1.0$. $h_m = (0.45 + 1.0)\{16.30^2/[(2)(32.2)]\} = 5.98$ ft, $H = 7.13 + 5.98 = 13.11$ ft.

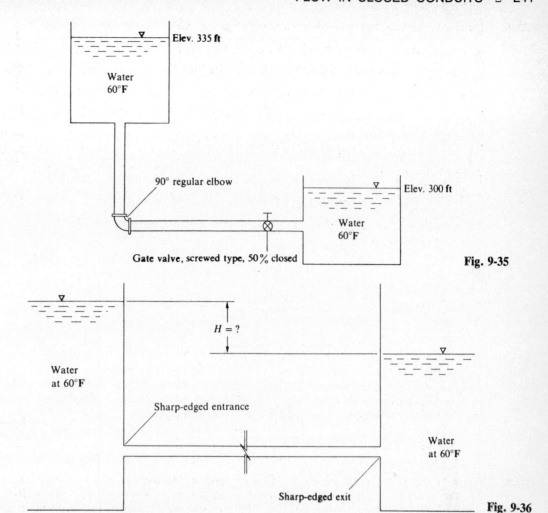

Fig. 9-35

Fig. 9-36

9.200 Repeat Prob. 9.87 by including losses due to a sharp-edged entrance, the exit, and a fully open flanged globe valve. By what percentage is the flow decreased?

$\blacksquare$ For entrance, $K_1 = 0.5$. For exit, $K_2 = 1.0$. For globe valve, $K_3 = 8.5$. From Prob. 9.87, $h_L = h_f + h_f =$ $[(f)(L/d) + K_1 + K_2 + K_3](v^2/2g)$, $100 = \{f[7000/(\frac{5}{100})] + 0.5 + 1.0 + 8.5\}\{v^2/[(2)(9.807)]\}$, $1961 = (140\,000f + 10.0)(v^2)$, $v = [1961/(140\,000f + 10.0)]^{1/2}$. Try $f = 0.02$: $v = \{1961/[(140\,000)(0.02) + 10.0]\}^{1/2} = 0.8354$ m/s, $N_R = \rho dv/\mu = (998)(\frac{5}{100})(0.8354)/(1.02 \times 10^{-3}) = 4.09 \times 10^4$. From Fig. A-5, $f = 0.022$. Try $f = 0.022$: $v = \{1961/[(140\,000)(0.022) + 10.0]\}^{1/2} = 0.7966$ m/s, $N_R = (998)(\frac{5}{100})(0.7966)/(1.02 \times 10^{-3}) = 3.90 \times 10^4$, $f = 0.022$ (O.K.); $Q = Av = [(\pi)(\frac{5}{100})^2/4](0.7966) = 0.00156$ m³/s, or 5.63 m³/h. This is $(5.64 - 5.63)/5.64 = 0.002$, or 0.2 percent less.

9.201 Repeat Prob. 9.95 by including losses due to a sharp entrance and a fully open screwed swing-check valve. By what percentage is the required tank level h increased?

$\blacksquare$ For sharp entrance, $K_1 = 0.5$. For swing-check valve, $K_2 = 5.1$. $h_m = (K_1 + K_2)(v^2/2g)$. From Prob. 9.95, $v = 14.67$ ft/s. $h_m = (0.5 + 5.1)\{14.67^2/[(2)(32.2)]\} = 18.71$ ft. From Prob. 9.95, $0 + 0 + h = 0 + 14.67^2/[(2)(32.2)] + 0 + (189.3 + 18.71)$, $h = 211.4$ ft. This is $(211.4 - 192.6)/192.6 = 0.098$, or 9.8 percent more.

9.202 The two reservoirs in Fig. 9-37 are connected by 20-ft-long wrought iron pipes joined abruptly. The entrance and exit are sharp-edged. The fluid is water at 20 °C. Including minor losses, compute the flow rate if reservoir 1 is 60 ft higher than reservoir 2.

$\blacksquare$

$$p_1/\gamma + v_1^2/2g + z_1 = p_2/\gamma + v_2^2/2g + z_2 + h_L \qquad h_L = h_f + h_m$$

$$h_f = (f)(L/D)(v^2/2g) = (f_a)[20/(\tfrac{1}{12})]\{v_a^2/[(2)(32.2)]\} + f_b[20/(\tfrac{2}{12})]\{v_b^2/[(2)(32.2)]\}$$

$$A_a v_a = A_b v_b \qquad [(\pi)(\tfrac{1}{12})^2/4](v_a) = [(\pi)(\tfrac{2}{12})^2/4](v_b) \qquad v_b = v_a/4$$

$$h_f = 3.727 f_a v_a^2 + (1.863)(f_b)(v_a/4)^2 = 3.727 f_a v_a^2 + 0.01164 f_b v_a^2$$

For sharp entrance, $K_1 = 0.5$. For sharp exit, $K_2 = 1.0$. For sudden expansion, $K_2 = (1 - 0.5)^2 = 0.5625$.
$h_m = (0.5 + 0.5625)\{v_a^2/[(2)(32.2)]\} + (1.0)\{v_b^2/[(2)(32.2)]\} = 0.01650v_a^2 + (0.01553)(v_a/4)^2 = 0.01747v_a^2$, $0 + 0 + 60 = 0 + 0 + 0 + [(3.727f_a v_a^2 + 0.1164f_b v_a^2) + 0.01747v_a^2]$. Try $f_a = f_b = 0.020$:

$$60 = (3.727)(0.020)(v_a^2) + (0.1164)(0.020)(v_a^2) + 0.01747v_a^2 \qquad v_a = 25.22 \text{ ft/s} \qquad v_b = 25.22/4 = 6.305 \text{ ft/s}$$

$$N_R = Dv/v \qquad (N_R)_a = (\tfrac{1}{12})(25.22)/(1.11 \times 10^{-5}) = 1.89 \times 10^5 \qquad (\epsilon/D)_a = 0.00015/(\tfrac{1}{12}) = 0.00180$$

From Fig. A-5, $f_a = 0.024$. $(N_R)_b = (\tfrac{2}{12})(6.305)/(1.11 \times 10^{-5}) = 9.47 \times 10^4$, $(\epsilon/D)_b = 0.00015/(\tfrac{2}{12}) = 0.000900$, $f_b = 0.022$. Try $f_a = 0.024$ and $f_b = 0.022$:

$$60 = (3.727)(0.024)(v_a^2) + (0.1164)(0.022)(v_a^2) + 0.01747v_a^2 \qquad v_a = 23.41 \text{ ft/s} \qquad v_b = 23.41/4 = 5.852 \text{ ft/s}$$

$$(N_R)_a = (\tfrac{1}{12})(23.41)/(1.11 \times 10^{-5}) = 1.76 \times 10^5 \qquad f_a = 0.024$$

$$(N_R)_b = (\tfrac{2}{12})(5.852)/(1.11 \times 10^{-5}) = 8.79 \times 10^4 \qquad f_b = 0.022$$

Therefore, $f_a = 0.024$ and $f_b = 0.022$ is O.K. $Q = Av = [(\pi)(\tfrac{1}{12})^2/4](23.41) = 0.128 \text{ ft}^3/\text{s}$.

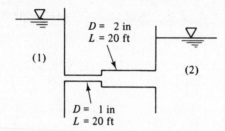

Fig. 9-37

9.203 Two reservoirs containing water at 20 °C are connected by 700 m of 18-cm cast iron pipe, including a sharp entrance, a submerged exit, a gate valve 75 percent open, two 1-m-radius bends, and six regular 90° elbows. If the flow rate is 0.15 m³/s, what is the difference in reservoir elevations?

$$\blacksquare \qquad p_1/\gamma + v_1^2/2g + z_1 = p_2/\gamma + v_2^2/2g + z_2 + h_L \qquad h_L = h_f + h_m \qquad h_f = (f)(L/D)(v^2/2g)$$

$$v = Q/A = 0.15/[(\pi)(\tfrac{18}{100})^2/4] = 5.895 \text{ m/s} \qquad N_R = Dv/v = (\tfrac{18}{100})(5.895)/(1.02 \times 10^{-6}) = 1.04 \times 10^6$$

$$\epsilon/D = 0.00026/(\tfrac{18}{100}) = 0.00144$$

From Fig. A-5, $f = 0.0217$. $h_f = 0.0217[700/(\tfrac{18}{100})]\{5.895^2/[(2)(9.807)]\} = 149.5 \text{ m}$. For sharp entrance, $K_1 = 0.5$. For exit, $K_2 = 1.0$. For gate valve 75 percent open, $K_3 = 0.3$. For bends, $K_4 = (2)(0.15) = 0.30$. For elbows, $K_5 = (6)(0.27) = 1.62$. $h_m = (0.5 + 1.0 + 0.3 + 0.30 + 1.62)\{5.895^2/[(2)(9.807)]\} = 6.6 \text{ m}$, $h_L = 149.5 + 6.6 = 156.1 \text{ m}$, $0 + 0 + z_1 = 0 + 0 + z_2 + 156.1$, $z_1 - z_2 = 156.1 \text{ m}$.

9.204 The system in Fig. 9-38 consists of 1200 m of 5-cm cast iron pipe, two 45° and four 90° flanged long-radius elbows, a fully open flanged globe valve, and a sharp exit into a reservoir. If the elevation at point 1 is 400 m, what gage pressure is required at point 1 to deliver 0.005 m³/s of water at 20 °C into the reservoir?

$$\blacksquare \qquad p_1/\gamma + v_1^2/2g + z_1 = p_2/\gamma + v_2^2/2g + z_2 + h_L \qquad h_L = h_f + h_m \qquad h_f = (f)(L/D)(v^2/2g)$$

$$v = Q/A = 0.005/[(\pi)(\tfrac{5}{100})^2/4] = 2.546 \text{ m/s} \qquad N_R = Dv/v = (\tfrac{5}{100})(2.546)/(1.02 \times 10^{-6}) = 1.25 \times 10^5$$

$$\epsilon/D = 0.00026/(\tfrac{5}{100}) = 0.00520$$

From Fig. A-5, $f = 0.0315$. $h_f = 0.0315[1200/(\tfrac{5}{100})]\{2.546^2/[(2)(9.807)]\} = 249.8 \text{ m}$. For 45° elbows, $K_1 = (2)(0.20) = 0.40$. For 90° elbows, $K_2 = (4)(0.30) = 1.20$. For the open valve, $K_3 = 8.5$. For exit, $K_4 = 1.0$. $h_m = (0.40 + 1.20 + 8.5 + 1.0)\{2.546^2/[(2)(9.807)]\} = 3.7 \text{ m}$, $h_L = 249.8 + 3.7 = 253.5 \text{ m}$, $p_1/9.79 + 2.546^2/[(2)(9.807)] + 400 = 0 + 0 + 500 + 253.5$, $p_1 = 3458 \text{ kN/m}^2$ gage.

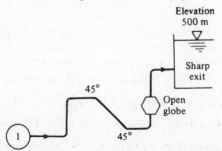

Fig. 9-38

9.205 The water pipe in Fig. 9-39 slopes upward at 30°. The pipe is 1 in in diameter and smooth. The flanged globe valve is fully open. If the mercury manometer shows a 7-in deflection, what is the flow rate?

$$p_1/\gamma + v_1^2/2g + z_1 = p_2/\gamma + v_2^2/2g + z_2 + h_L \qquad v_1^2/2g = v_2^2/2g \qquad h_L = h_f + h_m$$

Therefore, $p_1/\gamma + z_1 = p_2/\gamma + z_2 + h_f + h_m$,

$$(p_1 - p_2)/\gamma = z_2 - z_1 + h_f + h_m \tag{1}$$

From manometer,

$$p_1 - p_2 = (\gamma_{\text{Hg}} - \gamma_{\text{H}_2\text{O}})(\tfrac{7}{12}) + (\gamma_{\text{H}_2\text{O}})(z_2 - z_1) \tag{2}$$

Combining Eqs. (1) and (2), $[(\gamma_{\text{Hg}} - \gamma_{\text{H}_2\text{O}})/\gamma_{\text{H}_2\text{O}}](\tfrac{7}{12}) = h_f + h_m$, $h_f = (f)(L/D)(v^2/2g) = (f)[(10/\cos 30°)/(\tfrac{1}{12})]\{v^2/[(2)(32.2)]\} = 2.152fv^2$, $h_m = (K)(v^2/2g)$. For globe valve, $K = 13$. $h_m = 13\{v^2/[(2)(32.2)]\} = 0.2019v^2$, $\{[(13.6)(62.4) - 62.4]/62.4\}(\tfrac{7}{12}) = 2.152fv^2 + 0.2019v^2$, $v = \sqrt{7.350/(2.152f + 0.2019)}$. Try $f = 0.02$: $v = \sqrt{7.350/[(2.152)(0.02) + 0.2019]} = 5.478$ ft/s, $N_R = Dv/\nu = (\tfrac{1}{12})(5.478)/(1.05 \times 10^{-5}) = 4.35 \times 10^4$. From Fig. A-5, $f = 0.0217$. Try $f = 0.0217$: $v = \sqrt{7.350/[(2.152)(0.0217) + 0.2019]} = 5.437$ ft/s, $N_R = (\tfrac{1}{12})(5.437)/(1.05 \times 10^{-5}) = 4.32 \times 10^4$, $f = 0.0217$ (O.K.); $Q = Av = [(\pi)(\tfrac{1}{12})^2/4](5.437) = 0.0297$ ft³/s.

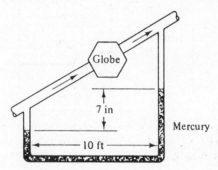

Fig. 9-39

9.206 A pipe system carries water from a reservoir and discharges it as a free jet, as shown in Fig. 9-40. How much flow is to be expected through a 200-mm steel commercial pipe with the fittings shown?

$$p_1/\gamma + v_1^2/2g + z_1 = p_2/\gamma + v_2^2/2g + z_2 + h_L \qquad h_L = h_f + h_m$$

$$h_f = (f)(L/D)(v^2/2g) = f[(60 + 20 + 60)/(\tfrac{200}{1000})]\{v_2^2/[(2)(9.807)]\} = 35.69fv_2^2$$

$$h_m = (K)(v^2/2g) = [0.05 + (2)(0.40)]\{v_2^2/[(2)(9.807)]\} = 0.04334v_2^2$$

$$0 + 0 + 30 = 0 + v_2^2/[(2)(9.807)] + 20 + (35.69fv_2^2 + 0.04334v_2^2) \qquad v_2 = \sqrt{10/(35.69f + 0.09432)}$$

Try $f = 0.014$: $v_2 = \sqrt{10/[(35.69)(0.014) + 0.09432]} = 4.103$ m/s, $N_R = Dv/\nu = (\tfrac{200}{1000})(4.103)/(0.0113 \times 10^{-4}) = 7.26 \times 10^5$, $\epsilon/D = 0.000046/(\tfrac{200}{1000}) = 0.000230$. From Fig. A-5, $f = 0.0152$. Try $f = 0.0152$: $v_2 = \sqrt{10/[(35.69)(0.0152) + 0.09432]} = 3.963$ m/s, $N_R = (\tfrac{200}{1000})(3.963)/(0.0113 \times 10^{-4}) = 7.01 \times 10^5$, $f = 0.0152$ (O.K.); $Q = Av = [(\pi)(\tfrac{200}{1000})^2/4](3.963) = 0.125$ m³/s.

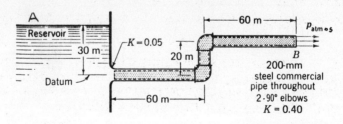

$\nu = .0113 \times 10^{-4}$ m²/s
$\rho = 999$ kg/m³ **Fig. 9-40**

9.207 A pipe system having a given centerline geometry as shown in Fig. 9-41 is to be chosen to transport a maximum of 1 ft³/s of oil from tank A to tank B. What is a pipe size that will do the job?

$$p_c/\gamma + v_c^2/2g + z_c = p_p/\gamma + v_p^2/2g + z_p + h_L \qquad h_L = h_f + h_m$$

$$h_f = (f)(L/D)(v^2/2g) = f[(100 + 130 + 300)/D]\{v_p^2/[(2)(32.2)]\} = 8.230fv_p^2/D$$

$$h_m = (K)(v^2/2g) = (0.05 + 0.5 + 0.5 + 1)\{v_p^2/[(2)(32.2)]\} = 0.03183v_p^2$$

$$(100)(144)/50 + 0 + 130 = (50)(144)/50 + v_p^2/[(2)(32.2)] + 0 + 8.230fv_p^2/D + 0.03183v_p^2$$

$$0.04736v_p^2 + 8.230fv_p^2/D = 274 \qquad v_p^2 = 274/(0.04736 + 8.230f/D) \qquad v_p = Q/A_p = 1/(\pi D^2/4) = 1.273/D^2$$

$$(1.273/D^2)^2 = 274/(0.04736 + 8.230f/D)$$

Try $f = 0.015$: $(1.273/D^2)^2 = 274/[0.04736 + (8.230)(0.015)/D]$. By trial and error, $D = 0.240$ ft.

$$v = 1.273/0.240^2 = 22.1 \text{ ft/s} \qquad N_R = \rho Dv/\mu = (\gamma/g)(Dv)/\mu = (50/32.2)(0.240)(22.1)/(50 \times 10^{-5}) = 1.65 \times 10^4$$

$$\epsilon/D = 0.00015/0.240 = 0.000625$$

From Fig. A-5, $f = 0.0285$. Try $f = 0.0285$: $(1.273/D^2)^2 = 274/[0.04736 + (8.230)(0.0285)/D]$. By trial and error, $D = 0.272$ ft.

$$v = 1.273/0.272^2 = 17.2 \text{ ft/s} \qquad N_R = (50/32.2)(0.272)(17.2)/(50 \times 10^{-5}) = 1.45 \times 10^4$$

$$f = 0.0285 \qquad \text{(O.K.)}$$

Hence, $D = 0.272$ ft, or 3.26 in.

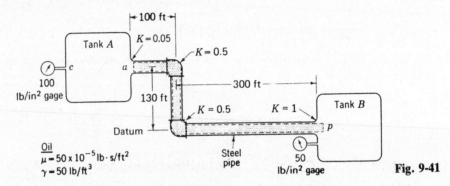

Fig. 9-41

9.208 What gage pressure p_1 is required to cause 5 ft³/s of water to flow through the system shown in Fig. 9-42? Assume that the reservoir is large. Neglect minor losses. Take $\nu = 2.11 \times 10^{-5}$ ft²/s.

$$p_1/\gamma + v_1^2/2g + z_1 = p_2/\gamma + v_2^2/2g + z_2 + h_L \qquad v_2 = Q/A_2 = 5/[(\pi)(\tfrac{6}{12})^2/4] = 25.46 \text{ ft/s}$$

$$h_L = h_f = (f)(L/D)(v^2/2g) \qquad N_R = dv/\nu = (\tfrac{6}{12})(25.46)/(2.11 \times 10^{-5}) = 6.03 \times 10^5$$

$$\epsilon/d = 0.00015/(\tfrac{6}{12}) = 0.00030$$

From Fig. A-5, $f = 0.016$.

$$h_L = 0.016[(500 + 150 + 200)/(\tfrac{6}{12})]\{25.46^2/[(2)(32.2)]\} = 273.8 \text{ ft}$$

$$(p_1)(144)/62.4 + 0 + 100 = 0 + 25.46^2/[(2)(32.2)] + 150 + 273.8 \qquad p_1 = 145 \text{ lb/in}^2 \text{ gage}$$

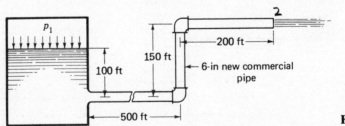

Fig. 9-42

9.209 In Prob. 9.208, take the diameter of the pipe to be the *nominal* diameter. For the entrance fitting, $r/d = 0.06$. Calculate the pressure p_1. The elbows are screwed elbows and there is now an open globe valve in the pipe system. Include minor losses.

Use $d = 6.065$ in

$$p_1/\gamma + v_1^2/2g + z_1 = p_2/\gamma + v_2^2/2g + z_2 + h_L \qquad v_2 = Q/A_2 = 5/[(\pi)(6.065/12)^2/4] = 24.92 \text{ ft/s}$$

$$h_L = h_f + h_m \qquad h_f = (f)(L/d)(v^2/2g) \qquad N_R = dv/\nu = (6.065/12)(24.92)/(2.11 \times 10^{-5}) = 5.97 \times 10^5$$

$$\epsilon/d = 0.00015/(6.065/12) = 0.000297$$

From Fig. A-5, $f = 0.016$. $h_f = 0.016[(500 + 150 + 200)/(6.065/12)]\{24.92^2/[(2)(32.2)]\} = 259.5$ ft, $h_m = (K)(v^2/2g)$. For entrance, $K_1 = 0.15$. For elbows, $K_2 = (2)(0.45) = 0.90$. For globe valve, $K_3 = 5.1$.

$$h_m = (0.15 + 0.90 + 5.1)\{24.92^2/[(2)(32.2)]\} = 59.3 \text{ ft} \qquad h_L = 259.5 + 59.3 = 318.8 \text{ ft}$$

$$(p_1)(144)/62.4 + 0 + 100 = 0 + 24.92^2/[(2)(32.2)] + 150 + 318.8 \qquad p_1 = 164 \text{ lb/in}^2 \text{ gage}$$

9.210 In Fig. 9-43, what pressure p_1 is needed to cause 100 L/s of water to flow into the device at a pressure p_2 of 40 kPa gage? The pipe is 150-mm commercial pipe. Take $v = 0.113 \times 10^{-5}$ m^2/s.

▌ $\quad p_A/\gamma + v_A^2/2g + z_A = p_B/\gamma + v_B^2/2g + z_B + h_L \qquad v_B = Q/A_B = (\frac{100}{1000})/[(\pi)(\frac{150}{1000})^2/4] = 5.659$ m/s

$$h_L = h_f + h_m \qquad h_f = (f)(L/d)(v^2/2g) \qquad N_R = dv/v = (\frac{150}{1000})(5.659)/(0.113 \times 10^{-5}) = 7.51 \times 10^5$$

$$\epsilon/d = 0.000046/(\tfrac{150}{1000}) = 0.000307$$

From Fig. A-5, $f = 0.016$.

$$h_f = 0.016[(325 + 160 + 260)/(\tfrac{150}{1000})]\{5.659^2/[(2)(9.807)]\} = 129.7 \text{ m}$$

$$h_m = (K)(v^2/2g) = (0.4 + 0.9 + 0.9 + 1)\{5.659^2/[(2)(9.807)]\} = 5.2 \text{ m} \qquad h_L = 129.7 + 5.2 = 134.9 \text{ m}$$

$$p_1/9.79 + 0 + 26 = 40/9.79 + 5.659^2/[(2)(9.807)] + 160 + 134.9 \qquad p_1 = 2689 \text{ kPa gage}$$

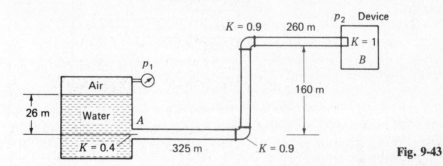

Fig. 9-43

9.211 What pressure p_1 is required in Fig. 9-44 to cause 1 ft^3/s of water to flow where pressure p_2 is 5 psig? Take $\mu = 2.11 \times 10^{-5}$ lb · s/ft^2 for water.

▌ $\quad p_A/\gamma + v_A^2/2g + z_A = p_B/\gamma + v_B^2/2g + z_B + h_L \qquad p_A/\gamma = (p_1)(144)/62.4 + (10)(0.8) = 2.308p_1 + 8.00$

$$v_B = Q/A_B = 1/[(\pi)(\tfrac{6}{12})^2/4] = 5.093 \text{ ft/s} \qquad h_L = h_f + h_m \qquad h_f = (f)(L/d)(v^2/2g)$$

$$N_R = \rho dv/\mu = (1.94)(\tfrac{6}{12})(5.093)/(2.11 \times 10^{-5}) = 2.34 \times 10^5 \qquad \epsilon/d = 0.00015/(\tfrac{6}{12}) = 0.00030$$

From Fig. A-5, $f = 0.0175$.

$$h_f = 0.0175[3000/(\tfrac{6}{12})]\{5.093^2/[(2)(32.2)]\} = 42.29 \text{ ft} \qquad h_m = (K)(v^2/2g) = (0.5 + 1)\{5.093^2/[(2)(32.2)]\} = 0.60 \text{ ft}$$

$$h_L = 42.29 + 0.60 = 42.89 \text{ ft}$$

$$(2.308p_1 + 8.00) + 0 + 20 = (5)(144)/62.4 + 5.093^2/[(2)(32.2)] + 0 + 42.89 \qquad p_1 = 11.6 \text{ psig}$$

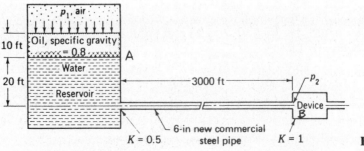

Fig. 9-44

9.212 What should be the flow through the system shown in Fig. 9-45? We have commercial steel pipe, 6 in in diameter.

▮ $p_A/\gamma + v_A^2/2g + z_A = p_B/\gamma + v_B^2/2g + z_B + h_L$ $0 + 0 + (100 + 100) = 0 + 0 + 100 + h_L$ $h_L = 100\,\text{ft} = h_f + h_m$

$$h_f = (f)(L/D)(v^2/2g) = f[(500 + 100 + 100)/(\tfrac{6}{12})]\{v^2/[(2)(32.2)]\} = 21.74fv^2$$

$$h_m = (K)(v^2/2g) = (0.4 + 0.8 + 0.8 + 1)\{v^2/[(2)(32.2)]\} = 0.04658v^2 \qquad h_L = 21.74fv^2 + 0.04658v^2$$

$$100 = 21.74fv^2 + 0.04658v^2 \qquad v = \sqrt{100/(21.74f + 0.04658)}$$

Try $f = 0.015$: $v = \sqrt{100/[(21.74)(0.015) + 0.04658]} = 16.38\,\text{ft/s}$, $N_R = Dv/\nu = (\tfrac{6}{12})(16.38)/(1.21 \times 10^{-5}) = 6.77 \times 10^5$, $\epsilon/D = 0.00015/(\tfrac{6}{12}) = 0.00030$. From Fig. A-5, $f = 0.016$. Try $f = 0.016$: $v = \sqrt{100/[(21.74)(0.016) + 0.04658]} = 15.92\,\text{ft/s}$, $N_R = (\tfrac{6}{12})(15.92)/(1.21 \times 10^{-5}) = 6.58 \times 10^5$, $f = 0.016$ (O.K.); $Q = Av = [(\pi)(\tfrac{6}{12})^2/4](15.92) = 3.13\,\text{ft}^3/\text{s}$.

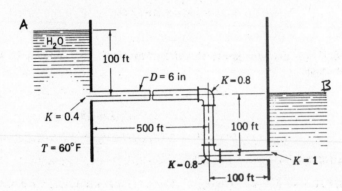

Fig. 9-45

9.213 A flow of 170 L/s is to go from tank A to tank B in Fig. 9-46. If $\nu = 0.113 \times 10^{-5}\,\text{m}^2/\text{s}$, what should the diameter be for the horizontal section of pipe?

▮ $p_A/\gamma + v_A^2/2g + z_A = p_B/\gamma + v_B^2/2g + z_B + h_L$ $0 + 0 + (35 + 35) = 0 + 0 + 16 + h_L$ $h_L = 54\,\text{m} = h_f + h_m$

$$v_1 = Q/A_1 = (\tfrac{170}{1000})/(\pi D^2/4) = 0.2165/D^2 \qquad v_2 = Q/A_2 = (\tfrac{170}{1000})/[(\pi)(\tfrac{150}{1000})^2/4] = 9.620\,\text{m/s}$$

$$h_f = (f)(L/D)(v^2/2g) = (f_1)(65/D)\{(0.2165/D^2)^2/[(2)(9.807)]\}$$
$$+ f_2[35/(\tfrac{150}{1000})]\{9.620^2/[(2)(9.807)]\} = 0.1553f_1/D^5 + 1101f_2$$

$$h_m = (K)(v^2/2g) = (0.4 + 0.9)\{(0.2165/D^2)^2/[(2)(9.807)]\}$$
$$+ 1.0\{9.620^2/[(2)(9.807)]\} = 0.003107/D^4 + 4.718$$

$$h_L = 0.1553f_1/D^5 + 1101f_2 + 0.003107/D^4 + 4.718$$

$$54 = 0.1553f_1/D^5 + 1101f_2 + 0.003107/D^4 + 4.718$$

Try $f_1 = f_2 = 0.015$: $54 = (0.1553)(0.015)/D^5 + (1101)(0.015) + 0.003107/D^4 + 4.718$.
By trial and error, $D = 0.154\,\text{m}$.

$$v_1 = 0.2165/0.154^2 = 9.129\,\text{m/s} \qquad N_R = Dv/\nu$$

$$(N_R)_1 = (0.154)(9.129)/(0.113 \times 10^{-5}) = 1.24 \times 10^6 \qquad (\epsilon/D)_1 = 0.000046/0.154 = 0.00030$$

From Fig. A-5, $f_1 = 0.0155$. $(N_R)_2 = (\tfrac{150}{1000})(9.620)/(0.113 \times 10^{-5}) = 1.28 \times 10^6$, $(\epsilon/D)_2 = 0.000046/(\tfrac{150}{1000}) = 0.00031$, $f_2 = 0.0155$. Try $f_1 = f_2 = 0.0155$: $54 = (0.1553)(0.0155)/D^5 + (1101)(0.0155) + 0.003107/D^4 + 4.718$. By trial and error, $D = 0.155\,\text{m}$.

$$v_1 = 0.2165/0.155^2 = 9.011\,\text{m/s} \qquad (N_R)_1 = (0.155)(9.011)/(0.113 \times 10^{-5}) = 1.24 \times 10^6$$

$$(\epsilon/D)_1 = 0.000046/0.155 = 0.00030 \qquad f_1 = 0.0155$$

$$(N_R)_2 = (\tfrac{150}{1000})(9.620)/(0.113 \times 10^{-5}) = 1.28 \times 10^6 \qquad f_2 = 0.0155 \qquad \text{(O.K.)}$$

Therefore, $D = 0.155\,\text{m}$, or 155 mm.

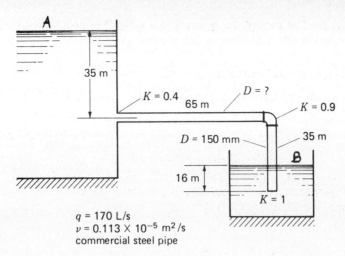

$q = 170$ L/s
$\nu = 0.113 \times 10^{-5}$ m²/s
commercial steel pipe

Fig. 9-46

9.214 In Prob. 9.208, if $p_1 = 200$ lb/in² gage, what should the inside pipe diameter be to transport 12 ft³/s of water? Neglect minor losses.

$$p_1/\gamma + v_1^2/2g + z_1 = p_2/\gamma + v_2^2/2g + z_2 + h_L \qquad v_2 = Q/A_2 = 12/(\pi d^2/4) = 15.28/d^2$$

$$h_L = h_f = (f)(L/d)(v^2/2g)$$

$$h_f = (f)[(500 + 150 + 200)/d]\{(15.28/d^2)^2/[(2)(32.2)]\} = 3082f/d^5$$

$$(200)(144)/62.4 + 0 + 100 = 0 + (15.28/d^2)^2/[(2)(32.2)] + 150 + 3082f/d^5 \qquad 411.5 = 3.625/d^4 + 3082f/d^5$$

Try $f = 0.015$: $411.5 = 3.625/d^4 + (3082)(0.015)/d^5$. By trial and error, $d = 0.652$ ft. $v = 15.28/0.652^2 = 35.94$ ft/s, $N_R = dv/v = (0.652)(35.94)/(2.11 \times 10^{-5}) = 1.11 \times 10^6$, $\epsilon/d = 0.00015/0.652 = 0.00023$. From Fig. A-5, $f = 0.015$ (O.K.). Therefore, $d = 0.652$ ft, or 7.82 in.

9.215 In Prob. 9.210, we found that for a flow of 100 L/s we needed a pressure p_1 of 2689 kPa gage. With this pressure p_1 and the given pressure on device B, what size pipe is needed to double the flow?

$$p_A/\gamma + v_A^2/2g + z_A = p_B/\gamma + v_B^2/2g + z_B + h_L \qquad v_B = Q/A_B = (2)(\tfrac{100}{1000})/(\pi d^2/4) = 0.2546/d^2$$

$$h_f = (f)(L/d)(v^2/2g) = f[(325 + 160 + 260)/d]\{(0.2546/d^2)^2/[(2)(9.807)]\} = 2.462f/d^5$$

$$h_m = (K)(v^2/2g) = (0.4 + 0.9 + 0.9 + 1)\{(0.2546/d^2)^2/[(2)(9.807)]\} = 0.01058/d^4$$

$$h_L = h_f + h_m = 2.462f/d^5 + 0.01058/d^4$$

$$2689/9.79 + 0 + 26 = 40/9.79 + (0.2546/d^2)^2/[(2)(9.807)] + 160 + 2.462f/d^5 + 0.01058/d^4$$

$$136.6 = 2.462f/d^5 + 0.01388/d^4$$

Try $f = 0.015$: $136.6 = (2.462)(0.015)/d^5 + 0.01388/d^4$. By trial and error, $d = 0.196$ m. $v = 0.2546/0.196^2 = 6.627$ m/s, $N_R = dv/v = (0.196)(6.627)/(0.113 \times 10^{-5}) = 1.15 \times 10^6$, $\epsilon/d = 0.000046/0.196 = 0.000235$. From Fig. A-5, $f = 0.015$ (O.K.). Therefore, $d = 0.196$ m, or 196 mm.

9.216 A 12-in-diameter pipe with a friction factor of 0.02 conducts fluid between two tanks at 10 fps. The entrance and exit conditions to and from the tank are flush with the wall of the tank. Find the ratio of the minor losses to the pipe friction loss if the length of the pipe is (a) 5 ft, (b) 100 ft, and (c) 2000 ft.

$$h_f = (f)(L/d)(v^2/2g) \qquad h_m = (K)(v^2/2g)$$

(a) $\qquad h_f = 0.02[(5/\tfrac{12}{12})](v^2/2g) = 0.10v^2/2g \qquad h_m = (0.5 + 1.0)(v^2/2g) = 1.5v^2/2g$

$$h_m/h_f = (1.5v^2/2g)/(0.10v^2/2g) = 15 \qquad (15:1)$$

(b) $\qquad h_f = 0.02[100/(\tfrac{12}{12})](v^2/2g) = 2.0v^2/2g \qquad h_m = (0.5 + 1.0)(v^2/2g) = 1.5v^2/2g$

$$h_m/h_f = (1.5v^2/2g)/(2.0v^2/2g) = 0.75 \qquad (0.75:1)$$

(c) $\qquad h_f = 0.02[2000/(\tfrac{12}{12})](v^2/2g) = 40v^2/2g \qquad h_m = (0.5 + 1.0)(v^2/2g) = 1.5v^2/2g$

$$h_m/h_f = (1.5v^2/2g)/(40v^2/2g) = 0.0375 \qquad (0.0375:1)$$

9.217 A smooth pipe 30 cm in diameter and 90 m long has a flush entrance and a submerged discharge. The velocity is 3 m/s. If the fluid is water at 15 °C, what is the total loss of head?

$$h_L = h_f + h_m \qquad h_f = (f)(L/d)(v^2/2g) \qquad N_R = dv/v = (\tfrac{30}{100})(3)/(1.16 \times 10^{-6}) = 7.76 \times 10^5$$

From Fig. A-5, $f = 0.0122$. $h_f = 0.0122[90/(\tfrac{30}{100})]\{3^2/[(2)(9.807)]\} = 1.68$ m, $h_m = (K)(v^2/2g) =$ $(0.5 + 1.0)\{3^2/[(2)(9.807)]\} = 0.69$ m, $h_L = 1.68 + 0.69 = 2.37$ m.

9.218 Suppose that the fluid in Prob. 9.217 were oil with a kinematic viscosity of 9.3×10^{-5} m²/s and a specific gravity of 0.925. What would be the head loss in meters of oil and in kilopascals?

$$h_L = h_f + h_m \qquad h_f = (f)(L/d)(v^2/2g) \qquad N_R = dv/v = (\tfrac{30}{100})(3)/(9.3 \times 10^{-5}) = 9.68 \times 10^3$$

From Fig. A-5, $f = 0.031$.

$$h_f = 0.031[90/(\tfrac{30}{100})]\{3^2/[(2)(9.807)]\} = 4.27 \text{ m} \qquad h_m = (K)(v^2/2g) = (0.5 + 1.0)\{3^2/[(2)(9.807)]\} = 0.69 \text{ m}$$

$$h_L = 4.27 + 0.69 = 4.96 \text{ m of oil} \qquad \Delta p = \gamma h_L = [(0.925)(9.79)](4.96) = 44.9 \text{ kPa}$$

9.219 A smooth pipe consists of 50 ft of 8-in pipe followed by 300 ft of 24-in pipe with an abrupt change of cross section at the junction. It has a flush entrance and a submerged discharge. If it carries water at 60 °F in the smaller pipe with a velocity of 18 fps, what is the total head loss?

$$h_L = h_f + h_m \qquad h_f = (f)(L/d)(v^2/2g) \qquad N_R = dv/v \qquad (N_R)_{d=8} = (\tfrac{8}{12})(18)/(1.21 \times 10^5) = 9.92 \times 10^5$$

From Fig. A-5, $(f)_{d=8} = 0.0117$.

$$A_1 v_1 = A_2 v_2 \qquad A_1 = [(\pi)(\tfrac{8}{12})^2/4] = 0.3491 \text{ ft}^2 \qquad A_2 = [(\pi)(\tfrac{24}{12})^2/4] = 3.142 \text{ ft}^2$$

$$(0.3491)(18) = (3.142)(v_2) \qquad v_2 = 2.000 \text{ ft/s} \qquad (N_R)_{d=24} = (\tfrac{24}{12})(2.000)/(1.21 \times 10^{-5}) = 3.31 \times 10^5$$

$$(f)_{d=24} = 0.0142 \qquad h_f = 0.0117[50/(\tfrac{8}{12})]\{18^2/[(2)(32.2)]\} + 0.0142[300/(\tfrac{24}{12})]\{2.000^2/[(2)(32.2)]\} = 4.55 \text{ ft}$$

$$h_m = (K)(v^2/2g)$$

For entrance, $K_1 = 0.5$. For abrupt change, $K_2 = (1 - A_1/A_2)^2 = (1 - 0.3491/3.142)^2 = 0.7901$. For exit, $K_3 = 1.0$. $h_m = (0.5 + 0.7901)\{18^2/[(2)(32.2)]\} + 1.0\{2.000^2/[(2)(32.2)]\} = 6.55$ ft, $h_L = 4.55 + 6.55 = 11.10$ ft.

9.220 A 6-in-diameter pipe ($f = 0.032$) of length 100 ft connects two reservoirs whose water-surface elevations differ by 10 ft. The pipe entrance is flush, and the discharge is submerged. (*a*) Compute the flow rate. (*b*) If the last 10 ft of pipe were replaced with a conical diffuser with a cone angle of 10°, compute the flow rate.

$$p_1/\gamma + v_1^2/2g + z_1 = p_2/\gamma + v_2^2/2g + z_2 + h_L$$

(*a*)
$$h_f = (f)(L/d)(v^2/2g) = 0.032[100/(\tfrac{6}{12})]\{v^2/[(2)(32.2)]\} = 0.09938v^2$$

$$h_m = (K)(v^2/2g) = (0.5 + 1.0)\{v^2/[(2)(32.2)]\} = 0.02329v^2$$

$$h_L = h_f + h_m = 0.09938v^2 + 0.02329v^2 = 0.1227v^2$$

$$0 + 0 + 10 = 0 + 0 + 0 + 0.1227v^2 \qquad v = 9.028 \text{ ft/s}$$

$$Q = Av = [(\pi)(\tfrac{6}{12})^2/4](9.028) = 1.77 \text{ ft}^3/\text{s}$$

(*b*) $h_f = 0.032[(100 - 10)/(\tfrac{6}{12})]\{v^2/[(2)(32.2)]\} = 0.08944v^2 \qquad h_m = (0.5 + 0.40)\{v^2/[(2)(32.2)]\} = 0.01398v^2$

$$h_L = 0.08944v^2 + 0.01398v^2 = 0.1034v^2 \qquad 0 + 0 + 10 = 0 + 0 + 0 + 0.1034v^2$$

$$v = 9.834 \text{ ft/s} \qquad Q = Av = [(\pi)(\tfrac{6}{12})^2/4](9.834) = 1.93 \text{ ft}^3/\text{s}$$

9.221 For two pipes in series with a diameter ratio of 1:2 and velocity of 20 fps in the smaller pipe, find the loss of head due to (*a*) sudden contraction and (*b*) sudden enlargement.

$$h_m = (K)(v^2/2g)$$

(*a*) $\qquad d/D = 0.5 \qquad K = 0.33 \qquad h_m = 0.33\{20^2/[(2)(32.2)]\} = 2.05$ ft

(*b*) $\qquad K = 0.55 \qquad h_m = 0.56\{20^2/[(2)(32.2)]\} = 3.48$ ft

9.222 In a 100-ft length of 4-in-diameter wrought iron pipe there are one open globe valve ($K = 10$), one 45° regular elbow ($K = 0.75$), and one pipe bend with a radius of curvature of 40 in ($K = 0.10$). The bend is 90°, and its length is not included in the 100 ft. No entrance or discharge losses are involved. If the fluid is water at 72 °F and the velocity is 6 fps, what is the total head loss?

$$h_L = h_f + h_m \qquad h_f = (f)(L/d)(v^2/2g)$$

$$N_R = dv/v = (\tfrac{4}{12})(6)/(1.02 \times 10^{-5}) = 1.96 \times 10^5 \qquad \epsilon/d = 0.00015/(\tfrac{4}{12}) = 0.00045$$

From Fig. A-5, $f = 0.0185$.

$$L = 100 + \tfrac{1}{4}[(2)(\pi)(\tfrac{40}{12})] = 105.2 \text{ ft} \qquad h_f = 0.0185[105.2/(\tfrac{4}{12})]\{6^2/[(2)(32.2)]\} = 3.26 \text{ ft}$$

$$h_m = (K)(v^2/2g) = (10 + 0.75 + 0.10)\{6^2/[(2)(32.2)]\} = 6.07 \text{ ft} \qquad h_L = 3.26 + 6.07 = 9.33 \text{ ft}$$

9.223 It has been found that with great care laminar flow can be maintained up to $N_R = 50\,000$. Compute the friction head per 100 ft of pipe for a Reynolds number of 50 000 if the flow is laminar. Consider two situations: one where the fluid is water at 60 °F, the other where the fluid is SAE10 oil at 150 °F ($v = 0.00016 \text{ ft}^2/\text{s}$). Pipe diameter is 2.0 in.

$$h_f = (f)(L/d)(v^2/2g) \qquad f = 64/N_R = 64/50\,000 = 0.00128 \qquad N_R = dv/v$$

For water:

$$50\,000 = (2.0/12)(v)/(1.21 \times 10^{-5}) \qquad v = 3.630 \text{ ft/s}$$

$$h_f = 0.00128[100/(2.0/12)]\{3.630^2/[(2)(32.2)]\} = 0.157 \text{ ft}$$

For oil:

$$50\,000 = (2.0/12)(v)/0.00016 \qquad v = 48.00 \text{ ft/s}$$

$$h_f = 0.00128[100/(2.0/12)]\{48.00^2/[(2)(32.2)]\} = 27.5 \text{ ft}$$

9.224 Repeat Prob. 9.223 if the flow is turbulent in a smooth pipe.

$h_f = (f)(L/d)(v^2/2g)$. From Fig. A-5, $f = 0.0207$.

For water: From Prob. 9.223, $v = 3.630 \text{ ft/s}$. $h_f = 0.0207[100/(2.0/12)]\{3.630^2/[(2)(32.2)]\} = 2.54 \text{ ft}$.

For oil: From Prob. 9.223, $v = 48.00 \text{ ft/s}$. $h_f = 0.0207[100/(2.0/12)]\{48.00^2/[(2)(32.2)]\} = 444 \text{ ft}$.

9.225 Repeat Prob. 9.223 if the flow is turbulent in a rough pipe with $\epsilon/d = 0.05$.

$h_f = (f)(L/d)(v^2/2g)$. From Fig. A-5, $f = 0.072$.

For water: From Prob. 9.223, $v = 3.630 \text{ ft/s}$. $h_f = 0.072[100/(2.0/12)]\{3.630^2/[(2)(32.2)]\} = 8.84 \text{ ft}$.

For oil: From Prob. 9.223, $v = 48.00 \text{ ft/s}$. $h_f = 0.072[100/(2.0/12)]\{48.00^2/[(2)(32.2)]\} = 1546 \text{ ft}$.

9.226 Water at 60 °F flows through 10 000 ft of 12-in-diameter pipe between two reservoirs whose water-surface elevation difference is 200 ft. Find the flow rate if $\epsilon = 0.0018$ in.

$$p_1/\gamma + v_1^2/2g + z_1 = p_2/\gamma + v_2^2/2g + z_2 + h_L \qquad 0 + 0 + 200 = 0 + 0 + 0 + h_L$$

$$h_L = 200 \text{ ft} = h_f = (f)(L/d)(v^2/2g) = f[10\,000/(\tfrac{12}{12})]\{v^2/[(2)(32.2)]\} = 155.3fv^2$$

$$200 = 155.3fv^2 \qquad v = 1.135/\sqrt{f}$$

Try $f = 0.03$: $v = 1.135/\sqrt{0.03} = 6.553 \text{ ft/s}$, $N_R = dv/v = (\tfrac{12}{12})(6.553)/(1.21 \times 10^{-5}) = 5.42 \times 10^5$, $\epsilon/d = 0.0018/12 = 0.00015$. From Fig. A-5, $f = 0.015$. Try $f = 0.015$: $v = 1.135/\sqrt{0.015} = 9.267 \text{ ft/s}$, $N_R = (\tfrac{12}{12})(9.267)/(1.21 \times 10^{-5}) = 7.66 \times 10^5$, $f = 0.0145$. Try $f = 0.0145$: $v = 1.135/\sqrt{0.0145} = 9.426 \text{ ft/s}$, $N_R = (\tfrac{12}{12})(9.426)/(1.21 \times 10^{-5}) = 7.79 \times 10^5$, $f = 0.0145$ (O.K.); $Q = Av = [(\pi)(\tfrac{12}{12})^2/4](9.426) = 7.40 \text{ ft}^3/\text{s}$.

9.227 Repeat Prob. 9.226 if ϵ is twenty times larger than in Prob. 9.226.

$v = 1.135/\sqrt{f}$ (from Prob. 9.226). Try $f = 0.03$: $v = 1.135/\sqrt{0.03} = 6.553 \text{ ft/s}$, $N_R = dv/v = (\tfrac{12}{12})(6.553)/(1.21 \times 10^{-5}) = 5.42 \times 10^5$, $\epsilon/d = (20)(0.0018)/12 = 0.00300$. From Fig. A-5, $f = 0.027$. Try $f = 0.027$: $v = 1.135/\sqrt{0.027} = 6.907 \text{ ft/s}$, $N_R = (\tfrac{12}{12})(6.907)/(1.21 \times 10^{-5}) = 5.71 \times 10^5$, $f = 0.027$ (O.K.); $Q = Av = [(\pi)(\tfrac{12}{12})^2/4](6.907) = 5.42 \text{ ft}^3/\text{s}$.

9.228 How large a wrought iron pipe is required to convey oil (s.g. = 0.9, $\mu = 0.0008$ lb · s/ft^2) from one tank to another at a rate of 1.0 cfs if the pipe is 3000 ft long and the difference in elevation of the free liquid surfaces is 40 ft?

❚
$$p_1/\gamma + v_1^2/2g + z_1 = p_2/\gamma + v_2^2/2g + z_2 + h_L \qquad 0 + 0 + 40 = 0 + 0 + 0 + h_L$$

$$h_L = 40 \text{ ft} = h_f = (f)(L/d)(v^2/2g) \qquad v = Q/A = 1.0/(\pi d^2/4) = 1.273/d^2$$

$$h_L = (f)(3000/d)\{(1.273/d^2)^2/[(2)(32.2)]\} = 75.49f/d^5 \qquad 40 = 75.49f/d^5 \qquad d = 1.135f^{1/5}$$

Try $f = 0.03$: $d = (1.135)(0.03)^{1/5} = 0.5629$ ft, $N_R = \rho dv/\mu = [(0.9)(1.94)](0.5629)(1.273/0.5629^2)/0.0008 = 4936$, $\epsilon/d = 0.00015/0.5629 = 0.00027$. From Fig. A-5, $f = 0.0375$. Try $f = 0.0375$: $d = (1.135)(0.0375)^{1/5} = 0.5886$ ft, $N_R = [(0.9)(1.94)](0.5886)(1.273/0.5886^2)/0.0008 = 4720$, $f = 0.038$. Try $f = 0.038$: $d = (1.135)(0.038)^{1/5} = 0.5901$ ft, $N_R = [(0.9)(1.94)](0.5901)(1.273/0.5901^2)/0.0008 = 4708$, $f = 0.038$ (O.K.). Therefore, $d = 0.5901$ ft, or 7.08 in.

9.229 If the diameter of a pipe is doubled, what effect does this have on the flow rate for a given head loss if the flow is laminar?

❚
$$h_L = (32)(v)(L/gd^2)(v) = (\text{constant})(v/d^2) \qquad v = kd^2 \qquad Q = Av = k'd^4$$

Thus, doubling the diameter will increase the flow rate by a factor of 2^4, or 16.

9.230 If the diameter of a pipe is doubled, what effect does this have on the flow rate for a given head loss if the flow is turbulent?

❚ If $f = $ constant (complete turbulence): $h_L = k_1 v^2/d$, $v = k_2 d^{1/2}$, $Q = Av = k_3 d^{5/2}$. Thus, doubling the diameter will increase the flow rate by a factor of $2^{5/2}$, or 5.66.

For smooth pipe with $N_R < 100\,000$: $f = 0.316/N_R^{0.25}$, $h_L = (f)(L/d)(v^2/2g) = (k_1)(v^{7/4}/d^{5/4})$, $v = k_2 d^{5/7}$, $Q = Av = k_3 d^{19/7}$. Thus, doubling the diameter will increase the flow rate by a factor of $2^{19/7}$, or 6.56.

9.231 A 15-cm-diameter pipeline 120 m long discharges a 5-cm-diameter jet of water into the atmosphere at a point which is 60 m below the water surface at intake. The entrance to the pipe is a projecting one, with $K = 0.9$, and the nozzle loss coefficient is 0.05. Find the flow rate and the pressure head at the base of the nozzle, assuming $f = 0.03$.

❚
$$p_1/\gamma + v_1^2/2g + z_1 = p_2/\gamma + v_2^2/2g + z_2 + h_L \qquad (1)$$

$$h_f = (f)(L/d)(v^2/2g) = 0.03[120/(\tfrac{15}{100})]\{v^2/[(2)(9.907)]\} = 1.224v^2$$

$$h_m = (K)(v^2/2g) = (0.9)\{v^2/[(2)(9.907)]\} + (0.05)\{v_{\text{jet}}^2/[(2)(9.907)]\} = 0.04589v^2 + 0.002549v_{\text{jet}}^2$$

$$h_L = h_f + h_m = 1.224v^2 + 0.04589v^2 + 0.002549v_{\text{jet}}^2 = 1.270v^2 + 0.002549v_{\text{jet}}^2$$

$$0 + 0 + 60 = 0 + v_{\text{jet}}^2/[(2)(9.807)] + 0 + 1.270v^2 + 0.002549v_{\text{jet}}^2$$

Since velocity varies with the square of the diameter, $v_{\text{jet}} = (\tfrac{15}{5})^2(v) = 9v$, $60 = (9v)^2/[(2)(9.807)] + 1.270v^2 + (0.002549)(9v)^2$, $v = 3.271$ m/s; $Q = Av = [(\pi)(\tfrac{15}{100})^2/4](3.271) = 0.0578$ m^3/s. Applying Eq. (1) to the nozzle, $p_1/9.79 + 3.271^2/[(2)(9.807)] + 0 = 0 + [(9)(3.271)]^2/[(2)(9.807)] + 0 + 0.05\{[(9)(3.271)]^2/[(2)(9.807)]\}$, $p_1 = 449$ kN/m^2.

9.232 A 2.0-m-diameter concrete pipe of length 1560 m for which $\epsilon = 1.5$ mm conveys water at 12 °C between two reservoirs at a rate of 8.0 m^3/s. What must be the difference in water-surface elevation between the two reservoirs? Consider minor losses at entrance and exit.

❚
$$p_1/\gamma + v_1^2/2g + z_1 = p_2/\gamma + v_2^2/2g + z_2 + h_L \qquad h_L = h_f + h_m \qquad h_f = (f)(L/d)(v^2/2g)$$

$$v = Q/A = 8.0/[(\pi)(2.0)^2/4] = 2.546 \text{ m/s} \qquad N_R = dv/\nu = (2.0)(2.546)/(1.24 \times 10^{-6}) = 4.11 \times 10^6$$

$$\epsilon/d = (1.5/1000)/2.0 = 0.00075$$

From Fig. A-5, $f = 0.0185$.

$$h_f = (0.0185)(1560/2.0)\{2.546^2/[(2)(9.807)]\} = 4.77 \text{ m}$$

$$h_m = (K)(v^2/2g) = (0.5 + 1.0)\{2.546^2/[(2)(9.807)]\} = 0.50 \text{ m}$$

$$h_L = 4.77 + 0.50 = 5.27 \text{ m} \qquad 0 + 0 + z_1 = 0 + 0 + z_2 + 5.27 \text{ m} \qquad z_1 - z_2 = 5.27 \text{ m}$$

9.233 A pipe with an average diameter of 62 in is 6272 ft long and delivers water to a powerhouse at a point 1375 ft lower in elevation than the water surface at intake. Assume $f = 0.025$. When the pipe delivers 300 cfs, what is its efficiency? What is the horsepower delivered to the plant?

$$v_2 = Q/A_2 = 300/[(\pi)(\tfrac{62}{12})^2/4] = 14.31 \text{ ft/s}$$

$$h_f = (f)(L/d)(v^2/2g) = 0.025[6272/(\tfrac{62}{12})]\{14.31^2/[(2)(32.2)]\} = 96.50 \text{ ft}$$

$$h_m = (K)(v_2/2g) = 0.50\{14.31^2/[(2)(32.2)]\} = 1.59 \text{ ft}$$

$$h_L = h_f + h_m = 96.50 + 1.59 = 98.09 \text{ ft}$$

Efficiency $= (1375 - 98.09)/1375 = 0.929$ or 92.9 percent (i.e., loss $= 7.1$ percent)

$$P = Q\gamma h_L = (300)(62.4)(1375 - 98.09) = 2.390 \times 10^7 \text{ ft} \cdot \text{lb/s} = (2.390 \times 10^7)/550 = 43\,450 \text{ hp}$$

9.234 Find the kilowatt loss in 1000 m of 50-cm-diameter pipe for which $\epsilon = 0.05$ mm when crude oil at 45 °C (s.g. $= 0.855$, $v = 4.4 \times 10^{-6}$ ft²/s) flows at 0.22 m³/s. Neglect minor losses.

$$h_f = (f)(L/d)(v^2/2g) \qquad v = Q/A = 0.22/[(\pi)(\tfrac{50}{100})^2/4] = 1.120 \text{ m/s}$$

$$N_R = dv/v = (\tfrac{50}{100})(1.120)/(4.4 \times 10^{-6}) = 1.27 \times 10^5 \qquad \epsilon/d = (0.05/1000)/(\tfrac{50}{100}) = 0.00010$$

From Fig. A-5, $f = 0.018$. $h_f = 0.018[1000/(\tfrac{50}{100})]\{1.120^2/[(2)(9.807)]\} = 2.302$ m, $P = Q\gamma h_f = 0.22[(0.855)(9.79)](2.302) = 4.24$ kW.

9.235 California crude oil, warmed until its kinematic viscosity is 0.0004 ft²/s, is pumped through a 2-in pipe ($\epsilon = 0.001$ in). Its specific weight is 59.8 lb/ft³. (*a*) At what maximum velocity would the flow still be laminar? (*b*) What would then be the loss in energy head in pounds per square inch per 1000 ft of pipe?

(*a*) Assume laminar flow exists for $N_R \le 2000$. $N_R = dv/v$, $2000 = (\tfrac{2}{12})(v)/0.0004$, $v = 4.80$ ft/s.

(*b*) $f = 64/N_R = \tfrac{64}{2000} = 0.032 \qquad h_f = (f)(L/d)(v^2/2g) = 0.032[1000/(\tfrac{2}{12})]\{4.80^2/[(2)(32.2)]\} = 68.69 \text{ ft}$

$$p = \gamma h_f = (59.8)(68.69)/144 = 28.5 \text{ psi per 1000 ft}$$

9.236 Repeat Prob. 9.235 if the velocity is three times the maximum velocity to ensure laminar flow as determined in Prob. 9.235.

$$v = (3)(4.80) = 14.40 \text{ ft/s} \qquad h_f = (f)(L/d)(v^2/2g)$$

$$N_R = dv/v = (\tfrac{2}{12})(14.40)/0.0004 = 6000 \qquad \epsilon/d = 0.001/2 = 0.00050$$

From Fig. A-5, $f = 0.036$.

$$h_f = 0.036[1000/(\tfrac{2}{12})]\{14.40^2/[(2)(32.2)]\} = 695.5 \text{ ft}$$

$$p = \gamma h_f = (59.8)(695.5)/144 = 289 \text{ psi per 1000 ft}$$

9.237 Water flows upward at 3 m/s through a vertical 15-cm-diameter pipe standing in a body of water with its lower end 0.9 m below the surface. Considering all losses and with $f = 0.025$, find the pressure at a point 3 m above the surface of the water.

$p_1/\gamma + v_1^2/2g + z_1 = p_2/\gamma + v_2^2/2g + z_2 + h_L$. Let point 1 be at the water surface and point 2 be 3 m above the water surface. $h_L = h_f + h_m$, $h_f = (f)(L/d)(v^2/2g) = 0.025[(3 + 0.9)/(\tfrac{15}{100})]\{3^2/[(2)(9.807)]\} = 0.298$ m, $h_m = (K)(v^2/2g)$. For entrance loss, assume $K = 0.8$. $h_m = 0.8\{3^2/[(2)(9.807)]\} = 0.367$ m, $h_L = 0.298 + 0.367 = 0.665$ m, $0 + 0 + 0 = p_2/9.79 + 3^2/[(2)(9.807)] + 3 + 0.665$, $p_2 = -40.4$ kN/m².

9.238 Work Prob. 9.237 if the flow is downward.

$$p_2/\gamma + v_2^2/2g + z_2 = p_1/\gamma + v_1^2/2g + z_1 + h_L \qquad h_L = h_f + h_m$$

$$h_f = 0.298 \text{ m} \qquad \text{(from Prob. 9.237)} \qquad h_m = (K)(v^2/2g)$$

For exit loss, assume $K = 1.0$.

$$h_m = 1.0\{3^2/[(2)(9.807)]\} = 0.459 \text{ m} \qquad h_L = 0.298 + 0.459 = 0.757 \text{ m}$$

$$p_2/9.79 + 3^2/[(2)(9.807)] + 3 = 0 + 0 + 0 + 0.757 \qquad p_2 = -26.5 \text{ kN/m}^2$$

9.239 A horizontal pipe 15 cm in diameter and for which $f = 0.025$ projects into a body of water 1 m below the surface. Considering all losses, find the pressure at a point 4 m from the end of the pipe if the velocity is 3 m/s and the flow is from the pipe into the body of water.

▮ $p_1/\gamma + v_1^2/2g + z_1 = p_2/\gamma + v_2^2/2g + z_2 + h_L$. Let point 1 be 4 m from the end of the pipe and point 2 be at the water surface. $h_L = h_f + h_m$, $h_f = (f)(L/d)(v^2/2g) = 0.025[(4)/(\frac{15}{100})]\{3^2/[(2)(9.807)]\} = 0.306$ m, $h_m = (K)(v^2/2g)$. For exit loss, $K = 1.0$. $h_m = 1.0\{3^2/[(2)(9.807)]\} = 0.459$ m, $h_L = 0.306 + 0.459 = 0.765$ m, $p_1/9.79 + 3^2/[(2)(9.807)] + 0 = 0 + 0 + 1 + 0.765$, $p_1 = 12.8$ kN/m^2.

9.240 Repeat Prob. 9.239 if the flow is from the body of water into the pipe.

▮
$$p_2/\gamma + v_2^2/2g + z_2 = p_1/\gamma + v_1^2/2g + z_1 + h_L \qquad h_L = h_f + h_m$$
$$h_f = 0.306 \text{ m} \qquad \text{(from Prob. 9.239)} \qquad h_m = (K)(v^2/2g)$$

For entrance, $K = 0.8$.

$$h_m = 0.8\{3^2/[(2)(9.807)]\} = 0.367 \text{ m} \qquad h_L = 0.306 + 0.367 = 0.673 \text{ m}$$
$$0 + 0 + 1 = p_1/9.79 + 3^2/[(2)(9.807)] + 0 + 0.673 \qquad p_1 = -1.29 \text{ kN/m}^2$$

9.241 A pipe runs from one reservoir to another, both ends of the pipe being under water. The intake is nonprojecting. The length of pipe is 500 ft, its diameter is 10.25 in, and the difference in water levels between the two reservoirs is 110 ft. If $f = 0.02$, what will be the pressure at a point 300 ft from the intake, the elevation of which is 120 ft lower than the surface of the water in the upper reservoir?

▮
$$p_1/\gamma + v_1^2/2g + z_1 = p_2/\gamma + v_2^2/2g + z_2 + h_L \qquad\qquad\qquad (1)$$

Let points 1 and 2 be at the water surface in the upper and lower reservoirs, respectively.

$$h_f = (f)(L/d)(v^2/2g) = 0.02[(500)/(10.25/12)]\{v^2/[(2)(32.2)]\} = 0.1818v^2$$
$$h_m = (K)(v^2/2g) = (0.5 + 1.0)\{v^2/[(2)(32.2)]\} = 0.02329v^2$$
$$h_L = h_f + h_m = 0.1818v^2 + 0.02329v^2 = 0.2051v^2 \qquad 0 + 0 + 110 = 0 + 0 + 0 + 0.2051v^2 \qquad v = 23.16 \text{ ft/s}$$

Now apply Eq. (1) between the upper reservoir (point 1) and the point 300 ft from the intake (point 2).

$$h_f = 0.02[(300)/(10.25/12)]\{23.16^2/[(2)(32.2)]\} = 58.51 \text{ ft} \qquad h_m = 0.5\{23.16^2/[(2)(32.2)]\} = 4.16 \text{ ft}$$
$$h_L = 58.51 + 4.16 = 62.67 \text{ ft} \qquad 0 + 0 + 120 = (p_2)(144)/62.4 + 23.16^2/[(2)(32.2)] + 0 + 62.67 \qquad p_2 = 21.1 \text{ lb/in}^2$$

9.242 A pipeline runs from one reservoir to another, both ends being under water, and the intake end is nonprojecting. The difference in water levels between the two reservoirs is 110 ft, and the length of pipe is 500 ft. (a) What is the discharge if the pipe is 10.25 in in diameter and $f = 0.022$? (b) When this same pipe is old, assume that the growth of tubercules has reduced the diameter to 9.5 in and that $f = 0.06$. What then will be the rate of discharge?

▮
$$p_1/\gamma + v_1^2/2g + z_1 = p_2/\gamma + v_2^2/2g + z_2 + h_L$$

(a)
$$h_f = (f)(L/d)(v^2/2g) = 0.022[(500)/(10.25/12)]\{v^2/[(2)(32.2)]\} = 0.2000v^2$$
$$h_m = (K)(v^2/2g) = (0.5 + 1.0)\{v^2/[(2)(32.2)]\} = 0.02329v^2$$
$$h_L = h_f + h_m = 0.2000v^2 + 0.02329v^2 = 0.2233v^2 \qquad 0 + 0 + 110 = 0 + 0 + 0 + 0.2233v^2 \qquad v = 22.19 \text{ ft/s}$$
$$Q = Av = [(\pi)(10.25/12)^2/4](22.19) = 12.7 \text{ ft}^3/\text{s}$$

(b) $\quad h_f = 0.06[(500)/(9.5/12)]\{v^2/[(2)(32.2)]\} = 0.5884v^2 \qquad h_m = (0.5 + 1.0)\{v^2/[(2)(32.2)]\} = 0.02329v^2$
$$h_L = 0.5884v^2 + 0.02329v^2 = 0.6117v^2 \qquad 0 + 0 + 110 = 0 + 0 + 0 + 0.6117v^2$$
$$v = 13.41 \text{ ft/s} \qquad Q = Av = [(\pi)(9.5/12)^2/4](13.41) = 6.60 \text{ ft}^3/\text{s}$$

9.243 A jet of water is discharged through a nozzle at a point 200 ft below the water level at intake. The jet is 4 in in diameter, and the loss coefficient of the nozzle is 0.15. If the pipeline is 12 in in diameter, 500 ft long, with a nonprojecting entrance, what is the pressure at the base of the nozzle? Assume $f = 0.015$.

$$p_1/\gamma + v_1^2/2g + z_1 = p_2/\gamma + v_2^2/2g + z_2 + h_L \tag{1}$$

$$h_f = (f)(L/d)(v^2/2g) = 0.015[500/(\tfrac{12}{12})]\{v^2/[(2)(32.2)]\} = 0.1165v^2$$

$$h_m = (K)(v^2/2g) = 0.5\{v^2/[(2)(32.2)]\} + (0.15)\{v_{\text{jet}}^2/[(2)(32.2)]\} = 0.007764v^2 + 0.002329v_{\text{jet}}^2$$

$$h_L = h_f + h_m = 0.1165v^2 + 0.007764v^2 + 0.002329v_{\text{jet}}^2 = 0.1243v^2 + 0.002329v_{\text{jet}}^2$$

$$0 + 0 + 200 = 0 + v_{\text{jet}}^2/[(2)(32.2)] + 0 + 0.1243v^2 + 0.002329v_{\text{jet}}^2$$

Since velocity varies with the square of diameter, $v_{\text{jet}} = (\tfrac{12}{4})^2(v) = 9v$, $200 = (9v)^2/[(2)(32.2)] + 0.1243v^2 + (0.002329)(9v)^2$, $v = 11.28$ ft/s. Applying Eq. (1) to the nozzle, $(p_1)(144)/62.4 + 11.28^2/[(2)(32.2)] + 0 = 0 + [(9)(11.28)]^2/[(2)(32.2)] + 0 + (0.15)\{[(9)(11.28)]^2/[(2)(32.2)]\}$, $p_1 = 78.9$ lb/in^2.

9.244 Compute the losses due to flow of 25 m^3/min of air, $p = 1$ atm, $T = 20\,°$C, through a sudden expansion from 300-mm pipe to 900-mm pipe. How much head would be saved by using a 10° conical diffuser?

For sudden expansion: $h_m = [1 - (D_1/D_2)^2]^2(v_1^2/2g)$, $v_1 = Q/A_1 = (\tfrac{25}{60})/[(\pi)(\tfrac{300}{1000})^2/4] = 5.895$ m/s, $h_m = [1 - (\tfrac{300}{900})^2]^2\{5.895^2/[(2)(9.807)]\} = 1.400$ m.

For conical diffuser: $h_m = 0.152[(v_1 - v_2)^2/2g]$, $v_2 = Q/A_2 = (\tfrac{25}{60})/[(\pi)(\tfrac{900}{1000})^2/4] = 0.655$ m/s, $h_m = 0.152\{(5.895 - 0.655)^2/[(2)(9.807)]\} = 0.213$ m.

Saving in head $= 1.400 - 0.213 = 1.187$ m or 1.187 N $\cdot$ m/N or 1.187 J/N

9.245 Calculate the value of H in Fig. 9-47 for 125 L/s of water at 15 °C through commercial steel pipe.

$$p_1/\gamma + v_1^2/2g + z_1 = p_2/\gamma + v_2^2/2g + z_2 + h_L \qquad h_L = h_f + h_m \qquad h_f = (f)(L/d)(v^2/2g)$$

$$v = Q/A = (\tfrac{125}{1000})/[(\pi)(\tfrac{30}{100})^2/4] = 1.768 \text{ m/s}$$

$$N_R = dv/\nu = (\tfrac{30}{100})(1.768)/(1.16 \times 10^{-6}) = 4.57 \times 10^5 \qquad \epsilon/d = 0.000046/(\tfrac{30}{100}) = 0.00015$$

From Fig. A-5, $f = 0.015$.

$$h_f = 0.015[30/(\tfrac{30}{100})]\{1.768^2/[(2)(9.807)]\} = 0.239 \text{ m}$$

$$h_m = (K)(v^2/2g) = (0.5 + 1.0)\{1.768^2/[(2)(9.807)]\} = 0.239 \text{ m}$$

$$h_L = 0.239 + 0.239 = 0.478 \text{ m} \qquad 0 + 0 + z_1 = 0 + 0 + z_2 + 0.478 \qquad z_1 - z_2 = H = 0.478 \text{ m}$$

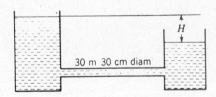

Fig. 9-47

9.246 In Fig. 9-47 for $H = 3$ m, calculate the discharge of oil, s.g. $= 0.8$, $\mu = 7$ cP, through smooth pipe.

$$p_1/\gamma + v_1^2/2g + z_1 = p_2/\gamma + v_2^2/2g + z_2 + h_L \qquad h_f = (f)(L/d)(v^2/2g) = (f)[30/(\tfrac{30}{100})]\{v^2/[(2)(9.807)]\} = 5.098fv^2$$

$$h_m = (K)(v^2/2g) = (0.5 + 1.0)\{v^2/[(2)(9.807)]\} = 0.07648v^2 \qquad h_L = h_f + h_m = 5.098fv^2 + 0.07648v^2$$

$$0 + 0 + 3 = 0 + 0 + 0 + 5.098fv^2 + 0.07648v^2 \qquad v = \sqrt{3/(5.098f + 0.07648)}$$

Try $f = 0.02$: $v = \sqrt{3/[(5.098)(0.02) + 0.07648]} = 4.100$ m/s, $N_R = \rho dv/\mu = [(0.8)(1000)](\tfrac{30}{100})(4.100)/[(7)(0.001)] = 1.41 \times 10^5$. From Fig. A-5, $f = 0.017$. Try $f = 0.017$: $v = \sqrt{3/[(5.098)(0.017) + 0.07648]} = 4.288$ m/s, $N_R = [(0.8)(1000)](\tfrac{30}{100})(4.288)/[(7)(0.001)] = 1.47 \times 10^5$, $f = 0.0165$. Try $f = 0.0165$: $v = \sqrt{3/[(5.098)(0.0165) + 0.07648]} = 4.322$ m/s, $N_R = [(0.8)(1000)](\tfrac{30}{100})(4.322)/[(7)(0.001)] = 1.48 \times 10^5$, $f = 0.0165$ (O.K.); $Q = Av = [(\pi)(\tfrac{30}{100})^2/4](4.322) = 0.306$ m^3/s.

9.247 If a valve is placed in the line in Prob. 9.246 and adjusted to reduce the discharge by one-half, what is K for the valve and what is its equivalent length of pipe at this setting?

∎ From Prob. 9.246, $H = 3 = [(f)(30)/(\frac{30}{100}) + 0.5 + 1.0 + K_{valve}]\{(4.322/2)^2/[(2)(9.807)]\}$, $K_{valve} = 11.10 - 100f$, $N_R = \rho dv/\mu = [(0.8)(1000)][(\frac{30}{100})(4.322/2)/[(7)(0.001)] = 7.41 \times 10^4$. From Fig. A-5, $f = 0.019$. $K_{valve} = 11.10 - (100)(0.019) = 9.2$, $L_e = Kd/f = (9.2)(\frac{30}{100})/0.019 = 145$ m.

9.248 A water line connecting two reservoirs at 70 °F has 5000 ft of 24-in-diameter steel pipe, three standard elbows ($K = 0.9$), a globe valve ($K = 10$), a re-entrant pipe entrance ($K = 1.0$), and a pipe exit ($K = 1.0$). What is the difference in reservoir elevations for 20 cfs?

∎ $$p_1/\gamma + v_1^2/2g + z_1 = p_2/\gamma + v_2^2/2g + z_2 + h_L \qquad h_L = h_f + h_m \qquad h_f = (f)(L/d)(v^2/2g)$$

$$v = Q/A = 20/[(\pi)(\tfrac{24}{12})^2/4] = 6.366 \text{ ft/s} \qquad N_R = dv/\nu = (\tfrac{24}{12})(6.366)/(1.05 \times 10^{-5}) = 1.21 \times 10^6$$

$$\epsilon/d = 0.00015/(\tfrac{24}{12}) = 0.000075$$

From Fig. A-5, $f = 0.013$.

$$h_f = 0.013[5000/(\tfrac{24}{12})]\{6.366^2/[(2)(32.2)]\} = 20.45 \text{ ft}$$

$$h_m = (K)(v^2/2g) = [(3)(0.9) + 10 + 1.0 + 1.0]\{6.366^2/[(2)(32.2)]\} = 9.25 \text{ ft}$$

$$h_L = 20.45 + 9.25 = 29.70 \text{ ft} \qquad 0 + 0 + z_1 = 0 + 0 + z_2 + 29.70 \qquad z_1 - z_2 = 29.70 \text{ ft}$$

9.249 For the conditions given in Prob. 9.248, determine the discharge if the difference in elevation is 40 ft.

∎ From Prob. 9.248, $40 = h_f + h_m = \{v^2/[(2)(32.2)]\}\{(f)[5000/(\tfrac{24}{12})] + [(3)(0.9) + 10 + 1.0 + 1.0]\}$, $v = \sqrt{2576/(2500f + 14.7)}$. Try $f = 0.013$: $v = \sqrt{2576/[(2500)(0.013) + 14.7]} = 7.388$ ft/s, $N_R = dv/\nu = (\tfrac{24}{12})(7.388)/(1.05 \times 10^{-5}) = 1.41 \times 10^6$, $\epsilon/d = 0.000075$ (from Prob. 9.248). From Fig. A-5, $f = 0.0125$. Try $f = 0.0125$: $v = \sqrt{2576/[(2500)(0.0125) + 14.7]} = 7.487$ ft/s, $N_R = (\tfrac{24}{12})(7.487)/(1.05 \times 10^{-5}) = 1.43 \times 10^6$, $f = 0.0125$ (O.K.); $Q = Av = [(\pi)(\tfrac{24}{12})^2/4](7.487) = 23.5$ ft^3/s.

9.250 What size commercial steel pipe is needed to convey 200 L/s of water at 20 °C a distance of 5000 m with a head drop of 4 m? The line connects two reservoirs, has a re-entrant ($K = 1.0$), a submerged outlet ($K = 1.0$), four standard elbows ($K = 0.9$), and a globe valve ($K = 10$).

∎ $$p_1/\gamma + v_1^2/2g + z_1 = p_2/\gamma + v_2^2/2g + z_2 + h_L \qquad 0 + 0 + 4 = 0 + 0 + 0 + h_L \qquad h_L = h_f + h_m$$

$$v = Q/A = (\tfrac{200}{1000})/(\pi d^2/4) = 0.2546/d^2$$

$$h_f = (f)(L/d)(v^2/2g) = (f)(5000/d)\{(0.2546/d^2)^2/[(2)(9.807)]\} = 16.52f/d^5$$

$$h_m = (K)(v^2/2g) = [1.0 + 1.0 + (4)(0.9) + 10]\{(0.2546/d^2)^2/[(2)(9.807)]\} = 0.05156/d^4$$

$$4 = 16.52f/d^5 + 0.05156/d^4$$

Try $f = 0.02$: $4 = (16.52)(0.02)/d^5 + 0.05156/d^4$. By trial and error, $d = 0.619$ m.

$$v = 0.2546/0.619^2 = 0.6645 \text{ m/s} \qquad N_R = dv/\nu = (0.619)(0.6645)/(1.02 \times 10^{-6}) = 4.03 \times 10^5$$

$$\epsilon/d = 0.000046/0.619 = 0.000074$$

From Fig. A-5, $f = 0.0145$. Try $f = 0.0145$: $4 = (16.52)(0.0145)/d^5 + 0.05156/d^4$, $d = 0.588$ m, $v = 0.2546/0.588^2 = 0.7364$ m/s, $N_R = (0.588)(0.7364)/(1.02 \times 10^{-6}) = 4.25 \times 10^5$, $f = 0.0145$. Therefore, $d = 0.588$ m, or 588 mm.

9.251 What is the equivalent length of 2-in-diameter pipe, $f = 0.022$, for (a) a re-entrant pipe entrance (use $K = 1.0$), (b) a sudden expansion from 2 in to 4 in diameter, and (c) a globe valve and a standard tee?

∎ $$L_e = KD/f$$

(a) $$L_e = (1.0)(\tfrac{2}{12})/0.022 = 7.58 \text{ ft}$$

(b) $$K = [1 - (D_1/D_2)^2]^2 = [1 - (\tfrac{2}{4})^2]^2 = 0.5625 \qquad L_e = (0.5625)(\tfrac{2}{12})/0.022 = 4.26 \text{ ft}$$

(c) $$K = 10 + 1.8 = 11.8 \qquad L_e = (11.8)(\tfrac{2}{12})/0.022 = 89.4 \text{ ft}$$

9.252 Find H in Fig. 9-48 for 200 gpm of oil ($\mu = 0.1$ P, $\gamma = 60$ lb/ft³) for the angle valve wide open ($K = 5.0$).

$$p_1/\gamma + v_1^2/2g + z_1 = p_2/\gamma + v_2^2/2g + z_2 + h_L \qquad h_L = h_f + h_m \qquad h_f = (f)(L/d)(v^2/2g)$$

$$Q = (200)(0.002228) = 0.4456 \text{ ft}^3/\text{s} \qquad v = Q/A = 0.4456/[(\pi)(\tfrac{3}{12})^2/4] = 9.078 \text{ ft/s}$$

$$\mu = 0.1[(0.1)(0.3048)^2/4.448] = 0.0002089 \text{ lb} \cdot \text{s/ft}^2$$

$$N_R = \rho dv/\mu = (\gamma/g)(dv)/\mu = (60/32.2)(\tfrac{3}{12})(9.078)/0.0002089 = 2.02 \times 10^4 \qquad \epsilon/d = 0.00015/(\tfrac{3}{12}) = 0.000600$$

From Fig. A-5, $f = 0.0275$.

$$h_f = 0.0275[210/(\tfrac{3}{12})]\{9.078^2/[(2)(32.2)]\} = 29.56 \text{ ft}$$

$$h_m = (K)(v^2/2g) = [0.5 + 5.0 + 1.0]\{9.078^2/[(2)(32.2)]\} = 8.32 \text{ ft}$$

$$h_L = 29.56 + 8.32 = 37.88 \text{ ft} \qquad 0 + 0 + z_1 = 0 + 0 + z_2 + 37.88 \qquad z_1 - z_2 = H = 37.88 \text{ ft}$$

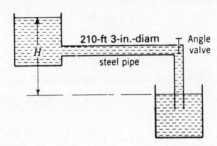

210-ft 3-in.-diam steel pipe — Angle valve

H

Fig. 9-48

9.253 Find K for the angle valve in Prob. 9.252 for a flow of 10 L/s at the same H.

$$v = Q/A = [(\tfrac{10}{1000})/0.3048^3]/[(\pi)(\tfrac{3}{12})^2/4] = 7.194 \text{ ft/s}$$

$$N_R = \rho dv/\mu = (\gamma/g)(dv)/\mu = (60/32.2)(\tfrac{3}{12})(7.194)/0.0002089 = 1.60 \times 10^4$$

$$\epsilon/d = 0.000600 \qquad \text{(from Prob. 9.252)}$$

From Fig. A-5, $f = 0.0285$. From Prob. 9.252, $H = [(f)(L/d) + K_1 + K_2 + K_3](v^2/2g)$, $37.88 = \{(0.0285)[210/(\tfrac{3}{12})] + 0.5 + K_2 + 1.0\}\{7.194^2/[(2)(32.2)]\}$, $K_2 = 21.7$.

9.254 What is the discharge through the system of Fig. 9-48 for water at 25 °C when $H = 8$ m?

$$p_1/\gamma + v_1^2/2g + z_1 = p_2/\gamma + v_2^2/2g + z_2 + h_L$$

$$h_f = (f)(L/d)(v^2/2g) = (f)[210/(\tfrac{3}{12})]\{v^2/[(2)(32.2)]\} = 13.04fv^2$$

$$h_m = (K)(v^2/2g) = (0.5 + 5.0 + 1.0)\{v^2/[(2)(32.2)]\} = 0.1009v^2 \qquad h_L = h_f + h_m = 13.04fv^2 + 0.1009v^2$$

$$0 + 0 + 8/0.3048 = 0 + 0 + 0 + 13.04fv^2 + 0.1009v^2 \qquad v = \sqrt{26.25/(13.04f + 0.1009)}$$

Try $f = 0.02$: $v = \sqrt{26.25/[(13.04)(0.02) + 0.1009]} = 8.519$ ft/s, $N_R = dv/\nu = (\tfrac{3}{12})(8.519)/(9.56 \times 10^{-6}) = 2.23 \times 10^5$, $\epsilon/d = 0.00015/(\tfrac{3}{12}) = 0.000600$. From Fig. A-5, $f = 0.019$. Try $f = 0.019$: $v = \sqrt{26.25/[(13.04)(0.019) + 0.1009]} = 8.677$ ft/s, $N_R = (\tfrac{3}{12})(8.677)/(9.56 \times 10^{-6}) = 2.27 \times 10^5$, $f = 0.019$ (O.K.); $Q = Av = [(\pi)(\tfrac{3}{12})^2/4](8.677) = 0.4259$ ft³/s $= (0.4259)(0.3048)^3 = 0.0121$ m³/s, or 12.1 L/s.

9.255 Find the discharge through the pipeline in Fig. 9-49 for $H = 10$ m, as shown. Use minor loss coefficients for the entrance, elbows, and globe valve of 0.5, 0.9 (each), and 10, respectively.

$$p_1/\gamma + v_1^2/2g + z_1 = p_2/\gamma + v_2^2/2g + z_2 + h_L$$

$$h_f = (f)(L/d)(v^2/2g) = f[(30 + 12 + 60)/(\tfrac{150}{1000})]\{v_2^2/[(2)(9.807)]\} = 34.67fv_2^2$$

$$h_m = (K)(v^2/2g) = [0.5 + (2)(0.9) + 10]\{v_2^2/[(2)(9.807)]\} = 0.6271v_2^2$$

$$h_L = h_f + h_m = 34.67fv_2^2 + 0.6271v_2^2 \qquad 0 + 0 + 10 = 0 + v_2^2/[(2)(9.807)] + 0 + 34.67fv_2^2 + 0.6271v_2^2$$

$$v = \sqrt{10/(34.67f + 0.6781)}$$

Try $f = 0.02$: $v = \sqrt{10/[(34.67)(0.02) + 0.6781]} = 2.700$ m/s, $N_R = dv/\nu = (\tfrac{150}{1000})(2.700)/(1.02 \times 10^{-6}) = 3.97 \times 10^5$, $\epsilon/d = 0.00026/(\tfrac{150}{1000}) = 0.00173$. From Fig. A-5, $f = 0.023$. Try $f = 0.023$: $v = \sqrt{10/[(34.67)(0.023) + 0.6781]} = 2.603$ m/s, $N_R = (\tfrac{150}{1000})(2.603)/(1.02 \times 10^{-6}) = 3.83 \times 10^5$, $f = 0.023$ (O.K.); $Q = Av = [(\pi)(\tfrac{150}{1000})^2/4](2.603) = 0.0460$ m³/s, or 46.0 L/s.

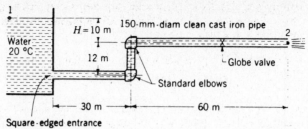

Fig. 9-49

9.256 Rework Prob. 9.255 to find H if $Q = 60$ L/s.

$$p_1/\gamma + v_1^2/2g + z_1 = p_2/\gamma + v_2^2/2g + z_2 + h_L \qquad h_L = h_f + h_m \qquad h_f = (f)(L/d)(v^2/2g)$$

$$v = Q/A = (\tfrac{60}{1000})/[(\pi)(\tfrac{150}{1000})^2/4] = 3.395 \text{ m/s} \qquad N_R = dv/v = (\tfrac{150}{1000})(3.395)/(1.02 \times 10^{-6}) = 4.99 \times 10^5$$

$$\epsilon/d = 0.00173 \qquad \text{(from Prob. 9.255)}$$

From Fig. A-5, $f = 0.0225$.

$$h_f = 0.0225[(30 + 12 + 60)/(\tfrac{150}{1000})]\{3.395^2/[(2)(9.807)]\} = 8.99 \text{ m}$$

$$h_m = (K)(v^2/2g) = [0.5 + (2)(0.9) + 10]\{3.395^2/[(2)(9.807)]\} = 7.23 \text{ m}$$

$$h_L = 8.99 + 7.23 = 16.22 \text{ m} \qquad 0 + 0 + z_1 = 0 + 3.395^2/[(2)(9.807)] + z_2 + 16.22 \qquad z_1 - z_2 = H = 16.81 \text{ m}$$

9.257 Assume that water at 10 °C is to be conveyed at 300 L/s through 500 m of commercial steel pipe with a total head drop of 6 m. Minor losses are $12v^2/2g$. Determine the required diameter.

$$N_R = R_5/D \tag{1}$$

$$f = R_7/[\ln (R_3/D + R_2/N_R^{0.9})]^2 \tag{2}$$

$$x = R_6 + R_4 D/f \tag{3}$$

$$D = (R_0)(x^{4.75} + R_1 x^{5.2})^{0.04} \tag{4}$$

where $R_0 = (0.66)(\epsilon^{1.25} Q^{9.5})^{0.04}$, $R_1 = v/\epsilon^{1.25} Q^{0.1}$, $R_2 = 5.74$, $R_3 = \epsilon/3.7$, $R_4 = K/gh_f$, $R_5 = 4Q/\pi v$, $R_6 = L/gh_f$, $R_7 = 1.325$, $\epsilon = 0.000046$ m, $Q = 300$ L/s, or 0.300 m³/s, $v = 1.30 \times 10^{-6}$ m²/s.

$$K = 12 \qquad h_f = 6 \text{ mm} \qquad R_0 = (0.66)(0.000046^{1.25} 0.300^{9.5})^{0.04} = 0.25351$$

$$R_1 = 1.30 \times 10^{-6}/(0.000046^{1.25} 0.300^{0.1}) = 0.38707 \qquad R_3 = 0.000046/3.7 = 1.2432 \times 10^{-5}$$

$$R_4 = 12/[(9.807)(6)] = 0.20394 \qquad R_5 = (4)(0.300)/[(\pi)(1.30 \times 10^{-6})] = 2.9382 \times 10^5$$

$$R_6 = 500/[(9.807)(6)] = 8.4973$$

Assume $D = 1$ m. Substituting into Eqs. (1), (2), (3), and (4),

$$N_R = 2.9382 \times 10^5/1 = 2.9382 \times 10^5 \qquad f = 1.325/\{\ln [(1.2432 \times 10^{-5})/1 + 5.74/(2.9382 \times 10^5)^{0.9}]\}^2 = 0.014938$$

$$x = 8.4973 + (0.20394)(1)/0.014938 = 22.150 \qquad D = (0.25351)[22.150^{4.75} + (0.38707)(22.150^{5.2})]^{0.04} = 0.47418 \text{ m}$$

Try $D = 0.47418$ m: $N_R = 2.9382 \times 10^5/0.47418 = 6.1964 \times 10^5$, $f = 1.325/\{\ln [(1.2432 \times 10^{-5})/0.47418 + 5.74/(6.1964 \times 10^5)^{0.9}]\}^2 = 0.014086$, $x = 8.4973 + (0.20394)(0.47418)/0.014086 = 15.363$, $D = (0.25351)[15.363^{4.75} + (0.38707)(15.363^{5.2})]^{0.04} = 0.44063$ m. Try $D = 0.44063$ m: $N_R = 2.9382 \times 10^5/0.44063 = 6.6682 \times 10^5$, $f = 1.325/\{\ln [(1.2432 \times 10^{-5})/0.44063 + 5.74/(6.6682 \times 10^5)^{0.9}]\}^2 = 0.014075$, $x = 8.4973 + (0.20394)(0.44063)/0.014075 = 14.882$, $D = (0.25351)[14.882^{4.75} + (0.38707)(14.882^{5.2})]^{0.04} = 0.43783$ m. Try $D = 0.43783$ m: $N_R = 2.9382 \times 10^5/0.43783 = 6.7108 \times 10^5$, $f = 1.325/\{\ln [(1.2432 \times 10^{-5})/0.43783 + 5.74/(6.7108 \times 10^5)^{0.9}]\}^2 = 0.014074$, $x = 8.4973 + (0.20394)(0.43783)/0.014074 = 14.842$, $D = (0.25351)[14.842^{4.75} + (0.38707)(14.842^{5.2})]^{0.04} = 0.43759$ m. Therefore, $D = 0.438$ m, or 438 mm.

9.258 One equation for determining the friction factor is $1/\sqrt{f} = 0.869 \ln (N_R\sqrt{f}) - 0.8$. Compare the smooth pipe curve on the Moody diagram (Fig. A-5) with results from the equation above for values of Reynolds number of 10^5, 10^6, and 10^7.

For $N_R = 10^5$, from Fig. A-5, $f = 0.0178$. From the equation, $1/\sqrt{f} = 0.869\{\ln [(10^5)(\sqrt{f})]\} - 0.8$. By trial and error, $f = 0.0183$. For $N_R = 10^6$, from Fig. A-5, $f = 0.0116$. From the equation, $1/\sqrt{f} = 0.869\{\ln [(10^6)(\sqrt{f})]\} - 0.8$. By trial and error, $f = 0.0116$. For $N_R = 10^7$, from Fig. A-5, $f = 0.0082$. From the equation, $1/\sqrt{f} = 0.869\{\ln [(10^7)(\sqrt{f})]\} - 0.8$. By trial and error, $f = 0.00810$.

9.259 An equation for determining the friction factor developed by Colebrook is $1/\sqrt{f} = -0.869 \ln[(\epsilon/D)/3.7 + 2.523/(N_R\sqrt{f})]$. Check the location of line $\epsilon/D = 0.0002$ on the Moody diagram (Fig. A-5) with the equation above for a Reynolds number of 10^5.

▎ From Fig. A-5, $f = 0.0190$. From the equation, $1/\sqrt{f} = -0.869 \ln\{0.0002/3.7 + 2.523/[(10^5)(\sqrt{f})]\}$. By trial and error, $f = 0.0192$.

9.260 Find the head loss in a pipeline consisting of 200 ft of 4-in steel pipe, a 90° bend on 24-in radius ($K = 0.15$), 4-in gate valve (fully open) ($K = 0.20$), 100 ft of 4-in steel pipe, expansion to 6 in with a 20° taper ($K = 0.4$), 300 ft of 6-in steel pipe, abrupt contraction to 3-in diameter ($K = 0.35$), and 50 ft of 3-in steel pipe. The discharge rate is 1.5 cfs.

▎ $h_f = (f)(L/d)(v^2/2g)$, $h_m = (K)(v^2/2g)$. For 200 ft of 4-in pipe: $v_1 = Q/A = 1.5/[(\pi)(\frac{4}{12})^2/4] = 17.19$ ft/s, $N_R = dv_1/\nu = (\frac{4}{12})(17.19)/(1.05 \times 10^{-5}) = 5.46 \times 10^5$, $\epsilon/d = 0.00015/(\frac{4}{12}) = 0.00045$. From Fig. A-5, $f = 0.0175$. $h_f = 0.0175[200/(\frac{4}{12})]\{17.19^2/[(2)(32.2)]\} = 48.2$ ft.

For bend: $h_m = 0.15\{17.19^2/[(2)(32.2)]\} = 0.7$ ft.

For gate valve: $h_m = 0.20\{17.19^2/[(2)(32.2)]\} = 0.9$ ft.

For 100 ft of 4-in pipe: $h_f = 0.0175[100/(\frac{4}{12})]\{17.19^2/[(2)(32.2)]\} = 24.1$ ft.

For expansion: $v_2 = 1.5/[(\pi)(\frac{6}{12})^2/4] = 7.639$ ft/s, $h_m = 0.4\{17.19 - 7.639)^2/[(2)(32.2)]\} = 0.6$ ft.

For 300 ft of 6-in pipe: $N_R = (\frac{6}{12})(7.639)/(1.05 \times 10^{-5}) = 3.64 \times 10^5$, $\epsilon/d = 0.00015/(\frac{6}{12}) = 0.00030$, $f = 0.0170$, $h_f = 0.0170[300/(\frac{6}{12})]\{7.639^2/[(2)(32.2)]\} = 9.2$ ft.

For abrupt contraction: $v_3 = 1.5/[(\pi)(\frac{3}{12})^2/4] = 30.56$ ft/s, $h_m = 0.35\{30.56^2/[(2)(32.2)]\} = 5.1$ ft.

For 50 ft of 3-in pipe: $N_R = (\frac{3}{12})(30.56)/(1.05 \times 10^{-5}) = 7.28 \times 10^5$, $\epsilon/d = 0.00015/(\frac{3}{12}) = 0.00060$, $f = 0.0180$, $h_f = 0.0180[50/(\frac{3}{12})]\{30.56^2/[(2)(32.2)]\} = 52.2$ ft, $h_L = 48.2 + 0.7 + 0.9 + 24.1 + 0.6 + 9.2 + 5.1 + 52.2 = 141.0$ ft.

9.261 Using the Darcy–Weisbach formula, find the head loss in 1000 ft of 6-ft-diameter smooth concrete pipe carrying 80 cfs of water at 50 °F.

▎
$$h_f = (f)(L/D)(v^2/2g) \qquad v = Q/A = 80/[(\pi)(6)^2/4] = 2.829 \text{ ft/s}$$
$$N_R = Dv/\nu = (6)(2.829)/(1.40 \times 10^{-5}) = 1.21 \times 10^6 \qquad \epsilon/D = 0.001/6 = 0.000167$$

From Fig. A-5, $f = 0.014$.

$$h_f = (0.014)(\tfrac{1000}{6})\{2.829^2/[(2)(32.2)]\} = 0.290 \text{ ft}$$

9.262 Solve Prob. 9.261 using the Manning formula.

▎
$$v = (1.486/n)(R)^{2/3}(s)^{1/2} = 2.829 \text{ ft/s} \qquad \text{(from Prob. 9.261)}$$
$$n = 0.013 \qquad \text{(from Table A-13)} \qquad R = D/4 = \tfrac{6}{4} = 1.500 \text{ ft}$$
$$2.829 = (1.486/0.013)(1.500)^{2/3}(s)^{1/2} \qquad s = 0.0003567 \qquad h_f = (0.0003567)(1000) = 0.357 \text{ ft}$$

9.263 Solve Prob. 9.261 using the Hazen–Williams formula.

▎
$$v = 1.318CR^{0.63}s^{0.54} = 2.829 \text{ ft/s} \qquad \text{(from Prob. 9.261)}$$
$$C = 120 \qquad \text{(from Table A-14)} \qquad R = 1.500 \text{ ft} \qquad \text{(from Prob. 9.262)}$$
$$2.829 = (1.318)(120)(1.500)^{0.63}(s)^{0.54} \qquad s = 0.0003618 \qquad h_f = (1000)(0.0003618) = 0.362 \text{ ft}$$

9.264 Using the Darcy–Weisbach formula, find the head loss in 100 ft of 3-ft-diameter welded steel pipe carrying 15 cfs of water at 60 °F.

▎
$$h_f = (f)(L/D)(v^2/2g) \qquad v = Q/A = 15/[(\pi)(3)^2/4] = 2.122 \text{ ft/s}$$
$$N_R = Dv/\nu = (3)(2.122)/(1.21 \times 10^{-5}) = 5.26 \times 10^5 \qquad \epsilon/D = 0.00015/3 = 0.000050$$

From Fig. A-5, $f = 0.0135$. $h_f = (0.0135)(\tfrac{100}{3})\{2.122^2/[(2)(32.2)]\} = 0.031$ ft.

9.265 Solve Prob. 9.264 using the Manning formula with $n = 0.012$.

$$v = (1.486/n)(R)^{2/3}(s)^{1/2} = 2.122 \text{ ft/s} \qquad \text{(from Prob. 9.264)}$$

$$R = D/4 = \tfrac{3}{4} = 0.7500 \text{ ft} \qquad 2.122 = (1.486/0.012)(0.7500)^{2/3}(s)^{1/2} \qquad s = 0.0004309$$

$$h_f = (0.0004309)(100) = 0.043 \text{ ft}$$

9.266 Solve Prob. 9.264 using the Hazen–Williams formula with $C = 120$.

$$v = 1.318CR^{0.63}s^{0.54} = 2.122 \text{ ft/s} \qquad \text{(from Prob. 9.264)}$$

$$R = 0.7500 \text{ ft} \qquad \text{(from Prob. 9.265)}$$

$$2.122 = (1.318)(120)(0.7500)^{0.63}(s)^{0.54} \qquad s = 0.0004769 \qquad h_f = (100)(0.0004769) = 0.048 \text{ ft}$$

9.267 A 36-in-diameter concrete pipe is 4000 ft long and has a head loss of 12.7 ft. Find the discharge capacity of water for this pipe by the Hazen–Williams formula

$$v = 1.318CR^{0.63}s^{0.54} = (1.318)(120)[(\tfrac{36}{12})/4]^{0.63}(12.7/4000)^{0.54} = 5.906 \text{ ft/s}$$

$$Q = Av = [(\pi)(\tfrac{36}{12})^2/4](5.906) = 41.7 \text{ ft}^3/\text{s}$$

9.268 Solve Prob. 9.267 using the Manning formula.

$$v = (1.486/n)(R)^{2/3}(s)^{1/2} = (1.486/0.013)[(\tfrac{36}{12})/4]^{2/3}(12.7/4000)^{1/2} = 5.317 \text{ ft/s}$$

$$Q = Av = [(\pi)(\tfrac{36}{12})^2/4](5.317) = 37.6 \text{ ft}^3/\text{s}$$

9.269 A 1-m-diameter new cast iron pipe is 845 m long and has a head loss of 1.11 m. Find the discharge capacity of water for this pipe by the Hazen–Williams formula.

$$v = 0.8492CR^{0.63}s^{0.54} = (0.8492)(130)(\tfrac{1}{4})^{0.63}(1.11/845)^{0.54} = 1.281 \text{ m/s}$$

$$Q = Av = [(\pi)(1)^2/4](1.281) = 1.01 \text{ m}^3/\text{s}$$

9.270 Solve Prob. 9.269 using the Manning formula.

$$v = (1.0/n)(R)^{2/3}(s)^{1/2} = (1.0/0.013)(\tfrac{1}{4})^{2/3}(1.11/845)^{1/2} = 1.106 \text{ m/s}$$

$$Q = Av = [(\pi)(1)^2/4](1.106) = 0.869 \text{ m}^3/\text{s}$$

9.271 A riveted steel pipe must transport 2.4 ft³/s of water a distance of 190 ft with a head loss of 2.7 ft. Find the necessary pipe diameter using the Hazen–Williams formula.

$$v = 1.318CR^{0.63}s^{0.54} = Q/A = 2.4/(\pi D^2/4) = 3.056/D^2$$

$$3.056/D^2 = (1.318)(110)(D/4)^{0.63}(2.7/190)^{0.54} \qquad D = 0.7695 \text{ ft} \quad \text{or} \quad 9.23 \text{ in}$$

9.272 A square concrete conduit must transport 4.0 m³/s of water a distance of 45 m with a head loss of 1.80 m. Find the necessary conduit size using the Hazen–Williams formula.

$v = 0.8492CR^{0.63}s^{0.54}$. Let $a = $ length of conduit side. $v = Q/A = 4.0/a^2$, $4.0/a^2 = (0.8492)(120)(a^2/4a)^{0.63}(1.80/45)^{0.54}$, $a = 0.788$ m.

9.273 Solve Prob. 9.267 using the Hazen–Williams pipe diagram.

Let $h_1 = $ unit head loss. $h_1 = 12.7/4000 = 0.003175$. From Fig. A-13 with $h_1 = 0.003175$ and $D = 3$ ft, or 36 in, $Q = 41.5$ ft³/s.

9.274 Solve Prob. 9.269 using the Hazen–Williams pipe diagram.

$h_1 = 1.11/845 = 0.001314$. From Fig. A-14 with $h_1 = 0.001314$ and $D = 1$ m, or 1000 mm, $Q = 0.91$ m³/s. This value of Q is for $C = 120$ (since the pipe diagram is for $C = 120$). Inasmuch as $C = 130$ for new cast iron (Table A-14), this value of Q must be adjusted. Since Q varies directly with C, $0.91/(Q)_{C=130} = \frac{120}{130}$, $(Q)_{C=130} = 0.99$ m³/s.

9.275 Water is flowing in a 500-mm-diameter new cast iron pipe at a velocity of 2.0 m/s. Using the Hazen–Williams pipe diagram, find the pipe friction loss per 100 m of pipe.

❙ $C = 130$ for new cast iron pipe. In order to use the pipe diagram (Fig. A-14) for which $C = 120$, the given velocity must be adjusted for a value of C of 120. Since velocity varies directly with C, $2.0/(v)_{C=120} = \frac{130}{120}$, $(v)_{C=120} = 1.85$ m/s. From Fig. A-14 with $v = 1.85$ m/s and $D = 500$ mm, $h_1 = 0.0067$ m/m, or 0.67 m per 100 m. $p = \gamma h = (9.79)(0.67) = 6.6$ kPa per 100 m.

9.276 A new cast iron pipe must carry 30 cfs of water at a head loss of 19.0 ft per mile of pipe length. Find the required pipe diameter using the Hazen–Williams pipe diagram.

❙ $h_1 = 19.0/5280 = 0.003598$. $C = 130$ for new cast iron pipe. In order to use the pipe diagram (Fig. A-13) for which $C = 120$, the given discharge must be adjusted for a value of C of 120. Since discharge varies directly with C, $30/(Q)_{C=120} = \frac{130}{120}$, $(Q)_{C=120} = 27.7$ cfs. With $Q = 27.7$ cfs and $h_1 = 0.003598$, a pipe diameter of 30 in is determined from Fig. A-13.

9.277 Solve Prob. 9.270 using the Manning pipe diagram with $n = 0.012$.

❙ $h_1 = 1.11/845 = 0.001314$. With $D = 1$ m, or 1000 mm, and $h_1 = 0.001314$, a discharge of 0.88 m³/s is determined from Fig. A-16. Inasmuch as Fig. A-16 is for $n = 0.013$ and n is 0.012 in this problem, this discharge must be adjusted. Since discharge varies inversely with n, $0.88/(Q)_{n=0.012} = 0.012/0.013$, $(Q)_{n=0.012} = 0.95$ m³/s.

9.278 A concrete pipe must carry 80 cfs of water at a head loss of 1.5 ft per 100 ft of pipe length. Find the required pipe diameter using the Manning pipe diagram.

❙ $h_1 = 1.5/100 = 0.015$. With $Q = 80$ cfs and $h_1 = 0.015$, a pipe diameter of about 36 in is determined from Fig. A-15.

9.279 Solve Prob. 9.278 for the same given conditions except that the pipe has an n value of 0.015.

❙
$$h_1 = 0.015 \quad \text{(from Prob. 9.278)}$$

The Manning pipe diagram (Fig. A-15) is for a value of n of 0.013. The given discharge (80 cfs) is for a value of n of 0.015. In order to use Fig. A-15, the given discharge must be adjusted for a value of n of 0.013. Since discharge varies inversely with n, $80/(Q)_{n=0.013} = 0.013/0.015$, $(Q)_{n=0.013} = 92.3$ cfs. With $Q = 92.3$ cfs and $h_1 = 0.015$, a pipe diameter of about 38 in is determined from Fig. A-15.

9.280 A capillary tube of inside diameter 6 mm connects tank A and open container B, as shown in Fig. 9-50. The liquid in A, B, and capillary CD is water having a specific weight of 9780 N/m³ and a viscosity of 0.0008 kg/(m · s). The pressure $p_A = 34.5$ kPa gage. Which direction will the water flow? What is the flow rate?

❙ Assume laminar flow from B to A.
$$\frac{p_B}{\gamma} + \frac{v_B^2}{2g} + z_B = \frac{p_A}{\gamma} + \frac{v_A^2}{2g} + z_A + h_L \qquad h_f = \frac{128\mu LQ}{\pi\gamma D^4} = \frac{(128)(0.0008)(4.3)(Q)}{(\pi)(9780)(\frac{6}{1000})^4} = 11\,058Q$$

$$0 + 0 + (1.4 + 4.3\sin 45°) = 34.5/9.79 + 0 + 1 + 11\,058Q$$

$$Q = -7.55 \times 10^{-6}\ \text{m}^3/\text{s} \quad \text{or} \quad -7.55 \times 10^{-3}\ \text{L/s}$$

Since Q is negative, the flow must be from A to B. Check for laminar flow: $v = Q/A = (7.55 \times 10^{-6})/[(\pi)(\frac{6}{1000})^2/4] = 0.2670$ m/s, $N_R = \rho Dv/\mu = (9780/9.807)(\frac{6}{1000})(0.2670)/0.0008 = 1997$ (laminar).

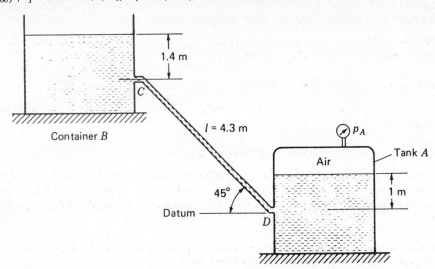

Container B

1.4 m

C

$l = 4.3$ m

p_A

Tank A

Air

1 m

45°

Datum

D

Fig. 9-50

9.281 An *equivalent length* of pipe is one whose head loss for the same value of flow is equal to that of some other system of different geometry for which it is equivalent. Consider a steel pipe of inside diameter 10.02 in having in it an open globe valve ($K = 4.8$) and four screwed 90° elbows ($K = 0.42$ each). The length of the pipe is 100 ft, and 5 ft^3/s of water at 60 °F flows through the pipe. What is the equivalent length of pipe with inside diameter 13.00 in?

▮ **For the given pipe:** $h_L = h_f + h_m$, $h_f = (f)(L/d)(v^2/2g)$, $v = Q/A = 5/[(\pi)(10.02/12)^2/4] = 9.131$ ft/s, $N_R = dV/\nu = (10.02/12)(9.131)/(1.21 \times 10^{-5}) = 6.30 \times 10^5$, $\epsilon/d = 0.00015/(10.02/12) = 0.00018$. From Fig. A-5, $f = 0.015$. $h_f = 0.015[100/(10.02/12)]\{9.131^2/[(2)(32.2)]\} = 2.33$ ft, $h_m = (K)(v^2/2g) = [4.8 + (4)(0.42)]\{9.131^2/[(2)(32.2)]\} = 8.39$ ft, $h_L = 2.33 + 8.39 = 10.72$ ft.

For the equivalent pipe: $v = 5/[(\pi)(13.00/12)^2/4] = 5.424$ ft/s, $N_R = (13.00/12)(5.424)/(1.21 \times 10^{-5}) = 4.86 \times 10^5$, $\epsilon/d = 0.00015/(13.00/12) = 0.000138$, $f = 0.015$, $10.72 = 0.015[L_e/(13.00/12)]\{5.424^2/[(2)(32.2)]\}$, $L_e = 1695$ ft.

9.282 In an air-conditioning system, there is a length of duct 200 ft transporting air at 50 °F at the rate of 8000 ft^3/min. The duct has a cross section of 2 ft by 1 ft and is made of galvanized iron. The pressure at the inlet of the duct is 2 psig. What is the pressure drop in millimeters of mercury over the length of the duct, if we hypothesize that the temperature remains very close to 50 °F and the pressure varies only slightly along the duct, i.e., we treat the flow as isothermal and incompressible? Consider that the flow is entirely turbulent.

▮
$$p_1/\gamma + v_1^2/2g + z_1 = p_2/\gamma + v_2^2/2g + z_2 + h_L \qquad h_L = h_f = (f)(L/d)(v^2/2g) \qquad N_R = \rho d v/\mu$$

$$\rho = p/RT = (2 + 14.7)(144)/[(1716)(460 + 50)] = 0.002748 \text{ slug/ft}^3 \qquad \gamma = \rho g = (0.002748)(32.2) = 0.08849 \text{ lb/ft}^3$$

$$d = D_h = 4A/p_w = 4[(2)(1)]/[(2)(2 + 1)] = 1.333 \text{ ft} \qquad v = Q/A = (\tfrac{8000}{60})/[(2)(1)] = 66.67 \text{ ft/s}$$

$$N_R = (0.002748)(1.333)(66.67)/(3.68 \times 10^{-7}) = 6.64 \times 10^5 \qquad \epsilon/d = 0.0005/1.333 = 0.000375$$

From Fig. A-5, $f = 0.0165$.

$$h_L = (0.0165)(200/1.333)\{66.67^2/[(2)(32.2)]\} = 170.9 \text{ ft}$$

$$p_1/0.08849 + v_1^2/2g + 0 = p_2/0.08849 + v_2^2/2g + 170.9$$

$$v_1^2/2g = v_2^2/2g \qquad p_1 - p_2 = 15.12 \text{ lb/ft}^2 \qquad \Delta p = \gamma h \qquad 15.12 = [(13.6)(62.4)](h)$$

$$h = 0.0178 \text{ ft} = (0.0178)(0.3048)(1000) = 5.43 \text{ mm of mercury}$$

9.283 In a heating system, there is a run of insulated duct of 50 m carrying air at a temperature of 35 °C at a pressure at the inlet of 100 kPa. The duct has a rectangular cross section of 650 mm by 320 mm. If there is a pressure drop from inlet to outlet of 5 mm of mercury, what is the volumetric flow? For such a small pressure drop in the duct, treat the flow as incompressible. Take $R = 287$ N $\cdot$ m/(kg $\cdot$ K). The duct is galvanized iron. Consider that the flow is entirely turbulent.

▮
$$p_1/\gamma + v_1^2/2g + z_1 = p_2/\gamma + v_2^2/2g + z_2 + h_L \qquad h_L = h_f = (f)(L/d)(v^2/2g) \qquad N_R = \rho d v/\mu$$

$$\rho = p/RT = (100)(1000)/[(287)(273 + 35)] = 1.131 \text{ kg/m}^3 \qquad \gamma = \rho g = (1.131)(9.807) = 11.09 \text{ N/m}^3$$

$$d = D_h = 4A/p_w = 4[(\tfrac{650}{1000})(\tfrac{320}{1000})]/[(2)(\tfrac{650}{1000} + \tfrac{320}{1000})] = 0.4289 \text{ m} \qquad h_L = (f)(50/0.4289)\{v^2/[(2)(9.807)]\} = 5.944 f v^2$$

$$p_1 - p_2 = \tfrac{5}{1000}[(13.6)(9.79)] = 0.6657 \text{ kN/m}^2 \quad \text{or} \quad 665.7 \text{ N/m}^2 \qquad p_1/\gamma - p_2/\gamma + v_1^2/2g + z_1 = v_2^2/2g + z_2 + h_L$$

$$v_1^2/2g = v_2^2/2g \qquad 665.7/11.09 + 0 = 0 + 5.944 f v^2 \qquad v = 3.178/\sqrt{f}$$

Try $f = 0.016$: $v = 3.178/\sqrt{0.016} = 25.12$ m/s, $N_R = (1.131)(0.4289)(25.12)/(1.88 \times 10^{-5}) = 6.48 \times 10^5$, $\epsilon/d = 0.00015/0.4289 = 0.000350$. From Fig. A-5, $f = 0.0165$. Try $f = 0.0165$: $v = 3.178/\sqrt{0.0165} = 24.74$ m/s, $N_R = (1.131)(0.4289)(24.74)/(1.88 \times 10^{-5}) = 6.38 \times 10^5$, $f = 0.0165$ (O.K.); $Q = Av = [(\tfrac{650}{1000})(\tfrac{320}{1000})](24.74) = 5.15 \text{ m}^3/\text{s}$.

9.284 Flowing through a pipe of 12 in diameter is 10 ft^3/s of water at 60 °F. The pipe is very smooth. Estimate the shear stress at the wall and the thickness of the viscous sublayer.

▮
$$v = Q/A = 10/[(\pi)(\tfrac{12}{12})^2/4] = 12.73 \text{ ft/s}$$

$$\tau_0 = 0.03325 \rho v^2 (\nu/Rv)^{1/4} = (0.03325)(1.94)(12.73)^2\{(1.21 \times 10^{-5})/[(\tfrac{6}{12})(12.73)]\}^{1/4} = 0.388 \text{ lb/ft}^2$$

$$\delta_1 = 5\nu/v_* = 5\nu\sqrt{\rho/\tau_0} = (5)(1.21 \times 10^{-5})\sqrt{1.94/0.388} = 0.000135 \text{ ft} \quad \text{or} \quad 0.00162 \text{ in}$$

9.285 Solve Prob. 9.284 using the following equation: $\tau_0 = (f/4)(\rho v^2/2)$. Compare values of δ_1.

$$N_R = dv/v = (\tfrac{12}{12})(12.73)/(1.21 \times 10^{-5}) = 1.05 \times 10^6$$

From Fig. A-5, $f = 0.0118$.

$$\tau_0 = (0.0118/4)[(1.94)(12.73)^2/2] = 0.464 \text{ lb/ft}^2$$

$$\delta_1 = 5v/v_* = 5v\sqrt{\rho/\tau_0} = (5)(1.21 \times 10^{-5})\sqrt{1.94/0.464} = 0.000124 \text{ ft} \quad \text{or} \quad 0.00148 \text{ in}$$

This value of δ_1 is $(0.00162 - 0.00148)/0.00162 = 0.086$, or 8.6 percent smaller than determined in Prob. 9.284.

9.286 What head H is needed in Fig. 9-51 to produce a discharge of $0.3 \text{ m}^3/\text{s}$? Use minor loss coefficients of 0.5 for the entrance and 0.12 for the diffusers.

$$p_A/\gamma + v_A^2/2g + z_A = p_B/\gamma + v_B^2/2g + z_B + h_L \qquad h_L = h_f + h_m \qquad h_f = (f)(L/d)(v^2/2g) \qquad N_R = dv/v$$

$$v_1 = Q/A_1 = 0.3/[(\pi)(\tfrac{200}{1000})^2/4] = 9.549 \text{ m/s} \qquad (N_R)_1 = (\tfrac{200}{1000})(9.549)/[(0.04)(0.1)/1000] = 4.77 \times 10^5$$

From Fig. A-5, $f_1 = 0.013$.

$$v_2 = Q/A_2 = 0.3/[(\pi)(\tfrac{300}{1000})^2/4] = 4.244 \text{ m/s} \qquad (N_R)_2 = (\tfrac{300}{1000})(4.244)/[(0.04)(0.1)/1000] = 3.18 \times 10^5$$

$$f_2 = 0.014 \qquad h_f = 0.013[30/(\tfrac{200}{1000})]\{9.549^2/[(2)(9.907)]\} + 0.014[60/(\tfrac{300}{1000})]\{4.244^2/[(2)(9.907)]\} = 11.64 \text{ m}$$

For entrance: $h_m = (K)(v^2/2g)$, $(h_m)_1 = 0.5\{9.549^2/[(2)(9.907)]\} = 2.32 \text{ m}$.

For first diffuser: $h_m = (K)(v_1 - v_2)^2/2g$, $(h_m)_2 = (0.12)(9.549 - 4.244)^2/[(2)(9.807)] = 0.17 \text{ m}$.

For second diffuser: $v_3 = Q/A_3 = 0.3/[(\pi)(\tfrac{450}{1000})^2/4] = 1.886 \text{ m/s}$, $(h_m)_3 = (0.12)(4.244 - 1.886)^2/[(2)(9.807)] = 0.03 \text{ m}$.

For exit: $(h_m)_4 = 1.0\{1.886^2/[(2)(9.807)]\} = 0.18 \text{ m}$,
$H = h_L = 11.64 + 2.32 + 0.17 + 0.03 + 0.18 = 14.34 \text{ m}$.

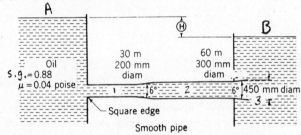

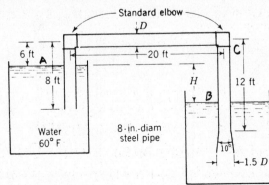

Fig. 9-51

9.287 Calculate the discharge through the siphon of Fig. 9-52 with the conical diffuser removed ($H = 4 \text{ ft}$). Use minor loss coefficients of 1.0 for both the entrance and exit and 0.9 for each elbow.

$$p_A/\gamma + v_A^2/2g + z_A = p_B/\gamma + v_B^2/2g + z_B + h_L$$

$$h_f = (f)(L/d)(v^2/2g) = f[(8 + 20 + 12)/(\tfrac{8}{12})]\{v^2/[(2)(32.2)]\} = 0.9317fv^2$$

$$h_m = (K)(v^2/2g) = [(1.0 + (2)(0.9) + 1.0]\{v^2/[(2)(32.2)]\} = 0.05901v^2 \qquad h_L = h_f + h_m = 0.9317fv^2 + 0.05901v^2$$

$$0 + 0 + 4 = 0 + 0 + 0 + 0.9317fv^2 + 0.05901v^2 \qquad v = \sqrt{4/(0.9317f + 0.05901)}$$

Try $f = 0.016$: $v = \sqrt{4/[(0.9317)(0.016) + 0.05901]} = 7.356 \text{ ft/s}$, $N_R = dv/v = (\tfrac{8}{12})(7.356)/(1.21 \times 10^{-5}) = 4.05 \times 10^5$, $\epsilon/d = 0.00015/(\tfrac{8}{12}) = 0.000225$. From Fig. A-5, $f = 0.016$ (O.K.). $Q = Av = [(\pi)(\tfrac{8}{12})^2/4](7.356) = 2.57 \text{ ft}^3/\text{s}$.

Fig. 9-52

9.288 Calculate the discharge in the siphon of Fig. 9-52 for $H = 8$ ft. What is the minimum pressure in the system? Use minor loss coefficients of 1.0 for both the entrance and exit, 0.9 for each elbow, and 0.15 for the diffuser.

I
$$p_A/\gamma + v_A^2/2g + z_A = p_B/\gamma + v_B^2/2g + z_B + h_L \tag{1}$$

$$h_L = h_f + h_m \qquad h_f = (f)(L/d)(v^2/2g) = f[(8 + 20 + 12)/(\tfrac{8}{12})]\{v_1^2/[(2)(32.2)]\} = 0.9317fv_1^2$$

$h_m = (K)(v^2/2g)$ for entrance, exit, and elbows. $h_m = (K)(v_1 - v_2)^2/2g$ for diffuser.

$$A_1v_1 = A_2v_2 \qquad [(\pi)(\tfrac{8}{12})^2/4](v_1) = \{(\pi)[(1.5)(\tfrac{8}{12})]^2/4\}(v_2) \qquad v_2 = 0.4444v_1$$

$$h_m = [1.0 + (2)(0.9)]\{v_1^2/[(2)(32.2)]\} + 0.15(v_1 - 0.4444v_1)^2/[(2)(32.2)]$$

$$+ 1.0\{(0.4444v_1)^2/[(2)(32.2)]\} = 0.04726v_1^2$$

$$h_L = 0.9317fv_1^2 + 0.04726v_1^2 \qquad 0 + 0 + 8 = 0 + 0 + 0 + 0.9317fv^2 + 0.04726v^2$$

$$v = \sqrt{8/(0.9317f + 0.04726)}$$

Try $f = 0.015$: $v = \sqrt{8/[(0.9317)(0.015) + 0.04726]} = 11.43$ ft/s, $N_R = dv/v = (\tfrac{8}{12})(11.43)/(1.21 \times 10^{-5}) = 6.30 \times 10^5$, $\epsilon/d = 0.00015/(\tfrac{8}{12}) = 0.000225$. From Fig. A-5, $f = 0.0155$. Try $f = 0.0155$: $v = \sqrt{8/[(0.9317)(0.0155) + 0.04726]} = 11.39$ ft/s, $N_R = (\tfrac{8}{12})(11.39)/(1.21 \times 10^{-5}) = 6.28 \times 10^5$, $f = 0.0155$ (O.K.); $Q = Av = [(\pi)(\tfrac{8}{12})^2/4](11.39) = 3.98$ ft^3/s. Minimum pressure would occur at C. Apply Eq. (1) between points A and C: $h_f = 0.0155[(8 + 20)/(\tfrac{8}{12})]\{11.39^2/[(2)(32.2)]\} = 1.311$ ft, $h_m = [1.0 + (2)(0.9)]\{11.39^2/[(2)(32.2)]\} = 5.641$ ft, $h_L = 1.311 + 5.641 = 6.952$ ft, $0 + 0 + 0 = (p_C)(144)/62.4 + 11.39^2/[(2)(32.2)] + 6 + 6.952$, $p_C = -6.49$ lb/in^2.

9.289 Find the discharge through the siphon of Fig. 9-53. What is the pressure at A? Estimate the minimum pressure in the system. Use minor loss coefficients of 1.0 for the entrance and 2.2 for the bend.

I Let v = velocity of water in the pipe and v_2 = velocity through the nozzle. Since velocity varies with the square of diameter, $v_2 = 4v$.

$$p_1/\gamma + v_1^2/2g + z_1 = p_2/\gamma + v_2^2/2g + z_2 + h_L \tag{1}$$

$$h_f = (f)(L/d)(v^2/2g) = f[(6 + 1.8 + 3.6)/(\tfrac{100}{1000})]\{v^2/[(2)(9.807)]\} = 5.812fv^2$$

$$h_m = (K)(v^2/2g) = (1.0 + 2.2)\{v^2/[(2)(9.807)]\} + (0.1)\{v_2^2/[(2)(9.807)]\}$$

$$= (1.0 + 2.2)\{v^2/[(2)(9.807)]\} + (0.1)\{(4v)^2/[(2)(9.807)]\} = 0.2447v^2$$

$$h_L = h_f + h_m = 5.812fv^2 + 0.2447v^2 \qquad 0 + 0 + (6 - 1.8) = 0 + (4v)^2/[(2)(9.807)]$$

$$+ (6 - 1.8 - 3.6) + 5.812fv^2 + 0.2447v^2 \qquad v = \sqrt{3.6/(5.812f + 1.060)}$$

Try $f = 0.03$: $v = \sqrt{3.6/[(5.812)(0.03) + 1.060]} = 1.708$ m/s, $N_R = \rho dv/\mu = [(0.8)(1000)](\tfrac{100}{1000})(1.708)/0.01 = 1.37 \times 10^4$. From Fig. A-5, $f = 0.028$. Try $f = 0.028$: $v = \sqrt{3.6/[(5.812)(0.028) + 1.060]} = 1.716$ m/s, $N_R = [(0.8)(1000)](\tfrac{100}{1000})(1.716)/0.01 = 1.37 \times 10^4$, $f = 0.028$ (O.K.); $Q = Av = [(\pi)(\tfrac{100}{1000})^2/4](1.716) = 0.0135$ m^3/s, or 13.5 L/s. To find pressure at A, apply Eq. (1) between points A and 2: $p_A/[(0.8)(9.79)] + 1.716^2/[(2)(9.807)] + \tfrac{150}{1000} = 0 + [(4)(1.716)]^2/[(2)(9.807)] + 0 + 0.1[(4)(1.716)]^2/[(2)(9.807)]$, $p_A = 18.3$ kN/m^2. Minimum pressure would occur at point 3. Apply Eq. (1) between points 1 and 3: $h_f = 0.028[6/(\tfrac{100}{1000})]\{1.716^2/[(2)(9.807)]\} = 0.2522$ m, $h_m = (1.0 + 2.2)\{1.716^2/[(2)(9.807)]\} = 0.4804$ m, $h_L = 0.2522 + 0.4804 = 0.7326$ m, $0 + 0 + 0 = p_3/[(0.8)(9.79)] + 1.716^2/[(2)(9.807)] + 1.8 + 0.7326$, $p_3 = -21.0$ kN/m^2.

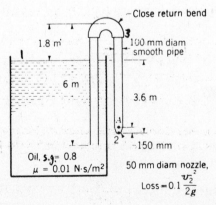

Close return bend
3
1.8 m
100 mm diam smooth pipe
6 m
3.6 m
150 mm
Oil, S.g. = 0.8
$\mu = 0.01$ N·s/m^2
50 mm diam nozzle,
Loss $= 0.1 \dfrac{v_2^2}{2g}$

Fig. 9-53

9.290 Find the velocity of water through the pipe in Fig. 9-54. What is the maximum height of point A for no cavitation? Barometer reading is 29.5 in of mercury. Use minor loss coefficients of 1.0 for both entrance and exit and 4.5 for the globe valve.

∎
$$p_1/\gamma + v_1^2/2g + z_1 = p_2/\gamma + v_2^2/2g + z_2 + h_L \tag{1}$$

$$h_f = (f)(L/d)(v^2/2g) = f[(100 + 75 + 100)/(\tfrac{8}{12})]\{v^2/[(2)(32.2)]\} = 6.405fv^2$$

$$h_m = (K)(v^2/2g) = (1.0 + 4.5 + 1.0)\{v^2/[(2)(32.2)]\} = 0.1009v^2$$

$$h_L = h_f + h_m = 6.405fv^2 + 0.1009v^2 \qquad 0 + 0 + 12 = 0 + 0 + 0 + 6.405fv^2 + 0.1009v^2$$

$$v = \sqrt{12/(6.405f + 0.1009)}$$

Try $f = 0.013$: $v = \sqrt{12/[(6.405)(0.013) + 0.1009]} = 8.072$ ft/s, $N_R = dv/v = (\tfrac{8}{12})(8.072)/(1.21 \times 10^{-5}) = 4.45 \times 10^5$. From Fig. A-5, $f = 0.0135$. Try $f = 0.0135$: $v = \sqrt{12/[(6.405)(0.0135) + 0.1009]} = 8.003$ ft/s, $N_R = (\tfrac{8}{12})(8.003)/(1.21 \times 10^{-5}) = 4.41 \times 10^5$, $f = 0.0135$ (O.K.). Now, to find the maximum height of point A, apply Eq. (1) between points 1 and A: $h_f = 0.0135[100/(\tfrac{8}{12})]\{8.003^2/[(2)(32.2)]\} = 2.014$ ft, $h_m = 0$, $h_L = 2.014 + 0 = 2.014$ ft. From Table A-1, $p_v = 36.5$ lb/ft^2. $p_{\text{atm}} = [(13.6)(62.4)](29.5/12) = 2086.2$ lb/ft^2. At cavitation, $p = (2086.2 - 36.5)/62.4 = 32.85$ ft (vacuum), or -32.85 ft, $0 + 0 + 0 = -32.85 + 0 + z_A + 2.014$, $z_A = 30.8$ ft. (z_A is the height of point A above the water surface in the left reservoir.)

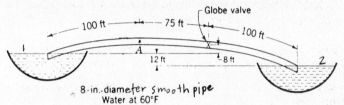

Fig. 9-54

9.291 What diameter smooth pipe is required to convey 8 L/s of kerosene ($v = 1.93 \times 10^{-6}$ m^2/s) at 32 °C, 150 m with a head of 5 m? There are a valve and other minor losses with total K of 7.6.

∎
$$p_1/\gamma + v_1^2/2g + z_1 = p_2/\gamma + v_2^2/2g + z_2 + h_L \qquad v = Q/A = (\tfrac{8}{1000})/(\pi D^2/4) = 0.01019/D^2$$

$$h_f = (f)(L/d)(v^2/2g) = (f)(150/D)\{(0.01019/D^2)^2/[(2)(9.807)]\} = 0.0007941f/D^5$$

$$h_m = (K)(v^2/2g) = 7.6\{(0.01019/D^2)^2/[(2)(9.807)]\} = 0.00004023/D^4$$

$$h_L = h_f + h_m = 0.0007941f/D^5 + 0.00004023/D^4 \qquad p_1/\gamma + z_1 - p_2/\gamma - z_2 = 5$$

$$v_1^2/2g = v_2^2/2g \qquad 5 = 0.0007941f/D^5 + 0.00004023/D^4$$

Try $f = 0.02$: $5 = (0.0007941)(0.02)/D^5 + 0.00004023/D^4$. By trial and error, $D = 0.0826$ m. $N_R = Dv/v = (0.0826)(0.01019/0.0826^2)/(1.93 \times 10^{-6}) = 6.39 \times 10^4$. From Fig. A-5, $f = 0.0195$. Try $f = 0.0195$: $5 = (0.0007941)(0.0195)/D^5 + 0.00004023/D^4$, $D = 0.0822$ m, $N_R = (0.0822)(0.01019/0.0822^2)/(1.93 \times 10^{-6}) = 6.42 \times 10^4$, $f = 0.0195$. Therefore, $D = 0.0822$ m, or 82.2 mm.

9.292 Find the value of the Hazen–Williams coefficient for the water flow in Prob. 9.130.

∎
$$v = 1.318CR^{0.63}s^{0.54} = Q/A = 3.0/[(\pi)(\tfrac{24}{12})^2/4] = 0.9549 \text{ ft/s}$$

$$0.9549 = (1.318)(C)[(\tfrac{24}{12})/4]^{0.63}(0.0003/1)^{0.54} \qquad C = 89.5$$

9.293 Find the value of the Hazen–Williams coefficient for the case where water flows at 0.18 m^3/s in a 60-cm-diameter pipeline with a head loss of 0.0012 m/m.

∎
$$v = 0.8492CR^{0.63}s^{0.54} = Q/A = 0.18/[(\pi)(\tfrac{60}{100})^2/4] = 0.6366 \text{ m/s}$$

$$0.6366 = (0.8492)(C)[(\tfrac{60}{100})/4]^{0.63}(0.0012)^{0.54} \qquad C = 93.6$$

9.294 In a field test of the 16-ft-diameter Colorado River aqueduct, Manning's n was found to have a value of 0.0132 when water at 50 °F was flowing at a Reynolds number of 10.5×10^6. Determine the average value of ϵ for this conduit.

∎
$$v = \left(\frac{1.486}{n}\right)(R^{2/3})(s^{1/2}) \qquad s = \left(\frac{nv}{1.486R^{2/3}}\right)^2 \qquad h_L = (f)\left(\frac{L}{d}\right)\left(\frac{v^2}{2g}\right) \qquad \frac{h_L}{L} = \left(\frac{f}{d}\right)\left(\frac{v^2}{2g}\right)$$

Equating s and h_L/L gives

$$\left(\frac{nv}{1.486R^{2/3}}\right)^2 = \left(\frac{f}{d}\right)\left(\frac{v^2}{2g}\right) \qquad \left[\frac{0.0132v}{(1.486)(\frac{16}{4})^{2/3}}\right]^2 = \left(\frac{f}{16}\right)\left\{\frac{v^2}{(2)(32.2)}\right\} \qquad f = 0.0128$$

From Fig. A-5 with $f = 0.0128$ and $N_R = 10.5 \times 10^6$, $\epsilon/d = 0.00015$; $\epsilon = (16)(0.00015) = 0.00240$ ft, or 0.0288 in.

9.295 Prepare a computer program that will determine for a single pipe with constant diameter either the flow rate or the required conduit diameter for closed conduit, incompressible flow. The program must be usable for problems in both the English system of units and the International system of units.

```
C     THIS PROGRAM DETERMINES THE FLOW RATE OR THE REQUIRED CONDUIT
C     DIAMETER FOR CLOSED CONDUIT, INCOMPRESSIBLE FLOW.  IT CAN BE USED
C     FOR PROBLEMS IN BOTH THE ENGLISH SYSTEM OF UNITS AND THE INTER-
C     NATIONAL SYSTEM OF UNITS.  THE APPLICATION OF THE PROGRAM IS
C     LIMITED TO CASES INVOLVING A SINGLE PIPE WITH CONSTANT DIAMETER.
C
C     THIS PROGRAM IS BASED ON THE SOLUTION OF THE BERNOULLI EQUATION;
C     ACCORDINGLY, CERTAIN DATA MUST BE ENTERED FOR EACH OF TWO POINTS
C     IN A PIPE SYSTEM.  HOWEVER, BOTH THE VELOCITY AT POINT 1 AND THE
C     VELOCITY AT POINT 2 ARE NOT CONSIDERED AS KNOWN VALUES INITIALLY.
C     IF EITHER VALUE OF VELOCITY IS ESSENTIALLY ZERO (SUCH AS AT A
C     RESERVOIR OR TANK SURFACE), ENTER 0 (ZERO) FOR THAT VELOCITY (OR
C     FOR BOTH VELOCITIES IF BOTH ARE ESSENTIALLY ZERO).  OTHERWISE,
C     ENTER 1.0 FOR ALL OTHER VELOCITIES.
C
C     INPUT DATA MUST BE SET UP AS FOLLOWS.
C
C     CARD 1    COLUMN 1       ENTER 0 (ZERO) OR BLANK IF ENGLISH SYSTEM
C                              OF UNITS IS TO BE USED.  ENTER 1 (ONE)
C                              IF INTERNATIONAL SYSTEM OF UNITS IS TO
C                              BE USED.
C               COLUMNS 2-79   ENTER TITLE, DATE, AND OTHER INFORMATION,
C                              IF DESIRED.
C     CARD 2    COLUMNS 1-10   ENTER NUMBER INCLUDING DECIMAL GIVING
C                              GAUGE PRESSURE AT POINT 1  (IN POUNDS
C                              PER SQUARE INCH OR KILOPASCALS).
C               COLUMNS 11-20  ENTER NUMBER INCLUDING DECIMAL GIVING
C                              GAUGE PRESSURE AT POINT 2  (IN POUNDS
C                              PER SQUARE INCH OR KILOPASCALS).
C               COLUMNS 21-30  ENTER 0 (ZERO) OR BLANK IF VELOCITY AT
C                              POINT 1 IS ESSENTIALLY ZERO; OTHERWISE,
C                              ENTER 1.0.
C               COLUMNS 31-40  ENTER 0 (ZERO) OR BLANK IF VELOCITY AT
C                              POINT 2 IS ESSENTIALLY ZERO; OTHERWISE,
C                              ENTER 1.0.
C               COLUMNS 41-50  ENTER NUMBER INCLUDING DECIMAL GIVING
C                              ELEVATION AT POINT 1 (IN FEET OR METERS).
C               COLUMNS 51-60  ENTER NUMBER INCLUDING DECIMAL GIVING
C                              ELEVATION AT POINT 2 (IN FEET OR METERS).
C               COLUMNS 61-70  ENTER NUMBER INCLUDING DECIMAL GIVING
C                              ACTUAL ENERGY ADDED (IN HORSEPOWER OR
C                              KILOWATTS).
C               COLUMNS 71-80  ENTER NUMBER INCLUDING DECIMAL GIVING
C                              ACTUAL ENERGY REMOVED (IN HORSEPOWER OR
C                              KILOWATTS).
C     CARD 3    COLUMNS 1-10   ENTER NUMBER INCLUDING DECIMAL GIVING
C                              TOTAL "MINOR LOSSES" (IN FEET OR METERS).
C               COLUMNS 11-20  ENTER NUMBER INCLUDING DECIMAL GIVING
C                              DIAMETER OF CONDUIT (IN INCHES OR
C                              MILLIMETERS).
C               COLUMNS 21-30  ENTER NUMBER INCLUDING DECIMAL GIVING
C                              LENGTH OF CONDUIT (IN FEET OR METERS).
C               COLUMNS 31-40  ENTER NUMBER INCLUDING DECIMAL GIVING
C                              KINEMATIC VISCOSITY OF FLUID (IN SQUARE
C                              FEET PER SECOND OR SQUARE METERS PER
C                              SECOND).
C               COLUMNS 41-50  ENTER NUMBER INCLUDING DECIMAL GIVING
C                              ROUGHNESS OF CONDUIT MATERIAL (IN FEET
C                              OR METERS).  ENTER A VALUE OF 0 (ZERO)
C                              FOR "SMOOTH" CONDUITS.
```

```
C                   COLUMNS 51-60    ENTER NUMBER INCLUDING DECIMAL GIVING
C                                    SPECIFIC (OR UNIT) WEIGHT OF FLUID (IN
C                                    POUNDS PER CUBIC FOOT OR KILO-
C                                    NEWTONS PER CUBIC METER).
C                   COLUMNS 61-70    ENTER NUMBER INCLUDING DECIMAL GIVING
C                                    FLOW RATE OF FLUID (IN CUBIC FEET PER
C                                    SECOND OR CUBIC METERS PER SECOND).
C       CARD 4      COLUMNS 1-24     ENTER TYPE OF FLUID
C                   COLUMNS 25-48    ENTER TYPE OF CONDUIT.
C
C ***********************************************************************
C *                                                                     *
C *   NOTE WELL....EITHER THE FLOW RATE (COLUMNS 61-70) OR THE          *
C *   CONDUIT DIAMETER (COLUMNS 11-20), WHICHEVER ONE IS TO BE DETER-   *
C *   MINED BY THIS PROGRAM, SHOULD BE LEFT BLANK ON CARD 3.            *
C *                                                                     *
C ***********************************************************************
C
C       MULTIPLE DATA SETS FOR SOLVING ANY NUMBER OF PROBLEMS MAY BE
C       INCLUDED FOR PROCESSING.
C
        REAL L
        DIMENSION TITLE(13),FLUID(4),PIPE(4)
        INTEGER UNITS
        COMMON F,ED,RN
        PI=3.14159265
1       READ(5,100,END=2)UNITS,TITLE
        FACTOR=12.0
        IF(UNITS.EQ.1)FACTOR=1000.0
100     FORMAT(I1,13A6)
        WRITE(6,101)TITLE
101     FORMAT('1',13A6,/////)
        READ(5,102)P1,P2,V1,V2,Z1,Z2,HA,HR,HM,D,L,VIS,E,SW,Q,FLUID,PIPE
102     FORMAT(8F10.0/7F10.0/8A6)
        P1SW=P1/SW*FACTOR**2
        IF(UNITS.EQ.1)P1SW=P1/SW
        G=32.2
        IF(UNITS.EQ.1)G=9.807
        P2SW=P2/SW*FACTOR**2
        IF(UNITS.EQ.1)P2SW=P2/SW
        FF=0.02
        IF(Q.GT.0.0001)GO TO 117
        IF(V1.GT.0.0001)V1=1.0/2.0/G
        IF(V2.GT.0.0001)V2=1.0/2.0/G
105     HF=FF*L/D*FACTOR/2.0/G
        HAT=550.*HA/SW
        IF(UNITS.EQ.1)HAT=HA/SW
        HRT=550.*HR/SW
        IF(UNITS.EQ.1)HRT=HR/SW
        Q=0.001
        VTRY=(Q/(PI*(D/FACTOR)**2/4.0))**2
        TRY1=P1SW+VTRY*V1+Z1+HAT/Q-HRT/Q-(P2SW+VTRY*V2+Z2+HF*VTRY)
116     Q=Q+0.001
        VTRY=(Q/(PI*(D/FACTOR)**2/4.0))**2
        TRY2=P1SW+VTRY*V1+Z1+HAT/Q-HRT/Q-(P2SW+VTRY*V2+Z2+HF*VTRY)
        IF(TRY1*TRY2)114,114,115
115     TRY1=TRY2
        GO TO 116
114     Q=Q-0.0005
        V=Q/(PI*(D/FACTOR)**2/4.0)
        RN=D/FACTOR*V/VIS
        ED=E/D*FACTOR
        CALL ROUGH
        DIFF=ABS(F-FF)
        IF(DIFF.LT.0.0001)GO TO 104
        FF=F
        GO TO 105
104     IF(V1.GT.0.0001)V1=V
        IF(V2.GT.0.0001)V2=V
        IF(UNITS.EQ.0)WRITE(6,106)P1,P2,Z1,Z2,HA,HR,HM,D,L,FLUID,PIPE,Q,
       *V1,V2
106     FORMAT(1X,'GIVEN DATA FOR A CIRCULAR CLOSED CONDUIT CARRYING INCOM
       *PRESSIBLE FLOW',//5X,'PRESSURE AT POINT 1 =',F7.1,' PSI',//5X,'PRE
```

```
      *SSURE AT POINT 2 =',F7.1,' PSI',//5X,'ELEVATION AT POINT 1 =',
      *F7.1,' FT',//5X,'ELEVATION AT POINT 2 =',F7.1,' FT',//5X,'ACTUAL E
      *NERGY ADDED BETWEEN POINTS 1 AND 2 =',F5.1,' HP',//5X,'ACTUAL ENER
      *GY REMOVED BETWEEN POINTS 1 AND 2 =',F5.1,' HP',//5X,'MINOR HEAD L
      *OSSES BETWEEN POINTS 1 AND 2 =',F5.1,' FT',//5X,'DIAMETER OF CONDU
      *IT =',F6.2,' IN',//5X,'LENGTH OF CONDUIT =',F8.1,' FT',//5X,'FLUID
      * FLOWING IS ',4A6,//5X,'CONDUIT MATERIAL IS ',4A6,////1X,'THE FLOW
      * RATE WILL BE',F7.1,' CU FT/S',//5X,'VELOCITY AT POINT 1 =',F6.2,
      *' FT/S',//5X,'VELOCITY AT POINT 2 =',F6.2,' FT/S')
       IF(UNITS.EQ.1)WRITE(6,107)P1,P2,Z1,Z2,HA,HR,HM,D,L,FLUID,PIPE,Q,
      *V1,V2
  107 FORMAT(1X,'GIVEN DATA FOR A CIRCULAR CLOSED CONDUIT CARRYING INCOM
      *PRESSIBLE FLOW',//5X,'PRESSURE AT POINT 1 =',F7.1,' KPA',//5X,'PRE
      *SSURE AT POINT 2 =',F7.1,' KPA',//5X,'ELEVATION AT POINT 1 =',
      *F7.1,' M ',//5X,'ELEVATION AT POINT 2 =',F7.1,' M ',//5X,'ACTUAL E
      *NERGY ADDED BETWEEN POINTS 1 AND 2 =',F5.1,' KW',//5X,'ACTUAL ENER
      *GY REMOVED BETWEEN POINTS 1 AND 2 =',F5.1,' KW',//5X,'MINOR HEAD L
      *OSSES BETWEEN POINTS 1 AND 2 =',F5.1,' M ',//5X,'DIAMETER OF CONDU
      *IT =',F7.1,' MM',//5X,'LENGTH OF CONDUIT =',F8.1,' M ',//5X,'FLUID
      * FLOWING IS ',4A6,//5X,'CONDUIT MATERIAL IS ',4A6,////1X,'THE FLOW
      * RATE WILL BE',F7.3,' CU M/S',//5X,'VELOCITY AT POINT 1 =',F6.2,
      *' M/S ',//5X,'VELOCITY AT POINT 2 =',F6.2,' M/S ')
       GO TO 1
  117 V1=V1*(Q/PI*4.0)
       V2=V2*(Q/PI*4.0)
  103 HF=FF*L*(Q/PI*4.0)**2/2.0/G
       HAT=550.*HA/SW/Q
       IF(UNITS.EQ.1)HAT=HA/SW/Q
       HRT=550.*HR/SW/Q
       IF(UNITS.EQ.1)HRT=HR/SW/Q
       D=0.001
       TRY1=HF/D**5+(V2**2/2.0/G)/D**4-(V1**2/2.0/G)/D**4-P1SW-Z1-HAT
      *+HRT*P2SW+Z2+HM
  110 D=D+0.001
       TRY2=HF/D**5+(V2**2/2.0/G)/D**4-(V1**2/2.0/G)/D**4-P1SW-Z1-HAT
      *+HRT+P2SW+Z2+HM
       IF(TRY1*TRY2)108,108,109
  109 TRY1=TRY2
       GO TO 110
  108 D=D-0.0005
       RN=D*Q/(PI*D**2/4.0)/VIS
       ED=E/D
       CALL ROUGH
       DIFF=ABS(F-FF)
       IF(DIFF.LT.0.0001)GO TO 111
       FF=F
       GO TO 103
  111 V1=V1/D**2
       V2=V2/D**2
       D=D*FACTOR
       IF(UNITS.EQ.0)WRITE(6,112)P1,P2,Z1,Z2,HA,HR,HM,Q,L,FLUID,PIPE,D,
      *V1,V2
  112 FORMAT(1X,'GIVEN DATA FOR A CIRCULAR CLOSED CONDUIT CARRYING INCOM
      *PRESSIBLE FLOW',//5X,'PRESSURE AT POINT 1 =',F7.1,' PSI',//5X,'PRE
      *SSURE AT POINT 2 =',F7.1,' PSI',//5X,'ELEVATION AT POINT 1 =',
      *F7.1,' FT',//5X,'ELEVATION AT POINT 2 =',F7.1,' FT',//5X,'ACTUAL E
      *NERGY ADDED BETWEEN POINTS 1 AND 2 =',F5.1,' HP',//5X,'ACTUAL ENER
      *GY REMOVED BETWEEN POINTS 1 AND 2 =',F5.1,' HP',//5X,'MINOR LOSSES
      * BETWEEN POINTS 1 AND 2 =',F5.1,' FT',//5X,'FLOW RATE =',F7.1,
      *' CU FT/S',//5X,'LENGTH OF CONDUIT =',F8.1,' FT',//5X,'FLUID FLOWI
      *NG IS ',4A6,//5X,'CONDUIT MATERIAL IS ',4A6,///1X,'THE CONDUIT DIA
      *METER REQUIRED WILL BE',F6.2,' IN',//5X,'VELOCITY AT POINT 1 =',
      *F6.2,' FT/S',//5X,'VELOCITY AT POINT 2 =',F6.2,' FT/S')
       IF(UNITS.EQ.1)WRITE(6,113)P1,P2,Z1,Z2,HA,HR,HM,Q,L,FLUID,PIPE,D,
      *V1,V2
  113 FORMAT(1X,'GIVEN DATA FOR A CIRCULAR CLOSED CONDUIT CARRYING INCOM
      *PRESSIBLE FLOW',//5X,'PRESSURE AT POINT 1 =',F7.1,' KPA',//5X,'PRE
      *SSURE AT POINT 2 =',F7.1,' KPA',//5X,'ELEVATION AT POINT 1 =',
      *F7.1,' M ',//5X,'ELEVATION AT POINT 2 =',F7.1,' M ',//5X,'ACTUAL E
      *NERGY ADDED BETWEEN POINTS 1 AND 2 =',F5.1,' KW',//5X,'ACTUAL ENER
      *GY REMOVED BETWEEN POINTS 1 AND 2 =',F5.1,' KW',//5X,'MINOR LOSSES
      * BETWEEN POINTS 1 AND 2 =',F5.1,' M ',//5X,'FLOW RATE =',F7.3,
      *' CU M/S ',//5X,'LENGTH OF CONDUIT =',F8.1,' M ',//5X,'FLUID FLOWI
```

```
         *NG IS ',4A6,//5X,'CONDUIT MATERIAL IS ',4A6,///1X,'THE CONDUIT DIA
         *METER REQUIRED WILL BE',F7.1,' MM',//5X,'VELOCITY AT POINT 1 =',
         *F6.2,' M/S ',//5X,'VELOCITY AT POINT 2 =',F6.2,' M/S ')
          GO TO 1
     2    STOP
          END
          SUBROUTINE ROUGH
          COMMON F,ED,RN
          IF(RN.LE.2000.0)F=64.0/RN
          IF(RN.LE.2000.0)RETURN
          IF(RN.LT.4000.0)WRITE(6,103)
    103   FORMAT(1X,'A REYNOLDS NUMBER IS IN THE CRITICAL ZONE, FOR WHICH TH
         *E FRICTION FACTOR IS UNCERTAIN.  HENCE, PROGRAM EXECUTION WAS TERM
         *INATED.')
          IF(RN.LT.4000.0)STOP
          F=0.006
          TRY1=1.0/SQRT(F)+2.0*ALOG10(ED/3.7+2.51/RN/SQRT(F))
    102   F=F+0.00001
          TRY2=1.0/SQRT(F)+2.0*ALOG10(ED/3.7+2.51/RN/SQRT(F))
          IF(TRY1*TRY2)100,100,101
    101   TRY1=TRY2
          GO TO 102
    100   F=F-0.000005
          RETURN
          END
```

9.296 For the data given in Prob. 9.65, find the flow rate at which water is being discharged from the pipe, utilizing the computer program of Prob. 9.295.

INPUT

```
1 2 3 4 5 6 7 8 9 10 11 12 13 14 15 16 17 18 19 20 21 22 23 24 25 26 27 28 29 30 31 32 33 34 35 36 37 38 39 40 41 42 43 44 45 46 47 48 49 50 51 52 53 54 55 56 57 58 59 60 61 62 63 64 65 66 67 68 69 70 71 72 73 74 75 76 77 78 79 80
0SAMPLE ANALYSIS ØF INCØMPRESSIBLE FLØW
                              1.0       150.5     98.4
           24.0      130.0    0.0000105 0.00085   62.4
WATER                    NEW CAST IRØN
```

OUTPUT

```
SAMPLE ANALYSIS OF INCOMPRESSIBLE FLOW


GIVEN DATA FOR A CIRCULAR CLOSED CONDUIT CARRYING INCOMPRESSIBLE FLOW

     PRESSURE AT POINT 1 =    0.0 PSI

     PRESSURE AT POINT 2 =    0.0 PSI

     ELEVATION AT POINT 1 =  150.5 FT

     ELEVATION AT POINT 2 =   98.4 FT

     ACTUAL ENERGY ADDED BETWEEN POINTS 1 AND 2 =  0.0 HP

     ACTUAL ENERGY REMOVED BETWEEN POINTS 1 AND 2 =  0.0 HP

     MINOR HEAD LOSSES BETWEEN POINTS 1 AND 2 =  0.0 FT

     DIAMETER OF CONDUIT = 24.00 IN

     LENGTH OF CONDUIT =   130.0 FT

     FLUID FLOWING IS WATER

     CONDUIT MATERIAL IS NEW CAST IRON


   THE FLOW RATE WILL BE  127.0 CU FT/S

     VELOCITY AT POINT 1 =  0.00 FT/S

     VELOCITY AT POINT 2 = 40.44 FT/S
```

9.297   For the data given in Prob. 9.66, find the pipe diameter, utilizing the computer program of Prob. 9.295.

INPUT

```
1 2 3 4 5 6 7 8 9 10 11 12 13 14 15 16 17 18 19 20 21 22 23 24 25 26 27 28 29 30 31 32 33 34 35 36 37 38 39 40 41 42 43 44 45 46 47 48 49 50 51 52 53 54 55 56 57 58 59 60 61 62 63 64 65 66 67 68 69 70 71 72 73 74 75 76 77 78 79 80
1SAMPLE ANALYSIS ØF INCØMPRESSIBLE FLØW
2.5                1.0        1.0        82.65      66.66
                   965.5      .000000406.00050      7.05           .10
GASØLINE                      (RØUGHNESS = 0.500 MM)
```

OUTPUT

```
SAMPLE ANALYSIS OF INCOMPRESSIBLE FLOW


GIVEN DATA FOR A CIRCULAR CLOSED CONDUIT CARRYING INCOMPRESSIBLE FLOW

     PRESSURE AT POINT 1 =    2.5 KPA

     PRESSURE AT POINT 2 =    0.0 KPA

     ELEVATION AT POINT 1 =   82.7 M

     ELEVATION AT POINT 2 =   66.7 M

     ACTUAL ENERGY ADDED BETWEEN POINTS 1 AND 2 =  0.0 KW

     ACTUAL ENERGY REMOVED BETWEEN POINTS 1 AND 2 =  0.0 KW

     MINOR LOSSES BETWEEN POINTS 1 AND 2 =  0.0 M

     FLOW RATE =  0.100 CU M/S

     LENGTH OF CONDUIT =   965.5 M

     FLUID FLOWING IS GASOLINE

     CONDUIT MATERIAL IS (ROUGHNESS = 0.500 MM)



THE CONDUIT DIAMETER REQUIRED WILL BE  257.5 MM

     VELOCITY AT POINT 1 =  1.92 M/S

     VELOCITY AT POINT 2 =  1.92 M/S
```

CHAPTER 10
Series Pipeline Systems

10.1 For a 12-in-diameter concrete pipe 12 000 ft long, find the diameter of a 1000-ft-long equivalent pipe.

    ▌ Assume a flow rate of 3.0 cfs. (The result should be the same regardless of the flow rate assumed.) From Fig. A-13, with $D = 12$ in and $Q = 3.0 \text{ ft}^3/\text{s}$, $h_1 = 0.0052$ ft/ft. Therefore, $h_f = (0.0052)(12\,000) = 62.40$ ft. For a 1000-ft-long equivalent pipe with the same head loss, $h_1 = 62.40/1000 = 0.06240$ ft/ft. From Fig. A-13, with $h_1 = 0.06240$ ft/ft and $Q = 3.0 \text{ ft}^3/\text{s}$, $D = 7.3$ in.

10.2 A 480-ft-long, 18-in-diameter concrete pipe and a 590-ft-long, 12-in-diameter concrete pipe are connected in series. Find the length of an equivalent pipe of 10 in diameter.

    ▌ Assume a flow rate of 5 cfs through the two given pipes. For the 18-in-diameter pipe, from Fig. A-13, $h_1 = 0.00180$ ft/ft. For the 12-in-diameter pipe, $h_1 = 0.0137$ ft/ft. The total head loss for both pipes is $h_f = (0.00180)(480) + (0.0137)(590) = 8.947$ ft. For a 10-in-diameter pipe with $Q = 5.0$ cfs, from Fig. A-13, $h_1 = 0.032$ ft/ft. Since the equivalent pipe must have the same head loss as that of the system it replaces (i.e., 8.947 ft), the required length of a 10-in-diameter equivalent pipe can now be determined by $0.032L = 8.947$, $L = 280$ ft. Note that the required length of pipe can be determined in a single computation as follows: $L = (0.00180)(480)/0.032 + (0.0137)(590)/0.032 = 280$ ft.

10.3 A 225-m-long, 300-mm-diameter concrete pipe and a 400-m-long, 500-mm-diameter concrete pipe are connected in series. Find the diameter of a 625-m-long equivalent pipe.

    ▌ Assume a flow rate of 0.1 m³/s. For the 300-mm-diameter pipe, from Fig. A-14, $h_1 = 0.0074$ m/m. For the 500-mm-diameter pipe, $h_1 = 0.00064$ m/m. The total head loss for both pipes is $h_f = (0.0074)(225) + (0.00064)(400) = 1.921$ m. For a 625-m-long equivalent pipe with this head loss, $h_1 = 1.921/625 = 0.00307$ m/m. From Fig. A-14, $D = 360$ mm.

10.4 Water flows at a rate of 0.020 m³/s from reservoir A to reservoir B through three concrete pipes connected in series, as shown in Fig. 10-1. Find the difference in water-surface elevations in the reservoirs. Neglect minor losses.

    ▌ $p_A/\gamma + v_A^2/2g + z_A = p_B/\gamma + v_B^2/2g + z_B + h_L$ $0 + 0 + z_A = 0 + 0 + z_B + h_L$ $h_L = h_f = z_A - z_B$

With $Q = 0.020$ m³/s and $D = 160$ mm, from Fig. A-14, $h_1 = 0.0082$ m/m. With $Q = 0.020$ m³/s and $D = 200$ mm, $h_1 = 0.0028$ m/m. With $Q = 0.020$ m³/s and $D = 180$ mm, $h_1 = 0.0046$ m/m. $H = h_f = (0.0082)(1000) + (0.0028)(1600) + (0.0046)(850) = 16.59$ m.

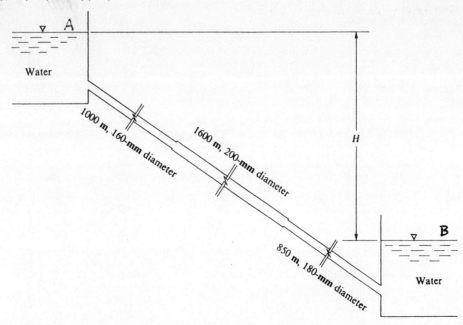

Fig. 10-1

10.5 Compute the flow rate of water through the three concrete pipes connected in series as shown in Fig. 10-2 by the equivalent length method.

▌ $h_f = z_A - z_B = 20$ ft/ft. Assume $Q = 1$ cfs. From Fig. A-13, $(h_1)_{12\text{-in}} = 0.00067$ ft/ft; $(h_1)_{10\text{-in}} = 0.0016$ ft/ft; $(h_1)_{8\text{-in}} = 0.00475$ ft/ft.

Consider a 10-in-diameter equivalent pipe: $L_e = 120 + (0.00475)(150)/0.0016 + (0.00067)(100)/0.0016 = 607$ ft. With $D = 10$ in and $h_1 = \frac{20}{607} = 0.0329$ ft/ft, from Fig. A-13, $Q = 5.1$ cfs.

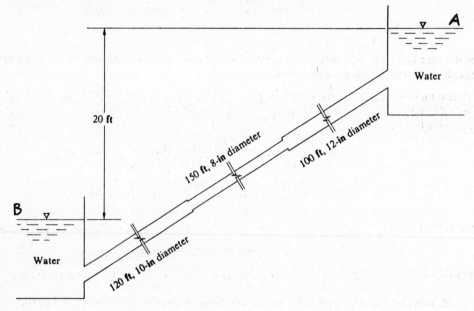

Fig. 10-2

10.6 Solve Prob. 10.5 by the equivalent diameter method.

▌ Using values obtained from Prob. 10.5, $h_f = (0.00067)(100) + (0.00475)(150) + (0.0016)(120) = 0.972$ ft. With $Q = 1$ cfs and $h_1 = 0.972/(100 + 150 + 120) = 0.00263$ ft/ft, $D = 8.9$ in. With $D = 8.9$ in and $h_1 = \frac{20}{370} = 0.0541$ ft/ft, $Q = 5.1$ cfs.

10.7 Two concrete pipes are connected in series. The flow rate of water through the pipes is 0.14 m³/s with a total friction loss of 14.10 m for both pipes. Each pipe has a length of 300 m. If one pipe has a diameter of 300 mm, what is the diameter of the other one? Neglect minor losses.

▌ **For first pipe:** With $Q = 0.14$ m³/s and $D = 300$ mm, $h_1 = 0.014$ m/m. $h_f = (0.014)(300) = 4.20$ m.
For second pipe: $h_f = 14.10 - 4.20 = 9.90$ m, $h_1 = 9.90/300 = 0.033$ m/m. With $Q = 0.14$ m³/s and $h_1 = 0.033$ m/m, $D = 250$ mm.

10.8 Three concrete pipes are connected in series, as shown in Fig. 10-3. Determine the length of an 8-in-diameter equivalent pipe.

▌ Assume $Q = 1$ cfs. $(h_1)_{8\text{-in}} = 0.00475$ ft/ft; $(h_1)_{6\text{-in}} = 0.0195$ ft/ft; $(h_1)_{10\text{-in}} = 0.0016$ ft/ft; $L_e = 1200 + (0.0195)(1000)/0.00475 + (0.0016)(2000)/0.00475 = 5979$ ft.

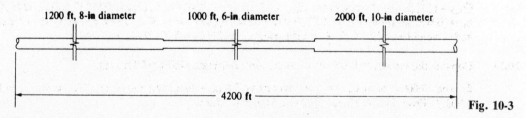

1200 ft, 8-in diameter   1000 ft, 6-in diameter   2000 ft, 10-in diameter

4200 ft

Fig. 10-3

10.9 For the three pipes in Fig. 10-3, determine the diameter of a 4200-ft-long equivalent pipe.

▌ Using values obtained from Prob. 10.8, for a 4200-ft-long equivalent pipe, $h_f = (0.00475)(1200) + (0.0195)(1000) + (0.0016)(2000) = 28.4$ ft, $h_1 = 28.4/4200 = 0.00676$ ft/ft. With $Q = 1$ cfs and $h_1 = 0.00676$ ft/ft, $D = 7.5$ in.

10.10 For three pipes in series, the total pressure drop is $p_A - p_B = 150\,000$ Pa, and the elevation drop is $z_A - z_B = 5$ m. The pipe data are

pipe	L, m	d, cm	ϵ, mm	ϵ/d
1	100	8	0.24	0.003
2	150	6	0.12	0.002
3	80	4	0.20	0.005

The fluid is water, $\rho = 1000$ kg/m^3, and $\nu = 1.02 \times 10^{-6}$ m^2/s. Calculate the flow rate in cubic meters per hour through the system. Neglect minor losses.

I The total head loss in the system is $h_L = (p_A - p_B)/\rho g + (z_A - z_B) = 150\,000/[(1000)(9.807)] + 5 = 20.30$ m $= h_f = (f)(L/d)(v^2/2g)$. From the continuity relation, $v_2 = (d_1/d_2)^2(v_1) = (\frac{8}{6})^2(v_1) = 1.778v_1$, $v_3 = (d_1/d_3)^2(v_1) = (\frac{8}{4})^2(v_1) = 4.000v_1$, and

$$(N_R)_2 = \left(\frac{v_2 d_2}{v_1 d_1}\right)(N_R)_1 = \left[\frac{(1.778v_1)(6)}{(v_1)(8)}\right](N_R)_1 = (1.333)(N_R)_1$$

$$(N_R)_3 = \left(\frac{v_3 d_3}{v_1 d_1}\right)(N_R)_1 = \left[\frac{(4.000v_1)(4)}{(v_1)(8)}\right](N_R)_1 = (2.000)(N_R)_1$$

$$20.30 = f_1[100/(\tfrac{8}{100})]\{v_1^2/[(2)(9.807)]\} + f_2[150/(\tfrac{6}{100})]\{(1.778v_1)^2/[(2)(9.807)]\} + f_3[80/(\tfrac{4}{100})]\{(4.000v_1)^2/[(2)(9.807)]\}$$

$$20.30 = (63.73f_1 + 402.9f_2 + 1631f_3)(v_1)^2$$

From Fig. A-5 from the fully rough regime, estimate $f_1 = 0.0262$, $f_2 = 0.0234$, and $f_3 = 0.0304$.

$$20.35 = [(63.73)(0.0262) + (402.9)(0.0234) + (1631)(0.0304)](v_1)^2 \qquad v_1 = 0.5791 \text{ m/s} \qquad N_R = dv/\nu$$

$$(N_R)_1 = (\tfrac{8}{100})(0.5791)/(1.02 \times 10^{-6}) = 4.54 \times 10^4$$

$$(N_R)_2 = (1.333)(4.54 \times 10^{-4}) = 6.05 \times 10^4 \qquad (N_R)_3 = (2.000)(4.54 \times 10^{-4}) = 9.08 \times 10^4$$

From Fig. A-5, $f_1 = 0.0288$, $f_2 = 0.0260$, and $f_3 = 0.0314$. $20.35 = [(63.73)(0.0288) + (402.9)(0.0260) + (1631)(0.0314)](v_1)^2$, $v_1 = 0.5660$ m/s; $Q = Av = [(\pi)(\tfrac{8}{100})^2/4](0.5660) = 0.002845$ m^3/s, or 10.2 m^3/h. An additional iteration (not shown) gives essentially the same result.

10.11 For a head loss of 5.0 ft/1000 ft, and using $C = 120$ for all pipes, how many 8-in pipes are equivalent to a 16-in pipe? To a 24-in pipe?

I From Fig. A-13, for $h_1 = 5.0/1000$, or 0.005, and $d = 8$ in, $Q = 1.0$ cfs. For $d = 16$ in, $Q = 6.6$ cfs. For $d = 24$ in, $Q = 17$ cfs. Thus it would take 6.6/1.0, or 6.6 eight-in pipes to be hydraulically equivalent to a 16-in pipe of the same relative roughness. Likewise, 17/1.0, or 17 eight-in pipes are equivalent to a 24-in pipe for a head loss of 5.0 ft/1000 ft, or for any other head loss condition.

10.12 A series piping system consists of 6000 ft of 20-in new cast iron pipe, 4000 ft of 16-in, and 2000 ft of 12-in pipe. Convert the system to an equivalent length of 16-in pipe.

I Assume a value of Q of 3.8 cfs. For new cast iron, $C = 130$. In order to use Fig. A-13, change Q_{130} to Q_{120}. $Q_{120} = (\tfrac{120}{130})(3.8) = 3.5$ cfs. From Fig. A-13, $(h_1)_1 = 0.00064$ ft/ft, $(h_1)_2 = 0.00187$ ft/ft, and $(h_1)_3 = 0.0070$ ft/ft. $h_L = (0.00064)(6000) + (0.00187)(4000) + (0.0070)(2000) = 25.32$ ft. The equivalent 16-in pipe must carry 3.8 cfs with a head loss of 25.32 ft ($C = 130$). Hence, $25.32/L_e = 0.00187$, $L_e = 13\,540$ ft.

10.13 Convert the system of Prob. 10.12 to an equivalent size pipe 12 000 ft long.

I The 12 000 ft of pipe, $C = 130$, must carry 3.8 cfs with a head loss of 25.32 ft. Hence, $h_1 = 25.32/12\,000 = 0.00211$. From Fig. A-13, using $Q = 3.5$ cfs, $d = 15.5$ in.

10.14 Suppose in Fig. 10-4 pipes 1, 2, and 3 are 300 m of 30-cm-diameter, 150 m of 20-cm-diameter, and 250 m of 25-cm-diameter, respectively, of new cast iron and are conveying water at 15 °C. If $h = 10$ m, find the rate of flow from A to B by the equivalent velocity method. Neglect minor losses.

I $(\epsilon/d)_1 = 0.00026/(\tfrac{30}{100}) = 0.000867 \qquad (\epsilon/d)_2 = 0.00026/(\tfrac{20}{100}) = 0.00130 \qquad (\epsilon/d)_3 = 0.00026/(\tfrac{25}{100}) = 0.00104$

From Fig. A-5, assume $f_1 = 0.019$, $f_2 = 0.021$, and $f_3 = 0.020$. $h_f = (f)(L/d)(v^2/2g) = 10 =$
$0.019[300/(\frac{30}{100})]\{v_1^2/[(2)(9.807)]\} + 0.021[150/(\frac{20}{100})]\{v_2^2/[(2)(9.807)]\} + 0.020[250/(\frac{25}{100})]\{v_3^2/[(2)(9.807)]\}$,
$10 = 0.9687v_1^2 + 0.8030v_2^2 + 1.020v_3^2$. From the continuity relation, $v_2 = (d_1/d_2)^2(v_1) = (\frac{30}{20})^2(v_1) = 2.250v_1$,
$v_3 = (d_1/d_3)^2(v_1) = (\frac{30}{25})^2(v_1) = 1.440v_1$, and

$$(N_R)_2 = \left(\frac{v_2 d_2}{v_1 d_1}\right)(N_R)_1 = \left[\frac{(2.250v_1)(20)}{(v_1)(30)}\right](N_R)_1 = (1.500)(N_R)_1$$

$$(N_R)_3 = \left(\frac{v_3 d_3}{v_1 d_1}\right)(N_R)_1 = \left[\frac{(1.440v_1)(25)}{(v_1)(30)}\right](N_R)_1 = (1.200)(N_R)_1$$

$$10 = 0.9687v_1^2 + (0.8030)(2.250v_1)^2 + (1.020)(1.440v_1)^2 \qquad v_1 = 1.183 \text{ m/s} \qquad N_R = dv/v$$

$$(N_R)_1 = (\tfrac{30}{100})(1.183)/(1.16 \times 10^{-6}) = 3.06 \times 10^5 \qquad (N_R)_2 = (1.500)(3.06 \times 10^5) = 4.59 \times 10^5$$

$$(N_R)_3 = (1.200)(3.06 \times 10^5) = 3.67 \times 10^5$$

From Fig. A-5, $f_1 = 0.021$, $f_2 = 0.021$, and $f_3 = 0.020$.

$$h_f = 10 = 0.021[300/(\tfrac{30}{100})]\{v_1^2/[(2)(9.807)]\} + 0.021[150/(\tfrac{20}{100})]\{v_2^2/[(2)(9.807)]\} + 0.020[250/(\tfrac{25}{100})]\{v_3^2/[(2)(9.807)]\}$$

$$10 = 1.071v_1^2 + 0.8030v_2^2 + 1.020v_3^2 = 1.071v_1^2 + (0.8030)(2.250v_1)^2 + (1.020)(1.440v_1)^2$$

$$v_1 = 1.174 \text{ m/s} \qquad Q = Av = [(\pi)(\tfrac{30}{100})^2/4](1.174) = 0.0830 \text{ m}^3/\text{s}$$

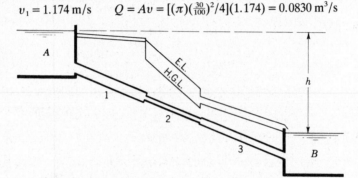

Fig. 10-4

10.15 Solve Prob. 10.14 by the equivalent length method.

▋ Using data from Prob. 10.14 and choosing a 30-cm pipe as the standard,

$$(L_e)_2 = (150)(0.021/0.020)(\tfrac{30}{20})^5 = 1196 \text{ m of 30-cm pipe}$$

$$(L_e)_3 = (250)(0.020/0.020)(\tfrac{30}{25})^5 = 622 \text{ m of 30-cm pipe}$$

$$(L_e)_{\text{total}} = 300 + 1196 + 622 = 2118 \text{ m of 30-cm pipe} \qquad h = 10 = 0.020[2118/(\tfrac{30}{100})]\{v^2/[(2)(9.807)]\}$$

$$v = 1.179 \text{ m/s} \qquad Q = [(\pi)(\tfrac{30}{100})^2/4](1.179) = 0.0833 \text{ m}^3/\text{s}$$

10.16 Suppose that in Fig. 10-4 pipes 1, 2, and 3 are 500 ft of 3.068-in, 200 ft of 2.067-in, and 400 ft of 2.469-in wrought iron pipe. With a total head loss of 20 ft from A to B, find the flow of water at 60 °F.

▋ $$(\epsilon/d)_1 = 0.00015/(3.068/12) = 0.000587 \qquad (\epsilon/d)_2 = 0.00015/(2.067/12) = 0.000871$$

$$(\epsilon/d)_3 = 0.00015/(2.469/12) = 0.000729$$

From Fig. A-5, assume $f_1 = 0.0172$, $f_2 = 0.0190$, and $f_3 = 0.0181$.

$$h_f = (f)(L/d)(v^2/2g) = 20 = 0.0172[500/(3.068/12)]\{v_1^2/[(2)(32.2)]\} + 0.0190[200/(2.067/12)]\{v_2^2/[(2)(32.2)]\}$$
$$+ 0.0181[400/(2.469/12)\{v_3^2/[(2)(32.2)]\}$$

$$20 = 0.5223v_1^2 + 0.3426v_2^2 + 0.5464v_3^2$$

From the continuity relation, $v_2 = (d_1/d_2)^2(v_1) = (3.068/2.067)^2(v_1) = 2.203v_1$, $v_3 = (d_1/d_3)^2(v_1) = (3.068/2.469)^2(v_1) = 1.544v_1$, and

$$(N_R)_2 = \left(\frac{v_2 d_2}{v_1 d_1}\right)(N_R)_1 = \left[\frac{(2.203v_1)(2.067)}{(v_1)(3.068)}\right](N_R)_1 = (1.484)(N_R)_1$$

$$(N_R)_3 = \left(\frac{v_3 d_3}{v_1 d_1}\right)(N_R)_1 = \left[\frac{(1.544v_1)(2.469)}{(v_1)(3.068)}\right](N_R)_1 = (1.243)(N_R)_1$$

$$20 = 0.5223v_1^2 + (0.3426)(2.203v_1)^2 + (0.5464)(1.544v_1)^2 \qquad v_1 = 2.395 \text{ ft/s}$$

$$N_R = dv/v \qquad (N_R)_1 = (3.068/12)(2.395)/(1.21 \times 10^{-5}) = 5.06 \times 10^4$$

$$(N_R)_2 = (1.484)(5.06 \times 10^4) = 7.51 \times 10^4 \qquad (N_R)_3 = (1.243)(5.06 \times 10^4) = 6.29 \times 10^4$$

From Fig. A-5, $f_1 = 0.0228$, $f_2 = 0.0225$, and $f_3 = 0.0225$.

$$h_f = 20 = 0.0228[500/(3.068/12)]\{v_1^2/[(2)(32.2)]\} + 0.0225[200/(2.067/12)]\{v_2^2/[(2)(32.2)]\}$$
$$+ 0.0225[400/(2.469/12)]\{v_3^2/[(2)(32.2)]\}$$
$$20 = 0.6924v_1^2 + 0.4057v_2^2 + 0.6792v_3^2 = 0.6924v_1^2 + (0.4057)(2.203v_1)^2 + (0.6792)(1.544v_1)^2 \qquad v_1 = 2.162 \text{ ft/s}$$
$$Q = Av = [(\pi)(3.068/12)^2/4](2.162) = 0.111 \text{ ft}^3/\text{s}$$

An additional iteration (not shown) produces an insignificant difference.

10.17 Two pipes connected in series are, respectively, 100 ft of 1-in ($\epsilon = 0.000005$ ft) and 500 ft of 6-in ($\epsilon = 0.0008$ ft). With a total head loss of 25 ft, find the flow of water at 60 °F.

▮ $\qquad (\epsilon/d)_1 = 0.000005/(\frac{1}{12}) = 0.0000600 \qquad (\epsilon/d)_2 = 0.0008/(\frac{6}{12}) = 0.00160$

From Fig. A-5, assume $f_1 = 0.0110$ and $f_2 = 0.0215$.

$$h_f = (f)(L/d)(v^2/2g) = 25 = 0.0110[100/(\tfrac{1}{12})]\{v_1^2/[(2)(32.2)]\} + 0.0215[500/(\tfrac{6}{12})]\{v_2^2/[(2)(32.2)]\}$$
$$25 = 0.2050v_1^2 + 0.3339v_2^2$$

From the continuity relation, $v_2 = (d_1/d_2)^2(v_1) = (\frac{1}{6})^2(v_1) = 0.02778v_1$ and

$$(N_R)_2 = \left(\frac{v_2 d_2}{v_1 d_1}\right)(N_R)_1 = \left[\frac{(0.02778v_1)(6)}{(v_1)(1)}\right](N_R)_1 = (0.1667)(N_R)_1$$
$$25 = 0.2050v_1^2 + (0.3339)(0.02778v_1)^2 \qquad v_1 = 11.04 \text{ ft/s} \qquad N_R = dv/v$$
$$(N_R)_1 = (\tfrac{1}{12})(11.04)/(1.21 \times 10^{-5}) = 7.60 \times 10^4 \qquad (N_R)_2 = (0.1667)(7.60 \times 10^4) = 1.27 \times 10^4$$

From Fig. A-5, $f_1 = 0.0195$ and $f_2 = 0.032$.

$$h_f = 25 = 0.0195[100/(\tfrac{1}{12})]\{v_1^2/[(2)(32.2)]\} + 0.032[500/(\tfrac{6}{12})]\{v_2^2/[(2)(32.2)]\}$$
$$25 = 0.3634v_1^2 + 0.4969v_2^2 = 0.3634v_1^2 + (0.4969)(0.02778v_1)^2 \qquad v_1 = 8.290 \text{ ft/s}$$
$$(N_R)_1 = (\tfrac{1}{12})(8.290)/(1.21 \times 10^{-5}) = 5.71 \times 10^4 \qquad (N_R)_2 = (0.1667)(5.71 \times 10^4) = 9.52 \times 10^3$$

From Fig. A-5, $f_1 = 0.0203$ and $f_2 = 0.033$.

$$h_f = 25 = 0.0203[100/(\tfrac{1}{12})]\{v_1^2/[(2)(32.2)]\} + 0.033[500/(\tfrac{6}{12})]\{v_2^2/[(2)(32.2)]\}$$
$$25 = 0.3783v_1^2 + 0.5124v_2^2 = 0.3783v_1^2 + (0.5124)(0.02778v_1)^2 \qquad v_1 = 8.125 \text{ ft/s}$$
$$Q = Av = [(\pi)(\tfrac{1}{12})^2/4](8.125) = 0.0443 \text{ ft}^3/\text{s}$$

An additional iteration (not shown) produces an insignificant difference.

10.18 Repeat Prob. 10.16 for the case where the fluid is an oil with s.g. $= 0.9$ and $\mu = 0.0008$ lb-s/ft^2.

▮ Using values from Prob. 10-16 and assuming $f = 0.03$ for all pipes $h_f = (f)(L/d)(v^2/2g)$.

$$20 = 0.03[500/(3.068/12)]\{v_1^2/[(2)(32.2)]\} + 0.03[200/(2.067/12)]\{v_2^2/[(2)(32.2)]\}$$
$$+ 0.03[400/(2.469/12)]\{v_3^2/[(2)(32.2)]\}$$
$$= 0.9110v_1^2 + 0.5409v_2^2 + 0.9056v_3^2$$
$$= 0.9110v_1^2 + (0.5409)(2.203v_1)^2 + (0.9056)(1.544v_1)^2$$
$$v_1 = 1.874 \text{ ft/s} \qquad N_R = \rho dv/\mu \qquad (N_R)_1 = [(0.9)(1.94)](3.068/12)(1.874)/0.0008 = 1046$$
$$(N_R)_2 = (1.484)(1046) = 1552 \qquad (N_R)_3 = (1.243)(1046) = 1300$$

Therefore, the flow is laminar and $h_L = (32)(\mu/\gamma)(L/d^2)(v)$.

$$20 = 32\{0.0008/[(0.9)(62.4)]\}[500/(3.068/12)^2](v_1) + 32\{0.0008/[(0.9)(62.4)]\}[200/(2.067/12)^2](2.203v_1)$$
$$+ 32\{0.0008/[(0.9)(62.4)]\}[400/(2.469/12)^2](1.544v_1)$$
$$v_1 = 1.183 \text{ ft/s} \qquad Q = Av = [(\pi)(3.068/12)^2/4](1.183) = 0.0607 \text{ ft}^3/\text{s}$$

10.19 A pipeline 240 m long discharges freely at a point 45 m below the water level at intake. The pipe projects into the reservoir. The first 150 m is 30 cm in diameter, and the remaining 90 m is 20 cm in diameter. Find the rate of discharge, assuming $f = 0.04$. If the junction of the two sizes of pipe is 35 m below the intake water-surface level, find the pressure head just above C and just below C, where C denotes the point of junction. Assume a sudden contraction at C. Use $K = 0.8$ for entrance, 0.24 for sudden contraction, and 1.0 for exit.

$$p_1/\gamma + v_1^2/2g + z_1 = p_2/\gamma + v_2^2/2g + z_2 + h_L \tag{1}$$

$$h_f = (f)(L/d)(v^2/2g) = 0.04[150/(\tfrac{30}{100})]\{v_1^2/[(2)(9.807)]\} + 0.04[90/(\tfrac{20}{100})]\{v_2^2/[(2)(9.807)]\} = 1.020v_1^2 + 0.9177v_2^2$$

$$h_m = (K)(v^2/2g) = 0.8\{v_1^2/[(2)(9.807)]\} + 0.24\{v_2^2/[(2)(9.807)]\} = 0.04079v_1^2 + 0.01224v_2^2$$

$$h_L = h_f + h_m = (1.020v_1^2 + 0.9177v_2^2) + (0.04079v_1^2 + 0.01224v_2^2) = 1.061v_1^2 + 0.9299v_2^2$$

$$A_1v_1 = A_2v_2 \qquad [(\pi)(\tfrac{30}{100})^2/4](v_1) = [(\pi)(\tfrac{20}{100})^2/4](v_2) \qquad v_1 = 0.4444v_2$$

$$h_L = (1.061)(0.4444v_2)^2 + 0.9299v_2^2 = 1.139v_2^2 \qquad 0 + 0 + 45 = 0 + v_2^2/[(2)(9.807)] + 0 + 1.139v_2^2$$

$$v_2 = 6.149 \text{ m/s} \qquad v_1 = (0.4444)(6.149) = 2.733 \text{ m/s} \qquad Q = A_2v_2 = [(\pi)(\tfrac{20}{100})^2/4](6.149) = 0.193 \text{ m}^3/\text{s}$$

To find the pressure head just above C, apply Eq. (1) between the water level at intake and a point just above C. $0 + 0 + 35 = p_2/\gamma + 2.733^2/[(2)(9.807)] + 0 + (1.020v_1^2 + 0.04079v_1^2)$, $35 = p_2/\gamma + 0.3808 + (1.020 + 0.04079)(2.733^2)$, $p_2/\gamma = 26.70$ m. To find the pressure head just below C, apply Eq. (1) between the water level at intake and a point just below C. $0 + 0 + 35 = p_2/\gamma + 6.149^2/[(2)(9.807)] + 0 + (1.020v_1^2 + 0.04079v_1^2 + 0.01224v_2^2)$, $35 = p_2/\gamma + 1.928 + [(1.020 + 0.04079)(2.733^2) + (0.01224)(6.149^2)]$, $p_2/\gamma = 24.69$ m.

10.20 Repeat Prob. 10.19 neglecting minor losses.

$$p_1/\gamma + v_1^2/2g + z_1 = p_2/\gamma + v_2^2/2g + z_2 + h_L \tag{1}$$

$$h_L = h_f + h_m \qquad h_m = 0$$

Using data from Prob. 10.19, $h_f = 1.020v_1^2 + 0.9177v_2^2$, $h_L = (1.020)(0.4444v_2)^2 + 0.9177v_2^2 = 1.119v_2^2$, $0 + 0 + 45 = 0 + v_2^2/[(2)(9.807)] + 0 + 1.119v_2^2$, $v_2 = 6.202$ m/s, $v_1 = (0.4444)(6.202) = 2.756$ m/s; $Q = A_2v_2 = [(\pi)(\tfrac{20}{100})^2/4](6.202) = 0.195 \text{ m}^3/\text{s}$. To find the pressure head just above C, apply Eq. (1) between the water level at intake and a point just above C. $0 + 0 + 35 = p_2/\gamma + 2.756^2/[(2)(9.807)] + 0 + 1.020v_1^2$, $35 = p_2/\gamma + 0.3873 + (1.020)(2.756^2)$, $p_2/\gamma = 26.87$ m. To find the pressure head just below C, apply Eq. (1) between the water level at intake and a point just below C. $0 + 0 + 35 = p_2/\gamma + 6.202^2/[(2)(9.807)] + 0 + 1.020v_1^2$, $35 = p_2/\gamma + 1.961 + (1.020)(2.756^2)$, $p_2/\gamma = 25.29$ m.

10.21 Three new cast iron pipes, having diameters of 30 in, 24 in, and 18 in, respectively, each 500 ft long, are connected in series. The 30-in pipe leads from a reservoir (flush entrance), and the 18-in pipe discharges into the air at a point 12 ft below the water surface in the reservoir. Assuming all changes in section to be abrupt, find the rate of discharge of water at 60 °F.

$$p_1/\gamma + v_1^2/2g + z_1 = p_2/\gamma + v_2^2/2g + z_2 + h_L \qquad h_L = h_f + h_m \qquad h_f = (f)(L/d)(v^2/2g)$$

Assume $f = 0.016$ for each pipe. $v_2 =$ velocity for 18-in pipe, velocity for 24-in pipe $= (\tfrac{18}{24})^2(v_2) = 0.5625v_2$, velocity for 30-in pipe $= (\tfrac{18}{30})^2(v_2) = 0.3600v_2$.

$$h_f = 0.016[500/(\tfrac{30}{12})]\{(0.3600v_2)^2/[(2)(32.2)]\} + 0.016[500/(\tfrac{24}{12})]\{(0.5625v_2)^2/[(2)(32.2)]\}$$
$$+ 0.016[500/(\tfrac{18}{12})]\{v_2^2/[(2)(32.2)]\} = 0.1089v_2^2$$

$$h_m = (K)(v^2/2g)$$

For entrance, take $K = 0.5$ (Fig. A-7). For sudden contractions, with $d/D = \tfrac{24}{30}$, or 0.80, $K = 0.15$ and with $d/D = \tfrac{18}{24}$, or 0.75, $K = 0.18$ (Fig. A-9). For exit, $K = 1.0$ (Fig. A-7).

$$h_m = (0.5)(0.3600v_2)^2 + (0.15)(0.5625v_2)^2 + 0.18v_2^2 = 0.2923v_2^2 \qquad h_L = 0.1089v_2^2 + 0.2923v_2^2 = 0.4012v_2^2$$

$$0 + 0 + 12 = 0 + v_2^2/[(2)(32.2)] + 0 + 0.4012v_2^2 \qquad v_2 = 5.366 \text{ ft/s} \qquad N_R = dv/v$$

$$(N_R)_{30\text{-in}} = \tfrac{30}{12}[(0.3600)(5.366)]/(1.21 \times 10^{-5}) = 3.99 \times 10^5 \quad (N_R)_{24\text{-in}} = \tfrac{24}{12}[(0.5625)(5.366)]/(1.21 \times 10^{-5}) = 4.99 \times 10^5$$

$$(N_R)_{18\text{-in}} = (\tfrac{18}{12})(5.366)/(1.21 \times 10^{-5}) = 6.65 \times 10^5 \qquad (\epsilon/d)_{30\text{-in}} = 0.00085/(\tfrac{30}{12}) = 0.000340$$

$$(\epsilon/d)_{24\text{-in}} = 0.00085/(\tfrac{24}{12}) = 0.000425 \qquad (\epsilon/d)_{18\text{-in}} = 0.00085/(\tfrac{18}{12}) = 0.000567$$

From Fig. A-5, $f_{30\text{-in}} = 0.0168$, $f_{24\text{-in}} = 0.0172$, and $f_{18\text{-in}} = 0.0176$.

$$h_f = 0.0168[500/(\tfrac{30}{12})]\{(0.3600v_2)^2/[(2)(32.2)]\} + 0.0172[500/(\tfrac{24}{12})]\{(0.5625v_2)^2/[(2)(32.2)]\}$$
$$+ 0.0176[500/(\tfrac{18}{12})]\{v_2^2/[(2)(32.2)]\} = 0.1190v_2^2$$

$$h_L = 0.1190v_2^2 + 0.2923v_2^2 = 0.4113v_2^2 \qquad 0 + 0 + 12 = 0 + v_2^2/[(2)(32.2)] + 0 + 0.4113v_2^2 \qquad v_2 = 5.302 \text{ ft/s}$$

An additional iteration (not shown) gives no significant change in v_2. $Q = Av = [(\pi)(\tfrac{18}{12})^2/4](5.302) = 9.37 \text{ ft}^3/\text{s}$.

10.22 In Fig. 10-5, $K_e = 0.5$, $L_1 = 300$ m, $D_1 = 600$ mm, $\epsilon_1 = 2$ mm, $L_2 = 240$ m, $D_2 = 1$ m, $\epsilon_2 = 0.3$ mm, $\nu = 3 \times 10^{-6}$ m²/s, and $H = 6$ m. Determine the discharge through the system.

❚ $$p_1/\gamma + v_1^2/2g + z_1 = p_2/\gamma + v_2^2/2g + z_2 + h_L \qquad h_L = h_f + h_m \qquad h_f = (f)(L/d)(v^2/2g)$$

Assume $f_1 = 0.026$ and $f_2 = 0.015$. $v_2 = [(\frac{600}{1000})/1]^2(v_1) = 0.3600v_1$, $h_f = 0.026[300/(\frac{600}{1000})]\{v_1^2/[(2)(9.807)]\} + (0.015)(240/1)\{(0.3600v_1)^2/[(2)(9.807)]\} = 0.6866v_1^2$.
For entrance: $h_m = (K)(v^2/2g) = 0.5\{v_1^2/[(2)(9.807)]\} = 0.02549v_1^2$.
For exit: $h_m = 1.0\{v_2^2/[(2)(9.807)]\} = 0.05098v_2^2 = (0.05098)(0.3600v_1)^2 = 0.006607v_1^2$.
For sudden contraction:

$$h_m = (v_1 - v_2)^2/2g = (v_1 - 0.3600v_1)^2/[(2)(9.807)]$$

$$= 0.02088v_1^2 (h_m)_{\text{total}} = 0.02549v_1^2 + 0.006607v_1^2 + 0.02088v_1^2 = 0.05298v_1^2$$

$$h_L = 0.6866v_1^2 + 0.05298v_1^2 = 0.7396v_1^2 \qquad 0 + 0 + 6 = 0 + 0 + 0 + 0.7396v_1^2 \qquad v_1 = 2.848 \text{ m/s}$$

$$v_2 = (0.3600)(2.848) = 1.025 \text{ m/s}$$

$$N_R = dv/\nu \qquad (N_R)_1 = (\tfrac{600}{1000})(2.848)/(3 \times 10^{-6}) = 5.70 \times 10^5 \qquad (N_R)_2 = (1)(1.025)/(3 \times 10^{-6}) = 3.42 \times 10^5$$

$$(\epsilon/d)_1 = \tfrac{2}{600} = 0.00333 \qquad (\epsilon/d)_2 = (0.3/1000)/1 = 0.000300$$

From Fig. A-5, $f_1 = 0.0265$ and $f_2 = 0.0168$.

$$h_f = 0.0265[300/(\tfrac{600}{1000})]\{v_1^2/[(2)(9.807)]\} + (0.0168)(240/1)\{(0.3600v_1)^2/[(2)(9.807)]\} = 0.7022v_1^2$$

$$h_L = 0.7022v_1^2 + 0.05298v_1^2 = 0.7552v_1^2 \qquad 0 + 0 + 6 = 0 + 0 + 0 + 0.7552v_1^2 \qquad v_1 = 2.819 \text{ m/s}$$

An additional iteration (not shown) gives no significant change in v_1. $Q = Av = [(\pi)(\tfrac{600}{1000})^2/4](2.819) = 0.797$ m³/s.

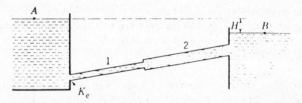

Fig. 10-5

10.23 Solve Prob. 10.22 by means of equivalent pipes.

❚ Expressing the minor losses in terms of equivalent lengths gives for pipe 1: $K_1 = 0.5 + [1 - (\tfrac{600}{1000})^2]^2 = 0.9096$, $(L_e)_1 = K_1 D_1/f_1 = (0.9096)(\tfrac{600}{1000})/0.026 = 20.99$ m; and for pipe 2: $K_2 = 1.0$, $(L_e)_2 = K_2 D_2/f_2 = (1.0)(1)/0.015 = 66.67$ m. The values of f_1 and f_2 are selected for the fully turbulent range as an approximation. The problem is now reduced to $300 + 20.99$, or 320.99 m of 600-mm pipe and $240 + 66.67$, or 306.67 m of 1-m pipe. Expressing the 1-m pipe in terms of an equivalent length of 600-m pipe, $L_e = (f_2/f_1)(L_2)(D_1/D_2)^5 = (0.015/0.026)(306.67)[(\tfrac{600}{1000})/1]^5 = 13.76$ m. By adding to the 600-mm pipe, the problem is reduced to finding the discharge through $320.99 + 13.76$, or 334.75 m of 600-mm pipe, $\epsilon_1 = 2$ mm, $H = 6$ m: $h_f = (f)(L/D)(v^2/2g)$, $6 = (f)[334.75/(\tfrac{600}{1000})]\{v^2/[(2)(9.807)]\}$, $v = 0.4593/\sqrt{f}$. Try $f = 0.026$: $v = 0.4593/\sqrt{0.026} = 2.848$ m/s, $N_R = Dv/\nu = (\tfrac{600}{1000})(2.848)/(3 \times 10^{-6}) = 5.70 \times 10^5$. From Fig. A-5, with $N_R = 5.70 \times 10^5$ and $\epsilon/D = 0.00333$, $f = 0.0265$. Try $f = 0.0265$: $v = 0.4593/\sqrt{0.0265} = 2.821$ m/s, $N_R = (\tfrac{600}{1000})(2.821)/(3 \times 10^{-6}) = 5.64 \times 10^5$, $f = 0.0265$ (O.K.); $Q = Av = [(\pi)(\tfrac{600}{1000})^2/4](2.821) = 0.798$ m³/s.

10.24 Two reservoirs are connected by three clean cast iron pipes in series: $L_1 = 300$ m, $D_1 = 200$ mm; $L_2 = 360$ m, $D_2 = 300$ mm; $L_3 = 1200$ m, $D_3 = 450$ mm. When $Q = 0.1$ m³/s of water at 20 °C, determine the difference in elevation of the reservoirs.

❚ $$v = Q/A \qquad v_1 = 0.1/[(\pi)(\tfrac{200}{1000})^2/4] = 3.183 \text{ m/s} \qquad v_2 = 0.1/[(\pi)(\tfrac{300}{1000})^2/4] = 1.415 \text{ m/s}$$

$$v_3 = 0.1/[(\pi)(\tfrac{450}{1000})^2/4] = 0.6288 \text{ m/s}$$

$$N_R = Dv/\nu \qquad (N_R)_1 = (\tfrac{200}{1000})(3.183)/(1.02 \times 10^{-6}) = 6.24 \times 10^5$$

$$(N_R)_2 = (\tfrac{300}{1000})(1.415)/(1.02 \times 10^{-6}) = 4.16 \times 10^5 \qquad (N_R)_3 = (\tfrac{450}{1000})(0.6288)/(1.02 \times 10^{-6}) = 2.77 \times 10^5$$

$$(\epsilon/D)_1 = 0.00026/(\tfrac{200}{1000}) = 0.00130 \qquad (\epsilon/D)_2 = 0.0026/(\tfrac{300}{1000}) = 0.000867 \qquad (\epsilon/D)_3 = 0.00026/(\tfrac{450}{1000}) = 0.000578$$

From Fig. A-5, $f_1 = 0.0215$, $f_2 = 0.020$, and $f_3 = 0.0185$. $H = h_f = (f)(L/D)(v^2/2g) = 0.0215[300/(\tfrac{200}{1000})]\{3.183^2/[(2)(9.807)]\} + 0.020[360/(\tfrac{300}{1000})]\{1.415^2/[(2)(9.807)]\} + 0.0185[1200/(\tfrac{450}{1000})]\{0.6288^2/[(2)(9.807)]\} = 20.10$ m.

10.25 Solve Prob. 10.24 by the method of equivalent lengths.

▮ Express pipes 2 and 3 in terms of pipe 1:

$$L_e = (f_2/f_1)(L_2)(D_1/D_2)^5$$

$$(L_e)_2 = (0.020/0.0215)(360)(\tfrac{200}{300})^5 = 44.10 \text{ m} \qquad (L_e)_3 = (0.0185/0.0215)(1200)(\tfrac{200}{450})^5 = 17.91 \text{ m}$$

$$(L_e)_{\text{total}} = 300 + 44.10 + 17.91 = 362.0 \text{ m}$$

$$H = h_f = (f)(L/D)(v^2/2g) = 0.0215[362.0/(\tfrac{200}{1000})]\{3.183^2/[(2)(9.807)]\} = 20.10 \text{ m}$$

10.26 Air at atmospheric pressure and 60 °F is carried through two horizontal pipes ($\epsilon = 0.06$ in) in series. The upstream pipe is 400 ft of 24 in diameter, and the downstream pipe is 100 ft of 36 in diameter. Estimate the equivalent length of 18-in ($\epsilon = 0.003$ in) pipe. Neglect minor losses.

▮ $$\epsilon/D_1 = 0.06/24 = 0.0025 \qquad \epsilon/D_2 = 0.06/36 = 0.00167$$

From Fig. A-5, assuming high Reynolds numbers, $f_1 = 0.025$ and $f_2 = 0.022$.

$$L_e = (0.025/f_3)(400)(\tfrac{18}{24})^5 + (0.022/f_3)(100)(\tfrac{18}{36})^5 = 2.442/f_3$$

$$\epsilon_3/D_3 = 0.003/18 = 0.000167 \qquad f_3 = 0.013 \qquad L_e = 2.442/0.013 = 188 \text{ ft}$$

10.27 What pressure drop, in inches of water is required for flow of 6000 ft³/min in Prob. 10.26? Include losses due to sudden expansion.

▮ $$h_L = h_f + h_m \qquad h_f = (f)(L/D)(v^2/2g)$$

From Prob. 10.26, $\epsilon/D_1 = 0.0025$, $\epsilon/D_2 = 0.00167$.

$$v_1 = Q/A_1 = (\tfrac{6000}{60})/[(\pi)(\tfrac{24}{12})^2/4] = 31.83 \text{ ft/s} \qquad v_2 = Q/A_2 = (\tfrac{6000}{60})/[(\pi)(\tfrac{36}{12})^2/4] = 14.15 \text{ ft/s}$$

$$N_R = Dv/v \qquad (N_R)_1 = (\tfrac{24}{12})(31.83)/(1.58 \times 10^{-4}) = 4.03 \times 10^5 \qquad (N_R)_2 = (\tfrac{36}{12})(14.15)/(1.58 \times 10^{-4}) = 2.69 \times 10^5$$

From Fig. A-5, $f_1 = 0.025$ and $f_2 = 0.022$.

$$h_f = 0.025[400/(\tfrac{24}{12})]\{31.83^2/[(2)(32.2)]\} + 0.022[100/(\tfrac{36}{12})]\{14.15^2/[(2)(32.2)]\} = 80.94 \text{ ft}$$

$$h_m = (v_1 - v_2)^2/2g = (31.83 - 14.15)^2/[(2)(32.2)] = 4.85 \text{ m} \qquad h_L = 80.94 + 4.85 = 85.79 \text{ ft of air}$$

$$(\gamma h)_{\text{air}} = (\gamma h)_{\text{H}_2\text{O}} \qquad \gamma = p/RT$$

$$\gamma_{\text{air}} = (14.7)(144)/[(53.3)(460 + 60)] = 0.07637 \text{ lb/ft}^3 \qquad (0.07637)(85.79) = 62.4h$$

$$h = 0.1050 \text{ ft} \quad \text{or} \quad 1.26 \text{ in of water}$$

10.28 Two pipes, $D_1 = 3$ in, $L_1 = 250$ ft, and $D_2 = 2$ in, $L_2 = 200$ ft, are laid in series with a total pressure drop of 3 lb/in². What is the flow rate of SAE30 oil ($\rho = 1.78$ slugs/ft³, $\mu = 0.00606$ lb-s/ft²) at 20 °C?

▮ $$h_f = \Delta p/\gamma = \Delta p/\rho g = (3)(144)/[(1.78)(32.2)] = 7.537 \text{ ft}$$

Assume laminar flow

$$h_f = \frac{128\mu LQ}{\pi \rho g D^4} \qquad 7.537 = \frac{(128)(0.00606)(250)(Q)}{(\pi)(1.78)(32.2)(\tfrac{3}{12})^4} + \frac{(128)(0.00606)(200)(Q)}{(\pi)(1.78)(32.2)(\tfrac{2}{12})^4} \qquad Q = 0.00541 \text{ ft}^3/\text{s}$$

Computation of N_R (not shown) indicates the flow is indeed laminar.

10.29 Convert the piping system shown in Fig. 10-6 to an equivalent length of 6-in pipe.

▮ $$p_A/\gamma + V_A^2/2g + z_A = p_M/\gamma + V_M^2/2g + z_M + h_L \qquad h_L = h_f + h_m$$

$$h_f = (f)(L/d)(V^2/2g) = 0.025[150/(\tfrac{12}{12})]\{V_{12}^2/[(2)(32.2)]\} + 0.020[100/(\tfrac{6}{12})]\{V_6^2/[(2)(32.2)]\}$$

$$= 0.05823V_{12}^2 + 0.06211V_6^2$$

$$h_m = (K)(V^2/2g) = [8.0 + (2)(0.5) + 0.7 + 1.0]\{V_{12}^2/[(2)(32.2)]\}$$

$$+ [0.7 + 6.0 + (2)(0.5) + 3.0 + 1.0]\{V_6^2/[(2)(32.2)]\}$$

$$= 0.1661V_{12}^2 + 0.1817V_6^2$$

$$h_L = 0.05823V_{12}^2 + 0.06211V_6^2 + 0.1661V_{12}^2 + 0.1817V_6^2 = 0.2243V_{12}^2 + 0.2438V_6^2$$

$$0 + 0 + h = 0 + 0 + 0 + 0.2243V_{12}^2 + 0.2438V_6^2 \qquad V_{12} = (\tfrac{6}{12})^2(V_6) = 0.2500V_6$$

$$h = (0.2243)(0.2500V_6)^2 + 0.2438V_6^2 = 0.2578V_6^2$$

For a 6-in equivalent pipe, $h = 0.020[L_e/(\tfrac{6}{12})]\{V_6^2/[(2)(32.2)]\} = 0.0006211L_eV_6^2$, $0.2578V_6^2 = 0.0006211L_eV_6^2$, $L_e = 415$ ft.

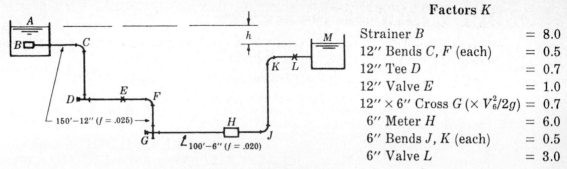

Factors K	
Strainer B	= 8.0
12″ Bends C, F (each)	= 0.5
12″ Tee D	= 0.7
12″ Valve E	= 1.0
12″ × 6″ Cross G ($\times V_6^2/2g$)	= 0.7
6″ Meter H	= 6.0
6″ Bends J, K (each)	= 0.5
6″ Valve L	= 3.0

Fig. 10-6

10.30 For the compound piping system described in Prob. 10.12, what flow will be produced for a total head loss of 70.0 ft?

▌ From Prob. 10.12, 13 540 ft of 16-in pipe is equivalent to the compound system. For a head loss of 70.0 ft, $h_1 = 70.0/13\,540 = 0.00517$. From Fig. A-13, $Q = 6.3$ cfs. This is for $C = 120$. For new cast iron pipe ($C = 130$), $Q = (\frac{130}{120})(6.3) = 6.8$ cfs.

10.31 For the system shown in Fig. 10-7, when the flow from reservoir A to main D is 3.25 mgd, the pressure at D is 20.0 psi. The flow to D must be increased to 4.25 mgd with the pressure at 40.0 psi. What size pipe, 5000 ft long, should be laid from B to C (shown dashed) parallel to the existing 12-in pipe to accomplish this?

▌ The elevation of reservoir A can be determined as follows: $Q = (3.25 \times 10^6)/[(7.48)(86\,400)] = 5.03$ cfs. From Fig. A-13, $(h_1)_{16} = 0.0038$ ft/ft, $(h_1)_{12} = 0.0150$ ft/ft. These values of h_1 are for $C = 120$; they must be adjusted for $C = 100$:

$$(h_1)_{16} = (\tfrac{120}{100})(0.0038) = 0.0046 \text{ ft/ft} \qquad (h_1)_{12} = (\tfrac{120}{100})(0.0150) = 0.018 \text{ ft/ft}$$

$$(h_f)_{16} = (0.0046)(8000) = 36.8 \text{ ft} \qquad (h_f)_{12} = (0.018)(5000) = 90.0 \text{ ft} \qquad (h_f)_{\text{total}} = 36.8 + 90.0 = 126.8 \text{ ft}$$

The hydraulic grade line drops 126.8 ft to an elevation of 46.2 ft above D (the equivalent of 20.0 psi). Thus reservoir A is 126.8 + 46.2, or 173.0 ft above point D. For a pressure of 40.0 psi, the elevation of the hydraulic grade line at D will be 92.4 ft above D, or the available head for the flow of 4.25 mgd will be 173.0 − 92.4, or 80.6 ft. In the 16-in pipe with $Q = 4.25$ mgd, or 6.58 cfs, $(h_1)_{16} = 0.0062$ ft/ft for $C = 120$.
For C = 100: $(h_1)_{16} = (\tfrac{120}{100})(0.0062) = 0.0074$ ft/ft, $(h_f)_{16} = (0.0074)(8000) = 59.2$ ft, $(h_f)_{B \text{ to } C} = 80.6 - 59.2 = 21.4$ ft.
For the existing 12-in pipe: $(h_1)_{12} = 21.4/5000 = 0.0043$ ft/ft, $Q_{12} = 2.7$ cfs for $C = 120$.
For $C = 100$, $Q_{12} = (\tfrac{100}{120})(2.7) = 2.3$ cfs. The flow in the new pipe must be 6.58 − 2.3, or 4.28 cfs with an available head (drop in the hydraulic grade line) of 21.4 ft from B to C. $h_1 = 21.4/5000 = 0.0043$ ft/ft. From Fig. A-13, $D = 14$ in, approximately.

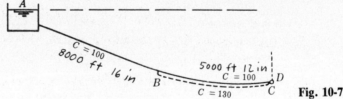

Fig. 10-7

CHAPTER 11
Parallel Pipeline Systems /

11.1 Figure 11-1 shows a looping pipe system. Pressure heads at points A and E are 70.0 m and 46.0 m, respectively. Compute the flow rate of water through each branch of the loop. Assume $C = 120$ for all pipes.

■ $(h_f)_{A-E} = 70.0 - 46.0 = 24.0$ m. From Fig. A-14, for pipe ABE, with $h_1 = 24.0/3000$, or 0.0080 m/m, and $D = 300$ mm, $Q_{ABE} = 0.105$ m³/s. For pipe ACE, with $h_1 = 24.0/1300$, or 0.0185 m/m, and $D = 200$ mm, $Q_{ACE} = 0.056$ m³/s. For pipe ADE, with $h_1 = 24.0/2600$, or 0.0092 m/m, and $D = 250$ mm, $Q_{ADE} = 0.070$ m³/s.

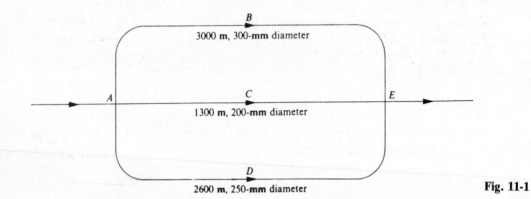

3000 m, 300-mm diameter

1300 m, 200-mm diameter

2600 m, 250-mm diameter

Fig. 11-1

11.2 A looping concrete pipe system is shown in Fig. 11-2. The total flow rate of water is 18.0 cfs. Determine the division of flow and the loss of head from point B to point E.

■ Assume a head loss of 20 ft from point B to point E. With $(h_1)_{BCE} = \frac{20}{5000}$, or 0.0040 ft/ft and $D = 15$ in, $Q_{BCE} = 4.75$ cfs. With $(h_1)_{BDE} = \frac{20}{4000}$, or 0.0050 ft/ft and $D = 18$ in, $Q_{BDE} = 8.60$ cfs. Fraction of flow through pipe $BCE = 4.75/(4.75 + 8.60) = 0.356$. Fraction of flow through pipe $BDE = 8.60/(4.75 + 8.60) = 0.644$. $Q_{BCE} = (18.0)(0.356) = 6.4$ cfs, $Q_{BDE} = (18.0)(0.644) = 11.6$ cfs. With $Q_{BCE} = 6.4$ cfs and $D = 15$ in, $(h_1)_{B-E} = 0.0070$ ft/ft. $(h_f)_{B-E} = (0.0070)(5000) = 35.0$ ft.

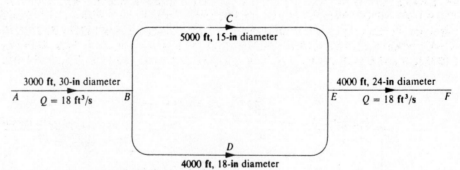

5000 ft, 15-in diameter

3000 ft, 30-in diameter

A $Q = 18$ ft³/s B

4000 ft, 24-in diameter

E $Q = 18$ ft³/s F

4000 ft, 18-in diameter

Fig. 11-2

11.3 The discharge of water in the concrete pipe looping system shown in Fig. 11-3 if 15.0 ft³/s. Compute the head loss from point A to point G.

■ With $Q_{A-B} = 15.0$ cfs and $D = 30$ in, $(h_1)_{A-B} = 0.00116$ ft/ft. With $Q_{FG} = 15.0$ cfs and $D = 24$ in, $(h_1)_{FG} = 0.00345$ ft/ft. $(h_f)_{AB} = (0.00116)(2500) = 2.90$ ft, $(h_f)_{FG} = (0.00345)(3000) = 10.35$ ft. Assume $(h_f)_{BF} = 30$ ft. With $D = 18$ in and $(h_1)_{BCF} = \frac{30}{1500} = 0.020$ ft/ft, $Q_{BCF} = 18.1$ cfs. With $D = 12$ in and $(h_1)_{BDF} = \frac{30}{1000} = 0.030$ ft/ft, $Q_{BDF} = 7.7$ cfs. With $D = 15$ in and $(h_1)_{BEF} = \frac{30}{2000} = 0.015$ ft/ft, $Q_{BEF} = 9.6$ cfs. Fraction of flow through pipe $BCF = 18.1/(18.1 + 7.7 + 9.6) = 0.511$, fraction of flow through pipe $BDF = 7.7/(18.1 + 7.7 + 9.6) = 0.218$, fraction of flow through pipe $BEF = 9.6/(18.1 + 7.7 + 9.6) = 0.271$; $Q_{BCF} = (15.0)(0.511) = 7.67$ cfs. With $Q_{BCF} = 7.67$ and $D = 18$ in, $(h_1)_{BCF} = 0.0040$ ft/ft. $(h_f)_{B-F} = (0.0040)(1500) = 6.00$ ft; $(h_f)_{A-G} = (h_f)_{AB} + (h_f)_{B-F} + (h_f)_{FG} = 2.90 + 6.00 + 10.35 = 19.25$ ft.

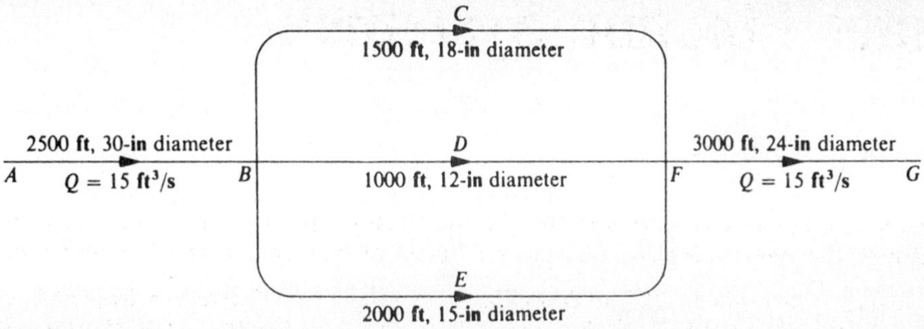

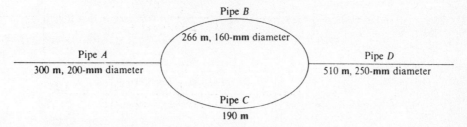

Fig. 11-3

11.4 If the flow rate of water through the pipe system shown in Fig. 11-4 is $0.050 \ \mathrm{m^3/s}$ under total head loss of 9.0 m, determine the diameter of pipe C. Assume a C coefficient of 120 for all pipes.

▮ With $Q_A = 0.050 \ \mathrm{m^3/s}$ and $D_A = 200$ mm, $(h_1)_A = 0.015$ m/m and $(h_f)_A = (0.015)(300) = 4.50$ m. With $Q_D = 0.050 \ \mathrm{m^3/s}$ and $D_D = 250$ mm, $(h_1)_D = 0.0049$ m/m and $(h_f)_D = (0.0049)(510) = 2.50$ m. $(h_f)_B = (h_f)_C = 9.0 - 4.50 - 2.50 = 2.00$ m. With $(h_1)_B = 2.00/266 = 0.0075$ m/m, $Q_B = 0.019 \ \mathrm{m^3/s}$. $Q_C = 0.050 - 0.019 = 0.031 \ \mathrm{m^3/s}$. With $Q_C = 0.031 \ \mathrm{m^3/s}$ and $(h_1)_C = 2.0/190 = 0.0105$ m/m, $D_C = 180$ mm.

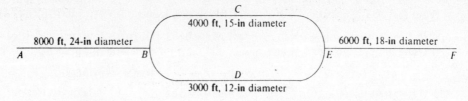

Pipe B
266 m, 160-mm diameter

Pipe A
300 m, 200-mm diameter

Pipe D
510 m, 250-mm diameter

Pipe C
190 m

Fig. 11-4

11.5 For the looping pipe system shown in Fig. 11-5, the head loss between points A and F is 160.0 ft of water. Determine the flow rate of water in the system. Assume $C = 120$ for all pipes.

▮ Assume a head loss of 12 ft from point B to point E. With $(h_1)_{BCE} = \frac{12}{4000} = 0.0030$ ft/ft and $D = 15$ in, $Q_{BCE} = 4.0$ cfs. With $(h_1)_{BDE} = \frac{12}{3000} = 0.0040$ ft/ft and $D = 12$ in, $Q_{BDE} = 2.6$ cfs. Fraction of flow through pipe $BCE = 4.0/(4.0 + 2.6) = 0.606$; fraction of flow through pipe $BDE = 2.6/(4.0 + 2.6) = 0.394$; $Q_{AB} = Q_{EF} = 4.0 + 2.6 = 6.6$ cfs. With $Q_{AB} = 6.6$ cfs and $D = 24$ in, $(h_1)_{AB} = 0.00075$ ft/ft; $(h_f)_{AB} = (0.00075)(8000) = 6.0$ ft. With $Q_{EF} = 6.6$ cfs and $D = 18$ in, $(h_1)_{EF} = 0.0030$ ft/ft; $(h_f)_{EF} = (0.0030)(6000) = 18.0$ ft. Fraction of total head loss through pipe $AB = 6.0/(6.0 + 12.0 + 18.0) = 0.167$, fraction of total head loss from B to $E = 12.0/(6.0 + 12.0 + 18.0) = 0.333$, fraction of total head loss through pipe $EF = 18.0/(6.0 + 12.0 + 18.0) = 0.500$. Actual head losses are, therefore, $(h_f)_{AB} = (160)(0.167) = 26.72$ ft, $(h_f)_{B-E} = (160)(0.333) = 53.28$ ft, $(h_f)_{EF} = (160)(0.500) = 80.00$ ft. With $(h_1)_{AB} = 26.72/8000 = 0.00334$ ft/ft and $D = 24$ in, $Q_{AB} = 14.5$ cfs. With $(h_1)_{EF} = 80.00/6000 = 0.0133$ ft/ft and $D = 18$ in, $Q_{EF} = 14.5$ cfs. $Q_{BCE} = (14.5)(0.606) = 8.8$ cfs, $Q_{BDE} = (14.5)(0.394) = 5.7$ cfs.

C
4000 ft, 15-in diameter

8000 ft, 24-in diameter

6000 ft, 18-in diameter

A   B   E   F

D
3000 ft, 12-in diameter

Fig. 11-5

11.6 A flow of 570 L/s is proceeding through the pipe network shown in Fig. 11-6. For a pressure of 690 kPa at node A, what pressure may be expected at node B? Neglect minor losses. Take $\rho = 1000 \ \mathrm{kg/m^3}$.

▮ $p_A/\gamma + v_A^2/2g + z_A = p_B/\gamma + v_B^2/2g + z_B + h_L$. Assume that a flow Q_1' of 170 L/s is proceeding through branch 1.

$$h_f = (f)(L/D)(v^2/2g) \qquad v_1 = Q_1' = (\tfrac{170}{1000})/[(\pi)(\tfrac{300}{1000})^2/4] = 2.405 \ \mathrm{m/s} \qquad N_R = Dv/\nu$$

$$(N_R)_1 = (\tfrac{300}{1000})(2.405)/(0.0113 \times 10^{-4}) = 6.38 \times 10^5 \qquad \epsilon/D_1 = 0.00026/(\tfrac{300}{1000}) = 0.00087$$

From Fig. A-5, $f = 0.0198$. $h_L = (h_f)_1 = 0.0198[600/(\tfrac{300}{1000})]\{2.405^2/[(2)(9.807)]\} = 11.68$ m. For branch 1,

$$p_A/\gamma + (v_1^2)_A/2g + 6 = p_B/\gamma + (v_1^2)_B/2g + 15 + 11.68 \qquad (v_1^2)_A/2g = (v_1^2)_B/2g \qquad [(p_A - p_B)/\gamma]_1 = 20.68 \ \mathrm{m}$$

Using this pressure head, which must be the same for each loop, for branch 2,

$$20.68 + (v_2^2)_A/2g + 6 = (v_2^2)_B/2g + 15 + (h_f)_2 \qquad (v_2^2)_A/2g = (v_2^2)_B/2g \qquad (h_f)_2 = 11.68 \text{ m}$$

$$11.68 = f_2[460/(\tfrac{470}{1000})]\{v_2^2/[(2)(9.807)]\} = 49.90 f_2 v_2^2 \qquad (\epsilon/D)_2 = 0.00026/(\tfrac{470}{1000}) = 0.00055$$

Estimate $f_2 = 0.018$.

$$11.68 = (49.90)(0.018)(v_2^2) \qquad v_2 = 3.606 \text{ m/s}$$

$$Q_2' = A_2 v_2 = [(\pi)(\tfrac{470}{1000})^2/4](3.606) = 0.6256 \text{ m}^3/\text{s} \quad \text{or} \quad 625.6 \text{ L/s}$$

Now the desired actual flows Q_1 and Q_2 may be computed so as to maintain the ratio Q_1'/Q_2' and to satisfy continuity so that $Q_1 + Q_2 = 570 \text{ L/s}$.

$$Q_1 = [170/(625.6 + 170)](570) = 121.8 \text{ L/s} \qquad Q_2 = [625.6/(625.6 + 170)](570) = 448.2 \text{ L/s}$$

$$v_1 = (121.8/1000)/[(\pi)(\tfrac{300}{1000})^2/4] = 1.723 \text{ m/s} \qquad v_2 = (448.2/1000)/[(\pi)(\tfrac{470}{1000})^2/4] = 2.583 \text{ m/s}$$

$$(N_R)_1 = (\tfrac{300}{1000})(1.723)/(0.0113 \times 10^{-4}) = 4.57 \times 10^5 \qquad (N_R)_2 = (\tfrac{470}{1000})(2.583)/(0.0113 \times 10^{-4}) = 1.07 \times 10^6$$

$$f_1 = 0.0198 \qquad f_2 = 0.0180 \qquad (h_f)_1 = 0.0198[600/(\tfrac{300}{1000})]\{1.723^2/[(2)(9.807)]\} = 5.99 \text{ m}$$

$$p_A/\gamma + (v_1^2)_A/2g + 6 = p_B/\gamma + (v_1^2)_B/2g + 15 + 5.99 \qquad [(p_A - p_B)/\gamma]_1 = 14.99 \text{ m}$$

$$(h_f)_2 = 0.0180[460/(\tfrac{470}{1000})]\{2.583^2/[(2)(9.807)]\} = 5.99 \text{ m}$$

$$p_A/\gamma + (v_2^2)_A/2g + 6 = p_B/\gamma + (v_2^2)_B/2g + 15 + 5.99$$

$$[(p_A - p_B)/\gamma]_2 = 14.99 \text{ m} \qquad [(690 - p_B)/9.79] = 14.99 \qquad p_B = 543 \text{ kPa}$$

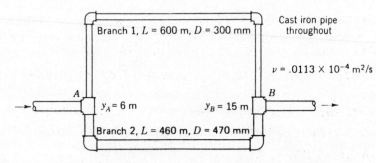

Branch 1, L = 600 m, D = 300 mm

Cast iron pipe throughout

$\nu = .0113 \times 10^{-4} \text{ m}^2/\text{s}$

A $y_A = 6 \text{ m}$ $y_B = 15 \text{ m}$ B

Branch 2, L = 460 m, D = 470 mm

Fig. 11-6

11.7 If 1 cfs of water flows into the system shown in Fig. 11-7 at A at a pressure of 100 psig, what is the pressure at B if one neglects minor losses? The pipe is commercial steel. Take $\mu = 2.11 \times 10^{-5} \text{ lb-s/ft}^2$.

❚ $$p_A/\gamma + v_A^2/2g + z_A = p_C/\gamma + v_C^2/2g + z_C + h_L.$$

Assume that a flow Q_1' of 0.4 ft³/s is proceeding through branch 1.

$$h_f = (f)(L/D)(v^2/2g) \qquad v_1 = Q_1'/A_1 = 0.4/[(\pi)(\tfrac{6}{12})^2/4] = 2.037 \text{ ft/s} \qquad N_R = \rho D v/\mu$$

$$(N_R)_1 = (1.94)(\tfrac{6}{12})(2.037)/(2.11 \times 10^{-5}) = 9.36 \times 10^4 \qquad \epsilon/D_1 = 0.00015/(\tfrac{6}{12}) = 0.00030$$

From Fig. A-5, $f = 0.0195$. $h_L = (h_f)_1 = 0.0195[(30 + 1300 + 30)/(\tfrac{6}{12})]\{2.037^2/[(2)(32.2)]\} = 3.417 \text{ ft}$.

For branch 1, $p_A/\gamma + (v_1^2)_A/2g + 0 = p_B/\gamma + (v_1^2)_B/2g + 0 + 3.417$, $(v_1^2)_A/2g = (v_1^2)_B/2g$, $[(p_A - p_B)/\gamma]_1 = 3.417$ ft. Using this pressure head, which must be the same for each loop, for branch 2,

$$3.417 + (v_2^2)_A/2g + 0 = (v_2^2)_B/2g + 0 + (h_f)_2 \qquad (v_2^2)_A/2g = (v_2^2)_B/2g \qquad (h_f)_2 = 3.417 \text{ ft}$$

$$3.417 = f_2[(30 + 1300 + 30)/(\tfrac{8}{12})]\{v_2^2/[(2)(32.2)]\} = 31.68 f_2 v_2^2 \qquad (\epsilon/D)_2 = 0.00015/(\tfrac{8}{12}) = 0.000225$$

Estimate $f_2 = 0.018$.

$$3.417 = (31.68)(0.018)(v_2^2) \qquad v_2 = 2.448 \text{ ft/s} \qquad Q_2' = A_2 v_2 = [(\pi)(\tfrac{8}{12})^2/4](2.448) = 0.8545 \text{ ft}^3/\text{s}$$

Now the desired actual flows Q_1 and Q_2 may be computed so as to maintain the ratio Q_1'/Q_2' and to satisfy continuity so that $Q_1 + Q_2 = 1 \text{ ft}^3/\text{s}$.

$$Q_1 = [0.4/(0.4 + 0.8545)](1) = 0.3189 \text{ ft}^3/\text{s} \qquad Q_2 = [0.8545/(0.4 + 0.8545)](1) = 0.6811 \text{ ft}^3/\text{s}$$

$$v_1 = (0.3189)/[(\pi)(\tfrac{6}{12})^2/4] = 1.624 \text{ ft/s} \qquad v_2 = (0.6811)/[(\pi)(\tfrac{8}{12})^2/4] = 1.951 \text{ ft/s}$$

$$(N_R)_1 = (1.94)(\tfrac{6}{12})(1.624)/(2.11 \times 10^{-5}) = 7.47 \times 10^4 \qquad (N_R)_2 = (1.94)(\tfrac{8}{12})(1.951)/(2.11 \times 10^{-5}) = 1.20 \times 10^5$$

$$f_1 = 0.0205 \qquad f_2 = 0.0190 \qquad (h_f)_1 = 0.0205[(30 + 1300 + 30)/(\tfrac{6}{12})]\{1.624^2/[(2)(32.2)]\} = 2.284 \text{ ft}$$

$$p_A/\gamma + (v_1^2)_A/2g + 0 = p_C/\gamma + (v_1^2)_C/2g + 0 + 2.284 \qquad [(p_A - p_B)/\gamma]_1 = 2.284 \text{ ft}$$

$$(h_f)_2 = 0.0190[(30 + 1300 + 30)/(\tfrac{8}{12})]\{1.951^2/[(2)(32.2)]\} = 2.291 \text{ ft}$$

$$p_A/\gamma + (v_2^2)_A/2g + 0 = p_C/\gamma + (v_2^2)_C/2g + 0 + 2.291 \qquad [(p_A - p_C)/\gamma]_2 = 2.291 \text{ ft}$$

$$[(100 - p_C)(144)/62.4] = 2.291 \text{ ft} \qquad p_C = 99.01 \text{ lb/in}^2$$

For pipe CB,

$$v = Q/A = 1/[(\pi)(\tfrac{8}{12})^2/4] = 2.865 \text{ ft/s} \qquad N_R = (1.94)(\tfrac{8}{12})(2.865)/(2.11 \times 10^{-5}) = 1.76 \times 10^5 \qquad f = 0.0178$$

$$h_f = 0.0178[3000/(\tfrac{8}{12})]\{2.865^2/[(2)(32.2)]\} = 10.21 \text{ ft} \qquad p_B = 99.01 - (10.21)(62.4)/144 = 94.6 \text{ lb/in}^2$$

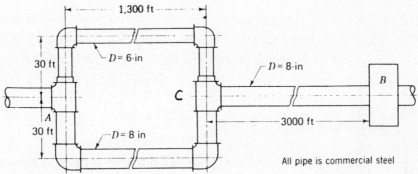

All pipe is commercial steel   **Fig. 11-7**

11.8   Do Prob. 11.7 for the case where we have nominal pipe sizes (6.065 in and 7.981 in diameters) and we have an open globe valve in the 8-in pipe just before reaching apparatus B. The fittings are all screwed fittings. Use $K = 0.45$ for a 6-in standard tee, 0.30 for elbows, and 4.8 for the globe valve.

$\blacksquare$ $p_A/\gamma + v_A^2/2g + z_A = p_C/\gamma + v_C^2/2g + z_C + h_L$. Assume that a flow Q_1' of 0.4 ft³/s is proceeding through branch 1.

$$h_L = h_f + h_m \qquad h_f = (f)(L/D)(v^2/2g) \qquad v_1 = Q_1'/A_1 = 0.4/[(\pi)(6.065/12)^2/4] = 1.994 \text{ ft/s}$$

$$N_R = \rho D v/\mu \qquad (N_R)_1 = (1.94)(6.065/12)(1.994)/(2.11 \times 10^{-5}) = 9.27 \times 10^4 \qquad \epsilon/D_1 = 0.00015/(\tfrac{6}{12}) = 0.00030$$

From Fig. A-5, $f = 0.0198$.

$$(h_f)_1 = 0.0198[(30 + 1300 + 30)/(6.065/12)]\{1.994^2/[(2)(32.2)]\} = 3.289 \text{ ft}$$

$$h_m = (K)(v^2/2g) \qquad (h_m)_1 = [(2)(0.30) + (2)(0.45)]\{1.994^2/[(2)(32.2)]\} = 0.093 \text{ ft}$$

$$(h_L)_1 = 3.289 + 0.093 = 3.382 \text{ ft}$$

For branch 1,

$$p_A/\gamma + (v_1^2)_A/2g + 0 = p_B/\gamma + (v_1^2)_B/2g + 0 + 3.382 \qquad (v_1^2)_A/2g = (v_1^2)_B/2g \qquad [(p_A - p_B)/\gamma]_1 = 3.382 \text{ ft}$$

Using this pressure head, which must be the same for each loop, for branch 2,

$$3.382 + (v_2^2)_A/2g + 0 = (v_2^2)_B/2g + 0 + (h_L)_2 \qquad (v_2^2)_A/2g = (v_2^2)_B/2g \qquad (h_L)_2 = 3.382 \text{ ft}$$

$$3.382 = f_2[(30 + 1300 + 30)/(7.981/12)]\{v_2^2/[(2)(32.2)]\} + [(2)(0.30) + (2)(0.45)]\{v_2^2/[(2)(32.2)]\}$$

$$= 31.75 f_2 v_2^2 + 0.02329 v_2^2$$

$$(\epsilon/D)_2 = 0.00015/(7.981/12) = 0.000226$$

Estimate $f_2 = 0.018$.

$$3.382 = (31.75)(0.018)(v_2^2) + 0.02329 v_2^2 \qquad v_2 = 2.385 \text{ ft/s} \qquad Q_2' = A_2 v_2 = [(\pi)(7.981/12)^2/4](2.385) = 0.8286 \text{ ft}^3/\text{s}$$

Now the desired actual flows Q_1 and Q_2 may be computed so as to maintain the ratio Q_1'/Q_2' and to satisfy continuity so that $Q_1 + Q_2 = 1$ ft³/s.

$$Q_1 = [0.4/(0.4 + 0.8286)](1) = 0.3256 \text{ ft}^3/\text{s} \qquad Q_2 = [0.8286/(0.4 + 0.8286)](1) = 0.6744 \text{ ft}^3/\text{s}$$

$$v_1 = (0.3256)/[(\pi)(6.065/12)^2/4] = 1.623 \text{ ft/s} \qquad v_2 = (0.6744)/[(\pi)(7.981/12)^2/4] = 1.941 \text{ ft/s}$$

$$(N_R)_1 = (1.94)(6.065/12)(1.623)/(2.11 \times 10^{-5}) = 7.54 \times 10^4$$

$$(N_R)_2 = (1.94)(7.981/12)(1.941)/(2.11 \times 10^{-5}) = 1.19 \times 10^5$$

$$f_1 = 0.0205 \qquad f_2 = 0.019$$

$$(h_f)_1 = 0.0205[(30 + 1300 + 30)/(6.065/12)]\{1.623^2/[(2)(32.2)]\} = 2.256 \text{ ft}$$

$$(h_m)_1 = [(2)(0.30) + (2)(0.45)]\{1.623^2/[(2)(32.2)]\} = 0.061 \text{ ft} \qquad (h_L)_1 = 2.256 + 0.061 = 2.317 \text{ ft}$$

$$p_A/\gamma + (v_1^2)_A/2g + 0 = p_C/\gamma + (v_1^2)_C/2g + 0 + 2.317 \qquad [(p_A - p_B)/\gamma]_1 = 2.317 \text{ ft}$$

$$(h_f)_2 = 0.019[(30 + 1300 + 30)/(7.981/12)]\{1.941^2/[(2)(32.2)]\} = 2.273 \text{ ft}$$

$$(h_m)_2 = [(2)(0.30) + (2)(0.45)]\{1.941^2/[(2)(32.2)]\} = 0.045 \text{ ft}$$

$$(h_L)_1 = 2.273 + 0.045 = 2.318 \text{ ft} \qquad p_A/\gamma + (v_2^2)_A/2g + 0 = p_C/\gamma + (v_2^2)_C/2g + 0 + 2.318$$

$$[(p_A - p_C)/\gamma]_2 = 2.318 \text{ ft} \qquad [(100 - p_C)(144)/62.4] = 2.318 \text{ ft} \qquad p_C = 99.00 \text{ lb/in}^2$$

For pipe CB,

$$v = Q/A = 1/[(\pi)(7.981/12)^2/4] = 2.878 \text{ ft/s} \qquad N_R = (1.94)(7.981/12)(2.878)/(2.11 \times 10^{-5}) = 1.76 \times 10^5$$

$$f = 0.0178 \qquad h_f = 0.0178[3000/(7.981/12)]\{2.878^2/[(2)(32.2)]\} = 10.33 \text{ ft}$$

$$h_m = 4.8\{2.878^2/[(2)(32.2)]\} = 0.62 \text{ ft} \qquad h_L = 10.33 + 0.62 = 10.95 \text{ ft}$$

$$p_B = 99.00 - (10.95)(62.4)/144 = 94.3 \text{ lb/in}^2$$

11.9 A two-branch pipe system (Fig. 11-8) is to deliver 400 L/s of water at 5 °C. The pressure at B is 20 kPa gage. What is the pressure at A? Neglect minor losses.

▌ $p_A/\gamma + v_A^2/2g + z_A = p_B/\gamma + v_B^2/2g + z_B + h_L$. Assume that a flow Q_1' of 200 L/s is proceeding through branch 1.

$$h_f = (f)(L/D)(v^2/2g) \qquad (v_1)_{250} = Q_1'/A_1 = (\tfrac{200}{1000})/[(\pi)(\tfrac{250}{1000})^2/4] = 4.074 \text{ m/s}$$

$$(v_1)_{200} = Q_1'/A_1 = (\tfrac{200}{1000})/[(\pi)(\tfrac{200}{1000})^2/4] = 6.366 \text{ m/s} \qquad N_R = Dv/v$$

$$[(N_R)_1]_{250} = (\tfrac{250}{1000})(4.074)/(1.52 \times 10^{-6}) = 6.70 \times 10^5 \qquad (\epsilon/D_1)_{250} = 0.000046/(\tfrac{250}{1000}) = 0.000184$$

From Fig. A-5, $(f_1)_{250} = 0.015$.

$$[(N_R)_1]_{200} = (\tfrac{200}{1000})(6.366)/(1.52 \times 10^{-6}) = 8.38 \times 10^5 \qquad (\epsilon/D_1)_{200} = 0.000046/(\tfrac{200}{1000}) = 0.00023 \qquad (f_1)_{200} = 0.015$$

$$h_L = 0.015[(70 + 100 + 50)/(\tfrac{250}{1000})]\{4.074^2/[(2)(9.807)]\} + 0.015[(100 + 50)/(\tfrac{200}{1000})]\{6.366^2/[(2)(9.807)]\} = 34.41 \text{ m}$$

For branch 1,

$$p_A/\gamma + (v_1^2)_A/2g + 0 = p_B/\gamma + (v_1^2)_B/2g + 0 + 34.41 \qquad (v_1^2)_A/2g = (v_1^2)_B/2g \qquad [(p_A - p_B)/\gamma]_1 = 34.41 \text{ m}$$

Using this pressure head, which must be the same for each loop, for branch 2,

$$34.41 + (v_2^2)_A/2g + 0 = (v_2^2)_B/2g + 0 + (h_f)_2 \qquad (v_2^2)_A/2g = (v_2^2)_B/2g \qquad (h_f)_2 = 34.41 \text{ m}$$

$$34.41 = f_2[(30 + 200)/(\tfrac{200}{1000})]\{v_2^2/[(2)(9.807)]\} = 58.63 f v_2^2 \qquad (\epsilon/D)_2 = 0.000046/(\tfrac{200}{1000}) = 0.00023$$

Estimate $f_2 = 0.015$.

$$34.41 = (58.63)(0.015)(v_2^2) \qquad v_2 = 6.255 \text{ m/s} \qquad Q_2' = A_2 v_2 = [(\pi)(\tfrac{200}{1000})^2/4](6.255) = 0.1965 \text{ m}^3/\text{s} \quad \text{or} \quad 196.5 \text{ L/s}$$

Now the desired actual flows Q_1 and Q_2 may be computed so as to maintain the ratio Q_1'/Q_2' and to satisfy continuity so that $Q_1 + Q_2 = 400 \text{ L/s}$.

$$Q_1 = [200/(200 + 196.5)](400) = 201.8 \text{ L/s} \qquad Q_2 = [196.5/(200 + 196.5)](400) = 198.2 \text{ L/s}$$

$$(v_1)_{250} = (201.8/1000)/[(\pi)(\tfrac{250}{1000})^2/4] = 4.111 \text{ m/s} \qquad (v_1)_{200} = (201.8/1000)/[(\pi)(\tfrac{200}{1000})^2/4] = 6.423 \text{ m/s}$$

$$v_2 = (198.2/1000)/[(\pi)(\tfrac{200}{1000})^2/4] = 6.309 \text{ m/s}$$

Since the assumed value of Q_1 is close to 201.8 L/s, the values of f of 0.015 should be close enough.

$$(h_f)_1 = 0.015[(70 + 100 + 50)/(\tfrac{250}{1000})]\{4.111^2/[(2)(9.807)]\} + 0.015[(100 + 50)/(\tfrac{200}{1000})]\{6.423^2/[(2)(9.807)]\} = 35.04 \text{ m}$$

$$p_A/\gamma + (v_1^2)_A/2g + 0 = p_B/\gamma + (v_1^2)_B/2g + 0 + 35.04 \qquad [(p_A - p_B)/\gamma]_1 = 35.04 \text{ m}$$

$$(h_f)_2 = 0.015[(30 + 200)/(\tfrac{200}{1000})]\{6.309^2/[(2)(9.807)]\} = 35.01 \text{ m}$$

$$p_A/\gamma + (v_2^2)_A/2g + 0 = p_B/\gamma + (v_2^2)_B/2g + 0 + 35.01 \qquad [(p_A - p_B)/\gamma]_2 = 35.01 \text{ m}$$

These values of $[(p_A - p_B)/\gamma]$ are close enough, so take an average value of $(35.04 + 35.01)/2$, or 35.02 m: $(p_A - 20)/9.79 = 35.02$, $p_A = 363 \text{ kPa}$.

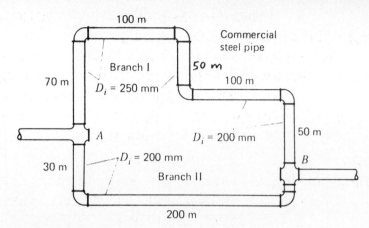

100 m

Commercial
steel pipe

Branch I

$D_i = 250$ mm

50 m

100 m

70 m

A

$D_i = 200$ mm

50 m

$D_i = 200$ mm

B

30 m

Branch II

200 m

Fig. 11-8

11.10 A flow of 800 L/s goes through the pipe system shown in Fig. 11-9. What is the pressure drop between A and B if the elevation of A is 100 m and of B is 200 m? Neglect minor losses. The water is at 5 °C.

❚
$$p_A/\gamma + v_A^2/2g + z_A = p_B/\gamma + v_B^2/2g + z_B + h_L$$

For branch I: Estimate $Q_1' = 250$ L/s, $h_f = (f)(L/D)(v^2/2g)$. Let $f = 0.014$: $v_1' = Q_1'/A_1 = (\frac{250}{1000})/[(\pi)(\frac{300}{1000})^2/4] = 3.537$ m/s, $h_f = 0.014[(300 + 500 + 300)/(\frac{300}{1000})]\{3.537^2/[(2)(9.807)]\} = 32.74$ m, $p_A/\gamma + v_A^2/2g + 100 = p_B/\gamma + v_B^2/2g + 200 + 32.74$, $v_A^2/2g = v_B^2/2g$, $(p_A - p_B)/\gamma = 132.74$ m.

For branch II: Let $f = 0.015$: $h_f = 0.015[500/(\frac{300}{1000})]\{(v_{II}')^2/[(2)(9.807)]\} = (1.275)(v_{II}')^2$, $p_A/\gamma + v_A^2/2g + 100 = p_B/\gamma + v_B^2/2g + 200 + (1.275)(v_{II}')^2$, $(p_A - p_B)/\gamma = 100 + (1.275)(v_{II}')^2$. Use $(p_A - p_B)/\gamma = 132.74$ m. $132.74 = 100 + (1.275)(v_{II}')^2$, $v_{II}' = 5.067$ m/s.

For branch III: Let $f = 0.013$: $h_f = 0.013[(400 + 500 + 400)/(\frac{300}{1000})]\{(v_{III}')^2/[(2)(9.807)]\} = (2.872)(v_{III}')^2$, $p_A/\gamma + v_A^2/2g + 100 = p_B/\gamma + v_B^2/2g + 200 + (2.872)(v_{III}')^2$, $(p_A - p_B)/\gamma = 100 + (2.872)(v_{III}')^2$. Use $(p_A - p_B)/\gamma = 132.74$ m. $132.74 = 100 + (2.872)(v_{III}')^2$, $v_{III}' = 3.376$ m/s. The new values of Q are $Q_1 = 250$ L/s; $Q_{II} = [(\pi)(\frac{300}{1000})^2/4](5.067) = 0.3582$ m³/s, or 358.2 L/s; $Q_{III} = [(\pi)(\frac{300}{1000})^2/4](3.376) = 0.2386$ m³/s, or 238.6 L/s. Find new values of Q: $Q_1 = [250/(250 + 358.2 + 238.6)](800) = 236.2$ L/s, $Q_{II} = [358.2/(250 + 358.2 + 238.6)](800) = 338.4$ L/s, $Q_{III} = [238.6/(250 + 358.2 + 238.6)](800) = 225.4$ L/s.

For branch I: $v_I = (236.2/1000)/[(\pi)(\frac{300}{1000})^2/4] = 3.342$ m/s, $N_R = Dv/\nu = (\frac{300}{1000})(3.342)/(1.52 \times 10^{-6}) = 6.60 \times 10^5$, $\epsilon/D = 0.000046/(\frac{300}{1000}) = 0.00015$. From Fig. A-5, $f = 0.0146$.

For branch II: $v_{II} = (338.4/1000)/[(\pi)(\frac{300}{1000})^2/4] = 4.787$ m/s, $N_R = (\frac{300}{1000})(4.787)/(1.52 \times 10^{-6}) = 9.45 \times 10^5$, $f = 0.0145$.

For branch III: $v_{III} = (238.6/1000)/[(\pi)(\frac{300}{1000})^2/4] = 3.375$ m/s, $N_R = (\frac{300}{1000})(3.375)/(1.52 \times 10^{-6}) = 6.66 \times 10^5$, $f = 0.0146$. Compute $(p_A - p_B)/\gamma$ for each branch.

For branch I: $h_f = 0.0146[(300 + 500 + 300)/(\frac{300}{1000})]\{3.342^2/[(2)(9.807)]\} = 30.48$ m, $p_A/\gamma + v_A^2/2g + 100 = p_B/\gamma + v_B^2/2g + 200 + 30.48$, $(p_A - p_B)/\gamma = 130.48$ m.

For branch II: $h_f = 0.0145[500/(\frac{300}{1000})]\{4.787^2/[(2)(9.807)]\} = 28.23$ m, $p_A/\gamma + v_A^2/2g + 100 = p_B/\gamma + v_B^2/2g + 200 + 28.23$, $(p_A - p_B)/\gamma = 128.23$ m.

For branch III: $h_f = 0.0146[(400 + 500 + 400)/(\frac{300}{1000})]\{3.375^2/[(2)(9.807)]\} = 36.74$ m, $p_A/\gamma + v_A^2/2g + 100 = p_B/\gamma + v_B^2/2g + 200 + 36.74$, $(p_A - p_B)/\gamma = 136.74$ m.

We could go through another cycle. Instead, we stop here and take an average value of $(p_A - p_B)/\gamma$ of $(130.48 + 128.23 + 136.74)/3$, or 131.82 m. $(p_A - p_B) = (131.82)(9.79) = 1291$ kPa gage.

11.11 In Fig. 11-10, $L_1 = 3000$ ft, $D_1 = 1$ ft, $\epsilon_1 = 0.001$ ft; $L_2 = 2000$ ft, $D_2 = 8$ in, $\epsilon_2 = 0.0001$ ft; $L_3 = 4000$ ft, $D_3 = 16$ in, $\epsilon_3 = 0.0008$ ft; $\rho = 2.00$ slugs/ft³, $\nu = 0.00003$ ft²/s, $p_A = 80$ psi, $z_A = 100$ ft, $z_B = 80$ ft. For a total flow of 12 cfs, determine the flow through each pipe and the pressure at B.

❚ $h_f = (f)(L/D)(v^2/2g)$. Assume $Q_1' = 3$ cfs.

$$v_1' = Q_1'/A_1 = 3/[(\pi)(1)^2/4] = 3.820 \text{ ft/s} \qquad N_R = Dv/\nu$$
$$(N_R)_1' = (1)(3.820)/0.00003 = 1.27 \times 10^5 \qquad \epsilon/D_1 = 0.001/1 = 0.0010$$

From Fig. A-5, $f_1' = 0.022$. $(h_f)_1' = (0.022)(\frac{3000}{1})\{3.820^2/[(2)(32.2)]\} = 14.95$ ft. For pipe 2, assume $f_2' = 0.020$.

$$14.95 = 0.020[2000/(\tfrac{8}{12})]\{v_2'^2/[(2)(32.2)]\} \qquad v_2' = 4.006 \text{ ft/s} \qquad (N_R)_2' = (\tfrac{8}{12})(4.006)/0.00003 = 8.90 \times 10^4$$
$$\epsilon_2/D_2 = 0.0001/(\tfrac{8}{12}) = 0.00015 \qquad f_2' = 0.019 \qquad 14.95 = 0.019[2000/(\tfrac{8}{12})]\{v_2'^2/[(2)(32.2)]\}$$
$$v_2' = 4.110 \text{ ft/s} \qquad Q_2' = [(\pi)(\tfrac{8}{12})^2/4](4.110) = 1.43 \text{ ft}^3/\text{s}$$

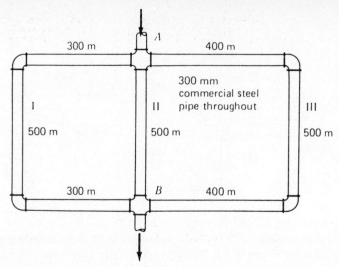

Fig. 11-9

For pipe 3, assume $f_3' = 0.020$.

$$14.95 = 0.020[4000/(\tfrac{16}{12})]\{v_3'^2/[(2)(32.2)]\} \qquad v_3' = 4.006 \text{ ft/s} \qquad (N_R)_2' = (\tfrac{16}{12})(4.006)/0.00003 = 1.78 \times 10^5$$

$$\epsilon_3/D_3 = 0.0008/(\tfrac{16}{12}) = 0.00060 \qquad f_3' = 0.020 \qquad Q_3' = [(\pi)(\tfrac{16}{12})^2/4](4.006) = 5.59 \text{ ft}^3/\text{s}$$

$$\Sigma Q = 3 + 1.43 + 5.59 = 10.02 \text{ ft}^3/\text{s} \qquad Q_1 = (3/10.02)(12) = 3.59 \text{ ft}^3/\text{s}$$

$$Q_2 = (1.43/10.02)(12) = 1.71 \text{ ft}^3/\text{s} \qquad Q_3 = (5.59/10.02)(12) = 6.69 \text{ ft}^3/\text{s}$$

Check the values of h_1, h_2, and h_3:

$$v_1 = 3.59/[(\pi)(1)^2/4] = 4.571 \text{ ft/s} \qquad (N_R)_1 = (1)(4.571)/0.00003 = 1.52 \times 10^5$$

$$f_1 = 0.021 \qquad (h_f)_1 = (0.021)(\tfrac{3000}{1})\{4.571^2/[(2)(32.2)]\} = 20.44 \text{ ft} \qquad v_2 = 1.71/[(\pi)(\tfrac{8}{12})^2/4] = 4.899 \text{ ft/s}$$

$$(N_R)_2 = (\tfrac{8}{12})(4.899)/0.00003 = 1.09 \times 10^5 \qquad f_2 = 0.018 \qquad (h_f)_2 = 0.018[2000/(\tfrac{8}{12})]\{4.899^2/[(2)(32.2)]\} = 20.12 \text{ ft}$$

$$v_3 = 6.69/[(\pi)(\tfrac{16}{12})^2/4] = 4.791 \text{ ft/s} \qquad (N_R)_3 = (\tfrac{16}{12})(4.791)/0.00003 = 2.13 \times 10^5$$

$$f_3 = 0.019 \qquad (h_f)_3 = 0.019[4000/(\tfrac{16}{12})]\{4.791^2/[(2)(32.2)]\} = 20.32 \text{ ft}$$

Since these values of h_f are close, accept the flow values determined above and take an average value of $(20.44 + 20.12 + 20.32)/3$, or 20.29 ft for calculating p_B: $p_A/\gamma + v_A^2/2g + z_A = p_B/\gamma + v_B^2/2g + z_B + h_L$, $(80)(144)/62.4 + v_A^2/2g + 100 = (p_B)(144)/62.4 + v_B^2/2g + 80 + 20.29$, $v_A^2/2g = v_B^2/2g$, $p_B = 79.9 \text{ lb/in}^2$.

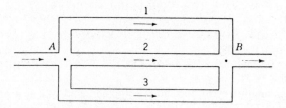

Fig. 11-10

11.12 Prepare a computer program to solve for the head drop and flow distribution through N pipes in parallel. Check it out with the data in Prob. 11.11.

```
10 REM B:                       SOLUTION OF PARALLEL PIPE PROBLEMS
20 DEFINT I,N:    DEF FNQ(D1,D2,D3,DH)=D1*DH*LOG(D2+D3/DH)
30 READ N,QQ,RHO,KVIS,PA,ZA,ZB,G,II
40 DATA 3,12.,2.,3E-5,80.,100.,80.,32.2,15
50 LPRINT:    LPRINT"N,QQ,RHO,KVIS=";N;QQ;RHO;KVIS
60 LPRINT:    LPRINT"PA,ZA,ZB,G,II=";PA;ZA;ZB;G;II:    LPRINT
70 FOR I=1 TO N:    READ L(I),D(I),EP(I):    NEXT I
80 DATA 3000.,1.,.001,2000.,.6667,.0001,4000.,1.3333,.0008
90 FOR I=1 TO N:    LPRINT"I,L,D,EPS=";I;L(I);D(I);EP(I):    NEXT I:    LPRINT
100 FOR I=1 TO N:    C=SQR(G*D(I)/L(I)):    E1(I)=-.9650001*D(I)^2*C
110 E2(I)=EP(I)/(3.7*D(I)):    E3(I)=1.784*KVIS/(D(I)*C):    NEXT I
120 X=.02*L(1)*QQ^2/(.7854*D(1)^5*2!*G)  ' MAX POSS. HEAD WITH FLOW THRU PIPE 1
130 HMA=SQR(X):    HMI=0!
140 FOR II=1 TO II:    S=0!:    HF=.5*(HMA+HMI)    ' START OF BISECTION METHOD
150 FOR I=1 TO N:    S=S+FNQ(E1(I),E2(I),E3(I),HF):    NEXT I
160 IF S-QQ>0! THEN HMA=HF ELSE HMI=HF
170 PRINT"S,QQ,HF=";S;QQ;HF
```

```
180 NEXT I1:   HFR=.5*(HMA+HMI):   HFF=HFR^2   ' FINAL VALUE OF HEAD DROP
190 FOR I=1 TO N:  LPRINT"I,HFF,Q=";I;HFF;FNQ(E1(I),E2(I),E3(I),HFR):  NEXT I
200 LPRINT: LPRINT"PB=";:  LPRINT USING" ###.### ";PA+(ZA-ZB-HFF)*RHO*G/144!

N,QQ,RHO,KVIS= 3  12  2  .00003

PA,ZA,ZB,G,II= 80  100  80  32.2  15

I,L,D,EPS= 1  3000   1      .001
I,L,D,EPS= 2  2000   .6667  .0001
I,L,D,EPS= 3  4000   1.3333 .0008

I,HFF,Q= 1  20.68776   3.576553
I,HFF,Q= 2  20.68776   1.710832
I,HFF,Q= 3  20.68776   6.713136

PB= 79.692
```

11.13 Two pipes are connected in parallel between two reservoirs: $L_1 = 2500$ m, $D_1 = 1.2$ m, $C = 100$; $L_2 = 2500$ m, $D_2 = 1$ m, $C = 90$. For a difference in elevation of 3.6 m, determine the total flow of water at 20 °C.

$$v = 0.8492CR^{0.63}s^{0.54}$$
$$v_1 = (0.8492)(100)(1.2/4)^{0.63}(3.6/2500)^{0.54} = 1.162 \text{ m/s}$$
$$v_2 = (0.8492)(90)(\tfrac{1}{4})^{0.63}(3.6/2500)^{0.54} = 0.9321 \text{ m/s}$$
$$Q = Av = [(\pi)(1.2)^2/4](1.162) + [(\pi)(1)^2/4](0.9321) = 2.05 \text{ m}^3/\text{s}$$

11.14 For 4.5 m³/s of flow in the system of Prob. 11.13, determine the difference in elevation of reservoir surfaces.

$$Q = Av = 0.8492ACR^{0.63}s^{0.54}$$
$$4.5 = 0.8492[(\pi)(1.2)^2/4](100)(1.2/4)^{0.63}s^{0.54} + 0.8492[(\pi)(1)^2/4](90)(\tfrac{1}{4})^{0.63}s^{0.54}$$
$$s = 0.006198 \text{ m/m} \qquad \Delta \text{ elevation} = (0.006198)(2500) = 15.5 \text{ m}$$

11.15 Three smooth tubes are connected in parallel: $L_1 = 40$ ft, $D_1 = \frac{1}{2}$ in; $L_2 = 60$ ft, $D_2 = 1$ in; $L_3 = 50$ ft, $D_3 = \frac{3}{4}$ in. For total flow of 30 gpm of oil, $\gamma = 55$ lb/ft³, $\mu = 0.65$ P, what is the drop in hydraulic grade line between junctions?

$$h_f = \frac{128\mu LQ}{\pi\gamma D^4}$$

Therefore, $L_1 Q_1/D_1^4 = L_2 Q_2/D_2^4 = L_3 Q_3/D_3^4$.

$$Q = Q_1 + Q_2 + Q_3 = (30)(0.002228) = 0.06684 \text{ ft}^3/\text{s} \qquad 40Q_1/[(\tfrac{1}{2})/12]^4 = 60Q_2/(\tfrac{1}{12})^4 \qquad Q_1 = 0.09375Q_2$$
$$60Q_2/(\tfrac{1}{12})^4 = 50Q_3/[(\tfrac{3}{4})/12]^4 \qquad Q_3 = 0.3797Q_2 \qquad 0.09375Q_2 + Q_2 + 0.3797Q_2 = 0.06684 \qquad Q_2 = 0.04536 \text{ ft}^3/\text{s}$$
$$\mu = \frac{(0.65)(0.1)(0.3048)^2}{4.448} = 0.001358 \text{ lb-s/ft}^2 \qquad h_f = \frac{(128)(0.001358)(60)(0.04536)}{(\pi)(55)(\tfrac{1}{12})^4} = 56.8 \text{ ft}$$

11.16 For $H = 12$ m in Fig. 11-11, find the discharge through each pipe. ($\mu = 8$ cP, s.g. $= 0.9$.)

$$h_f = (f)(L/D)(v^2/2g) \qquad (\epsilon/D)_1 = 0.006/(\tfrac{50}{1000}) = 0.120 \qquad f_1 = 0.114$$
$$(\epsilon/D)_2 = 0.009/(\tfrac{120}{1000}) = 0.075 \qquad f_2 = 0.088 \qquad (\epsilon/D)_1 = 0.012/(\tfrac{100}{1000}) = 0.120 \qquad f_3 = 0.114$$
$$(h_f)_1 = (h_f)_2 \qquad v_1 = Q_1/A_1 = Q_1/[(\pi)(\tfrac{50}{1000})^2/4] = 509.3Q_1 \qquad v_2 = Q_2/A_2 = Q_2/[(\pi)(\tfrac{120}{1000})^2/4] = 88.42Q_2$$
$$0.114[60/(\tfrac{50}{1000})]\{(509.3Q_1)^2/[(2)(9.807)]\} = 0.088[90/(\tfrac{120}{1000})]\{(88.42Q_2)^2/[(2)(9.807)]\}$$
$$Q_2 = 8.293Q_1 \qquad Q_3 = Q_1 + Q_2 = Q_1 + 8.293Q_1 = 9.293Q_1$$
$$v_3 = Q_3/A_3 = 9.293Q_1/[(\pi)(\tfrac{100}{1000})^2/4] = 1183Q_1 \qquad H = (h_f)_1 + (h_f)_3$$
$$12 = 0.114[60/(\tfrac{50}{1000})]\{(509.3Q_1)^2/[(2)(9.807)]\} + 0.114[120/(\tfrac{100}{1000})]\{(1183Q_1)^2/[(2)(9.807)]\}$$
$$Q_1 = 0.00102 \text{ m}^3/\text{s} \qquad Q_2 = (8.293)(0.00102) = 0.00846 \text{ m}^3/\text{s} \qquad Q_3 = (9.293)(0.00102) = 0.00948 \text{ m}^3/\text{s}$$
$$N_R = \frac{4Q\rho}{\pi D\mu} = \frac{(4)(0.00948)(0.9)(1000)}{(\pi)(\tfrac{100}{1000})(\tfrac{8}{1000})} = 13\,579$$

Hence, both f's are O.K.

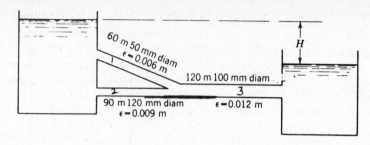

Fig. 11-11

11.17 Find H in Fig. 11-11 for 0.03 m^3/s flowing. ($\mu = 5$ cP, s.g. $= 0.9$.)

▌ $Q_3 = 0.03$ m^3/s. Q_1 and Q_2 will be in the same proportion as in Prob. 11.16. Hence, $Q_1 = (0.00102/0.00948)(0.03) = 0.00323$ m^3/s, $Q_2 = (8.293)(0.00323) = 0.0268$ m^3/s.

$$H = (h_f)_1 + (h_f)_3 \qquad h_f = (f)(L/D)(v^2/2g)$$

$$v_1 = Q_1/A_1 = 0.00323/[(\pi)(\tfrac{50}{1000})^2/4] = 1.645 \text{ m/s} \qquad v_3 = Q_3/A_3 = 0.03/[(\pi)(\tfrac{100}{1000})^2/4] = 3.820 \text{ m/s}$$

Using values of f from Prob. 11.16, $H = 0.114[60/(\tfrac{50}{1000})]\{1.645^2/[(2)(9.907)]\} + 0.114[120/(\tfrac{100}{1000})]\{3.820^2/[(2)(9.907)]\} = 120.6$ m.

11.18 Find the equivalent length of 300-mm-diameter clean cast iron pipe to replace the system of Fig. 11-12.

▌ $(\epsilon/D)_1 = 0.00026/(\tfrac{200}{1000}) = 0.0013 \qquad (\epsilon/D)_2 = 0.00026/(\tfrac{300}{1000}) = 0.00087 \qquad (\epsilon/D)_3 = 0.00026/(\tfrac{500}{1000}) = 0.00052$

$\qquad (\epsilon/D)_4 = 0.00026/(\tfrac{300}{1000}) = 0.00087 \qquad (\epsilon/D)_5 = 0.00026/(\tfrac{300}{1000}) = 0.00087$

Try $f_1 = 0.023$, $f_2 = 0.021$, $f_3 = 0.018$, $f_4 = 0.021$, and $f_5 = 0.021$: Assume a head loss in pipes 1 and 2 of 3 m.

$$h_f = (f)(L/D)(v^2/2g) \qquad 3 = 0.023[300/(\tfrac{200}{1000})]\{v_1^2/[(2)(9.807)]\} \qquad v_1 = 1.306 \text{ m/s}$$

$$Q_1 = A_1 v_1 = [(\pi)(\tfrac{200}{1000})^2/4](1.306) = 0.04103 \text{ m}^3/\text{s} \qquad 3 = 0.021[300/(\tfrac{300}{1000})]\{v_2^2/[(2)(9.807)]\} \qquad v_2 = 1.674 \text{ m/s}$$

$$Q_2 = A_2 v_2 = [(\pi)(\tfrac{300}{1000})^2/4](1.674) = 0.1183 \text{ m}^3/\text{s}$$

To find equivalent for 1 and 2, $v_e = (0.04103 + 0.1183)/[(\pi)(\tfrac{300}{1000})^2/4] = 2.254$ m/s, $3 = 0.021[L_e/(\tfrac{300}{1000})]\{2.254^2/[(2)(9.807)]\}$, $L_e = 165.5$ m (for pipes 1 and 2).
For pipe 3:

$$0.018\left(\frac{300}{\tfrac{500}{1000}}\right) \frac{\{Q/[(\pi)(\tfrac{500}{1000})^2/4]\}^2}{(2)(9.807)} = 0.021\left(\frac{L_e}{\tfrac{300}{1000}}\right) \frac{\{Q/[(\pi)(\tfrac{300}{1000})^2/4]\}}{(2)(9.807)}$$

$$L_e = 20.0 \text{ m} \qquad \text{(for pipe 3)}$$

Assume a head loss in pipes 4 and 5 of 3 m.

$$3 = 0.021[600/(\tfrac{300}{1000})]\{v_4^2/[(2)(9.807)]\} \qquad v_4 = 1.184 \text{ m/s}$$

$$Q_4 = [(\pi)(\tfrac{300}{1000})^2/4](1.184) = 0.08369 \text{ m}^3/\text{s} \qquad 3 = 0.021[800/(\tfrac{300}{1000})]\{v_5^2/[(2)(9.807)]\}$$

$$v_5 = 1.025 \text{ m/s} \qquad Q_5 = [(\pi)(\tfrac{300}{1000})^2/4](1.025) = 0.07245 \text{ m}^3/\text{s}$$

To find equivalent for 4 and 5,

$$v_e = (0.08369 + 0.07245)/[(\pi)(\tfrac{300}{1000})^2/4] = 2.209 \text{ m/s} \qquad 3 = 0.021[L_e/(\tfrac{300}{1000})]\{2.209^2/[(2)(9.807)]\}$$

$$L_e = 172.3 \text{ m} \qquad \text{(for pipes 4 and 5)} \qquad (L_e)_{\text{total}} = 165.5 + 20.0 + 172.3 = 357.8 \text{ m}$$

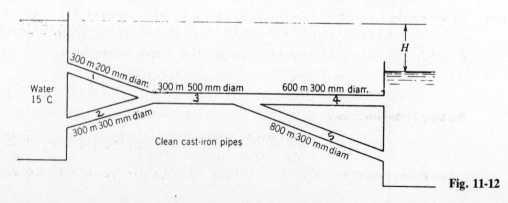

Fig. 11-12

11.19 With a velocity of 1 m/s in the 200-mm-diameter pipe of Fig. 11-12, calculate the flow through the system and the head H required.

❚ Using values from Prob. 11.18, $h_f = (f)(L/D)(v^2/2g)$, $(h_f)_1 = 0.023[300/(\frac{200}{1000})]\{1^2/[(2)(9.807)]\} = 1.759$ m, $(h_f)_2 = (h_f)_1 = 0.021[300/(\frac{300}{1000})]\{v_2^2/[(2)(9.807)]\} = 1.759$ m, $v_2 = 1.282$ m/s; $Q = Av = [(\pi)(\frac{200}{1000})^2/4](1) + [(\pi)(\frac{300}{1000})^2/4](1.282) = 0.122$ m^3/s. (This is the flow through the system.) $v_3 = Q_3/A_3 = 0.122/[(\pi)(\frac{500}{1000})^2/4] = 0.6213$ m/s, $(h_f)_3 = 0.018[300/(\frac{500}{1000})]\{0.6213^2/[(2)(9.807)]\} = 0.213$ m. The flow of 0.122 m^3/s will split in pipes 4 and 5 in the same proportions as determined in Prob. 11.18. Hence, $Q_4 = [0.08369/(0.08369 + 0.07245)](0.122) = 0.0654$ m^3/s, $v_4 = 0.0654/[(\pi)(\frac{300}{1000})^2/4] = 0.9252$ m/s, $(h_f)_4 = 0.021[600/(\frac{300}{1000})]\{0.9252^2/[(2)(9.807)]\} = 1.833$ m, $H = 1.759 + 0.213 + 1.833 = 3.805$ m.

11.20 Three pipes are in parallel with a total head loss of 20.3 m. The pipe data are

pipe	L, m	d, cm	ϵ, mm	ϵ/d
1	100	8	0.24	0.003
2	150	6	0.12	0.002
3	80	4	0.20	0.005

The fluid is water, $\rho = 1000$ kg/m^3, and $\nu = 1.02 \times 10^{-6}$ m^2/s. Calculate the total flow rate in cubic meters per hour, neglecting minor losses.

❚ $h_f = (f)(L/D)(v^2/2g)$, $(h_f)_1 = (h_f)_2 = (h_f)_3$. Guess fully rough flow in pipe 1: $f_1 = 0.0262$.

$$(h_f)_1 = 20.3 = 0.0262[100/(\tfrac{8}{100})]\{v_1^2/[(2)(9.807)]\} \qquad v_1 = 3.487 \text{ m/s}$$

$$N_R = Dv/\nu \qquad (N_R)_1 = (\tfrac{8}{100})(3.487)/(1.02 \times 10^{-6}) = 2.73 \times 10^5$$

From Fig. A-5, $f_1 = 0.0267$.

$$20.3 = 0.0267[100/(\tfrac{8}{100})]\{v_1^2/[(2)(9.807)]\} \qquad v_1 = 3.454 \text{ m/s}$$

$$Q_1 = A_1v_1 = [(\pi)(\tfrac{8}{100})^2/4](3.454) = 0.01736 \text{ m}^3/\text{s}$$

Guess $f_2 = 0.0234$:

$$(h_f)_2 = 20.3 = 0.0234[150/(\tfrac{6}{100})]\{v_2^2/[(2)(9.807)]\} \qquad v_2 = 2.609 \text{ m/s}$$

$$(N_R)_2 = (\tfrac{6}{100})(2.609)/(1.02 \times 10^{-6}) = 1.53 \times 10^5 \qquad f_2 = 0.0246 \qquad 20.3 = 0.0246[150/(\tfrac{6}{100})]\{v_2^2/[(2)(9.807)]\}$$

$$v_2 = 2.544 \text{ m/s} \qquad Q_2 = [(\pi)(\tfrac{6}{100})^2/4](2.544) = 0.007193 \text{ m}^3/\text{s}$$

Guess $f_3 = 0.0304$:

$$(h_f)_3 = 20.3 = 0.0304[80/(\tfrac{4}{100})]\{v_3^2/[(2)(9.807)]\} \qquad v_3 = 2.559 \text{ m/s}$$

$$(N_R)_3 = (\tfrac{4}{100})(2.559)/(1.02 \times 10^{-6}) = 1.00 \times 10^5$$

$$f_3 = 0.0313 \qquad 20.3 = 0.0313[80/(\tfrac{4}{100})]\{v_3^2/[(2)(9.807)]\} \qquad v_3 = 2.522 \text{ m/s}$$

$$Q_3 = [(\pi)(\tfrac{4}{100})^2/4](2.522) = 0.003169 \text{ m}^3/\text{s}$$

This is satisfactory convergence. Hence, $Q_{\text{total}} = 0.01736 + 0.007193 + 0.003169 = 0.02772$ m^3/s, or 99.8 m^3/h.

11.21 For the parallel pipe system of Fig. 11-13, each pipe is cast iron, and the pressure drop $p_1 - p_2 = 3$ psi. Compute the total flow rate between 1 and 2 if the fluid is SAE30 oil at 20 °C ($\rho = 1.78$ slugs/ft^3, $\mu = 0.00606$ lb-s/ft^2).

❚ $h_f = (p_1 - p_2)/\rho g = (3)(144)/[(1.78)(32.2)] = 7.537$ ft. Assume laminar flow.

$$h_f = \frac{128\mu L Q}{\pi \rho g D^4}$$

For the 3-in-diameter pipe:

$$7.537 = \frac{(128)(0.00606)(250)(Q_3)}{(\pi)(1.78)(32.2)(\tfrac{3}{12})^4} \qquad Q_3 = 0.02734 \text{ ft}^3/\text{s} \qquad N_R = \frac{\rho D v}{\mu}$$

$$v_3 = Q_3/A_3 = 0.02734/[(\pi)(\tfrac{3}{12})^2/4] = 0.5570 \text{ ft/s} \qquad (N_R)_3 = (1.78)(\tfrac{3}{12})(0.5570)/0.00606 = 41 \qquad \text{(laminar)}$$

For the 2-in-diameter pipe:

$$7.537 = \frac{(128)(0.00606)(200)(Q_2)}{(\pi)(1.78)(32.2)(\frac{2}{12})^4} \qquad Q_2 = 0.006750 \text{ ft}^3/\text{s} \qquad v_2 = 0.006750/[(\pi)(\tfrac{2}{12})^2/4] = 0.3094 \text{ ft/s}$$

$$(N_R)_3 = (1.78)(\tfrac{2}{12})(0.3094)/0.00606 = 15 \qquad \text{(laminar)} \qquad Q_{\text{total}} = 0.02734 + 0.006750 = 0.0341 \text{ ft}^3/\text{s}$$

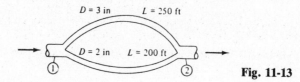

$D = 3$ in $\qquad L = 250$ ft

$D = 2$ in $\qquad L = 200$ ft

Fig. 11-13

11.22 For the pipe system of Fig. 11-14, all pipes are 6-cm-diameter wrought iron containing water at 20 °C. If the total flow rate from 1 to 2 is 0.01 m³/s, compute the total pressure drop $p_1 - p_2$. How does Q divide between the two parallel pipes? Neglect minor losses.

▌ For pipe c:

$$v_c = Q_C/A_C = 0.01/[(\pi)(\tfrac{6}{100})^2/4] = 3.537 \text{ m/s} \qquad N_R = dv/v$$

$$(N_R)_c = (\tfrac{6}{100})(3.537)/(1.02 \times 10^{-6}) = 2.08 \times 10^5 \qquad \epsilon/d = 0.000046/(\tfrac{6}{100}) = 0.00077$$

From Fig. A-5, $f_c = 0.020$.

$$h_f = (f)(L/d)(v^2/2g) \qquad (h_f)_c = 0.020[150/(\tfrac{6}{100})]\{3.537^2/[(2)(9.807)]\} = 31.89 \text{ m}$$

$$(\Delta p)_c = (\gamma)(h_f)_c = (9.79)(31.89) = 312 \text{ kPa}$$

For pipes a and b: $(h_f)_a = (h_f)_b$, $v_a = Q_a/A_a = Q_a/[(\pi)(\tfrac{6}{100})^2/4] = 353.7Q_a$, $v_b = Q_b/A_b = Q_b/[(\pi)(\tfrac{6}{100})^2/4] = 353.7Q_b$.

$$f_a[250/(\tfrac{6}{100})]\{(353.7Q_a)^2/[(2)(9.807)]\} = f_b[100/(\tfrac{6}{100})]\{(353.7Q_b)^2/[(2)(9.807)]\} \qquad (1)$$

If $f_a = f_b$, $Q_a = 0.6325Q_b$, $Q_a + Q_b = 0.6325Q_b + Q_b = 0.01$, $Q_b = 0.006126 \text{ m}^3/\text{s}$, $Q_a = (0.6325)(0.006126) = 0.003875 \text{ m}^3/\text{s}$; $(N_R)_a = \tfrac{6}{100}[(353.7)(0.003875)]/(1.02 \times 10^{-6}) = 8.06 \times 10^4$, $f_a = 0.0219$; $(N_R)_b = \tfrac{6}{100}[(353.7)(0.006125)]/(1.02 \times 10^{-6}) = 1.27 \times 10^5$, $f_b = 0.0208$. Substituting these values of f into Eq. (1), $0.0219[250/(\tfrac{6}{100})]\{353.7Q_a)^2/[(2)(9.807)]\} = 0.0208[100/(\tfrac{6}{100})]\{(353.7Q_b)^2/[(2)(9.807)]\}$, $Q_a = 0.6164Q_b$, $Q_a + Q_b = 0.6164Q_b + Q_b = 0.01$, $Q_b = 0.006187 \text{ m}^3/\text{s}$, $Q_a = (0.6164)(0.006187) = 0.003814 \text{ m}^3/\text{s}$.

$$(N_R)_a = \tfrac{6}{100}[(353.7)(0.003814)]/(1.02 \times 10^{-6}) = 7.94 \times 10^4 \qquad f_a = 0.0219 \qquad \text{(O.K.)}$$

$$(N_R)_b = \tfrac{6}{100}[(352.7)(0.006187)]/(1.02 \times 10^{-6}) = 1.29 \times 10^5 \qquad f_b = 0.0208 \qquad \text{(O.K.)}$$

Hence, $Q_a = 0.003814 \text{ m}^3/\text{s}$ and $Q_b = 0.006187 \text{ m}^3/\text{s}$.

$$(h_f)_a = (h_f)_b = 0.0219[250/(\tfrac{6}{100})]\{[(353.7)(0.003814)]^2/[(2)(9.807)]\} = 8.466 \text{ m}$$

$$(\Delta p)_{a,b} = (9.79)(8.466) = 83 \text{ kPa} \qquad (\Delta p)_{\text{total}} = 312 + 83 = 395 \text{ kPa}$$

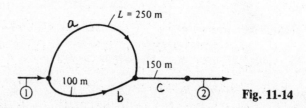

$L = 250$ m

a

150 m

100 m

c

b

Fig. 11-14

11.23 Modify Prob. 11.22 so that the flow rate is unknown but the total pressure drop is 900 kPa. Find the resulting flow rate Q.

▌ Using data from Prob. 11.22, guess $Q_a = 0.006 \text{ m}^3/\text{s}$: $Q_b = 0.006/0.6164 = 0.009734 \text{ m}^3/\text{s}$, $Q_c = 0.006 + 0.009734 = 0.015734 \text{ m}^3/\text{s}$, $v_c = Q_c/A_c = 0.015734/[(\pi)(\tfrac{6}{100})^2/4] = 5.565 \text{ m/s}$, $N_R = dv/v$, $(N_R)_c = (\tfrac{6}{100})(5.565)/(1.02 \times 10^{-6}) = 3.27 \times 10^5$, $\epsilon/d = 0.000046/(\tfrac{6}{100}) = 0.00077$. From Fig. A-5, $f_c = 0.0195$: $h_f = (f)(L/d)(v^2/2g)$, $(h_f)_c = 0.0195[150/(\tfrac{6}{100})]\{5.565^2/[(2)(9.807)]\} = 76.97 \text{ m}$.

For pipes a and b:

$$(h_f)_a = (h_f)_b \quad v_a = Q_a/A_a = 0.006/[(\pi)(\tfrac{6}{100})^2/4] = 2.122 \text{ m/s} \quad v_b = Q_b/A_b = 0.009734/[(\pi)(\tfrac{6}{100})^2/4] = 3.443 \text{ m/s}$$

$$(N_R)_a = (\tfrac{6}{100})(2.122)/(1.02 \times 10^{-6}) = 1.25 \times 10^5 \quad f_a = 0.0205$$

$$(N_R)_b = (\tfrac{6}{100})(3.443)/(1.02 \times 10^{-6}) = 2.03 \times 10^5 \quad f_b = 0.020$$

$$(h_f)_a = 0.0205[250/(\tfrac{6}{100})]\{2.122^2/[(2)(9.807)]\} = 19.61 \text{ m}$$

$$(h_f)_b = 0.020[100/(\tfrac{6}{100})]\{3.443^2/[(2)(9.807)]\} = 20.15 \text{ m}$$

Since these are fairly close, forego another iteration and take an average value of $(h_f)_{a,b}$:
$(h_f)_{a,b} = (19.61 + 20.15)/2 = 19.88 \text{ m}$, $(h_f)_{\text{total}} = 76.97 + 19.88 = 96.85 \text{ m}$, $(\Delta p)_{\text{total}} = (\gamma)(h_f)_{\text{total}} = (9.79)(96.85) =$
948 kPa. Since this is slightly larger than the required 900 kPa, reduce Q_a a small amount; e.g., try
$Q_a = 0.00585 \text{ m}^3/\text{s}$:

$$Q_b = 0.00585/0.6164 = 0.009491 \text{ m}^3/\text{s} \quad Q_c = 0.00585 + 0.009491 = 0.015341 \text{ m}^3/\text{s}$$

$$v_c = Q_c/A_c = 0.015341/[(\pi)(\tfrac{6}{100})^2/4] = 5.426 \text{ m/s} \quad (N_R)_c = (\tfrac{6}{100})(5.426)/(1.02 \times 10^{-6}) = 3.19 \times 10^5$$

$$f_c = 0.0195 \quad (h_f)_c = 0.0195[150/(\tfrac{6}{100})]\{5.426^2/[(2)(9.807)]\} = 73.18 \text{ m}$$

For pipes a and b:

$$v_a = Q_a/A_a = 0.00585/[(\pi)(\tfrac{6}{100})^2/4] = 2.069 \text{ m/s} \quad v_b = Q_b/A_b = 0.009491/[(\pi)(\tfrac{6}{100})^2/4] = 3.357 \text{ m/s}$$

$$(N_R)_a = (\tfrac{6}{100})(2.069)/(1.02 \times 10^{-6}) = 1.22 \times 10^5 \quad f_a = 0.0205$$

$$(N_R)_b = (\tfrac{6}{100})(3.357)/(1.02 \times 10^{-6}) = 1.97 \times 10^5 \quad f_b = 0.020$$

$$(h_f)_a = 0.0205[250/(\tfrac{6}{100})]\{2.069^2/[(2)(9.807)]\} = 18.64 \text{ m}$$

$$(h_f)_b = 0.020[100/(\tfrac{6}{100})]\{3.357^2/[(2)(9.807)]\} = 19.15 \text{ m}$$

Since these are fairly close, forego another iteration and take an average value of $(h_f)_{a,b}$:
$(h_f)_{a,b} = (18.64 + 19.15)/2 = 18.90 \text{ m}$, $(h_f)_{\text{total}} = 73.18 + 18.90 = 92.08 \text{ m}$, $(\Delta p)_{\text{total}} = (\gamma)(h_f)_{\text{total}} = (9.79)(92.08) =$
901 kPa (close enough). Hence, $Q = Q_c = 0.0153 \text{ m}^3/\text{s}$.

11.24 For the piping system of Fig. 11-15, all pipes are concrete with roughness of 0.03 in. If the flow rate is 18 cfs of
water at 20 °C, compute the pressure drop $p_1 - p_2$ and how the flow divides among the three parallel pipes.

▮ For pipe a:

$$v_a = Q_a/A_a = 18/[(\pi)(\tfrac{12}{12})^2/4] = 22.92 \text{ ft/s} \quad N_R = Dv/\nu$$

$$(N_R)_a = (\tfrac{12}{12})(22.92)/(1.08 \times 10^{-5}) = 2.12 \times 10^6 \quad \epsilon/D = 0.03/12 = 0.0025$$

From Fig. A-5, $f_a = 0.0249$. $h_f = (f)(L/D)(v^2/2g)$, $(h_f)_a = 0.0249[1000/(\tfrac{12}{12})]\{22.92^2/[(2)(32.2)]\} = 203.1 \text{ ft}$.

For pipes b, c, and d:

$$(h_f)_b = (h_f)_c = (h_f)_d \quad v_b = Q_b/A_b = Q_b/[(\pi)(\tfrac{8}{12})^2/4] = 2.865Q_b$$

$$v_c = Q_c/A_c = Q_c/[(\pi)(\tfrac{12}{12})^2/4] = 1.273Q_c \quad v_d = Q_d/A_d = Q_d/[(\pi)(\tfrac{15}{12})^2/4] = 0.8149Q_d$$

$$f_b\left(\frac{1500}{\tfrac{8}{12}}\right)\left[\frac{(2.865Q_b)^2}{(2)(32.2)}\right] = f_c\left(\frac{800}{\tfrac{12}{12}}\right)\left[\frac{(1.273Q_c)^2}{(2)(32.2)}\right] = f_d\left(\frac{1200}{\tfrac{15}{12}}\right)\left[\frac{(0.8149Q_d)^2}{(2)(32.2)}\right] \tag{1}$$

If $f_b = f_c = f_d$,

$$Q_b = 0.2649Q_c \quad Q_d = 1.426Q_c \quad Q_b + Q_c + Q_d = 0.2649Q_c + Q_c + 1.426Q_c = 18 \quad Q_c = 6.689 \text{ ft}^3/\text{s}$$

$$Q_b = (0.2649)(6.689) = 1.772 \text{ ft}^3/\text{s} \quad Q_d = (1.426)(6.689) = 9.539 \text{ ft}^3/\text{s}$$

$$(N_R)_b = \tfrac{8}{12}[(2.865)(1.772)]/(1.08 \times 10^{-5}) = 3.13 \times 10^5 \quad (\epsilon/D)_b = 0.03/8 = 0.0038 \quad f_b = 0.028$$

$$(N_R)_c = \tfrac{12}{12}[(1.273)(6.689)]/(1.08 \times 10^{-5}) = 7.88 \times 10^5 \quad (\epsilon/D)_c = 0.03/12 = 0.0025 \quad f_c = 0.025$$

$$(N_R)_d = \tfrac{15}{12}[(0.8149)(9.539)]/(1.08 \times 10^{-5}) = 9.00 \times 10^5 \quad (\epsilon/D)_d = 0.03/15 = 0.0020 \quad f_d = 0.0235$$

Substituting these values of f into Eq. (1),

$$0.028[1500/(\tfrac{8}{12})]\{(2.865Q_b)^2/[(2)(32.2)]\} = 0.025[800/(\tfrac{12}{12})]\{(1.273Q_c)^2/[(2)(32.2)]\}$$

$$= 0.0235[1200/(\tfrac{15}{12})]\{(0.8149Q_d)^2/[(2)(32.2)]\}$$

$$Q_b = 0.2504Q_c \quad Q_d = 1.471Q_c \quad Q_b + Q_c + Q_d = 0.2504Q_c + Q_c + 1.471Q_c = 18 \quad Q_c = 6.614 \text{ ft}^3/\text{s}$$

$$Q_b = (0.2504)(6.614) = 1.656 \text{ ft}^3/\text{s} \quad Q_d = (1.471)(6.614) = 9.729 \text{ ft}^3/\text{s}$$

$$(N_R)_b = \tfrac{8}{12}[(2.865)(1.656)]/(1.08 \times 10^{-5}) = 2.93 \times 10^5 \qquad f_b = 0.028 \qquad \text{(O.K.)}$$

$$(N_R)_c = \tfrac{12}{12}[(1.273)(6.614)]/(1.08 \times 10^{-5}) = 7.80 \times 10^5 \qquad f_c = 0.025 \qquad \text{(O.K.)}$$

$$(N_R)_d = \tfrac{15}{12}[(0.8149)(9.729)]/(1.08 \times 10^{-5}) = 9.18 \times 10^5 \qquad f_d = 0.0235 \qquad \text{(O.K.)}$$

Hence, $Q_b = 1.66 \text{ ft}^3/\text{s}$, $Q_c = 6.61 \text{ ft}^3/\text{s}$, and $Q_d = 9.73 \text{ ft}^3/\text{s}$.

$$(h_f)_b = (h_f)_c = (h_f)_d \qquad (h_f)_b = 0.028[1500/(\tfrac{8}{12})]\{[(2.865)(1.66)]^2/[(2)(32.2)]\} = 22.1 \text{ ft}$$

$$(h_f)_{\text{total}} = 203.1 + 22.1 = 225.2 \text{ ft} \qquad (\Delta p)_{\text{total}} = (\gamma)(h_f)_{\text{total}} = (62.4)(225.2) = 14\,052 \text{ lb/ft}^2 \quad \text{or} \quad 97.6 \quad \text{lb/in}^2$$

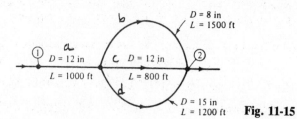

$D = 8$ in
$L = 1500$ ft
$D = 12$ in
$L = 1000$ ft
$D = 12$ in
$L = 800$ ft
$D = 15$ in
$L = 1200$ ft **Fig. 11-15**

11.25 For the system of Fig. 11-15 with gasoline at 20 °C ($v = 4.62 \times 10^{-6} \text{ ft}^2/\text{s}$, $\gamma = 42.5 \text{ lb/ft}^3$), compute the flow rate in all pipes if the pressure drop $p_1 - p_2$ is 35 psi. Neglect minor losses.

❚ Using data from Prob. 11.24, note that pipe a takes $203.1/225.2 = 0.9019$, or 90.19 percent of the total pressure drop. Hence, take $(\Delta p)_a = (35)(144)(0.9019) = 4546 \text{ lb/ft}^2$, $(h_f)_a = (\Delta p)_a/\gamma = 4546/42.5 = 107.0 \text{ ft}$, $h_f = (f)(L/d)(v^2/2g)$, $107.0 = f_a[1000/(\tfrac{12}{12})]\{v_a^2/[(2)(32.2)]\}$, $v_a = \sqrt{6.891/f_a}$. Try $f_a = 0.025$: $v_a = \sqrt{6.891/0.025} = 16.60 \text{ ft/s}$, $N_R = dv/v$, $(N_R)_a = (\tfrac{12}{12})(16.60)/(4.62 \times 10^{-6}) = 3.59 \times 10^6$, $(\epsilon/d)_a = 0.03/12 = 0.0025$. From Fig. A-5, $f_a = 0.0245$. Try $f_a = 0.0245$: $v_a = \sqrt{6.891/0.0245} = 16.77 \text{ ft/s}$, $(N_R)_a = (\tfrac{12}{12})(16.77)/(4.62 \times 10^{-6}) = 3.63 \times 10^6$, $f_a = 0.0245$ (O.K.); $Q_a = A_a v_a = [(\pi)(\tfrac{12}{12})^2/4](16.77) = 13.17 \text{ ft}^3/\text{s}$. This flow of $13.17 \text{ ft}^3/\text{s}$ for Q_a will divide among the three parallel pipes in the same proportions as found in Prob. 11.24. Hence,

$$Q_b = [1.66/(1.66 + 6.61 + 9.73)](13.17) = 1.215 \text{ ft}^3/\text{s} \qquad Q_c = [6.61/(1.66 + 6.61 + 9.73)](13.17) = 4.836 \text{ ft}^3/\text{s}$$

$$Q_d = [9.73/(1.66 + 6.61 + 9.73)](13.17) = 7.119 \text{ ft}^3/\text{s} \qquad (N_R)_b = (\tfrac{8}{12})(1.215)/(4.62 \times 10^{-6}) = 1.75 \times 10^5$$

$$(\epsilon/d)_b = 0.03/8 = 0.0038 \qquad f_b = 0.028 \qquad v_b = 1.215/[(\pi)(\tfrac{8}{12})^2/4] = 3.481 \text{ ft/s}$$

$$(h_f)_{b,c,d} = 0.028[1500/(\tfrac{8}{12})]\{3.481^2/[(2)(32.2)]\} = 11.85 \text{ ft} \qquad (\Delta p)_{b,c,d} = (42.5)(11.85) = 504 \text{ lb/ft}^2$$

$$(\Delta p)_{\text{total}} = 4546 + 504 = 5050 \text{ lb/ft}^2 \quad \text{or} \quad 35.1 \text{ lb/in}^2 \quad \text{(O.K.)}$$

Hence, $Q_a = 13.17 \text{ cfs}$, $Q_b = 1.21 \text{ cfs}$, $Q_c = 4.84 \text{ cfs}$, and $Q_d = 7.12 \text{ cfs}$.

11.26 In Fig. 11-16 all pipes are cast iron and $p_1 - p_2 = 35$ psi. Compute the total flow rate of water at 20 °C.

❚ $$h_f = (35)(144)/62.4 = 80.77 \text{ ft} = (f)(L/d)(v^2/2g)$$

For pipe c: $(h_f)_c = 80.77 = f_c[2200/(\tfrac{3}{12})]\{v_c^2/[(2)(32.2)]\}$, $v_c = \sqrt{0.5911/f_c}$. Try $f_c = 0.03$: $v_c = \sqrt{0.5911/0.03} = 4.439 \text{ ft/s}$, $N_R = dv/v$, $(N_R)_c = (\tfrac{3}{12})(4.439)/(1.08 \times 10^{-5}) = 1.03 \times 10^5$, $(\epsilon/d)_c = 0.00085/(\tfrac{3}{12}) = 0.0034$. From Fig. A-5, $f_c = 0.0283$. Try $f_c = 0.0283$: $v_c = \sqrt{0.5911/0.0283} = 4.570 \text{ ft/s}$, $(N_R)_c = (\tfrac{3}{12})(4.570)/(1.08 \times 10^{-5}) = 1.06 \times 10^5$, $f_c = 0.0283$ (O.K.); $Q_c = A_c v_c = [(\pi)(\tfrac{3}{12})^2/4](4.570) = 0.2243 \text{ ft}^3/\text{s}$.

For pipes a and b: $(h_f)_{a,b} = 80.77 = f_a[1000/(\tfrac{2}{12})]\{v_a^2/[(2)(32.2)]\} + f_b[1500/(\tfrac{4}{12})]\{v_b^2/[(2)(32.2)]\}$. From continuity, $v_b = v_a/4$. $80.77 = f_a[1000/(\tfrac{2}{12})]\{v_a^2/[(2)(32.2)]\} + f_b[1500/(\tfrac{4}{12})]\{(v_a/4)^2/[(2)(32.2)]\}$, $5202 = (v_a^2)(6000f_a + 281.2f_b)$. Try $f_a = f_b = 0.03$:

$$5202 = (v_a^2)[(6000)(0.03) + (281.2)(0.03)] \qquad v_a = 5.254 \text{ ft/s} \qquad (N_R)_a = (\tfrac{2}{12})(5.254)/(1.08 \times 10^{-5}) = 8.11 \times 10^4$$

$$(\epsilon/d)_a = 0.00085/(\tfrac{2}{12}) = 0.0051 \qquad f_a = 0.031 \qquad v_b = 5.254/4 = 1.314 \text{ ft/s}$$

$$(N_R)_b = (\tfrac{4}{12})(1.314)/(1.08 \times 10^{-5}) = 4.06 \times 10^4$$

$$(\epsilon/d)_b = 0.00085/(\tfrac{4}{12}) = 0.0026 \qquad f_b = 0.028$$

Try $f_a = 0.031$ and $f_b = 0.028$:

$$5202 = (v_a^2)[(6000)(0.031) + (281.2)(0.028)] \qquad v_1 = 5.180 \text{ ft/s}$$

$$(N_R)_a = (\tfrac{2}{12})(5.180)/(1.08 \times 10^{-5}) = 7.99 \times 10^4 \qquad f_a = 0.031 \qquad \text{(O.K.)}$$

$$v_b = 5.180/4 = 1.295 \text{ ft/s} \qquad (N_R)_b = (\tfrac{4}{12})(1.295)/(1.08 \times 10^{-5}) = 4.00 \times 10^4 \qquad f_b = 0.028 \qquad \text{(O.K.)}$$

$$Q_a = [(\pi)(\tfrac{2}{12})^2/4](5.180) = 0.1130 \text{ ft}^3/\text{s} \qquad Q_{\text{total}} = 0.2243 + 0.1130 = 0.3373 \text{ ft}^3/\text{s}$$

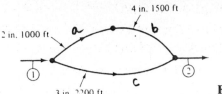

Fig. 11-16

11.27 In Fig. 11-16, all pipes are cast iron, and the total flow rate from 1 to 2 is 0.5 cfs of water at 20 °C. Compute the pressure drop $p_1 - p_2$ and the division of flow between the two paths.

▌ The flow will divide between the two paths in the same proportions as found in Prob. 11.26. Hence,

$$Q_c = [0.2243/(0.2243 + 0.1130)](0.5) = 0.3325 \text{ ft}^3/\text{s}$$

$$Q_a = Q_b = [0.1130/(0.2243 + 0.1130)](0.5) = 0.1675 \text{ ft}^3/\text{s}$$

$$h_f = (f)(L/d)(v^2/2g) \qquad v_c = Q_c/A_c = 0.3325/[(\pi)(\tfrac{3}{12})^2/4] = 6.774 \text{ ft/s} \qquad N_R = dv/v$$

$$(N_R)_c = (\tfrac{3}{12})(6.774)/(1.08 \times 10^{-5}) = 1.57 \times 10^5 \qquad (\epsilon/d)_c = 0.00085/(\tfrac{3}{12}) = 0.0034$$

From Fig. A-5, $f_c = 0.028$. $(h_f)_c = 0.028[2200/(\tfrac{3}{12})]\{6.774^2/[(2)(32.2)]\} = 175.6 \text{ ft}$, $p_1 - p_2 = (\gamma)(h_f)_c = (62.4)(175.6) = 10\,957 \text{ lb/ft}^2$, or 76.1 lb/in^2.

11.28 In Fig. 11-17, all three pipes are cast iron, Neglecting minor losses, compute the flow rate in each pipe for water at 20 °C.

▌ Assume $(h_f)_c = 20 \text{ ft}$.

$$\alpha = gd^3 h_f/Lv^2 \qquad N_R = -\sqrt{8\alpha} \log[(\epsilon/d)/3.7 + 2.51/\sqrt{2\alpha}]$$

$$\alpha_c = (32.2)(\tfrac{4}{12})^3(20)/[(1200)(1.08 \times 10^{-5})^2] = 1.704 \times 10^8 \qquad (\epsilon/d)_c = 0.00085/(\tfrac{4}{12}) = 0.00255$$

$$(N_R)_c = -\sqrt{(8)(1.704 \times 10^8)} \log[0.00255/3.7 + 2.51/\sqrt{(2)(1.704 \times 10^8)}] = 1.138 \times 10^5 \qquad N_R = 4Q/\pi vd$$

$$1.138 \times 10^5 = 4Q_c/[(\pi)(1.08 \times 10^{-5})(\tfrac{4}{12})] \qquad Q_c = 0.3218 \text{ ft}^3/\text{s} \qquad (h_f)_a = (h_f)_b = 50 - 20 = 30 \text{ ft}$$

$$\alpha_a = (32.2)(\tfrac{3}{12})^3(30)/[(1000)(1.08 \times 10^{-5})^2] = 1.294 \times 10^8 \qquad (\epsilon/d)_a = 0.00085/(\tfrac{3}{12}) = 0.00340$$

$$(N_R)_a = -\sqrt{(8)(1.294 \times 10^8)} \log[0.00340/3.7 + 2.51/\sqrt{(2)(1.294 \times 10^8)}] = 9.551 \times 10^4$$

$$9.551 \times 10^4 = 4Q_a/[(\pi)(1.08 \times 10^{-5})(\tfrac{3}{12})] \qquad Q_a = 0.2025 \text{ ft}^3/\text{s}$$

$$\alpha_b = (32.2)(\tfrac{2}{12})^3(30)/[(700)(1.08 \times 10^{-5})^2] = 5.477 \times 10^7 \qquad (\epsilon/d)_b = 0.00085/(\tfrac{2}{12}) = 0.00510$$

$$(N_R)_b = -\sqrt{(8)(5.477 \times 10^7)} \log[0.00510/3.7 + 2.51/\sqrt{(2)(5.447 \times 10^7)}] = 5.842 \times 10^4$$

$$5.842 \times 10^4 = 4Q_b/[(\pi)(1.08 \times 10^{-5})(\tfrac{2}{12})] \qquad Q_b = 0.0826 \text{ ft}^3/\text{s} \qquad Q_a = Q_b = 0.2025 + 0.0826 = 0.2851 \text{ ft}^3/\text{s}$$

Since this value of $Q_a + Q_b = 0.2851 \text{ ft}/\text{s}$ is not equal to $Q_c = 0.3218 \text{ ft}^3/\text{s}$, the assumed value of $(h_f)_c = 20 \text{ ft}$ is incorrect. Try $(h_f)_c = 17.2 \text{ ft}$:

$$\alpha_c = (32.2)(\tfrac{4}{12})^3(17.2)/[(1200)(1.08 \times 10^{-5})^2] = 1.466 \times 10^8$$

$$(N_R)_c = -\sqrt{(8)(1.466 \times 10^8)} \log[0.00255/3.7 + 2.51/\sqrt{(2)(1.466 \times 10^8)}] = 1.054 \times 10^5$$

$$1.054 \times 10^5 = 4Q_c/[(\pi)(1.08 \times 10^{-5})(\tfrac{4}{12})] \qquad Q_c = 0.2980 \text{ ft}^3/\text{s}$$

$$(h_f)_a = (h_f)_b = 50 - 17.2 = 32.8 \text{ ft} \qquad \alpha_a = (32.2)(\tfrac{3}{12})^3(32.8)/[(1000)(1.08 \times 10^{-5})^2] = 1.415 \times 10^8$$

$$(N_R)_a = -\sqrt{(8)(1.415 \times 10^8)} \log[0.00340/3.7 + 2.51/\sqrt{(2)(1.415 \times 10^8)}] = 9.997 \times 10^4$$

$$9.997 \times 10^4 = 4Q_a/[(\pi)(1.08 \times 10^{-5})(\tfrac{3}{12})] \qquad Q_a = 0.2120 \text{ ft}^3/\text{s}$$

$$\alpha_b = (32.2)(\tfrac{2}{12})^3(32.8)/[(700)(1.08 \times 10^{-5})^2] = 5.989 \times 10^7$$

$$(N_R)_b = -\sqrt{(8)(5.989 \times 10^7)} \log[0.00510/3.7 + 2.51/\sqrt{(2)(5.989 \times 10^7)}] = 6.115 \times 10^4$$

$$6.115 \times 10^4 = 4Q_b/[(\pi)(1.08 \times 10^{-5})(\tfrac{2}{12})] \qquad Q_b = 0.0864 \text{ ft}^3/\text{s} \qquad Q_a + Q_b = 0.2120 + 0.0864 = 0.2984 \text{ ft}^3/\text{s}$$

This value of $Q_a + Q_b = 0.2984 \text{ ft}^3/\text{s}$ is practically the same as $Q_c = 0.2980 \text{ ft}^3/\text{s}$. Hence, take $Q_a = 0.212 \text{ cfs}$, $Q_b = 0.086 \text{ cfs}$, $Q_c = 0.298 \text{ cfs}$.

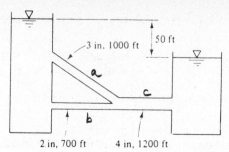

2 in, 700 ft   4 in, 1200 ft   **Fig. 11-17**

11.29 For the pipe system in Fig. 11-18, the pressure head at A is 120.0 ft of water and the pressure head at E is 72.0 ft of water. Assuming the pipes are in a horizontal plane, what are the flows in each branch of the loop? Neglect minor losses.

▌

$$h_f = 120.0 - 72.0 = 48.0 \text{ ft} \qquad Q = 1.318ACR^{0.63}s^{0.54}$$

$$Q_B = 1.318[(\pi)(\tfrac{12}{12})^2/4](100)[(\tfrac{12}{12})/4]^{0.63}(48.0/12\,000)^{0.54} = 2.19 \text{ ft}^3/\text{s}$$

$$Q_C = 1.318[(\pi)(\tfrac{8}{12})^2/4](100)[(\tfrac{8}{12})/4]^{0.63}(48.0/4000)^{0.54} = 1.37 \text{ ft}^3/\text{s}$$

$$Q_B = 1.318[(\pi)(\tfrac{10}{12})^2/4](100)[(\tfrac{10}{12})/4]^{0.63}(48.0/8000)^{0.54} = 1.69 \text{ ft}^3/\text{s}$$

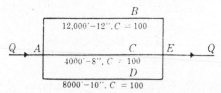

Fig. 11-18

11.30 In Fig. 11-18, if the total flow is 6.50 mgd, how much head loss occurs between A and E and how does Q divide in the loop?

▌ The total flow will divide in the loop in the same proportions as in Prob. 11.29. Hence, $Q_B =$ [2.19/(2.19 + 1.37 + 1.69)](6.50) = 2.71 mgd, or 4.20 cfs; $Q_C =$ [1.37/(2.19 + 1.37 + 1.69)](6.50) = 1.70 mgd, or 2.62 cfs; $Q_D =$ [1.69/(2.19 + 1.37 + 1.69)](6.50) = 2.09 mgd, or 3.24 cfs; $Q = 1.318ACR^{0.63}s^{0.54}$. For pipe B, $4.20 = 1.318[(\pi)(\tfrac{12}{12})^2/4](100)[(\tfrac{12}{12})/4]^{0.63}s^{0.54}$, $s = 0.01334$; $(h_f)_B = (0.01334)(12\,000) = 160$ ft. This is the head loss between A and E. It should, of course, be the same when computed by pipes C and D. To confirm this, for pipe C, $2.62 = 1.318[(\pi)(\tfrac{8}{12})^2/4](100)[(\tfrac{8}{12})/4]^{0.63}s^{0.54}$, $s = 0.04010$; $(h_f)_C = (0.04010)(4000) = 160$ ft. For pipe D, $3.24 = 1.318[(\pi)(\tfrac{10}{12})^2/4](100)[(\tfrac{10}{12})/4]^{0.63}s^{0.54}$, $s = 0.02005$; $(h_f)_B = (0.02005)(8000) = 160$ ft.

11.31 For the system shown in Fig. 11-19, what flow will occur when the drop in the hydraulic grade line from A to B is 200 ft?

▌ Assume $(h_f)_{WZ} = 30$ ft. Using Fig. A-13: For pipe 2, with $d = 12$ in and $h_1 = \tfrac{30}{5000} = 0.0060$, $Q_2 = 3.2$ ft^3/s. For pipe 3, with $d = 16$ in and $h_1 = \tfrac{30}{3000} = 0.0100$, $Q_3 = 9.0$ ft^3/s. For pipe 1, with $d = 24$ in and $Q = 3.2 + 9.0 = 12.2$ ft^3/s, $(h_1)_1 = 0.0024$ ft/ft; $(h_f)_{AW} = (0.0024)(10\,000) = 24.0$ ft. For pipe 4, with $d = 20$ in and $Q = 12.2$ ft^3/s, $(h_1)_4 = 0.0060$ ft/ft; $(h_f)_{ZB} = (0.0060)(8000) = 48.0$ ft, $(h_f)_{A-B} = 24.0 + 30 + 48.0 = 102.0$ ft. This value of $(h_f)_{A-B} = 24.0 + 30 + 48.0 = 102.0$ ft is not equal to the given value of 200 ft, but the actual head losses will be in the same proportions as those above. Hence, $(h_f)_{AW} = (24.0/102.0)(200) = 47.06$ ft. For pipe 1, with $d = 24$ in and $h_1 = 47.06/10\,000 = 0.0047$, $Q = 18$ cfs.

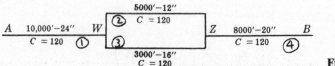

Fig. 11-19

11.32 In Prob. 11.31, what length of 20-in pipe with $C = 120$ is equivalent to section A–B?

▌ From Prob. 11.31, it is known that a flow of 18 cfs is produced when the drop in the hydraulic grade line from A to B is 200 ft. For 18 cfs in a 20-in pipe, from Fig. A-13, $h_1 = 0.012$ ft/ft, $0.012L_e = 200$, $L_e = 16\,700$ ft.

11.33 In Fig. 11-20, when pump YA delivers 5.00 cfs, find the pressure heads at A and B. Draw the hydraulic grade lines.

❚ Reduce loop BC to an equivalent pipe, 16 in in diameter, $C = 100$. By so doing, a single-size pipe of the same relative roughness is readily handled for all conditions of flow. Assume a drop in the grade line of 22 ft ftom B to C. $Q = 1.318AR^{0.63}s^{0.54}$.

For the 10-in pipe: $Q_{10} = 1.318[(\pi)(\frac{10}{12})^2/4](90)[(\frac{10}{12})/4]^{0.63}(22/10\,000)^{0.54} = 0.8843$ cfs.

For the 8-in pipe: $Q_8 = 1.318[(\pi)(\frac{8}{12})^2/4](100)[(\frac{8}{12})/4]^{0.63}(22/11\,000)^{0.54} = 0.5190$ cfs. $Q_{\text{total}} = 0.8843 + 0.5190 = 1.403$ cfs. For a 16-in-diameter equivalent pipe with $C = 100$, $1.403 = 1.318[(\pi)(\frac{16}{12})^2/4](100)[(\frac{16}{12})/4]^{0.63}s^{0.54}$, $s = 0.0004313$ ft/ft; $0.0004313L_e = 22$, $L_e = 51\,000$ ft. For a 16-in-diameter pipe from A to C with a length of $16\,000 + 51\,000 = 67\,000$ ft and carrying 5.00 cfs, $5.00 = 1.318[(\pi)(\frac{16}{12})^2/4](100)[(\frac{16}{12})/4]^{0.63}s^{0.54}$, $s = 0.004537$ ft/ft; $h_f = (0.004537)(67\,000) = 304.0$ ft. Thus the elevation of the hydraulic grade line at A is $217.0 + 304.0 = 521.0$ ft, as shown in the figure. The drop from A to B is $(0.004537)(16\,000) = 72.6$ ft and the elevation of the hydraulic grade line at B becomes $521.0 - 72.6 = 448.4$ ft. Pressure head at $A = 521.0 - 50.0 = 471.0$ ft, pressure head at $B = 448.4 - 50.0 = 398.4$ ft.

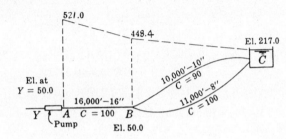

Fig. 11-20

11.34 In Fig. 11-21, which system has the greater capacity, $ABCD$ or $EFGH$? Use $C = 120$ for all pipes.

❚ Assume $Q = 3$ cfs in $ABCD$. Using Fig. A-13,

$$(h_1)_{AB} = 0.0014 \text{ ft/ft} \qquad (h_f)_{AB} = (0.0014)(9000) = 12.6 \text{ ft}$$
$$(h_1)_{BC} = 0.0053 \text{ ft/ft} \qquad (h_f)_{BC} = (0.0053)(6000) = 31.8 \text{ ft}$$

$(h_1)_{CD} = 0.013$ ft/ft $\quad (h_f)_{AB} = (0.013)(3000) = 39.0$ ft $\qquad (h_f)_{\text{total}} = 12.6 + 31.8 + 39.0 = 83.4$ ft   (for $ABCD$)

For $EFGH$, assume $(h_f)_{FG} = 24$ ft.

$$(h_1)_{FIG} = \tfrac{24}{5000} = 0.00480 \text{ ft/ft} \qquad (h_1)_{FJG} = \tfrac{24}{7000} = 0.00343 \text{ ft/ft} \qquad Q_{FIG} = 0.97 \text{ cfs} \qquad Q_{FJG} = 1.6 \text{ cfs}$$

Hence, pipe FIG carries $0.97/(0.97 + 1.6) = 0.3774$, or 37.74 percent of the flow and pipe FJG carries $1.6/(0.97 + 1.6) = 0.6226$, or 62.26 percent. For $Q = 3.0$ cfs in pipe EF,

$$(h_1)_{EF} = 0.00074 \text{ ft/ft} \qquad (h_f)_{EF} = (0.00074)(11\,000) = 8.1 \text{ ft} \qquad Q_{FIG} = (0.3774)(3.0) = 1.13 \text{ cfs}$$
$$(h_1)_{FIG} = 0.0060 \text{ ft/ft} \qquad (h_f)_{FIG} = (0.0060)(5000) = 30.0 \quad \text{ ft}$$

For $Q = 3.0$ cfs in pipe GH,

$(h_1)_{GH} = 0.013$ ft/ft $\qquad (h_f)_{GH} = (0.013)(2500) = 32.5$ ft $\qquad (h_f)_{\text{total}} = 8.1 + 30.0 + 32.5 = 70.6$ ft   (for $EFGH$)

Since $EFGH$ carries the assumed flow of 3.0 cfs with a lesser head loss than $ABCD$, it ($EFGH$) has the greater capacity.

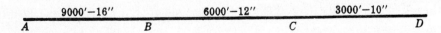

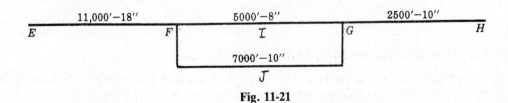

Fig. 11-21

11.35 Three pipes A, B, and C are interconnected as shown in Fig. 11-22. The pipe characteristics are as follows:

pipe	D, in	L, ft	f
A	6	2000	0.020
B	4	1600	0.032
C	8	4000	0.024

Find the rate at which water will flow in each pipe. Find also the pressure at point P. Neglect minor losses.

$$p_1/\gamma + v_1^2/2g + z_1 = p_2/\gamma + v_2^2/2g + z_2 + h_L \qquad (1)$$

$$h_L = h_f = (f)(L/d)(v^2/2g) = 0.020[2000/(\tfrac{6}{12})]\{v_A^2/[(2)(32.2)]\}$$

$$+ 0.024[4000/(\tfrac{8}{12})]\{v_C^2/[(2)(32.2)]\} = 1.242v_A^2 + 2.236v_C^2$$

$$0 + 0 + 200 = 0 + v_C^2/[(2)(32.2)] + 50 + 1.242v_A^2 + 2.236v_C^2$$

$$150 = 1.242v_A^2 + 2.252v_C^2 \qquad (2)$$

$$Q_A + Q_B = Q_C \qquad [(\pi)(\tfrac{6}{12})^2/4](v_A) + [(\pi)(\tfrac{4}{12})^2/4](v_B) = [(\pi)(\tfrac{8}{12})^2/4](v_C)$$

$$0.1963v_A + 0.08727v_B = 0.3491v_C \qquad (3)$$

$$(h_f)_A = (h_f)_B \qquad 0.020[2000/(\tfrac{6}{12})]\{v_A^2/[(2)(32.2)]\} = 0.032[1600/(\tfrac{4}{12})]\{v_B^2/[(2)(32.2)]\} \qquad v_B = 0.7217v_A$$

Substituting into Eq. (3), $0.1963v_A + (0.08727)(0.7217v_A) = 0.3491v_C$, $v_A = 1.346v_C$. Substituting into Eq. (2), $150 = (1.242)(1.346v_C)^2 + 2.252v_C^2$, $v_C = 5.772$ ft/s; $Q_C = A_Cv_C = [(\pi)(\tfrac{8}{12})^2/4](5.772) = 2.01$ ft^3/s, $v_A = (1.346)(5.772) = 7.769$ ft/s, $Q_A = [(\pi)(\tfrac{6}{12})^2/4](7.769) = 1.53$ ft^3/s. Substituting into Eq. (3), $(0.1963)(7.769) + 0.08727v_B = (0.3491)(5.772)$, $v_B = 5.614$ ft/s; $Q_B = [(\pi)(\tfrac{4}{12})^2/4](5.614) = 0.490$ ft^3/s. To find p_P, apply Eq. (1) between points P and 2. $(p_P)(144)/62.4 + v_P^2/2g + 120 = 0 + v_2^2/2g + 50 + 0.024[4000/(\tfrac{8}{12})]\{5.772^2/[(2)(32.2)]\}$, $v_P^2/2g = v_2^2/2g$, $p_P = 1.95$ lb/in^2.

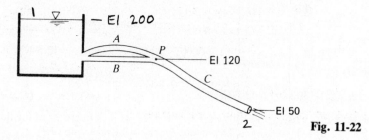

Fig. 11-22

11.36 Suppose that in Fig. 11-23 pipes 1, 2, and 3 are 500 ft of 2-in diameter, 300 ft of 3-in diameter, and 700 ft of 4-in diameter, respectively. The pipes are all smooth brass. When the total flow of crude oil at 80 °F (s.g. = 0.855) is 0.6 cfs, find the head loss from A to B and the flow in each pipe.

$$h_f = (f)(L/d)(v^2/2g) \qquad (h_f)_1 = (h_f)_2 = (h_f)_3$$

$$f_1[500/(\tfrac{2}{12})]\{v_1^2/[(2)(32.2)]\} = f_2[300/(\tfrac{3}{12})]\{v_2^2/[(2)(32.2)]\} = f_3[700/(\tfrac{4}{12})]\{v_3^2/[(2)(32.2)]\} \qquad (1)$$

If $f_1 = f_2 = f_3$,

$$46.58v_1^2 = 18.63v_2^2 = 32.61v_3^2 \qquad v_2 = 1.581v_1 \qquad v_3 = 1.195v_1 \qquad Q_{\text{total}} = Q_1 + Q_2 + Q_3 = A_1v_1 + A_2v_2 + A_3v_3$$

$$0.6 = [(\pi)(\tfrac{2}{12})^2/4](v_1) + [(\pi)(\tfrac{3}{12})^2/4](1.581v_1) + [(\pi)(\tfrac{4}{12})^2/4](1.195v_1)$$

$$v_1 = 2.945 \text{ ft/s} \qquad v_2 = (1.581)(2.945) = 4.656 \text{ ft/s} \qquad v_3 = (1.195)(2.945) = 3.519 \text{ ft/s} \qquad N_R = dv/v$$

From Fig. A-2, $v = 6.5 \times 10^{-5}$ ft^2/s, $(N_R)_1 = (\tfrac{2}{12})(2.945)/(6.5 \times 10^{-5}) = 7551$, $(N_R)_2 = (\tfrac{3}{12})(4.656)/(6.5 \times 10^{-5}) = 17\,908$, $(N_R)_3 = (\tfrac{4}{12})(3.519)/(6.5 \times 10^{-5}) = 18\,046$. For Reynolds numbers in this range the Blasius formula can be used to determine friction factors: $f = 0.316/N_R^{0.25}$.

$$f_1 = 0.316/7551^{0.25} = 0.0339 \qquad f_2 = 0.316/17\,908^{0.25} = 0.0273 \qquad f_3 = 0.315/18\,046^{0.25} = 0.0273$$

Substituting these values of f into Eq. (1),

$$0.0339[500/(\tfrac{2}{12})]\{v_1^2/[(2)(32.2)]\} = 0.0273[300/(\tfrac{3}{12})]\{v_2^2/[(2)(32.2)]\} = 0.0273[700/(\tfrac{4}{12})]\{v_3^2/[(2)(32.2)]\}$$

$$1.579v_1^2 = 0.5087v_2^2 = 0.8902v_3^2 \qquad v_2 = 1.762v_1 \qquad v_3 = 1.332v_1$$

$$0.6 = [(\pi)(\tfrac{2}{12})^2/4](v_1) + [(\pi)(\tfrac{3}{12})^2/4](1.762v_2) + [(\pi)(\tfrac{4}{12})^2/4](1.323v_1)$$

$$v_1 = 2.681 \text{ ft/s} \qquad v_2 = (1.762)(2.681) = 4.724 \text{ ft/s} \qquad v_3 = (1.332)(2.681) = 3.571 \text{ ft/s}$$

$$(N_R)_1 = (\tfrac{2}{12})(2.681)/(6.5 \times 10^{-5}) = 6874 \qquad (N_R)_2 = (\tfrac{3}{12})(4.724)/(6.5 \times 10^{-5}) = 18\,169$$

$$(N_R)_3 = (\tfrac{4}{12})(3.571)/(6.5 \times 10^{-5}) = 18\,313$$

$$f_1 = 0.316/6874^{0.25} = 0.0347 \qquad f_2 = 0.316/18\,169^{0.25} = 0.0272 \qquad f_3 = 0.316/18\,313^{0.25} = 0.0272$$

$$0.0347[500/(\tfrac{2}{12})]\{v_1^2/[(2)(32.2)]\} = 0.0272[300/(\tfrac{3}{12})]\{v_2^2/[(2)(32.2)]\}$$

$$= 0.0272[700/(\tfrac{4}{12})]\{v_3^2/[(2)(32.2)]\}$$

$$1.616v_1^2 = 0.5068v_2^2 = 0.8870v_3^2 \qquad v_2 = 1.786v_1 \qquad v_3 = 1.350v_1$$

$$0.6 = [(\pi)(\tfrac{2}{12})^2/4](v_1) + [(\pi)(\tfrac{3}{12})^2/4](1.786v_1) + [(\pi)(\tfrac{4}{12})^2/4](1.350v_1)$$

$$v_1 = 2.640 \text{ ft/s} \qquad v_2 = (1.786)(2.640) = 4.715 \text{ ft/s} \qquad v_3 = (1.350)(2.640) = 3.564 \text{ ft/s}$$

$$(N_R)_1 = (\tfrac{2}{12})(2.640)/(6.5 \times 10^{-5}) = 6769 \qquad (N_R)_2 = (\tfrac{3}{12})(4.715)/(6.5 \times 10^{-5}) = 18\,135$$

$$(N_R)_3 = (\tfrac{4}{12})(3.564)/(6.5 \times 10^{-5}) = 18\,277$$

$$f_1 = 0.316/6769^{0.25} = 0.0348 \quad \text{(O.K.)} \qquad f_2 = 0.316/18\,135^{0.25} = 0.0272 \quad \text{(O.K.)}$$

$$f_3 = 0.316/18\,277^{0.25} = 0.0272 \quad \text{(O.K.)}$$

$$Q_1 = [(\pi)(\tfrac{2}{12})^2/4](2.640) = 0.0576 \text{ ft}^3/\text{s} \qquad Q_2 = [(\pi)(\tfrac{3}{12})^2/4](4.715) = 0.231 \text{ ft}^3/\text{s}$$

$$Q_3 = [(\pi)(\tfrac{4}{12})^2/4](3.564) = 0.311 \text{ ft}^3/\text{s}$$

$$(h_f)_{AB} = (h_f)_1 = 0.0348[500/(\tfrac{2}{12})]\{2.640^2/[(2)(32.2)]\} = 11.3 \text{ ft}$$

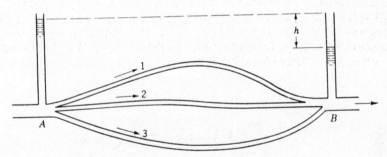

Fig. 11-23

11.37 Repeat Prob. 11.36 where the flow rate is 0.06 cfs.

$\blacksquare$ Assume this flow will be laminar. $h_f = (32v)(L/gD^2)(v)$. Since $(h_f)_1 = (h_f)_2 = (h_f)_3$,

$$L_1v_1/D_1^2 = L_2v_2/D_2^2 = L_3v_3/D_3^2 \qquad 500v_1/(\tfrac{2}{12})^2 = 300v_2/(\tfrac{3}{12})^2 = 700v_3/(\tfrac{4}{12})^2 \qquad v_2 = 3.750v_1 \qquad v_3 = 2.857v_1$$

$$Q_{\text{total}} = Q_1 + Q_2 + Q_3 = A_1v_1 + A_2v_2 + A_3v_3$$

$$0.06 = [(\pi)(\tfrac{2}{12})^2/4](v_1) + [(\pi)(\tfrac{3}{12})^2/4](3.750v_1) + [(\pi)(\tfrac{4}{12})^2/4](2.857v_1)$$

$$v_1 = 0.1318 \text{ ft/s} \qquad v_2 = (3.750)(0.1318) = 0.4942 \text{ ft/s} \qquad v_3 = (2.857)(0.1318) = 0.3766 \text{ ft/s} \qquad N_R = dv/v$$

From Fig. A-2, $v = 6.5 \times 10^{-5} \text{ ft}^2/\text{s}$;

$$(N_R)_1 = (\tfrac{2}{12})(0.1318)/(6.5 \times 10^{-5}) = 338 \qquad \text{(laminar)}$$

$$(N_R)_2 = (\tfrac{3}{12})(0.4942)/(6.5 \times 10^{-5}) = 1901 \qquad \text{(laminar)}$$

$$(N_R)_3 = (\tfrac{4}{12})(0.3766)/(6.5 \times 10^{-5}) = 1931 \qquad \text{(laminar)}$$

$$Q_1 = [(\pi)(\tfrac{2}{12})^2/4](0.1318) = 0.00288 \text{ ft}^3/\text{s} \qquad Q_2 = [(\pi)(\tfrac{3}{12})^2/4](0.4942) = 0.0243 \text{ ft}^3/\text{s}$$

$$Q_3 = [(\pi)(\tfrac{4}{12})^2/4](0.3766) = 0.0329 \text{ ft}^3/\text{s}$$

$$(h_f)_{AB} = (h_f)_1 = (32)(6.5 \times 10^{-5})\{500/[(32.2)(\tfrac{2}{12})^2]\}(0.1318) = 0.153 \text{ ft}$$

11.38 Suppose that in Fig. 11-23 pipes 1, 2, and 3 are copper tubing as follows: 80 m of 3 cm, 100 m of 4 cm, and 80 m of 5 cm, respectively. When the total flow of crude oil at 40 °C (s.g. = 0.855) is 5 L/s, find the head loss from A to B and the flow in each pipe.

▌ From Fig. A-2, $v = 4.8 \times 10^{-6}$ m²/s. Assume Reynolds numbers between 3000 and 100 000, in which case the Blasius equation can be used to determine f: $h_f = (f)(L/d)(v^2/2g)$, $f = 0.316/N_R^{0.25}$, $N_R = dv/v$. Therefore,

$$h_f = \left[\frac{0.316}{(dv/v)^{0.25}}\right]\left(\frac{L}{d}\right)\left(\frac{v^2}{2g}\right) = \frac{0.316 v^{0.25} L v^{1.75}}{2gd^{1.25}} \qquad h_f \propto \frac{Lv^{1.75}}{d^{1.25}} \qquad (h_f)_1 = (h_f)_2 = (h_f)_3$$

$$(80)(v_1^{1.75})/(\tfrac{3}{100})^{1.25} = (100)(v_2^{1.75})/(\tfrac{4}{100})^{1.25} = (80)(v_3^{1.75})/(\tfrac{5}{100})^{1.25} \qquad v_1 = 0.9250 v_2 \qquad v_3 = 1.332 v_2$$

$$Q_{total} = Q_1 + Q_2 + Q_3 = A_1 v_1 + A_2 v_2 + A_3 v_3$$

$$\tfrac{5}{1000} = [(\pi)(\tfrac{3}{100})^2/4](0.9250 v_2) + [(\pi)(\tfrac{4}{100})^2/4](v_2) + [(\pi)(\tfrac{5}{100})^2/4](1.332 v_2)$$

$$v_2 = 1.105 \text{ m/s} \qquad v_1 = (0.9250)(1.105) = 1.022 \text{ m/s} \qquad v_3 = (1.332)(1.105) = 1.472 \text{ m/s}$$

$$(N_R)_1 = (\tfrac{3}{100})(1.022)/(4.8 \times 10^{-6}) = 6.39 \times 10^3 \qquad (N_R)_2 = (\tfrac{4}{100})(1.105)/(4.8 \times 10^{-6}) = 9.21 \times 10^3$$

$$(N_R)_3 = (\tfrac{5}{100})(1.472)/(4.8 \times 10^{-6}) = 1.53 \times 10^4$$

Hence, use of the Blasius equation is O.K.

$$Q_1 = [(\pi)(\tfrac{3}{100})^2/4](1.022) = 0.00072 \text{ m}^3/\text{s} \quad \text{or} \quad 0.72 \text{ L/s}$$

$$Q_2 = [(\pi)(\tfrac{4}{100})^2/4](1.105) = 0.00139 \text{ m}^3/\text{s} \quad \text{or} \quad 1.39 \text{ L/s}$$

$$Q_3 = [(\pi)(\tfrac{5}{100})^2/4](1.472) = 0.00289 \text{ m}^3/\text{s} \quad \text{or} \quad 2.89 \text{ L/s}$$

$$(h_f)_{AB} = (h_f)_1 = \frac{(0.316)(4.8 \times 10^{-6})^{0.25}(80)(1.022)^{1.75}}{(2)(9.807)(\tfrac{3}{100})^{1.25}} = 5.02 \text{ m}$$

11.39 Repeat Prob. 11.38 for the case where the total flow rate is 0.40 L/s.

▌ Assume laminar flow, in which case the following equation can be used to determine head loss: $h_f = (32v)(L/gd^2)(v)$, $h_f \propto Lv/d^2$.

$$(h_f)_1 = (h_f)_2 = (h_f)_3 \qquad 80 v_1/(\tfrac{3}{100})^2 = 100 v_2/(\tfrac{4}{100})^2 = 80 v_3/(\tfrac{5}{100})^2 \qquad v_1 = 0.70310 v_2 \qquad v_3 = 1.953 v_2$$

$$Q_{total} = Q_1 + Q_2 + Q_3 = A_1 v_1 + A_2 v_2 + A_3 v_3$$

$$0.40/1000 = [(\pi)(\tfrac{3}{100})^2/4](0.7031 v_2) + [(\pi)(\tfrac{4}{100})^2/4](v_2) + [(\pi)(\tfrac{5}{100})^2/4](1.953 v_2)$$

$$v_2 = 0.07158 \text{ m/s} \qquad v_1 = (0.7031)(0.07158) = 0.05033 \text{ m/s} \qquad v_3 = (1.953)(0.07158) = 0.1398 \text{ m/s}$$

$$(N_R)_1 = (\tfrac{3}{100})(0.05033)/(4.8 \times 10^{-6}) = 315 \qquad \text{(laminar)}$$

$$(N_R)_2 = (\tfrac{4}{100})(0.07158)/(4.8 \times 10^{-6}) = 597 \qquad \text{(laminar)}$$

$$(N_R)_3 = (\tfrac{5}{100})(0.1398)/(4.8 \times 10^{-6}) = 1456 \qquad \text{(laminar)}$$

$$Q_1 = [(\pi)(\tfrac{3}{100})^2/4](0.05033) = 0.0000356 \text{ m}^3/\text{s} \quad \text{or} \quad 0.0356 \text{ L/s}$$

$$Q_2 = [(\pi)(\tfrac{4}{100})^2/4](0.07158) = 0.0000900 \text{ m}^3/\text{s} \quad \text{or} \quad 0.0900 \text{ L/s}$$

$$Q_3 = [(\pi)(\tfrac{5}{100})^2/4](0.1398) = 0.0002745 \text{ m}^3/\text{s} \quad \text{or} \quad 0.2745 \text{ L/s}$$

$$(h_f)_{AB} = (h_f)_1 = (32)(4.8 \times 10^{-6})\{80/[(9.807)(\tfrac{3}{100})^2]\}(0.05033) = 0.0701 \text{ m}$$

11.40 The pipes in the system shown in Fig. 11-24 are all cast iron. With a flow of 20 cfs, find the head loss from A to D. Neglect minor losses.

▌ $Q_n = A_n \sqrt{2gh_L/(f_n)(L_n/d_n)}$ or $Q_n = C_n\sqrt{h_L}$, where $C_n = A_n\sqrt{2gd_n/f_nL_n}$. Assume $f_1 = f_2 = f_3 = 0.019$.

$$C_2 = [(\pi)(\tfrac{14}{12})^2/4]\sqrt{(2)(32.2)(\tfrac{14}{12})/[(0.019)(3500)]} = 1.136$$

$$C_3 = [(\pi)(\tfrac{12}{12})^2/4]\sqrt{(2)(32.2)(\tfrac{12}{12})/[(0.019)(2500)]} = 0.9145$$

$$C_4 = [(\pi)(\tfrac{16}{12})^2/4]\sqrt{(2)(32.2)(\tfrac{16}{12})/[(0.019)(3000)]} = 1.714$$

$$Q_2 = [1.136/(1.136 + 0.9145 + 1.714)](20) = 6.035 \text{ ft}^3/\text{s}$$

$$Q_3 = [0.9145/(1.136 + 0.9145 + 1.714)](20) = 4.859 \text{ ft}^3/\text{s}$$

$$Q_4 = [1.714/(1.136 + 0.9145 + 1.714)](20) = 9.106 \text{ ft}^3/\text{s}$$

$$v_2 = Q_2/A_2 = 6.035/[(\pi)(\tfrac{14}{12})^2/4] = 5.645 \text{ ft/s} \qquad v_3 = 4.859/[(\pi)(\tfrac{12}{12})^2/4] = 6.187 \text{ ft/s}$$

$$v_4 = 9.106/[(\pi)(\tfrac{16}{12})^2/4] = 6.522 \text{ ft/s} \qquad N_R = dv/v$$

$$(N_R)_2 = (\tfrac{14}{12})(5.645)/(1.05 \times 10^{-5}) = 6.27 \times 10^5$$

$$(N_R)_3 = (\tfrac{12}{12})(6.187)/(1.05 \times 10^{-5}) = 5.89 \times 10^5$$

$$(N_R)_4 = (\tfrac{16}{12})(6.522)/(1.05 \times 10^{-5}) = 8.28 \times 10^5$$

$$(\epsilon/d)_2 = 0.00085/(\tfrac{14}{12}) = 0.00073 \qquad (\epsilon/d)_3 = 0.00085/(\tfrac{12}{12}) = 0.00085 \qquad (\epsilon/d)_4 = 0.00085/(\tfrac{16}{12}) = 0.00064$$

From Fig. A-5, $f_2 = 0.0185$, $f_3 = 0.0195$, and $f_4 = 0.0185$.

$$C_2 = [(\pi)(\tfrac{14}{12})^2/4]\sqrt{(2)(32.2)(\tfrac{14}{12})/[(0.0185)(3500)]} = 1.152$$
$$C_3 = [(\pi)(\tfrac{12}{12})^2/4]\sqrt{(2)(32.2)(\tfrac{12}{12})/[(0.0195)(2500)]} = 0.9027$$
$$C_4 = [(\pi)(\tfrac{16}{12})^2/4]\sqrt{(2)(32.2)(\tfrac{16}{12})/[(0.0185)(3000)]} = 1.737$$
$$Q_2 = [1.152/(1.152 + 0.9027 + 1.737)](20) = 6.076 \text{ ft}^3/\text{s}$$
$$Q_3 = [0.9027/(1.152 + 0.9027 + 1.737)](20) = 4.761 \text{ ft}^3/\text{s}$$
$$Q_4 = [1.737/(1.152 + 0.9027 + 1.737)](20) = 9.162 \text{ ft}^3/\text{s}$$
$$v_2 = Q_2/A_2 = 6.076/[(\pi)(\tfrac{14}{12})^2/4] = 5.684 \text{ ft/s} \qquad v_3 = 4.761/[(\pi)(\tfrac{12}{12})^2/4] = 6.062 \text{ ft/s}$$
$$v_4 = 9.162/[(\pi)(\tfrac{16}{12})^2/4] = 6.562 \text{ ft/s}$$
$$(N_R)_2 = (\tfrac{14}{12})(5.684)/(1.05 \times 10^{-5}) = 6.32 \times 10^5$$
$$(N_R)_3 = (\tfrac{12}{12})(6.062)/(1.05 \times 10^{-5}) = 5.77 \times 10^5$$
$$(N_R)_4 = (\tfrac{16}{12})(6.562)/(1.05 \times 10^{-5}) = 8.33 \times 10^5$$
$$f_2 = 0.0185 \qquad f_3 = 0.0195 \qquad \text{and} \qquad f_4 = 0.0185 \qquad \text{(O.K.)}$$
$$Q_{\text{total}} = \sqrt{h_L}(C_2 + C_3 + C_4) \qquad 20 = \sqrt{(h_L)_{BC}}(1.152 + 0.9027 + 1.737)$$
$$(h_L)_{BC} = 27.82 \text{ ft} \qquad h_L = h_f = (f)(L/d)(v^2/2g)$$
$$v_1 = 20/[(\pi)(\tfrac{24}{12})^2/4] = 6.366 \text{ ft/s} \qquad (N_R)_1 = (\tfrac{24}{12})(6.366)/(1.05 \times 10^{-5}) = 1.21 \times 10^6$$
$$(\epsilon/d)_1 = 0.00085/(\tfrac{24}{12}) = 0.00042 \qquad f_1 = 0.0168$$
$$(h_L)_1 = 0.0168[3000/(\tfrac{24}{12})]\{6.366^2/[(2)(32.2)]\} = 15.86 \text{ ft} \qquad v_5 = 20/[(\pi)(\tfrac{30}{12})^2/4] = 4.074 \text{ ft/s}$$
$$(N_R)_5 = (\tfrac{30}{12})(4.074)/(1.05 \times 10^{-5}) = 9.70 \times 10^5 \qquad (\epsilon/d)_5 = 0.00085/(\tfrac{30}{12}) = 0.00034 \qquad f_5 = 0.016$$
$$(h_L)_5 = 0.016[5000/(\tfrac{30}{12})]\{4.074^2/[(2)(32.2)]\} = 8.25 \text{ ft} \qquad (h_L)_{AD} = 15.86 + 27.82 + 8.25 = 51.93 \text{ ft}$$

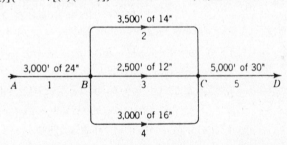

Fig. 11-24

11.41 In Prob. 11.40, what should be the diameter of a single pipe from B to C such that it replaces pipes 2, 3, and 4 without altering the capacity for the same head loss from A to D?

▎ $h_f = (f)(L/d)(v^2/2g)$. Assume $f = 0.016$:

$$v = Q/A = 20/(\pi d^2/4) = 25.46/d^2 \qquad 27.82 = (0.016)(2500/d)\{(25.46/d^2)^2/[(2)(32.2)]\} \qquad d = 1.707 \text{ ft}$$
$$N_R = dv/v = (1.707)(25.46/1.707^2)/(1.05 \times 10^{-5}) = 1.42 \times 10^6 \qquad \epsilon/d = 0.00085/1.707 = 0.00050$$

From Fig. A-5, $f = 0.017$. Try $f = 0.017$:

$$27.82 = (0.017)(2500/d)\{(25.46/d^2)^2/[(2)(32.2)]\} \qquad d = 1.727 \text{ ft}$$
$$N_R = (1.727)(25.46/1.727^2)/(1.05 \times 10^{-5}) = 1.40 \times 10^6 \qquad f = 0.017 \qquad \text{(O.K.)}$$

Therefore, $d = 1.727$ ft, or 20.7 in.

11.42 With the same pipe lengths, sizes, and connections as in Fig. 11-24, find the flow in each pipe if the head loss from A to D is 100 ft and if all pipes have $f = 0.018$. Also find the head losses from A to B, B to C, and C to D.

▎ $$h_f = (f)(L/d)(v^2/2g) = (fL/d)(Q^2/2gA^2) = (fL/2gd)[Q/(\pi d^2/4)]^2 = (8f/\pi^2 g)(LQ^2/d^5)$$

Hence,

$$h_f \propto LQ^2/d^5 \qquad (h_f)_2 = (h_f)_3 = (h_f)_4 \qquad 3500Q_2^2/(\tfrac{14}{12})^5 = 2500Q_3^2/(\tfrac{12}{12})^5 = 3000Q_4^2/(\tfrac{16}{12})^5$$
$$Q_2 = 1.243Q_3 \qquad Q_4 = 1.874Q_3 \qquad Q_1 = Q_5 = Q_2 + Q_3 + Q_4 = 1.243Q_3 + Q_3 + 1.874Q_3 = 4.117Q_3$$
$$(h_f)_{AD} = (h_f)_1 + (h_f)_3 + (h_f)_5$$

$$100 = \{(8)(0.018)/[(\pi)^2(32.2)]\}[(3000)(4.117Q_3)^2/(\tfrac{24}{12})^5 + (2500)(Q_3)^2/(\tfrac{12}{12})^5 + (5000)(4.117Q_3)^2/(\tfrac{30}{12})^5]$$

$$Q_3 = 6.67 \text{ ft}^3/\text{s} \qquad Q_2 = (1.243)(6.67) = 8.29 \text{ ft}^3/\text{s}$$

$$Q_4 = (1.873)(6.67) = 12.50 \text{ ft}^3/\text{s} \qquad Q_1 = Q_5 = (4.117)(6.67) = 27.46 \text{ ft}^3/\text{s}$$

$$(h_f)_{AB} = \{(8)(0.018)/[(\pi)^2(32.2)]\}(3000)(27.46)^2/(\tfrac{24}{12})^5 = 32.03 \text{ ft}$$

$$(h_f)_{BC} = \{(8)(0.018)/[(\pi)^2(32.2)]\}(2500)(6.67)^2/(\tfrac{12}{12})^5 = 50.40 \text{ ft}$$

$$(h_f)_{CD} = \{(8)(0.018)/[(\pi)^2(32.2)]\}(5000)(27.46)^2/(\tfrac{30}{12})^5 = 17.49 \text{ ft}$$

11.43 In Prob. 11.42, find the new head loss distributions and the percentage increase in the capacity of the system achieved by adding another 12-in pipe 2500 ft long between B and C.

❚ The additional pipe from B to C is identical to pipe 3. As in Prob. 11.42,

$$h_f = (8f/\pi^2 g)(LQ^2/d^5) \qquad Q_2 = 1.243Q_3 \qquad Q_4 = 1.874Q_3$$

$$Q_1 = Q_5 = Q_2 + 2Q_3 + Q_4 = 1.243Q_3 + 2Q_3 + 1.874Q_3 = 5.117Q_3 \qquad (h_f)_{AD} = (h_f)_1 + (h_f)_3 + (h_f)_5$$

$$100 = \{(8)(0.018)/[(\pi)^2(32.2)]\}[(3000)(5.117Q_3)^2/(\tfrac{24}{12})^5 + (2500)(Q_3)^2/(\tfrac{12}{12})^5 + (5000)(5.117Q_3)^2/(\tfrac{30}{12})^5]$$

$$Q_3 = 5.92 \text{ ft}^3/\text{s} \qquad Q_1 = Q_5 = (5.117)(5.92) = 30.29 \text{ ft}^3/\text{s}$$

$$(h_f)_{AB} = \{(8)(0.018)/[(\pi)^2(32.2)]\}(3000)(30.29)^2/(\tfrac{24}{12})^5 = 38.97 \text{ ft}$$

$$(h_f)_{BC} = \{(8)(0.018)/[(\pi)^2(32.2)]\}(2500)(5.92)^2/(\tfrac{12}{12})^5 = 39.70 \text{ ft}$$

$$(h_f)_{CD} = \{(8)(0.018)/[(\pi)^2(32.2)]\}(5000)(30.29)^2/(\tfrac{30}{12})^5 = 21.29 \text{ ft}$$

Increase in capacity $= (30.29 - 27.46)/27.46 = 0.103$, or 10.3 percent.

11.44 Compute the flow in each pipe of the system shown in Fig. 11-25, and determine the pressures at B and C. Pipe AB is 1000 ft long, 6 in in diameter, $f = 0.03$; pipe BC (upper) is 600 ft long, 4 in in diameter, $f = 0.02$; pipe BC (lower) is 800 ft long, 2 in in diameter, $f = 0.04$; pipe CD is 400 ft long, 4 in in diameter, $f = 0.02$. The elevations are reservoir surface 100 ft; A, 80 ft; B, 50 ft; C, 40 ft; and D, 25 ft. There is free discharge to the atmosphere at C. Neglect velocity heads. Neglect minor losses.

❚ $h_f = (f)(L/d)(v^2/2g)$. Let U denote the upper parallel pipe from B to C and L, the lower one. For B to C, $(h_f)_U = (h_f)_L$.

$$0.02[600/(\tfrac{4}{12})]\{v_U^2/[(2)(32.2)]\} = 0.04[800/(\tfrac{2}{12})]\{v_L^2/[(2)(32.2)]\} \qquad v_U = 2.309v_L$$

$$Q = Q_U + Q_L = A_U v_u + A_L v_L = [(\pi)(\tfrac{4}{12})^2/4](v_U) + [(\pi)(\tfrac{2}{12})^2/4](v_L) = [(\pi)(\tfrac{4}{12})^2/4](2.309v_L) + [(\pi)(\tfrac{2}{12})^2/4](v_L)$$

$$= 0.2233v_L \qquad Q = Q_U + A_L v_L = Q_U + [(\pi)(\tfrac{2}{12})^2/4](Q/0.2233)$$

$$Q_U = 0.9023Q \qquad Q_L = (1 - 0.9023)(Q) = 0.0977Q \qquad v_U = Q_U/A_U = 0.9023Q/[(\pi)(\tfrac{4}{12})^2/4] = 10.34Q$$

$$v_{AB} = Q/A_{AB} = Q/[(\pi)(\tfrac{6}{12})^2/4] = 5.093Q \qquad v_U = (10.34)(v_{AB}/5.093) = 2.030v_{AB} \qquad h_f = (h_f)_{AB} + (h_f)_U + (h_f)_{CD}$$

$$100 - 25 = 0.03[1000/(\tfrac{6}{12})]\{v_{AB}^2/[(2)(32.2)]\} + 0.02[600/(\tfrac{4}{12})]\{v_U^2/[(2)(32.2)]\} + 0.02[400/(\tfrac{4}{12})]\{v_{CD}^2/[(2)(32.2)]\}$$

$$[(\pi)(\tfrac{4}{12})^2/4](v_{AB}) = [(\pi)(\tfrac{4}{12})^2/4](v_{CD}) \qquad v_{CD} = 2.250v_{AB}$$

$$100 - 25 = 0.03[1000/(\tfrac{6}{12})]\{v_{AB}^2/[(2)(32.2)]\} + 0.02[600/(\tfrac{4}{12})]\{(2.030v_{AB})^2/[(2)(32.2)]\}$$

$$+ 0.02[400/(\tfrac{4}{12})]\{2.250v_{AB})^2/[(2)(32.2)]\}$$

$$v_{AB} = 3.827 \text{ ft/s} \qquad Q = Q_{AB} = [(\pi)(\tfrac{6}{12})^2/4](3.827) = 0.751 \text{ ft}^3/\text{s} \qquad Q_{CD} = Q_{AB} = 0.751 \text{ ft}^3/\text{s}$$

$$Q_U = (0.9023)(0.751) = 0.678 \text{ ft}^3/\text{s} \qquad Q_L = (0.0977)(0.751) = 0.073 \text{ ft}^3/\text{s}$$

$$p_1/\gamma + v_1^2/2g + z_1 = p_B/\gamma + v_B^2/2g + z_B + h_f$$

$$(h_f)_{AB} = 0.03[1000/(\tfrac{6}{12})]\{3.827^2/[(2)(32.2)]\} = 13.65 \text{ ft} \qquad 0 + 0 + 100 = (p_B)(144)/62.4 + 0 + 50 + 13.65$$

$$p_B = 15.75 \text{ lb/in}^2 \qquad p_C/\gamma + v_C^2/2g + z_C = p_D/\gamma + v_D^2/2g + z_D + h_f$$

$$(h_f)_{CD} = 0.02[400/(\tfrac{4}{12})]\{[(2.250)(3.827)]^2/[(2)(32.2)]\} = 27.63 \text{ ft}$$

$$(p_C)(144)/62.4 + 0 + 40 = 0 + 0 + 25 + 27.63 \qquad p_C = 5.47 \text{ lb/in}^2$$

11.45 Refer to Fig. 11-25. Compute the flow in each pipe and determine the pressures at B and C. Pipe AB is 500 m long, 20 cm in diameter, and $f = 0.03$; pipe BC (upper) is 400 m long, 10 cm in diameter, and $f = 0.02$; pipe BC (lower) is 300 m long, 15 cm in diameter, and $f = 0.025$; pipe CD is 800 m long, 30 cm in diameter, and $f = 0.018$. The elevations are reservoir water surface, 100 m; A, 80 m; B, 50 m; C, 40 m; and D, 15 m. There is free discharge to the atmosphere at D. Neglect velocity heads.

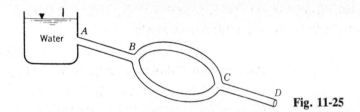

Fig. 11-25

▌ $h_f = (f)(L/d)(v^2/2g)$. Let U denote the upper parallel pipe from B to C and L, the lower one. For B to C, $(h_f)_U = (h_f)_L$.

$$0.02[400/(\tfrac{10}{100})]\{v_U^2/[(2)(9.807)]\} = 0.025[300/(\tfrac{15}{100})]\{v_L^2/[(2)(9.807)]\} \qquad v_U = 0.7906 v_L$$

$$Q = Q_U + Q_L = A_U v_U + A_L v_L = [(\pi)(\tfrac{10}{100})^2/4](v_U) + [(\pi)(\tfrac{15}{100})^2/4](v_L) = [(\pi)(\tfrac{10}{100})^2/4](0.7906 v_L) + [(\pi)(\tfrac{15}{100})^2/4](v_L)$$

$$Q = 0.02388 v_L \qquad Q = Q_U + A_L v_L = Q_U + [(\pi)(\tfrac{15}{100})^2/4](Q/0.02388)$$

$$Q_U = 0.2600 Q \qquad Q_L = (1 - 0.2600)(Q) = 0.7400 Q \qquad v_U = Q_U/A_U = 0.2600 Q/[(\pi)(\tfrac{10}{100})^2/4] = 33.10 Q$$

$$v_{AB} = Q/A_{AB} = Q/[(\pi)(\tfrac{20}{100})^2/4] = 31.83 Q \qquad v_U = (33.10)(v_{AB}/31.83) = 1.040 v_{AB}$$

$$h_f = (h_f)_{AB} + (h_f)_U + (h_f)_{CD}$$

$$100 - 15 = 0.03[500/(\tfrac{20}{100})]\{v_{AB}^2/[(2)(9.807)]\} + 0.02[400/(\tfrac{10}{100})]\{v_U^2/[(2)(9.807)]\}$$

$$+ 0.018[800/(\tfrac{30}{100})]\{v_{CD}^2/[(2)(9.807)]\}$$

$$[(\pi)(\tfrac{20}{100})^2/4](v_{AB}) = [(\pi)(\tfrac{30}{100})^2/4](v_{CD}) \qquad v_{CD} = 0.4444 v_{AB}$$

$$100 - 15 = 0.03[500/(\tfrac{20}{100})]\{v_{AB}^2/[(2)(9.807)]\} + 0.02[400/(\tfrac{10}{100})]\{(1.040 v_{AB})^2/[(2)(9.807)]\}$$

$$+ 0.018[800/(\tfrac{30}{100})]\{(0.4444 v_{AB})^2/[(2)(9.807)]\}$$

$$v_{AB} = 3.122 \text{ m/s} \qquad Q = Q_{AB} = [(\pi)(\tfrac{20}{100})^2/4](3.122) = 0.0981 \text{ m}^3/\text{s} \qquad Q_{CD} = Q_{AB} = 0.0981 \text{ m}^3/\text{s}$$

$$Q_U = (0.2600)(0.0981) = 0.0255 \text{ m}^3/\text{s} \qquad Q_L = (0.7400)(0.0981) = 0.0726 \text{ m}^3/\text{s}$$

$$p_1/\gamma + v_1^2/2g + z_1 = p_B/\gamma + v_B^2/2g + z_B + h_f$$

$$(h_f)_{AB} = 0.03[500/(\tfrac{20}{100})]\{3.122^2/[(2)(9.807)]\} = 37.27 \text{ m}$$

$$0 + 0 + 100 = p_B/9.79 + 0 + 50 + 37.27 \qquad p_B = 125 \text{ kPa}$$

$$p_C/\gamma + v_C^2/2g + z_C = p_D/\gamma + v_D^2/2g + z_D + h_f$$

$$(h_f)_{CD} = 0.018[800/(\tfrac{30}{100})]\{[(0.4444)(3.122)]^2/[(2)(9.807)]\} = 4.711 \text{ m}$$

$$p_C/9.79 + 0 + 40 = 0 + 0 + 15 + 4.711 \qquad p_C = -199 \text{ kPa}$$

The head loss in the parallel-pipe system is so great that it pulls the hydraulic grade line down below zero absolute pressure. Therefore, the sytem will not function.

11.46 Referring to Fig. 11-26, when the pump develops 25 ft of head, the velocity of flow in pipe C is 4 fps. Neglecting minor losses, find the flow rates in pipes A and B under these conditions. The pipe characteristics are as follows: pipe A, 4000 ft long, 2 ft diameter, and $f = 0.03$; pipe B, 4000 ft long, 1 ft in diameter, and $f = 0.03$; pipe C, 4000 ft long, 2 ft diameter, and $f = 0.02$.

▌ $h_f = (f)(L/d)(v^2/2g) \qquad (h_f)_A = (0.03)(\tfrac{4000}{2})\{v_A^2/[(2)(32.2)]\} = 0.9317 v_A^2 \qquad (h_L)_C = (h_f)_C - h_{\text{pump}}$

$(h_f)_C = (0.02)(\tfrac{4000}{2})\{4^2/[(2)(32.2)]\} = 9.94 \text{ ft} \qquad (h_L)_C = 9.94 - 25 = -15.06 \text{ ft}$

Hence the energy is greater at the right end of C and flow will be to the left in A.

$$0.9317 v_A^2 = 15.06 \qquad v_A = 4.020 \text{ ft/s} \qquad Q_A = A_A v_A = [(\pi)(2)^2/4](4.020) = 12.63 \text{ ft}^3/\text{s} \qquad \text{(to the left)}$$

$$Q_C = [(\pi)(2)^2/4](4) = 12.57 \text{ ft}^3/\text{s} \qquad \text{(to the right)}$$

$$Q_B = Q_A + Q_C = -12.63 + 12.57 = -0.06 \text{ ft}^3/\text{s} \qquad \text{(i.e., to the left)}$$

(The flow in pipe B is virtually zero.)

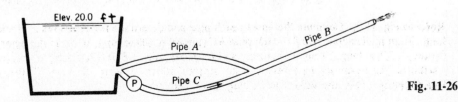

Fig. 11-26

11.47 In Prob. 11.46, find the elevation (El) of pipe B at discharge.

❚ $$\text{El}_B = \text{El}_{\text{reservoir surface}} + h_{\text{pump}} - \sum h_f = 20.0 + 25 - 9.94 = 35.1 \text{ ft}$$

11.48 Repeat Prob. 11.46 for the case where the velocity in pipe C is 5 fps with all other data remaining the same.

❚ $h_f = (f)(L/d)(v^2/2g)$ $\quad (h_f)_A = (0.03)(\frac{4000}{2})\{v_A^2/[(2)(32.2)]\} = 0.9317 v_A^2$ $\quad (h_L)_C = (h_f)_C - h_{\text{pump}}$
$(h_f)_C = (0.02)(\frac{4000}{2})\{5^2/[(2)(32.2)]\} = 15.53 \text{ ft}$ $\quad (h_L)_C = 15.53 - 25 = -9.47 \text{ ft}$

Hence the energy is greater at the right end of C and flow will be to the left in A.

$0.9317 v_A^2 = 9.47$ $\quad v_A = 3.188 \text{ ft/s}$ $\quad Q_A = A_A v_A = [(\pi)(2)^2/4](3.188) = 10.02 \text{ ft}^3/\text{s}$ $\quad$ (to the left)
$$Q_C = [(\pi)(2)^2/4](5) = 15.71 \text{ ft}^3/\text{s} \quad \text{(to the right)}$$
$$Q_B = Q_A + Q_C = -10.02 + 15.71 = 5.69 \text{ ft}^3/\text{s} \quad \text{(i.e., to the right)}$$

11.49 In Prob. 11.48, find the elevation of pipe B at discharge.

❚ $$\text{El}_B = \text{El}_{\text{reservoir surface}} + h_{\text{pump}} - \sum h_f - v_B^2/2g \quad\quad v_B = Q_B/A_B = 5.69[(\pi)(1)^2/4] = 7.245 \text{ ft/s}$$

$(h_f)_B = (0.03)(\frac{4000}{1})\{7.245^2/[(2)(32.2)]\} = 97.81 \text{ ft}$ $\quad\quad \sum h_f = 15.53 + 97.81 = 113.34 \text{ ft}$

$$\text{El}_B = 20.0 + 25 - 113.34 - 7.245^2/[(2)(32.2)] = -69.2 \text{ ft}$$

11.50 Refer to Fig. 11-26. Assume the water surface in the reservoir is at elevation 100 m. Pipes A, B, and C are all 800 m long, and they all have a diameter of 60 cm with $f = 0.025$. Neglecting minor losses, find the flow rate in all pipes under conditions where the pump develops 10 m of head when the velocity in pipe C is 3.0 m/s.

❚ $h_f = (f)(L/d)(v^2/2g)$ $\quad (h_f)_A = 0.025[800/(\frac{60}{100})]\{v_A^2/[(2)(9.807)]\} = 1.699 v_A^2$ $\quad (h_L)_C = (h_f)_C - h_{\text{pump}}$
$(h_f)_C = 0.025[800/(\frac{60}{100})]\{3.0^2/[(2)(9.807)]\} = 15.30 \text{ m}$ $\quad (h_L)_C = 15.30 - 10 = 5.30 \text{ ft}$
$$1.699 v_A^2 = 5.30 \quad v_A = 1.76 \text{ m/s}$$
$$Q_A = A_A v_A = [(\pi)(\tfrac{60}{100})^2/4](1.766) = 0.499 \text{ m}^2/\text{s} \quad \text{(to the right)}$$
$$Q_C = [(\pi)(\tfrac{60}{100})^2/4](3.0) = 0.848 \text{ m}^3/\text{s} \quad \text{(to the right)}$$
$$Q_B = Q_A + Q_C = 0.499 + 0.848 = 1.347 \text{ m}^3/\text{s} \quad \text{(to the right)}$$

11.51 In Prob. 11.50, find the elevation of pipe B at discharge.

❚ $$\text{El}_B = \text{El}_{\text{reservoir surface}} + h_{\text{pump}} - \sum h_f - v_B^2/2g \quad\quad v_B = Q_B/A_B = 1.347/[(\pi)(\tfrac{60}{100})^2/4] = 4.764 \text{ m/s}$$

$(h_f)_B = 0.025[800/(\frac{60}{100})]\{4.764^2/[(2)(9.807)]\} = 38.57 \text{ m}$ $\quad\quad \sum h_f = 15.30 + 38.57 = 53.87 \text{ m}$
$$\text{El}_B = 100 + 10 - 53.87 - 4.764^2/[(2)(9.807)] = 55.0 \text{ m}$$

11.52 Using $n = 0.013$ and neglecting minor losses, express the head loss through the pipe system of Fig. 11-27 in the form of $h_L = KQ^x$.

❚ $$h_L = \frac{n^2 v^2 L}{(1.486)^2 (R^{2/3})^2} = \frac{n^2 Q^2 L}{(1.486)^2 R^{4/3} A^2} \quad\quad (h_L)_{AB} = \frac{(0.013)^2(Q)^2(500)}{(1.486)^2[(\frac{24}{12})/4]^{4/3}[(\pi)(\frac{24}{12})^2/4]^2} = 0.009770 Q^2$$

$$(h_L)_{CD} = \frac{(0.013)^2(Q)^2(1000)}{(1.486)^2[(\frac{18}{12})/4]^{4/3}[(\pi)(\frac{18}{12})^2/4]^2} = 0.09063 Q^2 \quad\quad (h_L)_{BEC} = (h_L)_{BFC}$$

$$\frac{Q_1^2 L_1}{R_1^{4/3} A_1^2} = \frac{Q_2^2 L_2}{R_2^{4/3} A_2^2} \quad\quad \frac{(Q_1^2)(4000)}{[(\frac{18}{12})/4]^{4/3}[(\pi)(\frac{18}{12})^2/4]^2} = \frac{(Q_2^2)(5000)}{[(\frac{12}{12})/4]^{4/3}[(\pi)(\frac{12}{12})^2/4]^2}$$

$$Q_1 = 3.269 Q_2 \quad\quad Q = Q_1 + Q_2 = 3.269 Q_2 + Q_2 \quad\quad Q_2 = 0.2342 Q$$

$$(h_L)_{BFC} = \frac{(0.013)^2(Q_2)^2(5000)}{(1.486)^2[(\frac{12}{12})/4]^{4/3}[(\pi)(\frac{12}{12})^2/4]^2} = 3.939 Q_2^2$$

$(h_L)_{BFC} = (3.939)(0.2342 Q)^2 = 0.05485 Q^2$ $\quad\quad (h_L)_{AD} = 0.009770 Q^2 + 0.09063 Q^2 + 0.05485 Q^2 = 0.155 Q^2$

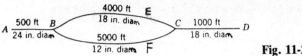

Fig. 11-27

11.53 A pipe system connects two reservoirs whose difference in elevation is 15 m. The pipe system consists of 300 m of 60-cm concrete pipe (pipe A), branching into 600 m of 30 cm (pipe B) and 45 cm (pipe C) in parallel, which join again to a single 60-cm line 1500 m long (pipe D). What would be the flow in each pipe? Assume $f = 0.030$ for all pipes.

$\blacksquare$

$$(h_f)_B = (h_f)_C \qquad h_f = (f)(L/D)(v^2/2g)$$

$$(f)[600/(\tfrac{30}{100})]\{v_B^2/[(2)(9.807)]\} = f[600/(\tfrac{45}{100})]\{v_C^2/[(2)(9.807)]\} \qquad v_C = 1.225v_B \qquad Q_A = Q_B + Q_C$$

$$[(\pi)(\tfrac{60}{100})^2/4](v_A) = [(\pi)(\tfrac{30}{100})^2/4](v_B) + [(\pi)(\tfrac{45}{100})^2/4](v_C) \qquad 0.2827v_A = 0.07069v_B + 0.1590v_C$$

$$0.2827v_A = 0.07069v_B + (0.1590)(1.225v_B) \qquad v_B = 1.065v_A$$

Convert the parallel pipes to a single equivalent 60-cm-diameter pipe.

$$f[600/(\tfrac{30}{100})]\{v_B^2/[(2)(9.807)]\} = f[L_e/(\tfrac{60}{100})]\{v_A^2/[(2)(9.807)]\}$$

$$102.0V_B^2 = 0.08497L_e v_A^2 \qquad (102.0)(1.065v_A)^2 = 0.08497L_e v_A^2 \qquad L_e = 1362 \text{ m} \qquad \Delta\text{El} = \sum h_f$$

$$15 = 0.030[(300 + 1362 + 1500)/(\tfrac{60}{100})]\{v_A^2/[(2)(9.807)]\} \qquad v_A = 1.364 \text{ m/s}$$

$$Q_A = A_A v_A = [(\pi)(\tfrac{60}{100})^2/4](1.364) = 0.386 \text{ m}^3/\text{s} \qquad Q_D = Q_A = 0.386 \text{ m}^3/\text{s} \qquad v_B = (1.065)(1.364) = 1.453 \text{ m/s}$$

$$Q_B = [(\pi)(\tfrac{30}{100})^2/4](1.453) = 0.103 \text{ m}^3/\text{s} \qquad Q_C = Q_A - Q_B = 0.386 - 0.103 = 0.283 \text{ m}^3/\text{s}$$

11.54 Refer to Fig. 11-27. Suppose $p_A/\gamma = 6.5$ ft, $p_D/\gamma = 20$ ft, and $z_A = z_D$. A pump in the 4000-ft pipe (flow from left to right) develops 30 ft of head. Find the flow rate in each pipe. Assume $n = 0.013$ for all pipes.

$\blacksquare$

$$Q_{AB} = Q_{BEC} - Q_{BFC} = Q_{CD} \tag{1}$$

$$6.5 - (h_f)_{AB} = p_B/\gamma \tag{2}$$

$$p_B/\gamma + h_{\text{pump}} - (h_f)_{BEC} = p_C/\gamma \tag{3}$$

$$p_C/\gamma - (h_f)_{BFC} = p_B/\gamma \tag{4}$$

$$p_C/\gamma - (h_f)_{CD} = 20 \tag{5}$$

$$h_f = n^2 v^2 L/(1.486R^{2/3})^2 \qquad v_{AB} = Q_{AB}/A_{AB} = Q_{AB}/[(\pi)(\tfrac{24}{12})^2/4] = 0.3183Q_{AB}$$

$$(h_f)_{AB} = (0.013)^2(0.3183Q_{AB})^2(500)/\{(1.486)[(\tfrac{24}{12})/4]^{2/3}\}^2 = 0.009769Q_{AB}^2 \tag{6}$$

$$v_{BEC} = Q_{BEC}/[(\pi)(\tfrac{18}{12})^2/4] = 0.5659Q_{BEC}$$

$$(h_f)_{BEC} = (0.013)^2(0.5659Q_{BEC})^2(4000)/\{(1.486)[(\tfrac{18}{12})/4]^{2/3}\}^2 = 0.3625Q_{BEC}^2 \tag{7}$$

$$v_{BFC} = Q_{BFC}/[(\pi)(\tfrac{12}{12})^2/4] = 1.273Q_{BFC}$$

$$(h_f)_{BFC} = (0.013)^2(1.273Q_{BFC})^2(5000)/\{(1.486)[(\tfrac{12}{12})/4]^{2/3}\}^2 = 3.938Q_{BFC}^2 \tag{8}$$

$$v_{CD} = Q_{CD}/[(\pi)(\tfrac{18}{12})^2/4] = 0.5659Q_{CD}$$

$$(h_f)_{CD} = (0.013)^2(0.5659Q_{CD})^2(1000)/\{(1.486)[(\tfrac{18}{12})/4]^{2/3}\}^2 = 0.09063Q_{CD}^2 \tag{9}$$

Substitute Eqs. (6), (7), (8), and (9) into Eqs. (2), (3), (4), and (5), respectively, and then solve simultaneously between Eqs. (2), (3), (4), and (5), introducing also Eq. (1). Two equations result, such as

$$Q_{BEC}^2 - 2Q_{BEC}Q_{BFC} - 37.40Q_{BFC}^2 = -132 \qquad Q_{BEC}^2 + 10.7Q_{BFC}^2 = 82$$

By trial and error, $Q_{BEC} = 6.25$ cfs and $Q_{BFC} = 2.00$ cfs. Hence, $Q_{AB} = 6.25 - 2.00 = 4.25$ cfs.

CHAPTER 12
Branching Pipeline Systems

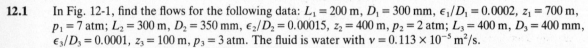

12.1 In Fig. 12-1, find the flows for the following data: $L_1 = 200$ m, $D_1 = 300$ mm, $\epsilon_1/D_1 = 0.0002$, $z_1 = 700$ m, $p_1 = 7$ atm; $L_2 = 300$ m, $D_2 = 350$ mm, $\epsilon_2/D_2 = 0.00015$, $z_2 = 400$ m, $p_2 = 2$ atm; $L_3 = 400$ m, $D_3 = 400$ mm, $\epsilon_3/D_3 = 0.0001$, $z_3 = 100$ m, $p_3 = 3$ atm. The fluid is water with $v = 0.113 \times 10^{-5}$ m²/s.

▌ We first calculate H_1, H_2, and H_3. Thus, $H_1 = z_1 + (7)(p_{atm})/\gamma = 700 + (7)(101\ 325)/9806 = 772.3$ m, $H_2 = z_2 + (2)(p_{atm})/\gamma = 400 + (2)(101\ 325)/9806 = 420.7$ m, $H_3 = z_3 + (3)(p_{atm})/\gamma = 100 + (3)(101\ 325)/9806 = 131.0$ m. Now estimate $H_J = 380$ m. We will hence use the continuity equation for test purposes. We can then say that $(772.3 - 380) = (1/g)f_1(200/0.300)(V_1^2/2)$, $(420.7 - 380) = (1/g)f_2(300/0.350)(V_2^2/2)$, $(380 - 131.0) = (1/g)f_3(400/0.400)(V_3^2/2)$. Estimating $f_1 = 0.014$, $f_2 = 0.013$, and $f_3 = 0.012$, we can compute velocities and the flows q_i. We get

$$V_1 = 28.72 \text{ m/s} \qquad q_1 = (\pi/4)(0.300^2)(28.72) = 2.03 \text{ m}^3/\text{s}$$

$$V_2 = 8.465 \text{ m/s} \qquad q_2 = (\pi/4)(0.350^2)(8.465) = 0.814 \text{ m}^3/\text{s}$$

$$V_3 = 20.18 \text{ m/s} \qquad q_3 = (\pi/4)(0.400^2)(20.18) = 2.536 \text{ m}^3/\text{s}$$

We see that $(q_1 + q_2) > q_3$. As a second estimate, we increase H_J and use more accurate friction factors. Using the preceding velocities, we now find a second set of friction factors: $(N_R)_1 = (28.72)(0.300)/(0.0113 \times 10^{-4}) = 7.625 \times 10^6$, $f_1 = 0.014$; $(N_R)_2 = (8.465)(0.350)/(0.0113 \times 10^{-4}) = 2.622 \times 10^6$, $f_2 = 0.0134$; $(N_R)_3 = (20.18)(0.400)/(0.0113 \times 10^{-4}) = 7.143 \times 10^6$, $f_3 = 0.012$. Suppose we next choose H_J to be 400 m. Thus, we still use the continuity equation. We get the following results: $V_1 = 27.98$ m/s, $q_1 = 1.977$ m³/s; $V_2 = 5.95$ m/s, $q_2 = 0.5721$ m³/s; $V_3 = 20.97$ m/s, $q_3 = 2.635$ m³/s.

Note that $(q_1 + q_2) < q_3$. We now *interpolate* to get the final result. That is, $H_J = 380$ m gave us a value $(q_1 + q_2) - q_3 = 0.308$, while $H_J = 400$ m gave us a value $(q_1 + q_2) - q_3 = -0.0859$. Hence we choose a final H_J to be $H_J = 380 + [0.308/(0.308 + 0.0859)](400 - 380) = 396$ m. For this we get $q_1 = 1.988$ m³/s $= 1988$ L/s, $q_2 = 0.6249$ m³/s $= 624.9$ L/s, $q_3 = 2.616$ m³/s $= 2616$ L/s. We come very close to satisfying the continuity equation so the above are the desired flows.

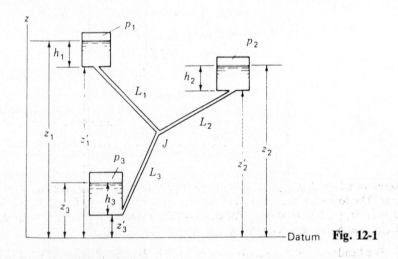

Datum **Fig. 12-1**

12.2 Consider the three interconnected reservoirs shown in Fig. 12-2. The following additional data apply: $L_1 = 2000$ m, $L_2 = 2300$ m, $L_3 = 2500$ m; $D_1 = 1$ m, $D_2 = 0.60$ m, $D_3 = 1.20$ m; $\epsilon_1/D_1 = 0.00015$, $\epsilon_2/D_2 = 0.001$, $\epsilon_3/D_3 = 0.002$. The water is at 5 °C. What is the volumetric flow through the pipes. Neglect minor losses.

▌ The heads H at the free surfaces if we use gage pressures are $H_1 = 120$ m, $H_2 = 100$ m, $H_3 = 30$ m. We estimate H_J as 80 m. The *continuity* equation we use now is $q_1 + q_2 = q_3$. Using the *first law for thermodynamics* for each branch we get:

$$(120 - 80) = f_1(\tfrac{2000}{1})(V_1^2/2g) \qquad (100 - 80) = f_2(2300/0.60)(V_2^2/2g) \qquad (80 - 30) = f_3(2500/1.20)(V_3^2/2g)$$

Estimate the f's: $f_1 = 0.013$, $f_2 = 0.020$, $f_3 = 0.023$. We get

$$V_1 = 5.494 \text{ m/s} \qquad q_1 = (\pi/4)(1^2)(5.494) = 4.315 \text{ m}^3/\text{s}$$
$$V_2 = 2.262 \text{ m/s} \qquad q_2 = (\pi/4)(0.60)^2(2.262) = 0.6396 \text{ m}^3/\text{s}$$
$$V_3 = 4.5247 \text{ m/s} \qquad q_3 = (\pi/4)(1.2)^2(4.5247) = 5.117 \text{ m}^3/\text{s}$$

Note

$$\left.\begin{array}{l} q_1 + q_2 = 4.955 \\ q_3 = 5.1173 \end{array}\right\} \quad \Delta = -0.1623$$

We will decrease H_J to 78 m. We get

$$(120 - 78) = 0.013(\tfrac{2000}{1})(V_1^2/2g) \qquad V_1 = 5.6297 \text{ m/s} \qquad q_1 = 4.4216 \text{ m}^3/\text{s}$$
$$(100 - 78) = 0.020(2300/0.6)(V_2^2/2g) \qquad V_2 = 2.3728 \text{ m/s} \qquad q_2 = 0.6709 \text{ m}^3/\text{s}$$
$$(78 - 30) = 0.023(2500/1.20)(V_3^2/2g) \qquad V_3 = 4.4333 \text{ m/s} \qquad q_3 = 5.0139 \text{ m}^3/\text{s}$$

$$\left.\begin{array}{l} q_1 + q_2 = 5.093 \\ q_3 = 5.014 \end{array}\right\} \quad \Delta = 0.079 \text{ m/s}$$

Now *interpolate*:

$$H_J = 80 - [0.1623/(0.079 + 0.1623)](2) = 78.6559 \text{ m}$$
$$V_1 = 5.586 \text{ m/s} \qquad q_1 = 4.387 \text{ m}^3/\text{s}$$
$$V_2 = 2.3372 \text{ m/s} \qquad q_2 = 0.6608 \text{ m}^3/\text{s}$$
$$V_3 = 4.4634 \text{ m/s} \qquad q_3 = 5.0480 \text{ m}^3/\text{s}$$

Final check:

$$q_1 + q_2 = 4.387 + 0.6608 = 5.048 \qquad q_3 = 5.0480 \qquad \text{O.K.}$$

Therefore,

$$q_1 = 4.39 \text{ m}^3/\text{s} \qquad q_2 = 0.661 \text{ m}^3/\text{s} \qquad q_3 = 5.048 \text{ m}^3/\text{s}$$

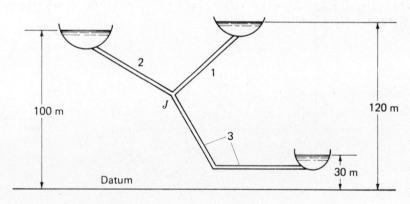

Fig. 12-2

12.3 Reservoirs 1 and 2 in Fig. 12-3a are connected to a tank which has on the free surface a pressure p of 50 lb/in^2 gage. The following data apply: $z_1 = 650$ ft, $L_1 = 2000$ ft, $D_1 = 3$ ft, $(\epsilon/D)_1 = 0.001$; $z_2 = 600$ ft, $L_2 = 2500$ ft $D_2 = 3.5$ ft, $(\epsilon/D)_2 = 0.002$; $z_3 = 50$ ft, $L_3 = 2200$ ft, $D_3 = 4$ ft, $(\epsilon/D)_3 = 0.002$. What are the flows in the pipes? Water is at 60 °F. Neglect minor losses.

▎ The heads for the free surfaces are $H = 650$ ft, $H_2 = 600$ ft, $H_3 = 50 + (50)(144)/62.4 = 165.4$ ft. We estimate H_J to be 400 ft. The *continuity* equation for this assumption is $q_1 + q_2 = q_3$. We can now say:

$$(650 - 400) = (1/g)f_1(\tfrac{2000}{3})(V_1^2/2) \qquad (600 - 400) = (1/g)f_2(2500/3.5)(V_2^2/2)$$
$$(400 - 165.4) = (1/g)f_3(\tfrac{2200}{4})(V_3^2/2) \qquad (1)$$

We make the following estimates of f: $f_1 = 0.02$, $f_2 = 0.023$, $f_3 = 0.023$. Solving for the velocities and flows:

$$V_1 = 34.75 \text{ ft/s} \qquad q_1 = 245 \text{ cfs} \qquad V_2 = 28.0 \text{ ft/s} \qquad q_2 = 269.4 \text{ cfs}$$
$$V_3 = 34.56 \text{ ft/s} \qquad q_3 = 434.3 \text{ cfs}$$

Look at *continuity*: $q_1 + q_2 = 245 + 269.4 = 514$ cfs, $q_3 = 434.3$ cfs. We raise H_J to 425 ft.

$$(650 - 425) = (1/g)(0.02)(\tfrac{2000}{3})(V_1^2/2) \qquad (600 - 425) = (1/g)(0.023)(2500/3.5)(V_2^2/2)$$

$$(425 - 165.4) = (1/g)(0.023)(\tfrac{2200}{4})(V_3^2/2)$$

We now get

$$V_1 = 32.96 \text{ ft/s} \qquad q_1 = 233 \text{ cfs} \qquad V_2 = 26.19 \text{ ft/s} \qquad q_2 = 252 \text{ cfs} \qquad V_3 = 36.35 \text{ ft/s} \qquad q_3 = 456 \text{ cfs}$$

Check continuity again: $q_1 + q_2 = 485$, $q_3 = 456$ ft^3/s. Now extrapolate in a linear manner.

		Δq	J
First:	$(q_1 + q_2) - (q_3) =$	79.7 $\cdots$	400
Second:	$(q_1 + q_2) - (q_3) =$	29.0 $\cdots$	425

See Fig. 12-3b. Choose $H = 437$ ft:

$$(650 - 437) = (1/g)(0.02)(\tfrac{2000}{3})(V_1^2/2) \qquad (600 - 437) = (1/g)(0.023)(2500/3.5)(V_2^2/2)$$

$$(437 - 165.4) = (1/g)(0.023)(\tfrac{2200}{4})(V_3^2/2)$$

$$V_1 = 32.07 \text{ ft/s} \qquad q_1 = 226.7 \text{ cfs} \qquad V_2 = 25.27 \text{ ft/s} \qquad q_2 = 243.2 \text{ cfs}$$

$$V_3 = 37.185 \text{ ft/s} \qquad q_3 = 467.3 \text{ cfs}$$

Check continuity: $q_1 + q_2 = 469.9$, $q_3 = 467.3$ ft^3/s. This is close enough. Therefore

$$q_1 = 227 \text{ cfs} \qquad q_2 = 243 \text{ cfs} \qquad q_3 = 467 \text{ cfs}$$

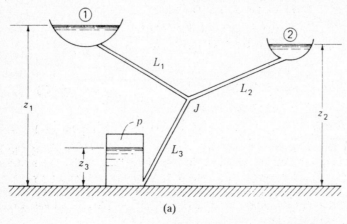
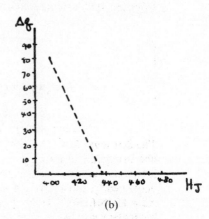

(a)   (b)

Fig. 12-3

12.4 In Fig. 12-4, valve F is partly closed, creating a head loss of 3.60 ft when the flow through the valve is 0.646 mgd (1.00 cfs). What is the length of 10-in pipe to reservoir A?

$$Q = 1.318 A C R^{0.63} s^{0.54} \qquad 1.00 = 1.318[(\pi)(\tfrac{12}{12})^2/4](80)[(\tfrac{12}{12})/4]^{0.63}(s_{DB})^{0.54}$$

$$s_{DB} = 0.001414 \text{ ft/ft}$$

$$(h_L)_{DB} = (0.001414)(1000) + 3.60 = 5.01 \text{ ft}$$

If El$_E = 0$, the grade line elevation at $B = 20 - 5.01 = 14.99$ ft.

$$s_{BE} = (14.99 - 0)/5000 = 0.002998 \text{ ft/ft}$$

$$Q_{BE} = 1.318[(\pi)(\tfrac{12}{12})^2/4](120)[(\tfrac{12}{12})/4]^{0.63}(0.002998)^{0.54} = 2.25 \text{ ft}^3/\text{s}$$

$$Q_{AB} = Q_{BE} - Q_{DB} = 2.25 - 1.00 = 1.25 \text{ ft}^3/\text{s}$$

$$1.25 = 1.318[(\pi)(\tfrac{10}{12})^2/4](100)[(\tfrac{10}{12})/4]^{0.63}(s_{AB})^{0.54}$$

$$s_{AB} = 0.003436 \text{ ft/ft} = (h_f)_{AB}/L_{AB} \qquad 0.003436 = (20 - 15 - 2)/L_{AB} \qquad L_{AB} = 873 \text{ ft}$$

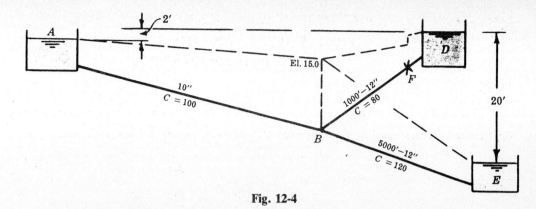

Fig. 12-4

12.5 For the constant elevations of the water surfaces shown in Fig. 12-5a, what flows will occur?

▮ Because the elevation of the hydraulic grade line at C cannot be computed (all flows unknown), the problem will be solved by successive trials. A convenient assumption is to choose the elevation of the hydraulic grade line at C at 190.0 ft. By so assuming, flow to or from reservoir B will be zero, reducing the number of calculations thereby. With this assumption,

$$Q = 1.318ACR^{0.63}s^{0.54} \qquad s_{AC} = (212.0 - 190.0)/8000 = 0.002750 \text{ ft/ft}$$

$$Q_{AB} = 1.318[(\pi)(\tfrac{24}{12})^2/4](100)[(\tfrac{24}{12})/4]^{0.63}(0.002750)^{0.54} = 11.1 \text{ ft}^3/\text{s}$$

$$s_{CD} = (190.0 - 100.0)/4000 = 0.02250 \text{ ft/ft}$$

$$Q_{CD} = 1.318[(\pi)(\tfrac{12}{12})^2/4](100)[(\tfrac{12}{12})/4]^{0.63}(0.02250)^{0.54} = 5.57 \text{ ft}^3/\text{s} \qquad \text{(away from } C\text{)}$$

Examination of these values of flow indicates that the grade line at C must be higher, thereby reducing the flow from A and increasing the flow to D as well as adding flow to B. In an endeavor to "straddle" the correct elevation at C, assume a value of 200.0 ft. Thus, for elevation at $C = 200.0$ ft,

$$s_{AC} = (212.0 - 200.0)/8000 = 0.001500 \text{ ft/ft}$$

$$Q_{AB} = 1.318[(\pi)(\tfrac{24}{12})^2/4](100)[(\tfrac{24}{12})/4]^{0.63}(0.001500)^{0.54} = 7.99 \text{ ft}^3/\text{s}$$

$$s_{CD} = (200.0 - 100.0)/4000 = 0.02500 \text{ ft/ft}$$

$$Q_{CD} = 1.318[(\pi)(\tfrac{12}{12})^2/4](100)[(\tfrac{12}{12})/4]^{0.63}(0.02500)^{0.54} = 5.90 \text{ ft}^3/\text{s} \qquad \text{(away from } C\text{)}$$

$$s_{CB} = (200.0 - 190.0)/4000 = 0.002500 \text{ ft/ft}$$

$$Q_{CB} = 1.318[(\pi)(\tfrac{16}{12})^2/4](120)[(\tfrac{16}{12})/4]^{0.63}(0.002500)^{0.54} = 4.35 \text{ ft}^3/\text{s} \qquad \text{(away from } C\text{)}$$

The flow away from C is $5.90 + 4.35$, or $10.25 \text{ ft}^3/\text{s}$ compared with the flow to C of $7.90 \text{ ft}^3/\text{s}$. Using Fig. 12-5b to obtain a guide regarding a reasonable third assumption, connect plotted points R and S. The line so drawn intersects the $(Q_{\text{to}} - Q_{\text{away}})$ zero abscissa at approximately $Q_{\text{to}} = 8.8 \text{ ft}^3/\text{s}$. Inasmuch as the values plotted do not vary linearly, use a flow to C slightly larger, say $9.2 \text{ ft}^3/\text{s}$. For $Q = 9.2 \text{ ft}^3/\text{s}$ to C (i.e., $Q_{AB} = 9.2 \text{ ft}^3/\text{s}$), $9.2 = 1.318[(\pi)(\tfrac{24}{12})^2/4](100)[(\tfrac{24}{12})/4]^{0.63}s^{0.54}$, $s_{AC} = 0.001948 \text{ ft/ft}$; $(h_f)_{AC} = (0.001948)(8000) = 15.6 \text{ ft}$. Then, the hydraulic grade line at C will be $212.0 - 15.6 = 196.4 \text{ ft}$.

$$s_{CD} = (196.4 - 100.0)/4000 = 0.02410 \text{ ft/ft}$$

$$Q_{CD} = 1.318[(\pi)(\tfrac{12}{12})^2/4](100)[(\tfrac{12}{12})/4]^{0.63}0.02410^{0.54} = 5.78 \text{ ft}^3/\text{s} \qquad \text{(away from } C\text{)}$$

$$s_{CB} = (196.4 - 190.0)/4000 = 0.001600 \text{ ft/ft}$$

$$Q_{CB} = 1.318[(\pi)(\tfrac{16}{12})^2/4](120)[(\tfrac{16}{12})/4]^{0.63}0.001600^{0.54} = 3.42 \text{ ft}^3/\text{s} \qquad \text{(away from } C\text{)}$$

The flow away from C is $5.78 + 3.42$, or $9.20 \text{ ft}^3/\text{s}$ compared with the flow to C of $9.2 \text{ ft}^3/\text{s}$; hence these values of Q_{AC}, Q_{CD}, and Q_{CB} must be O.K.

12.6 Three pipes connect three reservoirs as shown in Fig. 12-6 at these surface elevations: $z_1 = 20$ m, $z_2 = 100$ m, and $z_3 = 40$ m. The pipe data are

pipe	L, m	d, m	ϵ, mm	ϵ/d
1	100	8	0.24	0.003
2	150	6	0.12	0.002
3	80	4	0.20	0.005

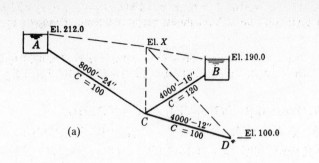

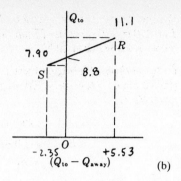

Fig. 12-5

The fluid is water, $\rho = 1000 \text{ kg/m}^3$, and $\nu = 1.02 \times 10^{-6} \text{ m}^2/\text{s}$. Find the resulting flow in each pipe, neglecting minor losses.

▌ As a first guess, take h_J equal to the middle reservoir height, $z_3 = h_J = 40$ m. This saves one calculation ($Q_3 = 0$) and enables us to get the lay of the land:

reservoir	h_J, m	$z_i - h_J$, m	f_i	V_i, m/s	Q_i, m³/h	L_i/d_i
1	40	−20	0.0267	−3.43	−62.1	1250
2	40	60	0.0241	4.42	45.0	2500
3	40	0		0	0	2000
					$\Sigma Q = -17.1$	

Since the sum of the flow rates toward the junction is negative, we guessed h_J too high. Reduce h_J to 30 m and repeat:

reservoir	h_J, m	$z_i - h_J$, m	f_i	V_i, m/s	Q_i, m³/h
1	30	−10	0.0269	−2.42	−43.7
2	30	70	0.0241	4.78	48.6
3	30	10	0.0317	1.76	8.0
					$\Sigma Q = 12.9$

This is positive ΣQ, and so we can linearly interpolate to get an accurate guess: $h_J \approx 34.3$ m. Make one final list:

reservoir	h_J, m	$z_i - h_J$, m	f_i	V_i, m/s	Q_i, m³/h
1	34.3	−14.3	0.0268	−2.90	−52.4
2	34.3	65.7	0.0241	4.63	47.1
3	34.3	5.7	0.0321	1.32	6.0
					$\Sigma Q = 0.7$

This is close enough; hence we calculate that the flow rate is 52.4 m³/h toward reservoir 3, balanced by 47.1 m³/h away from reservoir 1 and 6.0 m³/h away from reservoir 3.

One further iteration with this problem would give $h_J = 34.53$ m, resulting in $Q_1 = -52.8$, $Q_2 = 47.0$, and $Q_3 = 5.8$ m³/h, so that $\Sigma Q = 0$ to three-place accuracy. Pedagogically speaking, we would then be exhausted.

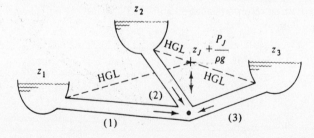

Fig. 12-6

12.7 For the three-reservoir system of Fig. 12-7, $z_1 = 30$ m, $L_1 = 80$ m, $z_2 = 130$ m, $L_2 = 150$ m, $z_3 = 70$ m, and $L_3 = 110$ m. All pipes are 25-cm-diameter concrete with roughness height of 0.5 mm. Compute the flow rate in all pipes for water at 20 °C.

❚ $h_f = (f)(L/d)(v^2/2g)$. Guess $z_a = 70$ m: $(h_f)_1 = z_1 - z_a = 30 - 70 = -40$ m, $\epsilon/d = (0.5/1000)/(\frac{25}{100}) = 0.00200$ (for all pipes), $N_R = dv/v$. Try $v_1 = 10$ m/s: $(N_R)_1 = (\frac{25}{100})(10)/(1.02 \times 10^{-6}) = 2.45 \times 10^6$. From Fig. A-5, $f_1 = 0.0235$.

$$40 = 0.0235[80/(\tfrac{25}{100})]\{v_1^2/[(2)(9.807)]\} \qquad v_1 = 10.21 \text{ m/s}$$

$$(N_R)_1 = (\tfrac{25}{100})(10.21)/(1.02 \times 10^{-6}) = 2.50 \times 10^6 \qquad f_1 = 0.0235 \qquad \text{(O.K.)}$$

$$Q_1 = A_1 v_1 = [(\pi)(\tfrac{25}{100})^2/4](10.21) = 0.5012 \text{ m}^3/\text{s} \qquad \text{(away from } a\text{)}$$

$$(h_f)_2 = z_2 - z_a = 130 - 70 = 60 \text{ m}$$

Try $v_2 = 10$ m/s:

$$(N_R)_2 = (\tfrac{25}{100})(10)/(1.02 \times 10^{-6}) = 2.45 \times 10^6 \qquad f_2 = 0.0235$$

$$60 = 0.0235[150/(\tfrac{25}{100})]\{v_2^2/[(2)(9.807)]\} \qquad v_2 = 9.136 \text{ m/s}$$

$$(N_R)_2 = (\tfrac{25}{100})(9.136)/(1.02 \times 10^{-6}) = 2.24 \times 10^6 \qquad f_1 = 0.0235 \qquad \text{(O.K.)}$$

$$Q_2 = [(\pi)(\tfrac{25}{100})^2/4](9.136) = 0.4485 \text{ m}^3/\text{s} \qquad \text{(toward } a\text{)}$$

$$(h_f)_3 = z_3 - z_a = 70 - 70 = 0 \text{ m}$$

Hence, $Q_3 = 0$; $Q_{\text{to } a} - Q_{\text{from } a} = 0.4485 - 0.5012 = -0.0527$ m^3/s. Hence z_a must be a little lower. Try $z_a = 69.5$ m: $(h_f)_1 = z_1 - z_a = 30 - 69.5 = -39.5$ m. Try $v_1 = 10$ m/s:

$$(N_R)_1 = (\tfrac{25}{100})(10)/(1.02 \times 10^{-6}) = 2.45 \times 10^6 \qquad f_1 = 0.0235$$

$$39.5 = (0.0235)[80/(\tfrac{25}{100})]\{v_1^2/[(2)(9.807)]\} \qquad v_1 = 10.15 \text{ m/s}$$

$$(N_R)_1 = (\tfrac{25}{100})(10.15)/(1.02 \times 10^{-6}) = 2.49 \times 10^6 \qquad f_1 = 0.0235 \qquad \text{(O.K.)}$$

$$Q_1 = [(\pi)(\tfrac{25}{100})^2/4](10.15) = 0.4982 \text{ m}^3/\text{s} \qquad \text{(away from } a\text{)}$$

$$(h_f)_2 = z_2 - z_a = 130 - 69.5 = 60.5 \text{ m}$$

Try $v_2 = 10$ m/s:

$$(N_R)_2 = (\tfrac{25}{100})(10)/(1.02 \times 10^{-6}) = 2.45 \times 10^6 \qquad f_2 = 0.0235$$

$$60.5 = 0.0235[150/(\tfrac{25}{100})]\{v_2^2/[(2)(9.807)]\} \qquad v_2 = 9.174 \text{ m/s}$$

$$(N_R)_2 = (\tfrac{25}{100})(9.174)/(1.02 \times 10^{-6}) = 2.25 \times 10^6 \qquad f_2 = 0.0235 \qquad \text{(O.K.)}$$

$$Q_2 = [(\pi)(\tfrac{25}{100})^2/4](9.174) = 0.4503 \text{ m}^3/\text{s} \qquad \text{(toward } a\text{)}$$

$$(h_f)_3 = z_3 - z_a = 70 - 69.5 = 0.5 \text{ m}$$

Try $v_3 = 1$ m/s:

$$(N_R)_3 = (\tfrac{25}{100})(1)/(1.02 \times 10^{-6}) = 2.45 \times 10^5 \qquad f_3 = 0.0244$$

$$0.5 = 0.0244[110/(\tfrac{25}{100})]\{v_3^2/[(2)(9.807)]\} \qquad v_3 = 0.9558 \text{ m/s}$$

$$(N_R)_3 = (\tfrac{25}{100})(0.9558)/(1.02 \times 10^{-6}) = 2.34 \times 10^5 \qquad f_3 = 0.0244 \qquad \text{(O.K.)}$$

$$Q_3 = [(\pi)(\tfrac{25}{100})^2/4](0.9558) = 0.0469 \text{ m}^3/\text{s} \qquad \text{(toward } a\text{)}$$

$$Q_{\text{to } a} - Q_{\text{from } a} = (0.4503 + 0.0469) - 0.4982 = -0.001 \text{ m}^3/\text{s}$$

Hence, $Q_1 = 0.498$ m^3/s from a, and $Q_2 = 0.450$ m^3/s and $Q_3 = 0.047$ m^3/s toward a.

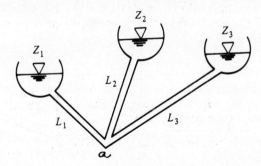

Fig. 12-7

12.8 The three pipes in Fig. 12-8 are cast iron: $d_1 = 7$ in, $L_1 = 2000$ ft; $d_2 = 5$ in, $L_2 = 1000$ ft; $d_3 = 8$ in, $L_3 = 1500$ ft. Compute the flow rate in all pipes for water at 20 °C.

▌ $h_f = (f)(L/d)(v^2/2g)$. Guess $z_J = 50$ ft: $(h_f)_1 = z_1 - z_a = 20 - 50 = -30$ m, $(\epsilon/d)_1 = 0.00085/(\frac{7}{12}) = 0.00146$, $N_R = dv/v$. Try $v_1 = 5$ ft/s: $(N_R)_1 = (\frac{7}{12})(5)/(1.02 \times 10^{-5}) = 2.86 \times 10^5$. From Fig. A-5, $f_1 = 0.0225$.

$$30 = 0.0225[2000/(\tfrac{7}{12})]\{v_1^2/[(2)(32.2)]\} \qquad v_1 = 5.004 \text{ ft/s} \qquad \text{(value of } f_1 \text{ O.K.)}$$
$$Q_1 = A_1 v_1 = [(\pi)(\tfrac{7}{12})^2/4](5.004) = 1.337 \text{ ft}^3/\text{s} \qquad \text{(away from } J)$$
$$(h_f)_2 = z_2 - z_J = 100 - 50 = 50 \text{ ft}$$

Try $v_2 = 7$ ft/s:

$$(N_R)_2 = (\tfrac{5}{12})(7)/(1.02 \times 10^{-5}) = 2.86 \times 10^5$$
$$(\epsilon/d)_2 = 0.00085/(\tfrac{5}{12}) = 0.00204 \qquad f_2 = 0.024$$
$$50 = 0.024[1000/(\tfrac{5}{12})]\{v_2^2/[(2)(32.2)]\} \qquad v_2 = 7.477 \text{ ft/s}$$
$$(N_R)_2 = (\tfrac{5}{12})(7.477)/(1.02 \times 10^{-5}) = 3.05 \times 10^5 \qquad f_2 = 0.024 \qquad \text{(O.K.)}$$
$$Q_2 = [(\pi)(\tfrac{5}{12})^2/4](7.477) = 1.020 \text{ ft}^3/\text{s} \qquad \text{(toward } J) \qquad (h_f)_3 = z_3 - z_J = 50 - 50 = 0 \text{ ft}$$

Hence, $Q_3 = 0$; $Q_{\text{to } J} - Q_{\text{from } J} = 1.020 - 1.337 = -0.317$ ft^3/s. Hence z_J must be a little lower. Try $z_J = 49.35$ ft: $(h_f)_1 = z_1 - z_a = 20 - 49.35 = -29.35$ m. Try $v_1 = 5$ ft/s:

$$(N_R)_1 = (\tfrac{7}{12})(5)/(1.02 \times 10^{-5}) = 2.86 \times 10^5 \qquad f_1 = 0.0225$$
$$29.35 = 0.0225[2000/(\tfrac{7}{12})]\{v_1^2/[(2)(32.2)]\} \qquad v_1 = 4.950 \text{ ft/s} \qquad \text{(value of } f_1 \text{ O.K.)}$$
$$Q_1 = [(\pi)(\tfrac{7}{12})^2/4](4.950) = 1.323 \text{ ft}^3/\text{s} \qquad \text{(away from } J)$$
$$(h_f)_2 = z_2 - z_J = 100 - 49.35 = 50.65 \text{ ft}$$

Try $v_2 = 7.5$ ft/s:

$$(N_R)_2 = (\tfrac{5}{12})(7.5)/(1.02 \times 10^{-5}) = 3.06 \times 10^5 \qquad f_2 = 0.024$$
$$50.65 = 0.024[1000/(\tfrac{5}{12})]\{v_2^2/[(2)(32.2)]\} \qquad v_2 = 7.525 \text{ ft/s} \qquad \text{(value of } f_2 \text{ O.K.)}$$
$$Q_2 = [(\pi)(\tfrac{5}{12})^2/4](7.525) = 1.026 \text{ ft}^3/\text{s} \qquad \text{(toward } J)$$
$$(h_f)_3 = z_3 - z_J = 50 - 49.35 = 0.65 \text{ ft}$$

Try $v_3 = 1$ ft/s:

$$(N_R)_3 = (\tfrac{8}{12})(1)/(1.02 \times 10^{-5}) = 6.54 \times 10^4$$
$$(\epsilon/d)_3 = 0.00085/(\tfrac{8}{12}) = 0.00128 \qquad f_3 = 0.024$$
$$0.65 = 0.024[1500/(\tfrac{8}{12})]\{v_3^2/[(2)(32.2)]\} \qquad v_3 = 0.8804 \text{ ft/s}$$
$$(N_R)_3 = (\tfrac{8}{12})(0.8804)/(1.02 \times 10^{-5}) = 5.75 \times 10^4 \qquad f_3 = 0.024 \qquad \text{(O.K.)}$$
$$Q_3 = [(\pi)(\tfrac{8}{12})^2/4](0.8804) = 0.307 \text{ ft}^3/\text{s} \qquad \text{(toward } J)$$
$$Q_{\text{to } J} - Q_{\text{from } J} = (1.026 + 0.307) - 1.323 = 0.010 \text{ ft}^3/\text{s}$$

Hence, $Q_1 = 1.323$ ft^3/s from J, and $Q_2 = 1.026$ ft^3/s and $Q_3 = 0.307$ ft^3/s toward J.

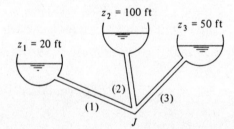

$z_2 = 100$ ft

$z_3 = 50$ ft

$z_1 = 20$ ft

(2)

(1)

(3)

J

Fig. 12-8

12.9 In Fig. 12-9 find the discharges for water at 20 °C with the following pipe data and reservoir elevations: $L_1 = 3000$ m, $D_1 = 1$ m, $\epsilon_1/D_1 = 0.0002$; $L_2 = 600$ m, $D_2 = 0.45$ m, $\epsilon_2/D_2 = 0.002$; $L_3 = 1000$ m, $D_3 = 0.6$ m, $\epsilon_3/D_3 = 0.001$; $z_1 = 30$ m, $z_2 = 18$ m, $z_3 = 9$ m.

▌ Assume $z_J + p_J/\gamma = 23$ m. Then $7 = f_1(\frac{3000}{1})(V_1^2/2g)$, $f_1 = 0.014$, $V_1 = 1.75$ m/s, $Q_1 = 1.380$ m^3/s; $5 = f_2(600/0.45)(V_2^2/2g)$, $f_2 = 0.024$, $V_2 = 1.75$ m/s, $Q_2 = 0.278$ m^3/s; $14 = f_3(1000/0.60)(V_3^2/2g)$, $f_3 = 0.020$, $V_3 =$

2.87 m/s, $Q_3 = 0.811$ m³/s; so that the inflow is greater than the outflow by $1.380 - 0.278 - 0.811 = 0.291$ m³/s. Assume $z_J + p_J/\gamma = 24.6$ m. Then $5.4 = f_1(\frac{3000}{1})(V_1^2/2g)$, $f_1 = 0.015$, $V_1 = 1.534$ m/s, $Q_1 = 1.205$ m³/s; $6.6 = f_2(600/0.45)(V_2^2/2g)$, $f_2 = 0.024$, $V_2 = 2.011$ m/s, $Q_2 = 0.320$ m³/s; $15.6 = f_3(1000/0.60)(V_3^2/2g)$, $f_3 = 0.020$, $V_3 = 3.029$ m/s, $Q_3 = 0.856$ m³/s. The inflow is still greater by 0.029 m³/s. By extrapolating linearly, $z_J + p_J/\gamma = 24.8$ m, $Q_1 = 1.183$, $Q_2 = 0.325$, $Q_3 = 0.862$ m³/s.

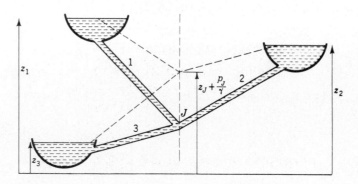

Fig. 12-9

12.10 Prepare a computer program to balance the flows in a system of N reservoirs connected by a common junction. Use the data of Prob. 12.9 to check out the program.

```
10 REM B:   N reservoirs with a common junction
20 DEFINT I,N: DEF FNQ(D1,D2,D3,DH)=D1*DH*LOG(D2+D3/DH)
30 READ N,G,KVIS,II:    DATA 3,9.806,1.007E-6,15
40 LPRINT:    LPRINT"N,G,KVIS,II=";N;G;KVIS;II
50 ZMA=-1000!:    ZMI=1000!
60 FOR I=1 TO N:    READ L(I),D(I),EP(I),Z(I)
70 LPRINT"L,D,EP,Z=";L(I);D(I);EP(I);Z(I)
80 NEXT I
90 DATA 3000.,1.,.0002,30.,600.,.45,.0009,18.,1000.,.6,.0006,9.
100 FOR I=1 TO N:    C=SQR(G*D(I)/L(I)):    E1(I)=-.9650001*D(I)^2*C
110 E2(I)=EP(I)/(3.7*D(I)):    E3(I)=1.784*KVIS/(D(I)*C)    ' FOR EQ. (5.8.15)
120 IF Z(I)>ZMA THEN ZMA=Z(I)                ' ELEV. OF HIGHEST RESERVOIR
130 IF Z(I)<ZMI THEN ZMI=Z(I)                ' ELEV. OF LOWEST RESERVOIR
140 NEXT I
150 FOR I1=1 TO 15:    HJUN=.5*(ZMA+ZMI):    S=0!    ' START BISECTION METHOD
160 FOR I=1 TO N    :HF=Z(I)-HJUN:    HFS=SQR(ABS(HF))
170 Q(I)=FNQ(E1(I),E2(I),E3(I),HFS)*SGN(HF)
180 S=S+Q(I):    PRINT"HF,Q,S=";HF;Q(I);S
190 NEXT I
200 IF S>0! THEN ZMI=HJUN ELSE ZMA=HJUN
210 NEXT I1
220 LPRINT"ELEVATION OF JUNCTION IS ";HJUN
230 LPRINT"PIPE DISCHARGES ARE (POSITIVE INTO THE JUNCTION)"
240 FOR I=1 TO N:    LPRINT"Q(";I;")=";:    LPRINT USING" ###.### ";Q(I);
250 NEXT I:    LPRINT

N,G,KVIS,II= 3  9.806001  1.007E-06  15
L,D,EP,Z= 3000  1  .0002  30
L,D,EP,Z= 600  .45  .0009  18
L,D,EP,Z= 1000  .6  .0006  9
ELEVATION OF JUNCTION IS  24.8801
PIPE DISCHARGES ARE (POSITIVE INTO THE JUNCTION)
Q( 1 )=   1.198 Q( 2 )=  -0.329 Q( 3 )=  -0.869
```

12.11 In Fig. 12-10, find the flow through the system when the pump is removed.

$$(\epsilon/D)_{AJ} = \tfrac{1}{200} = 0.0050 \qquad (\epsilon/D)_{BJ} = \tfrac{1}{200} = 0.0050 \qquad (\epsilon/D)_{CJ} = \tfrac{3}{300} = 0.0100$$

$$f_{AJ} = 0.032 \qquad f_{BJ} = 0.032 \qquad f_{CJ} = 0.038 \qquad h_f = (f)(L/D)(v^2/2g) = (f)(L/D)^5(8Q^2/g\pi^2)$$

$$30 - h_J = 0.032[1000/(\tfrac{200}{1000})^5]\{8Q_{AJ}^2/[(9.807)(\pi)^2]\} = 8265Q_{AJ}^2$$

$$27 - h_J = 0.032[300/(\tfrac{200}{1000})^5]\{8Q_{BJ}^2/[(9.807)(\pi)^2]\} = 2480Q_{BJ}^2$$

$$h_J - 17 = 0.038[(300+300)/(\tfrac{300}{1000})^5]\{8Q_{CJ}^2/[(9.807)(\pi)^2]\} = 775.5Q_{CJ}^2$$

Try $h_J = 21$ m:

$$30 - 21 = 8265Q_{AJ}^2 \qquad Q_{AJ} = 0.03300 \text{ m}^3/\text{s} \qquad 27 - 21 = 2480Q_{BJ}^2 \qquad Q_{BJ} = 0.04919 \text{ m}^3/\text{s}$$

$$21 - 17 = 775.5Q_{CJ}^2 \qquad Q_{CJ} = 0.07182 \text{ m}^3/\text{s}$$

$$Q_{\text{to }J} - Q_{\text{from }J} = 0.03300 + 0.04919 - 0.07182 = 0.01037 \text{ m}^3/\text{s}$$

Try $h_J = 22$ m:

$$30 - 22 = 8265Q_{AJ}^2 \quad Q_{AJ} = 0.03111 \text{ m}^3/\text{s} \quad 27 - 22 = 2480Q_{BJ}^2 \quad Q_{BJ} = 0.04490 \text{ m}^3/\text{s}$$

$$22 - 17 = 775.5Q_{CJ}^2 \quad Q_{CJ} = 0.08030 \text{ m}^3/\text{s}$$

$$Q_{\text{to }J} - Q_{\text{from }J} = 0.03111 + 0.04490 - 0.08030 = -0.00429 \text{ m}^3/\text{s}$$

Try $h_J = 21.7$ m:

$$30 - 21.7 = 8265Q_{AJ}^2 \quad Q_{AJ} = 0.03169 \text{ m}^3/\text{s} \quad 27 - 21.7 = 2480Q_{BJ}^2 \quad Q_{BJ} = 0.04623 \text{ m}^3/\text{s}$$

$$21.7 - 17 = 775.5Q_{CJ}^2 \quad Q_{CJ} = 0.07785 \text{ m}^3/\text{s}$$

$$Q_{\text{to }J} - Q_{\text{from }J} = 0.03169 + 0.04623 - 0.07785 = 0.00007 \text{ m}^3/\text{s}$$

Hence, $Q_{CJ} = 77.8$ L/s from J, and $Q_{AJ} = 31.7$ L/s and $Q_{BJ} = 46.2$ L/s toward J.

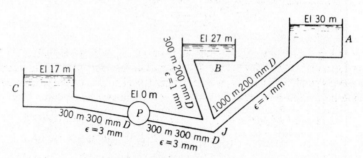

Fig. 12-10

12.12 Find the flow through the system of Fig. 12-11 for no pump in the system.

$$h_f = (f)(L/D)(v^2/2g) = (f)(L/D^5)(8Q^2/g\pi^2)$$

$$h_{J_1} - 80 = 0.025[(300 + 300)/(\tfrac{600}{1000})^5]\{8Q_A^2/[(9.807)(\pi)^2]\} = 15.945Q_A^2$$

$$h_{J_1} - 90 = 0.030[600/(\tfrac{450}{1000})^5]\{8Q_B^2/[(9.807)(\pi)^2]\} = 80.624Q_B^2$$

$$h_{J_2} - h_{J_1} = 0.030[900/(\tfrac{450}{1000})^5]\{8Q_{JJ}^2/[(9.807)(\pi)^2]\} = 120.94Q_{JJ}^2$$

$$100 - h_{J_2} = 0.030[300/(\tfrac{450}{1000})^5]\{8Q_C^2/[(9.807)(\pi)^2]\} = 40.312Q_C^2$$

$$90 - h_{J_2} = 0.030[300/(\tfrac{450}{1000})^5]\{8Q_D^2/[(9.807)(\pi)^2]\} = 40.312Q_D^2$$

Try $h_{J_1} = 84$ m:

$$84 - 80 = 15.945Q_A^2 \quad Q_A = 0.5009 \text{ m}^3/\text{s} \quad 84 - 90 = 80.624Q_B^2 \quad Q_B = -0.2728 \text{ m}^3/\text{s}$$

$$h_{J_2} - 84 = (120.94)(0.5009 - 0.2728)^2 \quad h_{J_2} = 90.29 \text{ m}$$

$$100 - 90.29 = 40.312Q_C^2 \quad Q_C = 0.4908 \text{ m}^3/\text{s}$$

$$90 - 90.29 = 40.312Q_D^2 \quad Q_D = -0.0848 \text{ m}^3/\text{s}$$

If the above values are correct, Q_{JJ} (i.e., $0.5009 - 0.2728 = 0.2281$ m³/s) must equal $Q_C + Q_D$ (i.e., $0.4908 - 0.0848 = 0.4060$ m³/s). Since they are not equal, try $h_{J_1} = 85$ m, but note that $Q_{JJ} < Q_C + Q_D$:

$$85 - 80 = 15.945Q_A^2 \quad Q_A = 0.5600 \text{ m}^3/\text{s} \quad 85 - 90 = 80.624Q_B^2 \quad Q_B = -0.2490 \text{ m}^3/\text{s}$$

$$h_{J_2} - 85 = (120.94)(0.5600 - 0.2490)^2 \quad h_{J_2} = 96.70 \text{ m}$$

$$100 - 96.70 = 40.312Q_C^2 \quad Q_C = 0.2861 \text{ m}^3/\text{s}$$

$$90 - 96.70 = 40.312Q_D^2 \quad Q_D = -0.4077 \text{ m}^3/\text{s}$$

If these values are correct, Q_{JJ} (i.e., $0.5600 - 0.2490 = 0.3110$ m³/s) must equal $Q_C + Q_D$ (i.e., $0.2861 - 0.4077 = -0.1216$ m³/s). Since this time, $Q_{JJ} > Q_C + Q_D$, the correct value of h_{J_1} must be between 84 m and 85 m. Try $h_{J_1} = 84.24$ m:

$$84.24 - 80 = 15.945Q_A^2 \quad Q_A = 0.5157 \text{ m}^3/\text{s} \quad 84.24 - 90 = 80.624Q_B^2 \quad Q_B = -0.2673 \text{ m}^3/\text{s}$$

$$h_{J_2} - 84.24 = (120.94)(0.5157 - 0.2673)^2 \quad h_{J_2} = 91.70 \text{ m}$$

$$100 - 91.70 = 40.312Q_C^2 \quad Q_C = 0.4538 \text{ m}^3/\text{s}$$

$$90 - 91.70 = 40.312Q_D^2 \quad Q_D = -0.2054 \text{ m}^3/\text{s}$$

If these values are correct, Q_{JJ} (i.e., $0.5157 - 0.2673 = 0.2484$ m³/s) must equal $Q_C + Q_D$ (i.e.,

$0.4538 - 0.2054 = 0.2484 \text{ m}^3/\text{s}$). Since they are equal, the correct values must be $Q_A = 516 \text{ L/s}$, $Q_B = 267 \text{ L/s}$, $Q_{JJ} = 248 \text{ L/s}$, $Q_C = 454 \text{ L/s}$, and $Q_D = 205 \text{ L/s}$.

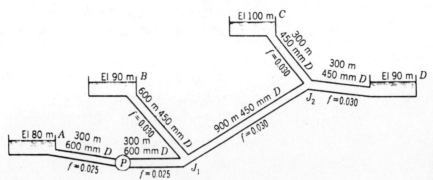

Fig. 12-11

12.13 Suppose that in Fig. 12-12, pipe 1 is 36-in smooth concrete, 5000 ft long; pipe 2 is 24-in cast iron, 1500 ft long; and pipe 3 is 18-in cast iron, 4000 ft long. The elevations of water surfaces in reservoirs A and B are 200 ft and 150 ft, respectively, and discharge Q_1 is 50 cfs. Find the elevation of the surface of reservoir C. Neglect minor losses and assume the energy line and hydraulic grade line are coincident.

$$h_f = (f)(L/d)(v^2/2g) \qquad (\epsilon/d)_1 = 0.001/(\tfrac{36}{12}) = 0.000333 \qquad N_R = dv/v$$
$$v_1 = Q_1/A_1 = 50/[(\pi)(\tfrac{36}{12})^2/4] = 7.074 \text{ ft/s}$$
$$(N_R)_1 = (\tfrac{36}{12})(7.074)/(1.05 \times 10^{-5}) = 2.02 \times 10^6$$

From Fig. A-5, $f_1 = 0.0157$. $(h_f)_1 = 0.0157[5000/(\tfrac{36}{12})]\{7.074^2/[(2)(32.2)]\} = 20.33 \text{ ft}$, $h_J = 200 - 20.33 = 179.67 \text{ ft}$, $(h_f)_2 = 179.67 - 150 = 29.67 \text{ ft}$. Assume $f_2 = 0.0162$:

$$29.67 = 0.0162[1500/(\tfrac{24}{12})]\{v_2^2/[(2)(32.2)]\} \qquad v_2 = 12.54 \text{ ft/s}$$
$$(N_R)_2 = (\tfrac{24}{12})(12.54)/(1.05 \times 10^{-5}) = 2.39 \times 10^6$$
$$(\epsilon/d)_2 = 0.00085/(\tfrac{24}{12}) = 0.000425 \qquad f_2 = 0.0162 \qquad \text{(O.K.)}$$
$$Q_2 = A_2 v_2 = [(\pi)(\tfrac{24}{12})^2/4](12.54) = 39.40 \text{ ft}^3/\text{s} \qquad Q_3 = 50 - 39.40 = 10.6 \text{ ft}^3/\text{s}$$
$$v_3 = 10.6/[(\pi)(\tfrac{18}{12})^2/4] = 5.998 \text{ ft/s} \qquad (N_R)_3 = (\tfrac{18}{12})(5.998)/(1.05 \times 10^{-5}) = 8.57 \times 10^5$$
$$(\epsilon/d)_3 = 0.00085/(\tfrac{18}{12}) = 0.000567 \qquad f_3 = 0.0177$$
$$(h_f)_3 = 0.0177[4000/(\tfrac{18}{12})]\{5.998^2/[(2)(32.2)]\} = 26.37 \text{ ft}$$
$$h_C = 179.67 - 26.37 = 153.30 \text{ ft}$$

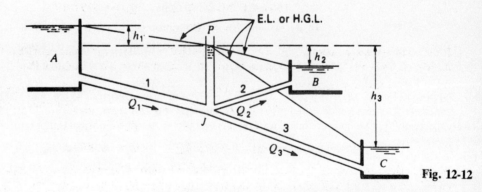

Fig. 12-12

12.14 With the sizes and lengths and materials of pipes given in Prob. 12.13, suppose that the surface elevations of reservoirs A and C are 200 ft and 125 ft, respectively, and discharge Q_2 is 20 cfs into reservoir B. Find the surface elevation of reservoir B.

$$h_f = (f)(L/d)(v^2/2g) \qquad v_2 = Q_2/A_2 = 20/[(\pi)(\tfrac{24}{12})^2/4] = 6.366 \text{ ft/s} \qquad N_R = dv/v$$
$$(N_R)_2 = (\tfrac{24}{12})(6.366)/(1.05 \times 10^{-5}) = 1.21 \times 10^6$$
$$(\epsilon/d)_2 = 0.000425 \qquad \text{(from Prob. 12.13)}$$

From Fig. A-5, $f_2 = 0.0168$. $(h_f)_2 = 0.0168[1500/(\frac{24}{12})]\{6.366^2/[(2)(32.2)]\} = 7.93$ ft.

$$Q_1 - Q_3 = 20 \tag{1}$$

$$(h_f)_1 + (h_f)_3 = 200 - 125 = 75 \text{ ft} \tag{2}$$

We need to find an h_f in line 1 or 3 such that both Eqs. (1) and (2) are satisfied. Try $(h_f)_3 = 64$ ft and $f_3 = 0.0175$:

$$64 = 0.0175[4000/(\tfrac{18}{12})]\{v_3^2/[(2)(32.2)]\} \qquad v_3 = 9.398 \text{ ft/s}$$

$$(N_R)_3 = \tfrac{18}{12}(9.398)/(1.05 \times 10^{-5}) = 1.34 \times 10^6$$

$$(\epsilon/d)_3 = 0.000567 \qquad \text{(from Prob. 12.13)}$$

$$f_3 = 0.0175 \quad \text{(O.K.)} \qquad Q_3 = A_3 v_3 = [(\pi)(\tfrac{18}{12})^2/4](9.398) = 16.61 \text{ ft}^3/\text{s}$$

$$Q_1 - 16.61 = 20 \qquad Q_1 = 36.61 \text{ ft}^3/\text{s}$$

$$v_1 = 36.61/[(\pi)(\tfrac{36}{12})^2/4] = 5.179 \text{ ft/s} \qquad (N_R)_1 = (\tfrac{36}{12})(5.179)/(1.05 \times 10^{-5}) = 1.48 \times 10^6$$

$$(\epsilon/d)_1 = 0.000333 \qquad \text{(from Prob. 12.13)} \qquad f_1 = 0.0158$$

$$(h_f)_1 = 0.0158[5000/(\tfrac{36}{12})]\{5.179^2/[(2)(32.2)]\} = 10.97 \text{ ft}$$

$$\text{El}_B = \text{El}_A - (h_f)_1 - (h_f)_2 = 200 - 10.97 - 7.93 = 181.1 \text{ ft}$$

12.15 Suppose, in Prob. 12.14, that discharge Q_2 is known to be 20 cfs from reservoir B. Find the elevation of the surface of reservoir B.

▮ As in Prob. 12.14, $(h_f)_2 = 7.93$ ft, but Eq. (1) becomes

$$Q_3 - Q_1 = 20 \tag{1}$$

Equation (2) is the same as in Prob. 12.14; that is,

$$(h_f)_1 + (h_f)_3 = 200 - 125 = 75 \text{ ft} \tag{2}$$

We need to find an h_f in line 1 or 3 such that both Eqs. (1) and (2) are satisfied. Try $(h_f)_3 = 77$ ft and $f_3 = 0.0175$:

$$77 = 0.0175[4000/(\tfrac{18}{12})]\{v_3^2/[(2)(32.2)]\} \qquad v_3 = 10.31 \text{ ft/s}$$

$$(N_R)_3 = (\tfrac{18}{12})(10.31)/(1.05 \times 10^{-5}) = 1.47 \times 10^6$$

$$(\epsilon/d)_3 = 0.000567 \qquad \text{(from Prob. 12.13)}$$

$$f_3 = 0.0175 \quad \text{(O.K.)}$$

$$Q_3 = A_3 v_3 = [(\pi)(\tfrac{18}{12})^2/4](10.31) = 18.22 \text{ ft}^3/\text{s} \qquad 18.22 - Q_1 = 20$$

$$Q_1 = -1.78 \text{ ft}^3/\text{s} \qquad \text{(i.e., flow is into reservoir } A)$$

$$v_1 = 1.78/[(\pi)(\tfrac{36}{12})^2/4] = 0.2518 \text{ ft/s}$$

$$(N_R)_1 = (\tfrac{36}{12})(0.2518)/(1.05 \times 10^{-5}) = 7.19 \times 10^4$$

$$(\epsilon/d)_1 = 0.000333 \qquad \text{(from Prob. 12.13)} \qquad f_1 = 0.0208$$

$$(h_f)_1 = 0.0208[5000/(\tfrac{36}{12})]\{0.2518^2/[(2)(32.2)]\} = 0.03 \text{ ft}$$

$$\text{El}_B = \text{El}_A + (h_f)_1 + (h_f)_2 = 200 + 0.03 + 7.93 = 208.0 \text{ ft}$$

12.16 Suppose in Fig. 12-12, that pipes 1, 2, and 3 are 900 m of 60 cm, 300 m of 45 cm, and 1200 m of 40 cm, respectively, of new welded steel pipe. The surface elevations of A, B, and C are 30 m, 18 m, and 0 m, respectively. Find the approximate water flow in all pipes. Assume a normal temperature.

▮ $(\epsilon/d)_1 = 0.000046/(\tfrac{60}{100}) = 0.0000767 \qquad (\epsilon/d)_2 = 0.000046/(\tfrac{45}{100}) = 0.0001022$

$$(\epsilon/d)_3 = 0.000046/(\tfrac{40}{100}) = 0.0001150$$

Assuming complete turbulence (high Reynolds numbers), friction factors for these values of (ϵ/d) will be $f_1 = 0.0115$, $f_2 = 0.0122$, and $f_3 = 0.0126$.

$$h_f = (f)(L/d)(v^2/2g) = (f)(L/d^5)(8Q^2/g\pi^2)$$

$$(h_f)_1 = 0.0115[900/(\tfrac{60}{100})^5]\{8Q_1^2/[(9.807)(\pi)^2]\} - 11.00Q_1^2$$

$$(h_f)_2 = 0.0122[300/(\tfrac{45}{100})^5]\{8Q_2^2/[(9.807)(\pi)^2]\} = 16.39Q_2^2$$

$$(h_f)_3 = 0.0126[1200/(\tfrac{40}{100})^5]\{8Q_3^2/[(9.807)(\pi)^2]\} = 122.0Q_3^2$$

$$30 - 11.00Q_1^2 = (p/\gamma + z)_P \tag{1}$$

[*Note:* $(p/\gamma + z)_P$ denotes $(p/\gamma + z)$ at point P.]

$$(p/\gamma + z)_P - 16.39Q_2^2 = 18 \qquad \text{or} \qquad 18 - 16.39Q_2^2 = (p/\gamma + z)_P \tag{2}$$

$$(p/\gamma + z)_P - 122.0Q_3^2 = 0 \tag{3}$$

$$Q_1 = Q_2 + Q_3 \qquad \text{or} \qquad Q_1 + Q_2 = Q_3 \tag{4}$$

Assume no flow in pipe 2, in which case $(p/\gamma + z)_P = 18$ m. $30 - 11.00Q_1^2 = 18$, $Q_1^2 = 1.091$; $18 - 122.0Q_3^2 = 0$, $Q_3^2 = 0.1475$. Since $Q_1 > Q_3$, flow must be into reservoir B, and $(p/\gamma + z)_P > 18$ m. By trial and error (not shown), $(p/\gamma + z)_P = 21.5$ m. $30 - 11.00Q_1^2 = 21.5$, $Q_1 = 0.879$ m^3/s; $21.5 - 16.39Q_2^2 = 18$, $Q_2 = 0.462$ m^3/s; $21.5 - 122.0Q_3^2 = 0$, $Q_3 = 0.420$ m^3/s. Check Eq. (4): $0.879 = 0.462 + 0.420$, $0.879 = 0.882$ (close enough). Check the Reynolds numbers for normal temperature:

$$N_R = dv/\nu \qquad v_1 = 0.879/[(\pi)(\tfrac{60}{100})^2/4] = 3.109 \text{ m/s}$$

$$(N_R)_1 = (\tfrac{60}{100})(3.109)/(1.02 \times 10^{-6}) = 1.83 \times 10^6$$

$$v_2 = 0.462/[(\pi)(\tfrac{45}{100})^2/4] = 2.905 \text{ m/s}$$

$$(N_R)_2 = (\tfrac{45}{100})(2.905)/(1.02 \times 10^{-6}) = 1.28 \times 10^6$$

$$v_3 = 0.420/[(\pi)(\tfrac{40}{100})^2/4] = 3.342 \text{ m/s}$$

$$(N_R)_1 = (\tfrac{40}{100})(3.342)/(1.02 \times 10^{-6}) = 1.31 \times 10^6$$

From Fig. A-5, $f = 0.0135$ approximately for all pipes; hence, the above flows should be multiplied by $(0.012/0.0135)$, giving $Q_1 = (0.012/0.0135)(0.879) = 0.781$ m^3/s, $Q_2 = (0.012/0.0135)(0.462) = 0.411$ m^3/s, $Q_3 = (0.012/0.0135)(0.420) = 0.373$ m^3/s. Further refinement of these approximations is not justified.

12.17 Suppose that, in Fig. 12-12, pipe 1 is 1500 ft of 12-in new cast iron pipe, pipe 2 is 800 ft of 6-in wrought iron pipe, and pipe 3 is 1200 ft of 8-in wrought iron pipe. The water surface of reservoir B is 20 ft below that of A, while junction J is 35 ft below the surface of A. In place of reservoir C, pipe 3 leads away to some other destination but its elevation at C is 60 ft below A. When the pressure head at J is 25 ft, find the flow in each pipe and the pressure head at C.

$$\blacksquare \qquad (h_f)_1 = 35 - 25 = 10 \text{ ft} \qquad (h_f)_2 = 20 - h_1 = 20 - 10 = 10 \text{ ft} \qquad h_f = (f)(L/d)(v^2/2g)$$

Assume $f_1 = 0.019$:

$$10 = 0.019[1500/(\tfrac{12}{12})]\{v_1^2/[(2)(32.2)]\} \qquad v_1 = 4.754 \text{ ft/s} \qquad N_R = dv/\nu$$

$$(N_R)_1 = (\tfrac{12}{12})(4.754)/(1.05 \times 10^{-5}) = 4.53 \times 10^5 \qquad (\epsilon/d)_1 = 0.00085/(\tfrac{12}{12}) = 0.00085$$

From Fig. A-5, $f_1 = 0.0198$. Try $f_1 = 0.0198$:

$$10 = 0.0198[1500/(\tfrac{12}{12})]\{v_1^2/[(2)(32.2)]\} \qquad v_1 = 4.657 \text{ ft/s}$$

$$(N_R)_1 = (\tfrac{12}{12})(4.657)/(1.05 \times 10^{-5}) = 4.44 \times 10^5 \qquad f_1 = 0.0198 \quad \text{(O.K.)}$$

$$Q_1 = A_1 v_1 = [(\pi)(\tfrac{12}{12})^2/4](4.657) = 3.658 \text{ ft}^3/\text{s}$$

Assume $f_2 = 0.015$:

$$10 = 0.015[800/(\tfrac{6}{12})]\{v_2^2/[(2)(32.2)]\} \qquad v_2 = 5.180 \text{ ft/s}$$

$$(N_R)_2 = (\tfrac{6}{12})(5.180)/(1.05 \times 10^{-5}) = 2.47 \times 10^5 \qquad (\epsilon/d)_2 = 0.00015/(\tfrac{6}{12}) = 0.00030 \qquad f = 0.0175$$

Try $f_2 = 0.0175$:

$$10 = 0.0175[800/(\tfrac{6}{12})]\{v_2^2/[(2)(32.2)]\} \qquad v_2 = 4.796 \text{ ft/s}$$

$$(N_R)_2 = (\tfrac{6}{12})(4.796)/(1.05 \times 10^{-5}) = 2.28 \times 10^5 \qquad f_2 = 0.0175 \quad \text{(O.K.)}$$

$$Q_2 = [(\pi)(\tfrac{6}{12})^2/4](4.796) = 0.942 \text{ ft}^3/\text{s} \qquad Q_3 = Q_1 - Q_2 = 3.658 - 0.942 = 2.716 \text{ ft}^3/\text{s}$$

$$v_3 = Q_3/A_3 = 2.716/[(\pi)(\tfrac{8}{12})^2/4] = 7.781 \text{ ft/s} \qquad (N_R)_3 = (\tfrac{8}{12})(7.781)/(1.05 \times 10^{-5}) = 4.94 \times 10^5$$

$$(\epsilon/D)_3 = 0.00015/(\tfrac{8}{12}) = 0.000225 \qquad f_3 = 0.016$$

$$(h_f)_3 = 0.016[1200/(\tfrac{8}{12})]\{7.781^2/[(2)(32.2)]\} = 27.08 \text{ ft} \qquad p_C/\gamma = 60 - 10 - 27.08 - 7.781^2/[(2)(32.2)] = 22.0 \text{ ft}$$

12.18 In the reservoir system of Fig. 12-13, $z_A = 200$ ft, $z_B = 230$ ft, $z_C = 120$ ft, $z_D = 140$ ft, $BD = 3000$ ft of 4-in cast iron pipe, $AD = 2000$ ft of 1-in steel pipe, and $DC = 500$ ft of 6-in cast iron pipe. Using $f = 0.025$ and neglecting minor losses, determine the flow in each pipe.

▮ $h_f = (f)(L/d)(v^2/2g)$. Let $p_D/\gamma = -15$ ft:

$$(h_f)_{AD} = 200 - 140 - (-15) = 0.025[2000/(\tfrac{1}{12})]\{v_{AD}^2/[(2)(32.2)]\} \qquad v_{AD} = 2.837 \text{ ft/s}$$

$$(h_f)_{BD} = 230 - 140 - (-15) = 0.025[3000/(\tfrac{4}{12})]\{v_{BD}^2/[(2)(32.2)]\} \qquad v_{BD} = 5.482 \text{ ft/s}$$

$$(h_f)_{CD} = 120 - 140 - (-15) = 0.025[500/(\tfrac{6}{12})]\{v_{CD}^2/[(2)(32.2)]\} \qquad v_{CD} = 3.589 \text{ ft/s} \quad (\text{away from } D)$$

If these velocities are correct, $Q_{AD} + Q_{BD}$ must equal Q_{CD}: $Q_{AD} + Q_{BD} = [(\pi)(\tfrac{1}{12})^2/4](2.837) + [(\pi)(\tfrac{4}{12})^2/4](5.482) = 0.4939 \text{ ft}^3/\text{s}$, $Q_{CD} = [(\pi)(\tfrac{6}{12})^2/4](3.589) = 0.7047 \text{ ft}^3/\text{s}$. Since they are unequal, try $p_D/\gamma = -17$ ft:

$$(h_f)_{AD} = 200 - 140 - (-17) = 0.025[2000/(\tfrac{1}{12})]\{v_{AD}^2/[(2)(32.2)]\} \qquad v_{AD} = 2.875 \text{ ft/s}$$

$$(h_f)_{BD} = 230 - 140 - (-17) = 0.025[3000/(\tfrac{4}{12})]\{v_{BD}^2/[(2)(32.2)]\} \qquad v_{BD} = 5.534 \text{ ft/s}$$

$$(h_f)_{CD} = 120 - 140 - (-17) = 0.025[500/(\tfrac{6}{12})]\{v_{CD}^2/[(2)(32.2)]\} \qquad v_{CD} = 2.789 \text{ ft/s} \quad (\text{away from } D)$$

$$Q_{AD} + Q_{BD} = [(\pi)(\tfrac{1}{12})^2/4](2.875) + [(\pi)(\tfrac{4}{12})^2/4](5.534) = 0.4986 \text{ ft}^3/\text{s}$$

$$Q_{CD} = [(\pi)(\tfrac{6}{12})^2/4](2.789) = 0.5476 \text{ ft}^3/\text{s}$$

$Q_{AD} + Q_{BD}$ is still not equal to Q_{CD}; try $p_D/\gamma = -17.4$ ft:

$$(h_f)_{AD} = 200 - 140 - (-17.5) = 0.025[2000/(\tfrac{1}{12})]\{v_{AD}^2/[(2)(32.2)]\} \qquad v_{AD} = 2.884 \text{ ft/s}$$

$$(h_f)_{BD} = 230 - 140 - (-17.5) = 0.025[3000/(\tfrac{4}{12})]\{v_{BD}^2/[(2)(32.2)]\} \qquad v_{BD} = 5.547 \text{ ft/s}$$

$$(h_f)_{CD} = 120 - 140 - (-17.5) = 0.025[500/(\tfrac{6}{12})]\{v_{CD}^2/[(2)(32.2)]\} \qquad v_{CD} = 2.538 \text{ ft/s} \quad (\text{away from } D)$$

$$Q_{AD} + Q_{BD} = [(\pi)(\tfrac{1}{12})^2/4](2.884) + [(\pi)(\tfrac{4}{12})^2/4](5.547) = 0.4998 \text{ ft}^3/\text{s}$$

$$Q_{CD} = [(\pi)(\tfrac{6}{12})^2/4](2.538) = 0.4983 \text{ ft}^3/\text{s}$$

Since $Q_{AD} + Q_{BD}$ is close enough to Q_{CD}, take

$$Q_{AD} = [(\pi)(\tfrac{1}{12})^2/4](2.884) = 0.015 \text{ ft}^3/\text{s} \qquad (\text{toward } D)$$

$$Q_{BD} = [(\pi)(\tfrac{4}{12})^2/4](5.547) = 0.483 \text{ ft}^3/\text{s} \qquad (\text{toward } D)$$

$$Q_{CD} = 0.498 \text{ ft}^3/\text{s} \qquad (\text{away from } D)$$

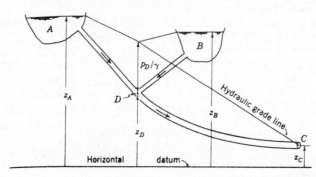

Fig. 12-13

12.19 Three reservoirs A, B, and C whose water-surface elevations are 50 ft, 40 ft, and 22 ft, respectively, are interconnected by a pipe system with a common junction D at elevation 25 ft. The pipes are as follows: from A to junction, $L = 800$ ft, $d = 3$ in; from B to junction, $L = 500$ ft, $d = 10$ in; from C to junction, $L = 1000$ ft, $d = 4$ in. Assume $f = 0.02$ for all pipes and neglect minor losses and velocity heads. Determine whether the flow is into or out of reservoir B.

▮ Assume no flow in B.

$$h_f = (f)(L/d)(v^2/2g) \qquad (h_f)_A = 0.02[800/(\tfrac{3}{12})]\{v_A^2/[(2)(32.2)]\} = 0.9938v_A^2$$

$$(h_f)_C = 0.02[1000/(\tfrac{4}{12})]\{v_C^2/[(2)(32.2)]\} = 0.9317v_C^2$$

$$A_A v_A = A_C v_C \qquad [(\pi)(\tfrac{3}{12})^2/4](v_A) = [(\pi)(\tfrac{4}{12})^2/4](v_C)$$

$$v_A = 1.778v_C \qquad (h_f)_A/(h_f)_C = 0.9938v_A^2/0.9317v_C^2 = (0.9938)(1.778v_C)^2/0.9317v_C^2 = 3.372$$

Hence, $(h_f)_A = [3.372/(3.372 + 1)](50 - 22) = 21.60$ ft, $\text{El}_D = 50 - 21.60 = 28.40$ ft. Since El_D (28.40 ft) is below the surface elevation of reservoir B (40 ft), flow is out of reservoir B.

13.1 Develop the expression utilized to study flows in a pipe network (Hardy Cross method).

▮ The method of attack, developed by Professor Hardy Cross, consists of assuming flows throughout the network, and then balancing the calculated head losses (LH). In the simple looping pipe system shown in Fig. 13-1, for the correct flow in each branch of the loop,

$$\text{LH}_{ABC} = \text{LH}_{ADC} \quad \text{or} \quad \text{LH}_{ABC} - \text{LH}_{ADC} = 0 \tag{1}$$

In order to utilize this relationship, the flow formula to be used must be written in the form $\text{LH} = kQ^n$. For the Hazen–Williams formula, this expression is $\text{LH} = kQ^{1.85}$.

But, since we are assuming flows Q_o, the correct flow Q in any pipe of a network can be expressed as $Q = Q_o + \Delta$, where Δ is the correction to be applied to Q_o. Then, using the binomial theorem, $kQ^{1.85} = k(Q_o + \Delta)^{1.85} = k(Q_o^{1.85} + 1.85Q_o^{1.85-1}\Delta + \cdots)$. Terms beyond the second can be neglected since Δ is small compared with Q_o.

For the loop above, substituting in expression (1) we obtain $k(Q_o^{1.85} + 1.85Q_o^{0.85}\Delta) - k(Q_{o'}^{1.85} + 1.85Q_{o'}^{0.85}\Delta) = 0$, $k(Q_o^{1.85} - Q_{o'}^{1.85}) + 1.85k(Q_o^{0.85} - Q_{o'}^{0.85})\Delta = 0$. Solving for Δ,

$$\Delta = -k(Q_o^{1.85} - Q_{o'}^{1.85})/[1.85k(Q_o^{0.85} - Q_{o'}^{0.85})] \tag{2}$$

In general, we may write for a more complicated loop,

$$\Delta = -\sum kQ_o^{1.85}/(1.85 \sum kQ_o^{0.85}) \tag{3}$$

But $kQ_o^{1.85} = \text{LH}$ and $kQ_o^{0.85} = (\text{LH})/Q_o$. Therefore

$$\Delta = -\sum (\text{LH})/[1.85 \sum (\text{LH}/Q_o)] \quad \text{for each loop of a network} \tag{4}$$

In utilizing expression (4), care must be exercised regarding the sign of the numerator. Expression (1) indicates that clockwise flows may be considered as producing clockwise losses, and counterclockwise flows, counterclockwise losses. This means that the minus sign is assigned to all counterclockwise conditions in a loop, namely flow Q and lost head LH. Hence to avoid mistakes, this sign notation must be observed in carrying out a solution. On the other hand, the denominator of (4) is always positive.

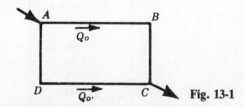

Fig. 13-1

13.2 In Fig. 13-2, for $Q = 11.7$ mgd total, how much flow occurs in each branch of the loop, using the Hardy Cross method?

▮ Values of Q_{12} and Q_{16} are assumed to be 4.0 mgd and 7.7 mgd, respectively. The tabulation below is prepared, (note the -7.70 mgd), the values of S calculated by Fig. A-17, then $\text{LH} = S \times L$, and LH/Q_o can be calculated. Note that the large $\sum \text{LH}$ indicates that the Q's are not well-balanced. (The values were assumed deliberately to produce this large $\sum \text{LH}$, to illustrate the procedure.)

D	L	assumed Q_o mgd	S, ft/1000 ft	LH, ft	LH/Q_o	Δ	Q_1
12 in	5000 ft	4.00	19.5	97.5	24.4	−0.85	3.15
16 in	3000 ft	−7.70	−16.3	−48.9	6.4	−0.85	−8.55
		$\mid\sum\mid = 11.70$		$\sum = +48.6$	30.8		11.70

$$\Delta = -\sum \text{LH}/[1.85 \sum (\text{LH}/Q)] = -(+48.6)/[1.85(30.8)] = -0.85 \text{ mgd}$$

Then the Q_1 values become $(4.00 - 0.85) = 3.15$ mgd and $(-7.70 - 0.85) = -8.55$ mgd. Repeating the calculation produces

S	LH	LH/Q_1	Δ	Q_2
12.5	62.5	19.80	−0.06	3.09
−19.8	−59.4	6.95	−0.06	−8.61
	$\Sigma = +3.1$	26.75		11.70

No further calculation is necessary, since the slide rule or chart cannot be read to the accuracy of 0.06 mgd. Ideally, Σ LH should equal zero, but this goal is seldom attained.

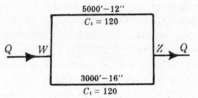

Fig. 13-2

13.3 For the pipe network giving pipe diameters and lengths and external flows entering and leaving the network as shown in Fig. 13-3a, find the flow rate in each pipe in the network.

❚ The first iteration in carrying out the Hardy Cross analysis is given in the table below. A detailed explanation corresponding to steps in carrying out the procedure is given after the table.

Iteration No. 1

Loop I

(1) pipe no.	(2) length, ft	(3) diameter, in	(4) flow rate (Q), ft³/s	(5) unit head loss, ft/ft	(6) head loss (h_f), ft	(7) h_f/Q s/ft²
given	given	given	estimated or assumed	from pipe diagram	(2) × (5)	(6)/(4)
1	2000	18	7.00	0.0034	6.800	0.971
2	900	24	3.50	0.000235	0.212	0.061
3	2800	12	−2.19	−0.0029	−8.120	3.708
4	1100	24	−7.00	−0.00085	−0.935	0.134
					−2.043	4.874

Loop II

(1) pipe no.	(2) length, ft	(3) diameter, in	(4) flow rate (Q), ft³/s	(5) unit head loss, ft/ft	(6) head loss (h_f), ft	(7) h_f/Q, s/ft²
given	given	given	estimated or assumed	from pipe diagram	(2) × (5)	(6)/(4)
5	2200	12	3.50	0.0070	15.400	4.400
6	750	18	−3.14	−0.00078	−0.585	0.186
7	2600	24	−5.69	−0.00057	−1.482	0.260
2	900	24	−3.50	−0.00023	−0.207	0.059
					13.126	4.905

Step 1: Columns (1), (2), and (3) in the table above record given data. Flow rates recorded in column (4) are the initial flow rate estimates. These are listed on the sketch of Fig. 13-3*b*, but they are not "given" data. The flow rate of 14.00 cfs entering joint *A* was estimated to separate such that 7.00 cfs goes through pipe 1 and an equal amount goes through pipe 4. At joint *F*, since 7.00 cfs enters the joint from pipe 4 and 4.81 cfs leaves the joint externally, the flow rate in pipe 3 has to be 2.19 cfs (that is, 7.00 cfs − 4.81 cfs) in the direction away from joint *F* in order to satisfy the principle that the total flow entering a joint must be equal to the total flow leaving that joint. Remaining flow rates were estimated in a similar manner, making sure that for each joint the total flow entering equaled the total flow leaving. It should be noted that clockwise flows in each loop (such as in pipes 1 and 5) are indicated as positive, while counterclockwise flows (such as in pipes 3 and 6) are indicated as negative. It should particularly be noted that the flow in pipe 2 is clockwise with respect to loop I but counterclockwise with respect to loop II; hence, it is indicated as positive when listed in loop I and negative when listed in loop II.

Step 2: Unit head losses in column (5) are determined from Fig. A-13, based on diameters [column (3)] and flow rates [column (4)]. For example, pipe 1 has a diameter of 18 in and flow rate of 7.00 cfs; hence, the unit head loss is determined from Fig. A-13 to be 0.0034 ft/ft. Head losses in column (6) are determined by multiplying pipe lengths [column (2)] by unit head losses [column (5)]. For example, pipe 1 has a length of 2000 ft and unit head loss of 0.0034 ft/ft; hence, the head loss is (2000 ft)(0.0034 ft/ft), or 6.800 ft. It should be noted that head losses are positive if their corresponding flow rates are positive and negative if their corresponding flow rates are negative. The h_f/Q fractions in column (7) are determined by dividing head losses [column (6)] by flow rates [column (4)]. For example, pipe 1 has a head loss of 6.800 ft and a flow rate of 7.00 cfs; hence, h_f/Q is 6.800/7.00, or 0.971 s/ft^2.

Step 3: Algebraic sums of head losses are observed from the table to be −2.043 ft in loop I and 13.126 ft in loop II. Since these are not zero, the original estimated flows are not correct.

Step 4: A flow rate correction (ΔQ) can be computed for each loop: $\Delta Q = -\sum h_f/[n \sum (h_f/Q)]$. (Since the Hazen–Williams formula is being used, $n = 1.85$.) $(\Delta Q)_{\text{loop I}} = -(-2.043)/[(1.85)(4.874)] = 0.23$ cfs, $(\Delta Q)_{\text{loop II}} = -(13.126)/[(1.85)(4.905)] = -1.45$ cfs.

Step 5: Adjusted flow rates for each pipe are determined by adding flow rate corrections to the previous rate for each pipe. These are as follows:

Loop I

pipe	old Q, cfs	ΔQ, cfs	new Q, cfs
1	7.00	+0.23	7.23
2	3.50	+0.23 + 1.45	5.18
3	−2.19	+0.23	−1.96
4	−7.00	+0.23	−6.77

Loop II

pipe	old Q, cfs	ΔQ, cfs	new Q, cfs
5	3.50	−1.45	2.05
6	−3.14	−1.45	−4.59
7	−5.69	−1.45	−7.14
2	−3.50	−1.45 − 0.23	−5.18

These adjusted flow rates are shown in Fig. 13-3*c*. It should be noted that the flow rate in pipe 2 was adjusted using flow rate corrections for both loops, since this pipe is common to both. It should be particularly noted that the sign of the flow rate correction for loop II was reversed when it was applied in loop I (and vice versa).

Step 6: Return to step 2 and repeat the entire procedure using adjusted flow rates. This is given in the table below.

Iteration No. 2

Loop I

(1) pipe no.	(2) length, ft	(3) diameter, in	(4) flow rate (Q), ft³/s	(5) unit head loss, ft/ft	(6) head loss (h_f), ft	(7) h_f/Q, s/ft²
given	given	given	estimate or assumed	from pipe diagram	(2) × (5)	(6)/(4)
1	2000	18	7.23	0.0036	7.200	0.996
2	900	24	5.18	0.00047	0.423	0.082
3	2800	12	−1.96	−0.0024	−6.720	3.429
4	1100	24	−6.77	−0.00080	−0.880	0.130
					0.023	4.637

Loop II

(1) pipe no.	(2) length, ft	(3) diameter, in	(4) flow rate (Q), ft³/s	(5) unit head loss, ft/ft	(6) head loss (h_f), ft	(7) h_f/Q s/ft
given	given	given	estimate or assumed	from pipe diagram	(2) × (5)	(6)/(4)
5	2200	12	2.05	0.0026	5.720	2.790
6	750	18	−4.59	−0.00156	−1.170	0.255
7	2600	24	−7.14	−0.00088	−2.288	0.320
2	900	24	−5.18	−0.00047	−0.423	0.082
					1.839	3.447

Since the algebraic sums of head losses are not both zero, new flow rate corrections must be computed. $(\Delta Q)_{\text{loop I}} = -(0.023)/[(1.85)(4.637)] = 0.00$ cfs, $(\Delta Q)_{\text{loop II}} = -(1.839)/[(1.85)(3.447)] = -0.29$ cfs. Revised flow rates for each pipe are determined using these flow rate corrections. These are as follows:

Loop I

pipe	old Q, cfs	ΔQ, cfs	new Q, cfs
1	7.23	0.00	7.23
2	5.18	0.00 + 0.29	5.47
3	−1.96	0.00	−1.96
4	−6.77	0.00	−6.77

Loop II

pipe	old Q, cfs	ΔQ, cfs	new Q, cfs
5	2.05	−0.29	1.76
6	−4.59	−0.29	−4.88
7	−7.14	−0.29	−7.43
2	−5.18	−0.29	−5.47

These adjusted flow rates are shown in Fig. 13-3d. Return to step 2 and repeat the entire procedure using these revised flow rates. This is given in the table below.

Iteration No. 3

Loop I

(1) pipe no.	(2) length, ft	(3) diameter, in	(4) flow rate (Q), ft³/s	(5) unit head loss, ft/ft	(6) head loss (h_f), ft	(7) h_f/Q, s/ft²
given	given	given	estimate or assumed	from pipe diagram	(2) × (5)	(6)/(4)
1	2000	18	7.23	0.0036	7.200	0.996
2	900	24	5.47	0.00053	0.477	0.087
3	2800	12	−1.96	−0.0024	−6.720	3.429
4	1100	24	−6.77	−0.00080	−0.880	0.130
					0.777	4.642

Loop II

(1) pipe no.	(2) length, ft	(3) diameter, in	(4) flow rate (Q), ft³/s	(5) unit head loss, ft/ft	(6) head loss (h_f), ft	(7) h_f/Q, s/ft²
given	given	given	estimate or assumed	from pipe diagram	(2) × (5)	(6)/(4)
5	2200	12	1.76	0.0019	4.180	2.375
6	750	18	−4.88	−0.00175	−1.312	0.269
7	2600	24	−7.43	−0.00094	−2.444	0.329
2	900	24	−5.47	−0.00053	−0.477	0.087
					−0.053	3.060

Since the algebraic sums of head losses are not both zero, new flow rate corrections must be computed. $(\Delta Q)_{\text{loop I}} = -(0.077)/[(1.85)(4.642)] = -0.01$ cfs, $(\Delta Q)_{\text{loop II}} = -(-0.053)/[(1.85)(3.060)] = 0.01$ cfs. Revised flow rates for each pipe are determined using these flow rate corrections. These are as follows:

Loop I

pipe	old Q, cfs	ΔQ, cfs	new Q, cfs
1	7.23	−0.01	7.22
2	5.47	−0.01 − 0.01	5.45
3	−1.96	−0.01	−1.97
4	−6.77	−0.01	−6.78

Loop II

pipe	old Q, cfs	ΔQ, cfs	new Q, cfs
5	1.76	+0.01	1.77
6	−4.88	+0.01	−4.87
7	−7.43	+0.01	−7.42
2	−5.47	+0.01 + 0.01	−5.45

These adjusted flow rates are shown in Fig. 13-3e. It would be appropriate to return to step 2 and repeat the entire procedure using these revised flow rates. However, an additional iteration (not shown) indicates that the next flow rate corrections would be no greater than 0.01 cfs, and further computation would appear to be wasted effort. Hence, the "new Q" values just before this paragraph are taken to be the correct flow rates for these pipes.

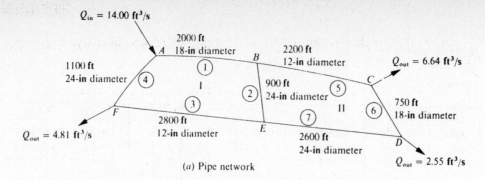

(a) Pipe network

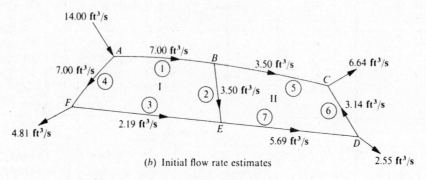

(b) Initial flow rate estimates

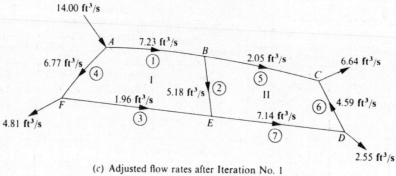

(c) Adjusted flow rates after Iteration No. 1

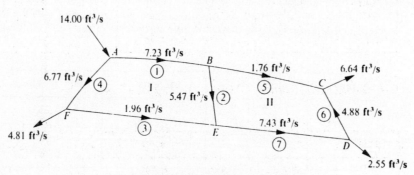

(d) Adjusted flow rates after Iteration No. 2

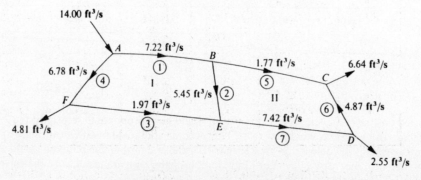

(e) Adjusted flow rates after Iteration No. 3

Fig. 13-3

13.4 For the pipe network shown in Fig. 13-4, find the rate of flow of water in each pipe. Assume $C = 120$ for all pipes.

▌

pipe	D, in	L, ft	$Q_{assumed}$, cfs	h_1, ft/ft	h_f, ft	h_f/Q	ΔQ	Q_{new}
AB	18	1500	8.0	0.0044	6.60	0.825	−0.11	7.89
BC	12	500	3.0	0.0053	2.65	0.883	−0.11	2.89
CF	15	1500	−4.0	−0.00295	−4.42	1.105	2.23	−1.77
FA	15	500	−7.0	−0.0083	−4.15	0.593	−0.11	−7.11
					0.68	3.406		
FC	15	1500	4.0	0.00295	4.42	1.105	−2.23	1.77
CD	10	500	7.0	0.059	29.50	4.214	−2.34	4.66
DE	12	1500	2.0	0.0025	3.75	1.875	−2.34	−0.34
EF	12	500	−3.0	−0.0053	−2.65	0.883	−2.34	−5.34
					35.02	8.077		

$$\Delta Q_{\mathrm{I}} = -\sum h_f/[n\sum (h_f/Q)] = -0.68/[(1.85)(3.406)] = -0.11 \text{ cfs} \qquad \Delta Q_{\mathrm{II}} = -35.02/[(1.85)(8.077)] = -2.34 \text{ cfs}$$

pipe	h_1	h_f	h_f/Q	ΔQ	Q_{new}
AB	0.0042	6.30	0.798	−0.66	7.23
BC	0.0049	2.45	0.848	−0.66	2.23
CF	0.00067	−1.00	0.565	0.08	−1.69
FA	0.0086	−4.30	0.605	−0.66	−7.77
		3.45	2.816		
FC	0.00067	1.00	0.565	−0.08	1.69
CD	0.028	14.00	3.000	−0.74	3.92
DE	0.000095	−0.14	0.412	−0.74	−1.08
EF	0.015	−7.50	1.404	−0.74	−6.08
		7.36	5.381		

$$\Delta Q_{\mathrm{I}} = -3.45/[(1.85)(2.816)] = -0.66 \text{ cfs} \qquad \Delta Q_{\mathrm{II}} = -7.36/[(1.85)(5.381)] = -0.74 \text{ cfs}$$

pipe	h_1	h_f	h_f/Q	ΔQ	Q_{new}
AB	0.0036	5.40	0.747	−0.20	7.03
BC	0.00295	1.48	0.664	−0.20	2.03
CF	−0.00060	−0.90	0.533	−0.20	−1.89
FA	−0.0100	−5.00	0.644	−0.20	−7.97
		0.98	2.588		
FC	0.00060	0.90	0.533	0.20	1.89
CD	0.0200	10.00	2.551	0.00	3.92
DE	−0.00080	−1.20	1.111	0.00	−1.08
EF	−0.0195	−9.75	1.604	0.00	−6.08
		−0.05	5.799		

$$\Delta Q_{\mathrm{I}} = -0.98/[(1.85)(2.588)] = -0.20 \text{ cfs} \qquad \Delta Q_{\mathrm{II}} = -(-0.05)/[(1.85)(5.799)] = 0.00 \text{ cfs}$$

pipe	h_1	h_f	h_f/Q	ΔQ	Q_{new}
AB	0.0034	5.10	0.725	0.02	7.05
BC	0.0026	1.30	0.640	0.02	2.05
CF	−0.00075	−1.11	0.587	0.03	−1.86
FA	−0.0108	−5.40	0.678	0.02	−7.95
		−0.11	2.630		
FC	0.00074	1.11	0.587	−0.03	1.86
CD	0.0200	10.00	2.551	−0.01	3.91
DE	−0.00080	−1.20	1.111	−0.01	−1.09
EF	−0.0195	−9.75	1.604	−0.01	−6.09
		0.16	5.853		

$$\Delta Q_{\rm I} = -(-0.11)/[(1.85)(2.630)] = 0.02 \text{ cfs} \qquad \Delta Q_{\rm II} = -0.16/[(1.85)(5.853)] = -0.01 \text{ cfs}$$

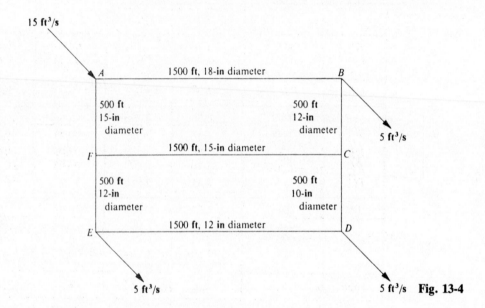

15 ft³/s

A — 1500 ft, 18-in diameter — B

500 ft
15-in
diameter

500 ft
12-in
diameter

5 ft³/s

F — 1500 ft, 15-in diameter — C

500 ft
12-in
diameter

500 ft
10-in
diameter

E — 1500 ft, 12 in diameter — D

5 ft³/s

5 ft³/s   5 ft³/s   **Fig. 13-4**

13.5 The pipe network shown in Fig. 13-5 represents a spray rinse system. Find the flow rate of water in each pipe. Assume $C = 120$ for all pipes.

◼

pipe	D, mm	L, m	$Q_{assumed}$, m³/s	h_1, m/m	h_f, m	h_f/Q	ΔQ	Q_{new}
AB	300	600	0.200	0.027	16.20	81.0	0.011	0.211
BG	250	400	0.100	0.0175	7.00	70.0	−0.003	0.097
GH	300	600	−0.100	−0.0074	−4.44	44.4	0.011	−0.089
HA	250	400	−0.200	−0.064	−25.60	128.0	0.011	−0.189
					−6.84	323.4		
BC	300	600	0.100	0.0074	4.44	44.4	0.014	0.114
CF	250	400	0.050	0.0049	1.96	39.2	0.014	0.064
FG	300	600	−0.100	−0.0074	−4.44	44.4	0.014	−0.086
GB	250	400	−0.100	−0.0175	−7.00	70.0	0.003	−0.097
					−5.04	198.0		

pipe	D, mm	L, m	$Q_{assumed}$, m³/s	h_1, m/m	h_f, m	h_f/Q	ΔQ	Q_{new}
CD	300	600	0.050	0.0020	1.20	24.0	0.000	0.050
DE	250	400	0.050	0.0049	1.96	39.2	0.000	0.050
EF	300	600	−0.050	−0.0020	−1.20	24.0	0.000	−0.050
FC	250	400	−0.050	−0.0049	−1.96	39.2	−0.014	−0.064
					0.00	126.4		

$$\Delta Q_{\mathrm{I}} = -\sum h_f / [n \sum (h_f/Q)] = -(-6.84)/[(1.85)(323.4)] = 0.011$$

$$\Delta Q_{\mathrm{II}} = -(-5.04)/[(1.85)(198.0)] = 0.014 \qquad \Delta Q_{\mathrm{III}} = -0.00/[(1.85)(126.4)] = 0.00$$

pipe	h_1	h_f	h_f/Q	ΔQ	Q_{new}
AB	0.0295	17.70	83.9	0.004	0.215
BG	0.017	6.80	70.1	0.001	0.098
GH	−0.0059	−3.54	39.8	0.004	−0.085
HA	−0.058	−23.20	122.8	0.004	−0.185
		−2.24	316.6		
BC	0.0095	5.70	50.0	0.003	0.117
CF	0.0079	3.16	49.4	−0.002	0.062
FG	−0.0056	−3.36	39.1	0.003	−0.083
GB	−0.017	−6.80	70.1	−0.001	−0.098
		−1.30	208.6		
CD	0.0020	1.20	24.0	0.005	0.055
DE	0.0049	1.96	39.2	0.005	0.055
EF	−0.0020	−1.20	24.0	0.005	−0.045
FC	−0.0079	−3.16	49.4	0.002	−0.062
		−1.20	136.6		

$$\Delta Q_{\mathrm{I}} = -(-2.24)/[(1.85)(316.6)] = 0.004 \qquad \Delta Q_{\mathrm{II}} = -(-1.30)/[(1.85)(208.6)] = 0.003$$

$$\Delta Q_{\mathrm{III}} = -(-1.20)/[(1.85)(136.6)] = 0.005$$

pipe	h_1	h_f	h_f/Q	ΔQ	Q_{new}
AB	0.031	18.60	86.5	0.000	0.215
BG	0.0172	6.88	70.2	−0.003	0.095
GH	0.0055	−3.30	38.8	0.000	−0.085
HA	0.056	−22.40	121.1	0.000	−0.185
		−0.22	316.6		
BC	0.010	6.00	51.3	0.003	0.120
CF	0.0075	3.00	48.4	0.002	0.064
FG	0.0052	−3.12	37.6	0.003	−0.080
GB	0.0172	−6.88	70.2	0.003	−0.095
		−1.00	207.5		
CD	0.0024	1.44	26.2	0.001	0.056
DE	0.0059	2.36	42.9	0.001	0.056
EF	0.0017	−1.02	22.7	0.001	−0.044
FC	0.0075	−3.00	48.4	−0.002	−0.064
		−0.22	140.2		

$$\Delta Q_{\mathrm{I}} = -(-0.22)/[(1.85)(316.6)] = 0.000 \qquad \Delta Q_{\mathrm{II}} = -(-1.00)/[(1.85)(207.5)] = 0.003$$

$$\Delta Q_{\mathrm{III}} = -(-0.22)/[(1.85)(140.2)] = 0.001$$

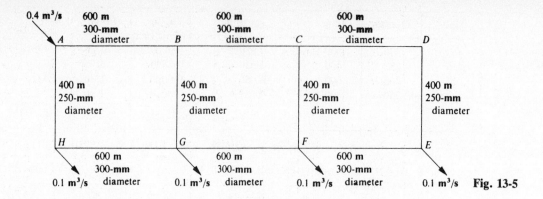

Fig. 13-5

13.6 The pipe network shown in Fig. 13-6 represents a small industrial park. Find the flow rate of water in each pipe. Assume $C = 120$ for all pipes.

pipe	D, mm	L, m	$Q_{assumed}$, m³/s	h_1, m/m	h_f, m	h_f/Q	ΔQ	Q_{new}
AB	500	1000	0.250	0.0034	3.40	13.6	0.028	0.278
BE	400	1200	0.120	0.0026	3.12	26.0	−0.005	0.115
EF	300	1000	−0.130	−0.012	−12.00	92.3	0.063	−0.067
FA	600	1200	−0.250	−0.0014	−1.68	6.7	0.028	−0.222
					−7.16	138.6		
BC	500	1000	0.130	0.00102	1.02	7.8	0.033	0.163
CD	400	1200	0.030	0.00020	0.24	8.0	0.033	0.063
DE	400	1000	−0.100	−0.0018	−1.8	18.0	0.050	−0.050
EB	400	1200	−0.120	−0.0026	−3.12	26.0	0.005	−0.115
					−3.66	59.8		
ED	400	1000	0.100	0.0018	1.80	18.0	−0.050	0.050
DI	300	1200	0.080	0.0048	5.76	72.0	−0.017	0.063
IH	300	1000	−0.020	−0.00037	−0.37	18.5	−0.017	−0.037
HE	300	1200	−0.050	−0.0020	−2.40	48.0	0.018	−0.032
					4.79	156.5		
FE	300	1000	0.130	0.012	12.00	92.3	−0.063	0.067
EH	300	1200	0.050	0.0020	2.40	48.0	−0.018	0.032
HG	400	1000	−0.020	−0.000095	−0.10	5.0	−0.035	−0.055
GF	400	1200	−0.120	−0.0026	−3.12	26.0	−0.035	−0.155
					11.18	171.3		

$$\Delta Q_{\mathrm{I}} = -\sum h_f/[n \sum (h_f/Q)] = -(-7.16)/[(1.85)(138.6)] = 0.028 \qquad \Delta Q_{\mathrm{II}} = -(-3.66)/[(1.85)(59.8)] = 0.033$$

$$\Delta Q_{\mathrm{III}} = -4.79/[(1.85)(156.5)] = -0.017 \qquad \Delta Q_{\mathrm{IV}} = -11.18/[(1.85)(171.3)] = -0.035$$

pipe	h_1	h_f	h_f/Q	ΔQ	Q_{new}
AB	0.0043	4.30	15.5	−0.013	0.265
BE	0.0023	2.76	24.0	−0.020	0.095
EF	−0.0034	−3.40	50.7	−0.018	−0.085
FA	−0.0011	−1.32	5.9	−0.013	−0.235
		2.34	96.1		

pipe	h_1	h_f	h_f/Q	ΔQ	Q_{new}
BC	0.0016	1.60	9.8	0.007	0.170
CD	0.00078	0.94	14.9	0.007	0.070
DE	−0.00050	−0.50	10.0	0.015	−0.035
EB	−0.0023	−2.76	24.0	0.020	−0.095
		−0.72	58.7		
ED	0.00050	0.50	10.0	−0.015	0.035
DI	0.0031	3.72	59.0	−0.008	0.055
IH	−0.00115	−1.15	31.1	−0.008	−0.045
HE	−0.00087	−1.04	32.5	−0.013	−0.045
		2.03	132.6		
FE	0.0034	3.40	50.7	0.018	0.085
EH	0.00087	1.04	32.5	0.013	0.045
HG	−0.00060	−0.60	10.9	0.005	−0.050
GF	−0.0041	−4.92	31.7	0.005	−0.150
		−1.08	125.8		

$$\Delta Q_{\mathrm{I}} = -2.34/[(1.85)(96.1)] = -0.013 \qquad \Delta Q_{\mathrm{II}} = -(-0.72)/[(1.85)(58.7)] = 0.007$$

$$\Delta Q_{\mathrm{III}} = -2.03/[(1.85)(132.6)] = -0.008 \qquad \Delta Q_{\mathrm{IV}} = -(-1.08)/[(1.85)(125.8)] = 0.005$$

pipe	h_1	h_f	h_f/Q	ΔQ	Q_{new}
AB	0.0039	3.90	14.7	0.006	0.271
BE	0.00165	1.98	20.8	0.012	0.107
EF	−0.0055	−5.50	64.7	0.014	−0.071
FA	−0.00125	−1.50	6.4	0.006	−0.229
		−1.12	106.6		
BC	0.0017	1.70	10.0	−0.006	0.164
CD	0.00094	1.13	16.1	−0.006	0.064
DE	−0.000265	−0.26	7.4	−0.008	−0.043
EB	−0.00165	−1.98	20.8	−0.012	−0.107
		0.59	54.3		
ED	0.000265	0.26	7.4	0.008	0.043
DI	0.0024	2.88	52.4	0.002	0.057
IH	−0.0017	−1.70	37.8	0.002	−0.043
HE	−0.0017	−2.04	45.3	0.010	−0.035
		−0.60	142.9		
FE	0.0055	5.50	64.7	−0.014	0.071
EH	0.0017	2.04	45.3	−0.010	0.035
HG	−0.00051	−0.51	10.2	−0.008	−0.058
GF	−0.0039	−4.68	31.2	−0.008	−0.158
		2.35	151.4		

$$\Delta Q_{\mathrm{I}} = -(-1.12)/[(1.85)(106.6)] = 0.006 \qquad \Delta Q_{\mathrm{II}} = -0.59/[(1.85)(54.3)] = -0.006$$

$$\Delta Q_{\mathrm{III}} = -(-0.60)/[(1.85)(142.9)] = 0.002 \qquad \Delta Q_{\mathrm{IV}} = -2.35/[(1.85)(151.4)] = -0.008$$

pipe	h_1	h_f	h_f/Q	ΔQ	Q_{new}
AB	0.0040	4.00	14.8	−0.006	0.265
BE	0.0021	2.52	23.6	−0.009	0.098
EF	−0.0039	−3.90	54.9	−0.008	−0.079
FA	−0.0012	−1.44	6.3	−0.006	−0.235
		1.18	99.6		
BC	0.0016	1.60	9.8	0.003	0.167
CD	0.00080	0.96	15.0	0.003	0.067
DE	−0.00038	−0.38	8.8	0.006	−0.037
EB	−0.0021	−2.52	23.6	0.009	−0.098
		−0.34	57.2		
ED	0.00038	0.38	8.8	−0.006	0.037
DI	0.0026	3.12	54.7	−0.003	0.054
IH	−0.0015	−1.50	34.9	−0.003	−0.046
HE	−0.00105	−1.26	36.0	−0.005	−0.040
		0.74	134.4		
FE	0.0039	3.90	54.9	0.008	0.079
EH	0.00105	1.26	36.0	0.005	0.040
HG	−0.00067	−0.67	11.6	0.002	−0.056
GF	−0.0041	−4.92	31.1	0.002	−0.156
		−0.43	133.6		

$$\Delta Q_{\mathrm{I}} = -1.18/[(1.85)(99.6)] = -0.006 \qquad \Delta Q_{\mathrm{II}} = -(-0.34)/[(1.85)(57.2)] = 0.003$$

$$\Delta Q_{\mathrm{III}} = -0.74/[(1.85)(134.4)] = -0.003 \qquad \Delta Q_{\mathrm{IV}} = -(-0.43)/[(1.85)(133.6)] = 0.002$$

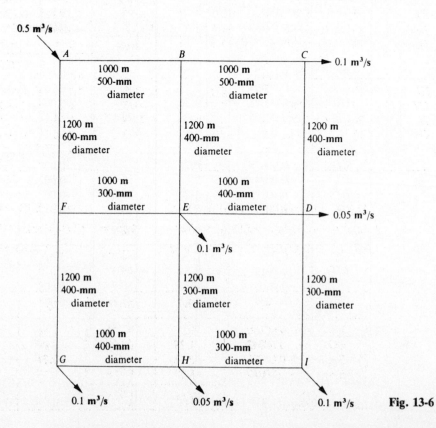

Fig. 13-6

13.7   Compute the flow rate of water in each pipe of the network shown in Fig. 13-7. Assume $C = 120$ for all pipes.

pipe	D, in.	L, ft	$Q_{assumed}$, cfs	h_1, ft/ft	h_f, ft	h_f/Q	ΔQ	Q_{new}
AB	24	2000	5.0	0.00045	0.90	0.180	0.29	5.29
BG	15	3000	1.0	0.00022	0.66	0.660	0.42	1.42
GH	18	2000	−2.5	−0.00051	−1.02	0.408	0.34	−2.16
HA	24	3000	−5.0	−0.00045	−1.35	0.270	0.29	−4.71
					−0.81	1.518		
BC	24	2000	4.0	0.00030	0.60	0.150	−0.13	3.87
CF	15	3000	1.0	0.00022	0.66	0.660	0.53	1.53
FG	18	2000	−1.0	−0.000095	−0.19	0.190	−0.48	−1.48
GB	15	3000	−1.0	−0.00022	−0.66	0.660	−0.42	−1.42
					0.41	1.660		
CD	15	2000	3.0	0.0017	3.40	1.133	−0.66	2.34
DE	12	3000	1.0	0.00067	2.01	2.010	−0.66	0.34
EF	18	2000	−0.3	−0.000010	−0.02	0.067	−0.33	−0.63
FC	15	3000	−1.0	−0.00022	−0.66	0.660	−0.53	−1.53
					4.73	3.870		
FE	18	2000	0.3	0.000010	0.02	0.067	0.33	0.63
EL	12	3000	1.3	0.0011	3.30	2.538	−0.33	0.97
LK	15	2000	−1.7	−0.00059	−1.18	0.694	−0.33	−2.03
KF	15	3000	−0.2	−0.000011	−0.03	0.150	−0.68	−0.88
					2.11	3.449		
GF	18	2000	1.0	0.000095	0.19	0.190	0.48	1.48
FK	15	3000	0.2	0.000011	0.03	0.150	0.68	0.88
KJ	18	2000	−1.5	−0.00019	−0.38	0.253	0.35	−1.15
JG	15	3000	−1.0	−0.00022	−0.66	0.660	0.40	−0.60
					−0.82	1.253		
HG	18	2000	2.5	0.00051	1.02	0.408	−0.34	2.16
GJ	15	3000	1.0	0.00022	0.66	0.660	−0.40	0.60
JI	24	2000	−0.5	−0.0000063	−0.01	0.020	−0.05	−0.55
IH	18	3000	−2.5	−0.00050	−1.50	0.600	−0.05	−2.55
					0.17	1.688		

$$\Delta Q_{\mathrm{I}} = -(-0.81)/[(1.85)(1.518)] = 0.29 \qquad \Delta Q_{\mathrm{II}} = -0.41/[(1.85)(1.660)] = -0.13$$

$$\Delta Q_{\mathrm{III}} = -4.73/[(1.85)(3.870)] = -0.66 \qquad \Delta Q_{\mathrm{IV}} = -2.11/[(1.85)(3.449)] = -0.33$$

$$\Delta Q_{\mathrm{V}} = -(-0.82)/[(1.85)(1.253)] = 0.35 \qquad \Delta Q_{\mathrm{VI}} = -0.17/[(1.85)(1.688)] = -0.05$$

pipe	h_1	h_f	h_f/Q	ΔQ	Q_{new}
AB	0.00050	1.00	0.189	−0.10	5.19
BG	0.00043	1.29	0.908	0.00	1.42
GH	−0.00039	−0.78	0.361	−0.31	−2.47
HA	−0.00040	−1.20	0.255	−0.10	−4.81
		0.31	1.713		

pipe	h_1	h_f	h_f/Q	ΔQ	Q_{new}
BC	0.00029	0.58	0.150	−0.10	3.77
CF	0.00050	1.50	0.980	0.07	1.60
FG	−0.00019	−0.38	0.257	0.04	−1.44
GB	−0.00043	−1.29	0.908	0.00	−1.42
		0.41	2.295		
CD	0.0011	2.20	0.940	−0.17	2.17
DE	0.000090	0.27	0.794	−0.17	0.17
EF	−0.000039	−0.08	0.127	−0.20	−0.83
FC	−0.00050	−1.50	0.980	−0.07	−1.60
		0.89	2.841		
FE	0.000039	0.08	0.127	0.20	0.83
EL	0.00064	1.92	1.979	0.03	1.00
LK	−0.00082	−1.64	0.808	0.03	−2.00
KF	−0.00018	−0.54	0.614	0.17	−0.71
		−0.18	3.528		
GF	0.00019	0.38	0.257	−0.04	1.44
FK	0.00018	0.54	0.614	−0.17	0.71
KJ	−0.00012	−0.24	0.209	−0.14	−1.29
JG	−0.000090	−0.27	0.450	−0.35	−0.95
		0.41	1.530		
HG	0.00039	0.78	0.361	0.31	2.47
GJ	0.000090	0.27	0.450	0.35	0.95
JI	−0.0000075	−0.02	0.036	0.21	−0.34
IH	−0.00053	−1.59	0.624	0.21	−2.34
		−0.56	1.471		

$$\Delta Q_{\mathrm{I}} = -0.31/[(1.85)(1.713)] = -0.10 \qquad \Delta Q_{\mathrm{II}} = -0.41/[(1.85)(2.295)] = -0.10$$

$$\Delta Q_{\mathrm{III}} = -0.89/[(1.85)(2.841)] = -0.17 \qquad \Delta Q_{\mathrm{IV}} = -(-0.18)/[(1.85)(3.528)] = 0.03$$

$$\Delta Q_{\mathrm{V}} = -0.41/[(1.85)(1.530)] = -0.14 \qquad \Delta Q_{\mathrm{VI}} = -(-0.56)/[(1.85)(1.471)] = 0.21$$

pipe	h_1	h_f	h_f/Q	ΔQ	Q_{new}
AB	0.00048	0.96	0.185	0.00	5.19
BG	0.00043	1.29	0.908	0.11	1.53
GH	−0.00050	−1.00	0.405	0.08	−2.39
HA	−0.00042	−1.26	0.262	0.00	−4.81
		−0.01	1.760		
BC	0.00027	0.54	0.143	−0.11	3.66
CF	0.00053	1.59	0.994	−0.05	1.55
FG	−0.00018	−0.36	0.250	−0.17	−1.61
GB	−0.00043	−1.29	0.908	−0.11	−1.53
		0.48	2.295		
CD	0.00097	1.94	0.894	−0.06	2.11
DE	0.000025	0.08	0.471	−0.06	0.11
EF	−0.000066	−0.13	0.157	−0.04	−0.87
FC	−0.00053	−1.59	0.994	0.05	−1.55
		0.30	2.156		

pipe	h_1	h_f	h_f/Q	ΔQ	Q_{new}
FE	0.000066	0.13	0.157	0.04	0.87
EL	0.00067	2.01	2.010	−0.02	0.98
LK	−0.00081	−1.62	0.810	−0.02	−2.02
KF	−0.00012	−0.36	0.507	−0.08	−0.79
		0.16	3.484		
GF	0.00018	0.36	0.250	0.17	1.61
FK	0.00012	0.36	0.507	0.08	0.79
KJ	−0.00015	−0.30	0.233	0.06	−1.23
JG	−0.00020	−0.60	0.632	0.14	−0.81
		−0.18	1.622		
HG	0.00050	1.00	0.405	−0.08	2.39
GJ	0.00020	0.60	0.632	−0.14	0.81
JI	−0.0000031	−0.01	0.029	−0.08	−0.42
IH	−0.00045	−1.35	0.577	−0.08	−2.42
		0.24	1.643		

$$\Delta Q_{\mathrm{I}} = -(-0.01)/[(1.85)(1.760)] = 0.00 \qquad \Delta Q_{\mathrm{II}} = -0.48/[(1.85)(2.295)] = -0.11$$

$$\Delta Q_{\mathrm{III}} = -0.30/[(1.85)(2.516)] = -0.06 \qquad \Delta Q_{\mathrm{IV}} = -0.16/[(1.85)(3.484)] = -0.02$$

$$\Delta Q_{\mathrm{V}} = -(-0.18)/[(1.85)(1.622)] = 0.06 \qquad \Delta Q_{\mathrm{VI}} = -0.24/[(1.85)(1.643)] = -0.08$$

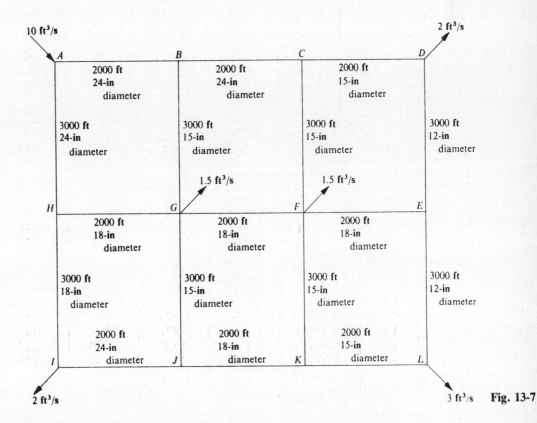

Fig. 13-7

13.8 Water flows through the piping system shown in Fig. 13-8 with certain measured flows indicated on the sketch. Determine the flows throughout the network. Use $C = 120$ throughout.

line	D, in	L, ft	assumed Q_1, mgd	S, ft/1000 ft	LH, ft	$\dfrac{LH}{Q_1}$	Δ	Q_2
AB	20	3000	4.0	1.62	4.86	1.22	+0.31	4.31
BE	16	4000	1.0	0.37	1.48	1.48	+0.31 − (0.13) = +0.18	1.18
EF	16	3000	−2.0	−1.33	−3.99	2.00	+0.31 − (0.50) = −0.19	−2.19
FA	24	4000	−6.0	−1.41	−5.64	0.94	+0.31	−5.69
					$\Sigma = -3.29$	5.64		
BC	20	3000	3.0	0.95	2.85	0.95	+0.13	3.13
CD	16	4000	2.0	1.33	5.32	2.66	+0.13	2.13
DE	12	3000	−1.5	−3.15	−9.45	6.30	+0.13 − (−0.12) = +0.25	−1.25
EB	16	4000	−1.0	−0.37	−1.48	1.48	+0.13 − (0.31) = −0.18	−1.18
					$\Sigma = -2.76$	11.39		
FE	16	3000	2.0	1.33	3.99	2.00	+0.50 − (+0.31) = +0.19	2.19
EH	12	4000	1.0	1.48	5.92	5.92	+0.50 − (−0.12) = +0.62	1.62
HG	16	3000	−2.0	−1.33	−3.99	2.00	+0.50	−1.50
GF	16	4000	−4.0	−4.85	−19.40	4.85	+0.50	−3.50
					$\Sigma = -13.48$	14.77		
ED	12	3000	1.5	3.15	9.45	6.30	−0.12 − (0.13) = −0.25	1.25
DI	12	4000	1.0	1.48	5.92	5.92	−0.12	0.88
IH	12	3000	−1.0	−1.48	−4.44	4.44	−0.12	−1.12
HE	12	4000	−1.0	−1.48	−5.92	5.92	−0.12 − (0.50) = −0.62	−1.62
					$\Sigma = +5.01$	22.58		

$$\Delta_{\mathrm{I}} = -(-3.29)/[1.85(5.64)] = +0.31 \qquad \Delta_{\mathrm{II}} = -(-2.76)/[1.85(11.39)] = +0.13$$

$$\Delta_{\mathrm{III}} = -(-13.48)/[1.85(14.77)] = +0.50 \qquad \Delta_{\mathrm{IV}} = -(+5.01)/[1.85(22.58)] = -0.12$$

For line *EF* in loop I, its net Δ term is $(\Delta_{\mathrm{I}} - \Delta_{\mathrm{III}})$ or $[+0.31 - (+0.50)] = -0.19$. It will be observed that the Δ for loop I is combined with that of the Δ for loop III since the line *EF* occurs in each loop. In similar fashion, for line *FE* in loop III, the net Δ term is $(\Delta_{\mathrm{III}} - \Delta_{\mathrm{I}})$ or $[+0.50 - (+0.31)] = +0.19$. Note that the net Δ's have the same magnitude but *opposite signs*. This can readily be understood since flow in *EF* is counterclockwise for loop I, whereas flow in *FE* in loop III is clockwise.

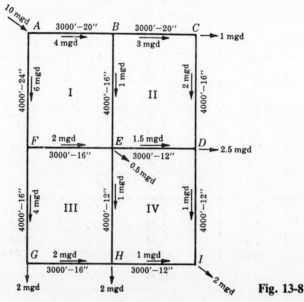

Fig. 13-8

In determining the values of Q_2 for the second calculation, $Q_{AB} = (4.00 + 0.31) = 4.31$ mgd whereas $Q_{EF} = (-2.00 - 0.19) = -2.19$ mgd and $Q_{FA} = (-6.00 + 0.31) = -5.69$ mgd.

line	Q_2	S	LH	LH/Q	Δ
AB	4.31	1.86	5.58	1.29	+0.20
BE	1.18	0.51	2.04	1.72	+0.20 + negl = +0.20
EF	-2.19	-1.57	-4.71	2.15	+0.20 - (-0.06) = +0.26
FA	-5.69	-1.28	-5.12	0.90	+0.20
			$\Sigma = -2.21$	6.06	
BC	3.13	1.02	3.06	0.98	negl
CD	2.13	1.48	5.92	2.79	negl
DE	-1.25	-2.28	-6.84	5.50	negl - 0.19 = -0.19
EB	-1.18	-0.51	-2.04	1.72	negl - 0.20 = -0.20
			$\Sigma = +0.10$	10.99	
FE	2.19	1.57	4.71	2.15	-0.06 - 0.20 = -0.26
EH	1.62	3.65	14.60	9.02	-0.06 - 0.19 = -0.25
HG	-1.50	-0.79	-2.37	1.58	-0.06
GF	-3.50	-3.75	-15.00	4.28	-0.06
			$\Sigma = +1.94$	17.03	
ED	1.25	2.28	6.84	5.42	+0.19 + negl = +0.19
DI	0.88	1.18	4.72	5.38	+0.19
IH	-1.12	-1.83	-5.49	4.90	+0.19
HE	-1.62	-3.65	-14.60	9.02	+0.19 - (-0.06) = +0.25
			$\Sigma = -8.53$	24.72	

line	Q_3	S	LH	LH/Q	Δ	Q_4
AB	4.51	2.02	6.06	1.34	-0.02	4.49
BE	1.39	0.68	2.72	1.95	-0.02 - 0.12 = -0.14	1.25
EF	-1.93	-1.25	-3.75	1.94	-0.02 - 0.12 = -0.14	-2.07
FA	-5.49	-1.20	-4.80	0.88	-0.02	-5.51
			$\Sigma = +0.23$	6.11		
BC	3.12	1.02	3.06	0.98	+0.12	3.24
CD	2.12	1.49	5.96	2.81	+0.12	2.24
DE	-1.45	-2.97	-8.91	6.15	+0.12 + 0.02 = +0.14	-1.31
EB	-1.39	-0.68	-2.72	1.95	+0.12 + 0.02 = +0.14	-1.25
			$\Sigma = -2.61$	11.89		
FE	1.93	1.25	3.75	1.94	+0.12 + 0.02 = +0.14	2.07
EH	1.37	2.68	10.72	7.83	+0.12 + 0.02 = +0.14	1.51
HG	-1.56	-0.84	-2.52	1.62	+0.12	-1.44
GF	-3.56	-3.90	-15.60	4.38	+0.12	-3.44
			$\Sigma = -3.65$	15.77		
ED	1.45	2.97	8.91	6.15	-0.02 - 0.12 = -0.14	1.31
DI	1.07	1.68	6.72	6.28	-0.02	1.05
IH	-0.93	-1.31	-3.93	4.23	-0.02	-0.95
HE	-1.37	-2.68	-10.72	7.83	-0.02 - 0.12 = -0.14	-1.51
			$\Sigma = +0.98$	24.49		

13.9 For the piping system of Prob. 13.8, if the elevation at point A is 200.0 ft and the pressure head is 150.0 ft and the elevation at I is 100.0 ft, find the pressure head at I.

▌ The elevation of the hydraulic grade line at A is $(200.0 + 150.0) = 350.0$. The lost head to I can be calculated by any route from A to I, adding the losses in the usual manner, i.e., in the direction of flow. Using $ABEHI$ we obtain $LH_{A\,to\,I} = (6.06 + 2.72 + 10.72 + 3.93) = 23.43$ ft. As a check, using $ABEDI$, LH $= (6.06 + 2.72 + 8.91 + 6.72) = 24.41$ ft. Using a value of 24 ft, the elevation of the hydraulic grade line at $I = (350.0 - 24.0) = 326.0$ ft. Hence the pressure head at $I = (326.0 - 100.0) = 226.0$ ft.

13.10 If the flows into and out of a two-loop pipe system are as shown in Fig. 13-9, determine the flow in each pipe. The K values for each pipe were calculated from the pipe and minor loss characteristics and from an assumed value of f.

▌ As a first step, assume flow in each pipe such that continuity is satisfied at all junctions. Calculate ΔQ for each loop, make corrections to the assumed Q's and repeat several times until the ΔQ's are quite small. As a final step the values of f for each pipe should be checked against the Moody diagram and modified, if necessary.

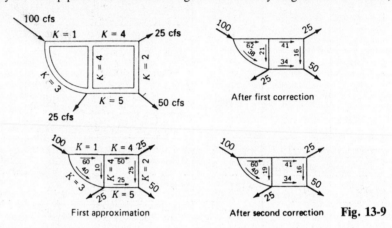

Fig. 13-9

	left loop		right loop	
$\sum KQ_0^n$	$\sum \lvert KnQ_0^{n-1}\rvert$	$\sum KQ_0^n$	$\sum \lvert KnQ_0^{n-1}\rvert$
$1 \times 60^2 = 3\,600$ $4 \times 10^2 = 400$ ⎯⎯⎯ $4\,000\downarrow$ $3 \times 40^2 = 4\,800\uparrow$ ⎯⎯⎯ $800\uparrow$	$1 \times 2 \times 60 = 120$ $4 \times 2 \times 10 = 80$ ⎯⎯⎯ $3 \times 2 \times 40 = 240$ ⎯⎯⎯ 440	$4 \times 50^2 = 10\,000$ $2 \times 25^2 = 1\,250$ ⎯⎯⎯ $11\,250\downarrow$ $4 \times 10^2 = 400$ $5 \times 25^2 = 3\,125$ ⎯⎯⎯ $3\,525\uparrow$ ⎯⎯⎯ $7\,725\downarrow$	$4 \times 2 \times 50 = 400$ $2 \times 2 \times 25 = 100$ ⎯⎯⎯ $4 \times 2 \times 10 = 80$ $5 \times 2 \times 25 = 250$ ⎯⎯⎯ 830
$\Delta Q_1 = \dfrac{800}{440} \approx 2\downarrow$		$\Delta Q_1 = \dfrac{7\,725}{830} \approx 9\uparrow$	
$1 \times 62^2 = 3\,844$ $4 \times 21^2 = 1\,764$ ⎯⎯⎯ $5\,608\downarrow$ $3 \times 38^2 = 4\,332\uparrow$ ⎯⎯⎯ $1\,276\downarrow$	$1 \times 2 \times 62 = 124$ $4 \times 2 \times 21 = 168$ ⎯⎯⎯ $3 \times 2 \times 38 = 228$ ⎯⎯⎯ 520	$4 \times 41^2 = 6\,724$ $2 \times 16^2 = 512$ ⎯⎯⎯ $7\,236\downarrow$ $4 \times 21^2 = 1\,764$ $5 \times 34^2 = 5\,780$ ⎯⎯⎯ $7\,544\uparrow$ ⎯⎯⎯ $308\uparrow$	$4 \times 2 \times 41 = 328$ $2 \times 2 \times 16 = 64$ ⎯⎯⎯ $4 \times 2 \times 21 = 168$ $5 \times 2 \times 34 = 340$ ⎯⎯⎯ 900
$\Delta Q_2 = \dfrac{1\,276}{520} \approx 2\uparrow$		$\Delta Q_2 = \dfrac{308}{900} \approx 0$	

13.11 Solve the pipe network shown in Fig. 13-10, using four approximations, to find the flow in each pipe. For simplicity, take $n = 2.0$ and use the values of f for complete turbulence, as given in Fig. A-5. All pipe is cast iron.

▌ With $V = Q/(\pi D^2/4)$: $h_L = f(L/D)(16Q^2/\pi^2 D^4)/2g$, that is, $h_L = KQ^n$ with $K = 16fL/(2\pi^2 gD^5) = 0.025fL/D^5$, $n = 2$. For complete turbulence (given), $f = \text{const}$, $K = \text{const}$ (for a given pipe). Table A-9 tor cast iron: $\epsilon = 0.00085$ ft. Thus for pipe ab, $\epsilon/D = 0.00085$. Figure A-5 for complete turbulence: $f = 0.0189$; $K = 0.476$. Similarly $K = 1.606$ for ac and be, $K = 2.01$ for cd and de, $K = 1.722$ for cf, dg, and eh, $K = 0.620$ for fg and gh.

First Approximation and Correction

loop A			
pipe	K	KQ_0^n	$\lvert KnQ_0^{n-1}\rvert$
ab	0.476	0.476	0.952
be	1.606	1.606	3.21
ed	2.01	0.502	2.01
		2.584	
ac	1.606	1.606	3.21
cd	2.01	0.502	2.01
		2.108	11.39

$$\Delta Q = (2.584 - 2.108)/11.39 = 0.0418$$

loop B				loop C			
pipe	K	KQ_0^n	$\lvert KnQ_0^{n-1}\rvert$	pipe	K	KQ_0^n	$\lvert KnQ_0^{n-1}\rvert$
cd	2.01	0.502	2.01	eh	7.22	1.804	7.22
cf	7.22	1.804	7.22	ed	2.01	0.502	2.01
			9.22				9.22

$$\Delta Q = (1.804 - 0.502)/9.22 = 0.1412 \qquad \Delta Q = (1.804 - 0.502)/9.22 = 0.1412$$

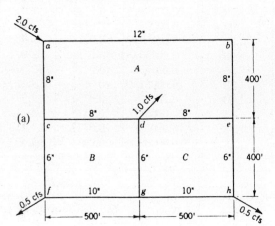

(a)

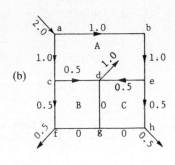

Initial flow assumptions:

(b)

Resulting final flows:

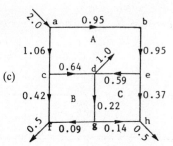

(c)

Fig. 13-10

pipe	Q(2d approximation), cfs	Q(4th and final approx), cfs
ab	$1.00 - 0.04 = 0.96$	0.95
be	$1.00 - 0.04 = 0.96$	0.95
ed	$0.50 - 0.04 + 0.14 = 0.60$	0.59
ac	$1.00 + 0.04 = 1.04$	1.06
cd	$0.50 + 0.04 + 0.14 = 0.68$	0.64
dg	$0 + 0.14 + 0.14 = 0.28$	0.22
gf	$0 + 0.14 = 0.14$	0.09
cf	$0.50 - 0.14 = 0.36$	0.42
eh	$0 + 0.14 = 0.14$	0.37
gh	$0.50 - 0.14 = 0.36$	0.14

13.12 If in Prob. 13.11 the pressure head at a is 100 ft, find the pressure head at d (which might represent a fire demand, for example) neglecting velocity heads.

▮ From the final approximations in Prob. 13.11,

In pipe ac $\qquad\qquad h_L = KQ^2 = 1.606 \times 1.06^2 = 1.805$ ft

In pipe cd, $\qquad\qquad h_L = 2.01 \times 0.64^2 \qquad = \underline{0.823}$ ft

$\qquad\qquad\qquad\qquad\qquad\qquad\qquad\qquad\qquad 2.628$ ft

In pipes $ac + cd$, $h_L = p_a/\gamma = p_a/\gamma - (h_L)_{ad} = 100 - 2.63 = 97.37$ ft.

13.13 Solve the pipe network shown in Fig. 13-11 using five trials. The 12-in and 16-in pipes are of average cast iron, while the 18-in and 24-in sizes are of average concrete ($\epsilon = 0.003$ ft). Assume $n = 2.0$, and use the values of f from Fig. A-5 for complete turbulence.

▮ $\qquad\qquad h_L = KQ^n = (f)(L/D)(V^2/2g) = (f)(L/D)(Q^2/[2g(\pi/4)^2 D^4]) = (fL/37.7D^5)Q^2$

Values of "f" for fully turbulent flow from Fig. A-5 ($n = 2$):

pipe	material	length, ft	diam., ft	e/D	f	D^5	$K = \dfrac{fL}{39.7D^5}$
ab	Avg. conc.	1000	1.50	0.00200	0.0233	7.59	0.0772
bc	Avg. conc.	500	1.50	0.00200	0.0233	7.59	0.0386
cd	Avg. conc.	500	1.50	0.00200	0.0233	7.59	0.0386
ef	New C.I.	500	1.00	0.00085	0.0188	1.00	0.237
fg	New C.I.	500	1.00	0.00085	0.0188	1.00	0.237
hi	Avg. conc.	1000	2.00	0.00150	0.0218	32.00	0.01715
ij	Avg. conc.	1000	2.00	0.00150	0.0218	32.00	0.01715
ah	Avg. conc.	1600	2.00	0.00150	0.0218	32.00	0.0274
be	New C.I.	800	1.33	0.00062	0.0175	4.21	0.0836
ei	Avg. conc.	800	1.50	0.00200	0.0233	7.59	0.0618
cf	New C.I.	800	1.33	0.00062	0.0175	4.21	0.0836
dg	New C.I.	800	1.33	0.00062	0.0175	4.21	0.0836
gi	Avg. conc.	800	1.50	0.00200	0.0233	7.59	0.0618

	loop A				loop B		
pipe	Q_0	KQ_0^2	$\|h_L/Q_0\|$	**pipe**	Q_0	KQ_0^2	$\|h_L/Q_0\|$
ha	3	0.247	0.0823	bc	2	0.1545	0.0772
ab	3	0.695	0.232	cf	2	0.335 ⌉	0.1673
						0.489 ↵	
be	1	0.0836 ⌉	0.0836				
		1.026 ↵					
hi	7	0.840	0.1200	be	1	0.0836	0.0836
ie	4	0.989 ↵	0.247	ef	2	0.947 ↵	0.473
		1.829 ⌋	0.765			1.030 ⌋	0.801

loop A	loop B
$\Delta Q = \dfrac{1.026 - 1.829}{2 \times 0.765} = -0.525\downarrow$	$\Delta Q = \dfrac{0.489 - 1.030}{2 \times 0.801} = -0.338\downarrow$

loop C				loop D			
pipe	Q_0	KQ_0^2	$|h_L/Q_0|$	pipe	Q_0	KQ_0^2	$|h_L/Q_0|$
cd	0	0.000	0.000	ie	4	0.989	0.247
gf	1	0.237 ⌉	0.237	ef	2	0.947 ⌉	0.473
		0.237 ↙				1.935 ⌡	
cf	2	0.335	0.1673	ij	3	0.1543	0.0514
gd	2	0.335 ↖	0.1673	jg	3	0.556	0.1854
		0.669 ⌡	0.571				
				gf	1	0.237 ↖	0.237
						0.947 ⌡	1.194

$\Delta Q = \dfrac{0.237 - 0.669}{2 \times 0.573} = -0.379\downarrow$	$\Delta Q = \dfrac{1.935 - 0.947}{2 \times 1.194} = +0.414\uparrow$

After five trials, the flows (cfs) within approximately 1 percent are $ab = 3.54$; $bc = 2.48$; $cd = 0.40$; $ef = 1.47$; $gf = 1.45$; $hi = 6.46$; $ij = 3.05$; $ha = 3.54$; $be = 1.06$; $ie = 3.41$; $cf = 2.08$; $gd = 1.60$; $jg = 3.05$.

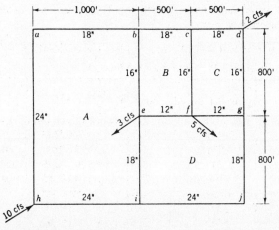

Fig. 13-11

13.14   If in Prob. 13.13 the pressure at h is 80 psi, find the pressure at f.

▮   Head loss from h to f (any path):

$$hi = 0.716 \text{ ft}$$
$$ie = 0.719 \text{ ft}$$
$$ef = \underline{0.511} \text{ ft}$$
$$1.946 \text{ ft}$$

$$p_f = 80 - (1.946)(62.4)/144 = 79.2 \text{ psi}$$

13.15   An 8-in cast iron pipe 1000 ft long forms one length of a pipe network. If the velocities to be encountered are assumed to fall within the range of 2 to 8 fps, derive an equation for the flow of water at 60 °F in this pipe in the form $h_f = KQ^n$.

▮ $$h_f = (f)(L/D)(v^2/2g) \qquad \epsilon/D = 0.00085/(\tfrac{8}{12}) = 0.00128 \qquad N_R = Dv/v$$

For $v = 2$ fps: $N_R = (\tfrac{8}{12})(2)/(1.21 \times 10^{-5}) = 1.10 \times 10^5$. From Fig. A-5, $f = 0.0230$: $h_f = 0.0230[1000/(\tfrac{8}{12})]\{2^2/[(2)(32.2)]\} = 2.143$ ft, $Q = Av = [(\pi)(\tfrac{8}{12})^2/4](2) = 0.6981$ ft³/s. For $v = 8$ fps: $N_R = (\tfrac{8}{12})(8)/(1.21 \times 10^{-5}) = 4.41 \times 10^5$, $f = 0.0213$, $h_f = 0.0213[1000/(\tfrac{8}{12})]\{8^2/[(2)(32.2)]\} = 31.75$ ft, $Q = Av = [(\pi)(\tfrac{8}{12})^2/4](8) = 2.793$ ft³/s. Given $h_f = KQ^n$, at 2 fps, $2.143 = (K)(0.6981)^n$. At 8 fps, $31.75 = (K)(2.793)^n$:

$$\log 2.143 = \log K + n \log 0.6981 \tag{1}$$

$$\log 31.75 = \log K + n \log 2.793 \tag{2}$$

Subtracting Eq. (*1*) from Eq. (*2*) gives $1.171 = 0.6022n$, $n = 1.945$. Substituting into Eq. (*1*), $\log 2.142 = \log K + (1.945)(\log 0.6981)$, $K = 4.309$; $h_f = 4.309Q^{1.945}$.

13.16 The distribution of flow through the network of Fig. 13-12 is desired for the inflows and outflows as given. For simplicity n has been given the value 2.0.

❚ The assumed distribution is shown in Fig. 13-12*a*. At the upper left the term $\sum rQ_0 |Q_0|^{n-1}$ is computed for the lower circuit number 1. Next to the diagram on the left is the computation of $\sum nr |Q_0|^{n-1}$ for the same circuit. The same format is used for the second circuit in the upper right of the figure. The corrected flow after the first step for the top horizontal pipe is determined as $15 + 11.06 = 26.06$ and for the diagonal as $35 + (-21.17) + (-11.06) = 2.77$. Figure 13-12*b* shows the flows after one correction and Fig. 13-12*c* the values after four corrections.

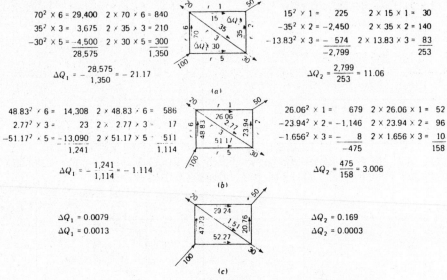

Fig. 13-12

13.17 Calculate the flow through each of the pipes of the network shown in Fig. 13-13 ($n = 2$).

❚

$1 \times 60^2 = 3600$	$2 \times 1 \times 60 = 120$	$2 \times 30^2 = 1800$	$2 \times 2 \times 30 = 120$
$3 \times 5^2 = 75$	$2 \times 3 \times 5 = 30$	$3 \times 5^2 = -75$	$2 \times 3 \times 5 = 30$
$2 \times 40^2 = -3200$	$2 \times 2 \times 40 = 160$	$1 \times 45^2 = -2025$	$2 \times 1 \times 45 = 90$
$\overline{475}$	$\overline{310}$	$\overline{-300}$	$\overline{240}$

$$\Delta Q = -\frac{475}{310} \approx -1.5 \qquad \Delta Q = \frac{300}{240} \approx 1$$

Fig. 13-13*b*

$1 \times 58.5^2 = 3422.25$	$2 \times 1 \times 58.5 = 117$	$2 \times 31^2 = 1922$	$2 \times 2 \times 31 = 124$
$3 \times 2.5^2 = 18.75$	$2 \times 3 \times 2.5 = 15$	$3 \times 2.5^2 = -19$	$2 \times 3 \times 2.5 = 15$
$2 \times 41.5^2 = -3444.5$	$2 \times 2 \times 41.5 = 166$	$1 \times 44^2 = -1936$	$2 \times 1 \times 44 = 88$
$\overline{-3.5}$	$\overline{298}$	$\overline{-33}$	$\overline{227}$

$$\Delta Q = \frac{3.5}{298} = 0.012 \qquad \Delta Q = \frac{33}{227} = 0.145$$

Fig. 13.13*c*

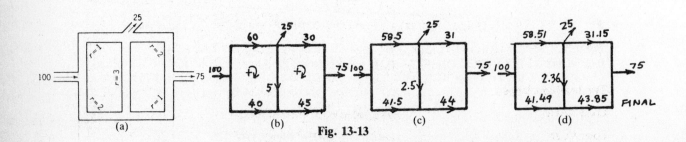

(a)   (b)   (c)   (d)

Fig. 13-13

13.18 Determine the flow through each line of Fig. 13-14 ($n = 2$).

$2 \times 19^2 =$ 722 $2 \times 2 \times 19 =$ 76 $1 \times 32^2 =$ 1024 $2 \times 1 \times 32 =$ 64 $2 \times 31^2 = -1922$ $2 \times 2 \times 31 = 124$ $\overline{\quad -176 \quad}$ $\overline{\quad 264 \quad}$ $\Delta Q = \dfrac{176}{264} = 0.67$	$4 \times 37^2 =$ 5476 $2 \times 4 \times 37 = 296$ $1 \times 32^2 = -1024$ $2 \times 1 \times 32 =$ 64 $3 \times 38^2 = -4332$ $2 \times 3 \times 38 = 228$ $\overline{\quad 120 \quad}$ $\overline{\quad 588 \quad}$ $\Delta Q = \dfrac{-120}{588} = -0.20$ ⎫⎬⎭ Fig. 13-14b
$2 \times 19.67^2 =$ 773.82 $2 \times 2 \times 19.67 = 78.68$ $1 \times 32.87^2 =$ 1080.44 $2 \times 1 \times 32.87 = 65.74$ $2 \times 30.33^2 = -1839.82$ $2 \times 2 \times 30.33 = 121.32$ $\overline{\quad 14.4 \quad}$ $\overline{\quad 265.74 \quad}$ $\Delta Q = -\dfrac{14.44}{265.74} = -0.05$	$4 \times 36.8^2 =$ 5417.0 $2 \times 4 \times 36.8 = 294.4$ $1 \times 32.87^2 = -1080.4$ $2 \times 1 \times 32.87 =$ 65.74 $3 \times 38.2^2 = -4377.7$ $2 \times 3 \times 38.2 = 229.2$ $\overline{\quad -41.12 \quad}$ $\overline{\quad 589.3 \quad}$ $\Delta Q = -\dfrac{41.12}{589.3} = 0.07$

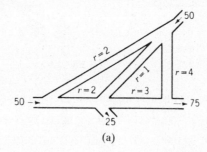

(a)

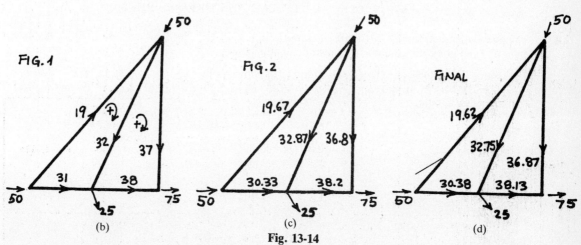

FIG. 1 (b) FIG. 2 (c) FINAL (d)

Fig. 13-14

13.19 In the five-pipe network of Fig. 13-15a, assume all pipes have a friction factor of 0.02. Find the flow rate in all pipes. The fluid is water at 20 °C.

Assume initial flows Q_1, Q_2, Q_3, Q_4, and Q_5, as shown in Fig. 13-15b.

$$h_f = (f)\left(\frac{L}{D}\right)\left(\frac{v^2}{2g}\right) = \frac{8fLQ^2}{\pi^2 gD^5} = KQ^2 \qquad K_1 = \frac{8f_1 L_1}{\pi^2 gD_1^5} = \frac{(8)(0.02)(3000)}{(\pi)^2(32.2)(\frac{6}{12})^5} = 48.33$$

$$K_2 = \frac{(8)(0.02)(4000)}{(\pi)^2(32.2)(\frac{8}{12})^5} = 15.29 \qquad K_3 = \frac{(8)(0.02)(5000)}{(\pi)^2(32.2)(\frac{9}{12})^5} = 10.61$$

$$K_4 = \frac{(8)(0.02)(4000)}{(\pi)^2(32.2)(\frac{8}{12})^5} = 15.29 \qquad K_5 = \frac{(8)(0.02)(3000)}{(\pi)^2(32.2)(\frac{3}{12})^5} = 1547$$

Two head-loss loops:

Loop ABC:
$$15.29Q_2^2 + 10.61Q_3^2 - 48.33Q_1^2 = 0 \qquad (1)$$

Loop BCD:
$$10.61Q_3^2 + 15.29Q_4^2 - 1547Q_5^2 = 0 \qquad (2)$$

Three junctions:

Junction A: $$Q_1 + Q_2 = 2.0 \tag{3}$$

Junction B: $$Q_2 = Q_3 + Q_5 \tag{4}$$

Junction C: $$Q_1 + Q_3 = Q_4 \tag{5}$$

Solving these simultaneous equations by trial and error gives $Q_1 = 0.81$ cfs, $Q_2 = 1.19$ cfs, $Q_3 = 0.99$ cfs, $Q_4 = 1.80$ cfs, $Q_5 = 0.20$ cfs.

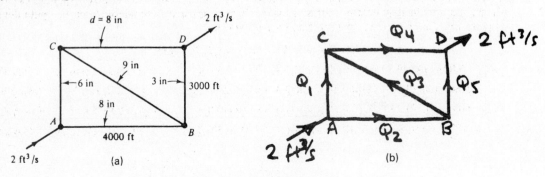

Fig. 13-15

13.20 If in Prob. 13.19 the pressure at A is 100 psig, find the pressures at B, C, and D.

❚ $p_B = p_A - \rho g K_2 Q_2^2 = (100)(144) - (1.94)(32.2)(15.29)(1.19)^2 = 13\,047$ lb/ft^2 or 90.6 lb/in^2

$p_C = p_A - \rho g K_1 Q_1^2 = (100)(144) - (1.94)(32.2)(48.33)(0.81)^2 = 12\,419$ lb/ft^2 or 86.2 lb/in^2

$p_D = p_C - \rho g K_4 Q_4^2 = (86.2)(144) - (1.94)(32.2)(15.29)(1.80)^2 = 9318$ lb/ft^2 or 64.7 lb/in^2

13.21 The pipe dimensions in Fig. 13-16a are the same as in Fig. 13-15a. The fluid is water at 20 °C. Assume that $f = 0.02$ in all pipes and compute the flow rate in all pipes.

❚ Assume initial flows Q_1, Q_2, Q_3, Q_4, and Q_5, as shown in Fig. 13-16b: $h_f = (f)(L/D)(v^2/2g) = 8fLQ^2/\pi^2 g D^5 = KQ^2$. From Prob. 13.19, $K_1 = 48.33$, $K_2 = 15.29$, $K_3 = 10.61$, $K_4 = 15.29$, and $K_5 = 1547$.

Two head-loss loops:

Loop ABC: $$15.29Q_2^2 + 10.61Q_3^2 - 48.33Q_1^2 = 0 \tag{1}$$

Loop BCD: $$10.61Q_3^2 + 15.29Q_4^2 - 1547Q_5^2 = 0 \tag{2}$$

Three junctions:

Junction A: $$Q_1 + Q_2 = 2.0 \tag{3}$$

Junction B: $$Q_2 = Q_3 + Q_5 + 1.0 \tag{4}$$

Junction C: $$Q_1 + Q_3 = Q_4 + 0.5 \tag{5}$$

Solving these simultaneous equations by trial and error gives $Q_1 = 0.725$ cfs, $Q_2 = 1.275$ cfs, $Q_3 = 0.226$ cfs, $Q_4 = 0.451$ cfs, $Q_5 = 0.049$ cfs.

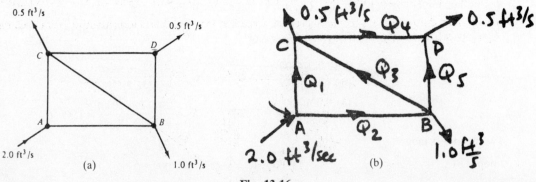

Fig. 13-16

13.22 If in Prob. 13.21 the pressure at A is 100 psig, find the pressures at B, C, and D.

$$p_B = p_A - \rho g K_2 Q_2^2 = (100)(144) - (1.94)(32.2)(15.29)(1.275)^2 = 12\,847\ \text{lb/ft}^2 \quad \text{or} \quad 89.2\ \text{lb/in}^2$$

$$p_C = p_A - \rho g K_1 Q_1^2 = (100)(144) - (1.94)(32.2)(48.33)(0.725)^2 = 12\,813\ \text{lb/ft}^2 \quad \text{or} \quad 89.0\ \text{lb/in}^2$$

$$p_D = p_C - \rho g K_4 Q_4^2 = (89.0)(144) - (1.94)(32.2)(15.29)(0.451)^2 = 12\,622\ \text{lb/ft}^2 \quad \text{or} \quad 87.7\ \text{lb/in}^2$$

13.23 In Fig. 13-17 four equal-sized pipes 30 m long and 6 cm diameter are joined at junction a. Pressures are known at each of the pipe ends: $p_1 = 900$ kPa, $p_2 = 300$ kPa, $p_3 = 700$ kPa, $p_4 = 100$ kPa. All pipes are cast iron, for which $f =$ approximately 0.0294 for this problem. Neglecting gravity and minor losses, compute the flow rate in all pipes when the pressure at the junction (p_a) is 500 kPa.

$$h_f = \Delta p / \gamma = 8fLQ^2/\pi^2 g D^5$$

$(900 - 500)/9.79 = (8)(0.0294)(30)(Q_1)^2/(\pi)^2(9.807)(\tfrac{6}{100})^5 \qquad Q_1 = 0.0209\ \text{m}^3/\text{s} \qquad$ (toward a)

$(500 - 100)/9.79 = (8)(0.0294)(30)(Q_4)^2/(\pi)^2(9.807)(\tfrac{6}{100})^5 \qquad Q_4 = 0.0209\ \text{m}^3/\text{s} \qquad$ (away from a)

$(500 - 300)/9.79 = (8)(0.0294)(30)(Q_2)^2/(\pi)^2(9.807)(\tfrac{6}{100})^5 \qquad Q_2 = 0.0148\ \text{m}^3/\text{s} \qquad$ (away from a)

$(700 - 500)/9.79 = (8)(0.0294)(30)(Q_3)^2/(\pi)^2(9.807)(\tfrac{6}{100})^5 \qquad Q_3 = 0.0148\ \text{m}^3/\text{s} \qquad$ (toward a)

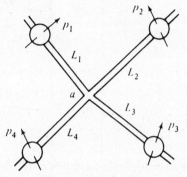

Fig. 13-17

13.24 Determine the flow in each pipe of the network shown in Fig. 13-18, using $f = 0.02$ throughout.

Taking $x = 2$, $h_L = f(L/D)(V^2/2g) = f(L/D)(1/2g)(4Q/\pi D^2)^2 = (8fL/\pi^2 g D^5)(Q^2) = KQ^2$. Hence $K = 0.81fL/gD^5$, and the K value for each pipe is

Diameter, in	3	4	5	6	7	8
K	1030	368	160	80.4	22.4	11.5

The assumed flows are indicated on the figure in parentheses. For loop $AEDB$,

$$\Delta_1 = -[(1030 \times 0.5^2) + (11.5 \times 0.1^2) - (22.4 \times 0.2^2) - (368 \times 0.7^2)]/$$
$$2[(1030 \times 0.5) + (11.5 \times 0.1) + (22.4 \times 0.2) + (368 \times 0.7)] = -0.05\ \text{cfs}$$

and for loop BDC,

$$\Delta_2 = -[(22.4 \times 0.2^2) + (80.4 \times 0.3^2) - (160 \times 0.5^2)]/2[(22.4 \times 0.2) + (80.4 \times 0.3) + (160 \times 0.5)] = +0.15\ \text{cfs}$$

The corrected flows appear on the figure below the first assumed flows. Recomputing Δ for each loop yields $\Delta_1 = +0.001$ cfs and $\Delta_2 = -0.001$ cfs.

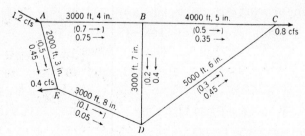

Fig. 13-18

13.25 What single pipe 1000 ft long would be the equivalent of a network $ABCD$ if $AB = 500$ ft of 8-in pipe, $BC = 500$ ft of 24-in pipe, $CD = 1000$ ft of 36-in pipe, and $DA = 800$ ft of 12-in pipe? Flow enters the system at A and leaves at C. Assume f is the same for all pipes.

❚ See Fig. 13-19.

$$(h_f)_{ABC} = (h_f)_{ADC} \qquad h_f = (f)(L/D)(v^2/2g)$$

$$f(500/\tfrac{8}{12})(V_8^2/2g) + f(\tfrac{500}{2})(V_{24}^2/2g) = f(\tfrac{800}{1})(V_{12}^2/2g) + f(\tfrac{1000}{3})(V_{36}^2/2g)$$

$$750V_8^2 + 250V_{24}^2 = 800V_{12}^2 + 333V_{36}^2 \qquad V_8 = 9V_{24} \qquad V_{12} = 9V_{36}$$

$$(750 \times 81 + 250)V_{24}^2 = (800 \times 81 + 333)V_{36}^2 \qquad V_{36} = 0.96V_{24}$$

$$D^2V = 2^2V_{24} + 3^2V_{36} = 4V_{24} + 9(0.96V_{24}) \qquad D^2V = 12.63V_{24}$$

$$(500/\tfrac{8}{12})V_8^2 + \tfrac{500}{2}V_{24}^2 = (1000/D)V^2$$

$$750V_8^2 + 250V_{24}^2 = 60\,900V_{24}^2 = (1000/D)V^2 \qquad (1)$$

But

$$V = 12.63V_{24}/D^2 \qquad (2)$$

Substituting (2) into (1) yields $D^5 = 2.63$. Hence $D = 1.21$ ft $= 14.6$ in.

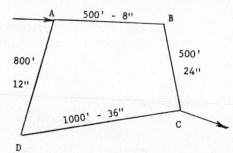

Fig. 13-19

13.26 The pipe network in Fig. 13-20 consists of pipes as follows: AB, 5000 ft, 12 in; BC, 3000 ft, 6 in; CD, 8000 ft, 24 in; DE, 7000 ft, 8 in; EA, 4000 ft, 10 in; BD, 7000 ft, 12 in. A flow of 4000 gpm enters the system at D, while outflow at the junctions is as follows: A, 500 gpm; B, 300 gpm; C, 1000 gpm; E, 2200 gpm. Find the flow in each pipe and the pressure at each junction if the head at D is 400 ft. Take $f = 0.023$.

❚ $$h_f = 8fLQ^2/\pi^2gD^5 = (8)(0.023)(LQ^2)/(\pi)^2(32.2)D^5 = 0.000579LQ^2/D^5$$

D^5	pipe	D, in	k
1.0	AB	12	2.89
0.031	BC	6	56.0
32	CD	24	0.143
0.132	DE	8	30.6
1.0	BD	12	4.05
0.402	AE	10	5.75

Loop 1:

$$\Delta_1 = [-(2.9)(3.62)^2 - (4.05)(3.40)^2 + (30.6)(2.43)^2 - (5.75)(2.5)^2]/$$
$$2[(2.9)(3.62) + (4.05)(3.40) + (30.6)(2.43) + (5.75)(2.5)] = -0.27$$

Loop 2:

$$\Delta_2 = -[-(56)(0.89)^2 - (0.143)(3.13)^2 + (4.05)(3.4)^2]/2[(56)(0.89) + (0.143)(3.13) + (4.05)(3.4)] = -0.01$$

$$\Delta_1 = [-(2.9)(3.89)^2 - (4.05)(3.66)^2 + (30.6)(2.16)^2 - (5.75)(2.77)^2]/$$
$$2[(2.9)(3.89) + (4.05)(3.66) + (30.6)(2.16) + (5.75)(2.77)] = 0$$

$$\Delta_2 = -(-45.3 - 1.4 + 54.1)/2(50.5 + 0.45 + 14.8) = -0.056$$

$$\Delta_1 = -(-44 - 52.5 + 142.5 - 44)/2(11.3 + 14.6 + 66.1 + 15.9) = -0.01$$

$$\Delta_2 = -[-(56)(0.96)^2 - (0.143)(3.20)^2 + (4.05)(3.6)^2]/2(53.6 + 0.46 + 14.6) = +0.01$$

$$\Delta_1 = -44.1 - 53.1 + 140.2 - 44.2 = -1.2 \qquad \Delta_2 = -50.5 - 1.46 + 53.2 = +1.2$$

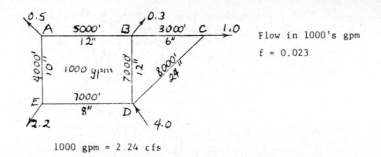

Flow in 1000's gpm

f = 0.023

1000 gpm = 2.24 cfs

First trial

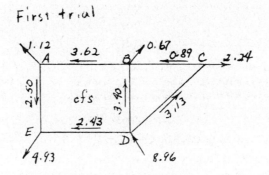

Second trial

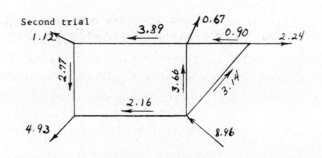

Third trial

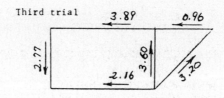

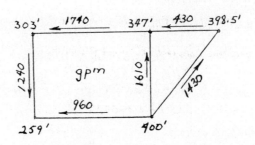

Fig. 13-20

13.27 The pipe network in Fig. 13-21 consists of the following pipes: *AB*, 3000 ft, 8 in; *BC*, 3000 ft, 12 in; *CD*, 10 000 ft, 36 in; *DE*, 8000 ft, 24 in; *EF*, 5000 ft, 6 in; *FA*, 4000 ft, 8 in; *BE*, 10 000 ft, 6 in; *BF*, 8000 ft, 12 in. Inflow at *D* = 6 cfs. Outflows are *A*, 2 cfs; *B*, 1 cfs; *E*, 1.5 cfs; *C*, 0.5 cfs; *F*, 1 cfs. Assume Manning *n* = 0.015, and find the flow in each pipe and the pressure at each junction if the pressure at *D* is 120 psi.

▮ $n = 0.015$ $\quad S = n^2 V^2 / 2.21 R^{4/3} = n^2 Q^2 / [2.21(D/4)^{4/3}(\pi^2 D^4/16)]$

$$= [(0.015)^2 Q^2 (16)(6.4)] / (2.21\pi^2 D^{16/3}) = 0.00106(Q^2/D^{16/3})$$

$$h_L = kQ^2 = (0.00106L/D^{16/3})(Q^2) \qquad k = 0.00106L/D^{16/3}$$

	D	$D^{16/3}$	$0.00106L$	k
AB	0.67	0.116	3.18	27.4
BC	1.00	1.0	3.18	3.2
CD	3.0	355	10.60	0.03
DE	2.0	40	8.48	0.21
EF	0.5	0.0245	5.30	216
FA	0.67	0.116	4.24	36.5
BF	1.0	1.0	8.48	8.48
BE	0.5	0.0245	10.60	432

Loop 1:

$$\Delta_1 = -\frac{-(27.4)(1.3)^2 + (8.48)(1.6)^2 + (36.5)(0.7)^2}{2[(27.4)(1.3) + (8.48)(1.6) + (36.5)(0.7)]} = +0.044$$

Loop 2:

$$\Delta_2 = -\frac{(216)(0.1)^2 - (8.48)(1.6)^2}{2[0 + (216)(0.1) + (8.48)(1.6)]} = +0.277$$

Loop 3:

$$\Delta_3 = -\frac{-(3.2)(3.9)^2 - (0.03)(4.4)^2 + (0.21)(1.6)^2}{2[(3.2)(3.9) + (0.03)(4.4) + (0.21)(1.6)]} = +1.88$$

$$\Delta_1 = -\frac{-43.5 + 15.8 + 20}{2[(27.4)(1.26) + (8.48)(1.36) + (36.5)(0.74)]} = +0.053$$

$$\Delta_2 = -\frac{-1110 + 31 - 15.6}{2[432(1.60) + 216(0.38) + (8.48)(1.36)]} = +0.695$$

$$\Delta_3 = -\frac{-13.1 - 0.19 + 2.56 + 1110}{2[(3.2)(2.02) + (0.03)(2.52) + (0.21)(3.48) + (432)(1.6)]} = -0.72$$

$$\Delta_1 = -\frac{-40 + 4.2 + 22.8}{2[(27.4)(1.21) + (8.48)(0.71) + (36.5)(0.79)]} = +0.096$$

$$\Delta_2 = -\frac{-14.1 + 251 - 4.2}{2[(432)(0.18) + (216)(1.08) + (8.48)(0.71)]} = -0.37$$

$$\Delta_3 = -\frac{-24.2 - 0.3 + 1.6 + 14.1}{2[(3.2)(2.74) + (0.03)(3.24) + (0.21)(2.76) + (432)(0.18)]} = +0.05$$

$$\Delta_1 = -\frac{-33.8 + 11.8 + 28.9}{2[(27.4)(1.11) + (8.48)(1.18) + (36.5)(0.89)]} = -0.05$$

$$\Delta_2 = -\frac{-155 + 109 - 11.8}{2[(432)(0.60) + (216)(0.71) + (8.48)(1.18)]} = +0.07$$

$$\Delta_3 = -\frac{-23 - 0.3 + 1.6 + 155}{2[(3.2)(2.69) + (0.03)(3.19) + (0.21)(2.81) + (432)(0.6)]} = -0.25$$

$$\Delta_1 = -\frac{-37 + 9.5 + 25.7}{2[(27.4)(1.16) + (8.5)(1.06) + (36.5)(0.84)]} = +0.01$$

$$\Delta_2 = -\frac{-33.8 + 132 - 9.6}{2[(432)(0.28) + (216)(0.78) + (8.5)(1.06)]} = -0.15$$

$$\Delta_3 = -\frac{-27.5 - 0.4 + 1.4 + 33.8}{2[(3.2)(2.94) + (0.03)(3.44) + (0.21)(2.56) + (432)(0.28)]} = -0.03$$

And after several more trials, final results are as shown on the last sketch of Fig. 13-21.

pressures (psi)	
A	87.5
B	104.5
C	119.8
D	120.0
E	119.5
F	98.0

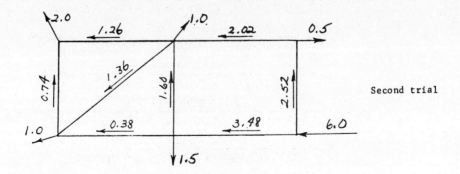

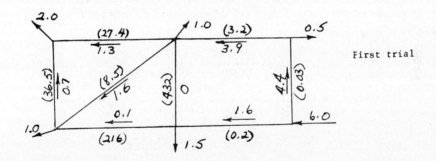

First trial

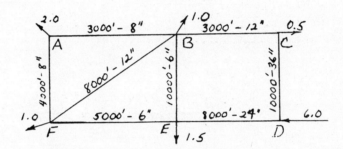

Second trial

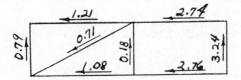

Third trial

Fig. 13-21

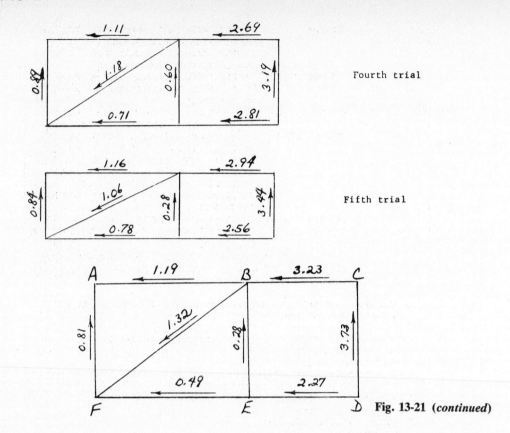

Fig. 13-21 (continued)

13.28 Prepare a computer program written in Fortran to determine the flow in each pipe in a pipe network by the Hardy Cross method. The program must be usable in both the English system of units and the International system of units.

```
C        THIS PROGRAM DETERMINES THE FLOW RATE IN EACH PIPE IN A PIPE NET-
C        WORK BY THE HARDY CROSS METHOD.  IT CAN BE USED FOR PROBLEMS IN
C        BOTH THE ENGLISH SYSTEM OF UNITS AND THE INTERNATIONAL SYSTEM OF
C        UNITS.  EACH LOOP MUST BE NUMBERED AND EACH INDIVIDUAL PIPE MUST
C        ALSO BE NUMBERED.
C
C        INPUT DATA MUST BE SET UP AS FOLLOWS.
C
C        CARD 1    COLUMN 1         ENTER 0 (ZERO) OR BLANK IF ENGLISH SYSTEM
C                                   OF UNITS IS TO BE USED.  ENTER 1 (ONE) IF
C                                   INTERNATIONAL SYSTEM OF UNITS IS TO BE
C                                   USED.
C                  COLUMN 2         ENTER 0 (ZERO) OR BLANK IF ONLY THE
C                                   RESULTS AFTER THE FINAL ITERATION ARE TO
C                                   BE PRINTED.  ENTER 1 (ONE) IF THE RESULTS
C                                   AFTER EACH ITERATION ARE TO BE PRINTED.
C                  COLUMNS 3-5      ENTER INTEGER NUMBER (RIGHT ADJUSTED)
C                                   TELLING HOW MANY LOOPS ARE IN THE PIPE
C                                   NETWORK.
C                  COLUMNS 6-80     ENTER TITLE, DATE, AND OTHER INFORMATION,
C                                   IF DESIRED.
C        CARD 2    COLUMNS 1-2      ENTER INTEGER NUMBER (RIGHT ADJUSTED)
C                                   TELLING HOW MANY PIPES ARE IN THE FIRST
C                                   LOOP.
C                  COLUMNS 3-4      ENTER INTEGER NUMBER (RIGHT ADJUSTED)
C                                   TELLING HOW MANY PIPES ARE IN THE SECOND
C                                   LOOP.
C                  COLUMNS 5-6      ENTER INTEGER NUMBER (RIGHT ADJUSTED)
C                                   TELLING HOW MANY PIPES ARE IN THE THIRD
C                                   LOOP.
C                  (CONTINUE PATTERN FOR ALL LOOPS.)
C        CARD 3    COLUMNS 1-3      ENTER INTEGER NUMBER (RIGHT ADJUSTED) TO
C                                   IDENTIFY AND TO REPRESENT A PARTICULAR
C                                   PIPE IN THE FIRST LOOP.
```

```
C                    COLUMNS 4-6      IF THIS PIPE IS IN COMMON WITH ANY OTHER
C                                     LOOP, ENTER INTEGER NUMBER (RIGHT AD-
C                                     JUSTED) IDENTIFYING THAT LOOP.  OTHER-
C                                     WISE, ENTER 0 (ZERO) OR BLANK.
C                    COLUMNS 7-16     ENTER NUMBER INCLUDING DECIMAL GIVING
C                                     DIAMETER OF THIS PIPE (IN INCHES OR
C                                     MILLIMETERS).
C                    COLUMNS 17-26    ENTER NUMBER INCLUDING DECIMAL GIVING
C                                     LENGTH OF THIS PIPE (IN FEET OR METERS).
C                    COLUMNS 27-36    ENTER NUMBER INCLUDING DECIMAL GIVING
C                                     INITIAL ESTIMATE OF FLOW RATE FOR THIS
C                                     PIPE (IN CUBIC FEET PER SECOND OR CUBIC
C                                     METERS PER SECOND).  (ENTER CLOCKWISE
C                                     FLOW RATES WITH RESPECT TO THIS LOOP AS
C                                     POSITIVE AND COUNTERCLOCKWISE FLOW RATES
C                                     AS NEGATIVE.)
C                    COLUMNS 37-46    ENTER NUMBER INCLUDING DECIMAL GIVING
C                                     HAZEN-WILLIAMS ROUGHNESS COEFFICIENT FOR
C                                     THIS PIPE.
C                    (ENTER ADDITIONAL CARDS LIKE CARD 3 FOR EACH REMAINING
C                    PIPE IN THE FIRST LOOP, THEN FOR EACH PIPE IN THE SECOND
C                    LOOP, ETC. FOR ALL LOOPS.)
C
C    MULTIPLE DATA SETS FOR SOLVING ANY NUMBER OF PROBLEMS MAY BE
C    INCLUDED FOR PROCESSING.
C
      DIMENSION TITLE(13),DIAM(100,10),Q(100,10),ROUGH(100,10),
     *       HLOSS(100,10),QDEL(100)
      REAL LENGTH(100,10)
      INTEGER UNITS,LOOPS,PIPES(100),PPLOOP,PIPENO(100,10),OLOOP(100,10)
1     READ(5,100,END=2)UNITS,IWRITE,LOOPS,TITLE
100   FORMAT(2I1,I3,12A6,A3)
      WRITE(6,105)TITLE
105   FORMAT('1',12A6,A3,////)
      COEFF=1.318
      FACTOR=12.0
      ERROR=.01
      IF(UNITS.EQ.1)COEFF=.8492
      IF(UNITS.EQ.1)FACTOR=1000.0
      IF(UNITS.EQ.1)ERROR=.001
      READ(5,101)(PIPES(J),J=1,LOOPS)
101   FORMAT(40I2)
      DO 200 J=1,LOOPS
      PPLOOP=PIPES(J)
      DO 200 K=1,PPLOOP
200   READ(5,102)PIPENO(J,K),OLOOP(J,K),DIAM(J,K),LENGTH(J,K),Q(J,K),
     *       ROUGH(J,K)
102   FORMAT(2I3,4F10.0)
      NIT=1
205   IF(IWRITE.EQ.1)WRITE(6,106)NIT
106   FORMAT(//38X,'ITERATION NO.',I3,/,38X,'========= === ==',//)
      IF(IWRITE.EQ.1.AND.UNITS.EQ.0)WRITE(6,107)
107   FORMAT(' LOOP NO.  PIPE NO.  DIAMETER (IN)  LENGTH (FT)  ROUGHNESS
     *  FLOW RATE (CFS)  HEAD LOSS (FT)',/' --------  --------  --------
     *-----  -----------  ---------  ---------------  --------------'/)
      IF(IWRITE.EQ.1.AND.UNITS.EQ.1)WRITE(6,108)
108   FORMAT(' LOOP NO.  PIPE NO.  DIAMETER (MM)  LENGTH ( M)  ROUGHNESS
     *  FLOW RATE (CMS)  HEAD LOSS ( M)',/' --------  --------  --------
     *-----  -----------  ---------  ---------------  --------------'/)
      DO 201 J=1,LOOPS
      IF(J.EQ.1)NIT=NIT+1
      SHLOSS=0.0
      SHQ=0.0
      PPLOOP=PIPES(J)
      DO 202 K=1,PPLOOP
      HLOSS(J,K)=(ABS(Q(J,K))*4.0**1.63/3.14159265/(DIAM(J,K)/FACTOR)
     *    **2.63/COEFF/ROUGH(J,K))**(1.0/.54)*LENGTH(J,K)
      IF(Q(J,K).LT.0.0)HLOSS(J,K)=-HLOSS(J,K)
      HQ=HLOSS(J,K)/Q(J,K)
      SHLOSS=SHLOSS+HLOSS(J,K)
      SHQ=SHQ+HQ
      IF(IWRITE.EQ.1)WRITE(6,109)J,PIPENO(J,K),DIAM(J,K),LENGTH(J,K),
     *     ROUGH(J,K),Q(J,K),HLOSS(J,K)
```

```
109 FORMAT(3X,I3,7X,I3,7X,F7.1,6X,F8.0,7X,F5.0,7X,F8.3,7X,F9.3)
202 CONTINUE
    QDEL(J)=-SHLOSS/1.85/SHQ
    IF(IWRITE.EQ.1.AND.UNITS.EQ.0)WRITE(6,110)SHLOSS,J,QDEL(J)
110 FORMAT(78X,'------------',/,78X,F9.3,/1X,'FLOW RATE ADJUSTMENT FOR
  * LOOP',I3,' = ',F7.3,' CFS',/)
    IF(IWRITE.EQ.1.AND.UNITS.EQ.1)WRITE(6,111)SHLOSS,J,QDEL(J)
111 FORMAT(78X,'------------',/,78X,F9.3,/1X,'FLOW RATE ADJUSTMENT FOR
  * LOOP',I3,' = ',F7.3,' CMS',/)
201 CONTINUE
    DO 203 J=1,LOOPS
    PPLOOP=PIPES(J)
    DO 203 K=1,PPLOOP
    Q(J,K)=Q(J,K)+QDEL(J)
    L=OLOOP(J,K)
    IF(OLOOP(J,K).NE.0)Q(J,K)=Q(J,K)-QDEL(L)
203 CONTINUE
    DO 204 J=1,LOOPS
    IF(ABS(QDEL(J)).GT.ERROR)GO TO 205
204 CONTINUE
    IF(IWRITE.EQ.1)GO TO 1
    IF(UNITS.EQ.0)WRITE(6,107)
    IF(UNITS.EQ.1)WRITE(6,108)
    DO 206 J=1,LOOPS
    PPLOOP=PIPES(J)
    DO 206 K=1,PPLOOP
    IF(K.NE.PPLOOP)WRITE(6,109)J,PIPENO(J,K),DIAM(J,K),LENGTH(J,K),
  *     ROUGH(J,K),Q(J,K),HLOSS(J,K)
    IF(K.EQ.PPLOOP)WRITE(6,112)J,PIPENO(J,K),DIAM(J,K),LENGTH(J,K),
  *     ROUGH(J,K),Q(J,K),HLOSS(J,K)
112 FORMAT(3X,I3,7X,I3,7X,F7.1,6X,F8.0,7X,F5.0,7X,F8.3,7X,F9.3,/)
206 CONTINUE
    NIT=NIT-1
    WRITE(6,104)NIT
104 FORMAT(//,1X,I3,' ITERATIONS WERE REQUIRED.')
    GO TO 1
2   STOP
    END
```

13.29 Use the computer program written for Prob. 13.28 to solve for the flow rate in each pipe of the network of Prob. 13.3.

Input

```
1 2 3 4 5 6 7 8 9 10 11 12 13 14 15 16 17 18 19 20 21 22 23 24 25 26 27 28 29 30 31 32 33 34 35 36 37 38 39 40 41 42 43 44 45 46 47 48 49 50 51 52 53 54 55 56 57 58 59 60 61 62 63 64 65 66 67 68 69 70 71 72 73 74 75 76 77 78 79 80
01   2SAMPLE PIPE NETWØRK ANALYSIS
 4 4
   1          18.     2000.     7.00     120.
   2    2     24.      900.     3.50     120.
   3          12.     2800.    -2.19     120.
   4          24.     1100.    -7.00     120.
   5          12.     2200.     3.50     120.
   6          18.      750.    -3.14     120.
   7          24.     2600.    -5.69     120.
   2    1     24.      900.    -3.50     120.
```

Output

SAMPLE PIPE NETWORK ANALYSIS

ITERATION NO.   1
========= === ==

LOOP NO.	PIPE NO.	DIAMETER (IN)	LENGTH (FT)	ROUGHNESS	FLOW RATE (CFS)	HEAD LOSS (FT)
1	1	18.0	2000.	120.	7.000	6.803
1	2	24.0	900.	120.	3.500	0.209
1	3	12.0	2800.	120.	-2.190	-7.978
1	4	24.0	1100.	120.	-7.000	-0.922

						-1.888

FLOW RATE ADJUSTMENT FOR LOOP 1 =   0.212 CFS

2	5	12.0	2200.	120.	3.500	14.936
2	6	18.0	750.	120.	-3.140	-0.578
2	7	24.0	2600.	120.	-5.690	-1.484
2	2	24.0	900.	120.	-3.500	-0.209

						12.665

FLOW RATE ADJUSTMENT FOR LOOP 2 = -1.435 CFS

ITERATION NO. 2
========= === ==

LOOP NO.	PIPE NO.	DIAMETER (IN)	LENGTH (FT)	ROUGHNESS	FLOW RATE (CFS)	HEAD LOSS (FT)
1	1	18.0	2000.	120.	7.212	7.190
1	2	24.0	900.	120.	5.147	0.427
1	3	12.0	2800.	120.	-1.978	-6.605
1	4	24.0	1100.	120.	-6.788	-0.870

						0.141

FLOW RATE ADJUSTMENT FOR LOOP 1 = -0.017 CFS

2	5	12.0	2200.	120.	2.065	5.624
2	6	18.0	750.	120.	-4.575	-1.160
2	7	24.0	2600.	120.	-7.125	-2.251
2	2	24.0	900.	120.	-5.147	-0.427

						1.786

FLOW RATE ADJUSTMENT FOR LOOP 2 = -0.286 CFS

ITERATION NO. 3
========= === ==

LOOP NO.	PIPE NO.	DIAMETER (IN)	LENGTH (FT)	ROUGHNESS	FLOW RATE (CFS)	HEAD LOSS (FT)
1	1	18.0	2000.	120.	7.196	7.159
1	2	24.0	900.	120.	5.416	0.469
1	3	12.0	2800.	120.	-1.994	-6.709
1	4	24.0	1100.	120.	-6.804	-0.874

						0.044

FLOW RATE ADJUSTMENT FOR LOOP 1 = -0.005 CFS

2	5	12.0	2200.	120.	1.779	4.267
2	6	18.0	750.	120.	-4.861	-1.298
2	7	24.0	2600.	120.	-7.411	-2.421
2	2	24.0	900.	120.	-5.416	-0.469

						0.079

FLOW RATE ADJUSTMENT FOR LOOP 2 = -0.014 CFS

ITERATION NO. 4
========= === ==

LOOP NO.	PIPE NO.	DIAMETER (IN)	LENGTH (FT)	ROUGHNESS	FLOW RATE (CFS)	HEAD LOSS (FT)
1	1	18.0	2000.	120.	7.190	7.149
1	2	24.0	900.	120.	5.425	0.470
1	3	12.0	2800.	120.	-2.000	-6.742
1	4	24.0	1100.	120.	-6.810	-0.876

						0.002

```
       FLOW RATE ADJUSTMENT FOR LOOP  1 =  -0.000 CFS
```

2	5	12.0	2200.	120.	1.765	4.206
2	6	18.0	750.	120.	-4.875	-1.305
2	7	24.0	2600.	120.	-7.425	-2.429
2	2	24.0	900.	120.	-5.425	-0.470

```
                                            -------------
                                                 0.001


       FLOW RATE ADJUSTMENT FOR LOOP  2 =  -0.000 CFS
```

13.30 Use the computer program written for Prob. 13.28 to solve for the flow rate in each pipe of the network shown in Fig. 13-22a.

❚ It is necessary to assume an initial value of flow rate for each pipe in the network. The values assumed for this example are shown in Fig. 13-22b.

Input

```
1 2 3 4 5 6 7 8 9 10 11 12 13 14 15 16 17 18 19 20 21 22 23 24 25 26 27 28 29 30 31 32 33 34 35 36 37 38 39 40 41 42 43 44 45 46 47 48 49 50 51 52 53 54 55 56 57 58 59 60 61 62 63 64 65 66 67 68 69 70 71 72 73 74 75 76 77 78 79 80
10   7SAMPLE PIPE NETWØRK ANALYSIS
 4 4 5 6 3 4 4
 1   0   1000.   1000.   -0.6    100.
 2   2    750.    925.    0.4    100.
 3   3    750.   1000.    1.15   100.
 4   0    750.    925.   -2.6    100.
10   0   1000.   1000.   -2.0    120.
11   0   1000.    925.   10.0    120.
 9   4    750.   1000.    4.0    120.
 2   1    750.    925.   -0.4    100.
 3   1    750.   1000.   -1.15   100.
 8   4    500.    650.    2.25   100.
 7   0    500.    400.   -0.75   100.
 6   0    500.    671.   -1.75   100.
 5   0    500.    350.   -2.75   100.
 9   2    750.   1000.   -4.0    120.
12   0   1000.    800.    5.0    120.
14   5   1000.    650.    2.0    120.
21   6    500.    800.    0.75   120.
22   0    500.   1000.   -2.0    120.
 8   3    500.    650.   -2.25   100.
13   0    750.    763.    2.0    120.
15   7    750.    400.    0.5    120.
14   4   1000.    650.   -2.0    120.
21   4    500.    800.   -0.75   120.
18   7    500.    125.    0.75   120.
19   0    500.    800.   -0.75   120.
20   0    500.    125.   -1.75   120.
15   5    750.    400.   -0.5    120.
16   0    750.    125.    0.5    120.
17   0    750.    400.   -0.5    120.
18   6    500.    125.   -0.75   120.
```

Output

```
SAMPLE PIPE NETWORK ANALYSIS
```

LOOP NO.	PIPE NO.	DIAMETER (MM)	LENGTH (M)	ROUGHNESS	FLOW RATE (CMS)	HEAD LOSS (M)
1	1	1000.0	1000.	100.	-0.532	-0.655
1	2	750.0	925.	100.	2.537	44.438
1	3	750.0	1000.	100.	0.211	0.481
1	4	750.0	925.	100.	-2.532	-44.269

continued

LOOP NO.	PIPE NO.	DIAMETER (MM)	LENGTH (M)	ROUGHNESS	FLOW RATE (CMS)	HEAD LOSS (M)
2	10	1000.0	1000.	120.	-4.068	-20.247
2	11	1000.0	925.	120.	7.932	64.482
2	9	750.0	1000.	120.	0.152	0.187
2	2	750.0	925.	100.	-2.537	-44.438
3	3	750.0	1000.	100.	-0.211	-0.481
3	8	500.0	650.	100.	1.478	82.694
3	7	500.0	400.	100.	0.258	1.998
3	6	500.0	671.	100.	-0.742	-23.884
3	5	500.0	350.	100.	-1.742	-60.446
4	9	750.0	1000.	120.	-0.152	-0.187
4	12	1000.0	800.	120.	6.780	41.703
4	14	1000.0	650.	120.	3.932	12.357
4	21	500.0	800.	120.	0.940	31.420
4	22	500.0	1000.	120.	-0.220	-2.676
4	8	500.0	650.	100.	-1.478	-82.694
5	13	750.0	763.	120.	1.848	14.537
5	15	750.0	400.	120.	-0.942	-2.191
5	14	1000.0	650.	120.	-3.932	-12.357
6	21	500.0	800.	120.	-0.940	-31.420
6	18	500.0	125.	120.	1.050	6.025
6	19	500.0	800.	120.	0.840	25.500
6	20	500.0	125.	120.	-0.160	-0.186
7	15	750.0	400.	120.	0.942	2.191
7	16	750.0	125.	120.	1.790	2.246
7	17	750.0	400.	120.	0.790	1.580
7	18	500.0	125.	120.	-1.050	-6.025

16 ITERATIONS WERE REQUIRED.

For illustrative purposes, the actual flow rates as determined by the computer program are shown on Fig. 13-22c.

13.31 Prepare a compute program written in Basic to determine the flow in each pipe in a pipe network by the Hardy Cross method.

```
10   'HARDY-CROSS LOOP BALANCING NETWORK PROGRAM,              NET.BAS
20   'U.S. CUSTOMARY(USC) OR SI UNITS(SI) MAY BE USED.
30   'HAZEN-WILLIAMS(HW) OR DARCY-WEISBACH(DW) EQUATION MAY BE USED FOR PIPES.
40   'DATA ENTRY VIA READ AND DATA STATEMENTS.
50   '*******READ DATA AND PRINT NETWORK INFORMATION****************************
60   DEFINT I,J,K,N
70   DIM ITYPE(1000),ELEM(500),IND(500),Q(100),H(100),S(20),IX(240)
80   FOR J=1 TO 100:          ITYPE(J)=5:          H(J)=-1000!:    NEXT J
90   READ TITLE$:             LPRINT:             LPRINT"   ";TITLE$
100  READ TT$,KK,TOL,VNU,DEFA  'NT$=USC OR SI, KK=NO. OF ITER., VNU=KIN. VISC.
110  'TOL=TOLERANCE IN ITERATION, DEFA=DEFAULT COEF.- EITHER C OR EPS
120  IF(TT$="SI" OR TT$="si") THEN GOTO 150
130  UNITS=4.727:             G=32.174:
140  LPRINT" US CUSTOMARY UNITS SPEC.,VISCOSITY IN FT^2/S=";VNU: GOTO 170
150  UNITS=10.674:            G=9.806:
160  LPRINT" SI UNITS SPEC.,VISCOSITY IN  M^2/S=";VNU
170  LPRINT" DESIRED TOLERANCE=";TOL;   " NO. OF ITERATIONS=";KK
180  LPRINT"   PIPE Q(CFS OR M^3/S)  L(FT OR M)    D(FT OR M)  HW C OR EPS"
190  READ NPI,TT$:            IF NPI=0 THEN 310
200  'NPI=NO. OF PIPES IN NETWORK, TT$="HW" OR "DW"
210  FOR I1=1 TO NPI:         READ I,QQ,L,D,X3:         IF X3=0! THEN X3=DEFA
220  'I=PIPE NO.,QQ=FLOW,L=LENGTH,D=DIAMETER,X3=C OR EPS
230  Q(I)=QQ:                 KP=4*(I-1)+1
240  IF TT$="HW" OR TT$="hw" THEN 280
250  ITYPE(I)=2:              ELEM(KP)=L/(2!*G*D^5*.7854^2):        'DW
260  ELEM(KP+1)=1!/(.7854*D*VNU): ELEM(KP+2)=X3/(3.7*D):
270  EX=2!:                   GOTO 290
280  ITYPE(I)=1:              EX=1.852:    ELEM(KP)=UNITS*L/(X3^EX*D^4.8704)  'HW
290  EN=EX-1!:                LPRINT"   ";I;:
300  LPRINT USING"  #####.#####";Q(I);L;D;X3:        NEXT I1
310  READ NPS,TT$:            IF NPS=0 THEN 360
320  'NPS=NO. OF PSEUDO ELEMENTS, TT$="PS"
330  FOR I1=1 TO NPS:         READ I,DH:       ITYPE(I)=3:    KP=4*(I-1)+1
```

```
340 ELEM(KP)=DH:              LPRINT"   ";I;" RESERVOIR ELEV. DIFFERENCE=";DH
350 NEXT I1
360 READ NPU,TT$:          IF NPU=0 THEN 480
370 'NPU=NO. OF PUMPS, TT$="PU"
380 FOR I1=1 TO NPU:       READ I,QQ,DQ,H1,H2,H3,H4:           ITYPE(I)=4
390 'I=PUMP NO.,QQ=FLOW,DQ=DEL Q, H1,H2,H3,H4=EQUALLY SPACED PTS ON PUMP CURVE
400 KP=4*(I-1)+1:                Q(I)=QQ:           ELEM(KP)=H1
410 ELEM(KP+3)=(H4-3!*(H3-H2)-H1)/(6!*DQ^3)
420 ELEM(KP+2)=(H3-2!*H2+H1)/(2!*DQ^2)-3!*ELEM(KP+3)*DQ
430 ELEM(KP+1)=(H2-H1)/DQ-ELEM(KP+2)*DQ-ELEM(KP+3)*DQ^2
440 LPRINT"   ";I;" PUMP CURVE, DQ=";DQ;" H=";H1,H2,H3,H4
450 LPRINT"      COEF. IN PUMP EQ.=";ELEM(KP);ELEM(KP+1);ELEM(KP+2);ELEM(KP+3)
460 NEXT I1
470 '*******READ LOOP INDEXING DATA AND BALANCE ALL LOOPS********************
480 READ NI,TT$:           IF NI=0 THEN 820
490 'NI=NO. OF ITEMS IN VECTOR IND, TT$="IND"
500 FOR I=1 TO NI:         READ IND(I):        NEXT I:        IND(NI+1)=0
510 LPRINT " IND=";;       FOR I=1 TO NI:      LPRINT IND(I);: NEXT I
520 FOR K=1 TO KK:         DDQ=0!:             IP=1
530 I1=IND(IP):            IF I1=0 THEN 780
540 DH=0!:                 HDQ=0!
550 FOR J=1 TO I1:         I=IND(IP+J):        IF I<0 THEN 570
560 IF I=0 THEN 710 ELSE GOTO 580
570 S(J)=-1!:              I=-I:               GOTO 590
580 S(J)=1!
590 NTY=ITYPE(I):          KP=4*(I-1)+1
600 ON NTY GOTO 610,620,680,690
610 R=ELEM(KP):            GOTO 660
620 REY=ELEM(KP+1)*ABS(Q(I)):                  IF REY<1! THEN REY=1!
630 IF REY<2000! THEN 640 ELSE GOTO 650
640 R=ELEM(KP)*64!/REY:    GOTO 660
650 R=ELEM(KP)*1.325/(LOG(ELEM(KP+2)+5.74/REY^.9))^2
660 DH=DH+S(J)*R*Q(I)*ABS(Q(I))^EN
670 HDQ=HDQ+EX*R*ABS(Q(I))^EN:                 GOTO 710
680 DH=DH+S(J)*ELEM(KP):                        GOTO 710
690 DH=DH-S(J)*(ELEM(KP)+Q(I)*(ELEM(KP+1)+Q(I)*(ELEM(KP+2)+Q(I)*ELEM(KP+3))))
700 HDQ=HDQ-(ELEM(KP+1)+2!*ELEM(KP+2)*Q(I)+3!*ELEM(KP+3)*Q(I)^2)
710 NEXT J
720 IF ABS(HDQ)<.0001 THEN HDQ=1!
730 DQ=-DH/HDQ:            DDQ=DDQ+ABS(DQ)
740 FOR J=1 TO I1:         I=ABS(IND(IP+J)):   IF ITYPE(I)=3 THEN 760
750 Q(I)=Q(I)+S(J)*DQ
760 NEXT J
770 IP=IP+I1+1:            GOTO 530
780 LPRINT:                LPRINT" ITERATION NO.";K;" SUM OF FLOW CORRECTIONS=";
790 LPRINT USING "###.#####";DDQ
800 IF DDQ<TOL THEN 820
810 NEXT K
820 LPRINT" ELEMENT FLOW":  FOR I=1 TO 100:   NTY=ITYPE(I)
830 ON NTY GOTO 840,840,850,840,850
840 LPRINT"   ";I;;        LPRINT USING"   ###.####";Q(I)
850 NEXT I
855 '******* DATA FOR PATH THRU SYSTEM TO COMPUTE HGL**********************
860 READ NU,TT$:           IF NU=0 THEN 80   'NU=NO. OF NODES WITH GIVEN HGL,TT$=NODES
865 FOR I=1 TO NU:         READ I1,H2:        H(I1)=H2:       NEXT I
870 READ NI,TT$:           IF NI=0 THEN 80   'NI=NO. OF ITEMS IN PATH,TT$="IX"
875 FOR I=1 TO NI:         READ IX(I):        NEXT I:         IX(NI+2)=0
880 LPRINT" IX=";;         FOR I=1 TO NI:     LPRINT IX(I);: NEXT I: LPRINT
885 IP=1
890 FOR J=1 TO 238 STEP 2: IF J=1 THEN K=IX(IP)
895 I=IX(IP+J):            N=IX(IP+J+1)
900 IF I<1 THEN SS=-1!:    I=-I:              GOTO 910
905 IF I=0 THEN 955 ELSE SS=1!
910 NTY=ITYPE(I):          KP=4*(I-1)+1
915 ON NTY GOTO 920,925,945,950,955
920 R=ELEM(KP):            GOTO 940
925 REY=ELEM(KP+1)*ABS(Q(I)):                  IF REY<1! THEN REY=1!
930 IF REY<2000! THEN R=ELEM(KP)*64!/REY:      GOTO 940
935 IF REY>=2000! THEN R=ELEM(KP)*1.325/(LOG(ELEM(KP+2)+5.74/REY^.9))^2
940 H(N)=H(K)-SS*R*Q(I)*ABS(Q(I))^EN:          GOTO 955
945 H(N)=H(K)-SS*ELEM(KP):                      GOTO 955
950 H(N)=H(K)+SS*(ELEM(KP)+Q(I)*(ELEM(KP+1)+Q(I)*(ELEM(KP+2)+Q(I)*ELEM(KP+3))))
955 IF IX(J+IP+3)=0 THEN 975
960 IF IX(J+IP+2)=0 THEN 970
965 K=N:                   NEXT J
970 IP=IP+J+3:             GOTO 890
975 LPRINT" JUNCTION HEAD"
980 FOR N=1 TO 100:        IF H(N)=-1000! THEN 990
985 LPRINT"   ";N;;        LPRINT USING" ###.###";H(N)
990 NEXT N
995 GOTO 80
```

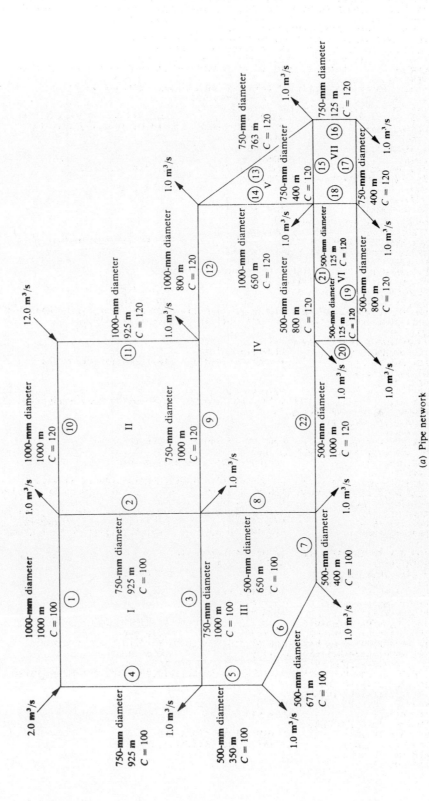

(a) Pipe network

Fig. 13-22

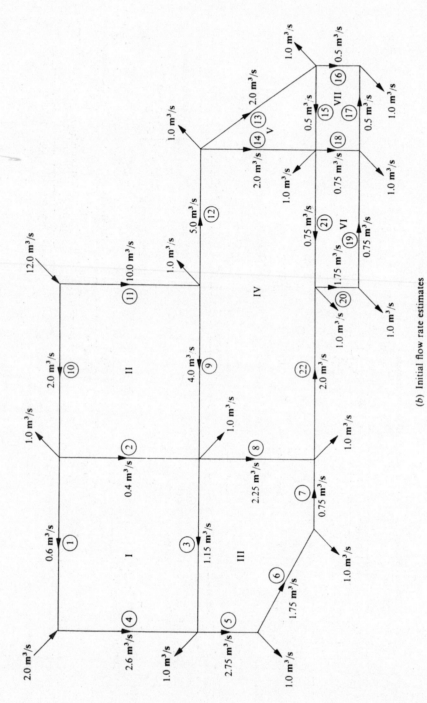

(b) Initial flow rate estimates

Fig. 13-22 (continued)

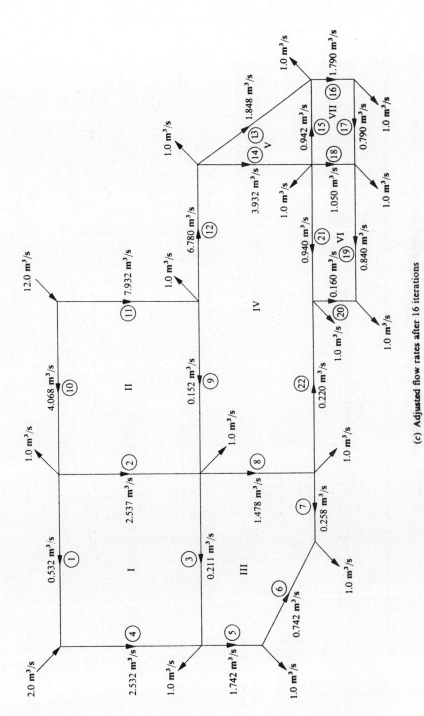

(c) Adjusted flow rates after 16 iterations

Fig. 13-22 (*continued*)

13.32 Use the computer program written for Prob. 13.31 to solve the network problem displayed in Fig. 13-23. The pump data are as follows:

Q, m³/s	0	0.03	0.06	0.09
H, m	30	29	26	20

Input

```
DATA EXAMPLE
DATA SI,30,.001,.000001,100.
DATA 5,HW
DATA 1,.12,600.,0.3,.0
DATA 2,.03,300.,.15,.0
DATA 3,.0,500.,.6,.0
DATA 4,.03,400.,.3,.0
DATA 5,.03,300.,.3,.0
DATA 2,PS
DATA 6,15.0
DATA 7,18.0
DATA 1,PU
DATA 8,.06,.03,30.,29.,26.,20.
DATA 16,IND
DATA 3,2,1,-3,3,4,-5,3,3,6,-4,-1,3,5,7,8
DATA 1,NODES
DATA 5,117.
DATA 9,IX
DATA 5,8,4,2,2,1,1,4,3
```

Output

```
    EXAMPLE
SI UNITS SPEC.,VISCOSITY IN  M^2/S= .000001
DESIRED TOLERANCE= .001  NO. OF ITERATIONS= 30
  PIPE Q(CFS OR M^3/S)  L(FT OR M)     D(FT OR M)   HW C OR EPS
   1       0.12000      600.00000      0.30000    100.00000
   2       0.03000      300.00000      0.15000    100.00000
   3       0.00000      500.00000      0.60000    100.00000
   4       0.03000      400.00000      0.30000    100.00000
   5       0.03000      300.00000      0.30000    100.00000
   6   RESERVOIR ELEV. DIFFERENCE= 15
   7   RESERVOIR ELEV. DIFFERENCE= 18
   8   PUMP CURVE, DQ= .03  H= 30         29         26         20
       COEF. IN PUMP EQ.= 30 -11.11112 -555.5555 -6172.841
IND= 3  2  1 -3  3  4 -5  3  3  6 -4 -1  3  5  7  8
ITERATION NO. 1  SUM OF FLOW CORRECTIONS=  0.1385

ITERATION NO. 2  SUM OF FLOW CORRECTIONS=  0.1040

ITERATION NO. 3  SUM OF FLOW CORRECTIONS=  0.0372

ITERATION NO. 4  SUM OF FLOW CORRECTIONS=  0.0034

ITERATION NO. 5  SUM OF FLOW CORRECTIONS=  0.0006
ELEMENT   FLOW
   1      0.143
   2     -0.034
   3      0.027
   4      0.080
   5      0.094
   8      0.087
IX= 5  8  4  2  2  1  1  4  3
JUNCTION HEAD
   1   137.811
   2   150.044
   3   135.044
   4   137.797
   5   117.000
```

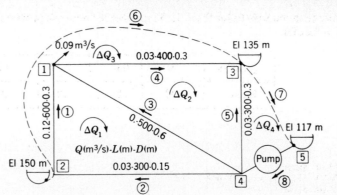

Fig. 13–23

CHAPTER 14
Flow in Open Channels

14.1 Water flows in a rectangular, concrete, open channel that is 12.0 m wide at a depth of 2.5 m. The channel slope is 0.0028. Find the water velocity and the flow rate.

$$v = (1.0/n)(R^{2/3})(s^{1/2}) \qquad n = 0.013 \qquad \text{(from Table A-13)}$$
$$R = A/p_w = (12.0)(2.5)/(2.5 + 12.0 + 2.5) = 1.765 \text{ m}$$
$$v = (1.0/0.013)(1.765)^{2/3}(0.0028)^{1/2} = 5.945 \text{ m/s} \qquad Q = Av = [(12.0)(2.5)](5.945) = 178 \text{ m}^3/\text{s}$$

14.2 Water flows in the symmetrical trapezoidal channel lined with asphalt shown in Fig. 14-1. The channel bottom drops 0.1 ft vertically for every 100 ft of length. What are the water velocity and flow rate?

$$v = (1.486/n)(R^{2/3})(s^{1/2}) \qquad n = 0.015 \qquad \text{(from Table A-13)}$$
$$R = A/p_w \qquad A = (16.0)(4.5) + (2)\{(4.5)[(3)(4.5)]/2\} = 132.8 \text{ ft}^2$$
$$p_w = 16.0 + 2\sqrt{(4.5)^2 + [(3)(4.5)]^2} = 44.46 \text{ ft} \qquad R = 132.8/44.46 = 2.987 \text{ ft}$$
$$s = 0.1/100 = 0.00100 \qquad v = (1.486/0.015)(2.987)^{2/3}(0.00100)^{1/2} = 6.498 \text{ ft/s}$$
$$Q = Av = (132.8)(6.498) = 863 \text{ ft}^3/\text{s}$$

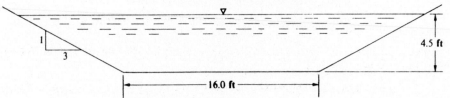

Fig. 14-1

14.3 Water is to flow at a rate of 30 m³/s in the concrete channel shown in Fig. 14-2. Find the required vertical drop of the channel bottom per kilometer of length.

$$A = (3.6)(2.0) + (4.0 - 2.0)[(1.6 + 3.6)/2] = 12.40 \text{ m}^2 \qquad v = (1.0/n)(R^{2/3})(s^{1/2}) = Q/A = 30/12.40 = 2.419 \text{ m/s}$$
$$p_w = 3.6 + 2.0 + \sqrt{(4.0 - 2.0)^2 + (3.6 - 1.6)^2} + 1.6 = 10.03 \text{ m} \qquad R = A/p_w = 12.40/10.03 = 1.236 \text{ m}$$
$$2.419 = (1.0/0.013)(1.236)^{2/3}(s)^{1/2} \qquad s = 0.000746$$

This slope represents a drop of the channel bottom of 0.000746 m per meter of length, or 0.746 m per kilometer of length.

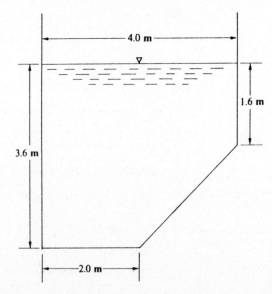

Fig. 14-2

14.4 Water flows in the triangular steel channel shown in Fig. 14-3 at a velocity of 2.9 ft/s. Find the depth of flow if the channel slope is 0.0015.

$$v = (1.486/n)(R^{2/3})(s^{1/2}) \qquad R = A/p_w = 2\{[(d)(d \tan 27.5°)/2]/(2d/\cos 27.5°)\} = 0.2309d$$
$$2.9 = (1.486/0.014)(0.2309d)^{2/3}(0.0015)^{1/2} \qquad d = 2.57 \text{ ft}$$

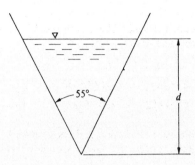

Fig. 14-3

14.5 After a flood had passed an observation station on a river, an engineer visited the site and, by locating flood marks, performing appropriate surveying, and doing necessary computations, determined that the cross-sectional area, wetted perimeter, and water-surface slope at the time of the peak flooding were 2960 m², 341 m, and 0.00076, respectively. The engineer also noted that the channel bottom was "earth with grass and weeds" ($n = 0.030$). Estimate the peak flood discharge.

$$v = (1.0/n)(R^{2/3})(s^{1/2}) = (1.0/0.030)(2960/341)^{2/3}(0.00076)^{1/2} = 3.881 \text{ m/s}$$
$$Q = Av = (2960)(3.881) = 11\,490 \text{ m}^3/\text{s}$$

14.6 A rectangular, concrete channel 50 ft wide is to carry water at a flow rate of 800 cfs. The channel slope is 0.00025. Find the depth of flow.

$$v = (1.486/n)(R^{2/3})(s^{1/2}) = Q/A = 800/50d = 16.00/d \qquad R = A/p_w = 50d/(50 + 2d)$$
$$16.00/d = (1.486/0.013)[50d/(50 + 2d)]^{2/3}(0.00025)^{1/2}$$

This equation is not readily solvable, but a trial-and-error solution (not shown here) reveals that $d = 3.92$ ft.

14.7 Prepare a computer program that will determine the depth of flow of water in a rectangular channel (as in Prob. 14.6).

```
C       THIS PROGRAM DETERMINES THE DEPTH OF FLOW FOR OPEN CHANNEL FLOW
C       IN RECTANGULAR SECTIONS.  IT CAN BE USED FOR PROBLEMS IN BOTH THE
C       ENGLISH SYSTEM OF UNITS AND THE INTERNATIONAL SYSTEM OF UNITS.
C
C       INPUT DATA MUST BE SET UP AS FOLLOWS.
C
C       CARD 1    COLUMN 1        ENTER 0 (ZERO) OR BLANK IF ENGLISH SYSTEM
C                                 OF UNITS IS TO BE USED.  ENTER 1 (ONE) IF
C                                 INTERNATIONAL SYSTEM OF UNITS IS TO BE
C                                 USED.
C                 COLUMNS 2-79    ENTER TITLE, DATE, AND OTHER INFORMATION,
C                                 IF DESIRED.
C       CARD 2    COLUMNS 1-10    ENTER NUMBER INCLUDING DECIMAL GIVING
C                                 WIDTH OF CHANNEL (IN FEET OR METERS).
C                 COLUMNS 11-20   ENTER NUMBER INCLUDING DECIMAL GIVING
C                                 FLOW RATE (IN CUBIC FEET PER SECOND OR
C                                 CUBIC METERS PER SECOND).
C                 COLUMNS 21-30   ENTER NUMBER INCLUDING DECIMAL GIVING
C                                 SLOPE.
C                 COLUMNS 31-40   ENTER NUMBER INCLUDING DECIMAL GIVING
C                                 MANNING N-VALUE.
C
C       MULTIPLE DATA SETS FOR SOLVING ANY NUMBER OF PROBLEMS MAY BE
C       INCLUDED FOR PROCESSING.
C
        DIMENSION TITLE(13)
        REAL N
        INTEGER UNITS
1       READ(5,100,END=2)UNITS,TITLE
```

```
100 FORMAT(I1,13A6)
    WRITE(6,105)TITLE
105 FORMAT('1',13A6,////)
    COEFF=1.486
    IF(UNITS.EQ.1)COEFF=1.0
    READ(5,101)W,Q,S,N
101 FORMAT(4F10.0)
    D=0.001
    TRY1=COEFF/N*(W*D/(W+2.0*D))**(2.0/3.0)*SQRT(S)-Q/W/D
104 D=D+0.001
    TRY2=COEFF/N*(W*D/(W+2.0*D))**(2.0/3.0)*SQRT(S)-Q/W/D
    IF(TRY1*TRY2)102,102,103
103 TRY1=TRY2
    GO TO 104
102 D=D-0.0005
    IF(UNITS.EQ.0)WRITE(6,106)W,Q,S,N,D
106 FORMAT(1X,'GIVEN DATA FOR AN OPEN CHANNEL FLOW IN A RECTANGULAR SE
   *CTION',//5X,'WIDTH =',F7.1,' FT',//5X,'FLOW RATE =',F7.1,' CU FT/S
   *',//5X,'SLOPE =',F10.7,//5X,'MANNING N-VALUE =',F6.3,////1X,'THE D
   *EPTH OF FLOW WILL BE',F7.2,' FT')
    IF(UNITS.EQ.1)WRITE(6,107)W,Q,S,N,D
107 FORMAT(1X,'GIVEN DATA FOR AN OPEN CHANNEL FLOW IN A RECTANGULAR SE
   *CTION',//5X,'WIDTH =',F7.1,' M ',//5X,'FLOW RATE =',F7.1,' CU M/S
   *',//5X,'SLOPE =',F10.7,//5X,'MANNING N-VALUE =',F6.3,////1X,'THE D
   *EPTH OF FLOW WILL BE',F7.2,' M')
    GO TO 1
2   STOP
    END
```

14.8 Solve Prob. 14.6 using the computer program of Prob. 14.7.

▮

Input

```
1 2 3 4 5 6 7 8 9 10 11 12 13 14 15 16 17 18 19 20 21 22 23 24 25 26 27 28 29 30 31 32 33 34 35 36 37 38 39 40 41 42 43 44 45 46 47 48 49 50 51 52 53 54 55 56 57 58 59 60 61 62 63 64 65 66 67 68 69 70 71 72 73 74 75 76 77 78 79 80
0SAMPLE ANALYSIS ØF ØPEN CHANNEL FLØW IN A RECTANGULAR SECTIØN
50.0        800.0       0.00025    0.013
```

Output

```
SAMPLE ANALYSIS OF OPEN CHANNEL FLOW IN A RECTANGULAR SECTION


GIVEN DATA FOR AN OPEN CHANNEL FLOW IN A RECTANGULAR SECTION

    WIDTH =    50.0 FT

    FLOW RATE =  800.0 CU FT/S

    SLOPE = 0.0002500

    MANNING N-VALUE = 0.013




THE DEPTH OF FLOW WILL BE   3.92 FT
```

14.9 A rectangular channel ($n = 0.016$) 20 m wide is to carry water at a flow rate of 30 m³/s at a slope of 0.00032. Find the depth of flow using the computer program of Prob. 14.7.

▮

Input

```
1 2 3 4 5 6 7 8 9 10 11 12 13 14 15 16 17 18 19 20 21 22 23 24 25 26 27 28 29 30 31 32 33 34 35 36 37 38 39 40 41 42 43 44 45 46 47 48 49 50 51 52 53 54 55 56 57 58 59 60 61 62 63 64 65 66 67 68 69 70 71 72 73 74 75 76 77 78 79 80
1SAMPLE ANALYSIS ØF ØPEN CHANNEL FLØW IN A RECTANGULAR SECTIØN
20.0        30.0        0.00032    0.016
```

Output

SAMPLE ANALYSIS OF OPEN CHANNEL FLOW IN A RECTANGULAR SECTION

GIVEN DATA FOR AN OPEN CHANNEL FLOW IN A RECTANGULAR SECTION

    WIDTH =   20.0 M

    FLOW RATE =   30.0 CU M/S

    SLOPE = 0.0003200

    MANNING N-VALUE = 0.016

THE DEPTH OF FLOW WILL BE 1.25 M

14.10 A corrugated metal pipe of 500 mm diameter flows half-full at a slope of 0.0050 (see Fig. 14-4). What is the flow rate for this condition?

$$v = (1.0/n)(R^{2/3})(s^{1/2}) = (1.0/0.024)[(\tfrac{500}{1000})/4]^{2/3}(0.0050)^{1/2} = 0.7366 \text{ m/s}$$
$$Q = Av = \{[(\pi)(\tfrac{500}{1000})^2/4]/2\}(0.7366) = 0.0723 \text{ m}^3/\text{s}$$

(*Note:* The hydraulic radius for both a circular cross section and a semicircular cross section is one-fourth the diameter.)

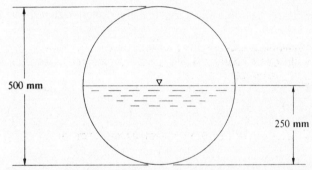

500 mm

250 mm

Fig. 14-4

14.11 A 24-in-diameter cast iron pipe on a $\frac{1}{400}$ slope carries water at a depth of 5.6 in, as shown in Fig. 14-5a. What is the flow rate?

$v = (1.486/n)(R^{2/3})(s^{1/2})$, $R = A/p_w$. The applicable area in this problem is the shaded area $(AECDA)$ in Fig. 14-5b: $AB = BC = 12$ in (both are radii), $BE = 12 - 5.6 = 6.4$ in. Therefore, $AE = EC = \sqrt{(12)^2 - (6.4)^2} = 10.15$ in and $\angle ABE = \angle EBC = \arccos(6.4/12) = 57.77°$, $(\text{Area})_{ABCDA} = [(\pi)(24)^2/4][(2)(57.77°)/360°] = 145.19$ in^2, $(\text{Area})_{ABEA} = (\text{Area})_{BCEB} = (6.4)(10.15)/2 = 32.48$ in^2, $(\text{Area})_{AECDA} = (\text{Area})_{ABCDA} - (2)(\text{Area})_{ABEA} = 145.19 - (2)(32.48) = 80.23$ in^2. The applicable wetted perimeter in this problem is the arc distance ADC in Fig. 14-5: $p_w = ADC = (\pi)(24)[(2)(57.77°)/360°] = 24.20$ in. $R = 80.23/24.20 = 3.315$ in, or 0.2763 ft, $v = (1.486/0.012)(0.2763)^{2/3}(\frac{1}{400})^{1/2} = 2.627$ ft/s, $Q = Av = (80.23/144)(2.627) = 1.46$ ft^3/s.

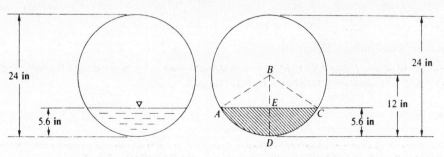

24 in

5.6 in

24 in

12 in

5.6 in

B

E

A C

D

(*a*)   (*b*)   **Fig. 14-5**

14.12 A 500-mm-diameter concrete pipe on a $\frac{1}{500}$ slope is to carry water at a flow rate of 0.040 m³/s. Find the depth of flow. See Fig. 14-6a.

■ $v = (1.0/n)(R^{2/3})(s^{1/2}) = Q/A = 0.040/A$ $0.040/A = (1.0/0.013)(R^{2/3})(\frac{1}{500})^{1/2}$ $AR^{2/3} = 0.01163$

Since $R = A/p_w$,

$$A^{5/3}/p_w^{2/3} = 0.01163 \tag{1}$$

Equation (1) contains two unknowns, A and p_w; however, both unknowns can be expressed in terms of the unknown depth of flow, d. The applicable area in this problem is the shaded area $(AECDA)$ in Fig. 14-6b: $AB = BC = 0.25$ m   (both are radii), $BE = 0.25 - d$. Therefore, $AE = CE = \sqrt{(0.25)^2 - (0.25 - d)^2}$, $\angle ABE = \angle EBC = \arccos[(0.25 - d)/0.25]$,

$$(\text{Area})_{ABCDA} = \left[\frac{(\pi)(0.50)^2}{4}\right]\left\{\frac{(2)\arccos[(0.25 - d)/0.25]}{360°}\right\} = (0.001091)\left(\arccos\frac{0.25 - d}{0.25}\right)$$

$$(\text{Area})_{ABEA} = (\text{Area})_{BCEB} = \frac{(0.25 - d)\sqrt{(0.25)^2 - (0.25 - d)^2}}{2}$$

$$(\text{Area})_{AECDA} = (\text{Area})_{ABCDA} - 2(\text{Area})_{ABEA} = (0.001091)\left(\arccos\frac{0.25 - d}{0.25}\right) - (2)\left[\frac{(0.25 - d)\sqrt{(0.25)^2 - (0.25 - d)^2}}{2}\right]$$

$$p_w = ADC = (\pi)(0.50)\left\{\frac{(2)\arccos[(0.25 - d)/0.25]}{360°}\right\} = (0.008727)\left(\arccos\frac{0.25 - d}{0.25}\right)$$

Therefore, substituting into Eq. (1),

$$\frac{[(0.001091)\{\arccos[(0.25 - d)/0.25]\} - (0.25 - d)\sqrt{(0.25)^2 - (0.25 - d)^2}]^{5/3}}{[(0.008727)\{\arccos[(0.25 - d)/0.25]\}]^{2/3}} = 0.01163 \frac{\text{m}^3}{\text{s}}$$

This equation is not readily solvable, but a trial-and-error solution (not shown here) reveals that $d = 0.166$ m or 166 mm.

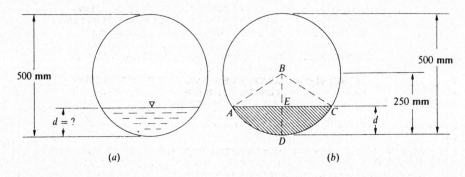

500 mm $d = ?$ (a)   B E A C D d 500 mm 250 mm   (b)   **Fig. 14-6**

14.13 Solve Prob. 14.11 utilizing Fig. A-18.

■ $s = \frac{1}{400} = 0.0025$ ft/ft. From Fig. A-15, $Q_{\text{full}} = 11.4$ ft³/s and $v_{\text{full}} = 3.6$ ft/s. These values of Q_{full} and v_{full} must be adjusted for an n value of 0.012 for the given cast iron pipe (Fig. A-15 is based on an n value of 0.013): $(Q_{\text{full}})_{n=0.012}/11.4 = 0.013/0.012$, $(Q_{\text{full}})_{n=0.012} = 12.4$ ft³/s, $(v_{\text{full}})_{n=0.012}/3.6 = 0.013/0.012$, $(v_{\text{full}})_{n=0.012} = 3.9$ ft/s, $d/d_{\text{full}} = 5.6/24 = 0.23$, or 23 percent. Enter the ordinate of Fig. A-18 with a value of d/d_{full} of 23 percent, move horizontally to the line marked "discharge," and then vertically downward to the abscissa to read $Q/Q_{\text{full}} = 12$ percent. In a similar manner using the "velocity" line, read $v/v_{\text{full}} = 63$ percent. Therefore, $Q = (0.12)(12.4) = 1.5$ ft³/s, $v = (0.63)(3.9) = 2.5$ ft/s.

14.14 Solve Prob. 14.12 utilizing Fig. A-18.

■ $s = \frac{1}{500} = 0.00200$. From Fig. A-16, $Q_{\text{full}} = 0.169$ m³/s: $Q/Q_{\text{full}} = 0.040/0.169 = 0.24$, or 24 percent. From Fig. A-18, $d/d_{\text{full}} = 34$ percent; $d = (0.34)(500) = 170$ mm.

14.15 A 30-in-diameter concrete storm sewer pipe must carry a flow rate of 9.0 cfs at a minimum velocity of 2.5 ft/s. Find the required slope and water depth.

■ $A = Q/v = 9.0/2.5 = 3.600$ ft² $A_{\text{full}} = [(\pi)(\frac{30}{12})^2/4] = 4.909$ ft²

$A/A_{\text{full}} = 3.600/4.909 = 0.73$ or   73 percent

From Fig. A-18, $d/d_{full} = 69$ percent and $R/R_{full} = 116$ percent.

$$d = (0.69)(30) = 20.7 \text{ in} \qquad v = (1.486/n)(R^{2/3})(s^{1/2})$$

$$2.5 = (1.486/0.013)\{(1.16)[(\tfrac{30}{12})/4]\}^{2/3}(s^{1/2}) \qquad s = 0.000734$$

14.16 A concrete pipe must carry water at a slope of 0.0075, at a velocity of 0.76 m/s, and at a depth of flow equal to one-tenth its diameter. What is the required pipe diameter?

$\blacksquare$ $\qquad v = (1.0/n)(R^{2/3})(s^{1/2}) \qquad 0.76 = (1.0/0.013)(R^{2/3})(0.0075)^{1/2} \qquad R = 0.03853 \text{ ft}$

From Fig. A-18 with $d/d_{full} = 10$ percent, $R/R_{full} = 25$ percent.

$$0.03853/R_{full} = 0.25 \qquad R_{full} = 0.1541 \text{ m} \qquad 0.1541 = d/4 \qquad d = 0.616 \text{ m} \quad \text{or} \quad 616 \text{ mm}$$

14.17 Prepare a computer program that will determine either the depth of flow or the flow rate for open channel flow in circular sections.

$\blacksquare$

```
C      THIS PROGRAM DETERMINES EITHER THE DEPTH OF FLOW OR THE FLOW RATE
C      FOR OPEN CHANNEL FLOW IN CIRCULAR SECTIONS.  IT CAN BE USED FOR
C      PROBLEMS IN BOTH THE ENGLISH SYSTEM OF UNITS AND THE INTERNATIONAL
C      SYSTEM OF UNITS.
C
C      INPUT DATA MUST BE SET UP AS FOLLOWS.
C
C      CARD 1    COLUMN 1        ENTER 0 (ZERO) OR BLANK IF ENGLISH SYSTEM
C                                OF UNITS IS TO BE USED.  ENTER 1 (ONE) IF
C                                INTERNATIONAL SYSTEM OF UNITS IS TO BE
C                                USED.
C                COLUMNS 2-79    ENTER TITLE, DATE, AND OTHER INFORMATION,
C                                IF DESIRED.
C      CARD 2    COLUMNS 1-10    ENTER NUMBER INCLUDING DECIMAL GIVING
C                                DIAMETER OF CHANNEL (IN INCHES OR MILLI-
C                                METERS).
C                COLUMNS 11-20   ENTER NUMBER INCLUDING DECIMAL GIVING
C                                DEPTH OF FLOW (IN INCHES OR MILLIMETERS).
C                COLUMNS 21-30   ENTER NUMBER INCLUDING DECIMAL GIVING
C                                SLOPE.
C                COLUMNS 31-40   ENTER NUMBER INCLUDING DECIMAL GIVING
C                                MANNING N-VALUE.
C                COLUMNS 41-50   ENTER NUMBER INCLUDING DECIMAL GIVING
C                                FLOW RATE (IN CUBIC FEET PER SECOND OR
C                                CUBIC METERS PER SECOND).
C
C ****************************************************************
C *                                                            *
C *  NOTE WELL....EITHER THE DEPTH OF FLOW (COLUMNS 11-20) OR THE   *
C *  FLOW RATE (COLUMNS 41-50), WHICHEVER ONE IS TO BE DETERMINED BY  *
C *  THIS PROGRAM, SHOULD BE LEFT BLANK ON CARD 2.              *
C *                                                            *
C ****************************************************************
C
C      MULTIPLE DATA SETS FOR SOLVING ANY NUMBER OF PROBLEMS MAY BE
C      INCLUDED FOR PROCESSING.
C
       DIMENSION TITLE(13)
       COMMON D,R,D1,DIAM,PI,FACTOR,AREA,WP
       REAL N
       INTEGER UNITS
       PI=3.14159265
     1 READ(5,100,END=2)UNITS,TITLE
   100 FORMAT(I1,13A6)
       WRITE(6,105)TITLE
   105 FORMAT('1',13A6,////)
       COEFF=1.486
       FACTOR=12.0
       IF(UNITS.EQ.1)COEFF=1.0
       IF(UNITS.EQ.1)FACTOR=1000.0
       READ(5,101)DIAM,D,S,N,Q
   101 FORMAT(5F10.0)
       R=DIAM/2.0
       IF(Q.GT.0.0001)GO TO 102
       D1=D
```

```
        CALL AREAWP
        HR=AREA/WP

        Q=AREA*COEFF/N*HR**(2.0/3.0)*SQRT(S)
        IF(UNITS.EQ.0)WRITE(6,103)DIAM,D,S,N,Q
103 FORMAT(1X,'GIVEN DATA FOR AN OPEN CHANNEL FLOW IN A CIRCULAR SECTI
    *ON'//5X,'DIAMETER =',F7.2,' IN',//5X,'DEPTH OF FLOW =',F7.2,' IN',
    *//5X,'SLOPE =',F10.7,//5X,'MANNING N-VALUE =',F6.3,////1X,'THE FLO
    *W RATE WILL BE',F8.3,' CU FT/S')
        IF(UNITS.EQ.1)WRITE(6,104)DIAM,D,S,N,Q
104 FORMAT(1X,'GIVEN DATA FOR AN OPEN CHANNEL FLOW IN A CIRCULAR SECTI
    *ON'//5X,'DIAMETER =',F7.1,' MM',//5X,'DEPTH OF FLOW =',F7.1,' MM',
    *//5X,'SLOPE =',F10.7,//5X,'MANNING N-VALUE =',F6.3,////1X,'THE FLO
    *W RATE WILL BE',F8.3,' CU M/S')
        GO TO 1
102 AWP=Q*N/COEFF/SQRT(S)
        D=0.01
        D1=D
        CALL AREAWP
        TRY1=AREA**(5.0/3.0)/WP**(2.0/3.0)-AWP
108 D=D+0.01
        IF(D.GT.DIAM)GO TO 112
        D1=D
        CALL AREAWP
        TRY2=AREA**(5.0/3.0)/WP**(2.0/3.0)-AWP
        IF(TRY1*TRY2)106,106,107
107 TRY1=TRY2
        GO TO 108
106 D=D-0.005
        IF(UNITS.EQ.0)WRITE(6,109)DIAM,Q,S,N,D
109 FORMAT(1X,'GIVEN DATA FOR AN OPEN CHANNEL FLOW IN A CIRCULAR SECTI
    *ON',//5X,'DIAMETER =',F7.2,' IN',//5X,'FLOW RATE =',F8.3,' CU FT/S
    *',//5X,'SLOPE =',F13.7,//5X,'MANNING N-VALUE =',F6.3,////1X,'THE D
    *EPTH OF FLOW WILL BE',F7.2,' IN')
        IF(UNITS.EQ.1)WRITE(6,110)DIAM,Q,S,N,D
110 FORMAT(1X,'GIVEN DATA FOR AN OPEN CHANNEL FLOW IN A CIRCULAR SECTI
    *ON',//5X,'DIAMETER =',F7.1,' MM',//5X,'FLOW RATE =',F8.3,' CU M/S
    *',//5X,'SLOPE =',F13.7,//5X,'MANNING N-VALUE =',F6.3,////1X,'THE D
    *EPTH OF FLOW WILL BE',F7.1,' MM')
        GO TO 1
112 WRITE(6,116)
116 FORMAT(1X,'THIS CIRCULAR CONDUIT CANNOT CARRY THIS GREAT A FLOW AS
    * OPEN CHANNEL FLOW.')
        GO TO 1
2   STOP
        END
        SUBROUTINE AREAWP
        COMMON D,R,D1,DIAM,PI,FACTOR,AREA,WP
        IF(D.GT.R)D1=DIAM-D
        ABCDA=DIAM**2/4.0*ARCOS((R-D1)/R)
        ABEA=(R-D1)*SQRT(R**2-(R-D1)**2)/2.0
        AREA=ABCDA-2.0*ABEA
        WP=DIAM*ARCOS((R-D1)/R)
        IF(D.GT.R)AREA=PI*DIAM**2/4.0-AREA
        IF(D.GT.R)WP=PI*DIAM-WP
        AREA=AREA/FACTOR**2
        WP=WP/FACTOR
        RETURN
        END
```

14.18   Solve Prob. 14.11 utilizing the computer program of Prob. 14.17.

❙ Input

```
1 2 3 4 5 6 7 8 9 10 11 12 13 14 15 16 17 18 19 20 21 22 23 24 25 26 27 28 29 30 31 32 33 34 35 36 37 38 39 40 41 42 43 44 45 46 47 48 49 50 51 52 53 54 55 56 57 58 59 60 61 62 63 64 65 66 67 68 69 70 71 72 73 74 75 76 77 78 79 80
OSAMPLE ANALYSIS ØF AN ØPEN CHANNEL FLØW IN A CIRCULAR SECTIØN
24.0      5.6      0.0025      0.012
```

Output

SAMPLE ANALYSIS OF AN OPEN CHANNEL FLOW IN A CIRCULAR SECTION

GIVEN DATA FOR AN OPEN CHANNEL FLOW IN A CIRCULAR SECTION

    DIAMETER = 24.00 IN

    DEPTH OF FLOW = 5.60 IN

    SLOPE = 0.0025000

    MANNING N-VALUE = 0.012

THE FLOW RATE WILL BE   1.463 CU FT/S

14.19   Solve Prob. 14.12 utilizing the computer program of Prob. 14.17.

❙ Input

1 2 3 4 5 6 7 8 9 10 11 12 13 14 15 16 17 18 19 20 21 22 23 24 25 26 27 28 29 30 31 32 33 34 35 36 37 38 39 40 41 42 43 44 45 46 47 48 49 50 51 52 53 54 55 56 57 58 59 60 61 62 63 64 65 66 67 68 69 70 71 72 73 74 75 76 77 78 79 80

1SAMPLE ANALYSIS ØF AN ØPEN CHANNEL FLØW IN A CIRCULAR SECTIØN
500.   0.002   0.013   0.040

Output

SAMPLE ANALYSIS OF AN OPEN CHANNEL FLOW IN A CIRCULAR SECTION

GIVEN DATA FOR AN OPEN CHANNEL FLOW IN A CIRCULAR SECTION

    DIAMETER = 500.0 MM

    FLOW RATE = 0.040 CU M/S

    SLOPE = 0.0020000

    MANNING N-VALUE = 0.013

THE DEPTH OF FLOW WILL BE 165.7 MM

14.20   An open channel is to be designed to carry $1.0 \text{ m}^3/\text{s}$ at a slope of 0.0065. The channel material has an n value of 0.011. Find the most efficient cross section for a semicircular section.

$$v = (1.0/n)(R^{2/3})(s^{1/2}) \qquad Q/A = (1.0/n)(A/p_w)^{2/3}(s^{1/2})$$
$$A^{5/3}/p_w^{2/3} = Qn/s^{1/2} = (1.0)(0.011)/0.0065^{1/2} = 0.1364$$
$$A = \pi d^2/8 \qquad p_w = \pi d/2 \qquad (\pi d^2/8)^{5/3}/(\pi d/2)^{2/3} = 0.1364 \qquad d = 0.951 \text{ m} \quad \text{or} \quad 951 \text{ mm}$$

(d is the diameter of the semicircular section; the depth of flow would, of course, be half of d.)

14.21   Find the most efficient cross section for Prob. 14.20 for a rectangular section.

❙ $A^{5/3}/p_w^{2/3} = 0.1364$ (from Prob. 14.20). The most efficient rectangular section has a width equal to twice its depth. Letting d = depth, $A = (d)(2d) = 2d^2$, $p_w = d + 2d + d = 4d$, $(2d^2)^{5/3}/(4d)^{2/3} = 0.1364$, $d = 0.434$ m, or 434 mm; width $= 2d = (2)(434) = 868$ mm.

14.22   Find the most efficient cross section for Prob. 14.20 for a triangular section.

❙ $A^{5/3}/p_w^{2/3} = 0.1364$ (from Prob. 14.20). The most efficient triangular section has a 90° angle and 1:1 side slopes (see Fig. 14-7). $A = (\frac{1}{2})(d\sqrt{2})(d\sqrt{2}) = d^2$, $p_w = (2)(d\sqrt{2}) = 2.828d$, $(d^2)^{5/3}/(2.828d)^{2/3} = 0.1364$, $d = 0.614$ m, or 614 mm; sides $= d\sqrt{2} = (614)(\sqrt{2}) = 868$ mm.

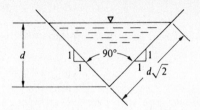

Fig. 14-7

14.23 Find the most efficient cross section for Prob. 14.20 for a trapezoidal section.

▮ $A^{5/3}/p^{2/3} = 0.1364$ (from Prob. 14.20). The most efficient trapezoidal section is half a regular hexagon (see Fig. 14-8). $A = (1.155d)(d) + (2)[(d)(d \tan 30°)/2] = 1.732d^2$, $p_w = (3)(1.155d) = 3.465d$, $(1.732d^2)^{5/3}/(3.465d)^{2/3} = 0.1364$, $d = 0.459$ m; sides and bottom: each $= 1.155d = (1.155)(0.459) = 0.530$ m.

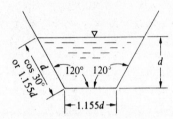

Fig. 14-8

14.24 For the same conditions given in Prob. 14.2, determine the status of flow (i.e., is it critical, subcritical, or supercritical?).

▮ $N_F = v/\sqrt{gd_m}$ $v = 6.498$ ft/s (from Prob. 14.2) $d_m = A/B$ $A = 132.8$ ft² (from Prob. 14.2)
$$B = (3)(4.5) + 16.0 + (3)(4.5) = 43.0 \text{ ft} \cdot \qquad d_m = 132.8/43.0 = 3.088 \text{ ft}$$
$$N_F = 6.498/\sqrt{(32.2)(3.088)} = 0.652$$

Since $N_F < 1.0$, the flow is subcritical.

14.25 The triangular channel ($n = 0.012$) shown in Fig. 14-9 is to carry water at a flow rate of 10 m³/s. Find the critical depth, critical velocity, and critical slope of the channel.

▮ $B/A^3 = g/Q^2$ $B = 6d_c$ $A = 2[(d_c)(3d_c)/2] = 3d_c^2$ $6d_c/(3d_c^2)^3 = 9.807/10^2$ $d_c = 1.178$ m
$$v_c = Q/A = 10/[(3)(1.178)^2] = 2.402 \text{ m/s} \qquad s_c = \{nv_c/[(1.0)(R_c^{2/3})]\}^2 \qquad R = A/p_w$$
$$R_c = [(3)(1.178)^2]/[(2)(\sqrt{10})(1.178)] = 0.5588 \text{ m} \qquad s_c = \{(0.012)(2.402)/[(1.0)(0.5588)^{2/3}]\}^2 = 0.00181$$

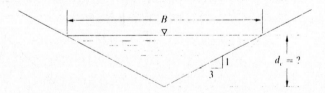

Fig. 14-9

14.26 The semicircular channel ($n = 0.010$) shown in Fig. 14-10 is to carry water at a depth of 1.0 ft. Find the velocity, slope, and discharge at the critical stage.

▮ $d_m = A/B = [(\frac{1}{2})(\pi)(2.0)^2/4]/2.0 = 0.7854 \text{ ft} \qquad v_c = \sqrt{gd_m} = \sqrt{(32.2)(0.7854)} = 5.029 \text{ ft/s}$
$$s_c = \{nv_c/[(1.486)(R_c^{2/3})]\}^2 = \{(0.010)(5.029)/[(1.486)(2.0/4)^{2/3}]\}^2 = 0.00289$$
$$Q = Av_c = [(\tfrac{1}{2})(\pi)(2.0)^2/4](5.029) = 7.90 \text{ ft}^3/\text{s}$$

14.27 A flow rate of 2.1 m³/s is to be carried in an open channel at a velocity of 1.3 m/s. Determine the dimensions of the channel cross section and required slope if the cross section is rectangular with depth equal to one-half the width. Use $n = 0.020$.

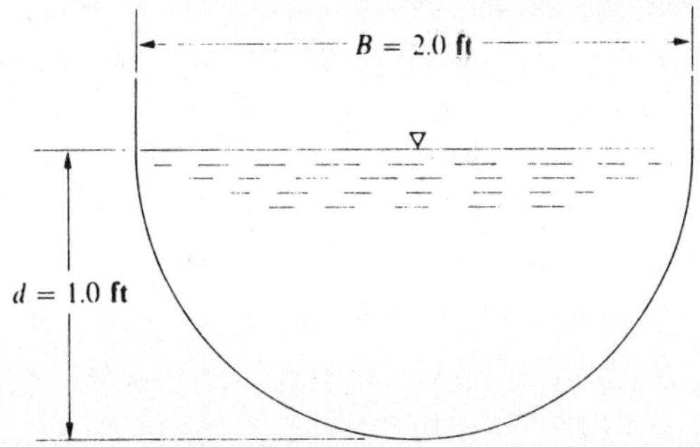

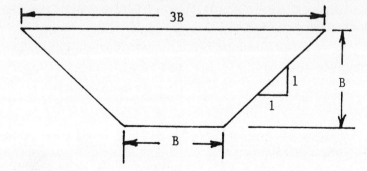

$B = 2.0$ ft

$d = 1.0$ ft

Fig. 14-10

▮ $Q = Av$. Let B = channel width. Then channel depth = $B/2$; $2.1 = [(B)(B/2)](1.3)$, $B = 1.797$ m. Hence, the required width is 1.797 m and depth is 1.797/2, or 0.898 m.

$$R = A/p_w = (1.797)(0.898)/[1.797 + (2)(0.898)] = 0.4491 \text{ m}$$
$$s = \{nv/[(1.0)(R^{2/3})]\}^2 = \{(0.020)(1.3)/[(1.0)(0.4491)^{2/3}]\}^2 = 0.00197$$

14.28 Repeat Prob. 14.27 if the depth must be equal to twice the width. Compare answers with Prob. 14.27.

▮ $Q = Av$. Let B = channel width. Then channel depth = $2B$; $2.1 = [(B)(2B)](1.3)$, $B = 0.899$ m. Hence, the required width is 0.899 m and depth is (2)(0.899), or 1.798 m.

$$R = A/p_w = (1.798)(0.899)/[0.899 + (2)(1.798)] = 0.3596 \text{ m}$$
$$s = \{nv/[(1.0)(R^{2/3})]\}^2 = \{(0.020)(1.3)/[(1.0)(0.3596)^{2/3}]\}^2 = 0.00264$$

The channel area is the same (neglecting round-off errors) but a steeper slope is required for the narrower channel.

14.29 Repeat Prob. 14.27 if the channel cross section is semicircular.

▮ $\quad Q = Av \qquad 2.1 = [(\tfrac{1}{2})(\pi d^2/4)](1.3) \qquad d = 2.028 \text{ m} \qquad r = 2.028/2 = 1.014 \text{ m}$
$$s = \{nv/[(1.0)(R^{2/3})]\}^2 = \{(0.020)(1.3)/[(1.0)(2.028/4)^{2/3}]\}^2 = 0.00167$$

14.30 Repeat Prob. 14.27 if the channel cross section is trapezoidal, with depth equal to the width of the channel bottom and side slopes of 1:1.

▮ $Q = Av$. Let depth and channel bottom width = B (see Fig. 14-11). Then surface width = $3B$; $A = (3B)(B) - (2)[(\tfrac{1}{2})(B)(B)] = 2B^2$, $2.1 = (2B^2)(1.3)$, $B = 0.899$ m.

$$R = A/p_w = (2)(0.899)^2/[0.899 + (2)(0.899)\sqrt{2}] = 0.4696 \text{ m}$$
$$s = \{nv/[(1.0)(R^{2/3})]\}^2 = \{(0.020)(1.3)/[(1.0)(0.4696)^{2/3}]\}^2 = 0.00185$$

$3B$

1
1
B

B

Fig. 14-11

14.31 For each of the channel cross sections shown in Fig. 14-12, compute the area, wetted perimeter, and hydraulic radius.

▮ **(a)** $A = \tfrac{1}{2}[(\pi)(4.0)^2/4] = 6.283 \text{ m}^2 \qquad p_w = \tfrac{1}{2}[(\pi)(4.0)] = 6.283 \text{ m} \qquad R = A/p_w = 6.283/6.283 = 1.000 \text{ m}$

  (b) $\qquad A = (5.0)(2.5) = 12.50 \text{ m}^2 \qquad p_w = 2.5 + 5.0 + 2.5 = 10.00 \text{ m} \qquad R = 12.50/10.00 = 1.250 \text{ m}$

  (c) $\qquad A = (5.0)(1.2) + (2)[(\tfrac{1}{2})(1.2)(1.2)] = 7.440 \text{ m}^2 \qquad p_w = 5.0 + (2)[(1.2)(\sqrt{2})] = 8.394 \text{ m}$
$$R = 7.440/8.394 = 0.886 \text{ m}$$

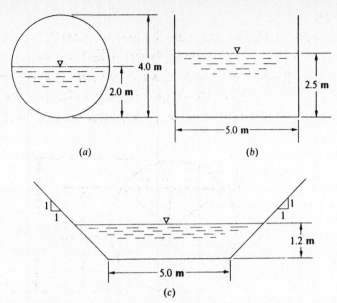

(a)   (b)

(c)   **Fig. 14-12**

14.32 Water is to flow in a rectangular flume at a rate of $1.42 \ m^3/s$ and at a slope of 0.0028. Determine the dimensions of the channel cross section if width must be equal to twice the depth. Use $n = 0.017$.

▌ $Q = (A)(1.0/n)(R^{2/3})(s^{1/2})$. Let B = channel width and $B/2$ = channel depth; $1.42 = [(B)(B/2)](1.0/0.017)[(B)(B/2)/(B/2 + B + B/2)]^{2/3}(0.0028)^{1/2}$, $B = 1.366$ m. Hence, required channel width = 1.366 m and depth = 1.366/2, or 0.683 m.

14.33 Rework Prob. 14.32, assuming width must be equal to the depth. Note which solution gives the smaller (and therefore more efficient) cross section.

▌ $Q = (A)(1.0/n)(R^{2/3})(s^{1/2})$. Let B = channel width and depth; $1.42 = [(B)(B)](1.0/0.017)[(B)(B)/(B + B + B)]^{2/3}(0.0028)^{1/2}$, $B = 0.981$ m. Hence, required channel width and depth are each 0.981 m. $A = (0.981)(0.981) = 0.962 \ m^2$. For Prob. 14.32, $A = (1.366)(0.683) = 0.933 \ m^2$. The cross section of Prob. 14.32 has the smaller cross-sectional area.

14.34 A rectangular channel ($n = 0.011$) 18 m wide is to carry water at a flow rate of 35 cfs. The slope of the channel is 0.00078. Determine the depth of flow.

▌ $$Q = (A)(1.0/n)(R^{2/3})(s^{1/2}) \qquad 35 = (18d)(1.0/0.011)[18d/(18 + 2d)]^{2/3}(0.00078)^{1/2}$$
$$d = 0.885 \ m \qquad \text{(by trial and error)}$$

14.35 The trapezoidal channel shown in Fig. 14-13 is laid on a slope of 0.00191. The channel must carry 60 cfs. Determine the depth of flow. Use $n = 0.015$.

▌ $$Q = (A)(1.486/n)(R^{2/3})(s^{1/2}) \qquad A = 4.0d + (2)[(d)(d)/2] = 4.0d + d^2$$
$$p_w = 4.0 + (2)(d\sqrt{2}) = 4.0 + 2d\sqrt{2} \qquad 60 = (4.0d + d^2)(1.486/0.015)[(4.0d + d^2)/(4.0 + 2d\sqrt{2})]^{2/3}(0.00191)^{1/2}$$
$$d = 2.00 \ ft \qquad \text{(by trial and error)}$$

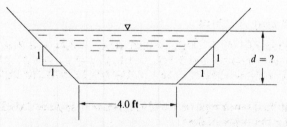

Fig. 14-13

14.36 A 36-in-diameter concrete pipe on a 0.0015 slope carries water at a depth of 26 in. Determine the flow rate for this pipe.

❚ See Fig. 14-14.

$$Q = (A)(1.486/n)(R^{2/3})(s^{1/2}) \qquad OE = 26 - \tfrac{36}{2} = 8 \text{ in} \quad \text{or} \quad 0.6667 \text{ ft} \qquad ∡COE = \arccos\,[8/(\tfrac{36}{2})] = 63.61°$$

$$EC = \sqrt{[(\tfrac{36}{2})/12]^2 - 0.6667^2} = 1.344 \text{ ft}$$

$$A = \{[360 - (2)(63.61)]/360\}[(\pi)(\tfrac{36}{12})^2/4] + (2)[(\tfrac{1}{2})(1.344)(0.6667)] = 5.467 \text{ ft}^2$$

$$p_w = \{[360 - (2)(63.61)]/360\}[(\pi)(\tfrac{36}{12})] = 6.094 \text{ ft}$$

$$Q = (5.467)(1.486/0.013)(5.467/6.094)^{2/3}(0.0015)^{1/2} = 22.5 \text{ ft}^3/\text{s}$$

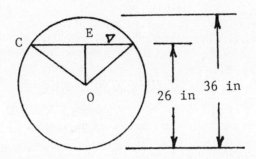

Fig. 14-14

14.37 Rework Prob. 14.36 using Fig. A-18.

❚ From Fig. A-15, $Q_{full} = 25.8 \text{ ft}^3/\text{s}$; $d/d_{full} = \tfrac{26}{36} = 0.722$, or 72.2 percent. From Fig. A-18, $Q/Q_{full} = 87.5$ percent; $Q = (0.875)(25.8) = 22.6 \text{ ft}^3/\text{s}$.

14.38 A sewer pipe, for which $n = 0.014$, is laid on a slope of 0.00018 and is to carry 2.76 m³/s when the pipe flows at 80 percent of full depth. Determine the required diameter of pipe.

❚ See Fig. 14-15.

$$Q = (A)(1.0/n)(R^{2/3})(s^{1/2}) \qquad OE = 0.80D - D/2 = 0.3000D \qquad \alpha = \arccos\,[0.30D/(D/2)] = 53.13°$$

$$CE = (0.3000D)(\tan 53.13°) = 0.4000D$$

$$A = \{[360 - (2)(53.13)]/360\}(\pi D^2/4) + 2[(\tfrac{1}{2})(0.3000D)(0.4000D)] = 0.6736D^2$$

$$p_w = \{[360 - (2)(53.13)]/360\}(\pi D) = 2.214D$$

$$2.76 = (0.6736D^2)(1.0/0.014)(0.6736D^2/2.214D)^{2/3}(0.00018)^{1/2} \qquad D = 2.32 \text{ m}$$

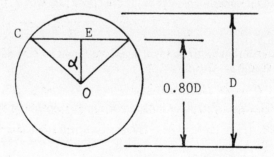

Fig. 14-15

14.39 Rework Prob. 14.38 using Fig. A-18.

❚ $D/D_{full} = 0.80$. From Fig. A-18, $Q/Q_{full} = 0.96$, $A/A_{full} = 0.84$, and $R/R_{full} = 1.21$: $Q_{full} = 2.76/0.96 = 2.88$ m³/s. Figure A-16 cannot be used for $Q_{full} > 1.0$ m³/s: $Q = (A)(1.0/n)(R^{2/3})(s^{1/2})$, $A = (0.84)(\pi D^2/4) = 0.6597D^2$, $R = (1.21)(D/4) = 0.3025D$, $2.76 = (0.6597D^2)(1.0/0.014)(0.3025D)^{2/3}(0.00018)^{1/2}$, $D = 2.34$ m.

14.40 A 72-in-diameter vitrified sewer pipe ($n = 0.014$) is laid on a slope of 0.00025 and carries wastewater at a flow rate of 50 cfs. What is the depth of flow?

❚ From Fig. A-15, $Q_{full} = 67 \text{ ft}^3/\text{s}$: $[(Q_{full})_{n=0.014}]/67 = 0.013/0.014$, $(Q_{full})_{n=0.014} = 62.2 \text{ ft}^3/\text{s}$; $Q/Q_{full} = 50/62.2 = 0.804$, or 80.4 percent. From Fig. A-18, $D/D_{full} = 69$ percent, $D = (0.69)(72) = 49.7$ in.

14.41 A 1.0-m-diameter pipe must carry a discharge of $0.40 \text{ m}^3/\text{s}$ at a velocity of 0.80 m/s. Determine the slope and the depth of water.

$$A = Q/v = 0.40/0.80 = 0.5000 \text{ m}^2 \qquad A_{\text{full}} = (\pi)(1.0)^2/4 = 0.7854 \text{ m}^2$$
$$A/A_{\text{full}} = 0.5000/0.7854 = 0.64 \quad \text{or} \quad 64 \text{ percent}$$

From Fig. A-18, $D/D_{\text{full}} = 0.63$ and $R/R_{\text{full}} = 1.12$.

$$D = (0.63)(1.0) = 0.630 \text{ m} \qquad v = (1.0/n)(R^{2/3})(s^{1/2}) \qquad R_{\text{full}} = 1.0/4 = 0.2500 \text{ m}$$
$$R = (1.12)(0.2500) = 0.2800 \text{ m} \qquad 0.80 = (1.0/0.013)(0.2800)^{2/3}(s^{1/2}) \qquad s = 0.000590$$

14.42 The trapezoidal channel of Fig. 14-16 is to carry 500 cfs of water. The maximum allowable velocity of flow is 3.0 fps to avoid scouring. Determine the depth of flow, d, and the width of the channel bottom, B, if the hydraulic radius of the channel is one-half the depth of flow. Also, determine the slope of the channel bottom. Use $n = 0.025$.

$$R = d/2 = A/p_w \qquad A = Bd + 2[(\tfrac{1}{2})(1.5d)(d)] = Bd + 1.5d^2 \qquad p_w = B + 2\sqrt{d^2 + (1.5d)^2} = B + 3.606d$$
$$d/2 = (Bd + 1.5d^2)/(B + 3.606d) \qquad B + 3.606d = 2B + 3.0d \qquad B = 0.606d \qquad A = Q/v$$
$$Bd + 1.5d^2 = 500/3.0$$

Substituting $B = 0.606d$, $(0.606d)(d) + 1.5d^2 = 166.7$, $d = 8.90$ ft; B = $(0.606)(8.90) = 5.39$ ft.

$$v = (1.486/n)(R^{2/3})(s^{1/2}) \qquad R = 8.90/2 = 4.45 \text{ ft} \qquad 3.0 = (1.486/0.025)(4.45)^{2/3}(s^{1/2}) \qquad s = 0.000348$$

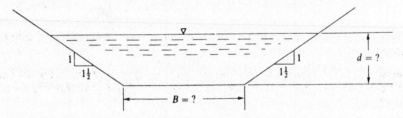

Fig. 14-16

14.43 An open channel to be made of concrete is to be designed to carry $1.5 \text{ m}^3/\text{s}$ at a slope of 0.00085. Find the most efficient cross section for a semicircular section.

$$v = (1.0/n)(R^{2/3})(s^{1/2}) \qquad Q/A = (1.0/n)(A/p_w)^{2/3}(s^{1/2})$$
$$A^{5/3}/p_w^{2/3} = Qn/s^{1/2} = (1.5)(0.013)/0.00085^{1/2} = 0.6688 \qquad A = \pi d^2/8$$
$$p_w = \pi d/2 \qquad (\pi d^2/8)^{5/3}/(\pi d/2)^{2/3} = 0.6688 \qquad d = 1.727 \text{ m}$$

(d is the diameter of the semicircular section; the depth of flow would, of course, be half of d.)

14.44 Find the most efficient cross section for Prob. 14.43 for a rectangular section.

$A^{5/3}/p_w^{2/3} = 0.6688$ (from Prob. 14.43). The most efficient rectangular section has a width equal to twice its depth. Letting $d = $ depth, $A = (d)(2d) = 2d^2$, $p_w = d + 2d + d = 4d$, $(2d^2)^{5/3}/(4d)^{2/3} = 0.6688$, $d = 0.789$ m; width $= 2d = (2)(0.789) = 1.578$ m.

14.45 Find the most efficient cross section for Prob. 14.43 for a triangular section.

$A^{5/3}/p_w^{2/3} = 0.6688$ (from Prob. 14.43). The most efficient triangular section has a 90° angle and 1:1 side slopes (see Fig. 14-7): $A = (\tfrac{1}{2})(d\sqrt{2})(d\sqrt{2}) = d^2$, $p_w = (2)(d\sqrt{2}) = 2.828d$, $(d^2)^{5/3}/(2.828d)^{2/3} = 0.6688$, $d = 1.115$ m; sides $= d\sqrt{2} = (1.115)(\sqrt{2}) = 1.577$ m.

14.46 Find the most efficient cross section for Prob. 14.43 for a trapezoidal section.

$A^{5/3}/p_w^{2/3} = 0.6688$ (from Prob. 14.43). The most efficient trapezoidal section is half a regular hexagon (see Fig. 14-8): $A = (1.155d)(d) + 2[(d)(d \tan 30°)/2] = 1.732d^2$, $p_w = (3)(1.155d) = 3.465d$, $(1.732d^2)^{5/3}/(3.465d)^{2/3} = 0.6688$, $d = 0.832$ m. Sides and bottom: each $= 1.155d = (1.155)(0.832) = 0.961$ m.

14.47 For the conditions given in Prob. 14.32, determine whether the flow is critical, subcritical, or supercritical.

$$v = Q/A = 1.42/[(1.366)(0.683)] = 1.522 \text{ m/s} \qquad N_F = v/\sqrt{gd_m} = 1.522/\sqrt{(9.807)(0.683)} = 0.588$$

Since $N_F < 1.0$, the flow is subcritical.

14.48 A rectangular channel with a width of 3.0 m and an n value of 0.014 is to carry water at a flow rate of 13.4 m³/s. Determine the critical depth, velocity, and channel slope.

■ $d_c = [(Q/B)^2/g]^{1/3} = [(13.4/3.0)^2/9.807]^{1/3} = 1.267$ m $\quad v_c = Q/A = 13.4/[(1.267)(3.0)] = 3.525$ m/s

$$R = A/p_w = (1.267)(3.0)/(1.267 + 3.0 + 1.267) = 0.6868 \text{ m}$$

$$s_c = \{nv_c/[(1.0)(R^{2/3})]\}^2 = \{(0.014)(3.525)/[(1.0)(0.6868)^{2/3}]\}^2 = 0.00402$$

14.49 The semicircular channel ($n = 0.013$) shown in Fig. 14-17 is to carry water while flowing full (i.e., at a depth of 1.5 ft). Determine the velocity, slope, and discharge when flow is critical.

■ $d_m = A/B = \frac{1}{2}[(\pi)(3.0)^2/4]/3.0 = 1.178$ ft $\quad v_c = \sqrt{gd_m} = \sqrt{(32.2)(1.178)} = 6.159$ ft/s

$$s_c = \{nv_c/[(1.486)(R^{2/3})]\}^2 = \{(0.013)(6.159)/[(1.486)(3.0/4)^{2/3}]\}^2 = 0.00426$$

$$Q = Av = \{\tfrac{1}{2}[(\pi)(3.0)^2/4]\}(6.159) = 21.8 \text{ ft}^3/\text{s}$$

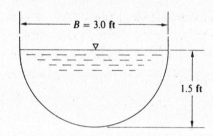

Fig. 14-17

14.50 Determine the dimensions of the most economical trapezoidal brick-lined ($n = 0.016$) channel to carry 200 m³/s with a slope of 0.0004.

■ The most economical trapezoidal channel has a cross section as shown in Fig. 14-8 and $R = d/2$ and $A = \sqrt{3}d^2$; $Q = (A)(1.0/n)(R^{2/3})(s^{1/2})$, $200 = (\sqrt{3}d^2)(1.0/0.016)(d/2)^{2/3}(0.0004)^{1/2}$, $d = 6.491$ m. Bottom width = $(1.155)(6.491) = 7.497$ m.

14.51 Determine the discharge for a trapezoidal channel with a bottom width of 8 ft and side slopes 1:1. The depth is 6 ft, and the slope of the bottom is 0.0009. The channel has a finished concrete lining ($n = 0.012$).

■ $A = (8)(6) + (2)[(\frac{1}{2})(6)(6)] = 84.00$ ft² $\quad p_w = 8 + (2)[(6)(\sqrt{2})] = 24.97$ ft

$$Q = (A)(1.486/n)(R^{2/3})(s^{1/2}) = (84.00)(1.486/0.012)(84.00/24.97)^{2/3}(0.0009)^{1/2} = 701 \text{ ft}^3/\text{s}$$

14.52 What depth is required for 4-m³/s flow in a rectangular planed-wood ($n = 0.012$) channel 2 m wide with a bottom slope of 0.002?

■ $\quad Q = (A)(1.0/n)(R^{2/3})(s^{1/2})$ $\quad 4 = (2d)(1.0/0.012)[2d/(2+2d)]^{2/3}(0.002)^{1/2}$

$$d^{5/3}/(2+2d)^{2/3} = 0.3381 \quad\quad d = 0.888 \text{ m} \quad \text{(by trial and error)}$$

14.53 A developer has been required by environmental regulatory authorities to line an open channel to prevent erosion. The channel is trapezoidal in cross section and has a slope of 0.0009. The bottom width is 10 ft and side slopes are 2:1 (horizontal to vertical). If he uses rubble ($\gamma_s = 135$ lb/ft³) for the lining, what is the minimum D_{50} of the rubble that can be used? The design flow is 1000 cfs. Assume the shear that rubble can withstand is described by $\tau = (0.040)(\gamma_s - \gamma)(D_{50})$ (lb/ft²), in which γ_s is the unit weight of rock and D_{50} is the average rock diameter in feet.

■ A Manning n of 0.03 is appropriate for rubble.

$$Q = (A)(1.486/n)(R^{2/3})(s^{1/2})$$

$$1000 = [(d)(10 + 2d)](1.486/0.03)\{(d)(10 + 2d)/[10 + (2)(\sqrt{5})(d)]\}^{2/3}(0.0009)^{1/2}$$

$$[(d)(10 + 2d)]^{5/3}/[10 + (2)(\sqrt{5})(d)]^{2/3} = 672.9 \quad\quad d = 8.63 \text{ ft} \quad \text{(by trial and error)} \quad\quad \tau_0 = \gamma Rs$$

$$R = 8.63[10 + (2)(8.63)]/[10 + (2)(\sqrt{5})(8.63)] = 4.841 \text{ ft} \quad\quad \tau_0 = (62.4)(4.841)(0.0009) = 0.2719 \text{ lb/ft}^2$$

To find the D_{50} size for incipient movement $\tau = \tau_0$ and $0.2719 = (0.040)(135 - 62.4)(D_{50})$, $D_{50} = 0.0936$ ft.

14.54 A rectangular channel is to carry 1.2 m³/s at a slope of 0.009. If the channel is lined with galvanized iron ($n = 0.011$), what is the minimum area of metal needed for each 100 m of channel? Neglect freeboard.

▌ For minimum area, $D = B/2$ and $R = B/4$.

$$Q = (A)(1.0/n)(R^{2/3})(s^{1/2}) \qquad 1.2 = [(B)(B/2)](1.0/0.011)(B/4)^{2/3}(0.009)^{1/2} \qquad B = 0.8754 \text{ m}$$

$$D = 0.8754/2 = 0.4377 \text{ m} \qquad A_{\text{metal}} = (0.4377 + 0.8754 + 0.4377)(100) = 175 \text{ m}^2 \text{ per } 100 \text{ m}$$

14.55 A trapezoidal channel, with side slopes 2 horizontal to 1 vertical ($m = 2$), is to carry 20 m³/s with a bottom slope of 0.0009. Determine the bottom width, depth, and velocity for the best hydraulic section ($n = 0.025$).

▌
$$Q = (A)(1.0/n)(R^{2/3})(s^{1/2})$$

$$p_w = 4d\sqrt{1 + m^2} - 2md = (4)(d)\sqrt{1 + d^2} - (2)(2)(d) = 4.944d = B + 2\sqrt{5}d$$

$$B = 0.4719d \qquad A = Bd + 2d^2 = (0.4719d)(d) + 2d^2 = 2.472d^2$$

$$20 = (2.472d^2)(1.0/0.025)(2.472d^2/4.944d)^{2/3}(0.0009)^{1/2} \qquad d = 2.433 \text{ m}$$

$$B = (0.4719)(2.433) = 1.148 \text{ m} \qquad v = Q/A = 20/[(2.472)(2.433)^2] = 1.367 \text{ m/s}$$

14.56 What radius semicircular corrugated-metal ($n = 0.022$) channel is needed to convey 2.5 m³/s a distance of 1000 m with a head loss of 2 m? Can you find another cross section that requires less perimeter?

▌
$$Q = (A)(1.0/n)(R^{2/3})(s^{1/2}) \qquad 2.5 = (\pi r^2/2)(1.0/0.022)(r/2)^{2/3}(\tfrac{2}{1000})^{1/2} \qquad r = 1.085 \text{ m}$$

No, the semicircular cross section requires the least perimeter.

14.57 Determine the best hydraulic trapezoidal section to convey 85 m³/s with a bottom slope of 0.002. The lining is finished concrete ($n = 0.012$).

▌
$$Q = (A)(1.0/n)(R^{2/3})(s^{1/2}) \qquad 85 = (A)(1.0/0.012)(A/p_w)^{2/3}(0.002)^{1/2} \qquad A^{5/3}/p_w^{2/3} = 22.81$$

The best trapezoidal section is half a regular hexagon (see Fig. 14-8) for which $A = 1.732d^2$ and $p_w = 3.465d$ (from Prob. 14.23). $(1.732d^2)^{5/3}/(3.465d)^{2/3} = 22.81$, $d = 3.127$ m. Sides and bottom: each $= 1.155d = (1.155)(3.127) = 3.612$ m.

14.58 Calculate the discharge in cubic feet per second through the channel and floodway of Fig. 14-18 for steady uniform flow, with $s = 0.0009$ and $y = 8$ ft.

▌
$$Q = (A)(1.0/n)(R^{2/3})(s^{1/2}) \qquad y = (8)(0.3048) = 2.438 \text{ m}$$

$$A_1 = (12)(5 + 2.438) + (2)(5 + 2.438)(5 + 2.438)/2 - (2.438)(2.438)/2 = 141.6 \text{ m}^2$$

$$(p_w)_1 = \sqrt{(5 + 2.438)^2 + (5 + 2.438)^2} + 12 + \sqrt{5^2 + 5^2} = 29.59 \text{ m}$$

$$Q_1 = (141.6)(1.0/0.025)(141.6/29.59)^{2/3}(0.0009)^{1/2} = 482.5 \text{ m}^3/\text{s}$$

$$A_2 = (120)(2.438) + (2.438)(2.438)/2 = 295.5 \text{ m}^2$$

$$(p_w)_2 = 120 + \sqrt{2.438^2 + 2.438^2} = 123.4 \text{ m} \qquad Q_2 = (295.5)(1.0/0.040)(295.5/123.4)^{2/3}(0.0009)^{1/2} = 396.7 \text{ m}^3/\text{s}$$

$$Q = Q_1 + Q_2 = 482.5 + 396.7 = 879.2 \text{ m}^3/\text{s} = 879.2/0.3048^3 = 31\,050 \text{ ft}^3/\text{s}$$

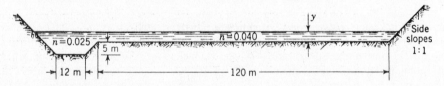

Fig. 14-18

14.59 For 25 000 cfs through the section of Fig. 14-18, find the depth of flow in the floodway (i.e., evaluate y) in feet when the slope of the energy grade line is 0.0004.

▌
$$Q = (A)(1.0/n)(R^{2/3})(s^{1/2}) \qquad A_1 = (12)(5 + y) + (2)(5 + y)(5 + y)/2 - (y)(y)/2 = y^2/2 + 22y + 85$$

$$(p_w)_1 = \sqrt{(5 + y)^2 + (5 + y)^2} + 12 + \sqrt{y^2 + y^2} = (5)(\sqrt{2}) + 12 + (2y)(\sqrt{2})$$

$$A_2 = 120y + (y)(y)/2 = 120y + y^2/2$$

$$(p_w)_2 = 120 + \sqrt{y^2 + y^2} = 120 + (y)(\sqrt{2}) \qquad Q = (25\,000)(0.3048)^3 = 707.9 \text{ m}^3/\text{s}$$

$$707.9 = (y^2/2 + 22y + 85)(1.0/0.025)\{(y^2/2 + 22y + 85)/[(5)(\sqrt{2}) + 12 + (2y)(\sqrt{2})]\}^{2/3}(0.0004)^{1/2}$$

$$+ (120y + y^2/2)(1.0/0.040)\{(120y + y^2/12)/[120 + (y)(\sqrt{2})]\}^{2/3}(0.0004)^{1/2}$$

$$y = 2.79 \text{ m} \quad \text{or} \quad 9.15 \text{ ft} \qquad \text{(by trial and error)}$$

14.60   What is the critical depth for flow of 1.5 m³/s per meter of width?

$$y_c = (q^2/g)^{1/3} = (1.5^2/9.807)^{1/3} = 0.612 \text{ m}$$

14.61   What is the critical depth for flow of 0.3 m³/s through the cross section of Fig. 14-19?

$$Q^2T/gA^3 = 1 \qquad T = (2)(y \tan 60°/2) = 1.155y$$
$$A = (y)(y \tan 60°/2) = 0.5774y^2 \qquad (0.3)^2(1.155y)/[(9.807)(0.5774y^2)^3] = 1 \qquad y = 0.560 \text{ m}$$

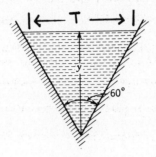

Fig. 14-19

14.62   Determine the critical depth for flow of 8.5 m³/s through a trapezoidal channel with a bottom width of 2.5 m and side slopes of 1 on 1.

$$Q^2T/gA^3 = 1 \qquad T = 2.5 + 2y \qquad A = By + my^2 = 2.5y + (1)(y)^2$$
$$(8.5)^2(2.5 + 2y)/[(9.807)(2.5y + y^2)^3] = 1 \qquad y = 0.928 \text{ m} \qquad \text{(by trial and error)}$$

14.63   Design a transition from a trapezoidal section, 8 ft bottom width and side slopes 1 on 1, depth 4 ft, to a rectangular section, 6 ft wide and 6 ft deep, for a flow of 250 cfs. The transition is to be 20 ft long, and the loss is one-tenth the difference between velocity heads. Show the bottom profile, and do not make any sudden changes in cross-sectional area.

$A_1 = 8 \times 4 + 4^2 = 48 \text{ ft}^2$, $A_2 = 36 \text{ ft}^2$, loss $= 0.1[(v_2^2/2g) - (v_1^2/2g)]$, and $y_1 + (v_1^2/2g) + z_1 = y_2 + (v_2^2/2g) + z_2 +$ loss. Assume a linear change in area, b, and T: $b = 8 - 2(x/L)$ and $T = 16 - 10(x/L)$. Hence: $A = (b + T)(y/2) = 48 - 12(x/L)$ and $y = 2\{[4 - (x/L)]/[2 - (x/L)]\}$.

x/L	A	v²/2g	loss	EGL	y	z
0	48	0.421		4.421	4.0	0
			0.008			
0.333	44	0.501		4.413	4.4	−0.488
			0.011			
0.667	40	0.607		4.402	5.001	−1.206
			0.014			
1.0	36	0.749		4.388	6.0	−2.361

The profile is shown in Fig. 14.20.

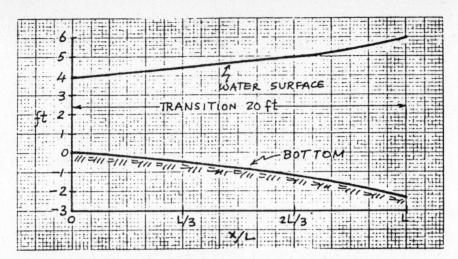

Fig. 14-20

14.64 A transition from a rectangular channel, 2.6 m wide and 2 m deep, to a trapezoidal channel, bottom width 4 m and side slopes 2 on 1, with depth 1.3 m has a loss four-tenths the difference between velocity heads. The discharge is 5.6 m³/s. Determine the difference between elevations of channel bottoms.

$$0 + v_1^2/2g + y_1 = \Delta z + v_2^2/2g + y_2 + (0.4)(v_1^2/2g - v_2^2/2g) \qquad A_1 = (2.6)(2) = 5.200 \text{ m}^2$$

$$A_2 = (4)(1.3) + (2)\{[(1.3)(2)](1.3)/2\} = 8.580 \text{ m}^2 \qquad v_1 = Q/A_1 = 5.6/5.200 = 1.077 \text{ m/s}$$

$$v_2 = 5.6/8.580 = 0.6527 \text{ m/s}$$

$$v_1^2/2g = 1.077^2/[(2)(9.807)] = 0.05914 \text{ m} \qquad v_2^2/2g = 0.6527^2/[(2)(9.807)] = 0.02172 \text{ m}$$

$$0 + 0.05914 + 2 = \Delta z + 0.02172 + 1.3 + (0.4)(0.05914 - 0.02172) \qquad \Delta z = 0.722 \text{ m}$$

14.65 A very wide gate (Fig. 14-21) admits water to a horizontal channel. Considering the pressure distribution hydrostatic at section 0, compute the depth at section 0 and the discharge per meter of width when $y = 1.0$ m.

$$d_0 = C_c y = (0.86)(1.0) = 0.86 \text{ m} \qquad p_1/\gamma + v_1^2/2g + z_1 = p_2/\gamma + v_2^2/2g + z_2 + h_L$$

$$0 + 0 + 6 = 0 + v_t^2/[(2)(9.807)] + 0.86 + 0 \qquad v_t = 10.04 \text{ m/s} \qquad v_a = C_v v_t = (0.96)(10.04) = 9.638 \text{ m/s}$$

$$Q = Av = [(0.86)(1)](9.638) = 8.29 \text{ m}^3/\text{s per meter of width}$$

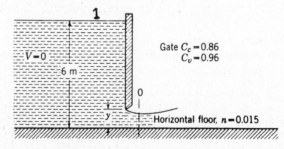

Fig. 14-21

14.66 A discharge of 4.5 m³/s occurs in a rectangular channel 1.83 m wide with $s = 0.002$ and $n = 0.012$. Find the normal depth for uniform flow and determine the critical depth. Is the flow subcritical or supercritical?

$$Q = (A)(1.0/n)(R^{2/3})(s^{1/2}) \qquad 4.5 = (1.83y_n)(1.0/0.012)[1.83y_n/(y_n + 1.83 + y_n)]^{2/3}(0.002)^{1/2}$$

$$y_n = 1.060 \text{ m} \qquad \text{(by trial and error)}$$

$$Q^2/g = A^3/B \qquad 4.5^2/9.807 = (1.83y_c)^3/1.83 \qquad y_c = 0.851 \text{ m}$$

Since $y_c < y_n$, the flow is subcritical.

14.67 What is the flow rate in a 48-in circular corrugated-metal ($n = 0.022$) pipe on a slope of 0.003 if the depth of flow is 19 in?

$$Q = (A)(1.486/n)(R^{2/3})(s^{1/2})$$

$$Q_{\text{full}} = [(\pi)(\tfrac{48}{12})^2/4](1.486/0.022)[(\tfrac{48}{12})/4]^{2/3}(0.003)^{1/2} = 46.49 \text{ ft}^3/\text{s} \qquad d/d_{\text{full}} = \tfrac{19}{48} = 0.40 \quad \text{or} \quad 40 \text{ percent}$$

From Fig. A-18, $Q/Q_{\text{full}} = 25$ percent, $Q = (0.25)(46.49) = 11.6 \text{ ft}^3/\text{s}$.

14.68 At what depth will 8.8 cfs flow in a 36-in-diameter concrete pipe on a slope of 0.004?

$$Q = (A)(1.486/n)(R^{2/3})(s^{1/2}) \qquad Q_{\text{full}} = [(\pi)(\tfrac{36}{12})^2/4](1.486/0.013)[(\tfrac{36}{12})/4]^{2/3}(0.004)^{1/2} = 42.18 \text{ ft}^3/\text{s}$$

$$Q/Q_{\text{full}} = 8.8/42.18 = 0.21 \quad \text{or} \quad 21 \text{ percent}$$

From Fig. A-18, $d/d_{\text{full}} = 32$ percent, $d = (0.32)(36) = 11.5$ in.

14.69 What is the flow rate in a 40-cm-diameter concrete pipe on a slope of 0.004 if the depth is 8.0 cm?

$$Q = (A)(1.0/n)(R^{2/3})(s^{1/2}) \qquad Q_{\text{full}} = [(\pi)(\tfrac{40}{100})^2/4](1.0/0.013)[(\tfrac{40}{100})/4]^{2/3}(0.004)^{1/2} = 0.1317 \text{ m}^3/\text{s}$$

$$d/d_{\text{full}} = 8.0/40 = 0.20 \quad \text{or} \quad 20 \text{ percent}$$

From Fig. A-18, $Q/Q_{\text{full}} = 10$ percent, $Q = (0.10)(0.1317) = 0.0132 \text{ m}^3/\text{s}$.

14.70 Repeat Prob. 14.68 for the case of a 36-in-diameter corrugated metal pipe ($n = 0.024$).

$$Q = (A)(1.486/n)(R^{2/3})(s^{1/2}) \qquad Q_{\text{full}} = [(\pi)(\tfrac{36}{12})^2/4](1.486/0.024)[(\tfrac{36}{12})/4]^{2/3}(0.004)^{1/2} = 22.85 \text{ ft}^3/\text{s}$$

$$Q/Q_{\text{full}} = 8.8/22.85 = 0.39 \quad \text{or} \quad 39 \text{ percent}$$

From Fig. A-18, $d/d_{\text{full}} = 43$ percent, $d = (0.43)(36) = 15.5$ in.

14.71 It is necessary to discharge 20 cfs in a cast iron ($n = 0.015$) pipe 36 in in diameter. What is the minimum possible slope which can be employed if the pipe is to flow at 0.8 depth?

$d/d_{\text{full}} = 0.8$. From Fig. A-18, $Q/Q_{\text{full}} = 96$ percent, $Q_{\text{full}} = 20/0.96 = 20.83 \text{ ft}^3/\text{s}$, $Q = (A)(1.486/n)(R^{2/3})(s^{1/2})$, $20.83 = [(\pi)(\tfrac{36}{12})^2/4](1.486/0.015)[(\tfrac{36}{12})/4]^{2/3}(s)^{1/2}$, $s = 0.00130$.

14.72 On what slope should one construct a 10-ft-wide rectangular channel ($n = 0.013$) so that critical flow will occur at a depth of 4.0 ft?

$$Q = \sqrt{A^3 g/B} = \sqrt{[(4)(10)]^3(32.2)/10} = 454.0 \text{ ft}^3/\text{s} = (A)(1.486/n)(R^{2/3})(s^{1/2})$$

$$454.0 = [(4)(10)](1.486/0.013)[(4)(10)/(4.0 + 10 + 4.0)]^{2/3}(s^{1/2}) \qquad s = 0.00340$$

14.73 Is the flow of Prob. 14.68 subcritical or supercritical?

For critical flow, $Q^2/g = A^3/B$. In a 36-in pipe flowing at a depth of 11.5 in, $B = 2.8$ ft and $A = 1.94 \text{ ft}^2$, approximately. $[8.8^2/32.2 = 2.40] < [1.94^3/2.8 = 2.6]$. Hence, the flow is barely subcritical.

14.74 Water flows steadily at 16.0 cfs in a very long triangular flume which has side slopes 1 on 1. The bottom of the flume is on a slope of 0.0040. At a certain section A, the depth of flow is 2.00 ft. Is the flow at this section subcritical or supercritical? At another section B, 200 ft downstream from A, the depth of flow is 2.50 ft. Approximately what is the value of Manning's n? Find y_n. Under what conditions can this flow occur?

$Q^2/g = A^3/B$. At critical flow, $A = (2)[(y_c)(y_c)/2] = y_c^2$, $16.0^2/32.2 = (y_c^2)^3/(2y_c)$, $y_c = 1.74$ ft. Since $y_c < 2.00$ ft, the flow at section A is subcritical.

$$L = (E_1 - E_2)/(S - S_0) \qquad v_A = Q/A_A = 16.0/2.00^2 = 4.000 \text{ ft/s} \qquad v_A^2/2g = 4.000^2/[(2)(32.2)] = 0.2484 \text{ ft}$$

$$v_B = 16.0/2.50^2 = 2.560 \text{ ft/s} \qquad v_B^2/2g = 2.560^2/[(2)(32.2)] = 0.1018 \text{ ft}$$

$$200 = [(2.00 + 0.2484) - (2.50 + 0.1018)]/(s - 0.0040) \qquad s = 0.00223 \qquad v = (1.486/n)(R^{2/3})(s^{1/2})$$

Apply this equation using average values of v and R:

$$R = A/p_w \qquad R_A = 2.00^2/[(2)(\sqrt{2.00^2 + 2.00^2})] = 0.7071 \text{ ft} \qquad R_B = 2.50^2/[(2)(\sqrt{2.50^2 + 2.50^2})] = 0.8839 \text{ ft}$$

$$(4.000 + 2.560)/2 = (1.486/n)[(0.7071 + 0.8839)/2]^{2/3}(0.00223)^{1/2} \qquad n = 0.018$$

$$Q = (A)(1.486/n)(R^{2/3})(s^{1/2}) \qquad 16.0 = (y_n^2)(1.486/0.018)\{y_n^2/[(2)(\sqrt{y_n^2 + y_n^2})]\}^{2/3}(0.004)^{1/2} \qquad y_n = 1.97 \text{ ft}$$

This is an M-1 curve (backwater from a dam).

14.75 Water flows at 8.5 m³/s in a 3.0-m-wide open channel of rectangular cross section. The bottom slope is adverse, i.e., it rises 0.2 m per 100 m in the direction of flow. If the water depth decreases from 2.10 m to 1.65 m in a 150 m length of channel, determine Manning's n.

❚
$$z_1 + y_1 + v_1^2/2g = z_2 + y_2 + v_2^2/2g + h_L$$
$$v_1 = Q/A_1 = 8.5/[(2.10)(3.0)] = 1.349 \text{ m/s} \qquad v_2 = 8.5/[(1.65)(3.0)] = 1.717 \text{ m/s}$$
$$0 + 2.10 + 1.349^2/[(2)(9.807)] = (0.2/100)(150) + 1.65 + 1.717^2/[(2)(9.807)] + h_L \qquad h_L = 0.0925 \text{ m}$$
$$v = (1.0/n)(R^{2/3})(s^{1/2}) = v_{avg} = (1.349 + 1.717)/2 = 1.533 \text{ m/s} \qquad R = A/p_w$$
$$R_1 = (2.10)(3.0)/(2.10 + 3.0 + 2.10) = 0.8750 \text{ m} \qquad R_2 = (1.65)(3.0)/(1.65 + 3.0 + 1.65) = 0.7857 \text{ m}$$
$$R_{avg} = (0.8750 + 0.7857)/2 = 0.8304 \text{ m} \qquad 1.533 = (1.0/n)(0.8304)^{2/3}(0.0925/150)^{1/2} \qquad n = 0.014$$

14.76 The flow in a long 12-ft-wide rectangular channel which has a constant bottom slope is 1000 cfs. A computation using the Manning equation indicates that the normal depth of flow for this flow rate is 7.0 ft. At a certain section A, the depth of flow in the channel is 3.0 ft. Will the depth of flow increase, decrease, or remain the same as one proceeds downstream from section A? Indicate a physical situation in which this type of flow will occur.

❚
$$y_n = 7.0 \text{ ft} \qquad y_c = (q^2/g)^{1/3} = [(\tfrac{1000}{12})^2/32.2]^{1/3} = 6.00 \text{ ft}$$

Hence, the depth of flow will increase. This is an M-3 curve (upstream of jump or downstream of sluice gate).

14.77 What would be the cross section of greatest hydraulic efficiency for a trapezoidal channel with side slopes of 1.5 horizontal to 1 vertical if the design discharge is 10.0 m³/s and the channel slope is 0.0005? Use $n = 0.025$.

❚ See Fig. 14-22: $R = A/p_w = (1.5y + x)(y)/\{x + (2)[\sqrt{y^2 + (1.5y)^2}]\}$. For greatest hydraulic efficiency, $R = y/2$. Hence, $(1.5y + x)(y)/[x + (2)\sqrt{y^2 + (1.5y)^2}] = y/2$, $x = 0.606y$.
$$Q = (A)(1.0/n)(R^{2/3})(s^{1/2}) \qquad A = (1.5y + x)(y) = (1.5y + 0.606y)(y) = 2.106y^2$$
$$10.0 = (2.106y^2)(1.0/0.025)(y/2)^{2/3}(0.0005)^{1/2} \qquad y = 2.22 \text{ m} \qquad x = (0.606)(2.22) = 1.35 \text{ m}$$

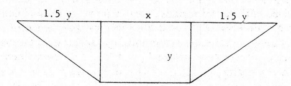

Fig. 14-22

14.78 An irrigation district plans to convey 500 cfs of water from a diversion dam to their distribution works by a canal which can be cut into a very tight clay ($n = 0.018$). Leakage is not considered serious. Between the two ends of the canal there is a drop of 100 ft in a distance of 10 000 ft. Design a suitable open-channel system to carry the water.

❚ $s_{max} = 100/10\,000 = 0.0100 \qquad A_{min} = Q/v_{max} = 500/4.0 = 125.0 \text{ ft}^2 \qquad Q = (A)(1.486/n)(R^{2/3})(s^{1/2})$

Use a trapezoidal channel with side slopes of $1:1$. Assume $v_{max} = 4.0$ fps. Try $b = 8$ ft and $y = 8$ ft:
$A = (8)(8 + 8) = 128.0 \text{ ft}^2$, $p_w = 8 + (2)(\sqrt{8^2 + 8^2}) = 30.63$ ft, $500 = (128.0)(1.486/0.018)(128.0/30.63)^{2/3}(s)^{1/2}$, $s = 0.000333$. Use $b = 8$ ft, $y = 8$ ft, side slopes $= 1:1$, $s = 000333$. This requires $100 - (0.000333)(10\,000) = 96.67$ ft of drop suitably spaced.

14.79 Given an open channel with a parabolic cross section ($x = 1.0$ m and $y = 1.0$ m in Fig. 14-23) on a slope of 0.02 with $n = 0.015$, find the normal depth and the critical depth for a flow rate of 2.0 m³/s.

❚ $Q = (A)(1.0/n)(R^{2/3})(s^{1/2})$. Equation of parabola: $x^2 = y$.
$$A = (2)(\tfrac{2}{3})(xy) = (\tfrac{4}{3})(y^{1/2}y) = 4y^{3/2}/3$$
$$p_w = (2x)[1 + (\tfrac{2}{3})(y/x)^2 - (\tfrac{2}{5})(y/x)^4 + \cdots] = (2y^{1/2})[1 + 2y/3 - 2y^2/5 + \cdots]$$

Try $y_n = 0.5$ m: $A = (4)(0.5)^{3/2}/3 = 0.4714 \text{ m}^2$, $p_w = (2)(0.5)^{1/2}[1 + (2)(0.5)/3 - (2)(0.5)^2/5 + \cdots] = 1.744$ m, $Q = (0.4714)(1.0/0.015)(0.4714/1.744)^{2/3}(0.02)^{1/2} = 1.86 \text{ m}^3/\text{s}$. Since this value of Q (1.86 m³/s) is slightly less than the given value (2.0 m³/s), try a slightly higher value of y_n, say 0.52 m: $A = (4)(0.52)^{3/2}/3 = 0.5000 \text{ m}^2$, $p_w = (2)(0.52)^{1/2}[1 + (2)(0.52)/3 - (2)(0.52)^2/5 + \cdots] = 1.786$ m, $Q = (0.5000)(1.0/0.015)(0.5000/1.786)^{2/3}(0.02)^{1/2} = 2.02 \text{ m}^3/\text{s}$. An additional iteration (not shown) indicates an appropriate value of y_n of 0.518 m. Critical depth occurs when $Q^2/g = A^3/B$: $2.0^2/9.807 = (4y_c^{3/2}/3)^3/[(2)(\sqrt{y_c})]$, $y_c = 0.766$ m.

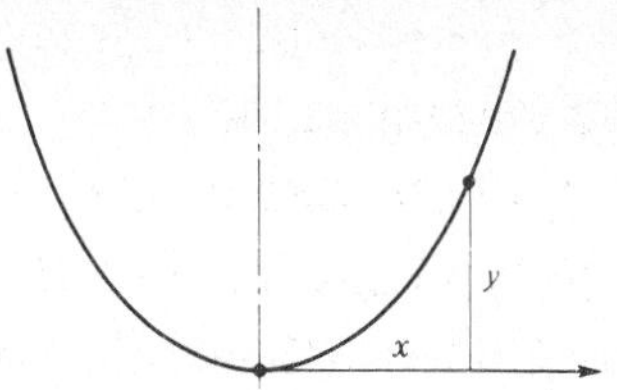

Fig. 14-23

14.80 Consider the frictionless open-channel flow of Fig. 14-24a, in which the water depth y_∞ is 2 ft and the volume flow rate per unit width $\dot{Q}/w$ is 5 ft²/s. Discuss what happens as the height of the obstacle on the channel floor is increased.

❚ We first find out whether the flow upstream of the obstacle is subcritical or supercritical by calculating the critical depth y_c, using $y_c = [(\dot{Q}/w)^2/g]^{1/3} = [(5)^2/32.17]^{1/3} = 0.9194$ ft. The upstream Froude number F_∞ is, then, $F_\infty = (y_c/y_\infty)^{3/2} = (0.9194/2)^{3/2} = 0.3117$. F_∞ is less than 1, showing that the upstream flow is subcritical.

Next we calculate the height of an obstacle which results in critical flow. The specific head of the flow upstream of the obstacle is $H_\infty = (\dot{Q}/w)^2/(2gy_\infty^2) + y_\infty = (5)^2/[(2)(32.17)(2)^2] + 2 = 2.097$ ft. The specific head of the flow at the critical point is the minimum specific head for that $\dot{Q}/w$: $H_{min} = \frac{3}{2}y_c = (\frac{3}{2})(0.9194) = 1.379$ ft. Then the height of an obstacle resulting in critical flow h_c is $h_c = H_\infty - H_{min} = 2.097 - 1.379 = 0.718$ ft.

Let us calculate the surface profile of a flow over an obstacle which is not high enough to cause critical flow, say an obstacle 0.5 ft high. Since the flow is subcritical, the water surface must dip or depress over the obstacle since dy/dx is opposite in sign to dh/dx. The water surface passes through a minimum depth right over the crest of the obstacle. The specific head of the flow over the crest is $H_{crest} = H_\infty - h_{crest} = 2.097 - 0.5 = 1.597$ ft. A dimensionless specific head of $H_{crest}/y_c = 1.597/0.9194 = 1.737$ corresponds to a dimensionless depth for subcritical flow of $y_{crest}/y_c = 1.52$. The depth of the flow over the crest is $y_{crest} = (1.52)(0.9194) = 1.4$ ft. The Froude number of the flow over the crest is $F_{crest} = (y_c/y_{crest})^{3/2} = (1/1.52)^{3/2} = 0.534$.

The local surface depth all along the obstacle is calculated in exactly the same way using the local height $h(x)$ of the obstacle. Figure 14-24a shows the water surface and Froude-number variation for this flow over a circular obstacle. The flow is symmetrical about the crest of the obstacle. Obstacles of the same crest height but different shape change the shape of the water-surface depression but not its minimum depth.

Suppose the height of the crest of the obstacle is increased to precisely $h_c = 0.718$ ft, the height at which critical flow occurs. The flow decreases in depth from y_∞ upstream of the obstacle to y_c at the crest with a corresponding increase in Froude number from $N_F = 0.3117$ to 1. Downstream of the crest there are two possible surface profiles. The first possibility is that the flow remains subcritical and its depth increases from y_c back to y_∞ as the Froude number decreases from 1 to $N_F = 0.3117$. This is the limiting case of subcritical flow. The flow is symmetrical about the crest. The second possibility is that the flow passes through the critical point and its Froude number continues to increase while its depth continues to decrease. The flow becomes supercritical. Downstream of the obstacle the flow has a new depth corresponding to the supercritical branch for the same specific head as upstream. Although the specific head is the same on either side of the obstacle, there is a different distribution between the kinetic and potential energy. Upstream of the obstacle the flow is subcritical, and most of the energy is potential, whereas downstream of the obstacle the flow is supercritical and a larger proportion of the energy is kinetic. The transition from subcritical to supercritical flow downstream of the critical point depends on whether the conditions downstream are favorable for maintaining supercritical flow.

Let us calculate the depth downstream of the obstacle for supercritical flow $y_{\infty,\,sup}$. The specific head is the same as upstream, $H_\infty = 2.097$ ft or $H/y_c = 2.281$. $y/y_c = 0.535$ is on the supercritical branch for this value of H/y_c. The downstream depth is then $y_{\infty,\,sup} = 0.535(0.9194) = 0.492$ ft. The Froude number there is $(N)_{\infty,\,sup} = (y_c/y_{\infty,sup})^{3/2} = (N)(1/0.535)^{3/2} = 2.56$.

The surface elevation undergoes its largest changes when the Froude number is in the vicinity of 1. At low subcritical Froude numbers $1 - N_F^2$ is nearly 1, and the decrease in surface elevation is about the same as the increase in obstacle elevation. As the Froude number in supercritical flow increases, $1 - N_F^2$ becomes increasingly more negative, causing the surface elevation to change less and less with changes in obstacle elevation.

Figure 14-24b shows the surface profile and Froude-number variation for subcritical and supercritical flow downstream of the crest.

What happens if the obstacle height is increased above h_c, the obstacle height which results in critical flow? The flow can no longer take place with the same values of y_∞ and $\dot{Q}/w$ as before. There must be an adjustment in either y_∞ or $\dot{Q}/w$ or both to raise H_∞ to at least the value which results in critical flow at the crest. This is the minimum specific head required to sustain the flow over the obstacle.

14.81 A large reservoir 5 m deep has a rectangular sluice gate 1 m wide. How does the volume flow rate change as the sluice gate is raised (Fig. 14-25)?

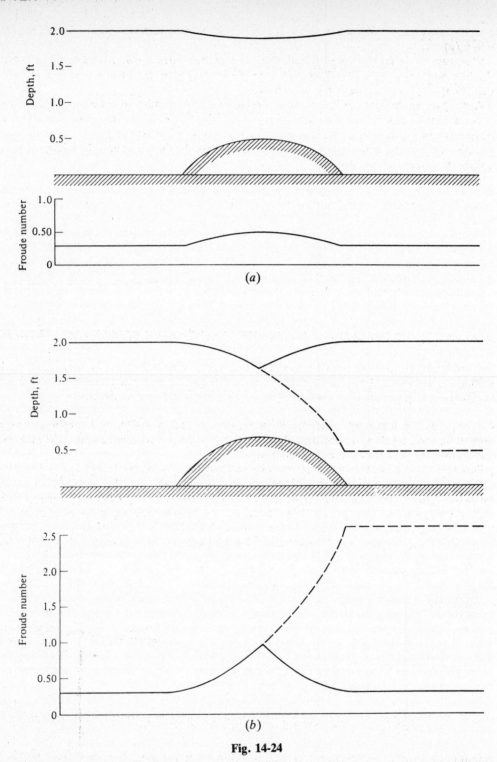

Fig. 14-24

▐ Apply the ideal Bernouilli equation in head form at the water surface and far enough up and downstream of the gate for the streamlines to be parallel: $(p_{atm}/\rho g) + (V_\infty^2/2g) + y_\infty = (p_{atm}/\rho g) + (V^2/2g) + y$. In terms of specific head $H_\infty = H$. This is a constant-specific-head flow. The sluice gate is not an obstacle which changes the specific head of the flow; it simply changes the distribution of energy between kinetic and potential.

Now we examine H_∞, $H_\infty = (V_\infty^2/2g) + y_\infty$. Assume that the reservoir is so large that y_∞ upstream of the gate does not change with changes in sluice-gate opening. Furthermore, V_∞ is very small. Then, $H_\infty \approx y_\infty$ and is the same for all sluice-gate openings. Thus, the outflow has the same specific head regardless of the height of the sluice-gate opening.

The flow rate is then given by $\dot{Q} = wy\sqrt{2g(y_\infty - y)}$. The critical depth y_c changes with the flow rate and consequently with y: $y_c = [(\dot{Q}/w)^2/g]^{1/3} = [2y^2(y_\infty - y)]^{1/3}$. The corresponding Froude number is $N_F = (y_c/y)^{3/2} = \sqrt{2[(y_\infty/y) - 1]}$.

What happens as the sluice gate is raised? First, consider a small opening which results in a y of, say, 0.5 m. The flow rate is $\dot{Q} = (1)(0.5)\sqrt{2(9.8)(5 - 0.5)} = 4.696$ m³/s and the corresponding Froude number is $N_F = \sqrt{2[(5/0.5) - 1]} = 4.243$. The flow is supercritical.

As the gate is raised further, y increases, causing the Froude number to decrease and the flow rate to increase. Finally, at a gate height which gives $y = \frac{2}{3}y_\infty = (\frac{2}{3})(5) = 3.333$ m, The Froude number is 1 and the maximum flow rate is passed: $\dot{Q}_{max} = w\sqrt{\frac{8}{27}gy_\infty^3} = (1)\sqrt{\frac{8}{27}(9.8)(5)^3} = 19.05$ m³/s.

If the gate is raised farther, the flow becomes subcritical, the Froude number passes through 1 and decreases, and the flow rate starts to decrease.

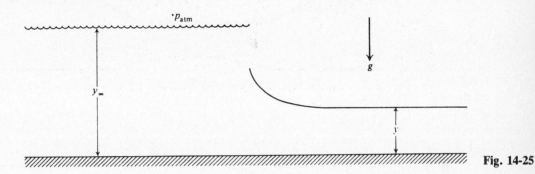

Fig. 14-25

14.82 Reconsider the flow leaving the sluice gate in Prob. 14-81 in which $y_\infty = 5$ m. The flow can be made subcritical by placing a step downstream of the gate. Suppose the flow leaving the sluice gate is $y_1 = 0.5$ m deep. Estimate the height of the step h required to cause a subcritical flow $y_3 = 2$ m deep over the step.

$\blacksquare$ The problem is solved by considering the flow in two parts (Fig. 14-26). First, a hydraulic jump occurs in front of the step, resulting in subcritical flow and a dissipation of mechanical energy. Then the level of the flow coming over the step decreases in a frictionless flow.

First we calculate the characteristics of the hydraulic jump. From Prob. 13.2 $(N_F)_1 = 4.243$. The depth ratio across the hydraulic jump is calculated from $y_2/y_1 = \frac{1}{2}(\sqrt{1 + 8(N_F)_1^2} - 1) = \frac{1}{2}[\sqrt{1 + (8)(4.243)^2} - 1] = 5.521$. The mechanical-energy dissipation across the hydraulic jump is $(h_f)_{1\to2}/y_1 = [(y_2/y_1) - 1]^3/(4y_2/y_1) = (5.521 - 1)^3/(4)(5.521) = 4.184$, $(h_f)_{1\to2} = 2.092$ m. The frictionless flow over the step is described by $H_2 = H_3 + h$. Now, $h_{t_2} = h_{t_1} - (h_f)_{1\to2}$, or $H_2 = H_1 - (h_f)_{1\to2}$. $H_1 = H_\infty = y_\infty = 5$ m. Then $H_2 = H_1 - (h_f)_{1\to2} = 5 - 2.092 = 2.908$ m, $H_3 = [(\dot{Q}/w)^2/2gy_3^2] + y_3 = \{(4.696)^2/[2(9.8)(2)^2]\} + 2 = 2.281$ m. Finally, $h = H_2 - H_3 = 2.908 - 2.281 = 0.627$ m.

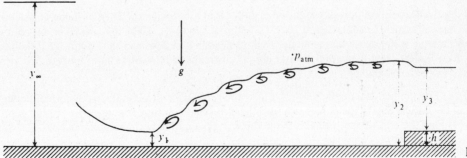

Fig. 14-26

14.83 A long channel with a rectangular cross section and an unfinished concrete surface ($n = 0.017$) is 35 ft wide and has a constant slope of 0.5°. What is the water depth when the channel carries 3500 cfs? Is the channel slope mild or steep?

$\blacksquare$ $\quad Q = (A)(1.486/n)(R^{2/3})(s^{1/2}) \qquad 3500 = (35d)(1.486/0.017)[35d/(35 + 2d)]^{2/3}(\tan 0.5°)^{1/2}$

$\quad d = 4.97$ ft $\quad$ (by trial and error) $\qquad d_c = (q^2/g)^{1/3} = [(\frac{3500}{35})^2/32.2]^{1/3} = 6.77$ ft

Since $d < d_c$, flow is supercritical and the channel slope is steep.

14.84 Find the depth for uniform flow in Fig. 14-27 when the flow rate is 225 cfs if $s = 0.0006$ and n is assumed to be 0.016. Compute the corresponding value of ϵ.

■ $$Q = (A)(1.486/n)(R^{2/3})(s^{1/2}) \qquad A = (y_0)(10 + 2y_0) \qquad p_w = 10 + (2)(\sqrt{5})(y_0)$$

$$225 = [(y_0)(10 + 2y_0)](1.486/0.016)\{[(y_0)(10 + 2y_0)]/[10 + (2)(\sqrt{5})(y_0)]\}^{2/3}(0.0006)^{1/2}$$

$$y_0 = 3.41 \text{ ft} \qquad \text{(by trial and error)} \qquad n = 0.093f^{1/2}R^{1/6}$$

$$R = 3.41[10 + (2)(3.41)]/[10 + (2)(\sqrt{5})(3.41)] = 2.272 \text{ ft} \qquad 0.016 = (0.093)(f^{1/2})(2.272)^{1/6} \qquad f = 0.0225$$

$$1/\sqrt{f} = 2\log(14.8R/\epsilon) \qquad 1/\sqrt{0.0225} = 2\log[(14.8)(2.272)/\epsilon] \qquad \epsilon = 0.0156 \text{ ft}$$

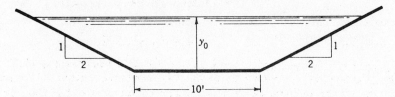

y_0

10'

Fig. 14-27

14.85 In Fig. 14-28, water flows uniformly at a steady rate of 14.0 cfs in a very long triangular flume which has side slopes of 1:1. The bottom of this flume is on a slope of 0.006, and $n = 0.012$. (**a**) Is the flow subcritical or supercritical? (**b**) Find the relation between $v_c^2/2g$ and y_c for this channel.

■ (**a**) $$Q = (A)(1.486/n)(R^{2/3})(s^{1/2}) \qquad A = (y)(2y)/2 = y^2 \qquad p_w = (2)(\sqrt{2})(y) = 2.828y$$
$$14.0 = (y^2)(1.486/0.012)(y^2/2.828y)^{2/3}(0.006)^{1/2} \qquad y = 1.49 \text{ ft} \qquad \text{(by trial and error)}$$
$$Q^2/g = A^3/B \qquad 14.0^2/32.2 = (y_c^2)^3/2y_c \qquad y_c = 1.65 \text{ ft}$$

Since $y < y_c$, flow is supercritical.

(**b**) $$v_c^2/g = A_c/B_c = y_c^2/2y_c = y_c/2 \qquad y_c^2/2g = y_c/4$$

Consequently, we see that the relation between $v^2/2g$ and y for critical-flow conditions depends on the geometry of the flow section. If the vertex angle of the triangle had been different, the relation would have been different.

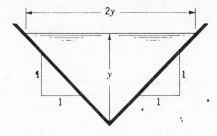

$2y$

y

Fig. 14-28

14.86 In Fig. 14-29, uniform flow of water occurs at 27 cfs in a 4-ft-wide rectangular flume at a depth of 2.00 ft. (**a**) Is the flow subcritical or supercritical? (**b**) If a hump of height $\Delta t = 0.30$ ft is placed in the bottom of the flume, calculate the water depth on the hump. Neglect head loss in flow over the hump. (**c**) If the hump height is raised to $\Delta z = 0.60$ ft, what then are the water depths upstream and downstream of the hump? Once again neglect head loss over the hump.

■ (**a**) First find critical depth: $y_c = (q^2/g)^{1/3} = [(\frac{27}{4})^2/32.2]^{1/3} = 1.12$ ft. Since the normal depth (2.00 ft) is greater than the critical depth, the flow is subcritical and the channel slope is mild.

(**b**) Find the critical hump height. Write the energy equation between sections 1 and 2, assume critical flow on the hump and apply continuity.

$$2.00 + (V_1^2/2g) = (\Delta z)_{\text{crit}} + 1.12 + (V_2^2/2g) \tag{1}$$
$$V_2 = 27/(4 \times 1.12) = 6.03 \text{ fps} \tag{2}$$
$$V_1 = 27/(4 \times 2) = 3.38 \text{ fps} \tag{3}$$

Substituging (2) and (3) in (1) gives $(\Delta z)_{\text{crit}} = 0.49$ ft. Thus the minimum-height hump that will produce critical depth on the hump is 0.49 ft.

Since the actual hump height, $\Delta z = 0.30$ ft, is less than the critical hump height, 0.49 ft, critical flow does not occur on the hump and there is no damming action.

Let us now find the depth y_2 on the hump:

Energy: $$2.00 + (V_1^2/2g) = 0.30 + y_2 + (V_2^2/2g) \tag{4}$$

Continuity: $$(4 \times 2)V_1 = 4y_2 V_2 = 27 \text{ cfs} \tag{5}$$

Eliminating V_1 and V_2 from Eqs. (4) and (5) gives three roots for y_2; $y_2 = 1.60$ ft, 0.82 ft, or a negative answer that has no physical meaning. Since the hump height is less than $(\Delta z)_{crit}$, the flow on the hump must be subcritical (that is, $y_2 > y_c$). Hence $y_2 = 1.60$ ft and the drop in the water surface on the hump $= 2.00 - (0.30 + 1.60) = 0.10$ ft.

(c) In this case the hump height $\Delta z = 0.60$ ft which is greater than the critical hump height. Hence critical depth ($y_c = 1.12$ ft) will occur on the hump. Writing the energy equation for this case, we have

$$y_1 + (V_1^2/2g) = 0.60 + 1.12 - (V_2^2/2g) \tag{6}$$

From continuity,
$$(4 \times y_1)V_1 = 27 \text{ cfs} \tag{7}$$

and, for critical flow,
$$V_2^2/2g = \tfrac{1}{2}y_2 = 0.56 \text{ ft} \tag{8}$$

Combining Eqs. (6), (7), and (8) gives $y_1 + [(\tfrac{27}{4})^2/(2g)y_1^2] = 2.28$ from which $y_1 = 2.12$ ft, 0.66 ft, or a negative answer which has no physical meaning. In this case, damming action occurs and the depth y, upstream of the hump, is 2.12 ft. On the hump the depth passes through critical depth of 1.12 ft and just downstream of the hump the depth will be 0.66 ft. The depth will then increase in the downstream direction following an M_2 water-surface profile until a hydraulic jump occurs to return the depth to the normal uniform flow depth of 2.00 ft.

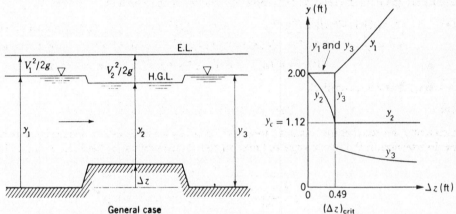

General case $(\Delta z)_{crit}$ Fig. 14-29

14.87   For the channel of Prob. 14.84, compute the "open-channel Reynolds number" assuming that water at 50 °F is flowing. Refer to Fig. A-5 to verify whether or not the flow is wholly rough. Determine ϵ from Fig. A-5 and compare it with the value computed in Prob. 14.84.

▌ $N_R = Rv/v$. From Prob. 14.84, $Q = 225$ cfs, $R = 2.272$ ft, $f = 0.0225$, and $A = 3.41[10 + (2)(3.41)] = 57.36$ ft²; $v = Q/A = 225/57.36 = 3.923$ ft/s.

Open channel: $N_R = (2.272)(3.923)/(1.40 \times 10^{-5}) = 6.37 \times 10^5$

Equivalent pipe: $N_R = (4)(6.37 \times 10^5) = 2.55 \times 10^6$

From Fig. A-5 with $N_R = 2.55 \times 10^6$ and $f = 0.0225$, $\epsilon/D = 0.0018$ and the flow is wholly rough: $\epsilon = 0.0018D = (0.0018)(4R) = 0.0018[(4)(2.272)] = 0.0164$ ft. This value of ϵ (0.0164) is close to the value of 0.0156 computed in Prob. 14.84.

14.88   Assuming the values of f versus R and ϵ/D given for pipes in Fig. A-5 to apply to open channels as well, find the rate of discharge of water at 60 °F in a 100-in-diameter smooth concrete pipe flowing half full, if the pipe is laid on a grade of 2 ft/mile. Note that D should be replaced by $4R$.

▌ $\epsilon/D = 0.001/(\tfrac{100}{12}) = 0.00012$. Try turbulent flow with $f = 0.0135$:

$$v = \sqrt{(8g/f)(Rs)} \qquad R = D/4 = (\tfrac{100}{12})/4 = 2.083 \text{ ft} \qquad s = \tfrac{2}{5280} = 0.0003788$$

$$v = \sqrt{[(8)(32.2)/0.0135][(2.083)(0.0003788)]} = 3.880 \text{ ft/s}$$

$$N_R = Dv/v = [(4)(2.083)](3.880)/(1.21 \times 10^{-5}) = 2.67 \times 10^6$$

From Fig. A-5, $f = 0.013$. Try $f = 0.013$:

$$v = \sqrt{[(8)(32.2)/0.013][(2.083)(0.0003788)]} = 3.954 \text{ ft/s}$$

$$N_R = [(4)(2.083)](3.954)/(1.21 \times 10^{-5}) = 2.72 \times 10^6$$

$$f = 0.013 \quad \text{(O.K.)} \qquad Q = Av = [(\tfrac{1}{2})(\pi)(\tfrac{100}{12})^2/4](3.954) = 108 \text{ ft}^3/\text{s}$$

14.89 For the channel of Prob. 14.84, compute the flow rate for depths of 1, 3, 5, and 7 ft.

▮ $Q = (A)(1.486/n)(R^{2/3})(s^{1/2})$. For $y_0 = 1$ ft, $A = 1[10 + (2)(1)] = 12.00$ ft², $p_w = 10 + (2)(\sqrt{5})(1) = 14.47$ ft, $Q = (12.00)(1.486/0.016)(12.00/14.47)^{2/3}(0.0006)^{1/2} = 24.1$ ft³/s. For $y_0 = 3$ ft, $A = 3[10 + (2)(3)] = 48.00$ ft², $p_w = 10 + (2)(\sqrt{5})(3) = 23.42$ ft, $Q = (48.00)(1.486/0.016)(48.00/23.42)^{2/3}(0.0006)^{1/2} = 176$ ft³/s. For $y_0 = 5$ ft, $A = 5[10 + (2)(5)] = 100.0$ ft², $p_w = 10 + (2)(\sqrt{5})(5) = 32.36$ ft, $Q = (100.0)(1.486/0.016)(100.0/32.36)^{2/3}(0.0006)^{1/2} = 483$ ft³/s. For $y_0 = 7$ ft, $A = 7[10 + (2)(7)] = 168.0$ ft², $p_w = 10 + (2)(\sqrt{5})(7) = 41.30$ ft, $Q = (168.0)(1.486/0.016)(168.0/41.30)^{2/3}(0.0006)^{1/2} = 974$ ft³/s.

14.90 Figure 14-30 shows a cross section of a canal forming a portion of the Colorado River Aqueduct, which is to carry 1600 cfs. The canal is lined with concrete, for which n is assumed to be 0.014. What must be the grade of the canal, and what will be the drop in elevation per mile?

▮
$$Q = (A)(1.486/n)(R^{2/3})(s^{1/2}) \qquad A = (10.2)(50.6 + 20)/2 = 360.1 \text{ ft}^2$$
$$p_w = 18.39 + 20 + 18.39 = 56.78 \text{ ft}$$
$$1600 = (360.1)(1.486/0.014)(360.1/56.78)^{2/3}(s)^{1/2} \qquad s = 0.000149$$
$$\text{Drop in elevation} = (0.000149)(5280) = 0.787 \text{ ft/mile}$$

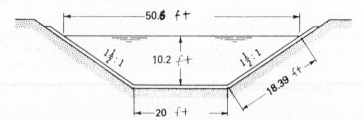

Fig. 14-30

14.91 If the flow in the canal of Prob. 14.90 were to decrease to 800 cfs, all other data, including the slope, being the same, what would be the depth of water?

▮
$$Q = (A)(1.486/n)(R^{2/3})(s^{1/2}) \qquad A = y[20 + (2)(1.5)(y) + 20]/2 = 20y + 1.5y^2$$
$$p_w = 20 + (2)[\sqrt{y^2 + (1.5y)^2}] = 20 + 3.606y$$
$$800 = (20y + 1.5y^2)(1.486/0.014)[(20y + 1.5y^2)/(20 + 3.606y)]^{2/3}(0.000149)^{1/2}$$
$$y = 7.10 \text{ ft} \qquad \text{(by trial and error)}$$

14.92 In Prob. 14.90 find the corresponding value of ϵ and compare it with values previously given for concrete pipe. Does it fall in the range given?

▮ $1/\sqrt{f} = 2\log(14.8R/\epsilon)$, $n = 0.093f^{1/2}R^{1/6}$. Therefore,

$$2\log(14.8R/\epsilon) = 0.093R^{1/6}/n \qquad R = 360.1/56.78 = 6.342 \text{ ft} \qquad \text{(from Prob. 14.90)}$$
$$2\log[(14.8)(6.342)/\epsilon] = (0.093)(6.342)^{1/6}/0.014 \qquad \epsilon = 0.00284 \text{ ft}$$

Table A-9 gives $\epsilon = 0.001$ ft to 0.01 ft for concrete; hence, the computed value of ϵ does fall within the given range.

14.93 What would be the capacity of the canal of Prob. 14.90 if the grade were to be 1.2 ft/mile?

▮ $Q = (A)(1.486/n)(R^{2/3})(s^{1/2})$. From Prob. 14.90, $A = 360.1$ ft², $n = 0.014$, $R = 360.1/56.78 = 6.342$ ft, $Q = (360.1)(1.486/0.014)(6.342)^{2/3}(1.2/5280)^{1/2} = 1974$ ft³/s.

14.94 Water flows uniformly in a 2-m-wide rectangular channel at a depth of 45 cm. The channel slope is 0.002 and $n = 0.014$. Find the flow rate.

▮
$$Q = (A)(1.0/n)(R^{2/3})(s^{1/2}) \qquad R = (2)(\tfrac{45}{100})/(\tfrac{45}{100} + 2 + \tfrac{45}{100}) = 0.3103 \text{ m}$$
$$Q = [(2)(\tfrac{45}{100})](1.0/0.014)(0.3103)^{2/3}(0.002)^{1/2} = 1.32 \text{ m}^3/\text{s}$$

14.95 At what depth will water flow in a 3-m-wide rectangular channel if $n = 0.017$, $s = 0.00085$, and $Q = 4$ m³/s?

▮
$$Q = (A)(1.0/n)(R^{2/3})(s^{1/2}) \qquad 4 = (3d)(1.0/0.017)[3d/(d + 3 + d)]^{2/3}(0.00085)^{1/2}$$
$$d = 1.07 \text{ m} \qquad \text{(by trial and error)}$$

14.96 Figure 14-31 shows a tunnel section on the Colorado River Aqueduct. The area of the water cross section is 191 ft², and the wetted perimeter is 39.1 ft. The flow is 1600 cfs. If $n = 0.013$ for its concrete lining, find the slope.

$$Q = (A)(1.486/n)(R^{2/3})(s^{1/2}) \qquad 1600 = (191)(1.486/0.013)(191/39.1)^{2/3}(s)^{1/2}$$
$$s = 0.000648 \quad \text{or} \quad 3.42 \text{ ft/mile}$$

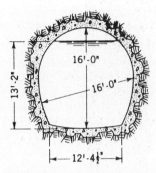

16'-0"

16'-0"

13'-2"

12'-4½"

Fig. 14-31

14.97 A monolithic concrete inverted siphon on the Colorado River Aqueduct is circular in cross section and is 16 ft in diameter. Obviously, it is completely filled with water, unlike the case of Prob. 14.96. If $n = 0.013$, find the slope of the hydraulic grade line for a flow of 1600 cfs using the Manning equation.

$$Q = (A)(1.486/n)(R^{2/3})(s^{1/2}) \qquad 1600 = [(\pi)(16)^2/4](1.486/0.013)(\tfrac{16}{4})^{2/3}(s)^{1/2}$$
$$s = 0.000763 \quad \text{or} \quad 4.03 \text{ ft/mile}$$

14.98 A 30-in-diameter pipe is known to have a Manning's n of 0.021. What is Manning's n for a 96-in-diameter pipe, having exactly the same ϵ as the smaller pipe?

$1/\sqrt{f} = 2 \log (14.8R/\epsilon)$, $n = 0.093f^{1/2}R^{1/6}$. Therefore, $2 \log (14.8R/\epsilon) = 0.093R^{1/6}/n$. For 30-in pipe: $R = (\tfrac{30}{12})/4 = 0.6250$ ft, $2 \log [(14.8)(0.6250)/\epsilon] = (0.093)(0.6250)^{1/6}/0.021$, $\epsilon = 0.08293$ ft. For 96-in pipe: $R = (\tfrac{96}{12})/4 = 2.000$ ft, $2 \log [(14.8)(2.000)/0.08293] = (0.093)(2.000)^{1/6}/n$, $n = 0.0204$.

14.99 For the channel shown in Fig. 14-32, if $a = 2$ m, $b = 5$ m, $d = 3$ m, $w = 25$ m, and $n = 0.014$, what slope is required so that the flow will be 30 m³/s when the depth of flow is 2.50 m?

$$Q = (A)(1.0/n)(R^{2/3})(s^{1/2}) \qquad A = (2.50)(5) + (2.50 - 2)[(25 - 5)/2]/2 = 15.00 \text{ m}^2$$
$$p_w = 2.50 + 5 + 2 + \sqrt{(2.50 - 2)^2 + [(25 - 5)/2]^2} = 19.51 \text{ m}$$
$$30 = (15.00)(1.0/0.014)(15.00/19.51)^{2/3}(s)^{1/2} \qquad s = 0.00111$$

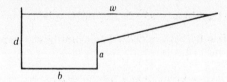

w

d

a

b

Fig. 14-32

14.100 Water flows in a rectangular flume 5 ft wide made of unplaned timber ($n = 0.013$). Find the necessary channel slope if the water flows uniformly at a depth of 2 ft and at 15 fps.

$$v = (1.486/n)(R^{2/3})(s^{1/2}) \qquad 15 = (1.486/0.013)[(5)(2)/(2 + 5 + 2)]^{2/3}(s)^{1/2} \qquad s = 0.0150$$

14.101 In Fig. 14-33, suppose the widths of A_1, A_2, and A_3 are 100 ft, 30 ft, and 200 ft and the total depths are 2 ft, 10 ft, and 3 ft, respectively. Compute the flow rate if $s = 0.0016$, $n_1 = n_3 = 0.04$, and $n_2 = 0.025$.

$$Q = (1.486/n)(R^{2/3})(s^{1/2})$$
$$Q_1 = [(100)(2)](1.486/0.04)[(100)(2)/(2 + 100)]^{2/3}(0.0016)^{1/2} = 466 \text{ ft}^3/\text{s}$$
$$Q_2 = [(30)(10)](1.486/0.025)\{(30)(10)/[(10 - 2) + 30 + (10 - 3)]\}^{2/3}(0.0016)^{1/2} = 2527 \text{ ft}^3/\text{s}$$
$$Q_3 = [(200)(3)](1.486/0.04)[(200)(3)/(3 + 200)]^{2/3}(0.0016)^{1/2} = 1836 \text{ ft}^3/\text{s}$$
$$Q = Q_1 + Q_2 + Q_3 = 466 + 2527 + 1836 = 4829 \text{ ft}^3/\text{s}$$

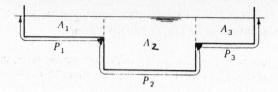

Fig. 14-33

14.102 Refer to Fig. 14-32. Find the flow rate at water depths of 1, 2, 3, 4, and 5 ft if $n = 0.020$ and $s = 0.0015$. The dimensions are as follows: $a = 3$ ft, $b = 6$ ft, $d = 5$ ft, and $w = 36$ ft.

∎ $Q = (1.486/n)(R^{2/3})(s^{1/2})$. For depth = 1 ft: $Q = [(1)(6)](1.486/0.020)[(1)(6)/(1 + 6 + 1)]^{2/3}(0.0015)^{1/2} = 14.3$ ft³/s. For depth = 2 ft: $Q = [(2)(6)](1.486/0.020)[(2)(6)/(2 + 6 + 2)]^{2/3}(0.0015)^{1/2} = 39.0$ ft³/s. For depth = 3 ft: $Q = [(3)(6)](1.486/0.020)[(3)(6)/(3 + 6 + 3)]^{2/3}(0.0015)^{1/2} = 67.9$ ft³/s. For depth = 4 ft: $A_1 = (6)(4) = 24.00$ ft², $A_2 = (4 - 3)[(36 - 6)/2]/2 = 7.50$ ft², $(p_w)_1 = 4 + 6 + 3 = 13.00$ ft, $(p_w)_2 = \sqrt{(4 - 3)^2 + [(36 - 6)/2]^2} = 15.03$ ft, $Q = (24.00)(1.486/0.020)(24.00/13.00)^{2/3}(0.0015)^{1/2} + (7.50)(1.486/0.020)(7.50/15.03)^{2/3}(0.0015)^{1/2} = 118$ ft³/s. For depth = 5 ft: $A_1 = (6)(5) = 30.00$ ft², $A_2 = (5 - 3)[(36 - 6)]/2 = 30.00$ ft², $(p_w)_1 = 5 + 6 + 3 = 14.00$ ft, $(p_w)_2 = \sqrt{(5 - 3)^2 + [(36 - 6)]^2} = 30.07$ ft, $Q = (30.00)(1.486/0.020)(30.00/14.00)^{2/3}(0.0015)^{1/2} + (30.00)(1.486/0.020)(30.00/30.07)^{2/3}(0.0015)^{1/2} = 230$ ft³/s.

14.103 In Fig. 14-34, with uniform flow in the wide open channel, $a = 2.80$ ft. Find b if $n = 0.020$.

∎ $v = (1.486/n)(R^{2/3})(s^{1/2})$. Working with a 1-ft width of channel,

$$v = (1.486/0.020)[(2 + 4 + 3)(1)/1]^{2/3}(s)^{1/2} \qquad s = 0.000009676 v^2 \qquad u_a^2/2g = 2.80 \text{ ft}$$

$$u_a = \sqrt{(2)(32.2)(2.80)} = 13.43 \text{ ft/s} \qquad u = v + (1/K)(\sqrt{gy_0 s})[1 + 2.3 \log (y/y_0)]$$

$$13.43 = v + (1/0.40)[\sqrt{(32.2)(2 + 4 + 3)(0.000009676 v^2)}][1 + 2.3 \log (2 + 4)/(2 + 4 + 3)] \qquad v = 12.45 \text{ ft/s}$$

$$u_b = 12.45 + (1/0.40)[\sqrt{(32.2)(2 + 4 + 3)(0.000009676)(12.45^2)}]\{1 + 2.3 \log [2/(4 + 3 + 2)]\} = 11.62 \text{ ft/s}$$

$$b = 11.62^2/[(2)(32.2)] = 2.10 \text{ ft}$$

Note: K in the equation above is the von Karman constant; it has a value of about 0.40 for clear water.

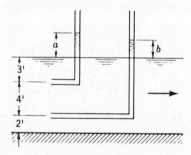

Fig. 14-34

14.104 Determine the depth below the surface (clear water) at which the velocity (u) is equal to the mean velocity (v). Also find the average of the velocities at 0.2 and 0.8 depths. Let $y_0 = 4$ ft, $s = 0.001$, and $n = 0.025$.

∎ $u = v + (1/K)(\sqrt{gy_0 s})[1 + 2.3 \log (y/y_0)]$. Where $u = v$, $v = v + (1/K)(\sqrt{gy_0 s})[1 + 2.3 \log (y/y_0)]$, $y/y_0 = 0.367$. Hence, the velocity (u) is equal to the mean velocity (v) when $y/y_0 = 0.367$, or at a depth below the surface of $1 - 0.367$, or 0.633 times the channel depth: $v = (1.486/n)(R^{2/3})(s^{1/2})$. Using a unit segment (width) of channel, $v = (1.486/0.025)[(1)(4)/1]^{2/3}(0.001)^{1/2} = 4.736$ ft/s. At 0.2 depth, $y = (0.8)(4) = 3.2$ ft; hence, $u = 4.736 + (1/0.40)[\sqrt{(32.2)(4)(0.001)}][1 + 2.3 \log (3.2/4)] = 5.433$ ft/s. At 0.8 depth, $y = (0.2)(4) = 0.8$ ft; hence, $u = 4.736 + (1/0.40)[\sqrt{(32.2)(4)(0.001)}][1 + 2.3 \log (0.8/4)] = 4.191$ ft/s, $u_{avg} = (5.433 + 4.191)/2 = 4.81$ ft/s. Note that u_{avg} is nearly equal to v.

14.105 Water flows uniformly in a very wide rectangular channel at a depth of 1.5 m $(s = 0.006$ and $n = 0.015)$. Calculate the velocities at values of 0.15, 0.3, 0.6, 0.9, and 1.5 m. Note the value of maximum velocity at the

∎ $v = (1.0/n)(R^{2/3})(s^{1/2})$. For a very wide channel, $R = y_0 = 1.5$ m: $v = (1.0/0.015)(1.5)^{2/3}(0.006)^{1/2} = 6.767$ m/s, $u = v + (1/K)(\sqrt{gy_0 s})[1 + 2.3 \log (y/y_0)]$. At $y = 0.15$ m, $u = 6.767 + (1/0.40)[\sqrt{(9.807)(1.5)(0.006)}][1 + 2.3 \log (0.15/1.5)] = 5.80$ m/s. At $y = 0.3$ m, $u = 6.767 + (1/0.40)[\sqrt{(9.807)(1.5)(0.006)}][1 + 2.3 \log (0.3/1.5)] = 6.32$ m/s. At $y = 0.6$ m, $u = 6.767 + (1/0.40)[\sqrt{(9.807)(1.5)(0.006)}][1 + 2.3 \log (0.6/1.5)] = 6.83$ m/s. At $y = $

0.9 m, $u = 6.767 + (1/0.40)[\sqrt{(9.807)(1.5)(0.006)}][1 + 2.3\log(0.9/1.5)] = 7.13$ m/s. At $y = 1.5$ m, $u = 6.767 + (1/0.40)[\sqrt{(9.807)(1.5)(0.006)}][1 + 2.3\log(1.5/1.5)] = 7.51$ m/s.

14.106 Consider a variety of rectangular sections all of which have a cross-sectional area of 20 ft². Tabulate hydraulic radii versus channel widths for a range of channel widths from 2 ft to 20 ft and note the depth:width ratio when R is maximum.

❚ $R = A/p_w = wd/(w + 2d)$. For $w = 2$ ft, $d = \frac{20}{2} = 10.00$ ft, $R = (2)(10.00)/[2 + (2)(10.00)] = 0.909$ ft. By similar computation,

w, ft	d, ft	R, ft
2	10.000	0.909
4	5.000	1.429
6	3.333	1.579
8	2.500	1.538
10	2.000	1.429
15	1.333	1.132
20	1.000	0.909

R_{max} appears from the table above to occur at a depth:width ratio around 3.333/6, or 0.556. In actuality, it occurs at a ratio of 0.5 (in this case, $w = 6.324$ ft and $d = 3.162$ ft).

14.107 Set up a general expression for the wetted perimeter p_w of a trapezoidal channel in terms of the cross-sectional area A, depth y, and angle of side slope ϕ. Then differentiate p_w with respect to y with A and ϕ held constant. From this, prove that $R = y/2$ for the section of greatest hydraulic efficiency (i.e., smallest p_w for a given A).

❚ Let B = bottom width.

$$A = By + (y)(y\tan\phi) = By + y^2\tan\phi \qquad B = A/y - y\tan\phi$$

$$p_w = B + 2y\sec\phi = A/y - y\tan\phi + 2y\sec\phi$$

$$dp_w/dy = -A/y^2 - \tan\phi + 2\sec\phi = -(By + y^2\tan\phi)/y^2 - \tan\phi + 2\sec\phi$$

Setting $dp_w/dy = 0$, $(By + y^2\tan\phi)/y^2 = 2\sec\phi - \tan\phi$, $B = 2y\sec\phi - 2y\tan\phi = (2y)(\sec\phi - \tan\phi)$.

$$R = \frac{A}{p_w} = \frac{By + y^2\tan\phi}{B + 2y\sec\phi} = \frac{(2y)(\sec\phi - \tan\phi)(y) + y^2\tan\phi}{(2y)(\sec\phi - \tan\phi) + 2y\sec\phi} = \frac{y}{2}$$

14.108 Prove that the most efficient triangular section is the one with a 90° vertex angle.

❚ See Fig. 14-35.

$$A = a^2\sin\phi\cos\phi \qquad p_w = 2a \qquad R = A/p_w$$

$$R = (a^2\sin\phi\cos\phi)/2a = (a/2)(\sin\phi\cos\phi) \qquad dR/d\phi = (a/2)(\cos^2\phi - \sin^2\phi) = 0$$

Hence, $\cos\phi = \sin\phi$; or $\phi = 45°$ and the vertex angle $= (2)(45)$, or 90°.

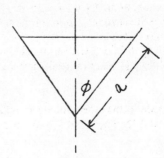

Fig. 14-35

14.109 The amount of water to be carried by a canal excavated in smooth earth ($n = 0.030$) is 10 m³/s. It has side slopes of 2:1 and the depth of water y is to be 1.5 m or less (see Fig. 14-36). If the channel slope is 45 cm/km, what must be the width at the bottom?

$$Q = (A)(1.0/n)(R^{2/3})(s^{1/2}) \qquad A = 1.5[b + b + (2)(1.5) + (2)(1.5)]/2 = 1.5b + 4.5$$

$$p_w = b + (2)\sqrt{1.5^2 + [(2)(1.5)]^2} = b + 6.708 \qquad s = (\tfrac{45}{100})/1000 = 0.000450$$

$$10 = (1.5b + 4.5)(1.0/0.030)[(1.5b + 4.5)/(b + 6.708)]^{2/3}(0.000450)^{1/2}$$

$$b = 6.05 \text{ m} \qquad \text{(by trial and error)}$$

Fig. 14-36

14.110 Refer to Fig. 14-36. If the discharge in the canal ($n = 0.030$) is to be 5 m³/s while the depth is 1.5 m and the velocity is not to exceed 45 m/min, what must be the width at the bottom and the drop in elevation per kilometer?

▮ $\qquad A = Q/v = 5/(\tfrac{45}{60}) = 6.667 \text{ ft}^2 \qquad 1.5[b + b + (2)(1.5) + (2)(1.5)]/2 = 6.667 \qquad b = 1.445 \text{ m}$

$$v = (1.0/n)(R^{2/3})(s^{1/2}) \qquad p_w = 1.445 + (2)\sqrt{1.445^2 + [(2)(1.445)]^2} = 7.907 \text{ m}$$

$$\tfrac{45}{60} = (1.0/0.030)(6.667/7.907)^{2/3}(s)^{1/2} \qquad s = 0.000636 \quad \text{or} \quad 0.636 \text{ m/km}$$

14.111 A rectangular flume of planed timber ($n = 0.012$) slopes 1 ft per 1000 ft. (**a**) Compute the rate of discharge if the width is 6 ft and the depth of water is 3 ft. (**b**) What would be the rate of discharge if the width were 3 ft and the depth of water 6 ft? (**c**) Which of the two forms would have the greater capacity and which would require less lumber?

▮ $$Q = (A)(1.486/n)(R^{2/3})(s^{1/2})$$

(**a**) $\qquad Q = [(6)(3)](1.468/0.012)[(6)(3)/(3 + 6 + 3)]^{2/3}(\tfrac{1}{1000})^{1/2} = 92.4 \text{ ft}^3/\text{s}$

(**b**) $\qquad Q = [(6)(3)](1.468/0.012)[(6)(3)/(6 + 3 + 6)]^{2/3}(\tfrac{1}{1000})^{1/2} = 79.6 \text{ ft}^3/\text{s}$

(**c**) $\qquad$ Lumber ratio $= (3 + 6 + 3)/(6 + 3 + 6) = 0.80 \qquad$ Flow ratio $= 92.4/79.6 = 1.16$

Hence, the first design provides 16 percent more flow capacity while requiring only 80 percent as much lumber.

14.112 What diameter of semicircular channel will have the same capacity as a rectangular channel of width 8 ft and depth 3 ft? Assume s and n are the same for both channels. Compare the lengths of the wetted perimeters.

▮ $Q = (A)(1.486/n)(R^{2/3})(s^{1/2})$. Since $Q_s = Q_r$ and $(1.486/n)(s^{1/2})$ is constant, $A_s R_s^{2/3} = A_r R_r^{2/3}$.

$$[(\pi d^2/4)/2](d/4)^{2/3} = [(8)(3)][(8)(3)/(3 + 8 + 3)]^{2/3} \qquad d = 7.57 \text{ ft}$$

$$(p_w)_s = (\pi)(7.57)/2 = 11.89 \text{ ft} \qquad (p_w)_r = 3 + 8 + 3 = 14.00 \text{ ft}$$

The semicircular channel has the smaller wetted perimeter.

14.113 Water flows uniformly in a circular concrete pipe ($n = 0.014$) of diameter 10 ft at a depth of 4 ft. Using Fig. A-18, determine the flow rate and the average velocity of flow ($s = 0.0003$).

▮ $\qquad Q = (A)(1.486/n)(R^{2/3})(s^{1/2}) \qquad Q_{\text{full}} = [(\pi)(10)^2/4](1.486/0.014)(\tfrac{10}{4})^{2/3}(0.0003)^{1/2} = 266 \text{ ft}^3/\text{s}$

$$v_{\text{full}} = Q_{\text{full}}/A_{\text{full}} = 266/[(\pi)(10)^2/4] = 3.39 \text{ ft/s} \qquad d/d_{\text{full}} = \tfrac{4}{10} = 0.40 \quad \text{or} \quad 40 \text{ percent}$$

From Fig. A-18, $Q/Q_{\text{full}} = 27$ percent and $v/v_{\text{full}} = 70$ percent: $Q = (0.27)(266) = 71.8 \text{ ft}^3/\text{s}$, $v = (0.70)(3.39) = 2.37 \text{ ft/s}$.

14.114 At what depth will water flow at 0.25 m³/s in a 100-cm-diameter concrete pipe on a slope of 0.004? (**a**) Assume $n = 0.013$. (**b**) Repeat with $n = 0.015$ and compare results.

▮ $$Q = (A)(1.0/n)(R^{2/3})(s^{1/2})$$

(**a**) $\qquad Q_{\text{full}} = [(\pi)(\tfrac{100}{100})^2/4](1.0/0.013)[(\tfrac{100}{100})/4]^{2/3}(0.004)^{1/2} = 1.516 \text{ m}^3/\text{s}$

$Q/Q_{\text{full}} = 0.25/1.516 = 0.16$, or 16 percent. From Fig. A-18, $d/d_{\text{full}} = 30$ percent: $d = (0.30)(100) = 30.0$ cm.

(**b**) $\qquad [(\pi)(\tfrac{100}{100})^2/4](1.0/0.015)[(\tfrac{100}{100})/4]^{2/3}(0.004)^{1/2} = 1.314 \text{ m}^3/\text{s}$

$Q/Q_{\text{full}} = 0.25/1.314 = 0.19$, or 19 percent. From Fig. A-18, $d/d_{\text{full}} = 33$ percent: $d = (0.33)(100) = 33.0$ cm. Increasing the n value increased the depth.

14.115 Evaluate the friction factor f for laminar flow in terms of the Reynolds number, and compare with the equation for pipe flow. (*Note:* Recall that for a wide channel the hydraulic radius is approximately equal to the depth.)

▮ $\quad N_R = (4R)(v)/v = 4y_0v/v \qquad q = (g/v)(y_0^3/3)(s) \qquad v = Q/A = q/y_0 = gsy_0^2/3v = [(8g/f)(R)(s)]^{1/2}$

Therefore, $[(8g/f)(R)(s)] = vgsy_0^2/3v$. With $R = y_0$, $f = 24v/y_0v = 96v/4y_0v = 96/N_R$. This compares with $f = 64/N_R$ for pipe flow.

14.116 Eastern lubricating oil (SAE30) at 90 °F ($v = 0.0014$ ft²/s) flows down a flat plate 10 ft wide. What is the maximum rate of discharge at which laminar flow will be ensured, assuming that the critical Reynolds number is 500? What should be the slope of the plate to secure a depth of 6 in at this flow rate?

▮ $\quad N_R = Rv/v = y_0q/vy_0 = q/v \qquad q = (500)(0.0014) = 0.7000$ ft³/s per ft $\qquad Q = (10)(0.7000) = 7.00$ ft³/s

$\quad q = (g/v)(y_0^3/3)(s) \qquad 0.7000 = (32.2/0.0014)[(\frac{6}{12})^3/3](s) \qquad s = 0.000730$

14.117 At what rate will water at 60 °F flow in a wide rectangular channel on a slope of 0.00015 if the depth is 0.01 ft? Assume laminar flow and justify that assumption by computing the Reynolds number.

▮ $\quad$ Assuming laminar flow, $q = (g/v)(y_0^3/3)(s) = [32.2/(1.21 \times 10^{-5})](0.01^3/3)(0.00015) = 1.33 \times 10^{-4}$ ft³/s/ft, $N_R = Rv/v = y_0v/v = (y_0)(q/y_0)/v = q/v = 1.33 \times 10^{-4}/(1.21 \times 10^{-5}) = 11$. Since $N_R < 500$, the assumption of laminar flow is justified.

14.118 At what rate will water at 15 °C flow in a wide, smooth, rectangular channel on a slope of 0.0003, if the depth is 8.0 mm? Assume laminar flow and justify that assumption by computing the Reynolds number.

▮ $\quad$ Assuming laminar flow, $q = (g/v)(y_0^3/3)(s) = [9.807/(1.16 \times 10^{-6})][(\frac{8}{1000})^3/3](0.0003) = 4.33 \times 10^{-4}$ (m³/s)/m, or 0.433 (L/s)/m; $N_R = Rv/v = y_0v/v = (y_0)(q/y_0)/v = q/v = 4.33 \times 10^{-4}/(1.16 \times 10^{-6}) = 373$ (laminar).

14.119 Water flows with a velocity of 4 fps and at a depth of 2 ft in a wide rectangular channel. Is the flow subcritical or supercritical? Find the alternate depth for the same discharge and specific energy.

▮ $\qquad\qquad v^2/2g = 4^2/[(2)(32.2)] = 0.2484$ ft $\qquad y/2 = \frac{2}{2} = 1.000$ ft

Since $0.2484 < 1.000$, the flow is subcritical.

$\qquad E = y + (1/2g)(q^2/y^2) = 2 + 0.2484 = 2.2484$ ft $\qquad q = (4)(2) = 8.000$ (ft²/s)/ft

$\qquad 2.2484 = y + \{1/[(2)(32.2)]\}(8.00^2/y^2) \qquad 2.2484y^2 - y^3 - 0.9938 = 0$

Since $y = 2$ is one known solution, divide by $(y - 2)$ to yield $y^2 - 0.2484y - 0.4968 = 0$, $y = [-(-0.2484) \pm \sqrt{(-0.2484)^2 - (4)(1)(-0.4968)}]/[(2)(1)] = 0.840$ ft.

14.120 Water flows down a wide rectangular channel of concrete ($n = 0.014$) laid on a slope of 0.002. Find the depth and rate of flow in SI units for critical conditions in this channel.

▮ $\qquad v = (1.0/n)(R^{2/3})(s^{1/2}) \qquad v_c = \sqrt{gy_c} \qquad \sqrt{(9.807)(y_c)} = (1.0/0.014)(y_c)^{2/3}(0.002)^{1/2}$

$\qquad y_c = 0.888$ m $\qquad q = y_cv_c = (y_c)(\sqrt{gy_c}) = (0.888)[\sqrt{(9.807)(0.888)}] = 2.62$ (m³/s)/m

14.121 Water is released from a sluice gate in a rectangular channel 5 ft wide such that the depth is 2 ft and the velocity is 15 fps. Find (**a**) the critical depth for this specific energy, (**b**) the critical depth for this rate of discharge, and (**c**) the type of flow and the alternate depth.

▮ (**a**) $\qquad E = y + v^2/2g = 2 + 15^2/[(2)(32.2)] = 5.494$ ft $\qquad y_c = (\frac{2}{3})(E) = (\frac{2}{3})(5.494) = 3.66$ ft

$\qquad$ (**b**) $\qquad q = yv = (2)(15) = 30.00$ ft³/s/ft $\qquad y_c = (q^2/g)^{1/3} = (30.00^2/32.2)^{1/3} = 3.03$ ft

$\qquad$ (**c**) Since $[y_0 = 2] < [y_c = 3.03]$, the flow is supercritical: $E = y + (1/2g)(q^2/y^2)$, $5.494 = y + \{1/[(2)(32.2)]\}(30.00^2/y^2)$, $5.494y^2 - y^3 - 13.98 = 0$. Since $y = 2$ is one known solution, divide by $(y - 2)$ to yield $y^2 - 3.494y - 6.988 = 0$, $y = [-(-3.494) \pm \sqrt{(-3.494)^2 - (4)(1)(-6.988)}]/[(2)(1)] = 4.92$ ft.

14.122 A flow of 100 cfs is carried in a rectangular channel 10 ft wide at a depth of 1.2 ft. If the channel is made of smooth concrete ($n = 0.012$), find the slope necessary to sustain uniform flow at this depth. What roughness coefficient would be required to produce uniform critical flow for the given rate of discharge on this slope?

▮ $\quad Q = (A)(1.486/n)(R^{2/3})(s^{1/2}) \qquad 100 = [(1.2)(10)][(1.486/0.012)][(1.2)(10)/(1.2 + 10 + 1.2)]^{2/3}(s)^{1/2}$

$$s = 0.00473 \qquad y_c = (q^2/g)^{1/3} \qquad y_c = [(\tfrac{100}{10})^2/32.2]^{1/3} = 1.459 \text{ ft}$$

$$v_c = \sqrt{gy_c} = \sqrt{(32.2)(1.459)} = 6.854 \text{ ft/s}$$

$$6.854 = (1.486/n)[(1.459)(10)/(1.459 + 10 + 1.459)]^{2/3}(0.00473)^{1/2} \qquad n = 0.0162$$

14.123 A long straight rectangular channel 3 m wide is observed to have a wavy surface at a depth of about 2 m. Estimate the rate of discharge.

▮ "Wavy surface" means the depth is near critical; hence y_c is approximately 2 m. $q_{max} = \sqrt{gy_c^3} = \sqrt{(9.807)(2)^3} = 8.858 \text{ m}^3/\text{s/m}$, $Q = (3)(8.858) = 26.6 \text{ m}^3/\text{s}$.

14.124 A thin rod is placed vertically in a stream which is 3 ft deep, and the resulting small disturbance wave is observed to make an angle of about 55° with the axis of the stream. Find the approximate velocity of the stream.

▮ $\quad \sin \beta = c/v \qquad c = \sqrt{gy} = \sqrt{(32.2)(3)} = 9.829 \text{ ft/s} \qquad \sin 55° = 9.829/v \qquad v = 12.0 \text{ ft/s}$

14.125 At a point in a shallow lake, the wave from a passing boat is observed to rise 1 ft above the undisturbed water surface. The observed speed of the wave is 10 mph. Find the approximate depth of the lake at this point.

▮ $\quad c = \sqrt{(g)(y + \Delta y)[(y + \Delta y/2)/y]} = (10)(5280)/3600 = 14.67 \text{ ft/s} \qquad 14.67 = \sqrt{(32.2)(y + 1)[(y + \tfrac{1}{2})/y]}$

$$215.2 = 32.2y + 48.30 + 16.1/y \qquad 32.2y^2 - 166.9y + 16.1 = 0$$

$$y = [-(-166.9) \pm \sqrt{(-166.9)^2 - (4)(32.2)(16.1)}]/[(2)(32.2)] = 5.08 \text{ ft}$$

(Another solution to the equation is 0.098 ft, but it is not a very practical solution to the problem.)

14.126 A rectangular channel 10 ft wide carries a flow of 200 cfs. Find the critical depth and critical velocity for this flow. Find also the critical slope if $n = 0.020$.

▮ $\quad y_c = (q^2/g)^{1/3} = [(\tfrac{200}{10})^2/32.2]^{1/3} = 2.32 \text{ ft} \qquad v_c = \sqrt{gy_c} = \sqrt{(32.2)(2.32)} = 8.64 \text{ ft/s}$

$$v = (1.486/n)(R^{2/3})(s^{1/2}) \qquad 8.64 = (1.486/0.020)[(2.32)(10)/(2.32 + 10 + 2.32)]^{2/3}(s)^{1/2} \qquad s = 0.00732$$

14.127 A flow of 10 cfs of water is carried in a 90° triangular flume built of planed timber ($n = 0.011$). Find the critical depth and critical slope.

▮ $Q^2/g = A^3/B$. If $y = $ depth, $A = y^2$ and $B = 2y$.

$$10^2/32.2 = (y^2)^3/2y \qquad y = y_c = 1.44 \text{ ft} \qquad Q = (A)(1.486/n)(R^{2/3})(s^{1/2})$$

$$R = A/p_w = 1.44^2/[(2)(\sqrt{1.44^2 + 1.44^2})] = 0.5091 \text{ ft} \qquad 10 = (1.44^2)(1.486/0.011)(0.5091)^{2/3}(s)^{1/2}$$

$$s = s_c = 0.00313$$

14.128 A trapezoidal canal with side slopes of 2:1 has a bottom width of 3 m and carries a flow of 20 m³/s. (a) Find the critical depth and critical velocity. (b) If the canal is lined with brick ($n = 0.015$), find the critical slope for the same rate of discharge.

▮ $Q^2/g = A^3/B$. If $y = $ depth, $A = 3y + 2y^2$ and $B = 3 + 4y$.

$$20^2/9.807 = (3y + 2y^2)^3/(3 + 4y) \qquad y = y_c = 1.25 \text{ m} \qquad \text{(by trial and error)}$$

$$Q = (A)(1.0/n)(R^{2/3})(s^{1/2}) \qquad R = A/p_w = [(3)(1.25) + (2)(1.25)^2]/\{3 + (2)\sqrt{1.25^2 + [(2)(1.25)]^2}\} = 0.8003 \text{ m}$$

$$20 = [(30)(1.25) + (2)(1.25)^2](1.0/0.015)(0.8003)^{2/3}(s)^{1/2} \qquad s = 0.00256$$

14.129 For a circular conduit with a diameter of 10 ft, compute the specific energy for a flow of 100 cfs at depths of 1, 3, 5, and 8 ft assuming $\alpha = 1.0$. At what depth is E least?

▮ $E = y + v^2/2g$. For $y = 1$ ft: $y/y_{full} = \tfrac{1}{10} = 0.10$, or 10 percent. From Fig. A-18, $A/A_{full} = 6$ percent. $A = [(\pi)(10)^2/4](0.06) = 4.71 \text{ ft}^2$, $v = Q/A = 100/4.71 = 21.2 \text{ ft/s}$, $E = 1 + 21.2^2/[(2)(32.2)] = 7.98 \text{ ft}$. For $y = 3$ ft:

$$y/y_{full} = \tfrac{3}{10} = 0.30 \quad \text{or} \quad 30 \text{ percent} \qquad A/A_{full} = 25 \text{ percent} \qquad A = [(\pi)(10)^2/4](0.25) = 19.6 \text{ ft}^2$$

$$v = Q/A = 100/19.6 = 5.10 \text{ ft/s} \qquad E = 3 + 5.10^2/[(2)(32.2)] = 3.40 \text{ ft}$$

For $y = 5$ ft:

$$y/y_{full} = \tfrac{5}{10} = 0.50 \quad \text{or} \quad 50 \text{ percent} \qquad A/A_{full} = 50 \text{ percent} \qquad A = [(\pi)(10)^2/4](0.50) = 39.3 \text{ ft}^2$$

$$v = Q/A = 100/39.3 = 2.54 \text{ ft/s} \qquad E = 5 + 2.54^2/[(2)(32.2)] = 5.10 \text{ ft}$$

For $y = 8$ ft:

$$y/y_{full} = \tfrac{8}{10} = 0.80 \quad \text{or} \quad 80 \text{ percent} \qquad A/A_{full} = 84 \text{ percent} \qquad A = [(\pi)(10)^2/4](0.84) = 66.0 \text{ ft}^2$$

$$v = Q/A = 100/66.0 = 1.52 \text{ ft/s} \qquad E = 8 + 1.52^2/[(2)(32.2)] = 8.04 \text{ ft}$$

By observing the above values, it appears that E_{min} occurs at a depth around 3 ft. Actually, it occurs at about 2.3 ft.

14.130 A circular conduit of well-laid (smooth) brickwork ($n = 0.012$) when flowing half full is to carry 400 cfs at a velocity of 10 fps. What is the necessary fall per mile? Will the flow be subcritical or supercritical?

$$\blacksquare \qquad A = Q/v = (\pi d^2/4)/2 \qquad \tfrac{400}{10} = (\pi d^2/4)/2 \qquad d = 10.09 \text{ ft} \qquad v = (1.486/n)(R^{2/3})(s^{1/2})$$

$$10 = (1.486/0.012)(10.09/4)^{2/3}(s)^{1/2} \qquad s = 0.001899 \quad \text{or} \quad 10.0 \text{ ft/mile}$$

$$A = [(\pi)(10.09)^2/4]/2 = 39.98 \text{ ft}^2 \qquad N_F = v/\sqrt{gy} = v/\sqrt{(g)(A/B)} = 10/\sqrt{(32.2)(39.98/10.09)} = 0.885$$

Since $N_F < 1.0$, the flow is subcritical.

14.131 Figure 14-37 describes the cross section of an open channel for which $s_0 = 0.02$ and $n = 0.015$. The sketch is drawn to the scale shown. When the flow rate is 100 cfs, find (a) the depth for uniform flow and (b) the critical depth.

$\blacksquare$ The cross-sectional area is found by planimetry and the wetted perimeter by use of dividers.

(a)

y_0, ft	A, ft	p_w, ft	R, ft	$Q = (A)(1.486/n)(R^{2/3})(s^{1/2})$, cfs
1	1.30	3.29	0.395	10
2	4.90	6.12	0.801	59
3	10.48	8.96	1.170	163

A plot of Q versus y_0 (not shown) indicates that $y_0 = 2.50$ ft for $Q = 100$ cfs.

(b)

y_c, ft	A, ft^2	$B(x_L + x_R)$, ft	$Q = (gA^3/B)^{1/2}$, cfs
1	1.30	2.60	5
2	4.90	4.60	29
3	10.48	6.55	75
4	18.30	9.10	147

A plot of Q versus y_c (not shown) indicates that $y_c = 3.35$ ft for $Q = 100$ cfs.

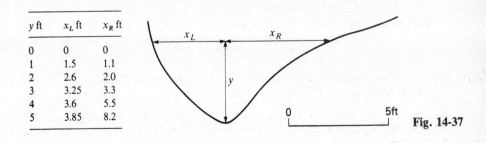

y ft	x_L ft	x_R ft
0	0	0
1	1.5	1.1
2	2.6	2.0
3	3.25	3.3
4	3.6	5.5
5	3.85	8.2

Fig. 14-37

14.132 Refer to Fig. 14-37 and replace feet dimensions with meters. Let the slope be 0.007 with $n = 0.015$. When the flow rate is 50 m^3/s, find (a) the depth for uniform flow and (b) the critical depth.

▮ The cross-sectional area is found by planimetry and the wetted perimeter by use of dividers.

(a)

y_0, m	A, m^2	p_w, m	R, m	$Q = (A)(1.0/n)(R^{2/3})(s^{1/2})$, m^3/s
1	1.30	3.29	0.395	4
2	4.90	6.12	0.801	24
3	10.48	8.96	1.170	65

A plot of Q versus y_0 (not shown) indicates that $y_0 = 2.70$ m for $Q = 50$ m^3/s.

(b)

y_c, m	A, m^2	$B(x_L + x_R)$, m	$Q = (gA^3/B)^{1/2}$, m^3/s
1	1.30	2.60	3
2	4.90	4.60	16
3	10.48	6.55	42
4	18.30	9.10	81

A plot of Q versus y_c (not shown) indicates that $y_c = 3.25$ m for $Q = 50$ m^3/s.

14.133 A rectangular channel 2 m wide carries 2 m^3/s of water in subcritical uniform flow at a depth of 0.80 m. A frictionless hump is to be installed across the bed. Find the critical hump height (i.e., the minimum hump that causes y_c on it).

▮ $$(\Delta z)_c = E_0 - E_{min} \qquad E = y + v^2/2g \qquad v = Q/A = 2/[(2)(0.80)] = 1.250 \text{ m/s}$$
$$E_0 = 0.80 + 1.250^2/[(2)(9.807)] = 0.8797 \text{ m}$$
$$E_{min} = (\tfrac{3}{2})(y_c) \qquad y_c = (q^2/g)^{1/3} = [(\tfrac{2}{2})^2/9.807]^{1/3} = 0.4672 \text{ m}$$
$$E_{min} = (\tfrac{3}{2})(0.4672) = 0.7008 \text{ m} \qquad (\Delta z)_c = 0.8798 - 0.7008 = 0.179 \text{ m}$$

14.134 Rework Prob. 14.86 ($y_0 = 2.0$ ft) for the case where the flow rate is 50 cfs.

▮ (a) $$q = Q/B = \tfrac{50}{4} = 12.50 \text{ (ft}^3\text{/s)/ft} \qquad y_c = (q^2/g)^{1/3} = (12.50^2/32.2)^{1/3} = 1.693 \text{ ft}$$

Since $y_0 > y_c$, the flow is subcritical (and slope is mild).

(b) Install hump with $\Delta z = 0.30$ ft. First find critical hump height:

$$E = y + (\tfrac{1}{2}g)(q^2/y^2) \qquad E_0 = 2.00 + \{1/[(2)(32.2)]\}(12.50^2/2.00^2) = 2.607 \text{ ft}$$
$$E_{min} = (\tfrac{3}{2})(y_c) = (\tfrac{3}{2})(1.693) = 2.540 \text{ ft} \qquad E_0 = \Delta z_{crit} + E_{min} \qquad 2.607 = \Delta z_{crit} + 2.540 \qquad \Delta z_{crit} = 0.067 \text{ ft}$$

Since $\Delta z > \Delta z_{crit}$, y_c occurs on the hump and damming action occurs. Hence, water depth on hump = 1.693 ft.

(c) Increase hump height to 0.60 ft. Still, $\Delta z > \Delta z_{crit}$, so y_c occurs on the hump. Let $y =$ upstream or downstream depth, with specific energy E.

$$E = \Delta z + E_{min} = y + (1/2g)(q^2/y^2) \qquad \Delta z + E_{min} = y + (1/2g)(q^2/y^2)$$
$$0.60 + 2.540 = y + \{1/[(2)(32.2)]\}(12.50^2/y^2) \qquad y = 2.84 \text{ ft} \quad \text{or} \quad 1.09 \text{ ft} \qquad \text{(by trial and error)}$$

This hump causes damming action, and the depths just upstream and downstream of the hump are 2.84 ft (subcritical) and 1.09 ft (supercritical), respectively.

14.135 A flow of 1.8 m^3/s is carried in a rectangular channel 1.6 m wide at a depth of 0.9 m. Will critical depth occur at a section where (a) a frictionless hump 15 cm high is installed across the bed, (b) a frictionless sidewall constriction (with no hump) reduces the channel width to 1.3 m, and (c) the hump and the sidewall constriction are installed together?

▮ (a) $$\Delta z_{crit} = E_0 - E_{min} \qquad E = y + (1/2g)(q^2/y^2) \qquad q = 1.8/1.6 = 1.125 \text{ (m}^3\text{/s)/m}$$
$$E_0 = 0.9 + \{1/[(2)(9.807)]\}(1.125^2/0.09^2) = 0.9797 \text{ m} \qquad E_{min} = (\tfrac{3}{2})(y_c) \qquad y_c = (q^2/g)^{1/3}$$
$$y_c = (1.125^2/9.807)^{1/3} = 0.5053 \text{ m} \qquad E_{min} = (\tfrac{3}{2})(0.5053) = 0.7580 \text{ m}$$
$$\Delta z_{crit} = 0.9797 - 0.7580 = 0.2217 \text{ m}$$
Since $[\Delta z = \tfrac{15}{100}] < [\Delta z_{crit} = 0.2217]$, y_c does not occur at the hump.

(b) $q = 1.8/1.3 = 1.385$ m $y_c = (1.385^2/9.807)^{1/3} = 0.5805$ m $E_{min} = (\tfrac{3}{2})(0.5805) = 0.8708$ m

Since $[E_{min} = 0.8708] < [E_0 = 0.9797]$, y_c does not occur at the constriction.

(c) $\Delta z_{crit} = E_0 - E_{min}$. With constriction, $E_{min} = 0.8708$ m; hence, $\Delta z_{crit} = 0.9797 - 0.8708 = 0.1089$ m. Since $[\Delta z = \tfrac{15}{100}] > [\Delta z_{crit} = 0.1089]$, y_c does occur at the hump with constriction.

14.136 Rework Prob. 14.86 ($y_0 = 2.0$ ft) for the case where the flow rate is 16 cfs.

$\blacksquare$ **(a)** $q = Q/B = \tfrac{16}{4} = 4.000$ ft^2/s/ft $y_c = (q^2/g)^{1/3} = (4.000^2/32.2)^{1/3} = 0.7921$ ft

Since $y_0 > y_c$, the flow is subcritical (and slope is mild).

(b) Install hump with $\Delta z = 0.30$ ft. First find critical hump height.

$$E = y + (\tfrac{1}{2g})(q^2/y^2)   E_0 = 2.00 + \{1/[(2)(32.2)]\}(4.000^2/2.00^2) = 2.062 \text{ ft}$$

$$E_{min} = (\tfrac{3}{2})(y_c) = (\tfrac{3}{2})(0.7921) = 1.188 \text{ ft}   E_0 = \Delta z_{crit} + E_{min}   2.062 = \Delta z_{crit} + 1.188   \Delta z_{crit} = 0.874 \text{ ft}$$

Since $\Delta z < \Delta z_{crit}$, y_c does not occur on the hump and damming action does not result.

$$E_h = E_0 - \Delta z   E = y + (1/2g)(q^2/y^2) = y_h + \{1/[(2)(32.2)]\}(4.000^2/y_h^2)$$

$$y_h + \{1/[(2)(32.2)]\}(4.000^2/y_h^2) = 2.062 - 0.30$$

By trial and error, $y_h = 1.67$ ft (subcritical) and 0.432 ft (supercritical). The flow does not pass through y_c, so it cannot become supercritical. Therefore, the water depth on the 0.30-ft hump is 1.67 ft.

(c) Increase hump height to 0.60 ft. Still, $\Delta z < \Delta z_{crit}$, so y_c does not occur on the hump and damming action does not result. $y_h + \{1/[(2)(32.2)]\}(4.000^2/y_h^2) = 2.062 - 0.60$. By trial and error $y_h = 1.32$ ft (subcritical) and 0.511 ft (supercritical). The flow does not pass through y_c, so it cannot become supercritical. Therefore, the water depth on the 0.60-ft hump is 1.32 ft.

14.137 A rectangular channel 40 ft wide carries 40 cfs of water in uniform flow at a depth of 2.80 ft. If a bridge pier 1 ft wide is placed in the middle of this channel, find the local change in the water-surface elevation. What is the minimum width of constricted channel that will not cause a rise in water surface upstream?

$\blacksquare$ $v_0 = Q/A = 40/[(2.80)(4)] = 3.571$ ft/s. At the constriction,

$$v = 40/[(4-1)(y)] = 13.33/y   E = y + v^2/2g = y + (13.33/y)^2/[(2)(32.2)] = y + 2.759/y^2$$

$$E_0 = 2.80 + 3.571^2/[(2)(32.2)] = 2.998 \text{ ft}   y + 2.759/y^2 = 2.998$$

By trial and error, $y = 2.59$ ft or 1.26 ft. Since supercritical conditions are impossible at the pier, $y = 2.59$ ft. Change in water depth $= 2.80 - 2.59 = 0.21$ ft drop. The minimum width of constricted channel that would produce critical flow is found by getting q_{max} for $E = 3.00$ ft:

$$y_c = (\tfrac{2}{3})(E) = (\tfrac{2}{3})(3.00) = 2.00 \text{ ft}   q_{max} = \sqrt{gy_c^3} = \sqrt{(32.2)(2.00)^3} = 16.05 \text{ ft}^3/\text{s/ft}$$

Minimum channel width $= 40/16.05 = 2.49$ ft. (A width less than this would produce backwater/damming action.)

14.138 A rectangular channel 10 ft wide carries 20 cfs in uniform flow at a depth of 0.90 ft. Find the local change in water-surface elevation caused by an obstruction 0.20 ft high on the floor of the channel.

$\blacksquare$ $y_c = (q^2/g)^{1/3} = [(\tfrac{20}{10})^2/32.2]^{1/3} = 0.499$ ft. Since $y_c < 0.90$, the flow is subcritical.

$$E = y + v^2/2g   v = Q/A = 20/[(0.90)(10)] = 2.222 \text{ ft/s}   E_1 = 0.9 + 2.222^2/[(2)(32.2)] = 0.9767 \text{ ft}$$

$$E_2 = E_1 - \Delta z = 0.9767 - 0.20 = 0.7767 \text{ ft}   E = y + (1/2g)(q^2/y^2)$$

$$E_2 = y_2 + \{1/[(2)(32.2)]\}[(\tfrac{20}{10})^2/y_2^2]   0.7767 = y_2 + 0.06211/y_2^2   y_2 = 0.610 \text{ ft} \quad \text{(by trial and error)}$$

Change in water-surface elevation $= 0.90 - (0.20 + 0.610) = 0.090$ ft (drop).

14.139 Suppose that the channel of Prob. 14.138 is so sloped that uniform flow of 20 cfs occurs at a depth of 0.30 ft. Find the local change in water-surface elevation caused by the 0.20-ft-high obstruction.

$\blacksquare$ $y_c = (q^2/g)^{1/3} = [(\tfrac{20}{10})^2/32.2]^{1/3} = 0.499$ ft. Since $y_c > 0.30$, the flow is supercritical.

$$E = y + v^2/2g   v = Q/A = 20/[(0.30)(10)] = 6.667 \text{ ft/s}   E_1 = 0.3 + 6.667^2/[(2)(32.2)] = 0.9902 \text{ ft}$$

$$E_2 = E_1 - \Delta z = 0.9902 - 0.20 = 0.7902 \text{ ft}   E = y + (1/2g)(q^2/y^2)   E_2 = y_2 + \{1/[(2)(32.2)]\}[(\tfrac{20}{10})^2/y_2^2]$$

$$0.7902 = y_2 + 0.06211/y_2^2   y_2 = 0.398 \text{ ft} \quad \text{(by trial and error)}$$

Change in water-surface elevation $= (0.20 + 0.398) - 0.30 = 0.298$ ft (rise).

14.140 Suppose that the depth of uniform flow of Prob. 14.137 is 0.90 ft. Find the change in water-surface elevation caused by the bridge pier. The flow rate is 40 cfs.

❚ $E = y + v^2/2g$ $v_0 = Q/A = 40/[(4)(0.90)] = 11.11$ ft/s $E_0 = 0.90 + 11.11^2/[(2)(32.2)] = 2.817$ ft

At constriction, $v = (\frac{40}{3})/y = 13.33/y$.

$$E = y + (13.33/y)^2/[(2)(32.2)] = y + 2.759/y^2 \qquad 2.817 = y + 2.759/y^2$$

$$y = 1.39 \text{ ft} \qquad \text{(by trial and error)}$$

Change in water-surface elevation $= 1.39 - 0.90 = 0.49$ ft (rise).

14.141 If 1.4 m³/s of water flows uniformly in a channel of width 1.8 m at a depth of 0.75 m, what is the change in water-surface elevation at a section contracted to a 1.2 m width with a 6-cm depression in the bottom?

❚ $y_c = (q^2/g)^{1/3}$ $(y_c)_2 = [(1.4/1.2)^2/9.807]^{1/3} = 0.5177$ m $E_2 = E_1 + \Delta z$ $E = y + v^2/2g$

$v_1 = Q/A = 1.4/[(0.75)(1.8)] = 1.037$ m/s $E_1 = 0.75 + 1.037^2/[(2)(9.807)] = 0.8048$ m

$$v_2 = 1.4/(1.2y_2) = 1.167/y_2$$

$$E_2 = y_2 + (1.167/y_2)^2/[(2)(9.807)] = y_2 + 0.06943/y_2^2 \qquad y_2 + 0.06943/y_2^2 = 0.8048 + \tfrac{6}{100}$$

$y_2 = 0.737$ m (subcritical) or 0.377 m (supercritical)   (by trial and error). y_2 cannot be less than $(y_c)_2$; hence, $y_2 = 0.737$ m. Change in water-surface elevation $= 0.75 - (0.737 - \tfrac{6}{100}) = 0.073$ m (drop).

14.142 A finished-concrete channel 8 ft wide has a slope of 0.5° and a water depth of 4 ft. Predict the uniform flow rate by (**a**) Manning's formula with $n = 0.012$ and (**b**) the friction-factor analysis with $\epsilon = 0.0032$ ft.

❚   (**a**) $R = A/p_w = (8)(4)/(4 + 8 + 4) = 2.000$ ft

$$Q = (A)(1.486/n)(R^{2/3})(s^{1/2}) = [(8)(4)](1.486/0.012)(2.000)^{2/3}(\tan 0.5°)^{1/2} = 588 \text{ ft}^3/\text{s}$$

(**b**) $C = (8g/f)^{1/2}$ $1/f^{1/2} = 2.0 \log(3.7 D_h/\epsilon)$ $D_h = 4R = (4)(2.000) = 8.000$ ft

$$1/f^{1/2} = (2.0)\log[(3.7)(8.000)/(0.0032)] \qquad f = 0.01589 \qquad C = [(8)(32.2)/0.01589]^{1/2} = 127.3$$

$$Q = CA(Rs)^{1/2} = (127.3)[(4)(8)][(2.000)(\tan 0.5°)]^{1/2} = 538 \text{ ft}^3/\text{s}$$

14.143 The asphalt-lined trapezoidal channel in Fig. 14-38 carries 300 cfs of water under uniform flow conditions when $s = 0.0015$. What is the normal depth? Use $n = 0.016$.

❚ $Q = (A)(1.486/n)(R^{2/3})(s^{1/2})$ $A = (\frac{1}{2})(6 + b_0)(y_n) = 6y_n + y_n^2 \cot 50°$ $p_w = 6 + 2y_n \csc 50°$

$$300 = (6y_n + y_n^2 \cot 50°)(1.486/0.016)[(6y_n + y_n^2 \cot 50°)/(6 + 2y_n \csc 50°)]^{2/3}(0.0015)^{1/2}$$

$$y_n = 4.58 \text{ ft} \qquad \text{(by trial and error)}$$

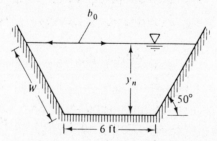

Fig. 14-38

14.144 What are the best dimensions for a rectangular brick ($n = 0.015$) channel designed to carry 5 m³/s of water in uniform flow with $s = 0.001$?

❚   The best dimensions are for width (b) equal to twice the depth (y): $Q = (A)(1.0/n)(R^{2/3})(s^{1/2})$,
$5 = [(y)(2y)](1.0/0.015)[(y)(2y)/(y + 2y + y)]^{2/3}(0.001)^{1/2}$, $y = 1.268$ m, $b = (2)(1.268) = 2.536$ m.

14.145 A wide rectangular clean-earth channel ($n = 0.022$) has a flow rate q of 50 cfs/ft (**a**) What is the critical depth? (**b**) What type of flow exists if $y = 3$ ft? (**c**) What is the critical slope?

❚   (**a**) $y_c = (q^2/g)^{1/3} = (50^2/32.2)^{1/3} = 4.27$ ft

(**b**) For $y < y_c$, the flow will be supercritical.

(**c**) $s_c = gn^2/(2.208y_c^{1/3}) = (32.2)(0.022)^2/[(2.208)(4.27)^{1/3}] = 0.00435$

14.146 The 50° triangular channel in Fig. 14-39 has a flow rate Q of 16 m³/s. Compute (**a**) y_c, (**b**) v_c, and (**c**) s_c if $n = 0.018$.

▌ (a)
$$gA^3 = b_0 Q^2 \qquad A = y^2 \cot 50° \qquad b_0 = 2y \cot 50°$$
$$(9.807)(y^2 \cot 50°)^3 = (2y \cot 50°)(16^2) \qquad y = y_c = 2.37 \text{ m}$$

(b)
$$v_c = Q/A = 16/[(2.37)^2(\cot 50)] = 3.39 \text{ m/s}$$

(c)
$$R = A/p_w = y^2 \cot 50°/(2y \csc 50°) = (2.37)^2(\cot 50°)/[(2)(2.37)(\csc 50°)] = 0.7617 \text{ m}$$
$$v = (1.0/n)(R^{2/3})(s^{1/2}) \qquad 3.39 = (1.0/0.018)(0.7617)^{2/3}(s)^{1/2} \qquad s = s_c = 0.00535$$

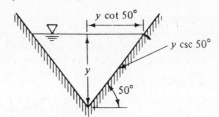

Fig. 14-39

14.147 The formula for shallow-water wave-propagation speed, $c_0 = \sqrt{gy}$, is independent of the physical properties of the liquid, i.e., density, viscosity, or surface tension. Does this mean that waves propagate at the same speed in water, mercury, gasoline, and glycerin? Explain.

▌ $c_0 = \sqrt{gy}$ is correct for any fluid except for viscosity and surface tension (very small wave) effects. It would be accurate for water, mercury, and gasoline but inaccurate for glycerin (too viscous).

14.148 A shallow-water wave 1 in (0.08333 ft) high propagates into still water of depth 2 ft. Compute the wave speed c and the velocity δv induced by the wave.

▌
$$c = \sqrt{gy(1 + \delta y/y)[1 + (\tfrac{1}{2})(\delta y/y)]} = \sqrt{(32.2)(2)(1 + 0.08333/2)[1 + (\tfrac{1}{2})(0.08333/2)]} = 8.28 \text{ ft/s}$$
$$\delta v = (c)(\delta y)/(y + \delta y) = (8.28)(0.08333)/(2 + 0.08333) = 0.331 \text{ ft/s}$$

14.149 Water flows rapidly in a flat wide channel 30 cm deep. Pebbles dropped successively in the water at the same spot create two circular ripples which are swept downstream as shown from above in Fig. 14-40. What is the current speed V?

▌ The centers of the circles move at current V; hence,
$$(x_0)(c_0) = 4V \qquad \text{(smaller circle)}$$
$$(x_0 + 4 + 6 + 9)(c_0) = 9V \qquad \text{(larger circles)}$$

Subtracting these equations, $19c_0 = 5V$, $V = 19c_0/5$, $c_0 = \sqrt{gy} = \sqrt{(9.807)(\tfrac{30}{100})} = 1.715$ m/s, $V = (19)(1.715)/5 = 6.52$ m/s.

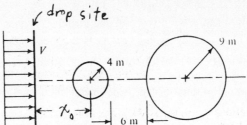

Fig. 14-40

14.150 Water flows in a wide flat channel 50 cm deep. Pebbles dropped successively at the same spot create two circular ripples as shown from above in Fig. 14-41. What is the current speed V?

▌ The centers of the circles move at current V. If the pebble drop site is at distance x_0 ahead of the small circle,
$$(x_0)(c_0) = 3V \qquad \text{(smaller circle)}$$
$$(x_0 + 4)(c_0) = 9V \qquad \text{(larger circle)}$$

Subtracting these equations, $4c_0 = 6V$, $V = 2c_0/3$, $c_0 = \sqrt{gy} = \sqrt{(9.807)(\tfrac{50}{100})} = 2.214$ m/s, $V = (2)(2.124)/3 = 1.48$ m/s.

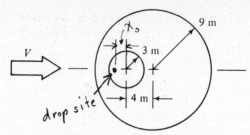

Fig. 14-41

14.151 Consider flow in a wide channel over a bump, as in Fig. 14-42. One can estimate the water-depth change or *transition* with frictionless flow. Use continuity and the Bernoulli equation to show that $dy/dx = -(dh/dx)/(1 - V^2/gy)$. Explain under what conditions the surface might rise above its upstream position y_0.

❚ From the Bernoulli equation, $p_0/\gamma + V_0^2/2g + y_0 = p_1/\gamma + V_1^2/2g + y_1 + h$, $0 + V_0^2/2g + y_0 = 0 + V_1^2/2g + y_1 + h$,

$$V\,dV/g + dy + dh = 0 \qquad (1)$$

From continuity, $V_0 y_0 = V_1 y_1 = \text{constant}$, $V\,dy + y\,dV = 0$, $dV = -V\,dy/y$. Substituting this value of dV into Eq. (1), $(V)(-V\,dy/y)/g + dy + dh = 0$, $dy/dx = -(dh/dx)/(1 - V^2/gy)$. If $dh/dx > 0$ (a bump) and $V^2 < gy$ (subcritical Froude number), dy/dx will be negative and the water level will drop across the bump. If $V^2 > gy$ (supercritical Froude number), the water level will rise.

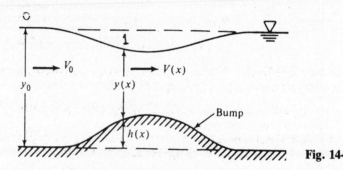

Fig. 14-42

14.152 In Fig. 14-42, suppose that $y_0 = 80$ cm and the flow rate is 1.2 m³/s per meter of channel width. If the bump has $h_{max} = 10$ cm, compute the velocity and water depth above the peak of the bump and comment about the Froude number effects and other details.

❚
$$p_0/\gamma + V_0^2/2g + y_0 = p_1/\gamma + V_1^2/2g + y_1 + h \qquad 0 + V_0^2/2g + y_0 = 0 + V_1^2/2g + y_1 + h_{max}$$
$$q = (V_0)(\tfrac{80}{100}) = V_1 y_1 = 1.2 \qquad V_0 = 1.500 \text{ m/s} \qquad y_1 = 1.2/V_1$$
$$1.500^2/[(2)(9.807)] + \tfrac{80}{100} = V_1^2/[(2)(9.807)] + 1.2/V_1 + \tfrac{10}{100} \qquad 0.05098V_1^3 - 0.8147V_1 + 1.2 = 0$$
$$V_1 = 1.91 \text{ m/s} \quad \text{(by trial and error)} \qquad y_1 = 1.2/1.91 = 0.628 \text{ m}$$
$$\text{Drop in water level} = \tfrac{80}{100} - 0.628 - \tfrac{10}{100} = 0.072 \text{ m} \qquad N_F = V/\sqrt{gy}$$

Over the bump, $N_F = 1.91/\sqrt{(9.807)(0.628)} = 0.770$ (subcritical).

14.153 Rework Prob. 14.152 for a flow rate of 3.6 m³/s per meter of channel width.

❚
$$p_0/\gamma + V_0^2/2g + y_0 = p_1/\gamma + V_1^2/2g + y_1 + h \qquad 0 + V_0^2/2g + y_0 = 0 + V_1^2/2g + y_1 + h$$
$$q = (V_0)(\tfrac{80}{100}) = V_1 y_1 = 3.6 \qquad V_0 = 4.500 \text{ m/s} \qquad y_1 = 3.6/V_1$$
$$4.500^2/[(2)(9.807)] + \tfrac{80}{100} = V_1^2/[(2)(9.807)] + 3.6/V_1 + \tfrac{10}{100} \qquad 0.05098V_1^3 - 1.732V_1 + 3.6 = 0$$
$$V_1 = 4.09 \text{ m/s} \quad \text{(by trial and error)} \qquad y_1 = 3.6/4.09 = 0.880 \text{ m}$$
$$\text{Rise in water level} = 0.880 - \tfrac{80}{100} + \tfrac{10}{100} = 0.180 \text{ m} \qquad N_F = V/\sqrt{gy}$$

Over bump, $N_F = 4.09/\sqrt{(9.807)(0.880)} = 1.39$ (supercritical).

14.154 Given the flow of a channel of large width b under a sluice gate, as shown in Fig. 14-43 and assuming frictionless steady flow with negligible upstream kinetic energy, derive a formula for the dimensionless ratio $Q^2/y_1^3 b^2 g$ as a function of the ratio y_2/y_1.

$\blacksquare$ From the Bernoulli equation, $y_1 + V_1^2/2g = y_2 + V_2^2/2g$. Assuming $V_1^2/2g$ to be negligible, $y_1 = y_2 + V_2^2/2g$. From continuity: $V_1 y_1 = V_2 y_2 = Q/b$, $V_2 = Q/by_2$, $y_1 = y_2 + (Q/by_2)^2/2g$. Multiplying by y_2^2/y_1^3, $y_2^2/y_1^2 = y_2^3/y_1^3 + Q^2/2gb^2y_1^3$, $Q^2/2gb^2y_1^3 = (y_2/y_1)^2 - (y_2/y_1)^3$.

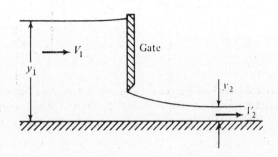

Fig. 14-43

14.155 For the sluice gate in Fig. 14-43, take the special case $y_1 = 90$ cm and $y_2 = 60$ cm, with $b = 10$ m. Compute flow rate Q if the upstream kinetic energy is (a) neglected and (b) considered.

$\blacksquare$ (a) If $V_1^2/2g$ is neglected, the equation developed in Prob. 14.154 applies: $Q^2/2gb^2y_1^3 = (y_2/y_1)^2 - (y_2/y_1)^3$,
$Q^2/[(2)(9.807)(10)^2(\frac{90}{100})^3] = [(\frac{60}{100})/(\frac{90}{100})]^2 - [(\frac{60}{100})/(\frac{90}{100})]^3$, $Q = 14.6$ m^3/s.
(b) From the Bernoulli equation,

$$y_1 + V_1^2/2g = y_2 + V_2^2/2g \qquad V_1 y_1 = V_2 y_2 \qquad (V_1)(\tfrac{90}{100}) = (V_2)(\tfrac{60}{100}) \qquad V_2 = 1.500V_1$$

$$\tfrac{90}{100} + V_1^2/[(2)(9.807)] = \tfrac{60}{100} + (1.500V_1)^2/[(2)(9.807)] \qquad V_1 = 2.170 \text{ m/s}$$

$$Q = Av = [(10)(\tfrac{90}{100})](2.170) = 19.5 \text{ m}^3/\text{s}$$

14.156 For laminar sheet draining, the flow may become turbulent if $N_R = 500$. If $s = 0.002$, what is the maximum sheet thickness y_0 to ensure laminar flow of water at 20 °C?

$$\blacksquare \qquad N_R = gy_0^3 s/3v^2 \qquad 500 = (9.807)(y_0^3)(0.002)/[(3)(1.02 \times 10^{-6})^2] \qquad y_0 = 0.00430 \text{ m}$$

14.157 A rectangular channel is 3 m wide and contains water 2 m deep. If the slope is 0.6° and the lining is unfinished concrete ($n = 0.014$), compute the discharge for uniform flow by the Manning formula.

$$\blacksquare \qquad Q = (A)(1.0/n)(R^{2/3})(s^{1/2}) = [(3)(2)](1.0/0.014)[(3)(2)/(2+3+2)]^{2/3}(\tan 0.6°)^{1/2} = 39.6 \text{ m}^3/\text{s}$$

14.158 Solve Prob. 14.157 by the Moody formula ($\epsilon = 0.0024$ m).

$$\blacksquare \qquad C = \sqrt{8g/f} \qquad 1/f^{1/2} = 2 \log (3.7D_h/\epsilon) \qquad D_h = 4R = (4)[(3)(2)/(2+3+2)] = 3.429 \text{ m}$$

$$1/f^{1/2} = 2 \log [(3.7)(3.429)/(0.0024)] \qquad f = 0.01804 \qquad C = \sqrt{(8)(9.807)/0.01804} = 65.95$$

$$Q = CA(Rs)^{1/2} = (65.95)[(3)(2)]\{[(3)(2)/(2+3+2)](\tan 0.6°)\}^{1/2} = 37.5 \text{ m}^3/\text{s}$$

14.159 A riveted-steel channel ($n = 0.015$) slopes at $1:400$ and has a V shape with an included angle of 70°. Find the water depth for uniform flow at 800 m^3/h.

$$\blacksquare \qquad Q = (A)(1.0/n)(R^{2/3})(s^{1/2}) \qquad A = (y)[y \tan (70°/2)] = 0.7002y^2 \qquad p_w = 2y/\cos (70°/2) = 2.442y$$

$$R = A/p_w = 0.7002y^2/2.442y = 0.2867y \qquad \tfrac{800}{3600} = (0.7002y^2)(1.0/0.015)(0.2867y)^{2/3}(\tfrac{1}{400})^{1/2} \qquad y = 0.566 \text{ m}$$

14.160 A trapezoidal channel similar to that of Fig. 14-38 has a bottom width of 10 ft, a side angle of 40°, and a depth of 3 ft. Compute the uniform flow discharge for a clean earth channel ($n = 0.022$) with $s = 0.0003$.

$$\blacksquare \qquad A = (10)(3) + (3)[3 \tan (90° - 40°)] = 40.73 \text{ ft}^2 \qquad p_w = 10 + (2)[3/\cos (90° - 40°)] = 19.33 \text{ ft}$$

$$Q = (A)(1.486/n)(R^{2/3})(s^{1/2}) = (40.73)(1.486/0.022)(40.73/19.33)^{2/3}(0.0003)^{1/2} = 78.3 \text{ ft}^3/\text{s}$$

14.161 For the channel of Prob. 14.160, compute the normal depth if the flow rate is 250 cfs.

$$\blacksquare \qquad Q = (A)(1.486/n)(R^{2/3})(s^{1/2}) \qquad A = 10y + (y)[y \tan (90° - 40°)] = 10y + 1.192y^2$$

$$p_w = 10 + (2)[y/\cos (90° - 40°)] = 10 + 3.111y$$

$$250 = (10y + 1.192y^2)(1.486/0.022)[(10y + 1.192y^2)/(10 + 3.111y)]^{2/3}(0.0003)^{1/2}$$

$$y = 5.63 \text{ ft} \qquad \text{(by trial and error)}$$

14.162 In smooth channels, the effective value of Manning's n can increase significantly with channel size. For roughness $\epsilon = 0.001$ ft, use the Moody formula to compute Manning's n for R equal to (a) 1 ft and (b) 8.75 ft.

$$C = \sqrt{8g/f} = (1.486/n)(R^{1/6}) \qquad n = 1.486R^{1/6}/C \qquad 1/f^{1/2} = 2\log(3.7D_h/\epsilon) \qquad D_h = 4R$$

(a)
$$D_h = (4)(1) = 4.000 \text{ ft} \qquad 1/f^{1/2} = 2\log[(3.7)(4.000)/0.001] \qquad f = 0.01438$$
$$C = \sqrt{(8)(32.2)/0.01438} = 133.8 \qquad n = (1.486)(1)^{1/6}/133.8 = 0.0111$$

(b)
$$D_h = (4)(8.75) = 35.00 \text{ ft} \qquad 1/f^{1/2} = 2\log[(3.7)(35.00)/0.001] \qquad f = 0.009566$$
$$C = \sqrt{(8)(32.2)/0.009566} = 164.1 \qquad n = (1.486)(8.75)^{1/6}/164.1 = 0.0130$$

14.163 A circular corrugated-metal ($n = 0.022$) storm drain is flowing half-full over a slope of 5 ft per mile. Estimate the discharge if the drain diameter is 7 m.

$$Q = (A)(1.486/n)(R^{2/3})(s^{1/2}) = [(\tfrac{1}{2})(\pi)(7)^2/4](1.486/0.022)(\tfrac{7}{4})^{2/3}(\tfrac{5}{5280})^{1/2} = 58.1 \text{ ft}^3/\text{s}$$

14.164 A trapezoidal aqueduct similar to that of Fig. 14-38 has a bottom width of 6 m, a side angle of 35°, and carries 50 m³/s of water at a depth of 3 ft. If $n = 0.014$, what is the required elevation drop?

$$Q = (A)(1.0/n)(R^{2/3})(s^{1/2}) \qquad A = (6)(3) + 3[3\tan(90° - 35°)] = 30.85 \text{ m}^2$$
$$p_w = 6 + (2)[3/\cos(90° - 35°)] = 16.46 \text{ m} \qquad 50 = (30.85)(1.0/0.014)(30.85/16.46)^{2/3}(s)^{1/2} \qquad s = 0.000223$$

14.165 For the aqueduct of Prob. 14.164, what will the water depth be if the slope is 0.0002 and the discharge is 25 m³/s?

$$Q = (A)(1.0/n)(R^{2/3})(s^{1/2}) \qquad A = 6y + y[y\tan(90 - 35)] = 6y + 1.428y^2$$
$$p_w = 6 + 2[y/\cos(90 - 35)] = 6 + 3.487y$$
$$25 = (6y + 1.428y^2)(1.0/0.014)[(6y + 1.428y^2)/(6 + 3.487y)]^{2/3}(0.0002)^{1/2}$$
$$y = 2.14 \text{ m} \qquad \text{(by trial and error)}$$

14.166 Uniform water flow in a wide brick channel ($n = 0.015$) of slope 0.02° moves over a 10-cm bump as in Fig. 14-44. A slight depression in water surface results. If the minimum water depth over the bump is 50 cm, compute the velocity over the bump and the flow rate per meter of width.

$$v = (1.0/n)(R^{2/3})(s^{1/2}) \qquad v_1 = (1.0/0.015)(y_1)^{2/3}(\tan 0.02°)^{1/2} = 1.246y_1^{2/3}$$

From continuity, $v_1 y_1 = v_2 y_2$, $(1.246y_1^{2/3})(y_1) = (v_2)(\tfrac{50}{100})$, $v_2 = 2.492y_1^{5/3}$. Applying the Bernoulli equation and neglecting bottom slope,

$$y_1 + v_1^2/2g = y_2 + v_2^2/2g + h_{max} \qquad y_1 + (1.246y_1^{2/3})^2/[(2)(9.807)] = \tfrac{50}{100} + (2.492y_1^{5/3})^2/[(2)(9.807)] + \tfrac{10}{100}$$
$$y_1 = 0.623 \text{ m} \qquad \text{(by trial and error)} \qquad v_1 = (1.246)(0.623)^{2/3} = 0.9089 \text{ m/s}$$
$$v_2 = (2.492)(0.623)^{5/3} = 1.13 \text{ m/s} \qquad q = v_1 y_1 = (0.9089)(0.623) = 0.566 \text{ (m}^3/\text{s)/m}$$

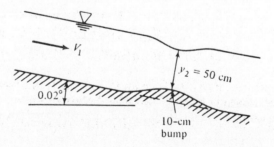

10-cm bump   **Fig. 14-44**

14.167 A triangular channel (similar to Fig. 14-39) has 45° sides and a slope of 0.0006 and is lined with asphalt ($n = 0.016$). What will the normal depth be when the discharge is (a) 150 cfs and (b) 15 cfs?

$$Q = (A)(1.486/n)(R^{2/3})(s^{1/2}) \qquad A = (y)(y\cot 45°) = 1.000y^2$$
$$p_w = (2y)(\csc 45°) = 2.828y \qquad R = A/p_w = 1.000y^2/2.828y = 0.3536y$$

(a)
$$150 = (1.000y^2)(1.486/0.016)(0.3536y)^{2/3}(0.0006)^{1/2} \qquad y = 6.24 \text{ ft}$$

(b)
$$15 = (1.000y^2)(1.486/0.016)(0.3536y)^{2/3}(0.0006)^{1/2} \qquad y = 2.63 \text{ ft}$$

14.168 A brick ($n = 0.015$) rectangular channel with $s = 0.002$ is designed to carry 230 cfs of water in uniform flow. There is an argument over whether the channel width should be 4 ft or 8 ft. Which design needs fewer bricks? By what percentage?

∎ $Q = (A)(1.486/n)(R^{2/3})(s^{1/2})$. For 4-ft width: $230 = (4y)(1.486/0.015)[4y/(y + 4 + y)]^{2/3}(0.002)^{1/2}$, $y = 9.31$ ft   (by trial and error), $p_w = 9.31 + 4 + 9.31 = 22.62$ ft. For 8-ft width: $230 = (8y)(1.486/0.015)[8y/(y + 8 + y)]^{2/3}(0.002)^{1/2}$, $y = 4.07$ ft   (by trial and error), $p_w = 4.07 + 8 + 4.07 = 16.14$ ft. Hence, the 8-ft-width design needs fewer bricks by $(22.62 - 16.14)/22.62 = 0.286$, or 28.6 percent.

14.169 In flood stage a natural channel often consists of a deep main channel plus two floodplains, as in Fig. 14-45. The floodplains are often shallow and rough. Suppose that $y_1 = 20$ ft, $y_2 = 5$ ft, $b_1 = 40$ ft, $b_2 = 100$ ft, $n_1 = 0.020$, $n_2 = 0.040$, with a slope of 0.0002. Estimate the discharge.

∎ Consider section 1 to be the main channel and section 2 to be each floodplain.

$$Q = (A)(1.486/n)(R^{2/3})(s^{1/2})$$

$$Q_1 = [(40)(20 + 5)](1.486/0.020)[(40)(20 + 5)/(20 + 40 + 20)]^{2/3}(0.0002)^{1/2} = 5659 \text{ ft}^3/\text{s}$$

$$Q_2 = [(100)(5)](1.486/0.040)[(100)(5)/(100 + 5)]^{2/3}(0.0002)^{1/2} = 744 \text{ ft}^3/\text{s}$$

$$Q = 5659 + (2)(744) = 7147 \text{ ft}^3/\text{s}$$

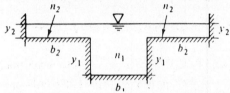

Fig. 14-45

14.170 For the flood stage in Fig. 14-45, compute the total flow rate if the main channel is 8 m by 8 m clean earth ($n = 0.022$) and each floodplain is 1-m-deep farm land ($n = 0.035$) 100 m wide. All channels have a slope of 0.1°.

∎ Consider section 1 to be the main channel and section 2 to be each floodplain.

$$Q = (A)(1.0/n)(R^{2/3})(s^{1/2})$$

$$Q_1 = [(8)(8 + 1)](1.0/0.022)[(8)(8 + 1)/(8 + 8 + 8)]^{2/3}(\tan 0.1°)^{1/2} = 284 \text{ m}^3/\text{s}$$

$$Q_2 = [(100)(1)](1.0/0.035)[(100)(1)/(100 + 1)]^{2/3}(\tan 0.1°)^{1/2} = 119 \text{ m}^3/\text{s}$$

$$Q = 284 + (2)(119) = 522 \text{ m}^3/\text{s}$$

14.171 A 1-m-diameter clay tile ($n = 0.014$) sewer pipe runs half-full on a slope of 0.2°. Compute the flow rate using the Manning formula.

∎ $$Q = (A)(1.0/n)(R^{2/3})(s^{1/2}) = [(\tfrac{1}{2})(\pi)(1)^2/4](1.0/0.014)(\tfrac{1}{4})^{2/3}(\tan 0.2°)^{1/2} = 0.658 \text{ m}^3/\text{s}$$

14.172 Solve Prob. 14.171 by the Moody method with $\epsilon = 0.0024$ m.

∎ $$1/f^{1/2} = 2 \log (3.7D/\epsilon) = 2 \log [(3.7)(1)/0.0024] \qquad f = 0.02460$$
$$C = \sqrt{8g/f} = \sqrt{(8)(9.807)/0.02469} = 56.47$$
$$Q = CA(Rs)^{1/2} = 56.47[(\tfrac{1}{2})(\pi)(1)^2/4][(\tfrac{1}{4})(\tan 0.2°)]^{1/2} = 0.655 \text{ m}^3/\text{s}$$

14.173 Four of the sewer pipes from Prob. 14.171 empty into a single finished-cement ($n = 0.012$) pipe, also sloping at 0.2°. If the large pipe is also to run at half-full, what should its diameter be?

∎ $$Q = (A)(1.0/n)(R^{2/3})(s^{1/2}) \qquad (4)(0.658) = [(\tfrac{1}{2})(\pi)(D)^2/4](1.0/0.012)(D/4)^{2/3}(\tan 0.2°)^{1/2} \qquad D = 1.59 \text{ m}$$

14.174 For the circular channel of Fig. 14-46, if $n = 0.016$, $D = 3$ m, and the slope is 0.2°, find the normal depth for a discharge of 20 m³/s.

∎ $$Q = (A)(1.0/n)(R^{2/3})(s^{1/2}) \qquad y = (D/2)(1 + \sin \theta) = (\tfrac{3}{2})(1 + \sin \theta)$$
$$A = (\tfrac{1}{2})(D/2)^2(\pi + 2\theta + \sin 2\theta) = (\tfrac{1}{2})(\tfrac{3}{2})^2(\pi + 2\theta + \sin 2\theta) = (1.125)(\pi + 2\theta + \sin 2\theta)$$
$$p_w = (D/2)(\pi + 2\theta) = (\tfrac{3}{2})(\pi + 2\theta)$$
$$20 = [(1.125)(\pi + 2\theta + \sin 2\theta)](1.0/0.016)[(1.125)(\pi + 2\theta + \sin 2\theta)/(\tfrac{3}{2})(\pi + 2\theta)]^{2/3}(\tan 0.2°)^{1/2}$$
$$\theta = 31.57° \qquad \text{(by trial and error)} \qquad y = (\tfrac{3}{2})(1 + \sin 31.57°) = 2.29 \text{ m}$$

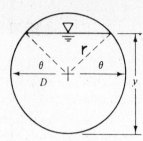

Fig. 14-46

14.175 A rectangular channel has $b = 3$ m and $y = 1$ m. If n and s are the same, what is the diameter of a semicircular channel that will have the same discharge? Compare the two wetted perimeters.

$$Q = (A)(1.0/n)(R^{2/3})(s^{1/2}) \qquad Q_1 = Q_2$$

$$[(3)(1)][1.0/n][(3)(1)/(1+3+1)]^{2/3}(s)^{1/2} = [(\tfrac{1}{2})(\pi)(D)^2/4](1.0/n)(D/4)^{2/3}(s)^{1/2} \qquad D = 2.67 \text{ m}$$

For rectangular channel, $p_w = 1 + 3 + 1 = 5$ ft. For semicircular channel, $p_w = (\pi)(2.67)/2 = 4.19$ m. p_w for semicircular channel is $(5 - 4.19)/5 = 0.162$, or 16.2 percent smaller.

14.176 A trapezoidal channel has $n = 0.022$ and $s = 0.0003$ and is made in the shape of a half-hexagon for maximum efficiency (see Fig. 14-47). What should be the length of the side of the hexagon if the channel is to carry 225 cfs of water?

$$Q = (A)(1.486/n)(R^{2/3})(s^{1/2}) \qquad A = (\tfrac{3}{2})(b^2)(\sin 60°) = 1.299b^2$$

$$225 = (1.299b^2)(1.486/0.022)(1.299b^2/3b)^{2/3}(0.0003)^{1/2} \qquad b = 8.03 \text{ ft}$$

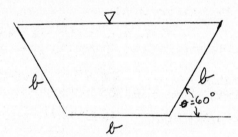

Fig. 14-47

14.177 What is the discharge of a semicircular channel of the same cross-sectional area and the same s and n as in Prob. 14.176?

$$A = (1.299)(8.03)^2 = 83.76 \text{ ft}^2 \qquad \text{(from Prob. 14.176)} \qquad 83.76 = (\tfrac{1}{2})(\pi)(d)^2/4 \qquad d = 14.60 \text{ ft}$$

$$Q = (A)(1.486/n)(R^{2/3})(s^{1/2}) = 83.76(1.486/0.022)(14.60/4)^{2/3}(0.0003)^{1/2} = 232 \text{ ft}^3/\text{s}$$

14.178 What are the most efficient dimensions for a planed-wood ($n = 0.012$) rectangular channel to carry 3.5 m³/s at $s = 0.0006$?

For most efficient rectangular section, width = twice the depth ($w = 2y$): $Q = (A)(1.0/n)(R^{2/3})(s^{1/2})$, $3.5 = [(y)(2y)][1.0/0.012][(y)(2y)/(y + 2y + y)]^{2/3}(0.0006)^{1/2}$, $y = 1.123$ m, $w = (2)(1.123) = 2.246$ m.

14.179 What is the most efficient depth for an asphalt ($n = 0.016$) trapezoidal channel, with sides sloping at 45°, to carry 4 m³/s at $s = 0.0004$?

For the most efficient trapezoidal section, $R = y/2$. $Q = (A)(1.0/n)(R^{2/3})(s^{1/2})$, $A = (y^2)[(2)(1 + \cot 45°)^{1/2} - \cot 45°] = 1.828y^2$, $4 = (1.828y^2)(1.0/0.016)(y/2)^{2/3}(0.0004)^{1/2}$, $y = 1.47$ m.

14.180 For the information given in Prob. 14.179, how much additional discharge is carried by a semicircle with the same area?

$$A = (1.828)(1.47)^2 = 3.950 \text{ m}^2 \qquad \text{(from Prob. 14.179)} \qquad 3.950 = (\tfrac{1}{2})(\pi)(D)^2/4 \qquad D = 3.172 \text{ m}$$

$$Q = (A)(1.0/n)(R^{2/3})(s^{1/2}) = (3.950)(1.0/0.016)(3.172/4)^{2/3}(0.0004)^{1/2} = 4.23 \text{ m}^3/\text{s}$$

The semicircular channel carries $(4.23 - 4)/4 = 0.058$, or 5.8 percent more than the trapezoidal channel.

14.181 Suppose that the side angles of the trapezoidal channel in Prob. 14.176 are reduced to $\theta = 20°$ to avoid earth slides. If the bottom flat width is 8 ft, what should the normal depth of this new channel be? Compare the wetted perimeter with that of Prob. 14.176.

$\blacksquare$ $\quad Q = (A)(1.486/n)(R^{2/3})(s^{1/2}) \qquad A = by + y^2 \cot 20° = 8y + 2.747y^2 \qquad p_w = b + 2y \csc 20° = 8 + 5.848y$

$$225 = (8y + 2.747y^2)(1.486/0.022)[(8y + 2.747y^2)/(8 + 5.848y)]^{2/3}(0.0003)^{1/2}$$

$$y = 4.68 \text{ ft} \qquad \text{(by trial and error)} \qquad p_w = 8 + (5.848)(4.68) = 35.4 \text{ ft}$$

p_w from Prob. 14.176 = $(3)(8.03) = 24.1$ ft. Hence, this trapezoidal section's ($\theta = 20°$) wetted perimeter is $(35.4 - 24.1)/24.1 = 0.469$, or 46.9 percent greater than the half-hexagon of Prob. 14.176.

14.182 A wide, clean-earth river ($n = 0.030$) has a flow rate $q = 135$ cfs/ft. What is the critical depth? If the actual depth is 15 ft, what is the Froude number of the river? Compute the critical slope by the Manning formula.

$\blacksquare$ $\qquad\qquad y_c = (q^2/g)^{1/3} = (135^2/32.2)^{1/3} = 8.27 \text{ ft} \qquad v = \frac{135}{15} = 9.000 \text{ ft/s}$

$$N_F = v/\sqrt{gy_c} = 9.000/\sqrt{(32.2)(8.27)} = 0.552 \qquad v_c = (1.486/n)(R_c^{2/3})(s_c^{1/2}) = \sqrt{gy_c} = \sqrt{(32.2)(8.27)} = 16.32 \text{ ft/s}$$

$$16.32 = (1.486/0.030)(8.27)^{2/3}(s_c)^{1/2} \qquad s_c = 0.00649$$

14.183 Find the critical slope in Prob. 14.182 by the Moody method, using $\epsilon = 0.8$ ft.

$\blacksquare$ $\qquad\qquad\qquad s_c = f/8 \qquad 1/f^{1/2} = 2 \log (3.7D/\epsilon)$

Using data from Prob. 14.182, $D = 4R = 4y_c = (4)(8.27) = 33.08$ ft, $1/f^{1/2} = 2 \log [(3.7)(33.08)/0.8]$, $f = 0.05238$, $s_c = 0.05238/8 = 0.00655$.

14.184 Find the critical depth of the brick channel in Prob. 14.168 for both the 4 ft and 8 ft widths. Are the normal flows subcritical or supercritical?

$\blacksquare$ $y_c = (Q^2/b^2g)^{1/3}$. Using data from Prob. 14.168,

(a) $y_c = \{230^2/[(4)^2(32.2)]\}^{1/3} = 4.68$ ft. Since $[y_n = 9.31 \text{ ft}] > [y_c = 4.68 \text{ ft}]$, the flow is subcritical.
(b) $y_c = \{230^2/[(8)^2(32.2)]\}^{1/3} = 2.95$ ft. Since $[y_n = 4.07 \text{ ft}] > [y_c = 2.95 \text{ ft}]$, the flow is subcritical.

14.185 A thumbnail piercing the surface of a channel flow creates a wedgelike wave of half-angle 25°, as seen from above in Fig. 14-48. If the water depth is 20 cm, what is the flow velocity?

$\blacksquare$ $\qquad\qquad \sin \theta = 1/N_F = \sqrt{gh}/v \qquad \sin 25° = \sqrt{(9.087)(\frac{20}{100})}/v \qquad v = 3.31 \text{ m/s}$

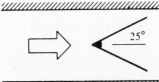

Fig. 14-48

14.186 Suppose that the wave of Fig. 14-48 is seen instead on the surface of water flowing half-full in a circular channel of diameter 75 cm. What is the flow rate if the surface is finished cement?

$\blacksquare$ Flow is supercritical.

$$v = N_F v_c \qquad N_F = \csc 25° = 2.37 \qquad A_c = (\tfrac{1}{2})(\pi)(\tfrac{75}{100})^2/4 = 0.2209 \text{ m}^2$$

$$v_c = (gA_c/b_0)^{1/2} = [(9.807)(0.2209)/(\tfrac{75}{100})]^{1/2} = 1.70 \text{ m/s} \qquad v = (2.37)(1.70) = 4.03 \text{ m/s}$$

$$Q = Av = (0.2209)(4.03) = 0.890 \text{ m}^3/\text{s}$$

14.187 Suppose that the wave of Fig. 14-48 is seen instead on the surface of water flowing full in a half-hexagon of side length 30 cm. What is the flow rate if the sides are of planed wood?

$\blacksquare$ Flow is supercritical.

$$v = N_F v_c \qquad N_F = \csc 25° = 2.37 \qquad v_c = (gA_c/b_0)^{1/2}$$

$$b_0 = 30 + (2)(30)(\cos 60°) = 60.00 \text{ cm} \qquad y = (30)(\sin 60°) = 25.98 \text{ cm}$$

$$A_c = (\tfrac{1}{2})(60.00 + 30)(25.98) = 1169 \text{ cm}^2 \quad \text{or} \quad 0.1169 \text{ m}^2 \qquad v_c = [(9.807)(0.1169)/(\tfrac{60}{100})]^{1/2} = 1.38 \text{ m/s}$$

$$v = (2.37)(1.38) = 3.27 \text{ m/s} \qquad Q = Av = (0.1169)(3.27) = 0.382 \text{ m}^3/\text{s}$$

14.188 For the given flow rate of 50 m³/s in the trapezoidal channel of Prob. 14.164, find the critical depth and the critical slope.

▮ $\quad A = (b_0 Q^2/g)^{1/3} \quad\quad b_0 = 6 + 2y_c \cot 35° = 6 + 2.856y_c \quad\quad A = (y_c)(6 + y_c \cot 35°) = 6y_c + 1.428y_c^2$

$\quad\quad 6y_c + 1.428y_c^2 = [(6 + 2.856y_c)(50^2)/9.807]^{1/3} \quad\quad y_c = 1.67 \text{ m} \quad\quad \text{(by trial and error)}$

$\quad v_c = (1.0/n)(R^{2/3})(s^{1/2}) \quad\quad A = (6)(1.67) + (1.428)(1.67)^2 = 14.00 \text{ m}^2 \quad\quad v_c = Q/A = 50/14.00 = 3.571 \text{ m/s}$

$\quad\quad p_w = 6 + (2)(1.67)/\sin 35° = 11.82 \text{ m} \quad\quad 3.571 = (1.0/0.014)(14.00/11.82)^{2/3}(s_c)^{1/2} \quad\quad s_c = 0.00199$

14.189 For the river flow of Prob. 14.182, find the depth y_2 that has the same specific energy as the given depth $y_1 = 15$ ft. These are called *conjugate depths*. What is $(N_F)_2$?

▮ $\quad\quad\quad E = y + v^2/2g \quad\quad v_1 = 9.000 \text{ ft/s} \quad\quad \text{(from Prob. 14.182)}$

$\quad E_1 = 15 + 9.000^2/[(2)(32.2)] = 16.26 \text{ ft} \quad\quad E_2 = E_1 = 16.26 = y_2 + (135/y_2)^2/[(2)(32.2)]$

$\quad\quad 16.26y_2^2 - y_2^3 - 283.0 = 0 \quad\quad y_2 = 5.02 \text{ ft} \quad\quad \text{(by trial and error)}$

$\quad\quad (N_F)_2 = v_2/\sqrt{gy_2} = (135/5.02)/\sqrt{(32.2)(5.02)} = 2.12 \quad\quad \text{(supercritical)}$

14.190 A clay-tile ($n = 0.014$) triangular channel has sides sloping at 60°. If the channel carries 10 m³/s, compute the (a) critical depth, (b) critical velocity, and (c) critical slope.

▮ (a) $\quad\quad gA_c^3 = b_0 Q^2 \quad\quad A_c = y_c^2 \cot 60° = 0.5774y_c^2 \quad\quad b_0 = 2y_c \cot 60° = 1.155y_c$

$\quad\quad (9.807)(0.5774y_c^2)^3 = (1.155y_c)(10)^2 \quad\quad y_c = 2.28 \text{ m}$

(b) $\quad\quad\quad v_c = Q/A_c = 10/[(0.5774)(2.28)^2] = 3.33 \text{ m/s}$

(c) $\quad\quad\quad v_c = (1.0/n)(R^{2/3})(s^{1/2})$

$\quad 3.33 = (1.0/0.014)\{(0.5774)(2.28)^2/[(2)(2.28)/\sin 60°]\}^{2/3}(s_c)^{1/2} \quad\quad s_c = 0.00460$

14.191 A riveted-steel ($n = 0.015$) triangular duct flows partly full as shown in Fig. 14-49. If the critical depth is 60 cm, compute (a) the critical flow rate and (b) the critical slope.

▮ (a) $\quad\quad b_0 = \{[0.8660 - (\frac{60}{100})]/0.8660\}(1) = 0.3072 \text{ m} \quad\quad A = (\frac{1}{2})(0.3072 + 1)(\frac{60}{100}) = 0.3922 \text{ m}^2$

$\quad\quad Q = A\sqrt{gA/b_0} = (0.3922)\sqrt{(9.807)(0.3922)/0.3072} = 1.39 \text{ m}^3/\text{s}$

(b) $\quad\quad Q = (A)(1.0/n)(R^{2/3})(s^{1/2}) \quad\quad p_w = 1 + (2)[(\frac{60}{100})/\sin 60°] = 2.386 \text{ m}$

$\quad\quad 1.39 = (0.3922)(1.0/0.015)(0.3922/2.386)^{2/3}(s)^{1/2} \quad\quad s = 0.0315$

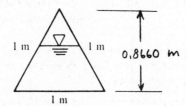

0.8660 m

1 m

Fig. 14-49

14.192 For the triangular duct of Prob. 14.191, if the critical flow rate is 1.0 m³/s, compute (a) the critical depth and (b) the critical slope.

▮ (a) $\quad\quad Q = A\sqrt{gA/b_0} \quad\quad b_0 = [(0.8660 - y_c)/0.8660](1) = 1 - 1.155y_c$

$\quad\quad A = \frac{1}{2}[(1 - 1.155y_c) + 1](y_c) = y_c - 0.5775y_c^2$

$\quad 1.0 = (y_c - 0.5775y_c^2)\sqrt{(9.807)(y_c - 0.5775y_c^2)/(1 - 1.155y_c)} \quad\quad y_c = 0.493 \text{ m} \quad\quad \text{(by trial and error)}$

(b) $\quad\quad Q = (A)(1.0/n)(R^{2/3})(s^{1/2}) \quad\quad p_w = 1 + (2)[0.493/\sin 60°] = 2.139 \text{ m}$

$\quad\quad A = 0.493 - (0.5775)(0.493)^2 = 0.3526 \text{ m}^2 \quad\quad 1.0 = (0.3526)(1.0/0.015)(0.3526/2.139)^{2/3}(s)^{1/2}$

$\quad\quad\quad\quad s = 0.200$

14.193 A rectangular channel 2 m wide and 1 m deep is found to have a critical slope of 0.008. Estimate (a) Manning's n, (b) the surface roughness height, and (c) the critical flow rate.

▮ (a) $\quad\quad s_c = gn^2 p_w/(1.0b_0 R^{1/3}) \quad\quad R = A/p_w = (2)(1)/(1 + 2 + 1) = 0.5000 \text{ m}$

$\quad\quad 0.008 = (9.807)(n)^2(1 + 2 + 1)/[(1.0)(2)(0.5000)^{1/3}] \quad\quad n = 0.0180$

(b) $n = 0.0382\epsilon^{1/6}$ $0.0180 = 0.0382\epsilon^{1/6}$ $\epsilon = 0.0109$ m   or   10.9 mm

(c) $Q_c = (A)\sqrt{gA/b_0} = [(2)(1)]\sqrt{(9.807)[(2)(1)]/2} = 6.26 \text{ m}^3/\text{s}$

14.194 Water flows in a rectangular unfinished-cement channel 10 m wide laid on a slope of 0.2°. Find the depth and discharge if the channel flow is critical.

▮ $s_c = gn^2p_w/(1.0b_0R^{1/3})$ $\tan 0.2° = (9.807)(0.014)^2(10 + 2y_c)/\{(1.0)(10)[10y_c/(10 + 2y_c)]^{1/3}\}$

$y_c = 0.196$ m   (by trial and error) $Q_c = (gA^3/b_0)^{1/3} = \{(9.807)[(10)(0.196)]^3/10\}^{1/3} = 1.95 \text{ m}^3/\text{s}$

14.195 A circular corrugated-metal ($n = 0.022$) channel 6 ft in diameter is found to have a Froude number of 0.5 in uniform half-full flow. Compute **(a)** the channel slope and **(b)** the flow rate.

▮ **(a)** $v_c = (gA/b_0)^{1/2} = \{(32.2)[(\frac{1}{2})(\pi)(6)^2/4]/6\}^{1/2} = 8.710 \text{ ft/s}$

$v = (1.486/n)(R^{2/3})(s^{1/2}) = v_c/2 = 8.710/2 = 4.355 \text{ ft/s}$

$4.355 = (1.486/0.022)(\frac{6}{4})^{2/3}(s)^{1/2}$ $s = 0.00242$

(b) $Q = Av = [(\frac{1}{2})(\pi)(6)^2/4](4.355) = 61.6 \text{ ft}^3/\text{s}$

14.196 Water is flowing in a finished-concrete, semicircular channel shown in Fig. 14-50 inclined at a slope of 0.0016. What is the flow rate if the flow is uniform?

▮ $A = (\frac{1}{2})(\pi)(10)^2 + (3)(10 + 10) = 217.1 \text{ ft}^2$ $R = A/p_w = 217.1/[(\pi)(10) + 3 + 3] = 5.802 \text{ ft}$

$Q = (A)(1.486/n)(R^{2/3})(s^{1/2}) = (217.1)(1.486/0.012)(5.802)^{2/3}(0.0016)^{1/2} = 3472 \text{ ft}^3/\text{s}$

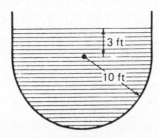

3 ft

10 ft

Fig. 14-50

14.197 In a planed-wood ($n = 0.012$) rectangular channel of width 4 m, water is flowing at the rate of 20 m³/s. If the slope of the channel is 0.0012, what is the depth of uniform flow?

▮ $Q = (A)(1.0/n)(R^{2/3})(s^{1/2})$ $20 = (4D)(1.0/0.012)[4D/(D + 4 + D)]^{2/3}(0.0012)^{1/2}$

$D = 1.80$ m   (by trial and error)

14.198 In a planed-wood rectangular channel of width 4 m, water is flowing at a rate of 5 m³/s. The slope of the channel is 0.0001. The roughness coefficient is 0.5 mm. What is the height h of the cross section of flow for normal, steady flow? The water has a temperature of 10 °C.

▮ $v = Q/A = (8gsR/f)^{1/2}$ $5/4y = \{(8)(9.807)(0.0001)[4y/(y + 4 + y)]/f\}^{1/2}$

Like problems in Chap. 9, this requires a trial-and-error solution whereby a value of the friction factor (f) is guessed and subsequent computations and work with Fig. A-5 are done to find the right combination of parameter values to satisfy the problem. Inasmuch as a number of trial-and-error problems of this type were presented in detail in Chap. 9, we shall save time and space in problems of this type in this chapter by "guessing" the correct value of f on the first try! Guess $f = 0.0132$.

$5/4y = \{(8)(9.807)(0.0001)[4y/(y + 4 + y)]/0.0132\}^{1/2}$ $y = 1.69$ m   (by trial and error)

$v = Q/A = 5/[(4)(1.69)] = 0.7396 \text{ m/s}$ $R = (4)(1.69)/(1.69 + 4 + 1.69) = 0.9160 \text{ m}$

$N_R = D_hv/v = (4R)(v)/v = (4)(0.9160)(0.7396)/(1.30 \times 10^{-6}) = 2.08 \times 10^6$

$\epsilon/D_h = \epsilon/4R = (0.5/1000)/[(4)(0.9160)] = 0.00014$

From Fig. A-5, $f = 0.0132$ (O.K.). Therefore, $y = 1.69$ m.

14.199 Do Prob. 14.196 using friction-factor formulas with $\epsilon = 0.0032$ ft. Assume a temperature of 60 °F.

▮ $v^*\epsilon/v = (\sqrt{gRs})(\epsilon)/v$. Using values from Prob. 14.196, $v^*\epsilon/v = [\sqrt{(32.2)(5.802)(0.0016)}](0.0032)/(1.21 \times 10^{-5}) = 145$. Since $v^*\epsilon/v > 100$, we are in the fully rough zone and $1/f^{1/2} = 2.16 - 2\log(\epsilon/R) = 2.16 -$

2 log $(0.0032/5.802)$, $f = 0.01328$, $v = (8gsR/f)^{1/2} = \{(8)(32.2)(0.0016)(5.802)/0.01328\}^{1/2} = 13.42$ ft/s, $Q = Av = (217.1)(13.42) = 2913$ ft^3/s.

14.200 A wide rectangular section has a flow of 70 cfs/ft. What is the critical depth? If the depth is 5 ft at a section, what is the Froude number there? Next, if we were to have normal, critical flow, what must the slope of the channel be? ($\epsilon = 0.0032$ ft.)

▌ $\qquad y_c = (q^2/g)^{1/3} = (70^2/32.2)^{1/3} = 5.34$ ft $\qquad N_F = v/\sqrt{gy} = \{70/[(5)(1)]\}/\sqrt{(32.2)(5)} = 1.10$

$\qquad\qquad s_c = f/8 \qquad N_R = D_h v/\nu \qquad D_h = 4R = (4)(5.34) = 21.36$ ft

$\qquad\qquad\qquad N_R = (21.36)\{70/[(5)(1)]\}/(1.21 \times 10^{-5}) = 2.47 \times 10^7$

$\qquad\qquad\qquad\qquad \epsilon/D_h = 0.0032/21.36 = 0.000150$

From Fig. A-5, $f = 0.0129$. $s_c = 0.0129/8 = 0.00161$.

14.201 For steady laminar flow in a thin sheet over a flat surface (see Fig. 14-51), $V_z = [(\gamma \sin \theta)/\mu][(3q\mu/\gamma \sin \theta)^{1/3}y - y^2/2]$, where q is the volumetric flow per unit width. What is the thickness t of such a flow of water at 5 °C for $\theta = 20°$? The volumetric flow $q = 0.0003$ (m^3/s)/m.

▌ When $y = t$, $dV_z/dy = 0$. $dV_z/dy = 0 = [(\gamma \sin \theta)/\mu][(3q\mu/\gamma \sin \theta)^{1/3} - t]$, $t = (3q\mu/\gamma \sin \theta)^{1/3}$:

$$t = \left\{\frac{(3)(0.0003)(1.52 \times 10^{-3})}{[(9.79)(1000)1(\sin 20°)]}\right\}^{1/3} = 0.000742 \text{ m} \quad \text{or} \quad 0.742 \text{ mm}$$

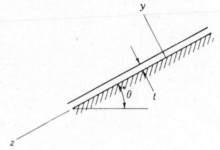

Fig. 14-51

14.202 A film of oil of thickness 0.002 ft moves at uniform speed down an inclined surface having an angle $\theta = 30°$ (see Fig. 14-51). What is the volume flow per unit width? ($\mu = 0.0003$ lb-s/ft^2 and $\gamma = 57$ lb/ft^3.)

▌ $\qquad\qquad\qquad t = (3q\mu/\gamma \sin \theta)^{1/3} \qquad$ (from Prob. 14.201)

$\qquad\qquad 0.002 = \{(3)(q)(0.0003)/[(57)(\sin 30°)]\}^{1/3} \qquad q = 0.000253$ cfs/ft

14.203 In Prob. 14.201, compute a Reynolds number as $N_R = vt/\nu$. We note that if this Reynolds number is larger than 500, we have turbulent flow rather than laminar flow. Is our laminar flow assumption valid? What is the limiting volumetric flow in Prob. 14.201 wherein the laminar flow assumption is valid? What is the film thickness for this case?

▌ $\quad v = q/t = 0.0003/0.000742 = 0.4043$ m/s $\qquad N_R = vt/\nu = (0.4043)(0.000742)/(1.52 \times 10^{-6}) = 197$

Since $N_R < 500$, the laminar flow assumption is O.K.

$\qquad\qquad N_R = vt/\nu = q/\nu \qquad 500 = q/(1.52 \times 10^{-6}) \qquad q = 0.0007600$ (m^3/s)/m

$\qquad\qquad\qquad t = (3q\mu/\gamma \sin \theta)^{1/3} \qquad$ (from Prob. 14.201)

$$t = \left\{\frac{(3)(0.0007600)(1.52 \times 10^{-3})}{[(9.79)(1000)](\sin 20°)}\right\}^{1/3} = 0.00101 \text{ m} \quad \text{or} \quad 1.01 \text{ mm}$$

14.204 Water at 5 °C is flowing in a finished-concrete rectangular channel of width 10 m with a slope of 0.001. The height of the water normal to the bed is constant, having the value of 1 m. What is the volume of flow for normal flow?

▌ $\quad Q = (A)(1.0/n)(R^{2/3})(s^{1/2}) = [(10)(1)](1.0/0.012)[(10)(1)/(1 + 10 + 1)]^{2/3}(0.001)^{1/2} = 23.3$ m^3/s

14.205 A wide rectangular channel dug from clean earth ($n = 0.025$) is to conduct a flow of 5 m^3/s per meter of width. The slope of the bed is 0.0015. What would be the depth of flow for normal flow?

▌ $\qquad\qquad s = (n/1.0)^2(q^2/y_N^{10/3}) \qquad 0.0015 = (0.025/1.0)^2(5^2/y_N^{10/3}) \qquad y_N = 2.02$ m

14.206 The triangular channel of Fig. 14-52 is made of corrugated steel ($n = 0.022$) and conducts 10 cfs at an elevation of 1000 ft to an elevation of 990 ft. What length L should the channel be for normal flow?

$$Q = (A)(1.486/n)(R^{2/3})(s^{1/2}) \qquad A = (2)(2\tan 45°) = 4.000 \text{ ft}^2$$
$$R = A/p_w = 4.000/[(2)(2/\cos 45°)] = 0.7071 \text{ ft}$$
$$10 = (4.000)(1.486/0.022)(0.7071)^{2/3}[(1000 - 990)/L]^{1/2} \qquad L = 4599 \text{ ft}$$

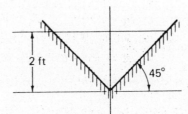

Fig. 14-52

14.207 What is the depth of normal flow and slope of a rectangular channel to conduct 5 m³/s of water a distance of 2000 m with a head loss of 15 m? The width of the channel is 2 m. The channel is made of brickwork ($n = 0.016$).

$$s = \tfrac{15}{2000} = 0.0075 \qquad Q = (A)(1.0/n)(R^{2/3})(s^{1/2})$$
$$5 = (2y_N)(1.0/0.016)[2y_N/(y_N + 2 + y_N)]^{2/3}(0.0075)^{1/2} \qquad y_N = 0.795 \text{ m} \qquad \text{(by trial and error)}$$

14.208 The asphalt-lined channel ($n = 0.016$) of Fig. 14-53 has a slope of 0.0017. The flow of water in the channel is 50 m³/s. What is the normal depth?

$$Q = (A)(1.0/n)(R^{2/3})(s^{1/2}) \qquad A = 10y_N - (\tfrac{10}{2})[(\tfrac{10}{2})(\tan 10°)] = 10y_N - 4.408$$
$$p_w = 2[y_N - (\tfrac{10}{2})(\tan 10°) + (\tfrac{10}{2})/\cos 10°] = 2y_N + 8.391 \qquad R = A/p_w = (10y_N - 4.408)/(2y_N + 8.391)$$
$$50 = (10y_N - 4.408)(1.0/0.016)[(10y_N - 4.408)/(2y_N + 8.391)]^{2/3}(0.0017)^{1/2}$$
$$y_N = 2.07 \text{ m} \qquad \text{(by trial and error)}$$

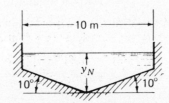

Fig. 14-53

14.209 Shown in Fig. 14-54 is a partially filled pipe. If Manning's n is 0.015, what slope is necessary for a steady flow of 50 cfs?

$$Q = (A)(1.486/n)(R^{2/3})(s^{1/2}) \qquad \theta = \arcsin \tfrac{2}{4} = 30.0°$$
$$A = \{[180 + (2)(30.0)]/360\}[(\pi)(8)^2/4] + (2)(4\cos 30.0°) = 40.44 \text{ ft}^2$$
$$p_w = \{[180 + (2)(30.0)]/360\}[(\pi)(8)] = 16.76 \text{ ft} \qquad R = A/p_w = 40.44/16.76 = 2.413 \text{ ft}$$
$$50 = (40.44)(1.486/0.015)(2.413)^{2/3}(s)^{1/2} \qquad s = 0.0000481$$

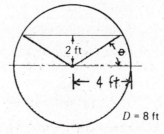

$D = 8$ ft **Fig. 14-54**

14.210 What is the flow in Fig. 14-55 when the slope of the channel is 0.0007 and the surface is that of clean excavated earth ($n = 0.025$)? The slopes on all inclined surfaces are 45°.

$$A = (3)(3)/2 + (3)(20) + (5)(3 + 3 + 5)/2 + (8)(3 + 5) + (5)(3 + 3 + 5)/2 + (3)(25) + (3)(3)/2 = 263.0 \text{ m}^2$$
$$p_w = \sqrt{3^2 + 3^2} + 20 + \sqrt{5^2 + 5^2} + 8 + \sqrt{5^2 + 5^2} + 25 + \sqrt{3^2 + 3^2} = 75.63 \text{ m} \qquad R = A/p_w = 263.0/75.63 = 3.477 \text{ m}$$
$$Q = (A)(1.0/n)(R^{2/3})(s^{1/2}) = (263.0)(1.0/0.025)(3.477)^{2/3}(0.0007)^{1/2} = 639 \text{ m}^3/\text{s}$$

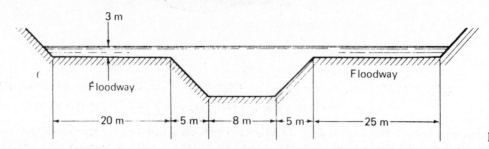

Fig. 14-55

14.211 The channel in Prob. 14.196 is to be replaced by a rectangular channel of width 16 ft. What is the ratio of cost of the concrete allowing 2 ft of freeboard (distance above the free surface) for the walls of the channels?

▮ From Prob. 14.196, $Q = 3472 \text{ ft}^3/\text{s}$, $s = 0.0016$, and $n = 0.012$.

$$Q = (A)(1.486/n)(R^{2/3})(s^{1/2}) \qquad 3472 = (16y)(1.486/0.012)[16y/(y + 16 + y)]^{2/3}(0.0016)^{1/2}$$
$$y = 14.65 \text{ ft} \qquad \text{(by trial and error)} \qquad (p_w)_1 = 16 + (2)(14.65) + (2)(2) = 49.30 \text{ ft}$$

For the channel of Prob. 14.196, $(p_w)_2 = (\pi)(10) + 3 + 3 + (2)(2) = 41.42 \text{ ft}$. Ratio of cost $= (p_w)_1/(p_w)_2 = 49.30/41.42 = 1.19$.

14.212 Do Prob. 14.204 using the friction-factor approach. Assume rough-zone flow and check at the conclusion of the problem whether this is appropriate.

▮ $R = A/p_w = (1)(10)/(1 + 10 + 1) = 0.8333 \text{ m} \qquad 1/f^{1/2} = 2.16 - 2 \log(\epsilon/R) = 2.16 - 2 \log(0.001/0.8333)$
$$f = 0.01562 \qquad v = (8gsR/f)^{1/2} = [(8)(9.807)(0.001)(0.8333)/0.01562]^{1/2} = 2.046 \text{ m/s}$$
$$Q = Av = [(1)(10)](2.046) = 20.5 \text{ m}^3/\text{s}$$
$$N_R = D_h v/v = (4R)(v)/v = [(4)(0.8333)](2.046)/(1.52 \times 10^{-6}) = 4.49 \times 10^6$$

For $f = 0.01562$ and $N_R = 4.49 \times 10^6$, we are in the rough zone according to Fig. A-5.

14.213 For a rectangular channel with a flow of 20 m³/s at a velocity of 5 m/s, what should width b and depth y be for the best hydraulic section?

▮ $A = Q/v = \frac{20}{5} = 4.000 \text{ m}^2$. For a rectangular section, the best hydraulic section has a width equal to twice the depth. Hence, $(b)(b/2) = 4.000$, $b = 2.828 \text{ m}$, $y = 2.828/2 = 1.414 \text{ m}$.

14.214 There is a steady flow at 5 m³/s at a slope of 0.002. What is the proper width b of a rectangular channel if we wish to minimize the wetted perimeter for the sake of economy of construction. The channel is made of unfinished concrete ($n = 0.0015$).

▮ $Q = (A)(1.0/n)(R^{2/3})(s^{1/2})$. For a rectangular section, the best hydraulic section has a width b equal to twice the depth. Hence, $A = (b)(b/2) = b^2/2$, $R = A/p_w = (b^2/2)/(b/2 + b + b/2) = b/4$, $5 = (b^2/2)(1.0/0.015)(b/4)^{2/3}(0.002)^{1/2}$, $b = 2.226 \text{ m}$.

14.215 A stream has a speed of about 16 ft/s and is 2 ft deep. If a thin obstruction such as a reed is present, what is the angle of the waves formed relative to the direction of motion of the stream?

▮ $$\sin \alpha = \sqrt{gy}/v = \sqrt{(32.2)(2)}/16 \qquad \alpha = 30.1°$$

14.216 A small boat is moving in shallow water where the depth is 2 m. A small bow wave is formed so that it makes an angle of 70° with the centerline of the boat (see Fig. 14-56). What is the speed of the boat?

▮ $$\sin \alpha = \sqrt{gy}/v \qquad \sin 70° = \sqrt{(9.807)(2)}/v \qquad v = 4.71 \text{ m/s}$$

14.217 A stone is thrown into a pond. A wave is formed, which has an amplitude of about 1 in and a speed of about 5 ft/s. Estimate the depth of the pond where these measurements are made.

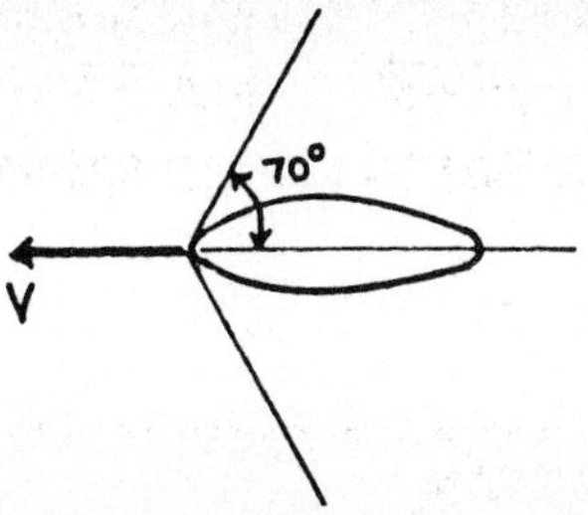

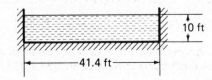

Fig. 14-56

$\blacksquare \qquad (g)(y + \Delta y/2)(y + \Delta y) = c^2 y \qquad 32.2[y + (\tfrac{1}{12})/2](y + \tfrac{1}{12}) = (5)^2 y \qquad y^2 - 0.6514y + 0.003472 = 0$

$$y = [-(-0.6514) \pm \sqrt{(-0.6514)^2 - (4)(1)(0.003472)}]/[(2)(1)] = 0.646 \text{ ft}$$

14.218 Consider a uniform flow in a channel of depth 0.4 m with a speed of 3 m/s. A small disturbance is created on the surface, forming a gravity wave. What is the difference in time at which an observer 10 m downstream from the disturbance first feels the wave as compared with an observer 10 m upstream of the disturbance? Observers and center of disturbance are positioned along a straight line.

$\blacksquare \qquad\qquad c = \sqrt{gy} = \sqrt{(9.807)(0.4)} = 1.981 \text{ m/s} \qquad t = d/v$

Downstream: $\qquad\qquad\qquad t = 10/(3 + 1.981) = 2.01 \text{ s}$

Upstream: $\qquad\qquad t = 10/(3 - 1.981) = 9.81 \text{ s} \qquad \Delta t = 9.81 - 2.01 = 7.80 \text{ s}$

14.219 A wide rectangular channel excavated from clean earth has a flow of 3 $(m^3/s)/m$. What is the critical depth and the minimum specific energy?

$\blacksquare \qquad\qquad y_c = (q^2/g)^{1/3} = (3^2/9.807)^{1/3} = 0.972 \text{ m} \qquad E_{\min} = (\tfrac{3}{2})(y_c) = (\tfrac{3}{2})(0.972) = 1.458 \text{ m}$

14.220 For the information given in Prob. 14.219, find the slope for critical normal flow. If $y = 3$ m at a section, what is the Froude number at this section for the given flow? Water is at 5 °C ($\epsilon = 0.037$ m).

$\blacksquare \qquad s_c = f/8 \qquad D_h = 4R = 4y_c = (4)(0.972) = 3.888 \text{ m} \qquad v = q/y_c = 3/0.972 = 3.086 \text{ m/s}$

$\qquad N_R = D_h v/\nu = (3.888)(3.086)/(1.52 \times 10^{-6}) = 7.89 \times 10^6 \qquad \epsilon/D_h = 0.037/3.888 = 0.0095$

From Fig. A-5, $f = 0.0375$.

$\qquad s_c = 0.0375/8 = 0.00469 \qquad v = 3/[(3)(1)] = 1.000 \text{ m/s} \qquad N_F = v/\sqrt{gy} = 1.000/\sqrt{(9.807)(3)} = 0.184$

14.221 A wide rectangular channel has a critical depth of 2 m and a critical slope of 0.001. What is the volume of flow for this condition? What is the depth of flow for normal flow with a value of flow of 5 $(m^3/s)/m$ with the above slope? Water is at 5 °C.

$\blacksquare \qquad\qquad y_c = (q^2/g)^{1/3} \qquad 2 = (q^2/9.807)^{1/3} \qquad q = 8.86 \text{ m}^3/\text{s/m}$

Assume rough-flow zone:

$$v = (8gsR/f)^{1/2} \qquad s_c = f/8 \qquad 0.001 = f/8 \qquad f = 0.008$$

$$5/[(1)(y)] = [(8)(9.807)(0.001)(y)/0.008]^{1/2} \qquad y = 1.366 \text{ m}$$

14.222 When the flow in a wide, finished-concrete, rectangular channel has a Froude number of unity, the depth is 1 m. What is the Froude number when the depth is 1.5 m for the same mass flow?

$\blacksquare \qquad\qquad y_c = (q^2/g)^{1/3} \qquad 1 = (q^2/9.807)^{1/3} \qquad q = 3.132 \ (m^3/s)/m$

$\qquad v = 3.132/1.5 = 2.088 \text{ m/s} \qquad N_F = v/\sqrt{gy} = 2.088/\sqrt{(9.807)(1.5)} = 0.544$

14.223 At a section in the rectangular channel shown in Fig. 14-57, the average velocity is 10 fps. Is the flow tranquil ($N_F < 1$) or shooting ($N_F > 1$)?

$\blacksquare \qquad q = (Av)/b = [(10)(41.4)](10)/41.4 = 100 \text{ cfs/ft} \qquad y_c = (q^2/g)^{1/3} = (100^2/32.2)^{1/3} = 6.77 \text{ ft}$

Since $y_c < 10$ ft, flow is tranquil.

Fig. 14-57

14.224 In Prob. 14.223, what is the specific energy? What other depth is possible for this energy?

    ❚ $E = q^2/2y^2g + y = 100^2/[(2)(10)^2(32.2)] + 10 = 11.55$ ft $11.55 = 100^2/[(2)(y)^2(32.2)] + y$

         $y^3 - 11.55y^2 + 155.3 = 0$ $y = 4.79$ ft   (by trial and error)

14.225 For 200 cfs of water flowing in a channel 10 ft wide, what is the minimum specific energy possible for this flow? What are the critical depth and critical velocity?

    ❚ $gA^3 = bQ^2$ $(32.2)(10y_c)^3 = (10)(200)^2$ $y_c = 2.316$ ft $E_{min} = A_c/2b + y_c$

         $E_{min} = (10)(2.316)/[(2)(10)] + 2.316 = 3.474$ ft

         $v_c = Q/A_c = 200/[(2.316)(10)] = 8.64$ ft/s

14.226 What is the critical depth for a rectangular finished-concrete channel of width 3 m? (The channel cannot be considered to be a wide channel.) What is the slope for critical normal flow? The flow is 2 m³/s. Water is at 5 °C ($\epsilon = 0.001$ m).

    ❚ $gA^3 = bQ^2$ $(9.807)(3y_c)^3 = (3)(2)^2$ $y_c = 0.357$ m $s_c = fp_w/8b$ $\epsilon/D_h = \epsilon/4R$

         $R = A/p_w = (3)(0.357)/(0.357 + 3 + 0.357) = 0.2884$ m

         $\epsilon/D_h = 0.001/[(4)(0.2884)] = 0.00087$

         $N_R = D_h v/v = (4R)(Q/A)/v = (4)(0.2884)\{2/[(3)(0.357)]\}/(1.52 \times 10^{-6}) = 1.42 \times 10^6$

     From Fig. A-5, $f = 0.019$. $s_c = (0.019)(0.357 + 3 + 0.357)/[(8)(3)] = 0.00294$.

14.227 What is the critical depth for a triangular cross section for a flow of 5 m³/s? The angle between sides is 60°.

    ❚ See Fig. 14-58.

         $gA^3 = bQ^2$ $b = (2)(y_c \tan 30°) = 1.155y_c$ $A = (y_c)(y_c \tan 30°) = 0.5774y_c^2$

         $(9.807)(0.5774y_c^2)^3 = (1.155y_c)(5)^2$ $y_c = 1.725$ m

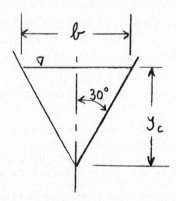

Fig. 14-58

14.228 What is the critical depth of a trapezoidal cross section for a flow of 10 m³/s? The width at the base is 3 m and angle θ at the sides is 60°.

    ❚ $gA^3 = bQ^2$ $b = 3 + (2)(y_c/\tan 60°) = 3 + 1.155y_c$

         $A = 3y_c + (y_c)(y_c/\tan 60°) = 3y_c + 0.5774y_c^2$

     $(9.807)(3y_c + 0.5774y_c^2)^3 = (3 + 1.155y_c)(10)^2$ $y_c = 0.976$ m   (by trial and error)

14.229 Shown in Fig. 14-59 is a partially filled pipe discharging 450 cfs. What is the critical depth?

    ❚ $gA^3 = bQ^2$ $b = (2)(4 \cos \theta) = 8 \cos \theta$

     $A = [(\pi)(8^2/4)]/2 + [(\pi)(8^2/4)](2\theta/360) + (\frac{8}{2})(\cos \theta)(\frac{8}{2})(\sin \theta)$

         $= 25.13 + 0.2793\theta + (16)(\cos \theta)(\sin \theta)$

     $(32.2)[25.13 + 0.2793\theta + (16)(\cos \theta)(\sin \theta)]^3 = (8 \cos \theta)(450)^2$

         $\theta = 20.5°$ (by trial and error)

         $y_c = \frac{8}{2} + (\frac{8}{2})(\sin 20.5°) = 5.40$ ft

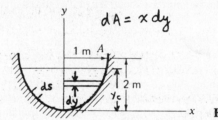

Fig. 14-59

14.230 What is the critical depth of the parabolic channel shown in Fig. 14-60 when there is a flow of $3 \text{ m}^3/\text{s}$? Position A on the channel has the coordinates shown in the figure.

▌ $gA^3 = bQ^2$, $x^2 = cy$. When $x = 1$, $y = 2$: $(1)^2 = (c)(2)$, $c = \frac{1}{2}$. Therefore, $x^2 = (\frac{1}{2})(y)$, or $y = 2x^2$.

$$A_c = (2) \int_0^{y_c} x \, dy = (2) \int_0^{y_c} (y/2)^{1/2} \, dy = 1.414[(\tfrac{2}{3})(y)^{3/2}]_0^{y_c}$$

$$= 1.414[(\tfrac{2}{3})(y_c)^{3/2}] = 0.9427 y_c^{3/2}$$

$$b = 2x = (2)(\sqrt{y_c/2}) = 1.414 y_c^{1/2}$$

$$(9.807)(0.9427 y_c^{3/2})^3 = (1.414 y_c^{1/2})(3)^2 \qquad y_c = 1.116 \text{ m}$$

y

$dA = x \, dy$

1 m A

ds

2 m

y_c

dy

x **Fig. 14-60**

14.231 In Prob. 14.230, what is the critical slope for normal flow? The friction factor is 0.015.

▌ $s_c = (f)(p_w)_c/8b$. To evaluate $(p_w)_c$, consider a differential length along the wetted perimeter (ds in Fig. 14-60): $ds = \sqrt{dx^2 + dy^2} = \sqrt{1 + (dy/dx)^2} \, dx$. From Prob. 14.230, when $y = y_c$, $x = \sqrt{y_c/2} = \sqrt{1.116/2} = 0.7470$ m.

$$(p_w)_c = \int ds = (2) \int_0^{0.7470} \sqrt{1 + (dy/dx)^2} \, dx \qquad y = 2x^2 \qquad dy/dx = 4x$$

$$(p_w)_c = (2) \int_0^{0.7470} \sqrt{1 + (4x)^2} \, dx = (8) \int_0^{0.7470} \sqrt{(\tfrac{1}{16} + x^2)} \, dx$$

$$= (\tfrac{8}{2})[(x)(\sqrt{\tfrac{1}{16} + x^2}) + \tfrac{1}{16}\{\log(x + \sqrt{\tfrac{1}{16} + x^2})\}]_0^{0.7470} = 2.400 \text{ m}$$

$$b = (2)(0.7470) = 1.494 \text{ m} \qquad s_c = (0.015)(2.400)/[(8)(1.494)] = 0.00301$$

14.232 At a section in the triangular channel of Fig. 14-61 the average velocity is 10 ft/s. Is the flow tranquil or shooting?

▌ $\tan(45°/2) = (b/2)/10 \qquad b = 8.284 \text{ ft} \qquad A = 10b/2 = (10)(8.282)/2 = 41.41 \text{ ft}^2$

$$Q = Av = (41.41)(10) = 414.1 \text{ ft}^3/\text{s} \qquad gA_c^3 = b_c Q^2$$

$$\tan(45°/2) = (b_c/2)/y_c \qquad b_c = 0.8284 y_c$$

$$A_c = y_c b_c/2 = (y_c)(0.8284 y_c)/2 = 0.4142 y_c^2$$

$$(32.2)(0.4142 y_c^2)^3 = (0.8284 y_c)(414.1)^2 \qquad y_c = 9.09 \text{ ft}$$

Since $[y = 10 \text{ ft}] > [y_c = 9.09 \text{ ft}]$, the flow is tranquil.

14.233 Water is moving in Fig. 14-62 with a velocity of 1 ft/s and a depth of 3 ft. It approaches a smooth rise in the channel bed of 1 ft. What should the estimated depth be after the rise? The channel is rectangular.

▌ $E = q^2/2y^2g + y \qquad Q = Av = (3b)(1) = 3b \qquad q = Q/b = 3b/b = 3 \text{ cfs/ft}$

$$E_1 = 3^2/[(2)(3)^2(32.2)] + 3 = 3.016 \text{ ft}$$

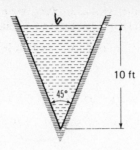

Fig. 14-61

Assuming no losses, $E_1 = E_2 + 1$, $3.016 = E_2 + 1$, $E_2 = 2.016$ ft; $y_c = (q^2/g)^{1/3} = (3^2/32.2)^{1/3} = 0.654$ ft. With surface elevation increasing, E must decrease. We must have one value of y which must be for tranquil flow. Hence, $2.016 = 3^2/[(2)(y_2)^2(32.2)] + y_2$, $y_2 = 1.98$ ft (by trial and error).

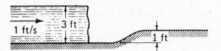

Fig. 14-62

14.234 In Fig. 14-63, a flow of 0.2 cfs flows over the rectangular channel of width 3 ft. If there is a smooth drop of 2 in, what is the elevation of the free surface above the bed of the channel after the drop? The velocity before the drop is 0.3 fps.

$$Q = Av \qquad 0.2 = (3y_1)(0.3) \qquad y_1 = 0.2222 \text{ ft}$$
$$y_c = (q^2/g)^{1/3} = [(0.2/3)^2/32.2]^{1/3} = 0.05168 \text{ ft}$$

Since $y_c < y_1$, we have tranquil flow: $E = q^2/2y^2g + y$, $E_1 = (0.2/3)^2/[(2)(0.2222)^2(32.2)] + 0.2222 = 0.2236$ ft. Assuming no losses, $E_2 = E_1 + \frac{2}{12} = 0.2236 + \frac{2}{12} = 0.3903$ ft. We must have one depth downstream greater than y_1: $0.3903 = (0.2/3)^2/[(2)(y_2)^2(32.2)] + y_2$, $y_2 = 0.390$ ft (by trial and error).

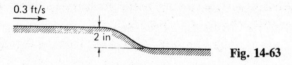

Fig. 14-63

14.235 Using the Powell equation, what quantity of liquid will flow in a smooth rectangular channel 2 ft wide, on a slope of 0.10 if the depth is 1.00 ft? Use $v = 0.00042$ ft^2/s.

▎ $C = -42 \log (C/N_R + \epsilon/R)$. For smooth channels, ϵ/R is small and can be neglected; hence,

$$C = 42 \log (N_R/C) \qquad v = C\sqrt{Rs} \qquad N_R = 4Rv/v = 4RC\sqrt{Rs}/v$$
$$N_R/C = (4)[(2)(1)/(1 + 2 + 1)]^{3/2}(0.010)^{1/2}/0.00042 = 336.7$$
$$C = (42)(\log 336.7) = 106.1$$
$$Q = CA\sqrt{Rs} = (106.1)[(2)(1)]\sqrt{[(2)(1)/(1 + 2 + 1)](0.010)} = 15.0 \text{ ft}^3/\text{s}$$

14.236 Determine C by the Powell equation for a 2 ft by 1 ft rectangular channel, if $v = 5.50$ fps, $\epsilon/R = 0.0020$, and $v = 0.00042$ ft^2/s.

▎ $$N_R = 4Rv/v = (4)[(2)(1)/(1 + 2 + 1)](5.50)/0.00042 = 26\,190$$
$$C = -42 \log (C/N_R + \epsilon/R) = (-42)[\log (C/26\,190 + 0.0020)] \qquad C = 95 \quad \text{(by trial and error)}$$

14.237 Show a correlation between roughness factor f and roughness factor n.

▎ Taking the Manning formula as a basis of correlation, $C = \sqrt{8g/f} = 1.486R^{1/6}/n$, $1/\sqrt{f} = 1.486R^{1/6}/(n\sqrt{8g})$, $f = 8gn^2/(2.208R^{1/3})$.

14.238 What is the average shear stress at the sides and bottom of a rectangular flume 12 ft wide, flowing 4 ft deep, and laid on a slope of 1.60 ft per 1000 ft?

▎ $$\tau_0 = \gamma Rs = 62.4[(4)(12)/(4 + 12 + 4)](1.60/1000) = 0.240 \text{ lb/ft}^2$$

14.239 What flow can be expected in a 4-ft-wide rectangular, concrete-lined channel laid on a slope of 4 ft in 10 000 ft, if the water flows 2 ft deep. Solve using the Manning formula with $n = 0.015$.

$\blacksquare$ $Q = (A)(1.486/n)(R^{2/3})(s^{1/2}) = [(4)(2)](1.486/0.015)[(4)(2)/(2 + 4 + 2]^{2/3}(4/10\,000)^{1/2} = 15.9 \text{ ft}^3/\text{s}$

14.240 Solve Prob. 14.239 using Kutter's C.

$\blacksquare$ $Q = AC\sqrt{Rs}$ $\qquad s = 4/10\,000 = 0.0004$ $\qquad R = A/p_w = (4)(2)/(2 + 4 + 2) = 1.000 \text{ ft}$

From Table A-15, with $s = 0.0004$, $R = 1.000$ ft, and $n = 0.015$, $C = 98$. $Q = [(4)(2)](98)\sqrt{(1.000)(0.0004)} = 15.7 \text{ ft}^3/\text{s}$.

14.241 In a hydraulics laboratory, a flow of 14.56 cfs was measured from a rectangular channel flowing 4 ft wide and 2 ft deep. If the slope of the channel was 0.00040, what is the roughness factor for the lining of the channel? Solve using Kutter's formula.

$\blacksquare$ $Q = AC\sqrt{Rs}$ $\quad 14.56 = [(4)(2)](C)\sqrt{[(4)(2)/(2 + 4 + 2)](0.00040)}$ $\quad C = 91$

From Table A-15, by interpolation, $n = 0.016$.

14.242 Solve Prob. 14.241 using Manning's formula.

$\blacksquare$ $$Q = (A)(1.486/n)(R^{2/3})(s^{1/2})$$
$$14.56 = [(4)(2)](1.486/n)[(4)(2)/(2 + 4 + 2)]^{2/3}(0.00040)^{1/2} \qquad n = 0.0163$$

14.243 On what slope should a 24-in vitrified sewer pipe be laid in order that 6.00 cfs will flow when the sewer is half-full? Use $n = 0.013$.

$\blacksquare$ $Q = (A)(1.486/n)(R^{2/3})$ $\quad 6.00 = [(\frac{1}{2})(\pi)(\frac{24}{12})^2/4](1.486/0.013)[(\frac{24}{12})/4]^{2/3}(s)^{1/2}$ $\quad s = 0.00281$

14.244 What would the slope be in Prob. 14.243 if the sewer flows full?

$\blacksquare$ $Q = (A)(1.486/n)(R^{2/3})(s^{1/2})$ $\quad 6.00 = [(\pi)(\frac{24}{12})^2/4](1.486/0.013)[(\frac{24}{12})/4]^{2/3}(s)^{1/2}$ $\quad s = 0.000703$

14.245 A trapezoidal channel, bottom width 20 ft and side slopes 1 to 1, flows 4 ft deep on a slope of 0.0009. For a value of $n = 0.025$, what is the uniform discharge?

$\blacksquare$ $$Q = (A)(1.486/n)(R^{2/3})(s^{1/2}) \qquad A = (20)(4) + (4)(4) = 96.00 \text{ ft}^2$$
$$p_w = 20 + (2)(4)(\sqrt{2}) = 31.31 \text{ ft}$$
$$Q = (96.00)(1.486/0.025)(96.00/31.31)^{2/3}(0.0009)^{1/2} = 361 \text{ ft}^3/\text{s}$$

14.246 Two concrete pipes ($C = 100$) must carry the flow from an open channel of half-square section 6 ft wide and 3 ft deep ($C = 120$). The slope of both structures is 0.00090. (**a**) Determine the diameter of the pipes. (**b**) Find the depth of water in the rectangular channel after it has become stabilized, if the slope is changed to 0.00160, using $C = 120$.

$\blacksquare$ (**a**) $\qquad Q_{\text{channel}} = Q_{\text{pipes}} \qquad (AC\sqrt{Rs})_{\text{channel}} = (2)(AC\sqrt{Rs})_{\text{pipe}}$

$[(3)(6)](120)\sqrt{[(3)(6)/(3 + 6 + 3)](0.00090)} = (2)[(\pi)(D)^2/4](100)\sqrt{(D/4)(0.00090)}$ $\qquad D = 4.08 \text{ ft}$

(**b**) $\qquad Q = [(3)(6)](120)\sqrt{[(3)(6)/(3 + 6 + 3)](0.00090)} = 79.36 \text{ ft}^3/\text{s}$

$79.36 = [(y)(6)](120)\sqrt{[(y)(6)/(y + 6 + y)](0.001600)}$ $\quad y = 2.39 \text{ ft}$ (by trial and error)

14.247 How deep will water flow at the rate of 240 cfs in a rectangular channel 20 ft wide, laid on a slope of 0.00010? Use $n = 0.0149$.

$\blacksquare$ $Q = (A)(1.486/n)(R^{2/3})(s^{1/2})$ $\quad 240 = (20y)(1.486/0.0149)[20y/(y + 20 + y)]^{2/3}(0.00010)^{1/2}$

$y = 5.27 \text{ ft}$ (by trial and error)

14.248 How wide must a rectangular channel be constructed in order to carry 500 cfs at a depth of 6 ft on a slope of 0.00040? Use $n = 0.010$.

$\blacksquare$ $Q = (A)(1.486/n)(R^{2/3})(s^{1/2})$ $\quad 500 = (6b)(1.486/0.010)[6b/(6 + b + 6)]^{2/3}(0.00040)^{1/2}$

$b = 13.1 \text{ ft}$ (by trial and error)

14.249 A rectangular channel carries 200 cfs. Find the critical depth and the critical velocity for (a) a width of 12 ft and (b) a width of 9 ft. (c) What slope will produce the critical velocity in (a) if $n = 0.020$?

    ▊ (a) $y_c = (q^2/g)^{1/3} = [(\frac{200}{12})^2/32.2]^{1/3} = 2.05$ ft $v_c = \sqrt{gy_c} = \sqrt{(32.2)(2.05)} = 8.13$ ft/s

     (b) $y_c = [(\frac{200}{9})^2/32.2]^{1/3} = 2.48$ ft $v_c = \sqrt{(32.2)(2.48)} = 8.94$ ft/s

     (c) $v = (1.486/n)(R^{2/3})(s^{1/2})$ $8.13 = (1.486/0.020)[(12)(2.05)/(2.05 + 12 + 2.05)]^{2/3}(s)^{1/2}$ $s = 0.00680$

14.250 A trapezoidal channel with side slopes of 2 horizontal to 1 vertical is to carry a flow of 590 cfs. For a bottom width of 12 ft, calculate the (a) critical depth and (b) critical velocity.

    ▊ (a) $Q^2/g = A^3/b$ $b = 12 + 4y$ $A = 12y + 2y^2$

     $590^2/32.2 = (12y + 2y^2)^3/(12 + 4y)$ $y = y_c = 3.46$ ft   (by trial and error)

     (b) $v_c = \sqrt{gA/b} = \sqrt{(32.2)[(12)(3.46) + (2)(3.46)^2]/[12 + (4)(3.46)]} = 9.03$ ft/s

14.251 A trapezoidal channel has a bottom width of 20 ft, side slopes of 1 to 1, and flows at a depth of 3.00 ft. For $n = 0.015$, and a discharge of 360 cfs, calculate (a) the normal slope, (b) the critical slope and critical depth for 360 cfs, and (c) the critical slope at the normal depth of 3.00 ft.

    ▊ (a) $Q = (A)(1.486/n)(R^{2/3})(s^{1/2})$ $A = (20)(3) + (3)(3) = 69.00$ ft^2 $p_w = 20 + (2)(\sqrt{3^2 + 3^2}) = 28.49$ ft

     $360 = (69.00)(1.486/0.015)(69.00/28.49)^{2/3}(s)^{1/2}$ $s = 0.000853$

     (b) $v_c = \sqrt{gA/b} = \sqrt{(32.2)(20y_c + y_c^2)/(20 + 2y_c)}$

Also,

$$v_c = Q/A_c = 360/(20y_c + y_c^2) \qquad \sqrt{(32.2)(20y_c + y_c^2)/(20 + 2y_c)} = 360/(20y_c + y_c^2)$$

$$y_c = 2.08 \text{ ft} \qquad \text{(by trial and error)}$$

$$360 = [(20)(2.08) + 2.08^2](1.486/0.015)\{[(20)(2.08) + 2.08^2]/[20 + (2)(\sqrt{2.08^2 + 2.08^2})]\}^{2/3}(s_c)^{1/2} \qquad s_c = 0.00291$$

     (c) $v_c = \sqrt{(32.2)[(20)(3.00) + 3.00^2]/[20 + (2)(3.00)]} = 9.24$ ft/s $v_c = (1.486/n)(R^{2/3})(s^{1/2})$

$$9.24 = (1.486/0.015)\{[(20)(3.00) + 3.00^2]/[20 + (2)(\sqrt{3.00^2 + 3.00^2})]\}^{2/3}(s_c)^{1/2} \qquad s_c = 0.00267$$

14.252 A rectangular channel, 30 ft wide, carries 270 cfs when flowing 3.00 ft deep. (a) What is the specific energy? (b) Is the flow subcritical or supercritical?

    ▊ (a) $v = Q/A = 270/[(3.00)(30)] = 3.00$ ft/s $E = v^2/2g + y = 3.00^2/[(2)(32.2)] + 3.00 = 3.14$ ft

     (b) $y_c = (q^2/g)^{1/3} = [(\frac{270}{30})^2/32.2]^{1/3} = 1.36$ ft

Since $[y_c = 1.36] < [y = 3.00]$, the flow is subcritical.

14.253 A trapezoidal channel has a bottom width of 20 ft and side slopes of 2 horizontal to 1 vertical. When the depth of water is 3.50 ft, the flow is 370 cfs. (a) What is the specific energy? (b) Is the flow subcritical or supercritical?

    ▊ (a) $A = (20)(3.50) + (3.50)[(2)(3.50)] = 94.50$ ft^2 $v = Q/A = 370/94.50 = 3.915$ ft/s

     $E = v^2/2g + y = 3.915^2/[(2)(32.2)] + 3.50 = 3.74$ ft

     (b) $Q^2/g = A_c^3/b$ $370^2/32.2 = \{20y_c + (y_c)[(2)(y_c)]\}^3/[20 + (2)(2y_c)]$ $y_c = 2.05$ ft   (by trial and error)

Since $[y_c = 2.05] < [y = 3.50$ ft], the flow is subcritical.

14.254 The discharge through a rectangular channel ($n = 0.012$) 15 ft wide is 400 cfs when the slope is 1 ft in 100 ft. Is the flow subcritical or supercritical?

    ▊ $y_c = (q^2/g)^{1/3} = [(\frac{400}{15})^2/32.2]^{1/3} = 2.81$ ft $Q = (A)(1.486/n)(R^{2/3})(s^{1/2})$

     $400 = [(15)(2.81)](1.486/0.012)[(15)(2.81)/(2.81 + 15 + 2.81)]^{2/3}(s_c)^{1/2}$ $s_c = 0.00226$

Since $[s_c = 0.00226] < [s = \frac{1}{100}]$, the flow is supercritical.

14.255 A rectangular channel, 10 ft wide, carries 400 cfs. (a) Tabulate depth of flow against specific energy for depths from 1 ft to 8 ft. (b) Determine the minimum specific energy. (c) What type of flow exists when the depth is 2 ft and when it is 8 ft? (d) For $C = 100$, what slopes are necessary to maintain the depths in (c)?

▌(a) $E = v^2/2g + y = (Q/A)^2/2g + y = (Q/10y)^2/2g + y$. For $y = 1$ ft, $E = \{400/[(10)(1)]\}^2/[(2)(32.2)] + 1 = 25.8$ ft. For succeeding depths,

y, ft	E, ft
1	25.8
2	8.21
3	5.76
4	5.55
5	5.99
6	6.69
7	7.51
8	8.39

(b) $y_c = (q^2/g)^{1/3} = [(\frac{400}{10})^2/32.2]^{1/3} = 3.676$ ft $\quad E_{min} = \{400/[(10)(3.676)]\}^2/[(2)(32.2)] + 3.676 = 5.51$ ft

(c) Since $[y = 2] < [y_c = 3.676]$, the flow is supercritical for a 2-ft depth. Since $[y = 8] > [y_c = 3.676]$, the flow is subcritical for an 8-ft depth.

(d) $Q = CA\sqrt{Rs}$. For a 2-ft depth: $400 = (100)[(2)(10)]\sqrt{[(2)(10)/(2 + 10 + 2)](s)}$, $s = 0.0280$. For an 8-ft depth: $400 = (100)[(8)(10)]\sqrt{[(8)(10)/(8 + 10 + 8)](s)}$, $s = 0.000812$.

14.256 A rectangular flume ($n = 0.012$) is laid on a slope of 0.0036 and carries 580 cfs. For critical-flow conditions, what width is required?

▌ $y_c = (q^2/g)^{1/3} = [(Q/b)^2/g]^{1/3}$. Try $b = 8.0$ ft:

$$y_c = [(580/8.0)^2/32.2]^{1/3} = 5.465 \text{ ft}$$

$Q = (A)(1.486/n)(R^{2/3})(s^{1/2}) = [(5.465)(8.0)](1.486/0.012)$

$$\times [(5.465)(8.0)/(5.465 + 8.0 + 5.465)]^{2/3}(0.0036)^{1/2} = 568 \text{ ft}^3/\text{s}$$

Try $b = 8.5$ ft:

$$y_c = [(580/8.5)^2/32.2]^{1/3} = 5.249 \text{ ft}$$

$$Q = [(5.249)(8.5)](1.486/0.012)[(5.249)(8.5)/(5.249 + 8.5 + 5.249)]^{2/3}(0.0036)^{1/2} = 586 \text{ ft}^3/\text{s}$$

Try $b = 8.33$ ft:

$$y_c = [(580/8.33)^2/32.2]^{1/3} = 5.320 \text{ ft}$$

$$Q = [(5.320)(8.33)](1.486/0.012)[(5.320)(8.33)/(5.320 + 8.33 + 5.320)]^{2/3}(0.0036)^{1/2} = 580 \text{ ft}^3/\text{s}$$

Hence, $b = 8.33$ ft.

14.257 For a constant specific energy of 6.60 ft, what maximum flow may occur in a rectangular channel 10.0 ft wide?

▌ $y_c = (\frac{2}{3})(E) = (\frac{2}{3})(6.60) = 4.40 \text{ ft} \qquad v_c = \sqrt{gy_c} = \sqrt{(32.2)(4.40)} = 11.90 \text{ ft/s}$

$$Q_{max} = A_c v_c = [(10.0)(4.40)](11.90) = 524 \text{ ft}^3/\text{s}$$

14.258 A rectangular channel, 20 ft wide, $n = 0.025$, flows 5 ft deep on a slope of 14.7 ft in 10 000 ft. A suppressed weir C, 2.45 ft high, is built across the channel ($m = 3.45$). Taking the elevation of the bottom of the channel just upstream from the weir to be 100.00 ft, estimate (using one reach) the elevation of the water surface at a point A, 1000 ft upstream from the weir. See Fig. 14-64.

▌ $Q = (A)(1.486/n)(R^{2/3})(s^{1/2}) = [(20)(5)](1.486/0.025)[(20)(5)/(5 + 20 + 5)]^{2/3}(14.7/10 000)^{1/2} = 509 \text{ ft}^3/\text{s}$

Calculate the new elevation of the water surface at B (before dropdown). Note that the flow is nonuniform since the depths, velocities, and areas are not constant after the weir is installed. Estimate a depth of 6 ft just upstream from the weir (i.e., at B). $v_{approach} = Q/A = 509/[(20)(6)] = 4.24$ ft. The applicable weir formula is $Q = mb[(H + v^2/2g)^{3/2} - (v^2/2g)^{3/2}]$.

$$509 = (3.45)(20)\left\{\left[H + \frac{4.24^2}{(2)(32.2)}\right]^{3/2} - \left[\frac{4.24^2}{(2)(32.2)}\right]^{3/2}\right\} \qquad H = 3.56 \text{ ft}$$

$$y_B = 3.56 + 2.45 = 6.01 \text{ ft} \qquad \text{(estimated depth of 6 ft O.K.)}$$

The new elevation at A must lie between $101.47 + 5 = 106.47$ ft and $101.47 + 6 = 107.47$ ft. Try an elevation of 106.90 ft:

$$(A_A)_{new} = (20)(106.90 - 101.47) = 108.6 \text{ ft}^2 \qquad (v_A)_{new} = Q/A = 509/108.6 = 4.69 \text{ ft/s}$$

$$v_{mean} = (4.24 + 4.69)/2 = 4.46 \text{ ft/s} \qquad y_A = (106.90 - 101.47) = 5.43 \text{ ft}$$

$$R_{mean} = \frac{\frac{1}{2}[(6)(20) + 108.6]}{\frac{1}{2}[(6 + 20 + 6) + (5.43 + 20 + 5.43)]} = 3.637 \text{ ft}$$

$$h_L = (vn/1.486R^{2/3})^2(L) = \{(4.46)(0.025)/[(1.486)(3.637)^{2/3}]\}^2(1000) = 1.01 \text{ ft}$$

Check by the Bernoulli equation: $v_A^2/2g + z_A = v_B^2/2g + z_B + h_L$, $4.69^2/[(2)(32.2)] + 106.90 = 4.24^2/[(2)(32.2)] + 106.00 + 1.01$, $107.24 = 107.29$ (approximately). Further refinement is not necessary. Hence, use an elevation of 106.90 ft at A.

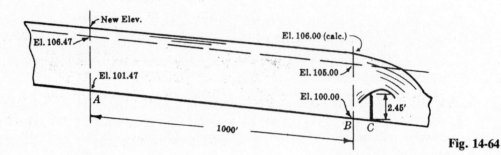

Fig. 14-64

14.259 Develop a formula for the length–energy–slope relationship for nonuniform flow problems similar to Prob. 14.258.

▮ Energy at 1 − head loss = energy at 2, $(z_1 + y_1 + v_1^2/2g) - h_L = (z_2 + y_2 + v_2^2/2g)$. Let s = slope of the energy line and s_0 = slope of the channel bottom: $s = h_L/L$, $h_L = sL$, $s_0 = (z_1 - z_2)/L$, $z_1 - z_2 = s_0L$. Therefore, $s_0L + (y_1 - y_2) + (v_1^2/2g - v_2^2/2g) = sL$, $L = [(y_1 + v_1^2/2g) - (y_2 + v_2^2/2g)]/(s - s_0) = (E_1 - E_2)/(s - s_0)$.

14.260 A rectangular flume ($n = 0.013$) is 6 ft wide and carries 66 cfs of water. At a certain section F, the depth is 3.20 ft. If the slope of the channel bed is constant at 0.000400, determine the distance from F where the depth is 2.70 ft. (Use one reach.)

▮ Assume the depth is upstream from F. Let subscript 2 refer to point F and subscript 1 to the other point.

$$L = [(y_1 + v_1^2/2g) - (y_2 + v_2^2/2g)]/(s - s_0) \qquad A_1 = (6)(2.70) = 16.20 \text{ ft}^2$$

$$v_1 = Q/A_1 = 66/16.20 = 4.074 \text{ ft/s}$$

$$R = A/p_w \qquad R_1 = 16.20/(2.70 + 6 + 2.70) = 1.421 \text{ ft}$$

$$A_2 = (6)(3.20) = 19.20 \text{ ft}^2 \qquad v_2 = Q/A_2 = 66/19.20 = 3.438 \text{ ft/s}$$

$$R_2 = 19.20/(3.20 + 6 + 3.20) = 1.548 \text{ ft} \qquad v_{mean} = (4.074 + 3.438)/2 = 3.756 \text{ ft/s}$$

$$R_{mean} = (1.421 + 1.548)/2 = 1.484 \text{ ft}$$

$$v = (1.486/n)(R^{2/3})(s^{1/2}) \qquad 3.756 = (1.486/0.013)(1.484)^{2/3}(s)^{1/2} \qquad s = 0.000638$$

$$L = \frac{\{2.70 + 4.074^2/[(2)(32.2)]\} - \{3.20 + 3.438^2/[(2)(32.2)]\}}{0.000638 - 0.000400} = -1789 \text{ ft}$$

The minus sign signifies that the section with the 2.70-ft depth is downstream from F, not upstream as assumed.

14.261 A rectangular channel, 40 ft wide, carries 900 cfs of water. The slope of the channel is 0.00283. At section 1 the depth is 4.50 ft and at section 2, 300 ft downstream, the depth is 5.00 ft. What is the average value of roughness factor n?

▮

$$L = [(y_1 + v_1^2/2g) - (y_2 + v_2^2/2g)]/(s - s_0) \qquad A_1 = (40)(4.50) = 180.0 \text{ ft}^2$$

$$v_1 = Q/A_1 = 900/180.0 = 5.000 \text{ ft/s}$$

$$R = A/p_w \qquad R_1 = 180.0/(4.50 + 40 + 4.50) = 3.673 \text{ ft}$$

$$A_2 = (40)(5.00) = 200.0 \text{ ft}^2 \qquad v_2 = Q/A = 900/200.0 = 4.500 \text{ ft/s}$$

$$R_2 = 200.0/(5.00 + 40 + 5.00) = 4.000 \text{ ft} \qquad v_{mean} = (5.000 + 4.500)/2 = 4.750 \text{ ft/s}$$

$$R_{\text{mean}} = (3.673 + 4.000)/2 = 3.836 \text{ ft}$$

$$300 = \frac{\{4.50 + 5.000^2/[(2)(32.2)]\} - \{5.00 + 4.500^2/[(2)(32.2)]\}}{s - 0.00283} \qquad s = 0.001409$$

$$v = (1.486/n)(R^{2/3})(s^{1/2}) \qquad 4.750 = (1.486/n)(3.836)^{2/3}(0.001409)^{1/2} \qquad n = 0.0288$$

14.262 A rectangular channel, 20 ft wide, has a slope of 1 ft per 1000 ft. The depth at section 1 is 8.50 ft and at section 2, 2000 ft downstream, the depth is 10.25 ft. If $n = 0.011$, determine the probable flow.

$\blacksquare \qquad E = y + v^2/2g + z \qquad E_1 = 8.50 + v_1^2/2g + (2000)(\frac{1}{1000}) \qquad E_2 = 10.25 + v_2^2/2g + 0$

$$s = \text{head loss}/L = [(10.50 - 10.25) + (v_1^2/2g - v_2^2/2g)]/2000$$

$$Q = (A)(1.486/n)(R^{2/3})(s^{1/2})$$

$$A_1 = (20)(8.50) = 170.0 \text{ ft}^2 \qquad R = A/p_w$$

$$R_1 = 170.0/(8.50 + 20 + 8.50) = 4.595 \text{ ft} \qquad A_2 = (20)(10.25) = 205.0 \text{ ft}^2$$

$$R_2 = 205.0/(10.25 + 20 + 10.25) = 5.062 \text{ ft} \qquad A_{\text{mean}} = (170.0 + 205.0)/2 = 187.5 \text{ ft}^2$$

$$R_{\text{mean}} = (4.595 + 5.062)/2 = 4.828 \text{ ft}$$

Assume $s = 0.000144$: $Q = (187.5)(1.486/0.011)(4.828)^{2/3}(0.000144)^{1/2} = 868.3 \text{ ft}^3/\text{s}$. Check on s:

$$v_1 = Q/A_1 = 868.3/170.0 = 5.108 \text{ ft/s} \qquad v_1^2/2g = 5.108^2/[(2)(32.2)] = 0.4052 \text{ ft}$$

$$v_2 = 868.3/205.0 = 4.236 \text{ ft/s}$$

$$v_2^2/2g = 4.236^2/[(2)(32.2)] = 0.2786 \text{ ft} \qquad s = [(10.50 - 10.25) + (0.4052 - 0.2786)]/2000 = 0.000188$$

This value of s (0.000188) does not equal the assumed value (0.000144); hence, try $s = 0.000210$: $Q = (187.5)(1.486/0.011)(4.828)^{2/3}(0.000210)^{1/2} = 1049 \text{ ft}^3/\text{s}$. Check on s:

$$v_1 = Q/A_1 = 1049/170.0 = 6.171 \text{ ft/s} \qquad v_1^2/2g = 6.171^2/[(2)(32.2)] = 0.5913 \text{ ft}$$

$$v_2 = 1049/205.0 = 5.117 \text{ ft/s}$$

$$v_2^2/2g = 5.117^2/[(2)(32.2)] = 0.4066 \text{ ft} \qquad s = [(10.50 - 10.25) + (0.5913 - 0.4066)]/2000 = 0.000217$$

This is close to the assumed value of s of 0.000210; hence, approximate $Q = 1050 \text{ ft}^3/\text{s}$.

14.263 A reservoir feeds a rectangular channel, 15 ft wide, $n = 0.015$, as shown in Fig. 14-65. At the entrance, the depth of water is 6.22 ft above the channel bottom. The flume is 800 ft long and drops 0.72 ft in this length. The depth behind a weir at the discharge end of the channel is 4.12 ft. Determine, using one reach, the capacity of the channel assuming the loss at the entrance to be $0.25v_1^2/2g$.

$\blacksquare \qquad p_A/\gamma + v_A^2/2g + z_A = p_1/\gamma + v_1^2/2g + z_1 + h_m \qquad 0 + 0 + 6.22 = 0 + v_1^2/2g + y_1 + 0.25v_1^2/2g$

$$L = [(v_2^2/2g + y_2) - (v_1^2/2g + y_1)]/(s_0 - s) \qquad s = (nv/1.486R^{2/3})^2$$

Solve these equations by successive trials until L approximates or equals 800 ft. Try $y_1 = 5.0$ ft:

$$6.22 = v_1^2/[(2)(32.2)] + 5.0 + 0.25v_1^2/[(2)(32.2)] \qquad v_1 = 7.928 \text{ ft/s}$$

$$q = y_1 v_1 = (5.0)(7.928) = 39.64 \text{ ft}^3/\text{s/ft} \qquad v_2 = q/y_2 = 39.64/4.12 = 9.621 \text{ ft/s}$$

$$v_{\text{mean}} = (7.928 + 9.621)/2 = 8.774 \text{ ft/s}$$

$$R_1 = (15)(5.0)/(5.0 + 15 + 5.0) = 3.000 \text{ ft} \qquad R_2 = (15)(4.12)/(4.12 + 15 + 4.12) = 2.659 \text{ ft}$$

$$R_{\text{mean}} = (3.000 + 2.659)/2 = 2.830 \text{ ft}$$

$$s = \{(0.015)(8.774)/[(1.486)(2.830)^{2/3}]\}^2 = 0.001960$$

$$L = \frac{\{(9.621)^2/[(2)(32.2)] + 4.12\} - \{(7.928)^2/[(2)(32.2)] + 5.0\}}{0.72/800 - 0.001960} = 395 \text{ ft}$$

Since L is not equal to 800 ft, try $y_1 = 5.21$ ft:

$$6.22 = v_1^2/[(2)(32.2)] + 5.21 + 0.25v_1^2/[(2)(32.2)] \qquad v_1 = 7.214 \text{ ft/s}$$

$$q = y_1 v_1 = (5.21)(7.214) = 37.58 \text{ ft}^3/\text{s/ft}$$

$$v_2 = q/y_2 = 37.58/4.12 = 9.121 \text{ ft/s} \qquad v_{\text{mean}} = (7.214 + 9.121)/2 = 8.168 \text{ ft/s}$$

$$R_1 = (15)(5.21)/(5.21 + 15 + 5.21) = 3.074 \text{ ft}$$

$$R_2 = (15)(4.12)/(4.12 + 15 + 4.12) = 2.659 \text{ ft} \qquad R_{mean} = (3.074 + 2.659)/2 = 2.866 \text{ ft}$$

$$s = \{(0.015)(8.168)/[(1.486)(2.866)^{2/3}]\}^2 = 0.001670$$

$$L = \frac{\{(9.121)^2/[(2)(32.2)] + 4.12\} - \{(7.214)^2/[(2)(32.2)] + 5.21\}}{0.72/800 - 0.001670} = 787 \text{ ft}$$

L is not exactly equal to 800 ft, but additional computations (not shown here) show that 5.21 ft is the best value of y_1 to the nearest hundredth of a foot. Hence, $Q = (37.58)(15) = 564 \text{ ft}^3/\text{s}$.

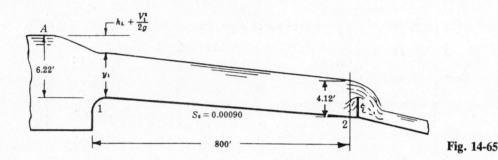

Fig. 14-65

14.264 A rectangular concrete channel 15 ft wide carries water as shown in Fig. 14-66. The channel bed slope is 0.0010. Find the theoretical rate of flow in the channel.

$$v_1^2/2g + d_1 + s_0L = v_2^2/2g + d_2 + sL \qquad v_1 = Q/[(15)(5.1)] = 0.01307Q$$

$$v_2 = Q/[(15)(3.9)] = 0.01709Q$$

$$s = (nv_m/1.486R_m^{2/3})^2 \qquad v_m = (0.01307Q + 0.01709Q)/2 = 0.01508Q \qquad R = A/p_w$$

$$R_1 = (15)(5.1)/(5.1 + 15 + 5.1) = 3.036 \text{ ft} \qquad R_2 = (15)(3.9)/(3.9 + 15 + 3.9) = 2.566 \text{ ft}$$

$$R_m = (3.036 + 2.566)/2 = 2.801 \text{ ft}$$

$$s = \{(0.013)(0.01508Q)/[(1.486)(2.801)^{2/3}]\}^2 = 4.408 \times 10^{-9}Q^2$$

$$(0.01307Q)^2/[(2)(32.2)] + 5.1 + (0.0010)(1000)$$

$$= (0.01709Q)^2/[(2)(32.2)] + 3.9 + (4.408 \times 10^{-9}Q^2)(1000) \qquad Q = 591 \text{ ft}^3/\text{s}$$

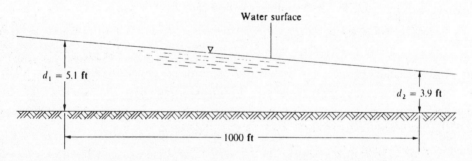

Fig. 14-66

14.265 Water flowing at the normal depth in a rectangular concrete channel that is 12.0 m wide encounters an obstruction, as shown in Fig. 14-67, causing the water level to rise above the normal depth at the obstruction and for some distance upstream. The water discharge is 126 m^3/s and the channel bottom slope is 0.00086. The depth of water just upstream from the obstruction (d_0) is 4.55 m. Find the distance upstream to the point where the water surface is at the normal depth.

$$d_c = (q^2/g)^{1/3} = [(126/12.0)^2/9.807]^{1/3} = 2.24 \text{ m} \qquad Q = (A)(1.0/n)(R^{2/3})(s^{1/2})$$

$$126 = (12.0d)(1.0/0.013)[12.0d/(d + 12.0 + d)]^{2/3}(0.00086)^{1/2}$$

$$2.256[12.0d/(d + 12.0 + d)]^{2/3} - 10.5/d = 0$$

$$d = 2.95 \text{ m} \qquad \text{(by trial and error)}$$

Since $d > d_c$, flow is subcritical, and computations should proceed upstream. The problem now is to determine the distance from the point where the depth is 4.55 m to the point upstream where the depth is 2.95 m. This will be done in ten equal depth increments of 0.16 m. The computations are given in the table below.

(1) d, m	(2) v, m/s $\dfrac{126}{12.0 \times (1)}$	(3) v_m, m/s	(4) $v^2/2g$, m $\dfrac{(2)^2}{2 \times 9.807}$	(5) R, m $\dfrac{12.0 \times (1)}{12.0 + 2 \times (1)}$	(6) R_m, m	(7) s $\left[\dfrac{0.013 \times (3)^2}{(6)^{2/3}}\right]^2$	(8) L, m $\dfrac{[(4)+(1)]_2 - [(4)+(1)]_1}{0.00086 - (7)}$
4.55	2.308		0.2716	2.588			
		2.350			2.562	0.0002662	−236
4.39	2.392		0.2917	2.535			
		2.437			2.508	0.0002946	−243
4.23	2.482		0.3141	2.481			
		2.531			2.453	0.0003272	−253
4.07	2.580		0.3394	2.425			
		2.633			2.396	0.0003654	−266
3.91	2.685		0.3676	2.367			
		2.743			2.338	0.0004098	−284
3.75	2.800		0.3997	2.308			
		2.863			2.277	0.0004626	−311
3.59	2.925		0.4362	2.246			
		2.993			2.214	0.0005246	−353
3.43	3.061		0.4777	2.182			
		3.136			2.150	0.0005989	−429
3.27	3.211		0.5257	2.117			
		3.294			2.083	0.0006893	−613
3.11	3.376		0.5811	2.048			
		3.468			2.013	0.0007997	−1580
2.95	3.559		0.6458	1.978			
							−4568 m

Hence, the answer to the problem is 4568 m.

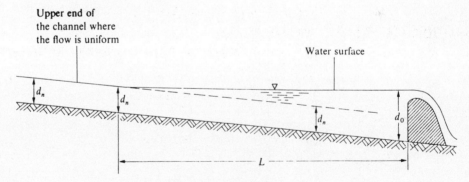

Upper end of the channel where the flow is uniform

Water surface

Fig. 14-67

14.266 Water flows in a rectangular concrete channel that is 5.0 ft wide, as shown in Fig. 14-68a, at a discharge of 16.5 cfs. Find the water-surface profile through the channel.

▌ $d_c = (q^2/g)^{1/3} = [(16.5/5.0)^2/32.2]^{1/3} = 0.70$ ft. In segment AB,

$$Q = (A)(1.486/n)(R^{2/3})(s^{1/2})$$
$$16.5 = (5.0d)(1.486/0.013)[5.0d/(d + 5.0 + d)]^{2/3}(0.00040)^{1/2}$$
$$2.286[5.0d/(d + 5.0 + d)]^{2/3} - 3.300/d = 0 \qquad d = 1.50 \text{ ft} \qquad \text{(by trial and error)}$$

Since $d > d_c$, the flow in segment AB is subcritical.
In segment BC,

$$16.5 = (5.0d)(1.486/0.013)[5.0d/(d + 5.0 + d)]^{2/3}(0.025)^{1/2}$$
$$18.07[5.0d/(d + 5.0 + d)]^{2/3} - 3.300/d = 0$$
$$d = 0.38 \text{ ft} \qquad \text{(by trial and error)}$$

Since $d < d_c$, the flow in segment BC is supercritical.

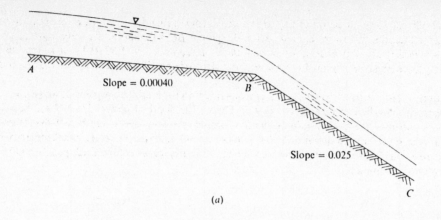

(a)

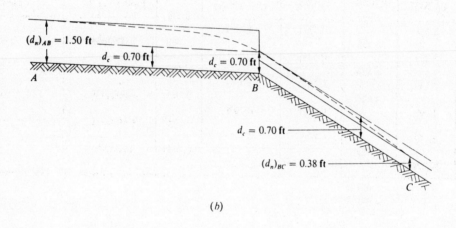

(b)

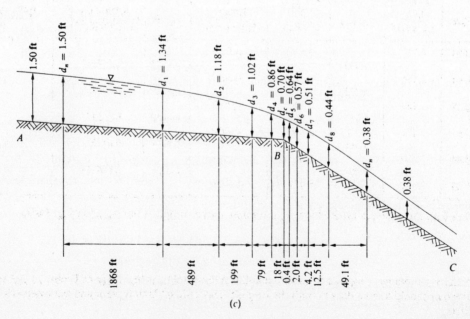

(c)

Fig. 14-68

Figure 14-68b shows the location of the critical depth (the dashed line), which is constant throughout, along with the normal depths of flow in segment AB and segment BC. Obviously, the water-surface profile cannot drop instantaneously at point B from the normal depth in segment AB (1.50 ft) to the normal depth in segment BC (0.38 ft); there must be a transition zone on both sides of point B as shown by the dashed line in Fig. 14-68b.

As a matter of fact, as the flow changes from subcritical to supercritical in going from segment AB to segment BC, it passes through the critical state at point B. Hence, the depth of flow at point B will be 0.70 ft. The problem now becomes one of determining the flow profile from the critical depth of 0.70 ft at point B upstream to the point where the normal depth of 1.50 ft is reached and downstream to the point where the normal depth of 0.38 ft is reached. These computations are carried out in the tables below. Each profile will be analyzed using five equal depth increments.

For Segment AB

(1) d, ft	(2) v, ft/s $\dfrac{16.5}{5.0 \times (1)}$	(3) v_m, ft/s	(4) $v^2/2g$, ft $\dfrac{(2)^2}{2 \times 32.2}$	(5) R, ft $\dfrac{5.0 \times (1)}{5.0 + 2 \times (1)}$	(6) R_m, ft	(7) s $\left[\dfrac{0.013 \times (3)}{1.486 \times (6)^{2/3}}\right]^2$	(8) L, ft $\dfrac{[(4)+(1)]_2 - [(4)+(1)]_1}{0.00040 - (7)}$
0.70	4.714		0.3451	0.5469			
		4.276			0.5934	0.0028063	−18
0.86	3.837		0.2286	0.6399			
		3.536			0.6822	0.0015934	−79
1.02	3.235		0.1625	0.7244			
		3.016			0.7630	0.0009985	−199
1.18	2.797		0.1215	0.8016			
		2.630			0.8370	0.0006711	−489
1.34	2.463		0.0942	0.8724			
		2.332			0.9050	0.0004755	−1868
1.50	2.200		0.0752	0.9375			

For Segment BC

(1) d, ft	(2) v, ft/s $\dfrac{16.5}{5.0 \times (1)}$	(3) v_m, ft/s	(4) $v^2/2g$, ft $\dfrac{(2)^2}{2 \times 32.2}$	(5) R, ft $\dfrac{5.0 \times (1)}{5.0 + 2 \times (1)}$	(6) R_m, ft	(7) s $\left[\dfrac{0.013 \times (3)}{1.486 \times (6)^{2/3}}\right]^2$	(8) L, ft $\dfrac{[(4)+(1)]_2 - [(4)+(1)]_1}{0.025 - (7)}$
0.70	4.714		0.3451	0.5469			
		4.935			0.5283	0.004364	0.4
0.64	5.156		0.4128	0.5096			
		5.473			0.4869	0.005985	2.0
0.57	5.789		0.5204	0.4642			
		6.130			0.4439	0.008493	4.2
0.51	6.471		0.6502	0.4236			
		6.986			0.3989	0.012720	12.5
0.44	7.500		0.8734	0.3741			
		8.092			0.3520	0.020164	49.1
0.38	8.684		1.1710	0.3299			

Based on the values computed above, the water-surface profile is illustrated in Fig. 14-68c.

14.267 Prepare a computer program to solve nonuniform flow problems like those of Probs. 14.265 and 14.266. The program should handle data in both the English Gravitational Unit System and the International System of Units.

```
C    THIS PROGRAM DETERMINES THE FLOW PROFILE FOR A NON-UNIFORM FLOW
C    IN A RECTANGULAR OPEN CHANNEL.  IT CAN BE USED FOR PROBLEMS IN
C    BOTH THE ENGLISH SYSTEM OF UNITS AND THE INTERNATIONAL SYSTEM OF
C    UNITS.
C
C    THE PROGRAM CONSIDERS TWO ADJACENT CHANNEL SEGMENTS.  CONSIDER
C    ONE SEGMENT TO GO FROM POINT "A" TO POINT "B" AND THE OTHER
C    SEGMENT TO GO FROM POINT "B" TO POINT "C" IN THE DOWNSTREAM
C    DIRECTION.  THE PROGRAM COMPUTES THE FLOW PROFILE IN SEGMENT AB
C    IN THE UPSTREAM DIRECTION (I.E., FROM "B" TO "A") BASED ON SUB-
C    CRITICAL FLOW IN SEGMENT AB AND THE FLOW PROFILE IN SEGMENT BC IN
C    THE DOWNSTREAM DIRECTION (I.E., FROM "B" TO "C") BASED ON SUPER-
C    CRITICAL FLOW IN SEGMENT BC.  THE PROGRAM CAN BE USED EITHER FOR
C    TWO SEGMENTS AS DESCRIBED ABOVE OR FOR A SINGLE SEGMENT WITH
C    COMPUTATIONS TO PROCEED UPSTREAM (SEGMENT AB) OR DOWNSTREAM (SEG-
C    MENT BC), AS DESIRED.  IN THE CASE OF A SINGLE SEGMENT, ENTER DATA
C    FOR SEGMENT AB IF FLOW IS SUBCRITICAL AND FOR SEGMENT BC IF FLOW
C    IS SUPERCRITICAL.
C
C    THIS PROGRAM IS BASED ON CONSTANT CHANNEL WIDTH, CONSTANT FLOW
C    RATE, AND CONSTANT MANNING N-VALUE THROUGHOUT AND ON A SEPARATE
C    CONSTANT CHANNEL SLOPE IN EACH OF SEGMENTS AB AND BC.
C
C    INPUT DATA MUST BE SET UP AS FOLLOWS.
C
C    CARD 1    COLUMN 1       ENTER 0 (ZERO) OR BLANK IF THE ENGLISH
C                             SYSTEM OF UNITS IS TO BE USED.  ENTER 1
C                             (ONE) IF THE INTERNATIONAL SYSTEM OF
C                             UNITS IS TO BE USED.
C              COLUMN 2       ENTER 1 (ONE) IF ONLY A SINGLE UPSTREAM
C                             COMPUTATION IS DESIRED.  (IN THIS CASE,
C                             ENTER DEPTHS AT POINTS A AND B AND SLOPE
C                             IN SEGMENT AB.  LEAVE DEPTH AT POINT C
C                             AND SLOPE IN SEGMENT BC BLANK.)  ENTER 2
C                             (TWO) IF ONLY A SINGLE DOWNSTREAM COM-
C                             PUTATION IS DESIRED.  (IN THIS CASE, ENTER
C                             DEPTHS AT POINTS B AND C AND SLOPE IN
C                             SEGMENT BC.  LEAVE DEPTH AT POINT A AND
C                             SLOPE IN SEGMENT AB BLANK.)  ENTER 3
C                             (THREE) IF COMPUTATIONS FOR BOTH SEGMENTS
C                             ARE DESIRED. (IN THIS CASE, ENTER DEPTHS
C                             AT POINTS A, B, AND C AND SLOPES IN SEG-
C                             MENTS AB AND BC.)
C              COLUMNS 3-5    ENTER INTEGER NUMBER (RIGHT ADJUSTED)
C                             GIVING NUMBER OF LENGTH INCREMENTS TO BE
C                             USED IN COMPUTING THE FLOW PROFILE IN
C                             EACH SEGMENT.
C              COLUMNS 6-80   ENTER TITLE, DATE, AND OTHER INFORMATION,
C                             IF DESIRED.
C    CARD 2    COLUMNS 1-10   ENTER NUMBER INCLUDING DECIMAL GIVING
C                             WIDTH OF RECTANGULAR CHANNEL (IN FEET OR
C                             METERS).
C              COLUMNS 11-20  ENTER NUMBER INCLUDING DECIMAL GIVING
C                             FLOW RATE (IN CUBIC FEET PER SECOND OR
C                             CUBIC METERS PER SECOND).
C              COLUMNS 21-30  ENTER NUMBER INCLUDING DECIMAL GIVING
C                             DEPTH AT POINT A (IN FEET OR METERS).
C                             THIS VALUE MAY BE LEFT BLANK, IN WHICH
C                             CASE THE "NORMAL DEPTH" IN SEGMENT AB
C                             WILL AUTOMATICALLY BE USED.
C              COLUMNS 31-40  ENTER NUMBER INCLUDING DECIMAL GIVING
C                             DEPTH AT POINT B (IN FEET OR METERS).
C                             THIS VALUE MAY BE LEFT BLANK, IN WHICH
C                             CASE THE "CRITICAL DEPTH" WILL AUTO-
C                             MATICALLY BE USED.
C              COLUMNS 41-50  ENTER NUMBER INCLUDING DECIMAL GIVING
C                             DEPTH AT POINT C (IN FEET OR METERS).
C                             THIS VALUE MAY BE LEFT BLANK, IN WHICH
C                             CASE THE "NORMAL DEPTH" IN SEGMENT BC
C                             WILL AUTOMATICALLY BE USED.
C              COLUMNS 51-60  ENTER NUMBER INCLUDING DECIMAL GIVING
C                             CHANNEL SLOPE IN SEGMENT AB.
C              COLUMNS 61-70  ENTER NUMBER INCLUDING DECIMAL GIVING
```

```
C                         CHANNEL SLOPE IN SEGMENT BC.
C             COLUMNS 71-80   ENTER NUMBER INCLUDING DECIMAL GIVING
C                         MANNING N-VALUE.
C
C      MULTIPLE DATA SETS FOR SOLVING ANY NUMBER OF PROBLEMS MAY BE
C      INCLUDED FOR PROCESSING.
C
       COMMON DMID,DUP,NSEG,Q,W,N,COEFF,G,SUP
       DIMENSION TITLE(13)
       REAL N
       INTEGER UNITS,CODE
1      READ(5,100,END=2)UNITS,CODE,NSEG,TITLE
100    FORMAT(2I1,I3,12A6,A3)
       WRITE(6,106)TITLE
106    FORMAT('1',12A6,A3,////)
       READ(5,101)W,Q,DUP,DMID,DDOWN,SUP,SDOWN,N
101    FORMAT(8F10.0)
       COEFF=1.486
       IF(UNITS.EQ.1)COEFF=1.0
       G=32.2
       IF(UNITS.EQ.1)G=9.807
       IF(DMID.LT.0.0001)DMID=((Q/W)**2/G)**(1.0/3.0)
       IF(CODE.EQ.2)GO TO 110
       IF(DUP.LT.0.0001)DUP=DNORM(W,Q,SUP,N,COEFF)
       IF(CODE.EQ.1)GO TO 109
110    IF(DDOWN.LT.0.0001)DDOWN=DNORM(W,Q,SDOWN,N,COEFF)
       IF(CODE.EQ.2)GO TO 105
109    X1=' UPST'
       X2='REAM '
       IF(UNITS.EQ.0)WRITE(6,107)X1,X2,W,Q,DUP,DMID,SUP,N
107    FORMAT(1X,'GIVEN DATA FOR ' ,2A5,' FLOW PROFILE FOR A RECTANGULAR
      *OPEN CHANNEL',//5X,'WIDTH OF CHANNEL =',F7.1,' FT',6X,'FLOW RATE O
      *F WATER =',F7.1,' CU FT/S',//5X,'DEPTH OF WATER AT UPSTREAM END OF
      * SEGMENT =',F7.2,' FT',//5X,'DEPTH OF WATER AT DOWNSTREAM END OF S
      *EGMENT =',F7.2,' FT',//5X,'SLOPE =',F10.7,20X,'MANNING N-VALUE =',
      *F6.3,///1X,'THE FLOW PROFILE WITHIN THE SEGMENT IS GIVEN IN THE TA
      *BLE BELOW'///4X,'DEPTH (FT)',10X,'VELOCITY (FT/S)',10X,'LENGTH OF
      *SUBSEGMENT (FT)')
       IF(UNITS.EQ.1)WRITE(6,108)X1,X2,W,Q,DUP,DMID,SUP,N
108    FORMAT(1X,'GIVEN DATA FOR ' ,2A5,' FLOW PROFILE FOR A RECTANGULAR
      *OPEN CHANNEL',//5X,'WIDTH OF CHANNEL =',F7.1,' M',6X,'FLOW RATE OF
      * WATER =',F7.1,' CU M/S',//5X,'DEPTH OF WATER AT UPSTREAM END OF S
      *EGMENT =',F7.2,' M',//5X,'DEPTH OF WATER AT DOWNSTREAM END OF SEGM
      *ENT =',F7.2,' M',//5X,'SLOPE =',F10.7,20X,'MANNING N-VALUE =',
      *F6.3,///1X,'THE FLOW PROFILE WITHIN THE SEGMENT IS GIVEN IN THE TA
      *BLE BELOW'///4X,'DEPTH (M)',10X,'VELOCITY (M/S)',10X,'LENGTH OF SU
      *BSEGMENT (M)')
       CALL LENGTH
       IF(CODE.EQ.1)GO TO 1
105    X1='DOWNS'
       X2='TREAM'
       IF(UNITS.EQ.0)WRITE(6,107)X1,X2,W,Q,DMID,DDOWN,SDOWN,N
       IF(UNITS.EQ.1)WRITE(6,108)X1,X2,W,Q,DMID,DDOWN,SDOWN,N
       DUP=DDOWN
       SUP=SDOWN
       CALL LENGTH
       GO TO 1
2      STOP
       END
       FUNCTION DNORM(W,Q,S,N,COEFF)
       D=0.001
       TRY1=COEFF/N*(W*D/(W+2.0*D))**(2.0/3.0)*SQRT(S)-Q/W/D
104    D=D+0.001
       TRY2=COEFF/N*(W*D/(W+2.0*D))**(2.0/3.0)*SQRT(S)-Q/W/D
       IF(TRY1*TRY2)102,102,103
103    TRY1=TRY2
       GO TO 104
102    DNORM=D-0.0005
       RETURN
       END
       SUBROUTINE LENGTH
       COMMON DMID,DUP,NSEG,Q,W,N,COEFF,G,SUP
       DINC=(DMID-DUP)/FLOAT(NSEG)
```

```
        D1=DMID
        TOTAL=0.0
        DO 102  J=1,NSEG
        D2=D1-DINC
        V1=Q/W/D1
        V2=Q/W/D2
        VMEAN=(V1+V2)/2.0
        V2G1=V1**2/2.0/G
        V2G2=V2**2/2.0/G
        HR1=W*D1/(W+2.0*D1)
        HR2=W*D2/(W+2.0*D2)
        HRMEAN=(HR1+HR2)/2.0
        SMEAN=(N*VMEAN/COEFF/HRMEAN**(2.0/3.0))**2
        SEGL=(V2G2+D2-(V2G1+D1))/(SUP-SMEAN)
        TOTAL=TOTAL+SEGL
        WRITE(6,103)D1,V1,SEGL
    103 FORMAT(1X,F10.2,11X,F12.3,/,55X,F8.1)
    102 D1=D2
        WRITE(6,103)D2,V2
        WRITE(6,104)TOTAL
    104 FORMAT(41X,'TOTAL LENGTH =',F8.1,//)
        RETURN
        END
```

14.268 Solve Prob. 14.265 using the computer program developed in Prob. 14.267.

❙ Input

```
1 2  3  4  5  6  7 8  9 1011 12131415161718192021 2223242526272829303132333435 3637 383940414243444546474849505152535455565758596061626364656667686970717273747576777879 80
11 10SAMPLE ANALYSIS ØF PRØFILE IN NØN-UNIFØRM FLØW
12.0       126.0               4.55               0.00086             0.013
```

Output

```
SAMPLE ANALYSIS OF PROFILE IN NON-UNIFORM FLOW


GIVEN DATA FOR  UPSTREAM  FLOW PROFILE FOR A RECTANGULAR OPEN CHANNEL

    WIDTH OF CHANNEL =    12.0 M     FLOW RATE OF WATER =   126.0 CU M/S

    DEPTH OF WATER AT UPSTREAM END OF SEGMENT =   2.95 M

    DEPTH OF WATER AT DOWNSTREAM END OF SEGMENT =   4.55 M

    SLOPE = 0.0008600                  MANNING N-VALUE = 0.013


THE FLOW PROFILE WITHIN THE SEGMENT IS GIVEN IN THE TABLE BELOW


    DEPTH (M)         VELOCITY (M/S)        LENGTH OF SUBSEGMENT (M)
      4.55              2.308
                                                  -235.2
      4.39              2.392
                                                  -242.8
      4.23              2.482
                                                  -252.6
      4.07              2.579
                                                  -265.6
      3.91              2.685
                                                  -283.6
      3.75              2.799
                                                  -309.9
      3.59              2.924
                                                  -351.8
      3.43              3.060
                                                  -427.9
      3.27              3.209
                                                  -607.7
      3.11              3.374
                                                 -1530.9
      2.95              3.556

                             TOTAL LENGTH = -4508.0
```

14.269 Solve Prob. 14.266 using the computer program developed in Prob. 14.267.

❚ Input

```
1 2 3 4 5 6 7 8 9 10 11 12 13 14 15 16 17 18 19 20 21 22 23 24 25 26 27 28 29 30 31 32 33 34 35 36 37 38 39 40 41 42 43 44 45 46 47 48 49 50 51 52 53 54 55 56 57 58 59 60 61 62 63 64 65 66 67 68 69 70 71 72 73 74 75 76 77 78 79 80
03   5SAMPLE ANALYSIS ØF PRØFILE IN NØN-UNIFØRM FLØW
5.0       16.5                                        0.00040   0.025    0.013
```

Output

SAMPLE ANALYSIS OF PROFILE IN NON-UNIFORM FLOW

GIVEN DATA FOR UPSTREAM FLOW PROFILE FOR A RECTANGULAR OPEN CHANNEL

WIDTH OF CHANNEL = 5.0 FT   FLOW RATE OF WATER = 16.5 CU FT/S

DEPTH OF WATER AT UPSTREAM END OF SEGMENT = 1.50 FT

DEPTH OF WATER AT DOWNSTREAM END OF SEGMENT = 0.70 FT

SLOPE = 0.0004000   MANNING N-VALUE = 0.013

THE FLOW PROFILE WITHIN THE SEGMENT IS GIVEN IN THE TABLE BELOW

DEPTH (FT)	VELOCITY (FT/S)	LENGTH OF SUBSEGMENT (FT)
0.70	4.736	
		-17.6
0.86	3.845	
		-78.9
1.02	3.236	
		-201.2
1.18	2.793	
		-501.2
1.34	2.457	
		-1986.4
1.50	2.193	

TOTAL LENGTH = -2785.2

GIVEN DATA FOR DOWNSTREAM FLOW PROFILE FOR A RECTANGULAR OPEN CHANNEL

WIDTH OF CHANNEL = 5.0 FT   FLOW RATE OF WATER = 16.5 CU FT/S

DEPTH OF WATER AT UPSTREAM END OF SEGMENT = 0.70 FT

DEPTH OF WATER AT DOWNSTREAM END OF SEGMENT = 0.38 FT

SLOPE = 0.0250000   MANNING N-VALUE = 0.013

THE FLOW PROFILE WITHIN THE SEGMENT IS GIVEN IN THE TABLE BELOW

DEPTH (FT)	VELOCITY (FT/S)	LENGTH OF SUBSEGMENT (FT)
0.70	4.736	
		0.5
0.63	5.208	
		1.9
0.57	5.783	
		4.5
0.51	6.501	
		11.0
0.44	7.423	
		46.1
0.38	8.650	

TOTAL LENGTH = 63.9

14.270 A rectangular channel 12.0 m wide is laid on a slope of 0.0028. The depth of flow at one section is 1.50 m, while the depth of flow at another section 500 ft downstream is 1.80 m. Determine the probable rate of flow, if $n = 0.026$.

❚
$$Q = A_m v_m = (A_m)(1.0/n)R_m^{2/3}s^{1/2} \qquad A_m = [(12.0)(1.50) + (12.0)(1.80)]/2 = 19.8 \text{ m}^2$$
$$R_m = [(12.0)(1.50)/(1.50 + 12.0 + 1.50) + (12.0)(1.80)/(1.80 + 12.0 + 1.80)]/2 = 1.29 \text{ m}$$
$$Q = (19.8)(1.0/0.026)(1.29)^{2/3}s^{1/2} = 902.44s^{1/2} \qquad (1)$$
$$v_1^2/2g + d_1 + s_0 L = v_2^2/2g + d_2 + sL \qquad v_1 = Q/A_1 = Q/[(12.0)(1.50)] = Q/18.0$$
$$v_2 = Q/A_2 = Q/[(12.0)(1.80)] = Q/21.6$$
$$(Q/18.0)^2/[(2)(9.807)] + 1.50 + (0.0028)(500) = (Q/21.6)^2/[(2)(9.807)] + 1.80 + 500s \qquad (2)$$

Substituting (1) into (2) gives $(902.44s^{1/2}/18.0)^2/[(2)(9.807)] + 1.50 + 1.40 = (902.44s^{1/2}/21.6)^2[(2)(9.807)] + 1.80 + 500s$, $460.8s = 1.10$, $s = 0.002387$; $Q = (902.44)(0.002387)^{1/2} = 44.1 \text{ m}^3/\text{s}$.

14.271 Solve Prob. 14.265 using five equal depth increments to determine the distance upstream to the point where the water surface is at the normal depth. Compare the answer with the one obtained in Prob. 14.265.

❚ Using data found in Prob. 14.265,

d	v	v_m	$v^2/2g$	R	R_m	s_g	L, m
4.55	2.308		0.2716	2.588			
		2.395			2.535	0.0002805	−479
4.23	2.482		0.3141	2.481			
		2.584			2.424	0.0003465	−519
3.91	2.685		0.3676	2.367			
		2.805			2.307	0.0004362	−593
3.59	2.925		0.4362	2.246			
		3.068			2.182	0.0005621	−774
3.27	3.211		0.5257	2.117			
		3.385			2.048	0.0007446	−1732
2.95	3.559		0.6458	1.978			
							−4097

The answer using five increments is $(4568 − 4097)/4568 = 0.103$, or 10.3 percent smaller than that using ten increments.

14.272 Water flows in a rectangular concrete channel similar to the one depicted in Fig. 14-68a. If the channel width is 3.0 ft and the discharge is 12.0 cfs, determine the water-surface profile throughout the channel shown.

❚
$$d_c = [(Q/B)^2/g]^{1/3} = [(12.0/3.0)^2/32.2]^{1/3} = 0.79 \text{ ft} \qquad Q = Av = (A)(1.486/n)(R^{2/3})(s^{1/2})$$
$$12.0 = (3.0d_{AB})(1.486/0.013)[3.0d_{AB}/(3.0 + 2d_{AB})]^{2/3}(0.00052)^{1/2}$$

By trial and error, $d_{AB} = 1.76$ ft. $12.0 = (3.0d_{BC})(1.486/0.013)[3.0d_{BC}/(3.0 + 2d_{BC})]^{2/3}(0.030)^{1/2}$. By trial and error, $d_{BC} = 0.42$ ft. Hence, the water-surface profile changes from the normal depth in segment AB (1.76 ft) to the critical depth at B (0.79 ft) to the normal depth in segment BC (0.42 ft).

For Segment *AB* (*B* to *A*)

d	v	v_m	$v^2/2g$	R	R_m	s_g	L, ft
0.79	5.063		0.398	0.517			
		4.572			0.555	0.003508	−17
0.98	4.082		0.259	0.593			
		3.736			0.626	0.001995	−81
1.18	3.390		0.178	0.660			
		3.155			0.688	0.001254	−196
1.37	2.920		0.132	0.716			
		2.734			0.742	0.0008516	−510
1.57	2.548		0.101	0.767			
		2.410			0.788	0.0006107	−1863
1.76	2.273		0.080	0.810			

For Segment *BC* (*B* to *C*)

d	v	v_m	$v^2/2g$	R	R_m	s_g	L, ft
0.79	5.063		0.398	0.517			
		5.310			0.502	0.005409	0.4
0.72	5.556		0.479	0.486			
		5.903			0.468	0.007340	2.1
0.64	6.250		0.607	0.449			
		6.634			0.431	0.010346	4.5
0.57	7.018		0.765	0.413			
		7.591			0.391	0.015425	13.0
0.49	8.163		1.035	0.369			
		8.844			0.349	0.024362	53.7
0.42	9.524		1.408	0.328			

14.273 In Fig. 14-69, 400 cfs flows through the transition. The rectangular section is 8 ft wide and $y_1 = 8$ ft. The trapezoidal section is 6 ft wide at the bottom with side slopes 1:1, and $y_2 = 7.5$ ft. Determine the rise z in the bottom through the transition.

$$v_1^2/2g + y_1 = v_2^2/2g + y_2 + z + E_1 \qquad v_1 = Q/A_1 = 400/[(8)(8)] = 6.250 \text{ ft/s}$$
$$v_2 = 400/[(6)(7.5) + (7.5)(7.5)] = 3.951 \text{ ft/s}$$
$$E_1 = (0.3)(v_1^2/2g - v_2^2/2g) = (0.3)\{6.250^2/[(2)(32.2)] - 3.951^2/[(2)(32.2)]\} = 0.109 \text{ ft}$$
$$6.250^2/[(2)(32.2)] + 8 = 3.951^2/[(2)(32.2)] + 7.5 + z + 0.109 \qquad z = 0.755 \text{ ft}$$

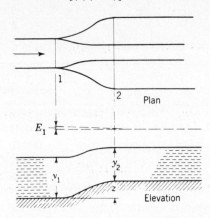

Plan

E_1

y_2

y_1

z

Elevation

Fig. 14-69

14.274 In a critical-depth meter 2 m wide with z = 0.3 m, the depth y_1 is measured to be 0.75 m. Find the discharge.

$q = (0.517)(g^{1/2})[y_1 - z + (0.55/g)(q^2/y_1^2)]^{3/2}$. Initially, neglect the last term in the equation above. $q = (0.517)(9.807)^{1/2}(0.75 - 0.3)^{3/2} = 0.489$ (m³/s)/m. Now, try $q = 0.500$ (m³/s)/m in the whole equation:

$q = (0.517)(9.807)^{1/2}[0.75 - 0.3 + (0.55/9.807)(0.500^2/0.75^2)]^{3/2} = 0.530 \ (m^3/s)/m$. Try $q = 0.536 \ (m^3/s)/m$: $q = (0.517)(9.807)^{1/2}[0.75 - 0.3 + (0.55/9.807)(0.536^2/0.75^2)]^{3/2} = 0.536 \ (m^3/s)/m$, $Q = (0.536)(2) = 1.07 \ m^3/s$.

14.275 At section 1 of a canal, the cross section is trapezoidal with bottom width $b_1 = 10$ m, depth $y_1 = 7$ m, and side slopes of 2 horizontal to 1 vertical. At section 2, 200 m downstream, the bottom is 0.08 m higher than at section 1, $b_2 = 15$ m, and side slopes are 3 horizontal to 1 vertical ($Q = 200 \ m^3/s$ and $n = 0.035$). Determine the depth of water at section 2.

❚
$$\Delta L = [(v_1^2 - v_2^2)/2g + y_1 - y_2]/(s - s_0) \qquad s = (nQ/1.0AR^{2/3})^2$$

Try $y = 6.92$ m:

$$A_1 = (10)(7) + (7)[(2)(7)] = 168.0 \ m^2 \qquad A_2 = (15)(6.92) + (6.92)[(3)(6.92)] = 247.5 \ m^2$$

$$A_{avg} = (168.0 + 247.5)/2 = 207.8 \ m^2 \qquad R = A/p_w \qquad (p_w)_1 = 10 + (2)(7)(\sqrt{2^2 + 1}) = 41.30 \ m$$

$$(p_w)_2 = 15 + (2)(6.92)(\sqrt{3^2 + 1}) = 58.77 \ m \qquad (p_w)_{avg} = (41.30 + 58.77)/2 = 50.04 \ m$$

$$R = 207.8/50.04 = 4.153 \ m$$

$$s = \{(0.035)(200)/[(1.0)(207.8)(4.153)^{2/3}]\}^2 = 0.0001700$$

$$v_1 = Q/A_1 = 200/168.0 = 1.190 \ m/s \qquad v_2 = 200/247.5 = 0.8081 \ m/s$$

$$\Delta L = \frac{(1.190^2 - 0.8081^2)/[(2)(9.807)] + 7 - 6.92}{0.0001700 - (-0.08/200)} = 209 \ m$$

This value of ΔL (209 m) is not equal to the given value (200 m). However, reworking the problem with $y_2 = 6.93$ m (not shown) yields a value of ΔL of 191 m. Thus y_2 must be approximately 6.925 m.

14.276 A trapezoidal channel, $b = 3$ m, side slopes of 1:1, $n = 0.014$, $s_0 = 0.001$, carries 28 m³/s. If the depth is 3 m at section 1, determine the water-surface profile for the next 700 m downstream.

❚ To determine whether the depth increases or decreases, the slope of the energy grade line must be computed.
$$A = (3)(3) + (3)(3) = 18.00 \ m^2 \qquad R = A/p_w = 18.00/[3 + (2)(3)(\sqrt{1^2 + 1^2})] = 1.567 \ m$$

$$s = (nQ/1.0AR^{2/3})^2 = \{(0.014)(28)/[(1.0)(18.00)(1.567)^{2/3}]\}^2 = 0.0002606$$

$$(Q^2/gA_c^3)(T_c) = \{28^2/[(9.807)(18.00)^3]\}(9) = 0.1234$$

Since $(Q^2/gA_c^3)(T_c) < 1.0$, the depth is above critical. With the depth greater than critical and the energy grade line less steep than the bottom of the channel, the specific energy is increasing. When the specific energy increases above critical, the depth of flow increases, y is then positive.

$$L = \int_{y_1}^{y_2} \frac{1 - Q^2 T/gA^3}{s_0 - (nQ/1.0AR^{2/3})^2} \, dy$$

$$L = \int_3^y \frac{1 - (28^2)(T)/[(9.807)(A)^3]}{0.001 - \{(0.014)(28)/[(1.0)(AR^{2/3})]\}^2} \, dy$$

$$L = \int_3^y \frac{1 - 79.94T/A^3}{0.001 - [0.1537/(A^2 R^{4/3})]} \, dy$$

The following table evaluates the terms of the integrand:

y	A	P	R	T	numerator	$10^6 \times$ denominator	$F(y)$	L
3	18	11.48	1.57	9	0.8766	739	1185	0
3.2	19.84	12.05	1.65	9.4	0.9038	799	1131	231.6
3.4	21.76	12.62	1.72	9.8	0.9240	843	1096	454.3
3.6	23.76	13.18	1.80	10.2	0.9392	876	1072	671.1
3.8	25.84	13.75	1.88	10.6	0.9509	901	1056	883.9

The integral $\int F(y) \, dy$ can be evaluated by plotting the curve and taking the area under it between $y = 3$ and the following values of y. As $F(y)$ does not vary greatly in this example, the average of $F(y)$ can be used for each reach (the trapezoidal rule); and when it is multiplied by Δy, the length of reach is obtained. Between $y = 3$ and $y = 3.2$, $[(1185 + 1131)/2](0.2) = 231.6$. Between $y = 3.2$ and $y = 3.4$, $[(1131 + 1096)/2](0.2) = 222.7$ and so on. Five points on it are known, so the water surface can be plotted.

14.277 After contracting below a slucie gate, water flows onto a wide horizontal floor with a velocity of 15 m/s and a depth of 0.7 m. Find the equation for the water-surface profile ($n = 0.015$).

$$x = -(\tfrac{3}{13})(1.0/nq)^2(y^{13/3} - y_1^{13/3}) + (3/4g)(1.0/n)^2(y^{4/3} - y_1^{4/3})$$

$$q = (0.7)(15) = 10.5 \ (\text{m}^3/\text{s})/\text{m}$$

$$x = -(\tfrac{3}{13})\{1.0/[(0.015)(10.5)]\}^2(y^{13/3} - 0.7^{13/3}) + \{3/[(4)(9.807)]\}(1.0/0.015)^2(y^{4/3} - 0.7^{4/3})$$

$$= -9.303y^{13/3} + 339.9y^{4/3} - 209.3$$

$$y_c = (q^2/g)^{1/3} = (10.5^2/9.807)^{1/3} = 2.240 \text{ m}$$

The depth must increase downstream, since the specific energy decreases, and the depth must move toward the critical value for less specific energy. The equation does not hold near the critical depth because of vertical accelerations that have been neglected in the derivation of gradually varied flow. If the channel is long enough for critical depth to be attained before the end of the channel, the high-velocity flow downstream from the gate may be drowned or a jump may occur. The water-surface calculation for the subcritical flow must begin with critical depth at the downstream end of the channel.

14.278 Prepare a computer program in BASIC to calculate the steady gradually varied water-surface profile in any prismatic rectangular, symmetric trapezoidal, or triangular channel.

```
10 REM B:PROFILES           WATER SURFACE PROFILES--STEADY STATE
20 ' WATER SURFACE PROFILE IN RECT, TRAPEZOIDAL, OR TRIANGULAR CHANNEL.
30 ' XL=LENGTH, B=BOT WIDTH, Z=SIDE SLOPE, RN=MANNING N,SO=BOT SLOPE,Q=FLOW.
40 ' YCONT=CONTROL DEPTH. IF YCONT=0 IN DATA, YCONT IS SET EQUAL TO YC.
50 ' IN SUBCRITICAL FLOW, CONTROL IS DOWNSTREAM AND DISTANCES ARE MEASURED
60 ' IN THE UPSTREAM DIRECTION
70 ' IN SUPERCRITICAL FLOW, CONTROL IS U.S. AND DISTANCES ARE MEASURED D.S.
80 LPRINT CHR$(14);"STEADY-STATE WATER-SURFACE PROFILES"
90 LPRINT CHR$(14);"    DATE=";DATE$;" TIME=";TIME$
100 DEF FNAREA(YY)=YY*(B+Z*YY): DEF FNPER(YY)=B+2!*YY*SQR(1'+Z^2)
110 DEF FNYCRIT(YY)=1!-Q^2*(B+2!*Z*YY)/(G*FNAREA(YY)^3)
120 DEF FNYNORM(YY)=1!-Q^2*CON/(FNAREA(YY)^3.333/FNPER(YY)^1.333)
130 DEFINT I:    DEF FNDL(YY)=FNYCRIT(YY)/(FNYNORM(YY)*SO)
140 DEF FNFPM(YY)=GAM*(YY^2*(.5*B+Z*YY/3!)+Q^2/(G*FNAREA(YY)))
150 ISI$="SI":    DEF FNENERGY(YY)=YY+Q^2/(2!*G*FNAREA(YY)^2)
160 READ IUNIT$,XL,B,Z,RN,SO,Q,YCONT: DATA "SI",200.,2.5,.8,.012,.025,25.,.0
170 IF IUNIT$=ISI$ THEN 190
180 GAM=62.4:G=32.2:CON=(RN/1.486)^2/SO:LPRINT"USC UNITS":LPRINT: GOTO 200
190 GAM=9802!:   G=9.806001:   CON=RN^2/SO:   LPRINT"SI UNITS":LPRINT
200 LPRINT" CHANNEL LENGTH=";XL;" DISCHARGE=";Q;" B=";B;" Z=";Z;"RN=";RN;
    " SO=";SO
210 ' DETERMINATION OF CRITICAL AND NORMAL DEPTHS
220 NN=30:   DN=0!:   UP=30!: YC=15: FOR I= 1 TO 15: IF FNYCRIT(YC)=0! THEN 250
230 IF FNYCRIT(YC)<0! THEN DN=YC ELSE UP=YC
240 YC=.5*(UP+DN):   NEXT I
250 IF YCONT=0! THEN YCONT=YC
260 IF SO<=0! THEN 320
270 UP=40!:   DN=0!:   YN=20!:   FOR I= 1 TO 15
280 X=FNYNORM(YN):   IF X<0! THEN DN=YN:   GOTO 300
290 IF X>0! THEN UP=YN ELSE GOTO 310
300 YN=.5*(UP+DN):   ' NEXT I
310 LPRINT:   LPRINT" NORMAL DEPTH=";YN;" CRITICAL DEPTH=";YC:   GOTO 330
320 YN=3!*YC:   LPRINT:   LPRINT" CRITICAL DEPTH=";YC
330 IF YN<YC THEN 410
340 ' MILD,ADVERSE,OR HORIZONTAL CHANNEL YN>YC
350 IF YCONT<YC THEN 390
360 ' SUBCRITICAL FLOW, YCONT>=YC
370 SIGN=-1!:   DY=(YCONT-YN)*.998/NN:   LPRINT:
    LPRINT"CONTROL IS DOWNSTREAM,DEPTH=";YCONT:   GOTO 460
380 ' SUPERCRITICAL FLOW
390 SIGN=1!:   DY=(YC-YCONT)/NN:   LPRINT:
    LPRINT"CONTROL IS UPSTREAM, DEPTH=";YCONT:   GOTO 460
400 ' STEEP CHANNEL,YN<YC
410 IF YCONT<=YC THEN 450
420 ' SUBCRITICAL FLOW, YCONT>YC
430 SIGN=-1!:   DY=(YCONT-YC)/NN:   LPRINT:
    LPRINT" CONTROL IS DOWNSTREAM, DEPTH=";YCONT:   GOTO 460
440 ' SUPERCRITICAL FLOW, YCONT<=YC
450 SIGN=1!:   NN=2*NN:   DY=(YN-YCONT)*.998/NN:   LPRINT:
    LPRINT" CONTROL IS UPSTREAM, DEPTH=";YCONT
460 SL=0!:   Y=YCONT:   E=FNENERGY(Y):   FM=FNFPM(Y):   LPRINT
470 LPRINT"    DISTANCE     DEPTH     ENERGY     F+M":   GOSUB 550
480 ' WATER-SURFACE PROFILE CALCULATION
490 FOR I=1 TO NN STEP 2:   Y2=YCONT+SIGN*DY*(I+1)
500 DX=DY*(FNDL(Y)+FNDL(Y2)+4!*FNDL(YCONT+SIGN*I*DY))/3!
510 SL=SL+DX:   IF SL>XL THEN 540
520 Y=Y2:   E=FNENERGY(Y):   FM=FNFPM(Y):   IF (I=NN-1) AND (SL<0!) THEN SL=XL
530 GOSUB 550:   NEXT I:   GOTO 160
540 Y=Y2-SIGN*2!*DY*(SL-XL)/DX:   E=FNENERGY(Y):   FM=FNFPM(Y):   SL=XL:
    GOSUB 550:   GOTO 160
550 LPRINT SPC(5) USING"   ####.#";SL;:   LPRINT USING"   ###.###";Y;E;:
    LPRINT USING"   ########.";FM:   RETURN
560 DATA "SI",600.,2.5,.8,.012,.0002,25.,.907
570 DATA "SI",600.,2.5,.8,.012,.0002,25.,2.
```

Input data include the specification of the system of units (SI or USC) in the first columns of the data, followed by the channel dimensions, discharge, and water-surface control depth. If the control depth is set to zero in data, it is automatically assumed to be the critical depth in the program. For subcritical flow the control is downstream, and distances are measured in the upstream direction. For supercritical flow the control depth is upstream, and distances are measured in the downstream direction.

The program begins with several line functions to compute the various variables and functions in the problem. After the necessary data input, critical depth is computed, followed by the normal-depth calculation if normal depth exists. The bisection method is used in these calculations. The type of profile is then categorized, and finally the water-surface profile, specific energy, and $F + M$ are calculated and printed. Simpson's rule is used in the integration for the water-surface profile.

The program begins with several line functions to compute the various variables and functions in the problem. After the necessary data input, critical depth is computed, followed by the normal-depth calculation if normal

14.279 A trapezoidal channel, $B = 2.5$ m, side slope $= 0.8$, has two bottom slopes. The upstream portion is 200 m long, $S_0 = 0.025$, and the downstream portion, 600 m long, $S_0 = 0.0002$, $n = 0.012$. A discharge of 25 m³/s enters at critical depth from a reservoir at the upstream end, and at the downstream end of the system the water depth is 2 m. Determine the water-surface profiles throughout the system, including jump location, using the computer program of Prob. 14.278.

❚ Three separate sets of data are included in the program and are needed to obtain the results used to plot the solution as shown in Fig. 14-70. The first set for the steep upstream channel has a control depth equal to zero since it will be automatically assumed critical depth in the program. The second set is for the supercritical flow in the mild channel. It begins at a control depth equal to the end depth from the upstream channel and computes the water surface downstream to the critical depth. The third set of data uses the 2-m downstream depth as the control depth and computes in the upstream direction. Computer output from the last two data sets are given below. The jump is located by finding the position of equal $F + M$ from the output of the last two data sets.

```
SI UNITS

 CHANNEL LENGTH= 600   DISCHARGE= 25   B= 2.5   Z= .8 RN= .012   SO= .0002

 NORMAL DEPTH= 3.190308   CRITICAL DEPTH= 1.780243

CONTROL IS UPSTREAM, DEPTH= .9070001

        DISTANCE        DEPTH       ENERGY         F+M
            0.0         0.907        4.630       225573
           25.8         0.965        4.160       211573
           51.2         1.023        3.786       199573
           76.0         1.082        3.487       189272
          100.1         1.140        3.247       180431
          123.3         1.198        3.054       172859
          145.5         1.256        2.900       166400
          166.5         1.315        2.777       160926
          186.0         1.373        2.679       156334
          204.0         1.431        2.603       152534
          220.0         1.489        2.544       149455
          233.9         1.547        2.500       147035
          245.4         1.606        2.469       145221
          254.1         1.664        2.448       143970
          259.6         1.722        2.437       143242
          261.5         1.780        2.433       143007
SI UNITS

 CHANNEL LENGTH= 600   DISCHARGE= 25   B= 2.5   Z= .8 RN= .012   SO= .0002

 NORMAL DEPTH= 3.190308   CRITICAL DEPTH= 1.780243

CONTROL IS DOWNSTREAM,DEPTH= 2

        DISTANCE        DEPTH       ENERGY         F+M
            0.0         2.000        2.474       146109
           33.0         2.079        2.504       148634
           80.5         2.158        2.541       151844
          145.4         2.238        2.583       155711
          231.3         2.317        2.630       160211
          343.0         2.396        2.681       165326
          486.6         2.475        2.734       171040
          600.0         2.524        2.769       174857
```

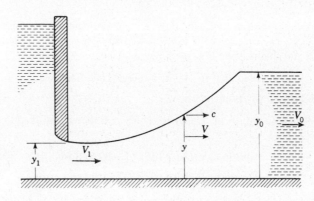

$Y_c = 1.780$ m

Energy grade line

$Y_n = 0.856$ m
$Q = 25$ m³/s

2.46 m

2.0 m

0.907 m   1.22 m $Y_n = 3.190$ m

Fig. 14-70

14.280 A rectangular channel 3 m wide and 2 m deep, discharging 18 m³/s, suddenly has the discharge reduced to 12 m³/s at the downstream end. Compute the height and speed of the surge wave.

$$(v_1 + c)(y_1) = (v_2 + c)(y_2) \qquad (\gamma/2)(y_1^2 - y_2^2) = (\gamma/g)(y_1)(v_1 + c)(v_2 + c - v_1 - c)$$
$$v_1 = Q/A_1 = 18/[(3)(2)] = 3 \text{ m/s} \qquad v_2 = 12/(3y_2) \qquad v_2 y_2 = 4 \text{ m}^2/\text{s}$$
$$(3 + c)(2) = 4 + cy_2 \qquad 6 = 4 + (c)(y_2 - 2)$$
$$(9.79/2)(2^2 - y_2^2) = (9.79/9.807)(2)(3 + c)(v_2 + c - 3 - c)$$

Eliminating c and v_2 gives $y_2^2 - 4 = (4/9.807)[2/(y_2 - 2) + 3](3 - 4/y_2)$, $y_2 = 2.75$ m (by trial and error), $v_2 = 4/2.75 = 1.455$ m/s. The height of the surge wave is $2.75 - 2 = 0.75$ m, and the speed of the wave is $c = 2/(y_2 - 2) = 2/(2.75 - 2) = 2.67$ m/s.

14.281 In Fig. 14-71 find the Froude number of the undisturbed flow such that the depth y_1 at the gate is just zero when the gate is suddenly closed. For $v_0 = 20$ ft/s, find the liquid-surface elevation.

It is required that $v_1 = 0$ when $y_1 = 0$ at $x = 0$ for any time after $t = 0$.

$$v = v_0 - (2)(\sqrt{g})(\sqrt{y_0} - \sqrt{y}) \qquad 0 = v_0 - (2)(\sqrt{g})(\sqrt{y_0} - \sqrt{0})$$
$$v_0 = (2)(\sqrt{gy_0}) \qquad N_F = v_0/\sqrt{gy_0} = 2$$
$$x = [v_0 - (2)(\sqrt{gy_0}) + (3)(\sqrt{gy})](t) \qquad y_0 = v_0^2/4g = 20^2/[(4)(32.2)] = 3.106 \text{ ft}$$
$$x = \{20 - (2)[\sqrt{(32.2)(3.106)}] + (3)(\sqrt{32.2y})\}(t) = 17.02ty^{1/2}$$

The liquid surface is a parabola with vertex at the origin and surface concave upward.

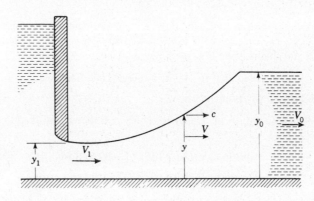

Fig. 14-71

14.282 In Fig. 14-71 the gate is partially closed at the instant $t = 0$ so that the discharge is reduced by 50 percent ($v_0 = 6$ m/s, $y_0 = 3$ m). Find v_1, y_1, and the surface profile.

The new discharge is

$$q = (6)(3)/2 = 9 = v_1 y_1 \qquad v = v_0 - (2)(\sqrt{g})(\sqrt{y_0} - \sqrt{y})$$
$$v_1 = 6 - (2)(\sqrt{9.807})(\sqrt{3} - \sqrt{y_1})$$
$$v_1 = 4.25 \text{ m/s} \quad \text{and} \quad y_1 = 2.11 \text{ m} \quad \text{(by trial and error)}$$
$$x = [v_0 - (2)(\sqrt{gy_0}) + (3)(\sqrt{gy})](t) = \{6 - (2)[\sqrt{(9.807)(3)}] + (3)[\sqrt{(9.807)(y)}]\}(t)$$
$$= [(9.39)(\sqrt{y}) - 4.85](t)$$

This holds for the range of values of y between 2.12 m and 3 m.

14.283 A discharge of 160 cfs occurs in a rectangular open channel 6 ft wide with $s_0 = 0.002$ and $n = 0.012$. If the channel ends in a free outfall, calculate the depth at the brink, y_n, and y_c. Determine the shape of the water-surface profile for a distance of 100 ft upstream from the brink.

∎ $Q = (A)(1.486/n)(R^{2/3})(s^{1/2})$ $160 = (6y_n)(1.486/0.012)[6y_n/(y_n + 6 + y_n)]^{2/3}(0.002)^{1/2}$

$$y_n = 3.5 \text{ ft} \quad \text{(by trial and error)}$$

$$Q^2/g = A^3/B \quad 160^2/32.2 = (6y_c)^3/6 \quad y_c = 2.81 \text{ ft}$$

Since $y_n > y_c$, the flow is subcritical and the water-surface profile is M_2 (see Fig. A-19). The depth at the outfall is approximately $0.7y_c = (0.7)(2.81) = 2.0$ ft. Critical depth occurs at about $4y_c = (4)(2.81) = 11$ ft upstream from the brink. Computations for the water-surface profile are given below.

y, ft	A, ft^2	$B+2y$, ft	R, ft	V, ft/s	$\frac{V^2}{2g}$, ft	$y+\frac{V^2}{2g}$, ft	$\Delta\left(y+\frac{V^2}{2g}\right)$, ft	V_{avg}, ft/s	R_{avg}, ft	S	$S-S_0$	x, ft	Σx,* ft
2.81	16.86	11.62	1.451	9.49	1.398	4.208							
							0.005	9.34	1.463	0.00341	0.00141	4	4
2.90	17.40	11.80	1.475	9.20	1.313	4.213							
							0.014	9.04	1.488	0.00312	0.00112	12	16
3.00	18.00	12.00	1.500	8.89	1.227	4.227							
							0.022	8.74	1.512	0.00284	0.00084	26	42
3.10	18.60	12.20	1.525	8.60	1.149	4.249							
							0.029	8.47	1.536	0.00262	0.00062	47	89
3.20	19.20	12.40	1.548	8.33	1.078	4.278							

* Summation x is measured from the point of critical depth 11 ft upstream from the brink.

The water-surface profile is sketched in Fig. 14-72.

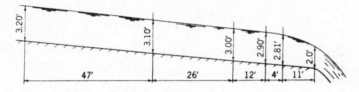

Fig. 14-72

14.284 Examine the flow conditions in a 10-ft-wide open rectangular channel of rubble masonry with $n = 0.017$ when the flow rate is 400 cfs. The channel slope is 0.020, and an ogee weir 5.0 ft high with $C_w = 3.8$ is located in the downstream end of the channel (see Fig. 14-73).

∎ $Q = (A)(1.486/n)(R^{2/3})(s^{1/2})$ $400 = (10y_n)(1.486/0.017)[10y_n/(y_n + 10 + y_n)]^{2/3}(0.020)^{1/2}$

$$y_n = 2.36 \text{ ft} \quad \text{(by trial and error)} \quad y_c = (q^2/g)^{1/3} \quad y_c = [(\tfrac{400}{10})^2/32.2]^{1/3} = 3.68 \text{ ft}$$

Since $y_n < y_c$, the flow is supercritical. The head required at the weir to discharge the given flow is found from the equation $Q = (C_w L)(h + V_0^2/2g)^{3/2}$:

$$V_0 = Q/A = 400/[(5+h)(10)] = 400/(50 + 10h) \quad 400 = (3.8)(10)\{h + [400/(50 + 10h)]^2/[(2)(32.2)]\}^{3/2}$$

$$h = 4.53 \text{ ft} \quad \text{(by trial and error)}$$

Hence, the depth of water just upstream of the weir is $5 + 4.53 = 9.53$ ft, which is greater than y_c. The flow at this point is subcritical, and a hydraulic jump must occur upstream. The depth y_2 after the jump is found from $y_2 = -y_1/2 + (y_1^2/4 + 2V_1^2 y_1/g)^{1/2}$, $V_1 = Q/A_1 = 400/[(2.36)(10)] = 16.95$ ft/s, $y_2 = -2.36/2 + [2.36^2/4 + (2)(16.95)^2(2.36)/32.2]^{1/2} = 5.42$ ft. The distance from the weir to the jump is determined by the equation $x = [(y_A + V_A^2/2g) - (y_B + V_B^2/2g)]/(s - s_0)$:

$$V_A = Q/A = 400/[(5.42)(10)] = 7.380 \text{ ft/s} \quad V_B = 400/[(9.53)(10)] = 4.197 \text{ ft/s}$$

$$s = (nV/1.486R^{2/3})^2 \quad V_m = (7.380 + 4.197)/2 = 5.788 \text{ ft/s}$$

$$(p_w)_A = 5.42 + 10 + 5.42 = 20.84 \text{ ft} \quad (p_w)_B = 9.53 + 10 + 9.53 = 29.06 \text{ ft}$$

$$R_m = [(5.42)(10)/20.84 + (9.53)(10)/29.06]/2 = 2.940 \text{ ft}$$

$$s = \{(0.017)(5.788)/[(1.486)(2.940)^{2/3}]\}^2 = 0.001041$$

$$x = \{5.42 + 7.380^2/[(2)(32.2)] - 9.53 - 4.197^2/[(2)(32.2)]\}/(0.001041 - 0.020) = 187 \text{ ft}$$

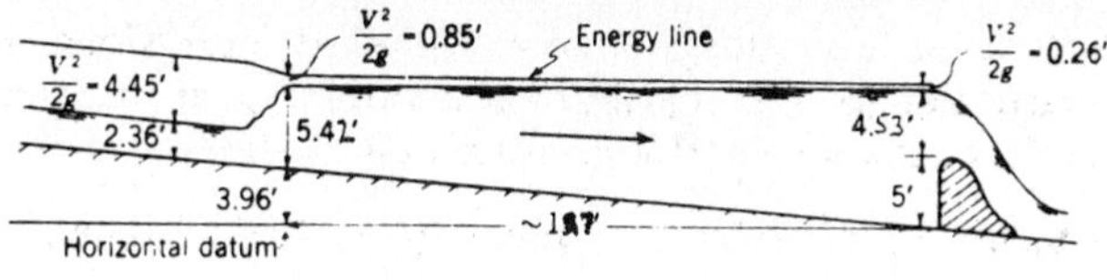

Fig. 14-73

14.285 A rectangular channel is 8 ft wide, has an 0.008 slope, discharge of 150 cfs, and $n = 0.014$. Find y_n and y_c. If the actual depth of flow is 5 ft, what type of profile exists?

$$Q = (A)(1.486/n)(R^{2/3})(s^{1/2}) \qquad 150 = (8y_n)(1.486/0.014)[8y_n/(y_n + 8 + y_n)]^{2/3}(0.008)^{1/2}$$

$$y_n = 1.74 \text{ ft} \quad \text{(by trial and error)} \qquad y_c = (q^2/g)^{1/3} = [(\tfrac{150}{8})^2/32.2]^{1/3} = 2.22 \text{ ft}$$

From Fig. A-19, this is an S_1 water-surface profile.

14.286 A rectangular channel is 3 m wide and ends in a free outfall. If the discharge is $10 \text{ m}^3/\text{s}$, slope is 0.0025, and $n = 0.016$, find y_n, y_c, and the water-surface profile for a distance of 150 m upstream from the outfall.

$$Q = (A)(1.0/n)(R^{2/3})(s^{1/2}) \qquad 10 = (3y_n)(1.0/0.016)[3y_n/(y_n + 3 + y_n)]^{2/3}(0.0025)^{1/2}$$

$$y_n = 1.34 \text{ m} \quad \text{(by trial and error)} \qquad y_c = (q^2/g)^{1/3} = [(\tfrac{10}{3})^2/9.807]^{1/3} = 1.04 \text{ m}$$

The depth at the outfall is approximately $0.7y_c = (0.7)(1.04) = 0.73$ m. Critical depth occurs at about $4y_c = (4)(1.04) = 4$ m upstream from the brink. Computations for the water-surface profile are given below $[s = (nv/1.0R^{2/3})^2, \; x = E/(s - s_0)]$.

y_n, m	v, m/s	$v^2/2g$, m	E, m	v_m, m/s	R_m, m	s	$\sum x$, m	x, m
1.04	3.205	0.524	1.546					
				3.06	0.630	0.00443	6	6
1.14	2.924	0.436	1.576					
				2.80	0.662	0.00346	33	39
1.24	2.688	0.368	1.608					
				2.58	0.694	0.00276	185	224
1.34	2.488	0.316	1.656					

14.287 A rectangular drainage channel is 15 ft wide and is to carry 500 cfs. The channel is lined with rubble masonry ($n = 0.017$) and has a bottom slope of 0.0015. It discharges into a stream which may reach a stage 10 ft above the channel bottom during floods. Calculate y_n, y_c, and the distance from the channel outlet to the point where normal depth would occur under this condition. Use one step.

$$Q = (A)(1.486/n)(R^{2/3})(s^{1/2}) \qquad 500 = (15y_n)(1.486/0.017)[15y_n/(y_n + 15 + y_n)]^{2/3}(0.0015)^{1/2}$$

$$y_n = 4.81 \text{ ft} \quad \text{(by trial and error)} \qquad y_c = (q^2/g)^{1/3} = [(\tfrac{500}{15})^2/32.2]^{1/3} = 3.26 \text{ ft}$$

$$\Delta x = [(y_1 + V_1^2/2g) - (y_2 + V_2^2/2g)]/(s - s_0) \qquad V_1 = Q/A = 500/[(4.81)(15)] = 6.930 \text{ ft/s}$$

$$V_2 = 500/[(10)(15)] = 3.333 \text{ ft/s}$$

$$s = (nV/1.486R^{2/3})^2 \qquad V_m = (6.930 + 3.333)/2 = 5.132 \text{ ft/s}$$

$$(p_w)_1 = 4.81 + 15 + 4.81 = 24.62 \text{ ft} \qquad (p_w)_2 = 10 + 15 + 10 = 35.00 \text{ ft}$$

$$R_m = [(4.81)(15)/24.62 + (10)(15)/35.00]/2 = 3.608 \text{ ft}$$

$$s = \{(0.017)(5.132)/[(1.486)(3.608)^{2/3}]\}^2 = 0.0006229$$

$$\Delta x = \{4.81 + 6.930^2/[(2)(32.2)] - 10 - 3.333^2/[(2)(32.2)]\}/(0.0006229 - 0.0015) = 5264 \text{ ft}$$

14.288 Solve Prob. 14.287 using three steps.

▌ Using data from Prob. 14.287, $s = (nv/1.486R^{2/3})^2$, $x = \Delta E/(s - s_0)$.

y_n, ft	v, ft/s	$v^2/2g$, ft	E, ft	v_m, ft/s	R_m, ft	s	x, ft
10	3.33	0.17	10.17				
				3.75	4.09	0.000285	1560
8	4.17	0.27	8.27				
				4.86	3.60	0.000559	1900
6	5.55	0.48	6.48				
				6.28	3.12	0.000112	2420
4.8	7.00	0.76	5.56				
							5880

14.289 A trapezoidal channel with a bottom width of 10 ft and side slopes of 2 horizontal to 1 vertical has a horizontal curve with a radius of 100 ft without superelevation. If the discharge is 800 cfs and the water surface at the inside of the curve is 5 ft above the channel bottom, find the water-surface elevation at the outside of the curve. Assume the flow is subcritical.

▌ See Fig. 14-74. $y_2 - y_1 = v^2B/gr$, $x - 5 = \{800/[(x + 5)(10 + x) - 25 - x^2]\}^2(10 + 10 + 2x)/[(32.2)(100)]$, $x = 5.53$ ft   (by trial and error).

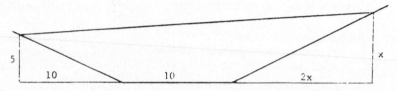

Fig. 14-74

14.290 A rectangular flume of planed timber ($n = 0.012$) is 1.5 m wide and carries 1.7 m³/s of water. The bed slope is 0.0006, and at a certain section the depth is 0.9 m. Find the distance (in one reach) to the section where the depth is 0.75 m. Is the distance upstream or downstream?

▌ $\Delta x = (y_1 + v_1^2/2g - y_2 - v_2^2/2g)/(s - s_0)$. Assume the 0.75 m depth is downstream.

$$v_1 = Q/A_1 = 1.7/[(0.9)(1.5)] = 1.259 \text{ m/s}$$
$$v_2 = Q/A_2 = 1.7/[(0.75)(1.5)] = 1.511 \text{ m/s} \qquad s = (nv/1.0R^{2/3})^2$$
$$v_m = (1.259 + 1.511)/2 = 1.385 \text{ m/s} \qquad (p_w)_1 = 0.9 + 1.5 + 0.9 = 3.30 \text{ m}$$
$$(p_w)_2 = 0.75 + 1.5 + 0.75 = 3.00 \text{ m}$$
$$R_m = [(0.9)(1.5)/3.30 + (0.75)(1.5)/3.00]/2 = 0.3920 \text{ m}$$
$$s = \{(0.012)(1.385)/[(1.0)(0.3920)^{2/3}]\}^2 = 0.0009628$$
$$\Delta x = \{0.9 + 1.259^2/[(2)(9.807)] - 0.75 - 1.511^2/[(2)(9.807)]\}/(0.0009628 - 0.0006) = 315 \text{ m}$$

Since Δx is positive, the 0.75 m depth is downstream, as assumed.

14.291 Suppose that the flume of Prob. 14.290 is now changed so that, with the same flow, the depth varies from 1.2 m at one section to 0.9 m at a section 300 m downstream. Find the new bed slope of the flume, using one reach. Sketch the flume, the energy grade line, and the water surface to assure that the answer is reasonable.

▌ $$\Delta x = (y_1 + v_1^2/2g - y_2 - v_2^2/2g)/(s - s_0) \qquad v_1 = Q/A_1 = 1.7/[(1.2)(1.5)] = 0.9444 \text{ m/s}$$
$$v_2 = Q/A_2 = 1.7/[(0.9)(1.5)] = 1.259 \text{ m/s} \qquad s = (nv/1.0R^{2/3})^2$$
$$v_m = (0.9444 + 1.259)/2 = 1.102 \text{ m/s}$$
$$(p_w)_1 = 1.2 + 1.5 + 1.2 = 3.90 \text{ m} \qquad (p_w)_2 = 0.9 + 1.5 + 0.9 = 3.30 \text{ m}$$
$$R_m = [(1.2)(1.5)/3.90 + (0.9)(1.5)/3.30]/2 = 0.4353 \text{ m}$$
$$s = \{(0.012)(1.102)/[(1.0)(0.4353)^{2/3}]\}^2 = 0.0005301$$
$$300 = \{1.2 + 0.9444^2/[(2)(9.807)] - 0.9 - 1.259^2/[(2)(9.807)]\}/(0.0005301 - s_0) \qquad s_0 = -0.000352$$

The minus sign indicates the slope is *adverse*. See Fig. 14-75.

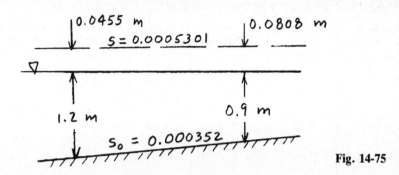

Fig. 14-75

14.292 Suppose that the slope of the flume in Prob. 14.290 is now increased to 0.01. With the same flow as before, find the depth 15 m downstream from a section where the flow is 0.45 m deep. Use only one reach. Is the flow subcritical or supercritical?

▮ $y_c = (q^2/g)^{1/3} = [(1.7/1.5)^2/9.807]^{1/3} = 0.508$ m. Since $y < y_c$, the flow is supercritical.

$$\Delta x = (y_1 + v_1^2/2g - y_2 - v_2^2/2g)/(s - s_0) \qquad v_1 = Q/A_1 = 1.7/[(0.45)(1.5)] = 2.519 \text{ m/s}$$
$$v_2 = Q/A_2 = 1.7/(1.5y_2) = 1.133/y_2 \qquad s = (nv/1.0R^{2/3})^2$$
$$v_m = (2.519 + 1.133/y_2)/2 = 1.260 + 0.5665/y_2$$
$$(p_w)_1 = 0.45 + 1.5 + 0.45 = 2.40 \text{ m} \qquad (p_w)_2 = y_2 + 1.5 + y_2 = 1.5 + 2y_2$$
$$R_m = [(0.45)(1.5)/2.40 + 1.5y_2/(1.5 + 2y_2)]/2 = 0.1406 + 0.75y_2/(1.5 + 2y_2)$$
$$s = \left\{ \frac{(0.012)(1.260 + 0.5665/y_2)}{1.0[0.1406 + 0.75y_2/(1.5 + 2y_2)]^{2/3}} \right\}^2$$
$$15 = \{0.45 + 2.519^2/[(2)(9.807)] - y_2 - (1.133/y_2)^2/[(2)(9.807)]\}/(s - 0.01)$$
$$y_2 = 0.383 \text{ m} \qquad \text{(by trial and error)}$$

14.293 A test on a rectangular glass flume 10 in wide yielded the following data on a reach of 30-ft length: with still water, $z_1 - z_2 = 0.009$ ft; with a measured flow of 0.1516 cfs, $y_1 = 0.361$ ft, $y_2 = 0.366$ ft. Find the value of the roughness coefficient n, using only one reach.

▮ $\quad Q = (A)(1.486/n)(R^{2/3})(s^{1/2}) \qquad A_1 = (0.361)(\tfrac{10}{12}) = 0.3008 \text{ ft}^2 \qquad A_2 = (0.366)(\tfrac{10}{12}) = 0.3050 \text{ ft}^2$
$$A_m = (0.3008 + 0.3050)/2 = 0.3029 \text{ ft}^2 \qquad (p_w)_1 = 0.361 + \tfrac{10}{12} + 0.361 = 1.555 \text{ ft}$$
$$(p_w)_2 = 0.366 + \tfrac{10}{12} + 0.366 = 1.565 \text{ ft}$$
$$R_m = (0.3008/1.555 + 0.3050/1.565)/2 = 0.1942 \text{ ft} \qquad x = (y_1 + v_1^2/2g - y_2 - v_2^2/2g)/(s - s_0)$$
$$v_1 = Q/A_1 = 0.1516/0.3008 = 0.5040 \text{ ft/s} \qquad v_2 = Q/A_2 = 0.1516/0.3050 = 0.4970 \text{ ft/s}$$
$$30 = \{0.361 + 0.5040^2/[(2)(32.2)] - 0.366 - 0.4970^2/[(2)(32.2)]\}/(s - 0.009/30) \qquad s = 0.0001370$$
$$0.1516 = (0.3029)(1.486/n)(0.1942)^{2/3}(0.0001370)^{1/2} \qquad n = 0.0117$$

14.294 A rectangular flume 10 ft wide is built of planed timber ($n = 0.012$) on a bed slope of 0.2 ft per 1000 ft ending in a free overfall. If the measured depth at the fall is 1.82 ft, find (a) the rate of flow and (b) the distance upstream from the fall to where the depth is 4 ft.

▮ (a) $\qquad\qquad\qquad y_{\text{brink}} = 0.72y_c \qquad 1.82 = 0.72y_c \qquad y_c = 2.528 \text{ ft}$
$$y_c = (q^2/g)^{1/3} \qquad 2.528 = (q^2/32.2)^{1/3} \qquad q = 22.81 \text{ cfs/ft} \qquad Q = (22.81)(10) = 228 \text{ ft}^3/\text{s}$$

(b) Between y_{brink} and y_c the flow is rapidly varying. So Manning's equation is not valid there and cannot be used there to determine Q. $s = (nV/1.486R^{2/3})^2$, $\Delta x = (y_1 + V_1^2/2g - y_2 - V_2^2/2g)/(s - s_0)$.

y, ft	V, fps	$V^2/2g$, ft	E, ft	A, ft²	P, ft	R, ft	R_m, ft	V_m, fps	S	Δx, ft
4.0	5.70	0.505	4.505	40	18.0	2.22				
							2.12	6.20	0.000915	568
3.4	6.71	0.698	4.095	34	16.8	2.02				
							1.949	7.15	0.001363	173
3.0	7.60	0.897	3.897	30	16.0	1.875				
							1.814	8.02	0.001888	53
2.7	8.44	1.108	3.808	27	15.4	1.753				
							1.717	8.73	0.00240	8
2.53	9.01	1.261	3.791	25.3	15.1	1.680				
									$4y_c = 10$	
y_b										$\sum(\Delta x) = 812$

14.295 A wide rectangular channel dredged in earth ($n = 0.035$) is laid on a slope of 10 ft/mi and carries a flow of 100 cfs/ft. Find the water depth 2 mi upstream of a section where the depth is 28.9 ft. Compute, using a single reach.

$$\Delta x = (y_1 + v_1^2/2g - y_2 - v_2^2/2g)/(s - s_0) \qquad v_1 = q/y_1 = 100/y_1$$
$$v_2 = q/y_2 = 100/28.9 = 3.460 \text{ ft/s}$$
$$v_m = (100/y_1 + 3.460)/2 = 50.0/y_1 + 1.730 \qquad s = (nv/1.486R^{2/3})^2$$
$$R_m = (y_1 + 28.9)/2 = y_1/2 + 14.45$$
$$s = \{(0.035)(50.0/y_1 + 1.730)/[(1.486)(y_1/2 + 14.45)^{2/3}]\}^2$$
$$(2)(5280) = \{y_1 + (100/y_1)^2/[(2)(32.2)] - 28.9 - 3.460^2/[(2)(32.2)]\}/(s - 10/5280)$$
$$y_1 = 11.7 \text{ ft} \qquad \text{(by trial and error)}$$

14.296 The slope of a stream of rectangular cross section is 0.0002, the width is 160 ft, and the value of the Chezy C is 78.3 ft$^{1/2}$/s. Find the depth for a uniform flow of 88.55 cfs per foot of width of the stream.

$$v_0 = q/y_0 = 88.55/y_0 \qquad v_0 = C\sqrt{Rs} = (78.3)\{\sqrt{[160y_0/(y_0 + 160 + y_0)](0.0002)}\}$$
$$88.55/y_0 = 78.3\{\sqrt{[160y_0/(y_0 + 160 + y_0)](0.0002)}\} \qquad y_0 = 20.0 \text{ ft} \qquad \text{(by trial and error)}$$

14.297 If in Prob. 14.296 a dam raises the water level so that at a certain distance upstream the increase is 5 ft, how far from this latter section will the increase be only 1 ft? Use reaches with 1-ft depth increments.

$$v = C\sqrt{Rs} \qquad \Delta x = (y_1 + v_1^2/2g - y_2 - v_2^2/2g)/(s - s_0)$$

y, ft	A (160y), ft²	P, ft	R (A/P), ft	V (q/y), fps	$V^2/2g$, ft	E, ft	ΔE $(E_1 - E_2)$, ft	V_m, fps	R_m, ft	S	Δx, ft
25	4000	210	19.05	3.54	0.195	25.195					
							-0.983	3.62	18.76	0.0001137	11 397
24	3840	208	18.46	3.69	0.211	24.211					
							-0.981	3.77	18.16	0.0001276	13 559
23	3680	206	17.86	3.85	0.230	23.230					
							-0.979	3.94	17.56	0.0001440	17 479
22	3520	204	17.26	4.03	0.252	22.252					
							-0.976	4.12	16.94	0.0001635	26 708
21	3360	202	16.63	4.22	0.276	21.276					
											$\sum(\Delta x) = 69\,143$

14.298 A portion of an outfall sewer is approximately a circular conduit 5 ft in diameter and with a slope of 1 ft in 1100 ft. It is of brick, for which $n = 0.013$. What would be its maximum capacity for uniform flow?

$$Q = (A)(1.486/n)(R^{2/3})(s^{1/2}) \qquad Q_{full} = [(\pi)(5)^2/4](1.486/0.013)(\tfrac{5}{4})^{2/3}(\tfrac{1}{1100})^{1/2} = 78.53 \text{ ft}^3/\text{s}$$

From Fig. A-18, $(Q_{max}/Q_{full}) = 1.06$. $Q_{max} = (78.53)(1.06) = 83.2 \text{ ft}^3/\text{s}$.

14.299 If the outfall sewer in Prob. 14.298 discharges 120 cfs with a depth at the end of 2.90 ft, how far back from the end must it become a pressure conduit? Proceeding from the mouth upstream, find by tabular solution the length of sewer that is not flowing full. Use three reaches with equal depth increments.

$$s = (nV/1.486R^{2/3})^2 \qquad \Delta x = (y_1 + v_1^2/2g - y_2 - v_2^2/2g)/(s - s_0)$$

y, ft	θ, deg	A, ft²	P, ft	R (A/P), ft	V (Q/A), fps	$V^2/2g$, ft	E, ft	ΔE (E_1-E_2), ft	V_m, fps	R_m, ft	S	Δx, ft
2.9	99.2	11.81	8.66	1.364	10.16	1.603	4.503					
								0.073	9.05	1.429	0.00387	24.7
3.6	116.1	15.13	10.13	1.494	7.93	0.976	4.576					
								0.417	7.31	1.504	0.00236	287
4.3	136.1	17.96	11.87	1.513	6.68	0.693	4.993					
								0.587	6.40	1.382	0.00202	527
5.0	180	19.63	15.71	1.250	6.11	0.580	5.580					$\Sigma(\Delta x) = 839$

The sewer must become a pressure conduit 839 ft back from the end.

14.300 For the channel of Prob. 14.131, find the distance between a section where the depth is 3.0 ft to another where the depth is 2.5 ft. Which section is upstream? Use only one reach.

Assume the 2.5 ft depth is upstream. Then $y_1 = 2.5$ ft and $y_2 = 3.0$ ft. From Prob. 14.131, $Q = 100$ cfs, $s = 0.02$, and $n = 0.015$.

$$\Delta x = (y_1 + v_1^2/2g - y_2 - v_2^2/2g)/(s - s_0) \qquad v_1 = Q/A_1 = 100/7.44 = 13.44 \text{ ft/s}$$
$$v_2 = Q/A_2 = 100/10.48 = 9.542 \text{ ft/s}$$
$$s = (nv/1.486R^{2/3})^2 \qquad v_m = (13.44 + 9.542)/2 = 11.49 \text{ ft/s} \qquad R = A/p_w$$
$$R_1 = 7.44/7.54 = 0.9867 \text{ ft}$$
$$R_2 = 10.48/8.96 = 1.170 \text{ ft} \qquad R_m = (0.9867 + 1.170)/2 = 1.078 \text{ ft}$$
$$s = \{(0.015)(11.49)/[(1.486)(1.078)^{2/3}]\}^2 = 0.01217$$
$$\Delta x = \{2.5 + 13.44^2/[(2)(32.2)] - 3.0 - 9.542^2/[(2)(32.2)]\}/(0.01217 - 0.02) = -114 \text{ ft}$$

Since Δx is negative, the assumption that the 2.5 ft depth was upstream is incorrect, and the 3.0 ft depth is upstream.

14.301 Analyze the water-surface profile in a long rectangular channel ($n = 0.013$). The channel is 10 ft wide, the flow rate is 400 cfs, and there is an abrupt change in slope from 0.0016 to 0.0150. Make a sketch showing normal depths, critical depths, and water-surface profile types.

$y_c = (q^2/g)^{1/3} = [(\frac{400}{10})^2/32.2]^{1/3} = 3.68$ ft, $Q = (A)(1.486/n)(R^{2/3})(s^{1/2})$. In upstream segment, $400 = (10y_n)(1.486/0.013)[10y_n/(y_n + 10 + y_n)]^{2/3}(0.0016)^{1/2}$, $y_n = 4.81$ ft (by trial and error). In downstream segment, $400 = (10y_n)(1.486/0.013)[10y_n/(y_n + 10 + y_n)]^{2/3}(0.0150)^{1/2}$, $y_n = 2.17$ ft (by trial and error). Thus, the flow is subcritical before the break and supercritical after the break. Critical depth occurs in the vicinity of the break in slope. The general profile is sketched in Fig. 14-76.

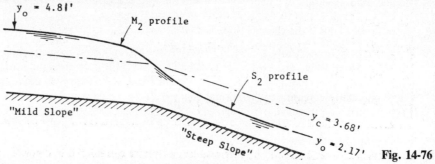

Fig. 14-76

14.302 Repeat Prob. 14.301 for the case where the flow rate is 150 cfs.

$y_c = (q^2/g)^{1/3} = [(\frac{150}{10})^2/32.2]^{1/3} = 1.92$ ft, $Q = (A)(1.486/n)(R^{2/3})(s^{1/2})$. In upstream segment, $150 = (10y_n)(1.486/0.013)[10y_n/(y_n + 10 + y_n)]^{2/3}(0.0016)^{1/2}$, $y_n = 2.38$ ft (bt trial and error). In downstream segment,

$150 = (10y_n)(1.486/0.013)[10y_n/(y_n + 10 + y_n)]^{2/3}(0.0150)^{1/2}$, $y_n = 1.13$ ft (by trial and error). Thus, the flow is subcritical before the break and supercritical after the break. Critical depth occurs in the vicinity of the break in slope. The general profile has the same characteristics as that of Prob. 14.301 (Fig. 14.76).

14.303 Repeat Prob. 14.302 for the case where the slope change is from 0.0016 to 0.0006. Compute the approximate distance upstream from the break to the point where normal depth occurs, using one reach.

▌ $y_c = 1.92$ ft (from Prob. 14.302). In upstream segment, $y_n = 2.38$ ft (from Prob. 14.302). In downstream segment, $150 = (10y_n)(1.486/0.013)[10y_n/(y_n + 10 + y_n)]^{2/3}(0.0006)^{1/2}$, $y_n = 3.36$ ft (by trial and error). Thus, the flow is subcritical before and after the break. A water surface on a mild slope cannot diverge from y_n in the upstream direction. Therefore, the depth at the break is 3.36 ft. See Fig. 14-77.

$$\Delta x = (y_1 + v_1^2/2g - y_2 - v_2^2/2g)/(s - s_0) \qquad v_1 = Q/A_1 = 150/[(2.38)(10)] = 6.303 \text{ ft/s}$$
$$v_2 = Q/A_2 = 150/[(3.36)(10)] = 4.464 \text{ ft/s} \qquad s = (nv/1.486R^{2/3})^2$$
$$v_m = (6.303 + 4.464)/2 = 5.384 \text{ ft/s}$$
$$(p_w)_1 = 2.38 + 10 + 2.38 = 14.76 \text{ ft} \qquad (p_w)_2 = 3.36 + 10 + 3.36 = 16.72 \text{ ft/s}$$
$$R_m = [(2.38)(10)/14.76 + (3.36)(10)/16.72]/2 = 1.811 \text{ ft}$$
$$s = \{(0.013)(5.384)/[(1.486)(1.811)^{2/3}]\}^2 = 0.001005$$
$$\Delta x = \{2.38 + 6.303^2/[(2)(32.2)] - 3.36 - 4.464^2/[(2)(32.2)]\}/(0.001005 - 0.0016) = 1130 \text{ ft}$$

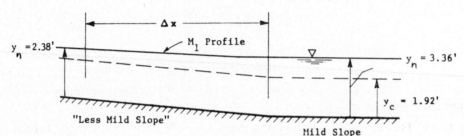

Fig. 14-77

14.304 In a 6-ft-wide rectangular channel ($s = 0.002$, $n = 0.013$), water flows at 250 cfs. A low dam (broad-crested weir) placed in the channel raises the water to a depth of 8.9 ft. Analyze the water-surface profile upstream from the dam.

▌ $y_c = (q^2/g)^{1/3} = [(\frac{250}{6})^2/32.2]^{1/3} = 3.78$ ft, $Q = (A)(1.486/n)(R^{2/3})(s^{1/2})$. In upstream segment, $250 = (6y_n)(1.486/0.013)[6y_n/(y_n + 6 + y_n)]^{2/3}(0.002)^{1/2}$, $y_n = 5.29$ ft (by trial and error). Since $y_n > y_c$, the channel slope is mild. Also, since $[y = 8.9$ ft$] > y_n > y_c$, upstream of the dam is an M_1 profile (see Fig. A-19) of great length with the depth gradually decreasing to the normal depth (5.29 ft). As damming action has occurred, critical depth (3.78 ft) occurs on the dam.

14.305 Solve Prob. 14.304 if the channel slope is 0.0005.

▌ $y_c = 3.78$ ft (from Prob. 14.304), $Q = (A)(1.486/n)(R^{2/3})(s^{1/2})$. In upstream segment, $250 = (6y_n)(1.486/0.013)[6y_n/(y_n + 6 + y_n)]^{2/3}(0.0005)^{1/2}$, $y_n = 9.42$ ft (by trial and error). Since $y_n > y_c$, the channel slope is mild. Insertion of a dam cannot lower the water surface below y_n to 8.9 ft, so this situation is impossible.

14.306 Solve Prob. 14.304 if the channel slope is 0.0008.

▌ $y_c = 3.78$ ft (from Prob. 14.304), $Q = (A)(1.486/n)(R^{2/3})(s^{1/2})$. In upstream segment, $250 = (6y_n)(1.486/0.013)[6y_n/(y_n + 6 + y_n)]^{2/3}(0.0008)^{1/2}$, $y_n = 7.71$ ft (by trial and error). Since $y_n > y_c$, the channel slope is mild. Also, since $[y = 8.9$ ft$] > y_n > y_c$, upstream of the dam is a long M_1 profile (see Fig. A-19) with the depth gradually decreasing to the normal depth (7.71 ft). As damming action has occurred, critical depth (3.78 ft) occurs on the dam.

14.307 Solve Prob. 14.304 if the channel slope is 0.005.

▌ $y_c = 3.78$ ft (from Prob. 14.304), $Q = (A)(1.486/n)(R^{2/3})(s^{1/2})$. In upstream segment, $250 = (6y_n)(1.486/0.013)[6y_n/(y_n + 6 + y_n)]^{2/3}(0.005)^{1/2}$, $y_n = 3.69$ ft (by trial and error). Since $y_n < y_c$, the channel slope is steep. Also, since $[y = 8.9$ ft$] > y_c > y_n$, upstream of the dam is an S_1 profile (see Fig. A-19), preceded by a hydraulic jump. Upstream of the jump is straight supercritical uniform flow with depth of 3.69 ft.

14.308 A rectangular channel changes in width from 4 ft to 6 ft, as shown in Fig. 14-78. Measurements indicate that $y_1 = 2.50$ ft and $Q = 50$ cfs. Determine the depth y_2, neglecting head loss.

❚
$$E_1 = E_2 + h_L \qquad E = y + v^2/2g \qquad v_1 = Q/A_1 = 50/[(4)(2.50)] = 5.000 \text{ ft/s}$$
$$E_1 = 2.50 + 5.000^2/[(2)(32.2)] = 2.888 \text{ ft} \qquad v_2 = Q/A_2 = 50/(6y_2) = 8.333/y_2$$
$$E_2 = y_2 + (8.333/y_2)^2/[(2)(32.2)] = y_2 + 1.078/y_2^2$$
$$h_L = 0 \qquad 2.888 = y_2 + 1.078/y_2^2 + 0 \qquad y_2 = 2.74 \text{ ft} \qquad \text{(by trial and error)}$$

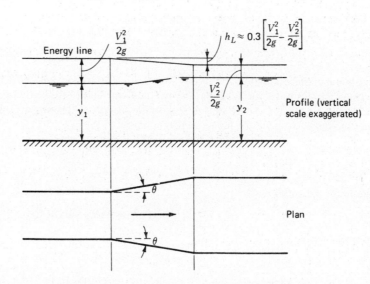

Fig. 14-78

14.309 Solve Prob. 14.308 considering head loss to be given as shown on Fig. 14-78.

❚ $h_L = (0.3)(v_1^2/2g - v_2^2/2g)$. Using data from Prob. 14.308, $h_L = 0.3\{5.000^2/[(2)(32.2)] - (8.333/y_2)^2/[(2)(32.2)]\} = 0.1165 - 0.3235/y_2^2$, $2.888 = y_2 + 1.078/y_2^2 + 0.1165 - 0.3235/y_2^2$, $y_2 = 2.67$ ft (by trial and error).

14.310 Consider a rectangular flume 4.5 m wide, built of unplaned planks ($n = 0.014$), leading from a reservoir in which the water surface is maintained constant at a height of 1.8 m above the bed of the flume at entrance (see Fig. 14-79). The flume is on a slope of 0.001. The depth 300 m downstream from the head end of the flume is 1.20 m. Assuming an entrance loss of $0.2V_1^2/2g$, find the flow rate for the given conditions.

❚ For a first approximate answer we shall consider the entire flume as one reach. The equations to be satisfied are

Energy at entrance:
$$y_1 + (1.2V_1^2/2g) = 1.80 \qquad (1)$$

Energy equation for the entire reach:

$$y_1 + (V_1^2/2g) = 1.20 + (V_2^2/2g) + (S - 0.001)L \qquad (2)$$

where S is given by
$$S = (nV_m/R_m^{2/3})^2 \qquad (3)$$

The procedure is to make successive trials of the upstream depth y_1. This determines corresponding values of V_1, q, V_2, V_m, R_m, and S. The trials are repeated until the value of Δx from Eq. (2) is close to 300 m. The solution is conveniently set in tabular form as follows:

Trial y_1, m	V_1, Eq. (1), m/s	$q = y_1 V_1$, m³/s/m	$V_2 = q/1.20$, m/s	V_m, m/s	R_{h_1}, m	R_{h_2}, m	R_{h_m}, m	S, Eq. (3)	Δx, Eq. (2), m
1.50	2.22	3.33	2.78	2.50	0.90	0.78	0.89	0.00143	358
1.48	2.29	3.39	2.82	2.56	0.89	0.78	0.835	0.00163	226

Thus $y_1 \approx 1.49$ m and the flow rate $Q = qB \approx 3.36 \times 4.5 \approx 15.1$ m³/s. The accuracy of the result, of course, depends on one's ability to select the correct value for Manning's n. If n was assumed to be 0.015, for example, rather than 0.014, the result would have been quite different. Also, a more accurate result can be obtained by dividing the flume into reaches in which the depth change is about 10 percent of the depth.

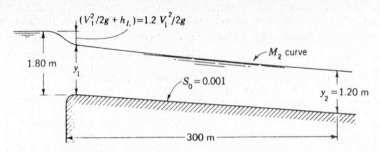

Fig. 14-79

14.311 A rectangular flume of planed timber ($n = 0.012$) 20 ft wide, 1000 ft long, with horizontal bed leads from a reservoir in which the still-water surface is 10 ft above the flume bed. Assume that the depth of the downstream end of the flume is fixed at 8 ft by some control section downstream. Allowing an entrance loss of 0.2 times the velocity head, find the flow rate in the flume using one reach.

▮ Energy at entrance:

$$y_1 + 1.2V_1^2/2g = 10 \qquad (1)$$

Energy equation for the entire reach:

$$y_1 + V_1^2/2g = 8 + V_2^2/2g + (S - 0)(L) \qquad (2)$$

where S is given by

$$S = (nV_m/1.486R_m^{2/3})^2 \qquad (3)$$

This problem involves a trial-and-error solution exactly like that of Prob. 14.310. After successive trials (not shown), the solution is reached with $y_1 = 8.82$ ft and $Q = qB = (70.25)(20) = 1405$ ft³/s.

14.312 Find the flow rate in the flume of Prob. 14.311 if it ends in a free fall, all other conditions remaining the same. The critical depth may be supposed to occur at about $6y_c$ back from the fall. Thus the length of the reach is $1000 - 6y_c$, and $y_2 = y_c$. As a first trial, y_c may be given a reasonable value, say 3 ft.

▮ Take the flume in one reach with y_1 and V_1 at the upstream end and y_c and V_c at the downstream end at a distance $L = (1000 - 6y_c)$ ft. Then, energy at entrance:

$$y_1 + 1.2V_1^2/2g = 10 \qquad (1)$$

Energy equation for the entire reach:

$$y_1 + V_1^2/2g = y_c + V_c^2/2g + (S - 0)(L) \qquad (2)$$

where S is given by

$$S = (nV_m/1.486R_m^{2/3})^2 \qquad (3)$$

First trial: Assume $y_1 = 7$ ft.

$$7 + 1.2V_1^2/[(2)(32.2)] = 10 \qquad V_1 = 12.69 \text{ ft/s} \qquad q = V_1 y_1 = (12.69)(7) = 88.83 \text{ cfs/ft}$$
$$y_c = (q^2/g)^{1/3} = (88.83^2/32.2)^{1/3} = 6.26 \text{ ft}$$
$$V_c = q/y_c = 88.83/6.26 = 14.19 \text{ ft/s} \qquad V_m = (12.69 + 14.19)/2 = 13.44 \text{ ft/s}$$
$$R_m = [(20)(7)/(7 + 20 + 7) + (20)(6.26)/(6.26 + 20 + 6.26)]/2 = 3.984 \text{ ft}$$
$$S = \{(0.012)(13.44)/[(1.486)(3.984)^{2/3}]\}^2 = 0.001865$$
$$7 + 12.69^2/[(2)(32.2)] = 6.26 + 14.19^2/[(2)(32.2)] + 0.001865L \qquad L = 61.1 \text{ ft}$$

But $L = 1000 - (6)(6.26) = 962$ ft. Since the computed value of L (61.1 ft) is not equal to the actual length of approximately 962 ft, the assumed value of y_1 of 7 ft is incorrect. As a second trial, y_1 should be increased. After successive trials (not shown), a value of $y_1 = 8.46$ ft is found to be appropriate. Then, $8.46 + 1.2V_1^2/[(2)(32.2)] = 10$, $V_1 = 9.091$ ft/s, $q = (9.091)(8.46) = 76.91$ cfs/ft, $Q = (76.91)(20) = 1538$ ft³/s.

14.313 At a certain section in a very smooth 6-ft-wide rectangular channel the depth is 3.00 ft when the flow rate is 160 cfs. Compute the distance to the section where the depth is 3.20 ft if $S_0 = 0.0020$ and $n = 0.012$.

$$\Delta x = (y_1 + V_1^2/2g - y_2 - V_2^2/2g)/(S - S_0) \qquad S = (nV_{avg}/1.486R_{avg}^{2/3})^2$$

The calculations are shown in the following table. The total distance is calculated to be 73 ft. The accuracy could be improved by taking more steps. In computing $S - S_0$, a slight error in the calculated value of S will introduce a sizeable error in the calculated value of x. Thus it is important that S be calculated as accurately as possible.

y, ft	A, ft^2	P $(6+2y)$, ft	R_h, ft	V, fps	$V^2/2g$, ft	$y + (V^2/2g)$	Numerator $\Delta[y+(V^2/2g)]$	V_{avg}, fps	R_{avg}, ft	S	Denominator $S - S_0$	Δx, ft
3.00	18.00	12.00	1.500	8.89	1.227	4.227						
							0.022	8.74	1.512	0.00284	0.00084	26
3.10	18.60	12.20	1.525	8.60	1.149	4.249						
							0.029	8.47	1.536	0.00262	0.00062	47
3.20	19.20	12.40	1.548	8.33	1.078	4.278						$\sum(\Delta x) = 73$

14.314 Assume the channel of Prob. 14.145 has a bottom slope of 0.0048. If $y_0 = 3$ ft at $x = 0$, how far along the channel $x = L$ does it take the depth to rise to $y_L = 4$ ft (see Fig. 14-80). Use increments of $y = 0.2$ ft and Manning's formula. Is the 4-ft-depth position upstream or downstream in Fig. 14-80?

$y_c = 4.27$ ft (from Prob. 14.145). Since $[y = 3$ ft$] < [y_c = 4.27$ ft$]$, the flow is supercritical; and the given channel slope (0.0048) is greater than s_c of 0.00435 (from Prob. 14.145). Therefore, we must be on an S profile. $Q = (A)(1.486/n)(R^{2/3})(s^{1/2})$. Since the channel is "wide", use $R = y_n$: $50 = [(1)(y_n)](1.486/0.022)(y_n)^{2/3}(0.0048)^{1/2}$, $y_n = 4.14$ ft. Thus y_0 and y_L are less than y_n, which is less than y_c and we *must* be on an S_3 curve. For the numerical solution, we tabulate $y = 3.0$ to 4.0 in intervals of 0.2, computing six values of $V = q/y$, $E = y + (V^2/2g)$, and $S = n^2V^2/2.208y^{4/3}$, from which S_{av} and Δx follow. The slope $S_0 = 0.0048$ is constant.

y, ft	$V = 50/y$	$E = y + (V^2/2g)$	S	S_{av}	Δx	$x = \sum \Delta x$
3.0	16.67	7.313	0.01407			0.0
				0.01271	40.7	
3.2	15.62	6.991	0.01135			40.7
				0.01031	42.3	
3.4	14.71	6.758	0.00927			83.0
				0.00847	44.4	
3.6	13.89	6.595	0.00766			127.4
				0.00703	48.0	
3.8	13.16	6.488	0.00640			175.4
				0.00590	56.4	
4.0	12.50	6.426	0.00539			$L = 231.8$ ft

For this depth increment of 0.2 ft, gradually-varied-flow theory predicts that a length of about 232 ft is required for the depth to rise from 3 to 4 ft in this supercritical flow. Using 10 increments by reducing Δy to 0.1 ft would give an estimate of $L = 235.2$ ft (calculations not shown). It should be clear that the $y = 4$ ft position is downstream.

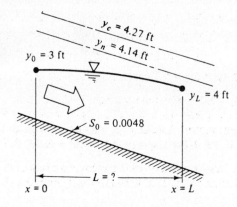

Fig. 14-80

14.315 At a section of a rectangular channel 10 ft wide with $n = 0.015$ and $s_0 = 0.001$, $Q = 300$ cfs. Compute the normal depth and the critical depth. Is the slope steep or mild? What type of gradually varied profile are we on if (**a**) $y = 2$ ft, (**b**) $y = 4$ ft, and (**c**) $y = 6$ ft?

> $$Q = (A)(1.486/n)(R^{2/3})(s^{1/2}) \qquad 300 = (10y_n)(1.486/0.015)[10y_n/(y_n + 10 + y_n)]^{2/3}(0.001)^{1/2}$$
> $$y_n = 5.15 \text{ ft} \quad \text{(by trial and error)} \qquad y_c = (q^2/g)^{1/3} = [(\tfrac{300}{10})^2/32.2]^{1/3} = 3.03 \text{ ft}$$

Since $y_c < y_n$, the slope is mild.
(**a**) If $y = 2$ ft, we have an M_3 curve (see Fig. A-19). (**b**) If $y = 4$ ft, we have an M_2 curve. (**c**) If $y = 6$ ft, we have an M_1 curve.

14.316 Repeat Prob. 14.315 if the slope is 0.007.

> $$Q = (A)(1.486/n)(R^{2/3})(s^{1/2}) \qquad 300 = (10y_n)(1.486/0.015)[10y_n/(y_n + 10 + y_n)]^{2/3}(0.007)^{1/2}$$
> $$y_n = 2.55 \text{ ft} \quad \text{(by trial and error)} \qquad y_c = (q^2/g)^{1/3} = [(\tfrac{300}{10})^2/32.2]^{1/3} = 3.03 \text{ ft}$$

Since $y_c > y_n$, the slope is steep.
(**a**) If $y = 2$ ft, we have an S_3 curve (see Fig. A-19). (**b**) If $y = 4$ ft, we have an S_1 curve. (**c**) If $y = 6$ ft, we have an S_1 curve.

14.317 Water flows in a wide clean-earth ($n = 0.022$) channel at $s_0 = 0.007$ and $q = 10 \text{ (m}^3/\text{s)/m}$. Is this a steep slope? For what range of depths will the flow be on a type 1, 2, or 3 curve?

> $$Q/B = (A/B)(1.0/n)(R^{2/3})(s^{1/2}) \qquad 10 = (y_n)(1.0/0.022)(y_n)^{2/3}(0.007)^{1/2}$$
> $$y_n = 1.79 \text{ m} \qquad y_c = (q^2/g)^{1/3} = (10^2/9.807)^{1/3} = 2.17 \text{ m}$$

Since $y_c > y_n$, the slope is steep. Flow will be S_1 for $[y > 2.17 \text{ m}]$ (see Fig. A-19), S_2 for $[1.79 \text{ m} < y < 2.17 \text{ m}]$, and S_3 for $[y < 1.79 \text{ m}]$.

14.318 Repeat Prob. 14.317 if the channel slope is 0.0015.

> $$Q/B = (A/B)(1.0/n)(R^{2/3})(s^{1/2}) \qquad 10 = (y_n)(1.0/0.022)(y_n)^{2/3}(0.0015)^{1/2}$$
> $$y_n = 2.84 \text{ m} \qquad y_c = (q^2/g)^{1/3} = (10^2/9.807)^{1/3} = 2.17 \text{ m}$$

Since $y_c < y_n$, the slope is mild. Flow will be M_1 for $[y > 2.84 \text{ m}]$ (see Fig. A-19), M_2 for $[2.17 \text{ m} < y < 2.84 \text{ m}]$, and M_3 for $[y < 2.17 \text{ m}]$.

14.319 Water flows at 125 cfs in an asphalt ($n = 0.016$) triangular channel with 45° sloping sides and $s_0 = 0.009$. Is the slope steep or mild? For what range of depths will the flow be a type 1, 2, or 3 curve?

> $$Q = (A)(1.486/n)(R^{2/3})(s^{1/2}) \qquad A = y_n y_n = y_n^2 \qquad p_w = 2y_n/(\cos 45°)$$
> $$R = A/p_w = y_n^2/[2y_n/(\cos 45°)] = 0.3536 y_n$$
> $$125 = (y_n^2)(1.486/0.016)(0.3536 y_n)^{2/3}(0.009)^{1/2} \qquad y_n = 3.51 \text{ ft}$$
> $$A_c = (bQ^2/g)^{1/3} \qquad y_c^2 = [(2y_c)(125)^2/32.2]^{1/3} \qquad y_c = 3.96 \text{ ft}$$

Since $y_c > y_n$, the slope is steep. Flow will be S_1 for $[y > 3.96 \text{ ft}]$ (see Fig. A-19), S_2 for $[3.51 \text{ ft} < y < 3.96 \text{ ft}]$, and S_3 for $[y < 3.51 \text{ ft}]$.

14.320 A circular riveted-steel ($n = 0.015$) duct is 2 m in diameter and laid on a slope of 4 ft per mile. The duct is half-full of water flowing at 2.5 m³/s. Is this a mild or steep slope? Is the flow on a type 1, 2, or 3 curve?

> $Q = (A)(1.0/n)(R^{2/3})(s^{1/2})$, $Q_{\text{full}} = [(\pi)(2)^2/4](1.0/0.015)(\tfrac{2}{4})^{2/3}(\tfrac{4}{5280})^{1/2} = 3.63 \text{ m}^3/\text{s}$, $Q/Q_{\text{full}} = 2.5/3.63 = 0.69$.
> From Fig. A-18, $D/D_{\text{full}} = 61$ percent. $y_n = (0.61)(2) = 1.22 \text{ m}$, $A_c = (bQ^2/g)^{1/3}$. Try $y_c = 0.70 \text{ m}$: $y_c/y_{\text{full}} = 0.70/2 = 0.35 \text{ m}$. From Fig. A-18, $A_c/A_{\text{full}} = 30$ percent.
> $$A_c = 0.30[(\pi)(2)^2/4] = 0.9425 \text{ m}^2 \qquad b = 2[\sqrt{1^2 - (1 - y_c)^2}] \quad \text{(see Fig. 14-81)}$$
> $$b = 2[\sqrt{1^2 - (1 - 0.70)^2}] = 1.908 \text{ m} \qquad 0.9425 = (1.908 Q^2/9.807)^{1/3} \qquad Q = 2.07 \text{ m}^3/\text{s}$$

Since the computed value of Q (2.07 m³/s) is not equal to the given value (2.5 m³/s), try $y_c = 0.75 \text{ m}$: $y_c/y_{\text{full}} = 0.75/2 = 0.375 \text{ m}$. From Fig. A-18, $A_c/A_{\text{full}} = 34$ percent.

> $$A_c = 0.34[(\pi)(2)^2/4] = 1.068 \text{ m}^2 \qquad b = 2[\sqrt{1^2 - (1 - 0.75)^2}] = 1.936 \text{ m}$$
> $$1.068 = (1.936 Q^2/9.807)^{1/3} \qquad Q = 2.48 \text{ m}^3/\text{s}$$

This value of Q is close enough to the given value of 2.5 m³/s so that y_c can be taken as 0.75 m. Since $[y_c = 0.75 \text{ m}] < [y = \frac{2}{2} = 1 \text{ m}] < [y_n = 1.22 \text{ m}]$, the slope is steep, an S_2 curve (see Fig. A-19).

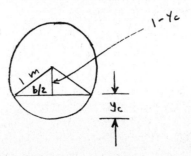

Fig. 14-81

14.321 Consider the gradual change from the profile beginning at point a in Fig. 14-82a, on a mild slope to a mild but steeper slope downstream. Sketch and label the curve $y(x)$ expected.

▌ See Fig. 14-82b.

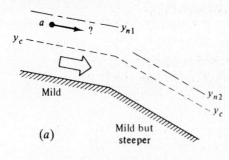

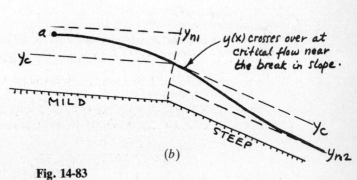

(a) (b)

Fig. 14-82

14.322 Consider the wide-channel flow in Fig. 14-83a, which changes from a mild to a steep slope. Beginning at point a, sketch the water-surface profile $y(x)$ which is expected for gradually varied flow.

▌ See Fig. 14-83b.

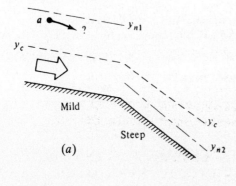

(a) (b)

Fig. 14-83

14.323 In Fig. 14-84a the channel slope changes from steep to less steep. Beginning at point a, sketch and label the expected surface curve $y(x)$ from gradually-varied-flow theory.

▌ See Fig. 14-84b.

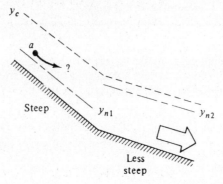

Fig. 14-84a

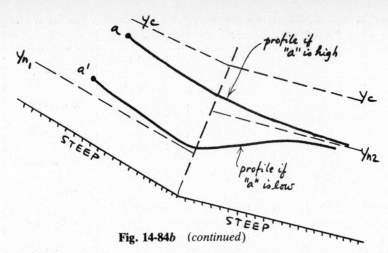

Fig. 14-84b (*continued*)

14.324 Repeat Prob. 14.314 to find the length L using only two depth increments of 0.5 ft each. Is the accuracy sufficient?

▌ Using data from Prob. 14.314,

y	V	E	S	S_{av}	Δx
3.0	16.67	7.313	0.01407		
				0.01125	100
3.5	14.29	6.669	0.00842		
				0.00691	115
4.0	12.50	6.426	0.00539		$L = 215$ ft

Using only two depth increments gives $(231.8 - 215)/231.8 = 0.07$, or 7 percent smaller length than that using five increments.

14.325 A wide-channel flow with $n = 0.018$ is proceeding up an adverse slope with $S_0 = -0.001$. Estimate from the gradually-varied-flow theory the distance required for the depth to drop from 2.5 m to 2.0 m if $q = 4$ (m³/s)/m.

▌ $y_c = (q^2/g)^{1/3} = (4^2/9.807)^{1/3} = 1.18$ m. Hence we have an A_2 profile (see Fig. A-19). Use $\Delta y = 0.125$ m.

y	$V = 4/y$	E	$S = n^2 V^2/y^{4/3}$	S_{av}	$\Delta x = \Delta E/(S_0 - S_v)$
2.5	1.600	2.6305	0.0002445		
				0.0002673	87.5
2.375	1.684	2.5196	0.0002900		
				0.0003187	82.3
2.25	1.778	2.4111	0.0003473		
				0.0003838	76.2
2.125	1.882	2.3056	0.0004202		
				0.0004673	69.3
2.0	2.000	2.2039	0.0005143		$\sum (\Delta x) = L \approx 315$ m

14.326 Figure 14-85 illustrates a free overfall or dropdown flow pattern, where a channel flow accelerates down a slope and falls freely over an abrupt edge. As shown, the flow reaches critical just before the overfall. Between y_c and the edge the flow is rapidly varied and does not satisfy gradually-varied-flow theory. Suppose that the flow rate is $q = 1.1$ (m³/s)/m and the surface is unfinished concrete ($n = 0.014$). Estimate the water depth 300 m upstream.

▌ $\quad q = (1.0/n)(y)(y^{2/3})(s^{1/2}) \qquad 1.1 = (1.0/0.014)(y_n)(y_n)^{2/3}(\tan 0.06°)^{1/2} \qquad y_n = 0.64$ m

$$y_c = (q^2/g)^{1/3} = (1.1^2/9.807)^{1/3} = 0.50 \text{ m}$$

Hence we have an M_2 profile (see Fig. A-19). Use $\Delta y = 0.02$ m.

y	V = 1.1/y	E	S	S_av	Δx
0.50	2.200	0.74669	0.0023904		
				0.0022439	1.2
0.52	2.115	0.74808	0.0020975		
				0.0019735	3.7
0.54	2.037	0.75149	0.0018495		
				0.0017439	7.4
0.56	1.964	0.75666	0.0016384		
				0.0015480	13.3
0.58	1.897	0.76333	0.0014575		
				0.0013796	24.0
0.60	1.833	0.77131	0.0013018		
				0.0012344	48.0
0.62	1.774	0.78044	0.0011670		
				0.0011367	55.2
0.63	1.746	0.78538	0.0011064		
				0.0010920	57.3
0.635	1.732	0.78795	0.0010776		
				0.0010692	70.9
0.638	1.724	0.78951	0.0010608		
				0.0010594	21.6
0.6385	1.723	0.78977	0.0010580		
					$\sum (\Delta x) \approx 300$ m

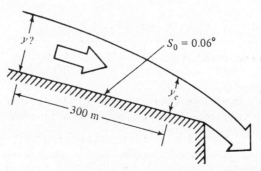

Fig. 14-85

14.327 The clean-earth ($n = 0.022$) channel in Fig. 14-86 is 6 m wide and slopes 0.3°. Water flows at 30 m³/s in the channel and enters a reservoir so that the channel depth is 3 m just before the entry. Assuming gradually varied flow, how far is the distance L to a point in the channel where $y = 2$ m? What type of curve is the water surface?

$$Q = (A)(1.0/n)(R^{2/3})(s^{1/2}) \qquad 30 = (6y_n)(1.0/0.022)[6y_n/(y_n + 6 + y_n)]^{2/3}(\tan 0.3°)^{1/2}$$

$$y_n = 1.52 \text{ m} \qquad \text{(by trial and error)} \qquad y_c = (q^2/g)^{1/3} = [(\tfrac{30}{6})^2/9.807]^{1/3} = 1.37 \text{ m}$$

Hence we have an M_1 profile (see Fig. A-19). Use $\Delta y = 0.25$ m.

y	V = 30/6y	E	R_h	S = n²V²/R_H^{4/3}	S_av	Δx
3.0	1.667	3.1416	1.500	0.0007830		
					0.0008859	51
2.75	1.818	2.9185	1.435	0.000988		
					0.0011345	52
2.50	2.000	2.7039	1.364	0.0012803		
					0.0014950	54
2.25	2.222	2.5017	1.286	0.0017096		
					0.0020409	57
2.0	2.500	2.3186	1.200	0.0023722		
						$L = \sum (\Delta x) \approx 214$ m

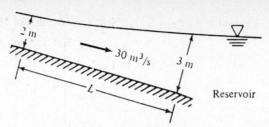

Fig. 14-86

14.328 Figure 14-87 shows a channel width change accompanied by a water-depth change, sometimes called a venturi flume, because measurements of y_1 and y_2 can be used to meter the flow rate. Suppose that $b_1 = 5$ m, $b_2 = 3$ m, $y_1 = 2$ m, and $y_2 = 1.5$ m. Assuming no losses, compute the flow rate Q.

▮
$$y_1 + v_1^2/2g = y_2 + v_2^2/2g \qquad Q = y_1 b_1 v_1 = y_2 b_2 v_2$$

Solve these equations simultaneously: $v_2 = \{(2g)(y_1 - y_2)/[1 - (y_2 b_2/y_1 b_1)^2]\}^{1/2}$, $Q = y_2 b_2 v_2 = \{(2g)(y_1 - y_2)/[(1/b_2^2 y_2^2) - (1/b_1^2 y_1^2)]\}^{1/2} = \{(2)(9.807)(2 - 1.5)/[1/(3^2)(1.5)^2 - 1/(5)^2(2)^2]\}^{1/2} = 15.8$ m³/s.

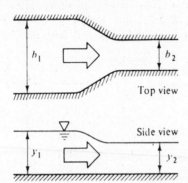

Top view

Side view

Fig. 14-87

14.329 In Prob. 14.328, keeping y_1, b_1, and b_2 the same, for what value of y_2 will the exit flow v_2 be critical? What will the critical flow rate be?

▮ $y_c = [(Q/b)^2/g]^{1/3}$. y_2 must equal y_c. $Q = \{(2g)(y_1 - y_2)/[(1/b_2^2 y_2^2) - (1/b_1^2 y_1^2)]\}^{1/2}$ (from Prob. 14.328). Try $y_2 = 1.5$ m: $Q = \{(2)(9.807)(2 - 1.5)/[1/(3^2)(1.5)^2 - 1/(5)^2(2)^2]\}^{1/2} = 15.78$ m³/s, $y_c = [(15.78/3)^2/9.807]^{1/3} = 1.41$ m. Since $y_c = 1.41$ m is not equal to the assumed $y_2 = 1.5$ m, try $y_2 = 1.42$ m: $Q = \{(2)(9.807)(2 - 1.42)/[1/(3^2)(1.42)^2 - 1/(5)^2(2)^2]\}^{1/2} = 15.88$ m³/s, $y_c = [(15.88/3)^2/9.807]^{1/3} = 1.42$ m (O.K.). Hence critical flow of 15.88 m³/s will occur at $y_2 = 1.42$ m.

14.330 In Prob. 14.328, keeping y_2, b_1, and b_2 the same, suppose the flow rate is 10 m³/s and y_1 is unknown. Find the proper value of y_1.

▮
$$Q = \{(2g)(y_1 - y_2)/[(1/b_2^2 y_2^2) - (1/b_1^2 y_1^2)]\}^{1/2} \qquad \text{(from Prob. 14.328)}$$
$$10 = \{(2)(9.807)(y_1 - 1.5)/[(1/(3^2)(1.5)^2 - 1/(5)^2(y_1)^2]\}^{1/2} \qquad y_1 = 1.68 \text{ m} \qquad \text{(by trial and error)}$$

There is a second solution, $y_1 = 1.36$ m, which causes a flow of 10 m³/s in the opposite direction.

14.331 In the flume transition of Fig. 14-87, suppose that $b_1 = 4$ ft, $b_2 = 3$ ft, $y_1 = 2$ ft, and $Q = 30$ cfs. Compute y_2 and v_2, assuming no losses.

▮
$$Q = \{(2g)(y_1 - y_2)/[(1/b_2^2 y_2^2) - (1/b_1^2 y_1^2)]\}^{1/2} \qquad \text{(from Prob. 14.328)}$$
$$30 = \{(2)(32.2)(2 - y_2)/[(1/(3^2)(y_2)^2 - 1/(4)^2(2)^2]\}^{1/2} \qquad y_2 = 1.64 \text{ ft} \qquad \text{(by trial and error)}$$

14.332 A flow of 35 m³/s flows along a trapezoidal concrete channel where (see Fig. 14-88) the base a is 4 m and β is 45°. If at section 1, the depth of the flow is 3 m, what is the water-surface profile up to a distance 600 m downstream. The channel is finished concrete and has a constant slope S_0 of 0.001.

▮
$$\Delta L = ([1 - (Q^2 b/gA^3)]/\{S_0 - (n/\kappa)^2[Q^2/(R_H^{4/3} A^2)]\}) \, \Delta y$$

We start with $y_1 = 3$ m. At this section we know that

$$A_1 = (3)(4) + \tfrac{1}{2}(3)(3)(2) = 21 \text{ m}^2$$
$$(R_H)_1 = A_1/(p_w)_1 = 21/[4 + (2)(3/0.707)] = 1.6818 \text{ m} \qquad (1)$$
$$b_1 = 4 + (2)(3) = 10 \text{ m}$$

We take $n = 0.012$ and $\kappa = 1.00$ and let $y_2 = 3.1$ m. Now we compute A_2, $(R_H)_2$, and b_2.*

$$A_2 = (3.1)(4) + \tfrac{1}{2}(3.1)(3.1)(2) = 22.01 \text{ m}^2$$

$$(R_H)_2 = 22.01/[4 + 2(3.1/0.707)] = 1.7236 \text{ m} \tag{2}$$

$$b_2 = 4 + (2)(3.1) = 10.20 \text{ m}$$

In the first interval, the average values of A, R_H, and b are

$$(A_{1-2})_{\text{av}} = 21.505 \text{ m}^2$$

$$[(R_H)_{1-2}]_{\text{av}} = 1.7027 \text{ m} \tag{3}$$

$$(b_{1-2})_{\text{av}} = 10.10 \text{ m}$$

$$(\Delta L)_{1-2} = \left\{ \frac{1 - (35^2)(10.10)/[(9.81)(21.505^3)]}{0.01 - (0.012/1)^2(35^2)/[(1.7027^{4/3})(21.505^2)]} \right\}(0.1) = 107.4 \text{ m}$$

We thus have two points on the free-surface profile. Next we compute A_3, R_{H_3}, and b_3 for $y = 3.2$ m. Using Eqs. (2) we now find the average values of these quantities in the interval 2–3. For instance, $(A_{2-3})_{\text{av}} = \tfrac{1}{2}\{(22.01) + [(3.2)(4) + (3.2^2)]\} = 22.525 \text{ m}^2$. We then proceed as indicated for the first interval.

The following table gives the results using six sections so that $L_{\text{total}} \approx 600$ m.

section	y, m	Δy, m	ΔL, m	L_{total}, m
1	3.0	0.1	0	0
2	3.1	0.1	107.4	107.4
3	3.2	0.1	106.9	214.3
4	3.3	0.1	105.6	319.9
5	3.4	0.1	104.7	424.6
6	3.5	0.1	104.1	528.7
7	3.6	0.1	103.8	632.5

We can now plot y versus L, that is, the second and last columns, starting with $y = 3$ for $L = 0$ m and going on to $y = 3.6$ m for $L = 632.5$ m. A smooth curve through these parts gives the approximate desired profile.

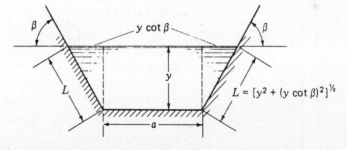

Fig. 14-88

14.333 Water enters a rectangular channel which is 3 ft wide at an average velocity of 0.8 fps and a depth of 3 in. The channel has an inclination α of 0.2°. If Manning's n for the surface is 0.012, estimate at what distance L along the channel the elevation will have risen to a depth of 4 in. Use $\Delta y = 0.2$ in in your numerical calculations.

∎

$$\Delta L = ([1 - (Q^2 b/g A^3)]/\{S_0 - (n/\kappa)^2[Q^2/(R_H^{4/3} A^2)]\}) \, \Delta y$$

$$S_0 = \tan 0.2° = 0.003499 \qquad R_H = by/(b + 2y) = y$$

Point 1:

$$V_1 = 0.8 \text{ ft/s} \qquad (F_R)_1 = V\sqrt{b}/\sqrt{gA} = (0.8)(\sqrt{3})/\sqrt{(g)(3)(\tfrac{1}{4})} = 0.282 \qquad \Delta y = 0.2/12 \text{ ft}$$

Point 2:

$$V_2 = [y_0(\text{in})/y_1(\text{in})]V_1 = (3/3.2)(0.8) = 0.750 \text{ ft/s} \qquad (F_R)_2 = (0.750)(\sqrt{3})/\sqrt{(g)(3)(3.2/12)} = 0.256$$

---

* We are assuming that the depth y relative to the channel is increasing. If we get a positive result for ΔL, we know that our assumption is correct. If ΔL is negative, then the depth must be decreasing along the channel flow. Use y_2 less than 3 m in that case.

1–2:

$$(V_{1-2})_{av} = (0.8 + 0.750)/2 = 0.775 \text{ ft/s} \qquad (F_{R,1-2})_{av} = (0.282 + 0.256)/2 = 0.269$$

$$(\Delta L)_{1-2} = \frac{1 - 0.269^2}{0.00349 - (0.012/1.486)^2[0.775^2/(3.1/12)^{4/3}]} \left(\frac{0.2}{12}\right) = 4.75 \text{ ft}$$

Point 3:

$$V_3 = (3/3.4)(0.8) = 0.706 \text{ ft/s} \qquad (F_R)_3 = 0.706/[(g(3.4/12)]^{1/2} = 0.234$$

2–3:

$$(V_{2-3})_{av} = (0.750 + 0.706)/2 = 0.728 \text{ ft/s} \qquad (F_{R,2-3})_{av} = (0.256 + 0.234)/2 = 0.245$$

$$(\Delta L)_{2-3} = \frac{1 - 0.245^2}{0.00349 - (0.012/1.486)^2[0.728^2/(3.3/12)^{4/3}]} \left(\frac{0.2}{12}\right) = 4.75 \text{ ft}$$

Point 4:

$$V_4 = (3/3.6)(0.8) = 0.667 \text{ ft/s} \qquad (F_R)_4 = 0.667/\sqrt{(g)(3.6/12)} = 0.215$$

3–4:

$$(V_{3-4})_{av} = (0.706 + 0.667)/2 = 0.686 \text{ ft/s} \qquad (F_{R,3-4})_{av} = (0.234 + 0.215)/2 = 0.224$$

$$(\Delta L)_{3-4} = \frac{1 - 0.224^2}{0.00349 - (0.012/1.486)^2[0.686^2/(3.5/12)^{4/3}]} \left(\frac{0.2}{12}\right) = 4.75 \text{ ft}$$

Point 5:

$$V_5 = (3/3.8)(0.8) = 0.632 \text{ ft/s} \qquad (F_R)_5 = 0.632/\sqrt{(g)(3.8/12)} = 0.198$$

4–5:

$$(V_{4-5})_{av} = (0.667 + 0.632)/2 = 0.650 \text{ ft/s} \qquad (F_{R,4-5})_{av} = (0.215 + 0.198)/2 = 0.206$$

$$(\Delta L)_{4-5} = \frac{1 - 0.206^2}{0.00349 - (0.012/1.486)^2[0.650^2/(3.7/12)^{4/3}]} \left(\frac{0.2}{12}\right) = 4.75 \text{ ft}$$

Point 6:

$$V_6 = (\tfrac{3}{4})(0.8) = 0.600 \text{ ft/s} \qquad (F_R)_6 = 0.600/[g(4.0/12)]^{1/2} = 0.183$$

5–6:

$$(V_{5-6})_{av} = (0.632 + 0.600)/2 = 0.616 \text{ ft/s} \qquad (F_{R,5-6})_{av} = (0.198 + 0.183)/2 = 0.190$$

$$(\Delta L)_{5-6} = \frac{1 - 0.190^2}{0.00349 - (0.012/1.486)^2[0.616^2/(3.9/12)^{4/3}]} \left(\frac{0.2}{12}\right) = 4.75 \text{ ft}$$

$$\text{Total length} = 23.75 \text{ ft}$$

14.334 In Prob. 14.332, compute the distance L in one calculation where the free surface has a depth of 3.6 m. Do not average.

▮

$$\Delta L = ([1 - (Q^2 b/gA^3)]/\{S_0 - (n/\kappa)^2[Q^2/(R_H^{4/3} A^2)]\}) \, \Delta y$$

$$A = (3)(4) + (3)(3) = 21.00 \text{ m}^2 \qquad p_w = 4 + (2)(\sqrt{3^2 + 3^2}) = 12.49 \text{ m}$$

$$R_H = 21.00/12.49 = 1.681 \text{ m} \qquad b = 3 + 4 + 3 = 10 \text{ m}$$

$$\Delta L = \left\{ \frac{1 - (35)^2(10)/(9.807)(21.00)^3}{0.001 - (0.012/1)^2[35^2/(1.681)^{4/3}(21.00)^2]} \right\}(3.6 - 3) = 649 \text{ m}$$

14.335 Do Prob. 14.334 using a linear average in the calculations. What is the percent error in not averaging?

▮

$$\Delta L = ([1 - (Q^2 b/gA^3)]/\{S_0 - (n/\kappa)^2[Q^2/(R_H^{4/3} A^2)]\}) \, \Delta y$$

At the section where $y = 3.6$ m,

$$A = (3.6)(4) + (3.6)(3.6) = 27.36 \text{ m}^2 \qquad p_w = 4 + (2)(\sqrt{3.6^2 + 3.6^2}) = 14.18 \text{ m}$$

$$R_H = 27.36/14.18 = 1.929 \text{ m} \qquad b = 3.6 + 4 + 3.6 = 11.20 \text{ m}$$

Using corresponding values where $y = 3$ m from Prob. 14.334,

$$A_{av} = (21.00 + 27.36)/2 = 24.18 \text{ m}^2 \qquad (R_H)_{av} = (1.681 + 1.929)/2 = 1.805 \text{ m}$$

$$b_{av} = (10 + 11.20)/2 = 10.60 \text{ m}$$

$$\Delta L = \left\{ \frac{1 - (35)^2(10.60)/(9.807)(24.18)^3}{0.001 - (0.012/1)^2[35^2/(1.805)^{4/3}(24.18)^2]} \right\}(3.6 - 3) = 630 \text{ m}$$

Error $= (649 - 630)/630 = 0.03$, or 3 percent.

14.336 A wide channel is made of finished concrete. It has a slope S_0 equal to 0.0003. The opening from a large reservoir to the channel is a sharp-edged sluice gate, as shown in Fig. 14-89. The coefficient of contraction C_c is 0.80 and the coefficient of friction $C_f = 0.85$. What is the approximate depth at the vena contracta and the flow q of fluid? How far from the vena contracta does the water increase depth by 0.030 m? Use one calculation with linear averages.

I

$$v_1^2/2g + y_1 = v_2^2/2g + y_2 \qquad y_2 = (0.30)(0.80) = 0.2400 \text{ m}$$

$$0 + 10 = v_2^2/[(2)(9.807)] + 0.2400 \qquad v_2 = 13.84 \text{ m/s}$$

$$q_{theoretical} = y_2 v_2 = (0.2400)(13.84) = 3.322 \text{ (m}^3\text{/s)/m} \qquad q_{actual} = (0.85)(3.322) = 2.82 \text{ (m}^3\text{/s)/m}$$

$$\Delta L = ([1 - (Q^2 b/gA^3)]/\{S_0 - (n/\kappa)^2[Q^2/(R_H^{4/3}A^2)]\}) \Delta y$$

$$\Delta y_{downstream} = 0.24 + 0.03 = 0.27 \text{ m}$$

$$\Delta L = \left\{ \frac{1 - (2.82)^2/(9.807)(0.27)^3}{0.0003 - (0.012/1)^2[2.82^2/(0.27)^{4/3}(0.27)^2]} \right\}(0.030) = 13.44 \text{ m}$$

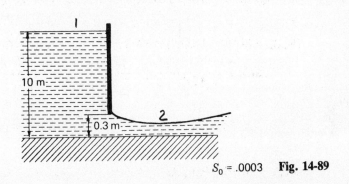

$S_0 = .0003$ **Fig. 14-89**

14.337 Water is moving at a speed of 4 m/s in a very wide horizontal channel at a depth of 1 m. If $n = 0.025$ for earth in good condition, at what distance x downstream will the depth increase to 1.1 m? Do problem analytically without the use of numerical methods.

I
$$x = (\kappa/n)^2 \left[\frac{3}{4g}(y^{4/3} - y_1^{4/3}) - \frac{3}{13q^2}(y^{13/3} - y_1^{13/3}) \right]$$

$$= \left(\frac{1}{0.025}\right)^2 \left\{ \left[\frac{3}{(4)(9.807)} \right](1.1^{4/3} - 1^{4/3}) - \left[\frac{3}{(13)(4)^2} \right](1.1^{13/3} - 1^{13/3}) \right\} = 4.78 \text{ m}$$

14.338 In a wide rectangular earth channel with weeds and stones, it is observed that the depth rises 0.2 m from a depth of 1 m in a distance of 6 m. What is the volumetric flow rate per unit length? Use $n = 0.035$.

I
$$x = \left(\frac{\kappa}{n}\right)^2 \left[\frac{3}{4g}(y^{4/3} - y_1^{4/3}) - \frac{3}{13q^2}(y^{13/3} - y_1^{13/3}) \right]$$

$$6 = \left(\frac{1}{0.035}\right)^2 \left\{ \left[\frac{3}{(4)(9.807)} \right](1.2^{4/3} - 1^{4/3}) - \left(\frac{3}{13q^2} \right)(1.2^{13/3} - 1^{13/3}) \right\}$$

$$q = 4.50 \text{ (m}^3\text{/s)/m}$$

14.339 In Prob. 14.196, for a flow of 5000 cfs at the geometry shown initially, at what distance downstream is the depth at an elevation of 14 ft? Use initial data and do not average. Do in one calculation.

∎ $$\Delta L = \left([1 - (Q^2 b/gA^3)]/\{S_0 - (n/\kappa)^2[Q^2/(R_H^{4/3}A^2)]\}\right)\Delta y$$

From Prob. 14.196, $A = 217.1\ \text{ft}^2$, $R_H = 5.802\ \text{ft}$, $n = 0.012$, $S_0 = 0.0016$, $b = 20\ \text{ft}$, and $y = 10\ \text{ft} + 3\ \text{ft} = 13\ \text{ft}$.

$$\Delta L = \left\{\frac{1 - (5000)^2(20)/(32.2)(217.1)^3}{0.0016 - (0.012/1.486)^2[5000^2/(5.802)^{4/3}(217.1)^2]}\right\}(14-13) = 301\ \text{ft}$$

14.340 Do Prob. 14.339 using a linear average over the interval. Compare results with ΔL in Prob. 14.339 not involving an average. Use one cycle of calculation. What is the error in not averaging in this problem?

∎ $$\Delta L = \left([1 - (Q^2 b/gA^3)]/\{S_0 - (n/\kappa)^2[Q^2/(R_H^{4/3}A^2)]\}\right)\Delta y$$

At $y = 13\ \text{ft}$, from Prob. 14.339, $A = 217.1\ \text{ft}^2$, $R_H = 5.802\ \text{ft}$, $n = 0.012$, $S_0 = 0.0016$, $b = 20\ \text{ft}$, and $y = 10 + 3 = 13\ \text{ft}$. At $y = 14\ \text{ft}$, $y = 10 + 4 = 14\ \text{ft}$.

$$A = (\tfrac{1}{2})(\pi)(10)^2 + (4)(10 + 10) = 237.1\ \text{ft}^2 \qquad R_H = 237.1/[(\pi)(10) + 4 + 4] = 6.015\ \text{ft}$$
$$(R_H)_{av} = (5.802 + 6.015)/2 = 5.908\ \text{ft} \qquad A_{av} = (217.1 + 237.1)/2 = 227.1\ \text{ft}^2$$
$$\Delta L = \left\{\frac{1 - (5000)^2(20)/(32.2)(227.1)^3}{0.0016 - (0.012/1.486)^2[5000^2/(5.908)^{4/3}(227.1)^2]}\right\}(14-13) = 240\ \text{ft}$$

Error $= (301 - 240)/240 = 0.25$, or 25 percent.

14.341 Water flows at a rate of 500 ft³/s through a rectangular section 10.0 ft wide from a "steep" slope to a "mild" slope creating a hydraulic jump, in the manner illustrated in Fig. 14-90. The upstream depth of flow (d_1) is 3.1 ft. Find the **(a)** downstream depth, **(b)** energy (head) loss in the hydraulic jump, and **(c)** upstream and downstream velocities.

∎ **(a)** $d_2 = (d_1/2)(\sqrt{1 + 8q^2/gd_1^3} - 1) = (3.1/2)\{\sqrt{1 + (8)(500/10.0)^2/[(32.2)(3.1)^3]} - 1\} = 5.7\ \text{ft}$

(b) $E_j = (d_2 - d_1)^3/4d_1 d_2 = (5.7 - 3.1)^3/[(4)(3.1)(5.7)] = 0.25\ \text{ft of water}$

(c) $v_1 = Q/A_1 = 500/[(3.1)(10.0)] = 16.1\ \text{ft/s} \qquad v_2 = 500/[(5.7)(10.0)] = 8.8\ \text{ft/s}$

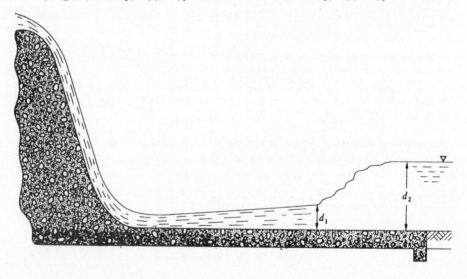

Fig. 14-90

14.342 Rework Prob. 14.341 for a flow rate of 20.0 m³/s, a channel width of 4.0 m, and an upstream depth of flow of 1.20 m.

∎ **(a)** $d_2 = (d_1/2)(\sqrt{1 + 8q^2/gd_1^3} - 1) = (1.20/2)\{\sqrt{1 + (8)(20.0/4.0)^2/[(9.807)(1.20)^3]} - 1\} = 1.55\ \text{m}$

(b) $E_j = (d_2 - d_1)^3/4d_1 d_2 = (1.55 - 1.20)^3/[(4)(1.20)(1.55)] = 0.006\ \text{m of water}$

(c) $v_1 = Q/A_1 = 20.0/[(1.20)(4.0)] = 4.17\ \text{m/s} \qquad v_2 = 20.0/[(1.55)(4.0)] = 3.23\ \text{m/s}$

14.343 Water flows over a concrete spillway into a rectangular channel 9.0 m wide through a hydraulic jump, in the manner illustrated in Fig. 14-90. The depths before and after the jump are 1.55 m and 3.08 m, respectively. Find the rate of flow in the channel.

❚ $q = \{[(d_1 + d_2)/2](gd_1d_2)\}^{1/2} = \{[(1.55 + 3.08)/2][(9.807)(1.55)(3.08)]\}^{1/2} = 10.41 \ (\text{m}^3/\text{s})/\text{m}$

$$Q = (10.41)(9.0) = 93.7 \ \text{m}^3/\text{s}$$

14.344 Rework Prob. 14.343 if the channel width is 20 ft and the depths of flow before and after the jump are 4.5 ft and 8.2 ft, respectively.

❚ $q = \{[(d_1 + d_2)/2](gd_1d_2)\}^{1/2} = \{[(4.5 + 8.2)/2][(32.2)(4.5)(8.2)]\}^{1/2} = 86.86 \ (\text{ft}^3/\text{s})/\text{ft}$

$$Q = (86.86)(20) = 1737 \ \text{ft}^3/\text{s}$$

14.345 A hydraulic jump occurs downstream from a 15-m-wide sluice gate. The depth is 1.5 m, and the velocity is 20 m/s. Determine (a) the Froude number and the Froude number corresponding to the conjugate depth, (b) the depth and velocity after the jump, and (c) the power dissipated by the jump.

❚ (a) $\qquad\qquad N_F = v/\sqrt{gy} \qquad (N_F)_1 = 20/\sqrt{(9.807)(1.5)} = 5.21$

$$(N_F)_2 = (2)(\sqrt{2})(N_F)_1/[\sqrt{1 + (8)(N_F)_1^2} - 1]^{3/2} = (2)(\sqrt{2})(5.21)/[\sqrt{1 + (8)(5.21)^2} - 1]^{3/2} = 0.288$$

(b) $\qquad\qquad (N_F)_2 = v_2/\sqrt{gy} \qquad v_2 y_2 = v_1 y_1 = (1.5)(20) = 30.00 \ \text{m}^2/\text{s}$

$$v_2^2 = (N_F)_2^2 g y_2 = (N_F)_2^2 (g)(30.00/v_2) = (0.288)^2(9.807)(30.00/v_2)$$

$$v_2 = 2.90 \ \text{m/s} \qquad y_2 = 30.00/2.90 = 10.34 \ \text{m}$$

(c) $\qquad E_j = (y_2 - y_1)^3/4y_1 y_2 = (10.34 - 1.5)^3/[(4)(1.5)(10.34)] = 11.13 \ \text{m of water}$

$$P = \gamma Q E_j \qquad Q = Av = [(15)(1.5)](20) = 450 \ \text{m}^3/\text{s} \qquad P = (9.79)(450)(11.13) = 49.0 \times 10^3 \ \text{kW}$$

14.346 Water flows at a rate of 16 m³/s at half critical depth in a trapezoidal channel with $b = 4$ m and $m = 0.4$ (see Fig. 14-91), before a hydraulic jump occurs. Prepare and execute a computer program to find the height after the jump and the energy loss in kilowatts.

```
❚ 10 REM B:EX124            JUMP IN A TRAPEZOIDAL CHANNEL
   20 DEFINT I:    DEF FNC(DY)=Q^2*(B+2!*M*DY)-G*(DY*(B+M*DY))^3  '
   30 DEF FNFM(DY)=.5*B*DY^2+M*DY^3/3!+Q^2/(G*DY*(B+M*DY))        '
   40 READ G,Q,B,M,GAM:    DATA 9.806,16.,4.,.4,9806.
   50 YMAX=16!:    YMIN=0!:    LPRINT:    LPRINT"G,Q,B,M,GAM=";G;Q;B;M;GAM
   60 FOR I=1 TO 15:    YC=.5*(YMAX+YMIN)
   70 IF FNC(YC)>0! THEN YMIN=YC ELSE YMAX=YC
   80 PRINT YMAX;YMIN;YC
   90 NEXT I:    LPRINT"Y1,YC=";.5*YC;YC
   100 Y1=.5*YC:    YMIN=YC:    YMAX=3!*YC:    FM=FNFM(Y1)
   110 FOR I=1 TO 15:    Y2=.5*(YMAX+YMIN)
   120 IF FNFM(Y2)-FM>0! THEN YMAX=Y2 ELSE YMIN=Y2
   130 PRINT"YMAX,YMIN,Y2=";YMAX;YMIN;Y2:    NEXT I
   140 A1=Y1*(B+M*Y1):    A2=Y2*(B+M*Y2):    V1=Q/A1:    V2=Q/A2
   150 LOSS=(V1^2-V2^2)/(2!*G)+Y1-Y2:    POWER=GAM*Q*LOSS/1000!
   160 LPRINT"Y1,Y2,V1,V2,LOSS,POWER=";
   170 LPRINT USING"###.### ";Y1;Y2;V1;V2;LOSS;POWER

   G,Q,B,M,GAM= 9.806001  16   4   .4   9806
   Y1,YC= .5661621  1.132324
   Y1,Y2,V1,V2,LOSS,POWER=  0.566   1.973   6.687   1.693   0.726 113.970
```

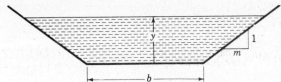

Fig. 14-91

14.347 A rectangular channel 10 ft wide carries 100 cfs in uniform flow at a depth of 1.67 ft. Suppose that an obstruction such as a submerged weir is placed across the channel, rising to a height of 6 in above the bottom. (a) Will this obstruction cause a hydraulic jump to form upstream? Why? (b) Find the water depth over the obstruction, and classify the surface profile, if possible, to be found upstream of the weir.

❚ (a) $y_c = (q^2/g)^{1/3} = [(\frac{100}{10})^2/32.2]^{1/3} = 1.459 \ \text{ft}$. Since $[y_0 = 1.67 \ \text{ft}] > [y_c = 1.459 \ \text{ft}]$, the slope is mild, flow is subcritical, and no jump can occur.

(b) $\qquad\qquad E = y + (1/2g)(q^2/y^2) \qquad E_0 = 1.67 + \{1/[(2)(32.2)]\}[(\frac{100}{10})^2/1.67^2] = 2.227 \ \text{ft}$

$$E_{min} = (\tfrac{3}{2})(y_c) = (\tfrac{3}{2})(1.459) = 2.188 \ \text{ft} \qquad \Delta z_{crit} = E_0 - E_{min} = 2.227 - 2.188 = 0.039 \ \text{ft}$$

Since $[\Delta z = 0.5 \ \text{ft}] > [\Delta z_{crit} = 0.039 \ \text{ft}]$, y_c occurs over the obstruction. Hence, the water depth over the obstruction = 1.459 ft. Let y_1 = water depth just upstream of the obstruction. Assuming no energy loss over the obstruction, $E_1 = E_{min} + \Delta z$, or $y_1 + \{1/[(2)(32.2)]\}[(\frac{100}{10})^2/y_1^2] = 2.188 + 0.5 = 2.688$, $y_1 = 2.42$ ft or 0.943 ft (by trial and error). There is no mild profile that enables normal flow to become supercritical downstream, so $y_1 = 0.943$ ft is not possible here. Thus $y_1 = 2.42$ ft. Upstream of y_1, the water surface tends asymptotically to $y_0 = 1.67$ ft.

14.348 Suppose that the slope and roughness of the channel in Prob. 14.347 are such that uniform flow of 100 cfs occurs at 1.00 ft. Consider an obstruction rising 4 in above the bottom of the channel. Will a hydraulic jump form upstream? As in Prob. 14.347, classify the surface profile found just upstream from the obstruction.

▮ $y_c = 1.459$ ft (from Prob. 14.347). Since $[y_0 = 1.000 \text{ ft}] < [y_c = 1.459 \text{ ft}]$, the slope is steep and for $y = y_0$, the flow is supercritical.

$$E = y + (1/2g)(q^2/y^2) \qquad E_0 = 1.000 + \{1/[(2)(32.2)]\}[(\tfrac{100}{10})^2/1.00^2] = 2.553 \text{ ft}$$

$$E_{min} = 2.188 \text{ ft} \qquad \text{(from Prob. 14.347)} \qquad \Delta z_{crit} = E_0 - E_{min} = 2.553 - 2.188 = 0.365 \text{ ft}$$

Since $[\Delta z = 4 \text{ in, or } 0.333 \text{ ft}] < [\Delta z_{crit} = 0.365 \text{ ft}]$, the obstruction is not sufficiently high to produce critical flow. Hence, upstream the flow is straight supercritical uniform flow, and a hydraulic jump cannot form.

14.349 In a rectangular channel 3 m wide with a flow of 5.65 m³/s the depth is 0.3 m. If a hydraulic jump is produced, what will be the depth after it? What will be the loss of energy?

▮ $$y_2 = (y_1/2)(\sqrt{1 + 8q^2/gy_1^3} - 1) = (0.3/2)\{\sqrt{1 + (8)(5.65/3)^2/[(9.807)(0.3)^3]} - 1\} = 1.410 \text{ m}$$
$$E_j = (y_2 - y_1)^3/4y_1 y_2 = (1.410 - 0.3)^3/[(4)(0.3)(1.410)] = 0.808 \text{ m}$$

14.350 The hydraulic jump may be used as a crude flowmeter. Suppose that in a horizontal rectangular channel 5 ft wide the observed depths before and after a hydraulic jump are 0.66 ft and 3.00 ft, respectively. Find the rate of flow and the head loss.

▮ $$q = \{[(d_1 + d_2)/2](gd_1d_2)\}^{1/2} = \{[(0.66 + 3.00)/2][(32.2)(0.66)(3.00)]\}^{1/2} = 10.80 \text{ (ft}^3\text{/s)/m}$$
$$Q = (10.80)(5) = 54.0 \text{ ft}^3\text{/s} \qquad E_j = (d_2 - d_1)^3/4d_1d_2 = (3.00 - 0.66)^3/[(4)(0.66)(3.00)] = 1.62 \text{ ft}$$

14.351 The tidal bore, which carries the tide into an estuary of a large river, is an example of an abrupt translatory wave, or moving hydraulic jump. Suppose such a bore is observed to rise to a height of 12 ft above the normal low-tide river depth of 8 ft. The speed of travel of the bore upstream is observed to be 15 mph. Find the approximate velocity of the undisturbed river. Does this represent subcritical or supercritical flow?

▮ $$q = \{[(d_1 + d_2)/2](gd_1d_2)\}^{1/2} = \{[(12 + 8 + 8)/2][(32.2)(12 + 8)(8)]\}^{1/2} = 268.6 \text{ cfs/ft}$$
$$v_1 = q/d_1 = 268.6/8 = 33.58 \text{ ft/s} \quad \text{(relative to the jump)} \qquad v_{jump} = (15)(\tfrac{5280}{3600}) = 22.00 \text{ ft/s}$$

The jump is moving upstream at a velocity of 22.00 ft/s while the river is moving into the jump at 33.58 ft/s. Therefore, the river is moving downstream at $33.58 - 22.00 = 11.58$ ft/s; $v^2/2g = 11.58^2/[(2)(32.2)] = 2.08$ ft. Since $[v^2/2g = 2.08 \text{ ft}] < [y/2 = \tfrac{8}{2} = 4.00 \text{ ft}]$, the flow is subcritical.

14.352 A hydraulic jump occurs in a triangular flume having side slopes 1:1. The flow rate is 15 cfs and the depth before the jump is 1.0 ft. Find the depth after the jump and the power loss in the jump.

▮ $Q^2/gA_1 + (h_c)_1A_1 = Q^2/gA_2 + (h_c)_2A_2$. For the triangular section, $h_c = y/3$, $A = y^2$, and $y_1 = 1.0$ ft.

$$15^2/[(32.2)(1.0)^2] + (1.0/3)(1.0)^2 = 15^2/[(32.2)(y_2)^2] + (y_2/3)(y_2)^2$$
$$y_2 = 2.67 \text{ ft} \qquad \text{(by trial and error)}$$
$$v_1 = Q/A_1 = 15/1.0^2 = 15.00 \text{ ft/s} \qquad v_2 = 15/2.67^2 = 2.104 \text{ ft}$$
$$E_j = (y_1 + v_1^2/2g) - (y_2 + v_2^2/2g) = 1.0 + 15.00^2/[(2)(32.2)] - 2.67 - 2.104^2/[(2)(32.2)] = 1.755 \text{ ft}$$
$$P = Q\gamma E_j = (15)(62.4)(1.755) = 1643 \text{ ft-lb/s} = \tfrac{1643}{550} = 2.99 \text{ hp}$$

14.353 A hydraulic jump occurs in a 5-m-wide rectangular channel carrying 6 m³/s on a slope of 0.005. The depth after the jump is 1.4 m. (a) What must be the depth before the jump? (b) What are the losses of energy and power in the jump?

▮ (a) $$y_1 = (y_2/2)(\sqrt{1 + 8q^2/gy_2^3} - 1) = (1.4/2)\{\sqrt{1 + (8)(\tfrac{6}{5})^2/[(9.807)(1.4)^3]} - 1\} = 0.137 \text{ m}$$

(b) $$E_j = (y_2 - y_1)^3/4y_1 y_2 = (1.4 - 0.137)^3/[(4)(0.137)(1.4)] = 2.626 \text{ m}$$
$$P = Q\gamma E_j = (6)(9.79)(2.626) = 154 \text{ kW}$$

14.354 Analyze the water-surface profile in a long rectangular channel with concrete lining ($n = 0.013$). The channel is 10 ft wide, the flow rate is 400 cfs, and there is an abrupt change in channel slope from 0.0150 to 0.0016. Find also the horsepower loss in the jump.

▮ $\qquad Q = (1.486/n)(A)(R^{2/3})(s^{1/2}) \qquad 400 = (1.486/0.013)(10y_{0_1})[10y_{0_1}/(10 + 2y_{0_1})]^{2/3}(0.015)^{1/2}$

By trial, $y_{0_1} = 2.17$ ft (normal depth on upper slope). Using a similar procedure, the normal depth y_{0_2} on the lower slope is found to be 4.80 ft. $y_c = (q^2/g)^{1/3} = [(\frac{400}{10})^2/32.2]^{1/3} = 3.68$ ft. Thus flow is supercritical ($y_{0_1} < y_c$) before break in slope and subcritical ($y_{0_2} > y_c$) after break, so a hydraulic jump must occur. $y_2 = (y_1/2)[-1 + \sqrt{1 + (8q^2/gy_1^3)}]$, $y_2' = (2.17/2)(-1 + \{1 + [8(40)^2/32.2(2.17)^3]\}^{1/2}) = 5.75$ ft. Therefore the depth conjugate to the upper-slope normal depth of 2.17 ft is 5.75 ft. This jump cannot occur because the normal depth y_{0_2} on the lower slope is less than 5.75 ft. $y_1' = (4.80/2)(-1 + \{1 + [8(40)^2/32.2(4.8)^3]\}^{1/2}) = 2.76$ ft. The lower conjugate depth of 2.76 ft will occur downstream of the break in slope. The location of the jump (i.e., its distance below the break in slope) may be found by $\Delta x = (E_1 - E_2)/(S - S_0)$, $E_1 = 2.17 + [(400/21.7)^2/64.4] = 7.45$ ft, $E_2 = 2.76 + [(400/27.6)^2/64.4] = 6.02$ ft, $V_m = (\frac{1}{2})[(400/21.7) + (400/27.6)] = 16.46$ fps, $R_m = (\frac{1}{2})[(21.7/14.34) + (27.6/15.52)] = 1.645$ ft, $S = (nv/1.486R^{2/3})^2 = [(0.013)(16.45)/1.49(1.645)^{2/3}]^2 = 0.01060$, $\Delta x = (7.45 - 6.02)/(0.0106 - 0.0016) = 160$ ft. Thus depth on the upper slope is 2.17 ft; downstream of the break the depth increases gradually (M_3 curve) to 2.76 ft over a distance of approximately 160 ft; then a hydraulic jump occurs from a depth of 2.76 ft to 4.80 ft; downstream of the jump the depth remains constant (i.e., normal) at 4.80 ft. HP loss = $\gamma Q h_{L_j}/550$ where $h_{L_j} = \Delta E$.

Before jump: $E_1' = 2.76 + [(400/27.6)^2/64.4] = 6.02$ ft. After jump: $E_{0_2} = 4.80 + [(400/48.0)^2/64.4] = 5.88$ ft. Hence HP loss = $62.4(400)(6.02 - 5.88)/550 = 6.35$.

14.355 A very wide rectangular channel with bed slope $S_0 = 0.0003$ and roughness $n = 0.020$ carries a steady flow of 50 cfs/ft of width. If a sluice gate is so adjusted as to produce a minimum depth of 1.5 ft in the channel, determine whether a hydraulic jump will occur downstream, and if so, find (using one reach) the distance from the gate to the jump.

▮ $\qquad q = (y)(1.486/n)(y^{2/3})(s^{1/2}) \qquad 50 = (y_0)(1.486/0.020)(y_0)^{2/3}(0.0003)^{1/2} \qquad y_0 = 8.99$ ft

$$y_c = (q^2/g)^{1/3} = (50^2/32.2)^{1/3} = 4.27 \text{ ft}$$

Since $y_c < y_0$, the slope is mild. Since $y = 1.5$ ft caused by the gate $< y_c < y_0$, downstream of the sluice gate is an M_3 profile, which must be followed by a hydraulic jump to enable the flow to return to normal depth.

$$y_1 = (y_2/2)(\sqrt{1 + 8q^2/gy_2^3} - 1) = (8.99/2)\{\sqrt{1 + (8)(50)^2/[(32.2)(8.99)^3]} - 1\} = 1.63 \text{ ft}$$

$$\Delta x = (E_1 - E_2)/(S - S_0) \qquad E = y + v^2/2g$$

For $y = 1.5$ ft, $v = q/y = 50/1.5 = 33.33$ ft/s, $E = 1.5 + 33.33^2/[(2)(32.2)] = 18.75$ ft. For $y = 1.63$ ft, $v = q/y = 50/1.63 = 30.67$ ft/s, $E = 1.63 + 30.67^2/[(2)(32.2)] = 16.24$ ft.

$$v_m = (33.33 + 30.67)/2 = 32.00 \text{ ft/s} \qquad R_m = (1.5 + 1.63)/2 = 1.565 \text{ ft}$$
$$S = (nv/1.486R^{2/3})^2 = \{(0.020)(32.00)/[(1.486)(1.565)^{2/3}]\}^2 = 0.1021$$
$$\Delta x = (18.75 - 16.24)/(0.1021 - 0.0003) = 24.7 \text{ ft}$$

14.356 A rectangular channel 10 ft wide carries 300 cfs in uniform flow at a depth of 4 ft. By how much should the outside wall be elevated above the inside wall for a bend of 40-ft radius to the centerline of the channel?

▮ $v = Q/A = 300/[(4)(10)] = 7.500$ ft/s, $N_F = v\sqrt{gy} = 7.500/\sqrt{(32.2)(4)} = 0.661$. Since $N_F < 1.0$, the flow is subcritical. $\Delta y = v^2 B/gr = (7.500)^2(10)/[(32.2)(40)] = 0.437$ ft.

14.357 Repeat Prob. 14.356 for the same conditions except that the normal depth is 2 ft.

▮ $v = Q/A = 300/[(2)(10)] = 15.00$ ft/s, $N_F = v/\sqrt{gy} = 15.00/\sqrt{(32.2)(2)} = 1.87$. Since $N_F > 1.0$, the flow is supercritical. $\Delta y = v^2 B/gr = (15.00)^2(10)/[(32.2)(40)] = 1.75$ ft. Because of (supercritical) wave action, maximum water depth at inside wall = y_0, and maximum water depth at outside wall = $(y_0 + \Delta y)$. Therefore, the required difference in wall elevations = 1.75 ft.

14.358 A rectangular channel 4 m wide carries 6 m³/s in uniform flow at a depth of 1.5 m. What will be the maximum difference in water-surface elevations between the inside and outside of a circular bend in this channel if the radius of the bend is 25 m?

▮ $v = Q/A = 6/[(4)(1.5)] = 1.000$ m/s, $N_F = v/\sqrt{gy} = 1.000/\sqrt{(9.807)(1.5)} = 0.261$. Since $N_F < 1.0$, the flow is subcritical. $\Delta y = v^2 B/gr = (1.000)^2(4)/[(9.807)(25)] = 0.0163$ m.

14.359 What is the capacity of a 4-ft by 4-ft concrete box culvert ($n = 0.013$) with a rounded entrance ($k_e = 0.05$, $C_d = 0.95$) if the culvert slope is 0.005, the length 120 ft, and the headwater level 6 ft above the culvert invert? Assume free outlet conditions. Neglect headwater and tailwater velocity heads.

❙ Headwater/$D = \frac{6}{4} = 1.5$. Since $1.5 > 1.2$, conditions are those of Fig. 14-92b or 14-92c. Assume case b.

$$R = A/p_w = (4)(4)/(4 + 4 + 4 + 4) = 1.000 \text{ ft}$$
$$(h_L)_{1-3} = (y_1 - y_3) + (z_1 - z_3) = y_1 - y_3 + s_0 L$$
$$\Delta h = (k_e + 29n^2 L/R^{4/3} + 1)(v^2/2g) \qquad y_1 - y_3 + s_0 L = (k_e + 29n^2 L/R^{4/3} + 1)(v^2/2g)$$
$$6 - 4 + (0.005)(120) = [0.05 + (29)(0.013)^2(120)/1.000^{4/3} + 1]\{v^2/[(2)(32.2)]\} \qquad v = 10.11 \text{ ft/s}$$
$$Q = Av = [(4)(4)](10.11) = 162 \text{ ft}^3/\text{s} = (A)(1.486/n)(R^{2/3})(s^{1/2})$$

Now find the depth y_0 which occurs with normal uniform flow at this flow rate: $162 = (4y_0)(1.486/0.013)[4y_0/(y_0 + 4 + y_0)]^{2/3}(0.005)^{1/2}$, $y_0 = 4.11$ ft (by trial and error). Since $[y_0 = 4.11 \text{ ft}] > [D = 4 \text{ ft}]$, the culvert flows full. Free discharge at the outlet is given; therefore, the preceding assumption and computations are valid, and $Q = 162 \text{ ft}^3/\text{s}$.

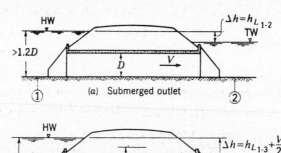

(a) Submerged outlet

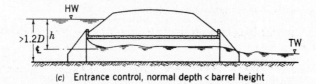

(b) Normal depth > barrel height

(c) Entrance control, normal depth < barrel height

Fig. 14-92

14.360 Solve Prob. 14.359 for the case where the tailwater elevation is 1 ft above the top of the box at the outlet.

❙ Headwater/$D = \frac{6}{4} = 1.5$. Since $1.5 > 1.2$, conditions are those of Fig. 14-92a with $y_2 = 5$ ft.

$$R = A/p_w = (4)(4)/(4 + 4 + 4 + 4) = 1.000 \text{ ft}$$
$$(h_L)_{1-3} = (y_1 - y_3) + (z_1 - z_3) = y_1 - y_3 + s_0 L$$
$$\Delta h = (k_e + 29n^2 L/R^{4/3} + 1)(v^2/2g) \qquad y_1 - y_3 + s_0 L = (k_e + 29n^2 L/R^{4/3} + 1)(v^2/g)$$
$$6 - 5 + (0.005)(120) = [0.05 + (29)(0.013)^2(120)/1.000^{4/3} + 1]\{v^2/[(2)(32.2)]\} \qquad v = 7.931 \text{ ft/s}$$
$$Q = Av = [(4)(4)](7.931) = 127 \text{ ft/s}$$

14.361 Solve Prob. 14.359 for the case where the tailwater elevation is 2 ft above the top of the box at the outlet.

❙ Headwater/$D = \frac{6}{4} = 1.5$. Since $1.5 > 1.2$, conditions are those of Fig. 14-92a with $y_2 = 6$ ft.

$$R = A/p_w = (4)(4)/(4 + 4 + 4 + 4) = 1.000 \text{ ft}$$
$$(h_L)_{1-3} = (y_1 - y_3) + (z_1 - z_3) = y_1 - y_3 + s_0 L$$
$$\Delta h = (k_e + 29n^2 L/R^{4/3} + 1)(v^2/2g) \qquad y_1 - y_3 + s_0 L = (k_e + 29n^2 L/R^{4/3} + 1)(v^2/2g)$$
$$6 - 6 + (0.0005)(120) = [0.05 + (29)(0.013)^2(120)/1.000^{4/3} + 1]\{v^2/[(2)(32.2)]\} \qquad v = 4.857 \text{ ft/s}$$
$$Q = Av = [(4)(4)](4.857) = 77.7 \text{ ft}^3/\text{s}$$

14.362 Repeat Prob. 14.359 for the case where the culvert slope is 0.04.

$\blacksquare$ Headwater/$D = \frac{6}{4} = 1.5$. Since $1.5 > 1.2$, conditions are those of Fig. 14-92b or 14-92c. Assume case b.

$$R = A/p_w = (4)(4)/(4 + 4 + 4 + 4) = 1.000 \text{ ft}$$
$$(h_L)_{1-3} = (y_1 - y_3) + (z_1 - z_3) = y_1 - y_3 + s_0 L$$
$$\Delta h = (k_e + 29n^2 L/R^{4/3} + 1)(v^2/2g) \qquad y_1 - y_3 + s_0 L = (k_e + 29n^2 L/R^{4/3} + 1)(v^2/2g)$$
$$6 - 4 + (0.04)(120) = [0.05 + (29)(0.013)^2(120)/1.000^{4/3} + 1]\{v^2/[(2)(32.2)]\} \qquad v = 16.35 \text{ ft/s}$$
$$Q = Av = [(4)(4)](16.35) = 262 \text{ ft}^3/\text{s} = (A)(1.486/n)(R^{2/3})(s^{1/2})$$

Now find the depth y_0 which occurs with normal uniform flow at this flow rate: $262 = (4y_0)(1.486/0.013)[4y_0/(y_0 + 4 + y_0)]^{2/3}(0.04)^{1/2}$, $y_0 = 2.63$ ft (by trial and error). Since $[y_0 = 2.63 \text{ ft}] < [D = 4 \text{ ft}]$, the culvert does not flow full, and case b cannot occur. Case c must occur, i.e., with entrance control. Hence, $Q = C_d A \sqrt{2gh} = (0.95)[(4)(4)]\sqrt{(2)(32.2)(4)} = 244 \text{ ft}^3/\text{s}$.

14.363 Repeat Prob. 14.359 for the case where the culvert slope is 0.08.

$\blacksquare$ Headwater/$D = \frac{6}{4} = 1.5$. Since $1.5 > 1.2$, conditions are those of Fig. 14-92b or 14-92c. Assume case b.

$$R = A/p_w = (4)(4)/(4 + 4 + 4 + 4) = 1.000 \text{ ft}$$
$$(h_L)_{1-3} = (y_1 - y_3) + (z_1 - z_3) = y_1 - y_3 + s_0 L$$
$$\Delta h = (k_e + 29n^2 L/R^{4/3} + 1)(v^2/2g) \qquad y_1 - y_3 + s_0 L = (k_e + 29n^2 L/R^{4/3} + 1)(v^2/2g)$$
$$6 - 4 + (0.08)(120) = [0.05 + (29)(0.013)^2(120)/1.000^{4/3} + 1]\{v^2/[(2)(32.2)]\} \qquad v = 21.35 \text{ ft/s}$$
$$Q = Av = [(4)(4)](21.35) = 342 \text{ ft}^3/\text{s} = (A)(1.486/n)(R^{2/3})(s^{1/2})$$

Now find the depth y_0 which occurs with normal uniform flow at this flow rate: $342 = (4y_0)(1.486/0.013)[4y_0/(y_0 + 4 + y_0)]^{2/3}(0.08)^{1/2}$, $y_0 = 2.47$ ft (by trial and error). Since $[y_0 = 2.47 \text{ ft}] < [D = 4 \text{ ft}]$, the culvert does not flow full, and case b cannot occur. Case c must occur, i.e., with entrance control. Hence, $Q = C_d A \sqrt{2gh} = (0.95)[(4)(4)]\sqrt{(2)(32.2)(4)} = 244 \text{ ft}^3/\text{s}$.

14.364 A culvert under a road must carry 4.3 m³/s. The culvert length will be 30 m and the slope will be 0.003. If the maximum permissible headwater level is 3.6 m above the culvert invert, what size corrugated-pipe culvert ($n = 0.025$) would you select? The outlet will discharge freely. Neglect velocity of approach. Assume square-edged entrance with $k_e = 0.5$ and $C_d = 0.65$.

$\blacksquare$ Assume $D < 3.0$ m, so that [headwater/D] $> [3.6/3.0 = 1.2]$ and the conditions are those of Fig. 14-92b or 14-92c. Assume case b.

$$(h_L)_{1-3} = (y_1 - y_3) + (z_1 - z_3) = y_1 - y_3 + s_0 L \qquad \Delta h = (k_e + 19.62n^2 L/R^{4/3} + 1)(v^2/2g)$$
$$y_1 - y_3 + s_0 L = (k_e + 19.62n^2 L/R^{4/3} + 1)(v^2/2g) \qquad v = Q/A = 4.3/(\pi D^2/4) = 5.475/D^2$$
$$3.6 - D + (0.003)(30) = [0.5 + (19.62)(0.025)^2(30)/(D/4)^{4/3} + 1]\{(5.475/D^2)^2/[(2)(9.807)]\}$$
$$D = 1.20 \text{ m} \qquad \text{(by trial and error)} \qquad Q = (A)(1.0/n)(R^{2/3})(s^{1/2})$$

Now find the diameter d_0 which just flows full with normal uniform flow: $4.3 = (\pi d_0^2/4)(1.0/0.025)(d_0/4)^{2/3}(0.003)^{1/2}$, $d_0 = 1.99$ m. Since $[d_0 = 1.99 \text{ m}] > [D = 1.20 \text{ m}]$, the culvert flows full, and free discharge at the outlet is given. Therefore, the above assumptions and analysis are valid, and $D = 1.20$ m. Use standard $D = 1.20$ m.

14.365 Repeat Prob. 14.364 for a culvert length of 100 m.

$\blacksquare$ Assume $D < 3.0$ m, so that [headwater/D] $> [3.6/3.0 = 1.2]$ and the conditions are those of Fig. 14-92b or 14-92c. Assume case b.

$$(h_L)_{1-3} = (y_1 - y_3) + (z_1 - z_3) = y_1 - y_3 + s_0 L \qquad \Delta h = (k_e + 19.62n^2 L/R^{4/3} + 1)(v^2/2g)$$
$$y_1 - y_3 + s_0 L = (k_e + 19.62n^2 L/R^{4/3} + 1)(v^2/2g) \qquad v = Q/A = 4.3/(\pi D^2/4) = 5.475/D^2$$
$$3.6 - D + (0.003)(100) = [0.5 + (19.62)(0.025)^2(100)/(D/4)^{4/3} + 1]\{(5.475/D^2)^2/[(2)(9.807)]\}$$
$$D = 1.41 \text{ m} \qquad \text{(by trial and error)} \qquad Q = (A)(1.0/n)(R^{2/3})(s^{1/2})$$

Now find the diameter d_0 which just flows full with normal uniform flow: $4.3 = (\pi d_0^2/4)(1.0/0.025)(d_0/4)^{2/3}(0.003)^{1/2}$, $d_0 = 1.99$ m. Since $[d_0 = 1.99 \text{ m}] > [D = 1.41 \text{ m}]$, the culvert flows full, and free discharge at the outlet is given. Therefore, the above assumptions and analysis are valid, and $D = 1.41$ m. Use standard $D = 1.50$ m.

14.366 Water flows in a wide channel at $q = 20$ cfs/ft, $y_1 = 1$ ft, and then undergoes a hydraulic jump. Compute y_2, v_2, $(N_F)_2$, E_j, the percentage dissipation, and the horsepower dissipated per unit width. What is the critical depth?

$$N_F = v/\sqrt{gy} \qquad v_1 = q/y_1 = \tfrac{20}{1} = 20.00 \text{ ft/s}$$
$$(N_F)_1 = 20.00/\sqrt{(32.2)(1)} = 3.52 \qquad E = y + v^2/2g$$
$$E_1 = 1 + 20.00^2/[(2)(32.2)] = 7.21 \text{ ft}$$
$$y_2 = (y_1/2)(\sqrt{1 + 8q^2/gy_1^3} - 1) = (\tfrac{1}{2})\{\sqrt{1 + (8)(20)^2/[(32.2)(1)^3]} - 1\} = 4.51 \text{ ft}$$
$$v_2 = q/y_2 = 20/4.51 = 4.43 \text{ ft/s} \qquad (N_F)_2 = 4.43/\sqrt{(32.2)(4.51)} = 0.368$$
$$E_j = (y_2 - y_1)^3/4y_1y_2 = (4.51 - 1)^3/[(4)(4.51)(1)] = 2.40 \text{ ft}$$
$$\text{Percent dissipation} = E_j/E_1 = 2.40/7.21 = 0.333 \quad \text{or} \quad 33.3 \text{ percent}$$
$$P_{\text{dissipated}} = Q\gamma E_j = (20)(62.4)(2.40) = 2995 \text{ ft-lb/s} = \tfrac{2995}{550} = 5.45 \text{ hp/ft}$$
$$y_c = (q^2/g)^{1/3} = (20^2/32.2)^{1/3} = 2.32 \text{ ft}$$

14.367 A wide-channel flow at depth 50 cm passes through a hydraulic jump and emerges at a depth of 2.8 m. Compute the velocities upstream and downstream of the jump. What is the critical depth for this flow?

$$2y_2/y_1 = -1 + [1 + 8(N_F)_1^2]^{1/2} \qquad (2)(2.8)/(\tfrac{50}{100}) = -1 + [1 + (8)(N_F)_1^2]^{1/2} \qquad (N_F)_1 = 4.30$$
$$N_F = v/\sqrt{gy} \qquad 4.30 = v_1/\sqrt{(9.807)(\tfrac{50}{100})} \qquad v_1 = 9.52 \text{ m/s}$$
$$v_1y_1 = v_2y_2 \qquad (9.52)(\tfrac{50}{100}) = (v_2)(2.8) \qquad v_2 = 1.70 \text{ m/s}$$
$$q = yv = (2.8)(1.70) = 4.76 \text{ (m}^3\text{/s)/m} \qquad y_c = (q^2/g)^{1/3} = (4.76^2/9.807)^{1/3} = 1.32 \text{ m}$$

14.368 The flow downstream of a hydraulic jump in a rectangular channel is 9 m deep and has a velocity of 3.5 m/s. What are the velocity and depth upstream of the jump? What is the critical depth for the flow?

$$2y_1/y_2 = -1 + [1 + 8(N_F)_2^2]^{1/2} \qquad N_F = v/\sqrt{gy} \qquad (N_F)_2 = 3.5/\sqrt{(9.807)(9)} = 0.373$$
$$2y_1/9 = -1 + [1 + (8)(0.373)^2]^{1/2} \qquad y_1 = 2.04 \text{ m} \qquad y_1v_1 = y_2v_2$$
$$2.04v_1 = (9)(3.5) \qquad v_1 = 15.4 \text{ m/s} \qquad q = yv = (2.04)(15.4) = 31.4 \text{ (m}^3\text{/s)/m}$$
$$y_c = (q^2/g)^{1/3} = (31.4^2/9.807)^{1/3} = 4.65 \text{ m}$$

14.369 Water in a horizontal channel accelerates smoothly over a bump and then undergoes a hydraulic jump, as in Fig. 14-93. If $y_1 = 1$ m and $y_3 = 40$ cm, estimate v_1, v_3, and y_4. Neglect friction.

$$E_1 = E_3 \qquad y_1 + v_1^2/2g = y_3 + v_3^2/2g \qquad 1 + v_1^2/[(2)(9.807)] = \tfrac{40}{100} + v_3^2/[(2)(9.807)]$$
$$y_1v_1 = y_3v_3 \qquad v_1 = y_3v_3/y_1 = (\tfrac{40}{100})(v_3)/1 = 0.4000v_3$$
$$1 + (0.4000v_3)^2/[(2)(9.807)] = \tfrac{40}{100} + v_3^2/[(2)(9.807)]$$
$$v_3 = 3.74 \text{ m/s} \qquad v_1 = (0.4000)(3.74) = 1.50 \text{ m/s}$$
$$2y_4/y_3 = -1 + [1 + 8(N_F)_3^2]^{1/2} \qquad N_F = v/\sqrt{gy}$$
$$(N_F)_3 = 3.74/\sqrt{(9.807)(\tfrac{40}{100})} = 1.89 \qquad 2y_4/(\tfrac{40}{100}) = -1 + [1 + (8)(1.89)^2]^{1/2} \qquad y_4 = 0.888 \text{ m}$$

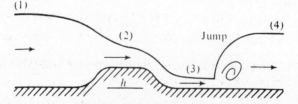

Fig. 14-93

14.370 For the conditions given in Prob. 14.369, estimate the bump height h.

$y_1 + v_1^2/2g = y_2 + v_2^2/2g + h$. Using data from Prob. 14.369,

$$1 + 1.50^2/[(2)(9.807)] = y_2 + v_2^2/[(2)(9.807)] + h$$
$$y_1v_1 = y_2v_2 \qquad (1)(1.50) = y_2v_2 \qquad v_2 = 1.50/y_2$$

Since flow over the bump is critical,

$$v_2 = v_c = \sqrt{gy_2} \qquad 1.50/y_2 = \sqrt{9.807 y_2} \qquad y_2 = 0.612 \text{ m} \qquad v_2 = 1.50/0.612 = 2.451 \text{ m/s}$$

$$1 + 1.50^2/[(2)(9.807)] = 0.612 + 2.451^2/[(2)(9.807)] + h \qquad h = 0.196 \text{ m}$$

14.371 Repeat Prob. 14.369 if $y_1 = 2$ ft and $y_3 = 6$ in.

▮ $\qquad E_1 = E_3 \qquad y_1 + v_1^2/2g = y_3 + v_3^2/2g \qquad 2 + v_1^2/[(2)(32.2)] = \frac{6}{12} + v_3^2/[(2)(32.2)]$

$$y_1 v_1 = y_3 v_3 \qquad v_1 = y_3 v_3/y_1 = (\tfrac{6}{12})(v_3)/2 = 0.2500 v_3$$

$$2 + (0.2500 v_3)^2/[(2)(32.2)] = \tfrac{6}{12} + v_3^2/[(2)(32.2)]$$

$$v_3 = 10.15 \text{ ft/s} \qquad v_1 = (0.2500)(10.15) = 2.54 \text{ ft/s} \qquad 2y_4/y_3 = -1 + [1 + 8(N_F)_3^2]^{1/2}$$

$$N_F = v/\sqrt{gy} \qquad (N_F)_3 = 10.15/\sqrt{(32.2)(\tfrac{6}{12})} = 2.53$$

$$2y_4/(\tfrac{6}{12}) = -1 + [1 + (8)(2.53)^2]^{1/2} \qquad y_4 = 1.56 \text{ ft}$$

14.372 For the conditions given in Prob. 14-371, estimate the bump height h.

▮ $y_1 + v_1^2/2g = y_2 + v_2^2/2g + h$. Using data from Prob. 14-371,

$$2 + 2.54^2/[(2)(32.2)] = y_2 + v_2^2/[(2)(32.2)] + h$$

$$y_1 v_1 = y_2 v_2 \qquad (2)(2.54) = y_2 v_2 \qquad v_2 = 5.08/y_2$$

Since flow over the bump is critical,

$$v_2 = v_c = \sqrt{gy_2} \qquad 5.08/y_2 = \sqrt{32.2 y_2} \qquad y_2 = 0.929 \text{ ft} \qquad v_2 = 5.08/0.929 = 5.468 \text{ ft/s}$$

$$2 + 2.54^2/[(2)(32.2)] = 0.929 + 5.468^2/[(2)(32.2)] + h \qquad h = 0.707 \text{ ft}$$

14.373 At the bottom of a spillway is a hydraulic jump with water depths 2 ft upstream and 11 ft downstream. If the spillway is 100 ft wide, how much horsepower is being dissipated in the jump? What percentage dissipation occurs?

▮ $\qquad E_j = (y_2 - y_1)^3/4y_1 y_2 = (11 - 2)^3/[(4)(11)(2)] = 8.284 \text{ ft} \qquad P = Q\gamma E_j \qquad N_F = v/\sqrt{gy}$

$$2y_2/y_1 = -1 + [1 + 8(N_F)_1^2]^{1/2} \qquad (2)(11)/2 = -1 + [1 + (8)(N_F)_1^2]^{1/2} \qquad (N_F)_1 = 4.23$$

$$4.23 = v_1/\sqrt{(32.2)(2)} \qquad v_1 = 33.95 \text{ ft/s} \qquad Q = A_1 v_1 = [(100)(2)](33.95) = 6790 \text{ ft}^3/\text{s}$$

$$P = (6790)(62.4)(8.284) = 3.510 \times 10^6 \text{ ft-lb/s} = (3.510 \times 10^6)/550 = 6382 \text{ hp}$$

$$E_1 = y_1 + v_1^2/2g = 2 + 33.95^2/[(2)(32.2)] = 19.90 \text{ ft}$$

$$\text{Percentage dissipation} = 8.284/19.90 = 0.416 \quad \text{or} \quad 41.6 \text{ percent}$$

14.374 At one point in a rectangular channel 8 ft wide, the depth of flow is 3 ft and the flow rate is 450 cfs. If a hydraulic jump occurs, will it be upstream or downstream of this point?

▮ $\qquad v_c = \sqrt{gy} = \sqrt{(32.2)(3)} = 9.83 \text{ ft/s} \qquad v = Q/A = 450/[(8)(3)] = 18.75 \text{ ft/s}$

Since $v > v_c$, the flow is supercritical, and a hydraulic jump would have to occur downstream.

14.375 A *bore* is a hydraulic jump which propagates upstream into a slower-moving or still fluid. Suppose a still fluid is 3 m deep and the water behind a bore is 5 m deep. What will the propagation speed of the bore be?

▮ $\qquad N_F = v/\sqrt{gy} \qquad 2y_2/y_1 = -1 + [1 + 8(N_F)_1^2]^{1/2} \qquad (2)(5)/3 = -1 + [1 + (8)(N_F)_1^2]^{1/2}$

$$(N_F)_1 = 1.49 \qquad 1.49 = v_1/\sqrt{(9.807)(3)} \qquad v_1 = c_{\text{bore}} = 8.08 \text{ m/s}$$

14.376 Repeat Prob. 14.375 assuming that the water upstream is flowing toward the bore at 5 m/s ground speed. What will the ground speed of propagation of the bore be?

▮ $v_1 = c_{\text{bore}} + 5$. v_1 will be the same as in Prob. 14.375; hence, $v_1 = 8.08$ m/s. $8.08 = c_{\text{bore}} + 5$, $c_{\text{bore}} = 3.08$ m/s.

14.377 Consider the flow under the sluice gate of Fig. 14-94. If $y_1 = 10$ ft and all losses are neglected except the dissipation in the jump, calculate y_2 and y_3 and the percentage dissipation. The channel is horizontal and wide.

▮ $\qquad y_1 + V_1^2/2g = y_2 + V_2^2/2g \qquad 10 + 2^2/[(2)(32.2)] = y_2 + V_2^2/[(2)(32.2)]$

$$V_1 y_1 = V_2 y_2 \qquad (2)(10) = V_2 y_2 \qquad y_2 = 20/V_2$$

$$10 + 2^2/[(2)(32.2)] = 20/V_2 + V_2^2/[(2)(32.2)] \qquad V_2 = 24.4 \text{ ft/s} \qquad \text{(by trial and error)}$$

$$y_2 = 20/24.4 = 0.820 \text{ ft} \qquad 2y_3/y_2 = -1 + [1 + 8(N_F)_2^2]^{1/2} \qquad N_F = V/\sqrt{gy}$$

$$(N_F)_2 = 24.4/\sqrt{(32.2)(0.820)} = 4.75 \qquad 2y_3/0.820 = -1 + [1 + (8)(4.75)^2]^{1/2} \qquad y_3 = 5.11 \text{ ft}$$

$$E_j = (y_3 - y_2)^3/4y_3\,y_2 = (5.11 - 0.820)^3/[(4)(5.11)(0.820)] = 4.71 \text{ ft}$$

$$E_2 = y_2 + V_2^2/2g = 0.820 + 24.4^2/[(2)(32.2)] = 10.06 \text{ ft}$$

Percentage dissipation $= 4.71/10.06 = 0.468$ or 46.8 percent

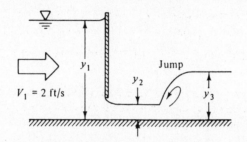

Fig. 14-94

14.378 If bottom friction is included in the sluice-gate flow of Fig. 14-94, the depths y_1, y_2, and y_3 will vary with x. What type of solution curves will we have in regions 1, 2, and 3?

▮ See Fig. 14-95.

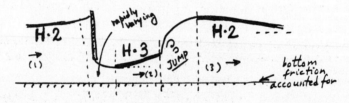

Fig. 14-95

14.379 A 10-cm-high bump in a wide horizontal water channel creates a hydraulic jump just upstream and the flow pattern in Fig. 14-96. Neglecting losses except in the jump, for the case $y_3 = 30$ cm, estimate V_4, y_4, V_1, and y_1.

▮ Since $y_4 < y_3$, assume flow over the hump is critical.

$$V = \sqrt{gy} \qquad V_3 = \sqrt{(9.807)(\tfrac{30}{100})} = 1.715 \text{ m}$$

$$E_2 = E_3 = E_4 = y_3 + V_3^2/2g + h = E_4 = \tfrac{30}{100} + 1.715^2/[(2)(9.807)] + \tfrac{10}{100} = E_4$$

$$= 0.5500 \text{ m} \qquad E_2 = y_2 + V_2^2/2g = 0.5500 \text{ m}$$

$$q = V_2\,y_2 = V_3\,y_3 = V_4\,y_4 = (1.715)(\tfrac{30}{100}) = 0.5145 \text{ m}^2/\text{s} \qquad V_2 = 0.5145/y_2$$

$$y_2 + (0.5145/y_2)^2/[(2)(9.807)] = 0.5500$$

$$y_2 = 0.495 \text{ m} \qquad \text{(by trial and error)} \qquad V_2 = 0.5145/0.495 = 1.039 \text{ m/s}$$

$$N_F = V/\sqrt{gy} \qquad (N_F)_2 = 1.039/\sqrt{(9.807)(0.495)} = 0.472 \qquad \text{(subcritical)}$$

$$2y_1/y_2 = -1 + [1 + 8(N_F)_2^2]^{1/2} \qquad 2y_1/0.495 = -1 + [1 + (8)(0.472)^2]^{1/2} \qquad y_1 = 0.165 \text{ m}$$

$$V_1 = q/y_1 = 0.5145/0.165 = 3.12 \text{ m/s} \qquad E_3 = E_4 = 0.5500 \text{ m} \qquad y_4 + V_4^2/2g = 0.5500 \text{ m}$$

$$V_4\,y_4 = 0.5145 \text{ m}^2/\text{s} \qquad V_4 = 0.5145/y_4 \qquad y_4 + (0.5145/y_4)^2/[(2)(9.807)] = 0.5500 \text{ m}$$

$$y_4 = 0.195 \text{ m} \qquad \text{(by trial and error)} \qquad V_4 = 0.5145/0.195 = 2.64 \text{ m/s}$$

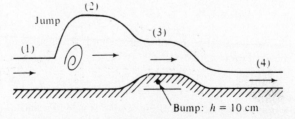

Bump: $h = 10$ cm   **Fig. 14-96**

14.380 Water flows in a wide channel at $q = 10 \ (\text{m}^3/\text{s})/\text{m}$ and $y_1 = 1.25$ m. If the flow undergoes a hydraulic jump, compute (*a*) y_2, (*b*) V_2, (*c*) $(N_F)_2$, (*d*) E_j, (*e*) the percentage dissipation, (*f*) the power dissipated per unit width, and (*g*) the temperature rise due to dissipation if $c_p = 4200 \ \text{J/(kg·K)}$.

▌ (a) $V_1 = q/y_1 = 10/1.25 = 8.00 \text{ m/s}$ $N_F = V/\sqrt{gy}$ $(N_F)_1 = 8.00/\sqrt{(9.807)(1.25)} = 2.285$

$2y_2/y_1 = -1 + [1 + 8(N_F)_1^2]^{1/2}$ $2y_2/1.25 = -1 + [1 + (8)(2.285)^2]^{1/2}$ $y_2 = 3.46 \text{ m}$

(b) $V_1 y_1 = V_2 y_2$ $(8.00)(1.25) = (V_2)(3.46)$ $V_2 = 2.89 \text{ m/s}$

(c) $(N_F)_2 = 2.89/\sqrt{(9.807)(3.46)} = 0.496$

(d) $E_j = (y_2 - y_1)^3/4y_2 y_1 = (3.46 - 1.25)^3/[(4)(3.46)(1.25)] = 0.624 \text{ m}$

(e) $E_1 = y_1 + v_1^2/2g = 1.25 + 8.00^2/[(2)(9.807)] = 4.51 \text{ m}$

Percentage dissipation $= E_j/E_1 = 0.624/4.51 = 0.138$ or   13.8 percent

(f) $P = q\gamma E_j$ $P_{\text{dissipated}} = (10)(9.79)(0.624) = 61.1 \text{ kW/m}$

(g) $P = \dot{m}c_p \, \Delta T = \rho q c_p \, \Delta T$ $(61.1)(1000) = (1000)(10)(4200)(\Delta T)$ $\Delta T = 0.0015 \text{ K}$

14.381 Water flows over a spillway into a stilling basin 20 m wide. Before the jump the water has a depth of 1 m and a velocity of 18 m/s. Determine (a) the initial Froude number before the jump, (b) the depth of flow after the jump, and (c) the head loss in the jump.

▌ (a) $N_F = V/\sqrt{gy}$ $(N_F)_1 = 18/\sqrt{(9.807)(1)} = 5.75$

(b) $y_2 = [-y_1 + \sqrt{y_1^2 + (8Q^2/gb^2)(1/y_1)}]/2$ $Q = A_1 V_1 = [(20)(1)](18) = 360 \text{ ft}^3/\text{s}$

$y_2 = \{-1 + \sqrt{1^2 + [(8)(360)^2/(9.807)(20)^2](\frac{1}{1})}\}/2 = 7.64 \text{ m}$

(c) $E_j = [y_2^3 - y_1^3 + y_1 y_2(y_1 - y_2)]/4y_1 y_2 = [7.64^3 - 1^3 + (1)(7.64)(1 - 7.64)]/[(4)(1)(7.64)] = 12.9 \text{ m}$

14.382 A rectangular channel has a width of 10 ft and a flow of 10 cfs. The depth is 3 in. Assume that a hydraulic jump occurs. What will the elevation of the free surface be after the jump, and what is the loss in kinetic energy?

▌ $y_2 = [-y_1 + \sqrt{y_1^2 + (8Q^2/gb^2)(1/y_1)}]/2$

$= \{-\frac{3}{12} + \sqrt{(\frac{3}{12})^2 + [(8)(10)^2/(32.2)(10)^2][1/(\frac{3}{12})]}\}/2 = 0.389 \text{ ft}$ or   4.67 in

$KE = v^2/2$ $v_1 = Q/A_1 = 10/[(\frac{3}{12})(10)] = 4.00 \text{ ft/s}$ $(KE)_1 = 4.00^2/2 = 8.00 \text{ ft-lb/slug}$

$v_2 = Q/A_2 = 10/[(0.389)(10)] = 2.57 \text{ ft/s}$ $(KE)_2 = 2.57^2/2 = 3.30 \text{ ft-lb/slug}$

$(KE)_{\text{loss}} = 8.00 - 3.30 = 4.70 \text{ ft-lb/slug}$

14.383 Water flows in a rectangular, finished-concrete channel and goes over a dam, as shown in Fig. 14-97. After the dam, the water goes into a stilling basin at which there is a hydraulic jump. The channel has a slope of 0.003 and the depth 50 m upstream of the dam is 4 m. What is the depth of flow just before reaching the dam? The volumetric flow is 18 m³/s and the width of the channel is 6 m. Do the problem the simplest way without averaging.

▌ $\Delta L = ([1 - (Q^2 b/gA^3)]/\{S_0 - (n/\kappa)^2[Q^2/(R_H^{4/3}A^2)]\}) \, \Delta y$

$A = (4)(6) = 24.00 \text{ m}^2$ $R_H = A/p_w = 24.00/(4 + 6 + 4) = 1.714 \text{ m}$

$50 = \left\{ \dfrac{1 - (18^2)(6)/(9.807)(24.00)^3}{0.003 - (0.012/1.0)^2[18^2/(1.714)^{4/3}(24.00)^2]} \right\}(\Delta y)$

$\Delta y = 0.150 \text{ m}$

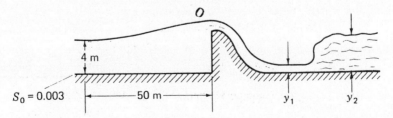

O

4 m

$S_0 = 0.003$ |————— 50 m —————| y_1 y_2

Fig. 14-97

14.384 Work Prob. 14.383 more accurately using linear averages.

▌ $\Delta L = ([1 - (Q^2 b/gA^3)]/\{S_0 - (n/\kappa)^2[Q^2/(R_H^{4/3}A^2)]\}) \, \Delta y$

Using data from Prob. 14.383, at the dam $y = 4 + 0.150 = 4.150$ m.

$$A = (4.150)(6) = 24.90 \text{ m}^2 \qquad R_H = A/p_w = 24.90/(4.150 + 6 + 4.150) = 1.741 \text{ m}$$

$$A_{av} = (24.00 + 24.90)/2 = 24.45 \text{ m}^2 \qquad (R_H)_{av} = (1.714 + 1.741)/2 = 1.728 \text{ m}$$

$$50 = \left\{ \frac{1 - (18^2)(6)/(9.897)(24.45)^3}{0.003 - (0.012/1.0)^2[18^2/(1.728)^{4/3}(24.45)^2]} \right\}(\Delta y)$$

$$\Delta y = 0.150 \text{ m} \qquad \text{(virtually the same as in Prob. 14.383)}$$

14.385 In Probs. 14.383 and 14.384, we computed the depth before the dam to be 4.150 m. If we neglect friction over the dam, what is the depth y_2 at the hydraulic jump (see Fig. 14-97)?

▌ Neglecting friction,

$$E_1 = E_0 \qquad E = y + q^2/2y^2 g \qquad E_0 = 4.150 + (\tfrac{18}{6})^2/[(2)(4.150)^2(9.807)] = 4.177 \text{ m}$$

$$E_1 = y_1 + (\tfrac{18}{6})^2/[(2y_1^2)(9.807)] = 4.177 \text{ m} \qquad y_1 = 0.346 \text{ m} \qquad \text{(by trial and error)}$$

$$y_2 = [-y_1 + \sqrt{y_1^2 + (8Q^2/gb^2)(1/y_1)}]/2$$

$$= \{-0.346 + \sqrt{0.346^2 + [(8)(18)^2/(9.807)(6)^2](1/0.346)}\}/2 = 2.137 \text{ m}$$

14.386 Water is moving in a rectangular channel to have a Froude number of $\sqrt{10}$. The channel is 5 m in width. The depth of flow is 1 m. If the water undergoes a hydraulic jump, what is the Froude number after the jump and what percentage of the mechanical energy of the flow is dissipated from the jump?

▌
$$N_F = v/\sqrt{gy} \qquad \sqrt{10} = v/\sqrt{(9.807)(1)} \qquad v = 9.903 \text{ m/s}$$

$$Q = Av = [(5)(1)](9.903) = 49.52 \text{ m}^3/\text{s}$$

$$y_2 = [-y_1 + \sqrt{y_1^2 + (8Q^2/gb^2)(1/y_1)}]/2$$

$$= \{-1 + \sqrt{1^2 + [(8)(49.52)^2/(9.807)(5)^2](1/1)}\}/2 = 4.000 \text{ m}$$

$$v_2 = Q/A_2 = 49.52/[(4.000)(5)] = 2.476 \text{ m/s} \qquad (N_F)_2 = 2.476/\sqrt{(9.807)(4.000)} = 0.395$$

$$E_j = [y_2^3 - y_1^3 + y_1 y_2(y_1 - y_2)]/4y_1 y_2$$

$$= [4.000^3 - 1^3 + (1)(4.000)(1 - 4.000)]/[(4)(1)(4.000)] = 3.188 \text{ m}$$

$$E = y + v^2/2g \qquad E_1 = 1 + 9.903^2/[(2)(9.807)] = 6.000 \text{ m}$$

$$\text{Percentage dissipated} = E_j/E_1 = 3.188/6.000 = 0.531 \quad \text{or} \quad 53.1 \text{ percent}$$

14.387 Water in a rectangular channel is seen to go through a hydraulic jump where the depth jumps from 3 m to 7 m. The width of the channel is 10 m. What is the volume flow rate? What are the final and initial Froude numbers?

▌
$$y_2 = [-y_1 + \sqrt{y_1^2 + (8Q^2/gb^2)(1/y_1)}]/2 \qquad 7 = \{-3 + \sqrt{3^2 + [8Q^2/(9.807)(10)^2](\tfrac{1}{3})}\}/2$$

$$Q = 321 \text{ m}^3/\text{s}$$

$$N_F = v/\sqrt{gy} \qquad v_1 = Q/A_1 = 321/[(10)(3)] = 10.70 \text{ m/s} \qquad (N_F)_1 = 10.70/\sqrt{(9.807)(3)} = 1.97$$

$$v_2 = Q/A_2 = 321/[(10)(7)] = 4.586 \text{ m/s} \qquad (N_F)_2 = 4.586/\sqrt{(9.807)(7)} = 0.553$$

14.388 Water flows in a rectangular concrete channel and undergoes a hydraulic jump such that 60 percent of its mechanical energy is to be dissipated. If the volume flow rate is 100 m³/s and the width of the channel is 5 m, what must the Froude number be just before the jump? Set up the proper equations but do not actually solve.

▌
$$N_F = v/\sqrt{gy} \qquad E = y + v^2/2g \qquad v_1 = Q/A_1 = 100/(5y_1) = 20.00/y_1$$

$$E_1 = y_1 + (20.00/y_1)^2/[(2)(9.807)] = y_1 + 20.39/y_1^2 \qquad E_j = [y_2^3 - y_1^3 + y_1 y_2(y_1 - y_2)]/4y_1 y_2$$

$$0.60E_1 = (0.60)(y_1 + 20.39/y_1^2) = [y_2^3 - y_1^3 + y_1 y_2(y_1 - y_2)]/4y_1 y_2 \tag{1}$$

$$y_2 = [-y_1 + \sqrt{y_1^2 + (8Q^2/gb^2)(1/y_1)}]/2 = \{-y_1 + \sqrt{y_1^2 + [(8)(100)^2/(9.807)(5)^2](1/y_1)}\}/2 \tag{2}$$

Solve Eqs. (1) and (2) to find y_1 and y_2. Then, $v_1 = Q/A_1 = 100/5y_1$, $(F_R)_1 = (100/5y_1)/\sqrt{(9.807)(y_1)} = 6.386y^{-3/2}$.

14.389 Water is flowing from a spillway into a stilling basin, as shown in Fig. 14-98. The elevation y_A ahead of the spillway is 10 m. The width of the rectangular channel is 8 m. If the stilling basin dissipates 60 percent of the mechanical energy, what is the volume flow rate? Set up simultaneous equations but do not actually solve.

▮
$$E = y + v^2/2g \qquad E_1 = 10 + \{Q/[(10)(8)]\}^2/(2g) = 10 + Q^2/[(80)^2(2g)]$$

$$10 + Q^2/[(80)^2(2g)] = y_1 + Q^2/[(8y_1)^2(2g)] \tag{1}$$

$$E_j = [y_2^3 - y_1^3 + y_1 y_2(y_1 - y_2)]/4 y_1 y_2$$
$$= 0.60\{10 + Q^2/[(80)^2(2g)]\} = [y_2^3 - y_1^3 + y_1 y_2(y_1 - y_2)]/4 y_1 y_1 \tag{2}$$

$$y_2 = [-y_1 + \sqrt{y_1^2 + (8Q^2/gb^2)(1/y_1)}]/2 = [-y_1 + \sqrt{y_1^2 + [8Q^2/(9.807)(8)^2](1/y_1)}]/2 \tag{3}$$

Solve Eqs. (1), (2), and (3) to find y_1, y_2, and Q.

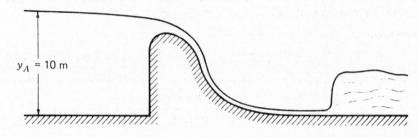

$y_A = 10$ m

Fig. 14-98

14.390 Water having a steady known volumetric flow rate Q is moving in a rectangular channel at supercritical speed up an adverse slope. It undergoes a hydraulic jump as shown in Fig. 14-99. If we know y_B and y_A at positions L_1 and L_2 apart, how do we approximately locate the position of the hydraulic jump, i.e., how do we get L_1 and L_2? The channel has a known value of n. Explain the simplest method. The width is b.

▮ Use the equation $\Delta L = ([1 - (Q^2 b/gA^3)]/\{S_0 - (n/\kappa)^2[Q^2/(R_H^{4/3}A^2)]\}) \Delta y$ and guess at a value L_1 and solve for $(\Delta y)_A$. Now compute $y_1 = y_A + (\Delta y)_A$ before the jump. Again go to the equation above and take L_2 and again solve for $(\Delta y)_B$ in the flow upstream of the jump: $y_2 = y_B - (\Delta y)_B$. Now insert y_1 and y_2 into the jump equation $y_2 = [-y_1 + \sqrt{y_1^2 + (8Q^2/gb^2)(1/y_1)}]/2$ to see whether this equation is satisfied for the given volumetric flow Q. If not, go back and choose a different value of L_1 proceeding in this way until the jump equation is satisfied.

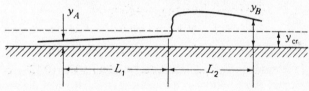

y_A y_B y_{cr} L_1 L_2

Fig. 14-99

14.391 The distance L_1 in Fig. 14-99 is 200 m and y_A is half the critical value. The channel is finished concrete with an adverse slope of -0.005. What is y_B at a distance $L_2 = 100$ m from the hydraulic jump? Solve in the simplest manner to get an approximate result. The flow rate is 200 m³/s and the width of the channel is 7 m.

▮
$$y_c = (q^2/g)^{1/3} = [(\tfrac{200}{7})^2/9.807]^{1/3} = 4.366 \text{ m} \qquad y_A = 4.366/2 = 2.183 \text{ m}$$
$$\Delta L = ([1 - (Q^2 b/gA^3)]/\{S_0 - (n/\kappa)^2[Q^2/(R_H^{4/3}A^2)]\}) \Delta y$$

$$A_1 = (7)(2.183) = 15.28 \text{ m}^2 \qquad R_H = A/p_w \qquad (R_H)_1 = 15.28/(2.183 + 7 + 2.183) = 1.344 \text{ m}$$

$$200 = \left\{ \frac{1 - (200)^2(7)/[(9.807)(15.28)^3]}{-0.005 - (0.012/1)^2[200^2/(1.344)^{4/3}(15.28)^2]} \right\}(\Delta y)_A \qquad (\Delta y)_A = 0.618 \text{ m}$$

$$y_1 = 2.183 + 0.618 = 2.801 \text{ m}$$

$$y_2 = [-y_1 + \sqrt{y_1^2 + (8Q^2/gb^2)(1/y_1)}]/2 = \{-2.801 + \sqrt{2.801^2 + [(8)(200)^2/(9.807)(7)^2](1/2.801)}\}/2 = 6.435 \text{ m}$$

$$A_2 = (7)(6.435) = 45.04 \text{ m}^2 \qquad (R_H)_2 = 45.04/(6.435 + 7 + 6.435) = 2.267 \text{ m}$$

$$100 = \left\{ \frac{1 - (200)^2(7)/[(9.807)(45.04)^3]}{-0.005 - (0.012/1)^2[200^2/(2.267)^{4/3}(45.04)^2]} \right\}(\Delta y)_B \qquad (\Delta y)_B = -0.866 \text{ m}$$

$$(\Delta y)_B = y_B - y_2 \qquad -0.866 = y_B - 6.435 \qquad y_B = 5.569 \text{ m}$$

14.392 For a rectangular channel, develop an expression for the relation between the depths before and after a hydraulic jump. Refer to Fig. 14-100.

▮ For the free body between sections 1 and 2, considering a unit width of channel and unit flow q, $P_1 = \gamma \bar{h} A = \gamma(\tfrac{1}{2}y_1)y_1 = \tfrac{1}{2}\gamma y_1^2$ and, similarly, $P_2 = \tfrac{1}{2}\gamma y_2^2$. From the principle of impulse and momentum, $\Delta P_x \, dt = \Delta$ linear momentum $= (W/g)(\Delta V_x)$, $\tfrac{1}{2}\gamma(y_2^2 - y_1^2) \, dt = (\gamma q \, dt/g)(V_1 - V_2)$.

Since $V_2 y_2 = V_1 y_1$ and $V_1 = q/y_1$, the above equation becomes

$$q^2/g = \tfrac{1}{2} y_1 y_2 (y_1 + y_2) \tag{1}$$

Since $q^2/g = y_c^3$,

$$y_c^3 = \tfrac{1}{2} y_1 y_2 (y_1 + y_2) \tag{2}$$

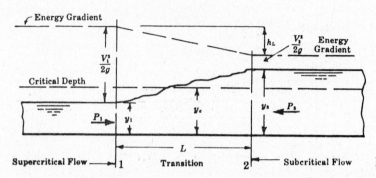

Fig. 14-100

14.393 A rectangular channel, 20 ft wide, carries 400 cfs and discharges onto a 20-ft-wide apron with no slope with a mean velocity of 20 fps. What is the height of the hydraulic jump? What energy is lost in the jump?

❚ $\quad q^2/g = y_1 y_2 (y_1 + y_2)/2 \qquad y_1 = q/v_1 = (\tfrac{400}{20})/20 = 1.00 \text{ ft} \qquad (\tfrac{400}{20})^2/32.2 = (1.00 y_2)(1.00 + y_2)/2$

$\quad 0.500 y_2^2 + 0.500 y_2 - 12.42 = 0 \qquad y_2 = 4.51 \text{ ft} \qquad \text{Height of jump} = 4.51 - 1.00 = 3.51 \text{ ft}$

$\quad E_j = (y_2 - y_1)^3 / 4 y_1 y_2 = (4.51 - 1.00)^3 / [(4)(4.51)(1.00)] = 2.40 \text{ ft}$

$\quad \text{Energy lost} = Q\gamma E_j = (400)(62.4)(2.40) = 59\,900 \text{ ft-lb/s}$

14.394 A rectangular channel, 16 ft wide, carries a flow of 192 cfs. The depth of water on the downstream side of a hydraulic jump is 4.20 ft. What is the upstream depth? What is the loss of head?

❚ $\qquad q^2/g = y_1 y_2 (y_1 + y_2)/2 \qquad (\tfrac{196}{16})^2/32.2 = (y_1)(4.20)(y_1 + 4.20)/2$

$\qquad 2.10 y_1^2 + 8.82 y_1 - 4.472 = 0 \qquad y_1 = 0.457 \text{ ft}$

$\qquad E_j = (y_2 - y_1)^3 / 4 y_1 y_2 = (4.20 - 0.457)^3 / [(4)(0.457)(4.20)] = 6.83 \text{ ft}$

14.395 After flowing over the concrete spillway of a dam, 9000 cfs then passes over a level concrete apron ($n = 0.013$). The velocity of the water at the bottom of the spillway is 42.0 ft/s and the width of the apron is 180 ft. Conditions will produce a hydraulic jump, the depth in the channel below the apron being 10.0 ft. In order that the jump be contained on the apron, (**a**) how long should the apron be built? (**b**) How much energy is lost from the foot of the spillway to the downstream side of the jump?

❚ Refer to Fig. 14-101.

(**a**) $\qquad q^2/g = y_1 y_2 (y_1 + y_2)/2 \qquad (\tfrac{9000}{180})^2/32.2 = 10 y_2 (10 + y_2)/2$

$\qquad 5 y_2^2 + 50 y_2 - 77.64 = 0 \qquad y_2 = 1.37 \text{ ft}$

$\quad y_1 = q/V_1 = (\tfrac{9000}{180})/42.0 = 1.19 \text{ ft} \qquad L = [(V_2^2/2g + y_2) - (V_1^2/2g + y_1)]/(S_0 - S)$

$\qquad V_2 = q/y_2 = (\tfrac{9000}{180})/1.37 = 36.50 \text{ ft/s} \qquad S = (nV/1.486 R^{2/3})^2$

$\qquad V_{av} = (42.0 + 36.40)/2 = 39.20 \text{ ft/s}$

$\quad R = A/p_w \qquad R_1 = (180)(1.19)/(1.19 + 180 + 1.19) = 1.174 \text{ ft}$

$\qquad R_2 = (180)(1.37)/(1.37 + 180 + 1.37) = 1.349 \text{ ft}$

$\quad R_{av} = (1.174 + 1.349)/2 = 1.262 \text{ ft} \qquad S = \{(0.013)(39.20)/[(1.486)(1.262)^{2/3}]\}^2 = 0.08623$

$\quad L = \{36.50^2/[(2)(32.2)] + 1.37 - 42.0^2/[(2)(32.2)] - 1.19\}/(0 - 0.08623) = 75.7 \text{ ft}$

The length of the jump L_3 from B to C is from $4.3 y_3$ to $5.2 y_3$. Assuming the conservative value of $5.0 y_3$, $L_3 = (5.0)(10.0) = 50 \text{ ft}$. Hence, $L_{ABC} = 76 + 50 = 126 \text{ ft}$ (approximately).

(**b**) $\qquad E = y + V^2/2g \qquad E_A = 1.19 + 42.0^2/[(2)(32.2)] = 28.58 \text{ ft}$

$\qquad V_3 = Q/A_3 = 9000/[(10.0)(180)] = 5.00 \text{ ft/s}$

$\quad E_C = 10.0 + 5.00^2/[(2)(32.2)] = 10.39 \text{ ft} \qquad E_{lost} = 28.58 - 10.39 = 18.19 \text{ ft}$

$\qquad P = Q\gamma E \qquad P_{lost} = (9000)(62.4)(18.19) = 10.22 \times 10^6 \text{ ft-lb/s}$

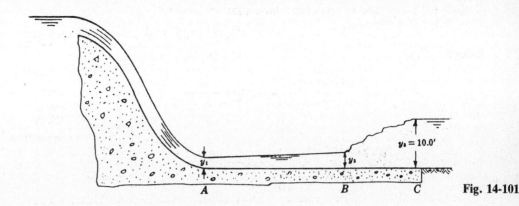

Fig. 14-101

14.396 Determine the elevation of the spillway apron if $q = 50$ cfs/ft, $h = 9$ ft, $D = 63$ ft, and the spillway crest is at elevation 200.0 ft.

$$(\pi)_1(\pi_3 - \pi_2)(\pi_3 + 1)^{1/2} + 0.353 = \sqrt{\tfrac{1}{8} + (2.828)(\pi_1)(\pi_3 + 1)^{3/2}}$$

$$\pi_1 = g^{1/2}h^{3/2}/q = (32.2)^{1/2}(9)^{3/2}/50 = 3.06$$

$$\pi_2 = D/h = \tfrac{63}{9} = 7.00 \qquad \pi_3 = d/h = d/9$$

$$(3.06)(d/9 - 7.00)(d/9 + 1)^{1/2} + 0.353 = \sqrt{\tfrac{1}{8} + (2.828)(3.06)(d/9 + 1)^{3/2}}$$

$d = 77.9$ ft   (by trial and error)   Elevation of spillway apron $= 200.0 - 77.9 = 122.1$ ft

14.397 Establish the equation for flow over a broad-crested weir assuming no lost head. See Fig. 14-102.

    At the section where critical flow occurs, $q = V_c y_c$. But $y_c = V_c^2/g = \tfrac{2}{3}E$, and $V_c = \sqrt{g(\tfrac{2}{3}E_c)}$. Hence the theoretical value of flow q becomes $q = \sqrt{g(\tfrac{2}{3}E_c)} \times \tfrac{2}{3}E_c = 3.09E_c^{3/2}$. However, the value of E_c is difficult to measure accurately, because the critical depth is difficult to locate. The practical equation becomes $q = CH^{3/2} \approx 3H^{3/2}$. The weir should be calibrated in place to obtain accurate results.

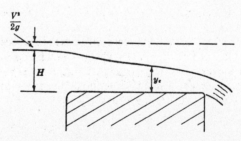

Fig. 14-102

14.398 Develop an expression for a critical-flow meter (Fig. 14-103).

    An excellent method of measuring flow in open channels is by means of a constriction. The measurement of the critical depth is not required. The depth y_1 is measured a short distance upstream from the constriction. The raised floor should be about $3y_c$ long and of such height as to have the critical velocity occur on it.

    For a rectangular channel of constant width, the Bernoulli equation is applied between sections 1 and 2, in which the lost head in accelerated flow is taken as one-tenth of the difference in velocity heads, i.e., $y_1 + (V_1^2/2g) - (\tfrac{1}{10})[(V_c^2/2g) - (V_1^2/2g)] = [y_c + (V_c^2/2g) + z]$, which neglects the slight drop in the channel bed between 1 and 2. Recognizing that $E_c = y_c + V_c^2/2g$, we rearrange as follows: $[y_1 + (1.10V_1^2/2g)] = z + 1.0E_c + (\tfrac{1}{10})(\tfrac{1}{3}E_c)$, $[y_1 - z + (1.10V_1^2/2g)] = 1.033E_c = (1.033)(\tfrac{3}{2}\sqrt[3]{q^2/g})$, or

$$q = (2.94)(y_1 - z + 1.10V_1^2/2g)^{3/2} \tag{1}$$

Since $q = V_1 y_1$, $\qquad\qquad q = (2.94)(y_1 - z + 0.0171q^2/y_1^2)^{3/2} \tag{2}$

14.399 Consider a rectangular channel 10 ft wide with the critical-flow meter of Fig. 14-103 having dimension $z = 1.10$ ft. If the measured depth y_1 is 2.42 ft, what is the discharge?

    $q = (2.94)(y_1 - z + 0.0171q^2/y_1^2)^{3/2} = (2.94)(2.42 - 1.10 + 0.0171q^2/2.42^2)^{3/2}$. As a first approximation, neglect the last term involving q. $q = (2.94)(2.42 - 1.10)^{3/2} = 4.46$ cfs/ft. Try $q = 4.80$ cfs/ft: $q = 2.94[2.42 - 1.10 + (0.0171)(4.80)^2/2.42^2]^{3/2} = 4.80$ cfs/ft (O.K.), $Q = (4.80)(10) = 48.0$ ft³/s.

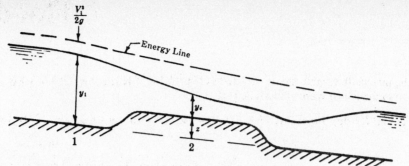

Fig. 14-103

14.400 In a 100-ft-wide rectangular channel, the depth upstream from a hydraulic jump is 4 ft. What flow rate is required in order for the downstream depth to be (**a**) twice the upstream depth and (**b**) ten times the upstream depth?

▮

$$q = \{[(d_1 + d_2)/2](gd_1d_2)\}^{1/2}$$

(**a**) $\quad d_2 = (2)(4) = 8\ \text{ft} \quad q = \{[(4 + 8)/2][(32.2)(4)(8)]\}^{1/2} = 78.63\ \text{cfs/ft}$

$$Q = (100)(78.63) = 7863\ \text{ft}^3/\text{s}$$

(**b**) $\quad d_2 = (10)(4) = 40\ \text{ft} \quad q = \{[(4 + 40)/2][(32.2)(4)(40)]\}^{1/2} = 336.7\ \text{cfs/ft}$

$$Q = (100)(336.7) = 33\,670\ \text{ft}^3/\text{s}$$

15.1 If the channel width for the stream depicted in Fig. 15-1 is 30 ft at a stage of 4 ft, what would be the velocity of a translatory monoclinal wave of small height?

    ▌ $u = (1/B)(dQ/dy)$. From Fig. 15-1, $dQ/dy = 340$ cfs/ft when $y = 4$ ft. $u = (\frac{1}{30})(340) = 11.33$ ft/s.

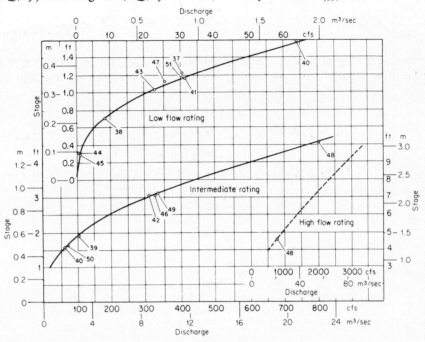

Fig. 15-1

15.2 Given the stage-discharge relation defined by $Q = (k)(g - a)^b$, where Q is flow rate, g is gage height, and a, b, and k are station constants, if $a = 0.055$, $b = 2.50$, and $k = 10.25$, compute the velocity of a monoclinal wave of small height. Assume a channel width of 20 m and a stage of 1.3 m.

    ▌ $u = (1/B)(dQ/dy)$ $Q = (10.25)(y - 0.055)^{2.50}$ $dQ/dy = (2.50)(10.25)(y - 0.055)^{1.50}$

For $y = 1.3$ m, $dQ/dy = (2.50)(10.25)(1.3 - 0.055)^{1.59} = 35.60$ (m³/s)/m, $u = (\frac{1}{20})(35.60) = 1.78$ m/s.

15.3 If the channel width for the stream depicted in Fig. 15-1 is 10 m at a stage of 1 m, what would be the velocity of a translatory monoclinal wave of small height?

    ▌ $u = (1/B)(dQ/dy)$. From Fig. 15-1, $dQ/dy = 26.6$ (m³/s)/m when $y = 1.0$ m. $u = (\frac{1}{10})(26.6) = 2.66$ m/s.

15.4 Given the hydrographs tabulated below (cubic meters per second), find the 12 hourly storage values for the reach in cubic meters. Ignore local inflow.

date	hour*	inflow	outflow	date	hour*	inflow	outflow
1	M	36	58	7	N	196	341
2	N	43	46		M	153	272
	M	121	42	8	N	124	218
3	N	346	61		M	101	180
	M	575	149	9	N	84	150
4	N	717	326		M	71	124
	M	741	536	10	N	60	104
5	N	612	674		M	52	86
	M	440	681	11	N	46	73
6	N	328	560		M	41	62
	M	251	437	12	N	37	52

    * M = midnight; N = noon.

459

▮ $[(I_1 + I_2)/2]t - [(O_1 + O_2)/2]t = S_2 - S_1$. Taking minimum storage equal to zero, and noting that $t = 0.5$ day $= 43\ 200$ s, computations are carried out as indicated in the following table. The last two columns apply to Prob. 15.8.

day	hour	I, m³/s	O, m³/s	$(I_1 + I_2)/2$	$(O_1 + O_2)/2$	$\bar{I} - \bar{O}$, $(\Delta s/t)$	$\sum (\bar{I} - \bar{O})$, (S/t)	$S \times 10^{-5}$, m³	$xI + (1-x)O$ $x = 0.2$	$x = 0.3$
1	M	36	58				13	6	54	51
				39	52	−13				
2	N	43	46				0	0	45	45
	M	121	42	82	44	38	38	16	58	66
				234	52	182				
3	N	346	61				220	95	118	147
				460	105	355				
	M	575	149				575	248	234	277
				646	238	408				
4	N	717	326				983	425	404	443
				729	431	298				
	M	741	536				1281	553	577	598
				676	605	71				
5	N	612	674				1352	584	662	655
				526	678	−152				
	M	440	681				1200	518	633	609
				384	620	−236				
6	N	328	560				964	416	514	490
				290	498	−208				
	M	251	437				756	327	400	381
				224	389	−165				
7	N	196	341				591	255	312	298
				174	306	−132				
	M	153	272				459	198	248	236
				138	245	−107				
8	N	124	218				352	152	199	190
				112	199	−87				
	M	101	180				265	114	164	156
				92	165	−73				
9	N	84	150				192	83	137	130
				78	137	−59				
	M	71	124				133	57	113	108
				66	114	−48				
10	N	60	104				85	37	95	91
				56	95	−39				
	M	52	86				46	20	79	76
				49	80	−31				
11	N	46	73				15	6	68	65
				44	68	−24				
	M	41	62				−9	−4	58	56

15.5 Given the following inflow data to a reservoir and the reservoir relationships of Fig. 15-2, route the flood through the reservoir. The initial outflow from the reservoir is 1.7 m³/s, and the initial value of $2S/t + 0$ is 9.0 m³/s.

day	hour	inflow, m³/s
1	Noon	2.0
	Midnight	5.2
2	Noon	10.1
	Midnight	12.2
3	Noon	8.5
	Midnight	4.7
4	Noon	2.3

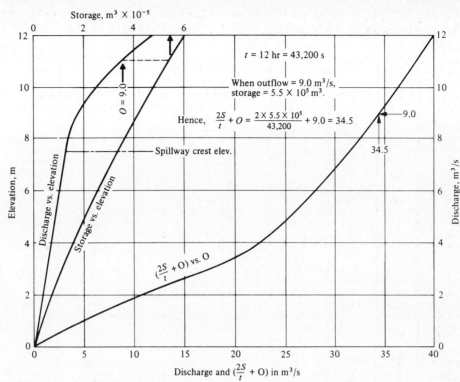

Storage, $m^3 \times 10^{-5}$

$t = 12$ hr $= 43,200$ s

When outflow $= 9.0$ m^3/s,
storage $= 5.5 \times 10^5$ m^3.

Hence, $\dfrac{2S}{t} + O = \dfrac{2 \times 5.5 \times 10^5}{43,200} + 9.0 = 34.5$

Spillway crest elev.

$\left(\dfrac{2S}{t} + O\right)$ vs. O

Discharge and $\left(\dfrac{2S}{t} + O\right)$ in m^3/s

Fig. 15-2

$$I_1 + I_2 + (2S_1/t - O_1) = 2S_2/t + O_2 \qquad 2S_1/t - O_1 = 2S_2/t + O_2 - 2O_2$$

$$2S_1/t - O_1 = 9.0 - (2)(1.7) = 5.6 \text{ m}^3/\text{s} \qquad 2.0 + 5.2 + 5.6 = 2S_2/t + O_2 \qquad 2S_2/t + O_2 = 12.8 \text{ m}^3/\text{s}$$

From Fig. 15-2, $O_2 = 2.3$ m^3/s. Succeeding computations are carried out in the same manner with the results as tabulated below.

day	hour	I	$(2S/t) - O$	$(2S/t) + O$	O
1	Noon	2.0	5.6	9.0	1.7
	Midnight	5.2	8.2	12.8	2.3
2	Noon	10.1	14.9	23.5	4.3
	Midnight	12.2	16.2	37.2	10.5
3	Noon	8.5	16.3	36.9	10.3
	Midnight	4.7	16.3	29.5	6.6
4	Noon	2.3		23.3	4.2

All tabular values are in cubic meters per second.

15.6 A small reservoir has an area of 300 acres at spillway level, and the banks are essentially vertical for several feet above spillway level. The spillway is 15 ft long and has a coefficient of 3.75. Taking the inflow hydrograph of Prob. 15.4 as the inflow to the reservoir (cubic feet per second), compute the maximum pool level and maximum discharge to be expected if the reservoir is initially at the spillway level at midnight on the first.

⬛ The discharge over the spillway is ($C = 3.75$ and $L = 15$ ft): $q = CLy^{3/2} = (3.75)(15)y^{3/2} = 56.25y^{3/2}$ cfs. The storage is equal to the reservoir area multiplied by the height of water above the spillway crest: $S = 300y$ acre-ft $= 151y$ cfs-days. From these equations for S and q, and noting that $t = 0.5$ day:

y, ft	q, cfs	S, cfs-day	$(2S/t) + O$, cfs
1	56	151	660
2	159	302	1370
3	292	453	2100
4	450	604	2870
5	629	755	3650

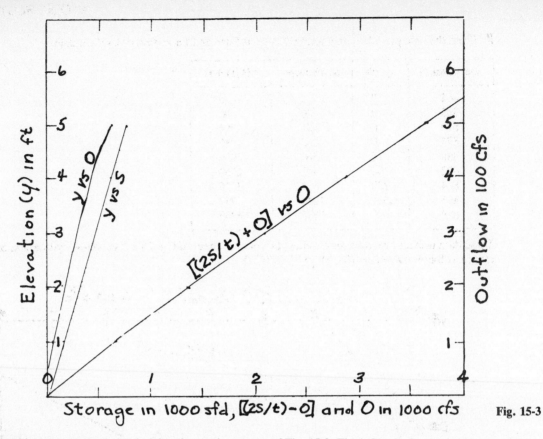

Fig. 15-3

These data provide the basis for deriving the routing curves of Fig. 15-3. The inflow hydrograph of Prob. 11.4 is routed as follows: $I_1 + I_2 + [(2S_1/t) - O_1] = (2S_2/t) + O_2$.

day	hour	I, cfs	$(2S_1/t) - O$	$(2S_2/t) + O$	O, cfs
1	M	36	0	0	0
2	N	43	71	79	4
	M	121	207	235	14
3	N	346	560	674	57
	M	575	1127	1481	177
4	N	717	1709	2419	355
	M	741	2137	3167	515
5	N	612	2310	3490	590
	M	440	2242	3362	560
6	N	328		3010	480

The peak flow is approximately 590 cfs and the maximum pool elevation is $y = (590/56.25)^{0.6667} = 4.79$ ft.

15.7 Tabulated below are the elevation-storage and elevation-discharge data for a small reservoir. Taking the inflow hydrograph of Prob. 15.4 as the reservoir inflow (cubic feet per second), and assuming the pool elevation to be 875 ft at midnight on the first, find the maximum pool elevation and peak outflow rate.

elevation, ft	storage, acre-ft	discharge, ft³/s	elevation, ft	storage, acre-ft	discharge, ft³/s
862	0	0	882	1220	100
865	40	0	884	1630	230
870	200	0	886	2270	394
875	500	0	888	3150	600
880	1000	0			

▮ From the data provided, and noting that $t = 0.5$ day and 1 acre-foot = 0.504 cfs-day:

elevation, ft	q, cfs	S, cfs-days	$(2S/t) + O$
862	0	0	0
865	0	20	80
870	0	101	404
875	0	252	1010
880	0	504	2020
882	100	615	2560
884	230	822	3520
886	394	1140	4950
888	600	1590	6960

These data provide the basis for deriving the routing curves of Fig. 15-4. The inflow hydrograph of Prob. 11.4 is routed as follows: $I_1 + I_2 + [(2S_1/t) - O_1] = (2S_2/t) + O_2$.

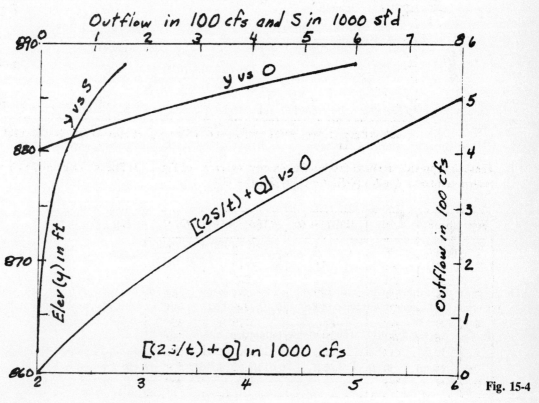

Fig. 15-4

day	hour	I, cfs	$(2S_1/t) - O$	$(2S_2/t) + O$	O, cfs
1	M	36	1010	1010	0
2	N	43	1089	1089	0
	M	121	1253	1253	0
3	N	346	1720	1720	0
	M	575	2415	2641	113
4	N	717	3201	3707	253
	M	741	3933	4659	363
5	N	612	4428	5286	429
	M	440	4580	5480	450
6	N	328	4474	5348	437
	M	251		5053	405

The peak flow is approximately 450 cfs and the maximum pool elevation is 886.6 ft.

15.8 Find the Muskingum K and x for the flood of Prob. 15.4 (flows in cubic meters per second) for use in the Muskingum method of flood routing.

▌ The computation of weighted flows ($x = 0$, $x = 0.2$, and $x = 0.3$) appears in the storage table of Prob. 15.4, and the plots appear in Fig. 15-5. It will be seen that increasing the value of x from 0.2 to 0.3 reverses the "loop" in the lower range of flow. Even so, a slight additional increase in x might provide a slightly better fit, without materially affecting K. Take $K = 26.7$ h, $x = 0.3$, and $t = 12$ h.

15.9 For the flood data of Prob. 15.8, determine the Muskingum routing coefficients for $t = 12$ h and $t = 24$ h.

▌
$$c_0 = (Kx - 0.5t)/(K - Kx + 0.5t) \qquad c_1 = (Kx + 0.5t)/(K - Kx + 0.5t)$$
$$c_2 = (K - Kx - 0.5t)/(K - Kx + 0.5t)$$

For $t = 12$ h, with $K = 26.7$ h and $x = 0.3$,

$$c_0 = -\frac{(26.7)(0.3) - (0.5)(12)}{26.7 - (26.7)(0.3) + (0.5)(12)} = -0.081 \qquad c_1 = \frac{(26.7)(0.3) + (0.5)(12)}{26.7 - (26.7)(0.3) + (0.5)(12)} = 0.567$$

$$c_2 = \frac{26.7 - (26.7)(0.3) - (0.5)(12)}{26.7 - (26.7)(0.3) + (0.5)(12)} = 0.514$$

For $t = 24$ h, with $K = 26.7$ h and $x = 0.3$,

$$c_0 = -\frac{(26.7)(0.3) - (0.5)(24)}{26.7 - (26.7)(0.3) + (0.5)(24)} = 0.130 \qquad c_1 = \frac{(26.7)(0.3) + (0.5)(24)}{26.7 - (26.7)(0.3) + (0.5)(24)} = 0.652$$

$$c_2 = \frac{26.7 - (26.7)(0.3) - (0.5)(24)}{26.7 - (26.7)(0.3) + (0.5)(24)} = 0.218$$

The 24-h routing period is to be preferred, since negative coefficients lead to instability when flow changes rapidly.

15.10 For a river reach with $K = 27$ h and $x = 0.2$, determine the Muskingum routing coefficients for $t = 12$ h.

▌
$$c_0 = (Kx - 0.5t)/(K - Kx + 0.5t) \qquad c_1 = (Kx + 0.5t)/(K - Kx + 0.5t)$$
$$c_2 = (K - Kx - 0.5t)/(K - Kx + 0.5t)$$

$$c_0 = -\frac{(27)(0.2) - (0.5)(12)}{27 - (27)(0.2) + (0.5)(12)} = 0.022 \qquad c_1 = \frac{(27)(0.2) + (0.5)(12)}{27 - (27)(0.2) + (0.5)(12)} = 0.413$$

$$c_2 = \frac{27 - (27)(0.2) - (0.5)(12)}{27 - (27)(0.2) + (0.5)(12)} = 0.565$$

15.11 Taking the outflow hydrograph of Prob. 15.4 as the inflow (cubic feet per second) to a reach, and using the routing coefficients determined in Prob. 15.10, find the peak outflow using the Muskingum method of routing.

▌ $O_2 = c_0 I_2 + c_1 I_1 + c_2 O_1$. The routing computations are as follows:

day	hour	I, cfs	$0.022I_2$	$0.413I_1$	$0.565O_1$	O, cfs
1	M	58				60 (est)
2	N	46	1	24	34	59
	M	42	1	19	33	53
3	N	61	1	17	30	48
	M	149	3	25	27	55
4	N	326	7	62	31	100
	M	536	12	135	56	203
5	N	674	15	221	115	351
	M	681	15	278	198	491
6	N	560	12	281	277	570
	M	437	10	231	322	563
7	N	341	8	180	318	506
	M	272	6	140	286	432

The routed peak flow is 570 cfs.

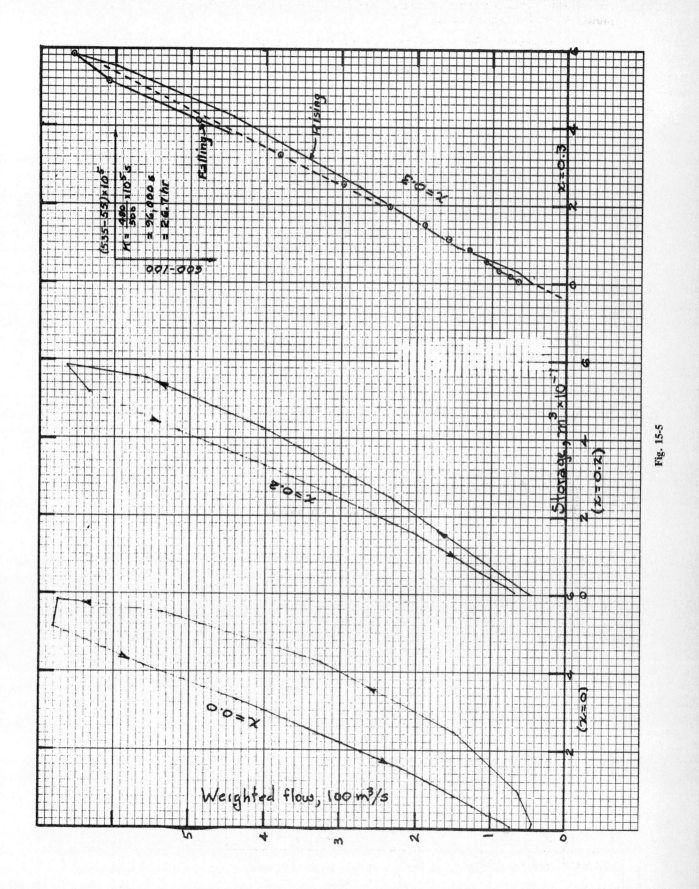

Fig. 15-5

15.12 For the inflow hydrograph given below for a reach, route the flood through the reach by the Muskingum method. Take $K = 11$ h, $t = 6$ h, and $x = 0.13$; hence, $c_0 = 0.124$, $c_1 = 0.353$, and $c_2 = 0.523$.

day	hour	I, cfs
1	6 a.m.	10
	Noon	30
	6 p.m.	68
	Midnight	50
2	6 a.m.	40
	Noon	31
	6 p.m.	23

Assume an initial outflow of 10 cfs.

▮
$$O_2 = c_0 I_2 + c_1 I_1 + c_2 O_1$$

day	hour	I	$c_0 I_2$	$c_1 I_1$	$c_2 O_1$	O
1	6 a.m.	10	—	—	—	10
	Noon	30	3.7	3.5	5.2	12.4
	6 p.m.	68	8.4	10.6	6.5	25.5
	Midnight	50	6.2	24.0	13.3	43.5
2	6 a.m.	40	5.0	17.7	22.7	45.4
	Noon	31	3.8	14.1	23.7	41.6
	6 p.m.	23	2.9	10.9	21.8	35.6

All tabular values are in cubic feet per second.

15.13 A reservoir of rectangular cross section is 52 ft deep at the dam and 20 ft deep 15 000 ft upstream. Compute the time for an infinitesimally small, shallow wave initiated at the dam to reach the point where the water is 20 ft deep. Assume channel slope is constant and flow rate is zero.

▮ See Fig. 15-6.

$$u = v_1 \pm \sqrt{gy} \qquad v_1 = 0 \qquad u = \sqrt{gy} = \sqrt{(g)(d + mx)} = \sqrt{a + bx}$$

$$t_2 - t_1 = \int_{x_1}^{x_2} \frac{dx}{\sqrt{a + bx}} = \left[\frac{(2)(\sqrt{a + bx})}{b} \right]_{x_1}^{x_2} = \frac{(2)(\sqrt{a + bx_2} - \sqrt{a + bx_1})}{b} \qquad b = \frac{(g)(y_2 - y_1)}{(x_2 - x_1)}$$

$$\sqrt{a + bx_2} = u_2 = \sqrt{gy_2} \qquad \sqrt{a + bx_1} = u_1 = \sqrt{gy_1} \qquad t_2 - t_1 = (2)(\sqrt{gy_2} - \sqrt{gy_1})/[(g)(y_2 - y_1)/(x_2 - x_1)]$$

$$t_2 - t_1 = (2)[\sqrt{(32.2)(20)} - \sqrt{(32.2)(52)}]/[(32.2)(20 - 52)/15\,000] = 452.5 \text{ s} \quad \text{or} \quad 7.54 \text{ min}$$

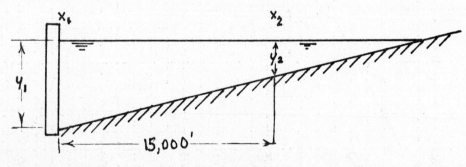

Fig. 15-6

15.14 Show the trajectory of the wave in Prob. 15.13 on the xt plane.

▮ From Prob. 15.13, $t - t_1 = (2)(\sqrt{gy} - \sqrt{gy_1})/[(g)(y - y_1)/(x - x_1)]$. Let $x_1 = 0$ and $t_1 = 0$: $t = (2)[\sqrt{32.2y} - \sqrt{(32.2)(52)}]/[(32.2)(y - 52)/x]$, $y = 52 - (x)(52 - 20)/15\,000$. For $x = 1000$ ft, $y = 52 - (1000)(52 - 20)/15\,000 = 49.87$ ft, $t = (2)[\sqrt{(32.2)(49.87)} - \sqrt{(32.2)(52)}]/[(32.2)(49.87 - 52)/1000] = 25$ s. For additional

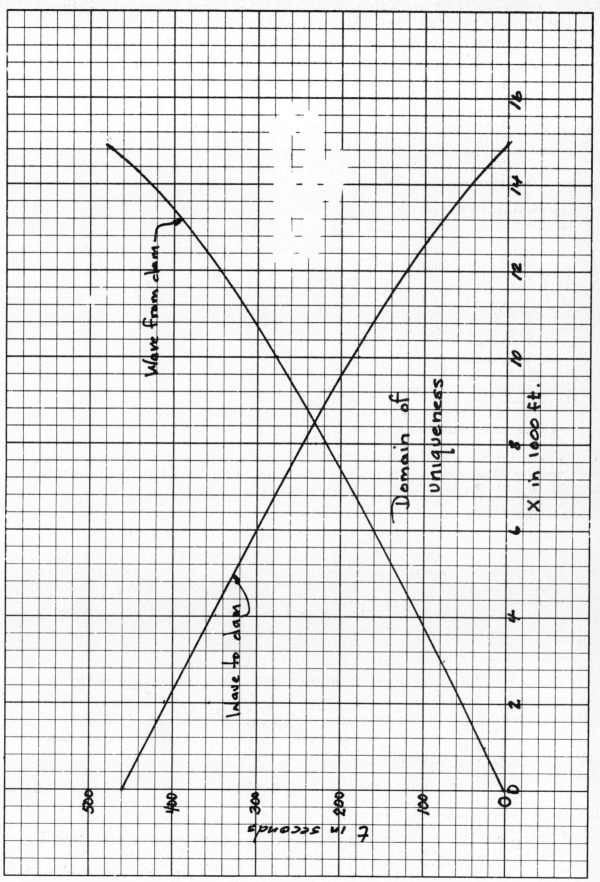

Fig. 15-7

values of x,

x, ft	t, s	x, ft	t, s
0	0	8 000	215
1 000	25	9 000	245
2 000	50	10 000	276
3 000	76	11 000	309
4 000	102	12 000	342
5 000	129	13 000	377
6 000	156	14 000	414
7 000	186	15 000	453

See plot in Fig. 15-7 on p. 467.

15.15 Suppose that another infinitesimally shallow wave is initiated where the depth is 20 ft at the same time as the wave in Prob. 15.13. Compute the point on the xt plane where the two waves will meet. Draw the trajectories and show the domain of uniqueness on the xt plane.

▮ For a wave starting at $y = 20$ ft, $t = (2)[\sqrt{32.2y} - \sqrt{(32.2)(20)}]/[(32.2)(y - 20)/(15\,000 - x)]$. For $x = 1000$ ft, $y = 49.87$ ft (from Prob. 15.14), $t = (2)[\sqrt{(32.2)(49.87)} - \sqrt{(32.2)(20)}]/[(32.2)(49.87 - 20)/(15\,000 - 1000)] = 428$ s. For additional values of x,

x, ft	t, s	x, ft	t, s
0	453	8 000	238
1 000	428	9 000	207
2 000	403	10 000	176
3 000	377	11 000	144
4 000	350	12 000	110
5 000	323	13 000	75
6 000	296	14 000	38
7 000	267	15 000	0

See plot in Fig. 15-7. Waves meet at about 8350 ft at $t = 225$ s.

15.16 A uniform rectangular channel 3 m wide with $n = 0.015$ and slope of 0.0004 is flowing at a depth of 1.6 m. A sudden gate opening increases the depth to 1.9 m. What are the celerity of the resulting abrupt wave and the wave velocity?

▮
$$c = \sqrt{(g)(J_2 - J_1)/[(A_1)(1 - A_1/A_2)]} \qquad J = Ad/2 \qquad A_1 = (3)(1.6) = 4.80 \text{ m}^2$$
$$J_2 = (4.80)(1.6/2) = 3.840 \text{ m}^3$$
$$A_2 = (3)(1.9) = 5.70 \text{ m}^2 \qquad J_1 = (5.70)(1.9/2) = 5.415 \text{ m}^3$$
$$c = \sqrt{(9.807)(5.415 - 3.840)/[(4.80)(1 - 4.80/5.70)]} = 4.51 \text{ m/s}$$
$$u = v_1 + c \qquad v = (1.0/n)(R^{2/3})(s^{1/2}) \qquad v_1 = (1.0/0.015)[4.80/(1.6 + 3 + 1.6)]^{2/3}(0.0004)^{1/2} = 1.12 \text{ m/s}$$
$$u = 1.12 + 4.51 = 5.63 \text{ m/s}$$

CHAPTER 16
Flow of Compressible Fluids

16.1 For air at a temperature of 30 °C and pressure of 470 kPa abs, find the specific weight of the air.

$$\gamma = p/RT = 470/[(29.3)(30 + 273)] = 0.0529 \text{ kN/m}^3$$

16.2 Find the mass density of helium at a temperature of 39 °F and a pressure of 26.9 psig when the atmospheric pressure is 14.9 psig.

$$\rho = p/RT = (26.9 + 14.9)(144)/[(12\,420)(39 + 460)] = 0.000971 \text{ slug/ft}^3$$

16.3 Methane at 22 °C flows through a pipeline at a velocity of 416.4 m/s. Is the flow subsonic, sonic, supersonic, or hypersonic?

$N_M = v/\sqrt{kgRT} = 416.4/\sqrt{(1.32)(9.807)(52.9)(22 + 273)} = 0.93$. Since $N_M < 1.0$, the flow is subsonic.

16.4 Chlorine gas at 51 °F flows through a pipeline. Find the velocity at which the flow will be sonic.

$$N_M = v/\sqrt{kgRT} \qquad 1.0 = v/\sqrt{(1.34)(32.2)(21.8)(51 + 460)} \qquad v = 693 \text{ ft/s}$$

16.5 Compute the change in internal energy and the change in enthalpy of 15 kg of air if its temperature is raised from 20 to 30 °C. The initial pressure is 95 kN/m² abs.

$\Delta i = c_v(T_2 - T_1) = (716)(30 - 20) = 7160 \text{ N-m/kg}$ Change in internal energy $= (\Delta i)(15) = 107\,400$ N-m

$\Delta h = c_p(T_2 - T_1) = 1003(10) = 10\,030 \text{ N-m/kg}$ Change in enthalpy $= (\Delta h)(15) = 150\,000$ N-m

16.6 Suppose that 15 kg of air ($T_1 = 20$ °C) of Prob. 16.5 were compressed isentropically to 40 percent of its original volume. Find the final temperature and pressure, the work required, and the changes in internal energy and enthalpy.

The following relations apply: $pv = RT$ and $pv^k = $ constant, where $k = 1.40$ for air. $pv^k = pv(v^k/v) = (RT/v)v^k = RTv^{k-1} = $ constant. Since $R = $ constant, $Tv^{k-1} = $ constant, and $T_2 = T_1(v_1/v_2)^{k-1} = (273 + 20)(1.0/0.4)^{1.40-1} = 422$ K $= 149$ °C, $pv/T = R = $ constant, $p_1 = 95$ kN/m² abs (from Prob. 16.5), $p_1v_1/T_1 = p_2v_2/T_2$ and $v_2 = 0.4v_1$, $95v_1/293 = [p_2(0.4v_1)/422]$, $p_2 = 342$ kN/m² abs. Since this is an isentropic process, the work required is equal to the change in internal energy. This can be confirmed by computing the values of the pressure and corresponding volumes occupied by the gas during the isentropic process, plotting a pressure-vs.-volume curve, and finding the area under the curve and thereby determining the work done on the fluid. Thus the work required is

$$\int_{s_1}^{s_2} F\, ds = \int_{s_1}^{s_2} \left(\frac{F}{A}\right) A\, ds = \int_{\text{vol } 1}^{\text{vol } 2} p\, d(\text{vol})$$

or $\Delta i = c_v(T_2 - T_1) = 716(422 - 293) = 92\,400$ N-m/kg, $(\Delta i)(15 \text{ kg}) = 1\,385\,000$ N-m $= $ work required $= $ change in internal energy, $\Delta h = c_p(T_2 - T_1) = 1003(129) = 129\,400$ N-m/kg, $(\Delta h)(15 \text{ kg}) = 1\,941\,000$ N-m $= $ change in enthalpy.

16.7 Using the data of Prob. 16.6 compute $\Delta(p/\rho)$ and thus show that $\Delta h = \Delta i + \Delta(p/\rho)$.

From Prob. 16.6, $\Delta h = 129\,400$ N-m/kg, $\Delta i = 92\,400$ N-m/kg, $R = 287$ N-m/(kg-K); $p_1/\rho_1 = RT_1 = (287)(293) = 84\,100$ N-m/kg; $p_2/\rho_2 = RT_2 = (287)(422) = 121\,100$ N-m/kg; $\Delta(p/\rho) = p_2/\rho_2 - p_1/\rho_1 = 37\,000$ N-m/kg. $\Delta i + \Delta(p/\rho) = 92\,000 + 37\,000 = 129\,400$ which checks accurately with the value of Δh.

16.8 Using the data of Prob. 16.6, determine the work done in compressing the air by finding the area under a pressure-vs.-volume curve. Compute and tabulate volumes and pressures using volume increments which are 10 percent of the original volume.

$R = 287$ N-m/(kg-K), $k = 1.4$. Initially, given: $T_1 = 273 + 20 = 293$ K; $p_1 = 95$ kN/m²; $v_1 = RT_1/p_1 = (287)(293)/95\,000 = 0.885$ m³/kg. After compression, from Prob. 16.6 $T = T_1(v_1/v)^{1-k} = (293)(v_1/v)^{0.4}$; $v = v_1/(v_1/v)$; $V = mv = 15v$; $p = RT/v$.

469

v_1/v	T, K	v, m³/kg	V, m³	p, N/m²
$\frac{100}{100}$	293	0.885	13.28	95 000
$\frac{100}{90}$	306	0.797	11.95	110 100
$\frac{100}{80}$	320	0.708	10.62	129 900
$\frac{100}{70}$	338	0.619	9.29	156 600
$\frac{100}{60}$	359	0.531	7.97	194 100
$\frac{100}{50}$	387	0.443	6.64	251 000
$\frac{100}{40}$	423	0.354	5.31	343 000

These data are plotted in Fig. 16-1. Measure or calculate area A: Work done in compression $= A = 1.395 \times 10^6$ N-m (by calculation).

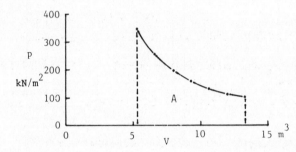

Fig. 16-1

16.9 Compute the change in enthalpy of 500 lb of oxygen if its temperature is increased from 120 °F to 155 °F.

$\blacksquare$ $c_p = 5437$ ft-lb/(slug-°R) $\Delta h = c_p(T_2 - T_1) = (5437)(155 - 120) = 190\,000$ ft-lb/slug

Five hundred pounds is equivalent to 15.53 slugs ($m = W/g$). Thus change in enthalpy $= (190\,000)(15.53) = 2.95 \times 10^6$ ft-lb.

16.10 Suppose the 500 lb of oxygen of Prob. 16.9 was compressed isentropically to 80 percent of its original volume.. Find the final temperature and pressure, the work required, and the change in enthalpy. Assume $T_1 = 120$ °F and $p_1 = 200$ psia.

$\blacksquare$ $pv^k = $ constant and $pv = RT$. Eliminating p: $(RT/v)V^k = $ constant $= RTv^{k-1}$. Since $R = $ constant also, $Tv^{k-1} = $ constant; $k = 1.4$. From which: $T_2 = T_1(v_1/v_2)^{k-1} = (120 + 460)(1/0.8)^{0.4} = 634.2$ °R $= 174.2$ °F. $p_1v_1/T_1 = p_2v_2/T_2$; so $200v_1/580 = p_2(0.8v_1)/634$; $p_2 = 273$ psia; $c_v = 3883$, $c_p = 5437$ ft²/(s²-°R). Work required $=$ change of internal energy (as isentropic). Work required $= (\Delta i)(W) = c_v(\Delta T)(500/32.2)$ where $\Delta T = 174.5 - 120 = 54.2$ °F. Work required $= (3883)(54.2)(500/32.2) = 3.27 \times 10^6$ ft-lb; increase in enthalpy $= (5437)(54.2)(500/32.2) = 4.68 \times 10^6$ ft-lb.

16.11 Determine the sonic velocity in air at sea level and at elevations 5000, 10 000, 20 000, and 30 000 ft. Assume standard atmosphere (Table A-7).

$\blacksquare$ $c = (kRT)^{1/2}$. In all cases $k = 1.4$ and $R = 1715$ ft-lb/(slug-°R). From Table A-7, temperatures $= 59$ °F, 41.2 °F, 23.4 °F, -12.3 °F, and -47.8 °F. Substitute for $T = (460 + $ temperature) and compute the resulting c's.

elevation ft	temperature °F	c, fps
0	59	1 116
5 000	41.2	1 097
10 000	23.4	1 077
20 000	−12.3	1 037
30 000	−47.8	995

16.12 Repeat Prob. 16.11 for sea level, 2000, and 10 000 m, expressing the answers in SI units.

$\blacksquare$ $c = (kRT)^{1/2}$, $k = 1.4$, and $R = 287$ N-m/(kg-K) $= 287$ m²/(s²-K); $g = 9.81$ m/s². From Table A-8: temperatures $= 15.0$, 2.0, and -49.9 °C. Substitute for $T = (273 + $ temp. °C) and compute the resulting c's.

elevation, m	temperature, °C	c, m/s
0	15.0	340
2 000	2.0	332
10 000	−49.9	299

16.13 Find the stagnation pressure and temperature in nitrogen flowing at 600 fps if the pressure and temperature in the undisturbed flow field are 100 psia and 200 °F, respectively.

$c = \sqrt{kRT} = \sqrt{(1.40)(1774)(460 + 200)} = 1280$ ft/s $\qquad N_M = v/c = \frac{600}{1280} = 0.469$

$\rho_0 = kp_0/c_0^2 = (1.40)[(100)(144)]/1280^2 = 0.0123$ slug/ft^3

$p_s = p_0 + (\rho_0)(v_0^2/2)[1 + (N_M)_0^2/4 + \cdots] = (100)(144) + (0.0123)(600^2/2)(1 + 0.469^2/4)$

$= 16\,736$ lb/ft^2 or 116 lb/in^2

$(c_p/g)(T_1) + V_1^2/2g = (c_p/g)(T_s)$

$(6210/32.2)(460 + 200) + 600^2/[(2)(32.3)] = (6210/32.2)(T_s)$ $\qquad T_s = 689$ °R or 229 °F

16.14 Find the stagnation pressure and temperature in air flowing at 88 fps if the pressure and temperature in the undisturbed flow field are 14.7 psia and 50 °F, respectively.

$c = \sqrt{kRT} = \sqrt{(1.40)(1716)(460 + 50)} = 1107$ ft/s $\qquad N_M = v/c \qquad (N_M)_0 = \frac{88}{1107} = 0.0795$

$(c_p/g)(T_0) + V_0^2/2g = (c_p/g)(T_s)$

$(6000/32.2)(460 + 50) + 88^2/[(2)(32.2)] = (6000/32.2)(T_s)$ $\qquad T_s = 511$ °R or 51 °F

$p_s = p_0 + (\rho_0)(v_0^2/2)[1 + (N_M)_0^2/4 + \cdots]$ $\qquad \rho_0 = p_0/RT_0 = (14.7)(144)/[(1716)(460 + 50)] = 0.00242$ slug/ft^3

$p_s = 14.7 + [(0.00242)(88^2/2)/144](1 + 0.0795^2/4) = 14.77$ psia

16.15 Air flows past an object at 600 fps. Determine the stagnation pressures and temperatures in the standard atmosphere at elevations of sea level, 5000 ft, and 30 000 ft.

From Table A-7: at sea level, 59 °F, $\rho = 0.00238$ slug/ft^3; at 5000 ft, 41.2 °F, $\rho = 0.00205$ slug/ft^3; at 30 000, −47.8 °F, $\rho = 0.00089$ slug/ft^3.

$c = \sqrt{kRT}$ $\qquad [R = 1716$ lb-ft/(slug-°R)] $\qquad N_M = v_0/c$ $\qquad$ (where $v_0 = 600$ fps)

$T_s = T_0 + (V_0^2/2c_p)$ $\qquad$ [where $c_p = 6000$ lb-ft/(slug-°R)]

$p_s = p_0 + [(\rho_0)(v_0^2/2)/144][1 + (N_M)_0^2/4 + \cdots]$

elevation, ft	p_0, psi	c, fps	N_M	T_s, °F	p_s, psi
0	14.70	1 116	0.538	89.0	17.89
5 000	12.24	1 097	0.547	71.2	14.99
30 000	4.37	995	0.603	−17.8	5.58

16.16 Repeat Prob. 16.15 for an air speed of 200 m/s and elevation of sea level, 2000 m, and 10 000 m.

From Table A-8: at 2000 m, 2 °C, $p = 79.50$ kN/m^2, $\rho = 1.007$ kg/m^3.

$c = \sqrt{kRT} = \sqrt{(1.40)(287)(2 + 273)} = 332$ m/s $\qquad N_M = v_0/c = \frac{200}{332} = 0.602$

$T_s = T_0 + (V_0^2/2c_p) = 2 + 200^2/[(2)(1003)] = 22$ °C

$p_s = p_0 + (\rho_0)(v_0^2/2)[1 + (N_M)_0^2/4 + \cdots] = 79.50 + (1.007)(200^2/2)(1 + 0.602^2/4)/(1000) = 101.5$ kN/m^2 abs

Similarly, at 10 000 m, $T_s = -29.9$ °C and $p_s = 35.6$ kN/m^2 abs. At sea level, $T_s = 34.9$ °C and $p_s = 106.1$ kN/m^2 abs.

16.17 Air at 250 psia is moving at 500 fps in a high-pressure wind tunnel at a temperature of 100 °F. Find the stagnation pressure and temperature. Note the magnitude of the sonic velocity for the 250-psia 100 °F air.

■ $\quad\quad c = \sqrt{kRT} = \sqrt{(1.40)(1716)(460 + 100)} = 1160 \text{ ft/s} \quad\quad N_M = v/c = \frac{500}{1160} = 0.431$

Since $[v_0 = 500 \text{ ft/s}] < [c = 1160 \text{ ft/s}]$,

$$p_s/p_0 = [1 + (v_0/c_0)^2(k-1)/2]^{k/(k-1)} \quad\quad p_s/250 = [1 + (\tfrac{500}{1160})^2(1.40 - 1)/2]^{1.40/(1.40-1)}$$

$$p_s = 284 \text{ psia} \quad\quad c_p T_0 + v_0^2/2g = c_p T_s \quad\quad (6000/32.2)(460 + 100) + 500^2/[(2)(32.2)] = (6000/32.2)(T_s)$$

$$T_s = 581 \,°\text{R} \quad \text{or} \quad 121\,°\text{F}$$

16.18 Show that the equation $p_s = p_0 + (\rho_0)(v_0^2/2)[1 + (N_M)_0^2/4 + \cdots]$ results from the binomial expansion of the equation $p_s/p_0 = [1 + (v_0/c)^2(k-1)/2]^{k/(k-1)}$.

■ $p_s/p_0 = \{1 + (v_0/c)^2[(k-1)/2]\}^{k/(k-1)}$; substitute $k = \rho_0 c_0^2/p_0$:

$$\frac{p_s}{p_0} = \left[1 + N_M^2\left(\frac{\rho_0 c_0^2 - p_0}{2p_0}\right)\right]^{\rho_0 c_0^2/(\rho_0 c_0^2 - p_0)}$$

Expanding by the binomial theorem: $(a + b)^n = a^n + (na^{n-1})(b) + [n(n-1)/2!]a^{n-1}b^2 + \cdots$ and then simplifying leads to $p_s = p_0 + (\rho_0 v_0^2/2)[1 + (N_M^2/4) + \cdots]$ Q.E.D.

16.19 Compute the value of R from the values of k and c_p for air.

■ $\quad\quad\quad\quad R = [(k-1)/k](c_p) = [(1.40 - 1)/1.40](6000) = 1714 \text{ ft-lb/(slug-°R)}$

16.20 Compute the enthalpy change in 5 kg of oxygen when the initial conditions are $p_1 = 130 \text{ kPa abs}$ and $T_1 = 10\,°\text{C}$, and the final conditions are $p_2 = 500 \text{ kPa abs}$ and $T_2 = 95\,°\text{C}$ $[c_p = 0.917 \text{ J/(kg-K)}]$.

■ $h_2 = h_1 = (c_p)(T_2 - T_1) \quad\quad H_2 - H_1 = (5)(c_p)(T_2 - T_1) = (5)(0.917)[(273 + 95) - (273 + 10)] = 390 \text{ kJ}$

16.21 A cylinder containing 2 kg nitrogen at 0.14 MPa abs and 5 °C is compressed isentropically to 0.30 MPa abs. Find the final temperature and the work required $[c_v = 0.741 \text{ kJ/(kg-K)}]$.

■ $T_2 = (T_1)(p_2/p_1)^{(k-1)/k} = (273 + 5)(0.30/0.14)^{(1.40-1)/1.40} = 346 \text{ K}$, or 73 °C. From the principle of conservation of energy, the work done on the gas must equal its increase in internal energy, since there is no heat transfer in an isentropic process; that is, $u_2 - u_1 = (c_v)(T_2 - T_1) = $ work per kilogram: Work $= (2)(0.741)[346 - (273 + 5)] = $ 101 kJ.

16.22 If 3.0 slugs of air are involved in a reversible polytropic process in which the initial conditions $p_1 = 12 \text{ psia}$ and $T_1 = 60\,°\text{F}$ change to $p_2 = 20 \text{ psia}$ and volume $V = 1011 \text{ ft}^3$, determine the (*a*) formula for the process and (*b*) work done on the air.

■ (*a*) $\rho_1 = p_1/RT_1 = (12)(144)/(53.3)(32.17)(460 + 60) = 0.00194 \text{ slug/ft}^3$. R was converted to foot-pounds per slug and degree Rankine by multiplying by 32.17. Also, $\rho_2 = \frac{3}{1011} = 0.002967 \text{ slug/ft}^3$; $p_1/\rho_1^n = p_2/\rho_2^n$, $n = [\ln(p_2/p_1)]/[\ln(\rho_2/\rho_1)] = [\ln(\frac{20}{12})]/[\ln(0.002967/0.00194)] = 1.20$; hence $p/\rho^{1.2} = $ const describes the polytropic process.

(*b*) Work of expansion is

$$W = \int_{V_1}^{V_2} p \, dV$$

This is the work done by the gas on its surroundings. Since $p_1 V_1^n = p_2 V_2^n = pV^n$, by substituting into the integral,

$$W = p_1 V_1^n \int_{V_1}^{V_2} \frac{dV}{V^n} = \frac{p_2 V_2 - p_1 V_1}{1 - n} = \frac{mR}{1 - n}(T_2 - T_1)$$

if m is the mass of gas. $V_2 = 1011 \text{ ft}^3$ and $V_1 = V_2(p_2/p_1)^{1/n} = (1011)(\frac{20}{12})^{1/1.2} = 1547 \text{ ft}^3$. Then $W = [(20)(144)(1011) - (12)(144)(1548)]/(1 - 1.2) = -1\,184\,000 \text{ ft-lb}$. Hence, the work done on the gas is $1\,184\,000$ ft-lb.

16.23 For Prob. 16.22, find the (*a*) amount of heat transfer and (*b*) entropy change.

■ (*a*) From the first law of thermodynamics the heat added minus the work done by the gas must equal the increase in internal energy; that is, $Q_H - W = U_2 - U_1 = c_v m(T_2 - T_1)$. First $T_2 = p_2/\rho_2 R = (20)(144)/(0.002965)(53.3)(32.17) = 566\,°\text{R}$. Then $Q_H = -(1\,184\,000/778) + (0.17)(32.17)[3(566 - 520)] = -761 \text{ Btu}$ and 761 Btu was transferred from the mass of air.

(b) $s_2 - s_1 = c_v \ln [(p_2/p_1)(\rho_1/\rho_2)^k]$ $s_2 - s_1 = (0.171) \ln [(\frac{20}{12})(0.00194/0.002967)^{1.4}] = -0.01436$ Btu/(lbm-°R)

and $S_2 - S_1 = -0.01436(3)(32.17) = -1.386$ Btu/°R

A rough check on the heat transfer can be made using an average temperature $T = (520 + 566)/2 = 543$ and by remembering that the losses are zero in a reversible process: $Q_H = T(S_2 - S_1) = (543)(-1.386) = -753$ Btu.

16.24   Carbon tetrachloride has a bulk modulus of elasticity of 1.124 GPa and a density of 1593 kg/m³. What is the speed of sound in the medium?

$$c = \sqrt{K/\rho} = \sqrt{(1.124 \times 10^9)/1593} = 840 \text{ m/s}$$

16.25   What is the speed of sound in dry air at sea level when $T = 68$ °F and in the stratosphere when $T = -67$ °F?

$c = \sqrt{kRT}$. At sea level, $c = \sqrt{(1.40)(1716)(460 + 68)} = 1126$ ft/s. In the stratosphere, $c = \sqrt{(1.40)(1716)(460 - 67)} = 972$ ft/s.

16.26   If 3 kg of a perfect gas, molecular weight 36, had its temperature increased 2 K when 6.4 kJ of work was done on it in an insulated constant-volume chamber, determine c_v and c_p.

$$c_v = \Delta u/\Delta t = 6.4/[(4.187)(2)(3)] = 0.255 \text{ kcal/(kg-K)} \quad (4.187 \text{ kJ} = 1 \text{ kcal}) \quad c_p = c_v + R$$
$$R = 8312/M = 8312/36 = 230.9 \text{ N-m/(kg-K)} = (230.9/4.187)/1000 = 0.055 \text{ kcal/(kg-K)}$$
$$c_p = 0.255 + 0.055 = 0.310 \text{ kcal/(kg-K)}$$

16.27   A gas of molecular weight 48 has $c_p = 1.558$ kJ/(kg-K). What is c_v for this gas?

$$R = (\tfrac{8312}{48})/1000 = 0.173 \text{ kJ/(kg-K)} \quad c_v = c_p - R = 1.558 - 0.173 = 1.385 \text{ kJ/(kg-K)}$$

16.28   Calculate the specific heat ratio k for Probs. 16.26 and 16.27.

$k = c_p/c_v$. For Prob. 16.26: $k = 0.310/0.255 = 1.216$. For Prob. 16.27: $k = 1.558/1.385 = 1.125$.

16.29   The enthalpy of a gas is increased by 0.4 Btu/(lbm-°R) when heat is added at constant pressure and the internal energy is increased by 0.3 Btu/(lbm-°R) when the volume is maintained constant and heat is added. Calculate the molecular weight.

$$R = 1545/M = c_p - c_v \quad c_p = 0.4 \text{ Btu/(lbm-°R)} \quad c_v = 0.3 \text{ Btu/(lbm-°R)}$$
$$R = (0.4 - 0.3)(778) = 77.8 \text{ ft-lb/(lbm-°R)} \quad 77.8 = 1545/M \quad M = 19.86$$

16.30   Calculate the enthalpy change of 2 kg carbon monoxide from $p_1 = 16$ kPa abs and $T_2 = 5$ °C to $p_2 = 30$ kPa abs and $T_2 = 170$ °C.

$$H = mc_p \Delta T = (2)(0.249)(170 - 5) = 82.2 \text{ kcal}$$

16.31   Calculate the entropy change in Prob. 16.30.

$$\Delta S = mc_v \ln [(T_2/T_1)^k(p_2/p_1)^{1-k}] \quad T_2 = 170 + 273 = 443 \text{ K} \quad T_1 = 5 + 273 = 278 \text{ K}$$
$$\Delta S = (2)(0.178)\{\ln [(\tfrac{443}{278})^{1.40}(\tfrac{30}{16})^{1-1.40}]\} = 0.143 \text{ kcal/K}$$

16.32   In an isentropic process, 1 kg of oxygen with a volume of 150 L at 20 °C has its absolute pressure doubled. What is the final temperature?

$$T_2/T_1 = (p_2/p_1)^{(k-1)/k} \quad T_2/(20 + 273) = (2)^{(1.40-1)/1.40} \quad T_2 = 357 \text{ K} \quad \text{or} \quad 84 \text{ °C}$$

16.33   Work out the expression for density change with temperature for a reversible polytropic process.

$$p/\rho^n = \text{constant} \quad p/\rho T = \text{constant} \quad \rho^{n-1}/T = \text{constant} \quad T_1/T_2 = (\rho_1/\rho_2)^{n-1}$$

16.34   Hydrogen at 60 psia and 30 °F has its temperature increased to 120 °F by a reversible polytropic process with $n = 1.20$. Calculate the final pressure.

$$T_2/T_1 = (p_2/p_1)^{(n-1)/n} \quad (120 + 460)/(30 + 460) = (p_2/60)^{(1.20-1)/1.20} \quad p_2 = 165 \text{ psia}$$

16.35   A gas has a density decrease of 10 percent in a reversible polytropic process when the temperature decreases from 45 °C to 5 °C. Compute the exponent n for the process.

▋ From Prob. 16.33, $T_1/T_2 = (\rho_1/\rho_2)^{n-1}$, $(45 + 273)/(5 + 273) = (1/0.9)^{n-1}$, $\log 1.144 = (n-1)\log 1.111$, $n = 2.28$.

16.36 A projectile moves through water at 80 °F at 2000 fps. What is the Mach number?

▋ $$c = \sqrt{K/\rho} = \sqrt{(322\,000)(144)/1.93} = 4902 \text{ ft/s} \qquad N_M = v/c = \tfrac{2000}{4902} = 0.408.$$

16.37 If an airplane travels at 1350 km/h at sea level, $p = 101$ kPa abs, $T = 20$ °C, and at the same speed in the stratosphere where $T = -55$ °C, how much greater is the Mach number in the latter case?

▋ $(N_M)_2/(N_M)_1 = c_1/c_2 = \sqrt{T_1/T_2} = \sqrt{(273 + 20)/(273 - 55)} = 1.159$. Thus, $(N_M)_2$ is 15.9 percent greater than $(N_M)_1$.

16.38 What is the speed of sound through hydrogen at 80 °F?

▋ $$c = \sqrt{kRT} = \sqrt{(1.40)(24\,649)(80 + 460)} = 4317 \text{ ft/s}$$

16.39 Argon ($M = 39.944$, $k = 1.67$) flows through a tube such that its initial condition is $p_1 = 250$ psia and $\rho_1 = 1.16$ lbm/ft³ and its final condition is $p_2 = 30$ psia and $T_2 = 265$ °F. Estimate the (a) initial temperature, (b) final density, (c) change in enthalpy, and (d) change in entropy.

▋ (a) $$R = 49\,720/39.944 = 1245 \text{ ft}^2/(\text{s}^2\text{-°R}) \qquad \rho_1 = 1.16/32.2 = 0.03602 \text{ slug/ft}^3$$
$$T_1 = p_1/R_1\rho_1 = (250)(144)/[(1245)(0.03602)] = 803 \text{ °R} \quad \text{or} \quad 343 \text{ °F}$$

(b) $$\rho_2 = p_2/RT_2 = (30)(144)/[(1245)(460 + 265)] = 0.00479 \text{ slug/ft}^3$$

(c) $$c_p = kR/(k-1) = (1.67)(1245)/(1.67 - 1) = 3103 \text{ ft}^2/(\text{s}^2\text{-°R})$$
$$h_2 - h_1 = (c_p)(T_2 - T_1) = (3103)[(460 + 265) - 803] = -242\,000 \text{ ft}^2/\text{s}^2$$

(d) $$s_2 - s_1 = c_p \ln(T_2/T_1) - R \ln(p_2/p_1) = (3103)\{\ln[(460 + 265)/803]\}$$
$$- (1245)\{\ln[(30)(144)/(250)(144)]\} = 2323 \text{ ft}^2/(\text{s}^2\text{-°R})$$

16.40 Estimate the speed of sound of carbon monoxide [$k = 1.40$, $R = 297$ N-m/(kg-K)] at 200 kPa and 300 °C.

▋ $$c = \sqrt{kRT} = \sqrt{(1.40)(297)(300 + 273)} = 488 \text{ m/s}$$

16.41 A gas flows adiabatically through a duct. At section 1, $p_1 = 200$ psia, $T_1 = 500$ °F, and $V_1 = 250$ fps, while farther downstream $V_2 = 1100$ fps and $p_2 = 40$ psia. Calculate T_2 and $s_2 - s_1$ if the gas is air [$c_p = 6010$ ft²/(s²-°R)].

▋ $$c_p T_1 + v_1^2/2 = c_p T_2 + v_2^2/2 \qquad (6010)(500) + 250^2/2 = 6010 T_2 + 1100^2/2 \qquad T_2 = 405 \text{ °F}$$
$$s_2 - s_1 = c_p \ln(T_2/T_1) - R \ln(p_2/p_1) = (6010)\{\ln[(405 + 460)/(500 + 460)]\} - (1716)[\ln(\tfrac{40}{200})] = 2136 \text{ ft}^2/(\text{s}^2\text{-°R})$$

16.42 Rework Prob. 16.41 if the gas is argon [$c_p = 3103$ ft²/(s²-°R), $k = 1.67$, $R = 1245$ ft²/(s²-°R)].

▋ $$c_p T_1 + v_1^2/2 = c_p T_2 + v_2^2/2 \qquad (3103)(500) + 250^2/2 = 3103 T_2 + 1100^2/2 \qquad T_2 = 315 \text{ °F}$$
$$s_2 - s_1 = c_p \ln(T_2/T_1) - R \ln(p_2/p_1) = (3103)\{\ln[(315 + 460)/(500 + 460)]\} - (1245)[\ln(\tfrac{40}{200})] = 1339 \text{ ft}^2/(\text{s}^2\text{-°R})$$

16.43 Solve Prob. 16.41 if the gas is steam. Assume an ideal gas, with $M = 18.02$ and $k = 1.33$.

▋ $$c_p T_1 + v_1^2/2 = c_p T_2 + v_2^2/2 \qquad R = 49\,720/M = 49\,720/18.02 = 2759 \text{ ft}^2/(\text{s}^2\text{-°R})$$
$$c_p = kR/(k-1) = (1.33)(2759)/(1.33 - 1) = 11\,120 \text{ ft}^2/(\text{s}^2\text{-°R})$$
$$(11\,120)(500) + 250^2/2 = 11\,120 T_2 + 1100^2/2 \qquad T_2 = 448 \text{ °F}$$
$$s_2 - s_1 = c_p \ln(T_2/T_1) - R \ln(p_2/p_1) = (11\,120)\{\ln[(448 + 460)/(500 + 460)]\} - (2759)[\ln(\tfrac{40}{200})] = 3821 \text{ ft}^2/(\text{s}^2\text{-°R})$$

16.44 Solve Prob. 16.41 if the gas is steam. Assume a real gas and use the steam tables.

▋ $h_1 + v_1^2/2 = h_2 + v_2^2/2$. With $p_1 = 200$ psi and $T_1 = 500$ °F, $h_1 = 1269.0$ Btu/lbm and $s_1 = 1.6242$ Btu/(lbm-°F) (from the steam tables). $1269.0 + 250^2/[(2)(32.2)(778)] = h_2 + 1100^2/[(2)(32.2)(778)]$, $h_2 = 1246.1$ Btu/lbm. With $p_2 = 40$ psi and $h_2 = 1246.1$ Btu/lbm, $T_2 = 420$ °F and $s_2 = 1.7720$ Btu/(lbm-°F) (from the steam tables). $s_2 - s_1 = 1.7720 - 1.6242 = 0.148$ Btu/(lbm-°F).

16.45 If 6 kg of oxygen in a closed tank at 250 kPa is heated from 100 °C to 350 °C, calculate the new pressure, the heat added, and the change in entropy.

$$p_2/p_1 = T_2/T_1 \qquad p_2/250 = (350 + 273)/(100 + 273) \qquad p_2 = 418 \text{ kPa} \qquad Q = mc_v \, \Delta T$$

$$c_p = kR/(k-1) = (1.40)(260)/(1.40-1) = 910 \text{ m}^2/(\text{s}^2\text{-K}) \qquad c_v = c_p/k = 910/1.40 = 650 \text{ m}^2/(\text{s}^2\text{-K})$$

$$Q = (6)(650)(350 - 100) = 975\,000 \text{ J}$$

$$s_2 - s_1 = mc_v \ln (T_2/T_1) = (6)(650)\{\ln [(350 + 273)/(100 + 273)]\} = 2001 \text{ J/K}$$

16.46   Steam at 400 °F and 60 psia is compressed isentropically to 100 psia. What is the new temperature? Assume an ideal gas with $k = 1.33$.

$$T_2/T_1 = (p_2/p_1)^{(k-1)/k} \qquad T_2/(400 + 460) = \left(\tfrac{100}{60}\right)^{(1.33-1)/1.33} \qquad T_2 = 976 \text{ °R} \quad \text{or} \quad 516 \text{ °F}$$

16.47   Solve Prob. 16.46 assuming a real gas and using the steam tables.

With $p_1 = 60$ psia and $T_1 = 400$ °F, $s_1 = 1.7134$ Btu/(lbm-°F) (from the steam tables). With $p_2 = 100$ psia and $s_2 = s_1 = 1.7134$ Btu/(lbm-°F), $T_2 = 509$ °F (from the steam tables).

16.48   Carbon dioxide ($k = 1.30$, $M = 44.01$) enters a constant-area duct at 400 °F, 100 psia, and 500 fps. Farther downstream the properties are $V_2 = 1000$ fps and $T_2 = 900$ °F. Compute (a) p_2, (b) the heat added between sections, (c) the entropy change between sections, and (d) the mass flow per unit area.

(a) $\qquad p_2/p_1 = (T_2/T_1)(V_1/V_2) \qquad p_2/100 = [(900 + 460)/(400 + 460)]\left(\tfrac{500}{1000}\right) \qquad p_2 = 79.1 \text{ psia}$

(b) $\quad q = (c_p)(T_2 - T_1) + (v_2^2 - v_1^2)/2 \qquad c_p = kR/(k-1) \qquad R = 49\,720/M = 49\,720/44.01 = 1130 \text{ ft}^2/(\text{s}^2\text{-°R})$

$$c_p = (1.30)(1130)/(1.30 - 1) = 4897 \text{ ft}^2/(\text{s}^2\text{-°R})$$

$$q = (4897)(900 - 400) + (1000^2 - 500^2)/2 = 2.834 \times 10^6 \text{ ft-lb/slug} = 2.824 \times 10^6/[(32.2)(778)] = 113 \text{ Btu/lbm}$$

(c) $\quad s_2 - s_1 = c_p \ln (T_2/T_1) - R \ln (p_2/p_1) = (4897)\{\ln [(900 + 460)/(400 + 460)]\} - (1130)[\ln (79.1/100)]$

$$= 2509 \text{ ft-lb/(slug-°R)}$$

(d) $\qquad \dot{m} = \rho_1 A_1 v_1 \qquad \rho_1 = p_1/RT_1 = (100)(144)/[(1130)(400 + 460)] = 0.01482 \text{ slug/ft}^3$

$$\dot{m}/A_1 = (0.01482)(500) = 7.41 \text{ (slugs/s)/ft}^2$$

16.49   Steam enters a duct at $p_1 = 50$ psia, $T_1 = 360$ °F, $V_1 = 200$ fps, and leaves at $p_2 = 100$ psia, $T_2 = 800$ °F, and $V_2 = 1200$ fps. How much heat was added?

$q = (h_2 - h_1) + (v_2^2 - v_1^2)/2$. From the steam tables, $h_1 = 1214.9$ Btu/lbm and $h_2 = 1429.7$ Btu/lbm.
$q = (1429.7 - 1214.9) + [(1200^2 - 200^2)/2]/[(32.2)(778)] = 243$ Btu/lbm.

16.50   The internal energy of a hypothetical perfect gas is given as $u = T^{1/2}/50 + 100$. Determine c_v and c_p. Take $R = 50$ ft-lb/(lbm-°F).

$$c_v = du/dt = (\tfrac{1}{2})(T^{-1/2})/50 = 0.0100T^{-1/2} \text{ Btu/(lbm-°R)}$$

$$c_p = R + c_v = \tfrac{50}{778} + 0.0100T^{-1/2} = 0.0643 + 0.0100T^{-1/2} \text{ Btu/(lbm-°R)}$$

16.51   Air at 15 °C and 101 325 Pa is compressed to a pressure of 345 000 Pa. If the compression is adiabatic and reversible, what is the final specific volume? How much work is done per kilogram of the gas?

$$p_1/p_2 = [(V_s)_2/(V_s)_1]^k \qquad V_s = RT/p \qquad (V_s)_1 = (287)(15 + 273)/101\,325 = 0.8158 \text{ m}^3/\text{kg}$$

$$101\,325/345\,000 = [(V_s)_2/(0.8158)]^{1.40} \qquad (V_s)_2 = 0.340 \text{ m}^3/\text{kg} \qquad W = \int p \, dv$$

Since $p(V_s)^k = \text{constant}$,

$$W = \int_{0.8158}^{0.340} \frac{\text{constant}}{(V_s)^k} \, dv = (\text{constant})\left[\frac{1/(V_s)^{k-1}}{1-k}\right]_{0.8158}^{0.340}$$

$$= (\text{constant})\left[\frac{(1/0.340)^{1.40-1} - (1/0.8158)^{1.40-1}}{1 - 1.40}\right] = (-1.137)(\text{constant})$$

$$\text{Constant} = (101\,325)(0.8158)^{1.40} = 76\,196 \qquad W = (-1.137)(76\,196) = -86\,600 \text{ N-m/kg}$$

16.52   Do Prob. 16.51 for an isothermal compression.

▮ $(p_1)(V_s)_1 = (p_2)(V_s)_2$. Using data from Prob. 16.51, $(101\,325)(0.8158) = (345\,000)(V_s)_2$, $(V_s)_2 = 0.240\,\text{m}^3/\text{kg}$.

$$W = \int p\,dv = \text{constant}\int_{0.8158}^{0.240}\frac{dV_s}{V_s} = \text{constant}\,[\ln V_s]_{0.8158}^{0.240} = (\text{constant})\left(\ln\frac{0.240}{0.8158}\right) = (\text{constant})(-1.224)$$

$$\text{Constant} = (101\,325)(0.8158) = 82\,661 \qquad W = (82\,661)(-1.224) = -101\,000\,\text{N-m/kg}$$

16.53 An airplane is capable of attaining a flight Mach number of 0.8. When it is flying at an altitude of 1000 ft in standard atmosphere, what is the ground speed if the air is not moving relative to the ground? What is the ground speed if the plane is at an altitude of 35 000 ft in standard atmosphere?

▮ At altitude 1000 ft, $c = 1113\,\text{ft/s}$, $v_{\text{plane}} = (0.8)(1113) = 890\,\text{ft/s}$. At altitude 35 000 ft, $c = 973\,\text{ft/s}$, $v_{\text{plane}} = (0.8)(973) = 778\,\text{ft/s}$.

16.54 Do the first part of Prob. 16.53 if the air is moving at 60 mph directly opposite to the direction of flight.

▮ $\qquad\qquad\qquad\qquad v_{\text{plane}} = 890 - (60)(5280)/3600 = 802\,\text{ft/s}$

16.55 What is the value of k for standard atmosphere at an altitude of 30 000 ft?

▮ $\qquad\qquad c = \sqrt{kp/\rho} \qquad 995 = \sqrt{(k)(628)/(0.374)(0.00238)} \qquad k = 1.40$

16.56 Suppose that a plane is moving horizontally relative to the ground at a speed of twice the velocity of sound (340.5 m/s) and that the air is moving in the opposite direction at a speed of one-half the velocity of sound relative to the ground. What is the Mach angle?

▮ Relative to the ground: $V_{\text{source}} = (2)(340.5) = 681.0\,\text{m/s}$, $V_{\text{air}} = 340.5/2 = 170.2\,\text{m/s}$. Velocity of source relative to the air is $681.0 + 170.2 = 851.2\,\text{m/s}$. Hence, $N_M = 851.2/340.5 = 2.500$, $\sin\alpha = 1/2.500$, $\alpha = 23.6°$.

16.57 Suppose that a cruise missile under test is moving horizontally at $N_m = 2$ in the atmosphere at an elevation of 305 m above the earth's surface. How long does it take for an observer to hear the disturbance from the instant when it is directly overhead? Assume standard atmosphere.

▮ See Fig. 16-2. $\sin\alpha = 1/2$, $\alpha = 30.0°$; $\tan 30° = 305/d$, $d = 528.3\,\text{m}$. Speed of shell is $(2)(340) = 680\,\text{m/s}$. $t = d/V = 528.3/680 = 0.777\,\text{s}$.

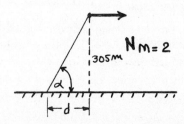

Fig. 16-2

16.58 Suppose in Prob. 16.57 that an observer in a plane is moving in the same direction as the missile at a speed of one-half the speed of sound at an elevation of 305 m above the missile. What is the time elapsed between the instant when the missile is directly below and the instant when the observer hears the sound? Neglect change of c from 305 m to 610 m elevation.

▮ See Fig. 16-3. Using data from Prob. 16.57, $680t = 528.3 + \left(\frac{680}{4}\right)(t)$, $t = 1.04\,\text{s}$.

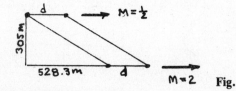

Fig. 16-3

16.59 A certain aircraft flies at the same Mach number regardless of its altitude. It flies 100 km/h slower at 10 000 m standard altitude than at sea level. What is its Mach number?

▮ $c = \sqrt{kRT}$. At sea level, $c_1 = \sqrt{kRT_1} = 340.3\,\text{m/s}$. At 10 000 m, $c_2 = \sqrt{kRT_2} = 299.4\,\text{m/s}$, $\Delta V = (N_M)(c_2 - c_1)$, $(100)(1000)/3600 = (N_M)(340.3 - 299.4)$, $N_M = 0.679$.

16.60 At 250 °C and 1 atm, compute the speed of sound of (a) air, (b) oxygen, (c) hydrogen, (d) steam, (e) carbon monoxide, and (f) $^{238}UF_6$ ($k = 1.06$).

▌ $$c = \sqrt{kRT}$$

(a) $c = \sqrt{(1.40)(287)(250 + 273)} = 458$ m/s   (b) $c = \sqrt{(1.40)(260)(250 + 273)} = 436$ m/s

(c) $c = \sqrt{(1.41)(4127)(250 + 273)} = 1745$ m/s   (d) $c = \sqrt{(1.33)(461)(250 + 273)} = 566$ m/s

(e) $c = \sqrt{(1.40)(297)(250 + 273)} = 466$ m/s   (f) $M = 238 + (6)(19) = 352$

$$R = 8313/M = \tfrac{8313}{352} = 23.62 \text{ N-m}/(\text{kg-K}) \qquad c = \sqrt{(1.06)(23.62)(250 + 273)} = 114 \text{ m/s}$$

16.61 Assuming that water follows the liquid equation of state $K = (n)(B + 1)(p_a)(\rho/\rho_a)^n$ with $n = 7$ and $B = 3000$, compute the bulk modulus and speed of sound at (a) 1 atm and (b) 1100 atm (the deepest part of the ocean).

▌ (a) At 1 atm (101 350 Pa), $\rho = \rho_a$, $K = (7)(3000 + 1)(101\,350)(1)^7 = 2.13 \times 10^9$ Pa, $c = \sqrt{K/\rho}$
$= \sqrt{(2.13 \times 10^9)/1025} = 1442$ m/s.

(b) At 1100 atm, $\rho/\rho_a = [(1100 + 3000)/3001]^{1/7} = 1.046$, $K = (K_{atm})(1.046)^7 = (2.13 \times 10^9)(1.046)^7$
$= 2.92 \times 10^9$ Pa, $c = \sqrt{(2.92 \times 10^9)/[(1025)(1.046)]} = 1650$ m/s.

16.62 The measured value of the speed of sound of water at 20 °C and 9000 atm is 2650 m/s.* Compare this with the value computed by the analysis of Prob. 16.61.

▌ $c = \sqrt{K/\rho}$, $K = (n)(B + 1)(p_a)(\rho/\rho_a)^n$, $n = 7$, $B = 3000$, $\rho_a = 998$ kg/m³. From Prob. 16.61, 9000 = $(3001)(\rho/998)^7 - 3000$, $\rho = 1216.5$ kg/m³; $K = (7)(3000 + 1)(101\,350)(1216.5/998)^7 = 8.513 \times 10^9$ Pa, $c = \sqrt{(8.513 \times 10^9)/1216.5} = 2645$ m/s. This is $(2650 - 2645)/2650 = 0.0019$, or 0.19 percent less than the measured value.

16.63 Mercury at 1 atm has a bulk modulus of about 2.8×10^{10} Pa. It has also $n = 7$ in the equation of state of Prob. 16.61. What value of B in that equation best fits the measured bulk modulus? Estimate the bulk modulus and speed of sound of mercury at 2000 atm.

▌ From Prob. 16.61, $K_a = (n)(B + 1)(p_a) = 2.8 \times 10^{10} = (7)(B + 1)(101\,350)$, $B = 39\,466$. At 2000 atm,

$$p/p_a = 2000 = (39\,466 + 1)(\rho/\rho_a)^7 - 39\,466 \qquad \rho/\rho_a = 1.007$$

$$K = (K_a)(\rho/\rho_a)^7 = (2.8 \times 10^{10})(1.007)^7 = 2.94 \times 10^{10} \text{ Pa}$$

$$\rho = 1.007\rho_a = (1.007)(13\,550) = 13\,645 \text{ kg/m}^3 \qquad c = \sqrt{K/\rho} = \sqrt{(2.94 \times 10^{10})/13\,645} = 1468 \text{ m/s}$$

16.64 Why do (a) water and mercury and (b) aluminum and steel have nearly equal speeds of sound in spite of the fact that the second material of each pair is much heavier than the first? Can this behavior be predicted from molecular theory?

▌ In both cases, the lighter material has a proportionately smaller bulk modulus.

16.65 An airplane flies at 350 m/s through air at −20 °C and 40 kPa. Is the airplane supersonic?

▌ $c = \sqrt{kRT} = \sqrt{(1.40)(287)(273 - 20)} = 319$ m/s, $N_M = v/c = \tfrac{350}{319} = 1.1$. Since $N_M > 1.0$, it is supersonic.

16.66 A weak pressure wave (sound wave) with a pressure change $\Delta p = 40$ Pa propagates through air at 20 °C and 1 atm. Estimate the (a) velocity change across the wave, (b) density change, and (c) temperature change.

▌ (a) $\Delta p = \rho c\, \Delta v$ $40 = (1.205)(343)(\Delta v)$ $\Delta v = 0.0968$ m/s

(b) $\Delta \rho = (\rho + \Delta \rho)(\Delta v/c) = (1.205 + \Delta \rho)(0.0968/343) = 0.0000340$ kg/m³

(c) Approximately isentropic:

$$(T + \Delta T)/T = [(p + \Delta p)/p]^{(k-1)/k}$$

$$(273 + 20 + \Delta T)/(273 + 20) = [(101\,350 + 40)/101\,350]^{(1.40-1)/1.40} \qquad \Delta T = 0.033 \text{ °C}$$

16.67 Air at 65 °F flows isothermally through a 6-in-diameter pipe. The pressure at one section is 82 psia, and that at a section 550 ft downstream is 65 psia. If the pipe surface is "smooth," find the weight flow rate of the air.

* A. H. Smith and A. W. Larson, *J. Chem. Phys.*, vol. 22, p. 351, 1954.

$$p_1^2 - p_2^2 = (G^2RT/g^2A^2)[fL/D + 2\ln(p_1/p_2)] \qquad A = (\pi)(\tfrac{6}{12})^2/4 = 0.1963 \text{ ft}^2$$

Assume $f = 0.007$: $(82^2 - 65^2)(144)^2 = (G^2)(53.3)(65 + 460)/[(32.2)^2(0.1963)^2][(0.007)(550)/(\tfrac{6}{12}) + (2)(\ln\tfrac{82}{65})]$, $G = 95.2$ lb/s. This is the weight flow rate of the air if the assumed value of f of 0.007 is correct. This must be checked.

$$N_R = \rho D/\mu \qquad \rho = p/RT = (82)(144)/[(1716)(65 + 460)] = 0.01311 \text{ lb-s}^2/\text{ft}^4 \qquad G = \gamma Av$$

$$\gamma = \rho g = (0.01311)(32.2) = 0.4221 \text{ lb/ft}^3 \qquad 95.2 = (0.4221)(0.1963)(v) \qquad v = 1149 \text{ ft/s}$$

$$N_R = (0.01311)(\tfrac{6}{12})(1149)/(3.78 \times 10^{-7}) = 1.99 \times 10^7$$

$$f = 0.007 \qquad \text{(from Fig. A-5, using the "smooth pipes" line)}$$

Since this value of f is the same as the assumed value, the computed weight flow rate of 95.2 lb/s is taken as the correct value.

16.68   Air at 18 °C flows isothermally through a 300-mm-diameter pipe at a flow rate of 0.450 kN/s. The pressure at one section is 550 kPa abs, and the pipe surface is smooth. Find the pressure at a section 200 m downstream.

$$p_1^2 - p_2^2 = (G^2RT/g^2A^2)[fL/D + 2\ln(p_1/p_2)] \qquad A = (\pi)(\tfrac{300}{1000})^2/4 = 0.07069 \text{ m}^2 \qquad N_R = \rho Dv/\mu$$

$$\gamma = p/RT = 550/[(29.3)(273 + 18)] = 0.06451 \text{ kN/m}^3 \qquad \rho = \gamma/g = 0.06451/9.807 = 0.006578 \text{ kN-s}^2/\text{m}^4$$

$$G = \gamma Av \qquad 0.450 = (0.06451)(0.07069)(v) \qquad v = 98.68 \text{ m/s}$$

$$N_R = (0.006578)(\tfrac{300}{1000})(98.68)/(1.81 \times 10^{-8}) = 1.08 \times 10^7$$

From Fig. A-5, $f = 0.0080$. Substituting into the equation given above and neglecting temporarily the second term inside the brackets, $550^2 - p_2^2 = \{(0.450)^2(29.3)(273 + 18)/[(9.807)^2(0.07069)^2]\}[(0.0080)(200)/(\tfrac{300}{1000})]$, $p_2 = 532$ kPa. Substituting this value of p_2 into the term that was neglected and solving for the p_2 on the left side of the equation, $550^2 - p_2^2 = \{(0.450)^2(29.3)(273 + 18)/[(9.807)^2(0.07069)^2]\}[(0.0080)(200)/(\tfrac{300}{1000}) + (2)(\ln\tfrac{550}{532})]$, $p_2 = 532$ kPa   (O.K.).

16.69   Air at 100 °F flows isothermally through a 4-in-diameter pipe. Pressures at sections 1 and 2 are 120 psia and 80 psia, respectively. Section 2 is located 400 ft downstream from section 1. Determine the weight flow rate of the air. Assume the pipe has a smooth surface.

$$p_1^2 - p_2^2 = (G^2RT/g^2A^2)[fL/D + 2\ln(p_1/p_2)] \qquad A = (\pi)(\tfrac{4}{12})^2/4 = 0.08727 \text{ ft}^2$$

Assume $f = 0.007$: $(120^2 - 80^2)(144)^2 = (G^2)(53.3)(100 + 460)/[(32.2)^2(0.08727)^2][(0.007)(400)/(\tfrac{4}{12}) + (2)(\ln\tfrac{120}{80})]$, $G = 69.0$ lb/s. This is the weight flow rate of the air if the assumed value of f of 0.007 is correct. Further checking (not shown here but following the pattern of Prob. 16.67) indicates that the assumed value of f of 0.007 is correct, and therefore $G = 69.0$ lb/s.

16.70   Air at 85 °F flows isothermally through a 6-in-diameter pipe at a flow rate of 10 lb/s. The pipe surface is very smooth. If the pressure at one section is 70 psia, determine the pressure at a section 600 ft downstream from the first section.

$$p_1^2 - p_2^2 = (G^2RT/g^2A^2)[fL/D + 2\ln(p_1/p_2)] \qquad A = (\pi)(\tfrac{6}{12})^2/4 = 0.1963 \text{ ft}^2 \qquad N_R = \rho Dv/\mu$$

$$\gamma = p/RT = (70)(144)/[(53.3)(460 + 85)] = 0.3470 \text{ lb/ft}^3 \qquad \rho = \gamma/g = 0.3470/32.2 = 0.01078 \text{ slug/ft}^3$$

$$G = \gamma Av \qquad 10 = (0.3470)(0.1963)(v) \qquad v = 146.8 \text{ ft/s} \qquad N_R = (0.01708)(\tfrac{6}{12})(146.8)/(3.78 \times 10^{-7}) = 2.09 \times 10^6$$

From Fig. A-5, $f = 0.0103$. Substituting into the equation given above and neglecting temporarily the second term inside the brackets, $(70^2)(144)^2 - (p_2)^2(144)^2 = (10)^2(53.3)(460 + 85)/[(32.2)^2(0.1963)^2][(0.0103)(600)/(\tfrac{6}{12})]$, $p_2 = 69.69$ lb/in². Substituting this value of p_2 into the term that was neglected and solving for the p_2 on the left side of the equation, $(70^2)(144)^2 - (p_2)^2(144)^2 = \{(10)^2(53.3)(460 + 85)/[(32.2)^2(0.1963)^2]\} \times \{(0.0103)(600)/(\tfrac{6}{12}) + (2)[\ln(70/69.69)]\}$, $p_2 = 69.69$ lb/in²   (O.K.).

16.71   Prepare a computer program to solve for either the weight flow rate or the pressure at a section downstream for isothermal flow of a compressible fluid.

```
C     THIS PROGRAM DETERMINES EITHER THE FLOW RATE OR THE PRESSURE AT A
C     SECOND (DOWNSTREAM) POINT FOR CLOSED CONDUIT, COMPRESSIBLE FLOW.
C     IT IS BASED ON AN ISOTHERMAL FLOW ANALYSIS AND IS THEREFORE SUB-
C     JECT TO THE RESTRICTIONS AND/OR ASSUMPTIONS IMPLICIT IN ISOTHERMAL
C     FLOW.  IT CAN BE USED FOR PROBLEMS IN BOTH THE ENGLISH SYSTEM OF
```

```
C       UNITS AND THE INTERNATIONAL SYSTEM OF UNITS.  THE APPLICATION OF
C       THIS PROGRAM IS LIMITED TO CASES INVOLVING A SINGLE CONDUIT WITH
C       A CONSTANT DIAMETER.
C
C       INPUT DATA MUST BE SET UP AS FOLLOWS.
C
C       CARD 1     COLUMN 1        ENTER 0 (ZERO, OR BLANK IF ENGLISH SYSTEM
C                                  OF UNITS IS TO BE USED.  ENTER 1 (ONE) IF
C                                  INTERNATIONAL SYSTEM OF UNITS IS TO BE USED.
C                  COLUMNS 2-79    ENTER TITLE, DATE, AND OTHER INFORMATION,
C                                  IF DESIRED.
C       CARD 2     COLUMNS 1-10    ENTER NUMBER INCLUDING DECIMAL GIVING
C                                  ABSOLUTE PRESSURE AT UPSTREAM POINT (IN
C                                  POUNDS PER SQUARE INCH OR KILO-
C                                  PASCALS).
C                  COLUMNS 11-20   ENTER NUMBER INCLUDING DECIMAL GIVING
C                                  ABSOLUTE PRESSURE AT DOWNSTREAM POINT (IN
C                                  POUNDS PER SQUARE INCH OR KILO-
C                                  PASCALS).
C                  COLUMNS 21-30   ENTER NUMBER INCLUDING DECIMAL GIVING GAS
C                                  CONSTANT (IN FEET PER DEGREE-RANKINE OR
C                                  METERS PER DEGREE-KELVIN).
C                  COLUMNS 31-40   ENTER NUMBER INCLUDING DECIMAL GIVING
C                                  TEMPERATURE (IN DEGREES-FAHRENHEIT OR
C                                  DEGREES-CELSIUS).
C                  COLUMNS 41-50   ENTER NUMBER INCLUDING DECIMAL GIVING
C                                  DIAMETER OF CONDUIT (IN INCHES OR MILLI-
C                                  METERS).
C                  COLUMNS 51-60   ENTER NUMBER INCLUDING DECIMAL GIVING
C                                  LENGTH OF CONDUIT (IN FEET OR METERS).
C                  COLUMNS 61-70   ENTER NUMBER INCLUDING DECIMAL GIVING
C                                  ROUGHNESS (IN FEET OR METERS).  ENTER A
C                                  VALUE OF 0 (ZERO) FOR "SMOOTH" CONDUITS.
C                  COLUMNS 71-80   ENTER NUMBER INCLUDING DECIMAL GIVING
C                                  VISCOSITY OF FLUID (IN POUNDS-SECONDS
C                                  PER SQUARE FOOT OR KILONEWTONS-SECONDS PER
C                                  SQUARE METER).
C       CARD 3     COLUMNS 1-10    ENTER NUMBER INCLUDING DECIMAL GIVING
C                                  WEIGHT FLOW RATE (IN POUNDS PER
C                                  SECOND OR KILONEWTONS PER SECOND).
C                  COLUMNS 11-34   ENTER TYPE OF FLUID.
C                  COLUMNS 35-58   ENTER TYPE OF CONDUIT.
C
C ********************************************************************
C *                                                                *
C *    NOTE WELL....EITHER THE FLOW RATE (COLUMNS 1-10 ON CARD 3) OR *
C *    THE PRESSURE AT DOWNSTREAM POINT (COLUMNS 11-20 ON CARD 2),  *
C *    WHICHEVER ONE IS TO BE DETERMINED BY THIS PROGRAM, SHOULD BE *
C *    LEFT BLANK.                                                  *
C *                                                                *
C ********************************************************************
C
C       MULTIPLE DATA SETS FOR SOLVING ANY NUMBER OF PROBLEMS MAY BE
C       INCLUDED FOR PROCESSING.
C
        DIMENSION TITLE(13),FLUID(4),PIPE(4)
        COMMON F,ED,RN
        PI=3.14159265
        REAL L
1       READ(5,100,END=2)UNITS,TITLE
100     FORMAT(I1,13A6)
        WRITE(6,105)TITLE
105     FORMAT('1',13A6,//)
        READ(5,101)P1,P2,R,T,D,L,E,VIS,Q,FLUID,PIPE
101     FORMAT(8F10.0,/,F10.0,8A6)
        FACTOR=12.0
        IF(UNITS.EQ.1)FACTOR=1000.0
        PFACT=144.0
        IF(UNITS.EQ.1)PFACT=1.0
        TFACT=460.0
        IF(UNITS.EQ.1)TFACT=273.0
        G=32.2
        IF(UNITS.EQ.1)G=9.807
```

```
                A=PI*D**2/FACTOR**2/4.0
                SW=P1*PFACT/R/(T+TFACT)
                RHO=SW/G
                ED=E/D*FACTOR
                IF(Q.GT.0.0001) GO TO 102
                FF=0.02
     104        Q=SQRT((P1**2-P2**2)*PFACT**2*G*G*A*A/(R*(T+TFACT)*(FF*L/D*FACTOR+
               *2.0*ALOG(P1/P2))))
                V=Q/SW/A
                RN=RHO*D/FACTOR*V/VIS
                CALL ROUGH
                DIFF=ABS(F-FF)
                IF(DIFF.LT.0.0001)GO TO 103
                FF=F
                GO TO 104
     103  IF (UNITS.EQ.0)WRITE(6,106)P1,P2,D,E,R,T,L,FLUID,PIPE,Q
     106  FORMAT(1X,'GIVEN DATA FOR A COMPRESSIBLE FLOW IN A CIRCULAR CLOSED
           * CONDUIT',//5X,'PRESSURE AT POINT 1 =',F8.1,' PSI',//5X,'PRESSURE
           *AT POINT 2 =',F8.1,' PSI',//5X,'DIAMETER OF CONDUIT =',F7.2,
           *' IN',//5X,'ROUGHNESS =',F10.7,' FT',//5X,'GAS CONSTANT =',F7.1,
           *' FT/DEG-R',//5X,'TEMPERATURE =',F7.1,' DEG-F',//5X,'LENGTH OF CON
           *DUIT =',F8.1,' FT',//5X,'FLUID FLOWING IS ',4A6,//5X,'CONDUIT MATE
           *RIAL IS ',4A6,//1X,'THE FLOW RATE WILL BE',F7.1,' LB/S')
                IF (UNITS.EQ.1)WRITE(6,107)P1,P2,D,E,R,T,L,FLUID,PIPE,Q
     107  FORMAT(1X,'GIVEN DATA FOR A COMPRESSIBLE FLOW IN A CIRCULAR CLOSED
           * CONDUIT',//5X,'PRESSURE AT POINT 1 =',F8.1,' KPA',//5X,'PRESSURE
           *AT POINT 2 =',F8.1,' KPA',//5X,'DIAMETER OF CONDUIT =',F7.1,
           *' MM',//5X,'ROUGHNESS =',F10.7,' M ',//5X,'GAS CONSTANT =',F7.1,
           * ' M/DEG-K',//5X,'TEMPERATURE =',F7.1,' DEG-C',//5X,'LENGTH OF CON
           *DUIT =',F8.1,' M ',//5X,'FLUID FLOWING IS ',4A6,//5X,'CONDUIT MATE
           *RIAL IS ',4A6,//1X,'THE FLOW RATE WILL BE',F7.3,' KN/S')
                GO TO 1
     102        V=Q/SW/A
                RN=RHO*D/FACTOR*V/VIS
                CALL ROUGH
                P2=SQRT(P1**2-Q**2*R*(T+TFACT/G**2/A**2*F*L/D*FACTOR/PFACT**2)
     109  P22=SQRT(P1**2-Q**2*R*(T+TFACT)/G**2/A**2*(F*L/D*FACTOR+2.0*
           *ALOG(P1/P2))/PFACT**2)
                DIFF=ABS(P22-P2)
                IF(DIFF.LT.0.01)GO TO 108
                P2=P22
                GO TO 109
     108  IF(UNITS.EQ.0)WRITE(6,111)P1,D,E,R,T,L,Q,FLUID,PIPE,P2
     111  FORMAT(1X,'GIVEN DATA FOR A COMPRESSIBLE FLOW IN A CIRCULAR CLOSED
           * CONDUIT',//5X,'PRESSURE AT POINT 1 =',F8.1,' PSI',//5X,'DIAMETER
           *OF CONDUIT =',F7.2,' IN',//5X,'ROUGHNESS =',F10.7,' FT',//5X,'GAS
           *CONSTANT =',F7.1,' FT/DEG-R',//5X,'TEMPERATURE =',F7.1,' DEG-F',
           *//5X,'LENGTH OF CONDUIT =',F8.1,' FT',//5X,'FLOW RATE =',F7.1,
           *' LB/S ',//5X,'FLUID FLOWING IS ',4A6,//5X,'CONDUIT MATERIAL IS ',
           *4A6,//1X,'THE PRESSURE AT POINT 2 WILL BE',F8.1,' PSI')
                IF(UNITS.EQ.1)WRITE(6,110)P1,D,E,R,T,L,Q,FLUID,PIPE,P2
     110  FORMAT(1X,'GIVEN DATA FOR A COMPRESSIBLE FLOW IN A CIRCULAR CLOSED
           * CONDUIT',//5X,'PRESSURE AT POINT 1 =',F8.1,' KPA',//5X,'DIAMETER
           *OF CONDUIT =',F7.1,' MM',//5X,'ROUGHNESS =',F10.7,' M ',//5X,'GAS
           *CONSTANT =',F7.1,' M/DEG-K ',//5X,'TEMPERATURE =',F7.1,' DEG-C',
           *//5X,'LENGTH OF CONDUIT =',F8.1,' M ',//5X,'FLOW RATE =',F7.3,
           *' KN/S ',//5X,'FLUID FLOWING IS ',4A6,//5X,'CONDUIT MATERIAL IS ',
           *4A6,//1X,'THE PRESSURE AT POINT 2 WILL BE ',F8.1,'KPA')
                GO TO 1
     2          STOP
                END
                SUBROUTINE ROUGH
                COMMON F,ED,RN
                IF(RN.LE.2000.0)F=64.0/RN
                IF(RN.LE.2000.0)RETURN
                IF(RN.LT.4000.0)WRITE(6,103)
     103  FORMAT(1X,'A REYNOLDS NUMBER IS IN THE CRITICAL ZONE, FOR WHICH TH
           *E FRICTION FACTOR IS UNCERTAIN.  HENCE, PROGRAM EXECUTION WAS TERM
           *INATED.')
                IF(RN.LT.4000.0)STOP
                F=0.006
                TRY1=1.0/SQRT(F)+2.0*ALOG10(ED/3.7+2.51/RN/SQRT(F))
```

```
102 F=F+0.00001
    TRY2=1.0/SQRT(F)+2.0*ALOG10(ED/3.7+2.51/RN/SQRT(F))
    IF(TRY1*TRY2)100,100,101
101 TRY1=TRY2
    GO TO 102
100 F=F-0.000005
    RETURN
    END
```

16.72 Solve Prob. 16.67 using the computer program of Prob. 16.71.

❙ **Input**

```
1 2 3 4 5 6 7 8 9 1011121314151617181920212223242526272829303132333435363738394041424344454647484950515253545556575859606162636465666768697071727374757677787980
0SAMPLE ANALYSIS ØF CØMPRESSIBLE FLØW
82.0      65.0      53.3      65.0      6.0      550.0              .000000378
          AIR                      "SMØØTH"
```

Output

```
SAMPLE ANALYSIS OF COMPRESSIBLE FLOW

GIVEN DATA FOR A COMPRESSIBLE FLOW IN A CIRCULAR CLOSED CONDUIT

     PRESSURE AT POINT 1 =    82.0 PSI

     PRESSURE AT POINT 2 =    65.0 PSI

     DIAMETER OF CONDUIT =    6.00 IN

     ROUGHNESS = 0.0000000 FT

     GAS CONSTANT =   53.3 FT/DEG-R

     TEMPERATURE =   65.0 DEG-F

     LENGTH OF CONDUIT =   550.0 FT

     FLUID FLOWING IS AIR

     CONDUIT MATERIAL IS "SMOOTH"

THE FLOW RATE WILL BE   95.2 LB/S
```

16.73 Solve Prob. 16.68 using the computer program of Prob. 16.71.

❙ **Input**

```
1 2 3 4 5 6 7 8 9 1011121314151617181920212223242526272829303132333435363738394041424344454647484950515253545556575859606162636465666768697071727374757677787980
1SAMPLE ANALYSIS ØF CØMPRESSIBLE FLØW
550.0               29.3      18.0      300.0      200.0              .000000018
0.450     AIR                      "SMØØTH"
```

Output

```
SAMPLE ANALYSIS OF COMPRESSIBLE FLOW

GIVEN DATA FOR A COMPRESSIBLE FLOW IN A CIRCULAR CLOSED CONDUIT

     PRESSURE AT POINT 1 =   550.0 KPA

     DIAMETER OF CONDUIT =   300.0 MM

     ROUGHNESS = 0.0000000 M

     GAS CONSTANT =   29.3 M/DEG-K

     TEMPERATURE =   18.0 DEG-C

     LENGTH OF CONDUIT =   200.0 M
```

```
FLOW RATE =  0.450 KN/S

FLUID FLOWING IS AIR

CONDUIT MATERIAL IS "SMOOTH"

THE PRESSURE AT POINT 2 WILL BE   532.1 KPA
```

16.74 Air flows isothermally at 65 °F through a horizontal 10- by 14-in rectangular duct at 100 lb/s. If the pressure at a section is 80 psia, find the pressure at a second section 500 ft downstream from the first. Assume the duct surface is very smooth; hence the lowest curve of Fig. A-5 may be used to determine f.

∎ $(N_M)_1^2/(N_M)_2^2 = 1 - k(N_M)_1^2\{2 \ln [(N_M)_2/(N_M)_1] + f(L/D)\}$ $R_h = A/p_w = \frac{140}{48} = 2.92$ in $= 0.243$ ft

$R_1 = DV\rho/\mu = GD/\mu gA = G(4R)/\mu gA = [100(4 \times 0.243)]/[(3.78 \times 10^{-7})(32.2)\frac{140}{144}] = 8.2 \times 10^6$

$R_2 = R_1$ since $\rho_1 V_1 = \rho_2 V_2$ and $\mu_1 = \mu_2$. From Fig. A-5, $f = 0.0083$.

$\gamma_1 = pg/RT = [(80 \times 144)32.2]/[1715(460 + 65)] = 0.41$ lb/ft^3 $V_1 = G/\gamma_1 A = 100/[(0.41)\frac{140}{144}] = 250$ fps

$c = \sqrt{kRT} = \sqrt{(1.4)(1715)(525)} = 1123$ fps $(N_M)_1 = V_1/c = \frac{250}{1123} = 0.222$

The limiting value of $(N_M)_1$ is $1/\sqrt{1.4} = 0.845$. Therefore,

$(0.222)^2/(0.845)^2 = 1 - 1.4(0.222)^2\{2 \ln (0.845/0.222) + 0.0083[L/4(0.243)]\}$ $L = 1260$ ft

Thus the equation $p_1^2 - p_2^2 = (G^2RT/g^2A^2)[f(L/D) + 2 \ln (p_1/p_2)]$ applies for all values of $L < 1260$ ft. Substituting $L = 500$ ft and neglecting the usually small logarithmic term, $[(80)(144)]^2 - p_2^2 = [(100)^2(1715)(460 + 65)/(32.2)^2(\frac{140}{144})^2]\{(0.0083)[500/4(0.243)]\}$ from which $p_2 = 67.1$ psia. Substituting this value of p_2 and considering the logarithmic term yields $p_2 = 66.6$ psia. Repeating the process again will give a more accurate answer.

16.75 Refer to Prob. 16.74. Neglecting the logarithmic term, find the pressures at sections 100, 300, and 800 ft downstream of the section where the pressure is 80 psia.

∎ $p_1 - p_2 = (G^2RT/g^2A^2)[f(L/D) + 2 \ln (p_1/p_2)]$

$[(80)(144)]^2 - p_2^2 = [(100)^2 1715(460 + 65)/32.2^2(\frac{140}{144})^2]\{(0.0085)[L/4(0.243)]\}$

$p_2^2 = 1.327 \times 10^8 - 80\,300L$

At $L = 100$ ft, $p_2 = 11\,170$ psfa $= 77.5$ psia. At $L = 300$ ft, $p_2 = 10\,420$ psfa $= 72.4$ psia. At $L = 800$ ft, $p_2 = 8270$ psfa $= 57.5$ psia.

Express in terms of $p + \gamma(V^2/2g)$:

At $L = 100$ ft, $\gamma = 0.41(77.5/80) = 0.397$ pcf.

At $L = 300$ ft, $\gamma = 0.41(72.4/80) = 0.371$ pcf.

At $L = 800$ ft, $\gamma = 0.41(57.5/80) = 0.295$ pcf.

$\gamma_1 V_1 = \gamma_2 V_2 = \gamma_3 V_3$

At $L = 100$ ft, $V = 250(0.41/0.397) = 258$ fps. At $L = 300$ ft, $V = 250(0.41/0.371) = 276$ fps. At $L = 800$ ft, $V = 250(0.41/0.295) = 347$ fps.

L, ft	p, psia	V, fps	$V^2/2g$	$\gamma V^2/2g$, psi	$p + \gamma V^2/2g$, psia
0	80.0	250	970	2.76	82.8
100	77.5	258	1034	2.86	80.4
300	72.4	276	1183	3.05	75.5
800	57.5	347	1870	3.83	61.3

$\Delta p = 80 - 57.5 = 22.5$ psi over the first 800 ft of length. *Note:* Greater accuracy can be achieved by considering the logarithmic term.

16.76 Air flows isothermally through a long horizontal pipe of uniform diameter. At a section where the pressure is 100 psia, the velocity is 120 fps. Because of fluid friction the pressure at a distant point is 40 psia. (**a**) What is the increase in kinetic energy per pound of air? (b) What is the amount of thermal energy in Btu per pound of air that must be transferred in order to maintain the temperature constant? (c) Is this heat transferred to the air in the pipe or removed from it? (d) If the temperature of the air is 100 °F and the diameter of the pipe is 3 in find the total heat transferred in Btu per hour.

▌ pV = const for isothermal flow, therefore $V_2 = p_1 V_1/p_2 = 100(120)/40 = 300$ fps, $V_1^2/2g = 224$ ft-lb/lb, $V_2^2/2g = 1398$ ft-lb/lb.

(a) $V_1^2/2g + Q_B = V_2^2/2g$. Increase in KE of air = $1398 - 224 = 1174$ ft-lb/lb.
(b) Thermal energy/lb transferred = $Q_H = \Delta KE = \frac{1174}{788} = 1.509$ Btu/lb.
(c) KE increases, therefore heat is transferred *to* the air in the pipe.
(d)
$$\gamma_1 = gp_1/RT_1 = [(32.2)(100)(144)]/[(1715)(460 + 100)] = 0.483 \text{ pcf}$$
$$G = \gamma_1 V_1 A = (0.483)(120)[(\pi/4)(\tfrac{3}{12})^2] = 2.85 \text{ lb/s}$$
$$(1.509 \times 2.85) = 4.30 \text{ Btu/s} = 15\,480 \text{ Btu/h}$$

16.77 For the case of Prob. 16.74 with a duct length of 500 ft, compute the thermal energy (heat) that must be added to the fluid to maintain isothermal conditions.

▌ Since the flow is isothermal, $p_1/\rho_1 = p_2/\rho_2 = RT = $ constant; $p_1 = 80$ psia and $p_2 = 66.6$ psia. Thus $\rho_1/\rho_2 = 80/66.6 = 1.20$ and $V_2/V_1 = 1.20$ since $\rho V = $ constant from continuity.
So $V_2 = 1.20(250) = 300$ fps; $V_1^2/2g + Q_H = V_2^2/2g$, $Q_H = (300)^2/[(2)(32.2)] - (250)^2/[(2)(32.2)] = 427$ ft-lb/lb of air. Since $G = 100$ lb/s, the rate at which heat must be added to the fluid is $(100)(427) = 42\,700$ ft-lb/s. *Note:* if $Q_H > 427$ ft-lb/lb of air, $T_2 > T_1$, and if $Q_H < 427$ ft-lb/lb of air, $T_2 < T_1$.

16.78 Carbon dioxide flows isothermally at 100 °F through a horizontal 6-in-diameter pipe. At this temperature $\mu = 4.0 \times 10^{-7}$ lb-s/ft^2. The pressure changes from 150.0 to 140.0 psig in a 100-ft length of pipe. Determine the flow rate if the atmospheric pressure is 14.5 psia and ϵ for the pipe is 0.002 ft.

▌ At section 1 (inlet), $p_1 = 150.0$ psig + 14.5 psi = 164.5 psia, $R = 1123$ ft^2/(s^2-°R), $\gamma = gp/RT$, $\gamma_1 = [(32.2)(164.5)(144)]/[(1123)(460 + 100)] = 1.213$ pcf; $\epsilon/D = 0.002/0.5 = 0.004$. Assume $N_R > 10^6$, Fig. A-5: $f = 0.0285$. At section 2, $p_2 = 140.0$ psig + 14.5 psi = 154.5 psia.
$$p_1^2 - p_2^2 = (G^2 RT/g^2 A^2)[f(L/D) + 2 \ln (p_1/p_2)]$$
$$[(164.5)(144)]^2 - [(154.5)(144)]^2 = [G^2(1123)(460 + 100)]/[32.2^2(0.1963)^2][0.0285(100/0.5) + 2 \ln (164.5/154.5)]$$
$G = 26.8$ lb/s (based on the above assumption) $\qquad V_1 = G/\gamma_1 A = 26.8/[1.213(0.1963)] = 112.6$ fps
$$N_R = (DV\gamma/g)/\mu \approx [(\tfrac{6}{12})112.6(1.213/32.2)]/(4.0 \times 10^{-7}) = 5.30 \times 10^6$$

Hence assumption for N_R was O.K., and $G = 26.8$ lb/s.

16.79 Methane gas is to be pumped through a 24-in-diameter welded-steel pipe connecting two compressor stations 25 mi apart. At the upstream station the pressure is not to exceed 60 psia, and at the downstream station it is to be at least 20 psia. Determine the maximum possible rate of flow (in cubic feet per day at 60 °F and 1 atm). Assume isothermal flow at 60 °F.

▌ For methane: $R = 3100$ ft^2/(s^2-°R), $\mu = 2.40 \times 10^{-7}$ lb-s/ft^2. Absolute viscosity is independent of pressure over a wide range of pressures but kinematic viscosity varies with pressure: $\nu_1 = \mu/\rho_1 = \mu/(p_1/RT_1) = 2.40 \times 10^{-7}/[(60)(144)/(3100)(520)] = 4.48 \times 10^{-5}$ ft^2/s, $\epsilon = 0.00015$ ft; $\epsilon/D = 0.000075$. As a first trial, assume $N_R > 10^7$, in which case from Fig. A-5 $f = 0.0115$.
$$p_1^2 - p_2^2 = (G^2 RT/g^2 A^2)[f(L/D) + 2 \ln (p_1/p_2)]$$
$$[(60)(144)]^2 - [(20)(144)]^2 = \{G^2(3100)(460 + 60)/32.2^2[(\pi)(\tfrac{24}{12})^2/4]\}\{(0.0115)[(25)(5280)/2] + 2 \ln \tfrac{60}{20}\}$$
$$G = 23.5 \text{ lb/s}$$
$$N_R = GD/\mu g A = (23.5)(2)/(2.40 \times 10^{-7})[(32.2)(\pi)] = 1.936 \times 10^6$$

Thus, from Fig. A-5: $f = 0.0125$; initially assumed f was too low. Substituting $f = 0.0125$ gives $G = 22.7$ lb/s, $\gamma_1 = gp_1/RT_1 = (32.2)(60)(144)/[(3100)(520)] = 0.1726$ lb/ft^3, $Q = G/\gamma = 22.7/0.1726 = 131.5$ ft^3/s = 11.36×10^6 ft^3/day at 60 °F and 60 psia or $Q = (11.36 \times 10^6)(60/14.7) = 46.4 \times 10^6$ ft^3/day at 60 °F and 14.7 psia.

16.80 Air flows isothermally in a long pipe. At one section the pressure is 90 psia, the temperature is 80 °F, and the velocity is 100 fps. At a second section some distance from the first the pressure is 15 psia. Find the energy head loss due to friction, and determine the thermal energy that must have been added to or taken from the fluid between the two sections. The diameter of the pipe is constant.

∎ For isothermal flow: $\rho_1/\rho_2 = p_1/p_2 = \frac{90}{15} = 6$. But from continuity $\rho_1 A_1 V_1 = \rho_2 A_2 V_2$, thus $V_2 = 6V_1$, $V_1^2/2g + Q_H = V_2^2/2g$, $Q_H = 600^2/64.4 - 100^2/64.4 = 5430$ ft-lb/lb of air. For head loss: $dp/\gamma + d(V^2/2g) = -(2\tau \, ds/\gamma r) = -dh_L$, $\gamma = g\rho = gp/RT$; substituting for γ, $(RT/g)(dp/p) + d(V^2/2g) = -dh_L$. Integrating

$$h_L = (RT/g) \ln (p_1/p_2) + (V_1^2 - V_2^2)/2g = [1715(460 + 80)/32.2] \ln 6 + (100^2 - 600^2)/2(32.2)$$
$$= 51\,500 - 5430 = 46\,100 \text{ ft-lb/lb of air}$$

Note: The total energy at section 2 is greater than that at section 1 because external heat is added. Thus h_L represents a degradation of mechanical energy, but not a net loss of energy, because it is converted to another form of energy, namely internal heat.

16.81 Air is flowing at constant temperature through a 3-in-diameter horizontal pipe, $f = 0.02$. At the entrance, $V_1 = 300$ ft/s, $T = 120\,°F$, and $p_1 = 30$ psia. What is the maximum pipe length for this flow, and how much heat is transferred to the air per pound mass?

∎
$$L_{max} = (D/f)[(1 - kN_M^2)/kN_M^2 + \ln (kN_M^2)] \qquad N_M = V/\sqrt{kRT}$$
$$(N_M)_1 = 300/\sqrt{(1.40)(1716)(120 + 460)} = 0.254$$
$$L_{max} = [(\tfrac{3}{12})/0.02]\{[1 - (1.40)(0.254)^2]/(1.40)(0.254)^2 + \ln [(1.40)(0.254)^2]\} = 95.8 \text{ ft} \qquad q_H = (V_2^2 - V_1^2)/2g$$
$$V_2 = (N_M)_2(c) \qquad (N_M)_2 = 1/\sqrt{k} = 1/\sqrt{1.40} = 0.845$$
$$V_2 = (0.845)(1180) = 997.1 \text{ ft/s} \qquad q_H = (997.1^2 - 300^2)/[(2)(32.2)(778)] = 18.0 \text{ Btu/lbm}$$

16.82 Air at 15 °C flows through a 25-mm-diameter pipe at constant temperature. At the entrance $V_1 = 60$ m/s, and at the exit $V_2 = 90$ m/s ($f = 0.016$). What is the length of the pipe?

∎
$$(f/D)(\Delta L) = [(1 - kN_M^2)/kN_M^2 + \ln (kN_M^2)]_1 - [(1 - kN_M^2)/k_M^2 + \ln (kN_M^2)]_2 \qquad N_M = V/\sqrt{kRT}$$
$$(N_M)_1 = 60/\sqrt{(1.40)(287)(15 + 273)} = 0.176 \qquad (N_M)_2 = 90/\sqrt{(1.40)(287)(15 + 273)} = 0.265$$
$$[0.016/(\tfrac{25}{1000})](\Delta L) = \{[1 - (1.40)(0.176)^2]/(1.40)(0.176)^2 + \ln [(1.40)(0.176)^2]\}$$
$$- \{[1 - (1.40)(0.265)^2]/(1.40)(0.265)^2 + \ln [(1.40)(0.265)^2]\} \qquad \Delta L = 18.9 \text{ m}$$

16.83 If the pressure at the entrance of the pipe of Prob. 16.82 is 1.5 atm, what is the pressure at the exit and what is the heat transfer to the pipe per second?

∎
$$p^{*t}/p = (\sqrt{k})(N_M) \qquad p^{*t}/p_1 = (\sqrt{k})(N_M)_1 \qquad p^{*t}/p_2 = (\sqrt{k})(N_M)_2$$

Using data from Prob. 16.82,

$$p_2/p_1 = (N_M)_1/(N_M)_2 = 0.176/0.265 = 0.664 \qquad p_2 = (0.664)(1.5) = 0.996 \text{ atm}$$
$$q_H = (\dot{m})(V_2^2 - V_1^2)/2 \qquad \dot{m} = \rho_1 A_1 V_1 \qquad \rho_1 = p/RT = (1.5)(101\,310)/[(287)(15 + 273)] = 1.839 \text{ kg/m}^3$$
$$A_1 = (\pi)(\tfrac{25}{1000})^2/4 = 0.0004909 \text{ m}^2 \qquad \dot{m} = (1.839)(0.0004909)(60) = 0.05417 \text{ kg/s}$$
$$q_H = [(0.05417)(90^2 - 60^2)/2]/4187 = 0.0291 \text{ kcal/s}$$

16.84 Hydrogen enters a pipe from a converging nozzle at $N_M = 1$, $p = 2$ psia, and $T = 0\,°F$. Determine, for isothermal flow, the maximum length of pipe, in diameters, and the pressure change over this length ($f = 0.016$).

∎
$$L = (D/f)[(1 - kN_M^2)/kN_M^2 + \ln (kN_M^2)]$$
$$L/D = (1/0.016)\{[1 - (1.40)(1)^2]/(1.40)(1)^2 + \ln [(1.40)(1)^2]\} = 3.17$$
$$p^{*t}/p = (\sqrt{k})(N_M) \qquad p^{*t}/2 = (\sqrt{1.40})(1) \qquad p^{*t} = 2.366 \text{ psia} \qquad \Delta p = 2.366 - 2 = 0.366 \text{ psia}$$

16.85 Oxygen flows at constant temperature of 20 °C from a pressure tank, $p = 130$ atm, through 10 ft of 3-mm-ID tubing to another tank where $p = 110$ atm, $f = 0.016$. Determine the mass rate of flow.

∎
$$\dot{m} = \rho A v \qquad \rho = p/RT \qquad \rho_1 = (130)(101\,310)/[(260)(20 + 273)] = 172.9 \text{ kg/m}^3$$
$$A_1 = (\pi)(\tfrac{3}{1000})^2/4 = 0.000007069 \text{ m}^2 \qquad N_M = v/\sqrt{kRT}$$
$$(f/D)(\Delta L) = [(1 - kN_M^2)/kN_M^2 + \ln (kN_M^2)]_1 - [(1 - kN_M^2)/kN_M^2 + \ln (kN_M^2)]_2$$
$$p_2/p_1 = (N_M)_1/(N_M)_2 \qquad \text{(from Prob. 16.83)}$$

$$p_2/p_1 = \tfrac{110}{130} = 0.8462 \qquad (N_M)_2 = (N_M)_1/0.8462 = 1.182(N_M)_1$$

$$[0.016/(\tfrac{3}{1000})][(10)(0.3048)] = \{[1 - (1.40)(N_M)_1^2/(1.40)(N_M)_1^2 + \ln[(1.40)(N_M)_1^2]\}$$

$$- (\{1 - (1.40)[1.182(N_M)_1]^2\}/(1.40)[1.182(N_M)_1]^2 + \ln(1.40)[1.182(N_M)_1]^2)$$

$$(N_M)_1 = 0.111 \qquad 0.111 = v_1/\sqrt{(1.40)(260)(15 + 273)} \qquad v_1 = 35.94 \text{ m/s}$$

$$\dot{m} = (172.9)(0.000007069)(35.94) = 0.0439 \text{ kg/s}$$

16.86 In isothermal flow of nitrogen at 80 °F, 3 lbm/s is to be transferred 100 ft from a tank where $p = 200$ psia to a tank where $p = 160$ psia. What minimum size of tubing, $f = 0.016$, is needed?

$\blacksquare$
$$(f/D)(\Delta L) = [(1 - kN_M^2)/kN_M^2 + \ln(kN_M^2)]_1 - [(1 - kN_M^2)/k_M^2 + \ln(kN_M^2)]_2 \qquad N_M = v/c$$

$$\dot{m} = \rho_1 A_1 v_1 \qquad \rho = p/RT \qquad \rho_1 = (200)(144)/[(1773)(80 + 460)] = 0.03008 \text{ slug/ft}^3$$

$$\rho_2 = (\rho_1)(p_2/p_1) = (0.03008)(\tfrac{160}{200}) = 0.02406 \text{ slug/ft}^3 \qquad 3 = (0.03008)(32.17)(\pi D^2/4)(v_1) \qquad v_1 = 3.947/D^2$$

$$v_2 = (\tfrac{200}{160})(3.947/D^2) = 4.934/D^2 \qquad c = \sqrt{kRT} = \sqrt{(1.40)(55.2)(32.17)(80 + 460)} = 1159 \text{ ft/s}$$

$$(N_M)_1 = (3.947/D^2)/1159 = 0.003406/D^2 \qquad (N_M)_2 = (4.934/D^2)/1159 = 0.004257/D^2$$

$$(0.016/D)(100) = \{[1 - (1.40)(0.003406/D^2)^2]/(1.40)(0.003406/D^2)^2 + \ln[(1.40)(0.003406/D^2)^2]\}$$

$$- \{[1 - (1.40)(0.004257/D^2)^2]/(1.40)(0.004257/D^2)^2 + \ln[(1.40)(0.004257/D^2)^2]\}$$

$$D = 0.150 \text{ ft} \quad \text{or} \quad 1.80 \text{ in} \qquad \text{(by trial and error)}$$

16.87 Air enters a pipe of 1 in diameter at subsonic velocity and $p_1 = 30$ psia, $T_1 = 550$ °R. If the pipe is 10 ft long, $f = 0.025$, and the exit pressure $p_2 = 20$ psia, compute the mass flow for isothermal flow.

$\blacksquare$
$$G^2 = (\dot{m}/A)^2 = (p_1^2 - p_2^2)/\{RT[fL/D + 2\ln(p_1/p_2)]\}$$

$$fL/D + 2\ln(p_1/p_2) = [(0.025)(10)/\tfrac{1}{12}] + 2\ln\tfrac{30}{20} = 3.81$$

$$G^2 = [30(144)]^2 - [20(144)]^2/[(1717)(550)(3.81)] = 2.88 \text{ lb}^2\text{-s}^2/\text{ft}^6 \quad \text{or} \quad G = 1.70 \text{ lb-s/ft}^3 = 1.70 \text{ slugs/(s-ft}^2)$$

$$A = \tfrac{1}{4}\pi D^2 = \tfrac{1}{4}\pi(\tfrac{1}{12})^2 = 0.00545 \text{ ft}^2 \qquad \text{Hence} \qquad \dot{m} = GA = (1.70)(0.00545) = 0.00926 \text{ slug/s}$$

Check the Mach number at inlet and exit:

$$a_1 = a_2 \approx 49(550)^{1/2} = 1149 \text{ ft/s} \qquad \rho_1 = p_1/RT = (30)(144)/(1717)(550) = 0.00457 \text{ slug/ft}^3$$

$$\rho_2 = p_2/RT = (20)(144)/(1717)(550) = 0.00305 \text{ slug/ft}^3$$

Then

$$V_1 = G/\rho_1 = 1.70/0.00457 = 371 \text{ ft/s} \qquad (N_M)_1 = V_1/a_1 = \tfrac{371}{1149} = 0.323 \qquad V_2 = G/\rho_2 = 557 \text{ ft/s}$$

$$(N_M)_2 = \tfrac{557}{1149} = 0.484$$

Since these are well below choking, the solution is accurate.

16.88 Isentropic flow of nitrogen occurs in a 2-in-diameter pipe. At one point the velocity of flow, pressure, and specific weight are 409 fps, 85 psia, and 0.655 lb/ft³, respectively. At a second point a short distance away, the pressure is 83 psia. What is the velocity at the second point?

$\blacksquare$
$$(v_2^2 - v_1^2)/2g = (p_1/\gamma_1)[k/(k - 1)][1 - (p_2/p_1)^{(k-1)/k_1}]$$

$$(v_2^2 - 409^2)/[(2)(32.2)] = [(85)(144)/0.655][1.40/(1.40 - 1)][1 - (\tfrac{83}{85})^{(1.40-1)/1.40}] \qquad v_2 = 443 \text{ ft/s}$$

16.89 At one point on a streamline in an isentropic airflow, the velocity, pressure, and specific weight are 30.5 m/s, 350 kPa, and 0.028 kN/m³, respectively. At a second point on the streamline, the velocity is 150 m/s. Find the pressure at the second point.

$\blacksquare$
$$(v_2^2 - v_1^2)/2g = (p_1/\gamma_1)[k/(k - 1)][1 - (p_2/p_1)^{(k-1)/k}]$$

$$(150^2 - 30.5^2)/[(2)(9.807)] = (350/0.028)[1.40/(1.40 - 1)][1 - (p_2/350)^{(1.40-1)/(1.40)}] \qquad p_2 = 320 \text{ kPa}$$

16.90 Isentropic flow of oxygen occurs in a 100-m-diameter pipe. At one point the velocity of flow, pressure, and specific weight are 125 m/s, 450 kPa, and 0.058 kN/m³, respectively. At a second point a short distance away, the pressure is 360 kPa. What is the velocity at the second point?

$\blacksquare$
$$(v_2^2 - v_1^2)/2g = (p_1/\gamma)[k/(k - 1)][1 - (p_2/p_1)^{(k-1)/k}]$$

$$(v_2^2 - 125^2)/[(2)(9.807)] = (450/0.058)[1.40/(1.40 - 1)][1 - (\tfrac{360}{450})^{(1.40-1)/1.40}] \qquad v_2 = 220 \text{ m/s}$$

16.91   At one point on a streamline in an isentropic airflow, the velocity, pressure, and specific weight are 80 m/s, 405 kPa, and 0.046 kN/m³, respectively. At a second point on the streamline, the velocity is 165 m/s. Find the pressure at the second point.

▮
$$(v_2^2 - v_1^2)/2g = (p_1/\gamma)[k/(k-1)][1 - (p_2/p_1)^{(k-1)/k}]$$

$$(165^2 - 80^2)/[(2)(9.807)] = (405/0.046)[1.40/(1.40-1)][1 - (p_2/405)^{(1.40-1)/1.40}] \qquad p_2 = 358 \text{ kPa}$$

16.92   Prepare a computer program that determines the pressure either at point 1 or point 2 or the velocity at point 1 or point 2 for closed conduit, compressible, isentropic flow.

```
C       THIS PROGRAM DETERMINES EITHER THE PRESSURE AT POINT 1 OR
C       POINT 2 OR THE VELOCITY AT POINT 1 OR POINT 2 FOR CLOSED CONDUIT,
C       COMPRESSIBLE FLOW.  IT IS BASED ON AN ISENTROPIC FLOW ANALYSIS
C       AND IS THEREFORE SUBJECT TO THE RESTRICTIONS AND/OR ASSUMPTIONS
C       IMPLICIT IN ISENTROPIC FLOW.  IT CAN BE USED FOR PROBLEMS IN BOTH
C       THE ENGLISH SYSTEM OF UNITS AND THE INTERNATIONAL SYSTEM OF
C       UNITS.
C
C       INPUT DATA MUST BE SET UP AS FOLLOWS.
C
C       CARD 1    COLUMN 1           ENTER 0 (ZERO) OR BLANK IF ENGLISH SYSTEM
C                                    OF UNITS IS TO BE USED.  ENTER 1 (ONE) IF
C                                    INTERNATIONAL SYSTEM OF UNITS IS TO BE
C                                    USED.
C                 COLUMNS 2-79       ENTER TITLE, DATE, AND OTHER INFORMATION,
C                                    IF DESIRED.
C       CARD 2    COLUMNS 1-10       ENTER NUMBER INCLUDING DECIMAL GIVING
C                                    ABSOLUTE PRESSURE AT UPSTREAM POINT
C                                    (POINT 1) (IN POUNDS PER SQUARE
C                                    INCH OR KILOPASCALS).
C                 COLUMNS 11-20      ENTER NUMBER INCLUDING DECIMAL GIVING
C                                    ABSOLUTE PRESSURE AT DOWNSTREAM POINT
C                                    (POINT 2) (IN POUNDS PER SQUARE
C                                    INCH OR KILOPASCALS).
C                 COLUMNS 21-30      ENTER NUMBER INCLUDING DECIMAL GIVING
C                                    VELOCITY AT POINT 1 (IN FEET PER SECOND
C                                    OR METERS PER SECOND).
C                 COLUMNS 31-40      ENTER NUMBER INCLUDING DECIMAL GIVING
C                                    VELOCITY AT POINT 2 (IN FEET PER SECOND
C                                    OR METERS PER SECOND).
C                 COLUMNS 41-50      ENTER NUMBER INCLUDING DECIMAL GIVING
C                                    SPECIFIC (OR UNIT) WEIGHT AT POINT 1 (IN
C                                    POUNDS PER CUBIC FOOT OR KILO-
C                                    NEWTONS PER CUBIC METER).
C                 COLUMNS 51-60      ENTER NUMBER INCLUDING DECIMAL GIVING
C                                    SPECIFIC (OR UNIT) WEIGHT AT POINT 2 (IN
C                                    POUNDS PER CUBIC FOOT OR KILO-
C                                    NEWTONS PER CUBIC METER).
C                 COLUMNS 61-70      ENTER NUMBER INCLUDING DECIMAL GIVING
C                                    SPECIFIC HEAT RATIO.
C                 COLUMNS 71-80      ENTER TYPE OF FLUID.
C
C ***************************************************************
C *                                                             *
C *    NOTE WELL....EITHER THE PRESSURE AT POINT 1 (COLUMNS 1-10 ON   *
C *    CARD 2) OR AT POINT 2 (COLUMNS 11-20) OR THE VELOCITY AT POINT 1  *
C *    (COLUMNS 21-30) OR AT POINT 2 (COLUMNS 31-40), WHICHEVER ONE IS   *
C *    TO BE DETERMINED BY THIS PROGRAM, SHOULD BE LEFT BLANK.  IF   *
C *    EITHER THE PRESSURE OR THE VELOCITY AT POINT 1 IS BEING DETER-   *
C *    MINED, ENTER A VALUE FOR THE SPECIFIC (OR UNIT) WEIGHT OF THE   *
C *    FLUID AT POINT 2 (COLUMNS 51-60) AND LEAVE SPECIFIC (OR UNIT)   *
C *    WEIGHT OF THE FLUID AT POINT 1 (COLUMNS 41-50) BLANK.  IF EITHER  *
C *    THE PRESSURE OR THE VELOCITY AT POINT 2 IS BEING DETERMINED,   *
C *    ENTER A VALUE FOR THE SPECIFIC (OR UNIT) WEIGHT OF THE FLUID AT   *
C *    POINT 1 (COLUMNS 41-50) AND LEAVE SPECIFIC (OR UNIT) WEIGHT OF   *
C *    THE FLUID AT POINT 2 (COLUMNS 51-60) BLANK.                 *
C *                                                             *
C ***************************************************************
C
C       MULTIPLE DATA SETS FOR SOLVING ANY NUMBER OF PROBLEMS MAY BE
C       INCLUDED FOR PROCESSING.
```

```
C
      REAL K
      DIMENSION TITLE(13),FLUID(2)
1     READ(5,100,END=2)UNITS,TITLE
100   FORMAT(I1,13A6)
      WRITE(6,105)TITLE
105   FORMAT('1',13A6,//)
      READ(5,102)P1,P2,V1,V2,SW1,SW2,K,FLUID
102   FORMAT(7F10.0,2A5)
      G=32.2
      IF(UNITS.EQ.1)G=9.807
      PFACT=144.0
      IF(UNITS.EQ.1)PFACT=1.0
      IF(P1.LT.0.0001)GO TO 103
      IF(P2.LT.0.0001)GO TO 104
      IF(V1.LT.0.0001)GO TO 106
      V2=SQRT(2.0*G*P1*PFACT/SW1*K/(K-1.0)*(1.0-(P2/P1)**((K-1.0)/K))
     *+V1**2)
      IF(UNITS.EQ.0)WRITE(6,107)P1,P2,V1,SW1,FLUID,V2
107   FORMAT(1X,'GIVEN DATA FOR A COMPRESSIBLE FLOW IN A CIRCULAR CLOSED
     * CONDUIT',//5X,'PRESSURE AT POINT 1 =',F8.1,' PSI',//5X,'PRESSURE
     *AT POINT 2 =',F8.1,' PSI',//5X,'VELOCITY AT POINT 1 =',F7.1,' FT/S
     *',//5X,'SPECIFIC (OR UNIT) WEIGHT OF FLUID AT POINT 1 =',F8.3,
     * ' LB/CU FT',//5X,'FLUID FLOWING IS ',2A5,//1X,'THE VELOCITY AT PO
     *INT 2 WILL BE',F7.1,' FT/S')
      IF(UNITS.EQ.1)WRITE(6,108)P1,P2,V1,SW1,FLUID,V2
108   FORMAT(1X,'GIVEN DATA FOR A COMPRESSIBLE FLOW IN A CIRCULAR CLOSED
     * CONDUIT',//5X,'PRESSURE AT POINT 1 =',F8.1,' KPA',//5X,'PRESSURE
     *AT POINT 2 =',F8.1,' KPA',//5X,'VELOCITY AT POINT 1 =',F7.1,' M/S
     *',//5X,'SPECIFIC (OR UNIT) WEIGHT OF FLUID AT POINT 1 =',F8.3,
     *' KN/CU M ',//5X,'FLUID FLOWING IS ',2A5,//1X,'THE VELOCITY AT PO
     *INT 2 WILL BE',F7.1,' M/S')
      GO TO 1
103   P1=(((V2**2-V1**2)/2.0/G/(P2*PFACT)*SW2/K*(K-1.0)+1.0)*(P2*PFACT)
     ***((K-1.0)/K))**(K/(K-1.0)))/PFACT
      IF(UNITS.EQ.0)WRITE(6,109)P2,V1,V2,SW2,FLUID,P1
109   FORMAT(1X,'GIVEN DATA FOR A COMPRESSIBLE FLOW IN A CIRCULAR CLOSED
     * CONDUIT',//5X,'PRESSURE AT POINT 2 =',F8.1,' PSI',//5X,'VELOCITY
     *AT POINT 1 =',F7.1,' FT/S',//5X,'VELOCITY AT POINT 2 =',F7.1,' FT/
     *S',//5X,'SPECIFIC (OR UNIT) WEIGHT OF FLUID AT POINT 2 =',F8.3,
     * ' LB/CU FT',//5X,'FLUID FLOWING IS ',2A5,//1X'THE PRESSURE AT POI
     *NT 1 WILL BE ',F8.1,' PSI')
      IF(UNITS.EQ.1)WRITE(6,110)P2,V1,V2,SW2,FLUID,P1
110   FORMAT(1X,'GIVEN DATA FOR A COMPRESSIBLE FLOW IN A CIRCULAR CLOSED
     * CONDUIT',//5X,,'PRESSURE AT POINT 2 =',F8.1,' KPA',//5X,'VELOCITY
     *AT POINT 1 =',F7.1,' M/S ',//5X,'VELOCITY AT POINT 2 =',F7.1,' M/S
     * ',//5X,'SPECIFIC (OR UNIT) WEIGHT OF FLUID AT POINT 2 =',F8.3,
     *' KN/CU M ',//5X,'FLUID FLOWING IS ',2A5,//1X'THE PRESSURE AT POI
     *NT 1 WILL BE ',F8.1,' KPA')
      GO TO 1
104   P2=((1.0-(V2**2-V1**2)/2.0/G/(P1*PFACT)*SW1/K*(K-1.0))*(P1*PFACT)
     ***((K-1.0)/K))**(K/(K-1.0)))/PFACT
      IF(UNITS.EQ.0)WRITE(6,111)P1,V1,V2,SW1,FLUID,P2
111   FORMAT(1X,'GIVEN DATA FOR A COMPRESSIBLE FLOW IN A CIRCULAR CLOSED
     * CONDUIT',//5X,'PRESSURE AT POINT 1 =',F8.1,' PSI',//5X,'VELOCITY
     *AT POINT 1 =',F7.1,' FT/S',//5X,'VELOCITY AT POINT 2 =',F7.1,' FT/
     *S',//5X,'SPECIFIC (OR UNIT) WEIGHT OF FLUID AT POINT 1 =',F8.3,
     * ' LB/CU FT',//5X,'FLUID FLOWING IS ',2A5,//1X'THE PRESSURE AT POI
     *NT 2 WILL BE ',F8.1,' PSI')
      IF(UNITS.EQ.1)WRITE(6,112)P1,V1,V2,SW1,FLUID,P2
112   FORMAT(1X,'GIVEN DATA FOR A COMPRESSIBLE FLOW IN A CIRCULAR CLOSED
     * CONDUIT',//5X,'PRESSURE AT POINT 1 =',F8.1,' KPA',//5X,'VELOCITY
     *AT POINT 1 =',F7.1,' M/S ',//5X,'VELOCITY AT POINT 2 =',F7.1,' M/S
     * ',//5X,'SPECIFIC (OR UNIT) WEIGHT OF FLUID AT POINT 1 =',F8.3,
     *' KN/CU M ',//5X,'FLUID FLOWING IS ',2A5,//1X'THE PRESSURE AT POI
     *NT 2 WILL BE ',F8.1,' KPA')
      GO TO 1
106   V1=SQRT(V2**2-2.0*G*P2*PFACT/SW2*K/(K-1.0)*((P1/P2)**((K-1.0)/K)
     *-1.0))
      IF(UNITS.EQ.0)WRITE(6,113)P1,P2,V2,SW2,FLUID,V1
113   FORMAT(1X,'GIVEN DATA FOR A COMPRESSIBLE FLOW IN A CIRCULAR CLOSED
     * CONDUIT',//5X,'PRESSURE AT POINT 1 =',F8.1,' PSI',//5X,'PRESSURE
     *AT POINT 2 =',F8.1,' PSI',//5X,'VELOCITY AT POINT 2 =',F7.1,' FT/S
```

```
    *',//5X,'SPECIFIC (OR UNIT) WEIGHT OF FLUID AT POINT 2 =',F8.3,
    * ' LB/CU FT',//5X,'FLUID FLOWING IS ',2A5,//1X,'THE VELOCITY AT PO
    *INT 1 WILL BE',F7.1,' FT/S')
     IF(UNITS.EQ.1)WRITE(6,114)P1,P2,V2,SW2,FLUID,V1
 114 FORMAT(1X,'GIVEN DATA FOR A COMPRESSIBLE FLOW IN A CIRCULAR CLOSED
    * CONDUIT',//5X,'PRESSURE AT POINT 1 =',F8.1,' KPA',//5X,'PRESSURE
    *AT POINT 2 =',F8.1,' KPA',//5X,'VELOCITY AT POINT 2 =',F7.1,' M/S
    *',//5X,'SPECIFIC (OR UNIT) WEIGHT OF FLUID AT POINT 2 =',F8.3,
    *' KN/CU M  ',//5X,'FLUID FLOWING IS ',2A5,//1X,'THE VELOCITY AT PO
    *INT 1 WILL BE',F7.1,' M/S')
     GO TO 1
   2 STOP
     END
```

16.93 Solve Prob. 16.88 using the computer program of Prob. 16.92.

❚ Input

```
1 2 3 4 5 6 7 8 9 10 11 12 13 14 15 16 17 18 19 20 21 22 23 24 25 26 27 28 29 30 31 32 33 34 35 36 37 38 39 40 41 42 43 44 45 46 47 48 49 50 51 52 53 54 55 56 57 58 59 60 61 62 63 64 65 66 67 68 69 70 71 72 73 74 75 76 77 78 79 80
ØSAMPLE ANALYSIS ØF COMPRESSIBLE FLØW
85.0     83.0      409.0              0.655             1.40      NITRØGEN
```

Output

```
SAMPLE ANALYSIS OF COMPRESSIBLE FLOW



GIVEN DATA FOR A COMPRESSIBLE FLOW IN A CIRCULAR CLOSED CONDUIT

    PRESSURE AT POINT 1 =    85.0 PSI

    PRESSURE AT POINT 2 =    83.0 PSI

    VELOCITY AT POINT 1 =  409.0 FT/S

    SPECIFIC (OR UNIT) WEIGHT OF FLUID AT POINT 1 =   0.655 LB/CU FT

    FLUID FLOWING IS NITROGEN

THE VELOCITY AT POINT 2 WILL BE  442.5 FT/S
```

16.94 Solve Prob. 16.89 using the computer program of Prob. 16.92.

❚ Input

```
1 2 3 4 5 6 7 8 9 10 11 12 13 14 15 16 17 18 19 20 21 22 23 24 25 26 27 28 29 30 31 32 33 34 35 36 37 38 39 40 41 42 43 44 45 46 47 48 49 50 51 52 53 54 55 56 57 58 59 60 61 62 63 64 65 66 67 68 69 70 71 72 73 74 75 76 77 78 79 80
1SAMPLE ANALYSIS ØF COMPRESSIBLE FLØW
350.0           30.5     150.0   0.028          1.40      AIR
```

Output

```
SAMPLE ANALYSIS OF COMPRESSIBLE FLOW


GIVEN DATA FOR A COMPRESSIBLE FLOW IN A CIRCULAR CLOSED CONDUIT

    PRESSURE AT POINT 1 =    350.0 KPA

    VELOCITY AT POINT 1 =    30.5 M/S

    VELOCITY AT POINT 2 =  150.0 M/S

    SPECIFIC (OR UNIT) WEIGHT OF FLUID AT POINT 1 =   0.028 KN/CU M

    FLUID FLOWING IS AIR

THE PRESSURE AT POINT 2 WILL BE    320.2 KPA
```

16.95 Air at a pressure of 150 psia and a temperature of 100 °F expands in a suitable nozzle to 15 psia. (*a*) If the flow is frictionless and adiabatic and the initial velocity is negligible, find the final velocity. (*b*) Find the final temperature at the end of the expansion.

▐ (*a*) $\qquad V_1 \simeq 0 \qquad V_2^2/2g = (RT_1/g)[k/(k-1)][1 - (p_2/p_1)^{(k-1)/k}]$

$\qquad V_2^2/2g = [(1715)(560)/32.2](1.4/0.4)[1 - (\frac{15}{150})^{0.2860}] \qquad V_2 = 1800 \text{ fps}$

(*b*) $\qquad V_2^2 - V_1^2 = 2c_p(T_1 - T_2) \qquad 1800^2 - 0 = 2(6000)(560 - T_2) \qquad T_2 = 290 \text{ °R} = -170 \text{ °F}$

16.96 Derive the equation for isentropic flow: $(V_2^2 - V_1^2)/2g = (p_1/\gamma_1)[k/(k-1)][1 - (p_2/p_1)^{(k-1)/k}]$ by integrating the Euler equation.

▐ Euler equation: $\qquad\qquad\qquad dp/\gamma + (V\,dV)/g = 0 \qquad\qquad\qquad\qquad (1)$

$pv^k = \text{constant} = p/\rho^k$ because $v = 1/\rho$; so $p/\gamma^k = C$:

$p = C\gamma^k$: $\qquad\qquad\qquad\qquad\qquad dp = Ck\gamma^{k-1}\,d\gamma \qquad\qquad\qquad\qquad\qquad (2)$

Substitute (2) into (1) and integrate: $Ck\gamma^{k-2}\,d\gamma + (1/g)V\,dV = 0$.

$$\left[\frac{Ck\gamma^{k-1}}{k-1}\right]_{\gamma_1}^{\gamma_2} + \left[\frac{V^2}{2g}\right]_{V_1}^{V_2} = 0 \qquad \frac{V_2^2 - V_1^2}{2g} = \frac{k}{k-1}\left[\frac{p}{\gamma}\right]_2^1 = \frac{k}{k-1}\left(\frac{p_1}{\gamma_1} - \frac{p_2}{\gamma_2}\right)$$

Note that $p_1/\gamma_1 - p_2/\gamma_2 = p_1/\gamma_1[1 - (p_2/\gamma_2)/(p_1/\gamma_1)]$ where $\gamma_1 = (1/C)p_1^{1/k}$ and $\gamma_2 = (1/C)p_2^{1/k}$, $p_2\gamma_1/p_1\gamma_2 = (p_2^{1-(1/k)})/(p_1^{1-(1/k)}) = (p_2/p_1)^{(k-1)/k}$. Thus $(V_2^2 - V_1^2)/2g = (p_1/\gamma_1)[k/(k-1)][1 - (p_2/p_1)^{(k-1)/k}]$ Q.E.D.

16.97 Carbon dioxide flows isentropically. At a point in the flow the velocity is 50 fps and the temperature is 125 °F. At a second point on the same streamline the temperature is 80 °F. What is the velocity at the second point?

▐ For carbon dioxide: $k = 1.28$; $R = 1123$ ft-lb/(slug-°R). Flow is isentropic, therefore adiabatic. $V_2^2 - V_1^2 = [2k/(k-1)](RT_1)[1 - (T_2/T_1)]$, $V_2^2 - 50^2 = [2(1.28)/0.28](1123)(460 + 125)[1 - (460 + 80)/(460 + 125)]$; $V_2 = 682$ fps. Check to see if sonic velocity is exceeded: $c = (kRT)^{1/2} = [1.28(1123)(460 + 80)]^{1/2} = 881$ fps. Since $V_2 < c$, flow is subsonic, $V_2 = 682$ fps is correct.

16.98 Refer to Prob. 16.97. If the pressure at the first point were 20 psi, determine the pressure and temperature on the nose of a streamlined object placed in the flow at that point.

▐ $\rho_0 = p_0/RT_0 = (20)(144)/(1123)(585) = 0.00438$ slug/ft³ $\qquad c = (kRT_1)^{1/2} = [1.29(1123)(585)]^{1/2} = 921$ fps

$N_M = V/c = \frac{50}{921} = 0.0543 \qquad p_s = p_0 + \rho_0(V_0^2/2)(1 + N_M^2/4)(\frac{1}{144}) = 20.0 \text{ psia} \qquad (c_p/g)T_1 + V_1^2/2g = (c_p/g)T_2$

$\qquad (5132/32.2)(585) + (50^2/64.4) = (5132/32.2)T_s \qquad T_s = 585 \text{ °R} = 125 \text{ °F}$

Temperature increase is negligible.

16.99 From the equation $s_2 - s_1 = c_v \ln[(T_2/T_1)(\rho_1/\rho_2)^{k-1}]$ and the perfect gas law, derive the equation of state for isentropic flow.

▐ Since $s_2 - s_1 = c_v = 0$ for isentropic flow,

$\ln[(T_2/T_1)(\rho_1/\rho_2)^{k-1}] = 0 \qquad (T_2/T_1)(\rho_1/\rho_2)^{1-k} = 1 \qquad T_2/T_1 = (p_2/\rho_2)/(p_1/\rho_1)$ (from perfect gas law)

Eliminating T's, $p_2/\rho_2^k = p_1/\rho_1^k$.

16.100 Compute the enthalpy change per slug for helium from $T_1 = 0$ °F, $p_1 = 15$ psia to $T_2 = 140$ °F in an isentropic process [$c_p = 1.25$ Btu/(lbm-°R)].

▐ $\qquad\qquad\qquad H = mc_p\,\Delta T = (1)[(1.25)(32.17)](140 - 0) = 5630 \text{ Btu}$

16.101 Isentropic flow of air occurs at a section of a pipe where $p = 45$ psia, $T = 90$ °F, and $V = 537$ ft/s. An object is immersed in the flow, which brings the velocity to zero. What are the temperature and pressure at the stagnation point?

▐ $V^2/2 = [kR/(k-1)](T_0 - T) \qquad 537^2/2 = [(1.40)(1716)/(1.40 - 1)][T_0 - (90 + 460)] \qquad T_0 = 574 \text{ °R}$ or   114 °F

$\qquad T_2/T_1 = (p_2/p_1)^{(k-1)/k} \qquad 574/(90 + 460) = (p_2/45)^{(1.40-1)/1.40} \qquad p_2 = 52.3 \text{ psia}$

16.102 What is the Mach number for Prob. 16.101?

▮ $$N_M = V/\sqrt{kRT} = 537/\sqrt{(1.40)(1716)(90 + 460)} = 0.467$$

16.103 How do the temperature and the pressure at the stagnation point in isentropic flow compare with reservoir conditions?

▮ They are the same.

16.104 Air flows from a reservoir at 90 °C, 7 atm. Assuming isentropic flow, calculate the temperature, pressure, density, and velocity at a section where $N_M = 0.60$.

▮ $T_0/T = 1 + [(k - 1)/2]N_M^2$ $\qquad$ $(273 + 90)/T = 1 + [(1.40 - 1)/2](0.60)^2$ $\qquad$ $T = 339$ K or 66 °C

$\qquad$ $p_0/p = \{1 + [(k - 1)/2]N_M^2\}^{k/(k-1)}$ $\qquad$ $(7)(101\,310)/p = \{1 + [(1.40 - 1)/2](0.60)^2\}^{1.40/(1.40-1)}$

$\qquad\qquad$ $p = 556\,000$ Pa or 556 kPa abs $\qquad$ $\rho = p/RT = 556\,000/[(287)(339)] = 5.71$ kg/m^3

$\qquad\qquad$ $c = \sqrt{kRT} = \sqrt{(1.40)(287)(339)} = 369$ m/s $\qquad$ $V = cN_M = (369)(0.60) = 221$ m/s

16.105 Oxygen flows from a reservoir where $p_0 = 120$ psia, $T_0 = 90$ °F, to a 6-in-diameter section where the velocity is 600 fps. Calculate the mass rate of flow (isentropic) and the Mach number, pressure, and temperature in the 6-in section.

▮ $T = T_0 - [(k - 1)/2kR]V^2 = (90 + 460) - [(1.40 - 1)/(2)(1.40)(1554)](600)^2 = 517$ °R or 57 °F

$\qquad\qquad$ $N_M = V/\sqrt{kRT} = 600/\sqrt{(1.40)(1554)(517)} = 0.566$ $\qquad$ $\dot{m} = \rho A V$

$\qquad\qquad$ $\rho_0 = p_0/RT_0 = (120)(144)/[(1554)(90 + 460)] = 0.02022$ slug/ft^3

$\rho_0/\rho = \{1 + [(k - 1)/2]N_M^2\}^{1/(k-1)}$ $\qquad$ $0.02022/\rho = \{1 + [(1.40 - 1)/2](0.566)^2\}^{1/(1.40-1)}$ $\qquad$ $\rho = 0.01731$ slug/ft^3

$\qquad\qquad$ $\dot{m} = 0.01731[(\pi)(\tfrac{6}{12})^2/4](600) = 2.04$ slugs/s

$\qquad\qquad$ $p = \rho RT = (0.01731)(1554)(517) = 13\,907$ lb/ft^2 or 96.6 lb/in^2

16.106 Air flows adiabatically through a duct. At point 1 the velocity is 800 ft/s, $T_1 = 500$ °R, and $p_1 = 25$ lb/in^2 abs $= 3600$ lb/ft^2. Compute (a) T_0; (b) p_{01}; (c) ρ_0; (d) N_M; (e) V_{max}; (f) V^*. At point 2 further downstream $V_2 = 962$ ft/s, and $p_2 = 2850$ lb/ft^2. (g) What is the stagnation pressure p_{02}?

▮ For air take $k = 1.4$, $c_p = 6010$, and $R = 1717$ (BG units). With V_1 and T_1 known, we can compute T_0 without using ratios.

(a) $\qquad\qquad\qquad$ $T_{01} = T_1 + [\tfrac{1}{2}(V_1)^2/c_p] = 500 + [\tfrac{1}{2}(800)^2/6010] = 553$ °R

Then compute N_M from the known ratio T/T_0, using $(N_M)^2 = 5[(T_0/T) - 1]$:

(d) $\qquad\qquad\qquad$ $(N_M)_1^2 = 5(\tfrac{553}{500} - 1)$ $\qquad$ $(N_M)_1 = 0.73$

The stagnation pressure follows:

(b) $\qquad\qquad\qquad$ $p_{01} = p_1[1 + 0.2(N_M)_1^2]^{3.5} = 3600[1 + 0.2(0.73)^2]^{3.5} = 5130$ lb/ft^2

We need the density before we can compute stagnation density: $\rho_1 = p_1/RT_1 = 3600/[1717(500)] = 0.00419$ slug/ft^3. Then

(c) $\qquad\qquad\qquad$ $\rho_0 = \rho_1[1 + 0.2(N_M)_1^2]^{2.5} = 0.00419[1 + 0.2(0.73)^2]^{2.5} = 0.00540$ slug/ft^3

However, if we were clever, we could compute ρ_0 directly from p_0 and T_0: $\rho_0 = p_0/RT_0 = 5130/[1717(553)] = 0.00540$ slug/ft^3. The maximum velocity follows:

(e) $\qquad\qquad\qquad$ $V_{max} = (2c_p T_0)^{1/2} = [2(6010)(553)]^{1/2} = 2580$ ft/s

and the sonic velocity from $V^* = \{[2k/(k + 1)]RT_0\}^{1/2}$:

(f) $\qquad\qquad\qquad$ $V^* = \{[2(1.4)/(1.4 + 1)](1717)(553)\}^{1/2} = 1050$ ft/s

At point 2, the temperature is not given, but since we know it is adiabatic flow, $T_{02} = T_{01} = 553$ °R from solution (a). Thus we can compute T_2 from $T_2 = T_{02} - (\tfrac{1}{2}V_2^2/c_p) = 553 - [\tfrac{1}{2}(962)^2/6010] = 476$ °R (Trying to find T_2 from the Mach-number relations is a frustratingly laborious procedure.) The speed of sound a_2 thus equals $(kRT_2)^{1/2} \approx 49(476)^{1/2} = 1069$ ft/s, whence the Mach number $(N_M)_2 = V_2/a_2 = \tfrac{962}{1069} = 0.90$. Finally compute

(g) $\qquad\qquad\qquad$ $p_{02} = p_2[1 + 0.2(N_M)_2^2]^{3.5} = 28.50[1 + 0.2(0.9)^2]^{3.5} = 4820$ lb/ft^2 $\qquad$ (6 percent less than p_{01})

16.107 Air flows isentropically through a duct. At section 1 the area is 1 ft^2 and $V_1 = 600$ ft/s, $p_1 = 12\,000$ lb/ft^2, and

$T_1 = 850\,°R$. Compute (a) T_0, (b) $(N_M)_1$, (c) p_0, and (d) A^*. If $A_2 = 0.75\,\text{ft}^2$, compute $(N_M)_2$ and p_2 if V_2 is (e) subsonic or (f) supersonic.

▌ With V_1 and T_1 known, the energy equation gives

(a)
$$T_0 = T_1 + (V_1^2/2c_p) = 850 + [(600)^2/2(6010)] = 880\,°R$$

The sound speed $a_1 \approx 49(850)^{1/2} = 1429\,\text{ft/s}$; hence

(b)
$$(N_M)_1 = V_1/a_1 = \tfrac{600}{1429} = 0.42$$

With $(N_M)_1$ known $p_0/p_1 = [1 + 0.2(N_M)_1^2]^{3.5} = 1.129$. Hence

(c)
$$p_0 = 1.129 p_1 = 1.129(12\,000) = 13\,550\,\text{lb/ft}^2$$

Similarly, from $A/A^* = (1/N_M)[(1 + 0.2N_M^2)^3/1.728]$, $A_1/A^* = [1 + 0.2(0.42)^2]^3/[1.728(0.42)] = 1.529$. Hence

(d)
$$A^* = 1/1.529 = 0.654\,\text{ft}^2$$

This throat must actually be present in the duct to expand the subsonic $(N_M)_1$ to supersonic flow downstream.
Given $A_2 = 0.75\,\text{ft}^2$, we can compute $A_2/A^* = 0.75/0.654 = 1.147$. For subsonic flow, use $N_M \approx 1 - 0.88[\ln(A/A^*)]^{0.45}$:

(e) $\quad (N_M)_2 \approx 1 - 0.88(\ln 1.147)^{0.45} = 0.64 \quad$ whence $\quad p_2 = p_0/[1 + 0.2(0.64)^2]^{3.5} = 13\,550/1.317 = 10\,300\,\text{lb/ft}^2$

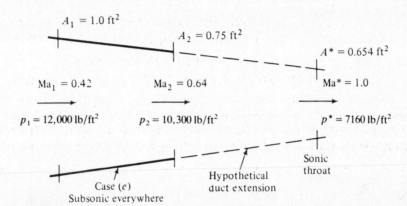

Fig. 16-4

For case (e) there is *no* sonic throat; i.e., the duct area has not decreased sufficiently to create supersonic flow. It may possibly do this further downstream, as illustrated in Fig. 16-4. If, on the other hand, the flow at section 2 is supersonic, we use $N_M = 1 + 1.2[(A/A^*) - 1]^{1/2}$:

(f) $\quad (N_M)_2 \approx 1 + 1.2(1.147 - 1)^{1/2} = 1.46 \quad$ whence $\quad p_2 = 13\,550/[1 + 0.2(1.46)^2]^{3.5} = 13\,550/3.46 = 3910\,\text{lb/ft}^2$

These answers are accurate within less than 1 percent. Note that the supersonic-flow pressure level is much less than the subsonic-flow condition at section 2 for the same duct area, and a sonic throat ($A^* = 0.654\,\text{ft}^2$) *must* have occurred between sections 1 and 2.

16.108 It is desired to expand air from $p_0 = 200\,\text{kPa}$ and $T_0 = 500\,\text{K}$ through a throat to an exit Mach number of 2.5. If the desired mass flow is 3 kg/s, compute (a) the throat area and the exit (b) pressure, (c) temperature, (d) velocity, and (e) area, assuming isentropic flow, with $k = 1.4$.

▌ (a) The throat area follows from the following equation, because the throat flow must be sonic to produce a supersonic exit: $A^* = [\dot{m}(RT_0)^{1/2}]/(0.6847 p_0) = \{3.0[287(500)]^{1/2}\}/[0.6847(200\,000)] = 0.00830\,\text{m}^2 = \tfrac{1}{4}\pi D^{*2}$ or $D_{\text{throat}} = 10.3\,\text{cm}$. With the exit Mach number known, the isentropic-flow relations give pressure and temperature:

(b)
$$p_e = p_0/[1 + 0.2(2.5)^2]^{3.5} = 200\,000/17.08 = 11\,700\,\text{Pa}$$

(c)
$$T_e = T_0/[1 + 0.2(2.5)^2] = 500/2.25 = 222\,\text{K}$$

(d) The exit velocity follows from the known Mach number and temperature $V_e = (N_M)_e(kRT_e)^{1/2} = 2.5[1.4(287)(222)]^{1/2} = 747\,\text{m/s}$.

(e) The exit area follows from the known throat area and exit Mach number and $A/A^* = (1/N_M)[(1 + 0.2N_M^2)^3/1.728]$: $A_e/A^* = [1 + 0.2(2.5)^2]^3/[1.728(2.5)] = 2.64$ or $A_e = 2.64A^* = 2.64(0.0083) = 0.0219\,\text{m}^2 = \tfrac{1}{4}\pi D_e^2$: $D_e = 16.7\,\text{cm}$.

16.109 A high-speed gas flow has $V = 800$ m/s, $p = 140$ kPa, and $T = 200\,°C$. Compute the pressure and temperature which the gas would achieve if brought isentropically to rest for air $[c_p = 1005$ m²/(s-K), $k = 1.40]$.

$$T_0 = T + V^2/2c_p = (200 + 273) + 800^2/[(2)(1005)] = 791\text{ K} \quad\text{or}\quad 518\,°C$$
$$p_0 = (p)(T_0/T)^{k/(k-1)} = 140[791/(200 + 273)]^{1.40/(1.40-1)} = 847\text{ kPa}$$

16.110 Rework Prob. 16.108 for argon $[c_p = 518$ m²/(s-K), $k = 1.67]$.

$$T_0 = T + V^2/2c_p = (200 + 273) + 800^2/[(2)(518)] = 1091\text{ K} \quad\text{or}\quad 818\,°C$$
$$p_0 = (p)(T_0/T)^{k/(k-1)} = (140)[1091/(200 + 273)]^{1.67/(1.67-1)} = 1124\text{ kPa}$$

16.111 Rework Prob. 16.108 for steam.

$h_0 = h + V^2/2$. From the steam tables, at 140 kPa and 200 °C, $h = 2873$ kJ/kg and $s = 7.703$ kJ/(kg-K): $h_0 = 2873 + (800^2/2)/1000 = 3193$ kJ/kg, $s_0 = 7.703$ kJ/(kg-K). From the steam tables, $T_0 = 362\,°C$ and $p_0 = 468$ kPa.

16.112 The Concorde airplane flies at $N_M = 2.2$ at 10 000 m standard altitude. Estimate the Celsius air temperature at the front stagnation point. At what Mach number would it fly to have a front stagnation temperature of 400 °C?

$$T = 273 - 49.9 = 223.1\text{ K} \qquad \text{(from Table A-8)}$$
$$T_0 = T\{1 + [(k - 1)/2](N_M)^2\} = 223.1\{1 + [(1.40 - 1)/2](2.2)^2\} = 439\text{ K} \quad\text{or}\quad 166\,°C$$
$$(400 + 273) = 223.1\{1 + [(1.40 - 1)/2](N_M)^2\} \qquad N_M = 3.17$$

16.113 Air flows at $V = 300$ m/s, $p = 100$ kPa, and $T = 150\,°C$. Compute the maximum velocity attainable by adiabatic expansion of this gas $[c_p = 1005$ m²/(s²-K)].

$$T_0 = T + V^2/2c_p = (150 + 273) + 300^2/[(2)(1005)] = 468\text{ K} \qquad V_{\max} = (2c_pT_0)^{1/2} = [(2)(1005)(468)]^{1/2} = 970\text{ m/s}$$

16.114 Rework Prob. 16.113 for helium flowing $[c_p = 5224$ m²/(s²-K)].

$$T_0 = T + V^2/2c_p = (150 + 273) + 300^2/[(2)(5224)] = 432\text{ K} \qquad V_{\max} = (2c_pT_0)^{1/2} = [(2)(5224)(432)]^{1/2} = 2125\text{ m/s}$$

16.115 Air expands isentropically through a duct from $p_1 = 125$ kPa and $T_1 = 100\,°C$ to $p_2 = 80$ kPa and $V_2 = 325$ m/s. Compute (a) T_2, (b) $(N_M)_2$, (c) T_0, (d) p_0, (e) V_1, and (f) $(N_M)_1$.

(a) $$T_2 = (T_1)(p_2/p_1)^{(k-1)/k} = (100 + 273)(\tfrac{80}{125})^{(1.40-1)/1.40} = 328\text{ K}$$

(b) $$N_M = V/c \qquad c = \sqrt{kRT} \qquad c_2 = \sqrt{(1.40)(287)(328)} = 363\text{ m/s} \qquad (N_M)_2 = \tfrac{325}{363} = 0.895$$

(c) $$(T_0)_1 = (T_0)_2 = (T_2)\{1 + [(k - 1)/2](N_M)_2^2\} \qquad T_0 = 328\{1 + [(1.40 - 1)/2](0.895)^2\} = 381\text{ K}$$

(d) $$(p_0)_1 = (p_0)_2 = (p_2)\{1 + [(k - 1)/2](N_M)_2^2\}^{k/(k-1)}$$
$$p_0 = 80\{1 + [(1.40 - 1)/2](0.895)^2\}^{1.40/(1.40-1)} = 135\text{ kPa}$$

(e) $$T_0 = T + V^2/2c_p \qquad 381 = (100 + 273) + V_1^2/[(2)(1005)] \qquad V_1 = 127\text{ m/s}$$

(f) $$c_1 = \sqrt{(1.40)(287)(100 + 273)} = 387\text{ m/s} \qquad (N_M)_1 = \tfrac{127}{387} = 0.328$$

16.116 Air flows isentropically in a duct. At section 1, $(N_M)_1 = 0.6$, $T_1 = 250\,°C$, and $p_1 = 300$ kPa. At section 2, $(N_M)_2 = 3.1$. Compute (a) $(p_0)_2$, (b) p_2, and (c) T_2.

(a) $$(p_0)_2 = (p_0)_1 = (p_1)\{1 + [k - 1)/2](N_M)_1^2\}^{k/(k-1)} = 300\{1 + [(1.40 - 1)/2](0.6)^2\}^{1.40/(1.40-1)} = 383\text{ kPa}$$

(b) $$p_2 = (p_0)_2/\{1 + [k - 1)/2](N_M)_2^2\}^{k/(k-1)} = 383/\{1 + [(1.40 - 1)/2](3.1)^2\}^{1.40/(1.40-1)} = 8.98\text{ kPa}$$

(c) $$(T_0)_2 = (T_0)_1 = (T_1)\{1 + [(k - 1)/2](N_M)_1^2\} = (250 + 273)\{1 + [(1.40 - 1)/2](0.6)^2\} = 561\text{ K}$$
$$T_2 = (T_0)_2/\{1 + [(k - 1)/2](N_M)_2^2\} = 561/\{1 + [(1.40 - 1)/2](3.1)^2\} = 192\text{ K}$$

16.117 Show that for isentropic flow of a perfect gas, if a pitot-static probe measures p_0, p, and T_0, the gas velocity can be calculated from $V^2 = 2c_pT_0[1 - (p/p_0)^{(k-1)/k}]$. What would be a source of error if a shock wave is formed in front of the probe?

$T = (T_0)(p/p_0)^{(k-1)/k} = T_0 - V^2/2c_p$. Solving for V^2 gives $V^2 = 2c_pT_0[1 - (p/p_0)^{(k-1)/k}]$. If a shock wave is formed (i.e., if V is supersonic), the probe measures $(p_0)_2$ inside the shock wave, which is less than stream stagnation pressure $(p_0)_1$.

16.118 At the exit of a nozzle flow, the air temperature is 50 °C and the velocity is 360 m/s. What is the Mach number and stagnation temperature there? If the flow is adiabatic, what is the Mach number upstream, where $T = 100$ °C?

$$N_M = V/c \qquad c = \sqrt{kRT} \qquad c_2 = \sqrt{(1.40)(287)(50 + 273)} = 360 \text{ m/s}$$

$$(N_M)_2 = \tfrac{360}{360} = 1.00 \qquad (T_0)_2 = T_2 + V_2^2/2c_p = (50 + 273) + 360^2/[(2)(1005)] = 387 \text{ K}$$

$$T_0/T_1 = 1 + [(k - 1)/2](N_M)_1^2 \qquad 387/(100 + 273) = 1 + [(1.40 - 1)/2](N_M)_1^2 \qquad (N_M)_1 = 0.433$$

16.119 The large compressed-air tank in Fig. 16-5 exhausts from a nozzle at an exit velocity of 235 m/s. The mercury manometer reads $h = 30$ cm. Assuming isentropic flow, compute the pressure in the tank and in the atmosphere. What is the exit Mach number?

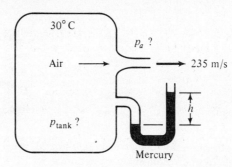

Fig. 16-5

$$N_M = V/c \qquad c = \sqrt{kRT} \qquad T_0 = t_{\text{exit}} + V_{\text{exit}}^2/2c_p \qquad (30 + 273) = T_{\text{exit}} + 235^2/[(2)(1005)] \qquad T_{\text{exit}} = 276 \text{ K}$$

$$c_{\text{exit}} = \sqrt{(1.40)(287)(276)} = 333 \text{ m/s} \qquad (N_M)_{\text{exit}} = \tfrac{235}{333} = 0.706$$

$$p_{\text{exit}} = (p_0)/\{1 + [(k - 1)/2](N_M)_{\text{exit}}^2\}^{k/(k-1)} \qquad p_0/p_{\text{exit}} = \{1 + [(1.40 - 1)/2](0.706)^2\}^{1.40/(1.40-1)} = 1.395$$

Since $\rho_{\text{Hg}} \gg \rho_0$, we can simply take $\rho_0 = 1.5 \text{ kg/m}^3$. $p_0 - p_{\text{exit}} = (\rho_{\text{Hg}} - \rho_0)gh = (13\,600 - 1.5)(9.807)(\tfrac{30}{100}) = 40\,000$ Pa. Then $p_0/p_{\text{exit}} - 1 = 1.395 - 1 = (p_0 - p_{\text{exit}})/p_{\text{exit}} = 40,000/p_{\text{exit}}$, $p_{\text{exit}} = 101\,300$ Pa, $p_0 = p_{\text{tank}} = 141\,300$ Pa.

16.120 Air flows adiabatically through a duct. At one section, $V_1 = 400$ fps, $T_1 = 200$ °F, and $p_1 = 35$ psia, while farther downstream $V_2 = 1100$ fps and $p_2 = 18$ psia. Compute (**a**) $(N_M)_2$, (**b**) U_{max}, and (**c**) $(p_0)_2/(p_0)_1$.

(a) $\qquad (N_M)_2 = V_2/c_2 \qquad c_2 = \sqrt{kRT_2} \qquad T_2 = (T_0)_2 - V_2^2/2c_p \qquad (T_0)_2 = (T_0)_1 = T_1 + V_1^2/2c_p$

$$(T_0)_2 = (200 + 460) + 400^2/[(2)(6010)] = 673 \text{ °R} \qquad T_2 = 673 - 1100^2/[(2)(6010)] = 572 \text{ °R}$$

$$c_2 = \sqrt{(1.40)(1716)(572)} = 1172 \text{ ft/s} \qquad (N_M)_2 = \tfrac{1100}{1172} = 0.939$$

(b) $\qquad\qquad U_{\text{max}} = \sqrt{2c_p T_0} = \sqrt{(2)(6010)(673)} = 2844 \text{ ft/s}$

(c) $\quad (p_0) = (p)\{[1 + (k - 1)/2](N_M)^2\}^{k/(k-1)} \qquad (N_M)_1 = V_1/\sqrt{kRT_1} = 400/\sqrt{(1.40)(1716)(200 + 460)} = 0.318$

$$(p_0)_1 = 35\{1 + [(1.40 - 1)/2](0.318)^2\}^{1.40/(1.40-1)} = 37.54 \text{ psia}$$

$$(p_0)_2 = 18\{1 + [(1.40 - 1)/2](0.939)^2\}^{1.40/(1.40-1)} = 31.78 \text{ psia}$$

$$(p_0)_2/(p_0)_1 = 31.78/37.54 = 0.847$$

16.121 Air flows isentropically from a reservoir, where $p = 300$ kPa and $T = 500$ K, to section 1 in a duct, where $A_1 = 0.2 \text{ m}^2$ and $V_1 = 600$ m/s. Compute (**a**) T_1, (**b**) $(N_M)_1$, (**c**) p_1, and (**d**) m. Is the flow choked?

(a) $\qquad\qquad T_1 = T_0 - V_1^2/2c_p = 500 - 600^2/[(12)(1005)] = 321 \text{ K}$

(b) $\qquad N_M = V/c \qquad c = \sqrt{kRT} \qquad c_1 = \sqrt{(1.40)(287)(321)} = 359 \text{ m/s} \qquad (N_M)_1 = \tfrac{600}{359} = 1.67$

(c) $\qquad p_1 = p_0/\{1 + [(k - 1)/2](N_M)_1^2\}^{k/(k-1)} = 300/\{1 + [(1.40 - 1)/2](1.67)^2\}^{1.40/(1.40-1)} = 63.6 \text{ kPa}$

(d) $\qquad\qquad \rho = p/RT \qquad \rho_1 = (63.6)(1000)/[(287)(321)] = 0.690 \text{ kg/m}^3$

$$\dot{m} = \rho_1 A_1 V_1 = (0.690)(0.2)(600) = 82.8 \text{ kg/s}$$

Flow is choked in order to generate a supersonic $(N_M)_1$.

16.122 Repeat Prob. 16.121 if the gas is argon [$k = 1.67$, $R = 208$ N-m/(kg-K), $c_p = 519$ m²/(s²–K)].

▌ (a)
$$T_1 = T_0 - V_1^2/2c_p = 500 - 600^2/[(2)(519)] = 153 \text{ K}$$

(b) $\quad N_M = V/c \qquad c = \sqrt{kRT} \qquad c_1 = \sqrt{(1.67)(208)(153)} = 231 \text{ m/s} \qquad (N_M)_1 = \frac{600}{231} = 2.60$

(c) $\quad p_1 = p_0/\{1 + [(k-1)/2](N_M)_1^2\}^{k/(k-1)} = 300/\{1 + [(1.67-1)/2](2.60)^2\}^{1.67/(1.67-1)} = 15.7 \text{ kPa}$

(d) $\qquad\qquad \rho = p/RT \qquad \rho_1 = (15.7)(1000)/[(208)(153)] = 0.493 \text{ kg/m}^3$
$$\dot{m} = \rho_1 A_1 V_1 = (0.493)(0.2)(600) = 59.2 \text{ kg/s}$$

Flow is choked in order to generate a supersonic $(N_M)_1$.

16.123 Sketch a passage which will **(a)** increase the pressure in a subsonic flow isentropically, **(b)** increase the pressure in a supersonic flow isentropically, **(c)** increase the Mach number in a supersonic flow isentropically, **(d)** decrease the Mach number in a subsonic flow isentropically.

▌ See Fig. 16-6.

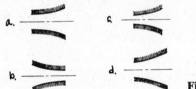

Fig. 16-6

16.124 Air at 28 °C flows from a large tank through a convergent nozzle, which has an exit diameter of 10 mm. Discharge is to the atmosphere where atmospheric pressure is 96.5 kPa abs. Air pressure inside the tank is 40.0 kPa gage. What is the weight flow rate of air through the nozzle?

▌
$$(p_2/p_1)_c = [2/(k+1)]^{k/(k-1)} = [2/(1.40+1)]^{1.40/(1.40-1)} = 0.5283$$

$p_2' = $ atmospheric pressure $= 96.5$ kPa (absolute) $\qquad p_1 = 40 + 96.5 = 136.5$ kPa (absolute)
$$p_2'/p_1 = 96.5/136.5 = 0.7070$$

Since the value of the ratio p_2'/p_1 of 0.7070 is greater than the value of the critical pressure ratio of 0.5283, flow through the nozzle will be subsonic and

$$G = A_2\sqrt{[2gk/(k-1)]p_1\gamma_1[(p_2/p_1)^{2/k} - (p_2/p_1)^{(k+1)/k}]} \qquad A_2 = (\pi)(0.010)^2/4 = 0.00007854 \text{ m}^2$$

$$\gamma = p/RT \qquad \gamma_1 = 136.5/[(29.3)(28+273)] = 0.01548 \text{ kN/m}^3 \qquad p_2 = p_2' = 96.5 \text{ kN/m}^2$$

$$G = (0.00007854)\sqrt{\frac{(2)(9.807)(1.40)}{1.40-1}(136.5)(0.01548)\left[\left(\frac{96.5}{136.5}\right)^{2/1.40} - \left(\frac{96.5}{136.5}\right)^{(1.40+1)/1.40}\right]}$$

$$= 0.000227 \text{ kN/s} \quad \text{or} \quad 0.227 \text{ N/s}$$

16.125 Air at 75 °F flows from a large tank through a convergent nozzle, which has an exit diameter of 1.5 in. The discharge is to the atmosphere, where atmospheric pressure is 14.0 psia. Air pressure inside the tank is 25.0 psig. What is the weight flow rate of air through the nozzle?

▌
$$(p_2/p_1)_c = 0.5283 \qquad \text{(as determined for air in Prob. 16.124)}$$

$p_2' = 14.0$ psia $\qquad p_1 = 25.0 + 14.0 = 39.0$ psia $\qquad p_2'/p_1 = 14.0/39.0 = 0.3590$

Since the value of the ratio p_2'/p_1 of 0.3590 is less than the value of the critical pressure ratio of 0.5283, flow through the nozzle will be sonic. However, the values substituted into the equation below for p_1 must be the pressure that makes the ratio p_2'/p_1 equal to the critical pressure ratio $(p_2/p_1)_c$.

$$G = (A_2 p_1/\sqrt{T_1})\sqrt{(gk/R)[2/(k+1)]^{(k+1)/(k-1)}} \qquad A_2 = (\pi)(1.5)^2/4 = 1.767 \text{ in}^2 = 0.01227 \text{ ft}^2$$

$$p_1 = p_2'/(p_2/p_1)_c = 14.0/0.5283 = 26.5 \text{ psia} \qquad T_1 = 75° + 460 = 535 °R$$

$$G = \{(0.01227)[(26.5)(144)]/\sqrt{535}\}\sqrt{[(32.2)(1.40)/53.3][2/(1.40+1)]^{(1.40+1)/(1.40-1)}} = 1.08 \text{ lb/s}$$

16.126 Air at 308 °C flows from a large tank through a convergent nozzle, which has an exit diameter of 20 mm. Discharge is to the atmosphere where atmospheric pressure is 95.0 kPa abs. Air pressure inside the tank is 50.0 kPa gage. What is the weight flow of air through the nozzle?

$$(p_2/p_1)_c = [2/(k + 1)]^{k/(k-1)} = [2/(1.40 + 1)]^{1.40/(1.40-1)} = 0.5283 \qquad p_2' = 95.0 \text{ kPa}$$

$$p_1 = 50 + 95.0 = 145.0 \text{ kPa} \qquad p_2'/p_1 = 95.0/145.0 = 0.6522$$

Since $[p_2'/p_1 = 0.6552] > [(p_2/p_1)_c = 0.5283]$, the flow through the nozzle will be subsonic:

$$G = A_2\sqrt{[(2gk)/(k-1)](p_1\gamma_1)[(p_2/p_1)^{2/k} - (p_2/p_1)^{(k+1)/k_1}]}$$

$$A_2 = (\pi)(0.02)^2/4 = 0.0003142 \text{ m}^2 \qquad \gamma = p/RT$$

$$\gamma_1 = 145.0/[(29.3)(30 + 273)] = 0.01633 \text{ kN/m}^3 \qquad p_2 = p_2' = 95.0 \text{ kN/m}^2$$

$$G = 0.0003142\sqrt{[(2)(9.807)(1.40)/(1.40-1)](145)(0.01633)(95.0/145.0)^{2/1.40} - (95.0/145.0)^{(1.40+1)/1.40}}$$

$$= 0.00100 \text{ kN/s} = 1.00 \text{ N/s}$$

16.127 Air at 60 °F flows from a large tank through a convergent nozzle, which has an exit diameter of 1.0 in. The discharge is to the atmosphere, where atmospheric pressure is 14.5 psia. Air pressure inside the tank is 30.0 psig. What is the weight flow rate of air through the nozzle?

▮ From Prob. 16.126, $(p_2/p_1)_c = 0.5283$, $p_2' = 14.5$ psia, $p_1 = 30.0 + 14.5 = 44.5$ psia, $p_2'/p_1 = 14.5/44.5 = 0.3258$. Since $[p_2'/p_1 = 0.3258] < [(p_2/p_1)_c = 0.5283]$, the flow through the nozzle will be sonic.

$$G = [A_2p_1/\sqrt{T_1}]\sqrt{(gk/R)[2/(k+1)]^{(k+1)/(k-1)}} \qquad A = (\pi)(\tfrac{1}{12})^2/4 = 0.005454 \text{ ft}^2$$

$$p_1 = p_2'/(p_2/p_1)_c = 14.5/0.5283 = 27.45 \text{ psia} \qquad T_1 = 60 + 460 = 520 °\text{R}$$

$$G = [(0.005454)(27.45)(144)/\sqrt{520}]\sqrt{[(32.2)(1.40)/53.3][2/(1.40+1)]^{(1.40+1)/(1.40-1)}} = 0.503 \text{ lb/s}$$

16.128 Air at 80 °F flows from a large tank through a converging nozzle of 2.0-in exit diameter. The discharge is to an atmospheric pressure of 13.5 psia. Determine the flow through the nozzle for pressures within the tank of 5, 10, 15, and 20 psig. Assume isentropic conditions. Plot G as a function of p_1. Assume that the temperature within the tank is 80 °F in all cases.

▮ From Prob. 16.126 the critical pressure ratio for air is $(p_2/p_1)_c = 0.528$. If the flow at the throat is subsonic, $p_2/p_1 > (p_2/p_1)_c$. Thus for subsonic flow at the throat, $p_2/p_1 > 0.528$ and $p_3 = p_2$. So $p_3/p_1 > 0.528$ and $p_1 < p_3/0.528$.

Since $p_3 = 13.5$ psia, the flow at the throat will be subsonic if $p_1 < 25.6$ psia (12.1 psig) and sonic if $p_1 > 25.6$ psia (12.1 psig).

To find the flow rate for conditions where $p_1 < 12.1$ psig (subsonic flow at throat), we use

$$G = A_2\sqrt{(2g)[k/(k-1)]p_1\gamma_1[(p_2/p_1)^{2/k} - (p_2/p_1)^{(k+1)/k}]} \qquad (1)$$

Substituting the appropriate value of p_1 into the equation and noting that for this condition $p_2 = p_3 = 13.5$ psia, we get: for $p_1 = 5$ psig (18.5 psia); $G = 1.20$ lb/s; for $p_1 = 10$ psig (23.5 psia), $G = 1.69$ lb/s. To find the

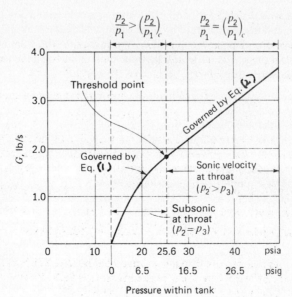

Pressure within tank   **Fig. 16-7**

flow rate for conditions where $p_1 > 12.1$ psig (sonic flow at the throat) we use

$$G' = g(A_2 p_1/\sqrt{T_1})\sqrt{(k/R)[2/(k+1)]^{(k+1)/(k-1)}} \tag{2}$$

We get: for $p_1 = 15$ psig (28.5 psia), $G' = 2.04$ lb/s; for $p_1 = 20$ psig (33.5 psia), $G' = 2.40$ lb/s. Substituting $p_1 = 25.6$ psia in Eq. (1) for subsonic flow gives $G = 1.84$ lb/s as does Eq. (2). This is the threshold point at which the flow in the throat changes from subsonic to sonic. When $p_1 = 25.6$ psia the flow rate as found from Eq. (2) is $G'_{p=25.6\,\text{psia}} = [32.2(0.0218)(144)(25.6)]/\sqrt{540}\sqrt{(1.4/1715)(2/2.4)^{2.4/0.4}} = 1.84$ lb/s. As p_1 is increased beyond 12.1 psig, sonic flow prevails at the throat and the flow rate increases linearly with $(p_1)_{\text{abs}}$ as indicated by Eq. (2). The variation of the flow rate with p_1 is shown in Fig. 16-7. Other information concerning various aspects of this problem can be found by applying the gas law ($pv = RT$ or $\gamma = gp/RT$), the equation of state ($pv^k = $ constant or $p/\gamma^k = $ constant), continuity ($G = \gamma A V$), and the energy equation. Applying these, for example, for the case where $p_1 = 5$ psig (18.5 psia) yields $\gamma_1 = 0.093$ lb/ft^3, $\gamma_2 = 0.0743$ lb/ft^3, $V_2 = 740$ fps, $T_2 = 491$ °R, and $p_2 = 13.5$ psia. Note that in this case $p_2 = p_3$.

16.129 Refer to Prob. 16.128. If the pressure in the tank is 5 psig, find G, p_2, and T_2.

▮ Air $\quad p_1 = 5$ psig $\qquad p_1 = 5 + p_{\text{atmos}} = 5 + 13.5 = 18.50$ psia

$$\gamma_1 = gp_1/RT_1 = (32.2)(18.50)(144)/[(17715)(540)] = 0.0926 \text{ lb/ft}^3$$

$$p_3/p_1 = 14.5/18.5 = 0.730 > 0.528 = (p_2/p_1)_c \text{ so } G = A_2\sqrt{2g[k/(k-1)]p_1\gamma_1[(p_2/p_1)^{2/k} - (p_2/p_1)^{(k+1)/k}]}$$

$$= (\pi/4)(\tfrac{2}{12})^2\sqrt{64.4(1.4/0.4)[(18.5)(144)(0.0926)][(13.5/18.5)^{2/1.4} - (13.5/18.5)^{2.4/1.4}]}$$

$$= 1.204 \text{ lb/s}.$$

$$V_2^2/2g = (p_2/\gamma_2)[k/(k-1)][(p_1/p_2)^{(k-1)/k} - 1] \tag{1}$$

$$G = \gamma_2 A_2 V_2 \tag{2}$$

$$\gamma_2 = gp_2/RT_2 \tag{3}$$

$$p_1/\gamma_1^k = p_2/\gamma_2^k \tag{4}$$

From (2)

$$\gamma_2 = G/A_2 V_2 = 1.204/(0.0218 V_2) = 55.0/V_2 \tag{5}$$

From (4) $p_2 = p_1(\gamma_2^k/\gamma_1^k) = [(18.5)(144)][\gamma_2^k/(0.0926)^{1.4}] = (2660\gamma_2^k)/0.0357 = 74\,500\gamma_2^k$. Substitute for γ_2 from (5): $p_2 = 74\,500(55/V_2)^{1.4} = 20.4 \times 10^6/V_2^{1.4}$. From (4): $p_1/p_2 = (\gamma_1/\gamma_2)^{1.4} = [0.0926/(55/V_2)]^{1.4} = 0.0001308 V_2^{1.4}$. Substitute into (1):

$$V_2^2 = (64.4)[(20.4 \times 10^6)(1.4/0.4)/(V_2^{1.4})(55/V_2)][0.0001308 V_2^{1.4})^{0.4/1.4} - 1]$$

$V_2^{2.4} = (83.6 \times 10^6)[0.0777 V_2^{0.4} - 1]$; by T and E, $V_2 = 741$ fps. From (2): $\gamma_2 = G/A_2 V_2 = 1.204/(0.0218)(741) = 0.0745$ lb/ft^3. From (4): $p_2 = (18.5)[(0.0742)^{1.4}/(0.0926)^{1.4}] = 13.57$ psia. From (3): $T_2 = gp_2/R\gamma_2 = (32.2)(13.57)(144)/[(1715)(0.0742)] = 492$ °R.

16.130 Air discharges from a tank through a converging-diverging nozzle with a 2.0-in-diameter throat. Within the tank the pressure is 50 psia and the temperature is 80 °F, while outside the tank the pressure is 13.5 psia. The nozzle is to operate with supersonic flow throughout its diverging section with a 13.5-psia pressure at its outlet. Find the required diameter of the nozzle outlet. Determine the flow rate and the velocities and temperatures at sections 2 and 3. Assume isentropic flow.

▮ The pressure at the throat must be such that sonic velocity will occur there. Hence $p_2 = p_c = 0.528 p_1 = 26.4$ psia.

The velocity at the outlet may be found from $V_3^2/2g = (RT_1/g)[k/(k-1)][1 - (p_3/p_1)^{(k-1)/k}] = [(1715)(540)/32.2](1.4/0.4)[1 - (13.5/50)^{0.4/1.4}] = 31\,400$ ft, $V_3 = 1420$ fps. The flow rate is computed from $G = (gA_2 p_1/\sqrt{T_1})\sqrt{(k/R)[2/(k+1)]^{(k+1)/(k-1)}}$, $G' = [(32.2)(0.0218)(50)(144)/\sqrt{540}]\sqrt{(1.4/1715)(2/2.4)^{2.4/0.4}} = 3.58$ lb/s. The temperature at 3 may be determined by using $V_3^2 - V_1^2 = 2c_p(T_1 - T_2)$, $(1420)^2 = 2(6000)(540 - T_3)$, $T_3 = 372$ °R $= -88$ °F. From the perfect-gas law $p_3/\gamma_3 = RT_3/g$. $\gamma_3 = (32.2)(13.5)(144)/[(1715)(372)] = 0.098$ lb/ft^3. Isentropic flow between 2 and 3 may be assumed, since the shock wave does not occur within that region. Thus $p_2/\gamma_2^{1.4} = p_3/\gamma_3^{1.4}$. $26.4/\gamma_2^{1.4} = 13.5/(0.098)^{1.4}$, $\gamma_2 = 0.158$ lb/ft^3. The velocity at 2 may now be computed: $V_2 = G/\gamma_2 A_2 = 3.58/[(0.158)(0.0218)] = 1040$ fps. The temperature at 2 results from $(1420)^2 - (1040)^2 = 2(6000)(T_2 - 372)$, $T_2 = 450$ °R $= -10$ °F. The area at 3 is computed from $A_3 = G/\gamma_3 V_3 = 3.58/[(0.098)(1420)] = 0.0257$ ft^2. Finally, $D_3 = 2.17$ in, the required outlet diameter.

Check for sonic velocity at throat: $c_2 = \sqrt{kRT_2} = \sqrt{(1.4)(1715)(450)} = 1040$ fps $= V_2$. With sonic velocity at the throat, if $D_3 < 2.17$ in, there will be supersonic flow throughout the tube and a shock wave will occur in the flow field downstream of the nozzle exit. If $D_3 > 2.17$ in, with sonic flow in the throat, in order to satisfy pressure conditions, a shock wave will occur in the tube somewhere between the throat and the nozzle exit.

16.131 Work Prob. 16.130 with all data the same except for the pressure within the tank, which is 100 psia rather than 50 psia.

❚ Pressure at throat: $p_2 = p_c = 0.528(100) = 52.8$ psia. Find the velocity at outlet from $V_3 = (2g)[k/(k - 1)](R/g)(T)[1 - (p_2/p_1)^{(k-1)/k}] = (64.4)(1.4/0.4)(1715/32.2)(540)[1 - (13.5/100)^{0.4/1.4}]$; $V_3 = 1680$ fps. By computation similar to that of Prob. 16.130, $G' = 7.20$ lb/s (exactly twice that of Prob. 16.130), $V_3^2 = [2k/(k - 1)](RT_1)[1 - (T_3/T_1)]$, $1680^2 = (2.8/0.4)(1715)(540)[1 - (T_3/540)]$; $T_3 = 305\,°\text{R} = -155\,°\text{F}$. $\gamma = pg/RT$; $\gamma_3 = (13.5)(144)(32.2)/[(1715)(305)] = 0.1197$ pcf. $p/\gamma^k = $ constant (isentropic), therefore, $5.28/\gamma_2^{1.4} = 13.5/(0.1197)^{1.4}$; $\gamma_2 = 0.317$ pcf. $V_2 = G/\gamma_2 A_2 = 7.20/[(0.317)(0.0218)] = 1042$ fps, $T_2 = 448\,°\text{R} = -12\,°\text{F}$, $A_3 = G/\gamma_3 V_3 = 7.20/[(0.1197)(1680)] = 0.0358$ ft^2 and $D = 2.56$ in.

16.132 Air within a tank at 120 °F flows isentropically through a 2-in-diameter convergent nozzle into a 14.2-psia atmosphere. Find the flow rate for air pressures within the tank of 5, 10, 20, 40, and 50 psia.

❚ For air ($k = 1.4$), $(p_2/p_1)_c = 0.528$; $p_2 = 14.2$ psia. If $p_2/p_1 > 0.528$, that is, $p_1 < 14.2/0.528 = 26.9$ psia, find G by

$$G = A_2\sqrt{(2g)[k/(k - 1)](p_1\gamma_1)[(p_2/p_1)^{2/k} - (p_2/p_1)^{(k+1)/k}]} \tag{1}$$

Otherwise $p_2/p_1 = 0.528$, that is, $p_1 = 26.9$ psia, find G' by

$$G' = g(A_2 p_1/\sqrt{T_1})\sqrt{(k/R)[2/(k + 1)]^{(k+1)/(k-1)}} \tag{2}$$

For $p_1 = 5 + 14.2 = 19.2$ psia (subsonic flow): $\gamma_1 = gp_1/RT_1 = (32.2)(19.2)(144)/[(1715)(580)] = 0.0895$ lb/ft^3, $G = 0.0218\sqrt{(2)(32.2)(1.4/0.4)(19.2)(144)(0.0895)[(14.2/19.2)^{1.43} - (14.2/19.2)^{1.71}]} = 1.193$ lb/s. For $p_1 \geq 26.9$ psia (sonic flow), G' is given by Eq. (2). For $p_1 = 20 + 14.2 = 34.2$ psia (sonic flow), $G' = (32.2)[(0.0218)(34.2)(144)/\sqrt{580}]\sqrt{(1.4/1715)(2/2.4)^{2.4/0.4}} = 2.37$ lb/s. If $p_1 < 14.7$ psia (negative gage pressure) reverse flow will occur.

p_1, psig	p_1, psia	T_1, °F	p_3, psia	p_2, psia	G or G', lb/s	
0	14.2	120	14.2	14.2	0	
5	19.2	120	14.2	14.2	0.0994	[Eq. (1)]
10	24.2	120	14.2	14.2	0.1237	[Eq. (2)]
12.7	26.9	120	14.2	14.2	1.867	[Eq. (2)]
20	34.2	120	14.2	18.07	2.37	[Eq. (2)]
40	54.2	120	14.2	28.6	3.76	[Eq. (2)]

16.133 Air flows at 150 °F from a large tank through a 1.5-in-diameter converging nozzle. Within the tank the pressure is 85 psia. Calculate the flow rate for external pressures of 10, 30, 50, and 70 psia. Assume isentropic conditions. Assume that the temperature within the tank is 150 °F in all cases. Compute also the temperature at the nozzle outlet for each condition.

❚ For air ($k = 1.4$), $(p_2/p_1)_c = 0.528$. If $p_3 > 0.528(85) = 44.9$ psia the flow is subsonic and $p_3 = p_2$. If $p_3 < 0.528(85) = 44.9$ psia the flow is sonic and $p_3 < p_2$. For sonic flows (i.e., $p_3 < 44.9$ psia), G' is given by

$$G' = (g)(A_2 p_1/\sqrt{T_1})\sqrt{(k/R)[2/(k + 1)]^{(k+1)/(k-1)}} \tag{1}$$
$$G' = [(32.2\pi)(1.5)^2/4][85/\sqrt{(460 + 150)}]\sqrt{(1.4/1715)(2/2.4)^{2.4/0.4}} = 3.24 \text{ lb/s}$$

For subsonic flows (i.e., $p_3 > 44.9$ psia), G is given by

$$G = A_2\sqrt{(2g)[k/(k - 1)](p_1\gamma_1)[(p_2/p_1)^{2/k} - (p_2/p_1)^{(k+1)/k}]} \tag{2}$$
$$\gamma_1 = gp_1/RT_1 = (32.2)(85)(144)/[(1715)(610)] = 0.378 \text{ pcf}$$

Substitute into Eq. (2): $p_2 = p_3$ (express in psf), $A_2 = 0.01227$ ft^2, $\gamma_1 = 0.378$ pcf, $p_1 = 85$ psia $= 12\,240$ psf, $k = 1.4$. For $p_2 < 44.9$ psia $= 6470$ psf, $G' = 3.35$ lb/s [Eq. (1)]. For $p_2 = p_3 = 50$ psia $= 7200$ psf, $G = 3.22$ lb/s [Eq. (2)]. For $p_2 = p_3 = 70$ psia $= 10\,000$ psf, $G = 2.49$ lb/s [Eq. (2)]. Determination of temperatures: For each case substitute the corresponding values of p_1 and p_2 into $T_2 = T_1 - (p_1 g/R\gamma_1)[1 - (p_2/p_1)^{(k-1)/k}]$. Also substitute

$T_1 = 150\,°F$, $\gamma_1 = 0.378$ pcf. Results are

p_1, psia	T_1, °F	p_3, psia	p_2, psia	G or G', lb/s	T_2, °F
85	150	10	44.9	3.35	49
85	150	30	44.9	3.35	49
85	150	50	50	3.22	64
85	150	70	70	2.49	117
85	150	85	85	0	150

16.134 Air enters a converging-diverging nozzle at a pressure of 120 psia and a temperature of 90 °F. Neglecting the entrance velocity and assuming a frictionless process, find the Mach number at the cross section where the pressure is 35 psia.

❚ $\quad T_2 = T_1(p_2/p_1)^{(k-1)/k}$ $\quad$ (from $pv^k = $ const $\quad$ and $\quad pv = RT$) $\quad T_2 = 550(\frac{35}{120})^{0.4/1.4} = 387\,°R$
$\quad V_2^2 = [2k/(k-1)](RT_1)[1 - (T_2/T_1)] = 2(1.40/0.40)(1715)(550)[1 - (\frac{387}{550})]$ $\quad V_2 = 139$ fps
$\quad c = (kRT_2)^{1/2} = [1.4(1715)387]^{1/2} = 964$ fps $\quad N_M = V/c = \frac{1399}{964} = 1.451$

16.135 Carbon dioxide within a tank at 40 psia and 80 °F discharges through a convergent nozzle into a 14.2-psia atmosphere. Find the velocity, pressure, and temperature at the nozzle outlet. Assume isentropic conditions.

❚ $\quad (p_2/p_1)_c = [2/(k+1)]^{k/(k-1)}$ $\quad k = 1.28$ $\quad (p_2/p_1)_c = (2/2.28)^{1.28/0.28} = 0.549$

If $p_3 > 0.549p_1 = 0.549(40) = 22.0$ psia, the flow is subsonic. But $p_3 = 14.2$ psia, which is < 22.0 psia; therefore, flow is sonic. The pressure at the nozzle outlet is 22.0 psia. $T_2 = T_1 - (p_1g/R\gamma_1)[1 - (p_2/p_1)^{(k-1)/k}]$. Noting that $p_1g/R\gamma_1 = T_1 = (460 + 80)\,°R$: $T_2 = 80 - 540[1 - (22.0/40)^{0.28/1.28}] = 13.80\,°F$, $V_2^2 = [2k/(k-1)](RT_1)[1 - (T_2/T_1)] = 2(1.28/0.28)(1123)(540)[1 - (\frac{474}{540})]$; $V_2 = 823$ fps.

16.136 In Prob. 16.135 if the pressure and temperature within the tank had been 20 psia and 100 °F, what would have been the velocity, pressure, and temperature at the nozzle outlet? Assume isentropic conditions.

❚ As for solution to Prob. 16.135: $(p_2/p_1)_c = 0.549$. If $p_3 > 0.549p_1 = 10.98$ psia, the flow is subsonic. $p_3 = 14.2$ psia > 10.98 psia, therefore flow is subsonic. With subsonic flow $p_2 = p_3 = 14.2$ psia, $T_2 = T_1 - (p_1g/R\gamma_1)[1 - (p_2/p_1)^{(k-1)/k}] = 100 - 560[1 - (14.2/20)^{0.219}] = 59.5\,°F$, $V_2^2 = [2k/(k-1)](RT_1)[1 - (T_2/T_1)] = (2)(1.28/0.28)(1123)(560)[1 - (519.5/560)]$; $V_2 = 645$ fps.

16.137 Air discharges from a large tank through a converging-diverging nozzle. The throat diameter is 3.0 in, and the exit diameter is 4.0 in. Within the tank the air pressure and temperature are 40 psia and 150 °F, respectively. Calculate the flow rate for external pressures of 39, 38, 36, and 30 psia. Assume no friction.

❚ $\gamma_1 = gp_1/RT_1 = (32.2)(40)(144)/[(1715)(610)] = 0.1773$ pcf. If sonic flow occurs,
$G' = (g)(A_2p_1/\sqrt{T})\sqrt{(k/R)[2/(k+1)]^{(k+1)/(k-1)}} = (32.2)[(0.0491)(40)(144)/\sqrt{460 + 150}]\sqrt{(1.4/1715)(2.0/2.4)^{2.4/0.4}} = 6.10$ lb/s. For subsonic flow:
$G = A_2\sqrt{(2g)[k/(k-1)](p_1\gamma_1)[(p_2/p_1)^{2/k} - (p_2/p_1)^{(k+1)/k}]} = (\pi/4)(\frac{4}{12})^2\sqrt{(2)(32.2)(1.4/0.4)(40)(144)(0.1773)[(p_3/40)^{2/1.4} - (p_3/40)^{2.4/1.4}]}$. By T and E using the above equation, $p_3 = p_2 = 36.75$ psia for $G = 6.10$ lb/s.

p_3, psia	G, lb/s	G', lb/s
40	0	
39	3.49	
38	4.87	
36.75	6.10	6.10
36	—	6.10
30	—	6.10

16.138 Air is to flow through a converging-diverging nozzle at 18 lb/s. At the throat the pressure, temperature, and velocity are to be 20 psia, 100 °F, and 500 fps, respectively. At the outlet the velocity is to be 200 fps. Determine the throat diameter. Assume isentropic flow.

$$G = \gamma_2 A_2 V^2 \qquad \gamma = p/RT \qquad \gamma_2 = (20)(144)/[(53.3)(100 + 460)] = 0.09649 \text{ lb/ft}^3$$
$$18 = (0.09649)(\pi D^2/4)(500) \qquad D = 0.6892 \text{ ft} \quad \text{or} \quad 8.27 \text{ in}$$

16.139 Air in a tank under a pressure of 140 psia and 70 °F flows out into the atmosphere through a 1.00-in-diameter converging nozzle. (a) Find the flow rate. (b) If a diverging section with an outlet diameter of 1.50 in were attached to the converging nozzle, what then would be the flow rate? Neglect friction.

∎ (a) $p_3/p_2 = 14.7/140 = 0.105 < 0.528$; so flow is sonic: Use $G' = (g)[A_2 p_1/\sqrt{T_1}] \times \sqrt{(k/R)[2/(k+1)]^{(k+1)/(k-1)}}$, $G' = (32.2)[(0.00545)(140)(144)/\sqrt{460 + 70}]\sqrt{(1.4/1715)(2/2.4)^{2.4/0.4}} = 2.54 \text{ lb/s}$.
(b) Attaching a diverging nozzle with sonic flow (and no change in Δp) will not change the flow rate.

16.140 Repeat Prob. 16.139 for the case where the air within the tank is at 20 psia. Assume all other data to be the same.

∎ (a) $p_3/p_1 = 14.7/20 = 0.735 > 0.528$; so flow is subsonic. $\gamma_1 = gp_1/RT_1 = (32.2)(20)(144)/[(1715)(530)] = 0.1020 \text{ pcf}$, $G = A_2\sqrt{(2g)[k/(k-1)](p_1\gamma_1)[(p_2/p_1)^{2/k} - (p_2/p_1)^{(k+1)/k}]} = (\pi/4)(\frac{1}{12})^2\sqrt{2(32.2)(1.4/0.4)(20)(144)(0.1020)(0.735^{2/1.4} - 0.735^{2.4/1.4})} = 0.327 \text{ lb/s}$.
(b) Attaching a diverging nozzle with subsonic flow (when the air behaves like an incompressible fluid) *will* increase the flow rate. It only changes A_2 in the equation above. Thus $G = 0.327(1.5/1.0)^2 = 0.736 \text{ lb/s}$ *providing* subsonic flow still occurs in the throat. Check this by computing G', which is G_{max} if p_1 is not increased. $G' = (g)(A_2 p_1/\sqrt{T_1})\sqrt{(k/R)[2/(k+1)]^{(k+1)/(k-1)}} = (32.2)(\pi/4)(1.0)^2(20/\sqrt{460 + 70})\sqrt{(1.4/1715)(2/2.4)^{2.4/0.4}} = 0.363 \text{ lb/s}$. Thus, with the 1.5-in diameter outlet the flow at the throat *has* become sonic, and $G = 0.363 \text{ lb/s}$.

16.141 Air discharges from a large tank through a converging-diverging nozzle with a 2.5-cm-diameter throat into the atmosphere. The gage pressure and temperature in the tank are 700 kN/m² and 40 °C, respectively; the barometric pressure is 995 mbar. (a) Find the nozzle-tip diameter required for p_3 to be equal to the atmospheric pressure. For this case, what are the flow velocity, sonic velocity, and Mach number at the nozzle exit?

∎ $p_{atm} = 995 \text{ mbar} = 99.5 \text{ kN/m}^2$ abs; $p_1 = 700 + 99.5 = 799.5 \text{ kN/m}^2$ abs. Assume sonic velocity at the throat:

$$V_3^2/2g = (RT/g)[k/(k-1)][1 - (p_2/p_1)^{(k-1)/k}] \qquad V_3^2/19.62 = [287(313)/9.81](1.4/0.4)[1 - (99.5/799.5)^{0.4/1.4}]$$
$$V_3 = 531 \text{ m/s}$$
$$G' = (g)(A_2 p_1/\sqrt{T_1})\sqrt{(k/R)[2/(k+1)]^{(k+1)/(k-1)}}$$
$$= [9.81(\pi)(0.0125^2)(799\,500)/\sqrt{273 + 40}]\sqrt{(1.4/287)(2/2.4)^{2.4/0.4}} = 8.80 \text{ N/s}$$
$$V_3^2 - V_1^2 = 2c_p(T_1 - T_3) \qquad 531^2 - 0 = 2(1003)(313 - T_3) \qquad T_3 = 171.9 \text{ K}$$
$$\gamma_3 = gp_3/RT_3 = (9.81)(99\,500)/[(287)(171.9)] = 19.78 \text{ N/m}^3$$
$$A_3 = G/\gamma_3 V_3 = 8.80/(19.78)(531) = 0.000838 \text{ m}^2 = 8.38 \text{ cm}^2$$
$$D_3 = 3.27 \text{ cm} \qquad c = [kRT]^{1/2} = [1.4(287)171.9]^{1/2} = 263 \text{ m/s} \qquad M_3 = V_3/c_3 = \tfrac{531}{263} = 2.02$$

Thus sonic velocity occurs at the throat and the shock wave is located downstream of the nozzle tip.

16.142 Helium enters a 100-mm-ID pipe from a converging-diverging nozzle at $N_M = 1.30$, $p = 14 \text{ N/m}^2$ abs, $T = 225 \text{ K}$. Determine for isothermal flow (a) the maximum length of pipe for no choking, (b) the downstream conditions, and (c) the length from the exit to the section where $N_M = 1.0$ ($f = 0.016$, $k = 1.66$).

∎ (a)
$$(f/D)L_{max} = [(1 - kN_M^2)/kN_M^2] + \ln(kN_M^2)$$
$$0.016 L_{max}/0.1 = \{[1 - 1.66(1.3)^2]/(1.66)(1.3)^2\} + \ln[1.66(1.3^2)]$$

from which $L_{max} = 2.425 \text{ m}$.

(b) $\qquad\qquad p^{*t}/p = \sqrt{k}N_M \qquad p^{*t} = p\sqrt{k}N_M = 14\sqrt{1.66}\ 1.3 = 23.45 \text{ kN/m}^2$ abs

The Mach number at the exit is $1/\sqrt{1.66} = 0.776$.

$$\int_V^{V^{*t}} \frac{dV}{V} = \frac{1}{2}\int_M^{1/\sqrt{k}} \frac{dN_M^2}{N_M^2} \qquad \text{or} \qquad \frac{V^{*t}}{V} = \frac{1}{\sqrt{k}N_M}$$

At the upstream section $V = N_M\sqrt{kRT} = 1.3\sqrt{(1.66)(2077)(225)} = 1145$ m/s and $V^{*'} = V/(\sqrt{k}N_M) = 1145/(\sqrt{1.66}\,1.3) = 683.6$ m/s.

(c) Substituting into the equation in (a) for $N_M = 1$, $(0.016/0.1)L'_{max} = [(1-1.66)/1.66] + \ln 1.66$ or $L'_{max} = 0.683$ m. $N_M = 1$ occurs 0.683 m from the exit.

16.143 Helium discharges from a $\frac{1}{2}$-in-diameter converging nozzle at its maximum rate for reservoir conditions of $p = 4$ atm, $T = 25\,°C$. What restrictions are placed on the downstream pressure? Calculate the mass flow rate and velocity of the gas at the nozzle.

$$\dot{m}_{max} = (A^*p_0/\sqrt{T_0})\sqrt{(k/R)[2/(k+1)]^{(k+1)/(k-1)}} \qquad A^* = (\pi)[(\tfrac{1}{2})(2.54)/100]^2/4 = 0.0001267\ m^2$$

$$p_0 = (4)(101\,310) = 405\,240\ Pa \qquad T_0 = 25 + 273 = 298\ K \qquad k = 1.66 \qquad R = 2077\ N\text{-}m/(kg\text{-}K)$$

$$\dot{m}_{max} = [(0.0001267)(405\,240)/\sqrt{298}]\sqrt{(1.66/2077)[2/(1.66+1)]^{(1.66+1)/(1.66-1)}} = 0.0473\ kg/s$$

$$V^* = \dot{m}_{max}/\rho^* A^* \qquad \rho^* = \rho_0(p^*/p_0)^{1/k} \qquad \rho_0 = p_0/RT_0 = 405\,240/[(2077)(298)] = 0.655\ kg/m^3$$

$$p^*/p_0 = [2/(k+1)]^{k/(k-1)} = [2/(1.66+1)]^{1.66/(1.66-1)} = 0.488 \qquad p^* = (0.488)(405\,240) = 197\,757\ Pa$$

$$\rho^* = 0.655(197\,757/405\,240)^{1/1.66} = 0.425\ kg/m^3$$

$$V^* = 0.0473/(0.425)(0.0001267) = 878\ m/s \qquad p_{downstream} < p^*$$

16.144 Air in a reservoir at 280 psia, $T = 290\,°F$, flows through a 2-in-diameter throat in a converging-diverging nozzle. For $N_M = 1$ at the throat, calculate p, ρ, and T there.

▮ For critical conditions at the throat, $T^* = 0.833$, $p^* = 0.528$, and $\rho^* = 0.632$.

$$T_{throat} = (0.833)(290 + 460) = 625\,°R \qquad p_{throat} = (0.528)(280) = 148\ psia$$

$$\rho = p/RT \qquad \rho_0 = (280)(144)/[(1716)(290 + 460)] = 0.0313\ slug/ft^3$$

$$\rho_{throat} = (0.632)(0.0313) = 0.0198\ slug/ft^3$$

16.145 What must be the velocity, pressure, density, temperature, and diameter at a cross section of the nozzle of Prob. 16.144 where $N_M = 2.4$?

▮
$$T_0/T = 1 + [(k-1)/2]N_M^2 \qquad (290 + 460)/T = 1 + [(1.40-1)/2](2.4)^2 \qquad T = 349\,°R$$

$$p_0/p = \{1 + [(k-1)/2]N_M^2\}^{k/(k-1)} \qquad 280/p = \{1 + [(1.40-1)/2](2.4)^2\}^{1.40/(1.40-1)} \qquad p = 19.2\ psia$$

$$\rho_0/\rho = \{1 + [(k-1)/2]N_M^2\}^{1/(k-1)} \qquad 0.0313/\rho = \{1 + [(1.40-1)/2](2.4)^2\}^{1/(1.40-1)} \qquad \rho = 0.00461\ slug/ft^3$$

$$N_M = V/\sqrt{kRT} \qquad 2.4 = V/\sqrt{(1.40)(1716)(349)} \qquad V = 2198\ ft/s$$

$$A/A^* = (1/N_M)[(5 + N_M^2)/6]^3 \qquad A/[(\pi)(\tfrac{2}{12})^2/4] = (1/2.4)[(5 + 2.4^2)/6]^3 \qquad A = 0.05243\ ft^2$$

$$\pi D^2/4 = 0.05243 \qquad D = 0.258\ ft \quad \text{or} \quad 3.10\ in$$

16.146 Nitrogen in sonic flow at a 25-mm-diameter throat section has a pressure of 50 kN/m² abs, $T = -20\,°C$. Determine the mass flow rate.

▮
$$\rho^* = p^*/RT^* = (50)(1000)/[(297)(273 - 20)] = 0.6654\ kg/m^3$$

$$V^* = \sqrt{kRT^*} = \sqrt{(1.40)(297)(273 - 20)} = 324.3\ m/s$$

$$\dot{m} = \rho^* A^* V^* = 0.6654[(\pi)(\tfrac{25}{1000})^2/4](324.3) = 0.106\ kg/s$$

16.147 What is the Mach number for Prob. 16.146 at a 40-mm-diameter section in supersonic and in subsonic flow?

▮
$$A/A^* = (1/N_M)[(5 + N_M^2)/6]^3 \qquad [(\pi)(40)^2/4]/[(\pi)(25)^2/4] = (1/N_M)[(5 + N_M^2)/6]^3$$

$$(N_M)_1 = 0.234 \qquad (N_M)_2 = 2.47$$

16.148 What diameter throat section is needed for critical flow of 0.5 lbm/s of carbon monoxide from a reservoir where $p = 300$ psia and $T = 100\,°F$?

▮
$$\dot{m} = \rho^* A^* V^* \qquad \rho_0 = p_0/RT_0 = (300)(144)/[(55.2)(100 + 460)] = 1.398\ lbm/ft^3$$

$$\rho^* = 0.634\rho_0 = (0.634)(1.398) = 0.8863\ lbm/ft^3 \qquad T^* = 0.833T_0 = (0.833)(100 + 460) = 466\,°R$$

$$V^* = \sqrt{kRT^*} = \sqrt{(1.40)[(55.2)(32.2)](466)} = 1077\ ft/s \qquad 0.5 = (0.8863)(A^*)(1077)$$

$$A^* = 0.0005238\ ft^2 \qquad \pi D^2/4 = 0.0005238 \qquad D = 0.02583\ ft \quad \text{or} \quad 0.31\ in$$

16.149 A supersonic nozzle is to be designed for airflow with $N_M = 3.5$ at the exit section, which is 200 mm in diameter and has a pressure of 7 kN/m^2 abs and temperature of $-85\,°C$. Calculate throat area and reservoir conditions.

$$A/A^* = (1/N_M)[(5 + N_M^2)/6]^3 \qquad A_{\text{exit}} = (\pi)(\tfrac{200}{1000})^2/4 = 0.03142 \text{ m}^2$$
$$0.03142/A^* = (1/3.5)[(5 + 3.5^2)/6]^3 \qquad A^* = 0.00463 \text{ m}^2$$
$$T_0 = (T)\{1 + [(k-1)/2]N_M^2\} = (273 - 85)\{1 + [(1.40 - 1)/2](3.5)^2\} = 649 \text{ K}$$
$$p_0 = (p)\{1 + [(k-1)/2]N_M^2\}^{k/(k-1)} = [(7)(1000)]\{1 + [(1.40 - 1)/2](3.5)^2\}^{1.40/(1.40-1)} = 534\,000 \text{ Pa}$$
$$\rho_0 = p_0/RT_0 = 534\,000/[(287)(649)] = 2.87 \text{ kg/m}^3$$

16.150 In Prob. 16.149, calculate the diameter of cross section for $N_M = 1.5$, 2.0, and 2.5.

Using data from Prob. 16.149,

$$\pi(D^*)^2/4 = 0.00463 \text{ m}^2 \qquad D^* = 0.07678 \text{ m} \qquad D = (D^*)(1/\sqrt{N_M})[(5 + N_M^2)/6]^{1.5}$$
$$N_M = 1.5 \qquad D = (0.07678)(1/\sqrt{1.5})[(5 + 1.5^2)/6]^{1.5} = 0.0833 \text{ m}$$
$$N_M = 2.0 \qquad D = (0.07678)(1/\sqrt{2.0})[(5 + 2.0^2)/6]^{1.5} = 0.0997 \text{ m}$$
$$N_M = 2.5 \qquad D = (0.07678)(1/\sqrt{2.5})[(5 + 2.5^2)/6]^{1.5} = 0.125 \text{ m}$$

16.151 Calculate the exit velocity and the mass rate of flow of nitrogen from a reservoir where $p = 4$ atm and $T = 25\,°C$, through a converging nozzle of 60 mm diameter discharging to the atmosphere.

$$T^* = 0.833T_0 = (0.833)(25 + 273) = 248 \text{ K} \quad V^* = \sqrt{kRT^*} = \sqrt{(1.40)(297)(248)} = 321 \text{ m/s} \quad \dot{m} = \rho^*A^*V^*$$
$$\rho^* = 0.634\rho_0 \qquad \rho_0 = p_0/RT_0 = (4)(101\,350)/[(297)(25 + 273)] = 4.580 \text{ kg/m}^3$$
$$\rho^* = (0.634)(4.580) = 2.904 \text{ kg/m}^3 \qquad A^* = (\pi)(\tfrac{60}{1000})^2/4 = 0.002827 \text{ m}^2$$
$$\dot{m} = (2.904)(0.002827)(321) = 2.64 \text{ kg/s}$$

16.152 A converging nozzle has a throat area of 1 in^2 and stagnation air conditions of 120 lb/in^2 abs and 600 °R. Compute the exit pressure and mass flow if the back pressure is (a) 90 lb/in^2 abs and (b) 45 lb/in^2 abs. Assume $k = 1.4$.

We are given $p_0 = 120$ lb/in^2 abs $= 17\,280$ lb/ft^2, $T_0 = 600$ °R. Hence $\rho_0 = p_0/RT_0 = 17\,280/(1717)(600) = 0.0168$ slug/ft^3, $p^*/p_0 = 0.5283$, or $p^* = 63.4$ lb/in^2 abs $= 9130$ lb/ft^2.
(a) Since $p_b = 90$ lb/in^2 abs $> p^*$, the flow is subsonic throughout. The throat Mach number is found from $p_e = p_b$ and

$$N_M^2 = 5[(p_0/p)^{2/7} - 1] \qquad (N_M)_e^2 = 5[(\tfrac{120}{90})^{2/7} - 1] = 0.4283 \qquad (N_M)_e = 0.6545$$
$$T_0/T_e = 1 + 0.2(N_M)_e^2 = 1 + 0.2(0.6545)^2 = 1.0857$$
$$\rho_0/\rho_e = [1 + 0.2(N_M)_e^2]^{2.5} = [1 + (0.2)(0.6545)^2]^{2.5} = 1.2281$$

Hence $T_e = 600/1.0857 = 553$ °R, so that $a_e \approx 49(553)^{1/2} = 1152$ ft/s. $\rho_e = 0.0168/1.2281 = 0.0137$ slug/ft^3. The mass flow is thus given by $\dot{m} = \rho_e A_e V_e = \rho_e A_e[(N_M)a_e] = 0.0137(\tfrac{1}{144})[0.6545(1152)] = 0.0715$ slug/s. The exit pressure equals the back pressure: $p_e = p_b = 90$ lb/in^2 abs.
(b) Since $p_b = 45$ lb/in^2 abs $< p^*$, the throat is choked. The exit pressure is sonic: $p_e = p^* = 63.4$ lb/in^2 abs, $\dot{m} = \dot{m}_{\max} = 0.6847p_0A_t/(RT_0)^{1/2} = 0.6847(17\,280)(\tfrac{1}{144})/[1717(600)]^{1/2} = 0.0810$ slug/s. Any back pressure less than or equal to $p^* = 63.4$ lb/in^2 abs would cause this same choked condition. Notice that the 50 percent increase in throat Mach number from 0.6545 to 1.0 increases the mass flux only 13 percent, from 0.0715 to 0.0810 slug/s.

16.153 A converging-diverging nozzle has a throat area of 0.002 m^2 and an exit area of 0.008 m^2. Air stagnation conditions are $p_0 = 1000$ kPa and $T_0 = 500$ K. Compute the exit pressure and mass flow for (a) design condition and the exit pressure and mass flow if (b) $p_b \approx 300$ kPa and (c) $p_b \approx 900$ kPa. Assume $\gamma = 1.4$.

(a) The design condition corresponds to supersonic isentropic flow at the given area ratio $A_e/A_t = 0.008/0.002 = 4.0$. We can find the design Mach number by $(N_M)_e = [216(A_1/A^*) - 254(A/A^*)^{2/3}]^{1/5}$, $(N_M)_{e,\text{ design}} \approx [216(4.0) - 254(4.0)^{2/3}]^{1/5} \approx 2.95$, $p_0/p_e = [1 + 0.2(N_M)_e^2]^{3.5}$, $p_0/p_e = [1 + 0.2(2.95)^2]^{3.5} = 34.1$ or $p_{e,\text{ design}} = 1000/34.1 = 29.3$ kPa. Since the throat is clearly sonic at design conditions, $\dot{m}_{\text{design}} = \dot{m}_{\max} = 0.6847p_0A_t/(RT_0)^{1/2} = 0.6847(10^6)(0.002)/[287(500)]^{1/2} = 3.61$ kg/s.

(b) For $p_b = 300$ kPa we are definitely far below the subsonic isentropic condition but we may even be below condition with a normal shock in the exit, where oblique shocks occur outside the exit plane. If it is, $p_e = p_{e, design} = 29.3$ kPa because no shock has yet occurred. To find out, compute condition F by assuming an exit normal shock with $(N_M)_1 = 2.95$, that is, the design Mach number just upstream of the shock. $p_2/p_1 = [1/(k+1)][2k(N_M)_1^2 - (k-1)] = (1/2.4)[2.8(2.95)^2 - 0.4] = 9.99$ or $p_2 = 9.99p_1 = 9.99p_{e, design} = 293$ kPa. Since this is less than the given $p_b = 300$ kPa, there is a normal shock just upstream of the exit plane. The exit flow is subsonic and equals the back pressure $p_e = p_b = 300$ kPa. Also $\dot{m} = \dot{m}_{max} = 3.61$ kg/s. The throat is still sonic and choked at its maximum mass flow.

(c) Finally, for $p_b = 900$ kPa, we compute $(N_M)_e$ and p_e. Again $A_e/A_t = 4.0$ for this condition, with a subsonic $(N_M)_e$: $(N_M)_e = [1 + 0.27(A/A^*)^{-2}]/(1.728)(A/A^*)$, $(N_M)_e(C) \approx [1 + 0.27/(4.0)^2]/(1.728)(4.0) = 0.147$. Then the isentropic exit-pressure ratio for this condition is $p_0/p_e = [1 + 0.2(0.147)^2]^{3.5} = 1.0152$ or $p_e = 1000/1.0152 = 985$ kPa. The given back pressure of 900 kPa is less than this value. Thus for this case there is a normal shock just downstream of the throat and the throat is choked: $p_e = p_b = 900$ kPa, $\dot{m} = \dot{m}_{max} = 3.61$ kg/s. For this large exit-area ratio the exit pressure would have to be larger than 985 kPa to cause a subsonic flow in the throat and a mass flow less than maximum.

16.154 Steam in a tank at 450 °F and 100 psia exhausts through a converging nozzle of throat area 0.1 in² to a 1-atm environment. Compute the initial mass flow rate. Assume an ideal gas with $k = 1.33$.

▌ It is sure to be choked since $p_0/p_e = 7$; therefore, $A_e = A^*$. $\dot{m} = \dot{m}_{max} = (k^{1/2})[2/(k+1)]^{(k+1)/2(k-1)}(A^*p_0/\sqrt{RT_0})$, $\dot{m} = 1.33^{1/2}[2/(1.33+1)]^{(1.33+1)/2(1.33-1)}\{(0.1/144)[(100)(144)]/\sqrt{(2759)(450+460)}\} = 0.00424$ slug/s.

16.155 Air in a tank at 700 kPa and 20 °C exhausts through a converging nozzle of throat area 0.65 cm² to a 1-atm environment. Compute the initial mass flow rate.

▌ It is sure to be choked since $p_0/p_e = 7$; therefore, $A_e = A^*$. $\dot{m} = \dot{m}_{max} = (k^{1/2})[2/(k+1)]^{(k+1)/2(k-1)}(A^*p_0/\sqrt{RT_0})$, $\dot{m} = 1.40^{1/2}[2/(1.40+1)]^{(1.40+1)/2(1.40-1)}\{(0.65/10\,000)[(700)(1000)]/\sqrt{(287)(20+273)}\} = 0.107$ kg/s.

16.156 At a point upstream of the throat of a converging-diverging nozzle the properties are $V_1 = 200$ m/s, $T_1 = 300$ K, and $p_1 = 125$ kPa. If the exit flow is supersonic, compute, from isentropic theory, $\dot{m}$ and A_1. The throat area is 35 cm².

▌ $A_1/A^* = (1 + 0.2N_M^2)^3/1.728N_M$ $c = \sqrt{kRT} = \sqrt{(1.40)(287)(300)} = 347$ m/s $N_M = V/c = \frac{200}{347} = 0.576$

Since the exit flow is supersonic, $A^* = A_{throat} = 35$ cm², or 0.00350 m².

$$A_1/0.00350 = [1 + (0.2)(0.576)^2]^3/[(1.728)(0.576)] \qquad A_1 = 0.00426 \text{ m}^2 \quad \text{or} \quad 42.6 \text{ cm}^2$$

$\rho_1 = p_1/RT = (125)(1000)/[(287)(300)] = 1.452$ kg/m³ $\dot{m} = \rho_1 A_1 V_1 = (1.452)(0.00526)(200) = 1.24$ kg/s

16.157 In transonic wind tunnel testing the small area decrease caused by model blockage can be significant. Suppose the test section Mach number is 1.08 and the section area is 1 m² for air with $T_0 = 20$ °C. Find, according to one-dimensional theory, the percentage change in test-section velocity caused by a model of cross section 0.005 (a 0.5 percent blockage).

▌ $T_1 = T_0/(1 + 0.2N_M^2) = (20 + 273)/[1 + (0.2)(1.08)^2] = 238$ K $c = \sqrt{kRT}$

$c_1 = \sqrt{(1.40)(287)(238)} = 309$ m/s $V_1 = c_1(N_M)_1 = (309)(1.08) = 334$ m/s   (unblocked)

$A_1/A^* = (1 + 0.2N_M^2)^3/1.728N_M$ $1/A^* = [1 + (0.2)(1.08)^2]^3/[(1.728)(1.08)]$ $A^* = 0.9949$ m²

If A_1 is blocked by the 0.5 percent model, $A_1 = 0.995$ m². $A_1/A^* = 0.995/0.9949 = 1.00010$. For the blocked condition,

$A_1/A^* = (1 + 0.2N_M^2)^3/1.728N_M$ $1.00010 = (1 + 0.2N_M^2)^3/1.728N_M$ $N_M = 1.011$

$T_1 = (20 + 273)/[1 + (0.2)(1.011)^2] = 243$ K $c_1 = \sqrt{(1.40)(287)(243)} = 312$ m/s

$V_1 = (312)(1.011) = 315$ m/s   (blocked)

Thus a 0.5 percent decrease in area causes a $(334 - 316)/334 = 0.054$, or 5.4 percent decrease in velocity for this transonic flow.

16.158 In flow of air in a converging-diverging duct with a supersonic exit, the throat area is 10 cm² and the throat pressure is 315 kPa. Find the pressure on either side of the throat where $A = 29$ cm².

▌ Supersonic exit means $A^* = A_{throat} = 10$ cm² and $p^* = 315$ kPa.

Subsonic side:

$$A_1/A^* = (1 + 0.2N_M^2)^3/1.728N_M \qquad \frac{29}{10} = (1 + 0.2N_M^2)^3/1.728N_M \qquad N_M = 0.204$$

$$p_1 = p_0/(1 + 0.2N_M^2)^{k/(k-1)} \qquad p_0 = p^*/[2/(k+1)]^{k/(k-1)} = 315/[2/(1.40+1)]^{1.40/(1.40-1)} = 596 \text{ kPa}$$

$$p_1 = 596/[1 + (0.2)(0.204)^2]^{1.40/(1.40-1)} = 579 \text{ kPa}$$

Subsonic side:

$$A_1/A^* = (1 + 0.2N_M^2)^3/1.728N_M \qquad \frac{29}{10} = (1 + 0.2N_M^2)^3/1.728N_M \qquad N_M = 2.60$$

$$p_1 = 596/[1 + (0.2)(2.60)^2]^{1.40/(1.40-1)} = 29.9 \text{ kPa}$$

16.159 A force $F = 2500$ N pushes a cylinder of diameter 12 cm through an insulated cylinder containing air at 20 °C, as in Fig. 16-8. The exit diameter is 3 mm and p_{atm}. Estimate (a) V_e, (b) V_p, and (c) m_e. (d) Under what conditions is the piston velocity independent of the force?

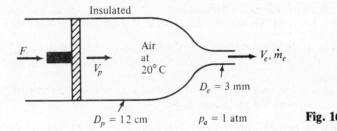

Insulated

Air at 20° C

$V_e, \dot{m}_e$

V_p

$D_e = 3$ mm

$D_p = 12$ cm $\qquad p_a = 1$ atm $\qquad$ **Fig. 16-8**

❙ (a) $\qquad V_e = \{[2k/(k+1)](RT)\}^{1/2} = \{[(2)(1.40)/(1.40+1)](287)(20+273)\}^{1/2} = 313 \text{ m/s}$

(b) $\qquad V_p = (V_e)(A_e/A_p) = (313)\{[(\pi)(\frac{3}{1000})^2/4]/[(\pi)(\frac{12}{100})^2/4]\} = 0.196 \text{ m/s}$

(c) $\dot{m}_e = \dot{m}_{max} = 0.6847 p_0 A^*/(RT_0)^{1/2} \qquad p_0 = F/A_p + p_{atm} = 2500/[(\pi)(\frac{12}{100})^2/4] + 101\,350 = 322\,400 \text{ Pa}$

$$A^* = (\pi)(\tfrac{3}{1000})^2/4 = 0.000007069 \text{ m}^2$$

$$\dot{m}_e = (0.6847)(322\,400)(0.000007069)/[(287)(20+273)]^{1/2} = 0.00538 \text{ kg/s}$$

(d) V_{piston} is independent of F (or p_1) if exit flow is choked.

16.160 Air flows steadily from a reservoir at 20 °C through a nozzle of exit area 20 cm² and strkies a vertical plate as in Fig. 16-9. The flow is subsonic throughout. A force of 135 N is required to hold the plate stationary. Compute (a) $(N_M)_e$, (b) V_e, and (c) p_0.

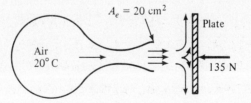

$A_e = 20$ cm²

Plate

Air 20° C

135 N

Fig. 16-9

❙ (a) Assume an ideal gas, $\rho_e V_e^2 = \gamma p_e (N_M)_e^2$, $\gamma p_e A_e (N_M)_e^2 = 135$, $(N_M)_e = \sqrt{135/[(1.40)(101\,350)(20/10\,000)]} = 0.690$.

(b) $\qquad V_e = c_e N_M \qquad T_e = T_0/(1 + 0.2N_M^2) = (20 + 273)/[1 + (0.2)(0.690)^2] = 268 \text{ K}$

$$c_e = \sqrt{kRT} = \sqrt{(1.40)(287)(268)} = 328 \text{ m/s} \qquad V_e = (328)(0.690) = 226 \text{ m/s}$$

(c) $\qquad p_0 = (p_e)\{1 + [(k-1)/2](N_M)_e^2\}^{k/(k-1)} = (101\,350)\{1 + [(1.40-1)/2](0.690)^2\}^{1.40/(1.40-1)}$

$$= 139\,000 \text{ Pa} \quad \text{or} \quad 139 \text{ kPa}$$

16.161 Air flows through a converging-diverging nozzle from a reservoir, where $p_0 = 400$ kPa. The throat area is 10 cm². A normal shock stands in the duct where $A = 15$ cm². Compute the pressure just downstream of this shock.

$$A/A^* = (1+0.2N_M^2)^3/1.728N_M = \tfrac{15}{10} = 1.50 \qquad 1.50 = (1+0.2N_M^2)^3/1.728N_M \qquad N_M = 1.85$$

$$P_1 = p_0/(1+0.2N_M^2)^{k/(k-1)} = 400/[1+(0.2)(1.85)^2]^{1.40/(1.40-1)} = 64.5 \text{ kPa}$$

$$p_2/p_1 = [1/(k+1)][2k(N_M)_1^2 - (k-1)] = [1/(1.40+1)][(2)(1.40)(1.85)^2 - (1.40-1)] = 3.826$$

$$p_2 = (3.826)(64.5) = 247 \text{ kPa}$$

16.162 Air in a large tank at 100 °C and 150 kPa exhausts to the atmosphere through a converging nozzle with a 5-cm² throat area. Compute the exit mass flow if the atmospheric pressure is 100 kPa.

▮
$$\dot{m} = \rho_e A_e V_e \qquad \rho_e = \rho_0/[1+(0.2)(N_M)_e^2]^{2.5}$$

$$\rho_0 = p_0/RT_0 = (150)(1000)/[(287)(100+273)] = 1.401 \text{ kg/m}^3$$

$$p_0/p_a = [1+(0.2)(N_M)_e^2]^{3.5} \qquad 1.5 = [1+(0.2)(N_M)_e^2]^{3.5} \qquad (N_M)_e = 0.784$$

$$\rho_e = 1.401/[1+(0.2)(0.784)^2]^{2.5} = 1.048 \text{ kg/m}^3$$

$$T_e = T/[1+0.2(N_M)_e^2] = (100+273)/[1+(0.2)(0.784)^2] = 322 \text{ K}$$

$$c_e = \sqrt{kRT_e} = \sqrt{(1.40)(287)(332)} = 365 \text{ m/s} \qquad V_e = (c_e)(N_M)_e = (365)(0.784) = 286 \text{ m/s}$$

$$\dot{m} = (1.048)(5/10\,000)(286) = 0.150 \text{ kg/s}$$

16.163 Air flows through a converging-diverging nozzle between two reservoirs, as shown in Fig. 16-10. A mercury manometer between the throat and the downstream reservoir reads $h = 20$ cm. Estimate the downstream reservoir pressure.

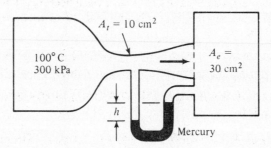

Fig. 16-10

▮ Guess critical flow in the throat: $p_t = p^* = 0.528p_0 = (0.528)(300) = 158.4$ kPa, or 158 400 Pa. Estimate $\rho_{\text{exit}} = 1.2 \text{ kg/m}^3$.
Manometer:

$$p^* - p_e = (\rho_M - \rho_a)(gh) \qquad 158\,400 - p_e = (13\,600 - 1.2)(9.807)(\tfrac{20}{100}) = 26\,670 \text{ Pa}$$

$$p_e = 131\,700 \text{ Pa} \qquad \text{(nonisentropic flow)}$$

16.164 Air in a tank at 120 kPa and 300 K exhausts to the atmosphere through a 5-cm²-throat converging nozzle at a rate of 0.12 kg/s. What is the atmospheric pressure? What is the maximum mass flow possible at low atmospheric pressure?

▮ $\dot{m}_{\max} = 0.6847 p_0 A^*/(RT_0)^{1/2} = (0.6847)[(120)(1000)](5/10\,000)/[(287)(300)]^{1/2} = 0.140$ kg/s. Therefore, the given $\dot{m} = 0.12$ kg/s is less than critical and we have a subsonic exit.

$$\dot{m} = \rho_e A_e V_e = (\rho_0)(\rho_e/\rho_0)(A_e)(N_M)_e(c_0)(c_e/c_0) = (\rho_0 A_e c_0)(N_M)_e[1+0.2(N_M)_e^2]^{-3}$$

$$\rho_0 = p_0/RT_0 = (120)(1000)/[(287)(300)] = 1.394 \text{ kg/m}^3 \qquad c_0 = \sqrt{kRT_0} = \sqrt{(1.40)(287)(300)} = 347.2 \text{ m/s}$$

$$0.12 = [(1.394)(5/10\,000)(347.2)](N_M)_e[1+0.2(N_M)_e^2]^{-3} \qquad (N_M)_e = 0.619 \qquad \text{(by trial and error)}$$

$$p_e = 120/[1+(0.2)(0.619)^2]^{3.5} = 92.7 \text{ kPa}$$

16.165 A supersonic nozzle with an exit area of 8 cm² discharges air at $N_M = 2.5$, $p = 101$ kPa, and $T = 300$ K. Compute (a) exit velocity, (b) p_0, (c) T_0, (d) throat area, and (e) mass flow.

▮ (a)
$$c_e = \sqrt{kRT} = \sqrt{(1.40)(287)(300)} = 347 \text{ m/s} \qquad V_e = c_e N_M = (347)(2.5) = 868 \text{ m/s}$$

(b)
$$p_0 = (p)(1+0.2N_M^2)^{3.5} = 101[1+(0.2)(2.5)^2]^{3.5} = 1726 \text{ kPa}$$

(c)
$$T_0 = (T)(1+0.2N_M^2) = 300[1+(0.2)(2.5)^2] = 675 \text{ K}$$

(d) At $(N_M)_e = 2.5$, $A_e/A^* = 2.637$, $A^* = 8/2.637 = 3.03 \text{ cm}^2$.

(e)
$$\dot{m} = \dot{m}_{max} = 0.6847 p_0 A^* / (RT_0)^{1/2}$$
$$\dot{m} = 0.6847[(1726)(1000)](3.03/10\,000)/[(287)(675)]^{1/2} = 0.814 \text{ kg/s}$$

16.166 A perfect gas (not air) expands isentropically through a supersonic nozzle with an exit area 5 times its throat area. The exit Mach number is 3.8. What is the specific-heat ratio of the gas? What might this gas be? If $p_0 = 300$ kPa, what is the exit pressure of the gas?

I
$$A_e/A^* = [1/(N_M)_e]\{[2 + (k-1)(N_M)_e^2]/(k+1)\}^{(k+1)/(2)(k-1)}$$
$$5 = (1/3.8)\{[2 + (k-1)(3.8)^2]/(k+1)\}^{(k+1)/(2)(k-1)} \qquad k = 1.67 \qquad \text{(by trial and error)}$$

Hence, the gas could be helium or argon. $p_e = p_0/\{1 + [(k-1)/2](N_M)_e^2\}^{k/(k-1)} = 300/\{1 + [(1.66 - 1)/2](3.8)^2\}^{1.66/(1.66-1)} = 3.66$ kPa.

16.167 Air with $p_0 = 300$ kPa and $T_0 = 500$ K flows through a converging–diverging nozzle with throat area of 1 cm^2 and exit area of 4 cm^2 into a receiver tank. The mass flow is 195.2 kg/h. For what range of receiver pressure is this mass flow possible?

I $\dot{m} = 195.2/3600 = 0.0542$ kg/s, $\dot{m}_{max} = 0.6847 p_0 A^*/(RT_0)^{1/2} = 0.6847[(300)(1000)](1/10\,000)/[(287)(500)]^{1/2} = 0.0542$ kg/s. Therefore the throat is choked, and p_b can lie anywhere in between $p_b = 0$ and p_e for isentropic subsonic exit: $A_e/A^* = \frac{4}{1} = 4.00$, $(N_M)_e = 0.147$, $(p_e)_{max} = p/[1 + (0.2)(N_M)_e^2]^{3.5} = 300/[1 + (0.2)(0.147)^2]^{3.5} = 296$ kPa. Thus the possible exit pressure range is $0 \le p_b \le 296$ kPa.

16.168 A spaceship rocket engine has a thrust of 1 000 000 lb when operating at its design point ($p_e = p_a$). If the chamber pressure and temperature are 500 psia and 5000 °R and the gas approximates air with $k = 1.4$, compute the throat diameter of the engine.

$$F = \rho_e A_e V_e^2 = k p_e A_e (N_M)_e^2 \qquad p_0/p_e = [1 + (0.2)(N_M)_e^2]^{3.5} \qquad 500/14.7 = [1 + (0.2)(N_M)_e^2]^{3.5} \qquad (N_M)_e = 2.95$$
$$1\,000\,000 = 1.4[(14.7)(144)](A_e)(2.95)^2 \qquad A_e = 38.77 \text{ ft}^2 \qquad \pi D_e^2/4 = 38.77 \qquad D_e = 7.03 \text{ ft}$$

At $(N_M)_e = 2.95$, $A_e/A^* = [1 + (0.2)(2.95)^2]^3/[(1.728)(2.95)] = 4.038$, $A^* = A_t = 38.77/4.038 = 9.601$ ft^2, $\pi D_t^2/4 = 9.601$, $D_t = 3.50$ ft.

16.169 A nozzle for an ideal rocket is to operate at a 15 250-m altitude in a standard atmosphere where the pressure is 11.60 kPa and is to give a 6.67-kN thrust when the chamber pressure is 1345 kPa and the chamber temperature is 2760 °C. What are the throat and exit areas and the exit velocity and temperature? Take $k = 1.4$ and $R = 355$ N-m/(kg-K) for this calculation. Take the exit pressure to be at ambient pressure.

I We have p/p_0 for the exit, $p/p_0 = 11.60/1345 = 0.008265$. From Table A-16, we see that $M_{exit} = 3.8$, and we have an area ratio $A_{exit}/A_{throat} = 8.95$. Finally, we get for the exit temperature T_e: $T_e/T_0 = 0.257$. Therefore, $T_e = (2760 + 273)(0.257) = 779$ K, and so $T_e = 779 - 273 = 506$ °C. We can also determine the exit velocity easily now. Thus $V_e = M_e c = M_e \sqrt{kRT_e}$. Therefore $V_e = 3.8\sqrt{(1.4)(355)(779)} = 2364$ m/s. To ascertain the *throat* and *exit* areas, we must consider the thrust. Using a control volume comprising the interior of the combustion chamber and nozzle and considering this control volume to be inertial we have from *linear momentum* considerations

$$6670 = (\rho_e V_e A_e) V_e = \rho_e A_e (2364^2) \qquad (1)$$

We can get ρ_e by using the *equation of state* at the exit conditions. Thus $p_e = \rho_e R T_e$. Therefore, $\rho_e = 11\,600/(355)(779) = 0.04195$ kg/m^3. Hence, going back to Eq. (1), we can say $A_e = 6670/(2364^2)(0.04195) = 0.02845$ m^2 and using the area ratio of 8.95 as determined earlier, we have for the throat area: $A^* = 0.0285/8.95 = 0.00318$ m^2.

16.170 The inlet velocity of an isentropic diffuser is 305 m/s and the undisturbed pressure and temperature are 34 500 Pa abs and 235 °C, respectively. If the pressure is increased by 30 percent at the exit of the diffuser, determine the exit velocity and temperature. Use tables.

I
$$V_1 = 305 \text{ m/s} \qquad p_1 = 34\,500 \text{ N/m}^2 \qquad T_1 = 508 \text{ K} \qquad p_2 = 44\,850 \text{ N/m}^2$$

Find (N_M) first:

$$(N_M)_1 = V_1/\sqrt{kRT_1} = 305/\sqrt{(1.4)(287)(508)} = 0.675 \qquad p/p_0 = 0.737$$
$$p_0 = 34\,500/0.737 = 46\,810 \text{ N/m}^2 \qquad p_2/p_0 = 44\,850/46\,810 = 0.958$$
$$(N_M)_2 = 0.249 \qquad T_2/T_0 = 0.988$$

But at section 1 $T_1/T_0 = 0.916$, $T_0 = 0.508/0.916 = 555$. Hence, $T_2 = (555)(0.988) = 548$ K. Also $V_2 = (0.249)\sqrt{(1.4)(287)(1)(548)}$, $V = 116.8$ m/s.

16.171 Air is kept in a tank at a pressure of 6.89×10^5 Pa abs and a temperature of 15 °C. If one allows the air to issue out in a one-dimensional isentropic flow, what is the greatest possible flow per unit area? What is the flow per unit area at the exit of the nozzle where $p = 101\,325$ Pa?

▌ $$p/p_0 = 101\,325/(6.89 \times 10^5) = 0.147 \qquad (N_M)_e = 1.91$$

For the largest G we go to the throat. $G^* = \sqrt{k/R}(p_0/\sqrt{T_0})[2/(k+1)]^{(k+1)/[2(k-1)]}$, $G^* = \sqrt{1.4/287}(6.89 \times 10^5/\sqrt{288})(2/2.4)^{2.4/0.8} = 1644$ kg/m²s. At the exit, $G = \sqrt{k/R}(p_0/\sqrt{T_0})(N_M/\{1 + [(k-1)/2]N_M^2\}^{(k+1)/[2(k-1)]}) = \sqrt{1.4/287}(6.89 \times 10^5/\sqrt{288})(1.91/\{1 + [(1.4-1)/2](1.91)^2\}^{2.4/0.8}) = 1047$ kg/m²s.

16.172 Determine the exit area and the exit velocity for isentropic flow of a perfect gas with $k = 1.4$ in the nozzle shown in Fig. 16-11. How small can we make the exit area and still have isentropic flow with the given conditions entering the nozzle? Use tables.

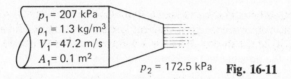

$p_1 = 207$ kPa
$\rho_1 = 1.3$ kg/m³
$V_1 = 47.2$ m/s
$A_1 = 0.1$ m²

$p_2 = 172.5$ kPa **Fig. 16-11**

▌ First find $(N_M)_1$: $N_M = V/c$, $c = \sqrt{kp/\rho} = \sqrt{(1.4)(207\,000)/1.3} = 472$ m/s, $(N_M)_1 = 47.2/472 = 0.1$. From Table A-16, $A/A^* = 5.82$, $A^* = 0.1/5.82 = 0.0172$ m². Also $p/p_0 = 0.993$, $p_0 = 207\,000/0.993 = 208\,460$ Pa. Now go to exit: $p/p_0 = 172\,500/208\,460 = 0.827$. Hence, $(N_M)_e = 0.53$ and $A_e/A^* = 1.29$, $A_e = (0.0172)(1.29) = 0.0222$ m². The smallest possible area is $A^* = 0.0172$ m².
At $N_M = 0.1$:

$$\rho_1/\rho_0 = 0.995 \qquad \rho_0 = 1/0.995 = 1.3/0.995 = 1.31 \text{ kg/m}^3$$

At $(N_M)_e = 0.53$ (exit):

$$\rho_2/\rho_0 = 0.873 \qquad \rho_2 = (0.873)(1.31) = 1.14 \text{ kg/m}^3$$

From continuity:

$$\rho_1 V_1 A_1 = \rho_2 V_2 A_2 \qquad (1.3)(47.2)(0.1) = (1.14)V_2(0.0222) \qquad V_2 = 242 \text{ m/s}$$

16.173 Determine the exit area in Prob. 16.172 for a nozzle efficiency of 90 percent. Take $c_p = 0.24$ Btu/(lbm-°F), $k = 1.4$, and $R = 287$ N-m/(kg-K). Give result in square meters.

▌ We shall need $(T_2)_{isen}$. To compute $(T_2)_{isen}$ we use the result that $(N_M)_1 = 0.1$ from Prob. 16.172. Hence, $T_1 = (T_0)(0.998)$, $T_0 = [207\,000/(1.3)(287)]0.998 = 555/0.998 = 556$ K, $T_1 = 555$ K. Since $(N_M)_2 = 0.53$ from Prob. 16.172 we have $T_2/T_0 = 0.947$, $T_2 = 527$ K, $\eta = (V_2^2/2)_{act}/[V_1^2/2 + c_p(T_1 - T_2)]_{isen}$, $0.9 = (V_2^2/2)_{act}/[47.2^2/2 + (1005)(555 - 527)]$, $(V_2)_{act} = 229$ m/s. We need $(\rho_2)_{act}$ now. For this we must evaluate $(T_2)_{act}$. If there were no "reheating" then $V_2^2/2 = 30\,259$ instead of $27\,233$. Thus we can expect an increase in exit temperature due to reheating of $(30\,259 - 27\,233)/(1005) = (\Delta T_2)$, $\Delta T_2 = \frac{3026}{1005} = 3.01$ K. Hence $(T_2)_{act} = 527 + 3 = 530$ K. Using the equation of state: $\rho_2 = 172\,500/(287)(530) = 1.134$ kg/m³. From *continuity* we get $w = (1.3)(0.1)(47.2) = (1.34)(229)(A_2)$, $A_2 = 0.0236$ m².

16.174 A rocket has an area ratio A_{exit}/A^* of 3.5 for the nozzle, and the stagnation pressure is 50×10^5 Pa abs. Fuel burns at the rate of 45 kg/s and the stagnation temperature is 2870 °C. What should the throat area and exit area be? Take $R = 355$ N-m/(kg-K) and $k = 1.4$.

▌ $$p_0 = 50 \times 10^5 \text{ Pa} \qquad A_e/A^* = 3.5 \qquad T_0 = 3143 \text{ K} \qquad \dot{m} = 45 \text{ kg/s}$$

Using Table A-16 we get for $A_e/A^* = 3.5$, $(N_M)_2 = 2.80$, $T_2/T_0 = 0.389$, $p_2/p_0 = 0.037$. Hence $T_2 = (0.389)(3143) = 1223$ K, $p_2 = (0.037)(50 \times 10^5) = 1.85 \times 10^5$ Pa. The *equation of state* gives us:

$$\rho = p/RT \qquad \rho_2 = 1.85 \times 10^5/(355)(1223) = 0.426 \text{ kg/m}^3$$
$$V = \sqrt{kRT}(N_M) \qquad V_2 = \sqrt{(1.4)(355)(1223)}(2.8) = 2183 \text{ m/s}$$

Use *continuity* next for A_2: $A_2 = 45/(0.426)(2183) = 0.0484$. Finally $A^* = 0.0494/3.5 = 0.01383$ m².

16.175 A nozzle expands air from a pressure $p_0 = 200 \, lb/in^2$ abs and temperature $T_0 = 100 \, °F$ to a pressure of $20 \, lb/in^2$ abs. If the mass flow w is $50 \, lbm/s$, what is the throat area and the exit area? Take $k = 1.4$ and $R = 53.3 \, ft\text{-}lb/(lbm\text{-}°R)$.

$\blacksquare$ $p_e/p_0 = \frac{20}{200} = 0.1$. From Table A-16:

$$(N_M)_e = 2.16 \qquad T_e/T_0 = 0.517 \qquad T_e = (560)(0.517) = 289.5 \, °R$$

$$\rho_e/\rho_0 = 0.192 \qquad A_e/A^* = 1.94$$

$$pv = RT \qquad \rho_e = p_e/RT_e = (20)(144)/[(53.3)(32.2)(289.5)] = 5.796 \times 10^{-3} \, slug/ft^3$$

$$V_e = (N_M)_e c_e \qquad c_e = \sqrt{kRT_e} = \sqrt{(1.4)(53.3)(32.2)(289.5)} = 834 \, ft/s$$

$$V_e = (2.16)(834) = 1801 \, ft/s$$

Continuity:

$$(50/g) = \rho_e V_e A_e = (5.796 \times 10^{-3})(1801)(A_e) \qquad A_e = 0.1488 \, ft^2$$

$$A^* = 0.1488/1.94 = 0.0767 \, ft^2$$

16.176 Determine the throat and the exit areas of an ideal rocket motor to give a static thrust of $6670 \, N$ at $6100\text{-}m$ altitude standard atmosphere if the chamber pressure is $1.035 \times 10^6 \, Pa$ abs and chamber temperature is $3315 \, °C$. Find the velocity at the throat. Take $k = 1.4$ and $R = 355 \, N\text{-}m/(kg\text{-}K)$. Assume that exit pressure is that of surroundings.

$\blacksquare$ $\qquad p_0 = 1.035 \times 10^6 \, Pa \qquad T_0 = 3588 \, K \qquad k = 1.4 \qquad R = 355 \, N\text{-}m/(kg\text{-}K)$

$$p = 46583 \, Pa \qquad \text{(see Table A-8)}$$

Find exit conditions first. Note: $p_e/p_0 = 46\,583/(1.035 \times 10^5) = 0.045$. Hence, from Table A-16, $(N_M)_e = 2.67$. Also $T_e/T_0 = 0.412$, $T_e = (0.412)(3588) = 1478 \, K$. Hence, $V_e = (2.67)\sqrt{(1.4)(355)(1478)} = 2290 \, m/s$, $\rho = p/RT$, $\rho_e = 46\,583/(355)(1478) = 0.0888 \, kg/m^3$. Now use the *momentum* equation. For steady flow and using gage pressures we get $6670 = \rho_e V_e^2 A_e$, $A_e = 6670/[(0.0888)(2290)^2] = 0.0143 \, m^2$. Also $A_e/A^* = 3.09$, $A^* = 0.0143/3.09 = 0.00463$. Finally we want V^*. For this we need T^*. From Table A-16: $T^*/T_0 = 0.833$, $T^* = (3588)(0.833) = 2988 \, K$. Hence, $V^* = \sqrt{(1.4)(355)(2988)} = 1220 \, m/s$.

16.177 Determine the exit area in Prob. 16.176 for a nozzle efficiency of 85 percent. Take $c_p = 1214 \, N\text{-}m/(kg\text{-}K)$.

$\blacksquare$ $\qquad \eta = 0.85 \qquad c_p = 1214 \, N\text{-}m/(kg\text{-}K) \qquad p_0 = 1.035 \times 10^6 \, Pa \qquad T_0 = 3588 \, K$

$$R = 355 \, N\text{-}m/(kg\text{-}K) \qquad p_e = 46\,583 \, Pa \qquad k = 1.4$$

$$\eta = (V_2^2/2)_{act}/[c_p(T_1 - T_2)_{isen}] \qquad (V_2^2/2)_{act} = (0.85)(1214)(3588 - 1478)$$

(see Prob. 16.176 for temperatures)

$$(V_2^2/2)_{act} = 2.177 \times 10^6 \qquad (V_2)_{act} = 2087 \, m/s$$

Next we get $(T_2)_{act}$. Use *first law*:

$$(V_2^2/2)_{act} = c_p[T_0 - (T_2)_{act}] - (2.177 \times 10^6/1214) + 3588 = (T_2)_{act} \qquad (T_2)_{act} = 1795 \, K$$

$$\rho = p/RT \qquad (\rho_2)_{act} = 46\,583/[(355)(1795)] = 0.07310 \, kg/m^3$$

Finally from *momentum* considerations we get $6670 = (\rho V_e^2 A_e)_{act}$, $(A_e) = 6670/[(0.07310)(2087)^2] = 0.021 \, m^2$.

16.178 Figure 16-12 is a convergent-divergent nozzle attached to a chamber (tank 1) where the pressure is $100 \, lb/in^2$ abs and the temperature is $200 \, °F$. The area of the throat is $3 \, in^2$ and A_1, where we happen to have a normal shock, is $4 \, in^2$. Finally A_e is $6 \, in^2$. What is the Mach number right after the shock wave? What is the Mach number at exit? Compute the stagnation pressure and actual pressure for the jet in tank 2. What is the stagnation temperature at exit? The fluid is air.

$\blacksquare$ **Data:**

$$T_0 = 660 \, °R \qquad p_0 = 100 \, psia \qquad A^* = 3 \, in^2 \qquad A_1 = 4 \, in^2 \qquad A_e = 6 \, in^2$$

Find $(N_M)_1$ to right of shock: Since $A/A^* = \frac{4}{3} = 1.333$, we have from Table A-16, $(N_M)_1 = 1.69$. $(N_M)_2$ from Table A-17 is 0.644. Find N_M at exit: For $N_M = 0.644$ Table A-16 gives $A_1/A^* = 1.16$. Therefore, $A^* = 4/1.16 = 3.45 \, m^2$. Now at exit: $A_e/A^* = 6/3.45 = 1.74$. From Table A-16: $(N_M)_e = 0.360$, $p_1/p_0 = 0.206$, $p_1 = (100)(0.206) = 20.6 \, psia$. From Table A-17: $p_2/p_1 = 3.165$, $p_2 = (3.165)(20.6) = 65.2$. From Table A-16: $p_2/p_0 = 0.759$, $p_0 = 652/0.759 = 86.0 \, psia$; $p_e/p_0 = 0.914$, $p_e = (86)(0.914) = 78.5 \, psia$, $T_0 = 200 \, °F$.

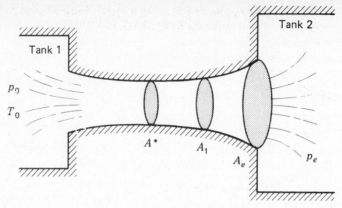

Fig. 16-12

16.179 A convergent nozzle has an exit area of 1.3×10^{-3} m². It permits flow of air to proceed from a large tank in which the pressure of the air is 138 000 Pa abs and the temperature is 20 °C. If the ambient pressure outside the tank is 101 325 Pa, what are the velocity of the flow on leaving the nozzle and the mass flow? Neglect friction.

▮ $\qquad p^*/p = 0.528 \qquad p^* = (138\,000)(0.528) = 72\,800$ N/m²

Hence fluid leaves subsonically and $p_j = 101\,325$ N/m²: $p_j/p_0 = 101\,325/138\,000 = 0.734$. Hence:

$$(N_M)_e = 0.68 \qquad T_e/T_0 = 0.915 \qquad T_e = 268 \text{ K}$$
$$V_e = c(N_M)_e = \sqrt{kRT_e}(N_M)_e = \sqrt{(1.4)(287)(268)}(0.68) = 223 \text{ m/s}$$
$$\rho = p/RT \qquad \rho_e = 101\,325/[(287)(268)] = 1.317 \text{ kg/m}^3$$
$$\dot{m} = \rho VA = (1.317)(223)(0.0013) = 0.382 \text{ kg/s}$$

16.180 In Prob. 16.179, suppose that you are changing the ambient pressure. What is the largest pressure that will permit the maximum flow through the nozzle? What are the maximum mass flow and temperature of the air leaving the nozzle? Neglect friction.

▮ Largest pressure is 72 800 N/m², i.e., the critical pressure. At $N_M = 1$:

$$T^*/T_0 = 0.833 \qquad T^* = (293)(0.833) = 244 \text{ K}$$
$$V = \sqrt{kRT} \qquad V_e = \sqrt{(1.4)(287)(244)} = 313 \text{ m/s}$$
$$\rho = p/RT \qquad \rho_e = 72\,800/[(287)(244)] = 1.040 \text{ kg/m}^3$$
$$\dot{m} = \rho VA = (1.040)(313)(1.3 \times 10^{-3}) = 0.423 \text{ kg/s}$$

16.181 A convergent-divergent nozzle with a throat area of 0.0013 m² and an exit area of 0.0019 m² is connected to a tank wherein air is kept at a pressure of 552 000 Pa abs and a temperature of 15 °C. If the nozzle is operating at design conditions, what should be the ambient pressure outside and the mass flow? What is the critical pressure? Neglect friction.

▮ For design operation we can say $A_e/A^* = 1.46$. From *isentropic* tables: $(N_M)_e = 1.82$, $p_e/p_0 = 0.169$, $p_e = (0.169)(552\,000) = 93\,300$ N/m². Also

$$T_e/T_0 = 0.602 \qquad T_e = (0.602)(288) = 173 \text{ K}$$
$$V_e = c_e(N_M)_e = \sqrt{kRT_e}(N_M)_e = \sqrt{(1.4)(287)(173)}(1.82) = 480 \text{ m/s} \qquad \rho = p/RT$$
$$\rho_e = 93\,300/[(287)(173)] = 1.879 \text{ kg/m}^3 \qquad \dot{m} = \rho VA = (1.879)(480)(0.0019) = 1.71 \text{ kg/s}$$

At throat $p^*/p_0 = 0.528$, $p^* = 291\,500$ N/m².

16.182 In Prob. 16.181, what is the ambient pressure at which a shock will first appear just inside the nozzle? What is the ambient pressure for the completely subsonic flow of maximum mass flow? Neglect friction.

▮ From previous problem we know that N_M just before the shock is 1.82. From the *normal shock* tables we have just downstream of the shock $(N_M)_e = 0.612$. Also $p_e/p_1 = 3.698$. Hence $p_e = (3.698)(93\,300) = 345\,000$ Pa. (*Note*: The pressure 93 300 Pa was taken from the previous problem.) For completely subsonic flow we assume isentropic flow and using $A_e/A^* = 1.46$ we have from the *isentropic* tables $p_e/p_0 = 0.874$. Hence $p_e = (552\,000)(0.874) = 482\,450$ Pa.

16.183 Suppose that you are given the data shown for the convergent-divergent nozzle in Fig. 16-13. A shock is present in the nozzle, as shown. Set up formulations by which one could proceed to ascertain the approximate position and strength of the shock.

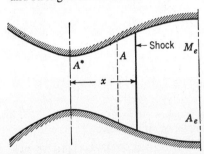

Fig. 16-13

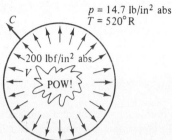

Fig. 16-14

1. First find $(N_M)_2$ as a function of ξ using isentropic data starting from exit conditions and working in.
2. Find $(N_M)_1$ as a function also of ξ using isentropic data starting from the throat and working downstream.
3. From these curves plot $(N_M)_2$ versus $(N_M)_1$.
4. Also plot $(N_M)_2$ versus $(N_M)_1$ using normal shock relations. Where they intersect gives the proper $(N_M)_1$ and $(N_M)_2$.
5. Now with either $(N_M)_1$ or $(N_M)_2$ go back to the isentropic data of step 2 or 1, respectively, and give ξ which is the distance of the shock from the exit.

16.184 Air flows through a 2-in constriction in a 3-in diameter pipeline. The pressure and temperature of the air in the pipeline are 108 psig and 105 °F, respectively, and the pressure in the constriction is 81 psig. Barometric pressure is 14.5 psia. Find the weight flow rate of the air in the pipeline.

$$G = [A_2/\sqrt{1 - (p_2/p_1)^{2/k}(A_2/A_1)^2}]\sqrt{[2gk/(k-1)](p_1\gamma_1)[(p_2/p_1)^{2/k} - (p_2/p_1)^{(k+1)/k}]}$$

$$A_2 = (\pi)(2)^2/4 = 3.142 \text{ in}^2 = 0.02182 \text{ ft}^2 \qquad p_2 = 81 + 14.5 = 95.5 \text{ psia}$$

$$p_1 = 108 + 14.5 = 122.5 \text{ psia} \qquad A_1 = (\pi)(3)^2/4 = 7.069 \text{ in}^2 = 0.04909 \text{ ft}^2$$

$$\gamma = p/RT \qquad \gamma_1 = (122.5)(144)/[(53.3)(105 + 460)] = 0.5858 \text{ lb/ft}^3$$

$$G = \frac{0.02182}{\sqrt{1 - (95.5/122.5)^{2/1.40}(0.02182/0.04909)^2}}\sqrt{\frac{(2)(32.2)(1.40)}{1.40 - 1}(122.5)(144)(0.5858)}$$

$$\times \sqrt{\left[\left(\frac{95.5}{122.5}\right)^{2/1.40} - \left(\frac{95.5}{122.5}\right)^{(1.40+1)/1.40}\right]} = 7.87 \text{ lb/s}$$

16.185 Air flows through a 1-in constriction in a 2-in-diameter pipeline. The pressure and temperature of the air in the pipeline are 100 psig and 102 °F, respectively, and the pressure in the constriction is 81 psig. Barometric pressure is 14.7 psia. Find the weight flow rate of the air in the pipeline.

$$G = [A_2/\sqrt{1 - (p_2/p_1)^{2/k}(A_2/A_1)^2}]\sqrt{[(2gk)/(k-1)](p_1\gamma_1)}\sqrt{(p_2/p_1)^{2/k} - (p_2/p_1)^{(k+1)/k}}$$

$$A_2 = (\pi)(\tfrac{1}{12})^2/4 = 0.005454 \text{ ft}^2 \qquad p_2 = 78 + 14.7 = 92.7 \text{ psia} \qquad p_1 = 100 + 14.7 = 114.7 \text{ psia}$$

$$A_1 = (\pi)(\tfrac{2}{12})^2/4 = 0.02182 \text{ ft}^2 \qquad \gamma = p/RT \qquad \gamma_1 = (114.7)(144)/[(53.3)(102 + 460)] = 0.5514 \text{ lb/ft}^3$$

$$G = [(0.005454)/\sqrt{1 - (92.7/114.7)^{2/1.4}(0.005454/0.02182)^2}]$$

$$\times \sqrt{[(2)(32.2)(1.40)/(1.40-1)](114.7)(144)(0.5514)}$$

$$\times \sqrt{[(92.7/114.7)^{2/1.40} - (92.7/114.7)^{(1.40+1)/1.40}]} = 1.67 \text{ lb/s}$$

16.186 A normal shock wave occurs in the flow of air where $p_1 = 10$ psia, $T_1 = 40$ °F, and $V_1 = 1400$ ft/s. Find p_2, V_2, and T_2.

$$p_2/p_1 = [2k(N_M)_1^2 - (k-1)]/(k+1) \qquad N_M = V/c$$

$$c = \sqrt{kRT} \qquad c_1 = \sqrt{(1.40)(1716)(40 + 460)} = 1096 \text{ ft/s}$$

$$(N_M)_1 = \tfrac{1400}{1096} = 1.28 \qquad p_2/10 = [(2)(1.40)(1.28)^2 - (1.40 - 1)]/(1.40 + 1) \qquad p_2 = 17.4 \text{ psia}$$

$$V_2/V_1 = [(k-1)(N_M)_1^2 + 2]/[(k+1)(N_M)_1^2] \qquad V_2/1400 = [(1.40-1)(1.28)^2 + 2]/[(1.40+1)(1.28)^2] \qquad V_2 = 945 \text{ ft/s}$$

$$\rho = p/RT \qquad \rho_1 V_1 = \rho_2 V_2 \qquad \rho_1 = (10)(144)/[(1716)(40 + 460)] = 0.001678 \text{ slug/ft}^3$$

$$(0.001678)(1400) = (\rho_2)(945) \qquad \rho_2 = 0.00249 \text{ slug/ft}^3$$

$$0.00249 = (17.4)(144)/(1716 T_2) \qquad T_2 = 586 \text{ °R} \quad \text{or} \quad 126 \text{ °F}$$

16.187 Rework Prob. 16.186 if $p_1 = 70 \text{ N/m}^2$ abs, $T_1 = 5\,°C$, and $V_1 = 425$ m/s.

▮ $\quad p_2/p_1 = [2k(N_M)_1^2 - (k-1)]/(k+1) \qquad N_M = V/c \qquad c = \sqrt{kRT} \qquad c_1 = \sqrt{(1.40)(287)(5+273)} = 334 \text{ m/s}$

$\quad (N_M)_1 = \frac{425}{334} = 1.27 \qquad p_2/70 = [(2)(1.40)(1.27)^2 - (1.40-1)]/(1.40+1) \qquad p_2 = 120 \text{ Pa}$

$\quad V_2/V_1 = [(k-1)(N_M)_1^2 + 2]/[(k+1)(N_M)_1^2] \qquad V_2/425 = [(1.40-1)(1.27)^2 + 2]/[(1.40+1)(1.27)^2]$

$$V_2 = 290 \text{ m/s}$$

$\quad \rho = p/RT \qquad \rho_1 V_1 = \rho_2 V_2 \qquad \rho_1 = 70/[(287)(5+273)] = 0.0008773 \text{ kg/m}^3$

$\quad (0.0008773)(425) = (\rho_2)(290) \qquad \rho_2 = 0.00129 \text{ kg/m}^3 \qquad 0.00129 = 120/(287 T_2) \qquad T_2 = 324 \text{ K} \quad \text{or} \quad 51\,°C$

16.188 The pressure, velocity, and temperature just upstream of a normal shock wave in air are 10 psia, 2200 fps, and 23 °F. Determine the pressure, velocity, and temperature just downstream of the wave.

▮ $\quad p_2 = p_1\{[2k(N_M)_1^2 - (k-1)]/(k+1)\} \qquad N_M = V/c \qquad c_1 = [kRT]^{1/2} = [1.4(1715)483]^{1/2} = 1080 \text{ fps}$

$\quad (N_M)_1 = \frac{2200}{1080} = 2.04 \qquad p_2 = (10)\{[2(1.4)(2.04)^2 - 0.4]/2.4\} = 46.9 \text{ psia} \qquad p_1 - p_2 = \gamma_1 V_1/g(V_2 - V_1)$

$\quad V_2 = V_1 + [(p_1-p_2)/\rho_1 V_1] = 2200 + \{[(10-46.9)144]/[(0.056/32.2)(2200)]\} = 811 \text{ fps}$

$\quad V_2^2 - V_1^2 = [2k/(k-1)]\{[p_1/(\gamma_1/g)] - (p_2/\rho_2)\}$

$\quad \gamma_1 = gp_1/RT_1 = (32.2)(10)(144)/[(1715)(483)] = 0.0560 \text{ pcf}$

$\quad 811^2 - 2200^2 = [2(1.4)/0.4]\{[(10)(144)/(0.56/32.2)][(46.9)(144)/\rho_2]\}$

$\quad \rho_2 = 0.00474 \text{ slug/ft}^3 = p_2/RT_2 \qquad T_2 = (46.9)(144)/[(1715)(0.00474)] = 831\,°R = 371\,°F$

16.189 Just downstream of a normal shock wave the pressure, velocity, and temperature are 52 psia, 400 fps, and 120 °F. Compute the Mach number upstream of the shock wave. Consider air and carbon dioxide.

▮ $$(N_M)_2^2 = [2 + (k-1)(N_M)_1^2]/[2k(N_M)_1^2 - (k-1)]$$

For air:

$$N_M = V/c \qquad c = \sqrt{kRT} \qquad c_2 = \sqrt{(1.40)(1716)(120+460)} = 1180 \text{ ft/s}$$

$$(N_M)_2 = \frac{400}{1180} = 0.339 \qquad 0.339^2 = [2 + (1.40-1)(N_M)_1^2]/[(2)(1.40)(N_M)_1^2 - (1.40-1)]$$

$$(N_M)_1^2 = -26.2 \qquad \text{(impossible)}$$

For carbon dioxide:

$$c_2 = \sqrt{(1.28)(1123)(120+460)} = 913 \text{ ft/s} \qquad (N_M)_2 = \frac{400}{913} = 0.438$$

$$0.438^2 = [2 + (1.28-1)(N_M)_1^2]/[(2)(1.28)(N_M)_1^2 - (1.28-1)] \qquad (N_M)_1 = 3.12$$

16.190 A schlieren photograph of a bullet shows a Mach angle of 30°. The air is at a pressure of 14 psia and 50 °F. Find the approximate speed of the bullet.

▮ $\quad N_M = V/c = 1/\sin\beta = 1/\sin 30° = 2.000 \qquad c = \sqrt{kRT} = \sqrt{(1.40)(1716)(50+460)} = 1107 \text{ ft/s}$

$$2.000 = V/1107 \qquad V = 2214 \text{ ft/s}$$

16.191 If a normal shock wave occurs in the flow of helium, $p_1 = 1$ psia, $T_1 = 40\,°F$, $V_1 = 4500$ ft/s, find p_2, ρ_2, V_2, and T_2.

▮ $\quad p_2 = [1/(k+1)][2\rho_1 V_1^2 - (k-1)p_1] \qquad \rho_1 = p_1/RT_1 = 1(144)/\{386[32.17(460+40)]\} = 0.0000232 \text{ slug/ft}^3$

$\quad p_2 = [1/(1.66+1)][2(0.0000232)(4500^2) - (1.66-1)(144)(1)] = 317 \text{ lb/ft}^2 \text{ abs} \qquad p_1 + \rho_1 V_1^2 = p_2 + \rho_1 V_1 V_2$

$\quad V_2 = V_1 - [(p_2-p_1)/\rho_1 V_1] = 4500 - [(317-144)(4500)(0.0000232)] = 2843 \text{ ft/s}$

$\quad \rho_2 = \rho_1(V_1/V_2) = (0.0000232)(\frac{4500}{2843}) = 0.0000367 \text{ slug/ft}^3$

$\quad T_2 = p_2/\rho_2 R = 317/\{(0.0000367)[(386)(32.17)]\} = 696\,°R \quad \text{or} \quad 236\,°F$

16.192 A shock wave occurs in a duct carrying air where the upstream Mach number is 2.0 and upstream temperature and pressure are 15 °C and 20 kPa abs, respectively. Calculate the Mach number, pressure, temperature, and velocity after the shock wave.

▮ From Table A-16, $(N_M)_2 = 0.577$, $p_2/p_1 = 4.500$, and $T_2/T_1 = 1.688$.

$$p_2 = (4.500)(20) = 90.0 \text{ kPa} \qquad T_2 = (1.688)(15+273) = 486 \text{ K} \quad \text{or} \quad 213\,°C$$

$$c_2 = \sqrt{kRT} = \sqrt{(1.40)(287)(486)} = 442 \text{ m/s} \qquad V_2 = c_2(N_M)_2 = (442)(0.577) = 255 \text{ m/s}$$

16.193 Show that entropy has increased across the shock wave of Prob. 16.192.

$\quad\blacksquare\quad \Delta s = c_v \ln \left[(T_2/T_1)^k (p_2/p_1)^{1-k} \right] = (0.171) \ln \left\{ [486/(15+273)]^{1.40} (90.0/20)^{1-1.40} \right\} = 0.0224 \text{ kcal/(kg-K)}$

16.194 Conditions immediately before a normal shock wave in airflow are $p_u = 6$ psia, $T_u = 100\,°F$, and $V_u = 1800$ ft/s. Find $(N_M)_u$, $(N_M)_d$, p_d, and T_d, where the subscript d refers to conditions just downstream from the shock wave.

$\quad\blacksquare\quad N_M = V/c \qquad c = \sqrt{kRT} = \sqrt{(1.40)(1716)(100+460)} = 1160 \text{ ft/s} \qquad (N_M)_u = \frac{1800}{1160} = 1.55$

From Table A-17, $(N_M)_d = 0.683$, $p_d/p_u = 2.65$, and $T_d/T_u = 1.356$. $p_d = (2.65)(6) = 15.9$ psia, $T_d = (1.356)(100+460) = 759\,°R$, or $299\,°F$.

16.195 For $A = 0.16$ ft² in Prob. 16.194, calculate the entropy increase across the shock wave in Btu/s and degrees Rankine.

$\quad\blacksquare\quad \Delta s = (0.171) \ln \left[(T_2/T_1)^k (p_2/p_1)^{1-k} \right] = (0.171) \ln \left\{ [759/(100+460)]^{1.40} (15.9/6)^{1-1.40} \right\} = 0.00613 \text{ Btu/(lbm-°R)}$

$$\Delta s = \dot m\, \Delta s$$

$$\dot m = \rho_u A V_u = (p_u/RT_u) A V_u = [(6)(144)/(1716)(100+460)](0.16)(1800) = 0.2589 \text{ slug/s}$$

$$\Delta s = (0.2589)(32.17)(0.00613) = 0.0511 \text{ Btu/(s-°R)}$$

16.196 An explosion in air, $\gamma = 1.4$, creates a spherical shock wave propagating radially into still air at standard conditions. At the instant shown in Fig. 16-14 (on p. 509) the pressure just inside the shock is 200 lb/in² abs. Estimate (a) the shock speed C and (b) the air velocity V just inside the shock.

$\quad\blacksquare\quad$ (a) In spite of the spherical geometry the flow across the shock moves normal to the spherical wavefront; hence the normal-shock relations apply. Fixing our control volume to the moving shock, we find that the proper conditions to use are $C = V_1$, $p_1 = 14.7$ lb/in² abs, $T_1 = 520\,°R$, $V = V_1 - V_2$, $p_2 = 200$ lb/in² abs. The speed of sound outside the shock is $c_1 \approx 49 T_1^{1/2} = 1117$ ft/s. We can find $(N_M)_1$ from the known pressure ratio across the shock: $p_2/p_1 = 200/14.7 = 13.61 = [1/(k+1)][2k(N_M)_1^2 - (k-1)]$, $13.61 = (1/2.4)[2.8(N_M)_1^2 - 0.4]$ or $(N_M)_1 = 3.436$. Then, by definition of the Mach number, $C = V_1 = (N_M)_1 c_1 = 3.436(1117) = 3840$ ft/s.

(b) $\qquad T_2/T_1 = [2 + (k-1)(N_M)_1^2]\{[2k(N_M)_1^2 - (k-1)]/(k+1)^2 (N_M)_1^2\}$

$\qquad = [2 + (0.4)(3.436)^2]\{[(2.8)(3.436)^2 - 0.4]/(2.4)^2(3.436)^2\} = 3.228$

$\qquad T_2 = 3.228 T_1 = 3.228(520) = 1679\,°R$

At such a high temperature we should account for non-perfect-gas effects or at least use the gas tables, but we won't. Here just estimate from the perfect-gas energy equation that $V_2^2 = 2c_p(T_1 - T_2) + V_1^2 = 2(6010)(520 - 1679) + (3840)^2 = 815\,500$ or $V_2 \approx 903$ ft/s. Notice that we did this without bothering to compute $(N_M)_2$, which equals 0.454, or $c_2 \approx 49 T_2^{1/2} = 2000$ ft/s.

Finally, the air velocity behind the shock is $V = V_1 - V_2 = 3840 - 903 \approx 2940$ ft/s. Thus a powerful explosion creates a brief but intense blast wind as it passes.

16.197 Air from a reservoir at 20 °C and 500 kPa flows through a duct and forms a normal shock downstream of a throat of area 10 cm². By an odd coincidence it is found that the stagnation pressure downstream of this shock exactly equals the throat pressure. What is the area where the shock wave stands?

$\quad\blacksquare\quad A/A^* = (1/N_M)[(1+0.2N_M^2)^3/1.728] \qquad p_1^* = (p_0)_1/(1+0.2N_M^2)^{3.5} = 500/[1+(0.2)(1)^2]^{3.5} = 264 \text{ kPa} = (p_0)_2$

$\qquad (p_0)_2/(p_0)_1 = \frac{264}{500} = 0.528$

From Table A-17, $(N_M)_1 = 2.43$, $A_1/10 = (1/2.43)\{[1+(0.2)(2.43)^2]^3/1.728\}$, $A_1 = 24.7$ cm².

16.198 Air passes through a normal shock with upstream conditions $V_1 = 800$ m/s, $p_1 = 100$ kPa, and $T_1 = 300$ K. What are the downstream conditions V_2 and p_2?

$\quad\blacksquare\quad p_2/p_1 = [2k(N_M)_1^2 - (k-1)]/(k+1) \qquad N_M = V/c \qquad c = \sqrt{kRT}$

$\qquad c_1 = \sqrt{(1.40)(287)(300)} = 347 \text{ m/s} \qquad (N_M)_1 = \frac{800}{347} = 2.305$

$\qquad p_2/100 = [(2)(1.40)(2.305)^2 - (1.40-1)]/(1.40+1) \qquad p_2 = 603 \text{ kPa}$

$V_2/V_1 = [(k-1)(N_M)_1^2 + 2]/(k+1)(N_M)_1^2 \qquad V_2/800 = [(1.40-1)(2.305)^2 + 2]/(1.40+1)(2.305)^2$

$\qquad\qquad V_2 = 259 \text{ m/s}$

16.199 Repeat Prob. 16.197 except this time let the odd coincidence be that the *static* pressure downstream of the shock exactly equals the throat pressure. What is the area where the shock stands?

❚

$$p_2 = p_{\text{throat}} = p_1^* = 264 \text{ kPa} \quad \text{(from Prob. 16.197)}$$
$$p_1 = (p_0)_1/(1 + 0.2N_M^2)^{3.5} = 500/(1 + 0.2N_M^2)^{3.5}$$
$$p_2/p_1 = [(k+1)(N_M)_1^2 - (k-1)]/(k+1) = [2.80(N_M)_1^2 - 0.40]/2.40$$

By trial and error, $(N_M)_1 = 2.15$, $p_1 = 50.6 \text{ kPa}$, $p_2 = 264 \text{ kPa}$, $A/A_1^* = 1.919$, $A_1 = (10)(1.919) = 19.2 \text{ cm}^2$.

16.200 An atomic explostion propagates into still air at 14.7 lb/in^2 abs and 520 °R. The pressure just inside the shock is 5000 lb/in^2 abs. Assuming $k = 1.4$, what is the speed C of the shock and the velocity V just inside the shock?

❚ $C = (N_M)_1(c_1)$ $p_2/p_1 = [2k(N_M)_1^2 - (k-1)]/(k+1)$ $5000/14.7 = [(2)(1.4)(N_M)_1^2 - (1.4-1)]/(1.4+1)$

$(N_M)_1 = 17.08$ $c = \sqrt{kRT}$ $c_1 = \sqrt{(1.4)(1716)(520)} = 1118 \text{ ft/s}$ $C = (17.08)(1118) = 19\,095 \text{ ft/s} = V_1$

$V_2/V_1 = [(k-1)(N_M)_1^2 + 2]/(k+1)(N_M)_1^2$ $V_2/19\,095 = [(1.4-1)(17.08)^2 + 2]/(1.4+1)(17.08)^2 = 0.1695$

$$V_2 = 3237 \text{ ft/s} \qquad V = C - V_2 = 19\,095 - 3237 = 15\,858 \text{ ft/s}$$

16.201 The normal-shock wave from an explosion propagates at 1500 m/s into still air at 20 °C and 101 kPa. What are the pressure, velocity, and temperature just inside the shock?

❚ $p_2/p_1 = [2k(N_M)_1^2 - (k-1)]/(k+1)$ $N_M = V/c$ $c = \sqrt{kRT} = \sqrt{(1.4)(287)(20+273)} = 343 \text{ m/s}$

$(N_M)_1 = \frac{1500}{343} = 4.37$ $p_2/101 = [(2)(1.4)(4.37)^2 - (1.4-1)]/(1.4+1)$ $p_2 = 2233 \text{ Pa}$

$$T_2/T_1 = [2 + (k-1)(N_M)_1^2]\{[(2k)(N_M)_1^2 - (k-1)]/(k+1)^2(N_M)_1^2\}$$

$$T_2/(20+273) = [2 + (1.4-1)(4.37)^2]\{[(2)(1.4)(4.37)^2 - (1.4-1)]/(1.4+1)^2(4.37)^2\}$$

$T_2 = 1363 \text{ K}$ $V_2/V_1 = [(k-1)(N_M)_1^2 + 2]/(k+1)(N_M)_1^2$ $V_2/1500 = [(1.4-1)(4.37)^2 + 2]/(1.4+1)(4.37)^2$

$$V_2 = 315 \text{ m/s} \qquad V_{\text{inside shock}} = C - V_2 = 1500 - 315 = 1185 \text{ m/s}$$

16.202 Air is moving at Mach number 3 in a duct and undergoes a normal shock. If the undisturbed pressure ahead of the shock is $69\,000 \text{ Pa}$ abs what is the increase in pressure after the shock? What is the loss in stagnation pressure across the shock?

❚ From *normal shock* tables (A-17), $(N_M)_2 = 0.475$, $p_2/p_1 = 10.333$, $(p_0)_2/(p_0)_1 = 0.328$. Hence $p_2 = (10.333)(69\,000) = 712\,980 \text{ Pa}$. Now go to the *isentropic* tables for $(p_0)_1$: $p_1/(p_0)_1 = 0.027$, $(p_0)_1 = 69\,000/0.027 = 2.56 \times 10^6 \text{ Pa}$. Hence $(p_0)_2 = (0.328)(2.56 \times 10^6) = 8.38 \times 10^5 \text{ Pa}$. Consequently, $\Delta p_0 = 2.56 \times 10^6 - 8.38 \times 10^5 = 1.72 \times 10^6 \text{ Pa}$.

16.203 An airplane having a diffuser designed for subsonic flight has a normal shock attached to the edge of the diffuser when the plane is flying at a certain Mach number. If at the exit of the diffuser the Mach number is 0.3, what must the flight Mach number be for the plane, assuming isentropic diffusion behind the shock? The inlet area is 0.25 m^2 and the exit area is 0.4 m^2.

❚ From *isentropic* tables (A-16), $A_e/A^* = 2.04$, $A^* = 0.4/2.04 = 0.196 \text{ m}^2$. At inlet, $A_1 = 0.25 \text{ m}^2$. Hence $A_1/A^* = 0.25/0.196 = 1.276$. From *isentropic* tables, $(N_M)_2 = 0.536$. This is the Mach number behind the inlet shock. From the *normal shock* tables we can now get $(N_M)_1$: $(N_M)_1 = 2.287$.

16.204 A normal shock forms ahead of the diffuser of a turbojet plane flying at Mach number 1.2. If the plane is flying at 35 000-ft altitude in standard atmosphere, what is the entering Mach number for the diffuser and the stagnation pressure.

❚ $(N_M)_1 = 1.2$. Hence from *normal shock* tables (A-17), $(N_M)_2 = 0.842$.

$$(p_0)_2/(p_0)_1 = 0.993 \tag{1}$$

Also, from *isentropic* tables (A-16) we have for $(N_M)_1 = 1.2$: $p_1/(p_0)_1 = 0.412$. Hence, $(p_0)_1 = 0.498/0.412 = 1210 \text{ psf}$, where we have used the standard atmosphere table. Now using Eq. (1) we get $(p_0)_2 = (0.993)(1210) = 1200 \text{ psf}$.

16.205 Consider a supersonic flow through a stationary duct wherein a stationary shock is present. The Mach number ahead of the shock is 2 and the pressure and temperature are $101\,300 \text{ Pa}$ abs and 40 °C, respectively. What is the velocity of propagation of the shock relative to the fluid ahead of the shock? The fluid is air.

$$(N_M)_1 = V_1/c_1 = V_1/\sqrt{kRT_1} \qquad 2 = V_1/\sqrt{(1.40)(287)(40 + 273)} \qquad V_1 = 709 \text{ m/s}$$

16.206   A jet plane is diving at supersonic speed at close to constant speed. There is a curved shock wave ahead of it. A static pressure gage near the nose of the plane measures 30.5 kPa abs. The ambient pressure and temperature of the atmosphere are 10 kPa and 245 K, respectively. What are the flight Mach number for the plane and its speed if one assumes that in front of the static pressure gage the shock wave is plane?

▐   Consider the reference of observation is from the plane. The ratio of the pressures across the shock wave is $p_2/p_1 = 30.5/10 = 3.05$, $(N_M)_1 = 1.66$. From *normal shock* tables (A-17) the plane is moving at a Mach number of 1.66. The velocity is determined next. The speed of sound in the undisturbed region ahead of the shock is $c = \sqrt{kRT} = \sqrt{(1.4)(287)(254)} = 319.5$ m/s. Therefore, $V = (1.66)(319.5) = 530$ m/s. Hence the velocity of the plane is 530 m/s.

16.207   A duct having a square cross section 0.300 m on a side has 25 kg of air per second flowing in it. The air, originally in a chamber where the temperature is 90 °C, has been insulated by the duct walls against heat transfer to the outside. The duct is operating in a choked condition. If the duct has a relative roughness of 0.002, determine the Mach number at a position 6 m from the exit of the duct.

▐   We may solve for N_M at this position by employing $[(1/M^2) - 1] + [(k + 1)/2] \ln [(k + 1)M^2/2\{1 + [(k - 1)/2]M^2\} = (fk/D_H)L$. To do this, we estimate f to be 0.024, from the Moody diagram (Fig. A-5). The hydraulic diameter D_H is $D_H = 4A/p_w = (4)(0.3^2)/(4)(0.3) = 0.3$ m. Using $k = 1.4$, we then have $[(1/N_M^2) - 1] + 1.2 \ln [1.2N_M^2/(1 + 0.2N_M^2)] = [(0.024)(1.4)/0.3](6) = 0.672$. Solving by trial and error, we get $N_M = 0.6$.

To check our friction factor, we must compute other conditions at this section of the pipe. The temperature T is determined from isentropic table A-16 for $N_M = 0.6$. Thus $T/T_0 = 0.933$. Therefore $T = 339$ K $= 66$ °C. We may determine the viscosity of air at that temperature to be 2.15×10^{-5} N-s/m^2 by making use of Fig. A-1, since viscosity does not depend greatly on the pressure. Noting that $G = \rho V = w/A = 25/0.300^2 = 277.8$ kg/(m^2-s), we get for N_R: $N_R = GD_H/\mu = (277.8)(0.300)/(2.15 \times 10^{-5}) = 3.88 \times 10^6$. Returning to the Moody diagram, we see that our choice of f is close enough not to require further computation and the desired Mach number is 0.6.

16.208   A constant-area duct having a circular cross-sectional area of 0.19 m^2 is operating in a choked condition. It is highly insulated against heat transfer, and the inside surface has a relative roughness of 0.002. At a distance 9 m from the end of the duct, what is the Mach number of 35 kg of airflow per second? The stagnation temperature is 95 °C. Perform one iteration. Fluid is air.

▐ $$[(1/N_M^2) - 1] + [(k + 1)/2] \ln ((k + 1)N_M^2/2\{1 + [(k - 1)/2]N_M^2\}) = (fk/D_H)L$$
$$\pi D^2/4 = 0.19 \text{ m}^2 \qquad D = 0.4918 \text{ m}$$

Guess at $f = 0.023$:

$$[(1/N_M^2) - 1] + [(1.40 + 1)/2] \ln ((1.40 + 1)N_M^2/(2)\{1 + [(1.40 - 1)/2]N_M^2\}) = [(0.023)(1.40)/0.4918](9)$$
$$N_M = 0.61 \qquad N_R = \rho DV/\mu \qquad \dot{m} = \rho AV \qquad V = cN_M \qquad c = \sqrt{kRT}$$

From Table A-16, $T/T_0 = 0.931$:

$$T = (0.931)(95 + 273) = 343 \text{ K} \qquad c = \sqrt{(1.40)(287)(343)} = 371 \text{ m/s} \qquad V = (371)(0.61) = 226 \text{ m/s}$$
$$35 = (\rho)(0.19)(226) \qquad \rho = 0.815 \text{ kg/m}^3 \qquad N_R = (0.815)(0.4918)(226)/(2.2 \times 10^{-5}) = 4.12 \times 10^6$$

From Fig. A-5 with $\epsilon/D = 0.002$, $f = 0.023$. Since this value is the same as the guessed value of f, $N_R = 0.61$.

16.209   In Prob. 16.208, determine where the Mach number is 0.5, and ascertain the pressure and temperature of the flow there.

▐   From Fig. A-18, $(fk/H)L = 1.5$, $[(0.023)(1.40)/0.4918]L = 1.5$, $L = 22.9$ m. From Fig. A-16, $T/T_0 = 0.952$:

$$T = (0.952)(95 + 273) = 350 \text{ K} \quad \text{or} \quad 77 \text{ °C} \qquad p = (G/N_M)(RT_0/k\{1 + [(k - 1)/2]N_M^2\})^{1/2}$$
$$p = [(35/0.19)/0.5]((287)(95 + 273)/(1.40)\{1 + [(1.40 - 1)/2](0.5)^2\})^{1/2} = 98\,750 \text{ N/m}^2$$

16.210   A constant-area duct is operating in the choked condition. The cross section is rectangular, having sides 6 by 4 ft, and the surface has a relative roughness of 0.0001. At 20 ft from the end of the duct, the pressure is 18 lb/in^2 abs. If there is no heat transferred through the walls, determine the Mach number and Reynolds number at this section of air flow. The exit pressure is that of ambient pressure of the surroundings, which is 14.7 lb/in^2.

▌ We first compute H.

$$H = 4A/p_w = (4)(24)/20 = 4.8 \text{ ft} \qquad p = (G/N_M)(RT_0/k\{1 + [(k-1)/2]N_M^2\})^{1/2}$$

$$(14.7)(144) = (G/1)\{1716T_0/(1.40)[1 + (0.4/2)(1)^2]\}^{1/2}$$

$$G_0^{1/2} = 66.2 \tag{1}$$

At position 20 ft from end we get $(18)(144) = (G/N_M)[1716T_0/1.4(1 + 0.2N_M^2)]^{1/2}$:

$$74.0 = (G/N_M)[T_0/(1 + 0.2N_M^2)]^{1/2} \tag{2}$$

We thus have two unknowns if we consider $GT_0^{1/2}$ and N_M as unknowns. Substituting for $GT_0^{1/2}$ from Eq. (1) into Eq. (2) we get $74.0 = 66.2/N_M(1 + 0.2N_M^2)^{1/2}$. Solve by trial and error to get $N_M = 0.838$. From Fig. A-72 we get $(fk/H)L = 0.08$, $f = (0.08)(4.8)/(1.4)(20) = 0.0137$. Now using $f = 0.0137$ and $\epsilon/D = 0.0001$, we get from the Moody diagram (Fig. A-5): $N_R = 8 \times 10^5$.

16.211 Determine the maximum length of 50-mm-ID pipe, $f = 0.02$ for flow of air, when the Mach number at the entrance to the pipe is 0.30.

▌
$$fL_{\max}/D = \tfrac{5}{7}\{[1/(N_M)_0^2 - 1]\} + \tfrac{6}{7}\ln\{6(N_M)_0^2/[(N_M)_0^2 + 5]\} \qquad k = 1.4$$
$$(0.02/0.05)L_{\max} = \tfrac{5}{7}[(1/0.3^2) - 1] + \tfrac{6}{7}\ln[6(0.30^2)/(0.30^2 + 5)]$$

from which $L_{\max} = 13.25$ m.

16.212 A 4.0-in-ID pipe, $f = 0.020$, has air at 14.7 psia and at $T = 60\,°\text{F}$ flowing at the upstream end with Mach number 3.0. Determine $L_{\max}$, p^*, V^*, T^*, and values of p_0', V_0', T_0', and L at $N_M = 2.0$.

▌ $fL_{\max}/D = \tfrac{5}{7}\{[1/(N_M)_0^2] - 1\} + \tfrac{6}{7}\ln\{6(N_M)_0^2/[(N_M)_0^2 + 5]\} \qquad (0.02/0.333)L_{\max} = \tfrac{5}{7}(\tfrac{1}{9} - 1) + \tfrac{6}{7}\ln[6(3^2)/(3^2 + 5)]$

from which $L_{\max} = 8.69$ ft. If the flow originated at $N_M = 2$, the length $L_{\max}$ is given by the same equation: $(0.02/0.333)L_{\max} = \tfrac{5}{7}(\tfrac{1}{4} - 1) + \tfrac{6}{7}\ln[6(2^2)/(2^2 + 5)]$ from which $L_{\max} = 5.08$ ft.
Hence, the length from the upstream section at $N_M = 3$ to the section where $N_M = 2$ is $8.69 - 5.08 = 3.61$ ft. The velocity at the entrance is $V = \sqrt{kRTN_M} = \sqrt{1.4(53.3)[32.17(460 + 60)]}(3) = 3352$ ft/s.

$$p^*/p_1 = (N_M)_0\sqrt{[(k-1)(N_M)_0^2 + 2]/(k+1)} \qquad V^*/V_0 = [1/(N_M)_0]\sqrt{[(k-1)(N_M)_0^2 + 2]/(k+1)}$$
$$T^*/T_0 = [(k-1)(N_M)_0^2 + 2]/(k+1) \qquad p^*/14.7 = 3\sqrt{[0.4(3^2) + 2]/2.4} = 4.583$$
$$V^*/3352 = \tfrac{1}{3}\sqrt{[0.4(3^2) + 2]/2.4} = 0.509 \qquad T^*/520 = [0.4(3^2) + 2]/2.4 = \tfrac{7}{3}$$

So $p^* = 67.4$ psia, $V^* = 1707$ ft/s, $T^* = 1213\,°\text{R}$. For $N_M = 2$ the same equations are now solved for p_0', V_0', and T_0': $67.4/p_0' = 2\sqrt{[0.4(2^2) + 2]/2.4} = 2.45$, $1707/V_0' = \tfrac{1}{2}\sqrt{[0.4(2^2) + 2]/2.4} = 0.6124$, $1213/T_0' = [0.4(2^2) + 2]/2.4 = \tfrac{3}{2}$. So $p_0' = 27.5$ psia, $V_0' = 2787$ ft/s, and $T_0' = 809\,°\text{R}$.

16.213 What length of 100-mm-diameter insulated duct, $f = 0.018$, is needed when oxygen enters at $N_M = 3.0$ and leaves at $N_M = 2.0$?

▌
$$fL/D = \tfrac{5}{7}\{[1/(N_M)_0^2] - [1/(N_M)^2]\} + \tfrac{6}{7}\ln\left([(N_M)_0/N_M]^2\{[(N_M)^2 + 5]/[(N_M)_0^2 + 5]\}\right)$$
$$0.018L/\tfrac{100}{1000} = \tfrac{5}{7}[(1/3.0^2) - (1/2.0^2)] + \tfrac{6}{7}\ln\{(3.0/2.0)^2[(2.0^2 + 5)/(3.0^2 + 5)]\} \qquad L = 1.206 \text{ m}$$

16.214 Air enters an insulated pipe at $N_M = 0.4$ and leaves at $N_M = 0.6$. What portion of the duct length is required for the flow to occur at $N_M = 0.5$?

▌
$$\frac{L_1}{L_2} = \frac{5\{[1(N_M)_0^2] - [1/(N_M)_1^2]\} + 6\ln\{[(N_M)_0/(N_M)_1]^2\{[(N_M)_1^2 + 5]/[(N_M)_0^2 + 5]\}\}}{5\{[1/(N_M)_0^2] - [1/(N_M)_2^2]\} + 6\ln\{[(N_M)_0/(N_M)_2]^2\{[(N_M)_2^2 + 5]/[(N_M)_0^2 + 5]\}\}}$$
$$= \frac{5[(1/0.4^2) - (1/0.5^2)] + 6\ln\{(0.4/0.5)^2[(0.5^2 + 5)/(0.4^2 + 5)]\}}{5[(1/0.4^2) - (1/0.6^2)] + 6\ln\{(0.4/0.6)^2[(0.6^2 + 5)/(0.4^2 + 5)]\}} = 0.682$$

Thus 68.2 percent of the duct length is required.

16.215 Determine the maximum length, without choking, for the adiabatic flow of air in a 110-mm-diameter duct, $f = 0.025$, when upstream conditions are $T = 50\,°\text{C}$, $V = 200$ m/s, and $p = 2$ atm. What are the pressure and temperature at the exit?

■

$$fL_{max}/D = \tfrac{5}{7}\{[1/(N_M)_0^2] - 1\} + \tfrac{6}{7}\ln\{6(N_M)_0^2/[(N_M)_0^2 + 5]\}$$

$$N_M = V/c = V/\sqrt{kRT} \qquad (N_M)_0 = 200/\sqrt{(1.40)(287)(50 + 273)} = 0.5552$$

$$0.025L_{max}/\tfrac{110}{1000} = \tfrac{5}{7}[(1/0.5552^2) - 1] + \tfrac{6}{7}\ln[(6)(0.5552)^2/(0.5552^2 + 5)] \qquad L_{max} = 3.077 \text{ m}$$

$$p^*/p_1 = (N_M)_0\sqrt{[(k-1)(N_M)_0^2 + 2]/(k+1)} \qquad p^*/(2)(101.3) = (0.5552)\sqrt{[(1.40-1)(0.5552)^2 + 2]/(1.40+1)}$$

$$p^* = 106 \text{ kPa abs} \qquad T^*/T = [(k-1)(N_M)_0^2 + 2]/(k+1)$$

$$T^*/(50 + 273) = [(1.40-1)(0.5552)^2 + 2]/(1.40+1) \qquad T^* = 286 \text{ K} \quad \text{or} \quad 13\,^\circ\text{C}$$

16.216 What minimum size insulated duct is required to transport nitrogen 1000 ft? The upstream temperature is 80 °F, and the velocity there is 200 fps ($f = 0.020$).

■

$$fL_{max}/D = \tfrac{5}{7}\{[1/(N_M)_0^2] - 1\} + \tfrac{6}{7}\ln\{6(N_M)_0^2/[(N_M)_0^2 + 5]\}$$

$$N_M = V/c = V/\sqrt{kRT} \qquad (N_M)_0 = 200/\sqrt{(1.40)(1776)(80 + 460)} = 0.1726$$

$$(0.020)(1000)/D = \tfrac{5}{7}[(1/0.1726^2) - 1] + \tfrac{6}{7}\ln[(6)(0.1726)^2/(0.1726^2 + 5)] \qquad D = 0.980 \text{ ft}$$

16.217 Find the upstream and downstream pressures in Prob. 16.216 for 3-lbm/s flow.

■

$$\rho = p/RT \qquad \dot{m} = \rho AV \qquad 3/32.2 = (\rho_u)[(\pi)(0.980)^2/4](200) \qquad \rho_u = 0.0006176 \text{ slug/ft}^3$$

$$0.0006176 = p_u/[(1776)(80 + 460)] \qquad p_u = 592.3 \text{ psfa} \quad \text{or} \quad 4.11 \text{ psia}$$

$$p^*/p_1 = (N_M)_0\sqrt{[(k-1)(N_M)_0^2 + 2]/(k+1)} \qquad p^*/592.3 = (0.1726)\sqrt{[(1.40-1)(0.1726)^2 + 2]/(1.40+1)}$$

$$p^* = p_d = 93.6 \text{ psfa} \quad \text{or} \quad 0.650 \text{ psia}$$

16.218 What is the maximum mass rate of flow of air from a reservoir, $T = 15\,^\circ\text{C}$, through 6 m of insulated 25-mm-diameter pipe, $f = 0.020$, discharging to atmosphere? ($p = 1$ atm).

■

$$\dot{m} = \rho_0 A V_0 \qquad \rho = p/RT \qquad p^*/p_0 = (N_M)_0\sqrt{[(k-1)(N_M)_0^2 + 2]/(k+1)}$$

$$fL_{max}/D = \tfrac{5}{7}\{[1/(N_M)_0^2] - 1\} + \tfrac{6}{7}\ln\{6(N_M)_0^2/[(N_M)_0^2 + 5]\}$$

$$(0.020)(6)/\tfrac{25}{1000} = \tfrac{5}{7}\{[1/(N_M)_0^2] - 1\} + \tfrac{6}{7}\ln\{6(N_M)_0^2/[(N_M)_0^2 + 5]\} \qquad (N_M)_0 = 0.311$$

$$101\,300/p_0 = 0.311\sqrt{[(1.40)(0.311)^2 + 2]/(1.40+1)} \qquad p_0 = 353\,000 \text{ Pa abs}$$

$$\rho_0 = 353\,000/[(287)(15 + 273)] = 4.27 \text{ kg/m}^3 \qquad V = (N_M)c = (N_M)\sqrt{kRT}$$

$$V_0 = (0.311)\sqrt{(1.40)(287)(15 + 273)} = 105.8 \text{ m/s} \qquad \dot{m} = (42.7)[(\pi)(\tfrac{25}{1000})^2/4](105.8) = 0.222 \text{ kg/s}$$

16.219 Air flows subsonically in an adiabatic 1-in-diameter duct. The average friction factor is 0.024. (*a*) What length of duct is necessary to accelerate the flow from $(N_M)_1 = 0.1$ to $(N_M)_2 = 0.5$? (*b*) What additional length will accelerate it to $(N_M)_3 = 1.0$? Assume $k = 1.4$.

■ (*a*) $f(\Delta L/D) = (fL^*/D)_1 - (fL^*/D)_2$. From Table A-24,

$$(fL^*/D)_{N_M=0.1} = 66.9216 \qquad (fL^*/D)_{N_M=0.5} = 1.0691$$

$$(0.024)(\Delta L)/(\tfrac{1}{12}) = 66.9216 - 1.0691 \qquad \Delta L = 229 \text{ ft}$$

(*b*) $\qquad f(\Delta L/D) = (fL^*/D)_{N_M=0.5} \qquad (0.024)(\Delta L)/(\tfrac{1}{12}) = 1.0691 \qquad \Delta L = 3.71 \text{ ft}$

16.220 For the duct flow of Prob. 16.219 assume that at $(N_M)_1 = 0.1$ we have $p_1 = 100$ lb/in² abs and $T_1 = 600\,^\circ$R. Compute at section 2 farther downstream [$(N_M)_2 = 0.5$] (*a*) p_2; (*b*) T_2; (*c*) V_2; and (*d*) $(p_0)_2$.

■ As preliminary information we can compute V_1 and $(p_0)_1$ from the given information: $V_1 = (N_M)_1 c_1 = 0.1[49(600)^{1/2}] = 120$ ft/s, $(p_0)_1 = p_1[1 + \tfrac{1}{2}(k-1)(N_M)_1^2]^{3.5} = 100[1 + 0.2(0.1)^2]^{3.5} = 100.7$ lb/in² abs. Now enter Table A-24, to find the following ratios:

section	N_M	p/p^*	T/T^*	V/V^*	p_0/p_0^*
1	0.1	10.9435	1.1976	0.1094	5.8218
2	0.5	2.1381	1.1429	0.5345	1.3399

Use these ratios to compute all properties downstream:

(a) $$p_2 = p_1(p_2/p^*)(p^*/p_1) = 100(2.1381/10.9435) = 19.5 \text{ lb/in}^2 \text{ abs}$$

(b) $$T_2 = T_1(T_2/T^*)(T^*/T_1) = 600(1.1429)/1.1976 = 573 \text{ °R}$$

(c) $$V_2 = V_1(V_2/V^*)(V^*/V_1) = 120(0.5345)/0.1094 = 586 \text{ ft/s}$$

(d) $$(p_0)_2 = (p_0)_1[(p_0)_2/p_0^*][p_0^*/(p_0)_1] = 100.7(1.3399)/5.8218 = 23.2 \text{ lb/in}^2 \text{ abs}$$

16.221 Air enters a duct of $L/D = 40$ at $V_1 = 200$ m/s and $T_1 = 300$ K. The flow at the exit is choked. What is the average friction factor in the duct for adiabatic flow?

▌ $$N_M = V/c = V/\sqrt{kRT} \qquad (N_M)_1 = 200/\sqrt{(1.40)(287)(300)} = 0.576$$

From Table A-24, $fL/D = 0.594$, $(f)(40) = 0.594$, $f = 0.0148$.

16.222 Air in a tank at $p_0 = 100$ psia and $T_0 = 520$ °R flows through a converging nozzle into pipe of 1 in diameter. What will be the mass flow through the pipe if its length is (a) 0 ft, (b) 1 ft? Assume $f = 0.025$ and the pressure outside the duct is negligibly small.

▌ (a) $$\dot{m} = 0.6847 p_0 A^*/(RT_0)^{1/2} \qquad \text{(nozzle is choked)}$$

$$= (0.6847)[(100)(144)][(\pi)(\tfrac{1}{12})^2/4]/[(1716)(520)]^{1/2} = 0.0569 \text{ slug/s}$$

(b) $$fL/D = (0.025)(1)/\tfrac{1}{12} = 0.300 \qquad \text{(section 2 is choked)} \qquad N_M = 0.6592 \qquad \dot{m} = \rho_1 A_1 V_1$$

$$\rho = p/RT \qquad p_1 = p_0/\{1 + [(k-1)/2](N_M)^2\}^{3.5} = 100/\{1 + [(1.40-1)/2](0.6592)^2\}^{3.5} = 74.7 \text{ psia}$$

$$T_1 = T_0/\{1 + [(k-1)/2](N_M)^2\} = 520/\{1 + [(1.40-1)/2](0.6592)^2\} = 478 \text{ °R}$$

$$\rho = (74.7)(144)/[(1716)(478)] = 0.01311 \text{ slug/ft}^3$$

$$V = (N_M)(c) = (N_M)\sqrt{kRT} = (0.6592)\sqrt{(1.40)(1716)(478)} = 706 \text{ ft/s}$$

$$\dot{m} = (0.01311)[(\pi)(\tfrac{1}{12})^2/4](706) = 0.0505 \text{ slug/s}$$

16.223 Hydrogen [$k = 1.41$ and $R = 4124$ m^2/(s^2·K)] enters a 5-cm-diameter pipe at $p_1 = 500$ kPa, $V_1 = 300$ m/s, and $T_1 = 20$ °C. The friction factor is 0.023. How long is the duct if the flow is choked? What is the exit pressure?

▌ $$fL/D = \{[1 - (N_M)_1^2]/k(N_M)_1^2\} + [(k+1)/2k] \ln \{(k+1)(N_M)_1^2/[2 + (k-1)(N_M)_1^2]\}$$

$$N_M = V/c = V/\sqrt{kRT} \qquad (N_M)_1 = 300/\sqrt{(1.41)(4124)(20 + 273)} = 0.231$$

$$0.023L/\tfrac{5}{100} = [(1 - 0.231^2)/(1.41)(0.231^2)]$$

$$+ [(1.41 + 1)/(2)(1.41)] \ln \{(1.41 + 1)(0.231)^2/[2 + (1.41 - 1)(0.231)^2]\} \qquad L = 22.2 \text{ m}$$

$$p_1/p^* = (1/N_M)\{(k+1)/[2 + (k-1)(N_M)_2]\}^{1/2}$$

$$500/p^* = (1/0.231)\{(1.41 + 1)/[2 + (1.41 - 1)(0.231)^2]\}^{1/2} \qquad p^* = p_2 = 106 \text{ kPa}$$

16.224 Air enters a 5-cm by 5-cm square duct at $V_1 = 900$ m/s and $T_1 = 300$ K. The friction factor is 0.018. For what length duct will the flow exactly decelerate to $N_M = 1.0$? If the duct length is 2 m, will there be a normal shock in the duct? If so, at what Mach number will it occur?

▌ $$N_M = V/c = V/\sqrt{kRT} \qquad (N_M)_1 = 900/\sqrt{(1.40)(287)(300)} = 2.59 \qquad \text{(supersonic entrance)}$$

From Table A-24, $fL^*/D = 0.451$, $0.018L^*/\tfrac{5}{100} = 0.451$, $L^* = 1.25$ m. If $L = 2$ m, which is larger than L^*, there will be a normal shock in the duct. By trial and error, the shock occurs at $N_M = 1.97$ where $fL/D = 0.296$, or $\Delta L_1 = (0.451 - 0.296)(\tfrac{5}{100})/0.018 = 0.43$ m. On the downstream side of the shock, $N_M = 0.582$ and $fL^*/D = 0.565$, so $\Delta L_2 = (0.565)(\tfrac{5}{100})/0.018 = 1.57$ m; $L_{duct} = \Delta L_1 + \Delta L_2 = 0.43 + 1.57 = 2.00$ m.

16.225 Air enters a 0.5-in-diameter pipe subsonically at $p_1 = 60$ psia and $T_1 = 600$ °R. The pipe length is 20 ft, $f = 0.022$, and the receiver pressure outside the pipe entrance is 20 psia. Compute the mass flow in the pipe, assuming isothermal flow.

▌ $$(\dot{m}/A)^2 = (p_1^2 - p_2^2)/RT[(fL/D) + 2 \ln (p_1/p_2)]$$

$$= \{[(60)(144)]^2 - [(20)(144)]^2\}/(1716)(600)\{[(0.022)(20)/(0.5/12)] + 2 \ln \tfrac{60}{20}\} = 5.052[\text{slugs}/(\text{s-ft}^2)]^2$$

$$\dot{m}/A = 2.248 \text{ slugs/(s-ft}^2) \qquad \dot{m} = (2.248)[(\pi)(0.5/12)^2/4] = 0.00307 \text{ slug/s}$$

16.226 Oxygen [$R = 260$ m^2/(s^2·K), $k = 1.40$, and $\mu = 2 \times 10^{-5}$ kg/(m-s)] enters a 120-m-long smooth pipe at 250 kPa and 60 °C. The pipe exits into a low-pressure reservoir. The desired mass flow is 0.25 kg/s. Using the Moody chart (Fig. A-5) to compute f, estimate the maximum pipe diameter to transport this flow and the resulting exit pressure and temperature.

▌ Guess $D = 0.05$ m

$$\dot{m} = \rho_1 A_1 V_1 \qquad \rho = p/RT \qquad \rho_1 = (250)(1000)/[(260)(60 + 273)] = 2.888 \text{ kg/m}^3$$
$$0.25 = (2.888)[(\pi)(0.05)^2/4](V_1) \qquad V_1 = 44.09 \text{ m/s}$$
$$N_M = V/c = V/\sqrt{kRT} \qquad (N_M)_1 = 44.09/\sqrt{(1.40)(260)(60 + 273)} = 0.1272$$

From Table A-24, $fL^*/D = 40.39$. $N_R = \rho DV/\mu = (2.888)(0.05)(44.09)/(2 \times 10^{-5}) = 3.18 \times 10^5$. From Fig. A-5, $f = 0.0143$. $(0.0143)(120)/D = 40.39$, $D = 0.0425$ m. Thus the guessed value of D of 0.05 m is too large. Additional iterations (not shown) yield a value of $D = 0.0485$ m at $(N_M)_1 = 0.135$.

16.227 Air enters a 1-in-diameter cast iron duct 20 ft long at $V_1 = 200$ fps, $p_1 = 40$ psia, and $T_1 = 520$ °R. Compute the exit pressure and mass flow using the Moody chart (Fig. A-5) to predict the friction factor.

▌
$$\dot{m} = \rho_1 A_1 V_1 \qquad \rho = p/RT \qquad \rho_1 = (40)(144)/[(1716)(520)] = 0.006455 \text{ slug/ft}^3$$
$$A_1 = (\pi)(\tfrac{1}{12})^2/4 = 0.005454 \text{ ft}^2 \qquad \dot{m} = (0.006455)(0.005454)(200) = 0.00704 \text{ slug/s}$$
$$N_M = V/c = V/\sqrt{kRT} \qquad (N_M)_1 = 200/\sqrt{(1.40)(1716)(520)} = 0.1789$$

From Table A-24, $(fL^*/D)_1 = 18.804$.
$$(fL^*/D)_2 = (fL^*/D)_1 - fL/D \qquad N_R = \rho DV/\mu$$
$$(N_R)_1 = (0.006455)(\tfrac{1}{12})(200)/(3.74 \times 10^{-7}) = 2.88 \times 10^5 \qquad \epsilon/D = 0.00085/(\tfrac{1}{12}) = 0.0102$$

From Fig. A-5, $f = 0.0384$.
$$(fL^*/D)_2 = 18.804 - (0.0384)(20)/(\tfrac{1}{12}) = 9.588 \qquad (N_M)_2 = 0.238$$
$$p_1/p^* = 6.104 \quad \text{and} \quad p_2/p^* = 4.578 \qquad p_2 = (4.578/6.104)(40) = 30.0 \text{ psia}$$

16.228 Air at 320 K is to be transported through a duct 40 m long. If $f = 0.025$, what is the minimum diameter of duct that can carry the flow without choking for $V_1 = $ (a) 50 m/s, (b) 100 m/s, and (c) 400 m/s. Assume adiabatic flow.

▌ (a)
$$N_M = V/c = V/\sqrt{kRT} \qquad (N_M)_1 = 50/\sqrt{(1.40)(287)(320)} = 0.139$$
From Fig. A-24, $(fL^*/D)_1 = 32.80$, $(0.025)(40)/D = 32.80$, $D = 0.0305$ m

(b)
$$(N_M)_1 = 100/\sqrt{(1.40)(287)(320)} = 0.279 \qquad (fL^*/D)_1 = 6.425$$
$$(0.025)(40)/D = 6.425 \qquad D = 0.156 \text{ m}$$

(c)
$$(N_M)_1 = 400/\sqrt{(1.40)(287)(320)} = 1.116 \qquad (fL^*/D)_1 = 0.0129$$
$$(0.025)(40)/D = 0.0129 \qquad D = 77.5 \text{ m}$$

16.229 Very cold air, to be used in an air-conditioning system of a test chamber, passes through a rectangular duct of cross-sectional area A ft^2 and of length L ft. It enters the duct at a temperature of T_1 °F at a pressure of p_1 psia. It is estimated that Q Btu per unit length per lbm will be transferred from the surroundings into the flow of air into the duct. If the exit temperature is to be T_2 °F at ambient pressure p_2, how much flow should there be? Set up equations only. Explain how you might go about solving the equations.

▌

equations	unknowns
(1) $\quad QL = c_p[(T_0)_2 - (T_0)_1]$	$(T_0)_1, (T_0)_2$
(2) $\quad c_1 = \sqrt{kRT_1}$	c_1
(3) $\quad c_2 = \sqrt{kRT_2}$	c_2
(4) $\quad \dot{m} = \rho_1 V_1 A_1 = (p_1/RT_1)V_1 A_1$	$\dot{m}, V_1$
(5) $\quad \dot{m} = (p_2/RT_2)V_2 A_2$	V_2
(6) $\quad (N_M)_1 = V_1/c_1$	$(N_M)_1$
(7) $\quad (N_M)_2 = V_2/c_2$	$(N_M)_2$
(8) $\quad T_2/(T_0)_2 = 1/\{1 + [(k-1)/2](N_M)_2^2\}$	
(9) $\quad T_1/(T_0)_1 = 1/\{1 + [(k-1)/2](N_M)_1^2\}$	

We thus have nine unknowns and nine equations. One way of proceeding is as follows. Solve for c_2 and c_3 immediately from Eqs. (2) and (3). Now guess at a "$\dot{m}$." Solve for V_2 from Eq. (5) . Get $(N_M)_2$ from Eq. (7) and then $(T_0)_2$ from Eq. (8). From Eq. (4) get V_1 then go to Eq. (6) for $(N_M)_1$ and finally to Eq. (9) for $(T_0)_1$. Finally go to Eq. (1) and see if the computed $(T_0)_1$ and $(T_0)_2$ satisfy the equation. If not, select another $\dot{m}$, etc.

16.230 Air at $V_1 = 300$ ft/s, $p = 40$ psia, $T = 60$ °F flows into a 4.0-in-diameter duct. How much heat transfer per unit mass is needed for sonic conditions at the exit? Determine pressure, temperature, and velocity at the exit and at the section where $N_M = 0.70$.

▋ $(N_M)_1 = V_1/\sqrt{kRT_1} = 300/\sqrt{1.4(53.3)(32.17)(460 + 60)} = 0.268$. The isentropic stagnation temperature at the entrance is $T_{01} = T_1\{1 + [(k - 1)/2](N_M)_1^2\} = 520[1 + 0.2(0.268^2)] = 527$ °R. The isentropic stagnation temperature at the exit is $T_0^* = T_0(1 + kN_M^2)^2/\{(k + 1)N_M^2[2 + (k - 1)N_M^2]\} = 527[1 + 1.4(0.268^2)]^2/\{(2.4 \times 0.268^2)[2 + 0.4(0.268^2)]\} = 1827$ °R. The heat transfer per slug of air flowing is $q_H = c_p(T_0^* - T_{01}) = 0.24[32.17(1827 - 527)] = 10\,037$ Btu/slug. The pressure at the exit is $p^* = p[(1 + kN_M^2)/(k + 1)] = (40/2.4)[1 + 1.4(0.268^2)] = 18.34$ psia and the temperature is $T^* = T[(1 + kN_M^2)/(k + 1)N_M]^2 = 520\{[1 + 1.4(0.268^2)]/2.4(0.268)\}^2 = 1522$ °R. At the exit, $V^* = c^* = \sqrt{kRT^*} = \sqrt{1.4(53.3)(32.17)(1522)} = 1911$ ft/s. At the section where $N_M = 0.7$, $p = p^*[(k + 1)/(1 + kN_M^2)] = 18.34(2.4)/[1 + 1.4(0.7^2)] = 26.1$ psia, $T = T^*[(k + 1)N_M/(1 + kN_M^2)]^2 = 1522\{2.4(0.7)/[1 + 1.4(0.7^2)]\}^2 = 1511$ °R and $V = N_M\sqrt{kRT} = 0.7\sqrt{1.4(53.3)(32.17)(1511)} = 1333$ ft/s.

16.231 In frictionless oxygen flow through a duct, the following conditions prevail at inlet and outlet: $V_1 = 300$ fps, $T_1 = 80$ °F, $(N_M)_2 = 0.5$. Find the heat added per slug and the pressure ratio p_1/p_2.

▋
$$p_1/p_2 = [1 + k(N_M)_2^2]/[1 + k(N_M)_1^2] \qquad N_M = V/c = V/\sqrt{kRT}$$
$$(N_M)_1 = 300/\sqrt{(1.40)(1554)(80 + 460)} = 0.277$$
$$p_1/p_2 = [1 + (1.40)(0.5)^2]/[1 + (1.40)(0.277)^2] = 1.22 \qquad q = c_p[(T_0)_2 - (T_0)_1]$$
$$(T_0) = T\{1 + [(k - 1)/2]N_M^2\} \qquad (T_0)_1 = (80 + 460)\{1 + [(1.40 - 1)/2](0.277)^2\} = 548 \text{ °R}$$
$$T_2 = T_1([(N_M)_2/(N_M)_1]\{[1 + k(N_M)_1^2]/[1 + k(N_M)_2^2]\})^2$$
$$= (80 + 460)((0.5/0.277)\{[1 + (1.40)(0.277)^2]/[1 + (1.40)(0.5^2)]\})^2 = 1184 \text{ °R}$$
$$(T_0)_2 = (1184)\{1 + [(1.40 - 1)/2](0.5)^2\} = 1243 \text{ °R} \qquad q = (0.219)(32.17)(1243 - 548) = 4896 \text{ Btu/slug}$$

16.232 In frictionless air, the flow through a 120-mm-diameter duct 0.15 kg/s enters at $T = 0$ °C and $p = 7$ kN/m² abs. How much heat can be added without choking the flow?

▋ $q_H = c_p[T_p^* - (T_0)_1] \qquad T_0^* = T_0(1 + kN_M^2)^2/(N_M)^2(k + 1)[2 + (k - 1)N_M^2] \qquad T_0 = T\{1 + [(k - 1)/2]N_M^2\}$
$N_M = V/c = V/\sqrt{kRT} \qquad \dot{m} = \rho AV \qquad \rho = p/RT = (7)(1000)/[(287)(0 + 273)] = 0.08934$ kg/m²
$0.15 = (0.08934)[(\pi)(\frac{120}{1000})^2/4](V) \qquad V = 148.5$ m/s $\qquad (N_M)_1 = 148.5/\sqrt{(1.40)(287)(0 + 273)} = 0.448$
$$(T_0)_1 = (0 + 273)\{1 + [(1.40 - 1)/2](0.448)^2\} = 284 \text{ K}$$
$$T_0^* = (284)[1 + (1.40)(0.448)^2]^2/(0.448)^2(1.40 + 1)[2 + (1.40 - 1)(0.448)^2] = 465 \text{ K}$$
$$q_H = (0.240)(465 - 284) = 43.4 \text{ kcal/kg}$$

16.233 Frictionless flow through a duct with heat transfer causes the Mach number to decrease from 2 to 1.75 ($k = 1.4$). Determine the temperature, velocity, pressure, and density ratios.

▋
$$p_1/p_2 = [1 + k(N_M)_2^2]/[1 + k(N_M)_1^2] = [1 + (1.40)(1.75)^2]/[1 + (1.40)(2)^2] = 0.801$$
$$T_1/T_2 = [(N_M)_1/(N_M)_2]^2(p_1/p_2)^2 = (2/1.75)^2(0.801)^2 = 0.838$$
$$\rho_2/\rho_1 = [(N_M)_1/(N_M)_2]\sqrt{T_1/T_2} = (2/1.75)\sqrt{0.838} = 1.05 \qquad V_1/V_2 = \rho_2/\rho_1 = 1.05$$

16.234 In Prob. 16.233 the duct is 2 in square, $p_1 = 15$ psia, and $V_1 = 2000$ fps. Calculate the mass rate of flow for air flowing.

▋
$$\dot{m} = \rho_1 A_1 V_1 \qquad \rho_1 = p_1/RT_1 \qquad RT_1 = V_1^2/kN_m^2$$
$$\rho_1 = p_1/(V_1^2/kN_M^2) = (15)(144)/\{2000^2/[(1.40)(2)^2]\} = 0.003024 \text{ slug/ft}^3$$
$$\dot{m} = (0.003024)[(\tfrac{2}{12})(\tfrac{2}{12})](2000) = 0.168 \text{ slug/s}$$

16.235 How much heat must be transferred per kilogram to cause the Mach number to increase from 2 to 2.8 in a frictionless duct carrying air? ($V_1 = 500$ m/s)

$$q_H = c_p[(T_0)_2 - (T_0)_1] \qquad T_0 = T\{1 + [(k-1)/2]N_M^2\} \qquad T = V^2/N_M^2 kR$$
$$T_1 = 500^2/(2)^2(1.40)(287) = 156 \text{ K}$$
$$(T_0)_1 = (156)\{1 + [(1.40-1)/2](2^2)\} = 281 \text{ K} \qquad T_1/T_2 = \{[(N_M)_1/(N_M)_2](p_1/p_2)\}^2$$
$$p_1/p_2 = [1 + k(N_M)_2^2]/[1 + k(N_M)_1^2] = [1 + (1.40)(2.8)^2]/[1 + (1.40)(2)^2] = 1.815$$
$$T_1/T_2 = [(2/2.8)(1.815)]^2 = 1.681 \qquad T_2 = 156/1.681 = 93 \text{ K}$$
$$(T_0)_2 = (93)\{1 + [(1.40-1)/2](2.8)^2\} = 239 \text{ K}$$
$$q_H = (0.240)(239 - 281) = -10.1 \text{ kcal/kg} \qquad (\text{i.e., cooling})$$

16.236 Oxygen at $V_1 = 525$ m/s, $p = 80$ kN/m^2 abs, and $T = -10$ °C flows in a 60-mm-diameter frictionless duct. How much heat transfer per kilogram is needed for sonic conditions at the exit?

$$q_H = c_p[(T_0)_2 - (T_0)_1] \qquad T_0 = T\{1 + [(k-1)/2]N_M^2\} \qquad N_M = V/c = V/\sqrt{kRT}$$
$$(N_M)_1 = 525/\sqrt{(1.40)(260)(-10+273)} = 1.70 \qquad (T_0)_1 = (-10+273)\{1 + [(1.40-1)/2](1.70)^2\} = 415 \text{ K}$$
$$T_2 = T_1\{[(N_M)_2/(N_M)_1](p_2/p_1)\}^2 \qquad p_1/p_2 = (1+k)/[1 + k(N_M)_1^2] = (1+1.40)/(1 + [1.40](1.70)^2) = 0.4756$$
$$T_2 = (-10+273)[1/1.70)(1/0.4756)]^2 = 402 \text{ K} \qquad (T_0)_2 = (402)\{1 + [(1.40-1)/2](1)^2\} = 482 \text{ K}$$
$$q_H = (0.219)(482 - 415) = 14.7 \text{ kcal/kg}$$

16.237 Apply the first law of thermodynamics, $q_H + p_1/\rho_1 + gz_1 + v_1^2/2 + u_1 = w_s + p_2/\rho_2 + gz_2 + v_2^2/2 + u_2$, to isothermal flow of a perfect gas in a horizontal pipeline, and develop an expression for the heat added per slug flowing.

■ For isothermal flow in a horizontal pipeline, $w = 0$, $z_1 = z_2$, $T_1 = T_2$, $p_1/\rho_1 = p_2/\rho_2$, $u_1 = u_2$. Hence, $q_H = (v_2^2 - v_1^2)/2$.

16.238 A fuel-mixture, assumed equivalent to air, enters a duct combustion chamber at $V_1 = 100$ m/s and $T_1 = 400$ K. What amount of heat addition in kilojoules per kilogram will cause the exit flow to be choked? What will be the exit Mach number and temperature if 1000 kJ/kg is added during combustion?

$$q_{\text{choke}} = c_p[T_0^* - (T_0)_1] \qquad T_0 = T + V^2/2c_p \qquad (T_0)_1 = 400 + [100^2/(2)(1005)] = 405 \text{ K}$$
$$N_M = V/c = V/\sqrt{kRT} \qquad (N_M)_1 = 100/\sqrt{(1.40)(287)(400)} = 0.249$$

From Table A-25, $(T_0)_1/T_0^* = 0.2558$. $T_0^* = 405/0.2558 = 1583$ K, $q_{\text{choke}} = (1005)(1583 - 405) = 1\,184\,000$ J/kg, or 1184 kJ/kg. For $q = 1000$ kJ/kg; $(T_0)_2 = (T_0)_1 + (q/c_p) = 405 + (1000/1.005) = 1400$ K, $(T_0)_2/T_0^* = \frac{1400}{1583} = 0.8844$. From Table A-25, $(N_M)_2 = 0.669$. $T_2 = (T_0)_2/\{1 + [(k-1)/2](N_M)_2^2\} = (1400)/\{1 + [(1.40-1)/2](0.669)^2\} = 1285$ K.

16.239 What happens to the inlet flow of Prob. 16.238 if the combustion yields 2500 kJ/kg heat addition and $(p_0)_1$ and $(T_0)_1$ remain the same? How much is the mass flow reduced?

$$(T_0)_2 = (T_0)_1 + (q/c_p)$$

Choking:

$$(T_0)_2 = T_0^* = 405 + (2500/1.005) = 2893 \text{ K}$$
$$(T_0)_1/T_0^* = \frac{405}{2893} = 0.1400$$

From Table A-25, new $(N_M)_1 = 0.1777$.

$$\dot{m}/A = \rho_1 V_1 = \rho_1 c_1 (N_M)_1 = \rho_0 c_0/\{1 + [(k-1)/2](N_M)_1^2\}^3$$
$$\frac{\dot{m} \text{ (new)}}{\dot{m} \text{ (old)}} = \frac{N_M \text{ (new)}}{N_M \text{ old)}} \left[\frac{1 + [(1.40-1)/2]N_M^2 \text{ (old)}}{1 + [(1.40-1)/2]N_M^2 \text{ (new)}}\right]^3 = \frac{0.1777}{0.249}\left[\frac{1 + (0.2)(0.249)^2}{1 + (0.2)(0.1777)^2}\right]^3 = 0.727$$

The mass flow is reduced by $1 - 0.727 = 0.273$, or 27.3 percent.

CHAPTER 17
Flow Measurement

17.1 Water is being discharged through a 3-in-diameter pipe directly into a container that has a volume of 27.0 ft^3. Find the rate and velocity of flow through the pipe if the time required to fill the container is 4 min 19.6 s.

■ $\quad Q = V/t = 27.0/[(4)(60) + 19.6] = 0.104 \text{ ft}^3/\text{s} \qquad v = Q/A = 0.104/[(\pi)(\frac{3}{12})^2/4] = 2.12 \text{ ft/s}$

17.2 Water is being discharged through a 6-in-diameter pipe directly into a container that has a volume of 200.0 ft^3. Find the rate and velocity of flow through the pipe if the time required to fill the container is 3 min 21.2 s.

■ $\quad Q = V/t = 200.0/[(3)(60) + 21.2] = 0.994 \text{ ft}^3/\text{s} \qquad v = Q/A = 0.994/[(\pi)(\frac{6}{12})^2/4] = 5.06 \text{ ft/s}$

17.3 A Pitot tube being used to determine the velocity of flow of water in a closed conduit indicates a difference between water levels in the Pitot tube and in the piezometer of 48 mm. What is the velocity of flow?

■ $\qquad\qquad v = \sqrt{2gh} = \sqrt{(2)(9.807)(\frac{48}{1000})} = 0.970 \text{ m/s}$

17.4 A Pitot tube being used to determine the velocity of flow of water in a closed conduit indicates a difference between water levels in the Pitot tube and in the piezometer of 58 mm. What is the velocity of flow?

■ $\qquad\qquad v = \sqrt{2gh} = \sqrt{(2)(9.807)(\frac{58}{1000})} = 1.07 \text{ m/s}$

17.5 An airplane uses a Pitot measuring device to determine its velocity. The instruments show a stagnation pressure (p_s) of 508 mm of mercury (mmHg), a static pressure (p_0) of 325 mmHg abs, and a stagnation temperature of $60 \,°\text{C}$. What is the plane's velocity?

■ $\qquad N_M = \sqrt{[2/(k-1)][(p_s/p_0)^{(k-1)/k} - 1]} = \sqrt{[2/(1.40-1)][(\frac{508}{325})^{(1.40-1)/1.40} - 1]} = 0.825$

Since $(N_M = 0.825) < 1.0$, flow is subsonic, and

$$v = \sqrt{[2kgRT/(k-1)][(p_s/p_0)^{(k-1)/k} - 1]} \qquad T = 60 + 273 = 333 \text{ K}$$
$$v = \sqrt{[(2)(1.40)(9.807)(29.3)(333)/(1.40-1)][(\frac{508}{325})^{(1.40-1)/1.40} - 1]} = 302 \text{ m/s}$$

17.6 An airplane uses a Pitot measuring device to determine its velocity. The instruments show a stagnation pressure (p_s) of 498 mmHg abs, a static pressure (p_0) of 318 mmHg abs, and a stagnation temperature of $75 \,°\text{C}$. What is the plane's velocity?

■ $\qquad N_M = \sqrt{[2/(k-1)][(p_s/p_0)^{(k-1)/k} - 1]} = \sqrt{[2/(1.40-1)][(\frac{498}{318})^{(1.40-1)/1.40} - 1]} = 0.827$

Since $(N_M = 0.827) < 1.0$, flow is subsonic, and

$$v = \sqrt{[2kgRT/(k-1)][(p_s/p_0)^{(k-1)/k} - 1]} \qquad T = 75 + 273 = 348 \text{ K}$$
$$v = \sqrt{[(2)(1.40)(9.807)(29.3)(348)/(1.40-1)][(\frac{498}{318})^{(1.40-1)/1.40} - 1]} = 309 \text{ m/s}$$

17.7 Air $(\gamma = 0.075 \text{ pcf})$ is flowing in Fig. 17-1. If $u = 13.0$ fps and $V_c = 15.0$ fps, determine the manometer reading.

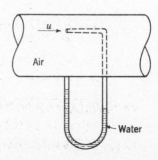

Fig. 17-1

$$u^2/2g = (\gamma_{H_2O}/\gamma_{air} - 1)(x) \qquad 13.0^2/[(2)(32.2)] = (62.4/0.075 - 1)(x)$$
$$x = 0.00316 \text{ ft} \quad \text{or} \quad 0.0379 \text{ in}$$

(Such a reading is too small for practical use.)

17.8 Repeat Prob. 17.7 for Fig. 17-2.

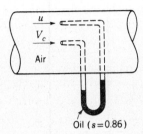

Oil (s=0.86) **Fig. 17-2**

$$V_c/2g - u^2/2g = (\gamma_{oil}/\gamma_{air} - 1)(x)$$
$$15.0^2/[(2)(32.2)] - 13.0^2/[(2)(32.3)] = [(0.86)(62.4)/0.075 - 1](x) \qquad x = 0.00122 \text{ ft} \quad \text{or} \quad 0.0146 \text{ in}$$

(As in Prob. 17.7, such a reading is too small for practical use.)

17.9 In Fig. 17-3, pressure gage A reads 10.0 psi, while pressure gage B reads 11.0 psi. Find the velocity if air at 50 °F is flowing. Atmospheric pressure is 26.8 inHg. Assume $C_I = 1.0$ and neglect compressibility effects.

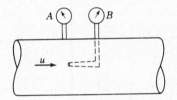

A B **Fig. 17-3**

$$p = (26.8/29.9)(14.7) + 10 = 23.18 \text{ psia} \qquad \gamma = p/RT = (23.18)(144)/[(53.3)(50 + 460)] = 0.1228 \text{ lb/ft}^3$$
$$u = C_I\sqrt{2g(p_s/\gamma - p_0/\gamma)} = (1.0)\sqrt{(2)(32.2)[(11.0)(144)/0.1228 - (10.0)(144)/0.1228]} = 275 \text{ ft/s}$$

17.10 In Prob. 17.9, if the two pressure gages were replaced by a differential manometer containing water, what would be the reading on the manometer (y)?

$$\Delta p/\gamma = (s.g._M/s.g._F - 1)(y) \qquad \Delta p/\gamma = (11.0)(144)/0.1228 - (10.0)(144)/0.1228 = 1173 \text{ ft}$$
$$1173 = (62.4/0.1228 - 1)(y) \qquad y = 2.313 \text{ ft} \quad \text{or} \quad 27.8 \text{ in}$$

17.11 In Fig. 17-3, assume kerosene (s.g. = 0.81) is flowing. The pressure gages at A and B read 65 Pa and 140 Pa, respectively. Find the velocity u, assuming $C_I = 1.0$.

$$\gamma = (0.81)(9.79) = 7.930 \text{ kN/m}^3 \quad \text{or} \quad 7930 \text{ N/m}^3$$
$$u = C_I\sqrt{2g(p_s/\gamma - p_0/\gamma)} = (1.00)\sqrt{(2)(9.807)(\tfrac{140}{7930} - \tfrac{65}{7930})} = 0.431 \text{ m/s}$$

17.12 The pitometer in Fig. 17-4 is connected to a mercury manometer, and the reading is 10.0 cm. The velocity is known to be 3.6 m/s. If carbon tetrachloride (s.g. = 1.59) is flowing, what is C_I for the measurement?

Fig. 17-4

$$u = C_I\sqrt{2g(p_s/\gamma - p_0/\gamma)} = C_I\sqrt{2g(\Delta p/\gamma)}$$
$$\Delta p/\gamma = (s.g._M/s.g._F - 1)(y) = (13.6/1.59 - 1)(10.0/100) = 0.7553 \text{ m}$$
$$3.6 = (C_I)\sqrt{(2)(9.807)(0.7553)} \qquad C_I = 0.935$$

17.13 In Fig. 17-4, suppose air at 50 °F is flowing. The pitometer is attached to a manometer containing a liquid (s.g. = 0.85). Find velocity u versus manometer readings assuming $C_I = 0.92$. Assume the air is at standard atmospheric pressure.

▮ From Table A-4, $\rho_{air} = 0.00242$ slug/ft³. $\rho_{liquid}/\rho_{air} = (0.85)(1.94)/0.00242 = 681.4$. Thus 1 ft of manometer reading is equivalent to 681.4 ft of air. $u = C_I\sqrt{2g(p_s/\gamma - p_0/\gamma)} = C_I\sqrt{2g(\Delta p/\gamma)}$. Letting y = manometer reading, $u = C_I\sqrt{2g(681.4y)}$. For $y = 1$ in, $u = (0.92)\sqrt{(2)(32.2)[(681.4)(\frac{1}{12})]} = 55.6$ ft/s. For $y = 5$ in, $u = (0.92)\sqrt{(2)(32.2)[(681.4)(\frac{5}{12})]} = 124$ ft/s. For $y = 10$ in, $u = (0.92)\sqrt{(2)(32.2)[(681.4)(\frac{10}{12})]} = 176$ ft/s. For $y = 20$ in, $u = (0.92)\sqrt{(2)(32.2)[(681.4)(\frac{20}{12})]} = 249$ ft/s. (At larger manometer readings, the effect of compressibility must be considered.)

17.14 A Pitot tube is placed in a pipe carrying water at 15 °C. The Pitot tube and a wall piezometer tube are connected to a water-mercury manometer which registers a differential of 7.5 cm. Assuming $C_I = 0.99$, what is the velocity approaching the tube?

▮ $$\Delta p/\gamma = (s.g._M/s.g._F - 1)(y) = (13.6/1 - 1)(7.5/100) = 0.9450 \text{ m}$$
$$u = C_I\sqrt{2g(p_s/\gamma - p_0/\gamma)} = C_I\sqrt{2g(\Delta p/\gamma)} = (0.99)\sqrt{(2)(9.807)(0.9450)} = 4.26 \text{ m/s}$$

17.15 Suppose that the fluids of Prob. 17.14 are reversed so that mercury is flowing in the pipe and water is the gage fluid (with the manometer now inverted). With the same gage differential, what would be the velocity of the mercury?

▮ $$\Delta p/\gamma = (1 - s.g._M/s.g._F)(y) = (1 - 1/13.6)(7.5/100) = 0.06949 \text{ m}$$
$$u = C_I\sqrt{2g(p_s/\gamma - p_0/\gamma)} = C_I\sqrt{2g(\Delta p/\gamma)} = (0.99)\sqrt{(2)(9.807)(0.06949)} = 1.16 \text{ m/s}$$

17.16 A Prandtl tube is placed on the centerline of a smooth 12-in-diameter pipe in which water at 80 °F is flowing. The reading on a differential manometer attached to this Prandtl tube is 10 in of carbon tetrachloride (s.g. = 1.59). Find the flow rate.

▮ $$V/u_{max} = 1/(1 + 1.33f^{1/2}) \qquad h = (1.59 - 1)(\tfrac{10}{12}) = 0.4917 \text{ ft}$$
$$u_{max} = \sqrt{2gh} = \sqrt{(2)(32.2)(0.4917)} = 5.627 \text{ ft/s}$$

Try $V = 5$ ft/s: $N_R = DV/\nu = (\frac{12}{12})(5)/(9.15 \times 10^{-6}) = 5.46 \times 10^5$. From Fig. A-5, $f = 0.013$. $V/5.627 = 1/[1 + (1.33)(0.013)^{1/2}]$, $V = 4.89$ ft/s. (f is still 0.013 for $V = 4.89$ ft/s.) $Q = AV = [(\pi)(\frac{12}{12})^2/4](4.89) = 3.84$ ft³/s.

17.17 A Pitot-static tube for which $C_I = 0.98$ is connected to an inverted U-tube containing oil (s.g. = 0.85). Water is flowing. What is the velocity if the manometer reading is 4.0 in?

▮ $$\Delta p/\gamma = (1 - s.g._M/s.g._F)(y) = (1 - 0.85/1)(4.0/12) = 0.0500 \text{ ft}$$
$$u = C_I\sqrt{2g(p_s/\gamma - p_0/\gamma)} = C_I\sqrt{2g(\Delta p/\gamma)} = (0.98)(\sqrt{(2)(32.2)(0.0500)}) = 1.76 \text{ ft/s}$$

17.18 A Pitot tube having a coefficient of 0.98 is used to measure the velocity of water at the center of a pipe, as shown in Fig. 17-5. What is the velocity?

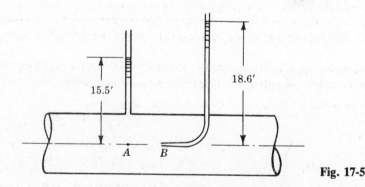

18.6'

15.5'

A B

Fig. 17-5

▮ $$u = C_I\sqrt{2g(p_s/\gamma - p_0/\gamma)} = (0.98)\sqrt{(2)(32.2)(18.6 - 15.5)} = 13.8 \text{ ft/s}$$

17.19 Air flows through a duct, and the Pitot-static tube measuring the velocity is attached to a differential gage containing water. If the deflection of the gage is 4 in, calculate the air velocity, assuming the specific weight of air is constant at 0.0761 lb/ft^3 and that the coefficient of the tube is 0.98.

$$\blacksquare \qquad (p_s/\gamma - p_0/\gamma) = (\tfrac{4}{12})(62.4)/0.0761 = 273.3 \text{ ft of air}$$

$$u = C_I\sqrt{2g(p_s/\gamma - p_0/\gamma)} = (0.98)\sqrt{(2)(32.2)(273.3)} = 130 \text{ ft/s}$$

17.20 Carbon tetrachloride (s.g. = 1.60) flows through a pipe. The differential gage attached to the Pitot-static tube shows a 3-in deflection of mercury. Assuming $C = 1.00$, find the velocity.

$$\blacksquare \qquad p_s - p_0 = (\tfrac{3}{12})(13.6 - 1.60)(62.4) = 187.2 \text{ lb/ft}^2$$

$$u = C_I\sqrt{2g(p_s/\gamma - p_0/\gamma)} = (1.00)\sqrt{(2)(32.2)\{187.2/[(1.60)(62.4)]\}} = 11.0 \text{ ft/s}$$

17.21 Water flows at a velocity of 4.65 fps. A differential gage ($C_I = 1.0$) which contains a liquid of specific gravity 1.25 is attached to the Pitot-static tube. What is the deflection of the gage fluid?

$$\blacksquare \quad u = C_I\sqrt{2g(p_s/\gamma - p_0/\gamma)} = C_I\sqrt{2g(\Delta p/\gamma)} \qquad 4.65 = (1.0)\sqrt{(2)(32.2)(\Delta p/\gamma)} \qquad \Delta p/\gamma = 0.3358 \text{ ft of water}$$

$$\Delta p/\gamma = (\text{s.g.}_M/\text{s.g.}_F - 1)(y) \qquad 0.3358 = [(1.25/1) - 1](y) \qquad y = 1.34 \text{ ft}$$

17.22 Develop the expression for measuring the flow of a gas with a Pitot tube.

$\blacksquare$ The flow from A to B in the figure of Prob. 17.18 may be considered adiabatic and with negligible loss. Using the Bernoulli equation, A to B, we obtain $\{[k/(k-1)](p_A/\gamma_A) + (V_A^2/2g) + 0\} - \text{negligible loss} = [k/(k-1)](p_A/\gamma_A)(p_B/p_A)^{(k-1)/k} + 0 + 0$ or

$$V_A^2/2g = [k/(k-1)](p_A/\gamma_A)[(p_B/p_A)^{(k-1)/k} - 1] \qquad (1)$$

The term p_B is the stagnation pressure. This expression (1) is usually rearranged, introducing the ratio of the velocity at A to the acoustic velocity c of the undisturbed fluid. The acoustic velocity $c = \sqrt{E/\rho} = \sqrt{kp/\rho} = \sqrt{kpg/\gamma}$. Combining with Eq. (1) above,

$$V_A^2 = [c^2/(k-1)][(p_B/p_A)^{(k-1)/k} - 1] \qquad \text{or} \qquad p_B/p_A = \{1 + [(k-1)/2](V_A/c)^2\}^{k/(k-1)} \qquad (2)$$

Expanding by the binomial theorem,

$$p_B/p_A = 1 + (k/2)(V_A/c)^2\{1 + \tfrac{1}{4}(V_A/c)^2 - [(k-2)/24](V_A/c)^4 + \cdots\} \qquad (3)$$

In order to compare this expression with the formula in Prob. 17.18, multiply through by p_A and replace kp_A/c^2 by ρ_A, obtaining

$$p_B = p_A + \tfrac{1}{2}\rho_A V_A^2\{1 + \tfrac{1}{4}(V_A/c)^2 - [(k-2)/24](V_A/c)^4 + \cdots\} \qquad (4)$$

The above expressions apply to all compressible fluids for ratios of V/c less than unity. For ratios over unity, shock wave and other phenomena occur, the adiabatic assumption is not sufficiently accurate, and the derivation no longer applies. The ratio V/c is called the *Mach number*.

The term in braces in (4) is greater than unity and the first two terms provide sufficient accuracy. The effect of compressibility is to increase the stagnation-point pressure over that of an incompressible fluid.

17.23 Air flowing under atmospheric conditions ($\gamma = 0.0763$ lb/ft^3 at 60 °F) at a velocity of 300 fps is measured by a Pitot tube. Calculate the error in the stagnation pressure by assuming the air to be incompressible.

$$\blacksquare \quad p_B = p_A + \rho V^2/2 = (14.7)(144) + (0.0763/32.2)(300)^2/2 = 2223 \text{ lb/ft}^2 \text{ abs}$$

Also:
$$p_B = p_A + \tfrac{1}{2}\rho_A V_A^2\{1 + \tfrac{1}{4}(V_A/c)^2 - [(k-2)/24](V_A/c)^4 + \cdots\}$$

$$c = \sqrt{kgRT} = \sqrt{(1.40)(32.2)(53.3)(60 + 460)} = 1118 \text{ ft/s}$$

$$p_B = (14.7)(144) + [(0.0763/32.2)(300)^2/2]\{1 + (\tfrac{1}{4})(\tfrac{300}{1118})^2 - [(1.40 - 2)/24](\tfrac{300}{1118})^4 + \cdots\} = 2225 \text{ lb/ft}^2 \text{ abs}$$

The error in the stagnation pressure is $(2225 - 2223)/2225 = 0.0009$, or 0.09 percent.

17.24 The difference between the stagnation pressure and the static pressure measured by a Pitot-static device is 412 lb/ft^2. The static pressure is 14.5 psia and the temperature in the air stream is 60 °F. What is the velocity of the air, assuming the air is (a) compressible and (b) incompressible.

▌ (a) Using Eq. (2) of Prob. 17.22,

$$p_B/p_A = \{1 + [(k-1)/2](V_A/c)^2\}^{k/(k-1)}$$
$$c = \sqrt{kgRT} = \sqrt{(1.40)(32.2)(53.3)(60+460)} = 1118 \text{ ft/s}$$
$$[(14.5)(144)+412]/(14.5)(144) = \{1 + [(1.40-1)/2](V_A/1118)^2\}^{1.40/(1.40-1)} \qquad V_A = 574 \text{ ft/s}$$

(b)
$$\gamma = p/RT = (14.5)(144)/[(53.3)(60+460)] = 0.07534 \text{ lb/ft}^3$$
$$V = \sqrt{2g(\Delta p/\gamma)} = \sqrt{(2)(32.2)(412/0.07534)} = 593 \text{ ft/s}$$

17.25 Air flows at 800 fps through a duct. At standard barometer, the stagnation pressure is −5.70 ft of water gage. The stagnation temperature is 145 °F. What is the static pressure in the duct?

▌ Using Eq. (2) of Prob. 17.22, $p_B/p_A = \{1 + [(k-1)/2](V_A/c)^2\}^{k/(k-1)}$. With two unknowns in the equation, assume a V/c ratio of 0.72. $(-5.70 + 34.0)(62.4)/p_A = \{1 + [(1.40-1)/2](0.72)^2\}^{1.40/(1.40-1)}$, $p_A = 1250 \text{ lb/ft}^2$ abs. Check the assumption using the adiabatic relation $T_B/T_A = (p_B/p_A)^{(k-1)/k}$:

$$(145+460)/T_A = [(-5.70+34.0)(62.4)/1250]^{(1.40-1)/1.40} \qquad T_A = 548 \text{ °R}$$
$$c = \sqrt{kgRT} = \sqrt{(1.40)(32.2)(53.3)(548)} = 1147 \text{ ft/s}$$
$$V/c = \tfrac{800}{1147} = 0.697 \qquad (-5.70+34.0)(62.4)/p_A = \{1 + [(1.40-1)/2](0.697)^2\}^{1.40/(1.40-1)}$$
$$p_A = 1277 \text{ lb/ft}^2 \text{ abs}$$

No further refinement is necessary.

17.26 A static tube indicates a static pressure that is 1 kPa too low when liquid is flowing at 2 m/s. Calculate the correction to be applied to the indicated pressure for the liquid flowing at 5 m/s.

▌ Since Δh is a function of V^2, $\Delta h_2/\Delta h_1 = V_2^2/V_1^2$, $\Delta p_2/\Delta p_1 = V_2^2/V_1^2$, $\Delta p_2/1 = 5^2/2^2$, $\Delta p_2 = 6.25$ kPa.

17.27 A simple Pitot tube is inserted into a small stream of flowing oil, $\gamma = 55 \text{ lb/ft}^3$, $\mu = 0.65$ P, $\Delta h = 1.5$ in, $h_0 = 5$ in. What is the velocity at point 1?

▌
$$v = \sqrt{2g\,\Delta h} = \sqrt{(2)(32.2)(1.5/12)} = 2.84 \text{ ft/s}$$

17.28 A stationary body immersed in a river has a maximum pressure of 69 kPa exerted on it at a distance of 5.4 m below the free surface. Calculate the river velocity at this depth.

▌
$$V^2/2g = p_t/\gamma - p_s/\gamma \qquad V^2/[(2)(9.807)] = 69/9.79 - 5.4 \qquad V = 5.69 \text{ m/s}$$

17.29 For Fig. 17-6 derive the equation for velocity at point 1.

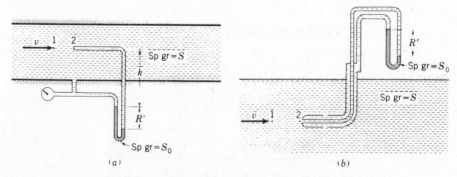

(a)   (b)   **Fig. 17-6**

▌ For Fig. 17-6a (Pitot tube and piezometer opening), $V_1^2/2g = (p_2 - p_1)/\gamma$.

Mamometer equation:

$$p_1(S/\gamma) + kS + R'S_0 - (k + R')S = p_2(S/\gamma) \qquad (p_2 - p_1)/\gamma = R'[(S_0 - S)/S]$$
$$V_1^2/2g = R'[(S_0 - S)/S] \qquad V_1 = \sqrt{2gR'(S_0/S - 1)}$$

For Fig. 17-6b (Pitot-static tube), we analyze this system in a manner similar to that for Fig. 17-6a and show that the same relations hold, but the uncertainty in the measurement of static pressure requires a corrective coefficient C_I to be applied so that $V_1 = C_I\sqrt{2gR'(S_0/S - 1)}$.

17.30 In Fig. 17-6a air is flowing ($p = 16$ psia, $T = 40$ °F) and water is in the manometer. For $R' = 1.2$ in, calculate the velocity of air.

$$V = \sqrt{2gR'(S_0/S - 1)} \qquad \rho = p/RT = (16)(144)/[(53.3)(40 + 460)] = 0.08645 \text{ lbm/ft}^3$$

$$S_0/S = 62.4/0.08645 = 721.8 \qquad V = \sqrt{(2)(32.2)(1.2/12)(721.8 - 1)} = 68.1 \text{ ft/s}$$

17.31 In Fig. 17-6 air is flowing ($p = 101$ kPa abs and $T_1 = 5$ °C) and mercury is in the manometer. For $R' = 200$ mm, calculate the velocity at 1 for air considered incompressible.

$$V = \sqrt{2gR'(S_{Hg}/S_{air} - 1)} \qquad \rho = p/RT \qquad \rho_{air} = (101)(1000)/[(287)(5 + 273)] = 1.266 \text{ kg/m}^3$$

$$S_{Hg}/S_{air} = (13.6)(1000)/1.266 = 10\,742 \qquad V = \sqrt{(2)(9.807)(\tfrac{200}{1000})(10\,742 - 1)} = 205 \text{ m/s}$$

17.32 Work Prob. 17.31 for isentropic compression of air between 1 and 2.

$$V_1^2/2 = c_p T_1[(p_2/p_1)^{(k-1)/k} - 1] \qquad (p_2 - p_1)/\gamma_{air} = R'[(S_{Hg} - S_{air})/S_{air}]$$

$$p_2 - p_1 = R'\gamma_{Hg} - R'\gamma_{air} \approx R'\gamma_{Hg} \qquad p_2 - 101 = (\tfrac{200}{1000})[(13.6)(9.79)] \qquad p_2 = 127.6 \text{ kPa}$$

$$V_1^2/2 = (1005)(5 + 273)[(127.6/101)^{(1.40-1)/1.40} - 1] \qquad V_1 = 196 \text{ m/s}$$

17.33 A Pitot-static tube directed into a 4-m/s water stream has a gage difference of 37 mm on a water-mercury differential manometer. Determine the coefficient for the tube.

$$V = C_I\sqrt{2gR'(S_0/S - 1)} \qquad 4 = C_I\sqrt{(2)(9.807)(\tfrac{37}{1000})[(13.6/1) - 1]} \qquad C_I = 1.32$$

17.34 A Pitot-static tube, $C_I = 1.12$, has a gage difference of 10 mm on a water-mercury manometer when directed into a water stream. Calculate the velocity.

$$V = C_I\sqrt{2gR'(S_0/S - 1)} = (1.12)\sqrt{(2)(9.807)(\tfrac{10}{1000})[(13.6/1) - 1]} = 1.76 \text{ m/s}$$

17.35 A Pitot-static tube of the Prandtl type has the following values of gage difference R' for the radial distance from the center of a 3-ft-diameter pipe:

r, ft	0.0	0.3	0.6	0.9	1.2	1.48
R', in	4.00	3.91	3.76	3.46	3.02	2.40

Water is flowing, and the manometer fluid has a specific gravity of 2.93. Calculate the discharge.

$V = C_I\sqrt{2gR'(S_0/S - 1)}$. For a Prandtl tube, $C_I = 1.0$

$$V_i = (1.0)\sqrt{(2)(32.2)(R'/12)[(2.93/1) - 1]} = 3.218\sqrt{R'} \qquad (R' \text{ in inches})$$

$$A_1 = (\pi)[(r_2 - r_1)/2]^2 \qquad A_i = (\pi)\{[(r_{i+1} + r_i)/2]^2 - [(r_i + r_{i-1})/2]^2\}$$

$$A_6 = (\pi)\{r_7^2 - [(r_5 + r_6)/2]^2\}$$

i	1	2	3	4	5	6	7	
r_i	0	0.3	0.6	0.9	1.2	1.48	1.5	ft
V_i	6.436	6.363	6.240	5.986	5.592	4.985	—	ft/s
A_i	0.071	0.565	1.131	1.696	2.177	1.334	—	ft^2

$$Q = \sum_{i=1}^{6} V_i A_i = (6.436)(0.071) + (6.363)(0.565) + (6.240)(1.131)$$

$$+ (5.986)(1.696) + (5.592)(2.177) + (4.985)(1.334) = 40.1 \text{ ft}^3/\text{s}$$

17.36 What would the gage difference be on a water-nitrogen manometer for flow of nitrogen at 200 m/s, using a Pitot-static tube? The static pressure is 175 kPa abs, and the corresponding temperature is 25 °C. True static pressure is measured by the tube.

$$p_2 = (p_1)(V_1^2/2c_pT_1 + 1)^{k/(k-1)} = (175)\{200^2/[(2)(1040)(25 + 273)] + 1\}^{1.40/(1.40-1)} = 217.8 \text{ kPa}$$
$$R' = (p_t - p_s)/\gamma = (217.8 - 175)/9.79 = 4.37 \text{ m}$$

17.37 Measurements in an air stream indicate that the stagnation pressure is 15 psia, the static pressure is 10 psia, and the stagnation temperature is 102 °F. Determine the temperature and velocity of the air stream.

▮ $$V_1^2/2 = c_pT_2[1 - (p_1/p_2)^{(k-1)/k}] = (6000)(102 + 460)[1 - (\tfrac{10}{15})^{(1.40-1)/1.40}]$$
$$V_1 = 859 \text{ ft/s} \qquad T_1 = T_2 - V^2/2c_p = (102 + 460) - 859^2/[(2)(6000)] = 501 \text{ °R} \quad \text{or} \quad 41 \text{ °F}$$

17.38 If 0.5 kg/s nitrogen flows through a 50-mm-diameter tube with stagnation temperature of 38 °C and undisturbed temperature of 20 °C, find the velocity and static and stagnation pressures.

▮ $$V_1 = \sqrt{2c_p(T_2 - T_1)} = \sqrt{(2)(1040)[(38 + 273) - (20 + 273)]} = 193 \text{ m/s}$$
$$p_1 = \rho_1 RT_1 = (\dot{m}/AV)RT_1$$
$$A = (\pi)(\tfrac{50}{1000})^2/4 = 0.001964 \text{ m}^2 \qquad p_1 = \{0.5/[(0.001964)(193)]\}(297)(20 + 273) = 114\,800 \text{ Pa}$$
$$p_2 = (p_1)(T_2/T_1)^{k/(k-1)} = (114\,800)[(38 + 273)/(20 + 273)]^{1.40/(1.40-1)} = 141\,400 \text{ Pa}$$

17.39 A Pitot tube is inserted into a flowing air stream at 15 °C and 110 kPa. The differential pressure reads 9 mm on an air-mercury manometer. What is the indicated air speed?

▮ $$\Delta p = (\rho_{\text{Hg}} - \rho_{\text{air}})gh = \rho_{\text{air}}V^2/2 \qquad \rho_{\text{air}} = p/RT = (110)(1000)/[(287)(15 + 273)] = 1.33 \text{ kg/m}^3$$
$$(13\,570 - 1.33)(9.807)(\tfrac{9}{1000}) = (1.33)(V^2/2) \qquad V = 42.4 \text{ m/s}$$

17.40 For the Pitot-static pressure arrangement of Fig. 17-7, compute the (*a*) centerline velocity, (*b*) pipe volume flow, and (*c*) wall shear stress. The manometer fluid is Meriam red oil (s.g. = 0.827).

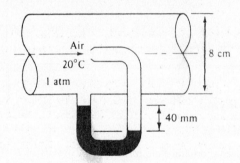

Fig. 17-7

▮ (*a*) $$p_0 - p = (\rho_{\text{oil}} - \rho_{\text{air}})gh = [(0.827)(998) - 1.20](9.807)(\tfrac{40}{1000}) = 323.3 \text{ Pa}$$
$$V_{C_L} = \sqrt{2\Delta p/\rho_{\text{air}}} = \sqrt{(2)(323.3)/1.20} = 23.2 \text{ m/s}$$

(*b*) $V_{\text{avg}} = V_{C_L}/(1 + 1.33\sqrt{f})$. Estimate $V_{\text{avg}} = 0.85V_{C_L} = (0.85)(23.2) = 19.7 \text{ m/s}$. $N_R = \rho DV/\mu = (1.20)(\tfrac{8}{100})(19.7)/(1.80 \times 10^{-5}) = 1.05 \times 10^5$. From Fig. A-5, $f = 0.0178$.

$$V_{\text{avg}} = 23.2/(1 + 1.33\sqrt{0.0178}) = 19.7 \text{ m/s} \quad \text{(O.K.)} \qquad Q = AV = [(\pi)(\tfrac{8}{100})^2/4](19.7) = 0.0990 \text{ m}^3/\text{s}$$

(*c*) $$\tau = (\tfrac{1}{8})(f\rho V_{\text{avg}}^2) = (\tfrac{1}{8})(0.0178)(1.20)(19.7)^2 = 1.04 \text{ Pa}$$

17.41 For the water flow at 20 °C of Fig. 17-8, use the manometer measurement to estimate the (*a*) centerline velocity and (*b*) volume flow in the 6-in-diameter pipe. Assume a smooth wall.

▮ (*a*) $V_{C_L} = \sqrt{2\Delta p/\rho_{\text{H}_2\text{O}}}$. Assume at first that static pressure at A = static pressure at B.

$$\Delta p = (p_0)_B - p_B = (p_0)_B - p_A = (\rho_{\text{Hg}} - \rho_{\text{H}_2\text{O}})gh = [(13.6 - 1)(1.94)](32.2)(\tfrac{2}{12}) = 131.2 \text{ lb/ft}^2$$
$$V_{C_L} = \sqrt{(2)(131.2)/1.94} = 11.6 \text{ ft/s}$$

(*b*) $V_{\text{avg}} = V_{C_L}/(1 + 1.33\sqrt{f})$. Estimate $V_{\text{avg}} = 0.87V = (0.87)(11.6) = 10.1 \text{ ft/s}$. $N_R = \rho DV/\mu = (1.94)(\tfrac{6}{12})(10.1)/(2.04 \times 10^{-5}) = 4.80 \times 10^5$. From Fig. A-5, $f = 0.0133$.

$$V_{\text{avg}} = 11.6/(1 + 1.33\sqrt{0.0133}) = 10.1 \text{ ft/s} \quad \text{(O.K.)}$$
$$Q = AV = [(\pi)(\tfrac{6}{12})^2/4](10.1) = 1.98 \text{ ft}^3/\text{s}$$

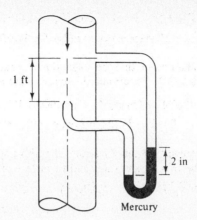

Mercury   **Fig. 17-8**

Check:

$$p_A - p_B = \rho g f (L/D)(V^2/2g) = (1.94)(32.2)(0.0133)[1/(\tfrac{6}{12})]\{10.1^2/[(2)(32.2)]\} = 2.6 \text{ lb/ft}^2$$

This gives a $2.6/131.2 = 0.020$, or 2.0 percent error in Δp.

17.42 A Pitot tube placed in a flow of helium ($\rho = 0.166$ kg/m^3) at 20 °C and 1 atm shows a helium-water differential manometer reading of 11 mm. What is the helium velocity? What will the manometer reading be if the helium velocity is 25 m/s?

❚
$$\Delta p = (\rho_{H_2O} - \rho_{He})gh = (998 - 0.166)(9.807)(\tfrac{11}{1000}) = 107.6 \text{ kPa}$$
$$V_{\mathbb{C}} = \sqrt{2\Delta p/\rho_{He}} = \sqrt{(2)(107.6)/0.166} = 36.0 \text{ m/s}$$

If $V_{\mathbb{C}} = 25$ m/s, $\Delta p = (\rho_{H_2O} - \rho_{He})gh = \rho V^2/2$, $(998 - 0.166)(9.807)(h) = (0.166)(25)^2/2$, $h = 0.0053$ m, or 5.3 mm.

17.43 A small airplane flying at 8000-m altitude uses a Pitot stagnation probe without a static tube. The measured stagnation pressure is 39 kPa. What is the indicated airplane speed in miles per hour and its probable uncertainty?

❚ $\Delta p = p_0 - p = \rho V^2/2$. From Table A-8, p at 8000 m = 35.65 kPa and 0.526 kg/m^3. $(39 - 35.65)(1000) = (0.526)(V)^2/2$, $V = 113$ m/s $= [(113/0.3048)/5280](3600) = 253$ mph. What is the uncertainty? Assume ρ_{air} and p_{air} have ±3 percent error and p_0 has ±1 percent. The worst case would be high ρ and p and low p_0.
$\Delta p = [(39)(0.99) - (35.65)(1.03)](1000) = [(0.526)(1.03)](V)^2/2$, $V = 83.5$ m/s. This is $(113 - 83.5)/113 = 0.26$, or 26 percent less. Thus, small errors in this case make for a large error in the velocity estimate.

17.44 An engineer who took college fluid mechanics on a pass-fail basis has placed the static pressure hole far upstream of the stagnation probe, as in Fig. 17-9, thus contaminating the Pitot measurement ridiculously with pipe friction losses. If the pipe flow is air at 20 °C and 1 atm and the manometer fluid is Meriam red oil (s.g. = 0.827), estimate the air centerline velocity for the given manometer reading. Assume a smooth-walled tube.

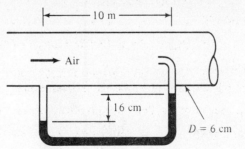

Fig. 17-9

❚
$$(p_0)_B - p_A = (\rho_{oil} - \rho_{air})gh = [(0.827)(998) - 1.20](9.807)(-\tfrac{16}{100}) = -1293 \text{ Pa}$$

Consider friction loss.

$$p_A - p_B = (f)(L/D)(V_{avg}^2/2)(\rho)$$
$$(p_0)_B - p_B = [(p_0)_B - p_A] + (p_A - p_B) = -1293 + (f)(L/D)(V_{avg}^2/2g)(\rho) = (\rho)(V_{\mathbb{C}})_B^2/2$$

Guess $f = 0.016$ and $V_{avg} = 0.85V_{\mathbb{C}}$.

$$-1293 + (0.016)[10/(\tfrac{6}{100})](V_{avg}^2/2)(1.20) = (1.20)(V_{avg}/0.85)^2/2 \qquad V_{avg} = 41.0 \text{ m/s}$$

$$N_R = \rho DV/\mu = (1.20)(\tfrac{6}{100})(41.0)/(1.81 \times 10^{-5}) = 1.63 \times 10^5$$

From Fig. A-5, $f = 0.016$ (O.K.).

$$V_{\mathbb{C}}/V_{avg} = 1 + 1.33f^{1/2} = 1 + (1.33)(0.016)^{1/2} = 1.168 \qquad V_{avg}/V_{\mathbb{C}} = 1/1.168 = 0.856 \qquad \text{(O.K.)}$$

Hence, $V_{avg} = 41.0$ m/s and $V_{\mathbb{C}} = (41.0)(1.168) = 47.9$ m/s.

17.45 The loss of head due to friction in an orifice nozzle, or tube, may be expressed as $h_L = kV^2/2g$, where V is the actual velocity of the jet. Compute k for the three tubes in Fig. 17-10.

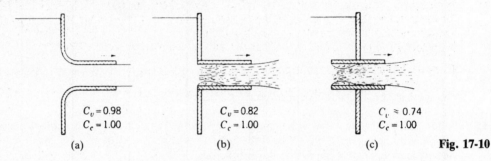

$C_v = 0.98$
$C_c = 1.00$

$C_v = 0.82$
$C_c = 1.00$

$C_v \approx 0.74$
$C_c = 1.00$

(a)   (b)   (c)   **Fig. 17-10**

∎ $\qquad\qquad h_L = kV^2/2g = (1/C_v^2 - 1)(V^2/2g) \qquad k = 1/C_v^2 - 1$

(a) $\qquad\qquad C_v = 0.98 \qquad k = (1/0.98^2) - 1 = 0.0412$

(b) $\qquad\qquad C_v = 0.82 \qquad k = (1/0.82^2) - 1 = 0.487$

(c) $\qquad\qquad C_v = 0.74 \qquad k = (1/0.74^2) - 1 = 0.826$

17.46 If the tubes in Prob. 17.45 and Fig. 17-10 discharge water under a head of 5 ft, compute the loss of head in each case.

∎ $\quad V = C_v\sqrt{2gh} \qquad h_L = kV^2/2g = (1/C_v^2 - 1)(V^2/2g) = (1/C_v^2 - 1)(C_v\sqrt{2gh})^2/2g = (1/C_v^2 - 1)(2hC_v^2/2)$

(a) $\qquad\qquad C_v = 0.98 \qquad h_L = [(1/0.98^2) - 1][(2)(5)(0.98)^2/2)] = 0.198 \text{ ft}$

(b) $\qquad\qquad C_v = 0.82 \qquad h_L = [(1/0.82^2) - 1][(2)(5)(0.82)^2/2)] = 1.64 \text{ ft}$

(c) $\qquad\qquad C_v = 0.74 \qquad h_L = [(1/0.74^2) - 1][(2)(5)(0.74)^2/2)] = 2.26 \text{ ft}$

17.47 The diverging tube shown in Fig. 17-11 discharges water when $h = 5$ ft. The area A is twice area A_0. Neglecting all friction losses, find the **(a)** velocity at the throat and **(b)** pressure head at the throat.

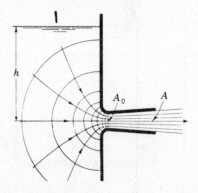

Fig. 17-11

∎ **(a)** $\qquad\qquad V = \sqrt{2gh} = \sqrt{(2)(32.2)(5)} = 17.94 \text{ ft/s}$

$$AV = A_0V_0 \qquad (2A_0)(17.94) = A_0V_0 \qquad V_0 = 35.9 \text{ ft/s}$$

(b) $\qquad p_0/\gamma = V^2/2g - V_0^2/2g = 17.94^2/[(2)(32.2)] - 35.9^2/[(2)(32.2)] = -15.0 \text{ ft}$

17.48 If the barometric pressure is 14.7 psia and the water temperature is 80 °F, what is the maximum value of h at which the tube will flow full, all other data being the same as in Prob. 17.47? What will happen if the value of h is made greater than this?

$$(p_0)_{min} = p_v = 73.5 \text{ lb/ft}^2 \quad \text{(from Table A-1)}$$
$$p_0/\gamma + V_0^2/2g = p_a/\gamma + V^2/2g \qquad V_0 = 2V \quad \text{(from Prob. 17.47)}$$
$$73.5/62.2 + (2V)^2/2g = (14.7)(144)/62.2 + V^2/2g \qquad V^2/2g = 10.95 \text{ ft}$$
$$h_{max} = V^2/2g = 10.95 \text{ ft}$$

If $h > 10.95$ ft, the jet will spring free from the diverging tube, so that $p_0 = p_a$, and discharge will decrease.

17.49 For a rounded entrance and tube flowing full as in the sketch for Prob. 17.47, $C_c = 1.0$ both for the throat and for the exit, and thus $C_v = C_d$ for both sections. For the throat, assume the value of C_v as given in Fig. 17-10a, and assume that for the tube as a whole the discharge coefficient applied to the exit end is 0.70. If $h = 5$ ft, find the velocity and the pressure head at the throat, and compare with Prob. 17.47.

$Q = AV = C_c A_0 (C_v \sqrt{2g\,\Delta h})$. Since $A = C_c A_0$,

$$V = 0.98\sqrt{(2)(32.2)(5)} = 17.59 \text{ ft/s} \qquad V_0 = 2V \quad \text{(from Prob. 17.47)}$$
$$V_0 = (2)(17.59) = 35.18 \text{ ft/s}$$
$$p_1/\gamma + v_1^2/2g + z_1 = p_0/\gamma + v_0^2/2g + z_0 + h_L$$
$$0 + 0 + 5 = p_0/\gamma + 35.18^2/[(2)(32.2)] + 0 + (1 - 0.98^2)(5) \qquad p_0/\gamma = -14.4 \text{ ft}$$

17.50 Suppose that the diverging tube shown in the figure for Prob. 17.47 is discharging water when $h = 2.5$ m. The area A is $1.8A_0$. Neglecting all friction losses, find the (a) velocity and (b) pressure head at the throat.

$$V = \sqrt{2gh} = \sqrt{(2)(9.807)(2.5)} = 7.002 \text{ m/s}$$

(a) $$AV = A_0 V_0 \qquad (1.8A_0)(7.002) = A_0 V_0 \qquad V_0 = 12.6 \text{ m/s}$$

(b) $$p_0/\gamma = V^2/2g - V_0^2/2g = 7.002^2/[(2)(9.807)] - 12.6^2/[(2)(9.807)] = -5.59 \text{ m}$$

17.51 If the tube of Prob. 17.50 is operating at standard atmospheric conditions at a 2000-m elevation, what would be the maximum value of h at which the tube will flow full?

$p_0/\gamma + V_0^2/2g = p_{atm}/\gamma + V^2/2g$. From Table A-8, at 2000-m elevation, $p_{atm} = 79.50$ kPa and $T = 2.0$ °C. From Table A-2, at 2 °C, $p_v = 0.735$ kPa and $\gamma = 9.81$ kN/m^3. Hence, $p_v/\gamma = 0.735/9.81 = 0.0749$ m, $V_0 = 1.8V$, $0.0749 + (1.8V)^2/2g = 79.50/9.81 + V^2/2g$, $V^2/2g = 3.58$ m. The tube will not flow full if $h > 3.58$ m.

17.52 Find the maximum theoretical head at which the Borda tube of Fig. 17-12 will flow full if the liquid is water at 80 °F and the barometer reads 28.4 inHg. Assume $C_d = 0.72$ for the tube flowing full.

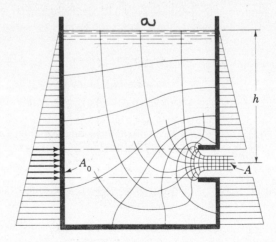

Fig. 17-12

$V = 0.72(2gh)^{1/2}$, $h = (V^2/2g)/0.72^2 = 1.929v^2/2g$. If $C_c = 0.52$ and $C_v = 0.98$ for the contracted throat section, $V_0^2/2g = (V^2/2g)/0.52^2 = (0.72^2/0.52^2)(h) = 1.917h$, $p_a/\gamma + V_a^2/2g + z_a = p_0/\gamma + V_0^2/2g + z_0 + h_L$. From Table A-1, at 80 °F, $p_v = 73.5$ lb/ft^2 and $\gamma = 62.2$ lb/ft^3. Hence, $(p_0/\gamma)_{min} = p_v/\gamma = 73.5/62.2 = 1.182$ ft,

$(28.4/12)(13.6) + 0 + h = 1.182 + 1.917h + (1 - 0.98^2)(h + 33.9)$. [$(h + 33.9)$ ft is the total absolute pressure head acting on the orifice.] $h = h_{max} = 31.0$ ft.

17.53 For the 4-in-diameter short tube shown in Fig. 17-13, **(a)** what flow of water at 75 °F will occur under a head of 30 ft? **(b)** What is the pressure head at section B? Use $C_v = 0.82$.

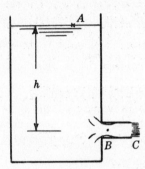

Fig. 17-13

▌ For a standard short tube, the stream contracts at B to about 0.62 of the area of the tube. The lost head from A to B has been measured at about 0.042 times the velocity head at B.

(a)
$$p_A/\gamma + V_A^2/2g + z_A = p_C/\gamma + V_C^2/2g + z_C + h_L$$
$$0 + 0 + 30 = 0 + V_C^2/[(2)(32.2)] + 0 + [(1/0.82^2) - 1]V_C^2/[(2)(32.2)] \qquad V_C = 36.04 \text{ ft/s}$$
$$Q = A_C V_C = [(\pi)(\tfrac{4}{12})^2/4](36.04) = 3.15 \text{ ft}^3/\text{s}$$

(b)
$$p_A/\gamma + V_A^2/2g + z_A = p_B/\gamma + V_B^2/2g + z_B + h_L \qquad V_B = V_C/0.62 = 36.04/0.62 = 58.13 \text{ ft/s}$$
$$h_L = 0.042 V_B^2/2g = (0.042)(58.13)^2/[(2)(32.2)] = 2.204 \text{ ft}$$
$$0 + 0 + 30 = p_B/\gamma + 58.13^2/[(2)(32.2)] + 0 + 2.204 \qquad p_B/\gamma = -24.7 \text{ ft of water}$$

17.54 In Prob. 17.53, what maximum head can be used if the tube is to flow full at the exit?

▌ As the head causing flow through the short tube is increased, the pressure head at B will become less and less. For steady flow (and with the tube full at the exit), the pressure head at B must not be less than the vapor pressure head for the liquid at the particular temperature. From Table A-1, for water at 75 °F, $p_v = 63.0$ lb/ft^2 and $\gamma = 62.3$ lb/ft^3; hence, $p_v/\gamma = 63.0/62.3 = 1.0$ ft abs at sea level (-33.0 ft gage). Applying the Bernoulli equation between A and B [see Prob. 17.53((b))], $C_c A V_B = A V_C = A C_v \sqrt{2gh}$, $V_B(C_v/C_c)\sqrt{2gh}$, $V_B^2/2g = (C_v/C_c)^2(h) = (0.82/0.62)^2(h) = 1.749h$, $0 + 0 + h = -33.0 + 1.749h + 0 + (0.042)(1.749h)$, $h = 40.1$ ft of water. Any head over this value will cause the stream to spring free of the sides of the tube. The tube will then function as an orifice.

17.55 A 4-ft-diameter tank contains oil of specific gravity 0.75. A 3-in-diameter short tube is installed near the bottom of the tank ($C = 0.85$). How long will it take to lower the level of the oil from 6 ft above the tube to 4 ft above the tube?

▌
$$t = t_2 - t_1 = (2A_T/CA_0\sqrt{2g})(h_1^{1/2} - h_2^{1/2}) \qquad A_T = (\pi)(4)^2/4 = 12.57 \text{ ft}^2$$
$$A_0 = (\pi)(\tfrac{3}{12})^2/4 = 0.04909 \text{ ft}^2$$
$$t = [(2)(12.57)/(0.85)(0.04909)\sqrt{(2)(32.2)}](6^{1/2} - 4^{1/2}) = 33.7 \text{ s}$$

17.56 A Borda mouthpiece 50 mm in diameter has a discharge coefficient of 0.51. What is the diameter of the issuing jet?

▌
$$C_c = (D_{jet}/D)^2 \qquad 1 = 2C_d C_v = 2C_v^2 C_c \qquad 1 = (2)(0.51)(C_v) \qquad C_v = 0.9804$$
$$1 = (2)(0.9804)^2(C_c) \qquad C_c = 0.5202 \qquad 0.5202 = (D_{jet}/50)^2 \qquad D_{jet} = 36.1 \text{ mm}$$

17.57 A reservoir of variable area is drained by a 150-mm-diameter short pipe with a valve attached. The valve is being adjusted so that the loss (in velocity heads) for the piping system is $K = 1.5 + 0.04t + 0.0001t^2$ with t in seconds. The reservoir area is given by $A = 4 + 0.1y + 0.01y^2$ m^2 where y is the elevation of the reservoir surface above the centerline of the valve. If $y = 20$ m at $t = 0$, determine y, A, K, and the discharge Q for 300 s. Solve using a computer program.

▮ $Q \, dt = -A \, dy$, $y + 0 + 0 = KV^2/2g = KQ^2/2gA_0^2$, $dy = -HQ/A$, and $Q = \sqrt{C_1 y/K}$ in which $C_1 = 2gA_0^2$. H is the time increment used in Runge–Kutta.

```
10 REM EXAMPLE DRAINAGE OF RESERVOIR—2ND ORDER RUNGE–KUTTA
20 DEF FNQ(YD,TD)=SQR(C1*YD/(K1+K2*TD+K3*TD^2))/(A1+A2*YD+A3*YD^2)
30 DEFINT I: READ A1,A2,A3,K1,K2,K3,Y,TMAX,PI,H,G,D
40 DATA 4.,.1,.01,1.5,.04,.0001,20!,300!,3.1416,1.,9.806001,.15
50 LPRINT: LPRINT"A1,A2,A3,K1,K2,K3=";A1;A2;A3;K1;K2;K3
60 LPRINT"Y,TMAX,PI,H,G,D=";Y;TMAX;PI;H;G;D: LPRINT
70 C1=2!*G*(.25*PI*D^2)^2: T=0!: I=0: A=A1+Y*(A2+A3*Y): Q=FNQ(Y,T)*A: K=K1
80 LPRINT"    T,s    Y,m    Q,m^3/s AREA,m^2    K": LPRINT
90 LPRINT USING"###.### ";T;Y;Q;A;K
100 IF T>= TMAX THEN STOP
110 I=I+1: U1=-H*FNQ(Y,T): U2=-H*FNQ(Y+U1,T+H): Y=Y+.5*(U1+U2): T=T+H
120 A=A1+A2*Y+A3*Y^2: Q=FNQ(Y,T)*A: K=K1+T*(K2+K3*T)
130 IF I MOD 30 = 0 THEN 90 ELSE GOTO 110
140 IF T<TMAX THEN 110 ELSE STOP
```

A1,A2,A3,K1,K2,K3= 4 .1 .01 1.5 .04 .0001
Y,TMAX,PI,H,G,D= 20 300 3.1416 1 9.806001 .15

T,s	Y,m	Q,m^3/s	AREA,m^2	K
0.000	20.000	0.286	10.000	1.500
30.000	19.265	0.206	9.638	2.790
60.000	18.687	0.164	9.361	4.260
90.000	18.202	0.137	9.133	5.910
120.000	17.778	0.119	8.938	7.740
150.000	17.401	0.105	8.768	9.750
180.000	17.060	0.094	8.616	11.940
210.000	16.748	0.085	8.480	14.310
240.000	16.460	0.077	8.355	16.860
270.000	16.191	0.071	8.241	19.590
300.000	15.941	0.066	8.135	22.500

17.58 A 2-in ISA flow nozzle is installed in a 3-in pipe carrying water at 72 °F. If a water-air manometer shows a differential of 2 in, find the flow.

▮ This is a trial-and-error type of solution. First assume a reasonable value of K. From Fig. A-20, for $D_2/D_1 = 0.67$, for the level part of the curve, $K = 1.06$. Then $Q = KA_2\sqrt{2g[(p_1/\gamma + z_1) - (p_2/\gamma + z_2)]}$, where $A_2 = (\pi/4)(2^2/144) = 0.0218 \text{ ft}^2$ and $\Delta(p/\gamma + z) = \frac{2}{12} = 0.167 \text{ ft}$. Thus $Q = (1.06)(0.0218)\sqrt{(64.4)(0.167)} = 0.0757 \text{ cfs}$. With this first determination of Q, $V_1 = Q/A = 0.0757/0.0492 = 1.54 \text{ fps}$. Then $D_1'' V_1 = (3)(1.54) = 4.62$. From Fig. A-20, $K = 1.04$ and $Q = (1.04/1.06)(0.0757) = 0.0743 \text{ cfs}$. No further correction is necessary.

17.59 The velocity of water in a 4-in-diameter pipe is 10 ft/s. At the end of the pipe is a nozzle whose velocity coefficient is 0.98. If the pressure in the pipe is 8 psi, what is the velocity in the jet? What is the diameter of the jet? What is the rate of discharge? What is the head loss?

▮ $p_1/\gamma + V_1^2/2g = p_2/\gamma + V_2^2/2g$. Since $p_2 = p_{atm} = 0$,

$$(V_2)_{ideal} = [2g(p_1/\gamma + V_1^2/2g)]^{1/2}$$

$$V_2 = C_v(V_2)_{ideal} = (0.98)\{(2)(32.2)[(8)(144)/62.4 + 10^2/(2)(32.2)]\}^{1/2} = 35.2 \text{ ft/s}$$

$$A_{jet} = Q/V \qquad Q = AV = [(\pi)(\tfrac{4}{12})^2/4](10) = 0.8727 \text{ ft}^3/\text{s} \qquad A_{jet} = 0.8727/35.2 = 0.02479 \text{ ft}^2$$

$$0.02479 = \pi D_{jet}^2/4 \qquad D_{jet} = 0.1777 \text{ ft} \quad \text{or} \quad 2.13 \text{ in}$$

$$h_L = h - V^2/2g \qquad h = \text{original head} = p/\gamma + V^2/2g + z$$

$$h = (8)(144)/62.4 + 10^2/[(2)(32.2)] + 0 = 20.01 \text{ ft} \qquad h_L = 20.01 - 35.2^2/[(2)(32.2)] = 0.770 \text{ ft}$$

17.60 A jet of water 3 in in diameter is discharged through a nozzle whose velocity coefficient is 0.96. If the pressure in the pipe is 12 psi and the pipe diameter is 8 in and if it is assumed that there is no contraction of the jet, what is the velocity at the tip of the nozzle? What is the rate of discharge?

▮ $V = \{C_v/[1 - C_v^2(D_2/D_1)^4]^{1/2}\}(2gp/\gamma)^{1/2} = \{0.96/[1 - (0.96)^2(\tfrac{3}{8})^4]^{1/2}\}\{(2)(32.2)[(12)(144)]/62.4\}^{1/2} = 40.9 \text{ ft/s}$

$$Q = AV = [(\pi)(\tfrac{3}{12})^2/4](40.9) = 2.01 \text{ ft}^3/\text{s}$$

17.61 The nozzle in Fig. 17-14 throws a stream of water vertically upward so that the power available in the jet at point 2 is 3.42 hp. If the pressure at the base of the nozzle, point 1, is 21.0 psi, find the (*a*) theoretical height to

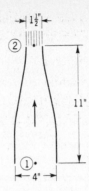

Fig. 17-14

which the jet will rise, **(b)** coefficient of velocity, **(c)** head loss between 1 and 2, and **(d)** theoretical diameter of the jet at a point 20 ft above point 2.

$$P = Q\gamma V_2^2/2g = A_2\gamma V_2^3/2g$$

$$(3.42)(550) = [(\pi)(1.5/12)^2/4](62.4)(V_2^3)/[(2)(32.2)] \quad V_2 = 54.08 \text{ ft/s}$$

(a) $\quad h_{\text{theor}} = V_2^2/2g = 54.08^2/[(2)(32.2)] = 45.4 \text{ ft}$

(b) $\quad C_v = V/V_i \quad V_i = \sqrt{2g(p_1/\gamma + V_1^2/2g + z_1 - z_2)}$

$$V_1^2/2g = (V_2^2/2g)(D_2/D_1)^4 = \{54.08^2/[(2)(32.2)]\}(1.5/4)^4 = 0.8981 \text{ ft}$$

$$V_i = \sqrt{(2)(32.2)[(21.0)(144)/62.4 + 0.8981 + 0 - \tfrac{11}{12}]} = 55.85 \text{ ft/s}$$

$$C_v = 54.08/55.85 = 0.968$$

(c) $\quad h_L = (1 - C_v^2)h = (1 - 0.968^2)\{55.85^2/[(2)(32.2)]\} = 3.05 \text{ ft}$

(d) $\quad V_2^2/2g = (V_3^2/2g)(D_3/D_2)^4 \quad V_3^2/2g = V_2^2/2g - z_3$

$$D_3 = D_2[(V_2^2/2g)/(V_3^2/2g)]^{1/4} = (1.5)\{[54.08^2/(2)(32.2)]/[54.08^2/(2)(32.2) - 20.0]\}^{1/4} = 1.73 \text{ in}$$

17.62 Suppose a 10-cm ISA flow nozzle is used in a 20-cm pipe to measure the flow of water at 40 °C. What would be the reading on a mercury manometer for the following flow rates: **(a)** 1.5 L/s, **(b)** 15 L/s, and **(c)** 150 L/s?

$$\Delta(p/\gamma + z) = R'(S_M/S_F - 1) = R'[(13.6/1) - 1] = 12.6R'$$

$$Q = KA_2\sqrt{2g\,\Delta(p/\gamma + z)} = KA_2\sqrt{(2g)(12.6R')}$$

Let $K = 1.0$: $Q = (1.0)[(\pi)(\tfrac{10}{100})^2/4]\sqrt{(2)(9.807)(12.6)R'} = 0.1235\sqrt{R'}$.

(a) $\quad 1.5/1000 = 0.1235\sqrt{R'} \quad R' = 0.000148 \text{ m} \quad \text{or} \quad 0.148 \text{ mm}$

(b) $\quad \tfrac{15}{1000} = 0.1235\sqrt{R'} \quad R' = 0.0148 \text{ m} \quad \text{or} \quad 14.8 \text{ mm}$

(c) $\quad \tfrac{150}{1000} = 0.1235\sqrt{R'} \quad R' = 1.475 \text{ m} \quad \text{or} \quad 1475 \text{ mm}$

Note: One should check the Reynolds number to confirm the value of K used.

17.63 A 6-in ISA nozzle is used to measure the flow of crude oil (s.g. $= 0.855$) at 15 °F. If a mercury manometer shows a reading of 3.5 in, what is the flow? Assume $D_2/D_1 = 0.70$, $\nu = 0.00034 \text{ ft}^2/\text{s}$.

$$\Delta(p/\gamma + z) = R'(S_M/S_F - 1) = R'[(13.6/0.855) - 1] = 14.91R'$$

$$Q = KA_2\sqrt{2g\,\Delta(p/\gamma + z)} = KA_2\sqrt{(2g)(14.91R')}$$

Let $K = 1.08$: $Q = (1.08)[(\pi)(\tfrac{6}{12})^2/4]\sqrt{(2)(32.2)(3.5/12)(13.6/0.855 - 1)} = 3.55 \text{ ft}^3/\text{s}$. Check N_R to confirm that the assumed K is applicable: $V = Q/A = 3.55/[(\pi)(\tfrac{6}{12})^2/4] = 18.08 \text{ ft/s}$, $N_R = DV/\nu = (\tfrac{6}{12})(18.08)/0.00034 = 2.66 \times 10^4$. From Fig. A-20, $K = 1.06$ approximately. $Q = (1.06/1.08)(3.55) = 3.48 \text{ ft}^3/\text{s}$.

17.64 Water flows through a 4-in pipe at the rate of 0.952 cfs and thence through a nozzle attached to the end of the pipe. The nozzle tip is 2 in in diameter and the coefficients of velocity and contraction for the nozzle are 0.950 and 0.930, respectively. What pressure head must be maintained at the base of the nozzle if atmospheric pressure surrounds the jet?

▮ Apply the Bernoulli equation between base of nozzle and jet.

$$p_b/\gamma + V_b^2/2g + z_b = p_j/\gamma + V_j^2/2g + z_j + h_L \qquad A_bV_b = A_jV_j = (C_cA_{2\text{-in}})V_j = 0.952$$

$$V_b = 0.952/A_b = 0.952/[(\pi)(\tfrac{4}{12})^2/4] = 10.91 \text{ ft/s}$$

$$V_j = 0.952/(C_cA_{2\text{-in}}) = 0.952/\{(0.930)[(\pi)(\tfrac{2}{12})^2/4]\} = 46.92 \text{ ft/s}$$

$$h_L = (1/C_v^2 - 1)(V_j^2/2g)$$

$$p_b/\gamma + 10.91^2/[(2)(32.2)] + 0 = 0 + 46.92^2/[(2)(32.2)] + 0 + (1/0.950^2 - 1)\{46.92^2/[(2)(32.2)]\}$$

$$p_b/\gamma = 36.0 \text{ ft of water}$$

17.65 A 4-in base diameter by 2-in tip diameter nozzle points downward and the pressure head at the base of the nozzle is 26.0 ft of water. The base of the nozzle is 3.0 ft above the tip and the coefficient of velocity is 0.962. Determine the horsepower in the jet of water.

▮ $P_j = Q\gamma H_j$. For a nozzle, unless C_c is given, it may be taken as unity. Hence, $V_j = V_t$.

$$p_b/\gamma + V_b^2/2g + z_b = p_t/\gamma + V_t^2/2g + z_t + h_L \qquad A_bV_b = A_tV_t$$

$$[(\pi)(4^2/4)](V_b) = [(\pi)(2^2/4)](V_t) \qquad V_b = V_t/4 \qquad h_L = (1/C_v^2 - 1)(V_t^2/2g)$$

$$26.0 + (V_t/4)^2/[(2)(32.2)] + 3.0 = 0 + V_t^2/[(2)(32.2)] + 0 + (1/0.962^2 - 1)\{V_t^2/[(2)(32.2)]\}$$

$$V_t = 42.83 \text{ ft/s} \qquad Q = A_tV_t = [(\pi)(\tfrac{2}{12})^2/4](42.83) = 0.9344 \text{ ft}^3/\text{s}$$

$$H_j = 0 + V_t^2/[(2)(32.2)] + 0 = 42.83^2/[(2)(32.2)] = 28.48 \text{ ft}$$

$$P_j = (0.9344)(62.4)(28.48) = 1661 \text{ ft-lb/s} = \tfrac{1661}{550} = 3.02 \text{ hp}$$

17.66 A nozzle with a 4-in-diameter tip is installed in a 10-in pipe. Medium fuel oil at 80 °F flows through the nozzle at the rate of 3.49 cfs. Assume the calibration of the nozzle is represented by curve $\beta = 0.40$ on Fig. A-21. Calculate the differential gage reading if a liquid of specific gravity 13.6 is the gage liquid.

▮
$$Q = A_4V_4 = A_4C\sqrt{[2g(p_A/\gamma - p_B/\gamma)]/[1 - (A_4/A_{10})^4]}$$

$$= A_4C\sqrt{[2g(p_A/\gamma - p_B/\gamma)]/[1 - (\tfrac{4}{10})^4]} \tag{1}$$

Figure A-21 indicates that C varies with Reynolds number: $V_4 = Q/A_4 = 3.49/[\tfrac{1}{4}\pi(\tfrac{4}{12})^2] = 40.0$ ft/s and $N_R = (40.0)(\tfrac{4}{12})/(3.65 \times 10^{-5}) = 365\,000$. Curve for $\beta = 0.40$ gives $C = 0.993$. Thus $3.49 = \tfrac{1}{4}\pi(\tfrac{4}{12})^2(0.993)\sqrt{[2g(p_A/\gamma - p_B/\gamma)]/[1 - (\tfrac{4}{10})^4]}$ and $(p_A/\gamma - p_B/\gamma) = 24.6$ ft of fuel oil. Differential gage principles produce, using specific gravity of the oil = 0.851, $24.6 = h[(13.6/0.851) - 1]$ and $h = 1.64$ ft (gage reading).

17.67 Derive an expression for the flow of a compressible fluid through a nozzle flowmeter and a Venturi meter.

▮ Since the change in velocity takes place in a very short period of time, little heat can escape and adiabatic conditions will be assumed. The Bernoulli theorem for compressible flow gives $\{[k/(k - 1)](p_1/\gamma_1) + (V_1^2/2g) + z_1\} - H_L = \{[k/(k - 1)](p_1/\gamma_1)(p_2/p_1)^{(k-1)/k} + (V_2^2/2g) + z_2\}$.

For a nozzle meter and for a horizontal Venturi meter, $z_1 = z_2$ and the lost head will be taken care of by means of the coefficient of discharge. Also, since $C_c = 1.00$, $W = \gamma_1A_1V_1 = \gamma_2A_2V_2$. Then upstream $V_1 = W/\gamma_1A_1$, downstream $V_2 = W/\gamma_2A_2$. Substituting and solving for W, $W^2/\gamma_2^2A_2^2 - W^2/\gamma_1^2A_1^2 = 2g[k/(k - 1)](p_1/\gamma_1)[1 - (p_2/p_1)^{(k-1)/k}]$ or (ideal) $W = [\gamma_2A_2/\sqrt{1 - (\gamma_2/\gamma_1)^2(A_2/A_1)^2}]\sqrt{[2gk/(k - 1)](p_1/\gamma_1)[1 - (p_2/p_1)^{(k-1)/k}]}$. It may be more practical to eliminate γ_2 under the radical. Since $\gamma_2/\gamma_1 = (p_2/p_1)^{1/k}$,

$$(\text{ideal}) \quad W = \gamma_2A_2\sqrt{[2gk/(k - 1)](p_1/\gamma_1)[1 - (p_2/p_1)^{(k-1)/k}]/[1 - (A_2/A_1)^2(p_2/p_1)^{2/k}]} \tag{1}$$

The true value of W in pounds per second is obtained by multiplying the right-hand side of the equation by coefficient C. From Eq. (1) of Prob. 17.66, the following equation may be written: $W = \gamma Q = [\gamma A_2C/\sqrt{1 - (A_2/A_1)^2}]\sqrt{2g(\Delta p/\gamma)}$ or $W = \gamma KA_2\sqrt{2g(\Delta p/\gamma)}$. The above equation can be expressed more generally so that it will apply both to compressible and incompressible fluids. An expansion (adiabatic) factor Y is introduced and the value of γ_1 at inlet is specified. The fundamental relation is then

$$W = \gamma_1KA_2Y\sqrt{2g(\Delta p/\gamma_1)} \tag{2}$$

For incompressible fluids, $Y = 1$. For compressible fluids, equate expressions (1) and (2) and solve for Y. By so doing,

$$Y = \sqrt{\{[1 - (A_2/A_1)^2]/[1 - (A_2/A_1)^2(p_2/p_1)^{2/k}]\}\{[k/(k - 1)][1 - (p_2/p_1)^{(k-1)/k}](p_2/p_1)^{2/k}/(1 - p_2/p_1)\}}$$

This expansion factor Y is a function of three dimensionless ratios. Table A-18 lists some typical values for nozzle flowmeters and for Venturi meters.

17.68 Air at a temperature of 80 °F flows through a 4-in pipe and through a 2-in flow nozzle. The pressure differential is 0.522 ft of oil, s.g. = 0.910. The pressure upstream from the nozzle is 28.3 psi gage. How many pounds per second are flowing for a barometric reading of 14.7 psi, assuming the air has constant density.

▌ $\gamma = p/RT$, $\gamma_1 = [(28.3 + 14.7)144]/[53.3(460 + 80)] = 0.215$ lb/ft^3. From differential gage principles, using pressure heads in feet of air, $\Delta p/\gamma_1 = 0.522[(\gamma_{oil}/\gamma_{air}) - 1] = 0.522\{[(0.910)(62.4)/0.215] - 1\} = 137$ ft of air.

Assuming $C = 0.980$ and using equation (1) of Prob. 17.66 after multiplying by γ_1, we have $W = \gamma_1 Q = (0.215)(\frac{1}{4}\pi)(\frac{2}{12})^2(0.980)\sqrt{2g(137)/[1 - (\frac{2}{4})^4]} = 0.445$ lb/s.

To check the value of C, find Reynolds number and use the appropriate curve on Fig. A-21 (here $\gamma_1 = \gamma_2$ and $v = 16.9 \times 10^{-5}$ ft^2/s at standard atmosphere).: $V_2 = W/A_2\gamma_2 = W/(\pi d_2^2/4)\gamma_2$. Then $N_R = V_2 d_2/v = 4W/\pi d_2 v\gamma_2 = 4(0.445)/[\pi(\frac{2}{12})(16.9 \times 14.7/43.0)10^{-5}(0.215)] = 274\,000$.

From Fig. A-21, $C = 0.986$. Recalculating, $W = 0.447$ lb/s.

Further refinement in calculation is not warranted inasmuch as the Reynolds number will not be changed materially, nor will the value of C read from Fig. A-21.

17.69 Find q for flow of water in a pipe of inside diameter $D = 100$ mm using a long-radius-type flow nozzle. The pipe is horizontal for this problem and the value of h for the manometer is 140 mm. The throat diameter of the nozzle is 60 mm. Take $\rho = 999$ kg/m^3 and $v = 1.12 \times 10^{-3}$ m^2/s.

▌
$$p_1 - p_2 = h(\gamma_{Hg} - \gamma_{H_2O}) = 0.140(13.6 - 1)(999)(9.81) = 17.300 \text{ kPa} \tag{1}$$
$$C_d = 0.99622 + 0.00059D - (6.36 + 0.13D - 0.24\beta^2)(1/\sqrt{N_R})$$
$$(C_d)_{noz} = [0.99622 + 0.00059(10.0/2.54)] - [6.36 + 0.13(10.0/2.54) - (0.24)(0.6^2)](1/\sqrt{N_R})$$
$$= 0.998 - 6.785(1/\sqrt{N_R}) \tag{2}$$
$$q_{act} = (C_d)_{noz}A_2\{2[(p_1 - p_2)/\rho]/[1 - (A_2/A_1)^2]\}^{1/2} = (C_d)_{noz}A_2\{2[(p_1 - p_2)/\rho]/(1 - \beta^4)\}^{1/2}$$
$$A_2V_2 = [0.998 - 6.785(1/\sqrt{N_R})](A_2)[2(17\,300/999)/(1 - 0.6^4)]^{1/2}$$

This equation becomes

$$V_2 = (0.998 - 6.785/\sqrt{\rho V_1 d/\mu})(6.31) = (0.998 - \{6.785/\sqrt{[(999)(V_1)(0.60)]/(1.12 \times 10^{-3})}\})(6.31) \tag{3}$$

From *continuity*,

$$V_2 = (\tfrac{100}{60})^2 V_1 \tag{4}$$

Therefore, substituting for V_2 in Eq. (3) using Eq. (4), $V_1 = [0.998 - (0.0293/\sqrt{\sqrt{V_1}})](2.77)$. Solving by trial and error we get $V_1 = 2.22$ m/s. The Reynolds number is then $N_R = [(999)(0.060)(2.22)/(1.12 \times 10^{-3})] = 1.189 \times 10^5$. We are well within the range of Eq. (2), so we can get q as $q = [(\pi)(0.100^2)/4](2.22) = 0.01744$ m^3/s.

17.70 Determine the flow through a 6-in-diameter water line that contains a 4-in-diameter flow nozzle. The mercury-water differential manometer has a gage difference of 10 in. Water temperature is 60 °F.

▌ $Q = CA_2\sqrt{2gR'(S_0/S_1 - 1)}$. From Fig. A-22, for $A_2/A_1 = (\frac{4}{6})^2 = 0.4444$, assume that the horizontal region of the curves applies. Hence, $C = 1.056$.

$$Q = (1.056)[(\pi)(\tfrac{4}{12})^2/4]\sqrt{(2)(32.2)(\tfrac{10}{12})[(13.6/1) - 1]} = 2.40 \text{ ft}^3/\text{s}$$
$$V = Q/A = 2.40/[(\pi)(\tfrac{6}{12})^2/4] = 12.22 \text{ ft/s}$$
$$N_R = \rho DV/\mu = (1.94)(\tfrac{6}{12})(12.22)/(2.35 \times 10^{-5}) = 5.04 \times 10^5$$

Figure A-22 shows the value of C to be correct; therefore, the discharge is 2.40 ft^3/s.

17.71 Air flows through an 80-mm-diameter ISA flow nozzle in a 120-mm-diameter pipe. $p_1 = 150$ kPa abs; $T_1 = 5$ °C; and a differential manometer with liquid, s.g. = 2.93, has a gage difference of 0.8 m when connected between taps. Calculate the mass rate of flow.

▌ $\dot{m} = \rho Q = \rho CYA_2\sqrt{2\,\Delta p/\rho}$ $\rho = p/RT = (150)(1000)/[(\tfrac{8312}{29})(5 + 273)] = 1.883$ kg/m^3
$$A_2/A_1 = (D_2/D_1)^2 = (\tfrac{80}{120})^2 = 0.4444$$

From Fig. A-22, C is assumed to be 1.055.

$$p_2 = p_1 - \gamma(\text{s.g.})\,\Delta h = 150 - (9.79)(2.93)(0.8) = 127.1 \text{ kPa}$$
$$p_2/p_1 = 127.1/150 = 0.847 \qquad D_2/D_1 = \tfrac{80}{120} = 0.667$$

From Fig. A-26, $Y = 0.89$.

$$A_2 = (\pi)(\tfrac{80}{1000})^2/4 = 0.005026 \text{ m}^2$$
$$\dot{m} = (1.883)(1.055)(0.89)(0.005026)\sqrt{(2)(150 - 127.1)(1000)/1.883} = 1.39 \text{ kg/s}$$
$$N_R = \rho DV/\mu \qquad V = \dot{m}/\rho A \qquad A_1 = (\pi)(\tfrac{120}{1000})^2/4 = 0.01131 \text{ m}^2$$
$$V_1 = 1.39/[(1.883)(0.01131)] = 65.27 \text{ m/s}$$
$$N_R = (1.883)(\tfrac{120}{1000})(65.27)/(1.74 \times 10^{-5}) = 8.48 \times 10^5$$

From Fig. A-22, assumed $C = 1.055$ is O.K.; hence, $\dot{m} = 1.39$ kg/s.

17.72 A 2.5-in-diameter ISA nozzle is used to measure flow of water at 40 °C in a 6-in-diameter pipe. What gage difference on a water-mercury manometer is required for 300 gpm?

▮ $\qquad R' = \Delta p/[(\gamma)(S_0/S_1 - 1)] \qquad \Delta p = (\rho/2)(Q/CA_2)^2 \qquad A_2/A_1 = (2.5/6)^2 = 0.1736$
$$V_1 = Q/A_1 = (300)(0.002228)/[(\pi)(\tfrac{6}{12})^2/4] = 3.404 \text{ ft/s}$$
$$(N_R)_1 = V_1 D_1/v_1 = (3.404)(\tfrac{6}{12})/(1.67 \times 10^{-5}) = 1.02 \times 10^5$$

From Fig. A-22, $C = 0.995$.

$$A_2 = (\pi)(2.5/12)^2/4 = 0.03409 \text{ ft}^2$$
$$\Delta p = (1.94/2)\{(300)(0.002228)/[(0.995)(0.03409)]\}^2 = 376.7 \text{ lb/ft}^2$$
$$R' = 376.7/\{(62.4)[(13.6/1) - 1]\} = 0.4791 \text{ ft} \quad \text{or} \quad 5.75 \text{ in}$$

17.73 We want to meter the volume flow of water ($\rho = 1000$ kg/m³, $v = 1.02 \times 10^{-6}$ m²/s) moving through a 200-mm-diameter pipe at an average velocity of 2.0 m/s. If the differential pressure gage selected reads accurately at $p_1 - p_2 = 50\,000$ Pa, what size meter should be selected for installing a long-radius flow nozzle? What would be the nonrecoverable head loss?

▮ Here the unknown is the β ratio of the meter. Since the discharge coefficient is a complicated function of β, iteration will be necessary. We are given $D = 0.2$ m and $V_1 = 2.0$ m/s. The pipe-approach Reynolds number is thus $N_R = V_1 D/v = (2.0)(0.2)/(1.02 \times 10^{-6}) = 392\,000$. The generalized formula gives the throat velocity as $V_t = V_1/\beta^2 = \alpha[2(p_1 - p_2)/\rho]^{1/2}$ where everything is known except α and β. Solve for β^2:

$$\beta^2 = (1/\alpha)(\rho V_1^2/2\,\Delta p)^{1/2} \tag{1}$$

With $V_1 = 2.0$ m/s and $\Delta p = 50\,000$ Pa, $\beta^2 = (1/\alpha)\{(1000)(2.0)^2/[2(50\,000)]\}^{1/2} = 0.2/\alpha$ or

$$\beta = 0.447/\alpha^{1/2} \tag{2}$$

The solution depends only on getting the proper flow coefficient. A good guess for flow-nozzle design is $\alpha = 1.0$. Iterate Eq. (2) and list the results:

α	β, Eq. (2)	C_d, Fig. A-23	$E = (1 - \beta^4)^{-1/2}$	$\alpha = EC_d$
1.0	0.447	0.9895	1.0206	1.0099
1.0099	0.445	0.9895	1.0202	1.0095
1.0095	0.445			

Convergence is rapid to $\beta = 0.445$, $d = \beta D = 89$ mm. The throat velocity is $2.0/(0.445)^2 = 10.1$ m/s; the throat head is $(10.1)^2/[2(9.81)] = 5.2$ m. From Fig. A-24 for the nozzle read $K_m \approx 0.7$. Then the nozzle loss is $h_m = 0.7(5.2 \text{ m}) = 3.6$ m.

17.74 Gasoline at 20 °C flows at 0.06 m³/s through a 15-cm pipe and is metered by a 9-cm long-radius flow nozzle. What is the expected pressure drop across the nozzle? ($\rho = 680$ kg/m³; $v = 4.29 \times 10^{-7}$ m²/s)

▮ $\qquad Q = C_d A_t[(2\,\Delta p/\rho)/(1 - \beta^4)]^{1/2} \qquad C_d = 0.9965 - 0.00653\beta^{1/2}[10^6/(N_R)_D]^{1/2}$
$$\beta = d/D = \tfrac{9}{15} = 0.600 \qquad N_R = DV/v \qquad V = Q/A = 0.06/[(\pi)(\tfrac{15}{100})^2/4] = 3.395 \text{ m/s}$$
$$(N_R)_D = (\tfrac{15}{100})(3.395)/(4.29 \times 10^{-7}) = 1.19 \times 10^6$$
$$C_d = 0.9965 - (0.00653)(0.600)^{1/2}[10^6/(1.19 \times 10^6)]^{1/2} = 0.9919$$
$$0.06 = (0.9919)[(\pi)(\tfrac{9}{100})^2/4][(2\,\Delta p/680)/(1 - 0.600^4)]^{1/2} \qquad \Delta p = 26\,760 \text{ Pa}$$

17.75 Ethyl alcohol at 20 °C flowing in a 6-cm-diameter pipe is metered through a 3-cm long-radius flow nozzle. If the measured pressure drop is 45 000 Pa, what is the estimated volume flow in cubic meters per hour? ($\rho = 789$ kg/m^3; $v = 1.52 \times 10^{-6}$ m^2/s)

    ■ $Q = C_d A_t [(2\,\Delta p/\rho)/(1 - \beta^4)]^{1/2}$. Estimate $C_d = 0.99$.

$$\beta = d/D = \tfrac{3}{6} = 0.500$$
$$Q = (0.99)[(\pi)(\tfrac{3}{100})^2/4]\{[(2)(45\,000)/789]/(1 - 0.500^4)\}^{1/2} = 0.007719 \text{ m}^3/\text{s}$$
$$C_d = 0.9965 - 0.00653\beta^{1/2}[10^6/(N_R)_D]^{1/2} \qquad N_R = DV/v$$
$$V = Q/A = 0.007719/[(\pi)(\tfrac{6}{100})^2/4] = 2.730 \text{ m/s}$$
$$(N_R)_D = (\tfrac{6}{100})(2.730)/(1.52 \times 10^{-6}) = 1.08 \times 10^5$$
$$C_d = 0.9965 - (0.00653)(0.500)^{1/2}[10^6/(1.08 \times 10^5)]^{1/2} = 0.9824$$
$$Q = (0.9824)[(\pi)(\tfrac{3}{100})^2/4]\{[(2)(45\,000)/789]/(1 - 0.500^4)\}^{1/2} = 0.007660 \text{ m}^3/\text{s} \quad \text{or} \quad 27.6 \text{ m}^3/\text{h}$$

17.76 Kerosene at 20 °C flows at 20 m^3/h in an 8-cm-diameter pipe. The flow is to be metered by an ISA 1932 flow nozzle so that the pressure drop is 7000 Pa. What is the proper nozzle diameter? ($\rho = 804$ kg/m^3; $v = 2.39 \times 10^{-6}$ m^2/s)

    ■ $Q = C_d A_t (\beta)^2 [(2\,\Delta p/\rho)/(1 - \beta^4)]^{1/2}$. Estimate $C_d = 0.99$.

$$\tfrac{20}{3600} = (0.99)[(\pi)(\tfrac{8}{100})^2/4](\beta)^2\{[(2)(7000)/804]/(1 - \beta^4)\}^{1/2}$$
$$\beta = 0.508 \qquad C_d = 0.9965 - 0.00653\beta^{1/2}[10^6/(N_R)_D]^{1/2}$$
$$N_R = DV/v \qquad V = Q/A = (\tfrac{20}{3600})/[(\pi)(\tfrac{8}{100})^2/4] = 1.105 \text{ m/s}$$
$$(N_R)_D = (\tfrac{8}{100})(1.105)/(2.39 \times 10^{-6}) = 3.70 \times 10^4$$
$$C_d = 0.9965 - (0.00653)(0.508)^{1/2}[10^6/(3.70 \times 10^4)]^{1/2} = 0.9723$$
$$\tfrac{20}{3600} = (0.9723)[(\pi)(\tfrac{8}{100})^2/4](\beta)^2\{[(2)(7000)/804]/(1 - \beta^4)\}^{1/2}$$
$$\beta = 0.513 \qquad d_{\text{nozzle}} = \beta D = (0.513)(8) = 4.10 \text{ cm}$$

17.77 Water flows from a large tank through an orifice and discharges to the atmosphere, as shown in Fig. 17-15. The coefficients of velocity and contraction are 0.96 and 0.62, respectively. Find the diameter and actual velocity in the jet and the discharge from the orifice.

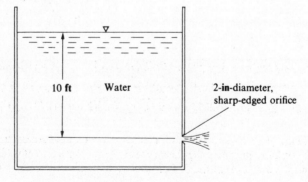

10 ft   Water   2-in-diameter, sharp-edged orifice

Fig. 17-15

    ■

$$C_c = a/A \qquad 0.62 = a/[(\pi)(\tfrac{2}{12})^2/4] \qquad a = 0.01353 \text{ ft}^2$$
$$(\pi)(D_{\text{jet}})^2/4 = 0.01353 \qquad D_{\text{jet}} = 0.1313 \text{ ft} \quad \text{or} \quad 1.58 \text{ in}$$

In order to determine the actual velocity in the jet, the discharge from the orifice will be determined next: $C = C_c C_v = (0.96)(0.62) = 0.595$, $Q = CA\sqrt{2gh} = (0.595)[(\pi)(\tfrac{2}{12})^2/4]\sqrt{(2)(32.2)(10)} = 0.329$ ft^3/s, $v = Q/a = 0.329/0.01353 = 24.3$ ft/s.

17.78 Oil discharges from a pipe through a sharp-crested, round orifice, as shown in Fig. 17-16. The coefficients of contraction and velocity are 0.62 and 0.98, respectively. Find the discharge from the orifice and the diameter and actual velocity in the jet.

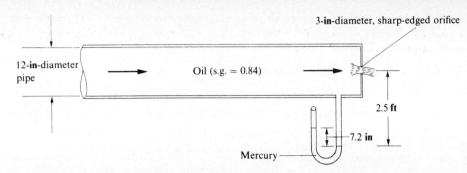

Fig. 17-16

∎ $Q = CA\sqrt{2g(p_1/\gamma)}/\sqrt{1 - C^2(D/D_p)^4}$ $C = C_v C_c = (0.98)(0.62) = 0.608$

$p_1/\gamma + 2.5 - (13.6/0.84)(7.2/12) = 0$ $p_1/\gamma = 7.21$ ft of oil

$Q = (0.608)[(\pi)(\tfrac{3}{12})^2/4]\sqrt{(2)(32.2)(7.21)}/\sqrt{1 - (0.608)^2(\tfrac{3}{12})^4} = 0.644$ ft³/s

$C_c = a/A$ $0.62 = a/[(\pi)(\tfrac{3}{12})^2/4]$ $a = 0.03043$ ft²

$(\pi)(D_{\text{jet}})^2/4 = 0.03043$ $D_{\text{jet}} = 0.1968$ ft or 2.36 in

$v = Q/a = 0.644/0.03043 = 21.2$ ft/s

17.79 Oil flows through a pipe as shown in Fig. 17-17. The coefficient of discharge for the orifice in the pipe is 0.63. What is the discharge of oil in the pipe?

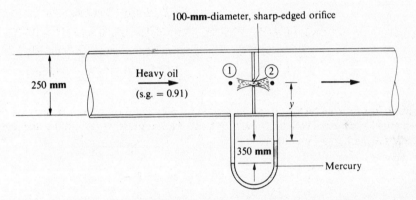

Fig. 17-17

∎ $Q = CA\sqrt{2g[(p_1 - p_2)/\gamma]}[1 + (C^2/2)(D/D_p)^4]$ $A = (\pi)(\tfrac{100}{1000})^2/4 = 0.007854$ m²

$p_1/\gamma + y + \tfrac{350}{1000} - (13.6/0.91)(\tfrac{350}{1000}) - y = p_2/\gamma$ $(p_1 - p_2)/\gamma = 4.881$ m of oil

$Q = (0.63)(0.007854)\sqrt{(2)(9.807)(4.881)}[1 + (0.63^2/2)(\tfrac{100}{250})^4] = 0.0487$ m³/s

17.80 The tank shown in Fig. 17-18a has the form of a frustum of a cone with dimensions as shown in the figure. The bottom of the tank contains an orifice that has a coefficient of discharge of 0.62. The tank contains water to its depth of 3.5 m. Find the diameter of the orifice needed to empty the tank in 8 min.

∎ $dt = A_s \, dh/CA\sqrt{2gh}$. Rearranging this equation gives

$$CA\sqrt{2gh} \, dt = A_s \, dh \tag{1}$$

In order to solve this problem, A_s must be expressed as a function of the water depth, h. This can be accomplished as follows: By observing Fig. 17-18b, it is evident that $y/(y + 3.5) = 1.5/3$, $y = 3.5$; $x/1.5 = (3.5 + h)/(3.5 + 3.5)$,

$$x = 0.7500 + 0.2143h \tag{2}$$

$$A = (\pi)(D)^2/4 = 0.7854D^2$$

$$A_s = \pi x^2 \tag{3}$$

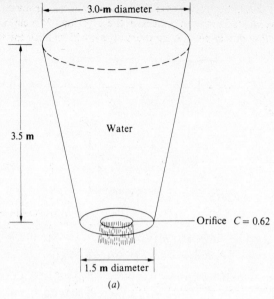

(a)

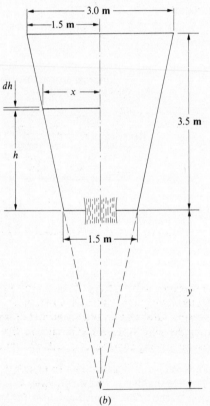

(b)

Fig. 17-18

Substituting Eq. (2) into Eq. (3) gives $A_s = (\pi)(0.7500 + 0.2143h)^2$. Therefore, substituting into Eq. (1),

$$(0.62)(0.7854D^2)\sqrt{(2)(9.807)}h^{1/2}\, dt = (\pi)(0.7500 + 0.2143h)^2\, dh$$

$$2.157D^2 h^{1/2}\, dt = (1.767 + 1.010h + 0.1443h^2)\, dh$$

$$2.157D^2 = (1.767h^{-1/2} + 1.010h^{1/2} + 0.1443h^{3/2})\, dh/dt$$

$$= \left[\int_0^{3.5}(1.767h^{-1/2} + 1.010h^{1/2} + 0.1443h^{3/2})\, dh\right] \Big/ \left(\int_0^{480} dt\right)$$

$$= [3.534h^{1/2} + 0.6733h^{3/2} + 0.05772h^{5/2}]_0^{3.5} / [t]_0^{480}$$

$$= 12.34/480$$

$$D = 0.109\,\text{m} \qquad \text{or} \qquad 109\,\text{mm}$$

17.81 A rectangular tank is divided by a partition into two chambers, as shown in Fig. 17-19a. A 6-in-diameter orifice, for which C is 0.65, is located near the bottom of the partition. At a certain time, the water level in chamber A is 10.0 ft higher than that in chamber B. Find the time it will take for the water surfaces in the two chambers to be at the same level.

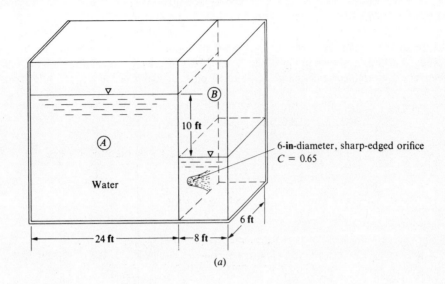

(a)

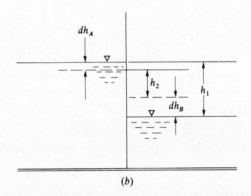

(b)

Fig. 17-19

▌ $dV = CA\sqrt{2gh}\ dt$. Let h = difference in elevations of the two water levels at any time, dh = change in h in time dt, and dV = volume of water flowing into chamber B in time dt. As illustrated in Fig. 17-19b, during a time dt, the water level in chamber A drops an amount indicated by dh_A while the level in chamber B rises an amount indicated by dh_B. During the same time, the difference in elevations of the two water levels decreases from h_1 to h_2. The volume of water leaving chamber A equals the volume entering chamber B. Hence,

$$(6)(24)(dh_A) = (6)(8)(dh_B) \tag{1}$$
$$dh_B = 3.00dh_A$$

Also from Fig. 17-19b,

$$dh = dh_A + dh_B \tag{2}$$

Substituting Eq. (1) into Eq. (2) gives $dh = dh_A + 3.00dh_A = 4.00dh_A$ or $dh_A = 0.250dh$, $dV = (0.250dh)(24)(6) = 36.0dh$, $A = (\pi)(0.5)^2/4 = 0.1963\ \text{ft}^2$, $36.0dh = (0.65)(0.1963)\sqrt{(2)(32.2)}h^{1/2}\ dt$, $35.16h^{-1/2}\ dh = dt$,

$$\int_0^t dt = 35.16\int_0^{10} h^{-1/2}\ dh \qquad [t]_0^t = 35.16[2h^{1/2}]_0^{10} \qquad t = 222\ \text{s}$$

17.82 Water flows from a large tank through an orifice and discharges to the atmosphere, as shown in Fig. 17-15, but the diameter of the orifice is 3 in and the head is 15 ft. The coefficients of velocity and contraction are 0.97 and 0.61, respectively. Find the diameter and actual velocity in the jet and the discharge from the orifice.

∎
$$C_c = a/A \qquad 0.61 = a/[(\pi)(\tfrac{3}{12})^2/4] \qquad a = 0.02994 \text{ ft}^2$$
$$(\pi)(D_{\text{jet}})^2/4 = 0.02994 \qquad D_{\text{jet}} = 0.1952 \text{ ft} \quad \text{or} \quad 2.34 \text{ in}$$

In order to determine the actual velocity in the jet, the discharge from the orifice will be determined next: $C = C_c C_v = (0.97)(0.61) = 0.592$, $Q = CA\sqrt{2gh} = (0.592)[(\pi)(\tfrac{3}{12})^2/4]\sqrt{(2)(32.2)(15)} = 0.903 \text{ ft}^3/\text{s}$, $v = Q/a = 0.903/0.02994 = 30.2 \text{ ft/s}$.

17.83 Oil with a specific gravity of 0.86 discharges from a 250-mm-diameter pipe through a 100-mm-diameter sharp-edged orifice, as shown in Fig. 17-20. Coefficients of velocity and contraction are 0.98 and 0.63, respectively. Determine the discharge of oil in the pipe, diameter of the jet, and velocity of oil in the jet.

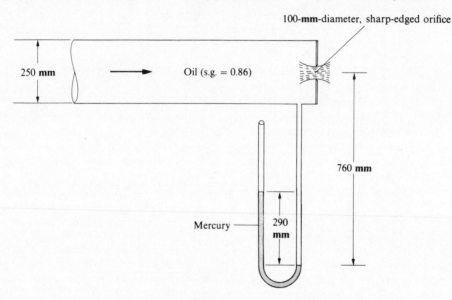

Fig. 17-20

∎
$$Q = CA\sqrt{2g(p_1/\gamma)}/\sqrt{1 - C^2(D/D_p)^4} \qquad C = C_v C_c = (0.98)(0.63) = 0.617$$
$$p_1/\gamma + \tfrac{760}{1000} - (13.6/0.86)(\tfrac{290}{1000}) = 0 \qquad p_1/\gamma = 3.826 \text{ m of oil}$$
$$Q = (0.617)[(\pi)(\tfrac{100}{1000})^2/4]\sqrt{(2)(9.807)(3.826)}/\sqrt{1 - (0.617)^2(\tfrac{100}{1000})^4} = 0.0420 \text{ m}^3/\text{s}$$
$$C_c = a/A \qquad 0.63 = a/[(\pi)(\tfrac{100}{1000})^2/4] \qquad a = 0.004948 \text{ m}^2$$
$$(\pi)(D_{\text{jet}})^2/4 = 0.004948 \qquad D_{\text{jet}} = 0.0794 \text{ m} \quad \text{or} \quad 79.4 \text{ mm}$$
$$v = Q/a = 0.0420/0.004948 = 8.49 \text{ m/s}$$

17.84 Oil flows in the pipe shown in Fig. 17-21. The orifice in the pipe has a coefficient of discharge of 0.64. Compute the discharge of the oil in the pipe.

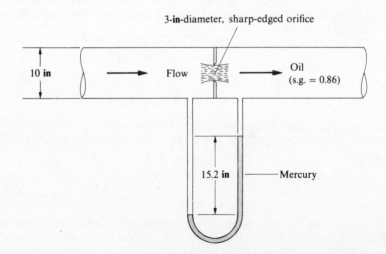

Fig. 17-21

$$Q = CA\sqrt{2g(p_1 - p_2)/\gamma}[1 + (C^2/2)(D/D_p)^4] \qquad A = (\pi)(\tfrac{3}{12})^2/4 = 0.04909 \text{ ft}^2$$

$$p_1/\gamma + y + 15.2/12 - (13.6/0.86)(15.2/12) - y = p_2/\gamma \qquad (p_1 - p_2)/\gamma = 18.76 \text{ ft of oil}$$

$$Q = (0.64)(0.04909)\sqrt{(2)(32.2)(18.76)}[1 + (0.64^2/2)(\tfrac{3}{10})^4] = 1.09 \text{ ft}^3/\text{s}$$

17.85 The water tank shown in Fig. 17-22a has the form of the frustum of a cone. The diameter of the top of the tank is 12 ft, while that at the bottom is 8 ft. The bottom of the tank contains a round, sharp-edged orifice, which has a diameter of 4 in. The discharge coefficient of the orifice is 0.60. If the tank is full at a depth of 10.0 ft, as shown in Fig. 17-22a, how long will it take to empty the tank?

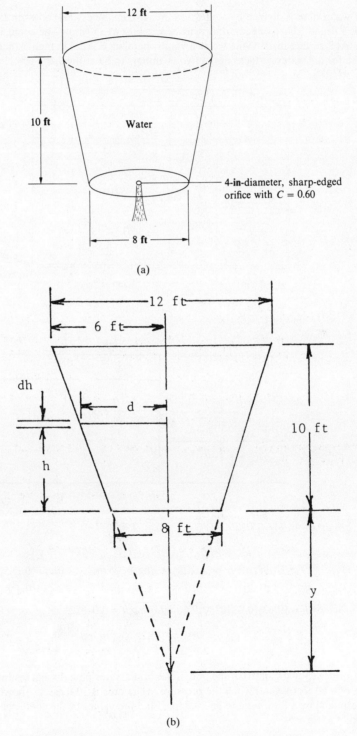

(a)

(b)

Fig. 17-22

▌ $dt = A_s \, dh/(CA\sqrt{2gh})$. From Fig. 17-22b,

$$y/(y + 10) = \tfrac{8}{12} \qquad y = 20 \text{ ft} \qquad d/12 = (20 + h)/30 \qquad d = 8 + 0.4h$$

$$A = (\pi)(\tfrac{4}{12})^2/4 = 0.08727 \text{ ft}^2 \qquad A_s = \pi d^2/4 = (\pi)(8 + 0.4h)^2/4 = 50.27 + 5.027h + 0.1257h^2$$

$$dt = (50.27 + 5.027h + 0.1257h^2) \, dh/[(0.60)(0.08727)\sqrt{(2)(32.2)(h)}] = (119.6h^{-1/2} + 11.96h^{1/2} + 0.2991h^{3/2}) \, dh$$

$$\int_0^t dt = \int_0^{10} (119.6h^{-1/2} + 11.96h^{1/2} + 0.2991h^{3/2}) \, dh$$

$$[t]_0^t = [239.2h^{1/2} + 7.973h^{3/2} + 0.1196h^{5/2}]_0^{10} \qquad t = 1046.4 \text{ s} = 17.44 \text{ min}$$

17.86 A rectangular water tank is divided by a partition into two chambers, as shown in Fig. 17-23a. In the bottom of the partition is a round, sharp-edged orifice with a diameter of 150 mm. The coefficient of discharge for the orifice is 0.62. At a certain instant, the water level in chamber B is 2.5 m higher than it is in chamber A. How long will it take for the water surfaces in the two chambers to be at the same level?

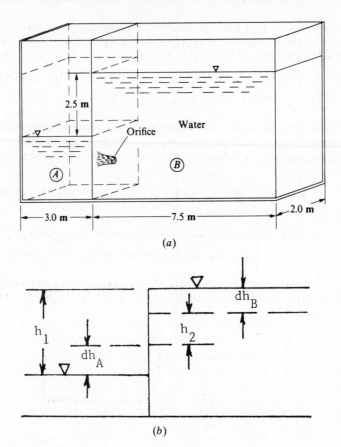

Fig. 17-23

▌ $dV = CA\sqrt{2gh} \, dt$. From Fig. 17-23b,

$$(2)(7.5)(dh_B) = (3)(2)(dh_A) \qquad dh_A = 2.5 \, dh_B$$

$$dh = dh_A + dh_B = 2.5 \, dh_B + dh_B = 3.5 \, dh_B \qquad dh_B = 0.2857 \, dh$$

$$dV = (dh_B)(7.5)(2.0) = (0.2857 \, dh)(15) = 4.286 \, dh \qquad A = (\pi)(0.150)^2/4 = 0.01767 \text{ m}^2$$

$$4.286 \, dh = (0.62)(0.01767)\sqrt{(2)(9.807)(h)} \, dt = 0.04852h \, dt \qquad dt = 88.33h^{-1/2} \, dh$$

$$\int_0^t dt = \int_0^{2.5} 88.33h^{-1/2} \, dh \qquad [t]_0^t = [176.66h^{1/2}]_0^{2.5} \qquad t = 279 \text{ s}$$

17.87 A 2-in circular orifice (not standard) at the end of a 3-in-diameter pipe discharges into the atmosphere a measured flow of 0.60 cfs of water when the pressure in the pipe is 10.0 psi, as shown in Fig. 17-24. The jet velocity is determined by a Pitot tube to be 39.2 fps. Find the values of the coefficients C_v, C_c, C_d. Find also the head loss for inlet to throat.

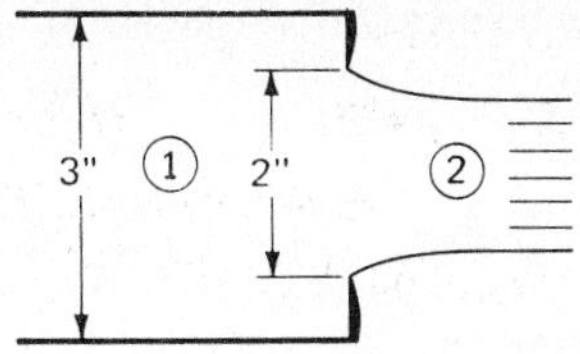

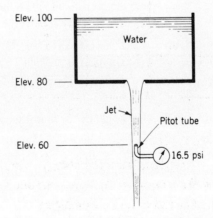

Fig. 17-24

▮ Define the inlet as section 1 and the throat as section 2. $p_1/\gamma = 10(144/62.4) = 23.1$ ft, $V_1 = Q/A_1 = 0.60/0.0491 = 12.22$ fps, $V_1^2/2g = 2.32$ ft. Express the ideal energy equation from 1 to 2 to determine the ideal velocity at 2: $(p_1/\gamma) + (V_1^2/2g) = V_2^2/2g$, $23.1 - 2.3 = V_2^2/2g$, $(V_2)_{\text{ideal}} = 40.4$ fps, $C_v = V/V_i = 39.2/40.4 = 0.97$. Area of jet: $A_2 = Q/V = 0.60/39.2 = 0.0153$ ft^2, $C_c = A_2/A_0 = 0.0153/0.0218 = 0.70$. Hence, $C_d = C_c C_v = 0.68$, $h_{L_{1-2}} = [(1/C_v^2) - 1][1 - (A_2/A_1)^2](V_2^2/2g) = \{[1/(0.97)^2] - 1\}[1 - (\frac{2}{3})^4](V_2^2/2g) = 0.051(V_2^2/2g) = 0.051[(39.2)^2/64.4] = 1.22$ ft. As a check, determine the actual velocity at 2 by expressing the real energy equation from 1 to 2: $(p_1/\gamma) + (V_2^2/2y) - h_{L_{1-2}} = V_2^2/2g$, $23.1 + 2.3 - 1.2 = V_2^2/2g$, $(V_2)_{\text{actual}} = 39.5$ fps, which is a good check. A better check would result if all numbers were carried out to more places.

17.88 Water issues from a circular orifice under a head of 40 ft. The diameter of the orifice is 4 in. If the discharge is found to be 479 ft^3 in 3 min, what is the coefficient of discharge? If the diameter at the vena contracta is measured to be 3.15 in, what is the coefficient of contraction and what is the coefficient of velocity?

▮ $$Q = C_d A_0 (2gh)^{1/2} \qquad 479/[(3)(60)] = (C_d)[(\pi)(\tfrac{4}{12})^2/4][(2)(32.2)(40)]^{1/2} \qquad C_d = 0.601$$
$$C_c = A/A_0 = [(\pi)(3.15)^2/4]/[(\pi)(4)^2/4] = 0.620 \qquad C_v = C_d/C_c = 0.601/0.620 = 0.969$$

17.89 A jet discharges from an orifice in a vertical plane under a head of 12 ft. The diameter of the orifice is 1.5 in, and the measured discharge is 0.206 cfs. The coordinates of the centerline of the jet are 11.54 ft horizontally from the vena contracta and 3.0 ft below the center of the orifice. Find the coefficients of discharge, velocity, and contraction.

▮ $$V = (x)(g/2z)^{1/2} = (11.54)\{32.2/[(2)(3.0)]\}^{1/2} = 26.73 \text{ ft/s} = (C_v)(2gh)^{1/2}$$
$$26.73 = (C_v)[(2)(32.2)(12)]^{1/2} \qquad C_v = 0.962$$
$$Q = C_d A_0 (2gh)^{1/2} \qquad 0.206 = (C_d)[(\pi)(1.5/12)^2/4][(2)(32.2)(12)]^{1/2} \qquad C_d = 0.604$$
$$C_v = C_d/C_c \qquad 0.962 = 0.604/C_c \qquad C_c = 0.628$$

17.90 In Fig. 17-25, the Pitot tube in a water jet at elevation 60 ft registers a pressure of 16.5 psi. The orifice at the bottom of the large open tank has a diameter of 1.00 in. Find C_c and C_v of the orifice. Neglect air resistance. The flow rate is 0.12 cfs.

Fig. 17-25

▮ $$V_3^2/2g = p_3/\gamma = (16.5)(144)/62.4 = 38.08 \text{ ft} \qquad V_3^2/[(2)(32.2)] = 38.08 \qquad V_3 = 49.52 \text{ ft/s}$$
$$A_3 = Q/V_3 = 0.12/49.52 = 0.002423 \text{ ft}^2$$
$$V_{\text{jet}}^2/[(2)(32.2)] + 80 = (16.5)(144)/62.4 + 60 \qquad V_{\text{jet}} = 34.12 \text{ ft/s}$$
$$V_{\text{ideal}} = (2gh)^{1/2} = [(2)(32.2)(100 - 80)]^{1/2} = 35.89 \text{ ft/s}$$
$$C_v = V_{\text{jet}}/V_{\text{ideal}} = 34.12/35.89 = 0.951$$
$$A_{\text{jet}} V_{\text{jet}} = C_c A_2 V_{\text{jet}} = A_3 V_3 \qquad (C_c)[(\pi)(\tfrac{1}{12})^2/4](34.12) = (0.002423)(49.52) \qquad C_c = 0.645$$

17.91 Water flows from one tank to an adjacent tank through a 50-mm sharp-edged orifice. The head of water on one side of the orifice is 2.0 m and that on the other is 0.4 m. Assuming $C_c = 0.62$ and $C_v = 0.95$, calculate the flow rate.

▮ $$Q = C_c C_v A (2gh)^{1/2} = (0.62)(0.95)[(\pi)(\tfrac{50}{1000})^2/4][(2)(9.807)(2.0 - 0.4)]^{1/2} = 0.00648 \text{ m}^3/\text{s}$$

17.92 Repeat Prob. 17.63 for a VDI orifice.

▮ $$Q = KA_2\sqrt{(2g)(14.91R')} \qquad \text{(from Prob. 17.63)}$$

Let $K = 0.70$ (see Fig. A-25): $Q = (0.70)[(\pi)(\tfrac{6}{12})^2/4]\sqrt{(2)(32.2)[(14.91)(3.5/12)]} = 2.30 \text{ ft}^3/\text{s}$. Check N_R to confirm that the assumed K is applicable.

$$V = Q/A = 2.30/[(\pi)(\tfrac{6}{12})^2/4] = 11.71 \text{ ft/s} \qquad N_R = DV/v = (\tfrac{6}{12})(11.71)/0.00034 = 1.72 \times 10^4$$
$$D_0/D_1 = 0.70 \qquad \text{(from Prob. 17.63)}$$

From Fig. A-25, $K = 0.73$, approximately, $Q = (0.73/0.70)(2.30) = 2.40 \text{ ft}^3/\text{s}$.

17.93 Helium, for which $k = 1.66$ and $R = 12\,400$ ft-lb/(slug-°R), is in a tank under a pressure of 50 psia and a temperature of 80 °F. It flows out through an orifice 0.5 in in diameter. For such an orifice, $C_v = 0.98$ and $C_c = 0.62$ for liquids. Find the rate of flow if the pressure into which the gas discharges is 40 psia. Assume $Y = 0.95$.

▮ $$G = CYA_2\sqrt{2g\gamma_1\{(p_1 - p_2)/[1 - (D_2/D_1)^4]\}} \qquad C = C_c C_v = (0.62)(0.98) = 0.608$$
$$Y = 0.95 \qquad \text{(from Fig. A-26)}$$
$$\gamma_1 = p_1/RT = (50)(144)/[(12\,400/32.2)(80 + 460)] = 0.03462 \text{ lb/ft}^3$$
$$G = (0.608)(0.95)[(\pi)(0.5/12)^2/4]\sqrt{(2)(32.2)(0.03462)[(50 - 40)(144)/(1 - 0)]} = 0.0446 \text{ lb/s}$$

17.94 For the data in Prob. 17.93, find the rate of discharge if $p_2 = p_c$.

▮ $$G = CYA_2\sqrt{2g\gamma_1\{(p_1 - p_2)/[1 - (D_2/D_1)^4]\}}$$
$$p_2 = p_c = 0.488p_1 = (0.488)(50) = 24.4 \text{ psia} \qquad Y = 0.86 \qquad \text{(from Fig. A-26)}$$

Using data from Prob. 17.93, for $D_0/D_1 = 1/\infty$,

$$G = (0.608)(0.86)[(\pi)(0.5/12)^2/4]\sqrt{(2)(32.2)(0.03462)[(50 - 24.4)(144)/(1 - 0)]} = 0.0646 \text{ lb/s}.$$

17.95 Air is in a tank under a pressure of 1400 kN/m² abs and a temperature of 40 °C. It flows out through an orifice having an area of 10 cm² into a space where the pressure is 550 kN/m² abs. Compute the rate of discharge, assuming $C_d = 0.60$.

▮ $$G = CYA_2\sqrt{2g\gamma_1\{(p_1 - p_2)/[1 - (D_2/D_1)^4]\}} \qquad Y = 0.823 \qquad \text{(from Fig. A-26)}$$
$$\gamma_1 = p_1/RT_1 = (1400)(1000)/[(29.3)(40 + 273)] = 152.7 \text{ N/m}^3$$
$$G = (0.60)(0.823)(10/10\,000)\sqrt{(2)(9.807)(152.7)[(1400 - 500)(1000)/(1 - 0)]} = 24.9 \text{ N/s}$$

17.96 Using the same data as in Prob. 17.95, what would be the flow if the air discharged into a space where the pressure is 105 kPa abs?

▮ $G = CYA_2\sqrt{2g\gamma_1\{(p_1 - p_2)/[1 - (D_2/D_1)^4]\}}$. With $p_2 = 105$ kN/m², $D_0/D_1 = 0$, $p_2/p_1 = \tfrac{105}{1400} = 0.075$, $Y = 0.73$, $G = (0.60)(0.73)(10/10\,000)\sqrt{(2)(9.807)(152.7)[(1400 - 105)(1000)/(1 - 0)]} = 27.3 \text{ N/s}$.

17.97 Air in a tank at 1500 kN/m² abs and 40 °C flows out through a 5.0-cm-diameter orifice into a space where the pressure is 500 kN/m² abs. Compute the rate of discharge assuming $C_d = 0.60$. Repeat for an external pressure of 1000 kN/m² abs.

▮ $p_2/p_1 = \tfrac{500}{1500} = 0.333$. Since $p_2/p_1 < 0.528$, sonic velocity occurs at the throat and $G = C_d g(A_2 p_1/\sqrt{T_1})\sqrt{(k/R)[2/(k + 1)]^{(k+1)/(k-1)}}$.

$$A_2 = (\pi)(5.0/100)^2/4 = 0.001964 \text{ m}^2$$
$$G = (0.60)(9.807)[(0.001964)(1500)/\sqrt{40 + 273}]\sqrt{(1.40/287)[2/(1.40 + 1)]^{(1.40+1)/(1.40-1)}} = 0.0396 \text{ kN/s}$$
$$p_2/p_1 = \tfrac{1000}{1500} = 0.667$$

Since $p_2/p_1 > 0.528$, $G = C_d Y A_2 \sqrt{2g\gamma_1\{(p_1 - p_2)/[1 - (D_2/D_1)^4]\}}$. For $D_0/D_1 = 0$, $p_2/p_1 = 0.667$, $Y = 0.91$.

$$\gamma_1 = p_1/RT_1 = 1500/[(29.3)(40 + 273)] = 0.1636 \text{ kN/m}^3$$

$$G = (0.60)(0.91)(0.001964)\sqrt{(2)(9.807)(0.1636)[(1500 - 1000)/(1 - 0)]} = 0.0430 \text{ kN/s}$$

17.98 A 4-in-diameter standard orifice discharges water under a 20.0-ft head as shown in Fig. 17-26. What is the flow?

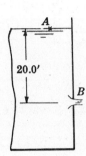

Fig. 17-26

▌ $Q = CA_0\sqrt{2gh}$. From Table A-19, $C = 0.594$, $Q = (0.594)[(\pi)(\frac{4}{12})^2/4]\sqrt{(2)(32.2)(20.0)} = 1.86 \text{ ft}^3/\text{s}$.

17.99 The actual velocity in the contracted section of a jet of liquid flowing from a 2-in-diameter orifice is 28.0 ft/s under a head of 15 ft. (**a**) What is the value of the coefficient of velocity? (**b**) If the measured discharge is 0.403 cfs, determine the coefficients of contraction and discharge.

▌ (**a**) $V_{\text{actual}} = C_v \sqrt{2gH}$ $28.0 = C_v \sqrt{(2)(32.2)(15)}$ $C_v = 0.901$

    (**b**) $Q_{\text{actual}} = CA\sqrt{2gH}$ $0.403 = (C)[(\pi)(\frac{2}{12})^2/4]\sqrt{(2)(32.2)(15)}$ $C = 0.594$

$$C = C_v C_c \qquad 0.594 = 0.901 C_c \qquad C_c = 0.659$$

17.100 Oil flows through a standard 1-in-diameter orifice under an 18.0-ft head at the rate of 0.111 cfs. The jet strikes a wall 5.00 ft away and 0.390 ft vertically below the centerline of the contracted section of the jet. Compute the coefficients.

▌ $Q = CA\sqrt{2gH}$ $0.111 = (C)[(\pi)(\frac{1}{12})^2/4]\sqrt{(2)(32.2)(18.0)}$ $C = 0.598$ $V = C_v\sqrt{2gH}$

Letting x and y represent the coordinates of the jet as measured, from kinematic mechanics, $x = Vt$, $y = gt^2/2$. Eliminating t gives

$$x^2 = (2V^2/g)(y) \qquad 5.00^2 = (2V^2/32.2)(0.390) \qquad V = V_{\text{actual}} = 32.13 \text{ ft/s}$$
$$32.13 = C_v\sqrt{(2)(32.2)(18.0)} \qquad C_v = 0.944 \qquad C = C_c C_v \qquad 0.598 = (C_c)(0.944) \qquad C_c = 0.633$$

17.101 The tank in Prob. 17.98 is closed and the air space above the water is under pressure, causing the flow to increase to 2.65 cfs. Find the pressure in the air space.

▌ $Q = CA_0\sqrt{2gH}$. Table A-19 indicates that C does not change appreciably at the range of head under consideration. Using $C = 0.593$, $2.65 = (0.593)[(\pi)(\frac{4}{12})^2/4]\sqrt{(2)(32.2)(20 + p/\gamma)}$, $p/\gamma = 20.72$ ft of water, $p' = \gamma h = (62.4)(20.72)/144 = 8.98 \text{ lb/in}^2$.

17.102 Oil of specific gravity 0.720 flows through a 3-in-diameter orifice whose coefficients of velocity and contraction are 0.950 and 0.650, respectively. What must be the reading of gage A in Fig. 17-27 in order for the power in the jet C to be 8.00 hp?

▌ $p_B/\gamma + V_B^2/2g + z_B = p_C/\gamma + V_C^2/2g + z_C + h_L$ $P_{\text{jet}} = \gamma Q H_{\text{jet}} = (\gamma)(C_c A_0 V_{\text{jet}})(0 + V_{\text{jet}}^2/2g + 0)$

$$(8.00)(550) = [(0.720)(62.4)](0.650)[(\pi)(\frac{3}{12})^2/4](V_{\text{jet}})\{V_{\text{jet}}^2/[(2)(32.2)]\}$$

$$V_{\text{jet}} = V_C = 58.25 \text{ ft/s} \qquad h_L = (1/C_v - 1)(V_{\text{jet}}^2/2g)$$

$$p_A/[(0.720)(62.4)] + 0 + 9.0 = 0 + 58.25^2/[(2)(32.2)] + 0 + (1/0.950 - 1)\{58.25^2/[(2)(32.2)]\}$$

$$p_A = 2087 \text{ lb/ft}^2 \quad \text{or} \quad 14.5 \text{ lb/in}^2$$

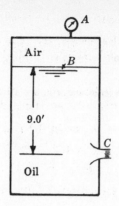

Fig. 17-27

17.103 Water at 100 °F flows at the rate of 0.525 cfs through a 4-in diameter orifice used in an 8-in pipe. What is the difference in pressure head between the upstream section and the contracted section (vena contracta section)?

▌ In Fig. A-27, it is observed that C' varies with the Reynolds number. Note that the Reynolds number must be calculated for the orifice cross section, not for the contracted section of the jet nor for the pipe section. This value is $N_R = V_o D_o/\nu = (4Q/\pi D_o^2)D_o/\nu = 4Q/\nu\pi D_o = 4(0.525)/[\pi(0.00000739)(\frac{4}{12})] = 272\,000$. Figure A-27 for $\beta = 0.500$ gives $C' = 0.604$.

Applying the Bernoulli theorem, pipe section to jet section, produces the general equation for incompressible fluids, as follows: $[(p_8/\gamma) + (V_8^2/2g) + 0] - [(1/C_v^2) - 1](V_{jet}^2/2g) = [(p_{jet}/\gamma) + (V_{jet}^2/2g) + 0]$ and $Q = A_8 V_8 = (C_c A_4)V_{jet}$.

Substituting for V_8 in terms of V_{jet} and solving, $V_{jet}^2/2g = C_v^2\{(p_8/\gamma - p_{jet}/\gamma)/[1 - C^2(A_4/A_8)^2]\}$ or $V_{jet} = C_v\sqrt{2g(p_8/\gamma - p_{jet}/\gamma)/[1 - C^2(D_4/D_8)^4]}$. Then $Q = A_{jet}V_{jet} = (C_c A_4)(C_v)\sqrt{2g(p_8/\gamma - p_{jet}/\gamma)/[1 - C^2(D_4/D_8)^4]} = CA_4\sqrt{2g(p_8/\gamma - p_{jet}/\gamma)/[1 - C^2(D_4/D_8)^4]}$.

More conveniently, for an orifice with velocity of approach and a contracted jet, the equation can be written

$$Q = [C'A_4/\sqrt{1 - (D_4/D_8)^4}]\sqrt{2g(\Delta p/\gamma)} \tag{1}$$

or

$$Q = KA_4\sqrt{2g(\Delta p/\gamma)} \tag{2}$$

where K is called the flow coefficient. The meter coefficient C' may be determined experimentally for a given ratio of diameter of orifice to diameter of pipe, or the flow coefficient K may be preferred.

Proceeding with the solution by substituting in Eq. (1),

$$0.525 = \frac{(0.604)[(\pi)(\frac{4}{12})^2/4]}{\sqrt{1 - (\frac{1}{2})^4}}\sqrt{(2)(32.2)\left(\frac{\Delta p}{\gamma}\right)} \qquad \frac{\Delta p}{\gamma} = 1.44 \text{ ft of water}$$

17.104 For the pipe orifice in Prob. 17.103, what pressure difference in pounds per square inch would cause the same quantity of turpentine at 68 °F (s.g. = 0.862, ν = 0.0000186 ft²/s) to flow?

▌

$$Q = [C'A_4/\sqrt{1 - (D_4/D_8)^4}]\sqrt{2g(\Delta p/\gamma)}$$

$$V = Q/A = 0.525/[(\pi)(\tfrac{4}{12})^2/4] = 6.016 \text{ ft/s} \qquad N_R = DV/\nu = (\tfrac{4}{12})(6.016)/0.0000186 = 1.08 \times 10^5$$

From Fig. A-27, for $\beta = 0.500$, $C' = 0.607$.

$$0.525 = \frac{(0.607)[(\pi)(\frac{4}{12})^2/4]}{\sqrt{1 - (\frac{1}{2})^4}}\sqrt{(2)(32.2)\left[\frac{\Delta p}{(0.862)(62.4)}\right]}$$

$$p = 76.9 \text{ lb/ft}^2 \quad \text{or} \quad 0.534 \text{ lb/in}^2$$

17.105 Determine the flow of water at 70 °F through a 6-in orifice installed in a 10-in pipeline if the pressure head differential for vena-contracta taps is 3.62 ft of water.

▌

$$Q = [C'A_6/\sqrt{1 - (D_6/D_{10})^4}]\sqrt{2g(\Delta p/\gamma)}$$

The value of C' cannot be found inasmuch as the Reynolds number cannot be computed. Referring to Fig. A-27, for $\beta = 0.600$, a value of C' will be assumed at 0.610. Using this assumed value,

$$Q = \frac{(0.610)[(\pi)(\frac{6}{12})^2/4]}{\sqrt{1 - (0.600)^4}}\sqrt{(2)(32.2)(3.62)} = 1.96 \text{ ft}^3/\text{s}$$

$$V = Q/A = 1.96/[(\pi)(\tfrac{6}{12})^2/4] = 9.982 \text{ ft/s} \qquad N_R = DV/\nu = (\tfrac{6}{12})(9.982)/0.00001059 = 4.71 \times 10^5$$

From Fig. A-27, for $\beta = 0.600$, $C' = 0.609$.

$$Q = \frac{(0.609)[(\pi)(\frac{6}{12})^2/4]}{\sqrt{1 - (0.600)^4}} \sqrt{(2)(32.2)(3.62)} = 1.96 \text{ ft}^3/\text{s}$$

17.106 Carbon dioxide discharges through a $\frac{1}{2}$-in hole in the wall of a tank in which the pressure is 110 psig and the temperature is 68 °F. What is the velocity in the jet (standard atmosphere)?

▮

$$\gamma_1 = p_1/RT_1 = (110 + 14.7)(144)/[(35.1)(68 + 460)] = 0.9689 \text{ lb/ft}^3$$
$$(p_2/p_1)_{\text{critical}} = [2/(k + 1)]^{k/(k-1)} = [2/(1.30 + 1)]^{1.30/(1.30-1)} = 0.546$$
$$p_{\text{atm}}/p_{\text{tank}} = 14.7/(110 + 14.7) = 0.118$$

Since the latter ratio is less than the critical pressure ratio, the pressure of the escaping gas is 0.546 times p_1; hence,

$$p_2 = (0.546)(110 + 14.7) = 68.1 \text{ psia} \qquad V_2 = c_2 = \sqrt{kgRT_2} \qquad T_2/T_1 = (p_2/p_1)^{(k-1)/k}$$
$$T_2/(68 + 460) = 0.546^{(1.30-1)/1.30} \qquad T_2 = 459 \text{ °R} \qquad V_2 = \sqrt{(1.30)(32.2)(35.1)(459)} = 821 \text{ ft/s}$$

17.107 Nitrogen flows through a duct in which changes in cross section occur. At a particular cross section the velocity is 1200 ft/s, the pressure is 12.0 psia, and the temperature is 90 °F. Assuming no friction losses and adiabatic conditions, **(a)** what is the velocity at a section where the pressure is 18.0 psia and **(b)** what is the Mach number at this section?

▮ **(a)**

$$V_2^2/2g - V_1^2/2g = [k/(k - 1)](p_1/\gamma_1)[1 - (p_2/p_1)^{(k-1)/k}]$$
$$\gamma_1 = p_1/RT_1 = (12.0)(144)/[(55.1)(90 + 460)] = 0.05702 \text{ lb/ft}^3$$
$$\{V_2^2/[(2)(32.2)]\} - \{1200^2/[(2)(32.2)]\} = [1.40/(1.40 - 1)][(12.0)(144)/0.05702]$$
$$\times [1 - (18.0/12.0)^{(1.40-1)/1.40}] \qquad V_2 = 775 \text{ ft/s}$$

(b)

$$N_M = V_2/c_2 = V_2/\sqrt{kgRT} \qquad T_2/T_1 = (p_2/p_1)^{(k-1)/k}$$
$$T_2/(90 + 460) = (18.0/12.0)^{(1.40-1)/1.40} \qquad T_2 = 618 \text{ °R}$$
$$N_M = 775/\sqrt{(1.40)(32.2)(55.1)(618)} = 0.626$$

17.108 Establish the formula to determine the time to lower the liquid level in a tank of constant cross section by means of an orifice. Refer to Fig. 17-28.

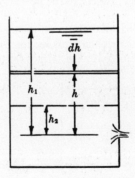

Fig. 17-28

▮ Inasmuch as the head is changing with time, we know that $\partial V/\partial t \neq 0$, i.e., we do not have steady flow. This means that the energy equation should be amended to include an acceleration term, which complicates the solution materially. As long as the head does not change too rapidly, no appreciable error will be introduced by assuming steady flow, thus neglecting the acceleration-head term.

Case A: With *no inflow* taking place, the instantaneous flow will be $Q = CA_0\sqrt{2gh}$ cfs. In time interval dt, the small volume dV discharged will be $Q\, dt$. In the same time interval, the head will decrease dh ft and the volume discharged will be the area of the tank A_T times dh. Equating these values, $(CA_0\sqrt{2gh})\, dt = -A_T\, dh$ where the negative sign signifies that h decreases as t increases. Solving for t yields

$$t = \int_{t_1}^{t_2} dt = (-A_T/CA_0\sqrt{2g}) \int_{h_1}^{h_2} h^{-1/2}\, dh$$

or
$$t = t_2 - t_1 = (2A_T/CA_0\sqrt{2g})(h_1^{1/2} - h_2^{1/2}) \qquad (1)$$

In using this expression, an average value of coefficient of discharge C may be used without producing significant error in the result. As h_2 approaches zero, a vortex will form and the orifice will cease to flow full. However, using $h_2 = 0$ will not produce serious error in most cases.

Equation (1) can be rewritten by multiplying and dividing by $(h_1^{1/2} + h_2^{1/2})$. There results

$$t = t_2 - t_1 = A_T(h_1 - h_2)/[\tfrac{1}{2}(CA_0\sqrt{2gh_1} + CA_0\sqrt{2gh_2})] \qquad (2)$$

Noting that the volume discharged in time $(t_2 - t_1)$ is $A_T(h_1 - h_2)$, this equation simplifies to

$$t = t_2 - t_1 = \text{volume discharged}/[\tfrac{1}{2}(Q_1 + Q_2)] = \text{volume discharged in ft}^3/\text{average flow } Q \text{ in ft}^3/\text{s} \qquad (3)$$

Case B. With a constant rate of inflow less than the flow through the orifice taking place,

$$-A_T\, dh = (Q_\text{out} - Q_\text{in})\, dt \qquad \text{and} \qquad t = t_2 - t_1 = \int_{h_1}^{h_2} \frac{-A_T\, dh}{Q_\text{out} - Q_\text{in}}$$

Should Q_in exceed Q_out, the head would increase, as would be expected.

17.109 The initial head on an orifice was 9 ft and when the flow was terminated the head was measured at 4 ft. Under what constant head H would the same orifice discharge the same volume of water in the same time interval? Assume coefficient C is constant.

❚ Volume under falling head = volume under constant head

$$\tfrac{1}{2}CA_0\sqrt{2g}\,(h_1^{1/2} + h_2^{1/2})(t) = CA_0\sqrt{2gH}\,(t)$$
$$\tfrac{1}{2}CA_0\sqrt{2g}\,(9^{1/2} + 4^{1/2})(t) = CA_0\sqrt{2g}\,\sqrt{H}\,(t) \qquad H = 6.25 \text{ ft}$$

17.110 A 75-mm-diameter orifice under a head of 4.88 m discharges 8900 N water in 32.6 s. The trajectory was determined by measuring $x_0 = 4.76$ m for a drop of 1.22 m. Determine C_v, C_c, C_d, the head loss per unit weight, and the power loss.

❚ The theoretical velocity V_{2t} is $V_{2t} = \sqrt{2gH} = \sqrt{2(9.806)(4.88)} = 9.783$ m/s. The actual velocity is determined from the trajectory. The time to drop 1.22 m is $t = \sqrt{2y_0/g} = \sqrt{2(1.22)/9.806} = 0.499$ s and the velocity is expressed by $x_0 = V_{2a}t$, $V_{2a} = 4.76/0.499 = 9.539$ m/s. Then $C_v = V_{2a}/V_{2t} = 9.539/9.783 = 0.975$. The actual discharge Q_a is $Q_a = 8900/[9806(32.6)] = 0.0278$ m^3/s. $C_d = Q_a/(A_0\sqrt{2gH}) = 0.0278/[\pi(0.0375^2)\sqrt{2(9.806)(4.88)}] = 0.643$, $C_c = C_d/C_v = 0.643/0.975 = 0.659$. The head loss is loss $= H(1 - C_v^2) = 4.88(1 - 0.975^2) = 0.241$ m-N/N. The power loss is $Q\gamma(\text{loss}) = 0.0278(9806)(0.241) = 65.7$ W.

17.111 A tank has a horizontal cross-sectional area of 2 m^2 at the elevation of the orifice, and the area varies linearly with elevation so that it is 1 m^2 at a horizontal cross section 3 m above the orifice. For a 100-mm-diameter orifice, $C_d = 0.65$, compute the time to lower the surface from 2.5 m to 1 m above the orifice.

❚
$$t = -\frac{1}{C_d A_0\sqrt{2g}} \int_{y_1}^{y_2} A_r y^{-1/2}\, dy \qquad A_r = 2 - \frac{y}{3}$$
$$t = -\frac{1}{(0.65)[(\pi)(\tfrac{100}{1000})^2/4]\sqrt{(2)(9.807)}} \int_{2.5}^{1} \left(2 - \frac{y}{3}\right)y^{-1/2}\, dy = -44.23[4y^{1/2} - \tfrac{2}{9}y^{3/2}]_{2.5}^{1} = 73.8 \text{ s}$$

17.112 Determine the equation for trajectory of a jet discharging horizontally from a small orifice with head of 6 m and velocity coefficient of 0.96. Neglect air resistance.

❚
$$x = V_2 t \qquad y = gt^2/2 \qquad V_2 = C_v\sqrt{2gH} \qquad t = x/V_2$$
$$y = (g)(x/V_2)^2/2 = (g)[x/(C_v\sqrt{2gH})]^2/2 = (9.807)\{x/[0.96\sqrt{(2)(9.807)(6)}]\}^2/2 = 0.0452x^2$$

17.113 An orifice of area 30 cm^2 in a vertical plate has a head of 1.1 m of oil, s.g. $= 0.91$. It discharges 6790 N of oil in 79.3 s. Trajectory measurements yield $x_0 = 2.25$ m, $y_0 = 1.23$ m. Determine C_v, C_c, and C_d.

❚ Letting subscript a denote "actual" and t, "theoretical,"

$$Q_a = W/\gamma = (6790/79.3)/[(0.91)(9790)] = 0.009611 \text{ m}^3/\text{s}$$
$$(V_2)_a = x\sqrt{g/2y} = (2.25)\sqrt{(9.807)/[(2)(1.23)]} = 4.492 \text{ m/s}$$
$$(V_2)_t = \sqrt{2gH} = \sqrt{(2)(9.807)(1.1)} = 4.645 \text{ m/s} \qquad C_v = (V_2)_a/(V_2)_t = 4.492/4.645 = 0.967$$
$$C_d = Q_a/[A(V_2)_t] = 0.009611/[(30/10\,000)(4.645)] = 0.690 \qquad C_d = C_d/C_v = 0.690/0.967 = 0.714$$

17.114 Calculate Y, the maximum rise of a jet from an inclined plate (Fig. 17-29), in terms of H and α. Neglect losses.

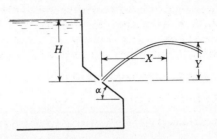

Fig. 17-29

┃
$$y = -gt^2/2 + (V \cos \alpha)(t) \qquad x = Vt \sin \alpha \qquad V = \sqrt{2gH}$$

Substituting for V and t,
$$y = -x^2/(4H \sin^2 \alpha) + x \cot \alpha \qquad dy/dx = -x/(2H \sin^2 \alpha) + \cot \alpha = 0$$
$$x = 2H \sin \alpha \cos \alpha \qquad y_{max} = H \cos^2 \alpha$$

17.115 In Fig. 17-29, for $\alpha = 45°$, $Y = 0.48H$. Neglecting air resistance of the jet, find C_v for the orifice.

┃ From Prob. 17.114 but with $V = C_v\sqrt{2gH}$, $Y = C_v^2 H \cos^2 \alpha$, $C_v = \sqrt{Y/H}/\cos \alpha = \sqrt{0.48H/H}/\cos 45° = 0.980$.

17.116 Show that the locus of maximum points of the jet of Fig. 17-29 is given by $X^2 = 4Y(H - Y)$ when losses are neglected.

┃ From Prob. 17.114, maximum point coordinates are given by $X = 2H \sin \alpha \cos \alpha$ and $Y = H \cos^2 \alpha$. Thus, $\cos \alpha = \sqrt{Y/H}$ and $\sin \alpha = \sqrt{1 - \cos^2 \alpha} = \sqrt{1 - Y/H}$. Hence, $X = 2H\sqrt{Y/H}\sqrt{1 - Y/H}$, $X^2 = 4H^2[Y/H - (Y/H)^2] = 4Y(H - Y)$.

17.117 A 3-in-diameter orifice discharges 64 ft³ of liquid, s.g. = 1.07, in 82.2 s under a 9-ft head. The velocity at the vena contract is determined by a Pitot-static tube with coefficient 1.0. The manometer liquid is acetylene tetrabromide, s.g. 2.96, and the gage difference is $R' = 3.35$ ft. Determine C_v, C_c, and C_d.

┃
$$V_a = C\sqrt{2gR'(S_0/S - 1)} = (1.0)\sqrt{(2)(32.2)(3.35)[(2.96/1.07) - 1]} = 19.52 \text{ ft/s}$$
$$V_t = \sqrt{2gH} = \sqrt{(2)(32.2)(9)} = 24.07 \text{ ft/s}$$
$$C_v = V_a/V_t = 19.52/24.07 = 0.811 \qquad Q_a = V/t = 64/82.2 = 0.779 \text{ ft}^3/\text{s}$$
$$C_d = Q_a/AV_t = 0.779/\{[(\pi)(\tfrac{3}{12})^2/4](24.07)\} = 0.659 \qquad C_c = C_d/C_v = 0.659/0.811 = 0.813$$

17.118 A 100-mm-diameter orifice discharges 44.6 L/s of water under a head of 2.75 m. A flat plate held normal to the jet downstream from the vena contracta requires a force of 320 N to resist impact of the jet. Find C_d, C_v, and C_c.

┃
$$F = \rho(V_2)_a Q_a \qquad (V_2)_a = F/\rho Q_a = (320)(1000)/[(1000)(44.6)] = 7.175 \text{ m/s}$$
$$C_v = (V_2)_a/\sqrt{2gH} = 7.175/\sqrt{(2)(9.807)(2.75)} = 0.977$$
$$C_d = Q_a/(A\sqrt{2gH}) = (44.6/1000)/\{[(\pi)(\tfrac{100}{1000})^2/4]\sqrt{(2)(9.907)(2.75)}\} = 0.773$$
$$C_c = C_d/C_v = 0.773/0.977 = 0.791$$

17.119 Compute the discharge from the tank shown in Fig. 17-30.

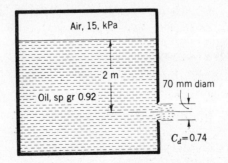

Fig. 17-30

┃
$$Q = C_d A\sqrt{2g(H + p_a/\gamma)} = (0.74)[(\pi)(\tfrac{70}{1000})^2/4]\sqrt{(2)(9.807)\{2 + 15/[(0.92)(9.79)]\}} = 0.0241 \text{ m}^3/\text{s}$$

17.120 For $C_v = 0.96$ in Fig. 17-30, calculate the losses in meter-newtons per newton and meter-newtons per second.

$$h_L = (H + p_a/\gamma)(1 - C_v^2) = \{2 + 15/[(0.92)(9.79)]\}(1 - 0.96^2) = 0.287 \text{ m-N/N}$$
$$Q = 0.0241 \text{ m}^3/\text{s} \quad \text{(from Prob. 17.119)}$$
$$\text{Losses} = Q\gamma h_L = (0.0241)[(0.92)(9.79)(1000)](0.287) = 62.3 \text{ m-N/s}$$

17.121 Calculate the discharge through the orifice of Fig. 17-31.

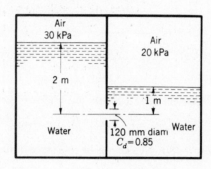

Fig. 17-31

$$H' = (2 + 30/9.79) - (1 + 20/9.79) = 2.021 \text{ m}$$
$$Q = C_d A\sqrt{2gH'} = (0.85)[(\pi)(\tfrac{120}{1000})^2/4]\sqrt{(2)(9.807)(2.021)} = 0.0605 \text{ m}^3/\text{s}$$

17.122 For $C_v = 0.93$ in Fig. 17-31, determine the losses in joules per newton and in watts.

$$H' = 2.021 \text{ m} \quad \text{(from Prob. 17.121)} \qquad h_L = H'(1 - C_v^2) = (2.021)(1 - 0.93^2) = 0.273 \text{ J/N}$$
$$Q = 0.0605 \text{ m}^3/\text{s} \quad \text{(from Prob. 17.121)} \qquad \text{Losses} = Q\gamma h_L = (0.0605)[(9.79)(1000)](0.273) = 161.7 \text{ W}$$

17.123 A 4-in-diameter orifice discharges 1.60 cfs of liquid under a head of 11.8 ft. The diameter of the jet at the vena contracta is found by calipering to be 3.47 in. Calculate C_v, C_d, and C_c.

$$V_2 = \sqrt{2gH} = \sqrt{(2)(32.2)(11.8)} = 27.57 \text{ ft/s} \qquad A_0 = (\pi)(\tfrac{4}{12})^2/4 = 0.08727 \text{ ft}^2$$
$$C_d = Q_a/V_2 A_0 = 1.60/[(27.57)(0.08727)] = 0.665$$
$$C_c = A_2/A_0 = [(\pi)(3.47/12)^2/4]/[(\pi)(\tfrac{4}{12})^2/4] = 0.753 \qquad C_v = C_d/C_c = 0.665/0.753 = 0.883$$

17.124 A 75-mm-diameter orifice, $C_d = 0.82$, is placed in the bottom of a vertical tank that has a diameter of 1.5 m. How long does it take to draw the surface down from 3 to 2.5 m?

$$t = (2A_r/C_d A_0\sqrt{2g})(y_1^{1/2} - y_2^{1/2}) \qquad A_r = (\pi)(1.5)^2/4 = 1.767 \text{ m}^2$$
$$A_0 = (\pi)(\tfrac{75}{1000})^2/4 = 0.004418 \text{ m}^2$$
$$t = \{(2)(1.767)/[(0.82)(0.004418)\sqrt{(2)(9.807)}]\}(3^{1/2} - 2.5^{1/2}) = 33.2 \text{ s}$$

17.125 Select the size of orifice that permits a tank of horizontal cross section 1.5 m^2 to have the liquid surface drawn down at the rate of 160 mm/s for 3.35-m head on the orifice ($C_d = 0.63$).

$$Q = (A_r)(\Delta y/\Delta t) = (1.5)(\tfrac{160}{1000}) = 0.2400 \text{ m}^3/\text{s}$$
$$A_0 = Q/(C_d\sqrt{2gH}) = 0.2400/[0.63\sqrt{(2)(9.807)(3.35)}] = 0.04700 \text{ m}^2$$
$$0.04700 = \pi D^2/4 \qquad D = 0.245 \text{ m} \quad \text{or} \quad 245 \text{ mm}$$

17.126 A 4-in-diameter orifice in the side of a 6-ft-diameter tank draws the surface down from 8 ft to 4 ft above the orifice in 83.7 s. Calculate the discharge coefficient.

$$t = [2A_r/(C_d A_0\sqrt{2g})](y_1^{1/2} - y_2^{1/2}) \qquad A_r = (\pi)(6)^2/4 = 28.27 \text{ ft}^2$$
$$A_0 = (\pi)(\tfrac{4}{12})^2/4 = 0.08727 \text{ ft}^2$$
$$83.7 = \{(2)(28.27)/[(C_d)(0.08727)\sqrt{(2)(32.2)}]\}(8^{1/2} - 4^{1/2}) \qquad C_d = 0.799$$

17.127 Select a reservoir of such size and shape that the liquid surface drops 1 m/min over a 3-m distance for flow through a 100-mm-diameter orifice ($C_d = 0.74$).

$$dy/dt = -Q/A_r = -(C_d A_0 \sqrt{2gy})/A_r$$

$$A_r = C_d A_0 \sqrt{2g}(\Delta t/\Delta y)y^{1/2} = (0.74)[(\pi)(\tfrac{100}{1000})^2/4]\sqrt{(2)(9.807)}(\tfrac{60}{1})y^{1/2} = 1.544y^{1/2}$$

$$\pi D^2/4 = 1.544y^{1/2} \qquad D = 1.402y^{1/4}$$

17.128 In Fig. 17-32 the truncated cone has an angle $\theta = 60°$. How long does it take to draw the liquid surface down from $y = 4$ m to $y = 1$ m?

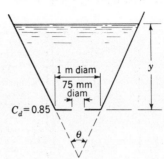

1 m diam

75 mm diam

$C_d = 0.85$

θ

Fig. 17-32

$$t = -\frac{1}{C_d A_0 \sqrt{2g}} \int_{y_1}^{y_2} \frac{A_r}{\sqrt{y}}\, dy \qquad A_r = \pi\left(r_0 + y \tan\frac{\theta}{2}\right)^2$$

$$t = \frac{\pi}{C_d A_0 \sqrt{2g}} \int_{y_2}^{y_1} \frac{[r_0 + y\tan(\theta/2)]^2}{\sqrt{y}}\, dy = \frac{\pi}{C_d A_0 \sqrt{2g}}\left[r_0^2(2\sqrt{y}) + 2r_0(\tfrac{2}{3}y^{3/2})\tan\frac{\theta}{2} + \tfrac{2}{5}y^{5/2}\tan^2\frac{\theta}{2}\right]_{y_2}^{y_1}$$

$$A_0 = (\pi)(\tfrac{75}{1000})^2/4 = 0.004418 \text{ m}^2$$

$$t = -\frac{\pi}{(0.85)(0.004418)\sqrt{(2)(9.807)}}\left[(\tfrac{1}{2})^2(2\sqrt{y}) + (2)(\tfrac{1}{2})(\tfrac{2}{3})y^{3/2}\tan\tfrac{60}{2} + \tfrac{2}{5}y^{5/2}\tan^2\tfrac{60}{2}\right]_4^1 = 1348 \text{ s} \quad \text{or} \quad 23.1 \text{ min}$$

17.129 Calculate the dimensions of a tank such that the surface velocity varies inversely as the distance from the centerline of an orifice draining the tank. When the head is 300 mm, the velocity of fall of the surface is 30 mm/s, orifice diameter is 12.5 mm, $C_d = 0.66$.

$$dy/dt = -Q/A_r = k/y \qquad k = y(dy/dt) = (\tfrac{300}{1000})(-\tfrac{30}{1000}) = -0.00900 \qquad -C_d A_0 \sqrt{2gy}/(\pi D^2/4) = k/y$$

$$D = y^{3/4}\sqrt{\frac{C_d A_0 \sqrt{2g}}{\pi/4}}\left(-\frac{1}{k}\right) = y^{3/4}\sqrt{\frac{(0.66)[(\pi)(12.5/1000)^2/4]\sqrt{(2)(9.807)}}{\pi/4}}\left(-\frac{1}{-0.00900}\right)$$

$$= 0.2253y^{3/4} \qquad \text{(all units in meters)}$$

17.130 Determine the time required to raise the right-hand surface of Fig. 17-33 by 2 ft.

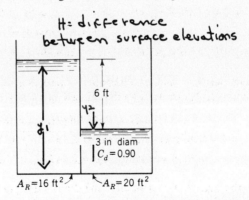

$H = $ difference between surface elevations

6 ft

y_2

3 in diam
$C_d = 0.90$

y_1

$A_R = 16$ ft^2 $A_R = 20$ ft^2 **Fig. 17-33**

$$A_1 y_1 + A_2 y_2 = C \qquad A_1\, dy_1 + A_2\, dy_2 = 0 \qquad H = y_1 - y_2$$

$$dH = dy_1 - dy_2 = -(1 + A_2/A_1)\, dy_2$$

$$dy_2 = -dH/(1 + A_2/A_1) \qquad Q\, dt = C_d A_0 \sqrt{2gH}\, dt = A_2\, dy_2 = -[A_2\, dH/(1 + A_2/A_1)]$$

Hence,

$$t = \frac{A_2}{C_d A_0 \sqrt{2g}(1 + A_2/A_1)} \int_{H_1}^{H_2} H^{-1/2}\, dH = \frac{2A_2(\sqrt{H_1} - \sqrt{H_2})}{C_d A_0 \sqrt{2g}(1 + A_2/A_1)} \qquad H_1 = 6 \text{ ft}$$

Rise of the right-hand surface by 2 ft will create a fall of the left-hand surface by $(2)(\frac{20}{16}) = 2.50$ ft. Thus, $H_2 = 6 - 2 - 2.5 = 1.5$ ft. $t = (2)(20)(\sqrt{6} - \sqrt{1.5})/\{(0.90)[(\pi)(\frac{3}{12})^2/4]\sqrt{(2)(32.2)}(1 + \frac{20}{16})\} = 61.4$ s.

17.131 How long does it take to raise the water surface of Fig. 17-34 by 2 m? The left-hand surface is a large reservoir of constant water-surface elevation.

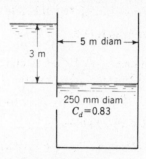

Fig. 17-34

▮ As in Prob. 17.130, $t = [2A_2(\sqrt{H_1} - \sqrt{H_2})]/[C_d A_0 \sqrt{2g}(1 + A_2/A_1)]$ where $A_1 = \infty$, $A_0 = (\pi)(\frac{250}{1000})^2/4 = 0.04909$ m², $t = [(2)(\pi 5^2/4)(\sqrt{3} - \sqrt{3 - 2})]/[(0.83)(0.04909)\sqrt{(2)(9.807)}(1 + 0)] = 159.3$ s.

17.132 Determine the discharge in a 300-mm-diameter line with a 160-mm-diameter VDI orifice for water at 20 °C when the gage difference is 300 mm on an acetylene tetrabromide (s.g. = 2.94, $v = 1.01 \times 10^{-6}$ m²/s)–water differential manometer.

▮ $$Q = CA_0\sqrt{2gR'(S_0/S_1 - 1)} \qquad A_0/A_1 = (\frac{160}{300})^2 = 0.284$$

From Fig. A-28, assume $C = 0.632$ (for high N_R).

$$Q = (0.632)[(\pi)(\frac{160}{1000})^2/4]\sqrt{(2)(9.807)(\frac{300}{1000})[(2.94/1) - 1]} = 0.0429 \text{ m}^3/\text{s}$$
$$N_R = DV/v = 4Q/\pi Dv = (4)(0.0429)/[(\pi)(\frac{300}{1000})(1.01 \times 10^{-6})] = 1.80 \times 10^5$$

From Fig. A-28, $C = 0.632$ as assumed is O.K.

17.133 A 10-mm-diameter VDI orifice is installed in a 25-mm-diameter pipe carrying nitrogen at $p_1 = 8$ atm, $T_1 = 50$ °C. For a pressure drop of 140 kPa across the orifice, calculate the mass flow rate.

▮ $$\dot{m} = CYA_0\sqrt{2\rho_1 \Delta p} \qquad A_0/A_1 = (\frac{10}{25})^2 = 0.1600$$

From Fig. A-28, assume $C = 0.63$.

$$k = 1.4 \qquad D_0/D_1 = 0.4 \qquad p_2/p_1 = [(8)(101.3) - 140]/[(8)(101.3)] = 0.827$$

From Fig. A-26, $Y = 0.95$.

$$\rho = p/RT = (8)(101.3)(1000)/[(297)(50 + 273)] = 8.448 \text{ kg/m}^3$$
$$\dot{m} = (0.63)(0.95)[(\pi)(\frac{10}{1000})^2/4]\sqrt{(2)(8.448)[(140)(1000)]} = 0.0723 \text{ kg/s} \qquad N_R = \rho DV/\mu$$
$$V = \dot{m}/\rho A = 0.0723/\{(8.448)[(\pi)(\frac{25}{1000})^2/4]\} = 17.43 \text{ m/s}$$
$$N_R = (8.448)(\frac{25}{1000})(17.43)/(2.15 \times 10^{-5}) = 1.71 \times 10^5$$

From Fig. A-28, $C = 0.61$. $\dot{m} = (0.61/0.63)(0.0723) = 0.0700$ kg/s.

17.134 Air at 1 atm, $T = 21$ °C, flows through a 1-m-square duct that contains a 500-mm-diameter square-edged orifice. With a head loss of 60 mm H_2O across the orifice, compute the flow in cubic meters per minute.

▮ $$\dot{m} = CYA_0\sqrt{2\rho \Delta p} \qquad A_0/A_1 = [(\pi)(\frac{500}{1000})^2/4]/[(1)(1)] = 0.196$$

From Fig. A-28, assume $C = 0.615$ (for high N_R).

$$p_2 = p_1 - \gamma h = 101\,300 - (9790)(\frac{60}{1000}) = 100\,700 \text{ Pa} \qquad p_2/p_1 = 100\,700/101\,300 = 0.994$$
$$D_0/D_1 = (\frac{500}{1000})/[(1)(1)] = 0.500 \qquad k = 1.4$$

From Fig. A-26, $Y = 0.995$.

$$\rho = p/RT = 101\,300/[(\tfrac{8312}{29})(21 + 273)] = 1.202 \text{ kg/m}^3$$
$$\dot{m} = (0.615)(0.995)[(\pi)(\tfrac{500}{1000})^2/4]\sqrt{(2)(1.202)(9790)(\tfrac{60}{1000})} = 4.515 \text{ kg/s}$$
$$Q = \dot{m}/\rho = 4.515/1.202 = 3.76 \text{ m}^3/\text{s} \quad \text{or} \quad 225 \text{ m}^3/\text{min}$$
$$N_R = \rho D_h V/\mu = (1.202)[(4)(1)]\{3.76/[(1)(1)]\}/(1.915 \times 10^{-5}) = 9.44 \times 10^5$$

From Fig. A-28, $C = 0.615$ as assumed is O.K.

17.135 A 6-in-diameter VDI orifice is installed in a 12-in-diameter oil line, $\mu = 6\,\text{cP}$, $\gamma = 52\,\text{lb/ft}^3$. An oil-air differential manometer is used. For a gage difference of 20 in determine the flow rate in gallons per minute.

$\blacksquare$ $\qquad Q = CA_0\sqrt{2gR'} \qquad A_0/A_1 = (\tfrac{6}{12})^2 = 0.250$

From Fig. A-28, assume $C = 0.625$ (for high N_R).

$$Q = (0.625)[(\pi)(\tfrac{6}{12})^2/4]\sqrt{(2)(32.2)(\tfrac{20}{12})} = 1.27 \text{ ft}^3/\text{s}$$
$$N_R = \rho DV/\mu = 4Q\rho/\pi D\mu = (4)(1.27)(52/32.2)/\{(\pi)(\tfrac{12}{12})[(\tfrac{6}{100})/479]\} = 2.08 \times 10^4$$

From Fig. A-28, $C = 0.63$ (O.K.) $Q = 1.27/0.002228 = 575$ gpm.

17.136 Solve Prob. 17.73 for an orifice with $D:\tfrac{1}{2}D$ taps (instead of a long-radius flow nozzle).

$\blacksquare$ A good initial guess for an orifice is $\alpha \approx 0.62$. From Eq. (2) of Prob. 17.73 compute $\beta \approx 0.447/(0.62)^{1/2} = 0.568$. From Fig. A-29 compute the discharge coefficient $C_d = 0.6064$ for $\beta = 0.568$ and $N_R = 392\,000$. Then $E = [1 - (0.568)^4]^{-1/2} = 1.0565$ and $\alpha = C_d E = 0.6407$. Iterate Eq. (2) again: $\beta = 0.447/(0.6407)^{1/2} = 0.558$, $C_d = 0.6061$, and $\alpha = 0.6378$. Stop. We have converged satisfactorily to a design value: $\beta \approx 0.56$, $d = \beta D = 112$ mm. The throat velocity is $V_t = V_1/\beta^2 = 2.0/(0.56)^2 = 6.38$ m/s. The throat head is $V_t^2/2g = (6.38)^2/[2(9.81)] = 2.07$ m. From Fig. A-24 for the orifice at $\beta = 0.56$, estimate $K_m \approx 1.7$. Then the nonrecoverable loss of the orifice will be $h_m = K_m(V_t^2/2g) \approx 1.7(2.07) = 3.5$ m.

17.137 Water at 20 °C in a 10-cm-diameter pipe flows through a 5-cm-diameter thin-plate orifice with $D:\tfrac{1}{2}D$ taps ($\beta = 0.5$). If the measured pressure drop is 65 kPa, what is the flow rate in cubic meters per hour?

$\blacksquare$ $Q = C_d A_t\sqrt{2\,\Delta p/[(\rho)(1 - \beta^4)]}$. Guess $C_d = 0.61$: $Q = (0.61)[(\pi)(\tfrac{5}{100})^2/4]\sqrt{(2)(65)(1000)/[(998)(1 - 0.5^4)]}$ $= 0.0141 \text{ m}^3/\text{s}$.

$$V = Q/A = 0.0141/[(\pi)(\tfrac{10}{100})^2/4] = 1.795 \text{ m/s}$$
$$N_R = \rho DV/\mu = (998)(\tfrac{10}{100})(1.795)/(1.02 \times 10^{-3}) = 1.76 \times 10^5$$

From Fig. A-29, $C_d = 0.605$. $Q = (0.605)[(\pi)(\tfrac{5}{100})^2/4]\sqrt{(2)(65)(1000)/[(998)(1 - 0.5^4)]} = 0.0140 \text{ m}^3/\text{s}$, or 50.4 m³/h.

17.138 Gasoline at 20 °C flows in a 5-cm-diameter pipe at 25 m³/h. If a 2-cm-diameter thin-plate orifice with corner taps is installed in the pipe, what will the measured pressure drop be? $\beta = 0.4$. Use $\rho = 680$ kg/m³ and $\mu = 0.000292$ kg/(m·s).

$\blacksquare$ $\qquad Q = C_d A_t\sqrt{2\,\Delta p/[(\rho)(1 - \beta^4)]} \qquad V = Q/A = (\tfrac{25}{3600})/[(\pi)(\tfrac{5}{100})^2/4] = 3.537 \text{ m/s}$

$$N_R = \rho DV/\mu = (680)(\tfrac{5}{100})(3.537)/0.000292 = 4.12 \times 10^5$$

For corner taps,

$$f(\beta) = 0.5959 + 0.0312\beta^{2.1} - 0.184\beta^8 = 0.5959 + (0.0312)(0.4)^{2.1} - (0.184)(0.4)^8 = 0.6003$$
$$C_d = f(\beta) + 91.71\beta^{2.5}N_R^{-0.75} = 0.6003 + (91.71)(0.4)^{2.5}(4.12 \times 10^5)^{-0.75} = 0.6009$$
$$\tfrac{25}{3600} = (0.6009)[(\pi)(\tfrac{2}{100})^2/4]\sqrt{2\,\Delta p/[(680)(1 - 0.4^4)]} \qquad \Delta p = 448\,000 \text{ Pa} \quad \text{or} \quad 448 \text{ kPa}$$

17.139 A 1-m-diameter cylindrical tank in Fig. 17-35 is initially filled with kerosene at 20 °C to a depth of 2 m. There is a 3-cm-diameter thin-plate orifice in the bottom. When the orifice is opened, what will the initial flow rate be in cubic meters per hour?

$\blacksquare$ $Q = C_d A_t\sqrt{2\,\Delta p/[(\rho)(1 - \beta^4)]}$, Δp across the orifice is $\rho gh(t)$. The orifice simulates "corner taps" as $\beta \to 0$, so $C_d \approx 0.596$. $Q_0 = C_d A_t\sqrt{2\rho g h_0/[(\rho)(1 - \beta^4)]} = (0.596)[(\pi)(\tfrac{3}{100})^2/4]\sqrt{(2)(\rho)(9.807)(2)/[(\rho)(1 - 0^4)]}$ $= 0.002639 \text{ m}^3/\text{s}$, or 9.50 m³/h (at $t = 0$).

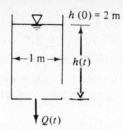

Fig. 17-35

17.140 For the conditions given in Prob. 17.139, how long will it take $h(t)$ to drop from 2 m to 1.4 m?

▮ $Q = -dV_{\text{tank}}/dt = -(A_0)(dh/dt)$. From Prob. 17.139,

$$Q_0 = (0.596)[(\pi)(\tfrac{3}{100})^2/4]\sqrt{(2)(\rho)(9.807)(h)/[(\rho)(1-0^4)]} = 0.001866h^{1/2}$$

$$A_0 = (\pi)(1)^2/4 = 0.7854 \text{ m}^2 \qquad (-0.7854)(dh/dt) = 0.001866h^{1/2}$$

$$\int_{2.0}^{1.5} \frac{dh}{h^{1/2}} = -\left(\frac{0.001866}{0.7854}\right)\int_0^t dt \qquad [2h^{1/2}]_{2.0}^{1.5} = -0.002376[t]_0^t$$

$$-0.002376t = (2)(1.5^{1/2} - 2.0^{1/2}) \qquad t = 159 \text{ s}$$

17.141 Water flowing in a 12-in-diameter pipe passes through a Venturi meter with a 6-in throat diameter. The measured pressure head differential is 150.5 in of water. Assuming a discharge coefficient of 0.98, find the flow rate of water through the pipe.

▮

$$Q = CA_2\sqrt{[2g(p_1 - p_2)/\gamma]/[1 - (A_2/A_1)^2]} \qquad A_2 = (\pi)(\tfrac{6}{12})^2/4 = 0.1963 \text{ ft}^2$$

$$A_1 = (\pi)(\tfrac{12}{12})^2/4 = 0.7854 \text{ ft}^2$$

$$Q = (0.98)(0.1963)\sqrt{[(2)(32.2)(150.5/12)]/[1 - (0.1963/0.7854)^2]} = 5.65 \text{ ft}^3/\text{s}$$

17.142 Oil is flowing upward through a Venturi meter as shown in Fig. 17-36. Assume a discharge coefficient of 0.984. What is the rate of flow of the oil?

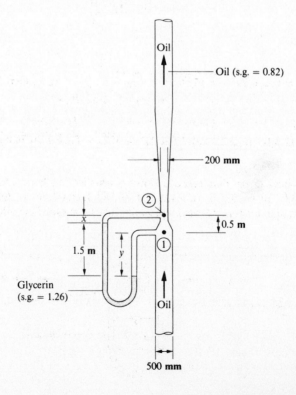

Fig. 17-36

$$Q = CA_2\sqrt{[2g(p_1 - p_2)/\gamma]/[1 - (A_2/A_1)^2]} \qquad A_2 = (\pi)(\tfrac{200}{1000})^2/4 = 0.03142 \text{ m}^2$$
$$A_1 = (\pi)(\tfrac{500}{1000})^2/4 = 0.1963 \text{ m}^2$$
$$p_1/\gamma + y - (1.26/0.82)(1.5) - x = p_2/\gamma \qquad x + 1.5 = 0.5 + y \qquad x = y - 1.0$$
$$p_1/\gamma + y - (1.26/0.82)(1.5) - (y - 1.0) = p_2/\gamma \qquad (p_1 - p_2)/\gamma = 1.305 \text{ m of oil}$$
$$Q = (0.984)(0.03142)\sqrt{(2)(9.807)(1.305)/[1 - (0.03142/0.1963)^2]} = 0.158 \text{ m}^3/\text{s}$$

17.143 Carbon dioxide at 20 °C flowing in a 200-mm-diameter pipe passes through a Venturi meter with a 100-mm throat diameter. Measured (gage) pressures are $p_1 = 22.0$ kPa and $p_2 = 5.2$ kPa. Atmospheric pressure is 762 mmHg, and the discharge coefficient is 0.98. Find the weight flow rate of carbon dioxide through the pipe.

$$G = C\gamma_2 A_2\sqrt{[2gk/(k - 1)](p_1/\gamma_1)[1 - (p_2/p_1)^{(k-1)/k}]/[1 - (A_2/A_1)^2(p_2/p_1)^{2/k}]} \qquad \gamma = p/RT$$
$$p_2 = 5.2 + (0.762)(9.79)(13.6) = 106.7 \text{ kPa} \qquad \gamma_2 = 106.7/[(19.3)(20 + 273)] = 0.01887 \text{ kN/m}^3$$
$$p_1 = 22.0 + (0.762)(9.79)(13.6) = 123.5 \text{ kPa} \qquad \gamma_1 = 123.5/[(19.3)(20 + 273)] = 0.02184 \text{ kN/m}^3$$
$$A_2 = (\pi)(\tfrac{100}{1000})^2/4 = 0.007854 \text{ m}^2 \qquad A_1 = (\pi)(\tfrac{200}{1000})^2/4 = 0.03142 \text{ m}^2$$
$$G = (0.98)(0.01887)(0.007854)$$
$$\times \sqrt{\frac{[(2)(9.807)(1.30)/(1.30 - 1)](123.5/0.02184)[1 - (106.7/123.5)^{(1.30-1)/1.30}]}{1 - (0.007854/0.03142)^2(106.7/123.5)^{2/1.30}}} = 0.0188 \text{ kN/s}$$

17.144 Water flows through a Venturi meter, as shown in Fig. 17-37. Determine the discharge coefficient of the Venturi meter if the discharge is determined to be 2.12 cfs.

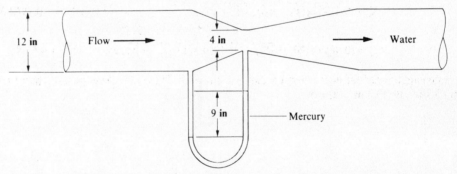

Fig. 17-37

$$Q = CA_2\sqrt{[2g(p_1 - p_2)/\gamma]/[1 - (A_2/A_1)^2]} \qquad A_2 = (\pi)(\tfrac{4}{12})^2/4 = 0.08727 \text{ ft}^2$$
$$A_1 = (\pi)(\tfrac{12}{12})^2/4 = 0.7854 \text{ ft}^2$$
$$p_1/\gamma + y + \tfrac{9}{12} - (13.6)(\tfrac{9}{12}) - y = p_2/\gamma \qquad (p_1 - p_2)/\gamma = 9.45 \text{ ft of water}$$
$$2.12 = (C)(0.08727)\sqrt{[(2)(32.2)(9.45)]/[1 - (0.08727/0.7854)^2]} \qquad C = 0.979$$

17.145 A Venturi meter having a throat diameter of 150 mm is installed in a horizontal 300-mm-diameter water main, as shown in Fig. 17-38. The coefficient of discharge is 0.982. Determine the difference in level of the mercury columns of the differential manometer attached to the Venturi meter if the discharge is 0.142 m³/s.

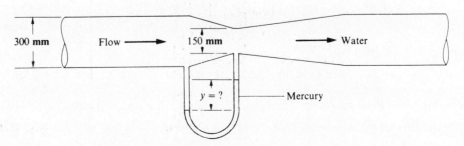

Fig. 17-38

▮
$$Q = CA_2\sqrt{[2g(p_1 - p_2)/\gamma]/[1 - (A_2/A_1)^2]} \qquad A_2 = (\pi)(0.150)^2/4 = 0.01767 \text{ m}^2$$
$$A_1 = (\pi)(0.300)^2/4 = 0.07069 \text{ m}^2$$
$$0.142 = (0.982)(0.01767)\sqrt{[(2)(9.807)(p_1 - p_2)/\gamma]/[1 - (0.01767/0.07069)^2]}$$
$$(p_1 - p_2)/\gamma = 3.201 \text{ m}$$
$$p_1/\gamma + x + y - 13.6y - x = p_2/\gamma \qquad (p_1 - p_2)/\gamma = 12.6y = 3.201 \qquad y = 0.254 \text{ m} = 254 \text{ mm}$$

17.146 Determine the weight flow rate when air at 20 °C and 700 kPa abs flows through a Venturi meter if the pressure at the throat of the meter is 400 kPa abs. The diameters at inlet and throat are 25 cm and 12.5 cm, respectively. Assume that $C = 0.985$.

▮
$$G = CYA_2\sqrt{2g\gamma_1\{(p_1 - p_2)/[1 - (D_2/D_1)^4]\}}$$
$$p_2/p_1 = \tfrac{400}{700} = 0.57 \qquad D_2/D_1 = 12.5/25 = 0.50 \qquad Y = 0.72 \qquad \text{(from Fig. A-26)}$$
$$\gamma = p/RT \qquad \gamma_1 = 700/[(29.3)(20 + 273)] = 0.08154 \text{ kN/m}^3$$
$$G = (0.985)(0.72)[(\pi)(12.5/100)^2/4]\sqrt{(2)(9.807)(0.08154)\{(700 - 400)/[1 - (12.5/25)^4]\}} = 0.197 \text{ kN/s}$$

17.147 Find the flow rate of water at 72 °F for the Venturi tube of Fig. 17-39 if the mercury manometer reads $y = 4$ in for the case where $D_1 = 8$ in, $D_2 = 4$ in, and $\Delta z = 1.5$ ft. Assume the discharge coefficients of Fig. A-30 are applicable. In addition, find the head loss from inlet to throat. Also, determine the total head loss across the meter. Assume a diverging cone angle of 10°.

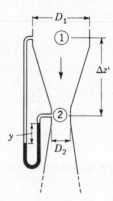

Fig. 17-39

▮
$$Q = \{CA_2/[1 - (D_2/D_1)^4]^{1/2}\}[2g \, \Delta(z + p/\gamma)]^{1/2}$$
$$(z_1 + p_1/\gamma) - (z_2 + p_2/\gamma) = (R')(S_M/S_F - 1) = (\tfrac{4}{12})[(13.6/1) - 1] = 4.20 \text{ ft}$$
$$A_2 = (\pi)(\tfrac{4}{12})^2/4 = 0.08727 \text{ ft}^2$$

From Fig. A-30, assume $C = 0.95$.
$$Q = \{(0.95)(0.08727)/[1 - (\tfrac{4}{8})^4]^{1/2}\}[(2)(32.2)(4.20)]^{1/2} = 1.41 \text{ ft}^3/\text{s}$$
$$V_2 = Q/A_2 = 1.41/0.08727 = 16.2 \text{ ft/s} \qquad D''V_2 = (4)(16.2) = 64.8$$

From Fig. A-30, $C = 0.985$. Since the curve is flat in this area, $C = 0.985$ is O.K.
$$Q = (0.985/0.95)(1.41) = 1.46 \text{ ft}^3/\text{s} \qquad V_2 = 1.46/0.08727 = 16.7 \text{ ft/s}$$
$$V_3 = V_2/4 = 16.7/4 = 4.18 \text{ ft/s}$$
$$(h_L)_{1-2} = (1/C^2 - 1)[1 - (D_2/D_1)^4](V_2^2/2g) = [(1/0.985^2) - 1][1 - (\tfrac{4}{8})^4]\{16.7^2/[(2)(32.2)]\} = 0.125 \text{ ft}$$
$$(h_L)_{2-3} = k'(V_2 - V_3)^2/2g$$

From Fig. A-31, $k' = 0.175$.
$$(h_L)_{2-3} = (0.175)(16.7 - 4.18)^2/[(2)(32.2)] = 0.426 \text{ ft} \qquad (h_L)_{\text{total}} = 0.125 + 0.426 = 0.551 \text{ ft}$$

17.148 Repeat Prob. 17.147 for the case of a horizontal Venturi tube (i.e., $\Delta z = 0$) with all other data the same.

▮ The answers are the same as for Prob. 17.147, since $R'(S_M/S_F - 1) = \Delta(p/\gamma + z)$.

17.149 Refer to Prob. 17.147. What would be the flow rate if the manometer fluid was carbon tetrachloride (s.g. = 1.59) with all other data remaining the same?

$$Q = \{CA_2/[1 - (D_2/D_1)^4]^{1/2}\}[2g\,\Delta(z + p/\gamma)]^{1/2}$$

Assume $C = 0.95$ (see Fig. A-30).

$$A_2 = 0.08727 \text{ ft}^2 \quad \text{(from Prob. 17.147)} \quad \Delta(p/\gamma + z) = (\tfrac{4}{12})[(1.59/1) - 1] = 0.1967 \text{ ft}$$
$$Q = \{(0.95)(0.08727)/[1 - (\tfrac{4}{8})^4]^{1/2}\}[(2)(32.2)(0.1967)]^{1/2} = 0.305 \text{ ft}^3/\text{s}$$
$$V_2 = Q/A_2 = 0.305/0.08727 = 3.49 \text{ ft/s} \quad D''V_2 = (4)(3.49) = 13.96$$

From Fig. A-30, $C = 0.985$.

$$Q = (0.985/0.95)(0.305) = 0.316 \text{ ft}^3/\text{s}$$

17.150 Find the flow rate of water for the Venturi tube shown in Fig. 17-39 if $D_1 = 80$ cm, $D_2 = 40$ cm, $\Delta z = 200$ cm, and $y = 15$ cm of mercury.

$$Q = \{CA_2/[1 - (D_2/D_1)^4]^{1/2}\}\sqrt{2gy(S_0/S_1 - 1)}. \text{ Try } C = 0.985.$$
$$Q = \{(0.985)[(\pi)(\tfrac{40}{100})^2/4]/[1 - (\tfrac{40}{80})^4]^{1/2}\}\sqrt{(2)(9.807)(\tfrac{15}{100})[(13.6/1) - 1]} = 0.778 \text{ m}^3/\text{s}$$
$$V_2 = Q/A_2 = 0.778/[(\pi)(\tfrac{40}{100})^2/4] = 6.19 \text{ m/s} \quad N_R = DV/v = (\tfrac{40}{100})(6.19)/10^{-6} = 2.48 \times 10^6$$

From Fig. A-30, $C = 0.985$. Hence, $Q = 0.778 \text{ m}^3/\text{s}$.

17.151 In Fig. 17-39 suppose $D_1 = 2$ in, $D_2 = 1$ in, and $\Delta z = 6$ in. Oil (s.g. = 0.90) with a kinematic viscosity of 0.0005 ft²/s is flowing. Determine the manometer reading y if mercury is used as the manometer fluid. The rate of flow is 0.10 cfs.

$$Q = \{CA_2/[1 - (D_2/D_1)^4]^{1/2}\}\sqrt{2gy(S_0/S - 1)}. \text{ Try } C = 0.96:$$
$$0.10 = \{(0.96)[(\pi)(\tfrac{1}{12})^2/4]/[1 - (\tfrac{1}{2})^4]^{1/2}\}\sqrt{(2)(32.2)(y)[(13.6/0.90) - 1]}$$
$$y = 0.3763 \text{ ft} \quad \text{or} \quad 4.52 \text{ in} \quad V = Q/A = 0.10/[(\pi)(\tfrac{1}{12})^2/4] = 18.33 \text{ ft/s}$$
$$N_R = DV/v = (\tfrac{1}{12})(18.33)/0.0005 = 3055$$

From Fig. A-30, $C = 0.91$. $y = (0.91/0.96)^2(4.52) = 4.06 \text{ in}$.

17.152 Assume that air at 70 °F and 100 psia flows through a Venturi tube and that the pressure at the throat is 60 psia. The inlet area is 0.60 ft², and the throat area is 0.15 ft². The tube coefficient is 0.98, and k is 1.4. Find the rate of discharge.

$$G = CA_t\sqrt{2g[k/(k-1)]p_1\gamma_1(p_2/p_1)^{2/k}\{[1 - (p_2/p_1)^{(k-1)/k}]/[1 - (A_t/A)^2(p_2/p_1)^{2/k}]\}}$$
$$p_1 = (100)(144) = 14\,400 \text{ lb/ft}^2 \quad \gamma = p/RT$$
$$\gamma_1 = 14\,400/[(53.3)(70 + 460)] = 0.5098 \text{ lb/ft}^3$$
$$G = (0.98)(0.15)\sqrt{(2)(32.2)\left(\frac{1.4}{1.4-1}\right)(14\,400)(0.5098)(\tfrac{60}{100})^{2/1.4}\left[\frac{1 - (\tfrac{60}{100})^{(1.4-1)/1.4}}{1 - (0.15/0.60)^2(\tfrac{60}{100})^{2/1.4}}\right]} = 49.1 \text{ lb/s}$$

17.153 Solve Prob. 17.152 by evaluating Y from Fig. A-26 and using it to find the rate of discharge.

$$G = CYA_t\sqrt{2g\gamma_1\{(p_1 - p_2)/[1 - (D_2/D_1)^4]\}}$$

For $D_2/D_1 = 0.5$ and $p_2/p_1 = \tfrac{60}{100} = 0.6$, from Fig. A-26, $Y = 0.74$.

$$G = (0.98)(0.74)(0.15)\sqrt{(2)(32.2)(0.5098)\{[(100 - 60)(144)]/(1 - 0.5^4)\}} = 48.9 \text{ lb/s}$$

17.154 What would be the value of Y and the rate of discharge for a square-edged orifice for the same data as in Prob. 17.152, assuming $C = 0.60$?

$$G = CYA_t\sqrt{2g\gamma_1\{(p_1 - p_2)/[1 - (D_0/D_1)^4]\}}$$

With $D_0/D_1 = 0.5$ and $p_2/p_1 = 0.6$, $Y = 0.875$ (from Fig. A-26).

$$G = (0.60)(0.875)(0.15)\sqrt{(2)(32.2)(0.5098)[(100 - 60)(144)/(1 - 0.5^4)]} = 35.4 \text{ lb/s}$$

17.155 What is the value of the throat velocity in Prob. 17.152?

$$V^2/2 = C_p T_1[1 - (p_2/p_1)^{(k-1)/k}] = (6000)(70 + 460)[1 - (\tfrac{60}{100})^{(1.40-1)/1.40}] \qquad V = 929 \text{ ft/s}$$

17.156 Air flows through a 15-cm by 7.5-cm Venturi meter. At the inlet the air temperature is 15 °C and the pressure is 140 kPa. Determine the flow rate if a mercury manometer reads 15 cm. Assume an atmospheric pressure of 101.3 kPa abs.

∎ $G = CYA_t\sqrt{2g\gamma_1\{(p_1 - p_2)/[1 - (D_0/D_1)^4]\}}$. From Fig. A-30, $C = 0.98$.

$$\gamma = p/RT \qquad \gamma_1 = (140 + 101.3)(1000)/[(29.3)(15 + 273)] = 28.60 \text{ N/m}^3$$
$$p_1 - p_2 = \gamma h = [(13.6)(9.79)](\tfrac{15}{100}) = 20.0 \text{ kN/m}^2$$
$$p_1 = 140 + 101.3 = 241.3 \text{ kN/m}^2 \qquad p_2 = 241.3 - 20.0 = 221.3 \text{ kN/m}^2$$

For $D_2/D_1 = 0.5$ and $p_2/p_1 = 221.3/241.3 = 0.917$, from Fig. A-26, $Y = 0.95$. $G = (0.98)(0.95)[(\pi)(0.075/2)^2]\sqrt{(2)(9.807)(28.60)[(20.0)(1000)/(1 - 0.5^4)]} = 14.2 \text{ N/s}$.

17.157 Natural gas, for which $k = 1.3$ and $R = 3100$ ft-lb/(slug-°R), flows through a Venturi tube with pipe and throat diameters of 12 in and 6 in, respectively. The initial pressure of the gas is 150 psia, and its temperature is 60 °F. If the meter coefficient is 0.98, find the rate of flow for a throat pressure of 100 psia.

∎ $G = CA_t\sqrt{2g[k/(k-1)]p_1\gamma_1(p_2/p_1)^{2/k}\{[1 - (p_2/p_1)^{(k-1)/k}]/[1 - (A_2/A_1)^2(p_2/p_1)^{2/k}]\}}$

$$A_t = (\pi)(\tfrac{6}{12})^2/4 = 0.1963 \text{ ft}^2 \qquad \gamma = p/RT$$
$$\gamma_1 = (150)(144)/[(3100/32.2)(60 + 460)] = 0.4315 \text{ lb/ft}^3$$

$$G = (0.98)(0.1963)$$
$$\times \sqrt{(2)(32.2)[1.3/(1.3 - 1)][(150)(144)](0.4315)(\tfrac{100}{150})^{2/1.3}\{[1 - (\tfrac{100}{150})^{(1.3-1)/1.3}]/[1 - (6^2/12^2)^2(\tfrac{100}{150})^{2/1.3}]\}}$$
$$= 69.0 \text{ lb/s}$$

17.158 Referring to Fig. 17-40, assume that liquid flows from A to C at the rate of 200 L/s and that the friction loss between A and B is negligible but that between B and C it is $0.1V_B^2/2g$. Find the pressure heads at A and C.

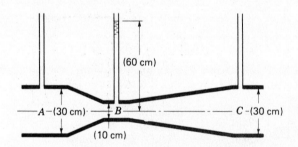

(60 cm)

A-(30 cm) ── B ·········· C -(30 cm)

(10 cm)   **Fig. 17-40**

∎
$$p_A/\gamma + V_A^2/2g + z_A = p_B/\gamma + V_B^2/2g + z_B + h_L$$
$$V_A = Q/A_A = 0.20/[(\pi)(\tfrac{30}{100})^2/4] = 2.83 \text{ m/s} \qquad V_B = 0.20/[(\pi)(\tfrac{10}{100})^2/4] = 25.5 \text{ m/s}$$
$$p_A/\gamma + 2.83^2/[(2)(9.807)] + 0 = \tfrac{60}{100} + 25.5^2/[(2)(9.807)] + 0 + 0 \qquad p_A/\gamma = 33.3 \text{ m}$$
$$p_B/\gamma + V_B^2/2g + z_B = p_C/\gamma + V_C^2/2g + z_C + h_L$$
$$\tfrac{60}{100} + 25.5^2/[(2)(9.807)] + 0 = p_C/\gamma + 2.83^2/[(2)(9.807)] + 0 + 0 + (0.1)\{25.5^2/[(2)(9.807)]\}$$
$$p_C/\gamma = 30.0 \text{ m}$$

17.159 Water flows through a 12-in by 6-in Venturi meter at the rate of 1.49 cfs and the differential gage is deflected 3.50 ft, as shown in Fig. 17-41. The specific gravity of the gage liquid is 1.25. Determine the coefficient of the meter.

∎ Applying the Bernoulli equation, A to B, ideal case, yields $(p_A/\gamma + V_{12}^2/2g + 0)$ − no lost head = $(p_B/\gamma + V_6^2/2g + 0)$ and $V_{12}^2 = (A_6/A_{12})^2 V_6^2$. Solving, $V_6 = \sqrt{2g(p_A/\gamma - p_B/\gamma)/[1 - (A_6/A_{12})^2]}$ (no lost head).

The true velocity (and hence the true value of flow Q) will be obtained by multiplying the ideal value by the coefficient C of the meter. Thus

$$Q = A_6 V_6 = A_6 C\sqrt{2g(p_A/\gamma - p_B/\gamma)/[1 - (A_6/A_{12})^2]} \qquad\qquad (1)$$

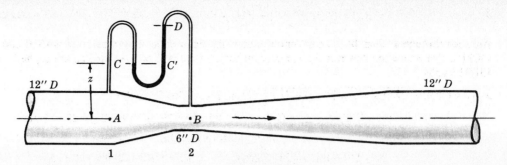

Fig. 17-41

To obtain the differential pressure head above, the principles of the differential gage must be used. $p_C = p_{C'}$, $(p_A/\gamma - z) = p_B/\gamma - (z + 3.50) + 1.25(3.50)$ or $p_A/\gamma - p_B/\gamma = 0.875$ ft.

Substituting in (1), $1.49 = \frac{1}{4}\pi(\frac{1}{2})^2 C\sqrt{2g(0.875)}/(1 - \frac{1}{16})$ and $C = 0.978$ (use 0.98).

17.160 Water flows upward through a vertical 12-in by 6-in Venturi meter whose coefficient is 0.980. The differential gage deflection is 3.88 ft of liquid of specific gravity 1.25, as shown in Fig. 17-42. Determine the flow in cfs.

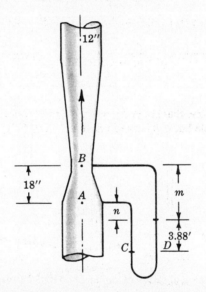

Fig. 17-42

▌ Reference to the Bernoulli equation in Prob. 17.159 indicates that, for this problem, $z_A = 0$ and $z_B = 1.50$ ft. Then $Q = CA_6\sqrt{2g[(p_A/\gamma - p_B/\gamma) - 1.50]/[1 - (\frac{1}{2})^4]}$.

Using the principles of the differential gage to obtain $\Delta p/\gamma$,

$$p_C/\gamma = p_D/\gamma \quad \text{(ft of water units)}$$
$$p_A/\gamma + (n + 3.88) = p_B/\gamma + m + 1.25(3.88)$$
$$(p_A/\gamma - p_B/\gamma) - (m - n) = 3.88(1.25 - 1.00)$$
$$(p_A/\gamma - p_B/\gamma) - 1.50 = 0.97 \text{ ft of water}$$

Substituting in the equation for flow, $Q = 0.980(\frac{1}{4}\pi)(\frac{1}{2})^2\sqrt{2g(0.97)/(1 - \frac{1}{16})} = 1.57$ cfs.

17.161 An 8-in by 4-in Venturi meter is used to measure the flow of carbon dioxide at 68 °F. The deflection of the water column in the differential gage is 71.8 in and the barometer reads 30.0 in of mercury. For a pressure at entrance of 18.0 psi abs, calculate the weight flow.

▌ The absolute pressure at entrance $= p_1 = (18.0)(144) = 2590$ psf abs and the specific weight γ_1 of the carbon dioxide is $\gamma_1 = 2590/[34.9(460 + 68)] = 0.1405$ lb/ft³. The pressure difference $= (71.8/12)(62.4 - 0.141) = 372$ psf and hence the absolute pressure at the throat $= p_2 = 2590 - 372 = 2220$ psf abs.

To obtain the specific weight γ_2, we use $p_2/p_1 = \frac{2220}{2590} = 0.858$ and $\gamma_2/\gamma_1 = (0.858)^{1/k}$. Thus $\gamma_2 = (0.1405)(0.858)^{1/1.30} = 0.1250$ lb/ft³. $W = \gamma_1 K A_2 Y\sqrt{2g(\Delta p/\gamma_1)}$ in lb/s. Using $k = 1.30$, $d_2/d_1 = 0.50$ and $p_2/p_1 =$

0.858, Y (Table A-18) = 0.909 by interpolation. Assuming $C = 0.985$, from Fig. A-32, and noting that $K = 1.032$, we have $W = (0.1405)[(1.032)(0.985)][\frac{1}{4}\pi(\frac{4}{12})^2](0.909)\sqrt{2g}(372/0.1405) = 4.67$ lb/s.

To check the assumed value of C, determine the Reynolds number and use the appropriate curve on Fig. A-32. $N_R = 4W/\pi d_2 v\gamma_2 = 4(4.67)/\{\pi(\frac{4}{12})[(9.1)(14.7)/18.0](10^{-5})](0.1250)\} = 1.92 \times 10^6$.

From Fig. A-32, $C = 0.984$. Recalculating, $W = 4.66$ lb/s.

17.162 A 10-cm-diameter pipe has a 5-cm-diameter, long constriction. If the mercury manometer indicates a column-height difference of 20 cm, estimate the flow rate of water passed by the pipe.

▌ We assume that the streamlines are parallel at locations 1, 2, and 3 (Fig. 17-43). Applying the Bernoulli equation with mechanical-energy loss for the sudden contraction between points 1 and 2 gives $p_1/\rho g + \bar{V}_1^2/2g - K_{SC}(\bar{V}_2^2/2g) = p_2/\rho g + \bar{V}_2^2/2g$. The Bernoulli equation with mechanical-energy loss for the sudden enlargement between points 2 and 3 gives $p_2/\rho g + \bar{V}_2^2/2g - K_{SE}(\bar{V}_2^2/2g) = p_3/\rho g + \bar{V}_3^2/2g$. The steady incompressible continuity equation gives $\bar{V}_1 = \bar{V}_3 = (d/D)^2\bar{V}_2$. The pressure change indicated by the manometer is $p_1 - p_3 = (p_1 - p_2) + (p_2 - p_3)$. Using the Bernoulli equation to eliminate the pressure differences $p_1 - p_2$ and $p_2 - p_3$ and continuity to eliminate $\bar{V}_1$ and $\bar{V}_3$, we get $p_1 - p_3 = (K_{SC} + K_{SE})\frac{1}{2}\rho\bar{V}_2^2$. But $\dot{Q} = (\pi d^2/4)\bar{V}_2$ and $p_1 - p_3 = \rho g(\text{s.g.}_{Hg} - 1)h$. Then $\dot{Q} = (\pi d^2/4)\sqrt{2gh(\text{s.g.}_{Hg} - 1)/(K_{SC} + K_{SE})}$. From Fig. A-33, $K_{SC} = 0.478$ and from $K_{SE} = (1 - A_1/A_2)^2$, $K_{SE} = (1 - 0.25)^2 = 0.5625$. Then $\dot{Q} = (\pi/4)(5)^2\sqrt{2(980)(20)(13.6 - 1)/(0.478 + 0.5625)}(\frac{1}{1000})$ = 13.5 L/s.

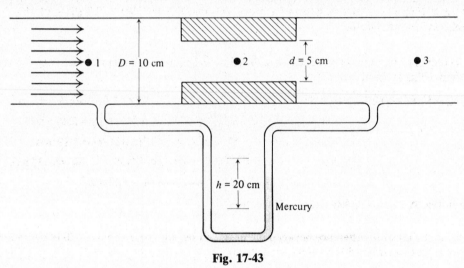

Fig. 17-43

17.163 Find q for flow of water in a pipe of inside diameter D of 100 mm using a Venturi meter with a machined convergent section. The throat diameter d of the Venturi meter is 60 mm. The Venturi meter is in a pipe section having an inclination θ (see Fig. 17-44) of 45°. The distance from 1 to 2 in the meter is 120 mm and the value of h for the manometer is 140 mm. Take $\rho = 999$ kg/m^3 and $\mu = 1.12 \times 10^{-3}$ kg/(m-s).

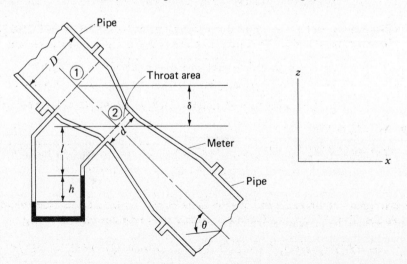

Fig. 17-44

❚ We note that $(C_d)_{ven}$ for this case is 0.995. We need next the value of $p_1 - p_2$. This is easily determined as follows from *manometry:* $p_1 + \gamma_{H_2O}(\delta + l + h) = p_2 + l\gamma_{H_2O} + h\gamma_{Hg}$, $p_1 - p_2 = h(\gamma_{Hg} - \gamma_{H_2O}) - \gamma_{H_2O}\delta = (0.140)(13.6 - 1)(999)(9.81) - (999)(9.81)(0.120)(0.707) = 16\,456$ Pa. We now compute V_2 in $q_{act} = (C_d)_{ven}A_2(2\{[(p_1 - p_2)/\rho] + (z_1 - z_2)g\}/[1 - (A_2/A_1)^2])^{1/2}$ by deleting A_2 in the formula. Thus, $V_2 = 0.995(2\{[16\,456/999] + (0.120)(0.707)(9.81)\}/[1 - (\frac{60}{100})^4]\})^{1/2}$ where we used diameters squared in place of areas. We get from the last equation and continuity $V_2 = 6.27$ m/s, $V_1 = (0.6^2)(6.27) = 2.26$ m/s. We must next compute N_R for the pipe: $N_R = \rho V_1 D/\mu = [(999)(2.26)(0.100)/(1.12 \times 10^{-3})] = 2.02 \times 10^5$. We see that the Reynolds number is within the proper range for the $(C_d)_{ven}$ we have used and so we have for q: $q = (V_2)(A_2) = (6.27)[(\pi)(0.060^2)/4] = 0.01773$ m³/s.

17.164 Consider in Prob. 17.163 that we have a flow of air at a temperature of 40 °C and a pressure of 200 kPa abs. The pipe as before is 100 mm in inside diameter, but the diameter in the Venturi is now 20 mm. The height h is 140 mmHg. What is the mass flow per hour?

❚ We can use the same C_d as in incompressible flow, which is 0.995. We next determine Y, so we can use $\dot{m} = \rho_1 C_d A_3\{2[(p_1 - p_2)/\rho_1]/[1 - (A_2/A_1)^2]\}^{1/2}Y$. We need p_2 for this purpose. Thus, $p_2 = p_1 - \Delta p = 200\,000 - (0.140)(9806)(13.6) = 181.3$ kPa. The ratio p_2/p_1 is then $p_2/p_1 = 181.3/200 = 0.907$. Noting that $A_2/A_1 = (d/D)^2 = (0.020/0.100)^2 = 0.04$, we go to Fig. A-34 to get Y: $Y = 0.95$. Finally, we need ρ_1. Using the equation of state of a perfect gas, $\rho_1 = p_1/RT_1 = 200\,000/[(287)(313)] = 2.2264$ kg/m³, $\dot{m} = \rho_1 C_d A_2\{2[(p_1 - p_2)/\rho_1]/[1 - (A_2/A_1)^2]\}^{1/2}Y = (2.2264)(0.995)[(\pi)(0.020^2)/4]\{2[(200\,000 - 181\,300)/2.2264]/[1 - (0.04^2)]\}^{1/2}(0.95) = 0.0858$ kg/s = 308.7 kg/h.

17.165 A 4-m by 2-m Venturi meter carries water at 25 °C. A water-air differential manometer has a gage difference of 60 mm. What is the discharge?

❚ $Q = C_v A_2\sqrt{2g(\Delta p/\gamma)/[1 - (D_2/D_1)^4]}$. Assume $N_R > 3 \times 10^5$. From Fig. A-35, $C_v = 0.983$.

$$Q = (0.983)[(\pi)(2)^2/4]\sqrt{(2)(9.807)(\tfrac{60}{1000})/[1 - (\tfrac{2}{4})^4]} = 3.46 \text{ m}^3/\text{s}$$

$$N_R = D_1 V_1/\nu \qquad V_2 = Q/A_2 = 3.46/[(\pi)(2)^2/4] = 1.10 \text{ m/s}$$

$$A_1 V_1 = A_2 V_2 \qquad [(\pi)(4)^2/4](V_1) = [(\pi)(2)^2/4](1.10) \qquad V_1 = 0.275 \text{ m/s}$$

$$N_R = (4)(0.275)/(9.02 \times 10^{-7}) = 1.22 \times 10^6$$

From Fig. A-35, $C_v = 0.983$. Hence, $Q = 3.46$ m³/s.

17.166 What is the pressure difference between the upstream section and throat of a 150-mm by 75-mm horizontal Venturi meter carrying 50 L/s of water at 48 °C?

❚
$$Q = C_v A_2\sqrt{2g(\Delta p/\gamma)/[1 - (D_2/D_1)^4]} \qquad V_1 = Q/A_1 = (\tfrac{50}{1000})/[(\pi)(\tfrac{150}{1000})^2/4] = 2.829 \text{ m/s}$$
$$N_R = V_1 D_1/\nu = (2.829)(\tfrac{150}{1000})/(5.70 \times 10^{-7}) = 7.44 \times 10^5$$

From Fig. A-35, $C_v = 0.984$.

$$\tfrac{50}{1000} = (0.984)[(\pi)(\tfrac{75}{1000})^2/4]\sqrt{(2)(9.807)(\Delta p/9.79)/[1 - (\tfrac{75}{150})^4]} \qquad \Delta p = 61.9 \text{ kN/m}^2$$

17.167 A 12-in by 6-in Venturi meter is mounted in a vertical pipe with the flow upward. 2500 gpm of oil, s.g. = 0.80, $\mu = 0.1$ P, flows through the pipe. The throat section is 6 in above the upstream section. What is $p_1 - p_2$?

❚
$$Q = C_v A_2\sqrt{2g[-h + (p_1 - p_2)/\gamma]/[1 - (D_2/D_1)^4]}$$
$$V_1 = Q/A_1 = (2500)(0.002228)/[(\pi)(\tfrac{12}{12})^2/4] = 7.092 \text{ ft/s}$$
$$N_R = \rho D_1 V_1/\mu = [(0.80)(1.94)](\tfrac{12}{12})(7.092)/(0.1/479) = 5.27 \times 10^4$$

From Fig. A-35, $C_v = 0.962$.

$$(2500)(0.002228) = (0.962)[(\pi)(\tfrac{6}{12})^2/4]\sqrt{(2)(32.2)\{-\tfrac{6}{12} + (p_1 - p_2)/[(0.80)(62.4)]\}/[1 - (\tfrac{6}{12})^4]}$$
$$p_1 - p_2 = 656.9 \text{ lb/ft}^2 \quad \text{or} \quad 4.56 \text{ lb/in}^2$$

17.168 Air flows through a Venturi meter in a 55-mm-diameter pipe having a throat diameter of 30 mm, $C_v = 0.97$. For $p_1 = 830$ kPa abs, $T_1 = 15$ °C, $p_2 = 550$ kPa abs, calculate the mass per second flowing.

$$\dot{m} = C_v Y A_2\sqrt{2\rho_1 \Delta p/[1 - (D_2/D_1)^4]} \qquad p_2/p_1 = \tfrac{550}{830} = 0.663 \qquad D_2/D_1 = \tfrac{30}{55} = 0.545$$

From Fig. A-26, $Y = 0.78$.

$$\rho = p/RT \qquad \rho_1 = (830)(1000)/[(287)(15 + 273)] = 10.04 \text{ kg/m}^3$$
$$\dot{m} = (0.97)(0.78)[(\pi)(\tfrac{30}{1000})^2/4]\sqrt{(2)(10.04)[(830 - 550)(1000)]/[1 - (\tfrac{30}{55})^4]} = 1.33 \text{ kg/s}$$

17.169 Oxygen, $p_1 = 40$ psia, $T_1 = 120\,°\text{F}$, flows through a 1-in by $\frac{1}{2}$-in Venturi meter with a pressure drop of 6 psi. Find the mass per second flowing.

▮ $\dot{m} = C_v Y A_2 \sqrt{(2\rho_1 \, \Delta p)/[1 - (D_2/D_1)^4]}$. Assume $C_v = 0.984$.

$$p_2/p_1 = (40 - 6)/40 = 0.850 \qquad D_2/D_1 = \tfrac{1}{2}/1 = 0.500$$

From Fig. A-26, $Y = 0.91$.

$$\rho = p/RT \qquad \rho_1 = (40)(144)/[(1551)(120 + 460)] = 0.006403 \text{ slug/ft}^3$$
$$\dot{m} = (0.984)(0.91)[(\pi)(0.5/12)^2/4]\sqrt{(2)(0.006403)[(6)(144)]/[1 - (0.5/1)^4]} = 0.00419 \text{ slug/s}$$
$$V_1 = \dot{m}/\rho_1 A_1 = 0.00419/\{(0.006403)[(\pi)(\tfrac{1}{12})^2/4]\} = 120 \text{ ft/s}$$
$$N_R = \rho_1 D_1 V_1/\mu = (0.006403)(\tfrac{1}{12})(120)/(4.18 \times 10^{-7}) = 1.53 \times 10^5$$

From Fig. A-35, $C_v = 0.984$ is O.K.

17.170 Solve Prob. 17.73 for a Venturi nozzle.

▮ For the Venturi guess $\alpha \approx 1.0$ and iterate Eq. (2) of Prob. 17.73: $\alpha = 1.0$, $\beta = 0.447$, $C_d = 0.9806$. This rapidly converges to give $\beta = 0.4468$, $d = \beta D = 89$ mm. The throat velocity is $2.0/(0.4468)^2 = 10.0$ m/s, and the throat head is $(10.0)^2/[2(9.81)] = 5.12$ m. From Fig. A-24 for the Venturi estimate $K_m \approx 0.15$; hence $h_m = 0.15(5.12 \text{ m}) = 0.8$ m.

17.171 It is planned to equip a 15-cm-diameter pipe carrying water at $20\,°\text{C}$ with a modern Venturi nozzle. In order for international standards to be met, what is the permissible range of flow rates and nozzle diameters?

▮ **Flow range:**

$$1.5 \times 10^5 < N_R < 2 \times 10^6 \qquad 1.5 \times 10^5 < 4Q\rho/\pi\mu D < 2 \times 10^6$$
$$1.5 \times 10^5 < (4)(Q)(998)/[(\pi)(1.02 \times 10^{-3})(\tfrac{15}{100})] < 2 \times 10^6 \qquad 0.0181 \text{ m}^3/\text{s} < Q < 0.241 \text{ m}^3/\text{s}$$

Size range:

$$0.316 < [\beta = d/D] < 0.775 \qquad 0.316 < d/15 < 0.775 \qquad 4.74 \text{ cm} < d < 11.62 \text{ cm}$$

17.172 For Prob. 17.171, what is the permissible range of pressure drops? Would compressibility be a problem for the highest pressure-drop condition?

▮ From Fig. A-36, $0.9236 < C_d < 0.9847$.

$$Q = (C_d \pi d^2/4)\sqrt{(2 \, \Delta p)/[\rho(1 - \beta^4)]} \qquad d = \beta D = \beta(\tfrac{15}{100}) = 0.15\beta$$
$$\Delta p = (1.598 \times 10^6)(1 - \beta^4)(Q/C_d\beta^2)^2$$

Smallest Δp occurs at largest β and smallest Q.

$$\Delta p = (1.598 \times 10^6)(1 - 0.775^4)\{0.0181/[(0.9847)(0.775)^2]\}^2 = 957 \text{ Pa}$$

Largest Δp occurs at smallest β and largest Q.

$$\Delta p = (1.598 \times 10^6)(1 - 0.316^4)\{0.241/[(0.9236)(0.316)^2]\}^2 = 1.08 \times 10^7 \text{ Pa}$$

Thus $957 \text{ Pa} < \Delta p < 1.08 \times 10^7$ Pa. The larger Δp is much smaller than the bulk modulus of water, which is 2.19×10^9 Pa, so compressibility is not important.

17.173 Light oil (s.g. $= 0.92$, $v = 10^{-5}$ ft^2/s) flows down a 6-in vertical pipe through a 3-in Venturi nozzle as in Fig. 17-45. If the mercury manometer reads a deflection of 4 in, what is the estimated flow rate?

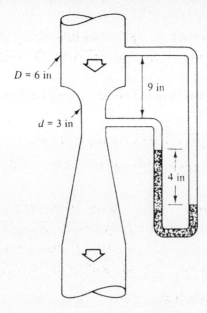

Fig. 17-45

$$Q = C_d A_t \sqrt{(2\,\Delta p)/[\rho(1 - \beta^4)]} \qquad C_d = 0.9858 - 0.196\beta^{4.5} \qquad \beta = d/D = \tfrac{3}{6} = 0.500$$
$$C_d = 0.9858 - (0.196)(0.500)^{4.5} = 0.9771$$
$$\Delta p = (\rho_M - \rho_0)gh = (\text{s.g.}_M - \text{s.g.}_0)\gamma h = (13.6 - 0.92)(62.4)(\tfrac{4}{12}) = 263.7 \text{ lb/ft}^2$$
$$Q = (0.9771)[(\pi)(\tfrac{3}{12})^2/4]\sqrt{(2)(263.7)/\{[(0.92)(1.94)](1 - 0.500^4)\}} = 0.852 \text{ ft}^3/\text{s}$$
$$N_R = 4Q/\pi\nu D = (4)(0.852)/(\pi)(10^{-5})(\tfrac{6}{12}) = 2.17 \times 10^5 \quad (\text{O.K.})$$

17.174 A Venturi meter (Fig. 17-46) is a carefully designed constriction whose pressure difference is a measure of the flow rate in a pipe. Using Bernoulli's equation for steady incompressible flow with no losses, show that flow rate Q is related to manometer reading h by $Q \approx [A_2/\sqrt{1 - (D_2/D_1)^4}]\sqrt{[2gh(\rho_M - \rho)/\rho]}$ where ρ_M is the density of the manometer fluid.

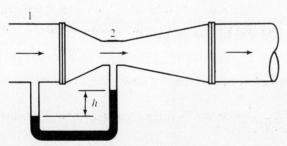

Fig. 17-46

Continuity: $\rho_1 A_1 V_1 = \rho_2 A_2 V_2$. Since $\rho_1 = \rho_2$, $V_2 = V_1(D_1/D_2)^2$.

Bernoulli (no losses): $p_1/\rho + V_1^2/2 + gz_1 = p_2/\rho + V_2^2/2 + gz_2$. Since $gz_1 = gz_2$, $V_2 = \sqrt{2(p_1 - p_2)/\{\rho[1 - (D_2/D_1)^4]\}}$, $Q = A_2 V_2$.

Manometer: $p_1 - p_2 = (\rho_M - \rho)gh$.
Combining these, $Q \approx [A_2/\sqrt{1 - (D_2/D_1)^4}]\sqrt{2gh(\rho_M - \rho)/\rho}$.

17.175 Water flows from a reservoir over a dam that is essentially a broad-crested weir. The dam is 20 ft long. How much higher will the water surface in the reservoir be above the top of the dam when the discharge over the dam is 1000 cfs?

$$Q = CLH^{1.5} \qquad 1000 = (3.09)(120)(H^{1.5}) \qquad H = 1.94 \text{ ft}$$

17.176 A sharp-crested, suppressed weir is under a head of 1.65 m. The weir is 2.5 m long and has a height of 4.50 m. What is the flow rate of water over the weir?

■ $$C = 0.611 + 0.075(H/H_w) = 0.611 + 0.075(1.65/4.50) = 0.6385$$
$$Q = C(\tfrac{2}{3})L\sqrt{2g}\, H^{1.5} = (0.6385)(\tfrac{2}{3})(2.50)\sqrt{(2)(9.807)}(1.65)^{1.5} = 9.99 \text{ m}^3/\text{s}$$

17.177 A sharp-crested, contracted, rectangular weir is to carry water at a flow rate of 12.75 cfs. What must the weir width be in order for the head of flow to be 1.50 ft?

■ $$Q = CL^{1.02}H^{1.47} \qquad 12.75 = (3.10)(L^{1.02})(1.50)^{1.47} \qquad L = 2.23 \text{ ft}$$

17.178 A sharp-crested, contracted, triangular weir with a weir angle of 80° carries water under a head of 0.955 m. Find the flow of water over the weir.

■ $$Q = C(\tfrac{8}{15})\sqrt{2g}\,[\tan(\theta/2)]H^{2.5} = (0.60)(\tfrac{8}{15})\sqrt{(2)(9.807)}\,[\tan(80°/2)](0.955)^{2.5} = 1.06 \text{ m}^3/\text{s}$$

17.179 Water flows over a Cipolletti weir 8.0 ft long under a head of 1.2 ft. What is the flow rate of water over the weir?

■ $$Q = CLH^{1.5} = (3.367)(8.0)(1.2)^{1.5} = 35.4 \text{ ft}^3/\text{s}$$

17.180 For the same conditions as given in Prob. 17.178 except that the weir is submerged and the downstream water surface is 0.450 m above the top of the weir, find the flow rate over the submerged weir.

■ $$Q = Q_1[1 - (d/h)^n]^{0.385} \qquad Q_1 = 1.06 \text{ m}^3/\text{s} \qquad \text{(from Prob. 17.178)}$$
$$Q = (1.06)[1 - (0.450/0.955)^{2.5}]^{0.385} = 0.995 \text{ m}^3/\text{s}$$

17.181 A sharp-crested, suppressed weir is under a head of 2.2 ft. The weir's length is 10.0 ft, and its height is 4.0 ft. Determine the flow rate of water over the weir.

■ $$C = 0.611 + 0.075(H/H_w) = 0.611 + 0.075(2.2/4.0) = 0.6522$$
$$Q = C(\tfrac{2}{3})L\sqrt{2g}\, H^{1.5} = (0.6522)(\tfrac{2}{3})(10)\sqrt{(2)(32.2)}\,(2.2)^{1.5} = 114 \text{ ft}^3/\text{s}$$

17.182 A sharp-crested, contracted, rectangular weir carries water at a flow rate of 1.81 m³/s. What must the length of the weir be in order for the head of flow to be 0.8 m?

■ $$Q = CL^{1.02}H^{1.47} \qquad 181 = (1.69)(L^{1.02})(0.8)^{1.47} \qquad L = 1.475 \text{ m}$$

17.183 A sharp-crested, contracted, triangular weir with a weir angle of 60° carries water under a head of 1.23 ft. Determine the flow rate of water over the weir.

■ $$Q = C(\tfrac{8}{15})\sqrt{2g}\,[\tan(\theta/2)]H^{2.5} = (0.60)(\tfrac{8}{15})\sqrt{(2)(32.2)}\,[\tan(60°/2)](1.23)^{2.5} = 2.49 \text{ ft}^3/\text{s}$$

17.184 The rate of flow of water over a sharp-crested, contracted, 45° triangular weir is 0.823 cfs. Determine the head of flow.

■ $$Q = C(\tfrac{8}{15})\sqrt{2g}\,[\tan(\theta/2)]H^{2.5} \qquad 0.823 = (0.60)(\tfrac{8}{15})\sqrt{(2)(32.2)}\,[\tan(45°/2)](H)^{2.5} \qquad H = 0.902 \text{ ft}$$

17.185 What length of a Cipolletti weir should be constructed in order that the head of flow will be 0.91 m when the rate of flow is 3.45 m³/s?

■ $$Q = CLH^{1.5} \qquad 3.45 = (1.859)(L)(0.910)^{1.5} \qquad L = 2.14 \text{ m}$$

17.186 Compute the rate of flow of water over a 4.5-ft-long Cipolletti weir under a head of 1.15 ft.

■ $$Q = CLH^{1.5} = (3.367)(4.5)(1.15)^{1.5} = 18.7 \text{ ft}^3/\text{s}$$

17.187 Water flows from a reservoir over a dam that is essentially a broad-crested weir. The dam is 45 m long, and the head of flow is 0.83 m. Compute the discharge of water over the dam.

■ $$Q = CLH^{1.5} = (1.71)(45)(0.83)^{1.5} = 58.2 \text{ m}^3/\text{s}$$

17.188 Given the same conditions as Prob. 17.181 except that the weir is submerged and the downstream water surface is 0.96 ft above the top of the weir, determine the flow rate of water over the weir.

■ $$Q = Q_1[1 - (d/h)^n]^{0.385} \qquad Q_1 = 114 \text{ ft}^3/\text{s} \qquad \text{(from Prob. 17.181)}$$
$$Q = (114)[1 - (0.96/2.2)^{1.5}]^{0.385} = 100 \text{ ft}^3/\text{s}$$

17.189 Flow is occurring in a rectangular channel at a velocity of 3 fps and depth of 1.0 ft. Neglecting the effect of velocity of approach, determine the height of a sharp-crested, suppressed weir that must be installed to raise the water depth upstream of the weir to 4 ft.

$$Q = 3.33LH^{3/2} = Av = (1.0L)(3) = 3.00L \qquad 3.00L = (3.33L)(H)^{3/2}$$
$$H = 0.93 \text{ ft} \qquad \text{Height of weir} = 4 - 0.93 = 3.07 \text{ ft}$$

17.190 A rectangular, sharp-crested weir 0.9 m high extends across a rectangular channel which is 2.4 m wide. When the head is 36 cm, find the rate of discharge by neglecting the velocity of approach.

$$Q = 1.8LH^{3/2} = (1.84)(2.4)(\tfrac{36}{100})^{3/2} = 0.954 \text{ m}^3/\text{s}$$

17.191 Suppose the rectangular weir of Prob. 17.190 is contracted at both ends. Find the rate of discharge for a head of 36 cm by the Francis formula. What would be the maximum value of H for which the Francis formula could be used?

$$Q = 1.84LH^{3/2} = (1.84)[2.4 - (2)(0.1)(\tfrac{36}{100})](\tfrac{36}{100})^{3/2} = 0.925 \text{ m}^3/\text{s}$$

Maximum head usable with Francis formula: $H_{max} = L/3 = 2.4/3 = 0.80$ m.

17.192 Develop a table of C_d versus P/H with H as a parameter. Use the Rehbock formula.

$C_d = 0.605 + 1/305H + 0.08H/P$. For $P/H = 0.5$ and $H = 0.2$, $C_d = 0.605 + 1/[(305)(0.2)] + 0.08/0.5 = 0.781$. Similarly, for additional values of P/H and H:

P/H	H = 0.2 ft	1.0 ft	5.0 ft
		C_d	
0.5	0.781	0.768	0.765
1.0	0.701	0.688	0.685
2.0	0.661	0.648	0.645
5.0	0.637	0.624	0.621
10.0	0.629	0.616	0.613

17.193 What is the rate of discharge of water over a 45° triangular weir when the head is 0.5 ft?

$Q = C_d(\tfrac{8}{15})\sqrt{2g}\,[\tan(\theta/2)](H^{5/2})$. From Fig. A-37, $C_d = 0.596$. $Q = (0.596)(\tfrac{8}{15})\sqrt{(2)(32.2)}\,[\tan(45°/2)](0.5)^{5/2} = 0.187 \text{ ft}^3/\text{s}$.

17.194 With the same head, what would be the increase in discharge obtained by doubling the notch angle in Prob. 17.193?

$Q = C_d(\tfrac{8}{15})\sqrt{2g}\,[\tan(\theta/2)](H^{5/2})$. From Fig. A-37, $C_d = 0.583$. $Q = (0.583)(\tfrac{8}{15})\sqrt{(2)(32.2)}\,[\tan(90°/2)](0.5)^{5/2} = 0.441 \text{ ft}^3/\text{s}$, increase $= (0.441 - 0.187)/0.187 = 1.36$, or 136 percent.

17.195 What would be the head for a discharge of 2.0 cfs of water over a 60° triangular weir? Use a value of C_d of 0.58.

$$Q = C_d(\tfrac{8}{15})\sqrt{2g}\,[\tan(\theta/2)](H^{5/2}) \qquad 2.0 = (0.58)(\tfrac{8}{15})\sqrt{(2)(32.2)}\,[\tan(60°/2)](H^{5/2}) \qquad H = 1.14 \text{ ft}$$

17.196 For the Cipolletti weir, derive the slope ($\tfrac{1}{4}$:1) of the sides of the trapezoid by setting the reduction in discharge due to contraction equal to the increase in discharge due to the triangular area added.

By the Francis formula, the decrease in flow due to two contractions is $Q = C_d(\tfrac{2}{3})\sqrt{2g}\,LH^{3/2}$, $\Delta Q = C_d(\tfrac{2}{3})\sqrt{2g}\,[(2)(0.1h)]H^{3/2}$. This deficiency is made up by the equivalent triangular weir of angle θ, for which $Q = C_d(\tfrac{8}{15})\sqrt{2g}\,[\tan(\theta/2)](H^{5/2})$. Equating these, $C_d(\tfrac{2}{3})\sqrt{2g}\,[(2)(0.1H)]H^{3/2} = C_d(\tfrac{8}{15})\sqrt{2g}\,[\tan(\theta/2)](H^{5/2})$. Assuming C_d is the same for both weirs, $\tan(\theta/2) = 0.2500$, or side slope is $\tfrac{1}{4}$:1.

17.197 Develop in general terms an expression for the percent of error in Q over a triangular weir if there is a small error in the measurement of the vertex angle. Assume there is no error in the weir coefficient. Compute the percent error in Q if there is a 2° error in the measurement of the total vertex angle of a triangular weir having a total vertex angle of 75°.

$$Q = C_d(\tfrac{8}{15})(2g)^{1/2}[\tan(\theta/2)]H^{5/2} \qquad \partial Q/\partial\theta = C_d(\tfrac{8}{15})(2g)^{1/2}H^{5/2}\{(\partial/\partial\theta)[\tan(\theta/2)]\}$$

$$\partial Q/Q = (\partial/\partial\theta)\{[\tan(\theta/2)]\,\partial\theta\}/\tan(\theta/2) = \sec^2(\theta/2)(\tfrac{1}{2})\,\partial\theta/[\tan(\theta/2)] = \partial\theta/[2\sin(\theta/2)\cos(\theta/2)]$$

For the case where $\partial\theta = 2°$ and $\theta = 75°$: $\theta/2 = 37.5°$, so that $\partial Q/Q = 2(\pi/180)/[2(0.609)0.793] = 0.0361 = 3.61$ percent.

17.198 A broad-crested weir rises 0.3 m above the bottom of a horizontal channel. With a measured head of 0.6 m above the crest, what is the rate of discharge per unit width?

$$Q = L\sqrt{g}\,(\tfrac{2}{3})^{3/2}E^{3/2} = L\sqrt{9.807}\,(\tfrac{2}{3})^{3/2}(0.6 + V_1^2/2g)^{3/2}$$

$$Q/L = q = (1.705)\{0.6 + V_1^2/[(2)(9.807)]\}^{3/2}$$

But $q = (0.6 + 0.3)V_1$: $V_1 = q/0.9$, $q = (1.705)\{0.6 + (q/0.9)^2/[(2)(9.807)]\}^{3/2}$, $q = 0.894$ m³/s/m (by trial and error).

17.199 A broad-crested weir of height 2.00 ft in a channel 5.00 ft wide has a flow over it of 9.50 cfs. What is the water depth just upstream of the weir?

$$q = 9.5/5 = 1.90 \text{ cfs/ft} \qquad y_2 = y_c = (q^2/g)^{1/3} = (1.90^2/32.2)^{1/3} = 0.4822 \text{ ft}$$

$$V_2 = q/y_2 = 1.90/0.4822 = 3.940 \text{ ft/s} \qquad y_1 + V_1^2/2g = 2 + 0.4822 + 3.940^2/[(2)(32.2)]$$

But $V_1 = 1.90/y_1$: $y_1 + (1.90/y_1)^2/[(2)(32.2)] = 2.723$, $y_1 = 2.72$ ft (by trial and error).

17.200 A 60° V-notch weir is used to measure a flow rate of 0.25 cfs. Assuming C_d is known precisely, compute the percentage of error in Q that would result from an error of 0.02 ft in head measurement.

$Q = C_d(\tfrac{8}{15})(2g)^{1/2}[\tan(\theta/2)]H^{5/2}$. For $\theta = 60°$, $C_d = 0.58$, and $Q = 0.25$ cfs: $H = 0.497$ ft. $\partial Q/\partial H = C_d(\tfrac{8}{15})(2g)^{1/2}[\tan(\theta/2)](\tfrac{5}{2})H^{3/2}$, $\partial Q/Q = \tfrac{5}{2}H^{3/2}(\partial H)/H^{5/2} = 5(\partial H)/2H = 5(0.02)/[2(0.497)] = 0.1006 = 10.06$ percent.

17.201 Work Prob. 17.200 for a rectangular weir with end contractions having a crest length of 2 ft.

$$Q = 3.33LH^{3/2} \qquad L = 2 \text{ ft} \qquad \text{and} \qquad Q = 0.25 \text{ cfs} \qquad H = 0.112 \text{ ft}$$

$$\partial Q/\partial H = C_w L(\tfrac{3}{2})H^{1/2} \qquad \partial Q/Q = \tfrac{3}{2}H^{1/2}(\partial H)/H^{3/2} = 3(\partial H)/2H = 3(0.02)/[2(0.112)] = 0.268 = 26.8 \text{ percent}$$

17.202 Derive the theoretical formula for flow over a rectangular weir. Refer to Fig. 17-47.

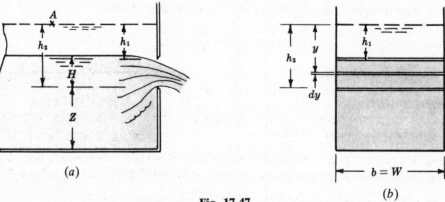

(a)

(b)

Fig. 17-47

Consider the rectangular opening in Fig. 17-47 to extend the full width W of the channel ($b = W$). With the liquid surface in the dashed position, application of the Bernoulli theorem between A and an elemental strip of height dy in the jet produces, for ideal conditions, $(0 + V_A^2/2g + y) -$ no losses $= (0 + V_{jet}^2/2g + 0)$ where V_A represents the average velocity of the particles approaching the opening.

Thus the ideal $V_{jet} = \sqrt{2g(y + V_A^2/2g)}$ and

$$\text{Ideal } dQ = dAV_{jet} = (b\,dy)V_{jet} = b\sqrt{2g}\left(y + \frac{V_A^2}{2g}\right)^{1/2}dy = b\sqrt{2g}\int_{h_1}^{h_2}\left(y + \frac{V_A^2}{2g}\right)^{1/2}dy$$

A weir exists when $h_1 = 0$. Let H replace h_2 and introduce a coefficient of discharge C to obtain the actual flow. Then

$$Q = Cb\sqrt{2g} \int_0^H \left(y + \frac{V_A^2}{2g}\right)^{1/2} dy = \tfrac{2}{3}Cb\sqrt{2g}\left[\left(H + \frac{V_A^2}{2g}\right)^{3/2} - \left(\frac{V_A^2}{2}\right)^{3/2}\right] = mb\left[\left(H + \frac{V_A^2}{2g}\right)^{3/2} - \left(\frac{V_A^2}{2g}\right)^{3/2}\right] \quad (1)$$

Notes:

(1) For a fully contracted rectangular weir the end contractions cause a reduction in flow. Length b is corrected to recognize this condition, and the formula becomes

$$Q = m(b - \tfrac{2}{10})[H + V_A^2/2g)^{3/2} - (V_A^2/2g)^{3/2}] \quad (2)$$

(2) For high weirs and most contracted weirs the velocity head of approach is negligible and

$$Q = m(b - \tfrac{2}{10}H)H^{3/2} \qquad \text{for contracted weirs} \quad (3)$$

$$\text{or} \qquad Q = mbH^{3/2} \qquad \text{for suppressed weirs} \quad (4)$$

(3) Coefficient of discharge C is not constant. It embraces the many complexities not included in the derivation, such as surface tension, viscosity, density, nonuniform velocity distribution, secondary flows, and possibly others.

17.203 Derive the theoretical formula for flow through a triangular-notched weir. Refer to Fig. 17-48.

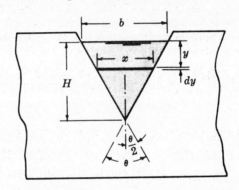

Fig. 17-48

▮ From Prob. 17.202, $V_{\text{jet}} = \sqrt{2g(y + \text{negligible } V^2/2g)}$ and ideal $dQ = dAV_{\text{jet}} = x\, dy\sqrt{2gy}$. By similar triangles, $x/b = (H - y)/H$ and $b = 2H \tan(\theta/2)$. Then

$$\text{Actual } Q = \left(\frac{b}{H}\right)C\sqrt{2g} \int_0^H (H - y)y^{1/2}\, dy$$

Integrating and substituting,

$$Q = \tfrac{8}{15}C\sqrt{2g}\, H^{5/2} \tan\left(\tfrac{1}{2}\theta\right) \quad (1)$$

A common V-notch is the 90° opening. Expression (1) then becomes $Q = 4.28CH^{5/2}$ where an *average* value of C is about 0.60 for heads above 1.0 ft.

17.204 During a test on an 8-ft suppressed weir which was 3 ft high, the head was maintained constant at 1.000 ft. In 38.0 s, 7600 gal of water were collected. Find weir factor m in Eqs. (1) and (4) of Prob. 17.202.

▮ (a) Change the measured flow to cfs. $Q = 7600/[(7.48)(38)] = 26.7$ cfs.
  (b) Check the velocity of approach. $V = Q/A = 26.7/[(8)(4)] = 0.834$ ft/s. Then $V^2/2g = (0.834)^2/2g = 0.018$ ft.
  (c) Using (1), $Q = mb[(H + V^2/2g)^{3/2} - (V^2/2g)^{3/2}]$ or $26.7 = (m)(8)[(1.000 + 0.011)^{3/2} - (0.0108)^{3/2}]$ and $m = 3.29$. Using (4), $Q = 26.7 = mbH^{3/2} = (m)(8)(1.000)^{3/2}$ and $m = 3.34$ (about 1.8 percent higher neglecting the velocity of approach terms).

17.205 Determine the flow over a suppressed weir 10.0 ft long and 4.00 ft high under a head of 3.000 ft. The value of m is 3.46.

▮ Since the velocity head term cannot be calculated, an approximate flow is $Q = mbH^{3/2} = 3.46(10)(3.000)^{3/2} = 178.8$ cfs. For this flow, $V = 178.8/[(10)(7)] = 2.554$ ft/s and $V^2/2g = 0.101$ ft. Using Eq. (1) of Prob. 17.202, $Q = 3.46(10)[(3.000 + 0.101)^{3/2} - (0.101)^{3/2}] = 188$ cfs.

This second calculation shows an increase of 9 cfs or about 5.0 percent over the first calculation. Further calculation will generally produce an unwarranted refinement, i.e., beyond the accuracy of the formula itself. However, to illustrate, the revised velocity of approach would be $V = 188/[(10)(7)] = 2.69$ ft/s, $V^2/2g = 0.112$ ft and $Q = 3.46(10)[(3.000 + 0.112)^{3/2} - (0.112)^{3/2}] = 189$ cfs.

17.206 A suppressed weir, 25.0 ft long, is to discharge 375.0 cfs into a channel. The weir factor $m = 3.42$. To what height Z (nearest $\frac{1}{100}$ ft) may the weir be built, if the water behind the weir must not exceed 6 ft in depth?

▮ Velocity of approach $V = Q/A = 375.0/[(25)(6)] = 2.50$ ft/s. Then $375.0 = (3.42)(25.0)\{[H + (2.50)^2/2g]^{3/2} - [(2.50)^2/2g]^{3/2}\}$ and $H = 2.59$ ft. Height of weir is $Z = 6.00 - 2.59 = 3.41$ ft.

17.207 A contracted weir, 4.00 ft high, is to be installed in a channel 8 ft wide. The maximum flow over the weir is 60.0 cfs when the total depth back of the weir is 7.00 ft. What length of weir should be installed if $m = 3.40$?

▮ Velocity of approach $V = Q/A = 60.0/[(8)(7)] = 1.071$ ft/s. It appears that the velocity head is negligible in this case. $Q = m(b - \frac{2}{10}H)(H)^{3/2}$, $60.0 = 3.40[b - \frac{2}{10}(3.00)](3.00)^{3/2}$, $b = 4.00$ ft long.

17.208 The discharge from a 6-in-diameter orifice, under a 10.0-ft head, $C = 0.600$, flows into a rectangular weir channel and over a contracted weir. The channel is 6 ft wide and, for the weir, $Z = 5.00$ ft and $b = 1.00$ ft. Determine the depth of water in the channel if $m = 3.35$.

▮ The discharge through the orifice is $Q = CA\sqrt{2gh} = (0.600)[\frac{1}{4}\pi(\frac{1}{2})^2]\sqrt{2g(10.0)} = 2.99$ cfs. For the weir, $Q = m(b - \frac{2}{10}H)H^{3/2}$ (velocity head neglected) or $2.99 = 3.35(1.00 - 0.20H)H^{3/2}$ and $H^{3/2} - 0.20H^{5/2} = 0.893$. By successive trials, $H = 1.09$ ft; and the depth $= Z + H = 5.00 + 1.09 = 6.09$ ft.

17.209 The discharge of water over a 45° triangular weir is 0.750 cfs. For $C = 0.580$, determine the head on the weir.

▮
$$Q = C(\tfrac{8}{15})\sqrt{2g}\,[\tan(\theta/2)]H^{5/2}$$
$$0.750 = (0.580)(\tfrac{8}{15})\sqrt{(2)(32.2)}\,[\tan(45°/2)]H^{5/2} \qquad H = 0.881 \text{ ft}$$

17.210 What length of Cipolletti weir should be constructed so that the head will be 1.54 ft when the discharge is 122 cfs?

▮
$$Q = 3.367bH^{3/2} \qquad 122 = (3.367)(b)(1.54)^{3/2} \qquad b = 19.0 \text{ ft}$$

17.211 Derive the expression for the time to lower the liquid level in a tank, lock, or canal, by means of a suppressed weir.

▮
$$Q\,dt = -A_T\,dH \qquad (mLH^{3/2})\,dt = -A_T\,dH$$
$$t = \int_{t_1}^{t_2} dt = \left(\frac{-A_T}{mL}\right)\int_{H_1}^{H_2} H^{-3/2}\,dH \qquad t = t_2 - t_1 = \left(\frac{2A_T}{mL}\right)(H_2^{-1/2} - H_1^{-1/2})$$

17.212 A rectangular flume, 50 ft long and 10 ft wide, feeds a suppressed weir under a head of 1.000 ft. If the supply to the flume is cut off, how long will it take for the head on the weir to decrease to 4 in? Use $m = 3.33$.

▮
$$t = (2A_T/mL)(H_2^{-1/2} - H_1^{-1/2}) \qquad \text{(from Prob. 17.211)}$$
$$= (2)[(50)(10)]/[(3.33)(10)][(\tfrac{4}{12})^{-1/2} - 1.000^{-1/2}] = 22.0 \text{ s}$$

17.213 The head on a sharp-crested, rectangular weir of height 2.0 ft and crest length 4.0 ft was incorrectly observed to be 0.38 ft when it was actually 0.40 ft. Determine the percentage error in the computed value of flow rate.

▮
$$Q = CLH^{3/2} \qquad \partial Q/\partial H = (\tfrac{3}{2})CLH^{1/2} \approx \Delta Q/\Delta H \qquad \Delta Q \approx \Delta H(\tfrac{3}{2})CLH^{1/2}$$
$$\Delta Q/Q \approx (\tfrac{3}{2})(\Delta H/H) = (\tfrac{3}{2})(0.02/0.40) = 0.075 \quad \text{or} \quad 7.5 \text{ percent}$$

17.214 It is desired to measure a discharge which may vary from 0.5 to 2.5 cfs with a relative accuracy of at least 0.5 percent through the entire range. A stage recorder which is accurate to the nearest 0.001 ft is available. What is the maximum width of rectangular sharp-crested weir that will satisfy these conditions? Assume the weir has end contractions and $C_w = 3.5$. Neglect velocity of approach.

▮
$$Q = 3.5Lh^{3/2} \qquad \partial Q/\partial h = 5.25Lh^{1/2} \approx \Delta Q/\Delta h$$
Thus $\Delta Q = 5.25Lh^{1/2}(\Delta h)$. $\Delta Q/Q = (5.25/3.50)(\Delta h/h) = 1.5(\Delta h/h)$, $0.005 = 1.5(0.001/h)$, $h = 0.0015/0.005 =$

0.3 ft. h must be ≥ 0.3 ft to acquire desired accuracy: $Q = 0.5 = 3.5L(0.3)^{3/2}$, $L \leq 0.865$ ft (0.93 ft if end contractions are considered).

17.215 What is the maximum permissible vertex angle of a V-notch weir that will satisfy the conditions of Prob. 17.214?

❚
$$Q = 4.28(0.58)h^{5/2}[\tan(\theta/2)] = 2.48h^{5/2}[\tan(\theta/2)]$$
$$\partial Q/\partial h = \tfrac{5}{2}(2.48)h^{3/2}[\tan(\theta/2)] = 6.2h^{3/2}[\tan(\theta/2)] \approx \Delta Q/\Delta h$$

Thus $\Delta Q = 6.2h^{3/2}[\tan(\theta/2)]\,\Delta h$; $\Delta Q/Q = 2.5(\Delta h/h)$, $0.005 = 2.5(0.001/h)$, $h \geq 0.5$ ft; $Q = 0.5 = 2.48h^{5/2}[\tan(\theta/2)] = 2.48(0.5)^{5/2}[\tan(\theta/2)]$, $\tan(\theta/2) = 0.5/[2.48(0.177)] = 1.14$, $\theta/2 \leq 48°45'$, $\theta < 97°30'$.

17.216 With a head of 0.15 m on a 60° V-notch weir, what error in the measured head will produce the same percentage error in the computed flow rate as an error of 1° in the vertex angle?

❚
$$Q = 2.48h^{5/2}[\tan(\theta/2)] \qquad \partial Q/\partial h = 6.2h^{3/2}[\tan(\theta/2)] \approx \partial Q/\partial h$$

Thus $\Delta Q = 6.2h^{3/2}[\tan(\theta/2)]\,\Delta h$; $\Delta Q/Q = 2.5(\Delta h/h)$, $\partial Q/\partial\theta = 2.48h^{5/2}[\sec^2(\theta/2)]\tfrac{1}{2} = 1.24h^{5/2}[\sec^2(\theta/2)] = \Delta Q/\Delta\theta$. Thus $\Delta Q = 1.24h^{5/2}[\sec^2(\theta/2)]\,\Delta\theta$; $\Delta Q/Q = [(\Delta\theta)\sec^2(\theta/2)]/(2\tan\theta)$. For $\theta = 30°$, $\sec^2(\theta/2)/\tan(\theta/2) = 2.31$, $\Delta Q = 2.5(\Delta h/0.15) = 2.31(1°)/2 = (2.31/2)(1/57.3)]$, $\Delta h = 0.0012$ m $= 0.12$ cm.

17.217 Tests on a 60°-notch weir yield the following values of head H on the weir and discharge Q:

H, ft	0.345	0.356	0.456	0.537	0.568	0.594	0.619	0.635	0.654	0.665
Q, cfs	0.107	0.110	0.205	0.303	0.350	0.400	0.435	0.460	0.490	0.520

Use the theory of least squares to determine the constants in $Q = CH^m$ for this weir.

❚ Taking the logarithm of each side of the equation, $\ln Q = \ln C + m \ln H$ or $y = B + mx$, it is noted that the best values of B and m are needed for a straight line through the data when plotted on log-log paper.

By the theory of least squares, the best straight line through the data points is the one yielding a minimum value of the sums of the squares of vertical displacements of each point from the line; or, from Fig. 17-49,

$$F = \sum_{i=1}^{i=n} s_i^2 = \sum [y_i - (B + mx_i)]^2$$

where n is the number of experimental points. To minimize F, $\partial F/\partial B$ and $\partial F/\partial m$ are taken and set equal to zero, yielding two equations in the two unknowns B and m: $\partial F/\partial B = 0 = 2\sum[y_i - (B + mx_i)](-1)$ from which

$$\sum y_i - nB - m\sum x_i = 0 \tag{1}$$

and $\partial F/\partial m = 0 = 2\sum[y_i - (B + mx_i)](-x_i)$ or

$$\sum x_i y_i - B\sum x_i - m\sum x_i^2 = 0 \tag{2}$$

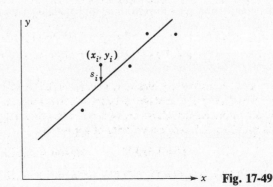

Fig. 17-49

Solving Eqs. (1) and (2) for m gives $m = (\sum x_i y_i/\sum x_i - \sum y_i/n)/(\sum x_i^2/\sum x_i - \sum x_i/n)$, $B = (\sum y_i - m \sum x_i)/n$. These equations are readily solved by hand calculator having the $\sum$ key, or a simple program can be written for the digital computer. The answer for the data of this problem is $m = 2.437$, $C = 1.395$.

17.218 A rectangular, sharp-crested weir 4 m long with end contractions suppressed is 1.3 m high. Determine the discharge when the head is 200 mm.

▌ Since the height of the weir is large compared to H, the upstream velocity head can be omitted; hence, $Q = 1.83LH^{1.5} = (1.83)(4)(\frac{200}{1000})^{1.5} = 0.655 \text{ m}^3/\text{s}$.

17.219 In Fig. 17-50, $L = 8$ ft, $P = 1.8$ ft, and $H = 0.80$ ft. Estimate the discharge over the weir ($C = 3.33$).

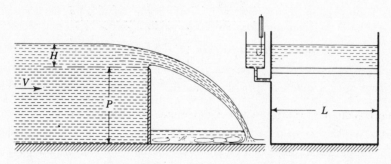

Fig. 17-50

▌ $Q = CL(H + \alpha V^2/2g)^{1.5}$ $V = Q/[(L)(P + H)]$ $Q = CL\{H + \alpha Q^2/[(2gL^2)(P + H)^2]\}^{1.5}$
$Q = (3.33)(8)\{0.80 + 1.4Q^2/[(2)(32.2)(8)^2(1.8 + 0.80)^2]\}^{1.5} = 19.8 \text{ ft}^3/\text{s}$ (by trial and error)

17.220 A rectangular, sharp-crested weir with end contractions is 1.5 m long. How high should it be placed in a channel to maintain an upstream depth of 2.25 m for 0.45 m³/s flow?

▌ $Q = 1.83L'H^{3/2}$ where $L' = L - 0.2H$ (due to end contractions).

$0.45 = (1.83)(1.5 - 0.2H)(H)^{3/2}$ $H = 0.31 \text{ m}$ $P = 2.25 - 0.31 = 1.94 \text{ m}$ (height of weir)

17.221 Determine the head on a 60° V-notch weir for a discharge of 170 L/s.

▌ $Q = C(\frac{8}{15})\sqrt{2g} [\tan (\theta/2)]H^{5/2}$ $\frac{170}{1000} = (0.58)(\frac{8}{15})\sqrt{(2)(9.807)} [\tan (60°/2)]H^{5/2}$ $H = 0.541 \text{ m}$

17.222 Tests on a 90° V-notch weir gave the following results: $H = 180$ mm, $Q = 19.4$ L/s, $H = 410$ mm, $Q = 150$ L/s. Determine the formula for the weir.

▌ Let $Q = AH^n$: $\frac{150}{1000} = A(\frac{410}{1000})^n$, $19.4/1000 = A(\frac{180}{1000})^n$. Dividing one equation by the other, $7.732 = 2.278^n$, $n = 2.48$. $\frac{150}{1000} = A(\frac{410}{1000})^{2.48}$, $A = 1.37$; $Q = 1.37H^{2.48}$.

17.223 A broad-crested weir 1.6 m high and 3 m long has a well-rounded upstream corner. What head is required for a flow of 2.85 m³/s?

▌ $Q = 1.67LH^{3/2}$ $2.85 = (1.67)(3)H^{3/2}$ $H = 0.687 \text{ m}$

17.224 A weir in a horizontal channel is 12 ft wide and 4 ft high. The upstream water depth is 5.2 ft. Estimate the discharge if the weir is sharp-crested.

▌ $C_w = 0.611 + 0.075H/Y = 0.611 + (0.075)(5.2 - 4)/4 = 0.6335$
$q = (\frac{2}{3})(C_w)(2g)^{1/2}H^{3/2} = (\frac{2}{3})(0.6335)[(2)(32.2)]^{1/2}(5.2 - 4)^{3/2} = 4.455 \text{ ft}^3/\text{s/ft}$ $Q = (12)(4.455) = 53.5 \text{ ft}^3/\text{s}$

17.225 Solve Prob. 17.224 if the weir is broad-crested.

▌ $C_w = 0.65/(1 + H/Y)^{1/2} = 0.65/[1 + (5.2 - 4)/4]^{1/2} = 0.5701$. Since everything else is the same as in Prob. 17.224, we simply scale down the sharp-crested result: $Q = (0.5701/0.6335)(53.5) = 48.1 \text{ ft}^3/\text{s}$.

17.226 A weir in a horizontal channel is 7 m wide and 1.2 m high. If the upstream depth is 2 m, estimate the channel discharge for a sharp-crested weir.

$$C_w = 0.611 + 0.075H/Y = 0.611 + (0.075)(2 - 1.2)/1.2 = 0.6610$$

$$q = (\tfrac{2}{3})(C_w)(2g)^{1/2}H^{3/2} = (\tfrac{2}{3})(0.6610)[(2)(9.807)]^{1/2}(2 - 1.2)^{3/2} = 1.396 \text{ m}^3/\text{s/m} \qquad Q = (7)(1.396) = 9.77 \text{ m}^3/\text{s}$$

17.227 Solve Prob. 17.226 if the weir is broad-crested.

▮ $C_w = 0.65/(1 + H/Y)^{1/2} = 0.65/[1 + (2 - 1.2)/1.2]^{1/2} = 0.5035$. Since everything else is the same as in Prob. 17.226, we simply scale down the sharp-crested result: $Q = (0.5035/0.6610)(9.77) = 7.44 \text{ m}^3/\text{s}$.

17.228 Show that the discharge of the V-notch weir in Fig. 17-51 is given by $Q = C_w(\tfrac{8}{15})(2g)^{1/2}(\tan \alpha)H^{5/2}$. Experiments show that C_w varies from 0.65 at $\alpha = 5°$ to 0.58 at $\alpha = 50°$.

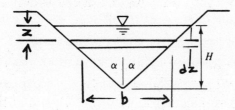

Fig. 17-51

▮ Assume V at any strip $= \sqrt{2gz}$, where z is measured down from the top.

$$Q = \int V \, dA = \int_0^H (2gz)^{1/2}b \, dz \qquad \text{where } b = b_0\left(1 - \frac{z}{H}\right)$$

$$= \int_0^H (2gz)^{1/2}\left[b_0\left(1 - \frac{z}{H}\right)\right] dz = (\tfrac{4}{15})(2g)^{1/2}\left(\frac{b_0}{H}\right)(H^{5/2})$$

$$b_0/H = 2\tan \alpha \qquad Q = (\tfrac{8}{15})(2g)^{1/2}(\tan \alpha)H^{5/2}$$

A correction factor C_w is added to account for end and edge effects and upstream kinetic energy; hence, $Q = C_w(\tfrac{8}{15})(2g)^{1/2}(\tan \alpha)H^{5/2}$.

17.229 Wastewater flows through a Parshall flume with a throat width of 3.0 ft at a depth of flow of 2.5 ft. Find the flow rate.

▮ $$Q = 4.0Bh^{1.522B^{0.026}} = (4.0)(3.0)(2.5)^{(1.522)(3.0)^{0.026}} = 50.4 \text{ ft}^3/\text{s}$$

17.230 Water is to flow through a proposed Parshall flume at a maximum flow rate of 30 m³/s. The width of the Parshall flume's throat is to be 10.0 m. What should be the channel depth entering the flume?

▮ $$Q = (2.293B + 0.474)h^{1.6} \qquad 30 = [(2.293)(10.0) + 0.474]h^{1.6} \qquad h = 1.17 \text{ m}$$

17.231 Wastewater flows through a Parshall flume with a throat width of 5.0 ft at a depth of flow of 3.2 ft. Determine the rate of flow of the wastewater.

▮ $$Q = 4.0Bh^{1.522B^{0.026}} = (4.0)(5.0)(3.2)^{(1.522)(5.0)^{0.026}} = 127 \text{ ft}^3/\text{s}$$

17.232 Water is to flow through a Parshall flume at a maximum flow rate of 25 m³/s. The width of the throat of the Parshall flume is 8.0 m. What will be the depth of flow at maximum flow rate?

▮ $$Q = (2.293B + 0.474)h^{1.6} \qquad 25 = [(2.293)(8.0) + 0.474]h^{1.6} \qquad h = 1.19 \text{ m}$$

17.233 The stilling-well depth-measurement scale of a Parshall flume with a 4-ft throat gives a head reading that is 0.05 ft too large. Compute the percentage errors in flow rate when the observed head readings are 0.5 ft and 2.0 ft.

▮ $$Q = 4Bh^{1.522B^{0.026}} \qquad B = 4 \text{ ft} \qquad \Delta h = 0.05 \text{ ft} \qquad e = 0.05 \text{ ft}$$

$$\% \text{ Error} = \{[(h + e)/h]^{1.522B^{0.026}} - 1\}(100) \qquad 4^{0.026} = 1.0367$$

$$\% \text{ Error} = [(0.55/0.50)^{1.578} - 1] = 0.162 \quad \text{or} \quad 16.2 \text{ percent}$$

If $h = 2.0$ ft, % Error $= [(2.05/2.00)^{1.578} - 1] = 0.0397$, or 3.97 percent.

17.234 It is desired to measure the flow in a canal which may carry between 25 cfs and 100 cfs. The flow is to be measured to an accuracy of 2 percent, and the available water-level recorder is accurate to 0.01 ft. The canal is on a very flat slope and the head loss in the measuring device should be as small as possible. What type and size of flow-measuring device would you recommend?

▮ Try an 8-ft-throat Parshall flume: $Q = 4Bh^{1.552B^{0.026}} = (4)(8)h^{1.522(8)^{0.026}} = 32h^{1.607}$, $dQ = 51.4h^{0.607} \, dh$. For $Q = 25$ cfs, $h = 0.858$ ft, $dQ/Q = (51.4)(0.858)^{0.607}(0.01)/25 = 0.0187 < 0.02$. For $Q = 100$ cfs, $h = 2.03$ ft, $dQ/Q = (51.4)(2.03)^{0.607}(0.01)/100 = 0.0079 < 0.02$. So a Parshall flume with an 8-ft throat provides the required accuracy.

17.235 A rectangular channel 1.8 m wide contains a sluice gate which extends across the width of the channel. If the gate produces free flow when it is open 0.12 m with an upstream depth of 1.05 m, find the rate of discharge, assuming $C_d = 0.60$ and $C_c = 0.62$.

▮ $$Q = C_d Ba[2g(y_1 - y_2 + V_1^2/2g)]^{1/2} \qquad y_2 = C_c a = (0.62)(0.12) = 0.0744 \text{ m}$$

Find a trial Q by neglecting the $V_1^2/2g$ term in the equation above.

$$Q = (0.60)(1.8)(0.12)[(2)(9.807)(1.05 - 0.0744 + 0)]^{1/2} = 0.567 \text{ m}^3/\text{s}$$

$$V_1 = Q/A_1 = 0.567/[(1.8)(1.05)] = 0.300 \text{ m/s} \qquad V_1^2/2g = 0.300^2/[(2)(9.807)] = 0.00459 \text{ m}$$

$$Q = (0.60)(1.8)(0.12)[(2)(9.807)(1.05 - 0.0744 + 0.00459)]^{1/2} = 0.568 \text{ m}^3/\text{s}$$

17.236 For the sluice gate shown in Fig. 17-52, if $C_v = 0.98$, what is the flow rate? If $C_c = 0.62$, what is the height of the opening?

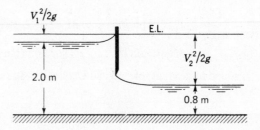

Fig. 17-52

▮ $V_2 = C_v[2g(y_1 - y_2 + V_1^2/2g)]^{1/2}$. Find a trial V_2 by neglecting the $V_1^2/2g$ term in the equation above.

$$V_2 = (0.98)[(2)(9.807)(2.0 - 0.8 + 0)]^{1/2} = 4.754 \text{ m/s} \qquad V_2 y_2 = V_1 y_1$$

$$(4.754)(0.8) = (V_1)(2.0) \qquad V_1 = 1.901 \text{ m/s}$$

$$V_1^2/2g = 1.901^2/[(2)(9.807)] = 0.184 \text{ m} \qquad V_2 = (0.98)[(2)(9.807)(2.0 - 0.8 + 0.184)]^{1/2} = 5.106 \text{ m/s}$$

Second trial:

$$(5.106)(0.8) = (V_1)(2.0) \qquad V_1 = 2.042 \text{ m/s}$$

$$V_1^2/2g = 2.042^2/[(2)(9.807)] = 0.213 \text{ m} \qquad V_2 = (0.98)[(2)(9.807)(2.0 - 0.8 + 0.213)]^{1/2} = 5.159 \text{ m/s}$$

Third trial:

$$(5.159)(0.8) = (V_1)(2.0) \qquad V_1 = 2.064 \text{ m/s}$$

$$V_1^2/2g = 2.064^2/[(2)(9.807)] = 0.217 \text{ m} \qquad V_2 = (0.98)[(2)(9.807)(2.0 - 0.8 + 0.217)]^{1/2} = 5.166 \text{ m/s}$$

Fourth trial:

$$(5.166)(0.8) = (V_1)(2.0) \qquad V_1 = 2.066 \text{ m/s} \qquad V_1^2/2g = 2.066^2/[(2)(9.807)] = 0.218 \text{ m}$$

$$V_2 = (0.98)[(2)(9.807)(2.0 - 0.8 + 0.218)]^{1/2} = 5.168 \text{ m/s} \qquad \text{(O.K.)}$$

$$q = y_2 V_2 = (0.8)(5.168) = 4.13 \text{ m}^3/\text{s/m} \qquad a = y_2 C_c = 0.8/0.62 = 1.29 \text{ m}$$

17.237 A disk meter has a volumetric displacement of 27 cm³ for one complete oscillation. Calculate the flow for 86.5 oscillations per minute.

$$Q = (\text{displacement per oscillation})(\text{number of oscillations per unit time})$$

$$= (27)(86.5) = 2336 \text{ cm}^3/\text{min} \quad \text{or} \quad 2.336 \text{ L/min}$$

17.238 A disk water meter with volumetric displacement of 40 cm³ per oscillation requires 470 oscillations per minute to pass 0.32 L/s and 3840 oscillations per minute to pass 2.57 L/s. Calculate the percent error, or slip, in the meter.

$$\text{Slip} = (Q_{actual} - Q_{indicated})/Q_{actual}$$

$$Q = (\text{displacement per oscillation})(\text{number of oscillations per unit time})$$

$Q_1 = (40)(470)/60 = 313.3 \text{ cm}^3/\text{s}$ or 0.3133 L/s $\text{Slip}_1 = (0.32 - 0.3133)/0.32 = 0.0209$ or   2.09 percent

$Q_2 = (40)(3840)/60 = 2560 \text{ cm}^3/\text{s}$ or 2.560 L/s $\text{Slip}_2 = (2.57 - 2.560)/2.57 = 0.0039$ or   0.39 percent

17.239 A volumetric tank 4 ft in diameter and 5 ft high was filled with oil in 16 min 32.4 s. What is the average discharge in gallons per minute?

$$Q = V/t = [(\pi)(4)^2/4](5)/[16 + 32.4/60)] = 3.799 \text{ ft}^3/\text{min} = (3.799)(7.48) = 28.4 \text{ gpm}$$

17.240 A weigh tank receives 75 N of liquid, s.g. = 0.86, in 14.9 s. What is the flow rate, in liters per minute?

$$V = W/\gamma = 75/[(0.86)(9.79)(1000)] = 0.008908 \text{ m}^3$$

$$Q = V/t = 0.008908/(14.9/60) = 0.0359 \text{ m}^3/\text{min}   \text{ or }   35.9 \text{ L/min}$$

17.241 A critical-depth meter 16 ft wide has a rise in the bottom of 2.0 ft. For an upstream depth of 3.52 ft, determine the flow through the meter.

$$q = 0.517 g^{1/2}(y_1 - z + 0.55 q^2/g y_1^2)^{3/2}$$
$$= (0.517)(32.2)^{1/2}\{3.52 - 2.0 + 0.55 q^2/[(32.2)(3.52)^2]\}^{3/2}$$
$$= (2.934)(1.52 + 0.001378 q^2)^{3/2}$$

$$q = 5.75 \text{ cfs/ft}   \text{ (by trial and error)}$$

$$Q = (5.75)(16) = 92.0 \text{ cfs}$$

CHAPTER 18
Dimensional Analysis and Similitude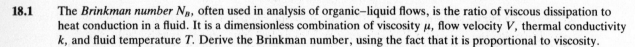

18.1 The *Brinkman number* N_B, often used in analysis of organic–liquid flows, is the ratio of viscous dissipation to heat conduction in a fluid. It is a dimensionless combination of viscosity μ, flow velocity V, thermal conductivity k, and fluid temperature T. Derive the Brinkman number, using the fact that it is proportional to viscosity.

I $N_B = f(\mu, V, k, T)$ and is proportional to μ.

$$N_B = \{1\} = \mu^1 V^a k^b T^c = \{M/LT\}^1 \{L/T\}^a \{ML/T^3\theta\}^b \{\theta\}^c = M^0 L^0 T^0 \theta^0$$

$$L: \quad -1 + a + b = 0 \qquad M: \quad 1 + b = 0 \qquad T: \quad -1 - a - 3b = 0 \qquad \theta: \quad -b + c = 0$$

Hence, $a = 2$ and $b = c = -1$. $N_B = \mu V^2 / kT$.

18.2 The velocity of sound c of a gas varies with pressure p and density ρ. Show by dimensional reasoning that the proper form must be $c = (\text{constant})(p/\rho)^{1/2}$.

I
$$c = f(p, \rho) = (\text{constant})p^a \rho^b \qquad \{L/T\} = \{M/LT^2\}^a \{M/L^3\}^b$$

$$M: \quad 0 = a + b \qquad L: \quad 1 = -a - 3b \qquad T: \quad -1 = -2a$$

Hence, $a = \frac{1}{2}$ and $b = -\frac{1}{2}$. $c = (\text{constant})(p/\rho)^{1/2}$.

18.3 The speed of propagation C of a capillary wave in deep water is known to be a function only of density ρ, wavelength λ, and surface tension σ. Find the proper functional relationship, completing it with a dimensionless constant. For a given density and wavelength, how does the propagation speed change if surface tension is doubled?

I
$$C = f(\rho, \lambda, \sigma) = (\text{constant})\rho^a \lambda^b \psi^c \qquad \{L/T\} = \{M/L^3\}^a \{L\}^b \{M/T^2\}^c$$

$$M: \quad 0 = a + c \qquad L: \quad 1 = -3a + b \qquad T: \quad -1 = -2c$$

Hence, $a = -\frac{1}{2}$, $b = -\frac{1}{2}$, $c = \frac{1}{2}$. $C = (\text{constant})(\sigma/\rho\lambda)^{1/2}$. If σ is doubled, C increases as $\sqrt{2}$, or 41 percent.

18.4 The excess pressure Δp inside a bubble is known to be a function of the surface tension and the radius. By dimensional reasoning determine how the excess pressure will vary if we double (a) the surface tension and (b) the radius.

I
$$\Delta p = f(\sigma, r) = (\text{constant})\sigma^a r^b \qquad \{F/L^2\} = \{F/L\}^a \{L\}^b$$

$$F: \quad 1 = a \qquad L: \quad -2 = -a + b$$

Hence, $a = 1$ and $b = -1$. $\Delta p = (\text{constant})(\sigma/r)$. (a) If σ is doubled, Δp is doubled. (b) If r is doubled, Δp is halved.

18.5 The pressure drop in a Venturi meter varies only with fluid density, pipe approach velocity, and the diameter ratio of the meter. A model Venturi meter tested in water at 20 °C shows a 5-kPa drop when the approach velocity is 4 m/s. A geometrically similar prototype meter is used to measure gasoline ($\rho = 680 \text{ kg/m}^3$) at 20 °C and a flow rate of 9 m^3/min. If the prototype pressure gage is most accurate at 15 kPa, what should the upstream pipe diameter be?

I
$$\Delta p = f(\rho, V, d/D) \qquad \Delta p / \rho V^2 = f(d/D) \qquad (\Delta p / \rho V^2)_{\text{model}} = (\Delta p / \rho V^2)_{\text{prototype}}$$

$$(\Delta p / \rho V^2)_{\text{model}} = (5)(1000) / [(998)(4)^2] = 0.3131$$

$$(\Delta p / \rho V^2)_{\text{prototype}} = (15)(1000) / (680V^2) = 22.06 / V^2 \qquad 0.3131 = 22.06 / V^2 \qquad V = 8.394 \text{ m/s}$$

$$Q = AV \qquad \tfrac{9}{60} = (\pi D_p^2 / 4)(8.394) \qquad D_p = 0.151 \text{ m} \quad \text{or} \quad 151 \text{ mm}$$

18.6 For two hydraulic machines to be homologous, they must (a) be geometrically similar, (b) have the same discharge coefficient when viewed as an orifice, $Q_1 / (A_1 \sqrt{2gH_1}) = Q_2 / (A_2 \sqrt{2gH_2})$, and (c) have the same ratio of peripheral speed to fluid velocity, $\omega D / (Q/A)$. Show that the scaling ratios can be expressed as $Q/ND^3 = \text{constant}$ and $H/(ND)^2 = \text{constant}$. N is the rotational speed.

▮ (1) From condition (b),

$$Q_1/(A_1\sqrt{H_1}) = Q_2/(A_2\sqrt{H_2}) \qquad (1)$$

From condition (c),

$$\omega_1 D_1 A_1/Q_1 = \omega_2 D_2 A_2/Q_2 \qquad (2)$$

$$(Q_1/\omega_1 D_1)(\omega_2 D_2/Q_2) = A_1/A_2$$

Geometrical similarity yields $A_1/A_2 = D_1^2/D_2^2 = (Q_1/\omega_1 D_1)(\omega_2 D_2/Q_2)$. Hence, $Q_1/\omega_1 D_1^3 = Q_2/\omega_2 D_2^3$.
   (2) From Eq. (1), $Q_1/A_1 = \sqrt{H_1/H_2}Q_2/A_2$. From Eq. (2), $Q_1/A_1 = (Q_2/A_2)(\omega_1 D_1/\omega_2 D_2)$. Hence, $\omega_1 D_1/\omega_2 D_2 = \sqrt{H_1/H_2}$, $H_1/(\omega_1 D_1)^2 = H_2/(\omega_2 D_2)^2$. (N is rotational speed in revolutions per minute; ω is in radians per second. They are equivalent for modeling purposes.)

18.7 Use the scaling ratios of Prob. 18.6 to determine the head and discharge of a 1:4 model of a centrifugal pump that produces 600 L/s at a 30 m head when turning at 240 rpm. The model operates at 1200 rpm.

▮ From Prob. 18.6, $H_1/(\omega_1 D_1)^2 = H_2/(\omega_2 D_2)^2$ or $H_m = H_p[(\omega_m/\omega_p)(D_m/D_p)]^2 = (30)[(\frac{1200}{240})(\frac{1}{4})]^2 = 46.88$ m; $Q_1/\omega_1 D_1 = Q_2/\omega_2 D_2$ or $Q_m = Q_p(\omega_m/\omega_p)(D_m/D_p)^3 = (600)(\frac{1200}{240})(\frac{1}{4})^3 = 46.88$ L/s.

18.8 For an ideal liquid, express the flow Q through an orifice in terms of the density of the liquid, the diameter of the orifice, and the pressure difference.

▮ $Q = f(\rho, p, d)$ or $Q = K\rho^a p^b d^c$. Then, dimensionally, $F^0 L^3 T^{-1} = (F^a T^{2a} L^{-4a})(F^b L^{-2b})(L^c)$ and $0 = a + b$, $3 = -4a - 2b + c$, $-1 = 2a$, from which $a = -\frac{1}{2}$, $b = \frac{1}{2}$, $c = 2$. Substituting, $Q = K\rho^{-1/2}p^{1/2}d^2$ or ideal $Q = Kd^2\sqrt{p/\rho}$. Factor K must be obtained by physical analysis and/or experimentation.
   For an orifice in the side of a tank under head h, $p = \gamma h$. To obtain the familiar orifice formula, let $K = \sqrt{2}(\pi/4)$. Then ideal $Q = \sqrt{2}(\pi/4)d^2\sqrt{\gamma h/\rho}$. But $g = \gamma/\rho$; hence ideal $Q = \frac{1}{4}\pi d^2\sqrt{2gh}$.

18.9 Determine the dynamic pressure exerted by a flowing incompressible fluid on an immersed object, assuming the pressure is a function of the density and the velocity.

▮ $p = f(\rho, V)$ or $p = K\rho^a V^b$. Then, dimensionally, $F^1 L^{-2} T^0 = (F^a T^{2a} L^{-4a})(L^b T^{-b})$ and $1 = a$, $-2 = -4a + b$, $0 = 2a - b$, from which $a = 1$, $b = 2$. Substituting, $p = K_\rho V^2$.

18.10 Develop the expression for lost head in a horizontal pipe for turbulent incompressible flow.

▮ For any fluid, the lost head is represented by the drop in the pressure gradient and is a measure of the resistance to flow through the pipe. The resistance is a function of the diameter of the pipe, the viscosity and density of the fluid, the length of the pipe, the velocity of the fluid, and the roughness K of the pipe. We may write $(p_1 - p_2) = f(d, \mu, \rho, L, V, K)$ or

$$(p_1 - p_2) = Cd^a \mu^b \rho^c L^d V^e (\epsilon/d)^f \qquad (1)$$

From experiment and observation, the exponent of the length L is unity. The value of K is usually expressed as a ratio of the size of the surface protuberances ϵ to the diameter d of the pipe, a dimensionless number. We may now write $F^1 L^{-2} T^0 = (L^a)(F^b T^b L^{-2b})(F^c T^{2c} L^{-4c})(L^e T^{-e})(L^f/L^f)$ and $1 = b + c$, $-2 = a - 2b - 4c + 1 + e + f - f$, $0 = b + 2c - e$ from which the values of a, b, and c may be determined in terms of e, or $c = e - 1$, $b = 2 - e$, $a = e - 3$. Substituting in (1), $(p_1 - p_2) = Cd^{e-3}\mu^{2-e}\rho^{e-1}L^1 V^e(\epsilon/d)^f$. Dividing the left side of the equation by w and the right side by its equivalent ρg, $(p_1 - p_2)/w =$ lost head $= C(\epsilon/d)^f L(d^{e-3}V^e \rho^{e-1}\mu^{2-e})/(\rho g)$, which becomes (introducing 2 in numerator and in denominator): lost head $= 2C(\epsilon/d)^f (L/d)(V^2/2g)(d^{e-2}V^{e-2}\rho^{e-2}/\mu^{e-2}) = K'(R_E^{e-2})(L/d)(V^2/2g) = f(L/d)(V^2/2g)$ (Darcy formula).

18.11 A copepod is a water crustacean approximately 1 mm in diameter. We want to known the drag force on the copepod when it moves slowly in fresh water. A scale model 100 times larger is made and tested in glycerin at $V = 30$ cm/s. The measured drag on the model is 1.3 N. For similar conditions, what are the velocity and drag of the actual copepod in water?

▮ Water (prototype): $\mu_p = 0.001$ kg/(m-s) $\rho_p = 999$ kg/m^3

   Glycerin (model): $\mu_m = 1.5$ kg/(m-s) $\rho_m = 1263$ kg/m^3

$(N_R)_m = \rho_m V_m L_m/\mu_m = (1263)(0.3)(0.1)/1.5 = 25.3$ $(C_F)_m = F_m/\rho_m V_m^2 L_m^2 = 1.3/[(1263)(0.3)^2(0.1)^2] = 1.14$

$(N_R)_p = (N_R)_m = 25.3 = 999 V_p(0.001)/0.001$ or $V_p = 0.0253$ m/s $= 2.53$ cm/s

$(C_F)_p = (C_F)_m = 1.14 = F_p/[999(0.0253)^2(0.001)^2]$ or $F_p = 7.31 \times 10^{-7}$ N

It would obviously be difficult to measure such a tiny drag force.

18.12 A 0.1-ft-diameter steel sphere ($\rho_s = 15.2$ slugs/ft^3) is dropped in water [$\rho = 1.94$ slugs/ft^3, $\mu = 0.000021$ slug/(ft-s)] until it reaches terminal velocity or zero acceleration. From the sphere data in Fig. A-38 compute the terminal velocity of the falling sphere in feet per second.

▮ At terminal velocity, the net weight of the sphere equals the drag; hence the drag is known in this problem: $D = W_{net} = (\rho_s - \rho)g(\pi/6)d^3 = (15.2 - 1.94)(32.2)(\pi/6)(0.1)^3 = 0.224$ lb. We can compute that portion of C_D and N_R which excludes the unknown velocity: $C_D = D/[\frac{1}{2}\rho(\pi/4)d^2V^2] = 0.224/[1.94(\pi/8)(0.1)^2V^2] = 29.4/V^2$, $N_R = \rho Vd/\mu = (1.94)(V)(0.1)/(2.1 \times 10^{-5}) = 9240V$. Now we will just have to guess an initial velocity V to get started on the iteration.

Guess $V = 1.0$ ft/s; then $N_R = 9240(1.0) = 9240$. From Fig. A-38 read $C_D \approx 0.38$; then $V \approx (29.4/C_D)^{1/2} = 8.8$ ft/s. Now try again with this new guess.

Guess $V = 8.8$ ft/s, $N_R = 9240(8.8) = 81\,000$. From Fig. A-38 read $C_D \approx 0.52$, $V \approx (29.4/0.52)^{1/2} = 7.5$ ft/s. One more try will give pretty good convergence.

Guess $V = 7.5$ ft/s, $N_R = 9240(7.5) = 69\,000$. From Fig. A-38 read $C_D \approx 0.51$, $V \approx (29.4/0.51)^{1/2} = 7.6$ ft/s. To the accuracy of the figure, $V_{term} \approx 7.6$ ft/s.

18.13 Repeat Prob. 18.12, using the regrouped chart, Fig. A-39.

▮ We must repeat the calculation of the net weight to establish that $D = W_{net} = 0.224$ lb. But now we can go directly to the new drag coefficient: $C_F' = D\rho/\mu^2 = (0.224)(1.194)/(2.1 \times 10^{-5})^2 = 9.85 \times 10^8$. Now enter Fig. A-39 and read $N_R = 70\,000$. Then the desired velocity is $V = \mu N_R/\rho d = (2.1 \times 10^{-5})(70\,000)/[(1.94)(0.1)] = 7.58$ ft/s.

18.14 A 6-cm-diameter sphere is tested in water at 20 °C and a velocity of 3 m/s and has a measured drag of 6 N. What will be the velocity and drag force of a 2-m-diameter weather balloon moving in air at 20 °C and 1 atm under similar conditions?

▮
$$(N_R)_m = (N_R)_p \qquad N_R = \rho DV/\mu \qquad (998)(\tfrac{6}{100})(3)/(1.02 \times 10^{-3}) = (1.20)(2)(V_b)/(1.81 \times 10^{-5})$$
$$V_b = 1.33 \text{ m/s}$$
$$(C_F)_m = (C_F)_p \qquad C_F = F/\rho V^2 D^2 \qquad 6/[(99)(3)^2(\tfrac{6}{100})^2] = F_b/[(1.20)(1.33)^2(2)^2] \qquad F_b = 1.58 \text{ N}$$

18.15 It is desired to measure the drag on an airplane whose velocity is 300 mph. Is it feasible to test a one-twentieth-scale model of a plane in a wind tunnel at the same pressure and temperature to determine the prototype drag coefficient?

▮
$$(N_R)_m = (N_R)_p \qquad N_R = LV/\nu \qquad L_m = L_p/20$$

Since $p_m = p_p$ and $T_m = T_p$, $\nu_m = \nu_p$. Therefore, $(L_p/20)(V_m)/\nu_p = (L_p)[(300)(5280)/3600]/\nu_p$, $V_m = 8800$ ft/s. Since this velocity is hypersonic, the proposed model is not feasible.

18.16 A one-twentieth-scale model of a submarine is tested at 200 ft/s in a wind tunnel using sea-level standard air. What is the prototype speed in sea water at 20 °C for dynamic similarity? If the model drag is 1.0 lb, what is the prototype drag?

▮
$$(N_R)_m = (N_R)_p \qquad N_R = \rho LV/\mu \qquad L_m = L_p/20$$
$$(0.00234)(L_p/20)(200)/(3.78 \times 10^{-7}) = (2.00)(L_p)(V_p)/(2.23 \times 10^{-5}) \qquad V_p = 0.690 \text{ ft/s}$$
$$(C_F)_m = (C_F)_p \qquad C_F = F/\rho V^2 L^2$$
$$1.0/[(0.00234)(200)^2(L_p/20)^2] = F_p/[(2.00)(0.690)^2(L_p)^2] \qquad F_p = 4.07 \text{ lb}$$

18.17 We wish to know the drag of a blimp which will move in air at 20 °C at 6 m/s. If a one-thirtieth-scale model is tested in water at 20 °C, what should the water velocity be? If the measured water drag on the model is 2700 N, what is the drag on the prototype blimp and the power required to propel it?

▮
$$(N_R)_m = (N_R)_p \qquad N_R = \rho LV/\mu \qquad L_m = L_p/30$$
$$(998)(L_p/30)(V_m)/(1.02 \times 10^{-3}) = (1.20)(L_p)(6)/(1.81 \times 10^{-5}) \qquad V_m = 12.2 \text{ m/s}$$
$$(C_F)_m = (C_F)_p \qquad C_F = F/\rho V^2 L^2 \qquad 2700/[(998)(12.2)^2(L_p/30)^2] = F_p/[(1.20)(6)^2(L_p)^2]$$
$$F_p = 707 \text{ N} \qquad P_p = F_p V_p = (707)(6) = 4242 \text{ N-m/s} \quad \text{or} \quad 4242 \text{ W}$$

18.18 A one-fifth-scale model automobile is tested in a wind tunnel in the same air properties as the prototype. The prototype velocity is 50 km/h. For dynamically similar conditions, the model drag is 350 N. What are the drag of the prototype automobile and the power required to overcome this drag?

$$ (N_R)_m = (N_R)_p \qquad N_R = \rho L V / \mu \qquad L_m = L_p/5 $$

$$ (\rho)(L_p/5)(V_m)/\mu = (\rho)(L_p)[(50)(1000)/3600]/\mu \qquad V_m = 69.44 \text{ m/s} $$

$$ (C_F)_m = (C_F)_p \qquad C_F = F/\rho V^2 L^2 \qquad 350/[(\rho)(69.44)^2(L_p/5)^2] = F_p/[(\rho)[(50)(1000)/3600]^2(L_p)^2] $$

$$ F_p = 350 \text{ N} \qquad P_p = F_p V_p = (350)[(50)(1000)/3600] = 4861 \text{ N-m/s} \quad \text{or} \quad 4861 \text{ W} $$

18.19 A model airplane has linear dimensions that are one-twentieth those of its prototype. If the plane is to fly at 400 mph, what must be the air velocity in the wind tunnel for the same Reynolds number if the air temperature and pressure are the same?

$$ (N_R)_m = (N_R)_p \qquad N_R = \rho L V / \mu \qquad L_m = L_p/20 \qquad (\rho)(L_p/20)(V_m)/\mu = (\rho)(L_p)(400)/\mu $$

$$ V_m = 8000 \text{ mph} $$

(This answer, which indicates a velocity about 10 times the speed of sound, illustrates why it is not generally possible to test model aircraft by the Reynolds model law in atmospheric wind tunnels.)

18.20 A model airplane has dimensions that are one-twentieth those of its prototype. It is desired to test it in a pressure wind tunnel at a speed the same as that of the prototype. If the air temperature is the same and the Reynolds number is the same, what must be the pressure in the wind tunnel relative to the atmospheric pressure?

$$ (N_R)_m = (N_R)_p \qquad N_R = LV/\nu \qquad L_m = L_p/20 \qquad (L_p/20)(V)\nu_m = (L_p)(V)\nu_p $$

$$ \nu_m/\nu_p = \tfrac{1}{20} = (\mu_m/\rho_m)/(\mu_p/\rho_p) = (\mu_m/\mu_p)(\rho_p/\rho_m) $$

For $T_m = T_p$, assume $\mu_m = \mu_p$; hence, $\rho_p/\rho_m = \gamma_p/\gamma_m = \tfrac{1}{20}$, $\gamma_p/\gamma_m = (p_p/RT)/(p_m/RT) = \tfrac{1}{20}$. Therefore, $p_m = 20p_p = 20$ atm.

18.21 What weight flow rate of air at 70 °F ($\mu = 3.82 \times 10^{-7}$ lb-s/ft²) at 50 psi in a 1-in-diameter pipe will give dynamic similarity to a 250-gpm flow of water at 60 °F in a 4-in-diameter pipe?

$$ (N_R)_{\text{air}} = (N_R)_{H_2O} \qquad N_R = DV/\nu \qquad \nu = \mu/\rho $$

$$ \rho_{\text{air}} = p/RT = (50)(144)/[(1716)(70+460)] = 0.007917 \text{ slug/ft}^3 $$

$$ \nu_{\text{air}} = (3.82 \times 10^{-7})/0.007917 = 4.825 \times 10^{-5} \text{ ft}^2/\text{s} $$

$$ V_{\text{air}} = Q/[(\pi)(\tfrac{1}{12})^2/4] = 183.3Q \qquad V_{H_2O} = Q/A = (250)(0.002228)/[(\pi)(\tfrac{4}{12})^2/4] = 6.383 \text{ ft/s} $$

$$ (\tfrac{1}{12})(183.3Q)/(4.825 \times 10^{-5}) = (\tfrac{4}{12})(6.383)/(1.21 \times 10^{-5}) \qquad Q = 0.5554 \text{ ft}^3/\text{s} $$

$$ \gamma = \rho g \qquad \gamma_{\text{air}} = (0.007917)(32.2) = 0.2549 \text{ lb/ft}^3 \qquad G = \gamma Q = (0.2549)(0.5554) = 0.142 \text{ lb/s} $$

18.22 A 1:30 scale model of a submarine is tested in a wind tunnel. It is desired to know the drag on the submarine when it is operating at 10 knots (16.9 fps) in ocean water at 40 °F. At what velocity should the object be tested in a wind tunnel containing air at 70 °F at atmospheric pressure?

$$ (N_R)_{\text{air}} = (N_R)_{H_2O} \qquad N_R = LV/\nu \qquad L_{\text{air}} = L_{H_2O}/30 $$

$$ (L_{H_2O}/30)(V_{\text{air}})/(1.64 \times 10^{-4}) = (L_{H_2O})(16.9)/(1.66 \times 10^{-5}) \qquad V_{\text{air}} = 5009 \text{ ft/s} $$

Since this velocity exceeds sonic velocity, the model will not operate properly to indicate prototype behavior.

18.23 A ship 600 ft long is to operate at a speed of 25 mph in ocean water whose viscosity is 1.2 cP and specific weight is 64 lb/ft³. What should be the kinematic viscosity of a fluid used with a model so that both the Reynolds number and the Froude number would be the same? Does such a liquid exist? Assume the model is 10 ft long.

$$ (N_R)_m = (N_R)_p \qquad N_R = LV/\nu $$

Let subscript r denote ratio of prototype to model.

$$ L_r V_r/\nu_r = 1 \qquad (N_F)_m = (N_F)_p \qquad N_F = V/(gL)^{1/2} \qquad V_r/(g_r L_r)^{1/2} = 1 \qquad L_r V_r/\nu_r = V_r/(g_r L_r)^{1/2} $$

$$ g_r = 1.0 \qquad L_r = \tfrac{600}{10} = 60.0 \qquad (60.0)(V_r)/\nu_r = V_r/[(1.0)(60.0)]^{1/2} \qquad \nu_r = 464.8 $$

Since $1\text{ cP} = 0.00002089\text{ lb-s/ft}^2$, $\mu_p = (1.2)(0.00002089) = 0.00002507\text{ lb-s/ft}^2$.

$$v_p = \mu_p/\rho = \mu_p/(\gamma/g) = 0.00002507/(64/32.2) = 0.00001261\text{ ft}^2/\text{s}$$
$$v_r = v_p/v_m = 0.00001261/v_m = 464.8 \qquad v_m = 2.71 \times 10^{-8}\text{ ft}^2/\text{s}$$

There is no such liquid available.

18.24 The valve coefficients $K = \Delta p/(\rho V^2/2)$ for a 600-mm-diameter valve are to be determined from tests on a geometrically similar 300-mm-diameter valve using atmospheric air at 80 °F ($v = 1.57 \times 10^{-5}\text{ m}^2/\text{s}$). The ranges of tests should be for flow of water at 70 °F ($v = 9.96 \times 10^{-7}\text{ m}^2/\text{s}$) at 1 m/s to 2.5 m/s. What ranges of airflows are needed?

∎ $\qquad N_R = DV/v \qquad (N_R)_{\min} = (\frac{600}{1000})(1)/(9.96 \times 10^{-7}) = 6.02 \times 10^5$
$$(N_R)_{\max} = (\tfrac{600}{1000})(2.5)/(9.6 \times 10^{-7}) = 1.51 \times 10^6$$

For air:

$$6.02 \times 10^5 = (\tfrac{300}{1000})(V_{\min})/(1.57 \times 10^{-5}) \qquad V_{\min} = 31.5\text{ m/s}$$
$$1.51 \times 10^6 = (\tfrac{300}{1000})(V_{\max})/(1.57 \times 10^{-5}) \qquad V_{\max} = 79.0\text{ m/s}$$
$$Q_{\min} = AV_{\min} = [(\pi)(\tfrac{300}{1000})^2/4](31.5) = 2.23\text{ m}^3/\text{s} \qquad Q_{\max} = AV_{\max} = [(\pi)(\tfrac{300}{1000})^2/4](79.0) = 5.58\text{ m}^3/\text{s}$$

18.25 A model of a Venturi meter has linear dimensions one-fifth those of the prototype. The prototype operates on water at 20 °C, and the model on water at 95 °C. For a throat diameter of 600 mm and a velocity at the throat of 6 m/s, what discharge is needed through the model for similitude?

∎ $(N_R)_m = (N_R)_p \qquad N_R = LV/v \qquad (L_p/5)(V_m)/(3.08 \times 10^{-7}) = (L_p)(6)/(1.02 \times 10^{-6}) \qquad V_m = 9.059\text{ m/s}$
$$Q_m = A_m V_m \qquad D_m = D_p/5 = \tfrac{600}{5} = 120\text{ mm} \qquad Q_m = [(\pi)(\tfrac{120}{1000})^2/4](9.059) = 0.102\text{ m}^3/\text{s}$$

18.26 The losses in a Y in a 1.2-m-diameter pipe system carrying gas ($\rho = 40\text{ kg/m}^3$, $\mu = 0.002\text{ P}$, $V = 25\text{ m/s}$) are to be determined by testing a model with water at 20 °C. The laboratory has a water capacity of 75 L/s. What model scale should be used, and how are the results converted into prototype losses?

∎ $(N_R)_m = (N_R)_p \qquad N_R = \rho DV/\mu \qquad V_m = Q_m/A_m = (\tfrac{75}{1000})/(\pi D_m^2/4) = 0.09549/D_m^2$
$$(998)(D_m)(0.09549/D_m^2)/(1.02 \times 10^{-3}) = (40)(1.2)(25)/[(0.002)(0.1)]$$
$$D_m = 0.01557\text{ m} \qquad D_m/D_p = 0.01557/1.2 = 1/77$$

Choose a model size equal to $\frac{1}{77}$ of the prototype size or smaller. $H_f = (f)(L/D)(V^2/2g)$. If roughness is to scale, $(\epsilon/D)_m = (\epsilon/D)_p$. Since $(N_R)_m = (N_R)_p$, $f_p = f_m$. Therefore, $(h_f)_p = (h_f)_m(V_p^2/V_m^2) = (h_f)_m(D_m v_p/D_p v_m)^2$. The actual ratio of losses depends on the chosen D ratio.

18.27 A 1:5 scale model of a water pumping station pumping system is to be tested to determine overall head losses. Air at 25 °C, 1 atm, is available. For a prototype velocity of 500 mm/s in a 4-m-diameter section with water at 15 °C, determine the air velocity and quantity needed and how losses determined from the model are converted into prototype losses.

∎ $(N_R)_m = (N_R)_p \qquad N_R = DV/v \qquad [(4)(\tfrac{1}{5})](V_m)/(1.56 \times 10^{-5}) = (4)(\tfrac{500}{1000})/(1.16 \times 10^{-6})$
$$V_m = 33.62\text{ m/s} \qquad Q_m = A_m V_m = \{(\pi)[(4)(\tfrac{1}{5})]^2/4\}(33.62) = 16.9\text{ m}^3/\text{s}$$

Losses are the same when expressed in velocity heads.

18.28 Water at 60 °F flows at 12.0 fps in a 6-in pipe. At what velocity must medium fuel oil at 90 °F ($v = 3.19 \times 10^{-5}\text{ ft}^2/\text{s}$) flow in a 3-in pipe for the two flows to be dynamically similar?

∎ $(N_R)_{H_2O} = (N_R)_{oil} \qquad N_R = DV/v \qquad (\tfrac{6}{12})(12.0)/(1.21 \times 10^{-5}) = (\tfrac{3}{12})(V_{oil})/(3.19 \times 10^{-5}) \qquad V_{oil} = 63.3\text{ ft/s}$

18.29 Air at 68 °F is to flow through a 24-in pipe at an average velocity of 6.00 fps. For dynamic similarity, what size pipe carrying water at 60 °F at 3.65 fps should be used?

∎ $(N_R)_{air} = (N_R)_{H_2O} \qquad N_R = DV/v \qquad (\tfrac{24}{12})(6.00)/(1.63 \times 10^{-4}) = (D/12)(3.65)/(1.21 \times 10^{-5}) \qquad D = 2.93\text{ in}$

(Would probably use a 3-in pipe.)

18.30 A 1:5 model of a submarine is to be tested in a towing tank containing salt water. If the submarine moves at 12.0 mph, at what velocity should the model be towed for dynamic similarity?

$$(N_R)_m = (N_R)_p \qquad N_R = LV/\nu \qquad (L/15)(V_m)/\nu = (L)(12.0)/\nu \qquad V_m = 180 \text{ mph}$$

18.31 A model of a torpedo is tested in a towing tank at a velocity of 80.0 fps. The prototype is expected to attain a velocity of 20.0 fps in water at 60 °F. (*a*) What model scale has been used? (*b*) What would be the model speed if tested in a wind tunnel under a pressure of 20 atm and at constant temperature 80 °F?

$$(N_R)_m = (N_R)_p \qquad N_R = DV/\nu$$

(*a*) $(L/x)(80.0)/\nu = (L)(20.0)/\nu$, $x = 4.00$. Hence, the scale model is 1:4.

(*b*) $\qquad v = \mu/\rho \qquad \rho_{air} = p/RT = (20)(14.7)(144)/[(1716)(80 + 460)] = 0.04569 \text{ slug/ft}^3$

$$\nu_{air} = (3.85 \times 10^{-7})/0.04569 = 8.426 \times 10^{-6} \text{ ft}^2/\text{s}$$

$$(L/4)(V_m)/(8.426 \times 10^{-6}) = (L)(20.0)/(1.21 \times 10^{-5}) \qquad V_m = 55.7 \text{ ft/s}$$

18.32 A centrifugal pump pumps medium lubricating oil at 60 °F ($\nu = 188 \times 10^{-5} \text{ ft}^2/\text{s}$) while rotating at 1200 rpm. A model pump, using air at 68 °F, is to be tested. If the diameter of the model is 3 times the diameter of the prototype, at what speed should the model run?

$(N_R)_m = (N_R)_p$, $N_R = DV/\nu$. Using the peripheral speeds (which equal radius times angular velocity in radians per second) as the velocities in Reynolds number, we obtain $(3D)[(3D/2)(\omega_m)]/(1.63 \times 10^{-4}) = (D)[(D/2)(\omega_p)]/(188 \times 10^{-5})$, $\omega_p = 103.8\omega_m$. Hence, model speed = $1200/103.8 = 11.6$ rpm.

18.33 An airplane wing of 3-ft chord is to move at 90 mph in air. A model of 3-in chord is to be tested in a wind tunnel with air velocity at 108 mph. For air temperature of 68 °F in each case, what should be the pressure in the wind tunnel?

$$(N_R)_m = (N_R)_p \qquad N_R = LV/\nu \qquad (\tfrac{3}{12})(108)/\nu_m = (3)(90)/(1.63 \times 10^{-4}) \qquad \nu_m = 1.63 \times 10^{-5} \text{ ft}^2/\text{s}$$

The pressure that produces this kinematic viscosity of air at 68 °F can be found by remembering that the absolute viscosity is not affected by pressure changes. The kinematic viscosity equals absolute viscosity divided by density. But density increases with pressure (temperature constant). $\nu = \mu/\rho$, $\nu_p/\nu_m = (1.63 \times 10^{-4})/(1.63 \times 10^{-5}) = 10$. Thus the density of air in the tunnel must be ten times standard (68 °F) air and the resulting pressure in the tunnel must be 10 atm.

18.34 A dam spillway is to be tested using Froude scaling with a one-thirtieth-scale model. The model flow has an average velocity of 0.6 m/s and a volume flow of 0.05 m³/s. What will the velocity and flow of the prototype be? If the measured force on a certain part of the model is 1.5 N, what will the corresponding force on the prototype be?

For Froude scaling,

$$V_p/V_m = 1/\sqrt{\alpha} \qquad \alpha = L_m/L_p = \tfrac{1}{30} \qquad V_p/0.6 = 1/\sqrt{\tfrac{1}{30}} \qquad V_p = 3.29 \text{ m/s}$$

$$Q_p/Q_m = (V_p/V_m)(L_p/L_m)^2 = (1/\sqrt{\alpha})(1/\alpha)^2 = (1/\alpha)^{5/2} \qquad Q_p/0.05 = [1/(\tfrac{1}{30})]^{5/2} \qquad Q_p = 246 \text{ m}^3/\text{s}$$

$$F_p/F_m = (\rho_p/\rho_m)(V_p/V_m)^2(L_p/L_m)^2 = (1)(1/\sqrt{\alpha})^2(1/\alpha)^2 = (1/\alpha)^3 \qquad F_p/1.5 = [1/(\tfrac{1}{30})]^3 \qquad F_p = 40\,500 \text{ N}$$

18.35 A prototype ship is 35 m long and designed to cruise at 11 m/s (about 21 knots). Its drag is to be simulated by a 1-m-long model pulled in a tow tank. For Froude scaling find (*a*) the tow speed, (*b*) the ratio of prototype to model drag, and (*c*) the ratio of prototype to model power.

For Froude scaling,

(*a*) $\qquad V_p/V_m = 1/\sqrt{\alpha} \qquad \alpha = L_m/L_p = \tfrac{1}{35} \qquad 11/V_m = 1/\sqrt{\tfrac{1}{35}} \qquad V_m = 1.86 \text{ m/s}$

(*b*) $\qquad F_m/F_p = (\rho_m/\rho_p)(V_m/V_p)^2(L_m/L_p)^2 = (1)(\sqrt{\alpha})^2(\alpha)^2 = \alpha^3 = (\tfrac{1}{35})^3 = 1/42\,875$

(*c*) $\qquad P_m/P_p = (F_m/F_p)(V_m/V_p) = (\alpha)^3(\sqrt{\alpha}) = \alpha^{3.5} = (\tfrac{1}{35})^{3.5} = 1/253\,652$

18.36 A ship 600 ft long is to operate at a speed of 25 mph. If a model is 10 ft long, what should be its speed to give the same Froude number? What is the value of the Froude number?

$$(N_F)_m = (N_F)_p \qquad N_F = V/(gL)^{1/2} \qquad V_m/[(g)(10)]^{1/2} = 25/[(g)(600)]^{1/2}$$

$$V_m = 3.23 \text{ mph} \quad \text{or} \quad 4.73 \text{ ft/s} \qquad N_F = 4.73/[(32.2)(10)]^{1/2} = 0.264$$

18.37 In a $1:40$ model of the flow over the crest of a spillway, the velocity at a particular point is 0.5 m/s. What velocity does this represent in the prototype? The force exerted on a certain area in the model is 0.12 N. What would be the force on the corresponding area in the prototype?

$$ (N_F)_m = (N_F)_p \qquad N_F = V/(gL)^{1/2} \qquad 0.5/[(g)(L_p/40)]^{1/2} = V_p/(gL_p)^{1/2} \qquad V_p = 3.16 \text{ m/s} $$

$$ (C_F)_m = (C_F)_p \qquad C_F = F/\rho V^2 L^2 \qquad 0.12/[(\rho)(0.5)^2(L_p/40)^2] = F_p/[(\rho)(3.16)^2(L_p)^2] \qquad F_p = 7669 \text{ N} $$

18.38 A $1:500$ model is constructed to study tides. What length of time in the model corresponds to a day in the prototype? Suppose the model could be transported to the moon and tested there. What then would be the time relationship between the model and prototype? g of earth equals six times g of moon.

$$ (N_F)_m = (N_F)_p \qquad N_F = V/(gL)^{1/2} \qquad [V/(gL)^{1/2}]_m = [V/(gL)^{1/2}]_p $$

$$ V_r = V_p/V_m = (gL)_r^{1/2} \qquad T_r = L_r/V_r = L_r(gL)_r^{1/2} \qquad T_p/T_m = [L/(gL)^{1/2}]_p/[L/(gL)^{1/2}]_m $$

On earth:

$$ 1/T_m = \{L_p/[(g)(L_p)]^{1/2}\}/\{(L_p/500)/[(g)(L_p/500)]^{1/2}\} \qquad T_m = 0.04472 \text{ day} \quad \text{or} \quad 1.07 \text{ h} $$

On the moon:

$$ 1/T_m = \{L_p/[(g_p)(L_p)]^{1/2}\}/\{L_p/500)/[(g_p/6)(L_p/500)]^{1/2}\} \qquad T_m = 0.1095 \text{ day} \quad \text{or} \quad 2.63 \text{ h} $$

18.39 A sectional model of a spillway 3 ft high is placed in a laboratory flume of 10-in width. Under a head of 0.375 ft the flow is 0.70 cfs. What flow does this represent in the prototype if the scale model is $1:25$ and the spillway is 650 ft long?

$$ (N_F)_m = (N_F)_p \qquad N_F = V/(gL)^{1/2} \qquad (N_F)_r = (N_F)_p/(N_F)_m = [V/(gL)^{1/2}]_p/[V/(gL)^{1/2}]_m = V_r/L_r^{1/2} = 1.00 $$

$$ V_r/25^{1/2} = 1.00 \qquad V_r = 5.00 \qquad Q_r = A_r V_r \qquad A_r = A_p/A_m = (650)[(0.375)(25)]/[(\tfrac{10}{12})(0.375)] = 19\,500 $$

$$ Q_r = A_r V_r = (19\,500)(5.00) = 97\,500 \qquad Q = (97\,500)(0.70) = 68\,250 \text{ ft}^3/\text{s} $$

18.40 The flow over a model spillway is 0.086 m³/s per meter of width. What flow does this represent in the prototype spillway if the model scale is $1:18$?

$$ q = Q/b = [(by)(V)]/b = yV \qquad q_r = y_r V_r \qquad y_r = L_r $$

$$ V_r/L_r^{1/2} = 1.00 \qquad \text{or} \qquad V_r = L_r^{1/2} \qquad \text{(from Prob. 18.39)} $$

$$ q_r = L_r L_r^{1/2} = L_r^{3/2} = 18^{3/2} = 76.37 \qquad q_p = (76.37)(0.086) = 6.57 \text{ m}^3/\text{s/m} $$

18.41 The velocity at a point in a model of a spillway for a dam is 1 m/s. For a ratio of prototype to model of $10:1$, what is the velocity at the corresponding point in the prototype under similar conditions?

$$ (N_F)_m = (N_F)_p \qquad N_F = V/(gL)^{1/2} \qquad 1/[(g)(L_p/10)]^{1/2} = V_p/(gL_p)^{1/2} \qquad V_p = 3.16 \text{ m/s} $$

18.42 The wave drag on a model of a ship is 16 N at a speed of 3 m/s. For a prototype 15 times as long, what will the corresponding speed and wave drag be if the liquid is the same in each case?

$$ (N_F)_m = (N_F)_p \qquad N_F = V/(gL)^{1/2} \qquad 3/[(g)(L_p/15)]^{1/2} = V_p/(gL_p)^{1/2} \qquad V_p = 11.62 \text{ m/s} $$

$$ C_D = F/(A\rho V^2/2) $$

At the same Froude number, C_D is equal for model and prototype; hence, $F_m/(A_m V_m^2/2) = F_p/(A_p V_p^2/2)$, $F_p = (F_m)(A_p/A_m)(V_p/V_m)^2 = (F_m)(L_p/L_m)^3 = (16)(15)^3 = 54\,000$ N, or 54.0 kN.

18.43 Oil of kinematic viscosity 50×10^{-5} ft²/s is to be used in a prototype in which both viscous and gravity forces dominate. A model scale of $1:5$ is also desired. What viscosity of model liquid is necessary to make both Froude number and Reynolds number the same in model and prototype?

$$ (N_F)_m = (N_F)_p \qquad N_F = V/(gL)^{1/2} \qquad V_r = (g_r L_r)^{1/2} = L_r^{1/2} \qquad (N_R)_m = (N_R)_p \qquad N_R = LV/\nu $$

$$ V_r = \nu_r/L_r \qquad L_r^{1/2} = \nu_r/L_r \qquad \nu_r = L_r^{3/2} = (\tfrac{1}{5})^{3/2} = 0.08944 $$

$$ \nu_m = (0.08944)(50 \times 10^{-5}) = 4.47 \times 10^{-5} \text{ ft}^2/\text{s} $$

18.44 A ship whose hull length is 460 ft is to travel at 25.0 fps. (**a**) Compute the Froude number. (**b**) For dynamic similarity, at what velocity should a $1:30$ model be towed through water?

❚ *(a)* $$N_F = V/(gL)^{1/2} = 25.0/[(32.2)(469)]^{1/2} = 0.205$$

(b) $\quad (N_F)_m = (N_F)_p \quad V_m/[(32.2)(\frac{460}{30})]^{1/2} = 25.0/[(32.2)(460)]^{1/2} \quad V_m = 4.56 \text{ ft/s}$

18.45 An airplane is designed to fly at 260 m/s at 10 000 m standard altitude. If a one-tenth-scale model is tested in a pressurized wind tunnel at 20 °C, what should the tunnel pressure be to scale both the Reynolds number and the Mach number correctly?

❚ At 10 000-m standard altitude, $\rho_p = 0.414 \text{ kg/m}^3$, $T_p = -49.9\,°C$, and $\mu_p = 1.46 \times 10^{-5}$ N-s/m^2 (from Table A-8).

$$(N_M)_m = (N_M)_p \qquad N_M = V/c \qquad V_m/c_m = V_p/c_p \qquad c \approx 20T^{1/2}$$
$$c_m \approx (20)(20 + 273)^{1/2} = 342.3 \text{ m/s} \qquad c_p \approx (20)(-49.9 + 273)^{1/2} = 298.7 \text{ m/s}$$
$$V_m/342.3 = 260/298.7 \qquad V_m = 298.0 \text{ m/s} \qquad (N_R)_m = (N_R)_p \qquad N_R = \rho L V/\mu$$
$$(\rho_m)(L_p/10)(298.0)/(1.81 \times 10^{-5}) = (0.414)(L_p)(260)/(1.46 \times 10^{-5}) \qquad \rho_m = 4.48 \text{ kg/m}^3$$
$$p_m = \rho_m R T_m = (4.48)(287)(20 + 273) = 377\,000 \text{ Pa} \quad \text{or} \quad 377 \text{ kPa}$$

18.46 One wishes to model the flow about a missile when traveling at 1000 mph through the atmosphere at elevation 10 000 ft. The model is to be tested in a wind tunnel at standard atmospheric conditions with air at 70 °F. What air speed in the wind tunnel is required for dynamic similarity?

❚ $$(N_M)_m = (N_M)_p \qquad N_M = V/c = V/(p/\rho)^{1/2}$$
$$V_m/(14.7/0.00223)^{1/2} = 1000/(10.11/0.001756)^{1/2} \qquad V_m = 1070 \text{ mph} \quad \text{or} \quad 1569 \text{ ft/s}$$

18.47 A model of a supersonic aircraft is tested in a variable-density wind tunnel at 1200 fps. The air is at 100 °F with a pressure of 18 psia. At what velocity should this model be tested to maintain dynamic similarity if the air temperature is raised to 120 °F and the pressure increased to 24 psia?

❚ $N_M = V/(p/\rho)^{1/2}$, $\rho = p/RT$. At 100 °F, $\rho = (18)(144)/[(53.3)(100 + 460)] = 0.08684 \text{ slug/ft}^3$, $N_M = 1200/[(18)(144)/0.08684]^{1/2} = 6.946$. At 120 °F, $\rho = (24)(144)/[(53.3)(120 + 460)] = 0.1118 \text{ slug/ft}^3$, $N_M = V/[(24)(144)/0.1118]^{1/2} = 6.946$, $V = 1221$ ft/s.

18.48 The flow about a ballistic missile which travels at 1500 fps through air at 60 °F and 14.7 psia is to be modeled in a high-speed wind tunnel with a 1:8 model. If the air in the wind tunnel test section has a temperature of 5 °F at a pressure of 11 psia, what velocity is required in the model test section?

❚ $$(N_M)_m = (N_M)_p \qquad N_M = V/(p/\rho)^{1/2}$$
$$\rho = p/RT \qquad \rho_m = (11)(144)/[(53.3)(5 + 460)] = 0.06391 \text{ slug/ft}^3$$
$$\rho_p = (14.7)(144)/[(53.3)(60 + 460)] = 0.07637 \text{ slug/ft}^3$$
$$V_m/[(11)(144)/0.06391]^{1/2} = 1500/[(14.7)(144)/0.07637]^{1/2} \qquad V_m = 1418 \text{ ft/s}$$

18.49 A prototype spillway has a characteristic velocity of 3 m/s and a characteristic length of 10 m. A small model is constructed using Froude scaling. What is the minimum scale ratio of the model that will ensure that its minimum Weber number is 100? Both flows use water at 20 °C.

❚ $$N_W = \rho V^2 L/\sigma \qquad (N_W)_p = (998)(3)^2(10)/0.0728 = 1\,233\,791$$

Froude scaling:

$$(N_W)_m/(N_W)_p = (\rho_m/\rho_p)(V_m/V_p)^2(L_m/L_p)/(\sigma_m/\sigma_p) = (1)(\sqrt{\alpha})^2(\alpha)(1) = \alpha^2 \ge 100/1\,233\,791$$
$$\alpha = 0.009003 \qquad L_p/L_m \le 1/\alpha \le 1/0.009003 = \tfrac{111}{1}$$

18.50 At low velocities (laminar flow), the volume flow Q through a small-bore tube is a function only of the tube radius R, the fluid viscosity μ, and the pressure drop per unit tube length dp/dx. Using the pi theorem, find an appropriate dimensionless relationship.

❚ Write the given relation and count variables: $Q = f(R, \mu, dp/dx)$ four variables ($n = 4$). Make a list of the dimensions of these variables:

Q	R	μ	dp/dx
$\{L^3 T^{-1}\}$	$\{L\}$	$\{ML^{-1}T^{-1}\}$	$\{ML^{-2}T^{-2}\}$

There are three primary dimensions (M, L, T), hence $j \le 3$. By trial and error we determine that R, μ, and dp/dx cannot be combined into a pi group. Then $j = 3$, and $n - j = 4 - 3 = 1$. There is only *one* pi group, which we find by combining Q in a power product with the other three: $\Pi_1 = R^a \mu^b (dp/dx)^c Q^1 = (L)^a (ML^{-1}T^{-1})^b (ML^{-2}T^{-2})^c (L^3 T^{-1}) = M^0 L^0 T^0$. Equate exponents:

Mass: $$b + c = 0$$

Length: $$a - b - 2c + 3 = 0$$

Time: $$-b - 2c - 1 = 0$$

Solving simultaneously, we obtain $a = -4$, $b = 1$, $c = -1$. Then $\Pi_1 = R^{-4} \mu^1 (dp/dx)^{-1} Q$ or $\Pi_1 = Q\mu/[R^4(dp/dx)] = \text{const}$. Since there is only one pi group, it must equal a dimensionless constant.

18.51 The capillary rise h of a liquid in a tube varies with tube diameter d, gravity g, fluid density ρ, surface tension σ, and the contact angle θ. **(a)** Find a dimensionless statement of this relation. **(b)** If $h = 3$ cm in a given experiment, what will h be in a similar case if diameter and surface tension are half as much, density is twice as much, and the contact angle is the same?

▌ **(a)** *Step 1* Write down the function and count variables $h = f(d, g, \rho, \sigma, \theta)$, $n = 6$ variables.

Step 2 List the dimensions (FLT):

h	d	g	ρ	¥ σ	θ
$\{L\}$	$\{L\}$	$\{LT^{-2}\}$	$\{FT^2L^{-4}\}$	$\{FL^{-1}\}$	None

Step 3 Find j. Several groups of three form no pi: σ, ρ, and g or ρ, g, and d. Therefore $j = 3$, and we expect $n - j = 6 - 3 = 3$ dimensionless groups. One of these is obviously θ, which is already dimensionless: $\Pi_3 = \theta$. If we chose carelessly to search for it using steps 4 and 5, we would still find $\Pi_3 = \theta$.

Step 4 Select j variables which do not form a pi group: ρ, g, d.

Step 5 Add one additional variable in sequence to form the pi's:

Add h: $$\Pi_1 = \rho^a g^b d^c h = (FT^2L^{-4})^a (LT^{-2})^b (L)^c (L) = F^0 L^0 T^0$$

Solve for $a = b = 0$, $c = -1$. Therefore $\Pi_1 = \rho^0 g^0 d^{-1} h = h/d$. Finally add Y, again selecting its exponent to be 1: $\Pi_2 = \rho^a g^b d^c \sigma = (FT^2L^{-4})^a (LT^{-2})^b (L)^c (FL^{-1}) = F^0 L^0 T^0$. Solve for $a = b = -1$, $c = -2$. Therefore $\Pi_2 = \rho^{-1} g^{-1} d^{-2} \sigma = \sigma/\rho g d^2$.

Step 6 The complete dimensionless relation for this problem is thus

$$h/d = F(\sigma/\rho g d^2, \theta) \tag{1}$$

This is as far as dimensional analysis goes. Theory, however, establishes that h is proportional to Y. Since Y occurs only in the second parameter, we can slip it outside $(h/d)_{\text{actual}} = (\sigma/\rho g d^2) F_1(\theta)$ or $h\rho g d/\sigma = F_1(\theta)$.

(b) We are given h_1 for certain conditions d_1, σ_1, ρ_1, and θ_1. If $h_1 = 3$ cm, what is h_2 for $d_2 = \frac{1}{2}d$, $\sigma_2 = \frac{1}{2}\sigma$, $\rho_2 = 2\rho_1$, and $\theta_2 = \theta_1$? We know the functional relation, Eq. (*1*), must still hold at condition 2: $h_2/d_2 = F(\sigma_2/\rho_2 g d_2^2, \theta_2)$. But $\sigma_2/\rho_2 g d_2^2 = \frac{1}{2}\sigma_1/2\rho_1 g(\frac{1}{2}d_1)^2 = \sigma_1/\rho_1 g d_1^2$. Therefore, functionally, $h_2/d_2 = F(\sigma_1/\rho_1 g d_1^2, \theta_1) = h_1/d_1$. We are given a condition 2 which is exactly similar to condition 1, and therefore a scaling law holds: $h_2 = h_1(d_2/d_1) = (3)(\frac{1}{2}d_1/d_1) = 1.5$ cm. If the pi groups had not been exactly the same for both conditions, we would have to know more about the functional relation F to calculate h_2.

18.52 Under laminar conditions, the volume flow Q through a small triangular-section pore of side length b and length L is a function of viscosity μ, pressure drop per unit length $\Delta p/L$, and b. Using the pi theorem, rewrite this relation in dimensionless form. How does the volume flow change if the pore size b is doubled?

▌ $$Q = f(\Delta p/L, \mu, b) \qquad \{L^3/T\} = \{M/L^2T^2\}\{M/LT\}\{L\}$$

$n = 4$, $j = 3$ (Q, μ, b do not make a Π), $n - j = 1$ Π expected. $\Pi_1 = Q^1(\Delta p/L)^a(\mu)^b(b)^c = \{L^3/T\}\{M/L^2T^2\}^a\{M/LT\}^b\{L\}^c = M^0 L^0 T^0$.

M: $a + b \quad = 0$

L: $3 - 2a - b + c = 0$

T: $-1 - 2a - b \quad = 0$

$a = -1$, $b = +1$, $c = -4$; $\Pi_1 = Q\mu/(\Delta p/L)b^4 = $ constant. If b is doubled, Q increases 2^4, or 16 times.

18.53 The power input P to a centrifugal pump is assumed to be a function of volume flow Q, impeller diameter D, rotational rate Ω, and the density ρ and viscosity μ of the fluid. Rewrite this as a dimensionless relationship.

▌ $P_{pump} = f(Q, D, \Omega, \rho, \mu)$. $n = 6$, $j = 3$, $n - j = 6 - 3 = 3$ Π's expected. $P/(\rho\Omega^3 D^5) = f(Q/\Omega D^3, \rho\Omega D^2/\mu)$.

18.54 The resistance force F of a surface ship is a function of its length L, velocity V, gravity g, and the density ρ and viscosity μ of the water. Rewrite in dimensionless form.

▌ $F_{ship} = f(L, V, g, \rho, \mu)$. $m = 6$, $j = 3$, $n - j = 6 - 3 = 3$ Π's expected. $F/(\rho V^2 L^2) = f(V^2/gL, \rho VL/\mu)$.

18.55 The torque M on an axial-flow turbine is a function of fluid density ρ, rotor diameter D, angular rotation rate Ω, and volume flow Q. Rewrite in dimensionless form. If it is known that M is proportional to Q for a particular turbine, how would M vary with Ω and D for that turbine?

▌ $M_{turbine} = f(\rho, D, \Omega, Q)$. $n = 5$, $j = 3$, $n - j = 5 - 3 = 2$ Π's expected. $M/(\rho\Omega^2 D^5) = f(Q/\Omega D^3)$. If $M \approx Q$, then $M \approx \Omega$ and $M \approx D^2$.

18.56 A *weir* is an obstruction in a channel flow that can be calibrated to measure flow rate. The volume flow Q varies with gravity g, weir width b, and upstream water height H above the weir crest. If it is known that Q is proportional to b, use the pi theorem to find a unique functional relationship $Q(g, b, H)$.

▌ $Q/b = f(g, H)$. $n = 3$, $j = 2(L, T)$, $n - j = 3 - 2 = 1$ Π expected. $(Q/b)/(g^{1/2}H^{3/2}) = $ constant.

18.57 The size of droplets produced by a liquid spray nozzle is thought to depend upon the nozzle diameter D, jet velocity U, and the properties of the liquid ρ, μ, and σ. Rewrite this relation in dimensionless form.

▌ $d = f(D, U, \rho, \mu, \sigma)$. $n = 6$, $j = 3$, $n - j = 6 - 3 = 3$ Π's expected. $d/D = f(\rho UD/\mu, \sigma/\rho U^2 D)$.

18.58 A certain fluid of specific gravity 0.92 in a tube of 3 cm diameter is found to have a capillary rise of 2 mm. (*a*) What will its capillary rise be in a 5-cm-diameter tube? (*b*) For what diameter will the rise be 1 cm?

▌ $h = f(\rho g, \sigma, D)$. $n = 4$, $j = 3$, $n - j = 4 - 3 = 1$ Π expected. $\rho ghD/\sigma = $ constant. For a given fluid (i.e., given ρg and σ), $hD = $ constant, or $h_m D_m = h_p D_p$.

(*a*) $(2)(3) = (h_p)(5)$ $h_p = 1.20$ mm

(*b*) $(2)(3) = [(1)(10)](D)$ $D = 0.600$ cm

18.59 An East Coast estuary has a tidal period of 12.42 h (the semidiurnal lunar tide) and tidal currents of approximately 80 cm/s. If a one-five-hundred-scale model is constructed with tides driven by a pump and storage apparatus, what should the period of the model tides be and what model current speeds are expected?

▌ $T_m/T_p = \sqrt{\alpha}$ $\alpha = L_m/L_p = \frac{1}{500}$ $T_m/12.42 = \sqrt{\frac{1}{500}}$ $T_m = 0.555$ h   or   33.3 min

$V_m/V_p = \sqrt{\alpha}$ $V_m/80 = \sqrt{\frac{1}{500}}$ $V_m = 3.58$ cm/s

18.60 Derive the expression for the drag on a submerged torpedo. The parameters are the size of the torpedo L, the velocity of the torpedo V, the viscosity of the water μ, and the density of the water ρ. The size of a torpedo may be represented by its diameter or its length.

▌ $F_D = f(L, V, \rho, \mu) = (\text{constant})L^a V^b \rho^c \mu^d$ $\{ML/T^2\} = \{L\}^a\{L/T\}^b\{M/L^3\}^c\{M/LT\}^d$

M: $1 = \quad c + d$

L: $1 = a + b - 3c - d$

T: $-2 = \quad -b \quad - d$

$c = 1 - d$ $b = 2 - d$ $a = 2 - d$

$F_D = (\text{constant})L^{2-d}V^{2-d}\rho^{1-d}\mu^d = (\text{constant})C\rho L^2 V^2(\mu/LV\rho)^d = (\text{constant})C\rho L^2 V^2(LV\rho/\mu)^{-d}$

Since $LV\rho/\mu = N_R$, $F_D = \phi(N_R)\rho L^2 V^2$.

18.61 Derive an expression for the drag on a surface vessel. Use the same parameters as in Prob. 18.60, and add the acceleration due to gravity g to account for the effect of wave action.

■ $F_D = f(L, V, \rho, \mu, g)$, $f'(F_D, L, V, \rho, \mu, g) = 0$, $n = 6$, $j = 3$, $n - j = 6 - 3 = 3$ Π's expected.

$$\Pi = \rho^a L^b V^c F_D^d \qquad \Pi_2 = \rho^a L^b V^c \mu^d \qquad \Pi_3 = \rho^a L^b V^c g^d$$
$$= F_D / L^2 V^2 \qquad = LV\rho/\mu = N_R \qquad = Lg/V^2 = 1/N_F^2$$
$$F_D = \phi(N_F, N_R)\rho L^2 V^2$$

18.62 The discharge through a horizontal capillary tube is thought to depend upon the pressure drop per unit length, the diameter, and the viscosity. Find the form of the equation.

■ The quantities are listed with their dimensions:

quantity	symbol	dimensions
Discharge	Q	$L^3 T^{-1}$
Pressure drop per length	$\Delta p/l$	$ML^{-2} T^{-2}$
Diameter	D	L
Viscosity	μ	$ML^{-1} T^{-1}$

Then $F(Q, \Delta p/l, D, \mu) = 0$. Three dimensions are used, and with four quantities there will be one Π parameter: $\Pi = Q^{x_1}(\Delta p/l)^{y_1} D^{z_1}\mu$. Substituting in the dimensions gives $\Pi = (L^3 T^{-1})^{x_1}(ML^{-2} T^{-2})^{y_1} L^{z_1} ML^{-1} T^{-1} = M^0 L^0 T^0$. The exponents of each dimension must be the same on both sides of the equation. With L first, $3x_1 - 2y_1 + z_1 - 1 = 0$ and similarly for M and T: $y_1 + 1 = 0$, $-x_1 - 2y_1 - 1 = 0$; from which $x_1 = 1$, $y_1 = -1$, $z_1 = -4$, and $\Pi = Q\mu/(D^4 \Delta p/l)$. After solving for Q, $Q = C(\Delta p/l)(D^4/\mu)$, from which dimensional analysis yields no information about the numerical value of the dimensionless constant C; experiment (or analysis) shows that it is $\pi/128$.

18.63 The discharge Q of a V-notch weir is some function of the elevation H of upstream liquid surface above the bottom of the notch. In addition, the discharge depends upon gravity and upon the velocity of approach V_0 to the weir. Determine the form of discharge equation.

■ A functional relation $F(Q, H, g, V_0, \phi) = 0$ is to be grouped into dimensionless parameters. ϕ is dimensionless; hence, it is one Π parameter. Only two dimensions are used, L and T. If g and H are the repeating variables, $\Pi_1 = H^{x_1} g^{y_1} Q = L^{x_1}(LT^{-2})^{y_1} L^3 T^{-1}$, $\Pi_2 = H^{x_2} g^{y_2} V_0 = L^{x_2}(LT^{-2})^{y_2} LT^{-1}$. Then

$$x_1 + y_1 + 3 = 0 \qquad x_2 + y_2 + 1 = 0$$
$$-2y_1 - 1 = 0 \qquad -2y_2 - 1 = 0$$

from which $x_1 = -\frac{5}{2}$, $y_1 = -\frac{1}{2}$, $x_2 = -\frac{1}{2}$, $y_2 = -\frac{1}{2}$, and $\Pi_1 = Q/(\sqrt{g}H^{5/2})$, $\Pi_2 = V_0/\sqrt{gH}$, $\Pi_3 = \phi$, or $f(Q/\sqrt{g}H^{5/2}), V_0/\sqrt{gH}, \phi) = 0$. This can be written $Q/(\sqrt{g}H^{5/2}) = f_1(V_0/\sqrt{gH}, \phi)$ in which both f and f_1 are unknown functions. After solving for Q, $Q = \sqrt{g}H^{5/2}f_1(V_0/\sqrt{gH}, \phi)$. Either experiment or analysis is required to yield additional information about the function f_1. If H and V_0 were selected as repeating variables in place of g and H, $\Pi_1 = H^{x_1} V_0^{y_1} Q = L^{x_1}(LT^{-1})^{y_1} L^3 T^{-1}$, $\Pi_2 = H^{x_2} V_0^{y_2} g = L^{x_2}(LT^{-1})^{y_2} LT^{-2}$. Then

$$x_1 + y_1 + 3 = 0 \qquad x_2 + y_2 + 1 = 0$$
$$-y_1 - 1 = 0 \qquad -y_2 - 2 = 0$$

from which $x_1 = -2$, $y_1 = -1$, $x_2 = 1$, $y_2 = -2$, and $\Pi_1 = Q/H^2 V_0$, $\Pi_2 = gH/V_0^2$, $\Pi_3 = \phi$ or $f(Q/H^2 V_0, gH/V_0^2, \phi) = 0$. Since any of the Π parameters can be inverted or raised to any power without affecting their dimensionless status, $Q = V_0 H^2 f_2(V_0/\sqrt{gH}, \phi)$. The unknown function f_2 has the same parameters as f_1, but it could not be the same function. The last form is not very useful, in general, because frequently V_0 may be neglected with V-notch weirs. This shows that a term of minor importance should not be selected as a repeating variable.

18.64 The losses $\Delta p/l$ in turbulent flow through a smooth horizontal pipe depend upon velocity V, diameter D, dynamic viscosity μ, and density ρ. Use dimensional analysis to determine the general form of the equation $F(\Delta p/l, V, D, \rho, \mu) = 0$.

▮ If V, D, and ρ are repeating variables, $\Pi_1 = V^{x_1}D^{y_1}\rho^{z_1}\mu = (LT^{-1})^{x_1}L^{y_1}(ML^{-3})^{z_1}ML^{-1}T^{-1}$

$$x_1 + y_1 - 3z_1 - 1 = 0$$
$$-x_1 \qquad\qquad -1 = 0$$
$$z_1 + 1 = 0$$

from which $x_1 = -1$, $y_1 = -1$, $z_1 = -1$, and $\Pi_2 = V^{x_2}D^{y_2}\rho^{z_2}(\Delta p/l) = (LT^{-1})^{x_2}L^{y_2}(ML^{-3})^{z_2}ML^{-2}T^{-2}$

$$x_2 + y_2 - 3z_2 - 2 = 0$$
$$-x_2 \qquad\qquad -2 = 0$$
$$z_2 + 1 = 0$$

From which $x_2 = -2$, $y_2 = 1$, and $z_2 = -1$. Then $\Pi_1 = \mu/VD\rho$, $\Pi_2 = (\Delta p/l)/(\rho V^2/D)$, $F[VD\rho/\mu, (\Delta p/l)/(\rho V^2/D)] = 0$, since the Π quantities can be inverted if desired. The first parameter. $VD\rho/\mu$, is the *Reynolds number* N_R. After solving for $\Delta p/l$ we have $\Delta p/l = f_1(N_R, \rho V^2/D)$. The usual formula is $\Delta p/l = f(N_R)(\rho V^2/2D)$ or, in terms of head loss, $\Delta h/l = f(N_R)(1/D)(V^2/2g)$.

18.65 A fluid-flow situation depends on the velocity V, the density ρ, several linear dimensions, l, l_1, l_2, pressure drop Δp, gravity g, viscosity μ, surface tension σ, and bulk modulus of elasticity K. Apply dimensional analysis to these variables to find a set of Π parameters: $F(V, \rho, l, l_1, l_2, \Delta p, g, \mu, \sigma, K) = 0$.

▮ As three dimensions are involved, three repeating variables are selected. For complete situations, V, ρ, and l are generally helpful. There are seven Π parameters: $\Pi_1 = V^{x_1}\rho^{y_1}l^{z_1}\Delta p$, $\Pi_2 = V^{x_2}\rho^{y_2}l^{z_2}g$, $\Pi_3 = V^{x_3}\rho^{y_3}l^{z_3}\mu$, $\Pi_4 = V^{x_4}\rho^{y_4}l^{z_4}\sigma$, $\Pi_5 = V^{x_5}\rho^{y_5}l^{z_5}K$, $\Pi_6 = l/l_1$, $\Pi = l/l_2$. By expanding the Π quantities into dimensions, $\Pi_1 = (LT^{-1})^{x_1}(ML^{-3})^{y_1}L^{x_1}ML^{-1}T^{-2}$

$$x_1 - 3y_1 + z_1 - 1 = 0$$
$$-x_1 \qquad\qquad -2 = 0$$
$$y_1 \qquad +1 = 0$$

from which $x_1 = -2$, $y_1 = -1$, $z_1 = 0$. $\Pi_2 = (LT^{-1})^{x_2}(ML^{-3})^{y_2}L^{z_2}LT^{-2}$

$$x_2 - 3y_2 + z_2 + 1 = 0$$
$$-x_2 \qquad\qquad -2 = 0$$
$$y_2 \qquad\qquad = 0$$

from which $x_2 = -2$, $y_2 = 0$, $z_2 = 1$. $\Pi_3 = (LT^{-1})^{x_3}(ML^{-3})^{y_3}L^{z_3}ML^{-1}T^{-1}$

$$x_3 - 3y_3 + z_3 - 1 = 0$$
$$-x_3 \qquad\qquad -1 = 0$$
$$y_3 \qquad +1 = 0$$

from which $x_3 = -1$, $y_3 = -1$, $z_3 = 1$. $\Pi_4 = (LT^{-1})^{x_4}(ML^{-3})^{y_4}L^{z_4}MT^{-2}$

$$x_4 - 3y_4 + z_4 \qquad = 0$$
$$-x_4 \qquad\qquad -2 = 0$$
$$y_4 \qquad +1 = 0$$

from which $x_4 = -2$, $y_4 = -1$, $z_4 = -1$. $\Pi_5 = (LT^{-1})^{x_5}(ML^{-3})^{y_5}L^{z_5}ML^{-1}T^{-2}$

$$x_5 - 3y_5 + z_5 - 1 = 0$$
$$-x_5 \qquad\qquad -2 = 0$$
$$y_5 \qquad +1 = 0$$

from which $x_5 = -2$, $y_5 = -1$, $z_5 = 0$. $\Pi_1 = \Delta p/\rho V^2$, $\Pi_2 = gl/V^2$, $\Pi_3 = \mu/Vl\rho$, $\Pi_4 = \sigma/V^2\rho l$, $\Pi_5 = K/\rho V^2$, $\Pi_6 = l/l_1$, $\Pi_7 = l/l_2$, and $f(\Delta p/\rho V^2, gl/V^2, \mu/Vl\rho, \sigma/V^2\rho l, K/\rho V^2, l/l_1, l/l_2) = 0$. It is convenient to invert some of the parameters and to take some square roots. $f_1(\Delta p/\rho V^2, V/\sqrt{gl}, Vl\rho/\mu, V^2 l\rho/\sigma, V/\sqrt{K/\rho}, l/l_1, l_1/l_2) = 0$. The first parameter, usually written $\Delta p/(\rho V^2/2)$, is the *pressure coefficient*; the second parameter is the *Froude* number N_F; the third is the *Reynolds* number N_R; the fourth is the *Weber* number N_W; and the fifth is the *Mach* number N_M. Hence, $f_1(\Delta p/\rho V^2, N_F, N_R, N_W, N_M, l/l_1, l/l_2) = 0$. After solving for pressure drop, $\Delta p = \rho V^2 f_2(N_F, N_R, N_W, N_M, l/l_1, l/l_2)$ in which f_1 and f_2 must be determined from analysis or experiment. By selecting other repeating variables, a different set of Π parameters could be obtained.

18.66 The thrust due to any one of a family of geometrically similar airplane propellers is to be determined experimentally from a wind-tunnel test on a model. Use dimensional analysis to find suitable parameters for plotting test results.

┃ The thrust F_T depends upon speed of rotation ω, speed of advance V_0, diameter D, air viscosity μ, density ρ, and speed of sound c. The function $F(F_T, V_0, D, \omega, \mu, \rho, c) = 0$ is to be arranged into four dimensionless parameters, since there are seven quantities and three dimensions. Starting first by selecting ρ, ω, and D as repeating variables, $\Pi_1 = \rho^{x_1}\omega^{y_1}D^{z_1}F_T = (ML^{-3})^{x_1}(T^{-1})^{y_1}L^{z_1}MLT^{-2}$, $\Pi_2 = \rho^{x_2}\omega^{y_2}D^{z_2}V_0 = (ML^{-3})^{x_2}(T^{-1})^{y_2}L^{z_2}LT^{-1}$, $\Pi_3 = \rho^{x_3}\omega^{y_3}D^{z_3}\mu = (ML^{-3})^{x_3}(T^{-1})^{y_3}L^{z_3}ML^{-1}T^{-1}$, $\Pi_4 = \rho^{x_4}\omega^{y_4}D^{z_4}c = (ML^{-3})^{x_4}(T^{-1})^{y_4}L^{z_4}LT^{-1}$. Writing the simultaneous equations in x_1, y_1, z_1, etc., as before and solving them gives $\Pi_1 = F_T/\rho\omega^2D^2$, $\Pi_2 = V_0/\omega D$, $\Pi_3 = \mu/\rho\omega D^2$, $\Pi_4 = c/\omega D$. Solving for the thrust parameter leads to $F_T/\rho\omega^2D^4 = f_1(V_0/\omega D, \rho\omega D^2/\mu, c/\omega D)$. Since the parameters can be recombined to obtain other forms, the second term is replaced by the product of the first and second terms, $VD\rho/\mu$, and the third term is replaced by the first term divided by the third term, V_0/c; thus $F_T/\rho\omega^2D^4 = f_2(V_0/\omega D, V_0D\rho/\mu, V_0/c)$. Of the dimensionless parameters, the first is probably of the most importance since it relates speed of advance to speed of rotation. The second parameter is a Reynolds number and accounts for viscous effects. The last parameter, speed of advance divided by speed of sound, is a Mach number, which would be important for speeds near or higher than the speed of sound. Reynolds effects are usually small, so that a plot of $F_T/\rho\omega^4$ against $V_0/\omega D$ should be most informative.

18.67 The variation Δp of pressure in static liquids is known to depend upon specific weight γ and elevation difference Δz. By dimensional reasoning determine the form of the hydrostatic law of variation of pressure.

┃
$$p = f(\gamma, \Delta z) = (\text{constant})\gamma^a(\Delta z)^b \qquad \{M/LT^2\} = \{M/L^2T^2\}^a\{L\}^b$$

M: $\qquad\qquad\qquad\qquad\qquad\qquad a \qquad\qquad 1$

L: $\qquad\qquad\qquad\qquad\qquad -2a + b = -1$

T: $\qquad\qquad\qquad\qquad\qquad -2a \qquad\;\; = -2$

$$a = 1 \qquad b = 1 \qquad \Delta p = (\text{constant})\gamma\Delta z$$

18.68 When viscous and surface-tension effects are neglected, the velocity V of efflux of liquid from a reservoir is thought to depend upon the pressure drop Δp of the liquid and its density ρ. Determine the form of expression for V.

┃
$$V = f(\Delta p, \rho) \qquad (\Delta p)^a\rho^b = (\text{constant})V \qquad \{M/LT^2\}^a\{M/L^3\}^b = L/T$$

M: $\qquad\qquad\qquad\qquad\qquad\quad a + b = 1$

L: $\qquad\qquad\qquad\qquad\qquad -a - 3b = 1$

T: $\qquad\qquad\qquad\qquad\qquad -2a \qquad = -1$

$$a = \tfrac{1}{2} \qquad b = -\tfrac{1}{2} \qquad V = (\text{constant})\sqrt{\Delta p/\rho}$$

18.69 The buoyant force F_b on a body is thought to depend upon its volume submerged V and the gravitational body force acting on the fluid. Determine the form of the buoyant-force equation.

┃
$$F_b = f(V, \rho g) \qquad V^a(\rho g)^b = (\text{constant})F_b \qquad \{L^3\}^a\{(M/L^3)(L/T^2)\}^b = \{ML/T^2\}$$

M: $\qquad\qquad\qquad\qquad\qquad\qquad\quad b = 1$

L: $\qquad\qquad\qquad\qquad\qquad 3a - 2b = 1$

T: $\qquad\qquad\qquad\qquad\qquad\qquad -2b = -2$

$$a = 1 \qquad b = 1 \qquad F_b = (\text{constant})V\rho g$$

18.70 In a fluid rotated as a solid about a vertical axis with angular velocity ω, the pressure rise p in a radial direction depends upon speed ω, radius r, and fluid density ρ. Obtain the form of equation for p.

┃
$$\Delta p = f(\omega, r, \rho) \qquad \omega^a r^b \rho^c = (\text{constant})\Delta p \qquad \{1/T\}^a\{L\}^b\{M/L^3\}^c = \{M/LT^2\}$$

M: $\qquad\qquad\qquad\qquad\qquad\qquad\qquad c = 1$

L: $\qquad\qquad\qquad\qquad\qquad\quad b - 3c = -1$

T: $\qquad\qquad\qquad\qquad\qquad -a \qquad\quad = -2$

$$a = 2 \qquad b = 2 \qquad c = 1 \qquad \Delta p = (\text{constant})\omega^2 r^2 \rho$$

18.71 The Mach number N_M for flow of a perfect gas in a pipe depends upon the specific-heat ratio k (dimensionless), the pressure p, the density ρ, and the velocity V. Obtain by dimensional reasoning the form of the Mach number expression.

▮ $N_M = f(k, \rho, p, V)$, $f'(N_M, k, \rho, p, V) = 0$; $n = 5$, $j = 3$, $n - j = 5 - 3 = 2$ Π's expected.

$$\Pi_1 = M \qquad \Pi_2 = k \qquad p^a\rho^b V = \Pi_3 \qquad \{M^a/L^a\}\{M^b/T^{2a}L^{3b}\}\{L/T\} = 1$$

M: $\qquad\qquad\qquad\qquad\qquad\qquad a + b \quad = 0$

L: $\qquad\qquad\qquad\qquad\qquad\qquad -a - 3b + 1 = 0$

T: $\qquad\qquad\qquad\qquad\qquad\qquad -2a \quad - 1 = 0$

$$a = -\tfrac{1}{2} \qquad b = \tfrac{1}{2} \qquad N_M = f(V/\sqrt{p/\rho}, k)$$

18.72 The moment exerted on a submarine by its rudder is to be studied with a 1:20 scale model in a water tunnel. If the torque measured on the water model is 5 N-m for a tunnel velocity of 15 m/s, what are the corresponding torque and speed for the prototype?

▮ $\qquad V_p/V_m = L_m/L_p \qquad V_p/15 = \tfrac{1}{20} \qquad V_p = 0.75$ m/s $\qquad T \propto \text{(force)(arm)} \propto (\rho V^2 A)(L)$

$\qquad T_r = \rho_r V_r^2 A_r L_r = (1)(1/L_r)^2 L_r^2 L_r = L_r \qquad T_p/T_m = L_p/L_m \qquad T_p/5 = \tfrac{20}{1} \qquad T_p = 100$ N-m

18.73 A rotary mixer is to be designed for stirring ethyl alcohol [$\rho = 789$ kg/m^3, $\mu = 1.20 \times 10^{-3}$ kg/(m-s)]. Tests with a one-fourth-scale model in SAE30 oil [$\rho = 917$ kg/m^2, $\mu = 0.29$ kg/(m-s)] indicate most efficient mixing at 1770 rpm. What should the speed of the prototype mixer be?

▮ $(N_R)_m = (N_R)_p$. Use Reynolds number scaling with "V" $= \Omega D$. $N_R = (\rho)(\Omega D)(D)/\mu = \rho\Omega D^2/\mu$, $(917)(1770)(D_p/4)^2/0.29 = (789)(\Omega_p)(D_p)^2/(1.20 \times 10^{-3})$, $\Omega_p = 0.532$ rpm.

18.74 A one-fifteenth-scale model of a parachute has a drag of 450 lb when tested at 20 ft/s in a water tunnel. If Reynolds number effects are negligible, estimate the terminal fall velocity at 5000 ft standard altitude of a parachutist using the prototype if chute and chutist together weight 160 lb. Neglect the drag coefficient of the woman.

▮ $(C_D)_m = (C_D)_p$ (if Reynolds number effects are negligible), $C_D = F/\rho V^2 D^2$.

$$\rho_p = 0.002048 \text{ slug/ft}^3 \qquad \text{(from Table A-7)} \qquad D_m = D_p/15$$

$$450/[(1.94)(20)^2(D_p/15)^2] = 160/[(0.002048)(V_p)^2(D_p)^2] \qquad V_p = 24.5 \text{ ft/s}$$

18.75 A one-twelfth-scale model of a weir has a measured flow rate of 2.2 cfs when the upstream water height is $H = 6.5$ in. Use the results of Prob. 18.56 to predict the prototype flow rate when $H = 3.6$ ft.

▮ $\qquad (Q/b)/(g^{1/2}H^{3/2}) = \text{constant} \qquad \text{(from Prob. 18.56)} \qquad [(Q/b)/(g^{1/2}H^{3/2})]_m = [(Q/b)/(g^{1/2}H^{3/2})]_p$

$\qquad b_m = b_p/12 \qquad [2.2/(b_p/12)]/[(32.2)^{1/2}(6.5/12)^{3/2}] = (Q_p/b_p)/[(32.2)^{1/2}(3.6)^{3/2}] \qquad Q_p = 452 \text{ ft}^3/\text{s}$

18.76 For model and prototype, show that, when gravity and inertia are the only influences, the ratio of flows Q is equal to the ratio of the length dimension to the five-halves power.

▮ $Q_m/Q_p = (L_m^3/T_m)/(L_p^3/T_p) = L_r^3/T_r$. The time ratio must be established for the conditions influencing the flow. Expressions can be written for the gravitation and inertia forces, as follows:

Gravity: $\qquad\qquad\qquad F_m/F_p = W_m/W_p = (\gamma_m/\gamma_p)(L_m^3/L_p^3) = \gamma_r L_r^3$

Inertia $\qquad\qquad\qquad F_m/F_p = M_m a_m/M_p a_p = (\rho_m/\rho_p)(L_m^3/L_p^3)(L_r/T_r^2) = \rho_r L_r^3(L_r/T_r^2)$

Equating the force ratios, $\gamma_r L_r^3 = \rho_r L_r^3(L_r/T_r^2)$, which, when solved for the time ratio, yields

$$T_r^2 = L_r(\rho_r/\gamma_r) = L_r/g_r \qquad\qquad\qquad (1)$$

Recognizing that the value of g_r is unity, substitution in the flow ratio expression gives

$$Q_r = Q_m/Q_p = L_r^3/L_r^{1/2} = L_r^{5/2} \qquad\qquad\qquad (2)$$

18.77 A spillway model is to be built to a scale of 1:25 across a flume which is 2 ft wide. The prototype is 37.5 ft high and the maximum head expected is 5.0 ft. What height of model and what head on the model should be used? If the flow over the model at 0.20 ft head is 0.70 cfs, what flow per foot of prototype may be expected;

$$L_m/L_p = \tfrac{1}{25} \qquad \text{Height of model} = (37.5)(\tfrac{1}{25}) = 1.50 \text{ ft}$$

$$\text{Head on model} = (5.0)(\tfrac{1}{25}) = 0.20 \text{ ft} \qquad Q_r = L_r^{5/2} \qquad \text{(from Prob. 18.76)}$$

$$Q_m/Q_p = (L_m/L_p)^{5/2} \qquad 0.070/Q_p = (\tfrac{1}{25})^{5/2} \qquad Q_p = 2188 \text{ ft}^3/\text{s}$$

$$b_m/b_p = \tfrac{1}{25} \qquad 2/b_p = \tfrac{1}{25} \qquad b_p = 50.0 \text{ ft} \qquad q_p = Q_p/b_p = 2188/50.0 = 43.8 \text{ ft}^3/\text{s}$$

18.78 For the model described in Prob. 18.77, if the model shows a measured hydraulic jump of 1.0 in, how high is the jump in the prototype? If the energy dissipated in the model at the hydraulic jump is 0.15 hp, what would be the energy dissipation in the prototype?

▮ Since $L_m/L_p = \tfrac{1}{25}$, height of jump = $(1.0)(25) = 25.0$ in, or 2.08 ft, $P_r = F_r L_r/T_r = (\gamma_r L_r^3)(L_r)/\sqrt{L_r/g_r}$. Since $g_r = \gamma_r = 1$, $P_r = L_r^{7/2}$, $0.15/P_p = (\tfrac{1}{25})^{7/2}$, $P_r = 11\,720$ hp.

18.79 A model of a reservoir is drained in 4 min by opening the sluice gate. The model scale is 1:225. How long should it take to empty the prototype?

▮ Since gravity is the dominant force, $Q_r = L_r^{5/2}$ (from Prob. 18-76), $Q_r = Q_m/Q_p = (L_m^3/t_m)/(L_p^3/t_p)$, $L_r^{5/2} = (L_r^3)(t_p/t_m)$, $t_p = t_m/L_r^{1/2} = (4)/(\tfrac{1}{225})^{1/2} = 60.0$ min.

18.80 A rectangular pier in a river is 4 ft wide by 12 ft long and the average depth of water is 9.0 ft. A model is built to a scale of 1:16. The velocity of flow of 2.50 fps is maintained in the model and the force acting on the model is 0.90 lb. What are the values of velocity in and force on the prototype?

▮ Since the gravity forces predominate, $V_m/V_p = \sqrt{L_r}$, $2.50/V_p = \sqrt{\tfrac{1}{16}}$, $V_p = 10.0$ ft/s. $F_m/F_p = \gamma_r L_r^3$, $0.90/F_p = (1.0)(\tfrac{1}{16})^3$, $F_p = 3686$ lb.

18.81 If a standing wave in the model of Prob. 18.80 is 0.16 ft high, what height of wave should be expected at the nose of the pier? What is the coefficient of drag resistance?

▮
$$V_m/V_p = \sqrt{L_m}/\sqrt{L_p} = \sqrt{h_m}/\sqrt{h_p} \qquad 2.50/10.0 = \sqrt{0.16}/\sqrt{h_p} \qquad h_p = 2.56 \text{ ft}$$
$$F_D = C_D\rho A V^2/2 \qquad 3686 = (C_D)(1.94)[(4)(9)](10.0)^2/2 \qquad C_D = 1.06$$

18.82 The measured resistance in fresh water of an 8-ft ship model moving at 6.50 fps was 9.60 lb. (**a**) What would be the velocity of the 128-ft prototype? (**b**) What force would be required to drive the prototype at this speed in salt water?

▮ (**a**) Since gravity forces predominate, $V_m/V_p = \sqrt{L_r}$, $6.50/V_p = \sqrt{\tfrac{8}{128}}$, $V_p = 26.0$ ft/s.

(**b**) $\qquad\qquad F_m/F_p = \gamma_r L_r^3 \qquad 9.60/F_p = (62.4/64.2)(\tfrac{1}{16})^3 \qquad F_p = 40\,460$ lb

19.1 A frictionless fluid column 2.18 m long has a speed of 2 m/s when $z = 0.5$ m. Find (a) the maximum value of z, (b) the maximum speed, and (c) the period.

▌ (a) $z = Z \cos \sqrt{2g/L} t$, $dz/dt = -3Z \sin \sqrt{2g/L} t$. If t_1 is the time when $z = 0.5$ and $dz/dt = 2$, $0.5 = Z \cos \sqrt{(2)(9.807)/2.18} t_1$, $0.5 = Z \cos 3.00 t_1$, $-2 = -3Z \sin 3.00 t_1$. Dividing the second equation by the first gives $1.333 = \tan 3t_1$, $t_1 = 0.309$ s; $0.5 = Z \cos[(3.00)(0.309)]$ (angle in radians), $Z = z_{max} = 0.833$ m.
(b) Maximum speed occurs when $\sin 3t = 1$, or $(dz/dt)_{max} = (-3)(0.833)(1) = 2.50$ m/s.
(c)   Period $= 2\pi\sqrt{L/2g} = 2\pi\sqrt{2.18/[(2)(9.807)]} = 2.09$ s

19.2 A 1.0-in-diameter U-tube contains oil, $v = 1 \times 10^{-4}$ ft²/s, with a total column length of 120 in. Applying air pressure to one of the tubes makes the gage difference 16 in. By quickly releasing the air pressure, the oil column is free to oscillate. Find the maximum velocity, the maximum Reynolds number, and the equation for position of one meniscus z, in terms of time.

▌ The assumption is made that the flow is laminar, and the Reynolds number will be computed on this basis. The constants m and n are $m = 16v/D^2 = (16 \times 10^{-4})/(\frac{1}{12})^2 = 0.2302$, $n = \sqrt{(16v/D^2)^2 - (2g/L)} = \sqrt{0.2302^2 - [2(32.2)/10]} = \sqrt{-6.387} = i2.527$, or $n' = 2.527$. The liquid will oscillate above and below $z = 0$. The oscillation starts from the maximum position, that is, $Z = 0.667$ ft. The velocity (fictitious) when $z = 0$ at time t_0 before the maximum is determined to be $V_0 = Z\sqrt{2g/L} \exp[(m/n') \tan^{-1}(n'/m)] = 0.667\sqrt{64.4/10} \exp[(0.2302/2.527) \tan^{-1}(2.527/0.2302)] = 1.935$ ft/s and $\tan n't_0 = n'/m$, $t_0 = (1/2.527) \tan^{-1}(2.527/0.2302) = 0.586$ s; $z = (V_0/n')e^{-mt} \sin(n't) = 0.766 \exp[-0.2302(t + 0.586)] \sin 2.527(t + 0.586)$ in which $z = Z$ at $t = 0$. The maximum velocity (actual) occurs for $t > 0$. Differentiating with respect to t to obtain the expression for velocity, $V = dz/dt = -0.1763 \exp[-0.2302(t + 0.586)] \sin 2.527(t + 0.586) + 1.935 \exp[-0.2302(t + 0.586)] \cos 2.527(t + 0.586)$. Differentiating again with respect to t and equating to zero to obtain maximum V produces $\tan 2.527(t + 0.586) = -0.1837$. The solution in the second quadrant should produce the desired maximum, $t = 0.584$ s. Substituting this time into the expression for V produces $V = -1.48$ ft/s. The corresponding Reynolds number is $VD/v = 1.48(\frac{1}{12} \times 10^4) = 1234$; hence the assumption of laminar resistance is justified.

19.3 A U-tube consisting of 500-mm-diameter pipe with $f = 0.03$ has a maximum oscillation (Fig. 19-1) of $z_m = 6$ m. Find the minimum position of the surface and the following maximum.

▌ $[1 + (fz_m/D)]e^{-fz_m/D} = [1 + (fz_{m+1}/D)]e^{-fz_{m+1}/D}$, $\{1 + [0.03(6)/0.5]\}e^{0.03(6)0.5} = (1 + 0.06z_{m+1})e^{-0.06z_{m+1}}$ or $(1 + 0.06z_{m+1})e^{-0.06z_{m+1}} = 0.9488$, which is satisfied by $z_{m+1} = -4.84$ m. Using $z_m = 4.84$ m in the previous equation, $(1 + 0.06z_{m+1})e^{-0.06z_{m+1}} = [1 + 0.06(4.84)]e^{-0.06(4.84)} = 0.9651$, which is satisfied by $z_{m+1} = -4.05$ m. Hence, the minimum water surface is $z = -4.84$ m and the next maximum is $z = 4.05$ m.

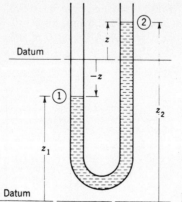

Fig. 19-1

19.4 The pipe of Prob. 19.3 is 1000 m long. By numerical integration of the equation

$$\frac{d^2z}{dt^2} + \frac{f}{2D}\frac{dz}{dt}\left|\frac{dz}{dt}\right| + \frac{2g}{L}z = 0$$

589

using the third-order Runge–Kutta method procedure, determine by a computer program the time to the first minimum z and the next maximum z and check the results of Prob. 19.3.

```
10 REM            EXAMPLE        TURBULENT OSCILLATION OF A U-TUBE.
20 DEFINT I:    DEF FNZ(ZZ,YY)=C1*YY*ABS(YY)+C2*ZZ    ' dy in Example 13.3
30 READ D,F,L,ZM,G,H,IPR:    DATA .5,.03,1000.,6.,9.806,.5,4
40 LPRINT:    LPRINT"D,F,L,ZM,G,H,IPR=";D;F;L;ZM;G;H;IPR
50 C1=-.5*F*H/D: C2=-2!*G*H/L: Y=0!: Z=6!                      ' Y IS DZ/DT
60 OZ=6!:        I=0:      EPS=.001
70 LPRINT:    LPRINT" T,sec      Z,m      V,m/s":    LPRINT
80 I=I+1:    OOZ=OZ:    OZ=Z      ' OOZ and OZ used to determine max and min Z
90 U11=FNZ(Z,Y):                         U12=H*Y
100  U21=FNZ(Z+.3333*U12,Y+.3333*U11):    U22=H*(Y+.3333*U11)
110 U31=FNZ(Z+.6667*U22,Y+.6667*U21):    U32=H*(Y+.6667*U21)
120 Y=Y+.25*U11+.75*U31:    Z=Z+.25*U12+.75*U32:    T=T+H
130 IF (OZ<OOZ--EPS) AND (OZ<Z-EPS) THEN LPRINT"ZMIN=";OZ;"TIME=";T-H
140 IF I MOD IPR=0 THEN LPRINT USING"###.###  ";T;Z;Y
150 IF (OZ>OOZ+EPS) AND (OZ>Z+EPS) THEN LPRINT"ZM=";OZ;"TIME=";T-H: STOP
    ELSE GOTO 80

D,F,L,ZM,G,H,IPR= .5  .03  1000  6  9.806001  .5  4

   T,sec      Z,m      V,m/s

   2.000     5.767    -0.231
   4.000     5.091    -0.438
   6.000     4.042    -0.602
   8.000     2.721    -0.709
  10.000     1.247    -0.755
  12.000    -0.257    -0.740
  14.000    -1.677    -0.671
  16.000    -2.912    -0.558
  18.000    -3.884    -0.409
  20.000    -4.533    -0.236
  22.000    -4.821    -0.050
ZMIN=-4.834092 TIME= 22.5
  24.000    -4.732     0.138
  26.000    -4.278     0.313
  28.000    -3.502     0.457
  30.000    -2.478     0.559
  32.000    -1.298     0.613
  34.000    -0.061     0.616
  36.000     1.136     0.574
  38.000     2.206     0.490
  40.000     3.075     0.374
  42.000     3.688     0.235
  44.000     4.006     0.082
ZM= 4.048564 TIME= 45
```

19.5   In Fig. 19-2 a valve is opened suddenly in the pipeline when $z_1 = 40$ ft. $L = 2000$ ft, $A_1 = 200$ ft^2, $A_2 = 300$ ft^2, $D = 3.0$ ft, $f = 0.024$, and minor losses are $3.50V^2/2g$. Determine the subsequent maximum negative and positive surges in the reservoir A_1.

The equivalent length of minor losses is $KD/f = 3.5(3)/0.024 = 438$ ft. Then $L_e = 2000 + 438 = 2438$ and $z_m = z_1 A_1/A = 40(200/(2.25\pi)) = 1132$ ft. The corresponding ϕ is $\phi = f(L_e/L)(z_m/D) = 0.024(\frac{2438}{2000})(\frac{1132}{3}) = 11.04$ and $F(\phi) = (1 + \phi)e^{-\phi} = (1 + 11.04)e^{-11.04} = 0.000193$, which is satisfied by $\phi \approx -1.0$. Then $F(\phi) = (1 + 1)e^{-1} = 0.736 = (1 + \phi)e^{-\phi}$ which is satisfied by $\phi = -0.593$. The values of z_m are, for $\phi = -1$. $z_m = \phi LD/fL_e = -1(2000)(3)/[0.024(2438)] = -102.6$ and, for $\phi = 0.593$, $z_m = 0.593(2000)(3)/[0.024(2438)] = 60.9$. The corresponding values of z_1 are $z_1 = z_m(A/A_1) = -102.6(2.25\pi/200) = -3.63$ ft and $z_1 = 60.9(2.25\pi/200) = 2.15$ ft.

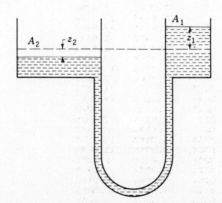

Fig. 19-2

19.6 In Fig. 19-3 the minor losses are $16V^2/2g$, $f = 0.030$, $L = 3000$ m, $D = 2.4$ m, and $H = 20$ m. Determine the time, after the sudden opening of a valve, for velocity to attain nine-tenths the final velocity.

$$H = f(L_e/D)(V_0^2/2g) \qquad L_e = 3000 + [16(2.4)/0.03] = 4280 \text{ m}$$
$$V_0 = \sqrt{2gHD/fL_e} = \sqrt{19.612(20)(2.4)/[0.030(4280)]} = 2.708 \text{ m/s}$$

Substituting $V = 0.9V_0$ into $t = (LV_0/2gH) \ln[(V_0 + V)/(V_0 - V)]$, $t = [3000(2.708)/19.612(20)] \ln(1.90/0.10) = 60.98$ s.

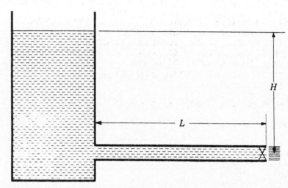

Fig. 19-3

19.7 In Fig. 19-4 the water surface changes according to the equation $H_{P_A} = H_0 + \Delta H \sin \omega t$ while the right end of the pipe contains a small orifice. The frequency of the waves is set at the natural period of the pipe, $4L/a$, which yields an ω of $2\pi/(4L/a)$. Determine the resulting motion of fluid in the pipe and the head fluctuations.

■ One reach is used. The upstream section is solved by $H_{P_i} = C_M + B_M Q_{P_i}$ together with the head, which is a known function of time. The downstream boundary condition solves the orifice equation together with $Q_{P_B}^2 = C_1 H_{P_A}$, $H_{P_B} = C_P - B_P Q_{P_B}$. The program and results are shown below.

```
10 REM ,                      TRANSIENT IN SINGLE-REACH PIPE.
20 DEFINT I:     READ F,L,D,A,DH,HO,IPR,PI,G,TM,CD,DORIF
30 DATA .018,600.,.5,1200.,3.,40.,1,3.1416,9.806,12.,.6,.03
40 AO=.25*PI*DORIF^2:    LPRINT"CD,DORIF,AO=";CD;DORIF;AO:   C1=(CD*AO)^2*2!*G
50 LPRINT"F,L,D,A,DH=";F;L;D;A;DH:    LPRINT"HO,IPR,PI,G,TM=";HO;IPR;PI;G;TM
60 PER=4!*L/A:    OM=2!*PI/PER:     AR=.25*PI*D^2
70 B=A/(G*AR):    R=.5*F*L/(G*D*AR^2):    QO=SQR(HO/(R+1!/C1))
80 T=0!:     DT=L/A:    I=0:    HA=HO:    HB=HO-R*QO^2:    QA=QO:    QB=QO
90 LPRINT"     T,s       HA,m      QA,l/s     HB,m      QB,l/s"
100 LPRINT USING"####.####  ";T;HA;1000!*QA;HB;1000!*QB
110 T=T+DT:    I=I+1:    IF T>TM THEN STOP
120 HPA=HO+DH*SIN(OM*T):    CM=HB-B*QB:    BM=B+R*ABS(QB):    QPA=(HPA-CM)/BM
130 CP=HA+B*QA:    BP=B+R*ABS(QA):    QPB=-.5*BP*C1+SQR((.5*BP*C1)^2+CP*C1)
140 HPB=CP-BP*QPB:    HA=HPA:    HB=HPB:    QA=QPA:    QB=QPB
150 IF I MOD IPR=0 THEN 100 ELSE GOTO 110
CD,DORIF,AO= .6    .03    7.0686E-04
F,L,D,A,DH= .018   600    .5    1200    3
HO,IPR,PI,G,TM= 40   1    3.1416    9.806001    12
     T,s       HA,m      QA,l/s     HB,m      QB,l/s
    0.0000    40.0000    11.8783    39.9960    11.8783
    0.5000    43.0000    16.6892    39.9960    11.8783
    1.0000    40.0000    11.8782    45.4994    12.6692
    1.5000    37.0000    -0.9676    39.9959    11.8783
    2.0000    40.0000    11.8784    29.9865    10.2851
    2.5000    43.0000    31.1506    39.9961    11.8783
    3.0000    39.9999    11.8780    53.8149    13.7783
    3.5000    37.0000   -13.1930    39.9958    11.8783
    4.0000    40.0001    11.8787    23.1428     9.0355
    4.5000    43.0000    40.8796    39.9963    11.8783
    5.0000    39.9999    11.8776    59.4364    14.4801
    5.5000    37.0000   -21.5050    39.9955    11.8782
    6.0000    40.0001    11.8792    18.5504     8.0895
    6.5000    43.0000    47.3015    39.9966    11.8784
    7.0000    39.9998    11.8771    63.1574    14.9265
    7.5000    37.0000   -27.0247    39.9952    11.8782
    8.0000    40.0002    11.8798    15.5371     7.4034
    8.5000    43.0000    51.4503    39.9970    11.8784
    9.0000    39.9998    11.8765    65.5652    15.2084
    9.5000    37.0000   -30.6034    39.9948    11.8781
   10.0000    40.0002    11.8804    13.6031     6.9273
   10.5000    43.0000    54.0776    39.9974    11.8785
   11.0000    39.9998    11.8759    67.0916    15.3844
   11.5000    37.0000   -32.8746    39.9944    11.8781
   12.0000    40.0003    11.8810    12.3853     6.6099
```

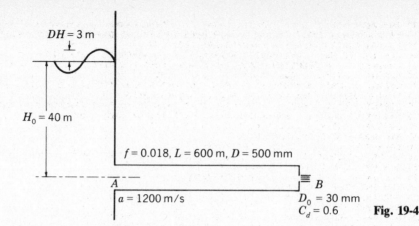

$DH = 3$ m

$H_0 = 40$ m

$f = 0.018, L = 600$ m, $D = 500$ mm

A

$a = 1200$ m/s

B

$D_0 = 30$ mm
$C_d = 0.6$

Fig. 19-4

19.8 Develop the necessary boundary-condition equations for the pump in Fig. 19-5. The pump is to be started with a linear speed rise to N_R in t_0 s. A check valve exists in the discharge pipe. The initial no-flow steady-state head on the downstream side of the check valve is H_C. For a steady flow of Q_0 there is a loss of ΔH_0 across the open check valve. Assume that the check valve opens when the pump has developed enough head to exceed H_C.

❚ The equation for the hydraulic grade line downstream from the pump and check valve (after the check valve is open) is $H_P = C_1 N^2 + C_2 N Q_P + C_3 Q_P^2 - Q_P^2 (\Delta H_0 / Q_0^2)$, $Q_P = \{(B_M - C_2 N)/[2(C_3 - \Delta H_0 / Q_0^2)]\}\{1 - [1 + 4(C_3 - \Delta H_0 / Q_0^2)^2 (C_M - C_1 N^2)/(B_M - C_2 N)^2]^{1/2}\}$. The equations for the boundary condition are $B_M = (a/gA) + R |Q_2|$, $C_M = H_2 - BQ_2$,

$$N = \begin{cases} N_R(t/t_0) & t \le t_0 \\ N_R & t > t_0 \end{cases}$$

If $C_1 N^2 > H_C$, Q_P is defined by the above equation for the quadratic, and $H_P = C_M + B_M Q_P$. If $C_1 N^2 < H_C$, $Q_P = 0$ and $H_P = H_C$.

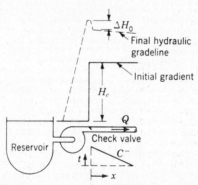

ΔH_0

Final hydraulic gradeline

Initial gradient

H_c

Q

Check valve

Reservoir

C^-

x

Fig. 19-5

19.9 The system of Fig. 19-6 initially has a valve opening $C_D A = 0.06$ m². At intervals of 5 s, $C_D A$ takes on the values 0.03, 0.01, 0.003, 0.001, 0.0005, 0.0002, 0.0, and remains closed. Using a computer program, calculate the transients of the system for 40 s after the valve starts to close.

❚ The data for this problem and some of the output are given below. The number of reaches was selected as $N = 4$. Hydraulic-grade-line elevations are given in meters and discharges in cubic meters per second.

```
10 REM B:WH   WATERHAMMER PROGRAM IN IBMPC BASICA FOR UPSTREAM RESERVOIR,
20 ' SINGLE PIPE, AND DOWNSTREAM VALVE. DARCY-WEISBACH FRICTION. F GIVEN.
30 ' A=WAVE SPEED, CVA=PRODUCT OF DISCHARGE COEF. AND AREA OF VALVE
40 ' OPENING, WITH VALUES GIVEN FOR INTERVALS DCV.
50 LPRINT "      BASIC WATERHAMMER PROGRAM"
60 LPRINT "       DATE=";DATE$;"   TIME=";TIME$
65 CLEAR:   DIM CVA(11):   DEFINT I,J,K
70 READ F,L,A,N,D,HRES,G,JPR,DCV,TMAX
80 DATA .022,4800.,1200.,4,2.,100.,9.806,1,5.,40.
90 LPRINT:   LPRINT"F,L,A,N,D=";F;L;A;N;D
100 LPRINT"HRES,G,JPR,DCV,TMAX=";HRES;G;JPR;DCV;TMAX
105 LPRINT"CVA=";
110 FOR I=1 TO 11:   READ CVA(I):   NEXT I:   LPRINT
115 FOR I=1 TO 11:   LPRINT CVA(I);:   NEXT I:   LPRINT
120 DATA .06,.03,.01,.003,.001,.0005,.0002,.0,.0,.0,.0
```

```
130 AR=.7854*D^2:    B=A/(G*AR):    NS=N+1:    DT=L/(A*N):    CV=CVA(1)
140 HP(1)=HRES:  J=0:  T=0!:  Q0=SQR(HRES/(F*L/(2!*G*D*AR^2)+1'/(2!*G*CV^2)))
150 H0=(Q0/CV)^2/(2!*G):   R=(HRES-H0)/(Q0^2*N)
160 FOR I=1 TO NS:  Q(I)=Q0:   H(I)=HRES-(I-1)*R*Q0^2:    NEXT I
170 LPRINT"    PIEZOMETER HEADS AND FLOWS ALONG THE PIPE"
180 LPRINT"  TIME    CV   X/L=    .0      .25      .5      .75     1."
190 LPRINT USING"###.###";T;:  LPRINT USING" #.#####";CV;:   LPRINT" H=";
200 LPRINT USING" ###.##";H(1);H(2);H(3);H(4);H(5)
210 LPRINT SPC(17);"Q=";:   LPRINT USING  ###.##";Q(1);Q(2);Q(3);Q(4);Q(5)
220 T=T+DT:   J=J+1:   K=FIX(T/DCV)+1:   IF T>TMAX THEN STOP
230 CV=CVA(K)+(T-(K-1)*DCV)*(CVA(K+1)-CVA(K))/DCV
240 REM DOWNSTREAM BOUNDARY CONDITION
250 CP=H(N)+B*Q(N):    BP=B+R*ABS(Q(N))
260 QP(NS)=-G*BP*CV^2+SQR((G*BP*CV^2)^2+2!*G*CV^2*CP):    HP(NS)=CP-BP*QP(NS)
270 REM UPSTREAM BOUNDARY CONDITION
280 QP(1)=(HP(1)-H(2)+B*Q(2))/(B+R*ABS(Q(2)))
290 REM INTERIOR SECTIONS
300 FOR I=2 TO N:    CP=H(I-1)+B*Q(I-1):    BP=B+R*ABS(Q(I-1))
310 CM=H(I+1)-B*Q(I+1):   BM=B+R*ABS(Q(I+1))
320 QP(I)=(CP-CM)/(BP+BM):   HP(I)=CP-BP*QP(I):    NEXT I
330 FOR I=1 TO NS:   H(I)=HP(I):   Q(I)=QP(I):   NEXT I
340 IF J MOD JPR=0 THEN 190 ELSE 220
```

```
F,L,A,N,D= .022  4800  1200   4  2
HRES,B,JPR,DCV,TMAX= 100  9.806001  1  5  36
CVA=
 .06  .03  .01  .003  .001  .0005  .0002  0  0  0  0
     PIEZOMETER HEADS AND FLOWS ALONG THE PIPE
   TIME    CV   X/L=     .0      .25      .5      .75      1.
  0.000 0.06000  H=   100.00   99.53    99.06    98.58    98.11
                 Q=     2.63    2.63     2.63     2.63     2.63
  1.000 0.05400  H=   100.00   99.53    99.06    98.58   105.14
                 Q=     2.63    2.63     2.63     2.63     2.45
  2.000 0.04800  H=   100.00   99.53    99.06   105.59   112.77
                 Q=     2.63    2.63     2.63     2.45     2.26
  3.000 0.04200  H=   100.00   99.53   106.03   113.18   121.07
                 Q=     2.63    2.63     2.45     2.26     2.05
  4.000 0.03600  H=   100.00  106.47   113.59   121.45   130.07
                 Q=     2.63    2.45     2.26     2.05     1.82
  5.000 0.03000  H=   100.00  114.00   121.83   130.41   139.85
                 Q=     2.28    2.26     2.05     1.82     1.57
  6.000 0.02600  H=   100.00  115.33   130.76   140.17   146.84
                 Q=     1.90    1.88     1.82     1.57     1.40
  7.000 0.02200  H=   100.00  116.76   133.63   147.14   154.25
                 Q=     1.48    1.46     1.40     1.40     1.21
  8.000 0.01800  H=   100.00  118.29   133.13   147.70   162.04
                 Q=     1.03    1.01     1.04     1.04     1.01
  9.000 0.01400  H=   100.00  116.37   132.36   148.02   157.80
                 Q=     0.54    0.61     0.64     0.65     0.78
 10.000 0.01000  H=   100.00  114.08   131.27   142.47   152.18
                 Q=     0.19    0.18     0.22     0.38     0.55
 11.000 0.00860  H=   100.00  114.90   124.19   135.44   139.90
                 Q=    -0.19   -0.20    -0.08     0.12     0.45
 12.000 0.00720  H=   100.00  110.11   119.07   121.62   126.02
                 Q=    -0.58   -0.45    -0.30    -0.02     0.36
 13.000 0.00580  H=   100.00  104.18   107.56   109.66   110.38
                 Q=    -0.70   -0.69    -0.38    -0.06     0.27
 14.000 0.00440  H=   100.00   97.46    94.78    96.32    99.66
                 Q=    -0.79   -0.64    -0.44    -0.09     0.19
 15.000 0.00300  H=   100.00   90.62    86.23    84.79    87.91
                 Q=    -0.57   -0.55    -0.35    -0.19     0.12
 16.000 0.00260  H=   100.00   88.78    80.63    77.83    73.65
                 Q=    -0.31   -0.28    -0.29    -0.13     0.10
 17.000 0.00220  H=   100.00   90.01    80.38    69.49    69.43
                 Q=     0.00   -0.05    -0.07    -0.01     0.08
 18.000 0.00180  H=   100.00   91.60    78.88    71.98    66.65
                 Q=     0.20    0.22     0.23     0.15     0.07
 30.000 0.00020  H=   100.00   96.69    93.19    93.90    94.64
                 Q=    -0.87   -0.72    -0.53    -0.21     0.01
 31.000 0.00016  H=   100.00   90.08    85.27    83.34    85.51
                 Q=    -0.64   -0.61    -0.43    -0.28     0.01
 32.000 0.00012  H=   100.00   88.59    80.24    76.88    72.20
                 Q=    -0.36   -0.34    -0.36    -0.21     0.00
 33.000 0.00008  H=   100.00   90.16    80.20    69.10    68.40
                 Q=    -0.05   -0.11    -0.13    -0.08     0.00
 34.000 0.00004  H=   100.00   91.61    79.03    71.72    66.12
                 Q=     0.15    0.16     0.18     0.09     0.00
 35.000 0.00000  H=   100.00   88.87    83.13    76.04    75.15
                 Q=     0.38    0.43     0.38     0.26     0.00
 36.000 0.00000  H=   100.00   91.52    85.88    86.56    86.02
                 Q=     0.72    0.60     0.51     0.29     0.00
```

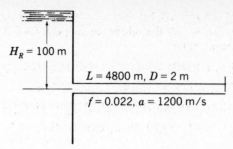

Fig. 19-6

19.10 Determine the period of oscillation of a U-tube containing $\frac{1}{2}$ L of water. The cross-sectional area is 2.4 cm². Neglect friction.

$\blacksquare$ $L = V/A = (\frac{1}{2})(1000)/2.4 = 208.3 \text{ cm}$ or   2.083 m

$$T = 2\pi\sqrt{L/2g} = (2)(\pi)\sqrt{2.083/[(2)(9.807)]} = 2.05 \text{ s}$$

19.11 A U-tube containing alcohol is oscillating with maximum displacement from an equilibrium position of 120 mm. The total column length is 1 m. Determine the maximum fluid velocity and the period of oscillation. Neglect friction.

$\blacksquare$ $z = Z \cos\sqrt{2g/L}t = (\frac{120}{1000}) \cos\sqrt{(2)(9.807)/1}t = 0.120 \cos 4.429t$ $dz/dt = -0.5315 \sin 4.429t$

Maximum velocity occurs when $\sin 4.429t = 1$, or $(dz/dt)_{max} = (-0.5315)(1) = -0.5315 \text{ m/s}$, $T = 2\pi\sqrt{L/2g} = (2)(\pi)\sqrt{1/[(2)(9.807)]} = 1.42 \text{ s}$.

19.12 A U-tube contains liquid oscillating with a velocity of 2 m/s at the instant the menisci are at the same elevation. Find the time to the instant the menisci are next at the same elevation and then determine the elevation. ($v = 1 \times 10^{-5} \text{ m}^2/\text{s}$, $D = 6$ mm, and $L = 750$ mm).

$\blacksquare$ $N_R = DV/v = (\frac{6}{1000})(2)/(1 \times 10^{-5}) = 1200$ (laminar flow)

$m = 16v/D^2 = (16)(1 \times 10^{-5})/(\frac{6}{1000})^2 = 4.444$ $n = \sqrt{(16v/D^2)^2 - 2g/L} = \sqrt{4.444^2 - (2)(9.807)/(\frac{750}{1000})} = 2.530i$

$$n' = 2.530 \qquad z = (V_0/n')(e^{-mt}) \sin(n't)$$

At next $z = 0$, $2.530t = \pi$. $t = 1.242 \text{ s}$, $V = dz/dt = V_0 e^{-mt}[\cos(n't) - (m/n') \sin(n't)]$. When $n't = \pi$, $V = (2)[e^{-(4.444)(1.242)}][\cos\pi - (m/n') \sin\pi] = 0.00802 \text{ m}$, or 0.802 cm/s.

19.13 A 10-ft-diameter horizontal tunnel has 10-ft-diameter vertical shafts spaced 1 mi apart. When valves are closed isolating this reach of tunnel, the water surges to a depth of 50 ft in one shaft when it is 20 ft in the other shaft. For $f = 0.022$, find the height of the next two surges.

$\blacksquare$ $z_m = (50 - 20)/2 = 15.00 \text{ ft}$ $[1 + (fz_m/D)]e^{-fz_m/D} = [1 + (fz_{m+1}/D)]e^{-fz_{m+1}/D}$

$$fz_m/D = (0.022)(15.00)/10 = 0.0330$$

$$(1 + 0.0330)e^{-0.0330} = [1 + (0.022z_{m+1}/10)]e^{-0.022z_{m+1}/10} \qquad z_{m+1} = 14.68 \text{ ft}$$

Repeat, using $z_m = 14.68 \text{ ft}$.

$$fz_m/D = (0.022)(14.68)/10 = 0.0323$$

$$(1 + 0.0323)e^{-0.0323} = [1 + (0.022z_{m+1}/10)]e^{-0.022z_{m+1}/10} \qquad z_{m+1} = 14.37 \text{ ft}$$

Hence, $z_1 = 14.68 \text{ ft}$ and $z_2 = 14.37 \text{ ft}$.

19.14 Two standpipes 6 m in diameter are connected by 900 m of 2.5-m-diameter pipe; $f = 0.020$ and minor losses are 4.5 velocity heads. One reservoir level is 9.0 m above the other one when a valve is rapidly opened in the pipeline. Find the maximum fluctuation in water level in the standpipe.

$\blacksquare$ $L_e = L + KD/f = 900 + (4.5)(2.5)/0.020 = 1462 \text{ m}$ $z_m = z_1(A_1/A) = (\frac{9}{2})(6/2.5)^2 = 25.92 \text{ m}$

$$\phi = (f)(L_e/L)(z_m/D) = (0.020)(\tfrac{1462}{900})(25.92/2.5) = 0.3368$$

$F(\phi) = (1 + \phi)e^{-\phi} = 0.954$, which is satisfied by $\phi = -0.275$.

$$z_m = \phi LD/fL_e = (-0.275)(900)(2.5)/[(0.020)(1462)] = -21.16 \text{ m}$$

$$z_1 = z_m(A/A_1) = (-21.16)(2.5/6)^2 = -3.67 \text{ m}$$

$$\text{Maximum fluctuation} = \tfrac{9}{2} - (-3.67) = 8.17 \text{ m}$$

19.15 A valve is quickly opened in a pipe 1200 m long, $D = 0.6$ m, with a 0.3-m-diameter nozzle on the downstream end. Minor losses are $4V^2/2g$, with V the velocity in the pipe, $f = 0.024$, and $H = 9$ m. Find the time to attain 95 percent of the steady-state discharge.

▌ $t = (LV/2gH)\ln[(V_0 + V)/(V_0 - V)]$ $H = (fL_e/D)(V^2/2g) + V_1^2/2g$ (V_1 = nozzle velocity)

$L_e = L + KD/f = 1200 + (4)(0.6)/0.024 = 1300$ m $V_1/V = (0.6/0.3)^2$ $V_1 = 4V$

$9 = [(0.024)(1300)/0.6]\{V^2/[(2)(9.807)]\} + (4V)^2/[(2)(9.807)]$ $V = 1.611$ m/s

$t = \{(1200)(1.611)/[(2)(9.807)(9)]\}\ln(1.95/0.05) = 40.1$ s

19.16 A globe valve ($K = 10$) at the end of a pipe 2000 ft long is rapidly opened. $D = 3.0$ ft, $f = 0.018$, minor losses are $2V^2/2g$, and $H = 75$ ft. How long does it take for the discharge to attain 80 percent of its steady-state value?

▌ $t = (LV/2gH)\ln[(V_0 + V)/(V_0 - V)]$ $V = \sqrt{2gHD/fL_e}$

$L_e = L + KD/f = 2000 + (10 + 2)(3)/0.018 = 4000$ ft

$V = \sqrt{(2)(32.2)(75)(3.0)/[(0.018)(4000)]} = 14.19$ ft/s

$t = \{(2000)(14.19)/[(2)(32.2)(75)]\}\ln(1.80/0.20) = 12.9$ s

19.17 Benzene ($K = 150\,000$ psi, s.g. $= 0.88$) flows through $\frac{3}{4}$-in-ID steel ($E = 30\,000\,000$ psi) tubing with $\frac{1}{8}$-in wall thickness. Determine the speed of a pressure wave.

▌ $a = \sqrt{(K/\rho)/[1 + (K/E)(D/e)]}$

$= \sqrt{\{[(150\,000)(144)]/[(0.88)(1.94)]\}/\{1 + (150\,000/30\,000\,000)[(\frac{3}{4})/(\frac{1}{8})]\}} = 3505$ ft/s

19.18 A valve is closed in 5 s at the downstream end of a 3000-m pipeline carrying water at 2 m/s ($a = 1000$ m/s). What is the peak pressure developed by the closure?

▌ $2L/a = (2)(3000)/1000 = 6.00$ s. Since $[2L/a = 6.00$ s$] > 5$ s, this is rapid closure. Then, $h = aV_0/g = (1000)(2)/9.807 = 203.9$ m, $p = \gamma h = (9.79)(203.9) = 1996$ kN/m^2.

19.19 Determine the length of pipe in Prob. 19.18 subject to the peak pressure.

▌ $x = L - at_c/2 = 3000 - (1000)(5)/2 = 500$ m

19.20 A valve is closed at the downstream end of a pipeline in such a manner that only one-third of the line is subjected to maximum pressure. During what proportion of the time $2L/a$ is it closed?

▌ $x = L - at_c/2$ $L/3 = L - at_c/2$ $2L/3 = at_c/2$ $t_c/(2L/a) = \frac{2}{3}$

19.21 A 10-mm-diameter U-tube contains oil, $v = 5$ m^2/s, with a total column length of 2 m. If the initial half-amplitude of displacement is 250 mm, prepare a computer program to find the first 10 maximum and minimum displacements and the times they occur.

```
▌ 10 REM                    LAMINAR OSCILLATION OF A U-TUBE.
  20 '  TO=ATN(N'/M)/N'
  30 '  VO=Z*N'/(EXP(-M*TO)*SIN(N'*TO))
  40 '  The angle THO is THO=N'*TO, and the
  50 '  general angle is TH=THO+I*PI, with T=TH/N'-TO.
  60 '  Z=VO/N'*EXP(-M*TH/N')*SIN(TH)
  70 DEFINT I: DEF FNZ(DT,DTH)=C1*EXP(-M*DT)*SIN(DTH)
  80 READ D,NU,L,G,DZ,PI: DATA .01,5E-6,2.,9.806,.25,3.1416
  90 LPRINT: LPRINT"D,NU,L,G,DZ,PI=";D;NU;L;G;DZ;PI: LPRINT
 100 M=16!*NU/D^2:CC=2!*G/L-M^2:IF CC>0! THEN NP=SQR(CC) ELSE
     LPRINT"NON-OSCILLATING CASE":STOP
 110 TO=ATN(NP/M)/NP:VO=NP*DZ*EXP(M*TO)/SIN(NP*TO):THO=NP*TO
 120 LPRINT"M,NP,TO,VO,THO=";M;NP;TO;VO;THO: LPRINT
 130 C1=VO/NP: FOR I=0 TO 10:TH=THO+I*PI:T=TH/NP:Z=FNZ(T,TH)
 140 LPRINT"T,Z,TH=";:LPRINT USING " ####.##### ";T-TO;Z;TH
 150 NEXT I
 160 LPRINT: LPRINT"MAXIMUM REYNOLD'S NUMBER,V*D/NU=";VO*D/NU


     D,NU,L,G,DZ,PI= .01  .000005  2  9.806001  .25  3.1416

     M,NP,TO,VO,THO= .8  3.02754  .4335071  1.107393  1.31246
```

```
T,Z,TH=     0.00000      0.25000      1.31246
T,Z,TH=     1.03767     -0.10900      4.45406
T,Z,TH=     2.07535      0.04752      7.59566
T,Z,TH=     3.11302     -0.02072     10.73726
T,Z,TH=     4.15070      0.00903     13.87886
T,Z,TH=     5.18837     -0.00394     17.02046
T,Z,TH=     6.22604      0.00172     20.16206
T,Z,TH=     7.26372     -0.00075     23.30366
T,Z,TH=     8.30139      0.00033     26.44526
T,Z,TH=     9.33907     -0.00014     29.58686
T,Z,TH=    10.37674      0.00006     32.72846

MAXIMUM REYNOLD'S NUMBER,V$D/NU= 2214.786
```

19.22 An equation for dealing with turbulent resistance is given by

$$\frac{d^2z}{dt^2} + \frac{f}{2D}\frac{dz}{dt}\left|\frac{dz}{dt}\right| + \frac{2g}{L} = 0$$

Put this equation in suitable form for solution by the third-order Runge-Kutta method.

❚ Let $y = dz/dt$, then $dy/dt = d^2z/dt^2$, and $dy/dt = -(f/2D)y\,|y| - (2g/L)z = F_1(y, z, t)$, $dz/dt = y = F_2(y, z, t)$. The two equations are solved simultaneously, from known initial conditions y_n, z_n, t_n: $u_{11} = hF_1(y_n, z_n, t_n) = h[-(f/2D)y_n\,|y_n| - (2g/L)z_n]$, $u_{12} = hF_2(y_n, z_n, t_n) = hy_n$, $u_{21} = hF_1[y_n + (u_{11}/3), z_n + (u_{12}/3), t_n + (h/3)] = h\{-(f/2D)[y_n + (u_{11}/3)]\,|y_n + (u_{11}/3)| - (2g/L)[z_n + (u_{12}/3)]\}$, $u_{22} = hF_2[y_n + (u_{11}/3), z_n + (u_{12}/3), t_n + (h/3)] = h[y_n + (u_{11}/3)]$, $u_{31} = hF_1(y_n + \frac{2}{3}u_{21}, z_n + \frac{2}{3}u_{22}, t_n + \frac{2}{3}h) = h[-(f/2D)(y_n + \frac{2}{3}u_{21})\,|y_n + \frac{2}{3}u_{21}| - (2g/L)(z_n + \frac{2}{3}u_{22})]$, $u_{32} = hF_2(y_n + \frac{2}{3}u_{21}, z_n + \frac{2}{3}u_{22}, t_n + \frac{2}{3}h) = h(y_n + \frac{2}{3}u_{21})$, $y_{n+1} = y_n + (u_{11}/4) + \frac{3}{4}u_{31}$, $z_{n+1} = z_n + (u_{12}/4) + \frac{3}{4}u_{32}$, $t_{n+1} = t_n + h$. The equations for simultaneous solution have been written for a general case as well as for the specific case of solution of the given equation.

19.23 Prepare a program to carry out the solution of Prob. 19.22 and apply it to the following case: $t = 0$, $z = 12$ ft, $V_0 = 0$, $L = 400$ ft, $f = 0.017$, $d = 2$ ft, $dt = 0.1$ s, and $t_{max} = 30$ s.

❚

```
10 REM                    TURBULENT OSCILLATION OF A U-TUBE
20 DEFINT I: DEF FNN(DY,DZ)=H$(-C1$DY$ABS(DY)-C2$DZ): DEF FNM(DY)=H$DY
30 READ Z,V,L,F,D,DT,TMAX,IPR,G: DATA 12.,,0,400.,.017,2.5,.1,30.,20,32.2
40 LPRINT: LPRINT"Z,V,L,F,D=";Z;V;L;F;D: LPRINT"DT,TMAX,IPR,G=";DT,TMAX,IPR,G
50 II=FIX(TMAX/DT)+1: H=DT: C1=.5$F/D: C2=2!$G/L: Y=0!: T=0!
60 LPRINT: LPRINT"        T              V              Z"
70 FOR I=0 TO II: IF I MOD IPR=0 THEN LPRINT USING "  ###.### ";T;Y;Z
80 U11=FNN(Y,Z):U12=FNM(Y)
90 U21=FNN(Y+U11/3!,Z+U12/3!):U22=FNM(Y+U11/3!)
100 U31=FNN(Y+2!$U21/3,Z+2!$U22/3!):U32=FNM(Y+2!$U21/3!)
110 Y=Y+.25$U11+.75$U31:Z=Z+.25$U12+.75$U32:T=T+H
120 NEXT I

Z,V,L,F,D= 12   0   400   .017   2.5
DT,TMAX,IPR,G= .1              30              20              32.2
```

T	V	Z
0.000	0.000	12.000
2.000	-3.434	8.354
4.000	-4.681	-0.240
6.000	-3.083	-8.433
8.000	0.308	-11.355
10.000	3.459	-7.356
12.000	4.418	0.982
14.000	2.702	8.495
16.000	-0.585	10.723
18.000	-3.466	6.425
20.000	-4.162	-1.647
22.000	-2.345	-8.513
24.000	0.834	-10.104
26.000	3.458	-5.552
28.000	3.911	2.244
30.000	2.011	8.492

19.24 A surge tank of area AR is to be used with a hydroelectric power plant. The turbines may change from full flow to zero flow in TC seconds, and similarly accept full flow in TC seconds. Assuming linear changes in flow to the turbines, determine the elevation of maximum and minimum water surfaces in the surge tank. The maximum flow is QM; elevation of reservoir is Z. Prepare a program using the Runge–Kutta second-order integration of the equations. Specifically, let $Z = 100$ m, $QM = 20$ m^3/s, $AR = 75$ m^2, and a pressure pipe $L = 10$ km, $D = 3$ m, $F = 0.016$, with $TC = 4$ s and $DT = H = 1$ s.

▮

```
10 REM                        SIMPLE SURGE TANK.
20 ' Continuity equation: DY/DT=(QIN-QOUT)/AR
30 ' Eq.Motion: GAM*AR*(Z-Y-R*Q*ABS(Q))=GAM*AR*(L/G)*DQ/DT
40 ' or DQ/DT=G*AP*(Z-Y-R*Q*ABS(Q))/L: in which Y=elev of water surface in
50 ' surge tank; AP=area of pressure pipe; R=head loss in pressure pipe
60 ' per unit discharge.
70 DEFINT I: DEF FNQ(DY,DQ)=C1*(Z-DY-R*DQ*ABS(DQ))
80 READ F,L,D,G,Z,AR,QM,TC,H,TM,IPR,ICLOSE
90 DATA .016,10000.,3.,9.806,100.,75.,20.,4.,1.,900.,60,0
100 LPRINT: LPRINT"F,L,D,G,Z=";F;L;D;G;Z: LPRINT
110 LPRINT"AR,QM,TC,H,TM,IPR,ICLOSE=";AR;QM;TC;H;TM;IPR;ICLOSE: LPRINT
120 AP=.7854*D^2: R=F*L/(2!*G*.7854^2*D^5): C1=G*AP*H/L: C2=H/AR: II=FIX(TM/H)+1

130 T=0!: IF ICLOSE=1 THEN Y=Z-R*QM^2: QT=QM: Q=QM ELSE Y=Z: QT=0!: Q=0!
140 LPRINT"      TIME       DISCHARGE       Y          QT"
150 YMAX=Z: YMIN=Z: LPRINT USING"  ###.###   ";T;Q;Y;QT: IF ICLOSE=0 THEN 270
160 FOR I=0 TO II:T=I*H:IF T<=TC THEN QT=QM*(1!-T/TC) ELSE QT=0!
170 U11=FNQ(Y,Q):U12=C2*(Q-QT):TT=T+H
180 IF TT<=TC THEN QT=QM*(1!-TT/TC) ELSE QT=0!
190 U21=FNQ(Y+U12,Q+U11):U22=C2*(Q+U11-QT)
200 Q=Q+.5*(U11+U21):Y=Y+.5*(U12+U22):T=T+H
210 IF Y>YMAX THEN YMAX=Y:TMAX=T
220 IF Y<YMIN THEN YMIN=Y:TMIN=T
230 IF (I+1) MOD IPR=0 THEN LPRINT USING"  ####.###   ";T;Q;Y;QT
240 NEXT I
250 GOTO 360
260 END
270 FOR I=1 TO II:T=I*H:IF T<=TC THEN QT=QM*T/TC ELSE QT=QM
280 U11=FNQ(Y,Q):U12=C2*(Q-QT):TT=T+H
290 IF TT<=TC THEN QT=QM*TT/TC ELSE QT=QM
300 U21=FNQ(Y+U12,Q+U11):U22=C2*(Q+U11-QT)
310 Q=Q+.5*(U11+U21):Y=Y+.5*(U12+U22):T=T+H
320 IF Y>YMAX THEN YMAX=Y:TMAX=T
330 IF Y<YMIN THEN YMIN=Y:TMIN=T
340 IF (I+1) MOD IPR=0 THEN LPRINT USING"  ####.###   ";T;Q;Y;QT
350 NEXT I
360 LPRINT"YMAX,TMAX=";:LPRINT USING"  ####.##   ";YMAX;TMAX
370 LPRINT"YMIN,TMIN=";:LPRINT USING"  ####.##   ";YMIN;TMIN
```

F,L,D,G,Z= .016 10000 3 9.806001 100

AR,QM,TC,H,TM,IPR,ICLOSE= 75 20 4 1 900 60 0

TIME	DISCHARGE	Y	QT
0.000	0.000	100.000	0.000
60.000	2.978	85.346	20.000
120.000	10.451	74.568	20.000
180.000	17.524	69.934	20.000
240.000	21.355	69.706	20.000
300.000	22.561	71.399	20.000
360.000	22.479	73.466	20.000
420.000	21.944	75.248	20.000
480.000	21.346	76.561	20.000
540.000	20.839	77.427	20.000
600.000	20.466	77.940	20.000
660.000	20.220	78.206	20.000
720.000	20.075	78.318	20.000
780.000	20.000	78.345	20.000
840.000	19.970	78.331	20.000
900.000	19.964	78.303	20.000
YMAX,TMAX=	100.00	0.00	
YMIN,TMIN=	69.44	212.00	

19.25 The open wedge-shaped tank in Fig. 19-7 has a length of 15 ft perpendicular to the sketch. It is drained with a 3-in-diameter pipe of length 10 ft whose discharge end is at elevation zero. The coefficient of loss at the pipe entrance is 0.50, the total of the bend loss coefficient is 0.20, and f for the pipe is 0.018. Find the time required to lower the water surface in the tank from elevation 8 ft to 5 ft. Neglect the possible change of f with N_R, and assume that the acceleration effects in the pipe are negligible.

▮ Energy equation from water surface to jet at discharge: $z - [0.5 + 0.2 + 0.018(10/0.25)](V^2/2g) = V^2/2g$, $z - 1.42(V^2/2g) = V^2/2g$, $V = 5.16z^{1/2}$; $Q_0 = AV = (\pi/4)(0.25)^2 5.16z^{1/2} = 0.253z^{1/2}$. The area of the water surface may be expressed as $A_s = 15b = 15Kz$. At the top of the tank, $A_s = 15(6) = 15K(10)$, $K = 0.6$. Thus $A_s = 15(0.6)z = 9z$.

$$t = \int_{z_1}^{z_2} \frac{A_s\, dz}{Q_i - Q_0} = \int_8^5 \frac{9z\, dz}{0 - 0.253z^{1/2}} = -\frac{9}{0.253} \int_8^5 z^{1/2}\, dz = -35.5[\tfrac{2}{3}z^{3/2}]_8^5 = 271\text{ s}$$

Note that if the pipe had discharged at an elevation other than zero, the integral would have been different, because the head on the pipe would then have been $z + h$, where h is the vertical distance of the discharge end of the pipe below (h positive) or above (h negative) point A of the figure.

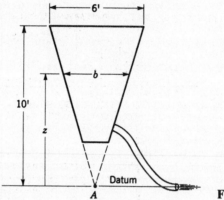

Fig. 19-7

19.26 Two large water reservoirs are connected to one another with a 10-cm-diameter pipe ($f = 0.02$) of length 15 m. The water-surface elevation difference between the reservoirs is 2.0 m. A valve in the pipe, initially closed, is suddenly opened. Determine the times required for the flow to reach $\frac{1}{4}$, $\frac{1}{2}$, and $\frac{3}{4}$ of the steady-state flow rate. Assume the water-surface elevations remain constant. Repeat for pipe lengths of 150 m and 1500 m with all other data remaining the same. In the first case $L/D = 15/0.10 = 150$, hence minor losses are significant. Assume square-edged entrance.

▮ Square-edge entrance: $k_L = 0.5$, $k = k_L + f(L/D) = 0.5 + 0.02(15/0.10) = 3.5$. For steady flow: $V_0 = \sqrt{2gH/(1+k)} = \sqrt{2(9.81)2/(1+3.5)} = 2.95$ m/s. For unsteady flow use $t = \{L/[(1+k)V_0]\} \ln[(V_0 + V)/(V_0 - V)] = [15/4.5(2.95)] \ln[(2.95 + V)/(2.95 - V)] = 1.129 \ln[(2.95 + V)/(2.95 - V)]$. For $Q = \frac{1}{4}Q_0$ substitute $V = \frac{1}{4}V_0$, etc.:

Q	V, m/s	$\dfrac{2.95 + V}{2.95 - V}$	ln	t, s
$0.25Q_0$	0.74	1.667	0.511	0.577
$0.50Q_0$	1.48	3.00	1.099	1.240
$0.75Q_0$	2.21	7.00	1.946	2.197

For the other two lengths the results are as follows:

Q	$L = 150$ m	$L = 1500$ m
$0.25Q_0$	2.18 s	7.04 s
$0.50Q_0$	4.69 s	15.15 s
$0.75Q_0$	8.30 s	26.84 s

19.27 In Fig. 19-8 the elasticity and dimensions of the pipe are such that the celerity of the pressure wave is 3200 fps. Suppose the pipe has a length of 2000 ft and a diameter of 4 ft. The flow rate is initially 30 cfs. Water is flowing.

Find (a) the water-hammer pressure for instantaneous valve closure; (b) the approximate water-hammer pressure at the valve if it is closed in 4.0 s; (c) the water-hammer pressure at the valve if it is manipulated so that the flow rate drops almost instantly from 30 to 10 cfs; (d) the maximum water-hammer pressure at a point in the pipe 300 ft from the reservoir if a 1.0-s valve closure reduces the flow rate from 10 cfs to zero.

$$V = Q/A = 30/(\pi 2^2) = 2.39 \text{ fps}$$

(a)
$$p_h = \rho c_P V = 1.94(3200)(2.39) = 14\,840 \text{ lb/ft}^2 = 103.0 \text{ psi}$$

(b)
$$p'_h \approx (\tfrac{4000}{3200}/4.0)p_h = (1.25/4.00)(103.0) = 32.2 \text{ psi}$$

(c) For this case of partial closure $\Delta p_h = -\rho c_P(\Delta V)$. $\Delta V = (10-30)/\pi 2^2 = -1.592$ fps, $\Delta p_h = -1.94(3200)(-1.592) = 9880 \text{ lb/ft}^2 = 68.6$ psi.

(d) If $2x_0/c_P = 1.0$ s, $x_0 = 1600$ ft, so that full water-hammer pressure will be developed in the pipe only in the region that is farther than 1600 ft from the reservoir.

   For this cases, at valve, $p_h = 1.94(3200)(2.39/3) = 4940 \text{ lb/ft}^2 = 34.3$ psi. At point 300 ft from reservoir: $p_h = 34.3(\tfrac{300}{1600}) = 6.43$ psi.

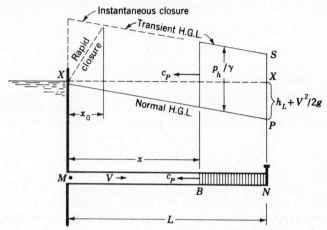

Fig. 19-8

19.28 Suppose a ship lock has vertical sides and that water enters or discharges through a conduit area A such that $Q = C_d A\sqrt{2gz}$, where z is the variable difference in level between the water surface in the lock and that outside. Prove that for the water level in the lock to change from z_1 to z_2 the time is $t = [2A_s/(C_dA\sqrt{2g})](z_1^{1/2} - z_2^{1/2})$. (*Note:* If the lock is being filled, the signs must be reversed.)

$$t = \int_{z_1}^{z_2} \frac{A_s\,dz}{Q_i - Q_o} \qquad Q_i = 0 \qquad Q_o = C_dA(2gz)^{1/2}$$

thus

$$t = \int_{z_1}^{z_2} \frac{A_s\,dz}{-C_dA(2gz)^{1/2}} = \frac{A_s}{C_dA(2g)^{1/2}} \int_{z_1}^{z_2} \frac{dz}{-z^{1/2}} = \frac{A_s}{C_dA(2g)^{1/2}}[-2z^{1/2}]_{z_1}^{z_2} = \frac{2A_s}{C_dA(2g)^{1/2}}(z_1^{1/2} - z_2^{1/2})$$

19.29 Suppose the lock of Prob. 19.28 is 300 ft long by 90 ft wide, and water enters through a conduit for which the discharge coefficient is 0.50. If the water surface in the lock is initially 36 ft below the level of the surface of the water upstream, how large must the conduit be if the lock is to be filled in 5 min?

$$t = [2A_s/(C_dA\sqrt{2g})](z_1^{1/2} - z_2^{1/2}) \qquad (5)(60) = \{(2)[(300)(90)]/[(0.50)(A)\sqrt{(2)(32.2)}]\}(36^{1/2} - 0^{1/2})$$

$$A = 269 \text{ ft}^2$$

19.30 (a) Suppose a reservoir has vertical sides and that initially there is a steady flow into it such that the height of the surface above the level of a spillway $(Q = C_WLH^{3/2})$ is z_1. If the inflow is suddenly cut off, prove that the time required for the water level to fall from z_1 to z_2 is $t = (2A_s/C_WL)(1/\sqrt{z_2} - 1/\sqrt{z_1})$. (*Note:* $z = H$.) (b) How long will it take theoretically for the outflow to cease entirely? What factors make this theoretical answer unrealistic?

(a)
$$t = \int_{z_1}^{z_2} \frac{A_s\,dz}{Q_i - Q_o} \qquad Q_i = 0 \qquad Q_o = C_WLz^{3/2}$$

$$= \int_{z_1}^{z_2} \frac{A_s\,dz}{-C_WLz^{3/2}} = \frac{A_s}{C_WL}\left[\frac{2}{z^{1/2}}\right]_{z_1}^{z_2} = \frac{2A_s}{C_WL}\left[\frac{1}{z_2^{1/2}} - \frac{1}{z_1^{1/2}}\right]$$

(b) As $z_2 \to 0$, $t \to \infty$, so theoretically the outflow would never cease. Factors which make this theoretical answer unrealistic include surface tension, surface ripples, evaporation, and leakage/seepage losses.

19.31 The crest of the overflow of a reservoir is 100 ft long, and the value of C_W is 3.45. For the range of levels considered here the area of the water surface is constant and is 700 000 ft². Initially, there is a flow into the reservoir at such a rate that the height of the water surface above that of the spillway crest is stabilized at 3 ft, and then the inflow is suddenly diverted. Find the length of time for the water surface to fall to a height of 1 ft above the level of the spillway.

❙

$$t = \int_{z_1}^{z_2} \frac{A_s \, dz}{Q_i - Q_o} \qquad Q_i = 0 \qquad Q_o = C_W L z^{3/2}$$

$$= \int_{z_1}^{z_2} \frac{A_s \, dz}{-C_W L z^{3/2}} = \frac{2A_s}{C_W L} \left[\frac{1}{z_2^{1/2}} - \frac{1}{z_1^{1/2}} \right]$$

$$= \frac{2(700\,000)}{3.45(100)} \left[\frac{1}{1^{1/2}} - \frac{1}{3^{1/2}} \right] = 1715 \text{ s} = 28.6 \text{ min}$$

19.32 The crest of the overflow spillway of a reservoir is 40 ft long, and the value of C_W is 3.50. The area of the water surface is assumed constant at 600 000 ft² for the range of heights considered here. Initially, the water surface is 3 ft below the level of the spillway crest. If suddenly there is turned into this reservoir a flow of 500 cfs, what will be the height of water in the reservoir for equilibrium? How long a time will be required for this height to be reached? How long a time will be required for the water surface to reach a height of 2 ft above the level of the spillway? (*Note:* This last can be solved by integration after substituting x^3 for $z^{3/2}$ and consulting integral tables. However, it will be easier to solve it graphically either by plotting and actually measuring the area under the curve or by computing the latter by some method, such as Simpson's rule.)

❙ $Q = C_W L z^{3/2}$, $Q_o = 3.50(40)z^{3/2} = 140z^{3/2}$. For equilibrium $Q_i = Q_o = 500$ cfs, from which $z = \left(\frac{500}{140}\right)^{2/3} = 2.34$ ft. The time required to reach this equilibrium height is theoretically infinite, but surface ripples and minor irregularities in flow cause practical equilibrium to be reached in a finite time. However, this does point out that true equilibrium is not quickly obtained. This should be borne in mind in experimental work. The time for the water level to reach the spillway crest = $3(600\,000)/500 = 3600$ s. The time for water level to rise from crest to 2 ft above it is

$$t = \int_{z_1}^{z_2} \frac{A_s \, dz}{Q_i - Q_o} = 600\,000 \int_0^2 \frac{dz}{500 - 140z^{3/2}}$$

Letting $z^{3/2} = x^3$, $\left(\frac{3}{2}\right)z^{1/2} \, dz = 3x^2 \, dx$ and $dz = (3x^2 \, dx)/[(\frac{3}{2})z^{1/2}] = (3x^2 \, dx)/[(\frac{3}{2})x] = 2x \, dx$.

$$t = 2(600\,000) \int_0^{2^{1/2}} \frac{x \, dx}{500 - 140x^3} = 1\,200\,000 \frac{(-\frac{500}{140})^{1/3}}{3(500)} \left[\frac{1}{2} \ln \frac{[x + (-\frac{500}{140})^{1/3}]^2}{(-\frac{500}{140})^{2/3} - (-\frac{500}{140})^{1/3}x + x^2} + 3^{1/2} \tan^{-1} \frac{2x - (-\frac{500}{140})^{1/3}}{(-\frac{500}{140})^{1/3}3^{1/2}} \right]_0^{2^{1/2}}$$

$$= 1\,200\,000 \left(\frac{-1.529}{1500} \right) \left[\frac{1}{2} \ln \frac{(x - 1.529)^2}{2.34 + 1.529x + x^2} + 3^{1/2} \tan^{-1} \frac{2x + 1.529}{-1.529(3)^{1/2}} \right]_0^{2^{1/2}} = 6830 \text{ s}$$

Total time = $3600 + 6830 = 10\,430$ s

19.33 Water enters a reservoir at such a rate that the height of water above the level of the spillway crest is 3 ft. The spillway ($Q = C_W L H^{3/2}$) is 100 ft long, and the value of C_W is 3.45. The area of the water surface for various water levels is as follows:

z, ft	A_s, ft²
3.00	860 000
2.50	830 000
2.00	720 000
1.50	590 000
1.25	535 000
1.00	480 000

If the inflow is suddenly reduced to 150 cfs, what will be the height of water for equilibrium? How long will it take, theoretically, for equilibrium to be attained? How long will it take for the level to drop from 3 ft to 1 ft above that of the spillway?

▮ $Q_o = 345z^{3/2}$; when $Q_o = Q_i = 150$ cfs, $z = 0.574$ ft and the time for equilibrium is theoretically infinite.

$$t = \int_{z_1}^{z_2} \frac{A_s\,dz}{Q_i - Q_o} = \int \frac{A_s\,dz}{345z^{3/2} - 150} \quad (z \text{ gets smaller})$$

z, ft	Δz, ft	A_s, ft²	$\bar{A}_s$, ft²	$\bar{A}_s(\Delta z)$, ft³	$Q_o = 345z^{3/2}$, cfs	$\bar{Q}_o$, cfs	$\bar{Q}_o - 150$, cfs	$\dfrac{A_s(\Delta z)}{\bar{Q}_o - 150}$
3.0		860 000			1793			
	0.5		845 000	422 500		1578	1428	296
2.5		830 000			1364			
	0.5		775 000	387 500		1170	1020	380
2.0		720 000			976			
	0.5		655 000	327 500		805	655	500
1.5		590 000			634			
	0.25		563 000	140 600		558	408	345
1.25		535 000			482			
	0.25		508 000	126 900		414	264	481
1.00		480 000			345			
								$\Sigma = 2002$ s

Total time for z to change from 3 ft to 1.0 ft is approximately 2000 s = 33.4 min.

19.34 Work Prob. 19.33 using the same numbers but changing feet to meters, square feet to square meters, and cubic feet per second to cubic meters per second.

▮ This problem is solved in the same manner as Prob. 19.33. $Q_o = 1.84(3.45/3.33)LH^{3/2} = 1.906(100)z^{3/2} = 190.6z^{3/2}$. When $Q_o = Q_i = 150$ m³/s, $z = (150/190.6)^{2/3} = 0.852$ m and the time for equilibrium is theoretically infinite. Columns 2 to 5 are omitted from the following solution table since they are numerically the same as those in Prob. 19.33.

z, m	$Q_o = 190.6z^{3/2}$, m³/s	$\bar{Q}_o$, m³/s	$\bar{Q}_o - 150$, m³/s	$\dfrac{A_s(\Delta z)}{\bar{Q}_o - 150}$
3.0	990			
		872	722	586
2.5	753			
		646	496	781
2.0	539			
		445	295	1112
1.5	350			
		308	158	890
1.25	266			
		229	79	1617
1.00	191			
				$\Sigma = 4986$ s

Total time for z to change from 3 m to 1.0 m is approximately 4990 s = 83.1 min.

19.35 Figure 19-9 shows a tank whose shape is the frustum of a cone with a 2-ft² orifice in the bottom. Assume $C_d = 0.62$. If the water level outside the tank is constant at section 2, how long will it take the water level in the tank to drop from section 1 to section 2? (*Note:* Diameter of tank = Ky, and $y = z + h_2$, where z is the variable distance between surface levels.)

▮ $D = ky$; $k = (44 - 20)/36 = \frac{2}{3}$; $A_s = (\pi/9)y^2$; $z = h_2 - y$; $y = z + h_2 = z + 30$; $Q_o = C_d A(2gz)^{1/2} = 0.62(2)(2gz)^{1/2}$.

$$t = \int_{z_1}^{z_2} \frac{A_s\,dz}{Q_i - Q_o} = \int_{36}^{0} \frac{(\pi/9)(z + 30)^2\,dz}{-(0.62)2(2gz)^{1/2}} = 791 \text{ s}$$

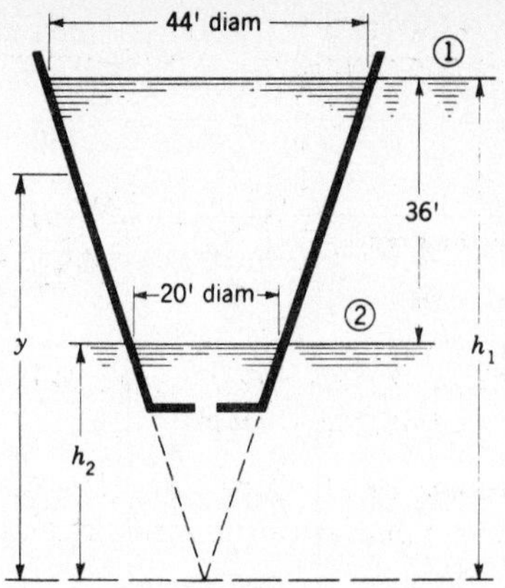

Fig. 19-9

19.36 If in Fig. 19-9 the water surface outside the tank is constant at section 1 and the tank is initially empty, how long will it take for the water level in the tank to rise from section 2 to section 1? Assume a 2-ft^2 orifice with $C_d = 0.62$.

▮ $z = h_1 = y$; $y = h_1 - z = 66 - z$; $Q_o = C_d A (2gz)^{1/2} = 0.62(2)(2gz)^{1/2}$.

$$t = \int_{z_1}^{z_2} \frac{A_s\, dz}{Q_i - Q_o} = \int_{36}^{0} \frac{(\pi/9)(66 - z)^2(-dz)}{(0.62)2(2gz)^{1/2}} = 1274 \text{ s}$$

19.37 A 1-in-diameter smooth brass pipe 1000 ft long drains an open cylindrical tank which contains oil having $\rho = 1.8$ slugs/ft^3, $\mu = 0.0006$ lb-s/ft^2. The pipe discharges at elevation 100 ft. Find the time required for the oil level to drop from elevation 120 ft to elevation 108 ft if the tank is 4 ft in diameter.

▮ Check if flow is always laminar. When $V = V_{max}$, the elevation of oil in the tank is at 120 ft, then $120 - f(L/D)(V^2/2g) = 100$; i.e., $20 = f(L/D)(V^2/2g)$. Assuming laminar flow: $f = 64/N_R$, $N_R = DV\rho/\mu$. Substituting: $20 = (64\mu/DV\rho)(L/D)(V^2/2g)$, $20 = \{64(0.0006)/[(\frac{1}{12})V(1.8)]\}(1000/\frac{1}{12})\{V^2/[2(32.2)]\} = 47.7V$, $V = V_{max} = 0.419$ fps; $N_{Rmax} = (\frac{1}{12})(0.419)(1.8)/0.0006 = 104.8$, $N_{Rmax} < 2000$, so the flow is always laminar. In general, from the above: elev $- 100 = 47.7V$, $V = (\text{elev} - 100)/47.7 = h/47.7$ (defining $h = \text{elev} - 100$); $Q = AV = 0.00545(h/47.7)$; $A_s\, dz = Q_i\, dt - Q_o\, dt$, $Q\, dt = -A\, dh$, $dt = -4\pi\, dh/[0.00545(h/47.7)] = -109\,900\, dh/h$.

$$\int_0^t dt = -109\,900 \int_{20}^{8} \frac{dh}{h} = -109\,900[\ln h]_{20}^{8} \qquad t = -109\,900(2.08 - 3.00) = 100\,700 \text{ s} = 28.0 \text{ h}$$

19.38 A large reservoir is being drained with a pipe system as shown in Fig. 19-10. Initially, when the pump is rotating at 200 rpm, the flow rate is 6.3 cfs. If the pump speed is increased instantaneously to 250 rpm, determine the flow rate as a function of time. Assume that the head h_p developed by the pump is proportional to the square of the rotative speed; that is, $h_p \propto n^2$.

▮ Initial condition neglecting minor losses: $z - h_{L_f} + h_p = V^2/2g$, $40 - 0.02(\frac{3000}{1})V^2/2g + h_p = V^2/2g$, where $V = Q/A = 6.3/0.785 = 8.02$ fps, from which $h_p = 20.9$ ft; $h_p' = 20.9(\frac{250}{200})^2 = 32.7$ ft.

New conditions: $40 - 60(V^2/2g) + 32.7 = V^2/2g + (L/g)(dV/dt)$. Eventually $dV/dt = 0$, $V^2/2g = 72.7/61 = 1.192$ ft, $V = 8.76$ fps when steady flow is eventually achieved.

In general: $72.7 - 61(V^2/2g) = (3000/32.2)(dV/dt)$, $dt = 98.4\, dV/(76.8 - V^2)$. Integrate the left side from zero to t; integrate the right side from 8.02 to V_t. Select various V_t's and find the corresponding t's:

$$\int dt = \frac{98.4}{17.53}\left[\ln\frac{8.76 + V}{8.76 - V}\right]_{8.02}^{V_t} = 5.61 \ln\frac{8.76 + V_t}{8.76 - V_t} - 17.51$$

V_t, fps	t, s	Q, cfs
8.02	0	6.30
8.15	1.112	6.40
8.40	4.14	6.60
8.65	10.79	6.79
8.70	14.10	6.83
8.74	19.84	6.86
8.76	∞	6.86

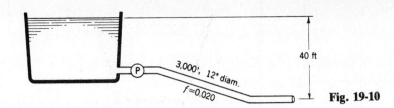

Fig. 19-10

19.39 (*a*) Repeat Prob. 19.38 with all data the same except use a 24-in-diameter pipe rather than a 12-in pipe. (*b*) Repeat also for the case of a 10-inch-diameter pipe.

▌ (*a*) **24-in pipe:** Initial conditions: $40 - 0.02(\frac{3000}{2})(V^2/2g) + h_p = V^2/2g$, $V = 8.02/4 = 2.01$ fps; $h_p = 31(V^2/2g) - 40 = -38.1$ ft. Thus a pump is not required; actually a head of 38.1 ft is available to drive a turbine when the flow rate is 6.3 cfs.

(*b*) **10-in pipe:** Initial conditions: $40 - 0.02(3000/0.833)(V^2/2g) + h_p = V^2/2g$ where $V = Q/A = 6.3/[\pi(\frac{5}{12})^2] = 11.55$ fps, $h_p = 73(V^2/2g) = 40 - 111.2$ ft; $h_p' = 111.2(\frac{250}{200})^2 = 173.8$ ft.

New conditions: $40 - 72(V^2/2g) + 173.8 = V^2/2g + (L/g)(dV/dt)$. Eventually $dV/dt = 0$, $V^2/2g = \frac{214}{73} = 2.93$ ft, $V = 13.73$ fps when steady flow is eventually achieved.

In general: $214 - 73(V^2/2g) = (3000/32.2)(dV/dt)$, $dt = 82.2\, dV/(188.6 - V^2)$.

$$\int_0^t dt = t = \int_{11.55}^{V_t} \frac{82.2\, dV}{(188.6 - V^2)} = \frac{82.2}{27.5}\left[\ln\frac{13.73 + V}{13.73 - V}\right]_{11.55}^{V_t} = 2.99 \ln\frac{13.73 + V_t}{13.73 - V_t} - 7.33$$

V_t, fps	t, s	Q, cfs
11.55	0	6.30
12.5	1.817	6.82
13.5	6.90	7.36
13.7	12.70	7.47
13.73	∞	7.49

19.40 Attached to the tank in Fig. 19-11 is a flexible 25-mm-diameter hose ($f = 0.015$) of length 60 m. The tank is hoisted in such a manner that $h = 6 + 0.9t$, where h is the head in meters and t is the time in seconds. (*a*) Find as accurately as you can the flow at $t = 10$ s. (*b*) Suppose h were decreasing at the same rate. What, then, would be the flow rate when $h = 15$ m?

▌ Find steady-state velocity for $t = 10$ s; then $h = 6 + 0.9(10) = 15$ m, $15 - 0.015(60/0.025)(V^2/2g) = V^2/2g$; $V = 2.82$ m/s at steady state. Now find the effect of changing head.

Given $h = 6 + 0.9t$, therefore $\partial h/\partial t = 0.9$ m/s, $h - 36(V^2/2g) = V^2/2g + (L/g)(dV/dt)$, $\partial h/\partial t = 37(2V/2g)(\partial h/\partial t) + (L/g)(\partial^2 V/\partial t^2)$. Neglect the last term (of higher order): $0.9 = 37(V/g)(\partial V/\partial t)$; $\partial V/\partial t = 0.9g/(37V) = 0.0846$ m/s per second.

(*a*) $15 = 37(V^2/2g) + (60/9.91)(0.0846)$; $37(V^2/2g) = 15 - 0.517 = 14.48$, $V = 2.77$ m/s, $Q = 0.001360$ m³/s.

(*b*) For decreasing h, one sign is reversed: $37(V^2/2g) = 15 + 0.517 = 15.52$, $V = 2.87$ m/s, $Q = 0.001408$ m³/s.

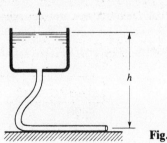

Fig. 19-11

19.41 Repeat Prob. 19.40 for the case of a 10-cm-diameter hose with all other data remaining the same.

▌ Find steady state for $t = 10$ s: $15 - 0.015(60/0.1)(V^2/2g) = V^2/2g$; $V = 5.42$ m/s at steady rate. Now find effect of changing head: $h = 10(V^2/2g) + (L/g)(dV/dt)$. Differentiating with respect to t and neglecting the higher-order term: $\partial h/\partial t = 0.9 = 10(2V/2g)(dV/dt)$; $\partial V/\partial t = 0.9g/(10V) = 0.1627$ m/s per second.

(a) $15 = 10(V^2/2g) + (60/9.81)(0.1627)$ $V = 5.24\,\text{m/s}$ $Q = 0.0412\,\text{m}^3/\text{s}$

(b) $15 = 10(V^2/2g) + (60/9.81)(-0.1627)$ $V = 5.60\,\text{m/s}$ $Q = 0.0440\,\text{m}^3/\text{s}$

19.42 In Prob. 19.40, suppose h was changed instantaneously from 6 m to 15 m. Under these conditions, find the flow rate at $t = 10\,\text{s}$.

▌ Initial condition where $h = 6\,\text{m}$: $6 - 0.015(60/0.025)(V^2/2g) = V^2/2g$; $V = 1.784\,\text{m/s}$ (initially). After head is changed to 15 m: $15 = 37(V^2/2g) + (L/g)(dV/dt)$. Eventually $dV/dt = 0$, $V = 2.82\,\text{m/s}$, $15 = [37/(2)(9.81)]V^2 + (60/9.81)(dV/dt)$; $15 = 1.886V^2 + 6.12(dV/dt)$; $7.95 = V^2 + 3.24(dV/dt)$; $dt = 3.24\,dV/(2.82^2 - V^2)$.

$$\int_0^t dt = \int_{1.784}^V \frac{3.24\,dV}{2.82^2 - V^2} = \frac{3.24}{5.64}\left[\ln\frac{2.81 + V}{2.82 - V}\right]_{1.785}^V \qquad t = 0.574\ln\frac{2.82 + V}{2.82 - V} - 0.857$$

V, m/s	t, s
1.784	0
2.4	0.589
2.7	1.341
2.8	2.37
2.82	∞

For all practical purposes $V = 2.82\,\text{m/s}$ at $t = 10\,\text{s}$ when $Q = (0.000491)2.82 = 0.001384\,\text{m}^3/\text{s}$.

19.43 A 4-in-diameter pipe of length 3000 ft drains a reservoir. The elevation difference between the reservoir water surface and the pipe outlet is 100 ft. The pipe entrance is flush. Initially there is no flow since there is a plug at the outlet. The plug is then removed. Tabulate Q versus t, assuming $f = 0.020$.

▌ $L/D = 3000/0.33 > 1000$, neglect minor losses, i.e., $k_L = 0$, $k = k_L + fL/D = 0 + 0.02(3000)/(\frac{4}{12}) = 180$, $V_0^2 = 2gH/(1 + k) = 2(32.2)100/(1 + 180)$; steady $V_0 = 5.96\,\text{fps}$.

$$t = [L/(1 + k)V_0]\ln[(V_0 + V)/(V_0 - V)] = [3000/181(5.96)]\ln[(V_0 + V)/(V_0 - V)]$$
$$= 2.78\ln[(5.96 + V)/(5.96 - V)]$$

V, fps	t, s	Q, cfs
0	0	0
1	0.941	0.0873
3	3.07	0.262
5	6.75	0.436
5.5	8.91	0.480
5.96	∞	0.521

19.44 Repeat Prob. 19.43 for the case where the pipe length is 300 ft rather than 3000 ft.

▌ $L/D = 300/0.33 = 900 < 1000$, minor losses are *not* negligible; $k_L = 0.5$ (flush entrance). $k = k_L + fL/D = 0.5 + 0.02(300)(\frac{4}{12}) = 18.5$, $V_0^2 = 2gH/(1 + k) = 2(32.2)100/(1 + 18.5)$; steady $V_0 = 18.17\,\text{fps}$.

$$t = [L/(1 + k)V_0]\ln[(V_0 + V)/(V_0 - V)] = [300/19.5(18.17)]\ln[(V_0 + V)/(V_0 - V)]$$
$$= 0.847\ln[(18.17 + V)/(18.17 - V)]$$

V, fps	t, s	Q, cfs
0	0	0
5	0.478	0.436
10	1.048	0.873
15	1.987	1.309
18	4.52	1.571
18.17	∞	1.586

19.45 A 15-cm-diameter pipe of length 500 m drains a reservoir. The elevation difference between the reservoir water surface and the pipe outlet is 60 m. Initially there is no flow because a valve at the pipe outlet is closed. The valve is suddenly opened; tabulate Q versus t assuming $f = 0.03$.

∎ $L/D = 500/0.15 > 1000$, neglect minor losses, i.e., $k_L = 0$, $k = k_L + fL/D = 0 + 0.03(500)/0.15 = 100$, $V_p^2 = 2gH/(1 + k) = 2(9.81)60/(1 + 100)$; steady-flow velocity $V_0 = 3.41$ m/s.

$$t = [L/(1 + k)V_0] \ln [(V_0 + V)/(V_0 - V)] = [500/101(3.41)] \ln [(V_0 + V)/(V_0 - V)]$$
$$= 1.450 \ln [(3.41 + V)/(3.41 - V)]$$

V, m/s	t, s	Q, m³/s	Q, L/s
0	0	0	0
1.0	0.875	0.01767	17.67
2.0	1.947	0.0353	35.3
3.0	3.97	0.0530	53.0
3.2	4.98	0.0565	56.5
3.3	5.91	0.0583	58.3
3.4	8.97	0.0600	60.0
3.41	∞	0.0603	60.3

19.46 A large water reservoir is drained by a pipeline that consists of 200 ft of 6-in-diameter pipe ($f = 0.030$) followed by 500 ft of 10-in-diameter pipe ($f = 0.020$). The point of discharge is 100 ft below the elevation of the reservoir water surface. A valve at the discharge end of the pipe is initially closed. It is then quickly opened. Derive an equation applicable to this situation, and tabulate flow rate versus time. Neglect minor losses.

∎ $h = f_1(L_1/D_1)(V_1^2/2g) - f_2(L_2/D_2)(V_2^2/2g) = V_2^2/2g + (L_1/g)(dV_1/dt) + (L_2/g)(dV_2/dt)$. By continuity: $V_1 = V_2(D_2/D_1)^2$; $dV_1 = dV_2(D_2/D_1)^2$. Eliminate V_1: $h - [f_1(L_1/D_1)(D_2/D_1)^4 + f_2(L_2/D_2) + 1](V_2^2/2g) = [L_1(D_2/D_1)^2 dV_2]/(g\,dt) + (L_2/g)(dV_2/dt)$. For steady-state condition when $dV_2/dt = 0$ and $V_2 = V_{20}$. Substituting known values yields steady-flow $V_{20} = 7.81$ fps. In general, rearranging: $dt = \{2[L_1(D_2/D_1)^2 + L_2]/[1 + f_1(L_1/D_1)(D_2/D_1)^4 + f_2(L_2/D_2)]\}[dV/(V_{20}^2 - V^2)]$; integrating

$$t = \{[L_1(D_2/D_1)^2 + L_2]/[1 + f_1(L_1/D_1)(D_2/D_1)^4 + f_2(L_2/D_2)]\}(1/V_{20}) \ln [(V_{20} + V_2)/(V_{20} - V_2)]$$
$$= 1.280 \ln [(7.81 + V_2)/(7.81 - V_2)]$$

V_1, fps	V_2, fps	t, s	Q, cfs
0	0	0	0
8.33	3	1.037	1.636
13.89	5	1.942	2.73
19.44	7	3.72	3.82
20.8	7.5	4.99	4.09
21.5	7.75	7.12	4.23
21.7	7.81	∞	4.26

19.47 A 10-in-diameter pipe ($f = 0.020$) of length 300 ft is connected to a reservoir. Entrance losses are negligible. At the discharge end of the pipe is a nozzle that produces a 4-in-diameter jet. The elevation difference between the jet and the water surface in the reservoir is 40 ft. The nozzle has a coefficient of velocity of 0.95. Initially, there is a tight-fitting plug in the nozzle, which is then removed. For this situation derive an equation and tabulate flow rate versus time. Assume that the liquid level in the reservoir does not drop.

∎ $h - k(V^2/2g) - [(1/C_V^2) - 1][1 - (D_j/D)^4](V_j^2/2g) = (V_j^2/2g) + (L/g)(dV/dt)$. But $V_j = V(A/A_j) = V(D/D_j)^2$, so $h = C(V^2/2g) + (L/g)(dV/dt)$, where $C = k + (D/D_j)^4\{1 + [(1/C_V^2) - 1][1 - (D_j/D)^4]\}$. For steady-state conditions when $dV/dt = 0$ and $V = V_0$: $h = CV_0^2/2g$, or $V_0 = (2gh/C)^{1/2}$. Eliminating h between the above two equations and rearranging: $dt = (2L/C)[dV/(V_0^2 - V^2)]$ and integrating $t = (L/CV_0) \ln [(V_0 + V)/(V_0 - V)]$. Given $k_L = 0$: $k = 0 + 0.020(300)/(\frac{10}{12}) = 7.2$. Given $C_V = 0.95$ and $D/D_j = \frac{10}{4}$: $C = 50.4$, $V_0 = [2(32.2)40/50.4]^{1/2} = 7.15$ fps, $t = [300/50.4(7.15)] \ln [(V_0 + V)/(V_0 - V)] = 0.833 \ln [(7.15 + V)/(7.15 - V)]$.

V_j, fps	V, fps	t, s	Q, cfs
0	0	0	0
12.5	2	0.479	1.091
25.0	4	1.053	2.18
37.5	6	2.03	3.27
43.8	7	3.78	3.82
44.4	7.1	4.69	3.87
44.7	7.15	∞	3.90

19.48 An open tank containing oil (s.g. = 0.85, $\mu = 0.0005$ lb-s/ft^2) is connected to a 2-in-diameter smooth pipe of length 3000 ft. The elevation drop from the liquid surface in the tank to the point of discharge is 15 ft. A valve on the discharge end of the pipe, initially closed, is then opened. Tabulate the ensuing flow rate versus time.

▎ $L/D = 3000/(\frac{2}{12}) = 18\,000 > 1000$, neglect minor losses. Assume laminar flow, neglect velocity head:
$h_L = 15 = 32(\mu/\gamma)(L/D^2)V$, $V = 15(0.85)(62.4)(0.1667)^2/[32(0.0005)(3000)] = 0.460$ fps, $N_R = DV\rho/\mu = 0.1667(0.46)(0.85)(1.94)/0.0005 = 253$. $N_R < 2000$, therefore flow *is* laminar, assumption was O.K. Assuming the velocity head is negligible: $h - 32(\mu/\gamma)(L/D^2)V = (L/g)(dV/dt)$, or $h - \alpha V = \beta(dV/dt)$, where $\alpha = 32(\mu/\gamma)(L/D^2) = 32(0.0005)3000/[(0.85)(62.4)(0.1667)^2] = 32.6$ and $\beta = L/g = 3000/32.2 = 93.2$.

$$dt = \frac{\beta\,dV}{h - \alpha V} = \frac{dV}{(h/\beta) - (\alpha/\beta)V}$$

$$t = \frac{-1}{\alpha/\beta}\ln\left[\left(\frac{h}{\beta}\right) - \left(\frac{\alpha}{\beta}\right)V\right]_0^V = -2.86\ln[0.1610 - 0.349V]_0^V = -2.86\ln\left(\frac{0.1610 - 0.349V}{0.1610}\right) = -2.86\ln(1 - 2.17V)$$

V, fps	t, s	Q, cfs
0	0	0
0.10	0.700	0.00218
0.20	1.630	0.00436
0.30	3.02	0.00655
0.40	5.81	0.00873
0.45	10.83	0.00982
0.460	∞	0.01004

Check velocity head assumption: $(V^2/2g)/h_L = (V^2/2g)/\alpha V = V/(2g\alpha) = V/2100$. As V varies from 0 to 0.460, $(V^2/2g)/h_L$ varies from 0 to 0.000219; therefore $V^2/2g < 0.022$ percent of h_L, velocity head *is* negligible (as assumed).

19.49 An open tank containing oil (s.g. = 0.82, $\nu = 0.002$ m^2/s) is connected to a 10-cm-diameter pipe of length 400 m. The elevation drop from the oil surface in the tank to the pipe outlet is 2.5 m. A valve at the end of the pipe, initially closed, is suddenly opened. Plot the ensuing flow rate as a function of time.

▎ $L/D = 400/0.1 = 4000 > 1000$, neglect minor losses. Assuming laminar flow for h_L, $H - 32\nu(L/gD^2)V = (L/g)(dV/dt)$; $2.5 - 32(0.002)[400/9.81(0.10)^2]V = (400/9.81)(dV/dt)$, $2.5 - 261V = 40.8\,dV/dt$; $dt = (40.8\,dV)/(2.5 - 261V) = (16.31\,dV)/(1 - 104.4V)$. For steady-state conditions when $dV/dt = 0$ and $V = V_0$: $2.5 - 261V_0 = 0$; steady-flow velocity $V_0 = 0.00958$ m/s. Integrating:

$$t = -(16.31/104.4)[\ln(1 - 104.4V)]_0^V = -0.1563\ln(1 - 104.4V)$$

V, m/s	t, s	Q, m^3/s	Q, L/s
0.001	0.01723	7.85×10^{-6}	0.00785
0.002	0.0366	15.71×10^{-6}	0.01571
0.004	0.0845	31.4×10^{-6}	0.0314
0.008	0.282	62.8×10^{-6}	0.0628
0.009	0.438	70.7×10^{-6}	0.0707
0.0093	0.552	73.0×10^{-6}	0.0730
0.00958	∞	75.2×10^{-6}	0.0752

$(N_R)_{max} = (0.1)0.00958/0.002 = 0.479$. $N_R < 2000$, so flow *is* laminar (as assumed) and answers are valid.

19.50   Find the celerity of a pressure wave in benzene (s.g. = 0.90, E_v = 150 000 psi) contained in a 6-in-diameter steel pipe having a wall thickness of 0.285 in.

$$c = (E_v/\rho)^{1/2} = [(150\,000)(144)/(0.9)(1.94)]^{1/2} = 3520 \text{ fps}$$
$$c_p = c/\sqrt{1 + (D/t)(E_v/E)} = 3520/\{1 + (6/0.285)[150\,000/(30 \times 10^6)]\}^{1/2} = 3350 \text{ fps}$$

19.51   (a) What is the celerity of a pressure wave in a 5-ft-diameter water pipe with 0.5-in steel walls? (b) If the pipe is 4000 ft long, what is the time required for a pressure wave to make the round trip from the valve?

   (a) $c = 4720$ fps in water
$$c_p = c/\sqrt{1 + (D/t)(E_v/E)} = 4720/\{1 + [(5)(12)/0.5][300\,000/(30 \times 10^6)]\}^{1/2} = 3182 \text{ fps}$$

   (b)
$$T_r = 2L/c_p = 2(4000)/3182 = 2.51 \text{ s}$$

19.52   In Prob. 19.51, (a) if the initial water velocity is 8 fps, what will be the rise in pressure at the valve if the time of closure is less than the time of a round trip? (b) If the valve is closed at such a rate that the velocity in the pipe decreases uniformly with respect to time and closure is completed in a time $t_c = 5L/c_P$, approximately what will be the pressure head at the valve when the first pressure unloading wave reaches the valve?

   (a)
$$p = (\gamma/g)c_pV = (62.4/32.2)(3182)8 = 49\,300 \text{ psf} = 791 \text{ ft}$$

   (b) If velocity changes uniformly from 8.0 fps to 0 in time $5L/c_p$, the change in V in $t' = T_r = 2L/c_p$ is $(\frac{2}{5})8.0 = 3.20$ fps. $\Delta p'/\gamma = c_p \Delta V/g = 3182(3.2)/32.2 = 316$ ft.

19.53   For the situation depicted in Prob. 19.27, find the water-hammer pressure at the valve if a flow of 80 cfs is reduced to 25 cfs in 3.0 s. Under these conditions what would be the maximum water-hammer pressures at points 500 ft and 1500 ft from the reservoir?

   $\Delta p'_h = (T_r/t_c)(\Delta p_h)$, $V_1 = 80/4\pi = 6.37$ fps, $V_2 = 25/4\pi = 1.989$ fps, $\Delta V = 4.38$ fps over 3.0 s; $\Delta p_h = \rho c_p(\Delta V) = 1.94(3200)(4.38) = 27\,200$ psf for instantaneous drop in flow rate. $T_r = 2L/c_p = \frac{4000}{3200} = 1.250$ s. At valve: $\Delta p'_h \approx (1.25/3.0)(27\,200) = 11\,333$ psf = 78.7 psi.
500 ft from reservoir: $\Delta p'_h = (\frac{500}{2000})78.7 = 19.7$ psi.
1500 ft from reservoir: $\Delta p'_h = (\frac{1500}{2000})78.7 = 59.0$ psi.

19.54   Water is flowing through a 30-cm-diameter welded steel pipe of length 2000 m that drains a reservoir under a head of 40 m. The pipe has a thickness of 8 mm. (a) If a valve at the end of the pipe is closed in 10 s, approximately what water-hammer pressure will be developed? (b) If the steady-state flow is instantaneously reduced to one-half its original value, what water-hammer pressure would you expect? Use $f = 0.013$.

$$h_f = (f)(L/D)(V^2/2g) \qquad 40 = 0.013(2000/0.3)V^2/2g \qquad V = 3.01 \text{ m/s}$$
$$c = 1440 \text{ m/s in water} \qquad c_p = c/\sqrt{1 + (D/t)(E_v/E)} = 1440/[1 + (30/0.8)(\tfrac{1}{100})]^{1/2} = 1228 \text{ m/s}$$

   (a) $p'_h = 2LV\rho/t_c = [2(2000)3.01(1000)/10] = 1204 \text{ kN/m}^2$

   (b) $p_h = \rho c_p V = (1)1228(3.01/2) = 1848 \text{ kN/m}^2$

19.55   In Fig. 19-12, the total length of pipe is 10 000 ft, its diameter is 36 in, and its thickness is $\frac{3}{4}$ in. Assume $E = 30\,000\,000$ psi and $E_v = 300\,000$ psi. If the initial velocity for steady flow is 10 fps and the valve at G is partially closed so as to reduce the flow to half of the initial velocity in 4 s, find (a) the maximum pressure rise from the water hammer; (b) the location of the point of maximum total pressure.

   (a) $c = 4720$ ft/s $c_p = c/\sqrt{1 + (D/t)(E_V/E)} = 4720/[1 + (36/\frac{3}{4})(300\,000/30\,000\,000)]^{1/2} = 3880 \text{ fps}$
$$\Delta p = (\gamma/g)c_p(\Delta V) = (62.4/32.2)(\tfrac{3880}{144})5 = 261 \text{ psi} \quad \text{or} \quad 602 \text{ ft}$$

   (b) $x_0 = c_p t_c/2 = 3880(4)/2 = 7760$ ft. Thus water-hammer pressure rise, at B: $\Delta p = 261(\frac{2000}{7760}) = 67.3$ psi; at E: $\Delta p = 261(100 \text{ percent}) = 261$ psi. Total pressure = static pressure plus water hammer. Total pressure at B: $p/\gamma = 600 + 67.3(144/62.4) = 755$ ft; at E: $p/\gamma = 120 + 261(144/62.4) = 722$ ft. Maximum total pressure occurs at point B.

19.56   Refer to Fig. 19-12, but take all the dimensions given in feet to be in meters instead. This 10-km-long pipe has a diameter of 1.5 m and a wall thickness of 25 mm. Assume $E = 200$ GPa and $E_v = 2$ GPa. The initial steady-flow velocity is 6 m/s. The valve at G is then partially closed so as to reduce the velocity to 1 m/s in 13 s. Find (a) the maximum pressure rise due to water hammer, and (b) the location of the point of maximum total pressure.

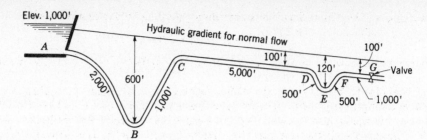

Fig. 19-12

∎ **(a)** $c_p = 1440 \text{ m/s}$ $c_p = c/\sqrt{1 + (D/t)(E_v/E)} = 1440/[1 + (\frac{1500}{25})(\frac{2}{200})]^{1/2} = 1138 \text{ m/s}$

$$\Delta p = \rho c_p \, \Delta V = 1000(1138)5 = 5690 \text{ kN/m}^2 \quad \text{or} \quad 580 \text{ m}$$

(b) $x_0 = c_p t_c/2 = 1138(13)/2 = 7400 \text{ m}$. Thus water-hammer pressure rise, at B: $\Delta p = 5690(\frac{2000}{7400}) = 1538 \text{ kN/m}^2$ or 156.8 m; at E: $\Delta p = 5690(100 \text{ percent}) = 5690 \text{ kN/m}^2$ or 174.0 m. Total pressure = static pressure plus water-hammer pressure. Total pressure at B: $p/\gamma = 600 \text{ m} + 156.8 \text{ m} = 756.8 \text{ m}$; at E: $p/\gamma = 120 \text{ m} + 580 \text{ m} = 700 \text{ m}$. Maximum total pressure occurs at point B.

19.57 Derive the equation $V^2 = (2gAD^2/LA_s f^2)[1 - (fA_s/AD)z] - Ce^{-(fA_s/AD)z}$, which expresses the relationship between velocity in a pipe and water-surface level in a tank over an interval from valve closure to the top of the first surge.

∎ Write dV/dt as $(dV/dz)(dz/dt)$ in $0 - f(L/D)(V^2/2g) = z + (L/g)(dV/dt)$ and substitute for dz/dt from $AV = A_s(dz/dt)$; $0 - (fL/D)(V^2/2g) = z + (L/g)(dV/dz)(AV/A_s)$ or $z + (fL/2gd)V^2 + (LA/2gA_s)[(2V\,dV)/dz] = 0$. Define $K = fL/2gD$; $M = (L/2g)(A/A_s)$; and $V^2 = x$ so that $2V\,dV = dx$. This gives $z + kx + (M\,dx)/dz = 0$ or $dx/dz + (K/M)x + (1/M)z = 0$. Now define $\alpha = K/M$, $\beta = 1/M$; then $dx/dz + \alpha Z + \beta z = 0$. Multiply by $e^{\alpha z}$: $(dx/dz)e^{\alpha z} + \alpha x e^{\alpha z} + \beta z e^{\alpha} = 0$, i.e., $d/dz(xe^{\alpha z}) + e^{\alpha z}\beta z = 0$ or $d(xe^{\alpha z}) + e^{\alpha z}\beta z\,dz = 0$. Integrate: $\int d(xe^{\alpha z}) + \int e^{\alpha z}\beta z\,dz = 0$, $xe^{\alpha z} + \beta(e^{\alpha z}/\alpha^2)(\alpha z - 1) + C = 0$ or $x = (\beta/\alpha^2)(1 - \alpha z) - Ce^{-\alpha z}$. But $\beta/\alpha^2 = M/K^2 = 2gAD^2/LA_s f^2$, $\alpha = K/M = fA_s/AD$, and $x = V^2$; therefore $V^2 = (2gAD^2/LA_s f^2)[1 - (fA_s/AD)z] - Ce^{-(fA_s/AD)z}$.

19.58 A 36-in steel pipe 3000 ft long supplies water to a small power plant. What height would be required for a simple surge tank 6 ft in diameter situated 50 ft upstream from the valve at a point where the centerline of the pipe is 120 ft below the water surface in the reservoir if the tank is to protect against instantaneous closure of a valve at the plant? The valve is 150 ft below reservoir level, and the discharge is 150 cfs. Take $f = 0.015$. The surge tank is not to overflow. Neglect all velocity heads and minor losses; in the surge tank (only) neglect fluid friction and inertial effects.

∎ $V^2 = (2gAD^2/LA_s f^2)[1 - (fA_s/AD)z] - Ce^{-(fA_s/AD)z}$

$= [2(32.2)/2950](\frac{3}{6})^2(3/0.015)^2[1 - (0.015/3)(\frac{6}{3})^2 z] - Ce^{-(0.015/3)(6/3)^2 z} = 218(1 - 0.02z) - Ce^{-0.02z}$

First find the constant of integration C, from initial (steady) conditions, when $Q_1 = 150 \text{ cfs}$, $V_1 = Q_1/A = 150/[(\pi/4)3^2] = 21.2 \text{ fps}$, and $-z_1 = (fL/D)(V_1^2/2g) = (0.015)(\frac{2950}{3})21.2^2/[(2)(32.2)] = 103.1 \text{ ft}$. Substituting these values $21.2^2 = 218[1 - 0.02(-103.1)] - Ce^{-0.02(-103.1)}$, from which $C = 27.7$. For the first surge, $z = z_{max}$ when $V = 0$, so $0.218(1 - 0.02z_{max}) - 27.7e^{-0.02z_{max}}$. [*Note:* The last (3rd) term is always negative. For a positive middle (2nd) term we need $0.002z_{max} < 1$, i.e., $z_{max} < 1/0.02 = 50$.] By trial and error: $z_{max} = 47.5 \text{ ft}$. Required height of surge tank $= (120 + 47.5) = 167.5 \text{ ft}$.

19.59 Repeat Prob. 19.58 for the case where the surge tank is to have a diameter of 10 ft.

∎ Following the procedure of Prob. 19.58: $V^2 = 78.67(1 - 0.0556z) - Ce^{-0.0556z}$. From initial (steady) conditions: $z_1 = -103.1 \text{ ft}$, $C = 0.255$. (Note the large change in C from Prob. 19.58.) For the first surge, we need $0.0556z_{max} < 1$, i.e., $z_{max} < 18.00 \text{ ft}$. By trial and error: $z_{max} = 17.98 \text{ ft}$. Required height of surge tank $= (120 + 17.98) = 138.0 \text{ ft}$.

19.60 Using the data of Prob. 19.58, find the diameter of a surge tank that will produce a surge requiring a tank height of 175 ft.

∎ As in Prob. 19.58 for initial (steady) conditions: $V_1 = 21.2 \text{ fps}$, $z_1 = -103.1 \text{ ft}$. Substituting these and other known values into $V^2 = (2gAD^2/LA_s f^2)[1 - (fA_s/AD)z] - Ce^{-(fA_s/AD)z}$, while writing A/A_s as $(D/D_s)^2$: $21.2^2 = 7860/D_s^2 + 450 - Ce^{0.0573D_s^2}$, that is,

$$0 = 7860/D_s^2 - Ce^{0.0573D_s^2} \tag{1}$$

When $z = z_{max} = 175 - 120 = 55$ ft, $V = 0$; substituting these and other known values

$$0 = 7860/D_s^2 - 240 - Ce^{-0.0306D_s^2} \qquad (2)$$

Eliminating the unknown C between Eqs. (1) and (2): $1 - 0.0306D_s^2 = e^{-0.0879D_s^2}$. By trial and error: $D_s = 5.52$ ft.

19.61   A 90-cm steel pipe ($f = 0.015$) of length 1000 m supplies water to a small power plant. What height would be required for a simple surge tank 3 m in diameter situated 10 m upstream from the valve at a point where the centerline of the pipe is 40 m below the water surface in the reservoir? Assume instantaneous closure of the valve. The valve is 50 m below reservoir level and the flow is 3.2 m³/s. The surge tank is not to overflow. Neglect all velocity heads and minor losses; in the surge tank (only) neglect fluid friction and inertial effects.

■ $V^2 = (2gAD^2/LA_s f^2)[1 - (fA_s/AD)z] - Ce^{-(fA_s/AD)z}$

$= [2(9.81)/900](0.9/3)^2(0.9/0.015)^2[1 - (0.015/0.9)(3/0.9)^2 z] - Ce^{-(0.015/0.9)(3/0.9)^2 z}$

$= 7.06(1 - 0.1852z) - Ce^{-0.1852z}$

First, find the constant of integration C from initial (steady) conditions when $Q_1 = 3.2$ m³/s, $V_1 = Q_1/A = 3.2/[(\pi/4)0.9^2] = 5.03$ m/s, and $-z_1 = (fL/D)(V_1^2/2g) = (0.015)(900/0.9)5.03^2/[(2)(9.81)] = 19.34$ m. Substituting these values, $5.03^2 = 7.06[1 - 0.1852(-19.34)] - Ce^{-0.1852(-19.34)}$, from which $C = 0.1965$. For the first surge, $z = z_{max}$ when $V = 0$, so $0 = 7.06(1 - 0.1852z_{max}) - 0.1965e^{-0.1852z_{max}}$. [*Note:* The last (3rd) term is always negative. For a positive middle (2nd) term we need $0.1852z_{max} < 1$, i.e., $z_{max} < 1./0.1852 = 5.40$ m.] By trial and error: $z_{max} = 5.34$ m. Required height of surge tank $= 40 + 5.34 = 45.34$ m.

19.62   A 1.25-m-diameter penstock ($f = 0.018$) 950 m long carries water at Q m³/s from a reservoir to a power plant. When the outlet valve is closed instantaneously, water rises in a 2-m-diameter surge tank immediately adjacent to the outlet valve. Determine the maximum allowable initial discharge Q so that the resulting surge will not rise more than 10 m above the reservoir water surface. Neglect all velocity heads and minor losses; in the surge tank (only) neglect fluid friction and inertial effects.

■ $V^2 = (2gAD^2/LA_s f^2)[1 - (fA_s/AD)z] - Ce^{-(fA_s/AD)z}$

$= [2(9.81)/950](1.25/2)^2(1.25/0.018)^2[1 - (0.018/1.25)(2/1.25)^2] - Ce^{-(0.018/1.25)(2/1.25)^2 z}$

$= 38.9(1 - 0.0369z) - Ce^{-0.0369z}$

For the first surge, $z = z_{max} = 10$ m when $V = 0$, so $0 = 38.9[1 - 0.0369(10)] - Ce^{-0.0369(10)}$; from which $C = 35.5$. For the initial conditions, when $V = $ steady-flow velocity $= V_1$, $z_1 = -h_L = -(fL/D)(V_1^2/2g)$. Substitute this expression for z_1 for V_1^2, and simplify: $V_1^2 = B[1 + (V_1^2/B)] - Ce^{V_1^2/B}$, where $B = 38.9$, $B = Ce^{V_1^2/B}$ or $\ln(B/C) = V_1^2/B$, $V_1 = [B\ln(B/C)]^{1/2} = [38.9\ln(38.9/35.5)]^{1/2} = 1.884$ m/s. So $Q = AV = (\pi/4)(1.25)^2(1.884) = 2.31$ m³/s.

CHAPTER 20
Pumps and Fans

20.1 A pump delivers $0.300 \text{ m}^3/\text{s}$ against a head of 200 m with a rotative speed of 2000 rpm. Find the specific speed.

$$N_s = 51.64 NQ^{0.5}/H^{0.75} = (51.64)(2000)(0.300)^{0.5}/200^{0.75} = 1064$$

20.2 A pump delivers $0.019 \text{ m}^3/\text{s}$ against a head of 16.76 m with a rotative speed of 1750 rpm. Find the specific speed.

$$N_s = 51.64 NQ^{0.5}/H^{0.75} = (51.64)(1750)(0.019)^{0.5}/16.76^{0.75} = 1504$$

20.3 A radial-flow pump must deliver 2000 gpm against a head of 950 ft. Find the minimum practical rotative speed.

■ $N_s = NQ^{0.5}/H^{0.75}$. From Fig. A-40, it is apparent that the minimum practical specific speed for a radial-flow pump is about 500. Hence, $500 = (N)(2000)^{0.5}/950^{0.75}$, $N = 1913$ rpm.

20.4 A radial-flow pump must deliver 300 gpm against a head of 30 ft. Find the operating rotative speed.

■ $N_s = NQ^{0.5}/H^{0.75}$. From Fig. A-40, $N_s = 2500$ for maximum efficiency of 92.5 percent for a radial-flow pump. Hence, $2500 = (N)(300)^{0.5}/30^{0.75}$, $N = 1850$ rpm.

20.5 An axial-flow pump is to be operated at a rotative speed of 2500 rpm against a head of 400 m. What flow rate will be delivered by the most efficient pump?

■ From Fig. A-40, it is apparent that the most efficient axial-flow pump has a specific speed of around 12 500. $N_s = 51.64 NQ^{0.5}/H^{0.75}$, $12\,500 = (51.64)(2500)(Q)^{0.5}/400^{0.75}$, $Q = 75.0 \text{ m}^3/\text{s}$.

20.6 A mixed-flow pump is to be operated at a rotative speed of 1500 rpm against a head of 15 m at maximum pump efficiency. Determine the flow rate the pump will deliver.

■ From Fig. A-40, it is apparent that the most efficient mixed-flow pump has a specific speed of around 6500. $N_s = 51.64 NQ^{0.5}/H^{0.75}$, $6500 = (51.64)(1500)(Q)^{0.5}/15^{0.75}$, $Q = 0.409 \text{ m}^3/\text{s}$.

20.7 A radial-flow pump operating at maximum efficiency is to deliver 260 gpm against a head of 129 ft at a rotative speed of 2100 rpm. Find the required number of stages (i.e., impellers).

■ From Fig. A-40, it is apparent that maximum efficiency for a radial-flow pump is about 93 percent at a specific speed of 2500. $N_s = NQ^{0.5}/H^{0.75}$, $2500 = (2100)(260)^{0.5}/H^{0.75}$, $H = 32.29$ ft. Since the given head is 129 ft, a total of $129/32.29$, or 4 stages will be needed.

20.8 A radial-flow pump operating at maximum efficiency is to deliver 400 gpm against a head of 191 ft at a rotative speed of 1920 rpm. Find the required number of stages (i.e., impellers).

■ From Fig. A-40, it is apparent that maximum efficiency for a radial-flow pump is about 93 percent at a specific speed of 2500. $N_s = NQ^{0.5}/H^{0.75}$, $2500 = (1920)(400)^{0.5}/H^{0.75}$, $H = 38.18$ ft; number of stages $= 191/38.18 = 5$.

20.9 The value of $(\text{NPSH})_{\text{min}}$ for a pump is given by the manufacturer as 20 ft. Water is being pumped from a reservoir at a rate of 25 cfs. The water level in the reservoir is 6.0 ft below the pump. Atmospheric pressure is 14.7 psia and water temperature is 40 °F. If the total head loss in the suction pipe is 4.0 ft, is the pump safe from cavitation effects?

■ $$\text{NPSH} = p_A/\gamma - z_s - h_L - p_v/\gamma = (14.7)(144)/62.4 - 6.0 - 4.0 - 18.5/62.4 = 23.6 \text{ ft}$$

Since $[\text{NPSH} = 23.6] > [(\text{NPSH})_{\text{min}} = 20]$, cavitation should not be a problem.

20.10 The $(\text{NPSH})_{\text{min}}$ for a pump is given by the manufacturer as 7.0 m. This pump is being used to pump water from a reservoir at a rate of $0.2832 \text{ m}^3/\text{s}$. The water level in the reservoir is 1.280 m below the pump. Atmospheric pressure is 98.62 kN/m^2 and water temperature is 20 °C. Assume total head loss in the suction pipe is 1.158 m of water. Determine whether or not the pump is safe from cavitation effects.

$$\text{NPSH} = p_A/\gamma - z_s - h_L - p_v/\gamma = 98.62/9.79 - 1.280 - 1.158 - 2.34/9.79 = 7.40 \text{ m}$$

Since $[\text{NPSH} = 7.40 \text{ m}] > [(\text{NPSH})_{\min} = 7.0 \text{ m}]$, cavitation should not be a problem.

20.11 A commercial pump is operating at 2150 rpm and delivers 1800 gpm against a head of 340 ft. Find the approximate efficiency of the pump.

$$N_s = NQ^{0.5}/H^{0.75} = (2150)(1800)^{0.5}/340^{0.75} = 1152$$

From Fig. A-41, with $N_s = 1152$ and $Q = 1800$ gpm, the approximate efficiency of the pump is determined to be 82 percent.

20.12 A pump operating at 1600 rpm delivers 0.189 m³/s against a head of 47.03 m. Determine the approximate efficiency of the pump.

$$N_s = 51.64NQ^{0.5}/H^{0.75} = (51.64)(1600)(0.189)^{0.5}/47.03^{0.75} = 2000$$

From Fig. A-41, with $N_s = 2000$ and $Q = 0.189$ m³/s, the approximate efficiency of the pump is determined to be 87 percent.

20.13 A centrifugal pump with a 700-mm-diameter impeller runs at 1800 rpm. The water enters without whirl, and $\alpha_2 = 60°$. The actual head produced by the pump is 17 m. Find its hydraulic efficiency when $V_2 = 6$ m/s.

▌ The theoretical head is $H = (u_2V_2 \cos \alpha_2)/g = [(\frac{1800}{60})(2\pi)(\frac{700}{1000})/2](6)(\cos 60°)/9.807 = 20.18$ m, $e_h = 17/20.18 = 0.842$, or 84.2 percent.

20.14 A centrifugal water pump has an impeller (Fig. 20-1a) with $r_2 = 12$ in, $r_1 = 4$ in, $\beta_1 = 20°$, $\beta_2 = 10°$. The impeller is 2 in wide at $r = r_1$ and $\frac{3}{4}$ in wide at $r = r_2$. For 1800 rpm, neglecting losses and vane thickness, determine **(a)** the discharge for shockless entrance when $\alpha_1 = 90°$, **(b)** α_2 and the theoretical head H, **(c)** the horsepower required, and **(d)** the pressure rise through the impeller.

▌ **(a)** The peripheral speeds are $u_1 = \frac{1800}{60}(2\pi)(\frac{1}{3}) = 62.8$ ft/s, $u_2 = 3u_1 = 188.5$ ft/s. The vector diagrams are shown in Fig. 20-1b. With u_1 and the angles α_1, β_1 known, the entrance diagram is determined, $V_1 = u_1 \tan 20° = 22.85$ ft/s; hence, $Q = 22.85(\pi)(\frac{2}{3})(\frac{2}{12}) = 7.97$ cfs.

(b) At the exit the radial velocity V_{r2} is $V_{r2} = 7.97(12)/[2\pi(0.75)] = 20.3$ ft/s. By drawing u_2 (Fig. 20-1b) and a parallel line distance V_{r2} from it, the vector triangle is determined when β_2 is laid off. Thus, $v_{u2} = 20.3 \cot 10° = 115$ ft/s, $V_{u2} = 188.5 - 115 = 73.5$ ft/s, $\alpha_2 = \tan^{-1}(20.3/73.5) = 15°26'$, $V_2 = 20.3 \csc 15°26' = 76.2$ ft/s; $H = (u_2V_2 \cos \alpha_2)/g = u_2V_{u2}/g = 188.5(73.5)/32.2 = 430$ ft.

(c) $$\text{Power} = Q\gamma H/550 = [7.97(62.4)(430)]/550 = 338 \text{ hp}$$

(d) By applying the energy equation from entrance to exit of the impeller, including the energy H added (elevation change across impeller is neglected), $H + (V_1^2/2g) + (p_1/\gamma) = (V_2^2/2g) + (p_2/\gamma)$ and $(p_2 - p_1)/\gamma = 430 + (22.85^2/64.4) - (76.2^2/64.4) = 348$ ft or $p_2 - p_1 = 348(0.433) = 151$ psi.

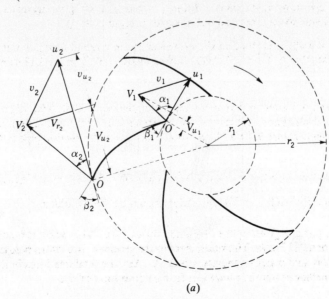

(a)   **Fig. 20-1**

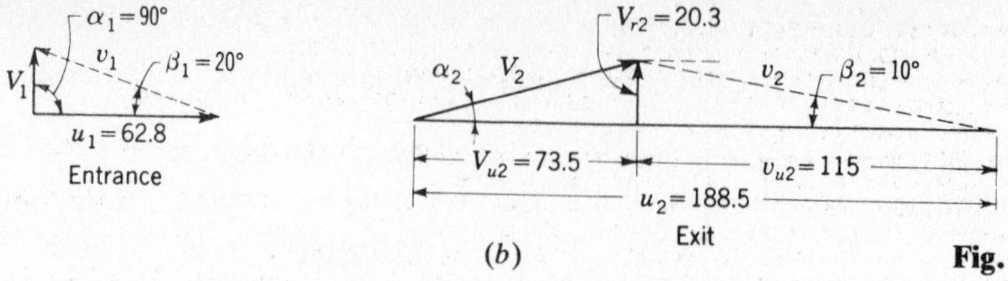

(b)

Fig. 20-1 *(continued)*

20.15 Tests of a pump model indicate a σ_c of 0.10. A homologous unit to be installed at a location where $p_a = 90$ kPa and $p_v = 3.5$ kPa is to pump water against a head of 25 m. The head loss from suction reservoir to pump impeller is 0.35 N·m/N. What is the maximum permissible suction head?

$$\sigma' = V_e^2/2gH = (p_a - p_v - \gamma z_s + h_l)/\gamma H$$

$$z_s = [(p_a - p_v)/\gamma] - \alpha'H + h_l = [(90\,000 - 3500)/9806] - 0.10(25) + 0.35 = 6.67 \text{ m}$$

20.16 A prototype test of a mixed-flow pump with a 72-in-diameter discharge opening, operating at 225 rpm, resulted in the following characteristics:

H, ft	Q, cfs	e, %	H, ft	Q, cfs	e, %	H, ft	Q, cfs	e, %
60	200	69	47.5	330	87.3	35	411	82
57.5	228	75	45	345	88	32.5	425	79
55	256	80	42.5	362	87.4	30	438	75
52.5	280	83.7	40	382	86.3	27.5	449	71
50	303	86	37.5	396	84.4	25	459	66.5

What size and synchronous speed (60 Hz) of homologous pump should be used to produce 200 cfs at 60-ft head at the point of best efficiency? Find the characteristic curves for this case.

Subscript 1 refers to the 72-in pump. For best efficiency $H_1 = 45$, $Q_1 = 345$, $e = 88$ percent. $H/N^2D^2 = H_1/N_1^2D_1^2$, $Q/ND^3 = Q/N_1D_1^3$, or $60/N^2D^2 = 45/[225^2(72^2)]$, $200/ND^3 = 345/[225(72^3)]$. After solving for N and D, $N = 366.7$ rpm, $D = 51.0$ in. The nearest synchronous speed (3600 divided by number of pairs of poles) is 360 rpm. To maintain the desired head of 60 ft, a new D is necessary. When its size is computed, $D = \sqrt{\frac{60}{45}(\frac{225}{360})}(72) = 52$ in, the discharge at best efficiency is $Q = Q_1ND^3/N_1D_1^3 = 345(\frac{360}{225})(\frac{52}{72})^3 = 208$ cfs, which is slightly more capacity than required. With $N = 360$ and $D = 52$, equations for transforming the corresponding values of H and Q for any efficiency can be obtained: $H = H_1(ND/N_1D_1)^2 = H_1[(\frac{360}{225})(\frac{52}{72})]^2 = 1.335H_1$ and $Q = Q_1(ND^3/N_1D_1^3) = Q_1(\frac{360}{225})(\frac{52}{72})^3 = 0.603Q_1$. The characteristics of the new pump are

H, ft	Q, cfs	e, %	H, ft	Q, cfs	e, %	H, ft	Q, cfs	e, %
80	121	69	63.5	200	87.3	46.7	248	82
76.7	138	75	60	208	88	43.4	257	79
73.4	155	80	56.7	219	87.4	40	264	75
70	169	83.7	53.5	231	86.3	36.7	271	71
66.7	183	86	50	239	84.4	33.4	277	66.5

The efficiency of the 52-in pump might be a fraction of a percent less than that of the 72-in pump, as the hydraulic radii of flow passages are smaller, so that the Reynolds number would be less.

20.17 Develop a program for calculating homologous pump characteristics and apply it to Prob. 20.16.

```
10 REM                        HOMOLOGOUS CHARACTERISTICS
20 DEFINT I: DIM H1(20),Q1(20),E(20),H(20),Q(20)
30 FOR I=1 TO 15: READ H1(I): NEXT I
40 DATA 60.,57.5,55.,52.5,50.,47.5,45.,42.5,40.,37.5,35.,32.5,30.,27.5,25.
50 LPRINT:LPRINT"H1=";: FOR I=1 TO 15:LPRINT USING"##.# ";H1(I);: NEXT I:LPRINT
60 FOR I=1 TO 15: READ Q1(I): NEXT I
70 DATA 200.,228.,256.,280.,303.,330.,345.,362.,382.,396.,411.,425.,438.,449.,45
9.
80 LPRINT:LPRINT"Q=";: FOR I=1 TO 15: LPRINT USING"###.# ";Q1(I);:NEXT I:LPRINT
90 FOR I=1 TO 15: READ E(I): NEXT I
100 DATA 69.,75.,80.,83.7,86.,87.3,88.,87.4,86.3,84.4,82.,79.,75.,71.,66.5
110 LPRINT:LPRINT"E=";: FOR I=1 TO 15: LPRINT USING"##.# ";E(I);: NEXT I:LPRINT
120 EE=0!: FOR I=1 TO 15: IF E(I)>EE THEN II=I: EE=E(I): NEXT I
```

```
130 LPRINT"II,E(I)=";II;EE
140 READ HH,QQ,NSYN,D1,N1:DATA 60.,200.,3600.,72.,225.
150 LPRINT"HH,QQ,NSYN,D1,N1=";HH;QQ;NSYN;D1;N1
160 D=((QQ/Q1(II))^2*H1(II)/HH)^.25*D1: N=N1*SQR(HH/H1(II))*D1/D
170 LPRINT"D,N=";D;N: I=FIX(NSYN/N)
180 NN1=FIX(NSYN/I): NN2=FIX(NSYN/(I+1)): LPRINT"N,NN1,NN2=";N;NN1;NN2
190 IF (NN1-N)<(N-NN2) THEN N=NN1 ELSE N=NN2
200 D=D1*N1*SQR(HH/H1(II))/N: QQ=(D/D1)^3*N*Q1(II)/N1: LPRINT"D,N,QQ=";D;N;QQ
210 C1=(N*D/(N1*D1))^2: C2=(D/D1)^3*N/N1
220 FOR I=1 TO 15: H(I)=C1*H1(I): Q(I)=C2*Q1(I): NEXT I
230 LPRINT: LPRINT"H=";: FOR I=1 TO 15:LPRINT USING" ###.#";H(I);:NEXT I:LPRINT
240 LPRINT"Q=";: FOR I=1 TO 15: LPRINT USING" ###.#";Q(I);: NEXT I: LPRINT


H1=60.0 57.5 55.0 52.5 50.0 47.5 45.0 42.5 40.0 37.5 35.0 32.5 30.0 27.5 25.0

Q=200.0 228.0 256.0 280.0 303.0 330.0 345.0 362.0 382.0 396.0 411.0 425.0 438.0
449.0 459.0

E=69.0 75.0 80.0 83.7 86.0 87.3 88.0 87.4 86.3 84.4 82.0 79.0 75.0 71.0 66.5
II,E(I)= 7  88
HH,QQ,NSYN,D1,N1= 60  200   3600  72  225
D,N= 51.01563  366.6749
N,NN1,NN2= 366.6749  400  360
D,N,QQ= 51.96153  360  207.4853

H= 80.0  76.7  73.3  70.0  66.7  63.3  60.0  56.7  53.3  50.0  46.7  43.3  40.0
   36.7  33.3
Q= 120.3 137.1 154.0 168.4 182.2 198.5 207.5 217.7 229.7 238.2 247.2 255.6 263.4
   270.0 276.0
```

20.18 Develop the characteristic curve for a homologous pump of the series of Prob. 20.16 for 18-in-diameter discharge and 1800 rpm.

∎ $Q = Q_1(N/N_1)(D/D_1)^3 = Q_1\frac{1800}{225}(1.5/6)^3 = 0.125Q_1$ $H = H_1(ND/N_1D_1)^2 = H_1[(1800)(1.5)/(225)(6)]^2 = 4H_1$

H_1	H	Q_1	Q
60	240	200	25.0
55	220	256	32.0
50	200	303	37.9
45	180	345	43.1
40	160	382	47.8
35	140	411	51.4
30	120	438	54.8
25	100	459	57.4

See curve plotted in Fig. 20-2.

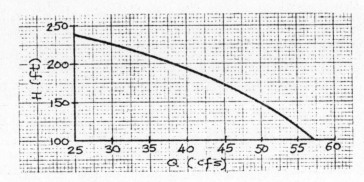

Fig. 20-2

20.19 Determine the size and synchronous speed of a pump homologous to the 72-in pump of Prob. 20.16 that will produce 3 m³/s at 100-m head at its point of best efficiency.

∎ $\qquad Q/ND^3 = Q_1/N_1D_1^3 \qquad Q_1 = 3/0.3048^3 = 105.9$ cfs $\qquad H_1 = 100/0.3048 = 328.1$ ft

$\qquad N_1D_1^3 = (Q_1/Q)ND^3 = (105.9/345)(225)(\frac{72}{12})^3 = 14\,918 \qquad H/N^2D^2 = H_1/N_1^2D_1^2$

$\qquad\qquad N_1^2D_1^2 = (H_1/H)N^2D^2 = (328.1/45)(225)^2(\frac{72}{12})^2 = 13\,288\,050$

$D_1 = 2.023$ ft, or 24.28 in, and $N_1 = 1801$ rpm. Use $N = 1800$ rpm (one pair of poles). Then, $D_1 = \sqrt{H_1/H}(N/N_1)D = \sqrt{328.1/45}(\frac{225}{1800})(6) = 2.03$ ft.

20.20 Construct a theoretical head-discharge curve for the following specifications of a centrifugal pump: $r_1 = 50$ mm, $r_2 = 100$ mm, $b_1 = 25$ mm, $b_2 = 20$ mm, $N = 1200$ rpm, and $\beta_2 = 30°$.

∎ $H = u_2^2/g - [(u_2 Q \cot \beta_2)/(2\pi r_2 b_2 g)]$ where $u_2 = N(2\pi/60)r_2 = 1200(2\pi/60)(0.10) = 12.566$ m/s, $H = (12.566^2/9.807) - \{[(12.566 \cot 30°)Q]/[2\pi(0.10)(0.02)(9.807)]\} = 16.10 - 176.61Q$. See curve plotted in Fig. 20-3.

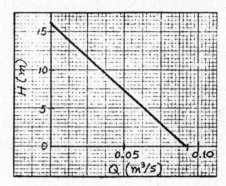

Fig. 20-3

20.21 A centrifugal pump (see Fig. 20-1a) has an impeller with dimensions $r_1 = 75$ mm, $r_2 = 160$ mm, $b_1 = 50$ mm, $b_2 = 30$ mm, $\beta_1 = \beta_2 = 30°$. For a discharge of 55 L/s and shockless entry to vanes, compute (a) the speed, (b) the head, (c) the torque, (d) the power, and (e) the pressure rise across the impeller. Neglect losses ($\alpha_1 = 90°$).

∎ $V_1 = V_{r1} = Q/2\pi r_1 b_1 = (\frac{55}{1000})/[(2)(\pi)(\frac{75}{1000})(\frac{50}{1000})] = 2.334$ m/s $u_1 = V_1/(\tan \beta_1) = 2.334/(\tan 30°) = 4.043$ m/s

(a) $N = (u_1/r_1)(60/2\pi) = [4.043/(\frac{75}{1000})](60/2\pi) = 514.8$ rpm

(b) $V_{r2} = Q/2\pi r_2 b_2 = (r_1 b_1/r_2 b_2)V_1 = [(75)(50)/(150)(30)](2.334) = 1.945$ m/s $u_2/u_1 = r_2/r_1 = \frac{150}{75} = 2$

   $u_2 = 2u_1 = (2)(4.043) = 8.086$ m/s $V_{u2} = u_2 - V_{r2} \cot \beta_2 = 8.086 - 1.945 \cot 30° = 4.717$ m/s

   $H = u_2 V_{u2}/g = (8.086)(4.717)/9.807 = 3.89$ m

(c) $T = \rho Q(r_2 V_{u2}) = (1000)(\frac{55}{1000})(\frac{150}{1000})(4.717) = 38.9$ N-m

(d) $P = T\omega = T(u_1/r_1) = (38.9)[4.043/(\frac{75}{1000})] = 2097$ W

(e) $V_2 = V_{r2}/(\sin \beta_2) = 1.945/(\sin 30°) = 3.89$ m/s

   $\Delta H = (8.086^2 - 4.043^2)/[(2)(9.807)] - (3.89^2 - 4.668^2)/[(2)(9.807)] = 2.84$ m   or $(2.84)(9.79) = 27.8$ kPa

20.22 A centrifugal pump with impeller dimensions $r_1 = 2$ in, $r_2 = 5$ in, $b_1 = 3$ in, $b_2 = 1.5$ in, $\beta_2 = 60°$ is to pump 5 cfs at a 64-ft head. Determine (a) β_1, (b) the speed, (c) the horsepower, and (d) the pressure rise across the impeller. Neglect losses and assume no shock at the entrance ($\alpha_1 = 90°$).

∎ $V_{r1} = Q/2\pi r_1 b_1 = 5/[(2)(\pi)(\frac{2}{12})(\frac{3}{12})] = 19.10$ ft/s

   $V_{2r} = Q/2\pi r_2 b_2 = (r_1 b_1/r_2 b_2)V_{r1} = [(2)(3)/(5)(1.5)](19.10) = 15.28$ ft/s

   $H = u_2 V_{u2}/g = u_2(u_1 - v_{r2} \cot \beta_2)$ $u_2^2 - u_2 V_{r2} \cot \beta_2 - gH = 0$

$u_2 = (V_{r2} \cot \beta_2 + \sqrt{V_{r2}^2 \cot^2 \beta_2 + 4gH})/2 = [15.28 \cot 60° + \sqrt{15.28^2 \cot^2 60° + (4)(32.2)(64)}]/2 = 50.0$ ft/s

   $u_1 = (r_1/r_2)u_2 = (\frac{2}{5})(50.0) = 20.0$ ft/s

(a) $\tan \beta_1 = V_{r1}/u_1 = 19.10/20.0 = 0.9550$ $\beta_1 = 43.68°$

(b) $N = u_2/r_2 = [50.0/(\frac{5}{12})](60/2\pi) = 1146$ rpm

(c) $P = Q\gamma H/550 = (5)(62.4)(64)/550 = 36.3$ hp

(d) $V_1 = V_{r1}/(\sin \beta_1) = 19.10/(\sin 43.68°) = 27.66$ ft/s $V_2 = V_{r2}/(\sin \beta_2) = 15.28/(\sin 60°) = 17.64$ ft/s

   $\Delta p = (\rho/2)(u_2^2 - u_1^2 - V_2^2 + V_1^2) = (1.94/2)(50.0^2 - 20.0^2 - 17.64^2 + 27.66^2) = 2477$ lb/ft² or 17.2 lb/in²

20.23 Select values of r_1, r_2, β_1, β_2, b_1, and b_2 of a centrifugal impeller to take 30 L/s of water from a 100-mm-diameter suction line and increase its energy by 15 m-N/N ($N = 1200$ rpm; $\alpha_1 = 90°$). Neglect losses.

▌ Select sample values $r_1 = 50$ mm, $r_2 = 100$ mm, $b_1 = 40$ mm, $b_2 = 25$ mm. Then find β_1 and β_2.

$$V_1 = Q/2\pi r_1 b_1 = (\tfrac{30}{1000})/[(2)(\pi)(\tfrac{50}{1000})(\tfrac{40}{1000})] = 2.387 \text{ m/s}$$

$$u_1 = \omega r_1 = [(2)(\pi)(1200)/60](\tfrac{50}{1000}) = 6.283 \text{ m/s} \qquad u_2 = (r_2/r_1)u_1 = (\tfrac{100}{50})(6.283) = 12.57 \text{ m/s}$$

Since $\alpha_1 = 90°$, $\tan \beta_1 = V_1/u_1 = 2.387/6.283 = 0.37991$, $\beta_1 = 20.8°$.

$$V_{u2} = Hg/u_2 = (15)(9.807)/12.57 = 11.70 \text{ m/s} \qquad V_{r2} = V_{r1}(r_1 b_1/r_2 b_2) = (2.387)[(50)(40)/(100)(25)] = 1.910 \text{ m/s}$$

$$\tan \beta_2 = V_{r2}/(u_2 - V_{u2}) = 1.910/(12.57 - 11.70) = 2.19540 \qquad \beta_2 = 65.5°$$

Note: It is best to choose $r_1 = 50$ mm (100-mm inlet diameter). Then r_2, b_2, and b_1 can be arbitrarily selected within reasonable limits.

20.24 A mercury–water differential manometer, $R' = 700$ mm, is connected from the 100-mm-diameter suction pipe to the 80-mm-diameter discharge pipe of a pump, as shown in Fig. 20-4. The centerline of the suction pipe is 300 mm below the discharge pipe. For $Q = 60$ L/s of water, calculate the head developed by the pump.

▌ $$V_1 = Q/A_1 = (\tfrac{60}{1000})/[(\pi)(\tfrac{100}{1000})^2/4] = 7.639 \text{ m/s} \qquad V_2 = V_1(D_1^2/D_2^2) = (7.639)(100^2/80^2) = 11.94 \text{ m/s}$$

The energy equation yields $p_1/\gamma + V_1^2/2g + H_p = p_2/\gamma + V_2^2/2g + h$, $H_p = \Delta p/\gamma + (V_2^2 - V_1^2)/2g + h$. The manometer equation yields $\Delta p/\gamma = R'(\text{s.g.} - 1) - h = (\tfrac{700}{1000})(13.6 - 1) - \tfrac{30}{100} = 8.52$ m of water, $H_p = 8.52 + (11.95^2 - 7.639^2)/[(2)(9.807)] + \tfrac{30}{1000} = 13.13$ m.

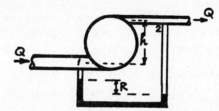

Fig. 20-4

20.25 An air blower is to be designed to produce pressure of 100 mmH$_2$O when operating at 3600 rpm; $r_2 = 1.1r_1$; $\beta_2 = \beta_1$; width of impeller is 100 mm; $\alpha_1 = 90°$. Find r_1. Assume a temperature of 30 °C.

▌ $Q = 2\pi r_1 b_1 V_1 = 2\pi r_2 b_2 V_2$. Since $r_2 = 1.1r_1$ and $b_1 = b_2$,

$$V_{r2} = V_{r1}/1.1 = V_1/1.1 \qquad u_2 = (r_2/r_1)u_1 = 1.1u_1$$

$$h_{air} = H_{H_2O}(\gamma_{H_2O}/\gamma_{air}) = (\tfrac{100}{1000})[(9.77)(1000)/11.4] = 85.70 \text{ m of air}$$

$$V_{u2} = gh_{air}/u_2 = (9.807)(85.70)/1.1u_1 = 764.1/u_1 \qquad u_2 = V_{u2} + V_{r2}\cot \beta$$

Since $\beta_1 = \beta_2 = \beta$ and $u_1 = V_1 \cot \beta$, $1.1u_1 = 764.1/u_1 + (V_1 \cot \beta)/1.1 = 764.1/u_1 + u_1/1.1$, $u_1 = 63.3$ m/s; $r_1 = u_1/\omega = 63.3/[(3600)(2\pi)/60] = 0.168$ m, or 168 mm.

20.26 In Prob. 20.25, when $\beta_1 = 30°$, calculate the discharge in cubic meters per minute.

▌ $$V_1 = u_1 \tan \beta_1 = 63.1 \tan 30° = 36.4 \text{ m/s}$$

$$Q = 2\pi r_1 b_1 V_1 = (2)(\pi)(0.168)(\tfrac{100}{1000})(36.4) = 3.842 \text{ m}^3/\text{s} \quad \text{or} \quad 231 \text{ m}^3/\text{min}$$

20.27 What is the cavitation index at a point in flowing water where $T = 20$ °C, $p = 14$ kPa, and the velocity is 12 m/s.

▌ $$\sigma' = (p - p_v)/(\rho V^2/2) = (14 - 2.34)(1000)/[(998)(12)^2/2] = 0.162$$

20.28 Two reservoirs A and B are connected with a long pipe which has characteristics such that the head loss through the pipe is expressible as $h_L = 20Q^2$, where h_L is in feet and Q is the flow rate in hundreds of gpm. The water-surface elevation in reservoir B is 35 ft above that in reservoir A. Two identical pumps are available for use to pump the water from A to B. The characteristic curve of the pump when operating at 1800 rpm is given in the table on page 616.

At the optimum point of operation the pump delivers 200 gpm at a head of 75 ft. Determine the specific speed N_s of the pump and find the rate of flow under the following conditions: (*a*) a single pump operating at 1800 rpm; (*b*) two pumps in series, each operating at 1800 rpm; (*c*) two pumps in parallel, each operating at 1800 rpm.

▌ The head-capacity curves for the pumping alternatives are plotted in Fig. 20-5 and so is the h_L versus Q curve for the pipe system. In this case $h = \Delta z + h_L = 35 + 20Q^2$. The answers are found at the points of

operation at 1800 rpm	
head, ft	flow rate, gpm
100	0
90	110
80	180
60	250
40	300
20	340

intersection of the curves. They are as follows: (*a*) single pump, 156 gpm; (*b*) two pumps in series, 224 gpm; (*c*) two pumps in parallel, 170 gpm.

If Δz had been greater than 100 ft, neither the single pump nor the two pumps in parallel would have delivered any water. If Δz had been -20 ft (i.e., with the water-surface elevation in reservoir B 20 ft below that in A), the flows would have been (*a*) 212 gpm; (*b*) 258 gpm; and (*c*) 232 gpm.

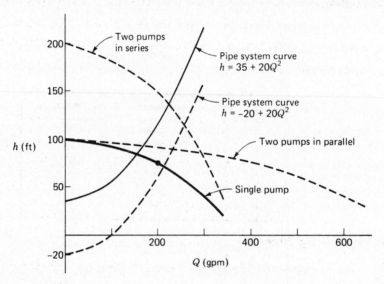

Fig. 20-5

20.29 The diameter of the discharge pipe of a pump is 6 in, and that of the intake pipe is 8 in. The pressure gage at discharge reads 30 psi, and the vacuum gage at intake reads 10 in of mercury. If $Q = 3.0$ cfs of water and the brake horsepower is 35.0, find the efficiency. The intake and the discharge are at the same elevation.

$$V_s = Q/A_s = 3.0/[(\pi)(\tfrac{8}{12})^2/4] = 8.594 \text{ ft/s} \qquad V_p = 3.0/[(\pi)(\tfrac{6}{12})^2/4] = 15.28 \text{ ft/s}$$

$$10 \text{ in of mercury vacuum} = -(10/29.9)(14.7) = -4.916 \text{ lb/ft}^2$$

$$H_p = p_d/\gamma + V_d^2/2g + z_d - (p_s/\gamma + V_s^2/2g + z_s) = (30)(144)/62.4 + 15.28^2/[(2)(32.2)] + 0$$
$$- \{(-4.916)(144)/62.4 + 8.594^2/[(2)(32.2)] + 0\} = 83.05 \text{ ft}$$

$$P = Q\gamma H_p = (3.0)(62.4)(83.05)/550 = 28.27 \text{ hp} \qquad \eta = 28.27/35.0 = 0.808 \quad \text{or} \quad 80.8 \text{ percent}$$

20.30 Under normal operation, a centrifugal pump with an impeller diameter of 2.84 in delivers 250 gpm of water at a head of 700 ft with an efficiency of 60 percent at 20 000 rpm. Compute the peripheral velocity u, specific speed N_s, and peripheral-velocity factor ϕ.

$$u = \omega r = [(20\,000)(2\pi)/60][(2.84/2)/12] = 248 \text{ ft/s} \qquad N_s = NQ^{0.5}/H^{0.75} = (20\,000)(250)^{0.5}/700^{0.75} = 2324$$

$$u = \phi\sqrt{2gh} \qquad 248 = (\phi)\sqrt{(2)(32.2)(700)} \qquad \phi = 1.168$$

20.31 Select the specific speed of the pump or pumps required to lift 15 cfs of water 375 ft through 10 000 ft of 3-ft-diameter pipe ($f = 0.020$). The pump rotative speed is to be 1750 rpm. Consider the following cases: single pump, two pumps in series, three pumps in series, two pumps in parallel, and three pumps in parallel.

$$Q = 15/0.002228 = 6732 \text{ gpm} \qquad V = Q/A = 15/[(\pi)(3)^2/4] = 2.122 \text{ ft/s}$$

$$h_f = (f)(L/D)(V^2/2g) = (0.020)(10\,000/3)\{2.122^2/[(2)(32.2)]\} = 4.7 \text{ ft}$$

$$h_p = 375 + 4.7 = 379.7 \text{ ft} \qquad N_s = NQ^{0.5}/H^{0.75}$$

case	gpm/pump	h_p/pump	N_s
Single pump	6732	379.7	1669
Two pumps in series	6732	189.8	2808
Three pumps in series	6732	126.6	3804
Two pumps in parallel	3366	379.7	1180
Three pumps in parallel	2244	379.7	964

20.32 The pump of Fig. 20-6a is placed in a 10-in-diameter pipe ($f = 0.020$), 1300 ft long, that is used to lift water from one reservoir to another. The difference in water-surface elevations between the reservoirs fluctuates from 20 ft to 100 ft. Plot a curve showing delivery rate versus water-surface-elevation difference. Plot also the corresponding efficiencies. The pump is operated at a constant speed of 1450 rpm. Neglect minor losses.

$$V = Q/A = Q/[(\pi)(\tfrac{10}{12})^2/4] = 1.833Q \qquad (Q \text{ in cfs})$$

$$h_p = \Delta z + (f)(L/D)(V^2/2g) = \Delta z + (0.020)[1300/(\tfrac{10}{12})]\{(1.833Q)^2/[(2)(32.2)]\} = \Delta z + 1.628Q^2$$

Plot the pump characteristic curve and pipe system curves for a variety of Δz's.

Δz, ft	Q, gpm	efficiency, %
20	2500	68
40	2320	76
60	2100	81
80	1860	84
100	1350	77
110	940	66

See plotted curves in Fig. 20-6b.

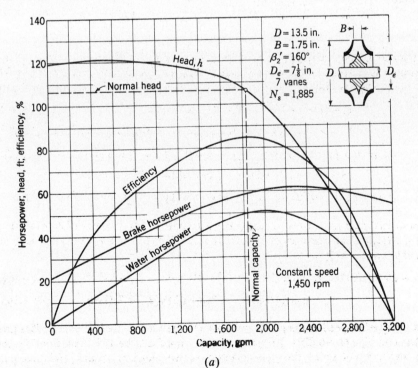

(a)

Fig. 20-6

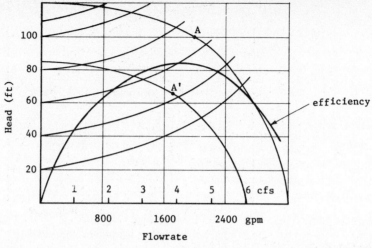

(b)   Fig. 20-6   (continued)

20.33   Repeat Prob. 20.32 for the case of the same pump operating at 1200 rpm. Assume efficiency pattern and values remain the same.

▮ The pump characteristic curve must be transformed using $Q \propto N$ and $h \propto N^2$. Thus point A in Fig. 20-6b transforms to A', where $Q = (2000)(\frac{1200}{1450}) = 1655$ gpm and $h = (100)(\frac{1200}{1450})^2 = 68.5$ ft, etc.

Δz, ft	Q, gpm	efficiency, %
20	2000	73
40	1750	82
60	1350	84
80	550	58
82.3	Shutoff	0

The transformed pump characteristic curve is shown in Fig. 20-6b. Note that the efficiency curve is transformed on the basis of corresponding flow rates. (The transformed efficiency curve is not shown in the figure.)

20.34   Repeat Prob. 20.32 for the case of a homologous pump whose diameter is 80 percent as large as the pump of Prob. 20.32. Assume efficiency pattern and values remain the same.

▮ In this case the pump characteristic curve is transformed using $Q \propto D^3$ and $h \propto D^2$. Thus point A in Fig. 20-6b for the solution of Prob. 20.32 transformations as follows: $Q = (2000)(0.80)^3 = 1024$ gpm, $h = (100)(0.80)^2 = 64.0$ ft.

Δz, ft	Q, gpm	efficiency, %
20	1400	60
40	1200	76
60	800	81
76.8	Shutoff	0

The transformed pump characteristic curve is shown in Fig. 20-7.

20.35   Repeat Prob. 20.32 for the case of a homologous pump with diameter 80 percent as large as the pump of Prob. 20.32 when operating at 1200 rpm. Assume efficiency pattern and values remain the same.

▮ In this case the pump characteristic curve is transformed using $Q \propto ND^3$ and $h \propto N^2D^2$. Thus point A in Fig. 20-6b for the solution of Prob. 20.32 transforms as follows: $Q = (2000)(\frac{1200}{1450})(0.80)^3 = 847$ gpm, $h = (100)(\frac{1200}{1450})^2(0.80)^2 = 43.8$ ft. Transforming the pump characteristic curve according to these rules and following

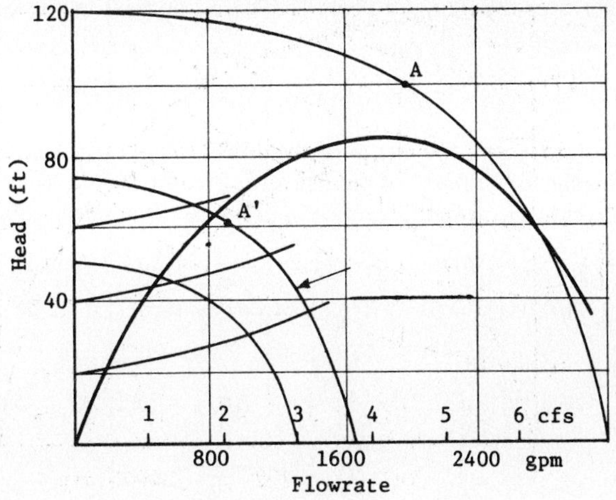

Fig. 20-7

the procedure as outlined in Prob. 20.34 gives

Δz, ft	Q, gpm	efficiency, %
20	1080	69
40	700	83
52.6	Shutoff	0

20.36 A pump with a critical value of σ_c of 0.10 is to pump against a head of 500 ft. The barometric pressure is 14.3 psia, and the vapor pressure of the water is 0.5 psia. Assume the friction losses in the intake piping are 5 ft. Find the maximum allowable height of the pump relative to the water surface at intake.

▌ $(z_s)_{max} = (p_0)_{abs}/\gamma - p_v/\gamma - \sigma_c h - h_L$ $p_0/\gamma = (14.3)(144)/62.4 = 33.0\,\text{ft}$

$p_v/\gamma = (0.5)(144)/62.4 = 1.154\,\text{ft}$ $(z_s)_{max} = 33.0 - 1.154 - (0.10)(500) - 5 = -23.2\,\text{ft}$ (submerged)

20.37 A boiler feed pump delivers water at 212 °F which it draws from an open hot well with a friction loss of 2 ft in the intake pipe between it and the hot well. Barometric pressure is 29 in of mercury and the value of σ_c for the pump is 0.10. What must be the elevation of the water surface in the hot well relative to that of the pump intake? The total pumping head is 240 ft.

▌ $(z_s)_{max} = (p_0)_{abs}/\gamma - p_v/\gamma - \sigma_c h - h_L$ $p_0 = (\frac{29}{12})(847.3/144) = 14.2\,\text{psia}$ $p_0/\gamma = (14.2)(144)/59.8 = 34.2\,\text{ft}$

$p_v/\gamma = (14.7)(144)/59.8 = 35.4\,\text{ft}$ $(z_s)_{max} = 34.2 - 35.4 - (0.10)(240) - 2 = -27.2\,\text{ft}$

Thus the water surface must be 27.2 ft above the pump suction.

20.38 Suppose a pump is to pump water at a head of 130 ft, the water temperature being 100 °F and the barometric pressure being 14.3 psia. At intake the pressure is a vacuum of 17 inHg and the velocity is 12 fps. What are the values of NPSH and σ?

▌ NPSH $= (p_s)_{abs}/\gamma + V_s^2/2g - p_v/\gamma$ $(p_s)_{abs} = p_0/\gamma + p_s/\gamma = (14.3)(144)/62.0 + (-\frac{17}{12})(847.3/62.0) = 13.85\,\text{ft}$

$V_s^2/2g = 12^2/[(2)(32.2)] = 2.24\,\text{ft}$ $p_v/\gamma = 135/62.0 = 2.18\,\text{ft}$

NPSH $= 13.85 + 2.24 - 2.18 = 13.91\,\text{ft}$ $\sigma = 13.91/130 = 0.107$

20.39 A pump is delivering 7500 gpm of water at 140 °F at a head of 240 ft, and the barometric pressure is 13.8 psia. Determine the reading on a pressure gage in inches of mercury vacuum at the suction flange when cavitation is incipient. Assume the suction pipe diameter equals 2 ft and neglect the effects of prerotation. Take $\sigma_c = 0.085$.

▌ $\sigma = [(p_s)_{abs}/\gamma + V_s^2/2g - p_v/\gamma]/h$. Let p = gage pressure at suction flange.

$(p_s)_{abs} = p_{atm} + p = 13.8 + p$ $V_s = Q/A_s = (7500)(0.002228)/[(\pi)(2)^2/4] = 5.319\,\text{ft/s}$

$p_v/\gamma = 416/61.4 = 6.775\,\text{ft}$ $0.085 = \{(13.8 + p)(144)/61.4 + 5.319^2/[(2)(32.2)] - 6.775\}/240$

$p = -2.400\,\text{lb/in}^2 = (-2.400)(29.9/14.7) = -4.88\,\text{inHg}$ or   4.88 inHg vacuum

20.40 If the maximum efficiency of the pump of Prob. 20.32 is 82 percent, approximately what would be the efficiency of the pump of Prob. 20.34?

$$(1 - \eta_1)/(1 - \eta_2) = (D_2/D_1)^{1/5} \qquad (1 - 0.82)/(1 - \eta_2) = (0.80/1.0)^{1/5} \qquad \eta_2 = 0.812 \quad \text{or} \quad 81.2 \text{ percent}$$

20.41 Figure 20-8 shows the dimensions and angles of the diffuser vanes of a centrifugal pump. The vane passages are 0.80 in wide perpendicular to the plane of the figure. If the impeller delivers water at the rate of 2.80 cfs under ideal and frictionless conditions, what is the rise in pressure through the diffuser?

$$V_1^2/2g = V_2^2/2g + \Delta p/\gamma \qquad V = Q/A \qquad A_1 = (2\pi)(\tfrac{6}{12})(\sin 15°)(0.80/12) = 0.05421 \text{ ft}^2$$

$$V_1 = 2.80/0.05421 = 51.65 \text{ ft/s} \qquad A_2 = (2\pi)(8.5/12)(\sin 28°)(0.80/12) = 0.1393 \text{ ft}^2$$

$$V_2 = 2.80/0.1393 = 20.10 \text{ ft/s}$$

$$51.65^2/[(2)(32.2)] = 20.10^2/[(2)(32.2)] + \Delta p/\gamma \qquad \Delta p/\gamma = 35.15 \text{ ft} \qquad \Delta p = (62.4)(35.15)/144 = 15.2 \text{ lb/in}^2$$

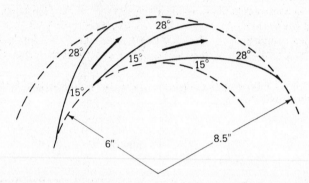

Fig. 20-8

20.42 Two pumps whose characteristics are given in Prob. 20.28 are to be used in parallel. They must develop a head $h = 35 + 20Q^2$ as in Prob. 20.28. One pump is to be operated at 1800 rpm. The speed of the other pump is to be gradually reduced until it no longer delivers water. At approximately what speed will this happen?

▮ From Prob. 20.28, with a single pump operating at 1800 rpm, the flow rate is 156 gpm and the head is 83.8 ft. If we add a pump in parallel to this pump with a shutoff head of 83.8 ft, the second pump will convey no flow. Thus, $h/h' = (n/n')^2$, $83.8/100 = (n/1800)^2$, $n = 1648$ rpm.

20.43 A pump is installed to deliver water from a reservoir of surface elevation zero to another of elevation 300 ft. The 12-in-diameter suction pipe ($f = 0.020$) is 100 ft long, and the 10-in-diameter discharge pipe ($f = 0.026$) is 5000 ft long. The pump characteristic at 1200 rpm is defined by $h_p = 375 - 24Q^2$ where h_p, the pump head, is in feet and Q is in cubic feet per second. Compute the rate at which this pump will deliver water under these conditions assuming the setting is low enough to avoid cavitation.

▮ $h_p = 375 - 24Q^2$. Also, since $h_f = (f)(L/D)(V^2/2g)$,

$$h_p = 300 + (0.020)[100/(\tfrac{12}{12})]V_{12}^2/2g + (0.026)(5000)(\tfrac{10}{12})V_{10}^2/2g \qquad V = Q/A$$

$$V_{12} = Q/[(\pi)(\tfrac{12}{12})^2/4] = 1.273Q \qquad V_{10} = Q/[(\pi)(\tfrac{10}{12})^2/4] = 1.833Q$$

$$h_p = 300 + (0.020)[100/(\tfrac{12}{12})]\{(1.273Q)^2/[(2)(32.2)]\}$$

$$+ (0.026)[5000/(\tfrac{10}{12})]\{(1.833Q)^2/[(2)(32.2)]\} = 300 + 8.189Q^2$$

Equating the two expressions for h_p gives $375 - 24Q^2 = 300 + 8.189Q^2$, $Q = 1.53$ ft³/s.

20.44 Repeat Prob. 20.43 determining the flow rate if two such pumps were installed in series. Repeat for two pumps in parallel.

▮ From Prob. 20.43, $h_p = 300 + 8.189Q^2$.

Two pumps in series:

$$h_p \text{ per pump} = (300 + 8.189Q^2)/2 = 150 + 4.094Q^2 = 375 - 24Q^2 \qquad Q = 2.83 \text{ ft}^3/\text{s}$$

Two pumps in parallel:

$$h_p \text{ per pump} = 300 + 8.189Q^2 = 375 - 24(Q/2)^2 \qquad Q = 2.30 \text{ ft}^3/\text{s}$$

20.45 A centrifugal pump with a 12-in-diameter impeller is rated at 600 gpm against a head of 80 ft when rotating at 1750 rpm. What would be the rating of a pump of identical geometric shape with a 6-in impeller? Assume pump efficiencies and rotative speeds are identical.

▌ $$h \propto D^2 \qquad h = (80)(\tfrac{6}{12})^2 = 20.0 \text{ ft} \qquad Q \propto D^3 \qquad Q = (600)(\tfrac{6}{12})^3 = 75.0 \text{ gpm}$$

20.46 Given the following data for a commercial centrifugal water pump: $r_1 = 4$ in, $r_2 = 7$ in, $\beta = 30°$, $\beta_2 = 20°$, speed = 1440 rpm. Estimate (**a**) design-point discharge, (**b**) water horsepower, and (**c**) head if $b_1 = b_2 = 1.75$ in.

▌ (**a**) The angular velocity is $\omega = 2\pi$ r/s $= 2\pi(\tfrac{1440}{60}) = 150.8$ rad/s. Thus the tip speeds are $u_1 = \omega r_1 = 150.8(\tfrac{4}{12}) = 50.3$ ft/s, and $u_2 = \omega r_2 = 150.8(\tfrac{7}{12}) = 88.0$ ft/s. From the inlet-velocity diagram, Fig. 20-9a, with $\alpha_1 = 90°$ for design point, we compute $V_{n1} = u_1 \tan 30° = 29.0$ ft/s, whence the discharge is $Q = 2\pi r_1 b_1 V_{n1} = 2\pi(\tfrac{4}{12})(1.75/12)(29.0) = 8.86$ ft³/s $= 3980$ gal/min. (The actual pump produces about 3500 gal/min.)

(**b**) The outlet radial velocity follows from Q: $V_{n2} = Q/(2\pi r_2 b_2) = 8.86/[2\pi(\tfrac{7}{12})(1.75/12)] = 16.6$ ft/s. This enables us to construct the outlet-velocity diagram as in Fig. 20-9b, given $\beta_2 = 20°$. The tangential component is $V_{t2} = u_2 - V_{n2} \cot \beta_2 = 88.0 - 16.6 \cot 20° = 42.4$ ft/s, $\alpha_2 = \tan^{-1}(16.6/42.4) = 21.4°$. The power is then computed with $V_{t1} = 0$ at the design point: $P_w = \rho Q u_2 V_{t2} = [(1.94)(8.8)(88.0)(42.4)]/550 = 117$ hp. (The actual pump delivers about 125 water horsepower, requiring 147 bhp at 85 percent efficiency.)

(**c**) Finally, the head is estimated from $H \approx P_w/\rho g Q = (117)(550)/[(62.4)(8.86)] = 116$ ft. (The actual pump develops about 140-ft head.)

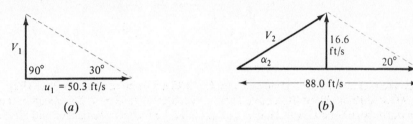

(a)   (b)   **Fig. 20-9**

20.47 The 32-in pump of Fig. 20-10 is to pump 24 000 gal/min of water at 1170 rpm from a reservoir whose surface is at 14.7 lb/in² abs. If head loss from reservoir to pump inlet is 6 ft, where should the pump inlet be placed to avoid cavitation for water at (**a**) 60 °F, $p_v = 0.26$ lb/in² abs, s.g. = 1.0 and (**b**) 200 °F, $p_v = 11.52$ lb/in² abs, s.g. = 0.9635?

▌ For either case read from Fig. 20-10 at 24 000 gal/min that the required NPSH is 40 ft.

(**a**) For this case $\rho g = 62.4$ lb/ft³. NPSH $\leq [(p_a - p_v)/\rho g] - Z_i - h_{fi}$ or 40 ft $\leq [(14.7 - 0.26)(144)/62.4] - Z_i - 6.0$ or $Z_i \leq 27.3 - 40 = -12.7$ ft. The pump must be placed at least 12.7 ft *below* the reservoir surface to avoid cavitation.

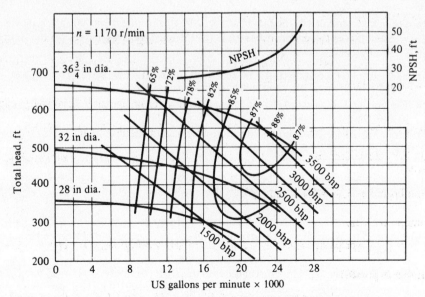

(From Ingersoll-Rand Corporation, Cameron Pump Division)   **Fig. 20-10**

(b) For this case $\rho g = 62.4(0.9635) = 60.1 \text{ lb/ft}^3$. $40 \text{ ft} \le [(14.7 - 11.52)(144)/60.1] - Z_i - 6.0$ or $Z_i \le 1.6 - 40 = 38.4 \text{ ft}$. The pump must now be placed at least 38.4 ft below the reservoir surface. These are unusually stringent conditions because a large, high-discharge pump requires a large NPSH.

20.48 A pump from the family of Fig. 20-11 has $D = 21$ in and $n = 1500$ rpm. Estimate (a) discharge, (b) head, (c) pressure rise, and (d) brake horsepower of this pump for water at 60 °F and best efficiency.

▌ (a) In BG units take $D = \frac{21}{12} = 1.75 \text{ ft}$ and $n = \frac{1500}{60} = 25 \text{ r/s}$. At 60 °F, ρ of water is 1.94 slugs/ft³. The BEP parameters are known from Fig. 20-11. The BEP discharge is thus $Q^* = C_Q \cdot nD^3 = 0.115(25)(1.75)^3 = 15.4 \text{ ft}^3/\text{s} = 6920 \text{ gal/min}$.

(b) Similarly, the BEP head is $H^* = C_H \cdot n^2 D^2/g = 5.0(25)^2(1.75)^2/32.2 = 297 \text{ ft water}$.

(c) Since we are not given elevation or velocity-head changes across the pump, we neglect them and estimate $\Delta p \approx \rho g H = 1.94(32.2)(297) = 18\,600 \text{ lb/ft}^2 = 129 \text{ lb/in}^2$.

(d) Finally, the BEP power is $P^* = C_p \cdot \rho n^3 D^5 = 0.65(1.94)(25)^3(1.75)^5/550 = 588 \text{ hp}$.

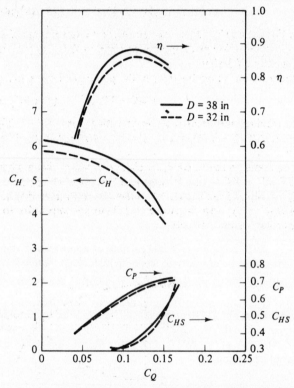

Fig. 20-11

20.49 We want to build a pump from the family of Fig. 20-11, which delivers 3000 gal/min water at 1200 rpm at best efficiency. Estimate (a) the impeller diameter, (b) the maximum discharge, (c) the shutoff head, and (d) the NPSH at best efficiency.

▌ (a) 3000 gal/min = 6.68 ft³/s and 1200 rpm = 20 r/s. At BEP we have $Q^* = C_Q \cdot nD^3 = 6.68 \text{ ft}^3/\text{s} = (0.115)(20)D^3$ or $D = [6.68/0.115(20)]^{1/3} = 1.43 \text{ ft} = 17.1 \text{ in}$.

(b) The max Q is related to Q^* by a ratio of capacity coefficients: $Q_{max} = Q^* C_{Q \cdot max}/C_{Q^*} \approx 3000(0.23)/0.115 = 6000 \text{ gal/min}$.

(c) From Fig. 20-11 we estimated the shutoff-head coefficient to be 6.0. Thus $H(0) \approx C_H(0)n^2 D^2/g = 6.0(20)^2(1.43)^2/32.2 = 152 \text{ ft}$.

(d) Finally, the NPSH at BEP is approximately NPSH$^* = C_{HS} \cdot n^2 D^2/g = 0.37(20)^2(1.43)^2/32.2 = 9.4 \text{ ft}$. Since this is a small pump, it will be less efficient than the pumps in Fig. 20-11, probably about 82 percent maximum.

20.50 We want to use a centrifugal pump from the family of Fig. 20-11 to deliver 100 000 gal/min of water at 60 °F with a head of 25 ft. What should be (a) the pump size and speed and (b) brake horsepower assuming operation at best efficiency?

▌ (a) Enter the known head and discharge into the BEP parameters: $H^* = 25 \text{ ft} = C_H \cdot n^2 D^2/g = 5.0n^2 D^2/32.2$, $Q^* = 100\,000 \text{ gal/min} = 222.8 \text{ ft}^3/\text{s} = C_Q \cdot nD^3 = 0.115nD^3$. The two unknowns are n and D. Solve simultaneously for $D = 12.4 \text{ ft}$, $n = 1.03 \text{ r/s} = 62 \text{ rpm}$.

(b) The most efficient horsepower is, then, $\text{bhp}^* \approx C_p \cdot \rho n^3 D^5 = 0.65(1.94)(1.03)^3(12.4)^5/550 = 734$ hp.
The solution to this problem is mathematically correct but results in a grotesque pump: an impeller more than 12 ft in diameter, rotating so slowly one can visualize oxen walking in a circle turning the shaft.

20.51 We want to use the 32-in pump of Fig. 20-10 at 1170 rpm to pump water at 60 °F from one reservoir to another 120 ft higher through 1500 ft of 16-in-ID pipe with friction factor $f = 0.030$. **(a)** What will the operating point and efficiency be? **(b)** To what speed should the pump be changed to operate the BEP?

▌ **(a)** For reservoirs the initial and final velocities are zero; thus the system head is $H_s = z_2 - z_1 +$
$(V^2/2g)(fL/D) = 120 + (V^2/2g)[0.030(1500)/\frac{16}{12}]$. From continuity in the pipe, $V = Q/A = Q/[\frac{1}{4}\pi(\frac{16}{12})^2]$, and so we substitute for V above to get

$$H_s = 120 + 0.269Q^2 \qquad Q \text{ in ft}^3/\text{s} \qquad (1)$$

Since Fig. 20-10 uses thousands of gallons per minute for the abscissa, we convert Q in Eq. (1) to this unit:
$$H_s = 120 + 1.335Q^2 \qquad Q \text{ in } 10^3 \text{ gal/min} \qquad (2)$$

We can plot Eq. (2) on Fig. 20-10 and see where it intersects the 32-in pump-head curve, as in Fig. 20-12. A graphical solution gives approximately $H \approx 430$ ft, $Q \approx 15\,000$ gal/min. The efficiency is about 82 percent, slightly off design.
An analytic solution is possible if we fit the pump-head curve to a parabola, which is very accurate:

$$H_{\text{pump}} \approx 490 - 0.26Q^2 \qquad Q \text{ in } 10^3 \text{ gal/min} \qquad (3)$$

Equations (2) and (3) must match at the operating point: $490 - 0.26Q^2 = 120 + 1.335Q^2$ or $Q^2 = (490 - 120)/(0.26 + 1.335) = 232$, $Q = 15.2 \times 10^3$ gal/min = 15 200 gal/min; $H = 490 - 0.26(15.2)^2 = 430$ ft.

(b) To move the operating point to BEP, we change n, which changes both $Q \propto n$ and $H \propto n^2$. From Fig. 20-10, at BEP, $H^* \approx 386$ ft; thus for any n, $H^* = 386(n/1170)^2$. Also read $Q^* \approx 20 \times 10^3$ gal/min; thus for any n, $Q^* = 20(n/1170)$. Match H^* to the system characteristics, Eq. (2), $H^* = 386(n/1170)^2 \approx 120 + 1.335[20(n/1170)]^2$, which gives $n^2 < 0$. Thus it is impossible to operate at maximum efficiency with this particular system and pump.

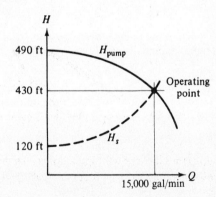

Fig. 20-12

20.52 Investigate extending Prob. 20.51 by using two 32-in pumps in parallel to deliver more flow. Is this efficient?

▌ Since the pumps are identical, each delivers $\frac{1}{2}Q$ at the same 1170 rpm speed. The system curve is the same, and the balance-of-head relation becomes $H = 490 - 0.26(\frac{1}{2}Q)^2 = 120 + 1.335Q^2$ or $Q^2 = (490 - 120)/(1.335 + 0.065)$, $Q = 16\,300$ gal/min. This is only 7 percent more than a single pump. Each pump delivers $\frac{1}{2}Q = 8150$ gal/min, for which the efficiency is only 60 percent. The total brake horsepower required is 3200, whereas a single pump used only 2000 bhp.

20.53 Suppose the elevation change in Prob. 20.51 is raised from 120 to 500 ft, greater than a single 32-in pump can supply. Investigate using 32-in pumps in series at 1170 rpm.

▌ Since the pumps are identical, the total head is twice as much and the constant 120 in the system-head curve is replaced by 500. The balance of heads becomes $H = 2(490 - 0.26Q^2) = 500 + 1.335Q^2$ or $Q^2 = (980 - 500)/(1.335 + 0.52)$, $Q = 16.1 \times 10^3$ gal/min. The operating head is $500 + 1.335(16.1)^2 = 846$ ft, or 97 percent more than a single pump in Prob. 20.50. Each pump is operating at 16.1×10^3 gal/min, which from Fig. 20-10 is 83 percent efficient, a pretty good match to the system. To pump at this operating point requires 4100 bhp, or about 2050 bhp for each pump.

20.54 A piston positive-displacement pump (PDP) has a 7-in diameter and a 3-in stroke. Its crankshaft rotates at 350 rpm. How many gallons per minute does it deliver at 100 percent volumetric efficiency?

▌ $V_{displaced} = [(\pi)(\frac{7}{12})^2/4](\frac{3}{12}) = 0.06681$ ft^3 $Q = (0.06681)(350) = 23.38$ ft^3/min $= (23.38)(7.48) = 175$ gpm

20.55 If the PDP of Prob. 20.54 delivers water against a total head of 30 ft, what horsepower is required at 75 percent efficiency?

▌ $P = Q\gamma E_p = (23.38/60)(62.4)(30) = 729.5$ ft-lb/s $P_{required} = (729.5/550)/0.75 = 1.77$ hp

20.56 A pump delivers 1500 L/min of water at 20 °C against a pressure rise of 300 kPa. Kinetic- and potential-energy changes are negligible. If the driving motor supplies 9 kW, what is the overall efficiency?

▌ $P = Q\gamma E_p = \Delta p \, Q = (300)(\frac{1500}{1000})/60 = 7.50$ kN-m/s   or   7.50 kW $\eta = 7.50/9 = 0.833$ or   83.3 percent

20.57 A 20-hp pump delivers 400 gpm of gasoline ($\gamma = 42.5$ lb/ft^3) at 20 °C with 80 percent efficiency. What head and pressure rise result across the pump?

▌ $P_{required} = Q\gamma E_p/\eta$ $(20)(550) = [(400)(0.002228)](42.5)(E_p)/0.80$ $E_p = 232$ ft

$$\Delta p = \gamma E_p = (42.5)(232) = 9860 \text{ lb/ft}^2 \text{ or } 68.5 \text{ lb/in}^2$$

20.58 A typical household basement sump pump provides a discharge of 10 gpm against a head of 20 ft. What is the minimum horsepower required to drive such a pump? Assume $N = 1750$ rpm.

▌ $N_s = NQ^{0.5}/H^{0.75} = (1750)(10)^{0.5}/20^{0.75} = 585$. From Fig. A-41, $\eta_{max} = 0.40$. $P_{required} = Q\gamma H/\eta = [(10)(0.002228)](62.4)(20)/0.40 = 69.5$ ft-lb/s, or 0.126 hp.

20.59 An 8-in model pump delivering water at 180 °F at 800 gpm and 2400 rpm begins to cavitate when the inlet pressure and velocity are 12 psia and 20 fps, respectively. Find the required NPSH of a prototype which is 4 times larger and runs at 1000 rpm.

▌ $\text{NPSH} = (p_i - p_v)/\gamma + V_i^2/2g$ $\text{NPSH}_{model} = [(12)(144) - 1086]/60.6 + 20^2/[(2)(32.2)] = 16.81$ ft

In prototype, $N = 1000$ rpm, $D = 32$ in. $\text{NPSH}_{proto} = (\text{NPSH}_{model})(N_p/N_m)^2(D_p/D_m)^2 = (16.81)(\frac{1000}{2400})^2(\frac{32}{8})^2 = 46.7$ ft.

20.60 Determine the specific speed of the pump whose operating characteristics are shown in Fig. 20–13. If this pump were operated at 1200 rpm, what head and discharge would be developed at rated capacity, and what power would be required?

▌ $N_s = NQ^{0.5}/H^{0.75}$. At its rated capacity of 700 gpm, this pump develops 120 ft of head when operating at 1450 rpm. Thus $N_s = (1450)(700)^{0.5}/120^{0.75} = 1058$. If the efficiency remains constant with the speed change, at 1200 rpm, $Q = (\frac{1200}{1450})(700) = 579$ gpm, $H = (\frac{1200}{1450})^2(120) = 82.2$ ft, $P = (\frac{1200}{1450})^3(26) = 14.7$ hp.

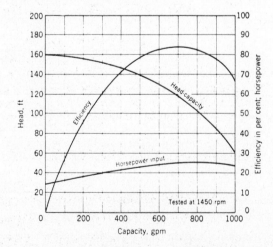

Fig. 20-13

20.61 A mixed-flow centrifugal pump ($N_s = 400$, $\sigma_c = 0.30$) is to develop 50 ft of head. Find the maximum permissible suction lift on the pump if the pump is at sea level and the water is at 80 °F.

▌ Neglecting head loss between reservoir and pump, $z_s \approx (p_{atm}/\gamma - p_v/\gamma) - \text{NPSH}$, $\text{NPSH} = \sigma_c h = (0.30)(50) = 15.0$ ft, $z_s \approx [(14.7)(144)/62.4 - 73.5/62.4] - 15.0 = 17.7$ ft.

20.62 A centrifugal pump driven by an electric motor lifts water through a total height of 150 ft from reservoir to discharge. The pump efficiency is 78 percent and the motor efficiency is 89 percent. The lift is through 1000 ft of 4-in-diameter pipe and the pumping rate is 300 gpm. If $f = 0.025$ and power costs 6 mills/kilowatthour, what is the cost of power for pumping a million gallons of water? An acre-ft?

▮
$$Q = (300)(0.002228) = 0.6684 \text{ ft}^3/\text{s} \qquad V = Q/A = 0.6684/[(\pi)(\tfrac{4}{12})^2/4] = 7.659 \text{ ft/s}$$

$$h_f = (f)(L/D)(V^2/2g) = (0.025)[1000/(\tfrac{4}{12})]\{7.659^2/[(2)(32.2)]\} = 68.3 \text{ ft} \qquad E_p = 150 + 68.3 = 218.3 \text{ ft}$$

$$P_{\text{H}_2\text{O}} = Q\gamma E_p = (0.6684)(62.4)(218.3) = 9105 \text{ ft-lb/s} = \tfrac{9105}{550} = 16.55 \text{ hp}$$

$$P_{\text{motor}} = 16.55/[0.78)(0.89)] = 23.84 \text{ hp} = (23.84)(550)(0.3048)(4.448)/1000 = 17.78 \text{ kW}$$

$$\text{Power cost} = (17.78)(1)(\tfrac{6}{1000}) = \$0.107/\text{h}$$

$$\text{Cost to pump } 1\,000\,000 \text{ gallons} = [(1\,000\,000/300)/60](0.107) = \$5.94$$

$$1 \text{ acre-ft} = (43\,560)(1)(7.48) \quad \text{or} \quad 325\,829 \text{ gal} \qquad \text{Cost to pump 1 acre-foot} = (325\,829/1\,000\,000)(5.94) = \$1.94$$

20.63 A model of a mixed-flow water pump has been tested in the laboratory to give performance curves as is shown in Fig. 20-14. What would be the total head ΔH delivered by a prototype pump with an impeller size of 1.2 m operating at a speed of 1750 rpm and delivering 1.300 m³/s flow? What is its mechanical efficiency? Consider that dynamic similarity can be achieved between model flow and prototype flow.

▮ We first compute the flow coefficient for the prototype flow: $(Q/ND^3)_p = 1.300/\{[(1750)(2\pi)/60](1.2^3)\} = 0.00411$. Examining Fig. 20-14 we see that the efficiency for this operating point is about 75 percent for the prototype. Also note that the head coefficient is about 0.17. Therefore we have for the head ΔH of the prototype $(g\,\Delta H)/N^2D^2 = 0.17$, $\Delta H = (0.17/9.81)[1750(2\pi/60)]^2(1.2^2) = 838 \text{ m}$.

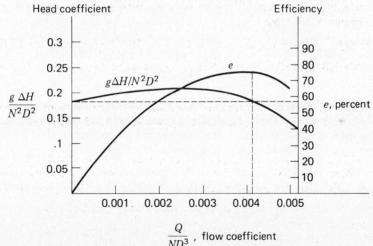

Fig. 20-14

20.64 A centrifugal blower compresses 1.5 lbm/s of air at a temperature of 60 °F and a pressure of 14.00 lb/in² abs to a pressure of 15.50 lb/in² abs. The impeller runs at a speed of 3550 rpm. The exit area of the blower casing is 0.15 ft² and the inlet area is 0.5 ft². What power is needed to run this blower if we have an efficiency of 80 percent for the blower? Neglect diffusion in the casing and consider that we do not have a diffuser. What torque is required?

▮
$$(dW_p/dt)_{\text{theo}} = \{[(V_2)^2_{\text{theo}} - V_1^2]/2 + c_p T_1[(p_2/p_1)^{(k-1)/k} - 1]\}\dot{m} \tag{1}$$

We need $(V_2)_{\text{theo}}$ and V_1.

$$p_1 = \rho_1 R T_1 \qquad \rho_1 = (14)(144)/[(53.3)(g)(520)] = 0.002259 \text{ slug/ft}^3 \qquad \rho_1/\rho_2 = (p_1/p_2)^{1/k}$$

$$\rho_2 = (0.002259)(15.5/14)^{1/1.4} = 0.002429 \text{ slug/ft}^3$$

$$\dot{m} = (1.5/g) = \rho_2(V_2)_{\text{theo}}A_2 \qquad 1.5/g = (0.002429)(V_2)_{\text{theo}}(0.15) \qquad (V_2)_{\text{theo}} = 127.85 \text{ ft/s}$$

Also $\dot{m} = 1.5/g = \rho_1 V_1 A_1 = (0.002259)(V_1)(0.5)$, $V_1 = 41.24 \text{ ft/s}$. Now go to Eq. (1):

$$(dW/dt)_{\text{theo}} = \{[(127.85^2 - 41.24^2)/2] + (0.24)(778)(g)(520)[(15.5/14)^{0.4/1.4} - 1]\}(1.5/g)/550 = 8.43 \text{ hp}$$

$$(dW/dt) = 8.43/0.80 = 10.54 \text{ hp} \qquad (T_s)[(3550)(2\pi)/60] = (10.54)(550) \qquad T_s = 15.6 \text{ ft-lb}$$

20.65 In Prob. 20.64, we found that it took 10.54 hp to do the job. If the impeller has an outside radius of 8 in, what should the blade angle β_2 be at exit? The width of the impeller is 1.5 in. The density ρ_2 found in the preceding solution is 0.002429 slug/ft³.

▌
$$(dW_p/dt) = (\omega/\eta)(r_2\dot{m})[(U_t)_2 + (V_r)_2 \cot \beta_2]$$

$$(10.54)(550) = [(3550)(2\pi)/(60)(0.80)](\tfrac{8}{12})(1.5/g)\{[(3550)(2\pi)/60](\tfrac{8}{12}) + (V_r)_2 \cot \beta_2\} \tag{1}$$

We need $(V_r)_2$. From *continuity*: $\dot{m} = \rho_2(V_r)_2(2\pi r_2)(b)$, $1.5/g = (0.002429)(V_r)_2(2\pi)(\tfrac{8}{12})(1.5/12)$, $(V_r)_2 = 36.63$ ft/s. Now go back to Eq. (1) to get β_2. $\beta_2 = 13.39°$.

20.66 An air compressor has a mass flow of 0.6 kg/s, taking the fluid from a pressure of 100 kPa at the inlet to a static pressure corresponding to 35 mm H₂O in a U-tube. The inlet temperature is 15 °C. The exit angle β_2 is 160 °. The inlet area is 7000 mm² and the exit area is 6000 mm². Neglect diffusion in the casing and consider that we do not have a diffuser. What is the speed of the compressor if the diameter of the rotor is 175 mm? The efficiency is 80 percent. The width of the impeller is 80 mm.

▌
$$dW_p/dm = [(V_2)^2_{theo} - V_1^2]/2 + c_p T_1[(p_2/p_1)^{(k-1)/k} - 1] \tag{1}$$

From *continuity*, $\rho_1 V_1 A_1 = \dot{m} = 0.6$ kg/s, $\rho_2 V_2 A_2 = 0.6$ kg/s, $V_2 = 0.6/[(1.2244)(6000 \times 10^{-6})] = 81.7$ m/s. Going back to Eq. (1)

$$(dW_p/dm)_{theo} = [(81.7^2 - 70.85^2)/2] + (0.24)(4187)(288)[(101\,668/100\,000)^{0.4/1.4} - 1] = 2199 \text{ N-m/kg}$$

$$(dW_p/dt) = (2199)(0.6)/0.80 = 1649 \text{ W} \quad \text{or} \quad 1.649 \text{ kW} \qquad T_s = r_2\dot{m}[(U_t)_2 + (V_{r_2})\cot \beta_2]$$

$$1649/\omega = (0.175/2)(0.6)[(\omega)(0.175/2) + (V_{r_2}) \cot 160°] \tag{2}$$

We need $(V_r)_2$ from *continuity*: $(1.2244)(V_{r_2})[(\pi)(0.175)](0.08) = 0.6$, $(V_r)_2 = 11.14$ m/s. Going back to Eq. (2), $\omega^2 - 349.8\omega - 3.590 \times 10^5 = 0$, $\omega = [349.8 \pm \sqrt{349.8^2 + (4)(3.590 \times 10^5)(1)}]/(2)(1) = 799$ rad/s = 7630 rpm. Also $\rho_1 = p_1/RT_1 = 100\,000/[(287)(273 + 15)] = 1.210$ kg/m³, $V_1 = 0.6/[(1.210)(7000 \times 10^{-6})] = 70.84$ m/s. Also $p_2 = 101\,325 + (0.035)(9806) = 101\,668$ N/m², $\rho_1/\rho_2 = (p_1/p_2)^{1/k} = (100\,000/101\,668)^{1/k} = 0.988$, $\rho_2 = (1.210)/(0.988) = 1.225$ kg/m³.

20.67 A performance chart is shown in Fig. 20-15 for a 3550-rpm blower from a bulletin of the Buffalo Forge Co. For a 21-in wheel having a flow of 3500 ft³/min, what is the efficiency and the required torque? The exit diameter is 8 in and the entrance diameter is 16 in. At the entrance $p = 14.7$ lb/in² and $T = 70$ °F. The capacity is given at the exit pressure. *Note:* The upper curves apply to the left ordinate and the lower curves apply to the right ordinate.

▌ From the chart:
$$\begin{cases} hp = 24 \\ (\Delta p)_{static} = 19 \text{ inH}_2\text{O} = (19)(62.4)/12 = 98.8 \text{ psf} \end{cases}$$

$$p_2 = (14.7)(144) + 98.8 = 2216 \text{ psf} \qquad \dot{m} = \rho_2 Q = [(0.002378)(\tfrac{2216}{2117})^{1/k}](\tfrac{3500}{60}) = 0.1433 \text{ slug/s}$$

$$V_2 = Q/A_2 = (\tfrac{3500}{60})/[\pi(\tfrac{8}{12})^2/4] = 167.1 \text{ ft/s}$$

Now get V_1.

$$\rho_1 V_1 A_1 = \dot{m} = 0.1433 \text{ slug/s} \qquad (0.002378)(V_1)(\pi)(\tfrac{16}{12})^2/4 = 0.1433 \qquad V_1 = 43.16 \text{ ft/s}$$

$$dW_p/dm = [(V_2^2/2) - (V_1^2/2)] + c_p T_1[(p_2/p_1)^{(k-1)/k} - 1]$$

$$dW_p/dt = (dW_p/dm)\dot{m} = (0.1433)\{[(167.1^2/2) - (43.16^2/2)]$$
$$+ (0.24)(778)(g)(460 + 70)[(\tfrac{2216}{2117})^{(1.4-1)/1.4} - 1]\}/550 = 14.31 \text{ hp}$$

$$e = [(24 - 14.31)/24](100) = 40.4\% \qquad T = (14.31)(550)/[3550(2\pi/60)] = 21.17 \text{ ft-lb}$$

20.68 A centrifugal pump takes 200 gal/min of water from a pressure of 13 lb/in² to a pressure p_2. The inside radius of the impeller is 3 in and the outside radius of the impeller is 0.8 ft. If the output of the pump is 26.5 hp running at a speed of 1750 rpm, what should the blade angle β_2 be? The efficiency is 80 percent. The width b of the vanes is 2 in. If the exiting fluid is at the same elevation as the incoming fluid, what is p_2? Neglect diffusion in casing. *Hint:* From $(V_r)_2$ get $(V_{rel})_2$. Then use the law of cosines to get V_2.

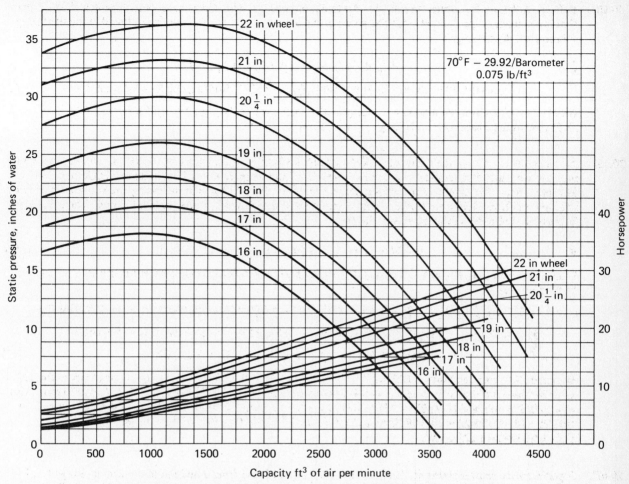

Fig. 20-15 (*Courtesy Buffalo Forge Company. Reprinted with permission.*)

$$T_s = \text{power}/\omega = r_2 \dot{m}[(U_t)_2 + (V_r)_2 \cot \beta_2] \qquad (1)$$

$$(1/0.80)\{(26.5)(550)/[(1750)(2\pi)/60]\} = (0.8)[(200)(231)/(1728)(60)](1.938)([(1750)(2\pi)/60](0.8)$$
$$+ \{[(200)(231)/(1728)(60)]/[(2\pi)(0.8)(\tfrac{2}{12})]\} \cot \beta_2)$$

$$\beta_2 = 168.7°$$

$$(dW_p/dm)_{\text{theo}} = (dW_p/dt)/\dot{m} = [(V_2^2)_{\text{theo}}/2 - V_1^2/2] + (p_2 - p_1)/\rho \qquad (2)$$
$$V_1 = [(200)(231)/(1728)(60)]/[(2\pi)(\tfrac{3}{12})(\tfrac{2}{12})] = 1.702 \text{ ft/s}$$

$$(V_2)_{\text{rel}} \sin(\pi - \beta_2) = (V_r)_2 \qquad (V_2)_{\text{rel}} = \{[(200)(231)/(1728)(60)]/[2\pi(0.8)(\tfrac{2}{12})]\}/[\sin(180° - 168.7°)] = 2.715 \text{ ft/s}$$

From law of cosines:

$$V_2^2 = (U_t)_2^2 + (V_{\text{rel}})_2^2 - 2(U_t)(V_{\text{rel}})_2 \cos(\pi - \beta_2)$$
$$= \{[(1750)(2\pi)/60](0.8)\}^2 + 2.715^2 - (2)\{[(1750)(2\pi)/60](0.8)\}(2.715 \cos 11.3°) = 20.721$$

Now go back to Eq. (2).

$$(1/0.80)(26.5)(550)/\{[(200)(231)/(1728)(60)](1.938)\} = [(20\,721/2) - (1.702^2/2)] + \Delta p/1.938$$

$$\Delta p = 20\,810 \text{ psf} = 144.5 \text{ psi} \qquad p_2 = 144.5 + 13 = 157.5 \text{ psia}$$

20.69 A centrifugal pump moves 10.0 L of water per second from a pressure of 100 kPa to a pressure of 550 kPa in one stage. The section inlet has a diameter of 50 mm and the discharge has a diameter of 40 mm. The impeller diameter is 0.4 m and β_2 is 150°. The width b of the impeller is 30 mm. What power is required if there is an efficiency of 65 percent? What speed should the pump run at for the above condition? The discharge pipe is 0.3 m above the inlet pipe. Use theoretical torque in computing ω.

∎ $(dW_p/dt)_{theo} = \dot{m}(\{[(V_2^2)_{theo} - V_1^2]/2\} + (p_2 - p_1)/\rho + g(z_2 - z_1))$ $\dot{m} = (0.0100)(1000) = 10$ kg/s

$V_1 = Q/A_1 = 0.0100/[(\pi)(0.05)^2/4] = 5.093$ m/s $V_2 = Q/A_2 = 0.01/[(\pi)(0.04)^2/4] = 7.958$ m/s

$dW_p/dt = (10)\{[(7.958^2 - 5.093^2)/2] + [(550 - 100)/1000] \times 10^3 + 9.81(0.3)\} = 4716$ N-m/s $= 4.716$ kW

$P = 4.716/0.65 = 7.26$ kW $T_s = r_2\dot{m}[(U_t)_2 + (V_r)_2 \cot \beta_2]$

$(V_r)_2 = Q/(2\pi r)(b) = 0.01/[(2\pi)(0.4/2)(0.03)] = 0.265$ m/s

$(U_t)_2 = (\omega)(0.2)$ $(T_2)_{theo} = \frac{4716}{60}$ $4716/\omega = (0.2)(0.01)(1000)[0.2\omega + (0.265) \cot 150°]$

$\omega^2 - 2.295\omega - 11\,790 = 0$ $\omega = [2.295 \pm \sqrt{(-2.295)^2 + (4)(11\,790)(1)}]/[(2)(1)] = 109.7$ rad/s $= 1048$ rpm

20.70 Shown in Fig. 20–16 are performance curves as found in a Buffalo Forge Co. bulletin for boiler feed pumps. It describes the operating characteristics of various multistage centrifugal pumps. The number designations 2×3, 4×4, etc., give the exit and the inlet diameters, respectively. For a 2×3 (four-stage) pump operating at 30 hp with a flow Q of 90 gal/min of water, compute the efficiency. What is the total required torque on the four rotors?

∎ $\Delta H(\text{output}) = 700$ ft $\Delta H(\text{input}) = [(\text{hp})(550)/Q\gamma] = (30)(550)/\{[(90)(231)/(1728)(60)]62.4\} = 1319$ ft

$\eta = \frac{700}{1319}(100) = 53$ percent   Torque $= (30)(550)/[(3500)(2\pi)/60] = 45$ ft-lb

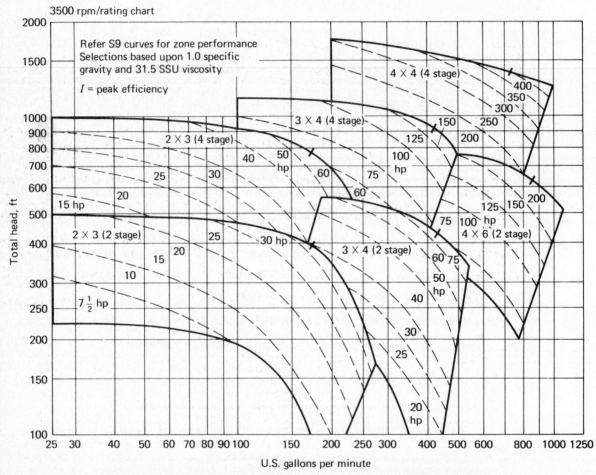

Fig. 20-16 (*Courtesy Buffalo Forge Company. Reprinted with permission.*)

20.71 Using the Buffalo Forge rating chart in Prob. 20.70, what centrifugal pump would you choose to move 100 gal/min of water from tank A to tank B in Fig. 20-17? The inlet and outlet pipes are steel pipes. Neglect head loss for the short inlet piping. Water is at 60 °F.

∎ Use the *first law* for pump and 2-in outlet pipe.

$$(1/g)(dW_p/dm) + (V_1^2/2g) + (p_1/\gamma) + 0 = (V_2^2/2g) + (p_2/\gamma) + 400 + [(h_L)/g] \tag{1}$$

For p_1 use *Bernoulli* between free surface of tank A and pump inlet. Use *gage* pressures.

$$(V_{fs}^2/2g) + (p_{atm}/\gamma) + 1 = (V_1^2/2g) + (p_1/\gamma) \qquad (2)$$

$$V_1 = [(100)(231)/(1728)(60)]/[(\pi)(3^2)/(4)(144)] = 4.539 \text{ ft/s} \qquad p_1/\gamma = 1 - (4.539^2/2g) = 0.680 \text{ ft}$$

$$V_2 = [(100)(231)/(1728)(60)]/[(\pi)(2^2)/(4)(144)] = 10.21 \text{ ft/s} \qquad N_R = (10.21)(\tfrac{2}{12})/(1.217 \times 10^{-5}) = 1.398 \times 10^5$$

$$\epsilon/D = 0.00015/(\tfrac{2}{12}) = 0.0009 \qquad f = 0.021 \qquad h_L = f(L/d)(V^2/2) = (0.021)(400/\tfrac{1}{6})(10.21^2/2) = 2627 \text{ ft}$$

Go back to Eq. (1).

$$(1/g)(dW_p/dm) + (4.539^2/2g) + 0.680 = (10.21^2/2g) + [(\gamma)(10)/\gamma] + 400 + (2627/g) = 492 \text{ ft}$$

Use 2×3(four-stage) 25-hp pump.

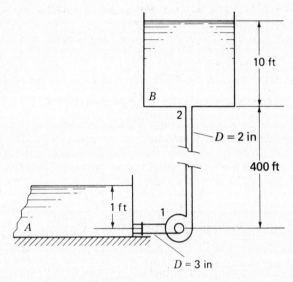

Fig. 20-17

20.72 A pump impeller, 12 in in diameter, discharges 5.25 cfs when running at 1200 rpm. The blade angle β_2 is $160°$ and the exit area A_2 is 0.25 ft². Assuming losses of $2.8(v_2^2/2g)$ and $0.38(V_2^2/2g)$, compute the efficiency of the pump (exit area A_2 is measured normal to v_2).

❚ The absolute and relative velocities at exit must be calculated first. Velocities u_2 and v_2 are $u_2 = r_2\omega = (\tfrac{6}{12})(2\pi)(\tfrac{1200}{60}) = 62.8 \text{ ft/s}$, $v_2 = Q/A_2 = 5.25/0.25 = 21.0 \text{ ft/s}$.

From the vector diagram shown in Fig. 20-18, the value of the absolute velocity at exit is $V_2 = 43.7 \text{ ft/s}$. Head furnished by impeller, $H' = (u_2^2/2g) - (v_2^2/2g) + (V_2^2/2g) = [(62.8)^2/2g] - [(21.0)^2/2g] + [(43.7)^2/2g] = 84.0 \text{ ft}$. Head delivered to water, $H = H' - \text{losses} = 84.0 - \{2.8[(21.0)^2/2g] + 0.38[(43.7)^2/2g]\} = 53.6 \text{ ft}$.

Efficiency $e = H/H' = 53.6/84.0 = 63.8$ percent. The value of H' might have been calculated by means of the commonly used expression $H' = (u_2/g)(u_2 + v_2 \cos \beta_2) = (62.8/g)[62.8 + 21.0(-0.940)] = 84.0 \text{ ft}$.

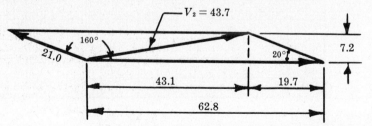

Fig. 20-18

20.73 A 6-in pump delivers 1300 gpm against a head of 90 ft when rotating at 1750 rpm. The head-discharge and efficiency curves are shown in Fig. 20-19. For a geometrically similar 8-in pump running at 1450 rpm and delivering 1800 gpm, determine (a) the probable head developed by the 8-in pump. (b) Assuming a similar efficiency curve for the 8-in pump, what power would be required to drive the pump at the 1800-gpm rate?

❚ (a) The homologous pumps will have identical characteristics at corresponding flows. Choose several rates of flow for the 6-in pump and read off the corresponding heads. Calculate the values of Q and H so that a curve for the 8-in pump can be plotted. One such calculation is detailed below and a table of values established by similar determinations.

Using the given 1300 gpm and the 90-ft head, we obtain from the speed relation $H_8 = (D_8/D_6)^2(N_8/N_6)^2 H_6 = (\frac{8}{6})^2(\frac{1450}{1750})^2 H_6 = 1.22 H_6 = 1.22(90) = 109.8$ ft. From the flow relation, $Q/D^3 N =$ constant, we obtain $Q_8 = (D_8/D_6)^3(N_8/N_6)Q_6 = (\frac{8}{6})^3(\frac{1450}{1750})Q_6 = 1.964 Q_6 = 1.964(1300) = 2550$ gpm.

Additional values, which have been plotted as the dashed line in Fig. 20-19, are as follows:

For 6-in pump at 1750 rpm			For 8-in pump at 1450 rpm		
Q, gpm	H, ft	Efficiency, %	Q, gpm	H, ft	Efficiency, %
0	124	0	0	151.6	0
500	119	54	980	145.5	54
800	112	64	1570	134.5	64
1000	104	68	1960	127.0	68
1300	90	70	2550	110.0	70
1600	66	67	3140	80.6	67

From the head-discharge curve, for $Q = 1800$ gpm the head is 130 ft.

(b) The efficiency of the 8-in pump would probably be somewhat higher than that of the 6-in pump at comparable rates of flow. For this case, the assumption is that the efficiency curves are the same at comparable rates of flow. The table above lists the values for the flows indicated. The figure gives the efficiency curve for the 8-in pump and, for the 1800 gpm flow, the value is 67 percent. Then $P = \gamma QH/550e = 62.4[1800/(60)(7.48)](130)/[550(0.67)] = 88.3$ horsepower required.

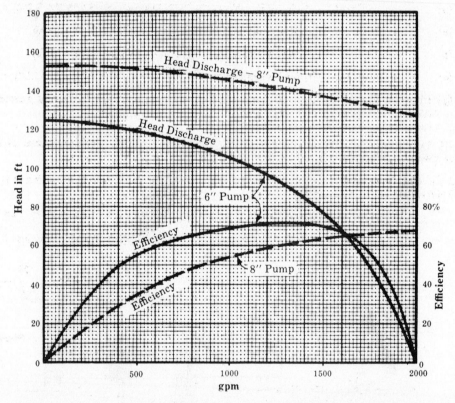

Fig. 20-19

20.74 In order to predict the behavior of a small oil pump, tests are to be made on a model using air. The oil pump is to be driven by a $\frac{1}{20}$-hp motor at 1800 rpm and a $\frac{1}{4}$-hp motor is available to drive the air pump at 600 rpm. Using the s.g. of oil at 0.912 and the density of air constant at 0.00238 slug/ft^3, what size model should be built?

▌ Using the power relation, we obtain prototype $[P/(\rho D^5 N^3)] = \text{model}[P/(\rho D^5 N^3)]$. Then $\frac{1}{20}/[0.912(62.4/32.2)D_p^5(1800)^3] = \frac{1}{4}/[0.00238 D_m^5(600)^3]$ and $D_m/D_p = \frac{10}{1}$. The model should be 10 times as large as the oil pump.

20.75 What is the power ratio of a pump and its $\frac{1}{5}$ scale model if the ratio of the heads is 4 to 1?

❚ For geometrically similar pumps, $P/(D^2H^{3/2})$ for pump = $P/(D^2H^{3/2})$ for model. Then $P_p/[(5D)^2(4H)^{3/2}] = P_m/(D^2H^{3/2})$ and $P_p = 25(4)^{3/2}P_m = 200P_m$.

20.76 Determine the thrust coefficient of a propeller which is 4 in in diameter, revolves at 1800 rpm, and develops a thrust of 2.50 lb in fresh water.

❚ Thrust coefficient = $F/(\rho N^2 D^4) = 2.50/[1.94(\frac{1800}{60})^2(\frac{4}{12})^4] = 0.116$. The coefficient is dimensionless when F is in pounds, N in revolutions/per second, and D in feet.

20.77 The power and thrust coefficients of an 8-ft-diameter propeller moving forward at 100 ft/s at a rotational speed of 2400 rpm are 0.068 and 0.095, respectively. **(a)** Determine the power requirement and thrust in air $(\rho = 0.00237 \text{ slug/ft}^3)$. **(b)** If the advance-diameter ratio for maximum efficiency is 0.70, what is the air speed for the maximum efficiency?

❚ **(a)** Power $P = C_P\rho N^3 D^5$ in ft-lb/s = $[0.068(0.00237)(\frac{2400}{60})^3(8)^5]/550 = 615$ horsepower.

Thrust $F = C_F\rho N^2 D^4$ in lb = $0.095(0.00237)(\frac{2400}{60})^2(8)^4 = 1476$ lb.

(b) Since $V/ND = 0.70$, $V = 0.70(\frac{2400}{60})(8) = 224$ ft/s.

20.78 An airplane flies at 180 mph in still air, $\gamma = 0.0750$ lb/ft^3. The propeller is 5.5 ft in diameter and the velocity of the air through the propeller is 320 ft/s. Determine **(a)** the slipstream velocity, **(b)** the thrust, **(c)** the horsepower input, **(d)** the horsepower output, **(e)** the efficiency, and **(f)** the pressure difference across the propeller.

❚ **(a)** $V = \frac{1}{2}(V_1 + V_4)$ $320 = \frac{1}{2}[180(\frac{5280}{3600}) + V_4]$ $V_4 = 376$ ft/s   (relative to fuselage)

(b)   Thrust $F = (\gamma/g)Q(V_4 - V_1) = (0.0750/32.2)[\frac{1}{4}\pi(5.5)^2(320)](376 - 264) = 1983$ lb

(c)   Power input $P_i = FV/550 = 1983(320)/550 = 1154$ hp

(d)   Power output $P_0 = FV_1/550 = 1983(264)/550 = 952$ hp

(e)   Efficiency $e = \frac{952}{1154} = 82.5$ percent or $e = 2V_1/(V_4 + V_1) = 2(264)/(376 + 264) = 82.5$ percent.

(f) Pressure difference = (thrust F)/[area $(\frac{1}{4}\pi D^2)$] = $1983/[\frac{1}{4}\pi(5.5)^2] = 83.5$ psf or pressure difference = $(0.0750/32.2)\{[(376)^2 - (264)^2]/2\} = 83.5$ psf.

20.79 Water is being pumped from the lower reservoir to the upper reservoir, as shown in Fig. 20-20. Find the pump horsepower required, assuming 75 percent pump efficiency. Neglect minor losses.

❚ $p_1/\gamma + V_1^2/2g + z_1 + E_p = p_2/\gamma + V_2^2/2g + z_2 + h_L$. From Fig. A-13, with $Q = 23$ cfs and $D = 24$ in, $h_1 = 0.0076$ ft/ft.

$$h_L = h_f = (0.0076)(3600) = 27.36 \text{ ft} \quad\quad 0 + 0 + 97.5 + E_p = 0 + 0 + 132.0 + 27.36$$

$$E_p = 61.86 \text{ ft} \quad\quad P = Q\gamma E_p/550 = (23)(62.4)(61.86)/550 = 161 \text{ hp}$$

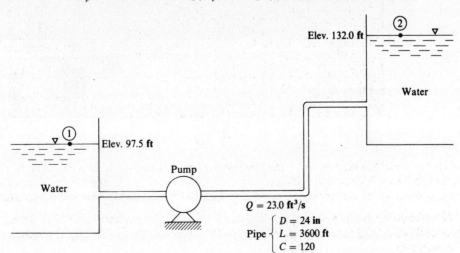

Elev. 132.0 ft

Water

Elev. 97.5 ft

Pump

Water

$Q = 23.0$ ft^3/s

Pipe $\begin{cases} D = 24 \text{ in} \\ L = 3600 \text{ ft} \\ C = 120 \end{cases}$

Fig. 20-20

20.80 Water is being pumped from a reservoir to the top of a hill, where it is discharged, as shown in Fig. 20-21. The pump, which is 70 percent efficient, is rated at 150 kW. Find the flow rate at which water is being discharged from the pipe. Neglect minor losses.

$$p_1/\gamma + V_1^2/2g + z_1 + E_p = p_2/\gamma + V_2^2/2g + z_2 + h_L \qquad E_p = P/Q\gamma = (150)(0.70)/[(Q)(9.79)] = 10.725/Q$$

$$V = Q/A \qquad V_2 = Q/[(\pi)(\tfrac{500}{1000})^2/4] = 5.093Q \qquad h_L = h_f = (f)(L/D)(V^2/2g)$$

Assume $f = 0.018$. $h_L = (0.018)[975/(\tfrac{500}{1000})]\{(5.093Q)^2/[(2)(9.807)]\} = 46.42Q^2$, $0 + 0 + 111.0 + 10.725/Q = 0 + (5.093Q)^2/[(2)(9.807)] + 150.2 + 46.42Q^2$, $Q = 0.254$ m³/s (by trial and error). This solution was based on the assumed value of f of 0.018. However, the value of f is dependent on the Reynolds number and relative roughness. Therefore a new value of f should be determined based on the computed value of Q of 0.254 m³/s.

$$\epsilon/D = 0.00030/(\tfrac{500}{1000}) = 0.00060 \qquad V = Q/A = 0.254/[(\pi)(\tfrac{500}{1000})^2/4] = 1.294 \text{ m/s}$$

$$N_R = DV/\nu = (\tfrac{500}{1000})(1.294)/(1.02 \times 10^{-6}) = 6.34 \times 10^5$$

From Fig. A-5, $f = 0.018$. Hence, the computed flow rate of 0.254 m³/s is correct.

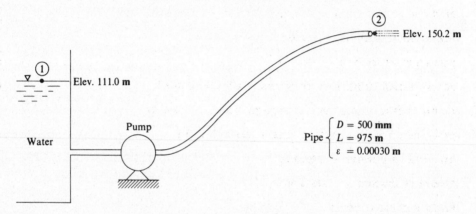

Fig. 20-21

20.81 Oil with a specific gravity of 0.87 is being pumped from a lower reservoir to an elevated tank as shown in Fig. 20–22. The pump in the system is 78 percent efficient and is rated at 185 kW. Determine the flow rate of the oil in the pipe if the total head loss from point 1 to point 2 is 12 m of oil.

$$p_1/\gamma + v_1^2/2g + z_1 + E_p = p_2/\gamma + v_2^2/2g + z_2 + h_L \qquad p_1/\gamma = p_2/\gamma = v_1^2/2g = 0 \qquad z_1 = 150 \text{ m}$$

$$p = Q\gamma E_p \qquad (0.78)(185) = (Q)[(0.87)(9.79)](E_p) \qquad E_p = 16.94/Q$$

$$v_2^2/2g = (Q/A)^2/2g = \{Q/[(\pi)(0.160)^2/4]\}^2/[(2)(9.807)] = 126.12Q^2$$

$z_2 = 200$ m $\qquad h_L = 12$ m $\qquad 0 + 0 + 150 + 16.94/Q = 0 + 126.12Q^2 + 200 + 12 \qquad 126.12Q^2 - 16.94/Q + 62 = 0$

By trial and error solution, $Q = 0.244$ m³/s.

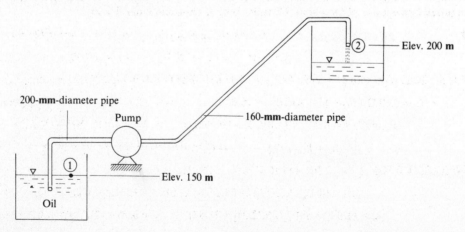

Fig. 20-22

20.82 For the conditions given in Prob. 20-80, find the flow rate at which water is being discharged from the pipe using the computer program written for Prob. 9.295.

I INPUT

```
( 2 ) 4 5 6 7 8 9 10 11 12 13 14 15 16 17 18 19 20 21 22 23 24 25 26 27 28 29 30 31 32 33 34 35 36 37 38 39 40 41 42 43 44 45 46 47 48 49 50 51 52 53 54 55 56 57 58 59 60 61 62 63 64 65 66 67 68 69 70 71 72 73 74 75 76 77 78 79 80
1SAMPLE ANALYSIS ØF INCØMPRESSIBLE FLØW
                            1.0        111.0       150.2      105.0
         500.0      975.0    .00000102 .00030     9.79
WATER                      (RØUGHNESS = 0.300 MM)
```

SAMPLE ANALYSIS OF INCOMPRESSIBLE FLOW

I OUTPUT

GIVEN DATA FOR A CIRCULAR CLOSED CONDUIT CARRYING INCOMPRESSIBLE FLOW

PRESSURE AT POINT 1 =   0.0 KPA

PRESSURE AT POINT 2 =   0.0 KPA

ELEVATION AT POINT 1 =   111.0 M

ELEVATION AT POINT 2 =   150.2 M

ACTUAL ENERGY ADDED BETWEEN POINTS 1 AND 2 =105.0 KW

ACTUAL ENERGY REMOVED BETWEEN POINTS 1 AND 2 =   0.0 KW

MINOR HEAD LOSSES BETWEEN POINTS 1 AND 2 =   0.0 M

DIAMETER OF CONDUIT =   500.0 MM

LENGTH OF CONDUIT =   975.0 M

FLUID FLOWING IS WATER

CONDUIT MATERIAL IS (ROUGHNESS = 0.300 MM)

THE FLOW RATE WILL BE   0.254 CU M/S

VELOCITY AT POINT 1 =   0.00 M/S

VELOCITY AT POINT 2 =   1.29 M/S

20.83 A pump draws water from a reservoir and then discharges it to an elevated tank, as shown in Fig. 20-23. The pipe's ends are squared-cornered (i.e., sharp-edged). There are two 90° bends as shown. The pipe roughness (ϵ) is 0.0084 ft and C is 120. If the rate of flow is 8.0 ft³/s and the efficiency of the pump is 75 percent, determine the required horsepower of the pump. Consider both friction and minor losses.

I
$$p_1/\gamma + v_1^2/2g + z_1 + E_p = p_2/\gamma + v_2^2/2g + z_2 + h_L \qquad p_1/\gamma = p_2/\gamma = v_1^2/2g = v_2^2/2g = 0$$

$$z_1 = 10\text{ ft} \qquad z_2 = 120\text{ ft} \qquad h_L = h_f + h_m$$

For the 12-in pipe with $Q = 8$ cfs, $(h_1)_{12} = 0.032$ ft/ft. For the 10-in pipe with $Q = 8$ cfs, $(h_1)_{10} = 0.074$ ft/ft.

$$h_f = (0.032)(40) + (0.074)(200) = 16.08\text{ ft} \qquad h_m = h_c + 2h_b + h_e \qquad h_c = K_c v_{10}^2/2g \qquad K_c = 0.45$$

$$v_{12} = Q/A = 8.0/[(\pi)(1.0)^2/4] = 10.19\text{ ft/s} \qquad h_c = (0.45)(10.19)^2/[(2)(32.2)] = 0.73\text{ ft}$$

$$h_b = K_b v_{10}^2/2g \qquad v_{10} = Q/A = 8.0/[(\pi)(0.833)^2/4] = 14.68\text{ ft/s}$$

With $\epsilon/D = 0.0084/(\frac{10}{12}) = 0.101$ and $R/D = \frac{20}{10} = 2.0$, $K_b = 0.6$.

$$h_b = (0.6)(14.68)^2/[(2)(32.2)] = 2.01\text{ ft} \qquad h_e = K_e v_{10}^2/2g \qquad K_e = 1.0$$

$$h_e = (1.0)(14.68)^2/[(2)(32.2)] = 3.35\text{ ft} \qquad h_m = 0.73 + (2)(2.01) + 3.35 = 8.10\text{ ft}$$

$$h_L = h_f + h_m = 16.08 + 8.10 = 24.18\text{ ft} \qquad 0 + 0 + 10 + E_p = 0 + 0 + 120 + 24.18 \qquad E_p = 134.18\text{ ft}$$

$$P = Q\gamma E_p = (8.0)(62.4)(134.18) = 66\,983\text{ ft-lb/s} \qquad \text{Required horsepower} = 66\,983/[(550)(0.75)] = 162\text{ hp}$$

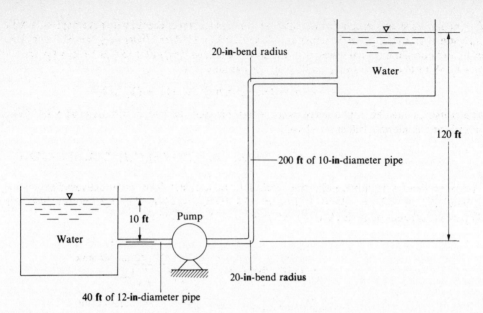

20-in-bend radius

Water

120 ft

200 ft of 10-in-diameter pipe

10 ft   Pump

Water

20-in-bend radius

40 ft of 12-in-diameter pipe

Fig. 20-23

20.84   Oil with a specific gravity of 0.86 is being pumped from a reservoir as shown in Fig. 20-24. The pressures at
points 1 and 2 are −4.0 psi and 43.0 psi, respectively. The rate of flow in the pipe is 0.50 ft³/s. The pump is rated
at 8 hp. Determine the efficiency of the pump. Neglect energy losses in the system.

▮ $p_1/\gamma + v_1^2/2g + z_1 + E_p = p_2/\gamma + v_2^2/2g + z_2 + h_L$ $p_1/\gamma = (-4)(144)/[(0.86)(62.4)] = -10.73$ ft

$$v_1^2/2g = v_2^2/2g \qquad z_1 = h_L = 0 \qquad p_2/\gamma = (43)(144)/[(0.86)(62.4)] = 115.38 \text{ ft}$$

$$z_2 = 3 \text{ ft} \qquad -10.73 + E_p = 115.38 + 3 \qquad E_p = 129.11 \text{ ft}$$

$$P = Q\gamma E_p = (0.50)[(0.86)(62.4)](129.11) = 3464 \text{ ft-lb/s} = \tfrac{3464}{550} = 6.30 \text{ hp}$$

Efficiency of pump = (6.30/8)(100) = 78.8 percent

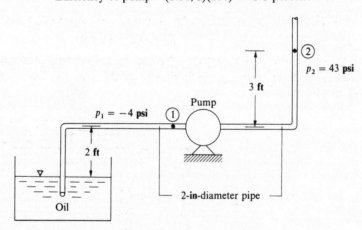

②
p_2 = 43 psi

3 ft

Pump

p_1 = −4 psi   ①

2 ft

2-in-diameter pipe

Oil

Fig. 20-24

20.85   A pump circulates water at the rate of 2000 gpm in a closed circuit holding 10 000 gal. The net head developed
by the pump is 300 ft and the pump efficiency is 90 percent. Assuming the bearing friction to be negligible and
that there is no loss of heat from the system, find the temperature rise in the water in 1 h.

▮ $P = Q\gamma h/\eta$ $Q = (2000)(0.002228) = 4.456 \text{ ft}^3/\text{s}$ $P_{\text{input}} = (4.456)(62.4)(300)/0.90 = 92\,685 \text{ ft-lb/s}$

Heat input in 1 h = (92 685)(3600) = 3.337×10^8 ft-lb = $mc\,\Delta T$

Mass heated = m = (10 000/7.48)(1.94) = 2594 slugs

c = 25 000 ft-lb/(slug-°R)   (2594)(25 000)(ΔT) = 3.337×10^8 ft-lb ΔT = 5.15 R° = 5.15 F°

20.86   The pump in Fig. 20-25 delivers water at 3 ft³/s to a machine at section 2, which is 20 ft higher than the reservoir
surface. The losses between 1 and 2 are given by $h_f = KV_2^2/2g$, where $K = 7.5$ is a dimensionless loss coefficient.
Find the power required to drive the pump if it is 80 percent efficient.

❙ The flow is steady except for the slow decrease in reservoir depth, which we neglect, taking $V_1 \approx 0$. We can compute V_2 from the given flow rate and diameter: $V_2 = Q/A_2 = 3/[(\pi/4)(\frac{3}{12})^2] = 61.1$ ft/s. Because of the solid walls and one-dimensional ports, the viscous work is zero. $(p_1/\rho g) + (V_1^2/2g) + z_1 = (p_2/\rho g) + (V_2^2/2g) + z_2 + h_s + h_f$. With $V_1 = z_1 = 0$ and $h_f = KV_2^2/2g$, we can solve for

$$h_s = [(p_1 - p_2)/\rho g] - z_2 - (1 + K)(V_2^2/2g) \qquad (1)$$

The pressures must be in pounds per square foot for consistent units, with $\rho g = 62.4$ lb/ft^3 for water. Introducing numerical values, we have

$$h_s = [(14.7 - 10)(144)/62.4] - 20 - [(1 + 7.5)(61.1)^2/2(32.2)] = -502 \text{ ft} \qquad (2)$$

The pump head is negative, indicating work done on the fluid. The power delivered is $P = \dot{m}w_s = \rho Qgh_s = (1.94)(3)(32.2)(-502) = -94\,100$ ft-lb/s, hp $= 94\,100/550 = 171$ hp. The input power thus required to drive the 80 percent efficient pump is $P_{input} = P/\text{efficiency} = 171/0.8 = 214$ hp.

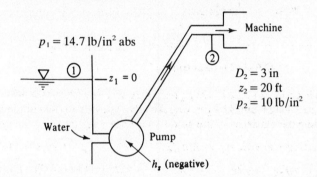

$p_1 = 14.7$ lb/in^2 abs

$z_1 = 0$

Water

Pump

h_s (negative)

Machine

$D_2 = 3$ in
$z_2 = 20$ ft
$p_2 = 10$ lb/in^2

Fig. 20-25

20.87 The pump in Fig. 20-26 draws water from a reservoir through a 12-cm-diameter pipe and discharges it at high velocity through a 5-cm-diameter nozzle. Total friction head loss is 7 m. If the pump delivers 35 kW of power to the water, what are the exit velocity V_e and the flow rate Q?

❙ $p_1/\gamma + V_1^2/2g + z_1 + E_p = p_2/\gamma + V_2^2/2g + z_2 + h_L \qquad 0 + 0 + 0 + E_p = 0 + V_e^2/[(2)(9.807)] + 2 + 7$

$$E_p = 0.05098V_e^2 + 9$$

Also, $P = Q\gamma E_p$, $Q = AV = [(\pi)(\frac{5}{100})^2/4](V_e) = 0.001964V_e$, $(35)(1000) = (0.001964V_e)[(9.79)(1000)](E_p)$, $E_p = 1820/V_e$. Equating expressions for E_p, $0.05098V_e^2 + 9 = 1820/V_e$, $V_e = 31.1$ m/s (by trial and error); $Q = (0.001964)(31.1) = 0.0611$ m^3/s.

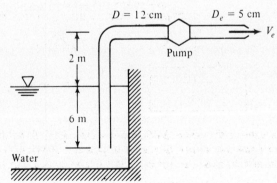

$D = 12$ cm $D_e = 5$ cm

V_e

2 m

Pump

6 m

Water

Fig. 20-26

20.88 A fireboat pump draws sea water (s.g. = 1.025) from a 6-in submerged pipe and discharges it as 120 fps through a 2-in nozzle, as in Fig. 20-27. Total head loss is 8 ft. If the pump is 70 percent efficient, how much horsepower is required to drive it?

❙ $\qquad P = Q\gamma E_p/\eta \qquad Q = AV = [(\pi)(\frac{2}{12})^2/4](120) = 2.618$ ft^3/s

$p_1/\gamma + V_1^2/2g + z_1 + E_p = p_2/\gamma + V_2^2/2g + z_2 + h_L \qquad 0 + 0 + 0 + E_p = 0 + 120^2/[(2)(32.2)] + 10 + 8$

$E_p = 241.6$ ft $P = \{(2.618)[(1.025)(62.4)](241.6)/550\}/0.70 = 105$ hp

20.89 The pump in Fig. 20-28 discharges water at 0.02 m³/s. Negelcting losses and elevation changes, what power is delivered to the water by the pump?

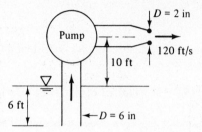

Fig. 20-27

▌ $p_1/\gamma + V_1^2/2g + z_1 + E_p = p_2/\gamma + V_2^2/2g + z_2 + h_L$ $V_1 = Q/A_1 = 0.02/[(\pi)(\frac{9}{100})^2/4] = 3.144$ m/s

$$V_2 = 0.02/[(\pi)(\frac{3}{100})^2/4] = 28.29 \text{ m/s}$$

$$120/9.79 + 3.144^2/[(2)(9.807)] + 0 + E_p = 400/9.79 + 28.29^2/[(2)(9.807)] + 0 + 0$$

$$E_p = 68.90 \text{ m} \qquad P = Q\gamma E_p = (0.02)(9.79)(68.90) = 13.5 \text{ kW}$$

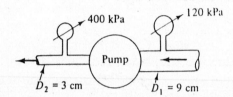

Fig. 20-28

20.90 What is the flow for the system shown in Fig. 20-29a? The pump has the characteristics shown in Fig. 20-29b. What is the power required?

▌ Express the first law for thermodynamics for pipe interior as shown:

$$(p_1/\rho) + (V_1^2/2) + gz_1 + gH = (p_2/\rho) + (V_2^2/2) + gz_2 + h \tag{1}$$

Bernoulii in left tank: Use gage pressures, $[(35)(144)/1.94] + 30g = (p_1/\rho) + (V^2/2) + 0$:

$$p_1/\rho = 3564 - V^2/2 \tag{2}$$

Hydrostatics in right-hand tank:

$$p_2/\rho = (1/\rho)[(10)(144) + 10.3(62.4)] = 1074 \tag{3}$$

Substitute Eqs. (2) and (3) into Eq. (1):

$$[3564 - (V^2/2)] + gH = 1074 + (g)(40) + f(1040/2.00)(V^2/2) + 2.3(V^2/2) \tag{4}$$

Assume $Q_1 = 80$ cfs. Then $V_1 = 80/[(\pi/4)(2.00)^2] = 25.5$ ft/s, $(N_R)_1 = [(25.5)(2.00)/(0.1217 \times 10^{-4})] = 4.19 \times 10^6$, $f_1 = 0.0118$. Solve for H_1 in Eq. (4): $H_1 = 57.9$ ft. The point (1) for H_1 and Q_1 is above Q curve. We take as a second estimate $Q_2 = 70$, $V_2 = 70/[(\pi/4)(2)^2] = 22.28$ ft/s, $(N_R)_2 = (22.28)(2)/(0.1217 \times 10^{-4}) = 3.66 \times 10^6$, $f_2 = 0.0118$, $H_2 = 35.4$ ft. New point is just above Q line. Third estimate can now be easily made: $Q_3 = 76$ cfs, $V_3 = 76/[(\pi/4)(2^2)] = 24.19$ ft/s, $(N_R)_3 = (24.19)(2)/(0.1217 \times 10^{-4}) = 3.98 \times 10^6$, $f_3 = 0.0118$, $H_3 = 48.4$ ft. We are close enough to the intersection so that we can say: $Q = 76$ cfs. The power needed for this operation is then Power $= g(48.4)(1.938)(76)/[(0.75)(550)] = 556$ hp.

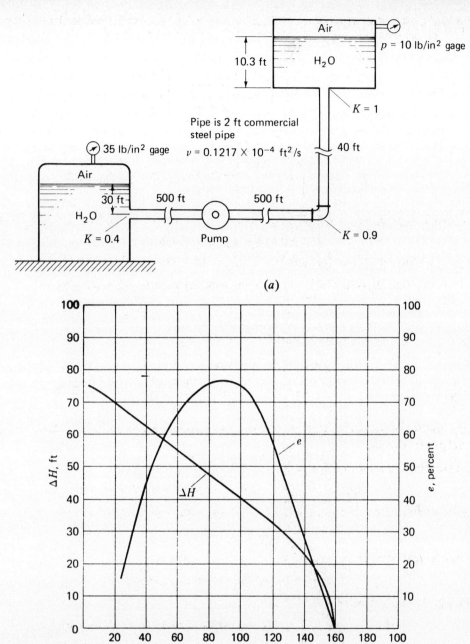

Fig. 20-29

CHAPTER 21
Turbines

21.1 A radial-flow turbine has the following dimensions: $r_1 = 1.6$ ft, $r_2 = 1.0$ ft, and $\beta_1 = 80°$. The width of the flow passage between the two sides of the turbine is 0.8 ft. At 300 rpm the flow rate through the turbine is 120 cfs. Find **(a)** the blade angle β_2 such that the water exits from the turbine in the radial direction, **(b)** the torque exerted by the water on the runner and the horsepower thus developed, **(c)** the head utilized by the runner and the power resulting therefrom. Assume that water enters and leaves the blades smoothly. Assume the blades are so thin that they do not occupy any of the available flow area.

▌ **(a)** At the outer periphery ($r_1 = 1.6$ ft): $u_1 = \omega r_1 = (2\pi/60)(300)(1.6) = 50.3$ fps. From continuity $Q = 120 = 2\pi r_1(z)V_{r_1} = 2\pi(1.6)(0.8)(V_{r_1})$, $V_{r_1} = 120/8.04 = 14.92$ fps $= v_1 \sin \beta_1 = v_1 \sin 80°$, $v_1 = 14.92/(\sin 80°) = 15.15$ fps, $v_1 \cos 80° = (15.15)(0.1736) = 2.63$ fps, $V_1 \cos \alpha_1 = u_1 + v_2 \cos \beta_1 = 50.3 + 2.6 = 52.9$ fps. (See Fig. 21-1a.) At the inner periphery ($r_2 = 1.0$ ft): $u_2 = \omega r_2 = (2\pi/60)(300)(1.0) = 31.4$ fps, $V_{r_2} = V_{r_1}(r_1/r_2) = 14.92(1.6/1.0) = 23.9$ fps. Because the water exits from the turbine in the radial direction, $\alpha_2 = 90°$ and $V_2 = V_{r_2}$. $\tan(180° - \beta_2) = V_2/u_2 = 23.9/31.4 = 0.760$. Thus $\beta_2 = 180° - 37.2° = 142.8°$, the required blade angle. (See Fig. 21-1b.)
(b) $T = \rho Q(r_1 V_1 \cos \alpha_1 - r_2 V_2 \cos \alpha_2) = 1.94(120)[(1.6)(52.9 - 0)] = 19\,700$ ft-lb and power $= T\omega = 19700(31.4) = 619\,000$ ft-lb/s $= 1125$ hp.
(c) $h'' = (u_1 V_1 \cos \alpha_1 - u_2 V_2 \cos \alpha_2)/g = [50.3(52.9) - 0]/32.2 = 82.6$ ft and power $= \gamma Q h'' = (62.4)(120)(82.6) = 619\,000$ ft-lb/s $= 1125$ hp.

Further calculations indicate that the absolute velocity of the water changed from $V_1 = 55.0$ fps at entry to $V_2 = 23.9$ fps at exit while the velocity of the water relative to the blades changed from $v_1 = 15.15$ fps at entry to $v_2 = 39.5$ fps at exit.

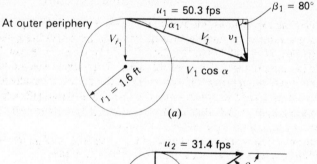

At outer periphery

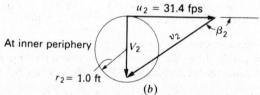

At inner periphery

Fig. 21-1

21.2 A 6-in-diameter pipe ($f = 0.020$) of length 1000 ft delivers water from a reservoir with a water-surface elevation of 500 ft to a nozzle at elevation 300 ft. The jet from the nozzle is used to drive a small impulse turbine. If the head loss through the nozzle can be expressed as $0.04\,V_j^2/2g$, find the jet diameter that will result in maximum power in the jet. Neglect the head loss at entrance to the pipe from the reservoir. Evaluate the power in the jet.

▌ Energy equation: $500 \times 0.02(1000/0.5)(V_p^2/2g) - 0.04(V_j^2/2g) = 300 + (V_j^2/2g)$. If we define the pipe diameter and velocity as D_p and V_p and the jet diameter and velocity as D_j and V_j, from continuity we get $A_p V_p = A_j V_j$, $D_p^2 V_p = D_j^2 V_j$.
Since the pipe diameter $D_p = 0.50$ ft, $0.25V_p = D_j^2 V_j$ and $V_p = 4D_j^2 V_j$. Substituting this expression for V_p in the energy equation gives $200 - (V_j^2/2g)(1.04 + 640D_j^4)$.
Assuming different values for D_j, we can compute corresponding values of V_j and Q, and then the jet power can be computed using $p_{\text{jet}} = [\gamma Q(V_j^2/2g)]/550$. The results are shown in the table on page 639.
Thus a 2-in-diameter jet is the optimum; it will have about 30 hp.
An alternative procedure for solving this problem is to set up an algebraic expression for the power of the jet, P_{jet}, as a function of the jet diameter, D_j, and differentiate P_{jet} with respect to D_j and equate to zero to find the value of D_j for which P_{jet} is a maximum.

D_j, in	D_j, ft	V_j, fps	A_j, ft²	$Q = A_j V_j$, cfs	P_{jet}, hp
1.0	0.083	111	0.0054	0.60	12.8
1.5	0.125	105	0.0122	1.28	24.2
2.0	0.167	91	0.0218	2.00	29.8
2.5	0.208	76	0.0338	2.57	26.2
3.0	0.250	60	0.0491	2.94	18.8
4.0	0.333	38	0.0873	3.29	8.4
6.0	0.500	18	0.197	3.49	1.9

21.3 A turbine is to operate at 400 rpm under a net head of 1320 ft. If a single 6-in-diameter water jet is used, find the specific speed of this machine assuming $C_v = 0.98$, $\phi = 0.45$, and $\eta = 0.85$. Find also the required pitch diameter of the wheel.

$$V = C_v\sqrt{2gh} = 0.98\sqrt{(64.4)(1320)} = 286 \text{ fps} \qquad Q = AV = (0.196)(286) = 56.0 \text{ cfs}$$

$$bhp = \eta(\gamma Qh/550) = 0.85[(62.4)(56)(1320)/550] = 7130$$

$$N_s = N\sqrt{bhp}/h^{5/4} = 400\sqrt{7130}/(1320)^{5/4} = 4.25 \qquad u = \phi\sqrt{2gh} = 0.45\sqrt{(64.4)(1320)} = 131.2 \text{ fps}$$

$$N = 400 \text{ rpm} = u(60)/\pi D = (131.2)(60)/\pi D \qquad D = 6.26 \text{ ft} = 75.2 \text{ in}$$

21.4 In lieu of the single impulse wheel of Prob. 21.3, suppose that three identical single-nozzle wheels are to be used, operating under the same head of 1320 ft. The total flow rate is to be 56.0 cfs. Determine the required specific speed of these turbines, their pitch diameter, the jet diameter, and the operating speed. Once again, assume $C_v = 0.98$, $\phi = 0.45$, and $\eta = 0.85$.

$$Q = \tfrac{56}{3} = 18.7 \text{ cfs} \qquad bhp = \tfrac{7130}{3} = 2377 \qquad N_s = N\sqrt{2377}/(1320^{5/4}) = N(0.00613)$$

$$N = 131.2(60)/\pi D = 2056/D$$

From the two preceding expressions it is apparent that the required N_s depends on the operating speed, as does D. Hence there are a number of possible answers. If we let the operating speed be 400 rpm (18-pole generator for 60-cycle electricity), $N_s = 400(0.00613) = 2.45$, $D = \tfrac{2506}{400} = 6.26$ ft = 75.2 in. Thus three $N_s = 2.45$ wheels of pitch diameter 75 in operating at 400 rpm would suffice. These wheels would have relatively small buckets. At such a low specific speed the optimum efficiency of impulse wheels is usually less than 0.85. An alternative solution, for example, would be to use an operating speed of 600 rpm (12-pole generator for 60-cycle electricity). For this case, $N_s = (600)(0.00613) = 3.68$, $D = \tfrac{2506}{600} = 4.18$ ft = 50.1 in. Thus three $N_s = 3.68$ wheels, of pitch diameter 50.1 in, operating at 600 rpm would suffice.

Let us now determine the required jet diameter for these 50.1-in-diameter wheels operating at 600 rpm: $Q = AV = (\pi D_j^2/4)(286) = 18.7$ cfs from which $D_j = 0.289$ ft = 3.46 in.

21.5 Refer to Prob. 21.3. Suppose a two-nozzle single-wheel installation were designed to operate under a head of 1320 ft with a total flow (for both nozzles) of 56 cfs. Determine the required specific speed of this turbine, its pitch diameter, and jet diameter for rotative speeds of 300, 400, and 600 rpm. C_v, ϕ, and η remains unchanged.

$$V_j = C_v(2gh)^{1/2} = (0.98)[(2)(32.2)(1320)]^{1/2} = 285.7 \text{ fps} \qquad Q_j = A_j V_j = (A_j)(285.7) = \tfrac{56}{2}$$

$$A_j = 0.09800 \text{ ft}^2 \qquad \pi D_j^2/4 = 0.09800 \qquad D_j = 0.3532 \text{ ft} \quad \text{or} \quad 4.24 \text{ in}$$

$$bp_j = (\eta)(\gamma Q_j h)/550 = (0.85)(62.4)(28.0)(1320)/550 = 3564 \text{ bhp per jet}$$

$$N_s = N(bhp)^{0.5}/h^{5/4} = (N)(3564)^{0.5}/1320^{5/4} = 0.007503N$$

$$D = 60\phi(2gh)^{1/2}/\pi N = (60)(0.45)[(2)(32.2)(1320)]^{1/2}/(\pi N) = 2506/N$$

N	N_s	pitch D, ft	pitch F, in
300	2.25	8.35	100.2
400	3.00	6.26	75.2
600	4.50	4.18	50.1

21.6 A series of vanes is acted on by a 3-in water jet having a velocity of 100 fps, $\alpha_1 = \beta_1 = 0°$. Find the required blade angle β_2 in order that the force acting on the vanes in the direction of the jet is 200 lb. Neglect friction. Solve for vane velocities of 85, 50, and 0 fps.

❚ From the velocity vector diagram of Fig. 21-2,

$$\Delta V_x = (u + v_2 \cos \beta_2) - V_1 \qquad Q = AV_1 = [(\pi)(\tfrac{3}{12})^2/4](100) = 4.909 \text{ ft}^3/\text{s}$$

$$F_x = \rho Q(-\Delta V) = -\rho Q[(u + v_2 \cos \beta_2) - V_1] \qquad 200 = -(1.94)(4.909)[(u + v_2 \cos \beta_2) - 100]$$

$$u + v_2 \cos \beta_2 = 79.0$$

u	$v_1 = v_2$	$v_2 \cos \beta_2$	$\cos \beta_2$	β_2
85	15	−6.0	−0.40000	113.6°
50	50	29.0	0.58000	54.5°
0	100	79.0	0.79000	37.8°

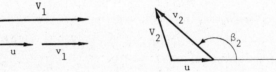

Fig. 21-2

21.7   Under a net head of 1 ft, the Pelton wheel for Fig. 21-3a discharges 0.286 cfs at full nozzle opening and the maximum power is 0.025 bhp for a value of $\phi = 0.465$. The corresponding brake torque is 3.69 ft-lb as shown in Fig. 21-3b. Assuming that the similarity laws apply precisely, determine the discharge, torque, power, and rotative speed of this wheel when it operates under a head of 1600 ft.

❚ On Fig. 21-3a, $h_1 = 1.0$ ft, $\phi = 0.465$, $Q_1 = 0.286$ cfs, $T_1 = 3.69$ ft-lb, $\text{bhp}_1 = 0.025$, and $r = 1$ ft.

$$u_1 = (\phi)(2gh_1)^{1/2} = (0.465)[(2)(32.2)(1.0)]^{1/2} = 3.732 \text{ fps} \qquad \omega_1 = u_1/r = 3.732/1 = 3.732 \text{ rad/s}$$

From the similarity laws, $Q \propto h^{1/2}$, $T \propto h$, $\omega \propto h^{1/2}$, and $\text{bhp} \propto h^{3/2}$. At $h_2 = 1600$ ft,

$$Q_2 = (0.286)(1600)^{1/2} = 11.4 \text{ ft}^3/\text{s} \qquad T_2 = (3.69)(1600) = 5904 \text{ ft-lb}$$

$$\omega_2 = (3.73)(1600)^{1/2} = 149.2 \text{ rad/s} \quad \text{or} \quad 1425 \text{ rpm} \qquad \text{bhp}_2 = (0.025)(1600)^{3/2} = 1600$$

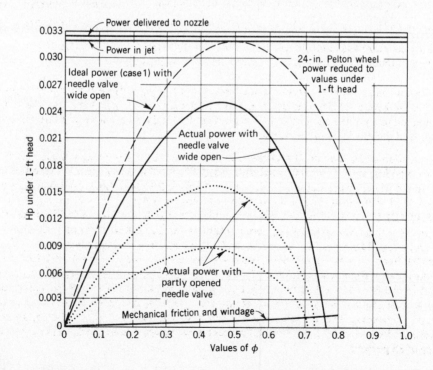

Fig. 21-3(a)

Relation between power and speed at constant head with maximum nozzle opening. (From tests made by F. G. Switzer and R. L. Daugherty.)

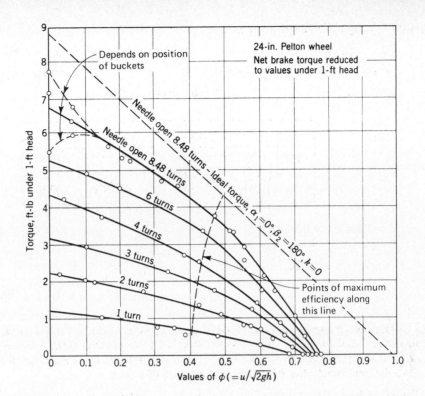

Fig. 21-3(b)

Relation between torque and speed at constant head. (From tests made by F. G. Switzer and R. L. Daugherty.)

21.8 A wheel and nozzle similar to that of Fig. 21-3a and b with a pitch diameter of 12 ft is used under a net head of 1600 ft. What are the torque, power, and rotative speed at point of best efficiency for full nozzle opening?

▋ In Fig. 21-3a, $\phi_e = 0.46$. In Fig. 21-3b, for $\phi = 0.46$, nozzle fully open, $T = 3.76$ ft-lb.

$$u = (\phi)(2gh)^{1/2} = (0.46)[(2)(32.2)(1)]^{1/2} = 3.691 \text{ fps} \qquad \omega = u/r = 3.691/1 = 3.691 \text{ rad/s}$$

$$\text{bhp} = Tu/550 = (3.76)(3.691)/550 = 0.02523 \qquad N = 60u/\pi D = (60)(3.691)/[(\pi)(2)] = 35.25 \text{ rpm}$$

For a 12-ft-diameter wheel and 1600 ft head, $T = (3.76)(\frac{12}{2})^3(1600) = 1.30 \times 10^6$ ft-lb, $P = (0.02523)(\frac{12}{2})^2(1600)^{3/2} = 58\,130$ hp, $N = (35.25)(1600)^{1/2}/6 = 235$ rpm.

21.9 Find the torque and power transferred to the buckets of an impulse wheel with $\alpha_1 = 0°$, $\beta_2 = 160°$, $k = 0.44$, $\phi = 0.46$, $C_v = 0.98$, a jet diameter of 10 in, and a pitch diameter of 10 ft. Find also, the hydraulic efficiency, and, expressed as a percentage of the total head, find the head loss in bucket friction, the energy head loss at discharge, and the head loss in the nozzle.

▋ For operation under the purely artificial value of 1-ft net head, $u = \phi\sqrt{2g} = 8.02\phi = 3.68$ fps, $V_1 = 0.98\sqrt{2g} = 7.86$ fps, and $\gamma Q = \gamma AV = (62.4)(0.545)(7.86) = 267$ lb/s. Hence, the expression for the torque exerted on the wheel by the water is $T = (r)(F_u) = 5(267/32.2)[1 + (0.940/1.2)](8.02)(C_v - \phi)$. With $C_v = 0.98$ and $\phi = 0.46$, $T = 309$ ft-lb. The power transferred from the water to the buckets is $F_u u = T\omega = T(u/r) = 309[(0.46)(8.02)/5] = 228$ ft-lb/s. The power input is $\gamma Qh = 267(1) = 267$ ft-lb/s. The hydraulic efficiency is $\frac{228}{267} = 0.85$. Note that the answers given here are for $h = 1$ ft. For other values of h, the flow rate, torque, and power can be found by adjusting the values for $h = 1$ ft according to similarity laws.

Under 1-ft head, $v_1 = V_1 - u = 7.86 - 3.68 = 4.18$ fps and $v_2 = v_1/\sqrt{1 + k} = 4.18/\sqrt{1.44} = 3.48$ fps. Hence the head loss in bucket friction is $0.44(3.48)^2/2g = 0.083$ ft, or 8.3 percent. $V_2\cos\alpha_2 = u + v_2\cos\beta_2 = 3.68 + 3.48\cos 160° = 3.68 - 3.27 = 0.41$ fps, $V_2\sin\alpha_2 = v_2\sin\beta_2 = 3.48\sin 160° = 1.19$ fps. Hence $\cot\alpha_2 = 0.41/1.19 = 0.344$ or $\alpha_2 = 71°$, and $V_2 = 1.19/0.945 = 1.26$ fps from which the energy head loss at discharge is $1.26^2/2g = 0.025$ ft, or 2.5 percent. The head loss in the nozzle is approximately $(1/C_v^2 - 1)V_1^2/2g$, about 4.0 percent; so the total hydraulic loss is $8.3 + 2.5 + 4.0 \approx 14.8$ percent, which gives a close check on the computed hydraulic efficiency of 85 percent.

21.10 A nozzle having a velocity coefficient of 0.98 discharges a jet 6 in in diameter under a head of 900 ft. As a simplifying assumption take $\alpha_1 = 0°$. The wheel diameter is 8 ft, $\beta_2 = 165°$, and it may be assumed that $k = 0.5$.

The mechanical efficiency of the wheel is 97 percent. What is the hydraulic efficiency? What is the gross efficiency? Assume $\phi = 0.46$.

▌ Nozzle efficiency $= 0.98^2 = 0.9604$, $\eta_h = (2)[1 - (\cos \beta_2/\sqrt{1+k})](C_v - \phi)(\phi) = (2)[1 - (\cos 165°/\sqrt{1+0.5})](0.98 - 0.46)(0.46) = 0.856$, or 85.6 percent. As C_v is included, η_h is here the hydraulic efficiency of both the wheel and the nozzle. $\eta = \eta_h \eta_m = (0.856)(0.97) = 0.830$, or 83.0 percent.

21.11 A 24-in laboratory Pelton wheel was tested under a head of 65.5 ft. With the nozzle open 6 turns of the needle, the net brake load at 275 rpm was 40 lb at a brake arm of 5.25 ft and the discharge was 1.897 cfs. Find values of brake horsepower and efficiency under operating conditions. What would be the values of torque and brake horsepower under 1-ft head?

▌ $\qquad T_2 = dW = (5.25)(40) = 210$ ft-lb $\qquad \omega = 2\pi n/60 = (2)(\pi)(275)/60 = 28.8$ rad/s

$$P_1 = T_2\omega = (210)(28.8) = 6048 \text{ ft-lb/s} \quad \text{or} \quad 11.0 \text{ hp}$$

$P_w = Q\gamma H = (1.897)(62.4)(65.5) = 7753$ ft-lb/s $\quad$ or $\quad$ 14.1 hp $\qquad \eta = 11.0/14.1 = 0.780$ $\quad$ or $\quad$ 78.0 percent

Under 1-ft head, $T_1 = (T_2)(h_1/h_2) = (210)(1/65.5) = 3.21$ ft-lb, $P_1 = (P_2)(h_1/h_2)^{3/2} = (11.0)(1/65.5)^{3/2} = 0.0208$ bhp.

21.12 On Prob. 21.11, at 275 rpm the bearing-friction and windage losses were found to be 0.2 hp. What percentage is this of the brake horsepower? What is the value of the mechanical efficiency? What is the value of the hydraulic efficiency?

▌ Bearing-friction and windage loss $= 0.2/11.0 = 0.018$, or 1.8 percent.

$$\eta_m = 100 - 1.8 = 98.2 \text{ percent} \qquad \eta_h = \eta/(\eta_v\eta_m) \qquad \eta_v = 1$$
$$\eta_h = 0.780/[(1)(0.982)] = 0.794 \quad \text{or} \quad 79.4 \text{ percent}$$

21.13 Find the approximate hydraulic efficiency of an impulse wheel for which the nozzle velocity coefficient is 0.97 and the bucket angle is 160°, if $\phi = 0.46$ and $k = 0.1$.

▌ $\qquad \eta_h = (2)[1 - (\cos \beta_2/\sqrt{1+k})](C_v - \phi)(\phi) = (2)[1 - (\cos 160°/\sqrt{1+0.1})](0.97 - 0.46)(0.46)$

$\qquad = 0.890$ $\quad$ or $\quad$ 89.0 percent

21.14 A double-overhung impulse-turbine installation is to develop 20 000 hp at 257 rpm under a net head of 1120 ft. Determine N_s, wheel-pitch diameter, and approximate jet diameter.

▌ $\qquad N_s = NP^{0.5}/H^{5/4} = (257)(20\,000/2)^{0.5}/1120^{5/4} = 3.97$

$$D = 153.2\phi\sqrt{h}/N = (153.2)(0.45)(\sqrt{1120})/257 = 8.98 \text{ ft}$$
$$D_j = D/12 = 8.98/12 = 0.7483 \text{ ft} \quad \text{or} \quad 8.98 \text{ in}$$

21.15 Repeat Prob. 21.14 for a single wheel with a single nozzle.

▌ $N_s = (257)(20\,000/1)^{0.5}/1120^{5/4} = 5.61$. Noting the computation of D in Prob. 21.14, D is unchanged; hence, $D_j = 8.98$ in.

21.16 Repeat Prob. 21.14 for a single wheel with four nozzles.

▌ $N_s = (257)(20\,000/4)^{0.5}/1120^{5/4} = 2.80$. Noting the computation of D in Prob. 21.14, D is unchanged; hence, $D_j = 8.98$ in.

21.17 An impulse turbine ($N_s = 5$) develops 100 000 hp under a head of 2000 ft. For 60-cycle electricity calculate the turbine speed, wheel diameter, and number of poles in the generator. Assume $\phi = 0.45$.

▌ $\qquad N_s = NP^{0.5}/H^{5/4} \qquad 5 = (N)(100\,000)^{0.5}/2000^{5/4} \qquad N = 211 \text{ rpm} = 7200/N_{\text{poles}}$

$211 = 7200/N_{\text{poles}} \qquad N_{\text{poles}} = 34 \qquad D = 153.2\phi\sqrt{h}/N = (153.2)(0.45)(\sqrt{2000})/212 = 14.54 \text{ ft}$

21.18 Rework Prob. 21.17 for a six-nozzle unit using the same N_s, bhp, and head.

▌ $\qquad N_s = NP^{0.5}/H^{5/4} \qquad 5 = (N)(100\,000/6)^{0.5}/2000^{5/4} \qquad N = 518 \text{ rpm} = 7200/N_{\text{poles}}$

$518 = 7200/N_{\text{poles}} \qquad N_{\text{poles}} = 14 \qquad D = 153.2\phi\sqrt{h}/N = (153)(0.45)(\sqrt{2000})/514 = 6.00 \text{ ft}$

21.19 It is desired to develop 15 000 bhp under a head of 1000 ft. Make any necessary assumptions, and estimate the diameter of the wheel required and the rotative speed.

▮ $N_s = NP^{0.5}/H^{5/4}$. Best efficiency occurs at $N_s < 4.5$. Assume $N_s = 4$: $4 = (N)(15\,000)^{0.5}/1000^{5/4}$, $N = 184$ rpm, $D = 153.2\phi\sqrt{h}/N$. Assume synchronous speed = 180 rpm and $\phi = 0.46$: $D = (153.2)(0.46)(\sqrt{1000})/180 = 12.38$ ft.

21.20 A 1:8 model of a 12-ft-diameter turbine is operated at 600 rpm under a net head of 54.0 ft. Under this mode of operation, the brake horsepower and Q of the model were observed to be 332 and 62 cfs, respectively. (**a**) From the above data, compute the specific speed of the model and the value of ϕ. (**b**) Calculate the efficiency and shaft torque of the model.

▮ (**a**)
$$N_s = NP^{0.5}/H^{5/4} = (600)(332)^{0.5}/54.0^{5/4} = 74.7$$
$$\phi = u/(2gh)^{1/2} = (2\pi N/60)r/(2gH)^{1/2} = [(2)(\pi)(600)/60][(\tfrac{12}{8})/2]/[(2)(32.2)(54.0)]^{1/2} = 0.799$$

(**b**)
$$\eta = P/\gamma QH = (332)(550)/[(62.4)(62)(54.0)] = 0.874 \quad \text{or} \quad 87.4 \text{ percent}$$
$$T = P/\omega = (332)(550)/[(2\pi)(600)/60] = 2906 \text{ ft-lb}$$

21.21 For the data of Prob. 21.20, (**a**) what would be the efficiency of the 12-ft-diameter prototype? (**b**) The prototype is to operate at 144 rpm under a net head of 200 ft. Find the horsepower output of the prototype and the flow rate.

▮ (**a**) $(1 - \eta)/(1 - \eta_1) = (D_2/D_1)^{1/5}$ $(1 - \eta)/(1 - 0.874) = (\tfrac{1}{8})^{1/5}$ $\eta = 0.917$ or 91.7 percent

(**b**) $N_s = NP^{0.5}/H^{5/4}$ $74.7 = 144P^{0.5}/200^{5/4}$ $P = 152\,000$ bhp
$$P = Q\gamma H/550 \quad 152\,000 = (Q)(62.4)(200)(0.917)/550 \quad Q = 7315 \text{ ft}^3/\text{s}$$

21.22 It is desired to install a single turbine that will develop 4200 hp under a head of 247 ft. If a turbine with $N_s \approx 25$ were selected, what rotative speed would you suggest for 50-cycle electricity? How many poles do you recommend for the generator?

▮
$$N_s = NP^{0.5}/H^{5/4} \quad 25 = (N)(4200)^{0.5}/247^{5/4} \quad N = 378 \text{ rpm}$$
For 50 cycles, $N_{\text{poles}} = 6000/N = \tfrac{6000}{378} = 16$.

21.23 The Francis turbine for which the test curves are shown in Figs. 21-4a and b has a 27-in-diameter runner and a maximum efficiency of 88 percent when discharging 38.8 cfs and developing 550 bhp at 600 rpm under a net head of 141.8 ft. Compute N_s, ϕ_e, and C_r, assuming $B/D = 0.15$.

▮
$$N_s = NP^{0.5}/H^{5/4} = (600)(550)^{0.5}/141.8^{5/4} = 28.8 \quad ND = 153.2\phi\sqrt{H}$$
$$(600)(\tfrac{27}{12}) = (153.2)(\phi_e)\sqrt{141.8} \quad \phi_e = 0.740$$
$$C_r = N_s^2/[(63\,800)(\phi_e^2)(B/D)(\eta)] = (28.8)^2/[(63\,800)(0.740)^2(0.15)(0.88)] = 0.1799 \quad V_r = Q/A_c$$
$$A_c = 0.95\pi(B/D)D^2 = (0.95)(\pi)(0.15)(\tfrac{27}{12})^2 = 2.266 \text{ ft}^2 \quad V_r = 38.8/2.266 = 17.12 \text{ fps}$$
$$C_r = V_r/(2gH)^{1/2} = 17.12/[(2)(32.2)(141.8)]^{1/2} = 0.1792 \quad \text{(a good check)}$$

21.24 If a turbine homologous to that in Prob. 21.23 were made with a runner diameter of 135 in, what would be its probable efficiency under the same head?

▮ $(1 - \eta_1)/(1 - \eta) = (D_2/D_1)^{1/5}$ $(1 - 0.88)/(1 - \eta) = (\tfrac{135}{27})^{1/5}$ $\eta = 0.913$ or 91.3 percent

21.25 The turbine of Prob. 21.23 has a horizontal shaft, and at the time of the test the centerline of the shaft was 12.67 ft above the surface of the water in the tailrace. The discharge edge of the runner at its highest point is 0.83 ft above the centerline of the shaft. If the temperature of the water were as high as 80 °F (vapor pressure = 0.5 psia) and the barometric pressure were 14.6 psia, what would be the value of the cavitation factor σ?

▮
$$\sigma = (p_{\text{atm}}/\gamma - p_v/\gamma - z_{\text{max}})/h \quad p_{\text{atm}}/\gamma = (14.6)(144)/62.4 = 33.69 \text{ ft}$$
$$p_v/\gamma = (0.5)(144)/62.4 = 1.15 \text{ ft} \quad z_{\text{max}} = 12.67 + 0.83 = 13.50 \text{ ft}$$
$$\sigma = (33.69 - 1.15 - 13.50)/141.8 = 0.134$$

21.26 If it is assumed that the critical value of the cavitation factor for the turbine in Prob. 21.23 is 0.06 and if the other data of Prob. 21.25 are used, what would be the maximum allowable height of the centerline of the shaft above the tailwater surface?

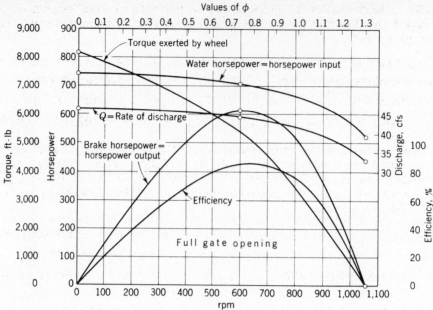

Fig. 21-4(a)
A certain Francis turbine operating at variable speed under constant head with constant gate opening. Runner diameter = 27 in, head = 140.5 ft.

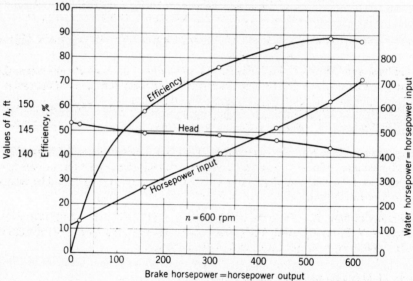

Fig. 21-4(b)
A certain Francis turbine operating at constant speed and variable gate opening. Runner diameter = 27 in.

$$z_{max} = p_{atm}/\gamma - p_v/\gamma - \sigma_c h = 33.69 - 1.15 - (0.06)(141.8) = 24.03 \text{ ft}$$
Height to centerline of shaft = 24.03 − 0.83 = 23.20 ft

21.27 In Prob. 21.26, assume all values the same except that the net head on the turbine is 400 ft. What would then be the maximum allowable height of the centerline of the shaft above the tailwater surface?

$$z_{max} = p_{atm}/\gamma - p_v/\gamma - \sigma_c h = 33.69 - 1.15 - (0.06)(400) = 8.54 \text{ ft}$$
Height to centerline of shaft = 8.54 − 0.83 = 7.71 ft

21.28 Find the maximum permissible head under which an axial-flow turbine ($N_s = 160$) can operate if it is set 5 ft below the tailwater. The installation is at elevation 3150 ft where $p_{atm} = 14.15$ psia, and the water temperature is 65 °F. Assume $\sigma_c = 0.90$.

$$z_{max} = p_{atm}/\gamma - p_v/\gamma - \sigma_c h \qquad -5 = (14.15)(144)/62.4 - 44.4/62.4 - 0.90h \qquad h = 41.0 \text{ ft}$$

21.29 In addition to the data of Prob. 21.23, the mechanical-friction losses in the Francis turbine were measured and found to be 2.7 hp. Assuming that the leakage is 1 percent of the measured discharge and that $\alpha_2 = 90°$, find the values of η_h, α_1, and β_1.

▌ $\eta_v = (Q - Q_L)/Q = (Q - 0.10Q)/Q = 0.990$ $\eta_m = bp/(bp + fp) = 550/(550 + 2.7) = 0.995$

$\eta_h = 0.88/[(0.99)(0.995)] = 0.893$ $C_1 = \cos \alpha_1 = 0.893/[(2)(0.74)] = 0.6054$ $\eta_h = 2\phi_e C_1 \cos \alpha_1$

$\tan \alpha_1 = (C_1 \sin \alpha_1)/(C_1 \cos \alpha_1) = C_r/(C_1 \cos \alpha_1) = 0.1799/0.6054 = 0.29716$ $\alpha_1 = 16.55°$

$\cot \beta_1 = (C_1 \cos \alpha_1 - \phi)/(C_1 \sin \alpha_1) = (C_1 \cos \alpha_1 - \phi)/C_r = (0.6054 - 0.74)/0.1799 = -0.74819$ $\beta_1 = 126.8°$

21.30 The Grand Coulee turbines have runner diameters of 197 in. The height of the guide vanes is 34.375 in. The diameter of the throat of the runner and also the diameter of the draft tube adjacent to the runner are 172 in. Each turbine is rated at 150 000 hp under a head of 330 ft at 120 rpm. At this power the efficiency is 89 percent and the absolute velocity of the water entering the runner is 77.2 fps. Compute the rated specific speed, ϕ, and C_1 for this full gate opening.

▌ $N_s = NP^{0.5}/H^{5/4} = (120)(150\,000)^{0.5}/330^{5/4} = 33.0$ $\phi = u_1/(2gH)^{1/2}$

$u_1 = r\omega = r(2\pi N/60) = [(\frac{197}{12})/2][(2)(\pi)(120)/60] = 103.1 \text{ ft/s}$

$\phi = 103.1/[(2)(32.2)(330)]^{1/2} = 0.707$ $C_1 = V_1/(2gh)^{1/2} = 77.2/[(2)(32.2)(330)]^{1/2} = 0.530$

21.31 For the data given in Prob. 21.30, find C_r, α_1, and β_1 for this full gate opening.

▌ $C_r = V_r/(2gh)^{1/2}$ $V_r = Q/A_c$ $A_c = 0.95\pi BD = (0.95)(\pi)(34.375/12)(\frac{197}{12}) = 140.4 \text{ ft}^2$

$V_r = 4500/140.4 = 32.05 \text{ fps}$ $C_r = 32.05/[(2)(32.2)(330)]^{1/2} = 0.220$

$\sin \alpha_1 = C_r/C_1 = 0.220/0.530 = 0.41509$ $\alpha_1 = 24.5°$

$\tan \beta_1 = (C_1 \sin \alpha_1)/(\phi - C_1 \cos \alpha_1) = (0.530)(\sin 24.5°)/[0.707 - (0.530)(\cos 24.5°)] = 0.97805$ $\beta_1 = 44.4°$

21.32 What is the least number of identical turbines that can be used at a powerhouse where the available head is 1200 ft and $Q = 1650$ cfs? Assume turbine efficiency is 90 percent and speed of operation 138.5 rpm.

▌ $bhp = \eta Q\gamma H_p/550 = (0.90)(1650)(62.4)(1200)/550 = 202\,000$. From Fig. A-42, $N_s \le 5.4$. Let x = number of identical turbines. $N_s \ge NP^{0.5}/H_p^{5/4}$, $5.4 = (138.5)(202\,000/x)^{0.5}/1200^{5/4}$, $x = 2.7$; use three turbines.

21.33 A single hydraulic turbine is to be selected for a power site with a net head of 100 ft. The turbine is to produce 25 000 hp at maximum efficiency. What speed (rpm) and diameter should this turbine have if a Francis turbine is selected? What is the highest "setting" (above or below tailwater) which should be recommended for this machine for it to run cavitation-free at its point of maximum efficiency?

▌ For a Francis turbine, $N_s = 20$ to 80; try $N_s = 50$: $N_s = NP^{0.5}/H^{5/4}$, $50 = N(25\,000)^{0.5}/100^{5/4}$, $N = 100$ rpm; $D = 153.2\phi\sqrt{H}/N$. From Fig. A-43 for $N_s = 50$, $\phi_e = 0.78$. $D = (153.2)(0.78)\sqrt{100}/100 = 11.95$ ft. From Fig. A-42 for $H = 100$ ft, $N_s = 50$: Set about 15 ft above tailwater.

21.34 Solve Prob. 21.33 if a propeller turbine is selected.

▌ For a propeller turbine, $N_s = 100$ to 250; try $N_s = 120$: $N_s = NP^{0.5}/H^{5/4}$, $120 = N(25\,000)^{0.5}/100^{5/4}$, $N = 240$ rpm; $D = 153.2\phi\sqrt{H}/N$. From Fig. A-43 for $N_s = 120$, $\phi_e = 1.63$. $D = (153.2)(1.63)\sqrt{100}/240 = 10.40$ ft. From Fig. A-42 for $H = 100$ ft, $N_s = 120$: Set about 12 ft below tailwater.

21.35 For 50-cycle electricity how may poles would you recommend for a generator which is connected to a turbine operating under a design head of 3000 ft with a flow of 80 cfs? Assume turbine efficiencies as given in Fig. A-44 and be sure the turbine is free of cavitation.

▌ For 50 cycles, $N_{poles} = 6000/N$. $N_s = NP^{0.5}/H^{5/4}$. From Fig. A-42, we must use $N_s \le 3$; use $N_s = 3$.

$$P = \eta Q\gamma H/550 = (0.84)(80)(62.4)(3000)/550 = 22\,872 \text{ hp}$$

$3 = (N)(22\,872)^{0.5}/3000^{5/4}$ $N = 440$ rpm $N_{poles} = \frac{6000}{440} = 13.6$

Use a 12- or 14-pole generator.

21.36 A turbine is to be installed where the net available head is 185 ft, and the available flow will average 900 cfs. What type of turbine would you recommend? Specify the operating speed and number of generator poles for

60-cycle electricity if a turbine with the highest tolerable specific speed that will safeguard against cavitation is selected. Assume the turbine is set 5 ft above tailwater. Assume turbine efficiency is 90 percent. Approximately what size of runner is required?

$$N_s = NP^{0.5}/H^{5/4} \qquad \text{bhp} = \eta Q\gamma H/550 = (0.90)(900)(62.4)(185)/550 = 17\,001$$

From Fig. A-42 for $H = 185$ ft and draft head $= +5$ ft, highest tolerable $N_s = 50$. $50 = (N)(17\,001)^{0.5}/185^{5/4}$, $N = 262$ rpm. With 60-cycle electricity, a 28-pole generator will operate at $\frac{7200}{28} = 257$ rpm. So use a 28-pole generator operating at 257 rpm. $N_s = (257)(17\,001)^{0.5}/185^{5/4} = 49.1$. Use a Francis turbine with $N_s = 49.1$. $D = 153.3\phi\sqrt{H}/N = (153.2)(0.78)(\sqrt{185})/257 = 6.32$ ft.

21.37 A turbine is to be installed at a point where the available head is 175 ft and the available flow will average 1000 cfs. What type of turbine would you recommend? Specify the operating speed and number of generator poles for 60-cycle electricity if a turbine with the highest tolerable specific speed to safeguard against cavitation is selected. Assume static draft head of 10 ft and 90 percent turbine efficiency. Approximately what size of turbine runner is required?

❚ $N_s = NP^{0.5}/H^{5/4}$. From Fig. A-42 for $H = 175$ ft and draft head $= +10$ ft, $N_s \leq 48$.

$$\text{bhp} = \eta Q\gamma H/550 = (0.90)(1000)(62.4)(175)/550 = 17\,869 \qquad 48 = (N)(17\,869)^{0.5}/175^{5/4} \qquad N = 229 \text{ rpm}$$

$N_{\text{poles}} = 7200/N = \frac{7200}{229} = 31$; use 32 poles. $32 = 7200/N$, $N = 225$ rpm; $N_s = (225)(17\,869)^{0.5}/175^{5/4} = 47.3$; use a Francis turbine. From Fig. A-43 for $N_s = 47.3$, $\phi_e = 0.77$. $D = 153.2\phi\sqrt{H}/N = (153.2)(0.77)(\sqrt{175})/225 = 6.94$ ft.

21.38 For the same conditions as given in Prob. 21.37, select a set of identical turbines to be operated in parallel. Specify the speed and size of the units.

$$N_s = NP^{0.5}/H^{5/4} \qquad N_s \leq 48 \qquad \text{bhp} = 17\,869/2 = 8934$$
$$48 = (N)(8934)^{0.5}/175^{5/4} \qquad N = 323 \qquad N_{\text{poles}} = 7200/N$$

$N_{\text{poles}} = \frac{7200}{323} = 22.3$; use 22 poles. $22 = 7200/N$, $N = 327$; $N_s = (327)(8934)^{0.5}/175^{5/4} = 48.6$. From Fig. A-43 for $N_s = 48.6$, $\phi_e = 0.75$. $D = 153.2\phi\sqrt{h}/N = (153.2)(0.75)(\sqrt{175})/327 = 4.65$ ft.

21.39 Water enters a rotating wheel with a relative velocity of 200 fps; $r_1 = 4.0$ ft and $N = 420$ rpm. There is no pressure drop in flow over the vanes. Assume $k = 0.2$. Find the relative velocity at discharge if (a) $r_2 = 3.0$ ft and (b) $r_2 = 5.0$ ft.

❚ With no pressure drop, $(V_1^2 - u_1^2)/2g - (V_2^2 - u_2^2)/2g = kV_2^2/2g$, $u = \pi DN/60$.

(a) $$u_1 = (\pi)[(2)(4)](420)/60 = 175.9 \text{ ft} \qquad u_2 = (\pi)[(2)(3)](420)/60 = 131.9 \text{ ft}$$
$$(200^2 - 175.9^2)/[(2)(32.2)] - (V_2^2 - 131.9^2)/[(2)(32.2)] = 0.2V_2^2/[(2)(32.2)] \qquad V_2 = 148 \text{ fps}$$

(b) $$u_2 = (\pi)[(2)(5)](420)/60 = 219 \text{ ft}$$
$$(200^2 - 175.9^2)/[(2)(32.2)] - (V_2^2 - 219.9^2)/[(2)(32.2)] = 0.2V_2^2/[(2)(32.2)] \qquad V_2 = 219 \text{ fps}$$

21.40 Water enters a rotating wheel which is so proportioned that the passages are completely filled. $Q = 400$ cfs, $a_1 = 10$ ft^2, $a_2 = 8$ ft^2, $r_1 = 1.5$ ft, $r_2 = 1.0$ ft, and $N = 540$ rpm. Assume $k = 0.2$. Find the drop in pressure head between entrance and exit.

❚ $$(p_1/\gamma - p_2/\gamma) + (V_1^2 - u_1^2)/2g - (V_2^2 - u_2^2)/2g = kV_2^2/2g \qquad V_1 = Q/a_1 = \frac{400}{10} = 40.0 \text{ fps}$$
$$V_2 = \frac{400}{8} = 50.0 \text{ fps} \qquad u = \pi DN/60$$
$$u_1 = (\pi)[(2)(1.5)](540)/60 = 84.82 \text{ fps} \qquad u_2 = (\pi)[(2)(1.0)](540)/60 = 56.55 \text{ fps}$$
$$(p_1/\gamma - p_2/\gamma) + (40.0^2 - 84.82^2)/[(2)(32.2)] - (50.0^2 - 56.55^2)/[(2)(32.2)] = (0.2)(50.0^2)/[(2)(32.2)]$$
$$p_1/\gamma - p_2/\gamma = 83.8 \text{ ft}$$

21.41 It is desired to develop 300 000 hp under a head of 49 ft and to operate at 60 rpm. (a) If turbines with a specific speed of approximately 150 are to be used, how many units will be required? (b) If Francis turbines with a specific speed of 80 were to be used, how many units would be required?

❚ $$N_s = NP^{0.5}/H^{5/4}$$

(a) $$150 = 60P^{0.5}/49^{5/4} \qquad P = 105\,044 \text{ bhp}$$

Number of units required $= 300\,000/105\,044 = 2.9$. Use three units.

(b) $$80 = 60P^{0.5}/49^{5/4} \qquad P = 29\,879 \text{ bhp}$$

Number of units required $= 300\,000/29\,879 = 10.0$.

21.42 Air $[R = 1715, c_p = 6003 \text{ ft-lb}/(\text{slug-}°R)]$ flows steadily, as shown in Fig. 21-5, through a turbine which produces 700 hp. For the inlet and exit conditions shown, estimate **(a)** the exit velocity V_2 and **(b)** the heat transferred $\dot Q$ in Btu per hour.

▮ **(a)** The inlet and exit densities can be computed from the perfect-gas law: $\rho_1 = p_1/RT_1 = 150(144)/[1715(460 + 300)] = 0.0166 \text{ slug/ft}^3$, $\rho_2 = p_2/RT_2 = 40(144)/[1715(460 + 35)] = 0.00679 \text{ slug/ft}^3$. The mass flow is determined by the inlet conditions $\dot m = \rho_1 A_1 V_1 = (0.0166)(\pi/4)(\frac{6}{12})^2(100) = 0.326 \text{ slug/s}$. Knowing mass flow, we compute the exit velocity

$$\dot m = 0.326 = \rho_2 A_2 V_2 = (0.00679)(\pi/4)(\tfrac{6}{12})^2 V_2 \quad \text{or} \quad V_2 = 245 \text{ fps}$$

(b) The steady-flow energy equation applies with $\dot W_v = 0$, $z_1 = z_2$, and $\hat h = c_p T$: $\dot Q - \dot W_s = \dot m(c_p T_2 + \frac{1}{2}V_2^2 - c_p T_1 - \frac{1}{2}V_1^2)$. Convert the turbine work to foot-pounds per second with the conversion factor 1 hp = 550 ft-lb/s. The turbine work is positive: $\dot Q - 700(550) = 0.326[6003(495) + \frac{1}{2}(245)^2 - 6003(760) - \frac{1}{2}(100)^2]$ or $\dot Q = -125\,000$ ft-lb/s. Convert this to British thermal units as follows: $\dot Q = -125\,000(\frac{3600}{778}) = -578\,000 \text{ Btu/h}$. The negative sign indicates that this heat transfer is a *loss* from the control volume.

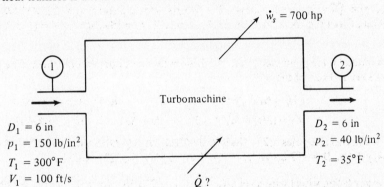

$\dot w_s = 700$ hp

Turbomachine

$D_1 = 6$ in
$p_1 = 150 \text{ lb/in}^2$
$T_1 = 300°F$
$V_1 = 100 \text{ ft/s}$

$D_2 = 6$ in
$p_2 = 40 \text{ lb/in}^2$
$T_2 = 35°F$

$\dot Q$?

Fig. 21-5

21.43 A hydroelectric power plant (Fig. 21-6) takes in 30 m³/s of water through its turbine and discharges it at $V_2 = 2$ m/s at atmospheric pressure. The head loss in the turbine and penstock system is $h_f = 20$ m. Estimate the power extracted by the turbine in megawatts.

▮ $(p_1/\rho g) + (V_1^2/2g) + z_1 = (p_2/\rho g) + (V_2^2/2g) + z_2 + h_s$. We neglect viscous work and heat transfer and take section 1 at the reservoir surface, where $V_1 \approx 0$ and $p_1 = p_a$. Section 2 is at the turbine outlet.

$$(p_a/\rho g) + \tfrac{1}{2}(0)^2 + 100 = (p_a/\rho g) + [\tfrac{1}{2}(2)^2/9.81] + 0 + h_f + h_s \qquad (1)$$

With h_f given as 20 m, we solve for $h_s = 100 - 20 - 0.2 = 79.8$ m or

$$w_s = h_s g = (79.8)(9.81) = 783 \text{ m}^2/\text{s}^2 = 783 \text{ N-m/kg} = 783 \text{ J/kg} \qquad (2)$$

The result is positive, as expected from our sign convention that work done by the fluid on a turbine is positive. The power extracted is $P = \dot m w_s = \rho Q w_s = (1000)(30)(783) = 23.5 \times 10^6 \text{ J/s} = 23.5 \times 10^6 \text{ W} = 23.5 \text{ MW}$. The turbine drives an electric generator which itself has losses of about 15 percent and probably generates about 20 MW of power.

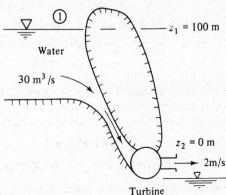

① Water

30 m³/s

$z_1 = 100$ m

$z_2 = 0$ m

2 m/s

Turbine **Fig. 21-6**

21.44 It is proposed to build a dam in a river where the flow rate is 10 m³/s and a 30-m drop in elevation can be achieved for flow through a turbine. If the turbine is 80 percent efficient, what is the maximum power that can be achieved?

$$P = \eta\gamma QH = (0.80)(9.79)(10)(30) = 2350 \text{ kN-m/s} \quad \text{or} \quad 2350 \text{ kW}$$

21.45 Steam enters a turbine at 400 psia, 600 °F, and 10 fps and is discharged at 100 fps and 20 psia saturated conditions. The mass flux is 2.0 lb/s, and the heat loss from the turbine is 6 Btu per pound of steam. Head losses and elevation changes are negligible. How much horsepower does the turbine develop?

$Q - W_s - W_v = \dot{m}(h_2 + V_2^2/2 - h_1 - V_1^2/2)$. 1 Btu/lbm = 25 000 ft-lb/slug; hence, $Q = (-6)(25\,000)(2.0/32.2) = -9332$ ft-lb/s. From the steam tables:

At 400 psia, 600 °F: $\qquad\qquad\qquad h_1 = 1306.6$ Btu/lbm

At 20 psia saturated: $\qquad\qquad\qquad h_2 = 1156.4$ Btu/lbm

$$-9332 - W_s - 0 = (2.0/32.2)[(1156.4)(25\,040) + 100^2/2 - (1306.6)(25\,000) - 10^2/2]$$
$$W_s = 224\,337 \text{ ft-lb/s} = 224\,337/550 = 408 \text{ hp}$$

21.46 Investigate the possibility of using a Pelton wheel similar to Fig. A-45 to deliver 30 000 bhp from a net head of 1200 ft.

From Fig. A-46, the most efficient Pelton wheel occurs at about $N_s \approx 4.5 = (N)(30\,000)^{1/2}/(1200)^{1.25}$ or $N = 183$ rpm. From Fig. A-45 the best operating point is $\phi \approx 0.47 = \pi D(3.06)/[2(32.2)(1200)]^{1/2}$ or $D = 13.6$ ft. This Pelton wheel is perhaps a little slow and a trifle large. You could reduce D and increase N by increasing N_s to, say, 6 or 7 and accepting the slight reduction in efficiency. Or you could use a double-hung, two-wheel configuration, each delivering 15 000 bhp, which changes D and N by the factor $2^{1/2}$.

Double wheel: $\qquad\qquad N = (183)2^{1/2} = 259$ rpm $\qquad D = 136/2^{1/2} = 9.6$ ft

21.47 For the data given in Prob. 21.46, investigate the possibility of using the Francis-turbine family of Fig. A-47.

The Francis wheel of Fig. A-47 must have $N_s = 29 = (N)(30\,000)^{1/2}/(1200)^{1.25}$ or $N = 1183$ rpm. Then the optimum power coefficient is $C_{P*} = 2.70 = P/\rho N^3 D^5 = 30\,000(550)/[(1.94)(\frac{1183}{60})^3 D^5]$ or $D = 3.33$ ft = 40 in. This is a faster speed than normal practice, and the casing would have to withstand 1200 ft of water or about 520 lb/in² internal pressure, but the 40-in size is extremely attractive. Francis turbines are now being operated at heads up to 1500 ft.

21.48 A lawn sprinkler can be used as a simple turbine. As shown in Fig. 21-7a, the flow enters normal to the paper in the center and splits evenly into $Q/2$ and V_{rel} leaving each nozzle. The arms rotate at angular velocity ω and do work on a shaft. Draw the velocity diagram for this turbine. Neglecting friction, find an expression for the power delivered to the shaft. Find the rotation rate for which the power is a maximum.

The velocity diagram for this turbine is shown in Fig. 21-7b.

$$P = \omega T = \rho Q(u_2 V_{t2} - u_1 V_{t1}) = \rho Qu(W - u) = \rho Q\omega R(V_{rel} - \omega R) \qquad dP/du = \rho Q(W - 2u) = 0$$
$$u = W/2 \quad \text{or} \quad \omega R = V_{rel}/2 \qquad P_{max} = \rho Q(\omega R)^2 = \rho Qu^2$$

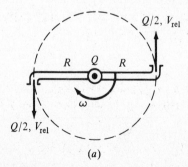

$$W = V_{rel} \qquad\qquad u = \omega R \qquad \textbf{Fig. 21-7}$$

(a)$\qquad\qquad\qquad\qquad\qquad\qquad\qquad\qquad$(b)

21.49 Apply the analysis of Prob. 21.48 to the specific case of $R = 15$ cm, $Q = 16$ m³/h, and water flow with a nozzle exit diameter of 1 cm. Compute the maximum power in watts which can be delivered.

$$P_{max} = \rho Q(\omega R)^2 = \rho Qu^2 \qquad u = V_{rel}/2 \qquad V_{rel} = (Q/2)/A_{exit} = [(\tfrac{16}{3600})/2]/[(\pi)(\tfrac{1}{100})^2/4] = 28.29 \text{ m/s}$$
$$u = 28.29/2 = 14.14 \text{ m/s} \qquad P_{max} = (998)(\tfrac{16}{3600})(14.14)^2 = 887 \text{ W}$$

21.50 Show that if the net head H is varied for a turbine operating at a given valve opening and efficiency, the speed will vary as $H^{1/2}$ and the output power as $H^{3/2}$.

■ At constant efficiency, gH/N^2D^2 is constant; and, since g and D are constant, $H \propto N^2$. Hence, $N \propto H^{1/2}$. Also at constant efficiency, $P/\rho N^3 D^5$ is constant; and, since ρ and D are constant, $P \propto N^3 \propto (H^{1/2})^3$. Hence, $P \propto H^{3/2}$.

21.51 Turbines are to be installed where the net head is 400 ft and the flow rate is 250 000 gal/min. Discuss the type, number, and size of turbine which might be selected if the generator selected is (a) 48-pole, 60 cycle ($n = 150$ rpm); and (b) 8-pole ($n = 900$ rpm). Why are at least two turbines desirable from a planning point of view?

■ Assume $\eta = 90$ percent. Select two turbines so that one will still be available for power if the other is shut down.

(a)
$$P = \eta\gamma QH = (0.90)(62.4)[(250\,000)(0.002228)](400)/550 = 22\,750 \text{ bhp}$$
$$N_s = NP^{0.5}/H^{5/4} = (150)(22\,750/2)^{0.5}/400^{5/4} = 8.9$$

Select two impulse turbines. $\phi = \pi ND/(2gH)^{1/2}$. Estimate $\phi = 0.47$. $0.47 = (\pi)(\frac{150}{60})(D)/[(2)(32.2)(400)]^{1/2}$, $D = 9.60$ ft.

(b) $N_s = (900)(22\,750/2)^{0.5}/400^{5/4} = 53.7$. Select two Francis turbines. $C_p^* = P/\rho N^3 D^5$. From Fig. A-47, $C_p^* = 2.6$. $2.6 = (22\,750/2)(550)/[(1.94)(\frac{900}{60})^3(D^5)]$, $D = 3.26$ ft.

21.52 Turbines at the Conowingo Plant on the Susquehanna River each develop 54 000 bhp at 82 rpm under a head of 89 ft. What type of turbines are these? Estimate the flow rate and impeller diameter.

■ $$N_s = NP^{0.5}/H^{5/4} = (82)(54\,000)^{0.5}/89^{5/4} = 69.7 \qquad \text{(Francis turbines)} \qquad P = \eta\gamma QH$$

From Fig. A-48, $\eta = 93$ percent. $(54\,000)(550) = (0.93)(62.4)(Q)(89)$, $Q = 5750 \text{ ft}^3/\text{s}$; $C_p^* = P/\rho N^3 D^5$. From Fig. A-47, $C_p^* = 2.6$. $2.6 = (54\,000)(550)/[(1.94)(82/60)^3(D^5)]$, $D = 18.7$ ft. (The actual Conowingo turbines have $D = 18.0$ ft.)

21.53 A certain turbine in Switzerland delivers 25 000 bhp at 500 rpm under a net head of 5330 ft. What type of turbine is this? Estimate the approximate discharge and size.

■ $$N_s = NP^{0.5}/H^{5/4} = (500)(25\,000)^{0.5}/5330^{5/4} = 1.74 \qquad \text{(impulse turbine)} \qquad P = \eta\gamma QH$$

From Fig. A-46, $\eta = 80$ percent. $(25\,000)(550) = (0.80)(62.4)(Q)(5330)$, $Q = 51.7 \text{ ft}^3/\text{s}$; $\phi = \pi ND/(2gH)^{1/2}$. Estimate $\phi = 0.47$. $0.47 = (\pi)(\frac{500}{60})(D)/[(2)(32.2)(5330)]^{1/2}$, $D = 10.5$ ft. (The actual turbines have $D = 10.9$ ft.)

21.54 A Pelton wheel of 12-ft pitch diameter operates under a net head of 2000 ft. Estimate the speed, power output, and flow rate for best efficiency if the nozzle exit diameter is 4 in.

■ $V_{\text{jet}} = C_v\sqrt{2gH}$. Take $C_v = 0.94$: $V_{\text{jet}} = (0.94)\sqrt{(2)(32.2)(2000)} = 337$ ft/s. $\phi = \pi ND/(2gH)^{1/2}$. Estimate $\phi = 0.47$: $0.47 = (\pi)(N/60)(12)/[(2)(32.2)(2000)]^{1/2}$, $N = 268$ rpm.

$$Q = AV = [(\pi)(\tfrac{4}{12})^2/4](337) = 29.4 \text{ ft}^3/\text{s} \qquad P_w = \rho Qu(V_{\text{jet}} - u)(1 - \cos\beta)$$

Assume $\beta = 165°$.

$$u = V_{\text{jet}}/2 = \tfrac{337}{2} = 168.5 \text{ fps} \qquad P_w = (1.94)(29.4)(168.5)(337 - 168.5)(1 - \cos 165°)/550 = 5788 \text{ hp}$$
$$N_s = NP^{0.5}/H^{5/4} = (268)(5788)^{0.5}/2000^{5/4} = 1.52$$

Hence, $\eta = 0.75$. $P = (0.75)(5788) = 4341$ hp.

21.55 It is planned to use the Francis-turbine family of Fig. A-47 at an installation with a head of 600 ft and a flow rate of 200 ft³/s. What are the proper impeller diameter and the optimum speed and power produced?

■ $N_s = NP^{0.5}/H^{5/4}$, $P^* = \eta\gamma QH$. Estimate $\eta = 87$ percent.

$$P^* = (0.87)(62.4)(200)(600)/550 = 11\,845 \text{ hp} \qquad N_s = 29 \qquad 29 = (N)(11\,845)^{0.5}/600^{5/4} \qquad N = 791 \text{ rpm}$$
$$Q_H^* = gH/N^2D^2 = 9.03 \qquad 9.03 = (32.2)(600)/[(\tfrac{791}{60})^2(D^2)] \qquad D = 3.51 \text{ ft}$$

21.56 One of the largest wind generators in operation today is the ERDA/NASA two-blade propeller HAWT in Sandusky, Ohio. The blades are 125 ft in diameter and reach maximum power in 19-mi/h winds. For this condition estimate (a) the power generated in kilowatts, (b) the rotor speed in revolutions per minute, and (c) the velocity V_2 behind the rotor.

▌ **(a)** $(C_p)_{max} = P/(\rho A V_1^3/2)$. From Fig. A-49, $(C_p)_{max} = 0.46$.

$$A = [(\pi)(125)^2/4] = 12\,272 \text{ ft}^2 \qquad V_1 = (19)(5280)/3600 = 27.87 \text{ fps}$$

$$0.46 = P/[(0.00233)(12\,272)(27.87)^3/2] \qquad P = 142\,367 \text{ ft-lb/s} = (142\,367)(0.3048)(4.448)/1000 = 193 \text{ kW}$$

(b) At P_{max}, $\omega r/V_1 = 5.5$. $(\omega)(\frac{125}{2})/27.87 = 5.5$, $\omega = 2.453$ rad/s, or 23.4 rpm.

(c) From ideal theory, $V_2 = V_1/3$, $V_2 = 27.87/3 = 9.29$ fps.

21.57 A hydroelectric plant (Fig. 21-8) has a difference in elevation from head water to tail water of $H = 50$ m and a flow $Q = 5$ m³/s of water through the turbine. The turbine shaft rotates at 180 rpm, and the torque in the shaft is measured to be $T = 1.16 \times 10^5$ N-m. Output of the generator is 2100 kW. Determine **(a)** the reversible power for the system, **(b)** the irreversibility, or losses, in the system, and **(c)** the losses and the efficiency in the turbine and in the generator.

▌ **(a)** The potential energy of the water is 50 m-N/N. Hence, for perfect conversion the reversible power is
$\gamma QH = (9806)(5)(50) = 2\,451\,500$ N-m/s $= 2451.5$ kW.

(b) The irreversibility, or lost power, in the system is the difference between the power into and out of the system, or $2451.5 - 2100 = 351.5$ kW.

(c) The rate of work by the turbine is the product of the shaft torque and the rotational speed:
$T\omega = (1.16 \times 10^5)[180(2\pi)/60] = 2186.5$ kW. The irreversibility through the turbine is then $2451.5 - 2186.5 = 265.0$ kW, or, when expressed as lost work per unit weight of fluid flowing,
$(265.0)(\frac{1000}{1})(\frac{1}{9806})(\frac{1}{5}) = 5.4$ m-N/N. The generator power loss is $2186.5 - 2100 = 86.5$ kW, or $86.5(1000)/[9806(5)] = 1.76$ m-N/N. Efficiency of the turbine η_t is $\eta_t = 100[(50 - 5.4)/50] = 89.2$ percent and efficiency of the generator η_g is $\eta_g = 100[(50 - 5.4 - 1.76)/(50 - 5.4)] = 96.1$ percent.

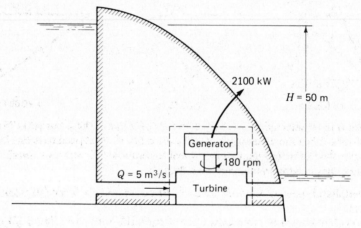

Fig. 21-8

21.58 A turbine has a velocity of 6 m/s at the entrance to the draft tube and a velocity of 1.2 m/s at the exit. For friction losses of 0.1 m and a tailwater 5 m below the entrance to the draft tube, find the pressure head at the entrance.

▌ $$p_1/\gamma = -z_s - V_1^2/2g + \text{losses} = -5 - 6^2/[(2)(9.807)] + 1.2^2/[(2)(9.807)] + 0.1 = -6.66 \text{ m}$$

21.59 The wicket gates of the propeller turbine of Fig. 21-9 are turned so that the flow makes an angle of 45° with a radial line at section 1, where the speed is 4.005 m/s. Determine the magnitude of tangential velocity component V_u over section 2.

▌ Since no torque is exerted on the flow between sections 1 and 2, the moment of momentum is constant and the motion follows the free-vortex law $V_u r = $ constant. At section 1, $V_{u1} = 4.005 \cos 45° = 2.832$ m/s, $V_{u1}r_1 = (2.832)(1.5/2) = 2.124$ m²/s. Across section 2, $V_{u2} = 2.124/r$. At the hub, $V_u = 2.124/(0.45/2) = 9.44$ m/s. At the outer edge, $V_u = 2.124/(1.2/2) = 3.54$ m/s.

21.60 Assuming uniform axial velocity over section 2 of Fig. 21-9 using the data of Prob. 21.59, determine the angle of the leading edge of the propeller at $r = 0.225$, 0.45, and 0.6 m for a propeller speed of 240 rpm.

▌ At $r = 0.225$ m, $u = \frac{240}{60}(2\pi)(0.225) = 5.66$ m/s, $V_u = 9.44$ m/s. At $r = 0.45$ m, $u = \frac{240}{60}(2\pi)(0.45) = 11.3$ m/s, $V_u = 4.72$ m/s. At $r = 0.6$ m, $u = \frac{240}{60}(2\pi)(0.6) = 15.06$ m/s, $V_u = 3.54$ m/s. The discharge through the turbine is, from section 1, $Q = (0.6)(1.5)(\pi)(4.005)(\cos 45°) = 8.01$ m³/s. Hence, the axial velocity at section 2 is $V_a = 8.01/[\pi(0.6^2 - 0.225^2)] = 8.24$ m/s. Figure 21-10 shows the initial vane angle for the three positions.

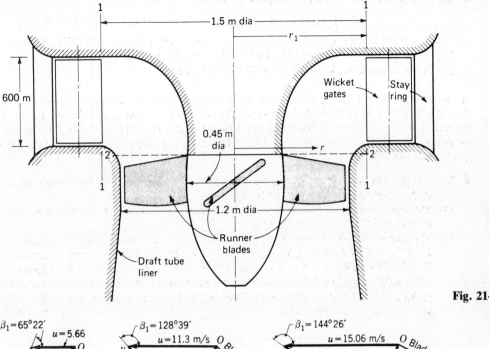

Fig. 21-9

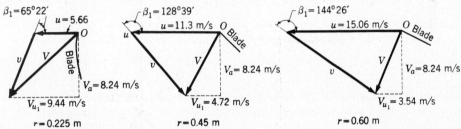

$\beta_1 = 65°22'$ $u = 5.66$ O
v V Blade
$V_a = 8.24$ m/s
$V_{u_1} = 9.44$ m/s
$r = 0.225$ m

$\beta_1 = 128°39'$ $u = 11.3$ m/s O Blade
u
v V $V_a = 8.24$ m/s
$V_{u_1} = 4.72$ m/s
$r = 0.45$ m

$\beta_1 = 144°26'$ $u = 15.06$ m/s O Blade
V
v $V_a = 8.24$ m/s
$V_{u_1} = 3.54$ m/s
$r = 0.60$ m

Fig. 21-10

21.61 A Pelton wheel is to be selected to drive a generator at 600 rpm. The water jet is 75 mm in diameter and has a velocity of 100 m/s. With the blade angle at 170°, the ratio of vane speed to initial jet speed at 0.47, and neglecting losses, determine (**a**) diameter of wheel to centerline of buckets (vanes), (**b**) power developed, and (**c**) kinetic energy per newton remaining in the fluid.

▌ (**a**) The peripheral speed of the wheel is $u = 0.47(100) = 47$ m/s. Then $\frac{600}{60}[2\pi(D/2)] = 47$ m/s or $D = 1.496$ m.

(**b**) The power, in kilowatts, is computed as $P = \rho Q u V_r (1 - \cos \theta) = (1000)(\pi/4)(0.075)^2(100)(47)(100 - 47)[1 - (-0.9848)](\frac{1}{1000}) = 2184$ kW.

(**c**) From Fig. 21.11, the absolute-velocity components leaving the vane are $V_x = (100 - 47)(-0.9848) + 47 = -5.2$ m/s, $V_y = (100 - 47)(0.1736) = 9.2$ m/s. The kinetic energy remaining in the jet is $(5.2^2 + 9.2^2)/[2(9.807)] = 5.69$ m-N/N.

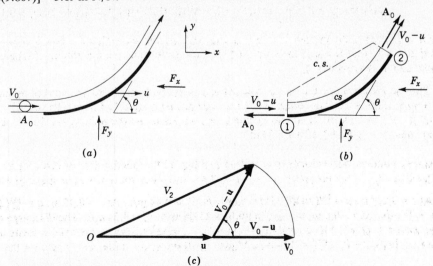

Fig. 21-11

21.62 A small impulse wheel is to be used to drive a generator for 60-Hz power. The head is 100 m, and the discharge is 40 L/s. Determine the diameter of the wheel at the centerline of the buckets and the speed of the wheel ($C_v = 0.98$). Assume efficiency of 80 percent.

▌ The power is $P = \gamma Q H_a \eta = 9806(0.040)(100)(0.80) = 31.38$ kW. By taking a trial value of N_s of 15, $N = N_s H_a^{5/4}/\sqrt{P} = 15(100^{5/4})/\sqrt{31.38} = 847$ rpm. For 60-Hz power the speed must be 3600 divided by the number of pairs of poles in the generator. For five pairs of poles the speed would be $\frac{3600}{5} = 720$ rpm, and for four pairs of poles it would be $\frac{3600}{4} = 900$ rpm. The closer speed 900 is selected. Then $N_s = N\sqrt{P}/H_a^{5/4} = 900\sqrt{31.38}/100^{5/4} = 15.94$. For $N_s = 15.94$, take $\phi = 0.448$, $u = \phi\sqrt{2gH_a} = 0.448\sqrt{2(9.807)(100)} = 19.84$ m/s and $\omega = \frac{900}{60}(2\pi) = 94.25$ rad/s. The peripheral speed u and D and ω are related: $u = \omega D/2$, $D = 2u/\omega = 2(19.84)/94.25 = 421$ mm. The diameter d of the jet is obtained from the jet velocity V_2; thus $V_2 = C_v\sqrt{2gH_a} = 0.98\sqrt{2(9.807)(100)} = 43.4$ m/s, $a = Q/V_2 = 0.040/43.4 = 9.22$ cm², $d = \sqrt{4a/\pi} = \sqrt{0.000922/0.7854} = 34.3$ mm, where a is the area of jet. Hence, the diameter ratio D/d is $D/d = 421/34.3 = 12.27$. The desired diameter ratio for the best efficiency is $D/d = 206/N_s = 206/15.94 = 12.92$ so the ratio D/d is satisfactory. The wheel diameter is 421 mm, and the speed is 900 rpm.

21.63 If 22 m³/s of water flowing through the fixed vanes of a turbine has a tangential component of 2 m/s at a radius of 1.25 m and the impeller, turning at 180 rpm, discharges in an axial direction, what torque is exerted on the impeller? Neglecting losses, what is the head on the turbine?

▌ $T = \rho Q[(rV_t)_{in} - (rV_t)_{out}] = \rho Q(r_1 V_{u1} - r_2 V_{u2}) = (1000)(22)[(1.25)(2) - 0] = 55\,000$ kg-m²/s² or 55 000 N-m

$$T\omega = Q\gamma H \qquad (55\,000)[(\tfrac{180}{60})(2\pi)] = (22)[(9.79)(1000)](H) \qquad H = 4.81 \text{ m}$$

21.64 A turbine model test with 260-mm-diameter impeller showed an efficiency of 90 percent. What efficiency could be expected from a 1.2-m-diameter impeller?

▌ $$\eta = 1 - (1 - \eta_1)(D_1/D)^{1/4} = 1 - (1 - 0.90)[(\tfrac{260}{1000})/1.2]^{1/4} = 0.932 \quad \text{or} \quad 93.2\%$$

21.65 A tangential turbine develops 7200 hp at 200 rpm under a head of 790 ft at an efficiency of 82 percent. (**a**) If the speed factor is 0.46 compute the wheel diameter, the flow, the unit speed, unit power, unit flow, and specific speed. (**b**) For this turbine, what would be the speed, power, and flow under a head of 529 ft? (**c**) For a turbine having the same design, what size of wheel should be used to develop 3800 hp under a 600-ft head and what would be its speed and rate of discharge? Assume no change in efficiency.

▌ (**a**) Since $\phi = D_1 N/(1840\sqrt{H})$, $D_1 = 1840\sqrt{790}(0.46)/200 = 119$ in. From horsepower output $= \gamma Q H/550$, $Q = (7200)(550)/[(62.4)(790)(0.82)] = 98.0$ cfs.

$N_u = ND_1/\sqrt{H} = (200)(119)/\sqrt{790} = 847$ rpm $\qquad P_u = P/D_1^2 H^{3/2} = 7200/[(119)^2(790)^{3/2}] = 0.0000229$ hp

$Q_u = Q/D_1^2\sqrt{H} = 98.0/[(119)^2\sqrt{790}] = 0.000246$ cfs $\qquad N_s = N\sqrt{P}/H^{5/4} = 200\sqrt{7200}/(790)^{5/4} = 4.05$ rpm

(**b**) $$\text{Speed } N = Nu\sqrt{H}/D_1 = 847\sqrt{529}/119 = 164 \text{ rpm}$$
$$\text{Power } P = P_u D_1^2 H^{3/2} = 0.0000229(119)^2(529)^{3/2} = 3946 \text{ hp}$$
$$\text{Flow } Q = Q_u D_1^2\sqrt{H} = 0.000246(119)^2\sqrt{529} = 80.1 \text{ cfs}$$

The above three quantities might have been obtained by noting that, for the same turbine (D_1 unchanged), the speed varies as $H^{1/2}$, the power varies as $H^{3/2}$, and Q varies as $H^{1/2}$. Thus $N = 200\sqrt{\tfrac{529}{790}} = 164$ rpm, $P = 7200(\tfrac{529}{790})^{3/2} = 3945$ hp, $Q = 98.0\sqrt{\tfrac{529}{790}} = 80.2$ cfs.
(**c**) From $P = P_u D_1^2 H^{3/2}$ we obtain $3800 = 0.0000229(D_1)^2(600)^{3/2}$ from which $D_1^2 = 11\,290$ and $D_1 = 106$ in. $N = N_u\sqrt{H}/D_1 = 847\sqrt{600}/106 = 196$ rpm, $Q = Q_u D_1^2\sqrt{H} = 0.000246(11\,290)\sqrt{600} = 68.0$ cfs.

21.66 An impulse wheel at best produces 125 hp under a head of 210 ft. By what percent should the speed be increased for a 290-ft head? Assuming equal efficiencies, what power would result?

▌ For the same wheel, the speed is proportional to the square root of the head. Thus $N_1/\sqrt{H_1} = N_2/\sqrt{H_2}$, $N_2 = N_1\sqrt{H_2/H_1} = N_1\sqrt{\tfrac{290}{210}} = 1.175N_1$. Thus, the speed should be increased 17.5 percent.

$N_s = NP^{0.5}/H^{5/4} \qquad N_1 P_1^{0.5}/H_1^{5/4} = N_2 P_2^{0.5}/H_2^{5/4} \qquad (N_1)(125)^{0.5}/210^{5/4} = (1.175N_1)(P_2)^{0.5}/290^{5/4} \qquad P_2 = 203$ hp

21.67 The reaction turbines at the Hoover Dam installation have a rated capacity of 115 000 hp at 180 rpm under a head of 487 ft. The diameter of each turbine is 11 ft and the discharge is 2350 cfs. Evaluate the speed factor, the unit speed, unit discharge and unit power, and the specific speed.

■ $\phi = D_1 N/1840\sqrt{H} = (11)(12)(180)/(1840\sqrt{487}) = 0.585$ $N_u = D_1 N/\sqrt{H} = (11)(12)(180)/\sqrt{487} = 1077$ rpm

$$Q_u = Q/D_1^2\sqrt{H} = 2350/[(132)^2\sqrt{487}] = 0.00611 \text{ cfs}$$

$$P_u = hp/D_1^2 H^{3/2} = 115\,000/[(132)^2(487)^{3/2}] = 0.000614 \text{ hp}   N_s = N_u\sqrt{P_u} = 1077\sqrt{0.000614} = 26.7$$

21.68 A single-stage impulse turbine has 5 lbm of combustion products at a speed of 6000 ft/s from the nozzles. The radius R to the nozzles is 2 ft. What is the most efficient speed ω of the turbine if $\alpha_1 = 20°$? If the turbine runs at 0.7 of this speed, what is the power developed by the turbine? What is β if $\beta_1 = \beta_2 = \beta$?

■ $$U_t = R\omega = \tfrac{1}{2}V_1 \cos \alpha_1   (2)(\omega) = \tfrac{1}{2}(6000) \cos 20°   \omega = 1410 \text{ rad/s} = 13\,460 \text{ rpm}$$

$$\text{Actual speed} = (0.7)(1410) = 987 \text{ rad/s}$$

$$dW_s/dt = 2\rho QU_t(V_1 \cos \alpha_1 - U_t) = (2)(5/g)(2)(987)[6000 \cos 20° - (2)(987)]/550 = 4084 \text{ hp}$$

Torque is found next:

$$M_z(987) = (4084)(550)   M_z = 2276 \text{ ft-lb}$$

$$(\mathbf{V}_{rel})_1 = \mathbf{V}_1 - \mathbf{U}_t = (6000)[\cos 20°\mathbf{j} + \sin 20°\mathbf{i}] - (2)(987)\mathbf{j} = 3664\mathbf{j} + 2052\mathbf{i} \text{ ft/s}$$

$$\beta = \tan^{-1}\tfrac{2052}{3664} = 29.25°$$

21.69 Three nozzles expand combustion products to a speed of 1800 m/s. The density of the fluid as it reaches the turbine blades is 0.46 kg/m³. If the turbine is to rotate at its most efficient speed of 12 000 rpm, what should the angle α_1 be for the three nozzles? The nozzles are at a distance R of 0.6 m from the axis of rotation. If the power desired is 6000 hp, what should the exit areas of the nozzles be and what should the blade angles β be? Take $\beta_1 = \beta_2$.

■ $$U_t = \tfrac{1}{2}V_1 \cos \alpha   (0.6)[(12\,000)(2\pi)/60] = \tfrac{1}{2}(1800) \cos \alpha_1   \alpha_1 = 33.1°$$

$$dW_s/dt = 2\rho QU_t(V_1 \cos \alpha - U_t)$$

$$(6000)(745.7) = (2)(0.46)(Q)(0.6)[(12\,000)(2\pi)/60]\{1800 \cos 33.1° - [(0.6)(12\,000)(2\pi)/60]\}$$

$$Q = 8.56 \text{ m}^3/\text{s}   3(A)(V_a) = 8.56   3(A)(1800 \sin 33.1°) = 8.56$$

$$A = 2.903 \times 10^{-3} \text{ m}^2 = 2903 \text{ mm}^2   M_z = (6000)(745.7)/[(12\,000)(2\pi)/60] = -2R\rho QV_a \cot \beta$$

$$3560 = (2)(0.6)(0.46)(8.56)(1800)(\sin 33.1°) \cot \beta   \cot \beta = 0.76646   \beta = 52.5°$$

21.70 If the effective head H for an axial-flow hydraulic turbine is 90 m and the volumetric flow Q is 20 m³/s, what should the speed ω of the rotor be? The radius R_o of the guide vanes is 1.5 m and the height b is 1 m. The exit flow is axial only and the guide vanes are at an angle α of 11.0°.

■ $g(\Delta H) = (U_t V_t)_2 - (U_t V_t)_1$. At any position r:

$$(9.81)(-90) = -(\omega)(r)(V_t)_1 \tag{1}$$

From *continuity*: $Q = (2)(\pi)(R_0)(b)(V_0) \sin \alpha$, $20 = (2\pi)(1.5)(1)(V_0) \sin 11°$, $V_0 = 11.12$ m/s. At any distance r:

$$V_t = (V_o R_o \cos \alpha)/r = [(11.12)(1.5) \cos 11°]/r = 16.37/r \tag{2}$$

Replace $(V_t)_1$ in Eq. (1) using Eq. (2): $(9.81)(90) = \omega r(16.37/r)$, $\omega = 53.9$ rad/s $= 515$ rpm.

21.71 The first stage of an axial-flow gas turbine has a hub-to-tip diameter ratio of 0.85 with the tip having a diameter of 4 ft. A flow of 50 lbm/s of combustion products flows through the machine. At the first stage, the pressure is 5 atm and the temperature is 1800 °R. If $\alpha_1 = 38°$ and $\beta_2 = 55°$ for the first stage and it is to develop 25 hp, at what speed should the rotor turn? Consider the gas constant R to be that of air for the working fluid.

■ Find V_a.

$$Q = V_a(\pi/4)\{4^2 - [(0.85)(4)]^2\} \tag{1}$$

Also we can say $Q = (50/g)/\rho$, $\rho = p/RT = (14.7)(144)(5)/[(533)(32.2)(1800)] = 0.003426$ slug/ft³, $Q = (50/g)/0.003426 = 453.2$ cfs. Go back to Eq. (1): $453.2 = V_a(\pi/4)\{4^2 - [(0.85)(4)]^2\}$, $V_a = 129.96$ ft/s. Compute ΔH for the first stage.

$$-\Delta H(Q\rho g) = 25(550)   \Delta H = -(25)(550)/[(453.2)(0.003426)(32.2)] = -275 \text{ ft}$$

$$\Delta Hg/U_t^2 = [1 - (V_a/U_t)(\cot \alpha_1 + \cot \beta_2)]   U_t^2 - (129.96)(\cot 38° + \cot 55°)U_t + (275)(32.2) = 0$$

$$U_t^2 - 257U_t + 8855 = 0   U_t = [257 \pm \sqrt{257^2 - (4)(8855)(1)}]/(2)(1)   (U_t)_1 = 216 \text{ ft/s}   (U_t)_2 = 41.0 \text{ fps}$$

$$U_t = R_{av}\omega   216 = (2 - 0.15)(\omega_1)   \omega_1 = 116.8 \text{ rad/s} = 1115 \text{ rpm}   41.0 = (2 - 0.15)(\omega_2)$$

$$\omega_2 = 22.16 \text{ rad/s} = 212 \text{ rpm}$$

21.72 Water flows from an upper reservoir to a lower one while passing through a turbine, as shown in Fig. 21-12. Find the power generated by the turbine. Neglect minor losses.

❚ $p_1/\gamma + V_1^2/2g + z_1 - E_t = p_2/\gamma + V_2^2/2g + z_2 + h_L$. From Fig. A-14, with $Q = 0.15$ m³/s and $D = 250$ mm, $h_1 = 0.037$ m/m.

$$h_L = h_f = (0.037)(100) = 3.70 \text{ m} \qquad 0 + 0 + 197.3 - E_t = 0 + 0 + 50.0 + 3.70$$

$$E_t = 143.6 \text{ m} \qquad P = Q\gamma E_t = (0.15)(9.79)(143.6) = 211 \text{ kN-m/s} \quad \text{or} \quad 211 \text{ kW}$$

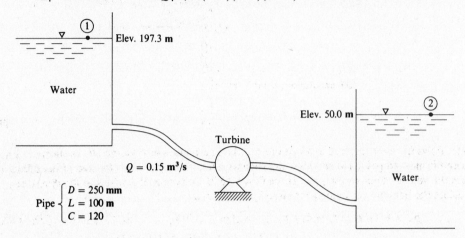

Elev. 197.3 m

Water

Elev. 50.0 m

Turbine

$Q = 0.15$ m³/s

Water

Pipe $\begin{cases} D = 250 \text{ mm} \\ L = 100 \text{ m} \\ C = 120 \end{cases}$

Fig. 21-12

21.73 Water is being discharged from a reservoir through a turbine, as shown in Fig. 21-13. What water-surface elevation is required in the reservoir in order for the turbine to generate 100 hp? Neglect minor losses.

❚ $p_1/\gamma + V_1^2/2g + z_1 - E_t = p_2/\gamma + V_2^2/2g + z_2 + h_L \qquad P = Q\gamma E_t \qquad 100 = (20)(62.4)(E_t)/550 \qquad E_t = 44.07$ ft

$$V_2 = Q/A_2 = 20/[(\pi)(\tfrac{36}{12})^2/4] = 2.829 \text{ ft/s}$$

From Fig. A-13, with $Q = 20$ ft³/s and $D = 36$ in, $h_1 = 0.00084$ ft/ft.

$$h_L = h_f = (0.00084)(270) = 0.23 \text{ ft} \qquad 0 + 0 + z_1 - 44.07 = 0 + 2.829^2/[(2)(32.2)] + 100 + 0.23$$

$$z_1 = 144.4 \text{ ft}$$

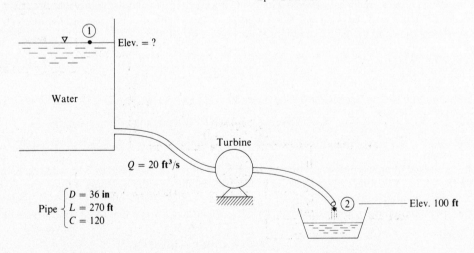

Elev. = ?

Water

Turbine

$Q = 20$ ft³/s

Pipe $\begin{cases} D = 36 \text{ in} \\ L = 270 \text{ ft} \\ C = 120 \end{cases}$

Elev. 100 ft

Fig. 21-13

21.74 The flow rate of water through the turbine shown in Fig. 21-14 is 0.20 m³/s, and the pressures at points 1 and 2 are 150 and −35.0 kPa, respectively. Determine the energy generated by the turbine.

❚

$$p_1/\gamma + v_1^2/2g + z_1 - E_t = p_2/\gamma + v_2^2/2g + z_2 + h_L \qquad p_1/\gamma = 150/9.79 = 15.32 \text{ m}$$

$$v_1 = Q/A_1 = 0.20/[(\pi)(0.25)^2/4] = 4.07 \text{ m/s}$$

$$v_1^2/2g = (4.07)^2/[(2)(9.807)] = 0.84 \text{ m} \qquad z_1 = 1.00 \text{ m} \qquad p_2/\gamma = -35/9.79 = -3.58 \text{ m}$$

$$v_2 = Q/A_2 = 0.20/[(\pi)(0.50)^2/4] = 1.02 \text{ m/s} \qquad v_2^2/2g = (1.02)^2/[(2)(9.807)] = 0.05 \text{ m} \qquad z_2 = 0 \qquad h_L = 0$$

$$15.32 + 0.84 + 1.0 - E_t = -3.58 + 0.05 + 0 + 0 \qquad E_t = 20.69 \text{ m} \qquad p = Q\gamma E_t = (0.20)(9.79)(20.69) = 40.5 \text{ kW}$$

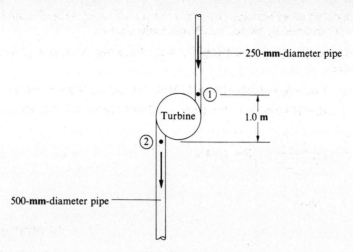

Fig. 21-14

21.75 Water flows from reservoir A to a turbine and then discharges into tailrace B, as shown in Fig. 21-15. The loss of head from A to point 1 is six times the velocity head in the pipe, and the loss of head from point 2 to B is 0.5 times the velocity head in the pipe. If the flow rate in the pipe is 30.0 ft^3/s and the diameter of the pipe is 24 in, calculate the horsepower being generated by the turbine.

$\blacksquare$ $\quad p_A/\gamma + v_A^2/2g + z_A - E_t = p_B/\gamma + v_B^2/2g + z_B + h_L \qquad p_A/\gamma = p_B/\gamma = v_A^2/2g = v_B^2/2g = 0$

$$v = Q/A = 30/[(\pi)(2.0)^2/4] = 9.55 \text{ ft/s}$$

$$v^2/2g = (9.55)^2/[(2)(32.2)] = 1.416 \text{ ft} \qquad h_L \text{ from } A \text{ to } 1 = (6)(1.416) = 8.50 \text{ ft}$$

$$h_L \text{ from } 2 \text{ to } B = (0.5)(1.416) = 0.71 \text{ ft}$$

$$0 + 0 + 210 - E_t = 0 + 0 + 0 + 8.50 + 0.71 \qquad E_t = 200.79 \text{ ft}$$

$$P = Q\gamma E_t = (30)(62.4)(200.79) = 375\,879 \text{ ft-lb/s} = 375\,879/550 = 683 \text{ hp}$$

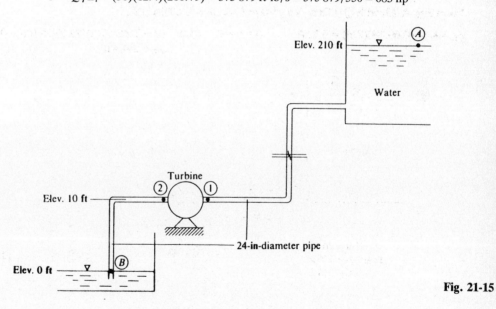

Fig. 21-15

21.76 Water is being discharged from a reservoir through a turbine, as shown in Fig. 21-16. Determine the elevation of the water surface in the reservoir that would be required in order for the turbine to generate 56 kW of power. Neglect minor losses. Assume $C = 120$ for all pipes.

$\blacksquare$ $\quad p_1/\gamma + v_1^2/2g + z_1 - E_t = p_2/\gamma + v_2^2/2g + z_2 + h_L \qquad p_1/\gamma = p_2/\gamma = v_1^2/2g = 0 \qquad p = Q\gamma E_t$

$56 = (0.28)(9.79)(E_t) \qquad E_t = 20.43 \text{ m} \qquad v_2^2/2g = (Q/A)^2/2g = \{0.28/[(\pi)(0.60)^2/4]\}^2/[(2)(9.807)] = 0.05 \text{ m}$

$$z_2 = 50 \text{ m}$$

With $Q = 0.28$ m^3/s and $D = 300$ m, $(h_1)_{300} = 0.050$ m/m. With $Q = 0.28$ m^3/s and $D = 600$ mm, $(h_1)_{600} = 0.0017$ m/m.

$$h_L = h_f = (0.050)(50) + (0.0017)(20) = 2.53 \text{ m} \qquad 0 + 0 + z_1 - 20.43 = 0 + 0.05 + 50 + 2.53 \qquad z_1 = 73.0 \text{ m}$$

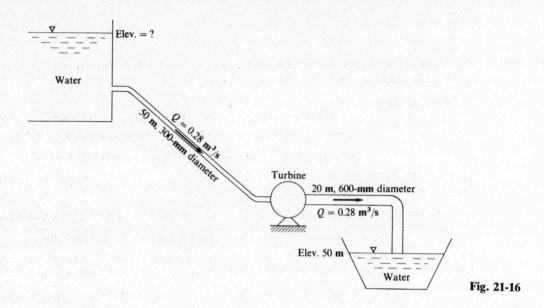

Fig. 21-16

22.1 Sketch the hydraulic grade line and the energy grade line for the pipe shown in Fig. 22-1a. Evaluate key points for the hydraulic grade line. The turbine is developing 50 kW. The water is at 5 °C.

┃ First compute q.

$$(V_1^2/2) + (p_1/\rho) + 0 - (dW_s/dm) = (V_2^2/2) + (p_2/\rho) + 0 + h_L \tag{1}$$

$$dW_s/dm = (dW_s/dt)(dt/dm) = (50\,000)/[V_2(\pi/4)(0.3^2)(1000)] = 707/V_2 \tag{2}$$

$$h = f_1(100/0.5)(V_1^2/2) + f_2(100/0.3)(V_2^2/2) + 0.05(V_1^2/2) + K(V_1^2/2) \tag{3}$$

Let $f_1 = 0.023$, $f_2 = 0.0258$: $K = 0.5[1 - (0.3/0.5)^2]/(0.3/0.5)^4 = 2.47$. Substitute the above results into Eq. (1) noting that $V_1 = (0.3/0.5)^2 V_2 = 0.36 V_2$,

$$[(0.36V_2)^2/2] + (p_1/\rho) - (707/V_2) = (V_2^2/2) + (0.023)(100/0.5)[(0.36V_2)^2/2] + (0.0258)(100/0.3)(V_2^2/2)$$
$$+ [(0.05)(0.36V_2)^2/2] + [2.47(0.36V_2)^2/2] \tag{4}$$

Use *Bernoulli* in tank: $100g = (p_1/\rho) + [(0.36V_2)^2/2] + 0$. Substitute into Eq. (4) and carry out the calculation: $100g - (707/V_2) = 5.26V_2^2$, $5.26V_2^3 - 100gV_2 + 707 = 0$. Solve by trial and error: $V_2 = 13.28$ m/s, $V_1 = 4.78$ m/s.

At A:

$$(H_{\text{Hyd}})_a = 100 - (0.05)(4.78^2/2g) - (0.023)(100/0.5)(4.78^2/2g) \qquad (H_{\text{Hyd}})_a = 94.6 \text{ m}$$

After contraction:

$$(H_{\text{Hyd}}) = -(2.47)(4.78^2/2g) + 94.6 = 91.7 \text{ m}$$

At B:

$$(H_{\text{Hyd}})_b = 91.7 - (0.0258)(50/0.3)(13.28^2/2g) \qquad (H_{\text{Hyd}})_b = 53.0 \text{ m}$$

After turbine (b'):

$$(H_{\text{Hyd}})_{b'} = 53.0 - [707/(13.28)(g)] = 47.6 \text{ m}$$

At C:

$$(H_{\text{Hyd}}) = 47.6 - (0.0258)(50/0.3)(13.28^2/2g) = 8.9 \text{ m}$$

This just equals the kinetic energy head at exit so solution is self-consistent. The hydraulic grade line is sketched in Fig. 22-1b.

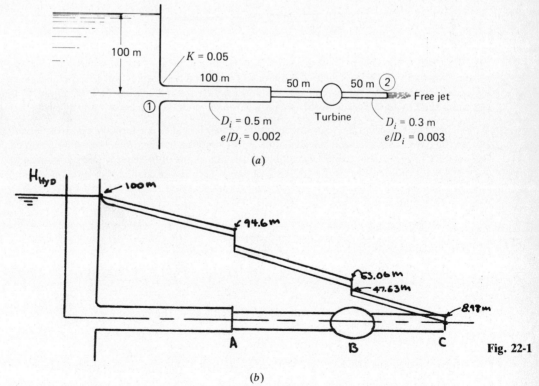

Fig. 22-1

22.2 At a point A in a horizontal 12-in pipe ($f = 0.020$) the pressure head is 200 ft. At a distance of 200 ft from A the 12-in pipe reduces suddenly to a 6-in pipe. At a distance of 100 ft from this sudden reduction the 6-in pipe ($f = 0.015$) suddenly enlarges to a 12-in pipe and point F is 100 ft beyond this change in size. For a velocity of 8.025 ft/s in the 12-in pipes, draw the energy and hydraulic grade lines. Refer to Fig. 22-2.

❚ The velocity heads are $V_{12}^2/2g = (8.025)^2/2g = 1.00$ ft and $V_6^2/2g = 16.0$ ft. The energy line drops in the direction of flow by the amount of the lost head. The hydraulic grade line (gradient) is below the energy line by the amount of the velocity head at any cross section. Note (in Fig. 22-2) that the *hydraulic grade line* can rise where a change (enlargement) in size occurs.

Tabulating the results to the nearest 0.1 ft,

	lost head in feet		elevation energy line	$V^2/2g$	elevation of hyd. gradient
at	**from**	**calculated**			
A	(Elevation 0.0)		201.0	1.0	200.0
B	A to B	$(0.020)(\frac{200}{1})(1) = 4.0$	197.0	1.0	196.0
C	B to C	$(K_c)(16) = (0.37)(16) = 5.9$	191.1	16.0	175.1
D	C to D	$(0.015)(100/\frac{1}{2})(16) = 48.0$	143.1	16.0	127.1
E	D to E	$\dfrac{(V_6 - V_{12})^2}{2g} = \dfrac{(32.1 - 8.0)^2}{64.4} = 9.0$	134.1	1.0	133.1
F	E to F	$(0.020)(\frac{100}{1})(1) = 2.0$	132.1	1.0	131.1

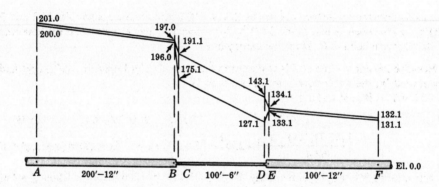

Fig. 22-2

22.3 The elevations of the energy line and hydraulic grade line at point G are 44.0 and 42.0 ft, respectively. For the system shown in Fig. 22-3. calculate **(a)** the power extracted between G and H if the energy line at H is at elevation 4.0 and **(b)** the pressure heads at E and F which are at elevation 20.0. **(c)** Draw, to the nearest 0.1 ft, the energy and hydraulic grade lines, assuming K for valve CD is 0.40 and $f = 0.010$ for the 6-in pipes.

❚ The flow must be from the reservoir since the energy line at G is below the reservoir level. GH is a turbine. Before the power extracted can be calculated, flow Q and the head extracted must be obtained.
(a) At G, $V_{12}^2/2g = 2.0$ (the difference between the elevations of the energy and hydraulic grade lines). Also $V_6^2/2g = (16)(2.0) = 32.0$ and $V_{24}^2/2g = \frac{1}{16}(2.0) = 0.13$ ft. To obtain Q, $V_{12} = 11.34$ ft/s and $Q = \frac{1}{4}\pi(1)^2(11.34) = 8.91$ cfs. h.p. $= \gamma QH_T/550 = 62.4(8.91)(44.0 - 4.0)/550 = 40.4$ extracted.
(b) F to G, datum zero

$$\text{(Energy at } F) - 0.030(\tfrac{100}{1})(2.0) = \text{(energy at } G = 44.0) \qquad \text{Energy at } F = 44.0 + 6.0 = 50.0 \text{ ft}$$

E to F, datum zero:

$$\text{(Energy at } E) - (45.4 - 11.3)^2/2g = \text{(energy at } F = 50.0) \qquad \text{Energy at } E = 50.0 + 18.0 = 68.0 \text{ ft}$$
$$z + V^2/2g$$

The pressure head at $E = 68.0 - (20 + 32) = 16.0$ ft water. The pressure head at $F = 50.0 - (20 + 2) = 28.0$ ft water.
(c) Working back from E:

Drop in energy line $DE = 0.010(25/\frac{1}{2})(32.0) = 16.0$ ft   Drop in energy line $CD = 0.40(32.0) = 12.8$ ft

Drop in energy line BC = same as $DE = 16.0$ ft   Drop in energy line $AB = 0.50(32.0) = 16.0$ ft

(Elevation at D − 16.0) = elevation at E of 68.0   Elev. D = 84.0

(Elevation at C − 12.8) = elevation at D of 84.0   Elev. C = 96.8

(Elevation at B − 16.0) = elevation at C of 96.8   Elev. B = 112.8

(Elevation at A − 16.0) = elevation at B of 112.8   Elev. A = 128.8

The hydraulic grade line is $V^2/2g$ below the energy line: 32.0 ft in the 6-in, 2.0 ft in the 12-in and 0.13 ft in the 24-in. The values are shown in Fig. 22-3.

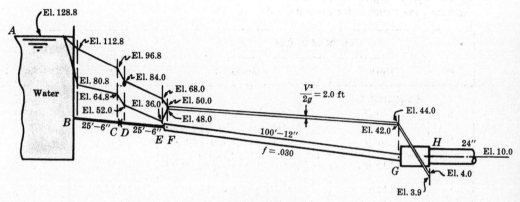

Fig. 22-3

22.4   The head extracted by turbine CR in Fig. 22-4 is 200 ft and the pressure at T is 72.7 psi. For losses of $2.0(V_{24}^2/2g)$ between W and R and $3.0(V_{12}^2/2g)$ between C and T, determine (a) how much water is flowing and (b) the pressure head at R. Draw the energy line.

❚   Because the energy line at T is at elevation $\{250.0 + [(72.7)(44)/62.4] + (V_{12}^2/2g)\}$ and well above the elevation at W, the water flows into the reservoir W.

(a) Using T to W, datum zero,

$$\underset{\text{at } T}{\underbrace{\left[\frac{(72.7)(144)}{62.4} + \frac{V_{12}^2}{2g} + 250\right]}} - \underset{T \text{ to } C \quad R \text{ to } W}{\underbrace{\left(3.0\frac{V_{12}^2}{2g} + 2.0\frac{V_{24}^2}{2g}\right)}} - \underset{H_T}{\underbrace{200}} = \underset{\text{at } W}{\underbrace{(0 + \text{negl} + 150)}}$$

Substituting $V_{24}^2 = \frac{1}{16}V_{12}^2$ and solving, $V_{12}^2/2g = 32.0$ ft or $V_{12} = 45.4$ ft/s. Then $Q = \frac{1}{4}\pi(1)^2(45.4) = 35.7$ cfs.

(b) Using R to W, datum R, $[(p_R/\gamma) + \frac{1}{16}(32.0) + 0] - 2(\frac{1}{16})(32.0) = (0 + \text{negl} + 50)$ and $p_R/\gamma = 52.0$ ft. The reader may check this pressure head by applying the Bernoulli equation between T and R.

To plot the energy line in the figure, evaluate the energy at the four sections indicated:

Elevation of energy line at T = 168.0 + 32.0 + 250.0 = 45.0

Elevation of energy line at C = 450.0 − 3(32.0) = 354.0

Elevation of energy line at R = 354.0 − 200.0 = 154.0

Elevation of energy line at W = 154.0 − 2($\frac{1}{16}$)(32) = 150.0

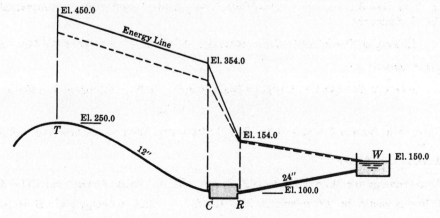

Fig. 22-4

22.5 A pipeline with a pump leads to a nozzle as shown in Fig. 22-5. Find the flow rate when the pump develops a head of 80 ft. Assume that the head loss in the 6-in-diameter pipe may be expressed by $h_L = 5V_6^2/2g$, while the head loss in the 4-in-diameter pipe is $h_L = 12V_4^2/2g$. Sketch the energy line and hydraulic grade line, and find the pressure head at the suction of the pump.

▌ Select the datum as the elevation of the water surface in the reservoir. Note from continuity that $V_6 = (\frac{3}{6})^2 V_3 = 0.25 V_3$ and $V_4 = (\frac{3}{4})^2 V_3 = 0.563 V_3$, where V_3 is the jet velocity. Writing an energy equation from the surface of the reservoir to the jet, $[z_1 + (p_1/\gamma) + (V_1^2/2g)] - h_{L_6} + h_p - h_{L_4} = z_3 + (p_3/\gamma) + (V_3^2/2g)$, $0 + 0 + 0 - 5(V_6^2/2g) + 80 - 12(V_4^2/2g) = 10 + 0 + (V_3^2/2g)$. Express all velocities in terms of V_3: $-[5(0.25V_3)^2/2g] + 80 - 12[(0.563V_3)^2/2g] = 10 + (V_3^2/2g)$, $V_3 = 29.7$ fps; $Q = A_3 V_3 = (\pi/4)(\frac{3}{12})^2(29.7) = 1.45$ cfs. Head loss in suction pipe: $h_L = 5(V_6^2/2g) = [5(0.25V_3)^2/2g] = 0.312V_3^2/2g = 4.3$ ft. Head loss in discharge pipe: $h_L = 12(V_4^2/2g) = 12(0.563V_3)^2/2g = 52.1$ ft, $V_3^2/2g = 13.7$ ft, $V_4^2/2g = 4.3$ ft, $V_6^2/2g = 0.86$ ft ≈ 0.9 ft. The energy line and hydraulic grade line are drawn on the figure to scale. Inspection of the figure shows that the pressure head on the suction side of the pump is $p_B/\gamma = 14.8$ ft. Likewise, the pressure head at any point in the pipe may be found if the figure is to scale.

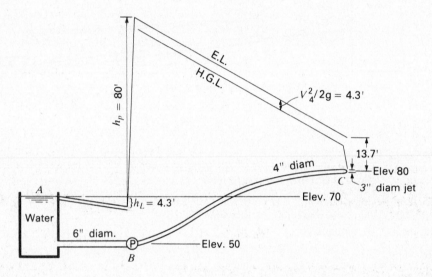

Fig. 22-5

22.6 A pump lifts water at the rate of 200 cfs to a height of 400 ft and the friction loss in the pipe is 30 ft. What is the horsepower required if the pump efficiency is 90 percent? Sketch the energy grade line and the hydraulic grade line.

▌ Water hp $= Q\gamma(\Delta z + h_L)/550 = (200)(62.4)(400 + 30)/550 = 9757$ Input power $= 9757/0.90 = 10\,840$

The energy and hydraulic grade lines are sketched in Fig. 22-6.

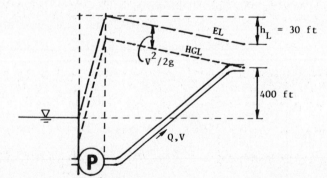

Fig. 22-6

22.7 The diameters of the suction and discharge pipes of a pump are 6 in and 4 in, respectively. The discharge pressure is read by a gage at a point 5 ft above the centerline of the pump, and the suction pressure is read by a gage 2 ft below the centerline. If the pressure gage reads 20 psi and the suction gage reads a vacuum of 10 inHg when gasoline (s.g. = 0.75) is pumped at the rate of 1.2 cfs, find the power delivered to the fluid. Sketch the energy grade line and the hydraulic grade line.

$$p_1/\gamma + z_1 + V_1^2/2g + E_p = p_2/\gamma + z_2 + V_2^2/2g + h_L \qquad V_1 = Q/A_1 = 1.2/[(\pi)(\tfrac{6}{12})^2/4] = 6.112 \text{ fps}$$

$$V_2 = 1.2/[(\pi)(\tfrac{4}{12})^2/4] = 13.75 \text{ fps} \qquad h_L = 0$$

$$-(\tfrac{10}{12})(13.6/0.75) - 2 + 6.112^2/[(2)(32.2)] + E_p = (20)(144)/[(0.75)(62.4)] + 5 + 13.75^2/[(2)(32.2)] + 0$$

$$E_p = 86.01 \text{ ft}$$

$$P = Q\gamma E_p = (1.2)[(0.75)(62.4)](86.01)/550 = 8.78 \text{ hp}$$

The energy and hydraulic grade lines are sketched in Fig. 22-7.

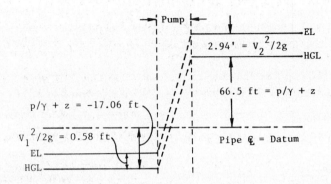

Fig. 22-7

22.8 Determine the elevation of hydraulic and energy grade lines at points A, B, C, D, and E of Fig. 22-8 ($z = 10$ ft).

I Solving for the velocity head is accomplished by applying the energy equation from the reservoir to E,
$10 + 60 + 0 + 0 = (V_E^2/2g) + 10 + 0 + \frac{1}{2}(V^2/2g) + 0.020(200/0.50)(V^2/2g) + 10(V^2/2g) + 0.10(V_E^2/2g)$. From the continuity equation, $V_E = 4V$. After simplifying, $60 = (V^2/2g)[16 + \frac{1}{2} + 8 + 10 + 16(0.1)] = 36.1(V^2/2g)$ and $V^2/2g = 1.66$ ft. Applying the energy equation for the portion from the reservoir to A gives $70 + 0 + 0 = (V^2/2g) + (p/\gamma) + z + 0.5(V^2/2g)$. Hence, the hydraulic grade line at A is

$$\frac{p}{\gamma} + z \bigg|_A = 70 - 1.5\frac{V^2}{2g} = 70 - 1.5(1.66) = 67.51 \text{ ft}$$

The energy grade line for A is $(V^2/2g) + z + (p/\gamma) = 67.51 + 1.66 = 69.17$ ft. For B, $70 + 0 + 0 = (V^2/2g) + (p/\gamma) + z + 0.5(V^2/2g) + 0.02(80/0.5)(V^2/2g)$ and

$$\frac{p}{\gamma} + z \bigg|_B = 70 - (1.5 + 3.2)(1.66) = 62.19 \text{ ft}$$

The energy grade line is at $62.19 + 1.66 = 63.85$ ft.
Across the valve the hydraulic grade line drops by $10V^2/2g$, or 16.6 ft. Hence, at C the energy and hydraulic grade lines are at 47.25 ft and 45.59 ft, respectively.
At point D, $70 = (V^2/2g) + (p/\gamma) + z + [10.5 + 0.02(200/0.50)](V^2/2g)$ and

$$\frac{p}{\gamma} + z \bigg|_D = 70 - 19.5(1.66) = 37.6 \text{ ft}$$

with the energy grade line at $37.6 + 1.66 = 39.26$ ft.
At point E the hydraulic grade line is 10 ft, and the energy grade line is $z + (V_E^2/2g) = 10 + 16(V^2/2g) = 10 + 16(1.66) = 36.6$ ft.

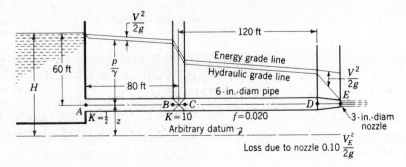

Fig. 22-8

22.9 A pump with a shaft input of 7.5 kW and an efficiency of 70 percent is connected in a water line carrying $0.1 \, m^3/s$. The pump has a 150-mm-diameter suction line and a 120-mm-diameter discharge line. The suction line enters the pump 1 m below the discharge line. For a suction pressure of $70 \, kN/m^2$, calculate the pressure at the discharge flange and the rise in the hydraulic grade line across the pump.

▌ If the energy added in meter-newtons per newton is symbolized by E, the fluid power added is $Q\gamma E = 7500(0.70)$ or $E = 7500(0.7)/[0.1(9806)] = 5.354$ m. Applying the energy equation from suction flange to discharge flange gives $(V_s^2/2g) + (p_s/\gamma) + 0 + 5.354 = (V_d^2/2g) + (p_d/\gamma) + 1$ in which the subscripts s and d refer to the suction and discharge conditions, respectively. From the continuity equation, $V_s = 0.1(4)/(0.15^2\pi) = 5.66$ m/s, $V_d = 0.1(4)/(0.12^2\pi) = 8.84$ m/s. Solving for p_d gives $p_d/\gamma = [5.66^2/2(9.806)] + (70\,000/9806) \div 5.354 - [8.84^2/2(9.806)] - 1 = 9.141$ m and $p_d = 89.6 \, kN/m^2$. The rise in hydraulic grade line is $[(p_d/\gamma) + 1] - (p_s/\gamma) = 9.141 + 1 - (70\,000/9806) = 3.002$ m. In this example much of the energy was added in the form of kinetic energy, and the hydraulic grade line rises only 3.002 m for a rise of energy grade line of 5.354 m.

22.10 Sketch the hydraulic and energy grade lines for Fig. 22-9a ($H = 8$ m).

▌
$$p_1/\gamma + V_1^2/2g + z_1 = p_2/\gamma + V_2^2/2g + z_2 + h_L \qquad h_L = h_f + h_m \qquad h_f = (f)(L/D)(V^2/2g)$$

$(\epsilon/D)_1 = 0.000046/(\frac{150}{1000}) = 0.000307$; try $f_1 = 0.015$. $(\epsilon/D)_2 = 0.000046/(\frac{300}{1000}) = 0.000153$; try $f_2 = 0.015$.

$$h_f = (0.015)[20/(\tfrac{150}{1000})]\{V_1^2/[(2)(9.807)]\} + (0.015)[(25+7)/(\tfrac{300}{1000})]\{(V_1/4)^2/[(2)(9.807)]\} = 0.1071V_1^2$$
$$h_m = KV^2/2g = [1 + (1-\tfrac{1}{4})^2 + 3.5/16 + \tfrac{1}{16}]\{V_1^2/[(2)(9.807)]\} = 0.09400V_1^2$$
$$h_L = 0.1071V_1^2 + 0.09400V_1^2 = 0.2011V_1^2 \qquad 0+0+8 = 0+0+0+0.2011V_1^2 \qquad V_1 = 6.307 \text{ m/s}$$
$$V_1^2/2g = 6.307^2/[(2)(9.807)] = 2.028 \text{ m} \qquad V_2 = 6.307/4 = 1.577 \text{ m/s} \qquad V_2^2/2g = 1.577^2/[(2)(9.807)] = 0.127 \text{ m}$$

Friction losses are $(f_1)(L/D)(v_1^2/2g) = 4.056$ m, $(f_2)(L/D)(v_2^2/2g) = 0.203$ m. Minor losses are $(V_1 - V_2)^2/2g = 1.140$ m, $3.5V_2^2/2g = 0.444$ m. The hydraulic and energy grade lines are sketched in Fig. 22-9b. [Further checking of the Reynolds number (not shown) indicates that the assumed values of f_1 and f_2 are acceptable.]

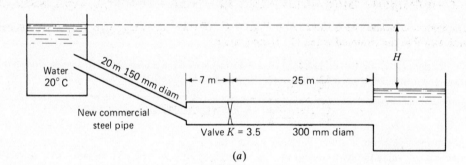

(a)

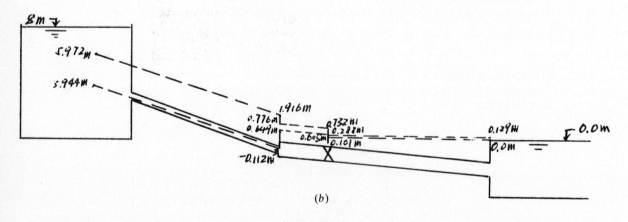

(b)

Fig. 22-9

22.11 Determine the slope of the hydraulic grade line for flow of atmospheric air through a rectangular 18-in by 6-in galvanized iron conduit ($V = 30$ fps).

▮ $h_f = (f)(L/D)(V^2/2g)$. Slope of the hydraulic grade line $= h_f/L = (f/D)(V^2/2g)$.

$$\epsilon = 0.0005 \text{ ft} \qquad R_h = (\tfrac{18}{12})(\tfrac{6}{12})/4 = 0.1875 \text{ ft} \qquad D = 4R_h = (4)(0.1875) = 0.7500 \text{ ft}$$

$$\epsilon/D = 0.0005/0.7500 = 0.000667 \qquad N_R = DV/v = (0.75)(30)/(1.69 \times 10^{-4}) = 1.33 \times 10^5$$

From Fig. A-5, $f = 0.021$. Slope of the hydraulic grade line $= (0.021/0.75)\{30^2/[(2)(32.2)]\} = 0.391$ ft/ft.

22.12 What size square conduit is needed to convey 300 L/s of water at 15 °C with a slope of the hydraulic grade line of 0.001? ($\epsilon = 1$ mm)

▮ Let d = width of each side of the square conduit.

$$R_h = d^2/4d = d/4 \qquad D = 4R_h = 4d/4 = d \qquad V = Q/A = (\tfrac{300}{1000})/d^2 = 0.300/d^2$$

$$N_R = DV/v = (d)(0.300/d^2)/(1.16 \times 10^{-6}) = 259\,000/d \qquad \epsilon/D = 0.001/d \qquad h_f = (f)(L/D)(V^2/2g)$$

$$h_f/L = 0.001 = (f/d)\{(0.300/d^2)^2/[(2)(9.807)]\} \qquad d^5 = 4.589f$$

Assume $f = 0.022$.

$$d^5 = (4.589)(0.022) \qquad d = 0.632 \text{ m} \qquad N_R = 259\,000/0.632 = 410\,000$$

From Fig. A-5, $f = 0.023$. $d^5 = (4.589)(0.023)$, $d = 0.638$ m.

22.13 Neglecting minor losses other than the valve, sketch the hydraulic grade line for Fig. 9-54. The globe valve has a loss coefficient of $K = 4.5$.

▮
$$p_1/\gamma + V_1^2/2g + z_1 = p_2/\gamma + V_2^2/2g + z_2 + h_L \qquad h_L = (f)(L/D)(V^2/2g) + KV^2/2g$$

Try $f = 0.013$.

$$h_L = (0.013)[(100 + 75 + 100)/(\tfrac{8}{12})]\{V^2/[(2)(32.2)]\} + (4.5)\{V^2/[(2)(32.2)]\} = 0.1531V^2$$

$$0 + 0 + 12 = 0 + 0 + 0 + 0.1531V^2 \qquad V = 8.853 \text{ fps}$$

$$12 - (0.013)[(100 + 75)/(\tfrac{8}{12})]\{8.853^2/[(2)(32.2)]\} = 7.85 \text{ ft} \qquad 4.5V^2/2g = (4.5)\{8.853^2/[(2)(32.2)]\} = 5.48 \text{ ft}$$

The hydraulic grade line is sketched in Fig. 22-10. [Further checking of the Reynolds number (not shown) indicates that the assumed value of f is acceptable.]

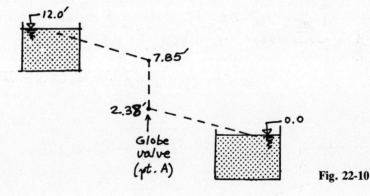

Fig. 22-10

CHAPTER 23
Forces Developed by Fluids in Motion

23.1 A hose and nozzle discharge a horizontal water jet against a nearby vertical plate, as shown in Fig. 23-1. The flow rate of water is 0.025 m³/s, and the diameter of the nozzle tip is 30 mm. Find the horizontal force necessary to hold the plate in place.

▌ $F = \rho Q(v_2 - v_1)$. The net external force acting on the fluid (F in the equation) is the horizontal force necessary to hold the plate in place (i.e., R in Fig. 23-1). Assume it acts toward the left, as shown in the figure, and this direction is taken to be positive. $v_1 = Q/A_1 = 0.025/[(\pi)(\frac{30}{1000})^2/4] = 35.37$ m/s, $v_2 = 0$, $R = (1000)(0.025)[0 - (-35.37)] = 884$ N   (leftward).

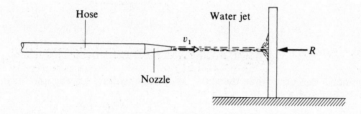

Fig. 23-1

23.2 Water flows from a large tank through an orifice of 3 in diameter and against a block, as shown in Fig. 23-2. The water jet strikes the block at the vena contracta. The block weighs 50 lb, and the coefficient of the friction between block and floor is 0.57. The orifice's coefficient of discharge (C) is 0.60, and its coefficient of contraction (C_c) is 0.62. What is the minimum height to which water must rise in the tank (y in Fig. 23-2) in order to start the block moving to the right?

▌ $F = \rho Q(v_2 - v_1)$. The force caused by the water striking the block must equal (or slightly exceed) the friction force between block and floor (F_f in Fig. 23-2). In other words, the net external force acting on the fluid (F in the equation above) is F_f when the block begins to move. Hence, $F = F_f = (0.57)(50) = 28.5$ lb, $-28.5 = (1.94)(Q)(0 - v_1)$, $Qv_1 = 14.69$. But $V_1 = Q/a$, where a is the area of the jet at its vena contracta.

$$a = (0.62)[(\pi)(\tfrac{3}{12})^2/4] = 0.03043 \text{ ft}^2 \qquad v_1 = Q/0.03043 = 32.86Q \qquad (Q)(32.86Q) = 14.69$$

$$Q = 0.6686 \text{ ft}^3/\text{s} = CA\sqrt{2gh}$$

$$0.6686 = (0.60)[(\pi)(\tfrac{3}{12})^2/4]\sqrt{(2)(32.2)(y-1.0)} \qquad y = 9.00 \text{ ft}$$

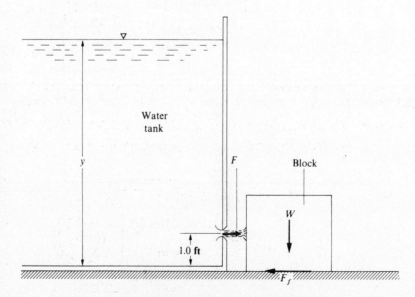

Fig. 23-2

23.3 A hose and nozzle discharge a horizontal water jet against a nearby vertical plate, as shown in Fig. 23-1. The flow rate of water is 0.043 m³/s, and the diameter of the nozzle tip is 50 mm. Find the horizontal force necessary to hold the plate in place.

$$F = \rho Q(v_2 - v_1) \qquad v_1 = Q/A_1 = 0.042/[(\pi)(\tfrac{50}{1000})^2/4] = 21.39 \text{ m/s}$$

$$v_2 = 0 \qquad R = (1000)(0.042)[0 - (-21.39)] = 898 \text{ N} \qquad \text{(leftward)}$$

23.4 In Fig. 23-3 a perfectly balanced weight and platform are supported by a steady water jet. If the total weight supported is 900 N, what is the proper jet velocity?

$$F = \rho Q(v_2 - v_1) \qquad Q = Av_1 = [(\pi)(\tfrac{5}{100})^2/4](v_1) = 0.001964v_1$$

$$900 = (1000)(0.001964v_1)(v_1) \qquad v_1 = 21.4 \text{ m/s}$$

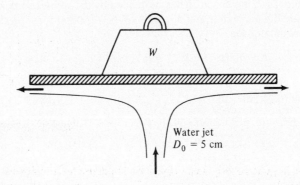

Water jet
$D_0 = 5$ cm

Fig. 23-3

23.5 The horizontal nozzle in Fig. 23-4 has $D_1 = 8$ in and $D_2 = 4$ in. The inlet pressure $p_1 = 50$ psia, and the exit velocity $V_2 = 72$ fps. Compute the force provided by the flange bolts to hold the nozzle on. Assume incompressible steady flow.

$$\Sigma F = \rho Q(V_2 - V_1) \qquad Q = AV = [(\pi)(\tfrac{4}{12})^2/4](72) = 6.283 \text{ ft}^3/\text{s} \qquad V_1 = (\tfrac{4}{8})^2(V_2) = (\tfrac{4}{8})^2(72) = 18.00 \text{ fps}$$

$$(50 - 15)[(\pi)(8)^2/4] - F_{\text{bolts}} = (1.94)(6.283)(72 - 18.00) \qquad F_{\text{bolts}} = 1101 \text{ lb}$$

$p_a = 15$ lb/in² abs

Open jet

Water

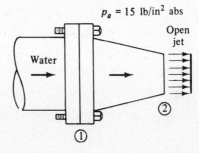

①   ②

Fig. 23-4

23.6 The water tank in Fig. 23-5 weighs 1000 N empty and contains 1 m³ of water at 20 °C. The entrance and exit pipes are identical, $D_1 = D_2 = 5$ cm, and both flow at 0.06 m³/s. What should the scale reading be?

$$\Sigma F_y = W - W_{\text{H}_2\text{O}} - W_{\text{tank}} = -\rho Q[(v_y)_2 - (v_y)_1] \qquad (v_y)_1 = Q/A = 0.06/[(\pi)(\tfrac{5}{100})^2/4] = 30.56 \text{ m/s}$$

$$W - (1)[(9.79)(1000)] - 1000 = -(1000)(0.06)(0 - 30.56) \qquad W = 12\,620 \text{ N}$$

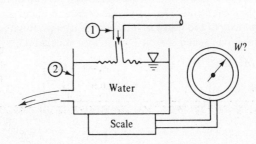

$W?$

Water

Scale

Fig. 23-5

23.7 A dredger is loading sand (s.g. = 2.65) onto a moored barge, as in Fig. 23-6. The sand leaves the dredger pipe at 5 fps with a weight flux of 800 lb/s. What is the tension on the mooring line caused by this sand loading?

$$F_x = \rho Q(v_2 - v_1)_x \qquad F_{\text{mooring line}} = [(2.65)(1.94)]\{800/[(2.65)(62.4)]\}(5\cos 30° - 0) = 108 \text{ lb}$$

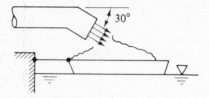

Fig. 23-6

23.8 The water jet in Fig. 23-7, moving at 50 ft/s, is divided by the splitter so that one-third of the water is diverted toward A. Calculate the magnitude and direction of the resultant force on this single stationary blade. Assume ideal flow in a horizontal plane.

$$F = \rho Q(v_2 - v_1) \qquad -F_x = \rho Q_A[(v_x)_2 - (v_x)_1] + \rho Q_B[(v_x)_2 - (v_x)_1]$$
$$F_y = \rho Q_A[(v_y)_2 - (v_y)_1] + \rho Q_B[(v_y)_2 - (v_y)_1]$$
$$Q = AV = (0.2)(50) = 10.0 \text{ ft}^3/\text{s} \qquad Q_A = 10.0/3 = 3.333 \text{ ft}^3/\text{s} \qquad Q_B = 10.0 - 3.333 = 6.667 \text{ ft}^3/\text{s}$$
$$-F_x = (1.94)(3.333)(-50\cos 60° - 50) + (1.94)(6.667)(50\cos 60° - 50) = 808 \text{ lb}$$
$$F_y = (1.94)(3.333)(-50\cos 30° - 0) + (1.94)(6.667)(50\cos 30° - 0) = 280 \text{ lb}$$
$$F_r = \sqrt{808^2 + 280^2} = 855 \text{ lb} \qquad \tan \alpha = \tfrac{280}{808} = 0.34653 \qquad \alpha = 19.1°$$

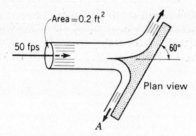

Area = 0.2 ft^2

50 fps

60°

Plan view

A

Fig. 23-7

23.9 For the conditions of Prob. 23.8, compute the magnitude and direction of the resultant force on the single blade if it is moving to the right at a velocity of 10 fps.

$$F = \rho Q(v_2 - v_1) \qquad -F_x = \rho Q_A[(v_x)_2 - (v_x)_1] + \rho Q_B[(v_x)_2 - (v_x)_1]$$
$$F_y = \rho Q_A[(v_y)_2 - (v_y)_1] + \rho Q_B[(v_y)_2 - (v_y)_1]$$
$$V = 50 - 10 = 40 \text{ ft/s} \qquad Q = AV = (0.2)(40) = 8.00 \text{ ft}^3/\text{s} \qquad Q_A = 8.00/3 = 2.667 \text{ ft}^3/\text{s}$$
$$Q_B = 8.00 - 2.667 = 5.333 \text{ ft}^3/\text{s}$$
$$-F_x = (1.94)(2.667)(-40\cos 60° - 40) + (1.94)(5.333)(40\cos 60° - 40) = 517 \text{ lb}$$
$$F_y = (1.94)(2.667)(-40\cos 30° - 0) + (1.94)(5.333)(40\cos 30° - 0) = 179 \text{ lb}$$
$$F_r = \sqrt{517^2 + 179^2} = 547 \text{ lb} \qquad \tan \alpha = \tfrac{179}{517} = 0.34623 \qquad \alpha = 19.1°$$

23.10 A locomotive tender running at 20 mph scoops up water from a trough between the rails, as shown in Fig. 23-8. The scoop delivers water at a point 8 ft above its original level and in the direction of motion. The area of the stream of water at entrance is 50 in^2. The water is everywhere under atmospheric pressure. Neglecting all losses, what is the absolute velocity of the water as it leaves the scoop? What is the force acting on the tender caused by this? What is the minimum speed at which water will be delivered to the point 8 ft above the original level?

$$F_u = \rho A_1 V_1 (\Delta V_u) \qquad u_1 = u_2 = -(20)(5280)/3600 = -29.33 \text{ ft/s} \qquad v_1 = 29.33 \text{ fps}$$

Relative to the moving car,

$$p_1/\gamma + v_1^2/2g + z_1 = p_2/\gamma + v_2^2/2g + z_2 \qquad 0 + 29.33^2/[(2)(32.2)] + 0 = 0 + v_2^2/[(2)(32.2)] + 8 \qquad v_2 = -18.58 \text{ fps}$$
$$V_2 = -29.33 - 18.58 = -47.91 \text{ fps} \qquad F_u = (1.94)(\tfrac{50}{144})(29.33)(47.91) = 947 \text{ lb}$$

At minimum speed, $V^2/2g = 8$ ft, since $v_2 = 0$.

$$V^2/[(2)(32.2)] = 8 \qquad V = 22.70 \text{ fps} \quad \text{or} \quad 15.5 \text{ mph}$$

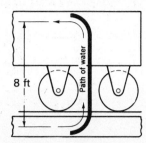

Fig. 23-8

23.11 Referring to Fig. 23-9, a 2-in-diameter stream of water strikes a 4-ft-square door which is at an angle of 30° with the stream's direction. The velocity of the water in the stream is 60.0 ft/s and the jet strikes the door at its center of gravity. Neglecting friction, what normal force applied at the edge of the door will maintain equilibrium?

❚ The force exerted by the door on the water will be normal to the door (no friction). Hence, since no forces act in the W direction in the figure, there will be no change in momentum in that direction. Thus, using W components, initial momentum $\pm 0 =$ final momentum, $+M(V \cos 30°) = +M_1 V_1 - M_2 V_2$, $(\gamma/g)(A_{\text{jet}} V)(V \cos 30°) = (\gamma/g)(A_1 V_1)V_1 - (\gamma/g)(A_2 V_2)V_2$. But $V = V_1 = V_2$ (friction neglected). Then $A_{\text{jet}} \cos 30° = A_1 - A_2$ and, from the equation of continuity, $A_{\text{jet}} = A_1 + A_2$. Solving, $A_1 = A_{\text{jet}}(1 + \cos 30°)/2 = A_{\text{jet}}(0.933)$ and $A_2 = A_{\text{jet}}(1 - \cos 30°)/2 = A_{\text{jet}}(0.067)$.

The stream divides as indicated and the momentum equation produces, for the X direction, $[(62.4/32.2)(\frac{1}{4}\pi)(\frac{1}{6})^2 60](60) - F_x(1) = [(62.4/32.2)(\frac{1}{4}\pi)(\frac{1}{6})^2(0.933)(60)](52.0) + [(62.4/32.2)(\frac{1}{4}\pi)(\frac{1}{6})^2(0.067)(60)](-52.0)$ and $F_x = 38.0$ lb.

Similarly, in the Y direction, $M(0) + F_y(1) = [(62.4/32.2)(0.0218)(0.933)(60)](30.0) + [(62.4/32.3)(0.0218)(0.067)(60)](-30.0)$ and $F_y = 65.9$ lb.

For the door as the free body, $\Sigma M_{\text{hinge}} = 0$ and $+38.0(1) + 65.9(2)(0.866) - P(4) = 0$ or $P = 38.0$ lb.

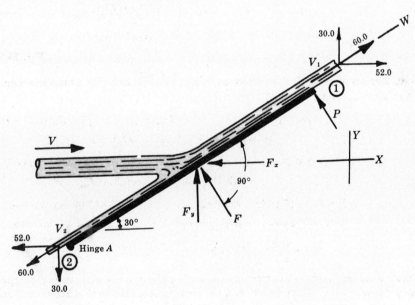

Fig. 23-9

23.12 The force exerted by a 1-in-diameter stream of water against a flat plate held normal to the stream's axis is 145 lb. What is the flow of water?

❚ $F = \rho Q(v_2 - v_1) \qquad v_1 = Q/A = Q/[(\pi)(\frac{1}{12})^2/4] = 183.3Q \qquad -145 = (1.94)(Q)(0 - 183.3Q) \qquad Q = 0.639 \text{ ft}^3/\text{s}$

23.13 Water flow in open channels can be controlled and measured with the sluice gate in Fig. 23-10. At a moderate distance upstream and downstream of the gate, sections 1 and 2, the flow is uniform and the pressure is hydrostatic. Derive an expression for the force F required to hold the gate as a function of ρ, V_1, g, h_1, and h_2. Neglect bottom friction in your analysis.

$$\Sigma F_x = \dot{m}(V_2 - V_1) \qquad -F + \gamma h_1(h_1 b)/2 - \gamma h_2(h_2 b)/2 = \rho h_1 b V_1[(h_1/h_2)(V_1) - V_1]$$

$$F = \gamma b(h_1^2 - h_2^2)/2 - \rho h_1 b V_1^2(h_1/h_2 - 1)$$

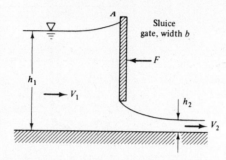

Fig. 23-10

23.14 Compute the force per unit width required to hold the gate in Fig. 23-10 if $h_1 = 3$ m, $h_2 = 0.5$ m, and $V_1 = 1.15$ m/s.

$$F = \gamma b(h_1^2 - h_2^2)/2 - \rho h_1 b V_1^2(h_1/h_2 - 1)$$
$$= (9.79)(1)(3^2 - 0.5^2)/2 - (\tfrac{998}{1000})(3)(1)(1.15)^2(3/0.5 - 1) = 23.0 \text{ kN} \qquad \text{(per meter width)}$$

23.15 Find the thrust developed when water is pumped through a 25-cm-diameter pipe in the bow of a boat at $V = 1.5$ m/s and emitted through a 15-cm-diameter pipe in the stern of the boat.

$$Q = AV = [(\pi)(\tfrac{25}{100})^2/4](1.5) = 0.07363 \text{ m}^3/\text{s} \qquad V_2 = (1.5)(\tfrac{25}{15})^2 = 4.167 \text{ m/s}$$

$$F = \rho Q(V_2 - V_1) = (988)(0.07363)(4.167 - 1.5) = 196 \text{ N}$$

23.16 A jet of water flowing freely in the atmosphere is deflected by a curved vane, as shown in Fig. 23-11a. If the water jet has a diameter of 1.5 in and a velocity of 25.5 fps, what is the force required to hold the vane in place?

▌ Forces acting on the water jet are shown in Fig. 23-11b.

$$F_x = \rho Q[(v_x)_2 - (v_x)_1] \qquad Q = AV = [(\pi)(1.5/12)^2/4](25.5) = 0.3129 \text{ ft}^3/\text{s}$$

$$R_x = (1.94)(0.3129)[0 - (-25.5)] = 15.48 \text{ lb} \qquad \text{(leftward)} \qquad F_y = \rho Q[(v_y)_2 - (v_y)_1]$$

$$R_y = (1.94)(0.3129)(25.5 - 0) = 15.48 \text{ lb} \qquad \text{(upward)} \qquad R = \sqrt{15.48^2 + 15.48^2} = 21.49 \text{ lb}$$

With equal x and y components, the direction of R is at a 45° angle (i.e., $\alpha = 45°$ in Fig. 23-11b).

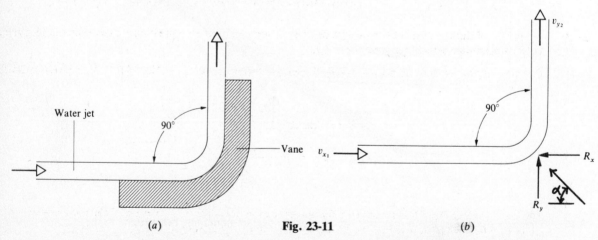

(a)   **Fig. 23-11**   (b)

23.17 A jet of water having a velocity of 10 m/s and a flow rate of 0.06 m³/s is being deflected by an angle chute, as shown in Fig. 23-12. Determine the reaction on the chute. Neglect the friction of the chute.

$$F_x = \rho Q[(v_x)_2 - (v_x)_1] \qquad R_x = (1000)(0.06)[-10 \sin 30° - (-10 \cos 30°)] = 219.6 \text{ N} \qquad \text{(leftward)}$$

$$F_y = \rho Q[(v_y)_2 - (v_y)_1] \qquad R_y = (1000)(0.06)[10 \sin 60° - (-10 \cos 60°)] = 819.6 \text{ N} \qquad \text{(upward)}$$

$$R = \sqrt{219.6^2 + 819.6^2} = 849 \text{ N} \qquad \alpha = \arctan(819.6/219.6) = 75.0°$$

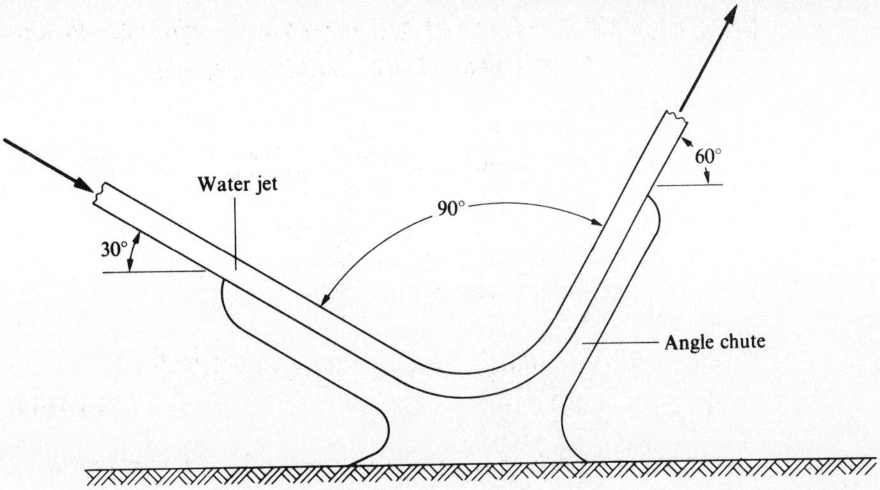

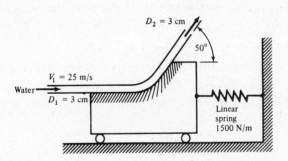

Fig. 23-12

23.18 The cart in Fig. 23-13 is supported by wheels and a linear spring as shown. The water jet is deflected 50° by the cart. Compute (a) the force on the wheels caused by the jet and (b) the spring deflection compared with its unstressed position.

▮ (a)
$$F_y = \rho Q[(v_y)_2 - (V_y)_1] \qquad Q = Av = [(\pi)(\tfrac{3}{100})^2/4](25) = 0.01767 \text{ m}^3/\text{s}$$
$$F_{\text{wheels}} = (998)(0.01767)(25 \sin 50° - 0) = 338 \text{ N} \qquad \text{(upward)}$$

(b)
$$F_x = \rho Q[(v_x)_2 - (V_x)_1] \qquad -F_{\text{spring}} = (998)(0.01767)(25 \cos 50° - 25) = -157 \text{ N}$$
$$F_{\text{spring}} = 157 \text{ N} \qquad \text{(leftward)} \qquad \Delta x_{\text{spring}} = \tfrac{157}{1500} = 0.105 \text{ m}$$

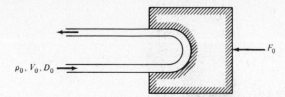

Fig. 23-13

23.19 In Fig. 23-14 the vane turns the water jet completely around. Find an expression for the maximum jet velocity V_0 if the maximum possible support force is F_0.

▮ $F = \rho Q(u_2 - u_1) \qquad F_0 = (\rho_0)(A_0 V_0)(-V_0 - V_0) = (\rho_0)[(\pi D_0^2/4)(V_0)](-V_0 - V_0) \qquad V_0 = \sqrt{2F_0/(\rho_0 D_0^2 \pi)}$

Fig. 23-14

23.20 The water tank in Fig. 23-15 stands on a frictionless cart and feeds a jet of diameter 4 cm and velocity 8 m/s, which is deflected 60° by a vane. Compute the tension in the supporting cable.

▮
$$F_x = \rho Q[(v_x)_2 - (v_x)_1] \qquad Q = Av = [(\pi)(\tfrac{4}{100})^2/4](8) = 0.01005 \text{ m}^3/\text{s}$$
$$T = (998)(0.01005)(8 \cos 60° - 0) = 40.1 \text{ N}$$

23.21 Water exits to the atmosphere ($p_{\text{atm}} = 101$ kPa) through a split nozzle as shown in Fig. 23-16. Duct areas are $A_1 = 0.01$ m^2 and $A_2 = A_3 = 0.005$ m^2. The flow rate is $Q_2 = Q_3 = 150$ m^3/h, and inlet pressure $p_1 = 140$ kPa abs. Compute the force on the flange bolts at section 1.

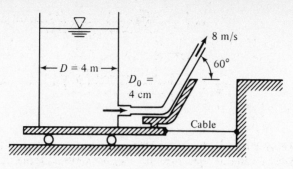

Fig. 23-15

∎ $\qquad \Sigma F_x = \rho Q[(v_x)_2 - (v_x)_1] \qquad -F_{\text{bolts}} + p_1 A_1 = 2\rho Q_2(-v_2 \cos 30°) - \rho Q_1 v_1$

$$v_2 = Q_2/A_2 = (\tfrac{150}{3600})/0.005 = 8.333 \text{ m/s} \qquad v_1 = (2)(\tfrac{150}{3600})/0.01 = 8.333 \text{ m/s}$$

$$-F_{\text{bolts}} + [(140 - 101)(1000)](0.01) = (2)(998)(\tfrac{150}{3600})(-8.333 \cos 30°) - (998)[(2)(\tfrac{150}{3600})](8.333) \qquad F_{\text{bolts}} = 1683 \text{ N}$$

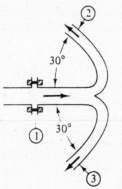

Fig. 23-16

23.22 A pump in a water tank directs a water jet at 30 ft/s and 0.4 ft³/s against a vane, as in Fig. 23-17. Compute the force F to hold the cart stationary if the jet follows path A.

∎ If the water follows path A, no momentum flux is created; thus, $F = 0$.

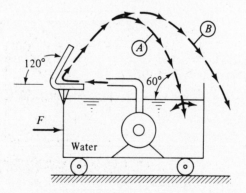

Fig. 23-17

23.23 Compute the force F in Prob. 23.22 if the fluid jet follows path B. The tank holds 35 ft³ of water at this instant.

∎ $F = \rho Q(v_2 - v_1)$. From "frictionless particle" concept, jet B will have constant horizontal velocity $v = v_{\text{jet}} \cos (180 - 120)°$. $F = (1.94)(0.4)[30 \cos (180 - 120)° - 0] = 11.6 \text{ lb}$.

23.24 The cart in Fig. 23-18 moves at constant velocity $V_0 = 35$ m/s and takes on water with a scoop 1.5 m wide which dips $h = 3$ cm into a pond. How much water is added to the cart per second and what is the drag force on the scoop?

∎ $\qquad \dot{m} = \rho A V_0 = (998)[(1.5)(\tfrac{3}{100})](35) = 1572 \text{ kg/s} \qquad F = \dot{m}(v_2 - v_1)$

The drag force F_D equals the thrust needed to keep the cart moving. $F_D = (1572)[0 - (-35)] = 55\,020 \text{ N}$.

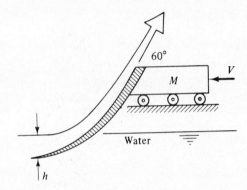

Fig. 23-18

23.25 Assume that a rocket sled is to be decelerated by a scoop, as shown in Fig. 23-19, which dips into water a distance h and deflects a water jet upward at 60°. Neglect air drag. If the sled and scoop weigh 2500 lb and $h = 4$ in, what will the acceleration of the sled be when $V = 500$ ft/s. The sled has zero thrust. The scoop has a width $b = 1$ ft into the paper.

$\blacksquare$ $\Sigma F_x - M(dv/dt) = \dot{m}_2 u_2 - \dot{m}_1 u_1 = \dot{m}(V_j \cos\theta - V_j)$. Since $V_j = V(t)$, $(dV/dt)_{\text{scoop}} = (\dot{m}V/M)(1 - \cos\theta)$, $\dot{m} = \rho V A = (1.94)(500)[(\frac{4}{12})(1)] = 323.3$ ft/s, $dV/dt = [(323.3)(500)/(2500/32.2)](1 - \cos 60°) = 1041$ ft/s^2.

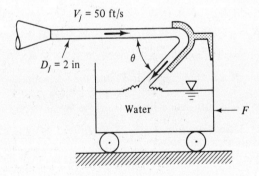

Fig. 23-19

23.26 A vane of turning angle $(180° - \theta)$ is mounted on a water tank, as shown in Fig. 23-20. It is struck by a 2-in-diameter, 50-fps water jet, which turns and falls off into the water tank without spilling. What is the force F required to hold the water tank stationary if $\theta = 90°$?

$\blacksquare$ $\Sigma F_x = \rho Q(v_2 - v_1)$ $\quad Q = Av = [(\pi)(\frac{2}{12})^2/4](50) = 1.091$ ft^3/s $\quad F = (1.94)(1.091)[0 - (-50)] = 106$ lb

23.27 Air $(1.2\ \text{kg/m}^3)$ flows in the 25-cm duct at 10 m/s. It is choked by a 90° cone in the exit, as shown in Fig. 23-21. Neglect wall friction and estimate the force of the airflow on the cone.

$\blacksquare$ $\quad\quad\quad \Sigma F_x = \dot{m}_2 u_2 - \dot{m}_1 u_1 \quad \dot{m}_1 = \rho A_1 V_1 = (1.2)[(\pi)(\frac{25}{100})^2/4](10) = 0.5890$ kg/s

Assume ρ is constant (somewhat unreasonable).

$A_2 = (\pi)[\frac{40}{100} + (\frac{1}{100})(\cos 45°)](\frac{1}{100}) = 0.01279$ m^2 $\quad \dot{m}_2 = \rho A_2 V_2 = (1.2)(0.01279)(V_2) = 0.5890$

$V_2 = 38.38$ m/s $\quad F = (0.5890)(38.38 \cos 45°) - (0.5890)(10) = 10.1$ N $\quad$ (rightward)

23.28 For the same situation and conditions as given in Prob. 23.16 except that the vane itself is moving to the right with a velocity of 10.0 ft/s, find the force exerted by the water on the vane.

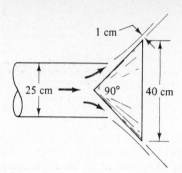

Fig. 23-21

▌ $F_x = \rho Q_r V_r(\cos \beta + 1)$ $V_r = 25.5 - 10.0 = 15.5 \text{ ft/s}$ $Q_r = A_r V_r = [(\pi)(1.5/12)^2/4](15.5) = 0.1902 \text{ ft}^3/\text{s}$

$$R_x = (1.94)(0.1902)(15.5)(\cos 90° + 1) = 5.719 \text{ lb} \quad F_y = \rho Q_r V_r(\sin \beta)$$

$$R_y = (1.94)(0.1092)(15.5)(\sin 90°) = 5.719 \text{ lb} \quad R = \sqrt{5.719^2 + 5.719^2} = 8.09 \text{ lb}$$

As in Prob. 23.16, R acts at an angle β of 45°.

23.29 Solve Prob. 23.28 if the "angle of the vane" is 30° (i.e., the 90° angle in Fig. 23-11b is 30° instead).

▌ $F_x = \rho Q_r V_r(\cos \beta + 1)$. Using data from Prob. 23.28, $R_x = (1.94)(0.1902)(15.5)(\cos 30° + 1) = 10.67 \text{ lb}$, $F_y = Q_r V_r(\sin \beta)$, $R_y = (1.94)(0.1902)(15.5)(\sin 30°) = 2.86 \text{ lb}$, $R = \sqrt{10.67^2 + 2.86^2} = 11.0 \text{ lb}$. The direction can be determined by referring to Fig. 23.22: $\alpha = \arctan(2.86/10.67) = 15.0°$.

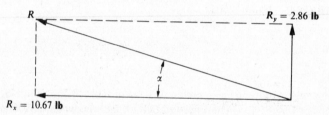

Fig. 23-22

23.30 A jet of water flowing freely in the atmosphere is deflected by a curved, 90° vane (see Fig. 23-11a). The vane is part of a turbine and is moving to the right with a velocity of 20 ft/s. The water jet has a diameter of 2 in and a velocity of 60 ft/s. Determine the force exerted by the water on the vane.

▌ $F_x = \rho Q_r V_r(\cos \beta + 1)$ $V_r = 60 - 20 = 40 \text{ ft/s}$ $Q_r = A_r V_r = [(\pi)(\frac{2}{12})^2/4](40) = 0.8727 \text{ ft}^3/\text{s}$

$$R_x = (1.94)(0.8728)(40)(\cos 90° + 1) = 67.72 \text{ lb} \quad F_y = Q_r V_r(\sin \beta)$$

$$R_y = (1.94)(0.8728)(40)(\sin 90°) = 67.72 \text{ lb} \quad R = \sqrt{67.72^2 + 67.72^2} = 95.8 \text{ lb}$$

Since R_x and R_y are equal, R acts at an angle of 45° with the horizontal.

23.31 Solve Prob. 23.30 if the "angle of the vane" is 60° (i.e., the 90° angle in Fig. 23-11b is 30° instead).

▌ $F_x = \rho Q_r V_r(\cos \beta + 1)$. Using data from Prob. 23.30, $R_x = (1.94)(0.8727)(40)(\cos 60° + 1) = 101.6 \text{ lb}$, $F_y = \rho Q_r V_r(\sin \beta)$, $R_y = (1.94)(0.8728)(40)(\sin 60°) = 58.65 \text{ lb}$, $R = \sqrt{101.6^2 + 58.65^2} = 117.3 \text{ lb}$. The direction can be determined by referring to Fig. 23-22: $\alpha = \arctan(58.65/101.6) = 30.0°$.

23.32 As shown in Fig. 23-23, a vane is curved so that the water jet is entirely reversed in direction. The vane is moving to the right at a velocity of 9.0 m/s, and the velocity of the water jet is 21.0 m/s. If the diameter of the water jet is 50 mm, find the force exerted by the vane on the water.

▌ $F_x = \rho Q_r V_r(\cos \beta + 1)$ $V_r = 21.0 - 9.0 = 12.0 \text{ m/s}$ $Q_r = A_r V_r = [(\pi)(\frac{50}{1000})^2/4](12.0) = 0.0236 \text{ m}^3/\text{s}$

$$R_x = (1000)(0.0236)(12.0)(\cos 0° + 1) = 566 \text{ N} \quad F_y = \rho Q_r V_r \sin \beta \quad R_y = (1000)(0.0236)(12.0)(\sin 0°) = 0$$

Hence, $R = 566 \text{ N}$ (acting horizontally).

23.33 A 2-in-diameter water jet with a velocity of 100 fps impinges on a single vane moving in the same direction (thus $F_x = F_u$) at a velocity of 60 fps. If $\beta_2 = 150°$ and friction losses over the vane are such that $v_2 = 0.9v_1$, compute the force exerted by the water on the vane. See Fig. 23-24a.

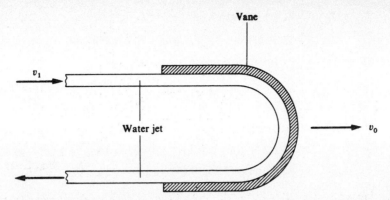

Fig. 23-23

▮ The velocity vector diagrams at entrance and exit to the vane are shown in Fig. 23-24b. Since $v_2 = (0.9)(40) = 36$ fps;

$$V_2 \sin \alpha_2 = v_2 \sin \beta_2 = (36)(0.5) = 18 \text{ fps} \tag{1}$$

$$V_2 \cos \alpha_2 = u + v_2 \cos \beta_2 = 60 + 36(-0.866) = 28.8 \text{ fps} \tag{2}$$

Solving (1) and (2) simultaneously yields $V_2 = 34$ fps, $\alpha_2 = 32°$. Hence $-F_x = \rho Q'(V_2 \cos \alpha_2 - V_1) = 1.94(0.0218)(100 - 60)(28.8 - 100) = -120.4$ lb. So $F_x = 120.4$ lb. The force of vane on water is to the left as assumed; hence force of water on vane is 120.4 lb to the right. $-F_y = \rho Q'(V_2 \sin \alpha_2 - 0) = 1.94(0.0218)(40)(-18) = -30.5$ lb. Thus $F_y = 30.5$ lb in the direction shown. The force of water on the vane is equal and opposite and thus 30.5 lb upward. If the blade were one of a series of blades, $-F_x = \rho Q(V_2 \cos \alpha_2 - V_1) = 1.94(0.0218)(100)(28.8 - 100) = -301$ lb.

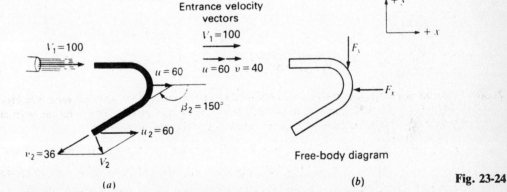

Fig. 23-24

23.34 A jet of water with an area of 3 in² and a velocity of 200 fps strikes a single vane which reverses it through 180° without friction loss. Find the force exerted if the vane moves (**a**) in the same direction as the jet with a velocity of 80 fps; (**b**) in a direction opposite to that of the jet with a velocity of 80 fps.

▮ $$F = \rho Q(v_2 - v_1)$$

(**a**) $v_1 = 200 - 80 = 120$ ft/s $Q = Av = (\frac{3}{144})(120) = 2.500$ ft³/s $F = (1.94)(2.500)[120 - (-120)] = 1164$ lb

(**b**) $v_1 = 200 - (-80) = 280$ ft/s $Q = (\frac{3}{144})(280) = 5.833$ ft³/s $F = (1.94)(5.833)[280 - (-280)] = 6337$ lb

23.35 The pipe bend shown in Fig. 23-25 is in a horizontal plane. Oil (s.g. = 0.88) enters the bend with a velocity of 3.50 m/s and a pressure of 280 kPa. Neglecting any energy losses in the bend, find the force required to hold the bend in place.

▮ $$\Sigma F_x = \rho Q[(v_x)_2 - (v_x)_1] \qquad p_1 A_1 + p_2 A_2 \cos 65° - R_x = \rho Q(-v_2 \cos 65° - v_1)$$
$$A_1 = (\pi)(\tfrac{200}{1000})^2/4 = 0.03142 \text{ m}^2$$

p_2 can be determined using the Bernoulli equation.

$$p_1/\gamma + v_1^2/2g + z_1 = p_2/\gamma + v_2^2/2g + z_2 \qquad A_2 v_2 = A_1 v_1$$

$$A_2 = (\pi)(\tfrac{80}{1000})^2/4 = 0.005027 \text{ m}^2 \qquad 0.005027 v_2 = (0.03142)(3.50) \qquad v_2 = 21.88 \text{ m/s}$$

$$280/[(0.88)(9.79)] + 3.50^2/[(2)(9.807)] + 0 = p_2/[(0.88)(9.79)] + 21.88^2/[(2)(9.807)] + 0$$

$$p_2 = 75.10 \text{ kN/m}^2 \qquad Q = A_1 v_1 = (0.03142)(3.50) = 0.1100 \text{ m}^3/\text{s}$$

$$[(280)(1000)](0.03142) + [(75.10)(1000)](0.005027)(\cos 65°) - R_x$$
$$= [(0.88)(1000)](0.1100)[(-21.88)(\cos 65°) - 3.50]$$

$$R_x = 10\,190 \text{ N} \quad \text{or} \quad 10.19 \text{ kN} \qquad \Sigma F_y = \rho Q[(v_y)_2 - (v_y)_1] \qquad -p_2 A_2 \sin 65° + R_y = \rho Q(v_2 \sin 65° - 0)$$

$$-[(75.10)(1000)](0.005027)(\sin 65°) + R_y = [(0.88)(1000)](0.1100)[(21.88)(\sin 65°) - 0]$$

$$R_y = 2262 \text{ N} \quad \text{or} \quad 2.26 \text{ kN} \qquad R = \sqrt{10.19^2 + 2.26^2} = 10.4 \text{ kN}$$

$$\alpha = \arctan(2.26/10.19) = 12.5° \qquad (\text{see Fig. 23-25}b)$$

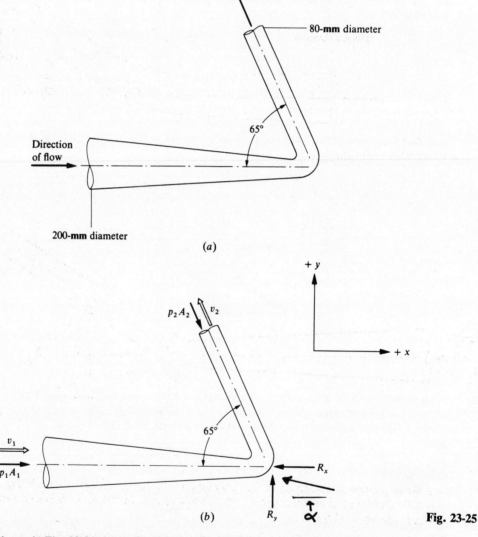

(a)

(b) R_y α **Fig. 23-25**

23.36 The pipe bend shown in Fig. 23.26a is in a horizontal plane. Oil with a specific gravity of 0.86 enters the reducing bend at section A with a velocity of 3.2 m/s and a pressure of 150 kPa. Determine the force required to hold the bend in place. Neglect any energy loss in the bend.

❚ External forces acting on the oil within the bend are shown in Fig. 23-26b.

$$A_1 v_1 = A_2 v_2 \qquad A_1 = \pi(0.150)^2/4 = 0.01767 \text{ m}^2 \qquad A_2 = \pi(0.100)^2/4 = 0.007854 \text{ m}^2$$

$$(0.01767)(3.2) = 0.007854 v_2 \qquad v_2 = 7.199 \text{ m/s} \qquad p_1/\gamma + v_1^2/2g + z_1 = p_2/\gamma + v_2^2/2g + z_2$$

$$150/[(0.86)(9.79)] + (3.2)^2/[(2)(9.807)] + 0 = p_2/[(0.86)(9.79)] + (7.199)^2/[(2)(9.807)] + 0 \qquad p_2 = 132.1 \text{ kN/m}^2$$

$$F_x = \rho Q(v_{x2} - v_{x1}) \qquad p_1 A_1 - p_2 A_2 \cos 30° - R_x = \rho Q(v_{x2} \cos 30° - v_{x1})$$

$$Q = Av = (0.01767)(3.2) = 0.05654 \text{ m}^3/\text{s}$$

$$(150\,000)(0.01767) - (132\,100)(0.007854) \cos 30° - R_x = (0.86)(1000)(0.05654)(7.199 \cos 30° - 3.2)$$

$$R_x = 1604 \text{ N}$$

$$F_y = \rho Q(v_{y2} - v_{y1}) \qquad -p_2 A_2 \sin 30° + R_y = \rho Q(v_{y2} \sin 30° - 0)$$

$$(-132\,100)(0.007854) \sin 30° + R_y = (0.86)(1000)(0.05654)(7.199 \sin 30°) \qquad R_y = 694 \text{ N}$$

$$R = \sqrt{(1604)^2 + (694)^2} = 1748 \text{ N} \qquad \alpha = \arctan \tfrac{694}{1604} = 23.4° \qquad \text{(see Fig. 23-26}b\text{)}$$

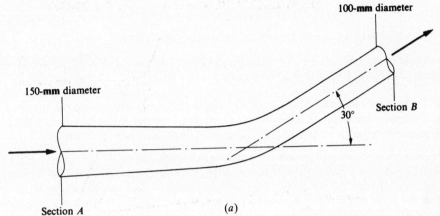

150-mm diameter

100-mm diameter

Section B

30°

Section A

(a)

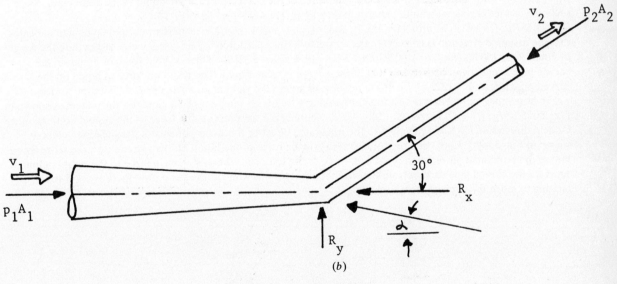

(b)

Fig. 23-26

23.37 Find the magnitude and direction of the resultant force on the compressor shown in Fig. 23-27. Air ($\gamma = 0.075 \text{ lb/ft}^3$) enters at A through a 3-ft^2 area at a velocity of 10 fps. Air is discharged at B through a 2-ft^2 area at a velocity of 12 fps.

$$F_x = \rho Q[(V_2)_x - (V_1)_x] \qquad Q = AV = (3)(10) = 30.0 \text{ ft}^3/\text{s}$$

$$F_x = (0.075/32.2)(30.0)(12 - 10 \cos 50°) = 0.389 \text{ lb} \qquad \text{(rightward)}$$

$$F_y = \rho Q[(V_2)_y - (V_1)_y] = (0.075/32.2)(30.0)[0 - (-10 \sin 50°)] = 0.535 \text{ lb} \qquad \text{(upward)}$$

Force of fluid on compressor is 0.389 lb leftward and 0.535 lb downward. $F = \sqrt{0.389^2 + 0.535^2} = 0.661 \text{ lb}$, $\alpha = \arctan (0.535/0.389) = 54.0°$ to the horizontal.

23.38 Repeat Prob. 23.37 for the case where a gas ($\gamma = 12.1 \text{ N/m}^3$) enters at A through a 60-cm-diameter pipe at 3.0 m/s and leaves at B through a 50-cm-diameter pipe at 4.0 m/s.

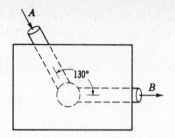

Fig. 23-27

$$F_x = \rho Q[(V_2)_x - (V_1)_x] \qquad Q = AV = [(\pi)(\tfrac{60}{100})^2/4](3) = 0.8482 \text{ m}^3/\text{s}$$

$$F_x = (12.1/9.807)(0.8482)(4.0 - 3.0\cos 50°) = 2.17 \text{ N} \qquad \text{(rightward)}$$

$$F_y = \rho Q[(V_2)_y - (V_1)_y] = (12.1/9.807)(0.8482)[0 - (-3.0\sin 50°)] = 2.41 \text{ N} \qquad \text{(upward)}$$

Force of fluid on compressor is 2.17 N leftward and 2.41 N downward. $F = \sqrt{2.17^2 + 2.41^2} = 3.24$ N, $\alpha = \arctan(2.41/2.17) = 48.0°$ to the horizontal.

23.39 In Fig. 23-28a is shown a curved pipe section of length 40 ft that is attached to the straight pipe section as shown. Determine the resultant force on the curved pipe, and find the horizontal component of the jet reaction. All significant data are given in the figure. Assume an ideal liquid with $\gamma = 55$ lb/ft³.

▌ The energy equation between sections 1 and 3 gives $[(30)(144)/55] + 35 = 20 + (V_3^2/2g)$ and $V_3 = 77.6$ fps (jet velocity); $Q = A_3 V_3 = 3.81$ cfs, $V_2 = Q/A_2 = 43.6$ fps. The energy equation between section 2 and 3 gives $10 + [p_2(144)/55] + [(43.6)^2/64.4] = 20 + [(77.6)^2/2g]$, and $p_2 = 28.3$ psi. The free-body diagram of the forces acting on the liquid contained in the curved pipe is shown in Fig. 23-28b. $p_2 A_2 - p_3 A_3 \cos 20° - (F_{N/L})_x = \rho Q(V_3 \cos 20° - V_2)$ where $(F_{N/L})_x$ represents the force of the curved pipe on the liquid in the x direction. Since section 3 is a jet in contact with the atmosphere, $p_3 = 0$. Thus $28.3[(\pi/4)(4^2)] - (F_{N/L})_x = [(1.94)(55/62.4)](3.81)[(77.6)(0.94) - 43.6]$, $356 - (F_{N/L})_x = 191$, $(F_{N/L})_x = +165$ lb where the plus sign indicates that the assumed direction is correct. In the y direction the $p_2 A_2$ force has no component. Estimating the weight of liquid W as 150 lb, $(F_{N/L})_y = \rho Q[(77.6)(0.342) - 0] + 150 = +323$ lb. The resultant force of liquid on the curved pipe is equal and opposite to that of the curved pipe on liquid. The resultant force of liquid on the curved pipe is $[(165)^2 + (323)^2]^{1/2} = 363$ lb downward and to the right at an angle of 62°56′ with the horizontal.

The horizontal jet reaction is best found by taking a free-body diagram of the liquid in the system as shown in Fig. 23-28c: $(F_{S/L})_x = \rho Q(V_3 \cos 20° - 0) = 475$ lb where $(F_{S/L})_x$ respresents the force of the system on the liquid in the x direction. $(F_{S/L})_x$ is equivalent to the integrated effect of the x components of the pressure vectors shown in Fig. 23-28c. Equal and opposite to $(F_{S/L})_x$ is the force of the liquid on the system, i.e., the jet reaction. Hence the horizontal jet reaction is a 475-lb force to the left. Thus there is a 165 lb force to the right tending to separate the curved pipe section from the straight pipe section, while at the same time there is a 475-lb force tending to move the entire system to the left.

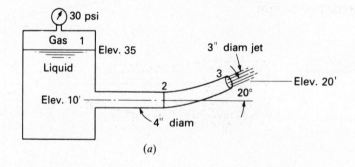

(a)

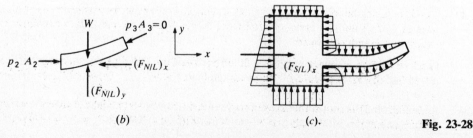

(b)   (c).   **Fig. 23-28**

23.40 A 24-in pipe carrying 31.4 cfs of oil (s.g. $= 0.85$) has a 90° bend in a horizontal plane. The loss of head in the bend is 3.50 ft of oil and the pressure at entrance is 42.5 psi. Determine the resultant force exerted by the oil on the bend.

▌ Referring to Fig. 23-29, the free-body diagram indicates the static and dynamic forces acting on the mass of oil in the bend. These forces are calculated as follows:

1. $$P_1 = p_1 A = (42.5)(\tfrac{1}{4}\pi)(24)^2 = 19\,200\,\text{lb}$$

2. $P_2 = p_2 A$, where $p_2 = p_1 -$ loss in psi, from the Bernoulli equation since $z_1 = z_2$ and $V_1 = V_2$. Then $P_2 = [42.5 - (0.85)(62.4)(3.50/144)](\tfrac{1}{4}\pi)(24)^2 = 18\,600\,\text{lb}$.

3. Using the impulse-momentum principle, and knowing that $V_1 = V_2 = Q/A = 10.0\,\text{ft/s}$, $MV_{x_1} + \Sigma$ (forces in X direction)(1) $= MV_{x_2}$, $19\,200 - F_x = [(0.85)(62.4)(31.4/32.2)](0 - 10.0) = -517\,\text{lb}$, and $F_x = +19\,720\,\text{lb}$, to the left on the oil.

4. Similarly, for $t = 1$ s, $MV_{y_1} + \Sigma$ (forces in Y direction)(1) $= MV_{y_2}$, $F_y - 18\,600 = [(0.85)(62.4)(31.4/32.2)](10.0 - 0) = +517\,\text{lb}$, and $F_y = +19\,120\,\text{lb}$, upward on the oil.

On the pipe bend, the resultant force R acts to the right and downward and is $R = \sqrt{(19\,720)^2 + (19\,120)^2} = 27\,470\,\text{lb}$ at $\theta_x = \tan^{-1}(19\,120/19\,720) = 44.1°$.

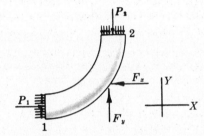

Fig. 23-29

23.41 The 24-in pipe of Prob. 23.40 is connected to a 12-in pipe by a standard *reducer* fitting. For the same flow of 31.4 cfs of oil, and a pressure of 40.0 psi, what force is exerted by the oil on the reducer, neglecting any lost head?

▌ Since $V_1 = 10.0$ fps, $V_2 = (\tfrac{2}{1})^2(10.0) = 40.0$ fps. Also, the Bernoulli equation between section 1 at entrance and section 2 at exit yields $\{(p_1/\gamma) + [(10)^2/2g] + 0\} -$ negligible lost head $= \{(p_2/\gamma) + [(40)^2/2g] + 0\}$. Solving, $p_2/\gamma = [(40.0)(144)/(0.85)(62.4)] + (100/2g) - (1600/2g) = 85.3$ ft of oil and $p_2' = 31.4$ psi.
    Figure 23-30 represents the forces acting on the mass of oil in the reducer: $P_1 = p_1 A_1 = (40.0)(\tfrac{1}{4}\pi)(24)^2 = 18\,100\,\text{lb}$ (to the right), $P_2 = p_2 A_2 = (31.4)(\tfrac{1}{4}\pi)(12)^2 = 3550\,\text{lb}$ (to the left).
    In the X direction the momentum of the oil is changed. Then $MV_{x_1} + \Sigma$ (forces in X direction)(1) $= MV_{x_2}$, $(18\,100 - 3550 - F_x)1 = [(0.85)(62.4)(31.4/32.2)](40.0 - 10.0)$, and $F_x = 13\,000\,\text{lb}$ acting to the left on the oil.
    The forces in the Y direction will balance each other and $F_y = 0$. Hence the force exerted by the oil on the reducer is 13 000 lb to the right.

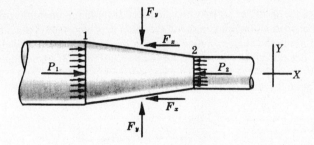

Fig. 23-30

23.42 A 45° reducing bend, 24-in diameter upstream, 12-in diameter downstream, has water flowing through it at the rate of 15.7 cfs under a pressure of 21.0 psi. Neglecting any loss in the bend, calculate the force exerted by the water on the reducing bend.

▌ $V_1 = 15.7/A_1 = 5.00$ ft/s and $V_2 = 20.0$ ft/s. The Bernoulli equation, section 1 to section 2, produces $\{[(21.0)(144)/62.4] + (25/2g) + 0\} -$ negligible lost head $= [(p_2/\gamma) + (400/2g) + 0]$ from which $p_2/\gamma = 42.6$ ft and $p_2' = 18.5$ psi.
    In Fig. 23-31 is shown the mass of water acted upon by static and dynamic forces: $P_1 = p_1 A_1 = (21.0)(\tfrac{1}{4}\pi)(24)^2 = 9500\,\text{lb}$, $P_2 = p_2 A_2 = (18.5)(\tfrac{1}{4}\pi)(12)^2 = 2090\,\text{lb}$, $P_{2_x} = P_{2_y} = (2090)(0.707) = 1478\,\text{lb}$.

In the X direction, $MV_{x_1} + \Sigma$ (forces in X direction)$(1) = MV_{x_2}$, $(9500 - 1478 - F_x)1 = [(62.4)(15.7/32.2)][(20.0)(0.707) - 5.00]$ and $F_x = 7740$ lb to the left.

In the Y direction, $(+F_y - 1478)1 = [(62.4)(15.7/32.2)][(20.0)(0.707) - 0]$ and $F_y = 1910$ lb upward.

The force exerted by the water on the reducing bend is $F = \sqrt{(7740)^2 + (1910)^2} = 7970$ lb to the right and downward, at an angle $\theta_x = \tan^{-1}\frac{1910}{7740} = 13°52'$.

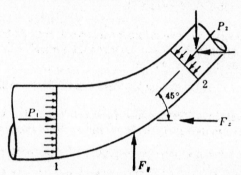

Fig. 23-31

23.43 Determine the reaction of a jet flowing through an orifice on the containing tank.

❚ In Fig. 23-32 a mass of liquid $ABCD$ is taken as a free body. The only horizontal forces acting are F_1 and F_2, which change the momentum of the water. $(F_1 - F_2)(1) = M(V_2 - V_1)$, where V_1 can be considered negligible. Reaction $F = F_1 - F_2 = (\gamma Q/g)V_2 = (\gamma A_2 V_2/g)V_2$. But $A_2 = c_c A_0$ and $V_2 = c_v \sqrt{2gh}$. Hence $F = [\gamma(c_c A_0)/g]c_v^2(2gh) = (c c_v)\gamma A_0(2h)$ (to the right on the liquid).

1. For average values of $c = 0.60$ and $c_v = 0.98$, the reaction $F = 1.176\gamma h A_0$. Hence the force acting to the left on the tank is about 18 percent more than the static force on a plug which would just fill the orifice.
2. For ideal flow (no friction, no contraction), $F = 2(\gamma h A_0)$. This force is equal to twice the force on a plug which would just fill the orifice.
3. For a nozzle ($c_c = 1.00$), the reaction $F = c_v^2 \gamma A(2h)$ where h would be the effective head causing the flow.

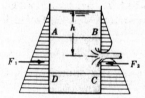

Fig. 23-32

23.44 The jets from a garden sprinkler are 1-in in diameter and are normal to the 2-ft radius. If the pressure at the base of the nozzles is 50 psi, what force must be applied on each sprinkler pipe, 1 ft from the center of rotation, to maintain equilibrium? (Use $c_v = 0.80$ and $c_c = 1.00$.)

❚ The reaction of the sprinkler jet may be calculated from the momentum principle. Inasmuch as the force which causes a change in momentum in the X direction acts along the X axis, no torque is exerted. We are interested, therefore, in the change in momentum in the Y direction. But the initial momentum in the Y direction is zero. The jet velocity $V_Y = c_v \sqrt{2gh} = 0.80\sqrt{2g[(50)(2.31) + \text{negligible velocity head}]} = 69.0$ ft/s. Thus $F_Y \, dt = M(V_Y) = [(62.4/32.2)(\frac{1}{4}\pi)(\frac{1}{12})^2(69.0) \, dt](-69.0)$ or $F_Y = -50.3$ lb downward on the water. Hence the force of the jet on the sprinkler is $+50.3$ lb upward. Then $\Sigma M_0 = 0$, $F(1) - 2(50.3) = 0$, $F = 100.6$ lb for equilibrium.

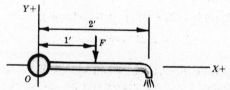

Fig. 23-33

23.45 A rocket engine burns 5.0 kg of fuel and oxidizer per second, and combustion gases exit the rocket at a velocity of 550 m/s relative to the rocket at a pressure approximately equal to the ambient air pressure. Find the thrust produced by the rocket engine.

❚
$$F = \rho Q v_2 = M v = (5.0)(550) = 2750 \text{ N}$$

23.46 A jet craft traveling at a velocity of 590 fps takes in air at a rate of 1.55 slugs/s. The air:fuel ratio is 25:1, and the exhaust velocity relative to the jet is 1950 fps. Find the thrust produced by the jet engine.

$$F = \rho_a Q_a(v_2 - v_1) + \rho_f Q_f v_2 \qquad \rho_a Q_a = M_a = 1.55 \text{ slugs/s}$$

Since the air:fuel ratio is 25:1, $\rho_f Q_f = 1.55/25 = 0.0620$ slug/s, $F = (1.55)(1950 - 590) + (0.0620)(1950) = 2229$ slug-ft/s². Since slug = lb-s²/ft, $F = 2229$ lb.

23.47 A rocket engine burns 8.5 kg of fuel and oxidizer per second, and combustion gases exit the rocket at a velocity of 565 m/s relative to the rocket at a pressure approximately equal to the surrounding atmospheric pressure. Find the thrust produced by the rocket engine.

$$F = \rho Q v_2 = Mv = (8.5)(565) = 4802 \text{ N}$$

23.48 A jet craft traveling at a velocity of 700 fps takes in air at a rate of 1.68 slugs/s. The air:fuel ratio is 20:1, and the exhaust velocity relative to the jet is 2105 fps. Find the thrust produced by the jet engine.

$$F = \rho_a Q_a(v_2 - v_1) + \rho_f Q_f v_2 \qquad \rho_a Q_a = M_a = 1.68 \text{ slugs/s}$$

Since the air:fuel ratio is 20:1, $\rho_f Q_f = 1.68/20 = 0.0840$ slug/s, $F = (1.68)(2105 - 700) + (0.0840)(2105) = 2537$ slug-ft/s². Since slug = lb-s²/ft, $F = 2537$ lb.

23.49 A jet engine is being tested in the laboratory. The engine consumes 50 lb of air per second and 0.5 lb of fuel per second. If the exit velocity of the gases is 1500 fps, what is the thrust?

$$F = W_{exit} V_2/g - W_1 V_1/g = (50 + 0.5)(1500)/32.2 - (50)(0)/32.2 = 2352 \text{ lb}$$

23.50 A jet engine operates at 600 fps and consumes air at the rate of 50.0 lb/s. At what velocity should the air be discharged in order to develop a thrust of 1500 lb?

$$F = 1500 = (50.0/32.2)(V_{exit} - 600) \qquad V_{exit} = 1566 \text{ fps}$$

23.51 A turbojet engine is tested in the laboratory under conditions simulating an altitude where atmospheric pressure is 785.3 lb/ft² abs, temperature is 429.5 °R, and specific weight is 0.0343 lb/ft³. If the exit area of the engine is 1.50 ft² and the exit pressure is atmospheric, what is the Mach number if the gross thrust is 1470 lb? Use $k = 1.33$.

$$F = W_{exit} V_{exit}/g = (\gamma A_{exit} V_{exit})(V_{exit})/g \qquad 1470 = (0.0343)(1.50)(V_{exit})^2/32.2 \qquad V_{exit} = 959 \text{ fps}$$

$$N_M = V_{exit}/\sqrt{kgRT} = 959/\sqrt{(1.33)(32.2)(53.3)(429.5)} = 0.969$$

23.52 A rocket device burns its propellant at a rate of 15.2 lb/s. The exhaust gases leave the rocket at a relative velocity of 3220 fps and at atmospheric pressure. The exhaust nozzle has an area of 50.0 in² and the gross weight of the rocket is 500 lb. At the given instant, 2500 hp is developed by the rocket engine. What is the rocket velocity?

$$P = FV_{rocket} \qquad F = W_{exit} V_{exit}/g = (15.2)(3220)/32.2 = 1520 \text{ lb}$$

$$(2500)(550) = 1520 V_{rocket} \qquad V_{rocket} = 905 \text{ fps}$$

23.53 A rocket motor is operating steadily, as shown in Fig. 23-34. The products of combustion flowing out the exhaust nozzle approximate a perfect gas with a molecular weight of 26. For the given conditions, calculate V_2.

$$\dot{m}_{exit} = \rho_{exit} A_{exit} V_{exit} = 0.5 + 0.1 = 0.6 \text{ slug/s} \qquad \rho = p/RT \qquad R = 49\,750/26 = 1913$$

$$\rho_{exit} = (15)(144)/[(1913)(1100 + 460)] = 0.0007238 \text{ slug/ft}^3 \qquad A_{exit} = (\pi)(5.5/12)^2/4 = 0.1650 \text{ ft}^2$$

$$0.6 = (0.0007238)(0.1650)(V_{exit}) \qquad V_{exit} = 5024 \text{ fps}$$

23.54 In contrast to the liquid rocket in Fig. 23-34, the solid-propellant rocket in Fig. 23-35 is self-contained and has no entrance ducts. Using a control-volume analysis for the conditions shown in Fig. 23–35, compute the rate of mass loss of the propellant, assuming that the exit gas has a molecular weight of 31.

$$\dot{m}_{propellant} = \rho_{exit} A_{exit} V_{exit} \qquad \rho = p/RT \qquad R = \tfrac{8313}{31} = 268$$

$$\rho_{exit} = (90)(1000)/[(268)(750)] = 0.4478 \text{ kg/m}^3 \qquad A_{exit} = (\pi)(\tfrac{18}{100})^2/4 = 0.02545 \text{ m}^2$$

$$\dot{m}_{propellant} = (0.4478)(0.02545)(1150) = 13.1 \text{ kg/s}$$

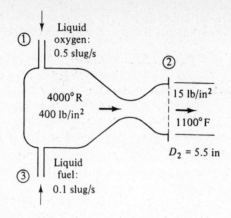

Liquid
oxygen:
0.5 slug/s

① ②

4000° R
400 lb/in²

15 lb/in²

1100° F

$D_2 = 5.5$ in

③ Liquid
fuel:
0.1 slug/s

Fig. 23-34

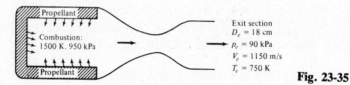

Propellant

Combustion:
1500 K, 950 kPa

Exit section
$D_e = 18$ cm
$p_e = 90$ kPa
$V_e = 1150$ m/s
$T_c = 750$ K

Propellant

Fig. 23-35

23.55 For the rocket engine of Prob. 23.53, compute the thrust if atmospheric pressure is 15 psia.

▌ Using data from Prob. 23.53, $F = \dot{m}_{exit}V_{exit} = (0.6)(5024) = 3014$ lb.

23.56 A rocket ship is moving at 9000 m/s in outer space, where gravity is negligible. It is desired to slow the rocket down to 8900 m/s by firing a retrorocket forward. The retrorocket burns 7 kg/s of fuel and oxidizer at an exhaust velocity of 1500 m/s relative to the rocket. If the initial mass of the rocket ship is 1600 kg, how long should the retrorocket burn, and how much fuel will be burned? Assume that the exhaust pressure is nearly equal to ambient conditions.

▌ $\int_{V_0}^{V_{final}} dV = \dot{m}V_{exit}\int_0^{t_{burn}}\dfrac{dt}{m_0 - \dot{m}t}$ $\quad V_{final} - V_0 = V_{exit}\ln\left(1 - \dfrac{\dot{m}t_{burn}}{m_0}\right)$ $\quad 8900 - 9000 = 1500\ln\left(1 - \dfrac{7t_{burn}}{1600}\right)$

$t_{burn} = 14.7$ s $\qquad m_{burned} = (7)(14.7) = 103$ kg

23.57 Determine the magnitude and direction of the resultant force exerted on the double nozzle of Fig. 23-36. Both nozzle jets have a velocity of 12 m/s. The axes of the pipes and both nozzles lie in a horizontal plane ($\gamma = 9.81$ kN/m³). Neglect friction.

▌ **Continuity:**

$$A_1V_1 = A_2V_2 + A_3V_3 \qquad [(\pi)(\tfrac{15}{100})^2/4](V_1) = [(\pi)(\tfrac{10}{100})^2/4](12) + [(\pi)(7.5/100)^2/4](12)$$

$$V_1 = 8.33 \text{ m/s} \qquad Q_1 = A_1V_1 = [(\pi)(\tfrac{15}{100})^2/4](8.33) = 0.147 \text{ m}^3/\text{s}$$

$$Q_2 = [(\pi)(\tfrac{10}{100})^2/4](12) = 0.094 \text{ m}^3/\text{s} \qquad Q_3 = [(\pi)(7.5/100)^2/4](12) = 0.053 \text{ m}^3/\text{s}$$

Energy equation:

$$(p_1/\gamma) + [8.33^2/2(9.81)] = 0 + [12^2/2(9.81)] \qquad p_1/\gamma = 3.80 \text{ m} \qquad p_1 = 37.3 \text{ kN/m}^2 \qquad p_1A_1 = 0.659 \text{ kN}$$

$$\sum F_x = p_1A_1 - (F_{N/L})_x = (\rho Q_2V_{2_x} + \rho Q_3V_{3_x}) - \rho Q_1V_{1_x} \qquad \rho = \gamma/g = 9.81/9.81 = 1.0 \text{ kN-s}^2/\text{m}^4 = 10^3 \text{ kg/m}^3$$

$$V_{2_x} = V_2\cos 15° = 12(0.966) = 11.6 \text{ m/s} \qquad V_{3_x} = V_3\cos 30° = 12(0.866) = 10.4 \text{ m/s} \qquad V_{1_x} = V_1 = 8.33 \text{ m/s}$$

$$0.659 - (F_{N/F})_x = 10^3(0.094)(11.6) + 10^3(0.053)(10.4) - 10^3(0.147)8.33 = 0.417 \text{ kN}$$

$$(F_{N/L})_x = 0.659 - 0.417 = 0.242 \text{ kN}$$

$$\sum F_y = (F_{N/L})_y = (\rho Q_2V_{2_y} + \rho Q_3V_{3_y}) - \rho Q_1V_{1_y} \qquad V_{2_y} = V_2\sin 15° = 12(0.259) = 3.1 \text{ m/s}$$

$$V_{3_y} = -V_3\sin 30° = -12(0.50) = -6.0 \text{ m/s} \qquad V_{1_y} = 0$$

$$(F_{N/L})_y = 10^3(0.094)(3.1) + 10^3(0.053)(-6.0) - 10^3(1.47)(0) = -0.027 \text{ kN}$$

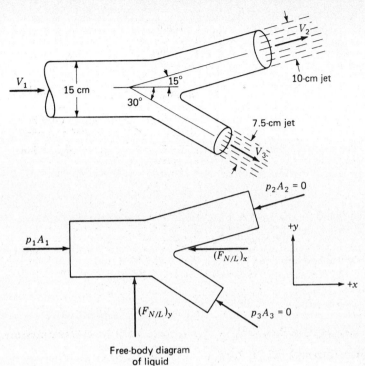

Free-body diagram
of liquid

Fig. 23-36

The minus sign indicates that the assumed direction of $(F_{N/L})_y$ was wrong. Therefore $(F_{N/L})_y$ acts in the negative y direction. Equal and opposite to $F_{N/L}$ is $F_{L/N}$: $(F_{L/N})_x = 0.242$ kN (in positive x direction), $(F_{L/N})_y = 0.027$ kN (in positive y direction).

23.58 In Fig. 23-37 suppose that $\theta = 30°$, $V_1 = 100$ fps, and the stream is a jet of water with an initial diameter of 2 in. Assume friction losses such that $V_2 = 95$ fps. Find the resultant force on the blade. Assume that flow occurs in a horizontal plane.

❚ This problem is best solved by taking a free-body diagram of the element of fluid in contact with the blade. The forces acting on this element are as shown in the sketch. The forces $(F_{B/W})_x$ and $(F_{B/W})_y$ represent the components of force of blade on water in the x and y directions. These forces include shear stresses tangential to the blade as well as pressure forces normal to the blade. $A = 0.0218$ ft^2, $-(F_{B/W})_x = \rho Q(V_{2_x} - V_{1_x}) = 1.94(0.0218)(100)[(0.866)(95) - 100] = 4.23(-17.7) = -74.9$ lb. Hence, $(F_{B/W})_x = +74.9$ lb. The plus sign indicates that the assumed direction of $(F_{B/W})_x$ was correct. $+(F_{B/W})_y = \rho Q(V_{2_y} - V_{1_y}) = 4.23[(0.50)(95) - 0] = 201$ lb. Thus the force of the blade on the fluid is the resultant of a 74.9 lb component to the left and a 201-lb component upward in the y direction. Equal and opposite to this is the force of the fluid on the blade (downward and to the right). The resultant force is 215 lb at an angle of 69.6° below the horizontal.

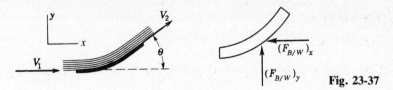

Fig. 23-37

23.59 On the right end of a 6-in diameter pipe is a nozzle which discharges a 2-in diameter jet. The pressure in the pipe is 57 psi, and the pipe velocity is 10 fps. The jet discharges into the air. If the fluid is water, what is the axial force exerted upon the nozzle? Find also the head loss in the nozzle.

❚

$$\sum F = \rho Q(v_2 - v_1) \qquad Q = A_1 v_1 = [(\pi)(\tfrac{6}{12})^2/4](10) = 1.963 \text{ ft}^3/\text{s} \qquad v_2 = (\tfrac{6}{2})^2(10) = 90.0 \text{ fps}$$

$$-F + (57)(144)[(\pi)(\tfrac{6}{12})^2/4] = (1.94)(1.963)(90.0 - 10) \qquad F = 1307 \text{ lb}$$

$$h_L = p_1/\gamma + v_1^2/2g - v_2^2/2g = (57)(144)/62.4 + 10^2/[(2)(32.2)] - 90.0^2/[(2)(32.2)] = 7.31 \text{ ft}$$

23.60 The absolute velocity of a jet of steam impinging upon the blades of a steam turbine is 4000 fps, and that leaving is 2800 fps. $\alpha_1 = 20°$, $\alpha_2 = 150°$, $u_1 = u_2 = 500$ fps, and $r_1 = r_2 = 0.5$ ft. Find the torque exerted on the rotor and the power delivered to it if $G = 0.4$ lb/s.

▌ $T = \rho Q(r_1 V_1 \cos \alpha_1 - r_2 V_2 \cos \alpha_2) = (0.4/32.2)[(0.5)(4000)(\cos 20°) - (0.5)(2800)(\cos 150°)] = 38.4$ ft-lb

$$P = FV = T\omega = (38.4)(500/0.5)/500 = 69.8 \text{ hp}$$

23.61 When a turbine runner is held so that it cannot rotate, the discharge under a head of 50 ft is found to be 29.5 cfs. $\alpha_1 = 35°$, $\beta_2 = 155°$, $r_1 = 0.70$ ft, $r_2 = 0.42$ ft, $A_1 = 0.837$ ft^2, $A_2 = 0.882$ ft^2. What is the value of the torque at zero speed? Neglect shock loss.

▌ $T = \rho Q(r_1 V_1 \cos \alpha_1 - r_2 V_2 \cos \beta_2)$ $\quad \rho Q = (29.5)(62.4)/32.2 = 57.17$ slugs/s

$$V_1 = Q/A_1 = 29.5/0.837 = 35.24 \text{ fps} \quad v_2 = 29.5/0.882 = 33.45 \text{ fps}$$

For stopped runner, $V_2 = v_2$ and $\alpha_2 = \beta_2 = 155°$. $T = (57.17)[(0.70)(35.24)(\cos 35°) - (0.42)(33.45)(\cos 155°)] = 1883$ ft-lb.

23.62 A 20-in-diameter household fan drives air ($\gamma = 0.076$ lb/ft^3) at a rate of 1.60 lb/s. Find the thrust exerted by the fan. What is the pressure difference on the two sides of the fan? Find the required horsepower to drive the fan. Neglect losses.

▌ $\gamma Q = \gamma A(\Delta V/2)$ $\quad 1.60 = (0.076)[(\pi)(\frac{20}{12})^2/4](\Delta V/2)$ $\quad \Delta V = 19.30$ fps

$$F = \rho Q(\Delta V) = (1.60/32.2)(19.30) = 0.959 \text{ lb} \quad \Delta p/\gamma = (\Delta V)^2/2g = 19.30^2/[(2)(32.2)] = 5.78 \text{ ft}$$

$$\Delta p = (0.076)(5.78) = 0.439 \text{ lb/ft}^2 \quad P = FV = F(\Delta V/2) = (0.959)(19.30/2)/550 = 0.0168 \text{ hp}$$

23.63 A fan sucks air from outside to inside a building through an 18-in-diameter duct. The density of the air is 0.0022 slug/ft^3. If the pressure difference across the two sides of the fan is 3.6 in of water, determine the flow rate of the air. What thrust must the fan support be designed to withstand?

▌ $F = A(\Delta p) = [(\pi)(\frac{18}{12})^2/4](62.4)(3.6/12) = 33.1 \text{ lb} \quad Q = A(\Delta V/2)$

$$F = \rho Q(\Delta V) = \rho[A(\Delta V/2)](\Delta V) \quad 33.1 = (0.0022)\{[(\pi)(\frac{18}{12})^2/4]/2\}(\Delta V)^2$$

$$\Delta V = 130.5 \text{ fps} \quad Q = [(\pi)(\frac{18}{12})^2/4](130.5/2) = 115 \text{ ft}^3/\text{s}$$

23.64 Find the thrust and efficiency of two 6.5-ft-diameter propellers through which flows a total of 20 000 cfs of air (0.072 lb/ft^3). The propellers are attached to an airplane moving at 150 mph through still air. Find also the pressure rise across the propellers and the horsepower input to each propeller. Neglect eddy losses.

▌ Velocity of air relative to airplane is $V_1 = 150$ mph $= (150)(44)/30 = 220$ fps. Velocity of air through the actuating disk is $V = V_1 + (\Delta V/2) = Q/A = (20\,000/2/[(\pi/4)(6.5)^2] = 301$ fps. Thus $\Delta V = 2(301 - 220) = 162$ fps, $F_T = \rho Q \, \Delta V = (0.072/32.2)(20\,000)(162) = 7240$ lb (total thrust of both propellers), $\eta = 1/[1 + \Delta V/2V_1] = 1/(1 + \frac{162}{440}) = 0.73 = 73$ percent. F_T on one propeller $= \frac{7240}{2} = 3620$ lb. But $F_T = (\Delta p)(A)$, thus $3620 = \Delta p(\pi/4)(6.5)^2$, $\Delta p = 109$ psf $= 0.758$ psi, hp/propeller $= \gamma Q(\Delta p/\gamma)/550 = Q(\Delta p)/550 = 10\,000(109)/550 = 1980$. Check: hp/propeller $= F_T(V_1 + \Delta V/2)/550 = 3620(301)/550 = 1980$.

23.65 This water passage (Fig. 23-38) is 10 ft wide normal to the figure. Determine the horizontal force acting on the shaded structure. Assume ideal flow.

▌ In free-surface flow such as this where the streamlines are parallel, the water surface is coincident with the hydraulic grade line. Writing an energy equation from the upstream section to the downstream section,

$$6 + (V_1^2/2g) = 3 + (V_2^2/2g) \tag{1}$$

From continuity,

$$6(10)V_1 = 3(10)V_2 \tag{2}$$

Substituting Eq. (2) into Eq. (1) yields $V_1 = 8.02$ fps, $V_2 = 16.04$ fps, $Q = A_1 V_1 = A_2 V_2 = 481$ cfs. Next take a free-body diagram of the element of water shown in the figure $F_1 - F_2 - F_x = \rho Q(V_2 - V_1)$, where F_x represents the force of the structure on the water in the horizontal direction. $F_1 = \gamma h_{c_1} A_1$ and $F_2 = \gamma h_{c_2} A_2$. Hence $62.4(3)(10)(6) - 62.4(1.5)(10)(3) - F_x = 1.94(481)(16.04 - 8.02)$ and $F_x = 940$ lb. The positive sign means that the assumed direction is correct. Hence the force of the water on the structure is equal and opposite, namely, 940 lb to the right.

The momentum principle will not permit one to obtain the vertical component of the force of the water on the shaded structure because the pressure distribution along the bottom of the channel is unknown. The pressure distribution along the boundary of the structure and along the bottom of the channel can be estimated by sketching a flow net and applying Bernoulli's principle. The horizontal and vertical components of the force can be found by computing the integrated effect of the pressure-distribution diagram.

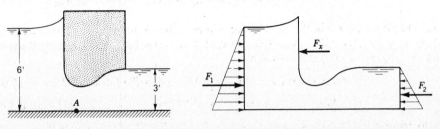

Fig. 23-38

23.66 Flow occurs over a spillway of constant section as shown in Fig. 23-39. Determine the horizontal force on the spillway per foot of spillway width (perpendicular to the spillway section). Assume ideal flow.

▌ $\sum F = \rho Q(v_2 - v_1)$. Energy consideration: $y_1 + v_1^2/2g = y_2 + v_2^2/2g$, $4 + v_1^2/[(2)(32.2)] = 0.6 + v_2^2/[(2)(32.2)]$. Continuity consideration: $y_1 v_1 = y_2 v_2$, $4v_1 = 0.6v_2$, $v_1 = 0.1500v_2$.

$$4 + (0.1500v_2)^2/[(2)(32.2)] = 0.6 + v_2^2/[(2)(32.2)] \qquad v_2 = 14.97 \text{ fps} \qquad v_1 = (0.1500)(14.97) = 2.246 \text{ fps}$$

$$(62.4)(4)(\tfrac{4}{2}) - (62.4)(0.6)(0.6/2) - F_x = (1.94)[(4)(2.246)](14.97 - 2.246) \qquad F_x = 266 \text{ lb/ft} \qquad \text{(rightward)}$$

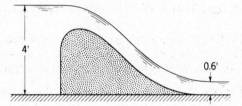

Fig. 23-39

CHAPTER 24
Dynamic Drag and Lift

24.1 A pitcher throws a baseball through air at 50 °F at a velocity of 92 mph. The diameter of the baseball is 2.82 in. Find the drag force on the baseball.

■ $F_D = C_D \rho(v^2/2)A$ $v = (92)(\frac{5280}{3600}) = 134.9$ ft/s $N_R = Dv/v = (2.82/12)(134.9)/(1.52 \times 10^{-4}) = 2.09 \times 10^5$

From Fig. A-50, $C_D = 0.42$. $F_D = (0.42)(0.00242)(134.9^2/2)[(\pi)(2.82/12)^2/4] = 0.401$ lb.

24.2 A flag pole 17 m high has the shape of a cylinder 100 mm in diameter, as shown in Fig. 24-1. Wind is blowing against the flag pole with a velocity of 15 m/s, and air temperature is 30 °C. Find the bending moment about the base (at ground level) of the flag pole. Neglect end effects.

■ $F_D = C_D \rho(v^2/2)A$ $N_R = Dv/v = (\frac{100}{1000})(15)/(1.60 \times 10^{-5}) = 9.38 \times 10^4$

From Fig. A-51, $C_D = 1.3$. $F_D = (1.3)(1.16)(15^2/2)[(17)(\frac{100}{1000})] = 288$ N. Wind acts uniformly along the flag pole; hence, the resultant drag force acts at the pole's midpoint as shown in Fig. 24-1. The bending moment about the base of the flag pole is, therefore, $M_{base} = (288)(8.5) = 2448$ N-m.

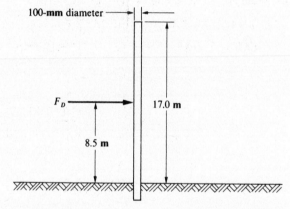

Fig. 24-1

24.3 A flat plate, 0.1 m by 0.1 m, moves at a velocity of 5.0 m/s perpendicular to the plate. Find the drag force on the plate if it is moving through (a) still air ($\rho = 1.200$ kg/m³) and (b) still water ($\rho = 1000$ kg/m³).

■ $F_D = C_D \rho(v^2/2)A$. From Fig. A-52 with $x/y = 0.1/0.1 = 1$, $C_D = 1.1$.

(a) $F_D = (1.1)(1.200)(5.0^2/2)[(0.1)(0.1)] = 0.165$ N

(b) $F_D = (1.1)(1000)(5.0^2/2)[(0.1)(0.1)] = 138$ N

24.4 A smooth, flat plate 8.0 ft wide and 80.0 ft long is being towed lengthwise (parallel to the plate) through still water at 70 °F at a velocity of 17.0 fps. What is the drag force on one side of the plate (i.e., the skin-friction drag)?

■ $F_D = C_D \rho(v^2/2)A$ $N_R = Dv/v = (80.0)(17.0)/(1.05 \times 10^{-5}) = 1.30 \times 10^8$

From Fig. A-53, $C_D = 0.0020$. $F_D = (0.0020)(1.93)(17.0^2/2)[(80.0)(8.0)] = 357$ lb.

24.5 A basketball is thrown through air at 50 °F at a velocity of 27 mph. The diameter of the basketball is 9.3 in. Determine the drag force on the basketball.

■ $F_D = C_D \rho(v^2/2)A$ $v = (27)(\frac{5280}{3600}) = 39.60$ ft/s $N_R = Dv/v = (9.3/12)(39.60)/(1.52 \times 10^{-4}) = 2.02 \times 10^5$

From Fig. A-50, $C_D = 0.41$. $F_D = (0.41)(0.00242)(39.60^2/2)[(\pi)(9.3/12)^2/4] = 0.367$ lb.

24.6 A silo with a height of 10 m above ground has the shape of a cylinder 3.5 m in diameter. Wind is blowing against the silo at a velocity of 8.3 km/h when air temperature is 30 °C. Determine the bending moment about the base of the silo.

▮ $F_D = C_D\rho(v^2/2)A$ $v = (8.3)(\frac{1000}{3600}) = 2.306$ m/s $N_R = Dv/v = (3.5)(2.306)/(1.60 \times 10^{-5}) = 5.04 \times 10^5$

From Fig. A-51, $C_D = 0.35$. $F_D = (0.35)(1.16)(2.306^2/2)[(3.5)(10)] = 37.8$ N, $M_{base} = (37.8)(\frac{10}{2}) = 189$ N-m.

24.7 A flat plate, 0.5 m by 2.0 m, moves at a velocity of 3.5 m/s perpendicular to the plate. Find the drag force on the plate if it is moving through **(a)** still air ($\rho = 1.200$ kg/m³) and **(b)** still water ($\rho = 1000$ kg/m³).

▮ $F_D = C_D\rho(v^2/2)A$. From Fig. A-52 with $x/y = 2.0/0.5 = 4.0$, $C_D = 1.2$.

(a) $F_D = (1.2)(1.200)(3.5^2/2)[(0.5)(2.0)] = 8.82$ N

(b) $F_D = (1.2)(1000)(3.5^2/2)[(0.5)(2.0)] = 7350$ N

24.8 Wind with a velocity of 35 mph is blowing perpendicularly against a 14-ft by 10-ft highway sign. If the mass density of the air is 0.00234 slug/ft³, compute the total drag force on the sign.

▮ $F_D = C_D\rho(v^2/2)A$. From Fig. A-52 with $x/y = 14/10 = 1.4$, $C_D = 1.12$. $v = (35)(\frac{5280}{3600}) = 51.33$ ft/s, $F_D = (1.12)(0.00234)(51.33^2/2)[(14)(10)] = 483$ lb.

24.9 A sign along an interstate is illustrated in Fig. 24-2. Assuming the sign can be analyzed as a flat plate, determine the wind velocity required to overturn the sign if it was designed to resist a total moment of 6000 lb-ft at its base. Use a mass density of air of 0.00234 slug/ft³.

▮ $F_D = C_D\rho(v^2/2)A$. From Fig. A-52 with $x/y = \frac{15}{9} = 1.67$, $C_D = 1.16$.

$$F_D = (1.16)(0.00234)(v^2/2)[(15)(9)] = 0.1832v^2 \qquad \sum M_{base} = 0$$

$$6000 - (6 + \tfrac{9}{2})(0.1832v^2) = 0 \qquad v = 55.85 \text{ ft/s} \quad \text{or} \quad 38.1 \text{ mph}$$

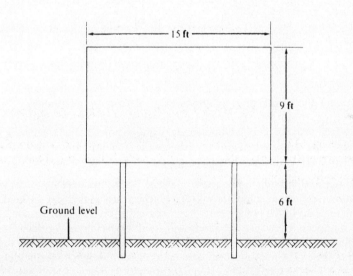

Fig. 24-2

24.10 A smooth, flat plate 2.0 m wide and 10.0 m long is being towed lengthwise (parallel to the plate) through still water at 20 °C at a velocity of 5.0 m/s. What is the drag force on one side of the plate (i.e., the skin-friction drag)?

▮ $F_D = C_D\rho(v^2/2)A$ $N_R = Dv/v = (10.0)(5.0)/(1.02 \times 10^{-6}) = 4.90 \times 10^7$

From Fig. A-53, $C_D = 0.0023$. $F_D = (0.0023)(998)(5.0^2/2)[(10.0)(2.0)] = 574$ N.

24.11 Find the friction drag on one side of a smooth flat plate 6 in wide and 18 in long, placed longitudinally in a stream of crude oil (s.g. $= 0.925$) at 60 °F flowing with undistributed velocity of 2 fps.

▮ From Fig. A-2, $v = 0.0010$ ft²/s. Then, at $x = L$, $N_R = LU/v = (1.5)(2)/0.0010 = 3000$, well within the laminar range; that is, $N_R < 500\,000$. $C_f = 1.328/\sqrt{N_R} = 1.328\sqrt{3000} = 0.0242$, $F_f = C_f\rho(V^2/2)BL = (0.0242)(0.925)(1.94)(2^2)(6)(18)/[(2)144] = 0.065$ lb. Find the thickness of the boundary layer and the shear stress at the trailing edge of the plate. $\delta/x = 4.91/\sqrt{N_R} = 4.91/\sqrt{3000} = 0.0896$, $\delta = (0.0896)(1.5) = 0.1344$ ft $= 1.61$ in. At $x = L$, $\tau_0 = 0.332(v\rho U/L)\sqrt{N_R} = (0.332)[(0.0010)(0.925)(1.94)(2)/1.5]\sqrt{3000} = 0.0435$ lb/ft².

24.12 **(a)** Find the frictional drag on the top and sides of a box-shaped moving van 8 ft wide, 10 ft high, and 35 ft long, traveling at 60 mph through air ($\gamma = 0.0725$ lb/ft^3) at 50 °F. Assume that the vehicle has a rounded nose so that the flow does not separate from the top and sides (see Fig. A-54b). Assume also that even though the top and sides of the van are relatively smooth there is enough roughness so that for all practical purposes a turbulent boundary layer starts immediately at the leading edge. **(b)** Find the thickness of the boundary layer and the shear stress of the trailing edge.

▮ **(a)** From Fig. A-2, for air at 50 °F, $v = 0.00015$ ft^2/s. Then $N_R = LU/v = (35)(88)/0.00015 = 20\,530\,000$. As $N_R > 10^7$, $C_f = 0.455/(\log N_R)^{2.58} = 0.455/(7.31)^{2.58} = 0.00269$, $F_f = C_f\rho(V^2/2)BL = (0.00269)(0.0725/32.2)[(88)^2/2](10 + 8 + 10)35 = 23.0$ lb.

(b) $$\delta/x = 0.377/N_R^{1/5} \qquad \delta_{35} = (35)(0.377)/[(205.3)^{1/5} \times 10] = 0.455 \text{ ft}$$
$$\tau_0 = 0.0587\rho(U^2/2)(v/Ux)^{1/5} \qquad (\tau_0)_{35} = (0.0587)(0.0725/32.2)(88^2/2)[0.00015/(88)(35)]^{1/5} = 0.0176 \text{ lb/ft}^2$$

24.13 A small submarine, which may be supposed to approximate a cylinder 10 ft in diameter and 50 ft long, travels submerged at 3 knots (5.06 fps) in sea water at 40 °F. Find the friction drag assuming no separation from the sides.

▮ $$F_f = C_f\rho(V^2/2)A \qquad N_F = DV/v$$

Viscosity of sea water $\approx$ viscosity of fresh water, $v = 0.0000166$ ft^2/s. Then $N_F = (50)(5.06)/0.0000166 = 1.52 \times 10^7$. From Fig. A-55, $C_f = 0.0028$ and $F_f = (0.0028)(64/32.2)[(5.06)^2/2](\pi)(10)(50) = 112$ lb. Find the value of the critical roughness for a point 1 ft from the nose of the submarine.

At $x = 1$ ft, $(N_F) = (5.06)(1)/0.0000166 = 305\,000$, $e_c = 26(v/V)(N_R)_x^{1/4} = [(26)(0.0000166)/5.06](305\,000)^{1/4} = 0.0020$ ft. Find the height of roughness at the midsection of the submarine which would class the surface as truly rough.

At $x = 25$ ft, $(N_R)_x = (5.05)(25)/0.0000166 = 7.61 \times 10^6$, $\tau_0/\rho = 0.0587(V^2/2)(v/Vx)^{1/5} = (0.0587)[(5.06)^2/2][1/(76.2 \times 10^5)^{1/5}] = 0.0316$ ft^2/s^2, $\delta_t = 60v/\sqrt{\tau_0/\rho} = (60)(0.0000166)/\sqrt{0.0316} = 0.0056$ ft.

24.14 Using the data of Prob. 24.12 determine the total drag exerted by the air on the van. Assume that $C_D \approx 0.45$ (see Fig. A-54).

▮ $$F_D = C_D\rho(V^2/2)A = 0.45(0.0725/32.2)[(88)^2/2](8)(10) = 314 \text{ lb}$$

Thus the pressure drag $= 314 - 23 = 291$ lb; in this case the pressure drag is responsible for about 93 percent of the total drag while the friction drag comprises only 7 percent of the total.

24.15 Find the "free-fall" velocity of an 8.5-in-diameter sphere (bowling ball) weighing 16 lb when falling through the following fluids under the action of gravity: through the standard atmosphere at sea level; through the standard atmosphere at 10 000-ft elevation; through water at 60 °F; through crude oil (s.g. = 0.925) at 60 °F.

▮ When first released the sphere will accelerate (Fig. 24-3a) because the forces acting on it are out of balance. This acceleration results in a buildup of velocity which causes an increase in the drag force. After a while the drag force will increase to the point where the forces acting on the sphere are in balance, as indicated in Fig. 24-3b. When that point is reached the sphere will attain a constant or terminal (free-fall) velocity. Thus for free-fall conditions, $\sum F_x = W - F_B - F_D = $ (mass)(acceleration) $= 0$ where W is the weight, F_B the buoyant force, and F_D the drag force. The buoyant force is equal to the unit weight of the fluid multiplied by the volume ($\pi D^3/6 = 0.186$ ft^3) of the sphere. The given data are approximately as follows:

fluid	γ, lb/ft^3	ρ, slugs/ft^3	$v = \mu/\rho$, ft^2/s	F_B, lb
Air (sea level)	0.0765	0.00238	1.57×10^{-4}	0.0142
Air (10 000 ft)	0.0564	0.00176	2.01×10^{-4}	0.0105
Water, 60 °F	62.4	1.94	1.22×10^{-5}	11.6
Oil, 60 °F	57.7	1.79	0.001	10.7

The detailed analysis for the sphere falling through the standard sea-level atmosphere is as follows:
$16 - 0.0142 - C_D\rho(V^2/2)A = 0$ where $\rho = 0.00238$ slug/ft^3 and $A = \pi(8.5/12)^2/4 = 0.394$ ft^2 or $15.986 = C_D(0.00238)(V^2/2)(0.394) = 0.00047C_DV^2$. A trial-and-error solution is required. Let $C_D = 0.2$, then $V = 412$ fps. $N_R = DV/v = (8.5/12)412/(1.57 \times 10^{-4}) = 1.86 \times 10^6$. The values of C_D and N_R check, hence $C_D = 0.2$ and $V = 412$ fps.

Following a similar procedure for the other three fluids gives the following free-fall velocities:

fluid	C_D	N_R	V_{fall}
Standard atmosphere at 10 000 ft	0.20	1.69×10^6	480 fps
Water at 60 °F	0.19	453 000	7.4 fps*
Crude oil (s.g. = 0.925) at 60 °F	0.39	4 390	6.2 fps

* In this instance the Reynolds number is 453 000 which, for the case of a sphere, generally indicates a turbulent boundary layer (Fig. A-56). This is very close to the point where the boundary layer changes from laminar to turbulent. If the water had been at a somewhat lower temperature and, hence, more viscous, a laminar boundary layer might have been present, in which case the free-fall velocity would have been only about 5.2 fps.

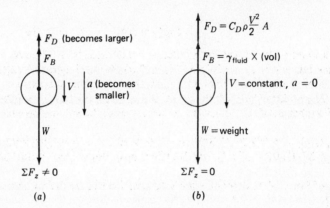

Fig. 24-3

24.16 For the critical Reynolds number of 500 000 for transition from laminar to turbulent flow in the boundary layer, find the corresponding critical Reynolds number for flow in a circular pipe. (*Hint:* Consider the boundary-layer thickness to correspond to the radius of the pipe in laminar flow, while the undisturbed velocity U of the boundary-layer theory represents the centerline velocity u_{max} of the pipe flow.)

∎ $\delta/x = 4.91/\sqrt{(N_R)_x} = 4.91/\sqrt{500\,000} = 0.00694$. Therefore, $N_R = 500\,000 = Ux/v = U\delta/0.00694v$. Since δ corresponds to pipe radius, $D/2$, U corresponds to $U_{max} = 2V$ for laminar flow. Then for pipe flow, $500\,000 = (2V)(D/2)/0.00694v$ or $DV/v = (N_R)_{pipe} = (500\,000)(0.00694) = 3470$.

24.17 Refer to the data of Prob. 24.12. Find the shear stress on the sides of the van at 5, 15, and 25 ft back from the front edge of the sides.

∎ $\tau_0 = 0.0587\rho(U^2/2)(v/Ux)^{1/5}$. At $x = 5$ ft, $\tau_0 = (0.0587)(0.0725/32.2)(88^2/2)\{0.00015/[(88)(5)]\}^{1/5} = 0.0260$ lb/ft^2. At $x = 15$ ft, $\tau_0 = (0.0587)(0.0725/32.2)(88^2/2)\{0.00015/[(88)(5)]\}^{1/5} = 0.0209$ lb/ft^2. At $x = 25$ ft, $\tau_0 = (0.0587)(0.0725/32.2)(88^2/2)\{0.00015/[(88)(25)]\}^{1/5} = 0.0189$ lb/ft^2.

24.18 It is well known that on the beach one can lie down to get out of the wind. Suppose the wind velocity 6 ft above the beach is 20 fps. Approximately what would be the wind velocity 0.5 ft and 1.0 ft above ground level?

∎ $u/U = (y/\delta)^{1/7}$. At 0.5 ft, $u/20 = (0.5/6)^{1/7}$, $u = 14.0$ fps. At 1.0 ft, $u/20 = (1.0/6)^{1/7}$, $u = 15.5$ fps.

24.19 A metal ball of diameter 1.0 ft and weight 90 lb is dropped from a boat into the ocean. Determine the maximum velocity the ball will achieve as it falls through the water. Properties of the ocean water: $\rho = 2.0$ slugs/ft^3, $\mu = 3.3 \times 10^{-5}$ lb-s/ft^2.

∎
$$F_D = W - F_b = C_D\rho(V^2/2)A \qquad W - (\gamma)(\pi D^3/6) = C_D\rho(V^2/2)(\pi D^2/4)$$
$$90 - [(62.4)(2.0/1.94)](\pi)(1.0)^3/6 = C_D(2.0)(V^2/2)(\pi)(1.0)^2/4 \qquad 90 - 33.68 = 0.7854C_DV^2 \qquad C_DV^2 = 71.71$$
$$N_R = DV\rho/\mu = (1.0)(V)(2.0)/(3.3 \times 10^{-5}) = 60\,606V$$

Try $C_D = 0.4$; $0.4V^2 = 71.71$, $V = 13.39$ ft/s, $N_R = (60\,606)(13.39) = 8.12 \times 10^5$. From Fig. A-56, $C_D = 0.20$. Try $C_D = 0.20$: $0.20V^2 = 71.71$, $V = 18.94$ ft/s, $N_R = (60\,606)(18.94) = 1.15 \times 10^6$, $C_D = 0.20$ (O.K.). Hence, $V = 18.94$ fps.

24.20 Suppose a well-streamlined automobile has a body form corresponding roughly to the airship hull of Fig. A-56, while a poorly streamlined car has a body approximating the $1:0.75$ oblate ellipsoid, each with a diameter of 6 ft. Find the power required to overcome air resistance in each of the two cases if the velocity is 60 mph (88 fps) through standard air at sea level.

$$P = F_D V \qquad F_D = C_D \rho (V^2/2) A \qquad N_R = DV/v = (6)(88)/0.0001564 = 3.38 \times 10^6$$

Assuming from Fig. A-56 that C_D remains constant for $N_R > 10^6$, C_D for well-streamlined automobile (airship hull) $= 0.04$ and C_D for poorly streamlined automobile ($1:0.75$ oblate ellipsoid) $= 0.20$. For well-streamlined automobile: $F_D = (0.04)(0.002378)(88^2/2)[(\pi)(6)^2/4] = 10.4$ lb, $P = (10.4)(88)/550 = 1.66$ hp. For poorly streamlined automobile: $F_D = (0.20)(0.002378)(88^2/2)[(\pi)(6)^2/4] = 52.1$ lb, $P = (52.1)(88)/550 = 8.34$ hp.

24.21 The drag coefficient for a hemispherical shell with the concave side upstream is approximately 1.33 if $N_R > 1000$. Find the diameter of a hemispherical parachute to provide a fall velocity no greater than that caused by jumping from a height of 2.5 m, if the total load is 900 N. Assume standard air at sea level.

In jump from 2.5-m wall,

$$V = \sqrt{2gH} = \sqrt{(2)(9.807)(2.5)} = 7.00 \text{ m/s} \qquad F_D = C_D \rho (V^2/2) A \qquad 900 = (1.33)(1.225)(7.00^2/2)(\pi D^2/4)$$
$$D = 5.36 \text{ m} \qquad N_R = DV/v = (5.36)(7.00)/(1.455 \times 10^{-5}) = 2.58 \times 10^6 \qquad \text{(O.K.)}$$

24.22 A regulation football is approximately 6.78 in in diameter and weighs 14.5 oz. Its shape is not greatly different from the prolate ellipsoid of Fig. A-56. Find the resistance when the ball is passed through still air (14.7 psia and 80 °F) at a velocity of 40 fps. Neglect the effect of spin about the longitudinal axis. What is the deceleration at the beginning of the trajectory?

$$F_D = C_D \rho (V^2/2) A \qquad N_R = DV/v = (6.78/12)(40)/(1.69 \times 10^{-4}) = 1.34 \times 10^5$$

From Fig. A-56, $C_D = 0.065$, $F_D = (0.065)(0.00228)(40^2/2)[(\pi)(6.78/12)^2/4] = 0.0297$ lb. At start of trajectory, $a = -F_D/m$. $a = -0.0297/[(14.5/16)/32.2)] = -1.06$ ft/s^2.

24.23 For the data given in Prob. 24.22, assuming no change in drag coefficient, find the percentage change in resistance if the air temperature is 20 °F rather than 80 °F.

$F_D = C_D \rho (V^2/2) A$. At $T = 20$ °F, $\rho = 0.00257$ slug/ft^3 (approximately, from Table A-4). $F_D = (0.065)(0.00257)(40^2/2)[(\pi)(6.78/12)^2/4] = 0.0335$ lb. Change in resistance $= (0.0335 - 0.0298)/0.0298 = 0.124$, or 12.4 percent increase.

24.24 What drag force is exerted at sea level by a 3-m-diameter braking parachute when the speed is 25 m/s? Assume $C_D = 1.20$. At what speed will the same braking force be exerted by this parachute at elevation 2000 m? Assume C_D remains constant.

$F_D = C_D \rho (V^2/2) A$. At sea level, $\rho_{air} = 1.225$ kg/m^3. $F_D = (1.20)(1.225)(25^2/2)[(\pi)(3)^2/4] = 3247$ N. At elevation 2000 m, $\rho_{air} = 1.007$ kg/m^3. $3247 = (1.20)(1.007)(V^2/2)[(\pi)(3)^2/4]$, $V = 27.6$ m/s.

24.25 Find the bending moment at the base of a cylindrical radio antenna 0.30 in in diameter extended to 6 ft in length on an automobile traveling through 60 °F air at 80 mph (117.3 fps).

$$F_D = C_D \rho (V^2/2) A \qquad N_R = DV/v = (0.30/12)(117.3)/(1.58 \times 10^{-4}) = 1.86 \times 10^4$$

From Fig. A-57, $C_D = 1.1$.

$$F_D = (1.1)(0.00237)(117.3^2/2)[(0.30/12)(6)] = 2.690 \text{ lb} \qquad M_{base} = (2.690)(\tfrac{6}{2}) = 8.07 \text{ lb/ft}$$

24.26 Approximately what frequency of oscillation is produced when a 60-mph (88-fps) wind blows across a 0.125-in-diameter wire at (a) standard sea level ($v = 1.57 \times 10^{-4}$ ft^2/s) and (b) standard atmosphere at 10 000-ft elevation ($v = 2.01 \times 10^{-4}$ ft^2/s)?

$$f \approx 0.20(V/D)(1 - 20/N_R) \qquad N_R = DV/v$$

(a) $\quad N_R = (0.125/12)(88)/(1.57 \times 10^{-4}) = 5839 \qquad f = (0.20)[88/(0.125/12)](1 - \tfrac{20}{5839}) = 1684$ cycles/s

(b) $\quad N_R = (0.125/12)(88)/(2.01 \times 10^{-4}) = 4561 \qquad f = (0.20)[88/(0.125/12)](1 - \tfrac{20}{4561}) = 1682$ cycles/s

24.27 Determine the rate of deceleration that will be experienced by the blunt-nosed projectile of Fig. A-58 when it is moving horizontally at 1000 mph (1467 fps). Assume standard sea-level atmosphere. The projectile has a diameter of 18 in and it weighs 600 lb.

▌ $F_D = C_D \rho (V^2/2)A$. At sea level, $c = 1116$ fps. $V/c = \frac{1467}{1116} = 1.31$. From Fig. A-58, $C_D = 1.22$.

$$F_D = (1.22)(0.002377)(1467^2/2)[(\pi)(\tfrac{18}{12})^2/4] = 5514 \text{ lb}$$
$$F_D = -a(W/g) \qquad 5514 = -a(600/32.2) \qquad a = -296 \text{ ft/s}^2$$

24.28 Solve Prob. 24.27 if the projectile is moving upward at an angle of 40° with the horizontal.

▌ In this case, $F_D - W \sin \theta = -a(W/g)$, $5514 - 600 \sin 40° = -a(600/32.2)$, $a = -275$ ft/s^2.

24.29 A sharp flat plate with $L = 1$ m and $b = 3$ m is immersed parallel to a stream of velocity 2 m/s. Find the drag on one side of the plate, and at the trailing edge find the thicknesses δ, δ^*, and θ for air, $\rho = 1.23$ kg/m^3 and $v = 1.46 \times 10^{-5}$ m^2/s.

▌ The airflow Reynolds number is $VL/v = (2.0)(1.0)/(1.46 \times 10^{-5}) = 137\,000$. Since this is less than 3×10^6, we assume that the boundary layer is laminar and the drag coefficient is $C_D = 1.328/N_R^{1/2} = 1.328/(137\,000)^{1/2} = 0.00359$. Thus the drag on one side in the airflow is $D = C_D \tfrac{1}{2} \rho U^2 bL = 0.00359(\tfrac{1}{2})(1.23)(2.0)^2(3.0)(1.0) = 0.0265$ N. The boundary-layer thickness at the end of the plate is $\delta/L = 5.0/N_R^{1/2} = 5.0/(137\,000)^{1/2} = 0.0135$ or $\delta = 0.0135(1.0) = 0.0135$ m $= 13.5$ mm. We find the other two thicknesses simply by ratios: $\delta^* = (1.721/5.0)\delta = 4.65$ mm, $\theta = \delta^*/2.59 = 1.80$ mm.

24.30 Solve Prob. 24.29 for water, $\rho = 1000$ kg/m^3 and $v = 1.02 \times 10^{-6}$ m^2/s.

▌ The water Reynolds number is $N_R = 2.0(1.0)/(1.02 \times 10^{-6}) = 1.96 \times 10^6$. This is rather close to the critical value of 3×10^6, so that a rough surface or noisy free stream might trigger transition to turbulence; but let us assume that the flow is laminar. The water drag coefficient is $C_D = 1.328/(1.96 \times 10^6)^{1/2} = 0.000949$ and $D = 0.000949(\tfrac{1}{2})(1000)(2.0)^2(3.0)(1.0) = 5.69$ N. The drag is 215 times more for water in spite of the higher Reynolds number and lower drag coefficient because water is 57 times more viscous and 813 times denser than air.

The boundary-layer thickness is given by $\delta/L = 5.0/(1.96 \times 10^6)^{1/2} = 0.00357$ or $\delta = 0.00357(1000$ mm$) = 3.57$ mm. By ratioing down we have $\delta^* = (1.721/5.0)\delta = 1.23$ mm, $\theta = \delta^*/2.59 = 0.47$ mm. The water layer is 3.8 times thinner than the air layer, which reflects the square root of the 14.3 ratio of air to water kinematic viscosity.

24.31 A hydrofoil 1.2 ft long and 6 ft wide is placed in a water flow of 40 fps, with $\rho = 1.99$ slugs/ft^3 and $v = 0.000011$ ft^2/s. Estimate the boundary-layer thickness at the end of the plate.

▌ The Reynolds number is $N_R = UL/v = (40)(1.2)/0.000011 = 4.36 \times 10^6$. Thus the trailing-edge flow is certainly turbulent. The maximum boundary-layer thickness would occur for turbulent flow starting at the leading edge. $\delta/L = 0.16/N_F^{1/7} = 0.16/(4.36 \times 10^6)^{1/7} = 0.018$ or $\delta = 0.018(1.2$ ft$) = 0.0216$ ft.

24.32 For the data given in Prob. 24.31, estimate the friction drag for turbulent smooth-wall flow from the leading edge.

▌ This is 7.5 times thicker than a fully laminar boundary layer at the same Reynolds number. For fully turbulent smooth-wall flow, the drag coefficient on one side of the plate is $C_D = 0.031/N_F^{1/7} = 0.031/(4.36 \times 10^6)^{1/7} = 0.00349$. Then the drag on both sides of the foil is approximately $D = 2C_D(\tfrac{1}{2}\rho U^2)bL = 2(0.00349)(\tfrac{1}{2})(1.99)(40)^2(6.0)(1.2) = 80.0$ lb.

24.33 Solve Prob. 24.32 for laminar-turbulent flow with $(N_R)_{\text{trans}} = 5 \times 10^5$.

▌ With a laminar leading edge and $(N_R)_{\text{trans}} = 5 \times 10^5$, $C_D = (0.031/N_R^{1/7}) - (1440/N_R) = 0.00349 - [1440/(4.36 \times 10^6)] = 0.00316$. The drag can be recomputed for this lower drag coefficient: $D = 2C_D(\tfrac{1}{2}\rho U^2)bL = 72.4$ lb.

24.34 Solve Prob. 24.32 for turbulent rough-wall flow with $\epsilon = 0.0004$ ft.

▌ For the rough wall, we calculate $L/\epsilon = 1.2/0.0004 = 3000$. From Fig. A-59 at $N_R = 4.36 \times 10^6$, this condition is just inside the fully rough regime. $C_D \approx [1.89 + 1.62 \log (L/\epsilon)]^{-2.5} = (1.89 + 1.62 \log 3000)^{-2.5} = 0.00644$ and the drag estimate is $D = 2C_D(\tfrac{1}{2}\rho U^2)bL = 148$ lb. This small roughness nearly doubles the drag. It is probable that the total hydrofoil drag is still another factor of 2 larger because of trailing-edge flow-separation effects.

24.35 A square 6-in piling is acted on by a water flow of 5 ft/s 20 ft deep, as shown in Fig. 24-4. Estimate the maximum bending exerted by the flow on the bottom of the piling.

▌ Assume sea water with $\rho = 1.99$ slugs/ft^3 and kinematic viscosity $\nu = 0.000011$ ft^2/s. With piling width of 0.5 ft, we have $N_R = VD/\nu = (5)(0.5)/0.000011 = 2.3 \times 10^5$. This is the range where Table A-20 applies. The worst case occurs when the flow strikes the flat side of the piling, $C_D \approx 2.1$. The frontal area is $A = Lh = (20)(0.5) = 10$ ft^2. The drag is estimated by $F = C_D(\frac{1}{2}\rho V^2 A) \approx 2.1(\frac{1}{2})(1.99)(5)^2(10) = 522$ lb. If the flow is uniform, the center of this force should be at approximately middepth. Therefore the bottom bending moment is $M_0 \approx FL/2 = 522(\frac{20}{2}) = 5220$ ft-lb. According to the flexure formula from strength of materials, the bending stress at the bottom would be $S = M_0 y/I = (5220)(0.25)/[\frac{1}{12}(0.5)^4] = 251\,000$ lb/ft$^2 = 1740$ lb/in^2 to be multiplied, of course, by the stress-concentration factor due to the built-in end conditions.

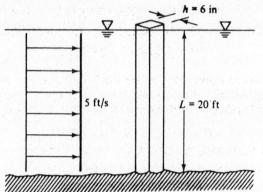

Fig. 24-4

24.36 A high-speed car with $m = 2000$ kg, $C_D = 0.3$, and $A = 1$ m^2, deploys a 2-m parachute to slow down from an initial velocity of 100 m/s (Fig. 24-5). Assuming constant C_D, brakes free, and no rolling resistance, calculate the distance and velocity of the car after 1, 10, 100, and 1000 s. For air, assume $\rho = 1.2$ kg/m^3 and neglect interference between the wake of the car and the parachute.

▌ Newton's law applied in the direction of motion gives $F_x = m(dV/dt) = -F_c - F_p = -\frac{1}{2}\rho V^2(C_{Dc}A_c + C_{Dp}A_p)$ where subscript c is the car and subscript p the parachute. This is of the form $dV/dt = -(K/m)V^2$, $K = \sum C_D A(\rho/2)$. Separate the variables and integrate

$$\int_{V_0}^{V} \frac{dV}{V^2} = -\frac{K}{m}\int_0^t dt$$

or $V_0^{-1} - V^{-1} = -(K/m)t$. Rearrange and solve for the velocity V: $V = V_0/[1 + (K/m)V_0 t]$, $K = (C_{Dc}A_c + C_{Dp}A_p)\rho/2$. We can integrate this to find the distance traveled; $S = (V_0/\alpha)\ln(1 + \alpha t)$, $\alpha = (K/m)V_0$. Now work out some numbers. From Table A-21, $C_{Dp} \approx 1.2$; hence $C_{Dp}A_c + C_{Dp}A_p = 0.3(1) + 1.2(\pi/4)(2)^2 = 4.07$ m^2. Then $(K/m)V_0 = \frac{1}{2}(4.07)(1.2)(100)/2000 = 0.122$ s$^{-1} = \alpha$. Now make a table of the results for V and S

t, s	1	10	100	1000
V, m/s	89	45	7.6	0.8
S, m	94	654	2100	3940

Air resistance alone will not stop a body completely. If you don't apply the brakes, you'll be halfway to the Yukon Territory and still going.

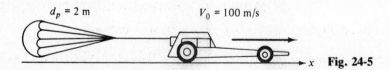

$d_p = 2$ m $V_0 = 100$ m/s   Fig. 24-5

24.37 A thin flat plate 45 cm by 90 cm is immersed in a stream of glycerin at 20 °C and a velocity of 7 m/s. Compute the viscous drag if the plate side parallel to the stream is the short side.

▌ $F_D = C_D\rho(V^2/2)A$ $N_R = \rho DV/\mu = (1258)(\frac{45}{100})(7)/1.49 = 2660$ (laminar)

$C_D = 1.328/\sqrt{N_R} = 1.328/\sqrt{2660} = 0.0257$ $(F_D)_{\text{total}} = (2)(0.0257)(1258)(7^2/2)[(\frac{90}{100})(\frac{45}{100})] = 642$ N

24.38 Solve Prob. 24.37 if the plate side parallel to the stream is the long side.

$$F_D = C_D \rho (V^2/2)A \qquad N_R = \rho DV/\mu = (1258)(\tfrac{90}{100})(7)/1.49 = 5319$$

$$C_D = 1.328/\sqrt{N_R} = 1.328/\sqrt{5319} = 0.0182 \qquad (F_D)_{\text{total}} = (2)(0.0182)(1258)(7^2/2)[(\tfrac{90}{100})(\tfrac{45}{100})] = 454 \text{ N}$$

24.39 A ship is 200 m long and has a wetted area of 8500 m². Estimate the power required to overcome friction for a smooth surface when the ship moves at 15 knots (7.72 m/s) in sea water at 20 °C.

$$P = F_D V \qquad F_D = C_D \rho (V^2/2)A \qquad C_D = 0.031/N_R^{1/7}$$

$$N_R = \rho DV/\mu = (1028)(200)(7.72)/(1.07 \times 10^{-3}) = 1.48 \times 10^9$$

$$C_D = 0.031/(1.48 \times 10^9)^{1/7} = 0.00152 \qquad F_D = (0.00152)(1028)(7.72^2/2)(8500) = 395\,800 \text{ N}$$

$$P = (395\,800)(7.72) = 3\,056\,000 \text{ W} \quad \text{or} \quad 3.056 \text{ MW}$$

24.40 A blimp approximates an ellipsoid 250 ft long and 50 ft in diameter, with a surface area of 31 370 ft². Estimate its skin-friction drag and the power required to overcome it when the blimp moves at 60 mph (88 fps) through still air at 68 °F and 12 psia ($\rho = 0.001906$ slug/ft³; $\mu = 3.76 \times 10^{-7}$ slug/ft-s).

$$P = F_D V \qquad F_D = C_D \rho (V^2/2)A \qquad C_D = 0.031/N_R^{1/7}$$

$$N_R = \rho DV/\mu = (0.001906)(250)(88)/(3.76 \times 10^{-7}) = 1.12 \times 10^8 \qquad C_D = 0.031/(1.12 \times 10^8)^{1/7} = 0.00220$$

$$F_D = (0.00220)(0.001906)(88^2/2)(31\,370) = 509 \text{ lb} \qquad P = (509)(88)/550 = 81.4 \text{ hp}$$

24.41 A hydrofoil 60 cm long and 3 m wide moves in water at a speed of 14 m/s. Using flat-plate theory, estimate the drag if $(N_R)_{\text{trans}} = 5 \times 10^5$ for a smooth wall.

$$F_D = C_D \rho (V^2/2)A \qquad C_D = 0.031/N_R^{1/7} - 1440/N_R$$

$$N_R = \rho DV/\mu = (998)(\tfrac{60}{100})(14)/(1.02 \times 10^{-3}) = 8.22 \times 10^6 \qquad \text{(turbulent)}$$

$$C_D = 0.031/(8.22 \times 10^6)^{1/7} - 1440/(8.22 \times 10^6) = 0.00301$$

$$(F_D)_{\text{total}} = (2)(0.00301)(998)(14^2/2)[(\tfrac{60}{100})(3)] = 1060 \text{ N}$$

24.42 Solve Prob. 24.41 for a rough wall, $\epsilon = 0.1$ mm.

$$F_D = C_D \rho (V^2/2)A \qquad L/\epsilon = (\tfrac{60}{100})/(0.1/1000) = 6000 \qquad \text{(fully rough)}$$

$$C_D = [1.89 + 1.62 \log (L/\epsilon)]^{-2.5} = (1.89 + 1.62 \log 6000)^{-2.5} = 0.00551$$

$$(F_D)_{\text{total}} = (2)(0.00551)(998)(14^2/2)[(\tfrac{60}{100})(3)] = 1940 \text{ N}$$

24.43 A jet airliner travels at 250 m/s at 10 000 m standard altitude. It has a smooth wing 7 m long and 55 m wide. Estimate the power required to overcome friction drag.

$$P = F_D V \qquad F_D = C_D \rho (V^2/2)A \qquad C_D = 0.031/N_R^{1/7}$$

$$N_R = \rho DV/\mu = (0.4125)(7)(250)/(1.49 \times 10^{-5}) = 4.84 \times 10^7 \qquad C_D = 0.031/(4.84 \times 10^7)^{1/7} = 0.00247$$

$$(F_D)_{\text{total}} = (2)(0.00247)(0.4125)(250^2/2)[(7)(55)] = 24\,500 \text{ N}$$

$$P = (24\,500)(250) = 6\,125\,000 \text{ W} \quad \text{or} \quad 6.125 \text{ MW}$$

24.44 For the data given in Prob. 24.43, if the wing is rough and requires 13 MW to overcome friction, estimate the wing roughness in millimeters.

$$P = F_D V \qquad (13)(1\,000\,000) = (F_D)(250) \qquad F_D = 52\,000 \text{ N} = C_D \rho (V^2/2)A$$

$$52\,000/2 = (C_D)(0.4125)(250^2/2)[(7)(55)] \qquad C_D = 0.00524$$

From Fig. A-59, $L/\epsilon = 7500$. $\epsilon = \tfrac{7}{7500} = 0.00093$ m, or 0.93 mm.

24.45 A flat barge 14 m wide and 40 m long moves at 3 knots (1.543 m/s) in sea water at 20 °C. Estimate the friction drag on the bottom of the barge for (*a*) a smooth wall and (*b*) wall roughness height of 3 mm.

$$F_D = C_D \rho (V^2/2)A$$

(*a*)
$$N_R = \rho DV/\mu = (1028)(40)(1.543)/(1.07 \times 10^{-3}) = 5.93 \times 10^7$$

$$C_D = 0.031/N_R^{1/7} = 0.031/(5.93 \times 10^7)^{1/7} = 0.00240 \qquad F_D = (0.00240)(1028)(1.543^2/2)[(14)(40)] = 1645 \text{ N}$$

(*b*) $L/\epsilon = 40/(\tfrac{3}{1000}) = 13\,333$. From Fig. A-59, $C_D = 0.00465$. $F_D = (0.00465)(1028)(1.543^2/2)[(14)(40)] = 3187$ N.

24.46 A torpedo 55 cm in diameter and 5 m long moves at 45 knots (23.15 m/s) in sea water at 20 °C. Estimate the power required to overcome friction drag if $(N_R)_{crit} = 5 \times 10^5$ and $\epsilon = 0.5$ mm.

$$\blacksquare \quad F_D = C_D \rho(V^2/2)A \qquad N_R = \rho DV/\mu = (1028)(5)(23.15)/(1.07 \times 10^{-3}) = 1.11 \times 10^8 \qquad \text{(roughness prevails)}$$

$$L/\epsilon = 5/(0.5/1000) = 10\,000$$

From Fig. A-59, $C_D = 0.00493$. $F_D = (0.00493)(1028)(23.15^2/2)[(\pi)(\frac{55}{100})(5)] = 11\,730$ N.

24.47 A chimney 2.5 m in diameter and 45 m high is exposed to sea-level storm winds at 20 m/s. Estimate the drag force and the bending moment about the bottom of the chimney $\rho = 1.2255$ kg/m³ and $\mu = 1.78 \times 10^{-5}$ kg/(m-s).

$$\blacksquare \qquad F_D = C_D \rho(V^2/2)A \qquad N_R = \rho DV/\mu = (1.2255)(2.5)(20)/(1.78 \times 10^{-5}) = 3.44 \times 10^6$$

Turbulent: $C_D \approx 0.3$. $F_D = (0.3)(1.2255)(20^2/2)[(2.5)(45)] = 8272$ N, $M_{bottom} = (8272)(\frac{45}{2}) = 186\,000$ N-m.

24.48 A logging boat tows a log 2.5 m in diameter and 25 m long at 4 m/s in fresh water at 20 °C. Estimate the power required if the axis of the log is parallel to the tow direction.

$$\blacksquare \qquad F_D = C_D \rho(V^2/2)A \qquad N_R = \rho DV/\mu = (998)(25)(4)/(1.02 \times 10^{-3}) = 9.78 \times 10^7$$

From Table A-21, $C_D \approx 1.05$ at $L/D \approx 10$. $F_D = (1.05)(998)(4^2/2)[(\pi)(2.5)^2/4] = 41\,200$ N.

24.49 A fishnet consists of 1-mm-diameter strings overlapped and knotted to form 1-cm by 1-cm squares. Estimate the drag of 1 m² of such a net when towed normal to its plane at 3 m/s in sea water. What horsepower is required to tow 500 ft² of this net?

$$\blacksquare \qquad F_D = C_D \rho(V^2/2)A \qquad N_R = \rho DV/\mu = (1028)(\tfrac{1}{1000})(3)/(1.07 \times 10^{-3}) = 2.88 \times 10^3$$

From Fig. A-60a, $C_D \approx 1.0$. For a single 1-cm strand: $F_D = (1.0)(1028)(3^2/2)[(\frac{1}{100})(\frac{1}{1000})] = 0.0463$ N. 1 m² of net contains 20 000 strands: $F_D = (0.0463)(20\,000) = 926$ N (the drag of 1 m² of net), 500 ft² = 46.45 m². For 500 ft² of net: $F_D = (926)(46.45) = 43\,013$ N, or 9670 lb; $P = F_D V = (9670)(3/0.3048)/550 = 173$ hp.

24.50 A filter may be idealized as an array of cylindrical fibers normal to the flow, as in Fig. 24-6a. Assuming that the fibers are uniformly distributed and have drag coefficients given by Fig. A-61a derive an approximate expression for the pressure drop Δp through a filter of thickness L.

$\blacksquare$ Let N = number of fibers per unit area of filter, D = fiber diameter, b = filter width, H = filter height, and L = filter length (see Fig. 24-6b). Each fiber has drag of $F_D = C_D \rho(V^2/2)Db$. Force balance on filter: $\Delta p Hb = \Sigma F_{fibers} = NHLC_D \rho(V^2/2)Db$, $\Delta p = NLC_D \rho(V^2/2)D$.

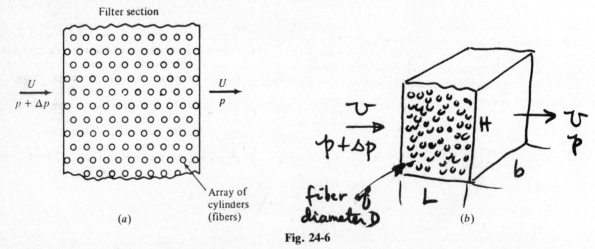

Fig. 24-6

24.51 Apply Prob. 24.50 to a filter consisting of 0.2-mm-diameter fibers packed 1000 per square centimeter of filter section in the plane of Fig. 24-6a. For sea-level standard air flowing at 1.2 m/s, estimate the pressure drop through a filter 3 cm thick.

$$\blacksquare \qquad \Delta p = NLC_D \rho(V^2/2)D \qquad N = 1000/\text{cm}^2 = 10\,000\,000/\text{m}^2$$

$$N_R = \rho DV/\mu = (1.2255)(0.2/1000)(1.2)/(1.78 \times 10^{-5}) = 16.5$$

From Fig. A-61a, $C_D \approx 2.8$. $\Delta p = (10\,000\,000)(\frac{3}{100})(2.8)(1.2255)(1.2^2/2)(0.2/1000) = 148$ N/m².

24.52 Assume that a radioactive dust particle approximates a sphere with a specific weight of 26 000 N/m³. Approximately how long will it take such a particle to settle to earth from an altitude of 10 km if the particle diameter is (**a**) 10^{-6} m and (**b**) 10^{-5} m?

▎ For creeping motion, $F_D = W_{net} = 3\pi\mu DV$. Assume average values of $\gamma_{air} = 11.8$ N/m³ and $\mu_{air} = 1.7 \times 10^{-5}$ kg/(m-s).

$$W = \gamma V \qquad W_{net} = (26\,000 - 11.8)(\pi D^3/6) = 13\,607 D^3$$
$$13\,607 D^3 = (3)(\pi)(1.7 \times 10^{-5})(D)(V) \qquad V = 8.493 = 10^7 D^2$$

(**a**) $\qquad D = 10^{-6}$ m $\qquad V = (8.493 \times 10^7)(10^{-6})^2 = 8.49 \times 10^{-5}$ m/s

$\qquad t = (10)(1000)/(8.49 \times 10^{-5}) = 1.18 \times 10^8$ s or 1363 days

(**b**) $\qquad D = 10^{-5}$ m $\qquad V = (8.493 \times 10^7)(10^{-5})^2 = 8.49 \times 10^{-3}$ m/s

$\qquad t = (10)(1000)/(8.49 \times 10^{-3}) = 1.18 \times 10^6$ s or 13.63 days

24.53 A water tower is approximated by a 15-m-diameter sphere mounted on a 1-m-diameter rod 20 m long. Estimate the bending moment at the base of the rod due to aerodynamic forces during hurricane winds of 40 m/s. Use 1.2255 kg/m³ for air.

▎ $F = C_D\rho(V^2/2)A$. Estimate $C_D = 0.2$ for sphere and 0.3 for rod.

$$F_{sphere} = (0.2)(1.2255)(40^2/2)[(\pi)(15^2)/4] = 34\,650 \text{ N} \qquad F_{rod} = (0.3)(1.2255)(40^2/2)[(1)(20)] = 5882 \text{ N}$$
$$M_{base} = (34\,650)(20 + \tfrac{15}{2}) + (5882)(\tfrac{20}{2}) = 1.01 \times 10^6 \text{ N-m}$$

24.54 In the great hurricane of 1938, winds of 85 mph blew over a boxcar in Providence, Rhode Island. The boxcar was 10 ft high, 40 ft long, and 6 ft wide, with a 3-ft clearance above tracks 4.8 ft apart. What wind speed would topple a boxcar weighing 40 000 lb?

▎ $F_D = C_D\rho(V^2/2)A$. From Table A-21, $C_D = 1.2$. The wind is, of course, blowing against the side of the boxcar; hence, $F_D = (1.2)(0.00237)(V^2/2)[(10)(40)] = 0.5688V^2$, $M_{overturning} = (0.5688V^2)(3 + \tfrac{10}{2}) = 4.550V^2$, $M_{righting} = (40\,000)(4.8/2) = 96\,000$ lb-ft, $4.550V^2 = 96\,000$, $V = 145.3$ ft/s, or 99.0 mph.

24.55 A long copper wire, 0.5 in in diameter, is stretched taut and is exposed to a wind of velocity 90.0 fps normal to the wire. Compute the drag force per foot of length.

▎ $\qquad F_D = C_D\rho(V^2/2)A \qquad N_R = DV/\nu = (0.5/12)(90.0)/(1.6 \times 10^{-4}) = 2.34 \times 10^4$

From Fig. A-62, $C_D = 1.30$. $F_D = (1.30)(0.00233)(90.0^2/2)[(0.5/12)(1)] = 0.511$ lb per foot.

24.56 Consider the area on one side of a moving van to be 600 ft². Determine the resultant force acting on the side of the van when the wind is blowing at 10 mph normal to the area (**a**) when the van is at rest and (**b**) when the van is moving at 30 mph normal to the direction of the wind. In (**a**) use $C_D = 1.30$, and in (**b**) use $C_D = 0.25$ and $C_L = 0.60$. ($\rho = 0.00237$ slug/ft³)

▎ (**a**) The force acting normal to the area $= C_D(\rho/2)AV^2$. Then Resultant force $=$ $1.30(0.00237/2)(600)[(10)(\tfrac{5280}{3600})]^2 = 199$ lb normal to area.

(**b**) It will be necessary to calculate the relative velocity of the wind with respect to the van. From kinetic mechanics, $V_{wind} = V_{wind/van} \leftrightarrow V_{van}$.

Figure 24-7 indicates this vector relationship, i.e., $OB = OA \leftrightarrow AB = 30.0 \leftrightarrow V_{w/v}$. Thus the relative velocity $= \sqrt{(30)^2 + (10)^2} = 31.6$ mph to the right and upward at an angle $\theta = \tan^{-1} \tfrac{10}{30} = 18.4°$.

The component of the resultant force normal to the relative velocity of wind with respect to van is Lift force $= C_L(\rho/2)AV^2 = 0.60(0.00237/2)(600)[(31.6)(\tfrac{5280}{3600})]^2 = 916$ lb normal to the relative velocity. The component of the resultant force parallel to the relative motion of wind to van is Drag force $= C_D(\rho/2)AV^2 = 0.25(0.00237/2)(600)[(31.6)(\tfrac{5280}{3600})]^2 = 382$ lb parallel to the relative velocity.

Referring to Fig. 24-7b, the resultant force $= \sqrt{(916)^2 + (382)^2} = 992$ lb at an angle $\alpha = \tan^{-1} \tfrac{916}{382} = 67.4°$. Hence the angle with the longitudinal axis (X axis) is $18.4° + 67.4° = 85.8°$.

24.57 A man weighing 170 lb is descending from an airplane using an 18-ft-diameter parachute. Assuming a drag coefficient of 1.00 and neglecting the weight of the parachute, what maximum terminal velocity will be attained?

▎ The forces on the parachute are the weight down and the drag force up. $F_D = C_D\rho(V^2/2)A$, $170 = (1.00)(0.00237)(V^2/2)[(\pi)(18)^2/4]$, $V = 23.7$ ft/s.

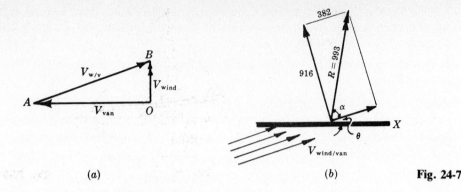

(a)   (b)   **Fig. 24-7**

24.58 A 1-in-diameter sphere of lead weighing 710 lb/ft^3 is moving downward in an oil at a constant velocity of 1.17 fps. Calculate the absolute viscosity of the oil if its specific gravity is 0.93.

$$F_D = (\gamma_{\text{lead}} - \gamma_{\text{oil}})(V_{\text{lead}}) = C_D\rho(V^2/2)A$$
$$[710 - (0.93)(62.4)][(\pi)(\tfrac{1}{12})^3/6] = C_D[(1.94)(0.93)](1.17^2/2)[(\pi)(\tfrac{1}{12})^2/4] \quad C_D = 29.3$$

From Fig. A-62, $N_R = 0.85$ for $C_D = 29.3$. $N_R = \rho DV/\mu$, $0.85 = [(1.94)(0.93)](\tfrac{1}{12})(1.17)/\mu$, $\mu = 0.207$ lb-s/ft^2.

24.59 A sphere, 0.5 in in diameter, rises in oil at the maximum velocity of 0.12 fps. What is the specific weight of the sphere if the density of the oil is 1.78 slugs/ft^3 and the absolute viscosity is 0.000710 lb-s/ft^2?

$$F_D = F_b - W = C_D\rho(V^2/2)(A) \qquad N_R = \rho DV/\mu = (1.78)(0.5/12)(0.12)/0.000710 = 12.5$$

From Fig. A-62, $C_D = 3.9$ for $N_R = 12.5$. $[(\pi)(0.5/12)^3/6][(1.78/1.94)(62.4)] - W = (3.9)(1.78)(0.12^2/2)[(\pi)(0.5/12)^2/4]$, $W = 0.002100$ lb; $\gamma = 0.002100/[(\pi)(0.5/12)^3/6] = 55.4$ lb/ft^3.

24.60 Measurements on a smooth sphere, 6 in in diameter, in an air stream (68 °F) gave a force for equilibrium equal to 0.250 lb. At what velocity was the air moving?

Total drag $= C_D\rho AV^2/2$, where $C_D =$ overall drag coefficient. Since neither Reynolds number nor C_D can be found directly, assume $C_D = 1.00$. Then $0.250 = C_D(0.00233)\tfrac{1}{4}\pi(\tfrac{1}{2})^2(V^2/2)$, $V^2 = 1093/C_D$, $V = 33.1$ ft/s. Calculate $N_R = Vd/\nu = 33.1(\tfrac{1}{2})/(16.0 \times 10^{-5}) = 103\,000$. From Fig. A-62, $C_D = 0.59$ (for spheres). Then $V^2 = 1093/0.59 = 1853$, $V = 43.0$ ft/s. Anticipating result, use $V = 44.0$ ft. Recalculate $N_R = Vd/\nu = 44.0(\tfrac{1}{2})/(16.0 \times 10^{-5}) = 137\,500$. From Fig. A-62, $C_D = 0.56$. Then $V^2 = 1093/0.56 = 1952$, $V = 44.2$ ft/s (satisfactory accuracy).

24.61 What should be the diameter of a sphere (s.g. $= 2.50$) in order that its freely falling velocity at 60 °F attains the acoustic velocity?

$F_D = W = C_D\rho(V^2/2)A$. From Fig. A-63, $C_D = 0.80$. $V = c = \sqrt{kgRT}$. For air at 60 °F, $c = \sqrt{(1.40)(32.2)(53.3)(460 + 60)} = 1118$ ft/s, $[(2.50)(62.4)][(\pi d^3/6)] = (0.80)(0.00237)(1118^2/2)(\pi d^2/4)$, $d = 11.4$ ft.

24.62 The fixed keel of a Columbia 22 sailboat is about 38 in long (see Fig. 24-8). Moving in Lake Ontario at a speed of 3 knots, what is the skin drag from the keel? The water is at 40 °F. Solve this problem using rectangular plate of length 38 in and width 24.5 in, which is the average width of the keel. Transition takes place at a Reynolds number of 10^6.

Compute the plate Reynolds number for the rectangular model of the keel.

$$N_R = Vl/\nu = [(3)(1.689)](24.5/12)/(1.664 \times 10^{-5}) = 6.22 \times 10^5 \tag{1}$$

We thus have a laminar boundary layer. We get the plate coefficient of drag using

$$C_f = 1.328/\sqrt{N_R} = 1.328/\sqrt{6.22 \times 10^5} = 1.684 \times 10^{-3} \tag{2}$$

Next we get the skin drag, realizing that there are two sides to be considered. Thus,

$$D = 2[(C_f)(\tfrac{1}{2}\rho V_0^2)(A)] = 2\{(1.684 \times 10^{-3})(\tfrac{1}{2})(1.940)[(3)(1.689)]^2[(38)(24.5)/144]\} = 0.542 \text{ lb} \tag{3}$$

24.63 Solve Prob. 24.62 using the actual dimensions of the keel as shown in Fig. 24-8. Compare answers and comment on the result.

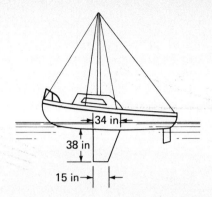

Fig. 24-8

For this purpose, consult Fig. 24-9 in which a keel with an infinitesimal slice dz is depicted. The length $l(z)$ of the slice is

$$l = \tfrac{15}{12} + (z/38)(19) = 1.250 + 0.5z \quad \text{ft} \tag{4}$$

with z in feet. Now let us first see if we have transition anywhere on the keel. Looking at the uppermost portion, we have for $(N_R)_{max}$

$$(N_R)_{max} = Vl_{max}/\nu = [(3)(1.689)](\tfrac{34}{12})/(1.664 \times 10^{-5}) = 8.63 \times 10^5 \tag{5}$$

We have only a laminar boundary layer. For an infinitesimal plate of length $l = 1.250 + 0.5z$ ft, $C_f = 1.328/(Vl/\nu)^{1/2} = 1.328\{[(3)(1.689)/(1.664 \times 10^{-5})](1.250 + 0.5z)\}^{-1/2}$. Now for the drag D,

$$D = 2\int_0^{38/12} C_f(\tfrac{1}{2}\rho V_0^2)(1.250 + 0.5z)(dz)$$

$$= 2\int_0^{38/12} 1.238\left[\frac{(3)(1.689)}{1.664 \times 10^{-5}}(1.250 + 0.5z)\right]^{-1/2}(\tfrac{1}{2})(1.940)[(3)(1.689)]^2(1.250 + 0.5z)dz$$

$$= 0.1199\int_0^{3.17} (1.250 + 0.5z)^{1/2} dz$$

Let $1.250 + 0.5z = \eta$; therefore $0.5dz = d\eta$, $dz = 2d\eta$. Hence,

$$D = \frac{0.1199}{0.5}\int_{1.250}^{2.833} \eta^{1/2} d\eta = \frac{0.1199}{0.5}\eta^{3/2}(\tfrac{2}{3})\Big|_{1.250}^{2.833} = 0.539 \text{ lb}$$

The averaging process in Prob. 24.62 gave a very good result when compared with the result above.

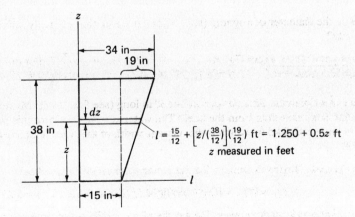

$$l = \tfrac{15}{12} + \left[z/(\tfrac{38}{12})\right](\tfrac{19}{12}) \text{ ft} = 1.250 + 0.5z \text{ ft}$$
z measured in feet

Fig. 24-9

24.64 The United States at one time in the thirties had three large dirigibles—the Los Angeles, the Graf Zeppelin, and the Akron. Two of them were destroyed by accidents. The largest was the Akron, having a length of 785 ft and a maximum diameter of 132 ft. Its maximum speed was 84 mph. The useful lift was 182 000 lb.

　　Moving at top speed, estimate the power needed to overcome skin friction, which is a significant part of the drag. Disregard effects of protrusion from engine cowlings, cabin region, etc. Assume that the surface is smooth. Take the critical Reynolds number to be 500 000. Consider the Akron at 10 000-ft standard atmosphere.

▮ We can make a reasonable estimate of the skin drag by "unwrapping" the outer surface of the Akron to form a flat plate. As a first step, we wish to calculate the plate Reynolds number. For this we find from the standard atmosphere table that $\rho = (0.7385)(0.002378) = 0.001756$ slug/ft^3, $T = 23.3$ °F. From the viscosity curves in Fig. A-1, we then find that $\mu = 3.7 \times 10^{-7}$ lb-s/ft^2. We can now compute $N_R = \rho dV/\mu$:

$$N_R = (0.001756)[(84)(\tfrac{5280}{3600})](785)/(3.7 \times 10^{-7}) = 4.59 \times 10^8 \tag{1}$$

We will use the Prandtl–Schlichting skin-friction formula with $A = 1700$:

$$C_f = [0.455/(\log N_R)^{2.58}] - (A/N_R) = \{0.455/[\log (4.59 \times 10^8)]^{2.58}\} - [1700/(4.59 \times 10^8)] = 1.730 \times 10^{-3} \tag{2}$$

Now going to $D = C_f(\tfrac{1}{2}\rho V_0^2)(bL)$, we have for the drag D owing to skin friction using the maximum diameter of 132 ft:

$$D = (1.730 \times 10^{-3})(\tfrac{1}{2})(0.001756)[84(\tfrac{5280}{3600})]^2(785)[(\pi)(132)] = 7505 \text{ lb} \tag{3}$$

The power needed then is

$$\text{Power} = (7505)(84)(\tfrac{5280}{3600})/550 = 1681 \text{ hp} \tag{4}$$

24.65 In Prob. 24.64, we computed the skin drag for the dirigible Akron using smooth flat-plate theory. Let us now evaluate the admissible roughness ϵ_{adm} for the results Prob. 24.64 to be valid. Then using a roughness of 0.05 in, recompute the skin-friction drag. What is the power needed to overcome this drag at the top speed of 84 mph?

▮ We can immediatly solve for the admissible roughness and hence the largest roughness that will still give hydraulically smooth flow. Using the plate Reynolds number from Prob. 24.64, we have

$$\epsilon_{adm} = L(100/N_R) = (785)[100/(4.59 \times 10^8)] = 1.710 \times 10^{-4} \text{ ft} = 0.00205 \text{ in} \tag{1}$$

For the dirigible, the actual roughness coefficient is 0.05 in. Now L/ϵ is $785/(0.05/12) = 1.884 \times 10^5$, so that on consulting Fig. A-64 we are clearly in the rough zone. We accordingly use

$$C_f = [1.89 + 1.62 \log (L/\epsilon)]^{-2.5} = \{1.89 + 1.62 \log [785/(0.05/12)]\}^{-2.5} = 0.002843 \tag{2}$$

For the skin drag, we have, using 0.001756 slug/ft^3 for ρ, $D = (0.002843)(\tfrac{1}{2})(0.001756)[(84)(\tfrac{5280}{3600})]^2(785)(\pi)(132) = 12\,333$ lb. The power needed to overcome skin friction is Power $= (12\,333)(84)(\tfrac{5280}{3600})/550 = 2763$ hp.

24.66 In Prob. 24.62, consider that the surface is rough with $\epsilon = 0.009$ in. Also, consider that transition takes place at $(N_R)_{crit} = 3.2 \times 10^6$. Calculate the skin drag on the keel as accurately as you can.

▮ From Prob. 24.62, we see that the plate Reynolds number ranges from 3.81×10^5 at the bottom of the keel to 8.63×10^5 at the top of the keel. The ratio L/ϵ ranges from 1.667×10^3 at the bottom to 3.778×10^3 at the top. Considering Fig. A-64, we see that at both extremes we are in the transition zone of flow. We accordingly will use $C_f = [0.031/(N_R)^{1/7}] - (A/N_R)$ for C_f. We will furthermore use strips of width dz (see Fig. 24-9) and of length $(1.250 + 0.5z)$ ft, with z measured in feet. We have for C_f for such a strip

$$C_f = \frac{0.031}{\{[(3)(1.689)(1.250 + 0.5z)]/(1.664 \times 10^{-5})\}^{1/7}} - \frac{1050}{\{[(3)(1.689)(1.250 + 0.5z)]/(1.664 \times 10^{-5})\}}$$

$$= 0.00510(1.250 + 0.5z)^{-1/7} - 0.00345(1.250 + 0.5z)^{-1}$$

The drag D then becomes

$$D = 2\int_0^{38/12} C_f(\tfrac{1}{2}\rho V_0^2)(1.250 + 0.5z) \, dz$$

$$= 2\int_0^{38/12} [(0.00510)(1.250 + 0.5z)^{-1/7} - (0.00345)(1.250 + 0.5z)^{-1}](\tfrac{1}{2})(1.490)[(3)(1.689)]^2(1.250 + 0.5z) \, dz$$

$$= \int_0^{3.17} [0.254(1.250 + 0.5z)^{6/7} - (0.1718)] \, dz$$

Integrating, we get

$$D = (0.254)(1.250 + 0.5z)^{13/7}\left(\frac{7}{13}\right)\left(\frac{1}{0.5}\right)\Big|_0^{3.17} - (0.1718)(3.17) = 0.936 \text{ lb}$$

We see that we get almost a doubling of the drag. In racing sailboats, every ounce of drag counts, so sailors carefully smooth the wetted surfaces of their boats.

24.67 We will consider the dirigible Akron once again for drag (see Prob. 24.64). This time we will use an ellipsoidal body of revolution from Table A-22 to represent the dirigible. Estimate the pressure drag.

▌ The coefficient of drag should correspond to an ellipsoid whose $L/D_{max} = \frac{785}{132} = 5.95$. For turbulent flow, we estimate C_D using simple interpolation from the 4:1 to the 8:1 ellipsoids:

$$C_D = 0.06 + [(5.95 - 4)/4](0.13 - 0.06) = 0.094 \tag{1}$$

Now the total drag is next computed using

$$D = C_D A(\rho V_0^2/2) = (0.094)[(\pi)(132^2)/4](\tfrac{1}{2})(0.001756)[84(\tfrac{5280}{3600})]^2 = 17\,143 \text{ lb} \tag{2}$$

We can now estimate the pressure drag on the Akron using the drag from Prob. 24.64. Thus considering hydraulically smooth flow, $D_{press} = 17\,143 - 7505 = 9638$ lb. The total power needed to move the Akron should be $P = (17\,143)[(84)(\tfrac{5280}{3600})]/550 = 3840$ hp. The Akron actually had 8450-hp diesel engines.

24.68 A well-streamlined car can have a drag coefficient as low as 0.45 compared with old-fashioned cars whose drag coefficient could be as high as 0.9 (see Fig. A-65). Consider a car with a drag coefficient of 0.45 moving at the speed of 100 km/h. The frontal area of the car is 2 m² and its mass with driver is 1300 kg. How far must it move to halve its speed if the engine is disengaged and the vehicle rolls freely along a straight flat road with negligible wind present. Neglect the rotational effects of the wheels but consider that the coefficient of rolling resistance, a, of the tires is 0.50 mm. The air temperature is 20 °C. The tire diameter is 450 mm.

▌ You will recall from your statics course that the resistance to rolling F_R is given as

$$F_R = Pa/r \tag{1}$$

where P = normal force on wheel, a = coefficient of rolling resistance, r = radius of wheel.

For this problem, we have for the total effect of the four wheels $F_R = [(1300)(9.81)](0.50 \times 10^{-3})/0.225 = 28.34$ N. The drag force from the air is next computed:

$$D = C_D(\tfrac{1}{2}\rho V^2)(A) = (0.45)(\tfrac{1}{2})(\rho)(V^2)(2) \tag{2}$$

For ρ we have $\rho = p/RT = 101\,325/[(287)(293)] = 1.205$ kg/m³. Hence we have for D

$$D = 0.542V^2 \tag{3}$$

We next express *Newton's* law for the car: $M(dV/dt) = F$.

$$(1300)(dV/dt) = -0.542V^2 - 28.34 \tag{4}$$

In order to separate variables with the distance x appearing, we may say, using the chain rule,

$$dV/dt = (dV/dx)(dx/dt) = V(dV/dx) \tag{5}$$

Now Eq. (4) becomes $1300V(dV/dx) = -(0.542V^2 + 28.34)$. Separating variables, we get

$$(1300V\,dV)/(0.542V^2 + 28.34) = -dx \tag{6}$$

Let $\eta = 0.542V^2 + 28.34$:

$$d\eta = 1.084V\,dV \tag{7}$$

Making the above substitution of variable, we get $(1300/1.084)(d\eta/\eta) = -dx$. Integrating, $1.199 \times 10^3 \ln \eta = -x + C$:

$$1.199 \times 10^3 \ln (0.542V^2 + 28.34) = -x + C \tag{8}$$

When $x = 0$, $V = 100(\tfrac{1000}{3600}) = 27.78$ m/s. We can determine C: $1.199 \times 10^3 \ln [(0.542)(27.78)^2 + 28.34] = C$, $C = 1.199 \times 10^3 \ln 446.6$. Equation (8) can now be written as

$$\ln [(0.542V^2 + 28.34)/446.6] = -x/(1.199 \times 10^3) \tag{9}$$

Let $V = 50(\tfrac{1000}{3600}) = 13.89$ m/s. We get for x

$$\ln \{[(0.542)(13.89)^2 + 28.34]/446.6\} = -x/(1.199 \times 10^3) \qquad x = 1453 \text{ m} \tag{10}$$

The vehicle moves a distance of 1453 m. Actually, by including the rotational energy of the wheels, the vehicle would move even farther.

24.69 A jet aircraft discharges solid particles of matter 10 μm in diameter, s.g. = 2.5, at the base of the stratosphere at 11 000 m. Assume the viscosity μ of air, in poises, to be expressed by $\mu = 1.78 \times 10^{-4} - 3.06 \times 10^{-9}y$ where y in

698 ☐ CHAPTER 24

meters is measured from sea level. Estimate the time for these particles to reach sea level. Neglect air currents and wind effects.

❚ Writing $U = -dy/dt$ in $U = (D^2/18\mu)(\gamma_s - \gamma)$ and recognizing the unit weight of air to be much smaller than the unit weight of the solid particles, one has $-dy/dt = (D^2/18)(\gamma_s/\mu)$:

$$\int_0^T dt = \int_{11\,000}^0 18(1.78 \times 10^{-4} - 3.06 \times 10^{-9}y)\left(\frac{0.1}{1}\right)\left[\frac{1}{(10 \times 10^{-6})^2}\right]\left[\frac{1}{2.5(9806)}\right] dy$$

$$T = (1/86\,400)[1.78y - (3.06 \times 10^{-5}y^2)/2]_0^{11\,000}(73.45) = 15.07 \text{ days}$$

24.70 How many 30-m-diameter parachutes ($C_D = 1.2$) should be used to drop a bulldozer weighing 45 000 N at a terminal speed of 10 m/s through air at 100 000 Pa abs at 20 °C?

❚ $$F_D = W = C_D\rho(V^2/2)A \qquad \rho = p/RT = 100\,000/[(287)(20 + 273)] = 1.189 \text{ kg/m}^3$$
$$W = (1.2)(1.189)(10^2/2)[(\pi)(30)^2/4] = 50\,430 \text{ N}$$

Thus, one parachute should be sufficient, with a factor of safety of 50 430/45 000 = 1.2.

24.71 An 0.8-m cubical box is placed on the luggage carrier on top of a station wagon. Estimate the additional power requirements for the vehicle to travel at (a) 80 km/h and (b) 110 km/h ($C_D = 1.1$ and $\rho = 1.2$ kg/m³).

❚ $$P = F_D V \qquad F_D = C_D\rho(V^2/2)A$$
(a) $$V = (80)(1000)/3600 = 22.22 \text{ m/s} \qquad F_D = (1.1)(1.2)(22.22^2/2)[(0.8)(0.8)] = 208.6 \text{ N}$$
$$P = (208.6)(22.22) = 4635 \text{ W} \quad \text{or} \quad 4.635 \text{ kW}$$
(b) $$V = (110)(1000)/3600 = 30.56 \text{ m/s} \qquad F_D = (1.1)(1.2)(30.56^2/2)[(0.8)(0.8)] = 394.5 \text{ N}$$
$$P = (394.5)(30.56) = 12\,060 \text{ W} \quad \text{or} \quad 12.06 \text{ kW}$$

24.72 A semitubular cylinder of 6-in radius with concave side upstream is submerged in water flowing 3 fps. Calculate the drag for a cylinder 24 ft long.

❚ $F_D = C_D\rho(V^2/2)A$. From Table A-23, $C_D = 2.3$. $F_D = (2.3)(1.94)(3^2/2)[(24)(6 + 6)/12] = 482$ lb.

24.73 A fully loaded, small aircraft weighing 5000 lb with a wing area (projected chord area) of 350 ft² is to take off at a horizontal velocity of 100 mph (146.7 fps). What is the necessary angle of attack (i.e., angle the wings make with the horizontal)? Assume the wings have the characteristics of the airfoil of Fig. A-66. Use $\rho_{air} = 0.00234$ slug/ft³.

❚ $$F_L = C_L\rho(v^2/2)A \qquad 5000 = C_L(0.00234)(146.7^2/2)(350) \qquad C_L = 0.567$$
From Fig. A-66, $\alpha = 6.0°$.

24.74 An aircraft weighing 1000.0 kN when empty has a wing area of 226 m². It is to take off at a velocity of 300 km/h and a 20° angle of attack. Assume $\rho_{air} = 1.20$ kg/m³. Also, assume the wing has the characteristics of the airfoil in Fig. A-66. What is the allowable weight of cargo?

❚ $F_L = C_L\rho(v^2/2)A$. From Fig. A-66, $C_L = 1.42$. $v = (300)(1000)/3600 = 83.33$ m/s, $F_L = (1.42)(1.20)(83.33^2/2)(226) = 1\,337\,000$ N, or 1337 kN. This value (1337 kN) represents the total weight that can be lifted. Since the aircraft weighs 1000.0 kN when empty, it can carry, in theory, 1337 − 1000 = 337 kN of cargo.

24.75 A fully loaded aircraft weighing 10 000 lb with a wing area of 450 ft² is to take off at a horizontal velocity of 125 mph (183.3 fps). What is the necessary angle of attack? Assume the wings have the characteristics of the airfoil of Fig. A-66. Use $\rho_{air} = 0.00234$ slug/ft³.

❚ $$F_L = C_L\rho(v^2/2)A \qquad 10\,000 = C_L(0.00234)(183.3^2/2)(450) \qquad C_L = 0.565$$
From Fig. A-66, $\alpha = 6.0°$.

24.76 An aircraft weighing 65.2 kN when empty has a wing area of 62.3 m². It is to take off at a velocity of 250 km/h and a 5.0° angle of attack. Assume $\rho_{air} = 1.20$ kg/m³. Also, assume the wing has the characteristics of the airfoil in Fig. A-66. What is the allowable weight of cargo?

❚ $F_L = C_L\rho(v^2/2)A$. From, Fig. A-66, $C_L = 0.50$. $v = (250)(1000)/3600 = 69.44$ m/s, $F_L = (0.50)(1.20)(69.44^2/2)(62.3) = 90\,000$ N, or 90.1 kN; $W_{cargo} = 90.1 - 65.2 = 24.9$ kN.

24.77 A 3-ft by 4-ft plate moves at 44 ft/s in still air at an angle of 12° with the horizontal. Using a coefficient of drag $C_D = 0.17$ and a coefficient of lift $C_L = 0.72$, determine **(a)** the resultant force exerted by the air on the plate, **(b)** the frictional force and **(c)** the horsepower required to keep the plate moving. (Use $\gamma = 0.0752\ \text{lb/ft}^3$.)

> **(a)**   Drag force $= C_D(\gamma/g)A(V^2/2)$ $F_D = 0.17(0.752/32.2)(12)[(44)^2/2] = 4.61\ \text{lb}$
>
>   Lift force $= C_L(\gamma/g)A(V^2/2)$ $F_L = 0.72(0.0752/32.2)(12)[(44)^2/2] = 19.5\ \text{lb}$
>
> Referring to Fig. 24-10, the resultant of the drag and lift components is $R = \sqrt{(4.61)^2 + (19.5)^2} = 20.0\ \text{lb}$ acting on the plate at $\theta_x = \tan^{-1}(19.5/4.61) = 76°42'$.
>
> **(b)** The resultant force might also have been resolved into a normal component and a frictional component (shown dotted in the figure). From the vector triangle, frictional component $= R\cos(\theta_x + 12°) = 20.0(0.0227) = 0.45\ \text{lb}$.
>
> **(c)** Horsepower $=$ (force in direction of motion)(velocity)$/550 = (4.61)(44)/550 = 0.369$

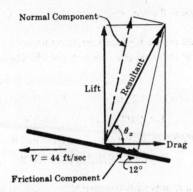

Normal Component

Lift

Resultant

θ_x

Drag

$V = 44\ \text{ft/sec}$

Frictional Component

12°

Fig. 24-10

24.78 If an airplane weighs 4000 lb and has a wing area of 300 ft², what *angle of attack* must the wings make with the horizontal at a speed of 100 mph? Assume the coefficient of lift varies linearly from 0.35 at 0° to 0.80 at 6° and use $\gamma = 0.0752\ \text{lb/ft}^3$ for air.

> For equilibrium in the vertical direction, $\sum Y = 0$. Hence, lift $-$ weight $= 0$, or weight $= C_L\gamma A(V^2/2g)$, $4000 = C_L(0.0752)(300)\{[(100)(\frac{5280}{3600})]^2/2g\}$, $C_L = 0.53$. By interpolation between 0° and 6°, angle of attack $= 2.4°$.

24.79 What wing area is required to support a 5000-lb plane when flying at an *angle of attack* of 5° at 88 ft/s? Use coefficients given in Prob. 24-78.

> From given data, $C_L = 0.725$ for 5° angle by interpolation. As in Prob. 24.78, weight $=$ lift, $5000 = 0.725(0.0752/32.2)A(88)^2/2$, $A = 763\ \text{ft}^2$.

24.80 A kite weighs 2.50 lb and has an area of 8.00 ft². The tension in the kite string is 6.60 lb when the string makes an angle of 45° with the horizontal. For a wind of 20 mph, what are the coefficients of lift and drag if the kite assumes an angle of 8° with the horizontal? Consider the kite essentially a flat plate and $\gamma_{\text{air}} = 0.0752\ \text{lb/ft}^3$.

> Figure 24-11 indicates the forces acting on the kite taken as a free body. The components of the tension are 4.66 lb each.
>
> From $\sum X = 0$, drag $= 4.66\ \text{lb}$. From $\sum Y = 0$, lift $= 4.66 + 2.50 = 7.16\ \text{lb}$. Drag force $= C_D\rho AV^2/2$, $4.66 = C_D(0.0752/32.2)(8.00)[(20)(\frac{5280}{3600})]^2/2$, $C_D = 0.58$. Lift force $= C_L\rho AV^2/2$, $7.16 = C_L(0.0752/32.2)(8.00)[(20)(\frac{5280}{3600})]^2/2$, $C_L = 0.89$.

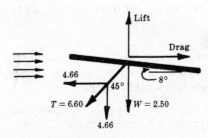

Lift

Drag

8°

4.66

45°

$T = 6.60$

$W = 2.50$

4.66

Fig. 24-11

24.81 If the minimum landing speed (V_{min}) of the Mustang racer is 1.3 times the stall speed, what is the minimum landing speed for this plane with flaps at 40°? The Mustang has a wing area of 233 ft^2 and a weight of 9500 lb. Take ρ to be 0.002378 slug/ft^3 and $(C_L)_{max}$ as 1.65.

$$V_{stall} = \sqrt{2W/[(\rho)(C_L)_{max}A]} = \sqrt{(2)(9500)/[(0.002378)(1.65)(233)]} = 144.2 \text{ ft/s} \quad \text{or} \quad 98.3 \text{ mph}$$
$$V_{min} = (1.3)(98.3) = 128 \text{ mph}$$

24.82 If the mean velocity along the top of a wing having a 2-m chord is 150 km/h (41.67 m/s) and that along the bottom of the wing is 120 km/h (33.33 m/s) when the wing moves through still air ($\gamma = 11.3$ N/m^3) at 128 km/h, estimate the lift per unit length of span.

$$\text{Lift} = (\Delta p)(c) \qquad p_{top}/\gamma + v_{top}^2/2g = p_{bottom}/\gamma + v_{bottom}^2/2g$$
$$p_{top}/11.3 + 41.67^2/[(2)(9.807)] = p_{bottom}/11.3 + 33.33^2/[(2)(9.807)]$$
$$p_{bottom} - p_{top} = \Delta p = 360.4 \text{ N/m}^2 \qquad \text{Lift} = (360.4)(2) = 721 \text{ N/m}$$

24.83 For a rectangular Clark Y airfoil of 6-ft chord by 36-ft span, find the value of the friction coefficient η if the angle of attack $\alpha = 5.4°$ when the wing is moving at 300 fps through standard atmosphere at altitude 10 000 ft. Find the weight which the wing will carry and the horsepower required to drive it.

From Fig. A-67, with $\alpha = 5.4°$, $C_L = 0.8$, $C_D = 0.047$. From Fig. A-68, for $B/c = 6$, $\tau = 0.175$. α_i(radians) = $[C_L/\pi(B/c)](1 + \tau) = [0.8/\pi(\frac{36}{6})](1 + 0.175) = 0.0499$ rad = 2.86°. From Fig. A-69, $\alpha_0 = \alpha - \alpha_i = 5.40 - 2.86 = 2.54°$ and since the angle of zero lift is $-5.6°$, $\alpha_0' = 2.54 + 5.6 = 8.14° = 0.1421$ rad, $\eta = C_L/2\pi\alpha_0' = 0.8/[(2\pi)(0.1421)] = 0.896$. The wing will support a weight equal to the lift force, $F_L = C_L\rho(V^2/2)(B)(c)$.

From Table A-7, at 10 000 ft, $\rho = 0.001756$ slug/ft^3. $F_L = (0.8)(0.001756)[(300)^2/2](36)(6) = 13\,650$ lb while $F_D = (0.047/0.8)(13\,650) = 802$ lb, horsepower required = $(802)(300)/550 = 437$ hp.

24.84 A cylinder 4 ft in diameter and 25 ft long rotates at 90 rpm with its axis perpendicular to an air stream with a wind velocity of 120 fps. The specific weight of the air is 0.0765 lb/ft^3. Assuming no slip between the cylinder and the circulatory flow, find (a) the value of the circulation; (b) the transverse or lift force; and (c) the position of the stagnation points.

(a) Peripheral velocity $v_t = 2\pi R n/60 = (2\pi)(2)(\frac{90}{60}) = 18.85$ fps, $\Gamma = 2\pi R v_t = (2\pi)(2)(18.85) = 237$ ft^2/s.

(b)
$$F_L = \rho B U \Gamma = (0.0765/32.2)(25)(120)(237) = 1690 \text{ lb}$$

(c)
$$\sin\theta_s = -\Gamma/4\pi R U = -237/(4\pi 2)(120) = -0.0786$$

Therefore $\theta_x = 184.5°$, 355.5°. Actually, the real circulation produced by surface drag of the rotating cylinder would be only about one-half of that obtained for the no-slip assumption.

24.85 A wing with a 20-m span and 60-m^2 "plan-form" area moves horizontally through the standard atmosphere at 10 000 m with a velocity of 800 km/h. If the wing supports 250 000 N, find (a) the required value of the lift coefficient, (b) the downwash velocity, assuming semielliptical distribution of lift over the span, and (c) the induced drag.

From Table A-8, $\rho = 0.414$ kg/m^3; $V = (800)(1000)/3600 = 222.2$ m/s.

(a)
$$F_L = C_L\rho(V^2/2)A \qquad 250\,000 = C_L(0.414)(222.2^2/2)(60) \qquad C_L = 0.408$$

(b)
$$V_i/V = C_L/[\pi(B^2/A)] \qquad V_i/222.2 = 0.408/[(\pi)(20^2/60)] \qquad V_i = 4.33 \text{ m/s}$$

(c)
$$(C_D)_i = C_L^2/[\pi(B^2/A)] = 0.408^2/[(\pi)(20^2/60)] = 0.00795$$
$$(F_D)_i = (C_D)_i(F_L/C_L) = (0.00795)(250\,000/0.408) = 4871 \text{ N}$$

24.86 A sailplane weighing 400 lb including its load has a 4-ft-chord by 24-ft-span wing of the Clark Y section. Assuming that its characteristics are the same as those for the larger wing of the same aspect ratio shown in Fig. A-67, find the angle of glide through standard air at 2000 ft which will give the greatest horizontal distance range. Neglect air forces on the fuselage and tail. (*Note:* The aspect ratio of 6 is here chosen for convenience in working the problem with the available data. Actually, the sailplane may be constructed with an aspect ratio of about twice this, so as to reduce drag to a minimum.)

From Fig. A-67 for maximum C_L/C_D: $\alpha = -0.1°$, $C_L = 0.4$, and $C_D = 0.019$. For maximum range, glide angle β is minimum. $\beta = \arctan(F_D/F_L) = \arctan(0.019/0.4) = 2.7°$.

24.87 A boat fitted with a hydrofoil weighs 5000 lb. At a velocity of 50 fps, what size hydrofoil is needed to support the boat? Use the lift characteristics in Fig. A-70 at an angle of attack of 4°.

▌ $F_L = C_L \rho (V^2/2) A$. From Fig. A-70, $C_L = 0.5$. $5000 = (0.5)(1.94)(50^2/2)A$, $A = 4.12 \text{ ft}^2$.

24.88 If the Mustang fighter plane weighs 42.7 kN, at what angle of attack should it be flown at a speed of 250 km/h? The planform area is 25 m². What power is required in horsepower to overcome wing drag? The temperature is 20 °C. The flaps are at zero degrees.

▌
$$F_L = C_L (\tfrac{1}{2}) \rho V^2 A \qquad 42\,700 = (C_L)(\tfrac{1}{2})(\rho)(250/3.6)^2(25) \qquad \rho = p/RT$$
$$\rho = 101\,325/[(287)(293)] = 1.205 \text{ kg/m}^3 \qquad C_L = 0.588 \qquad \alpha = 6° \qquad C_D = 0.04$$
$$D = (0.04)(\tfrac{1}{2})(1.205)(250/3.6)^2(25) = 2906 \text{ N} \qquad P = (2906)(250/3.6) = 202\,000 \text{ W} = 202 \text{ kW} = 202/0.7457 = 271 \text{ hp}$$

24.89 If the takeoff speed is about 1.3 times the stall speed for the Mustang, which weighs 42.7 kN, what is the takeoff distance for a constant thrust of 9 kN and a rolling resistance of 0.5 kN? The planform area is 25 m². The flaps are at 40°. The air is at 20 °C. The overall coefficient of drag for the plane is 0.20 and the frontal area is 15 m².

▌ The stall speed occurs at the condition of $(C_L)_{max}$ which for us is 1.64.
$$\rho = 101\,325/[(287)(293)] = 1.205 \text{ kg/m}^3$$
$$V_{stall} = \sqrt{(2)W/[(\rho)(C_L)_{max}A]} = \sqrt{(2)(42\,700)/[(1.205)(1.64)(25)]} = 41.6 \text{ m/s}$$

V at takeoff is $(1.3)(41.6) = 54.1 \text{ m/s}$. Newton's law is next:
$$9000 - 500 - (0.2)(\rho V^2/2)(15) = (42\,700/g)V(dV/dx)$$
$$8500 - 1.808V^2 = 4353V(dV/dx) \qquad (4353)(V\,dV)/(8500 - 1.808V^2) = dx$$

Let $8500 - 1.808V^2 = \eta$, $-3.62V\,dV = d\eta$, $V\,dV = (d\eta/3.62)$.
$$\int_{8500}^{3218} (4353)\frac{(-d\eta/3.62)}{\eta} = \int_0^{L_1} dx \qquad -\frac{4353}{3.62}\ln \eta \Big|_{8500}^{3218} = L \qquad L = 1168 \text{ m}$$

24.90 What weight can the Mustang plane have if it has a planform wing area of 233 ft² and is flying at an angle of attack of 3° at a speed of 210 mph? Air is at 60 °F. What power is needed for overcoming wing drag at this speed? The flaps are at zero degrees.

▌
$$C_L = 0.35 \qquad L = C_L(\tfrac{1}{2})\rho V^2 A = W = (0.35)(\tfrac{1}{2})(\rho)(V_0^2)(A)$$
$$\rho = (14.7)(144)/[(53.3)(g)(520)] = 0.00237 \text{ slug/ft}^3$$
$$W = (0.35)(\tfrac{1}{2})(0.00237)[(210)\tfrac{5280}{3600}]^2(233) = 9167 \text{ lb} = 4.58 \text{ tons}$$
$$D = (0.020)(\tfrac{1}{2})(0.00237)[(210)(\tfrac{5280}{3600})]^2(233) = 523.8 \text{ lb} \qquad \text{hp} = (D)(V)/550 = (523.8)(210)(\tfrac{5280}{3600})/550 = 293$$

24.91 A boat is fitted with hydrofoils having a total planform area of 1 m². The coefficient of lift is 1.5 when the boat is moving at 10 knots, which is the slowest speed for the hydrofoils to support the boat. The coefficient of drag is 0.6 for this case. What is the maximum weight of the boat to fulfill the minimal speed for hydrofoil support? What power is needed for this speed? Water is fresh water at 5 °C.

▌
$$W = F_L = C_L(\tfrac{1}{2})\rho V^2 A = (1.5)(\tfrac{1}{2})(1000)[(10)(0.5144)]^2(1) = 19\,846 \text{ N}$$
$$D = C_D(\tfrac{1}{2})\rho V^2 A = (0.6)(\tfrac{1}{2})(1000)[(10)(0.5144)]^2(1) = 7938 \text{ N}$$
$$p = DV = (7938)[(10)(0.5144)] = 40\,800 \text{ W} \quad \text{or} \quad 40.8 \text{ kW}$$

24.92 A fan used at the turn of the century and in some ice cream parlors today consists of flat wooden slats rotated at a small inclination of about 15°. At what speed ω would you have to rotate the fan (with four blades) to result in zero vertical force on the bearings supporting the wooden blades each of which weighs 1.5 lb? What torque is needed for this motion? *Hint:* Use Fig. A-71. Air is at 60 °F. Assume each slat is 10 in wide and 5 ft long and attached to an 8-in-diameter core.

▌From Fig. A-71 we get the following results for C_D and C_L: $C_D = 0.225$, $C_L = 0.79$. Consider one blade. Compute the lift:

$$L = C_L \int \left(\tfrac{1}{2}\rho V^2\right) dA = (0.79) \int_{0.333}^{5.333} \left(\tfrac{1}{2}\rho V_0^2\right)\left(\frac{10\,dr}{12}\right) \qquad \rho = \frac{(14.7)(144)}{(53.3)(g)(520)} = 0.002372 \text{ slug/ft}^3$$

$$L = 1.5 = (0.79)(\tfrac{1}{2})(0.002372)(\tfrac{10}{12}) \int_{0.333}^{5.333} (r\omega)^2\, dr \qquad 1.5 = (0.000781)\omega^2 \frac{r^3}{3}\Big|_{0.333}^{5.333} = 0.000260\omega^2(5.333^3 - 0.333^3)$$

$$\omega = 6.168 \text{ rad/s} = (6.168/2\pi)(60) = 58.9 \text{ rpm}$$

$$T = \left[(0.225)\int_{0.333}^{5.333} \tfrac{1}{2}\rho V^2 (dr)(\tfrac{10}{12})r \right] 4 = (0.225)(\tfrac{1}{2})(0.002372)(6.168)^2(\tfrac{10}{12})(4) \int_{0.333}^{5.333} r^3\, dr = 0.03384\frac{r^4}{4}\Big|_{0.333}^{5.333} = 6.84 \text{ N-m}$$

CHAPTER 25
Basic Hydrodynamics

25.1 Assuming ρ to be constant, do these flows satisfy continuity? *(a)* $u = -2y$, $v = 3x$; *(b)* $u = 0$, $v = 3xy$; *(c)* $u = 2x$, $v = -2y$. Continuity for incompressible fluids is satisfied if $\partial u/\partial y + \partial v/\partial y = 0$.

❙ *(a)* $\qquad\qquad [\partial(-2y)/\partial x] + [\partial(3x)/\partial y] = 0 + 0 = 0$ \qquad (continuity is satisfied)

(b) $\qquad\qquad [\partial(0)/\partial x] + [\partial(3xy)/\partial y] = 0 + 3x \neq 0$ \qquad (continuity is not satisfied)

(c) $\qquad\qquad [\partial(2x)/\partial x] + [\partial(-2y)/\partial y] = 2 - 2 = 0$ \qquad (continuity is satisfied)

Note: If *(b)* did indeed describe a flow field, the fluid must be compressible.

25.2 Determine whether these flows are rotational or irrotational. *(a)* $u = -2y$, $v = 3x$; *(b)* $u = 0$, $v = 3xy$; *(c)* $u = 2x$, $v = -2y$.

❙ If $(\partial v/\partial x) - (\partial u/\partial y) = 0$ (flow is irrotational):

(a) $\qquad\qquad [\partial(3x)/\partial x] - [\partial(-2y)/\partial y] = 3 + 2 \neq 0$ \qquad (flow is rotational)

(b) $\qquad\qquad [\partial(3xy)/\partial x] - [\partial(0)/\partial y] = 3y - 0 \neq 0$ \qquad (flow is rotational)

(c) $\qquad\qquad [\partial(-2y)/\partial x] - [\partial(2x)/\partial y] = 0 - 0 = 0$ \qquad (flow is irrotational)

25.3 Check these flows for continuity and determine the vorticity of each: *(a)* $v_t = 6r$, $v_r = 0$; *(b)* $v_t = 0$, $v_r = -5/r$.

❙ Applying the equations $(v_r/r) + (\partial v_r/\partial r) + (\partial v_t/r\,\partial\theta) = 0$, $\xi = (\partial v_t/\partial r) + (v_t/r) - (\partial v_r/r\,\partial\theta)$.

(a) $\qquad\qquad (0/r) + [\partial(0)/\partial r] + [\partial(6r)/r\,\partial\theta] = 0$ \qquad\qquad (continuity is satisfied)

$\xi = [\partial(6r)/\partial r] + (6r/r) - [\partial(0)/r\,\partial\theta] = 6 + 6 - 0 = 12$ \qquad (flow is rotational)

(b) $\quad -[(5/r)/r] + [\partial(-5r^{-1})/\partial r] + [\partial(0)/r\,\partial\theta] = -(5/r^2) + (5/r^2) + 0 = 0$ \qquad (continuity is satisfied)

$$\xi = \frac{\partial(0)}{\partial r} + \frac{0}{r} - \frac{\partial(-5/r)}{r\,\partial\theta} = 0 \qquad\qquad \text{(flow is rotational)}$$

25.4 A flow is defined by $u = 2x$ and $v = -2y$. Find the stream function and potential function for this flow and plot the flow net.

❙ Check continuity: $(\partial u/\partial x) + (\partial v/\partial y) = 2 - 2 = 0$. Hence continuity is satisfied and it is possible for a stream function to exist: $d\psi = -v\,dx + u\,dy = 2y\,dx + 2x\,dy$, $\psi = 2xy + C_1$. Check to see if the flow is irrotational: $(\partial v/\partial x) - (\partial u/\partial y) = 0 - 0 = 0$. Hence the flow is irrotational and a potential function exists: $d\phi = -u\,dx - v\,dy = -2x\,dx + 2y\,dy$, $\phi = -(x^2 - y^2) + C_2$. The location of lines of equal ψ can be found by substituting values of ψ into the expression $\psi = 2xy$. Thus for $\psi = 60$, $x = 30/y$. This line is plotted (in the upper right-hand quadrant) in Fig. 25-1. In a similar fashion lines of equal potential can be plotted. For example, for $\phi = 60$ we have $-(x^2 - y^2) = 60$ and $x = \pm\sqrt{y^2 - 60}$. This line is also plotted on the figure. The flow net depicts *flow in a corner*. Mathematically the net will plot symmetrically in all four quadrants.

25.5 Given a flow defined by $u = 3 + 2x$. If this flow satisfies continuity, what can be said about the density of the fluid?

❙ $\partial(\rho u)/\partial x = 0$. Therefore, $\rho u = \text{constant}$, or $\rho = \text{constant}/u$. Since the given equation has the velocity u increasing with x, the density ρ must decrease as x increases.

25.6 The flow of an incompressible fluid is defined by $u = 2$, $v = 8x$. Does a stream function exist for this flow? If so, determine the expression for the stream function.

❙ To find if a stream function exists, check for continuity. $\partial u/\partial x = 0$, $\partial v/\partial y = 0$, $w = 0$. Thus, $(\partial u/\partial x) + (\partial v/\partial y) + (\partial w/\partial z) = 0$ (continuity) is satisfied and a stream function exists.

$$d\psi = -v\,dx + u\,dy = -8x\,dx + 2\,dy \qquad \psi = -4x^2 + 2y$$

703

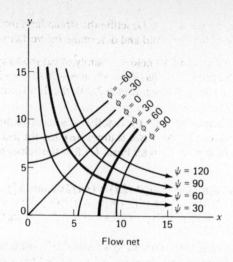

Flow net   **Fig. 25-1**

25.7   Plot the streamlines in the upper right-hand quadrant for the flow defined by $\psi = 1.5x^2 + y^2$ and determine the value of the velocity at $x = 4$, $y = 2$.

▌ $u = \partial\psi/\partial y = 2y$, $v = -\partial\psi/\partial x = -3x$. At $(4, 2)$ $u = 4$, $v = -12$, $V = \sqrt{4^2 + 12^2} = 12.65$. To plot streamlines, rewrite as $y = (\psi - 1.5x^2)^{1/2}$. Assume ψ, compute y's for different x's. See Fig. 25-2.

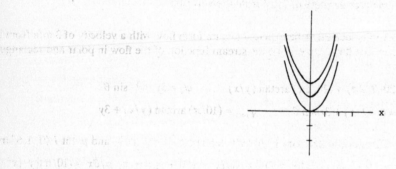

Fig. 25-2

25.8   The components of the velocities of a certain flow system are $u = -(Q/2\pi)[x/(x^2 + y^2)] + By + C$, $v = -A[y/(x^2 + y^2)] + Dx + E$. (**a**) Calculate a value of A consistent with continuous flow. (**b**) Sketch the streamlines for this flow system, assuming $B = C = D = E = 0$.

▌ (**a**) To satisy continuity 2-D flow: $(\partial u/\partial x) + (\partial v/\partial y) = 0$.

$$\partial u/\partial x = -(Q/2\pi)\{[(x^2 + y^2) - x(2x)]/(x^2 + y^2)^2\} = -(Q/2\pi)[(y^2 - x^2)/(x^2 + y^2)^2]$$
$$\partial v/\partial y = -A\{[(x^2 + y^2) - y(2y)]/(x^2 + y^2)^2\} = -A[(x^2 - y^2)/(x^2 + y^2)^2]$$

Substituting, $-Q/2\pi(y^2 - x^2) - A(x^2 - y^2) = 0$; thus $A = Q/2\pi$.

(**b**) $u = -(Q/2\pi)[x/(x^2 + y^2)]$ $v = -(Q/2\pi)[y/(x^2 + y^2)]$

$$d\psi = -v\,dx + u\,dy = (Q/2\pi)[(y\,dx)/(x^2 + y^2)] - (Q/2\pi)[(x\,dy)/(x^2 + y^2)]$$

Integrating, $\psi = (Q/2\pi)\tan^{-1}(x/y) - \tan^{-1}(y/x)$. See Fig. 25-3.

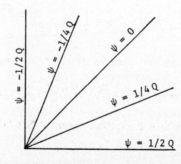

Fig. 25-3

25.9   A flow field is described by $\psi = x^2 - y$. Describe the streamlines for $\psi = 0$, 1, and 2. Derive an expression for the velocity at any point in the flow field and determine the vorticity of the flow.

▮   From the equation for ψ, the flow field is a family of parabolas symmetric about the y axis. The streamline $\psi = 0$ passes through the origin. $u = \partial u/\partial y = -1$, $v = -\partial u/\partial x = -2x$. Thus $V = (u^2 + v^2)^{1/2} = (1 + 4x^2)^{1/2}$. Vorticity: $\xi = (\partial v/\partial x) - (\partial u/\partial y) = -2$. (Since $\xi \neq 0$, the flow is rotational.)

25.10   A source of strength 8π is located at $(2, 0)$. Another source of strength 16π is located at $(-3, 0)$. For the combined flow field produced by these two sources: (**a**) find the location of the stagnation point; (**b**) describe the plotting of the $\psi = 0$, $\psi = 4\pi$, $\psi = 8\pi$ lines; (**c**) find the values of ψ at $(0, 2)$ and at $(3, -1)$; (**d**) find the velocity at $(-2, 5)$.

▮   (**a**) In general, $\psi = (q_1\theta_1/2\pi) + (q_2\theta_2/2\pi) = (8\pi/2\pi) \arctan[y/(x - 2)] + (16\pi/2\pi) \arctan[y/(x + 3)]$, $u = \partial\psi/\partial y = 4\{(x - 2)/[(x - 2)^2 + y^2]\} + 8\{(x + 3)/[(x + 3)^2 + y^2]\}$. Due to symmetry, stagnation point is on x axis ($y = 0$). If $y = 0$, $u = [4/(x - 2)] + [8/(x + 3)]$. At stagnation point, $u = 0 = [4/(x - 2)] + [8/(x + 3)]$; thus $x = \frac{1}{3}$.

  (**b**) ψ's at various points may be found by substituting coordinates of the point into the general expression for ψ (above). Lines of equal ψ may then be plotted by interpolation.

  (**c**) At $(0, 2)$: $\psi = 4 \arctan(2/-2) + 8 \arctan\frac{2}{3} = 14.13$. At $(3, -1)$: $\psi = 4 \arctan(-1/1) + 8 \arctan(-1/6) = 70.9$.

  (**d**) Differentiating ψ and simplifying: $u = \partial\psi/\partial y = 4\{(x - 2)/[(x - 2)^2 + y^2]\} + 8\{(x + 3)/[(x + 3)^2 + y^2]\}$. At $(-2, 5)$: $u = 4[-4/(16 + 25)] + 8[1/(1 + 25)] = -0.0826$. Similarly, $v = -\partial\psi/\partial x = 4\{y/[(x - 2)^2 + y^2]\} + 8\{y/[(x + 3)^2 + y^2]\}$. At $(-2, 5)$: $v = 4[5/(16 + 25)] + 8[5/(1 + 25)] = 2.03$. Adding vectorially, $V = (u^2 + v^2)^{1/2} = 2.03$, upwards at angle of $2°18'$ from the vertical.

25.11   A source discharging $20 \text{ m}^3/\text{s}$ per m is located at the origin and a uniform flow with a velocity of 3 m/s from left to right is superimposed on the source flow. Determine the stream function of the flow in polar and rectangular coordinates

▮
$$\psi_1 = 20(\theta/2\pi) = (20/2\pi) \arctan(y/x) \qquad \psi_2 = 3y = 3r \sin\theta$$
$$\psi_{\text{polar}} = (10\theta/\pi) + 3r \sin\theta \qquad \psi_{\text{rect}} = (10/\pi) \arctan(y/x) + 3y$$

25.12   Refer to Prob. 25.11. Find the difference in pressure head between point $A(-10 \text{ m}, 0)$ and point $B(0, 1.67 \text{ m})$.

▮   Differentiate ψ_{rect} from Prob. 25.11: $u = \partial\psi/\partial y = (10/\pi)[x/(x^2 + y^2)] + 3$; $v = -\partial\psi/\partial x = (10/\pi)[y/(x^2 + y^2)]$. At point A $(-10 \text{ m}, 0)$: $u = 2.68$ m/s, $v = 0$; $V_A = 2.68$ m/s. At point B $(0, 1.67 \text{ m})$: $u = 3.0$ m/s, $v = 1.9$ m/s; $V_B = \sqrt{3.0^2 + 1.9^2} = 3.55$ m/s. From Bernoulli's theorem: $p_A/\gamma - p_B/\gamma = V_B^2/2g - V_A^2/2g = (3.55^2 - 2.68^2)/[2(9.81)] = 0.276$ m.

25.13   Given is the two-dimensional flow described by $u = x^2 + 2x - 4y$, $v = -2xy - 2y$. (**a**) Does this satisfy continuity? (**b**) Compute the vorticity. (**c**) Plot the velocity vectors for $0 < x < 5$ and $0 < y < 4$ and sketch the general flow pattern. (**d**) Find the location of all stagnation points in the entire flow field. (**e**) Find the expression for the stream function.

▮   (**a**) $(\partial u/\partial x) + (\partial v/\partial y) = 2x + 2 - 2x - 2 = 0$; continuity is satisfied.

  (**b**) $\xi = (\partial v/\partial x) - (\partial u/\partial y) = -2y - 4 \neq 0$; flow is rotational.

  (**c**) See Fig. 25-4.

  (**d**) Stagnation points occur where both $u = 0$ and $v = 0$. $v = 0 = -2xy - 2y = -2y(x + 2)$; true when $y = 0$ or $x = -1$, $u = 0 = x^2 + 2x - 4y$. If $x = -1$, $y = -\frac{1}{4}$. If $y = 0$, $x = 0$ or -2. Thus there are three stagnation points: $(0, 0)$, $(-2, 0)$, and $(-1, -\frac{1}{4})$.

  (**e**)
$$u = \partial\psi/\partial y \qquad \psi = \int u\,\partial y = \int (x^2 + 2x - 4y)\,\partial y = x^2y + 2xy - 2y^2$$

$$v = -\partial\psi/\partial x \qquad \psi = -\int v\,\partial x = +\int (+2xy + 2y)\,\partial x = x^2y + 2xy$$

So $\psi = x^2y + 2xy - 2y^2$. Check: $\partial\psi/\partial y = u = x^2 + 2x - 4y$; $\partial\psi/\partial x = -v = 2xy + 2y$.

25.14   Given the stream function $\psi = 3x - 2y$. Is this a potential flow? Does it satisfy the Laplace equation?

▮ $\quad u = \partial\psi/\partial y = -2 \qquad v = -\partial\psi/\partial x = -3 \qquad \xi = (\partial v/\partial x) - (\partial u/\partial y) = 0$

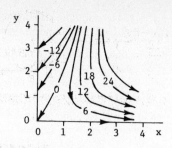

Fig. 25-4

Yes, this *is* potential flow, because $\xi = 0$.

$$u = -\partial\phi/\partial x \qquad v = -\partial\phi/\partial y \qquad -\partial\phi/\partial x = -2$$

$$\partial\phi/\partial y = -3 \qquad d\phi = 2\,dx + 3\,dy \qquad \phi = 2x + 3y$$

Therefore, by differentiation: $\partial^2\phi/\partial x^2 = 0$, $\partial^2\phi/\partial y^2 = 0$, and so it *does* satisfy the Laplace equation $(\partial^2\phi/\partial x^2) + (\partial^2\phi/\partial y^2) = 0$.

25.15 An ideal fluid flows in a two-dimensional 90° bend. The inner and outer radii of the bend are 0.4 and 1.4 ft. Sketch the flow net and estimate the velocity at the inner and outer walls of the bend if the velocity in the 1.0-ft-wide straight section is 10 fps. Develop an analytic expression for the stream function, in this case noting that $v_t = -\partial\psi/\partial r$ and $v_r = \partial\psi/r\,\partial\theta$. Determine the inner and outer velocities accurately.

$\blacksquare$ Velocity estimates: $V = -\partial\phi/\partial s$, $-\Delta\phi \approx V\,\Delta s \approx 10\,\Delta s_0$, $V \approx 10\,\Delta s_0/\Delta s$. From Fig. 25-5, $V_{\text{out}} \approx 10(\frac{5}{8}) = 6.25$ fps; $V_{\text{in}} \approx 10(5/2.5) = 20$ fps.

Accurate velocity determinations: $v_t(\Delta r) = \text{constant}$, Δr is proportional to r. $v_t(kr) = \text{constant}$, therefore $v_t = C/r$. Also $v_r = 0$. $\psi = \psi(r, \theta)$, so by chain rule of differentiation $d\psi = (\partial\psi/r\,\partial\theta)r\,d\theta + (\partial\psi/\partial r)\,dr = v_r r\,d\theta - v_t\,dr = 0 - v_t\,dr$. If $\psi_{\text{in}} = 0$, $\psi_{\text{out}} = \psi_{\text{in}} + dq = 0 + (1)(10) = 10$ cfs/ft. Integrating

$$\int_0^{10} d\psi = -\int_{0.4}^{1.4} v_t\,dr = -C\int_{0.4}^{1.4}\frac{dr}{r} \qquad \psi = 10 = -C[\ln r]_{0.4}^{1.4} = -C\ln\frac{1.4}{0.4} = -C\ln 3.5$$

and $C = -7.98$. Thus the analytic expression is $\psi = 7.98\ln r$, $v_t = C/r = -7.98/r$; $(v_t)_{r=0.4} = 19.9$ fps, $(v_t)_{r=1.4} = 5.7$ fps.

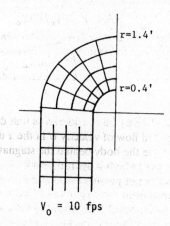

r=1.4'

r=0.4'

$V_0 = 10$ fps

Fig. 25-5

25.16 A cylindrical drum with a 2-ft radius is securely held in position in an open channel of rectangular section. The channel is 10 ft wide, and the flow rate is 240 cfs. Water flows beneath the drum as shown in Fig. 25-6a. Sketch the flow net, and determine from flow net measurements the pressure at the points indicated along the wetted drum surface. Neglect fluid friction. Sketch the pressure distribution, and by numerical integration determine an approximate value of the horizontal thrust on the cylinder.

$\blacksquare$ Sketch streamlines by eye and use continuity to determine velocities. High degree of accuracy is not expected. See Fig. 25-6b. At A, $p/\gamma = p_s/\gamma = V^2/2g = 6^2/2g = 0.599$ ft. At B, in contact with atmosphere ($p_{\text{atm}} = 0$ psig): $p/\gamma = 0$. At C, from flow net, $V \approx 7$ fps: $4 + 6^2/2g = (4 - 1.0) + p/\gamma + 7^2/2g$; $p/\gamma = 0.798$ ft. At D, from flow net, $V \approx 11.5$ fps: $4 + 6^2/2g = (4 - 1.73) + p/\gamma + 11.5^2/2g$; $p/\gamma = 0.235$ ft. See Fig. 25-6c. Area under projected p/γ diagram $\approx$ avg horizontal component = area/2.2 = 0.44 ft = 0.19 psi, horizontal thrust $\approx 0.19(2.2)(12)(10)(12) = 602$ lb.

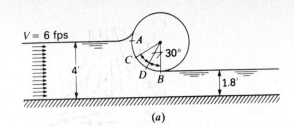

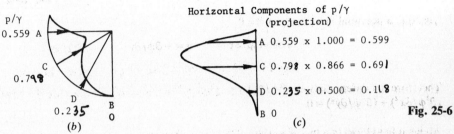

Fig. 25-6

25.17 For the two-dimensional flow of a frictionless incompressible fluid against a flat plate normal to the initial velocity, the stream function is given by $\psi = -2axy$, while its *conjugate function*, the velocity potential, is $\phi = a(x^2 - y^2)$ where a is a constant and the flow is symmetrical about the yz plane. By direct differentiation demonstrate that these functions satisfy the equation $(\partial^2\phi/\partial x^2) + (\partial^2\phi/\partial y^2) = 0$. Using a scale of 1 in = 1 unit of distance, plot the streamlines given by $\psi = \pm 2a$, $\pm 4a$, $\pm 6a$, $\pm 8a$, and the equipotential lines given by $\phi = 0$, $\pm 2a$, $\pm 4a$, $\pm 6a$, $\pm 8a$. Observe that this flow net also gives the ideal flow around an inside square corner.

▌ Given: $\psi = -2axy$; $\phi = a(x^2 - y^2)$. Differentiating: $\partial\psi/\partial y = -2ax$; $\partial\phi/\partial x = 2ax$; $-\partial\psi/\partial x = 2ay$; $\partial\phi/\partial y = a(-2y)$; $\partial^2\phi/\partial x^2 = 2a$; $\partial^2\phi/\partial y^2 = -2a$; thus the equation indicated is satisfied. See Fig. 25-7 for required plot.

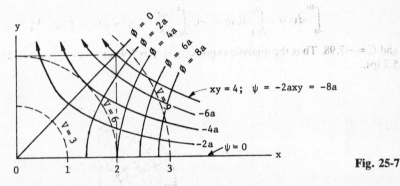

Fig. 25-7

25.18 The flow around the body of Fig. 25-8a may be considered as that due to the sum of two velocity potentials, $\phi_1 = -Ux$, representing an undisturbed flow of velocity U in the x direction, and $\phi_2 = -S \ln r$, representing the radial flow from a source located inside the body behind the stagnation point. To relate U and S, it is observed that the total flow $2\pi S$ from the source (which is hydrodynamically equivalent to the body itself) must be equal to the flow of the main stream which is not passing through the body of width b, or $2\pi S = Ub$. This gives $\phi_2 = -(Ub/2\pi) \ln r$. (**a**) The distance from the stagnation point to the source is determined by setting the radial velocity from the source, $v_r = -\partial\phi/\partial r$, equal and opposite to the undisturbed velocity U. Prove that that establishes the source at a distance $b/2\pi$ behind the stagnation point. The absolute velocity at any point of the field may be determined by the vector sum of the components U and v_r.

There follows an ingenious method of plotting the boundary of such a streamlined body, as shown in Fig. 25-8b. Suppose that the streamlines in the undisturbed flow are spaced a distance a apart, where $b/2a = n$, an integer. Next divide the upper half of the source into n radial sectors, each of angle α, that is, $n\alpha = \pi$. Then the undisturbed flow between the x axis and the first streamline is associated with the source flow in the first sector from the stagnation point. Thus the intersection of the first streamline with the first line must be a point on the boundary of the body, through which there can be no flow. Similarly, the intersection of the horizontal line at $2a$ with the radial line at 2α forms another point, and so on. Further streamlines can be plotted by connecting successive intersections of the original horizonal lines with the radial lines, recognizing that the same flow must exist between any adjacent pair of streamlines. Thus the intersection of a horizontal line ea above the axis with a radial line at $f\alpha$ from the stagnation point must lie on a streamline which is $(e - f)a$ distant from the axis in the undisturbed region, where e and f are integers. (**b**) Assume a value of $U = 20$ fps and a two-dimensional flow past a streamlined body for which $b = 36$ ft. Compute the distance from the source to the stagnation point and to the surface of the body at a radius 90° to the axis. What is the value of the source velocity at the latter point? (c) What is the magnitude of the velocity of the fluid along the surface at the 90° point? What is its direction relative to the axis?

▌ (a) Given: source potential, $\phi_2 = -(Ub/2\pi) \ln r$; $v_r = -\partial\phi/\partial r = Ub/2\pi r$. Setting $v_r = U$ for stagnation condition (on tip of nose) and solving for r, the desired distance behind the nose, $r = b/2\pi$.

(b) For $b = 36$ ft, distance to stagnation point $= -36/2\pi = -5.73$ ft. Since $b/2a = n = 6 = 36/2a =$ the number of radial sectors from half of source, streamline spacing $= a = \frac{36}{12} = 3$ ft. From Fig. 25–8b: At $90° = 3\alpha$, $r = 3a = 3(3) = 9$ ft. Source velocity at 90°, $r = 3a$: $v_r = Ub/2\pi r = (20)(36)/[2\pi(9)] = 12.73$ fps.

(c) At 90°, $V = (U^2 + v_r^2)^{1/2} = (20^2 + 12.73^2)^{1/2} = 23.7$ fps. Direction relative to axis is $\tan^{-1}(12.73/20) = 32.5°$.

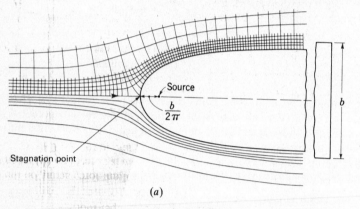

Stream lines in steady flow

Source

$\dfrac{b}{2\pi}$

Stagnation point

(a)

a

α

Source

(b)

Fig. 25-8

25.19 Find the distance to the surface of the body and the two velocities called for in the preceding problem for an angle of 30°.

▌ At 30°, $r = a/\sin 30° = 3/0.5 = 6.00$ ft.

$$v_r = \partial\phi/\partial r = Ub/2\pi r = 20(36)/[2\pi(6.0)] = 19.10 \text{ fps}$$
$$V^2 = U^2 + v_r^2 - 2Uv_r \cos \alpha = 20^2 + 19.10^2 - 2(20)19.10 \cos 30° \qquad V = 10.16 \text{ fps}$$

25.20 Superimpose a point source ($Q = 100$ cfs) on a rectilinear flow field ($U = 20$ fps). Plot the body contour at $\theta = 30, 60, 90, 120, 150, 180°$ using a scale of 1 in $= 1$ ft. Compute the velocities along the body contour at these points. Determine the pressures at these points assuming $\rho = 1.94$ slug/ft^3 with zero pressure in the undisturbed rectilinear flow field. What is the velocity and pressure in the combined flow field at the following points? *Hint:* Refer to Prob. 25.18.

(a) $\theta = 45°$, $r = 4.0$ ft   (b) $\theta = 90°$, $r = 2.0$ ft   (c) $\theta = 90°$, $r = 4.0$ ft   (d) $\theta = 135°$, $r = 2.0$ ft

▌ Given: $U = 20$ fps, $v_r = 100/2\pi r$ fps. Distance from source to stagnation point: $U = v_r$, $r = 5/2\pi = 0.796$ ft. Refer to Prob. 25.18; for upper half let $n = 6$, thus $\theta = 30°$. $q = Ua = 100/2n$; thus $a = 100/(2nU) = 0.417$ ft. See Fig. 25-9. Scale off radial distances to points a, b, c, d, e, and f; $v_x = U - v_r \cos \theta$, $v_y = v_r \sin \alpha$. Vector addition: $v = (v_x^2 + v_y^2)^{1/2}$.

point	r, ft	θ	v_r	v_x	v_y	v, fps
S	0.80	0	20.0	0	0	0
a	0.81	30°	19.64	2.99	9.90	10.34
b	0.96	60°	16.57	11.71	14.46	18.71
c	1.24	90°	12.83	20.0	12.83	23.8
d	1.90	120°	8.38	24.2	7.26	25.3
e	4.16	150°	3.83	23.3	1.915	23.4

$$p_0 + \rho V_0^2/2 = p + \rho V^2/2 \qquad 1.94(20^2/2) = 388 \text{ lb/ft}^2 = p + \rho V^2/2$$

Therefore, $p = 388 - \rho V^2/2$. Hence the pressures at 0°, 30°, etc., are 388, 284, 48.4, -161, -232, -143 psf, respectively. At other points in the flow field:

(a) $\quad\quad\quad v_x = [20 - 100/(2\pi)(4)]\cos 45°\quad\quad v_y = [100/(2\pi)(4)]\sin 45°\quad\quad v_x = 11.33\text{ fps}$

$\quad\quad\quad v_y = 2.81\text{ fps}\quad\quad\quad v = 11.67\text{ fps}\quad\quad\quad p = (\rho)(20^2)/2 - \rho V^2/2 = 256\text{ psf}$

(b) Similarly $v_x = 20.0\text{ fps}$, $v_y = 7.95\text{ fps}$, $v = 21.5\text{ fps}$, $p = -60.3\text{ psf}$.

(c) $\quad\quad\quad v_x = 20.0\text{ fps}\quad\quad\quad v_y = 4.98\text{ fps}\quad\quad\quad v = 20.6\text{ fps}\quad\quad\quad p = -23.6\text{ psf}$

(d) This point is inside the bounds of the body. Therefore, it is not in the flow field.

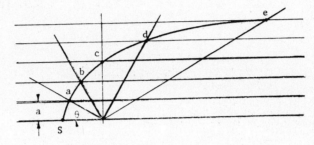

Fig. 25-9

Appendixes

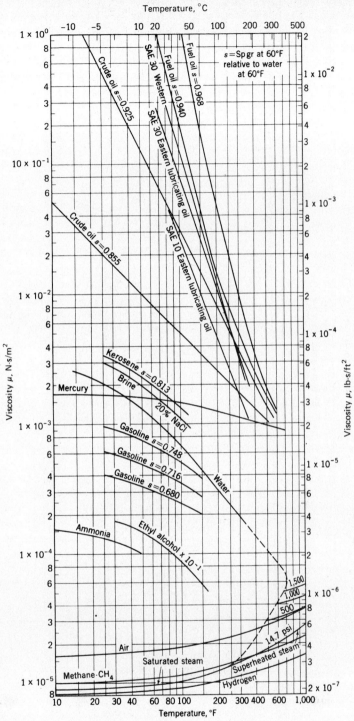

Fig. A-1

Absolute viscosity μ of fluids.

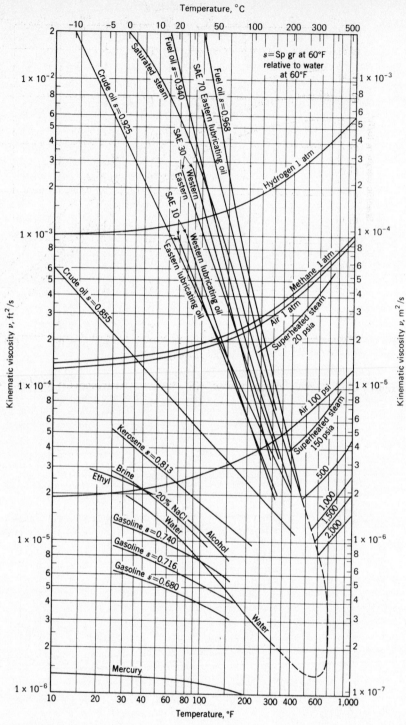

Fig. A-2

Kinematic viscosity ν of fluids.

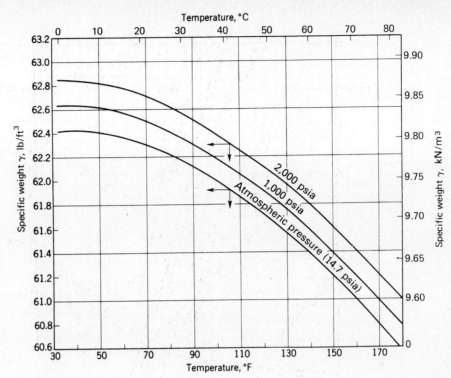

Fig. A-3

Specific weight γ of pure water as a function of temperature and pressure for condition where $g = 32.2\ \text{ft/s}^2$ $(9.81\ \text{m/s}^2)$.

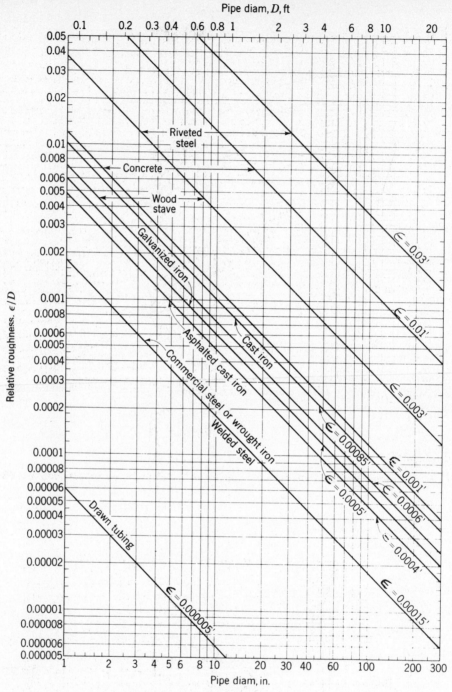

Fig. A-4

Roughness factors (ϵ expressed in feet) for commercial pipes.

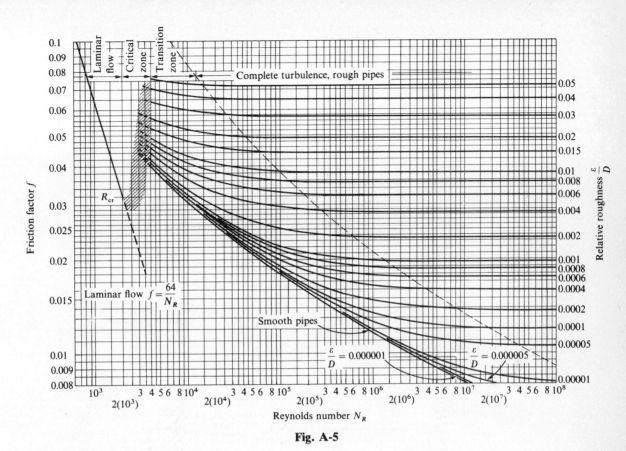

Fig. A-5

Moody diagram.

From Lewis F. Moody, "Friction Factors for Pipe Flows," *ASME Trans.*, vol. 66, pp. 671–684, 1944.

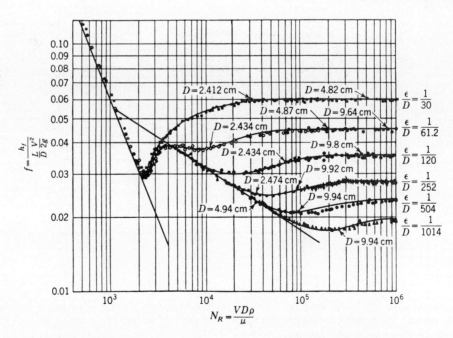

Fig. A-6

Nikuradse's sand-roughened-pipe tests.

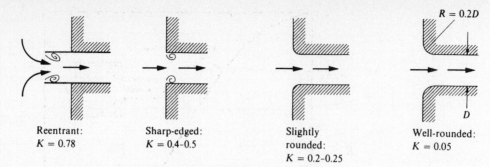

Reentrant:
$K = 0.78$

Sharp-edged:
$K = 0.4–0.5$

Slightly
rounded:
$K = 0.2–0.25$

Well-rounded:
$K = 0.05$

Exit losses: $K = 1.0$ for all shapes of exit (reentrant,
sharp-edged, slightly, or well-rounded)

Fig. A-7

Entrance and exit loss coefficients.

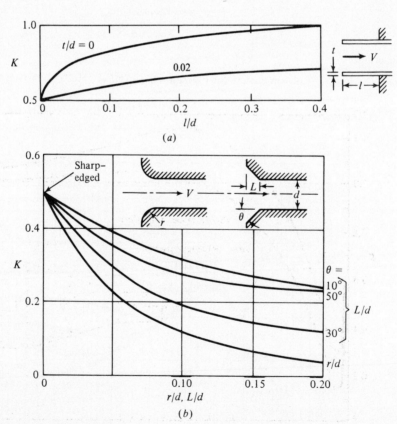

Entrance and exit loss coefficients: (*a*) reentrant inlets; (*b*) rounded and beveled inlets. Exit losses are $K \approx 1.0$ for all shapes of exit (reentrant, sharp, beveled, or rounded). (Adapted by permission from *ASHRAE Handbook of Fundamentals,* Atlanta, 1985.)

Fig. A-8

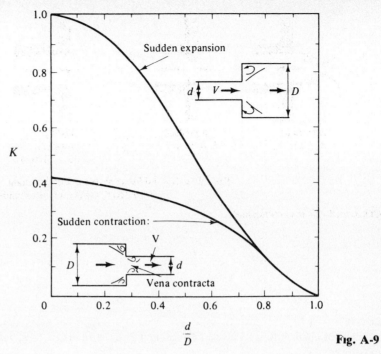

Fig. A-9

Sudden expansion and contraction losses. Note that the loss is based on velocity head in the small pipe.

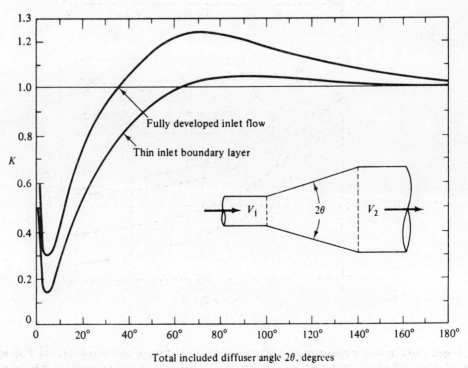

Fig. A-10

Flow losses in a gradual conical expansion region.

SUDDEN AND GRADUAL CONTRACTION

If: $\theta \lesssim 45°$ . . . . $K_2 =$ Formula 1

$45° < \theta \lesssim 180°$ . . . $K_2 =$ Formula 2

• **Formula 1**

$$K_2 = \frac{0.8\left(\sin\frac{\theta}{2}\right)(1 - \beta^2)}{\beta^4} = \frac{K_1}{\beta^4}$$

• **Formula 2**

$$K_2 = \frac{0.5\,(1 - \beta^2)\sqrt{\sin\frac{\theta}{2}}}{\beta^4} = \frac{K_1}{\beta^4}$$

$$\beta = \frac{d_1}{d_2}$$

$$\beta^2 = \left(\frac{d_1}{d_2}\right)^2 = \frac{a_1}{a_2}$$

Subscript 1 defines dimensions and coefficients with reference to the smaller diameter.
Subscript 2 refers to the larger diameter.

Fig. A-11

From "Flow of Fluids through Valves, Fittings, and Pipe," *Crane Co. Tech. Pap. 410*, New York, 1985.

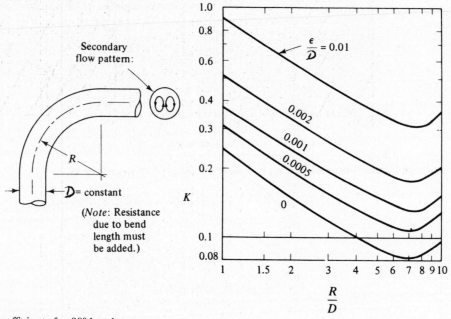

Resistance coefficients for 90° bends.

Fig. A-12

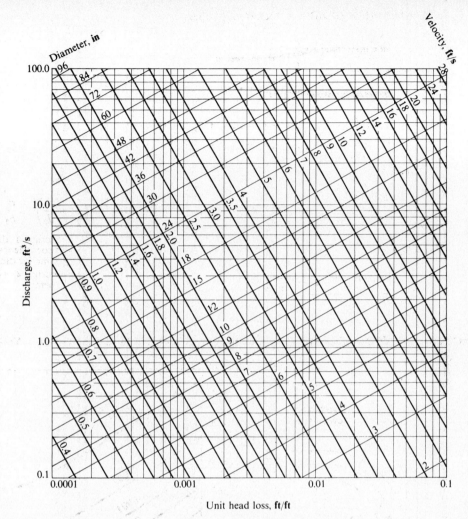

Fig. A-13

Pipe diagram: Hazen-Williams equation ($C = 120$); English Gravitational System.

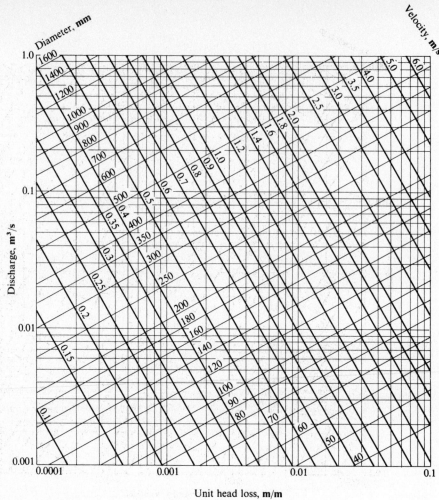

Pipe diagram: Hazen-Williams equation ($C = 120$); International System.

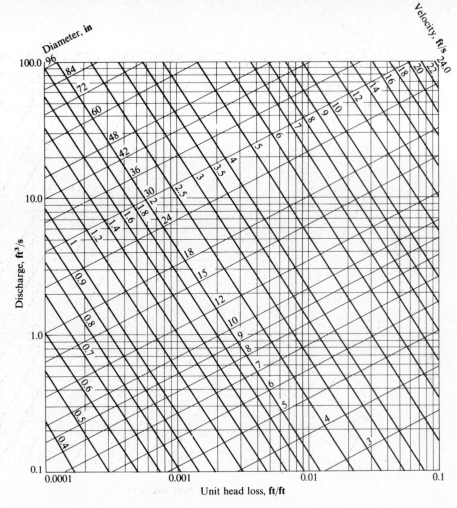

Fig. A-15

Pipe diagram: Manning equation ($n = 0.013$); English Gravitational System.

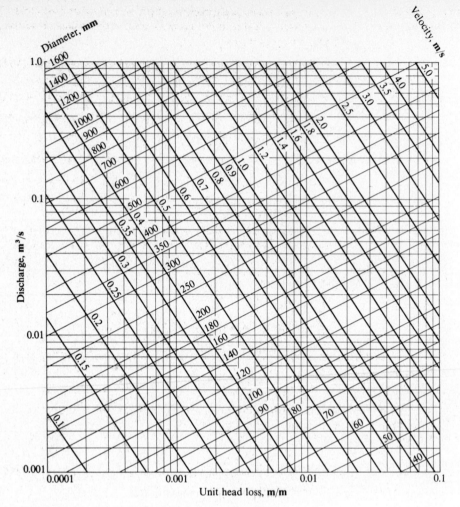

Fig. A-16

Pipe diagram: Manning equation ($n = 0.013$); International System.

FLOW CHART

HAZEN-WILLIAMS FORMULA, $C_1 = 100$

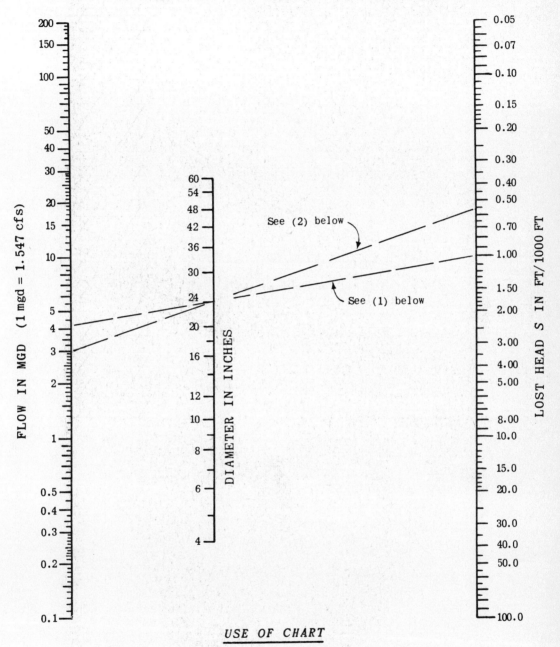

USE OF CHART

(1) Given $D = 24''$, $S = 1.0$ ft/1000 ft, $C_1 = 120$; find flow Q.

Chart gives $Q_{100} = 4.2$ mgd.

For $C_1 = 120$, $Q = (120/100)(4.2) = 5.0$ mgd.

(2) Given $Q = 3.6$ mgd, $D = 24''$, $C_1 = 120$; find Lost Head.

Change Q_{120} to Q_{100}: $Q_{100} = (100/120)(3.6) = 3.0$ mgd.

Chart gives $S = 0.55$ ft/1000 ft.

Fig. A-17

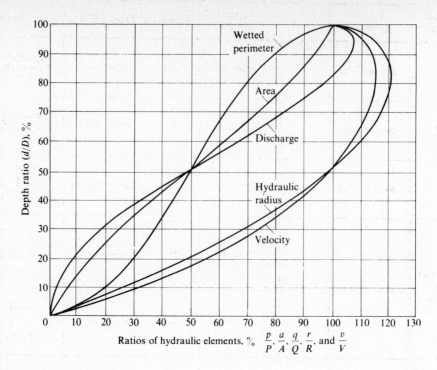

Fig. A-18

Hydraulic elements of a circular section.

Channel Slope	Depth Relations	$\left(\dfrac{dy}{dL}\right)$	Type of Profile	Symbol	Type of Flow	Form of Profile
Mild $0 < S < S_c$	$y > y_N > y_c$	$+$	Backwater	M_1	Subcritical	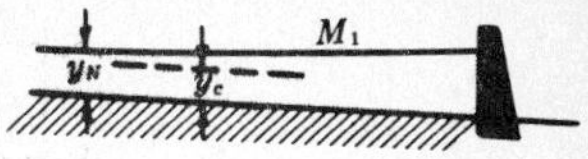
	$y_N > y > y_c$	$-$	Dropdown	M_2	Subcritical	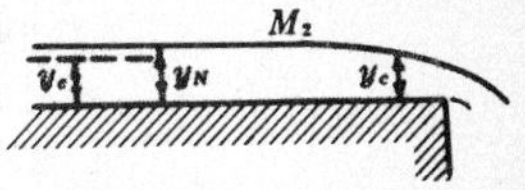
	$y_N > y_c > y$	$+$	Backwater	M_3	Supercritical	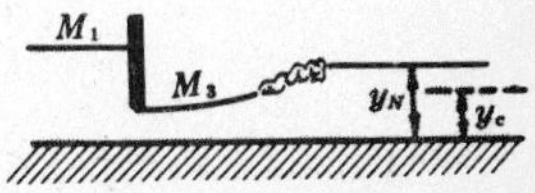
Horizontal $S = 0$ $y_N = \infty$	$y > y_c$	$-$	Dropdown	H_2	Subcritical	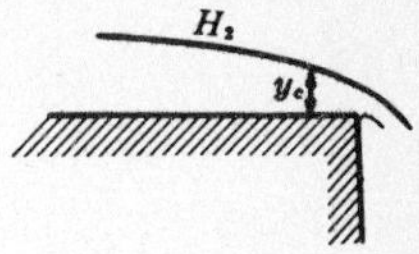
	$y_c > y$	$+$	Backwater	H_3	Supercritical	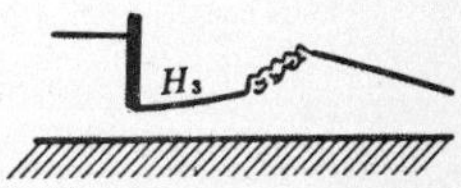
Critical $S_N = S_c$ $y_N = y_c$	$y > y_c = y_N$	$+$	Backwater	C_1	Subcritical	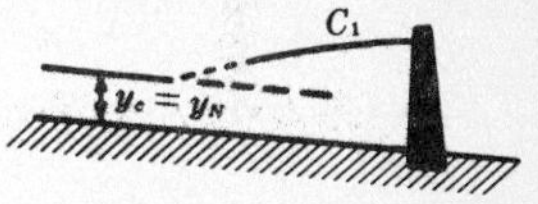
	$y_c = y = y_N$		Parallel to bed	C_2	Uniform, Critical	
	$y_c = y_N > y$	$+$	Backwater	C_3	Supercritical	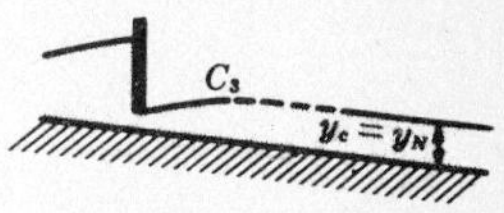
Steep $S > S_c > 0$	$y > y_c > y_N$	$+$	Backwater	S_1	Subcritical	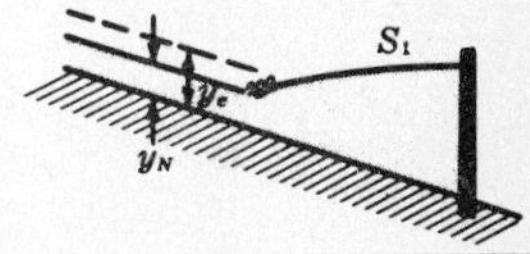
	$y_c > y > y_N$	$-$	Dropdown	S_2	Supercritical	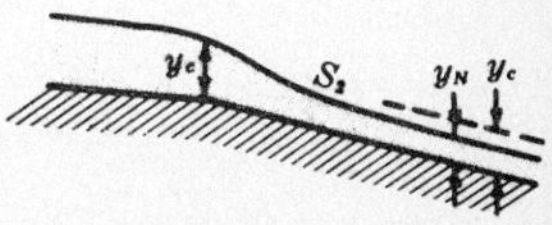
	$y_c > y_N > y$	$+$	Backwater	S_3	Supercritical	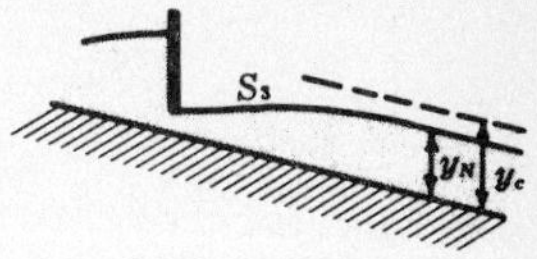
Adverse $S < 0$ $y_N = \infty$	$y > y_c$	$-$	Dropdown	A_2	Subcritical	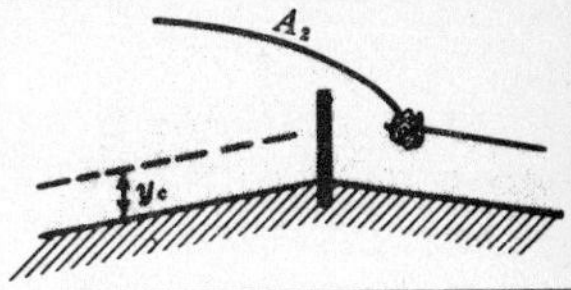
	$y_c > y$	$+$	Backwater	A_3	Supercritical	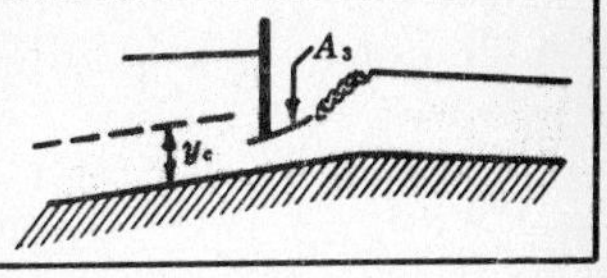

Fig. A-19

Types of profiles in open-channel flow.

Values of $(D_1'' V_1)$ for water at 72°F (diameter in inches x velocity in fps)

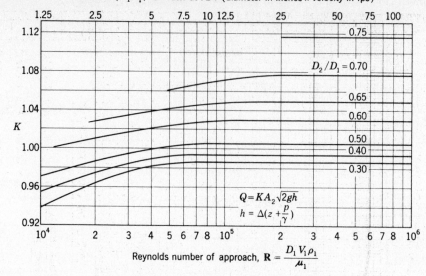

Reynolds number of approach, $\mathbf{R} = \dfrac{D_1 V_1 \rho_1}{\mu_1}$

Fig. A-20

Flow coefficients for ISA nozzle. (Adapted from *ASME Flow Measurement*, 1959.)

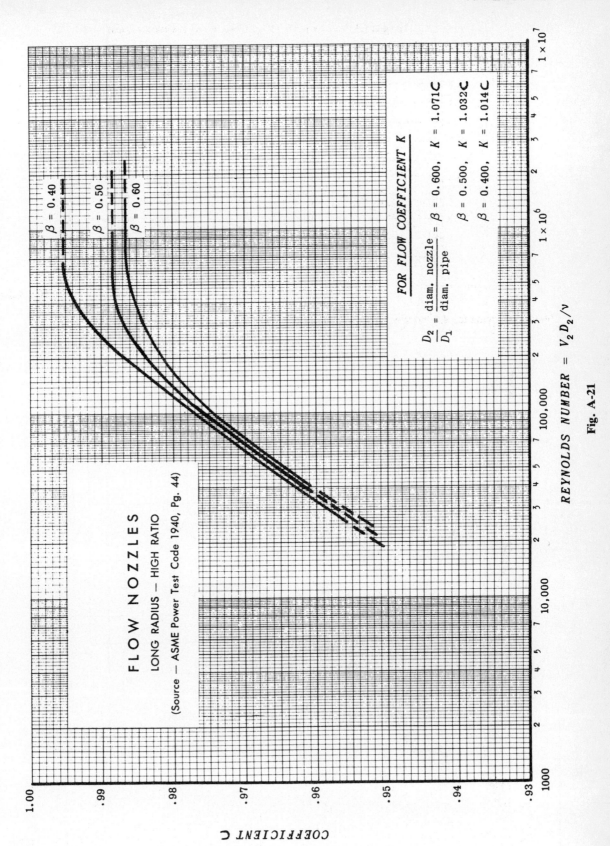

FLOW NOZZLES

LONG RADIUS — HIGH RATIO

(Source — ASME Power Test Code 1940, Pg. 44)

$\beta = 0.40$

$\beta = 0.50$

$\beta = 0.60$

FOR FLOW COEFFICIENT K

$\dfrac{D_2}{D_1} = \dfrac{\text{diam. nozzle}}{\text{diam. pipe}} = \beta = 0.600, \quad K = 1.071C$

$\beta = 0.500, \quad K = 1.032C$

$\beta = 0.400, \quad K = 1.014C$

REYNOLDS NUMBER $= V_2 D_2 / \nu$

Fig. A-21

COEFFICIENT C

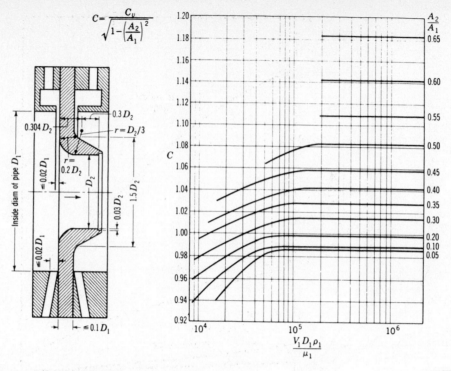

Fig. A-22

ISA (VDI) flow nozzle and discharge coefficients. (Ref. 11 in *NACA Tech. Mem. 952.*)

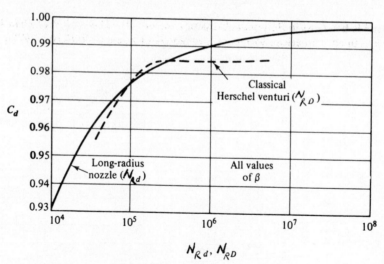

Fig. A-23

Discharge coefficient for long-radius nozzle and classical Herschel-type venturi.

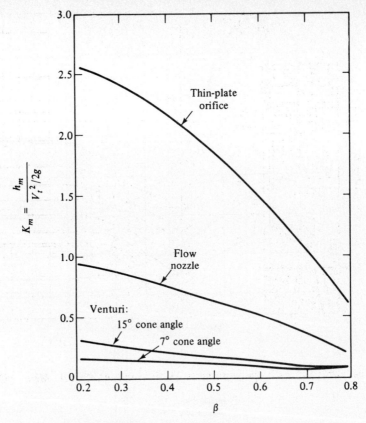

Fig. A-24

Nonrecoverable head loss in Bernoulli obstruction meters. [H. S. Bean (ed.), *Fluid Meters: Their Theory and Application,* 6th ed., American Society of Mechanical Engineers, New York, 1971.]

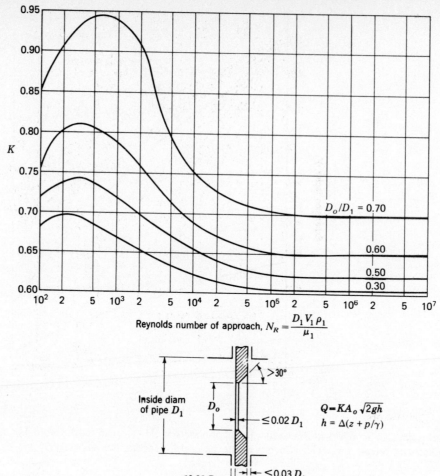

Fig. A-25

VDI orifice meter and flow coefficients for flange taps. (Adapted from *NACA Tech. Mem. 952.*)

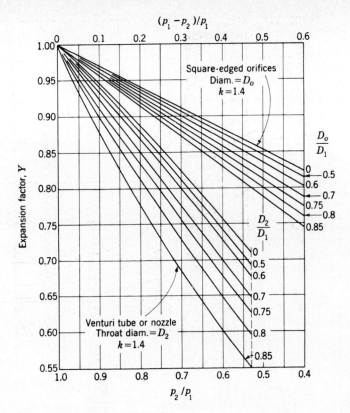

Fig. A-26

Expansion factors.

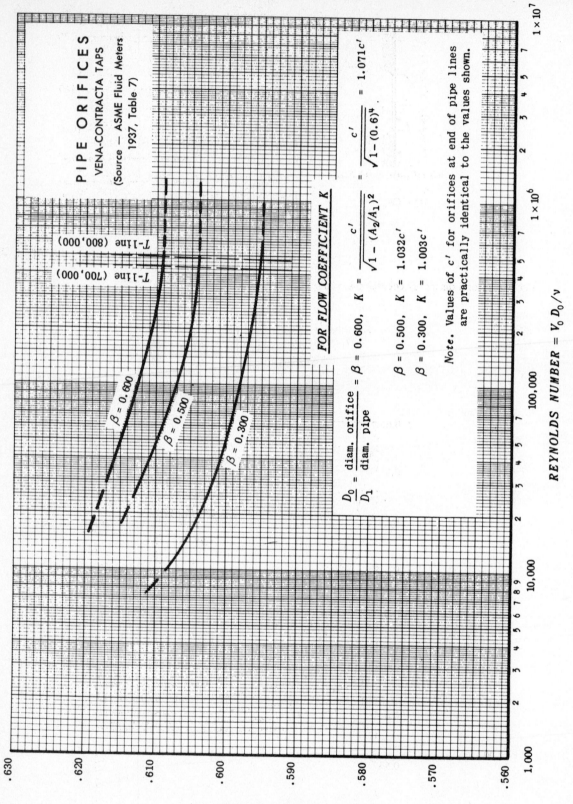

PIPE ORIFICES
VENA-CONTRACTA TAPS
(Source — ASME Fluid Meters
1937, Table 7)

FOR FLOW COEFFICIENT K

$$\frac{D_0}{D_1} = \frac{\text{diam. orifice}}{\text{diam. pipe}} = \beta = 0.600, \quad K = \frac{c'}{\sqrt{1 - (A_2/A_1)^2}} = \frac{c'}{\sqrt{1 - (0.6)^4}} = 1.071 c'$$

$$\beta = 0.500, \quad K = 1.032 c'$$

$$\beta = 0.300, \quad K = 1.003 c'$$

Note. Values of c' for orifices at end of pipe lines are practically identical to the values shown.

$\beta = 0.600$

$\beta = 0.500$

$\beta = 0.300$

T-line (800,000)

T-line (700,000)

COEFFICIENT c'

.630 .620 .610 .600 .590 .580 .570 .560

1,000 10,000 100,000 1×10^6 1×10^7

REYNOLDS NUMBER $= V_0 D_0 / \nu$

Fig. A-27

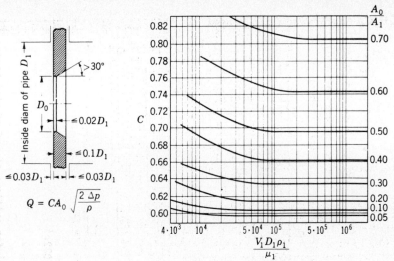

Fig. A-28

VDI orifice and discharge coefficients. (Ref. 11 in *NACA Tech. Mem. 952.*)

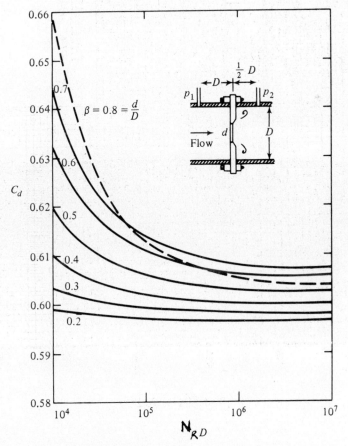

Fig. A-29

Discharge coefficient for a thin-plate orifice with $D: \frac{1}{2}D$.

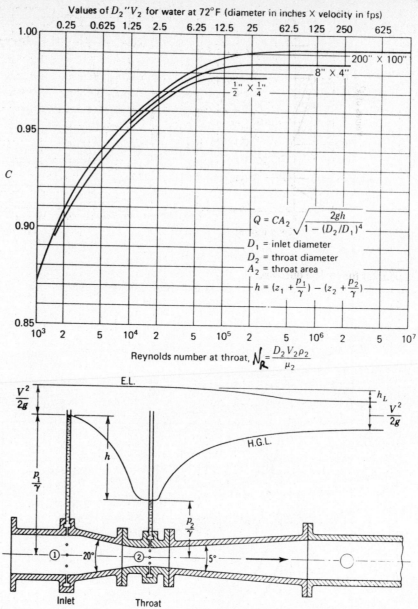

Values of $D_2''V_2$ for water at 72°F (diameter in inches × velocity in fps)

$$Q = CA_2 \sqrt{\frac{2gh}{1 - (D_2/D_1)^4}}$$

D_1 = inlet diameter
D_2 = throat diameter
A_2 = throat area
$h = (z_1 + \frac{p_1}{\gamma}) - (z_2 + \frac{p_2}{\gamma})$

Reynolds number at throat, $N_R = \frac{D_2 V_2 \rho_2}{\mu_2}$

Fig. A-30

Venturi meter with conical entrance and flow coefficients for $D_2/D_1 = 0.5$.

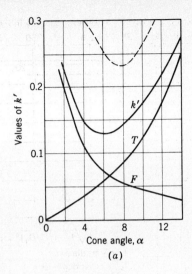

(a)

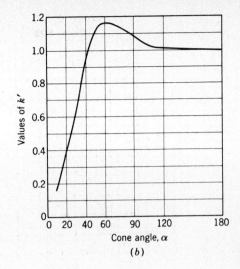

(b)

Fig. A-31

Loss coefficient for conical diffusers.

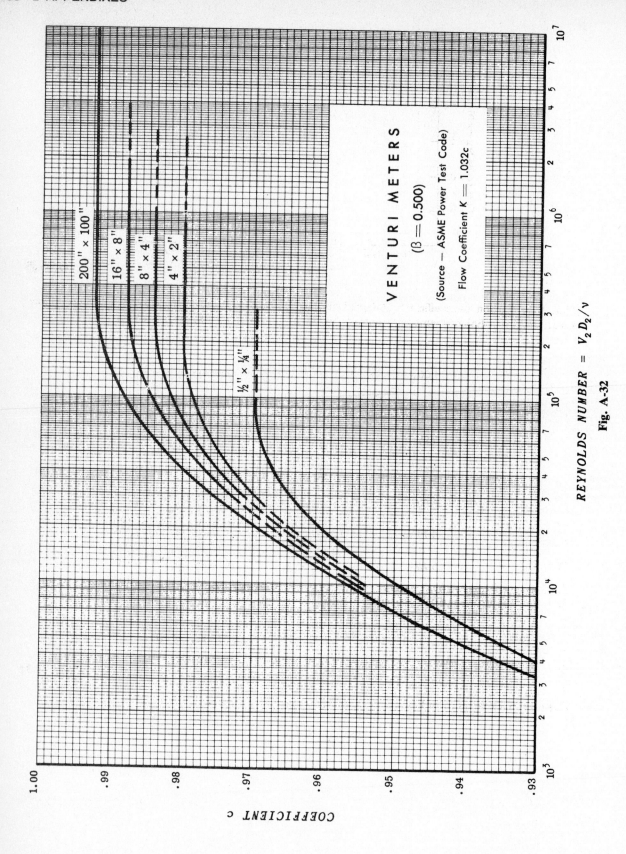

REYNOLDS NUMBER $= V_2 D_2/\nu$

Fig. A-32

VENTURI METERS

$(\beta = 0.500)$

(Source — ASME Power Test Code)

Flow Coefficient $K = 1.032c$

200" × 100"

16" × 8"

8" × 4"

4" × 2"

½" × ¼"

COEFFICIENT c

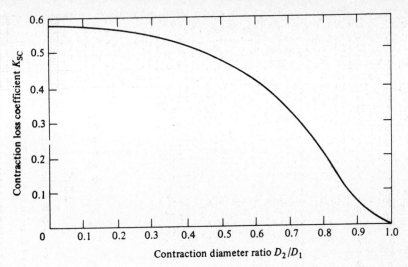

Fig. A-33

Sudden-contraction loss coefficient. (Adapted from R. P. Benedict, N. A. Carlucci, and S. D. Swetz, "Flow Losses in Abrupt Enlargements and Contractions," *J. Eng. Power,* vol. 48, January 1966.)

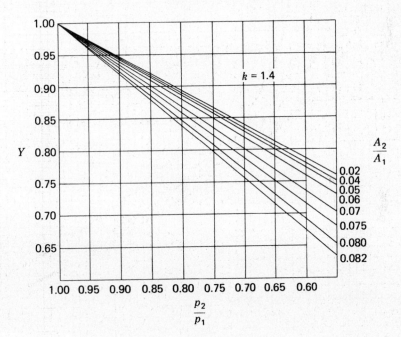

Fig. A-34

Compressibility factor Y for flow nozzles and Venturi meters. $k = 1.4$.

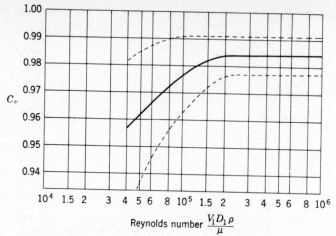

Fig. A-35

Coefficient C_v for venturi meters (*Fluid Meters: Their Theory and Application,* 5th ed., American Society of Mechanical Engineers, 1959.)

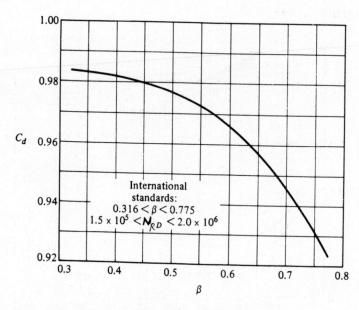

Fig. A-36

Discharge coefficient for a venturi nozzle.

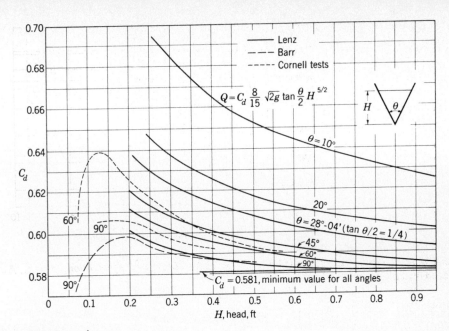

Fig. A-37

Coefficients for triangular weirs.

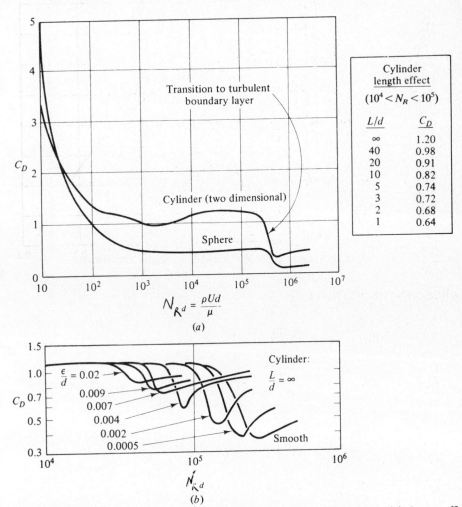

Fig. A-38

The proof of practical dimensional analysis: drag coefficients of a cylinder and sphere: (*a*) drag coefficient of a smooth cylinder and sphere (data from many sources); (*b*) increased roughness causes earlier transition to a turbulent boundary layer.

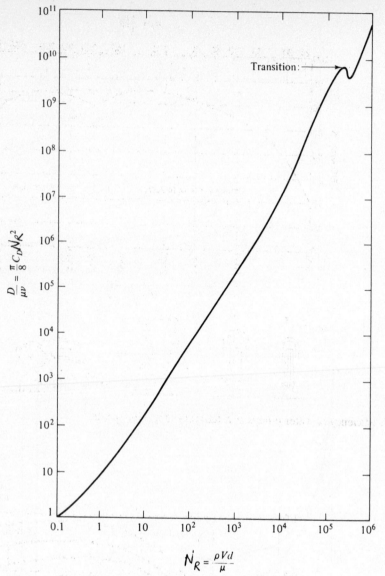

Fig. A-39

Crossplot of sphere-drag data from Fig. A-38*a* to isolate diameter and velocity.

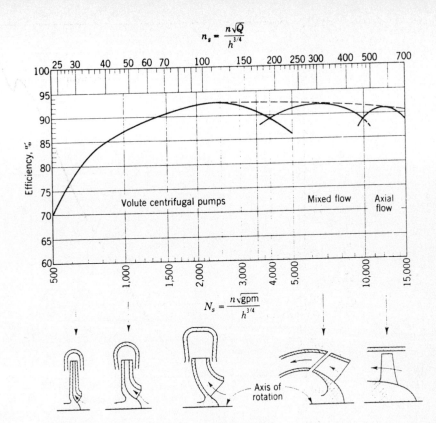

Fig. A-40

Optimum efficiency of water pumps as a function of specific speed.

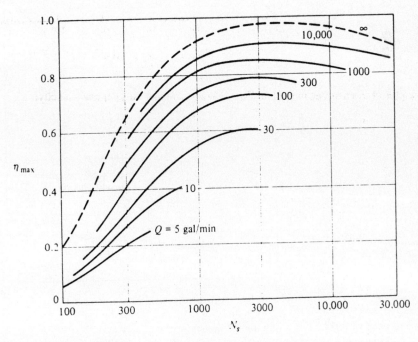

Fig. A-41

Optimum efficiency of pumps versus capacity and specific speed. (G. F. Wislicenus, *Fluid Mechanics of Turbomachinery*, 2d ed., McGraw-Hill, New York, 1965; I. J. Karassick et al., *Pump Handbook*, 2d ed., McGraw-Hill, New York, 1985.)

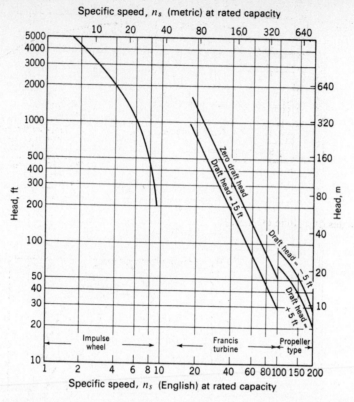

Fig. A-42

Recommended limits of specific speed for turbines under various effective heads at sea level with water temperature at 80 °F. (After Moody.)

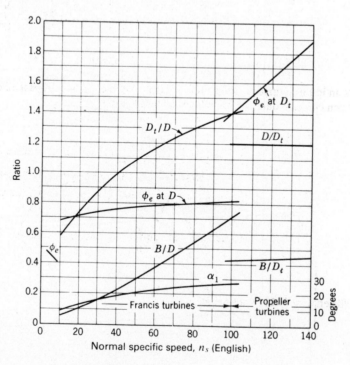

Fig. A-43

Characteristics of turbines as a function of specific speed.

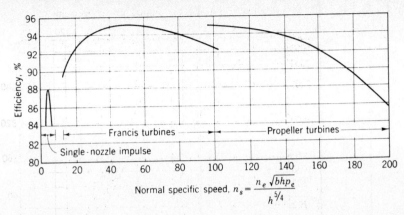

Fig. A-44

Optimum values of turbine efficiency.

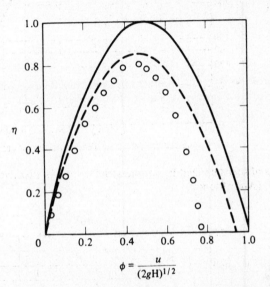

$$\phi = \frac{u}{(2gH)^{1/2}}$$

Fig. A-45

Efficiency of an impulse turbine: solid curve = ideal, $\beta = 180°$, $C_v = 1.0$; dashed curve = actual, $\beta = 160°$, $C_v = 0.94$; open circles = data, Pelton wheel, diameter = 2 ft.

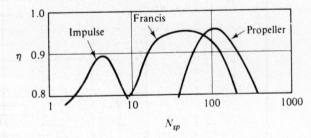

Fig. A-46

Optimum efficiency of turbine designs.

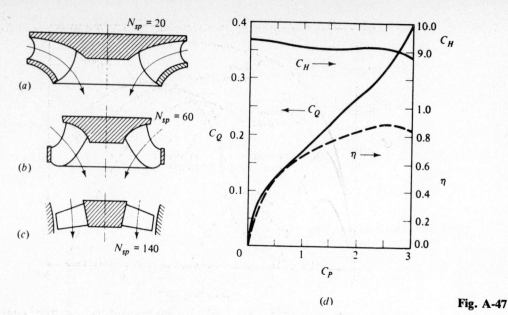

(d)

Fig. A-47

Reaction turbines: (a) Francis (radial type); (b) Francis (mixed-flow); (c) propeller (axial-flow); (d) performance curves for a Francis turbine, $n = 600$ r/min, $D = 2.25$ ft, $N_{sp} = 29$.

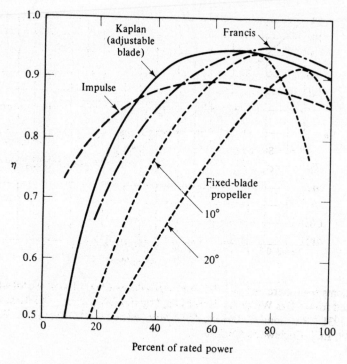

Fig. A-48

Efficiency versus power level for various turbine designs at constant speed and head.

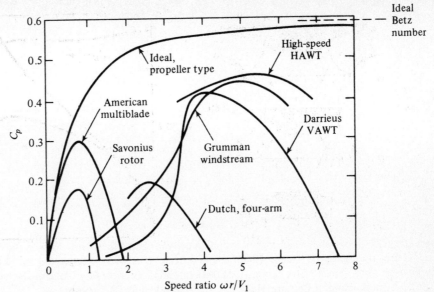

Fig. A-49

Estimated performance of various wind turbine designs as a function of blade-tip speed ratio. (Reprinted from *The Aeronautical Journal,* Vol. 85, No. 845, June 1981, by kind permission of The Royal Aeronautical Society.)

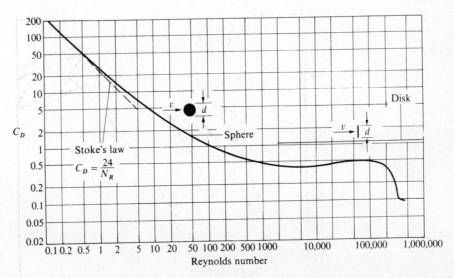

Fig. A-50

Drag coefficients for sphere and circular disk. Area in the drag relation is projected area normal to the stream. (Data adapted from "Das Widerstandsproblem," by F. Eisner, *Proc. 3d Intern. Congr. Appl. Mech.,* Stockholm, 1931.) Raymond C. Binder, *Fluid Mechanics,* 5e, © 1973, p. 132. Reprinted by permission of Prentice-Hall, Inc., Englewood Cliffs, N.J.

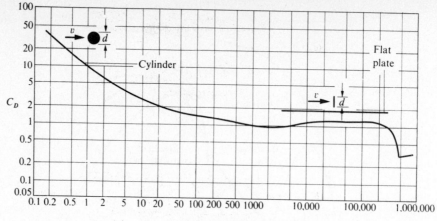

Fig. A-51

Drag coefficient for two-dimensional flow around a cylinder and a flat plate. Area in the drag relation is projected area normal to the stream. (Data adapted from "Das Widerstandsproblem," by F. Eisner, *Proc. 3d Intern. Congr. Appl. Mech.*, Stockholm, 1931.) Raymond C. Binder, *Fluid Mechanics*, 5e, © 1973, p. 134. Reprinted by permission of Prentice-Hall, Inc. Englewood Cliffs, N.J. 1973.

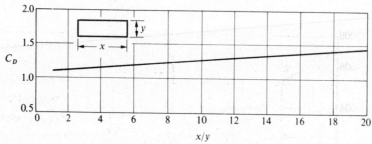

Fig. A-52

Drag coefficients for a flat plate of finite length normal to flow. Raymond C. Binder, *Fluid Mechanics*, 5e, © 1973, p. 135. Reprinted by permission of Prentice-Hall, Inc., Englewood Cliffs, N.J.

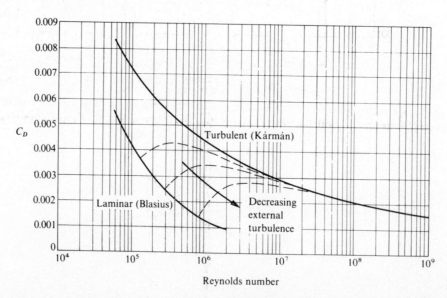

Fig. A-53

Skin-friction drag for smooth flat plates. T. v. Kármán, "Turbulence and Skin Friction," *J. Aeronaut. Sci.*, 1, no. 1 (January, 1934), © American Institute of Aeronautics and Astronautics, reprinted with permission.

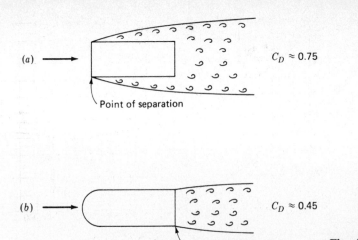

$(a) \longrightarrow$ $C_D \approx 0.75$

Point of separation

$(b) \longrightarrow$ $C_D \approx 0.45$

Point of separation   **Fig. A-54**

Plan view of flow about a motor vehicle (delivery van). (*a*) Blunt nose with separated flow along the entire side wall and a large drag coefficient $C_D = 0.75$. (*b*) Round nose with separation at the rear of the vehicle and smaller drag coefficient $C_D = 0.45$. (Adapted from H. Schlicting, *Boundary Layer Theory*, 4th ed., McGraw-Hill Book Co., New York, N.Y., 1960, p. 34).

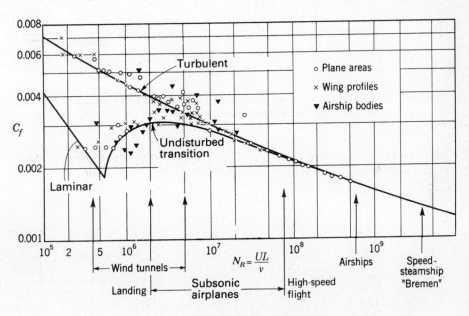

Fig. A-55

Drag coefficients for a smooth flat plate. (Adapted from *NACA Tech. Mem. 1218*, p. 117, 1949.)

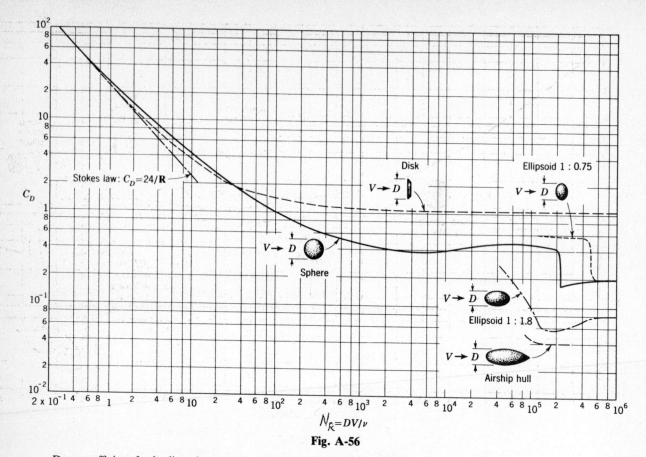

Fig. A-56

Drag coefficient for bodies of revolution. (Adapted from L. Prandtl, *Ergebnisse der aerodynamischen Versuchsanstalt zu Göttingen*, R. Oldenbourg, Munich and Berlin, 1923, p. 29; and F. Eisner, "Das Widerstandsproblem," *Proc. 3d Intern. Congr. Appl. Mech.*, 1930, p. 32.)

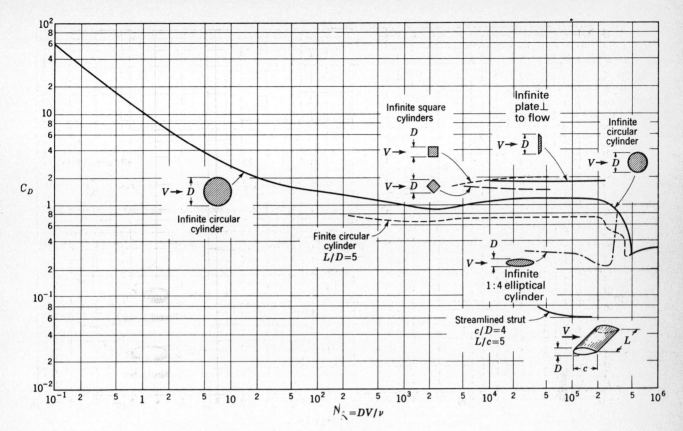

Fig. A-57

Drag coefficient for two-dimensional bodies. (Adapted from L. Prandtl, *Ergebnisse der aerodynamischen Versuchsanstalt zu Göttingen*, R. Oldenbourg, Munich and Berlin, 1923, p. 24; F. Eisner, "Das Widerstandsproblem," *Proc. 3d Intern. Congr. Appl. Mech.*, p. 32, 1930; A. F. Zahm, R. H. Smith, and G. C. Hill, "Point Drag and Total Drag of Navy Struts No. 1 Modified," *NACA Rept. 137*, p. 14, 1972; and W. F. Lindsey, "Drag of Cylinders of Simple Shapes," *NACA Rept. 619*, pp. 4–5, 1938.)

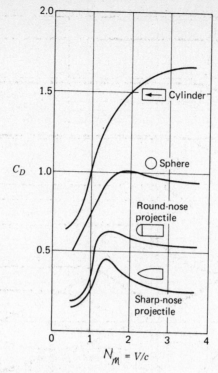

Fig. A-58

Drag coefficients as a function of Mach number.

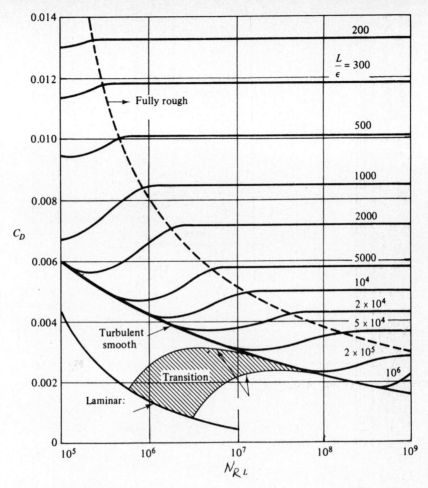

Fig. A-59

Drag coefficient of laminar and turbulent boundary layer on smooth and rough flat plates.

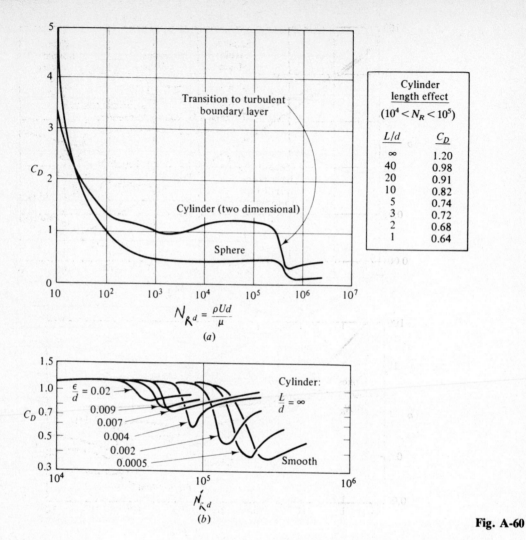

Cylinder length effect
$(10^4 < N_R < 10^5)$

L/d	C_D
∞	1.20
40	0.98
20	0.91
10	0.82
5	0.74
3	0.72
2	0.68
1	0.64

$$N_{R_d} = \frac{\rho U d}{\mu}$$

(a)

(b)

Fig. A-60

The proof of practical dimensional analysis: drag coefficients of a cylinder and sphere: (a) drag coefficient of a smooth cylinder and sphere (data from many sources); (b) increased roughness causes earlier transition to a turbulent boundary layer.

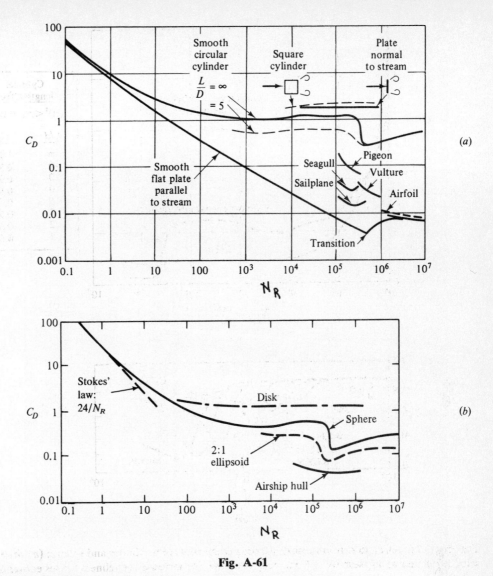

Fig. A-61

Drag coefficients of smooth bodies at low Mach numbers: (*a*) two-dimensional bodies; (*b*) three-dimensional bodies. Note the Reynolds-number independence of blunt bodies at high N_R.

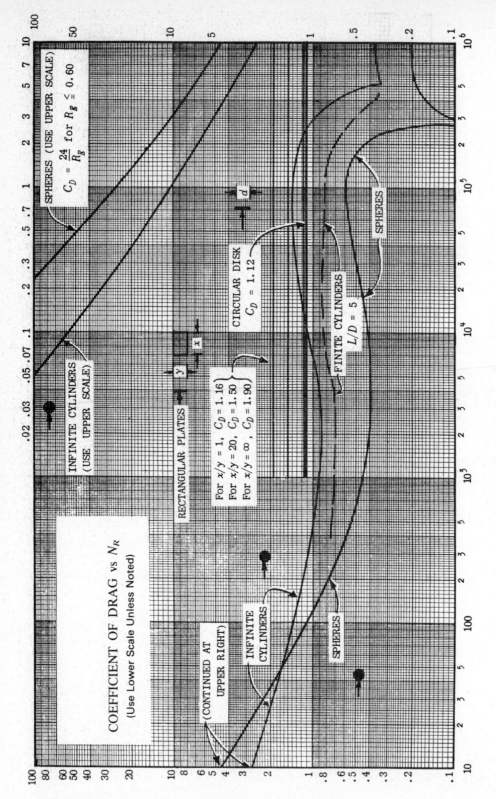

COEFFICIENT OF DRAG vs N_R
(Use Lower Scale Unless Noted)

SPHERES (USE UPPER SCALE)

$$C_D = \frac{24}{R_\beta} \text{ for } R_\beta \leq 0.60$$

INFINITE CYLINDERS
(USE UPPER SCALE)

RECTANGULAR PLATES

For $x/y = 1$, $C_D = 1.16$
For $x/y = 20$, $C_D = 1.50$
For $x/y = \infty$, $C_D = 1.90$

CIRCULAR DISK
$C_D = 1.12$

FINITE CYLINDERS
$L/D = 5$.

SPHERES

(CONTINUED AT
UPPER RIGHT)

INFINITE
CYLINDERS

SPHERES

COEFFICIENT OF DRAG C_D

REYNOLDS NUMBER (VD/ν)

Fig. A-62

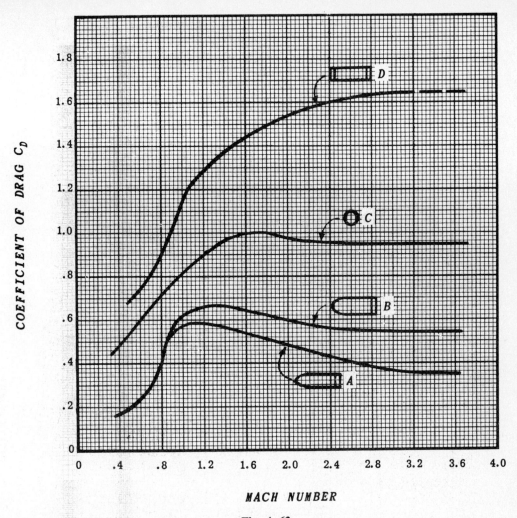

MACH NUMBER

Fig. A-63

Drag coefficients at supersonic velocities.

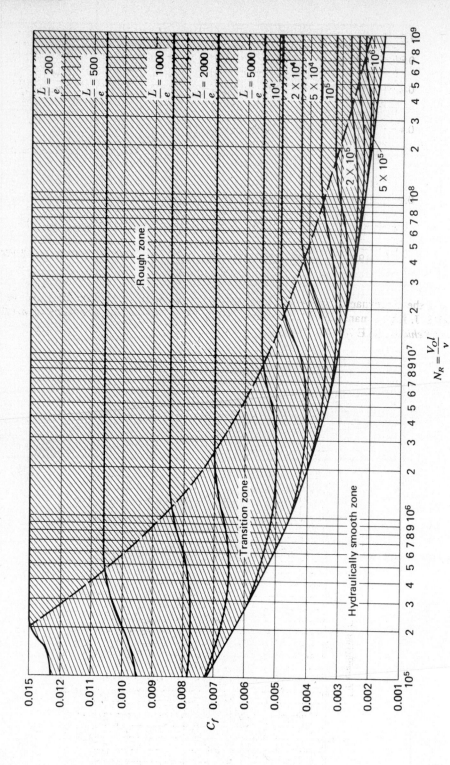

Fig. A-64

Three zones of flow for a rough plate.

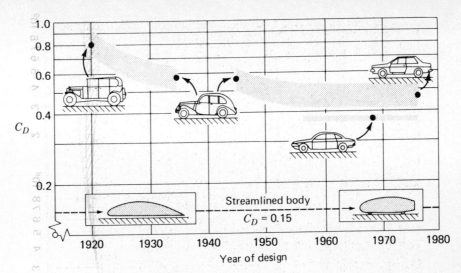

Fig. A-65

Time history of the aerodynamic drag of cars in comparison with streamlined bodies. (From W. H. Hucho, L. J. Janssen, and H. J. Emmelmann, *The Optimisation of Body Details—A Method For Reducing the Aerodynamic Drag of Road Vehicles,* SAE 760185, 1976.)

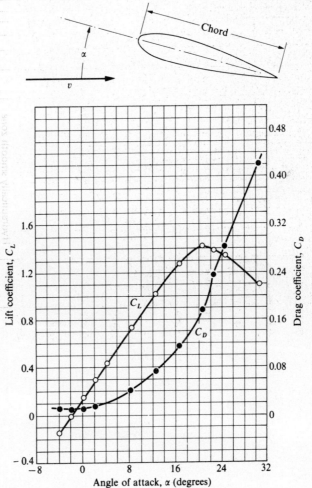

Fig. A-66

Lift and drag coefficients plotted against angle of attack for N.A.C.A. 2418 airfoil. Reynolds number = 3 060 000. (*NACA. Tech. Rept. No. 669.*)

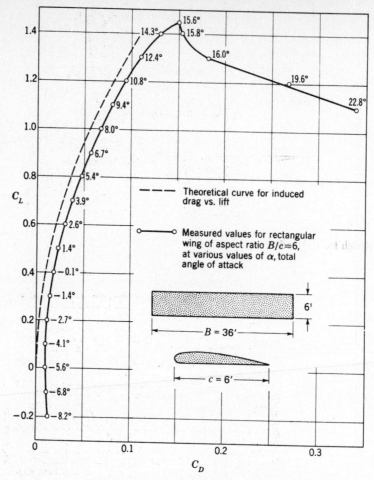

Fig. A-67

Polar diagram for rectangular Clark Y airfoil of 6-ft chord by 36-ft span. (Data from A. Silverstein, *NACA Rept. 502*, p. 15, 1934.)

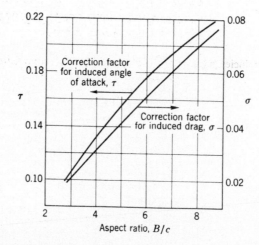

Fig. A-68

Correction factors for transforming rectangular airfoils from finite to infinite aspect ratio. (From A. Silverstein, *NACA Rept. 502*, Fig. 7, 1934.)

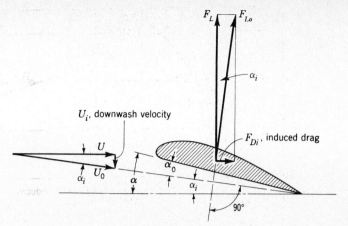

Fig. A-69

Definition sketch for induced drag.

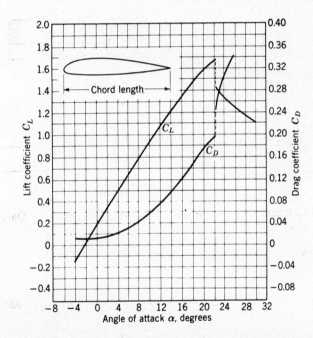

Fig. A-70

Typical lift and drag coefficients for an airfoil; C_L and C_D based on maximum projected wing area.

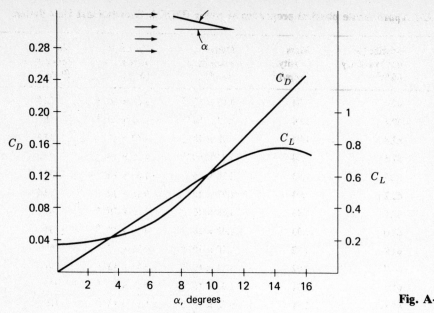

Fig. A-71

Coefficients of lift and drag for a flat plate at varying inclination α.

TABLE A-1 Approximate physical properties of water (English Gravitational Unit System)

Temperature (°F)	Specific (or unit) weight, γ (lb/ft³)	Mass density, ρ (slugs/ft³)	Dynamic viscosity, μ (lb-s/ft²)	Kinematic viscosity, ν (ft²/s)	Vapor pressure, p_v (lb/ft²)	Surface tension,† σ (lb/ft)	Bulk modulus of elasticity, K (lb/in²)
32	62.4	1.94	3.66×10^{-5}	1.89×10^{-5}	12.8	0.00518	293,000
40	62.4	1.94	3.23×10^{-5}	1.67×10^{-5}	18.5	0.00514	294,000
50	62.4	1.94	2.72×10^{-5}	1.40×10^{-5}	25.7	0.00509	305,000
60	62.4	1.94	2.35×10^{-5}	1.21×10^{-5}	36.5	0.00504	311,000
70	62.3	1.93	2.04×10^{-5}	1.05×10^{-5}	52.2	0.00500	320,000
80	62.2	1.93	1.77×10^{-5}	9.15×10^{-6}	73.5	0.00492	322,000
90	62.1	1.93	1.60×10^{-5}	8.29×10^{-6}	101	0.00486	323,000
100	62.0	1.93	1.42×10^{-5}	7.37×10^{-6}	135	0.00480	327,000
110	61.9	1.92	1.26×10^{-5}	6.55×10^{-6}	189	0.00473	331,000
120	61.7	1.92	1.14×10^{-5}	5.94×10^{-6}	251	0.00465	333,000
130	61.5	1.91	1.05×10^{-5}	5.49×10^{-6}	322	0.00460	334,000
140	61.4	1.91	9.60×10^{-6}	5.03×10^{-6}	416	0.00454	330,000
150	61.2	1.90	8.90×10^{-6}	4.68×10^{-6}	545	0.00447	328,000
160	61.0	1.90	8.30×10^{-6}	4.38×10^{-6}	693	0.00441	326,000
170	60.8	1.89	7.70×10^{-6}	4.07×10^{-6}	875	0.00433	322,000
180	60.6	1.88	7.23×10^{-6}	3.84×10^{-6}	1086	0.00426	318,000
190	60.4	1.88	6.80×10^{-6}	3.62×10^{-6}	1358	0.00419	313,000
200	60.1	1.87	6.25×10^{-6}	3.35×10^{-6}	1671	0.00412	308,000
210	59.9	1.86	5.95×10^{-6}	3.20×10^{-6}	2042	0.00405	301,000
212	59.8	1.86	5.89×10^{-6}	3.17×10^{-6}	2116	0.00404	300,000

† In contact with air.

TABLE A-2 Approximate physical properties of water (International System of Units)

Temperature (°C)	Specific (or unit) weight, γ (kN/m³)	Mass density, ρ (kg/m³)	Dynamic Viscosity, μ (N-s/m²)	Kinematic Viscosity, ν (m²/s)	Vapor Pressure, (kN/m²) (kPa)	Surface Tension,† σ (N/m)	Bulk modulus of elasticity, K (GN/m²) (GPa)
0	9.81	1000	1.75×10^{-3}	1.75×10^{-6}	0.611	0.0756	2.02
10	9.81	1000	1.30×10^{-3}	1.30×10^{-6}	1.23	0.0742	2.10
20	9.79	998	1.02×10^{-3}	1.02×10^{-6}	2.34	0.0728	2.18
30	9.77	996	8.00×10^{-4}	8.03×10^{-7}	4.24	0.0712	2.25
40	9.73	992	6.51×10^{-4}	6.56×10^{-7}	7.38	0.0696	2.28
50	9.69	988	5.41×10^{-4}	5.48×10^{-7}	12.3	0.0679	2.29
60	9.65	984	4.60×10^{-4}	4.67×10^{-7}	19.9	0.0662	2.28
70	9.59	978	4.02×10^{-4}	4.11×10^{-7}	31.2	0.0644	2.25
80	9.53	971	3.50×10^{-4}	3.60×10^{-7}	47.4	0.0626	2.20
90	9.47	965	3.11×10^{-4}	3.22×10^{-7}	70.1	0.0608	2.14
100	9.40	958	2.82×10^{-4}	2.94×10^{-7}	101.3	0.0589	2.07

† In contact with air.

TABLE A-3 Approximate physical properties of some common liquids at 1 atmosphere pressure and 20 °C (68 °F)

Liquid	Specific (or unit) weight, γ (lb/ft³)	(kN/m³)	Mass density, ρ (slugs/ft³)	(kg/m³)	Specific gravity	Dynamic viscosity, μ (lb-s/ft²)	(N-s/m²)	Vapor pressure (lb/ft²)	(kN/m²) (kPa)	Surface tension,† σ (lb/ft)	(N/m)
Ammonia	51.7	8.13	1.61	829	0.83	4.60×10^{-6}	2.20×10^{-4}	19,000	910	0.00146	0.0213
Benzene	54.8	8.62	1.70	879	0.88	1.36×10^{-5}	6.51×10^{-4}	210	10.1	0.00198	0.0289
Carbon tetrachloride	99.1	15.57	3.08	1,588	1.59	2.02×10^{-5}	9.67×10^{-4}	250	12.0	0.00185	0.0270
Ethanol	49.2	7.73	1.53	788	0.79	2.51×10^{-5}	1.20×10^{-3}	120	5.75	0.00156	0.0228
Gasoline	44.9	7.05	1.40	719	0.72	6.10×10^{-6}	2.92×10^{-4}	1,150	55.1		
Glycerin	78.5	12.34	2.44	1,258	1.26	3.11×10^{-2}	1.49	0.0003	0.000014	0.00434	0.0633
Kerosine	51.1	8.03	1.59	819	0.82	4.00×10^{-5}	1.92×10^{-3}	65	3.11	0.00190	0.0277
Mercury	847.3	133.1	26.34	13,570	13.6	3.25×10^{-5}	1.56×10^{-3}	0.000023	0.0000011	0.0352	0.514
Methanol	49.2	7.73	1.53	788	0.79	1.25×10^{-5}	5.98×10^{-4}	280	13.4	0.00155	0.0226
SAE 10 Oil	54.2	8.52	1.68	869	0.87	1.70×10^{-3}	8.14×10^{-2}	.....	.....	0.00250	0.0365
SAE 30 Oil	55.4	8.71	1.72	888	0.89	9.20×10^{-3}	4.40×10^{-1}	.....	.....	0.00240	0.0350
Water	62.3	9.79	1.94	998	1.00	2.09×10^{-5}	1.02×10^{-3}	48	2.34	0.00500	0.0728
Seawater	64.2	10.08	2.00	1,028	1.03	2.23×10^{-5}	1.07×10^{-3}	48	2.34	0.00500	0.0728

† In contact with air.

TABLE A-4 Approximate physical properties of air at standard atmospheric pressure (English Gravitational Unit System)

Temperature (°F)	Specific (or unit) weight, γ (lb/ft^3)	Mass density, ρ (slugs/ft^3)	Dynamic viscosity, μ (lb-s/ft^2)	Kinematic viscosity, ν (ft^2/s)
32	0.0808	0.00251	3.59×10^{-7}	1.43×10^{-4}
40	0.0794	0.00247	3.62×10^{-7}	1.46×10^{-4}
50	0.0779	0.00242	3.68×10^{-7}	1.52×10^{-4}
60	0.0763	0.00237	3.74×10^{-7}	1.58×10^{-4}
70	0.0750	0.00233	3.82×10^{-7}	1.64×10^{-4}
80	0.0735	0.00228	3.85×10^{-7}	1.69×10^{-4}
90	0.0723	0.00224	3.90×10^{-7}	1.74×10^{-4}
100	0.0709	0.00220	3.96×10^{-7}	1.80×10^{-4}
110	0.0696	0.00218	4.02×10^{-7}	1.84×10^{-4}
120	0.0684	0.00215	4.07×10^{-7}	1.89×10^{-4}
130	0.0674	0.00210	4.10×10^{-7}	1.95×10^{-4}
140	0.0663	0.00206	4.14×10^{-7}	2.01×10^{-4}
150	0.0652	0.00202	4.18×10^{-7}	2.06×10^{-4}
160	0.0641	0.00199	4.22×10^{-7}	2.12×10^{-4}
170	0.0631	0.00196	4.28×10^{-7}	2.18×10^{-4}
180	0.0621	0.00193	4.34×10^{-7}	2.25×10^{-4}
190	0.0612	0.00190	4.42×10^{-7}	2.32×10^{-4}
200	0.0602	0.00187	4.49×10^{-7}	2.40×10^{-4}
210	0.0594	0.00184	4.57×10^{-7}	2.48×10^{-4}
212	0.0592	0.00184	4.58×10^{-7}	2.50×10^{-4}

TABLE A-5 Approximate physical properties of air at standard atmospheric pressure (International System of Units)

Temperature (°C)	Specific (or unit) weight, γ (N/m^3)	Mass density, ρ (kg/m^3)	Dynamic viscosity, μ (N-s/m^2)	Kinematic viscosity, ν (m^2/s)
0	12.7	1.29	1.72×10^{-5}	1.33×10^{-5}
10	12.2	1.25	1.77×10^{-5}	1.42×10^{-5}
20	11.8	1.20	1.81×10^{-5}	1.51×10^{-5}
30	11.4	1.16	1.86×10^{-5}	1.60×10^{-5}
40	11.0	1.13	1.91×10^{-5}	1.69×10^{-5}
50	10.7	1.09	1.95×10^{-5}	1.79×10^{-5}
60	10.4	1.06	1.99×10^{-5}	1.89×10^{-5}
70	10.1	1.03	2.04×10^{-5}	1.99×10^{-5}
80	9.80	1.00	2.09×10^{-5}	2.09×10^{-5}
90	9.53	0.972	2.13×10^{-5}	2.19×10^{-5}
100	9.28	0.946	2.17×10^{-5}	2.30×10^{-5}

TABLE A-6 Approximate physical properties of some common gases at 1 atmosphere pressure and 30 °C (68 °F)

Gas	Specific (or unit) weight, γ		Mass density, ρ		Dynamic viscosity (μ)		Gas constant (**R**)			
	(lb/ft³)	(N/m³)	(slugs/ft³)	(kg/m³)	lb-s/ft²	kN-s/m²	ft/°R	m/K	lb-ft/slug-°R	N-m/kg-K
Air	0.0752	11.8	0.00234	1.20	3.78×10^{-7}	1.81×10^{-8}	53.3	29.3	1 716	287
Carbon dioxide	0.115	18.1	0.00357	1.84	3.10×10^{-7}	1.48×10^{-8}	35.1	19.3	1 130	189
Helium	0.0104	1.63	0.000323	0.166	4.11×10^{-7}	1.97×10^{-8}	385.7	212.0	12 420	2079
Hydrogen	0.00522	0.823	0.000162	0.0839	1.89×10^{-7}	9.05×10^{-9}	765.5	420.8	24 649	4127
Methane	0.0416	6.53	0.00129	0.666	2.80×10^{-7}	1.34×10^{-8}	96.2	52.9	3 098	519
Nitrogen	0.0726	11.4	0.00225	1.16	3.68×10^{-7}	1.76×10^{-8}	55.1	30.3	1 774	297
Oxygen	0.0830	13.0	0.00258	1.33	4.18×10^{-7}	2.00×10^{-8}	48.2	26.5	1 552	260

Gas	Specific heat ratio (k)	Specific heat (c_p)		Specific heat (c_v)		Molecular weight (M)
	Dimensionless ratio c_p/c_v	lb-ft/slug-°R	N-m/kg-K	lb-ft/slug-°R	N-m/kg-k	
Air	1.40	6 000	1 003	4 285	716	29.00
Carbon dioxide	1.30	5 132	858	4 009	670	44.00
Helium	1.66	31 230	5 220	18 810	3 143	4.00
Hydrogen	1.41	86 390	14 450	61 710	10 330	2.02
Methane	1.32	13 400	2 250	10 300	1 730	16.00
Nitrogen	1.40	6 210	1 040	4 437	743	28.00
Oxygen	1.40	5 437	909	3 883	649	32.00

TABLE A-7 The ICAO standard atmosphere in English units

Elevation above sea level, ft	Temp, °F	Absolute pressure, psia	Specific weight γ, lb/ft³	Density ρ, slugs/ft³	Viscosity $\mu \times 10^7$, lb·s/ft²
0	59.0	14.70	0.07648	0.002377	3.737
5,000	41.2	12.24	0.06587	0.002048	3.637
10,000	23.4	10.11	0.05643	0.001756	3.534
15,000	5.6	8.30	0.04807	0.001496	3.430
20,000	−12.3	6.76	0.04070	0.001267	3.325
25,000	−30.1	5.46	0.03422	0.001066	3.217
30,000	−47.8	4.37	0.02858	0.000891	3.107
35,000	−65.6	3.47	0.02367	0.000738	2.995
40,000	−69.7	2.73	0.01882	0.000587	2.969
45,000	−69.7	2.15	0.01481	0.000462	2.969
50,000	−69.7	1.69	0.01165	0.000364	2.969
60,000	−69.7	1.05	0.00722	0.000226	2.969
70,000	−69.7	0.65	0.00447	0.000140	2.969
80,000	−69.7	0.40	0.00277	0.000087	2.969
90,000	−57.2	0.25	0.00168	0.000053	3.048
100,000	−40.9	0.16	0.00102	0.000032	3.150

TABLE A-8 The ICAO standard atmosphere in SI units

Elevation above sea level, km	Temp, °C	Absolute pressure, kN/m², abs	Specific weight γ, N/m³	Density ρ, kg/m³	Viscosity $\mu \times 10^5$, N·s/m²
0	15.0	101.33	12.01	1.225	1.79
2	2.0	79.50	9.86	1.007	1.73
4	−4.5	60.12	8.02	0.909	1.66
6	−24.0	47.22	6.46	0.660	1.60
8	−36.9	35.65	5.14	0.526	1.53
10	−49.9	26.50	4.04	0.414	1.46
12	−56.5	19.40	3.05	0.312	1.42
14	−56.5	14.20	2.22	0.228	1.42
16	−56.5	10.35	1.62	0.166	1.42
18	−56.5	7.57	1.19	0.122	1.42
20	−56.5	5.53	0.87	0.089	1.42
25	−51.6	2.64	0.41	0.042	1.45
30	−40.2	1.20	0.18	0.018	1.51

TABLE A-9 Typical wall roughness values for commercial conduits

Material (new)	Roughness (ε) ft	m
Riveted steel	0.003–0.03	0.0009–0.009
Concrete	0.001–0.01	0.0003–0.003
Wood stave	0.0006–0.003	0.0002–0.0009
Cast iron	0.00085	0.00026
Galvanized iron	0.0005	0.00015
Asphalted cast iron	0.0004	0.0001
Commercial steel or wrought iron	0.00015	0.000046
Drawn brass or copper tubing	0.000005	0.0000015
Glass and plastic	"smooth"	"smooth"

From: Lewis F. Moody, "Friction Factors for Pipe Flow," *ASME Trans.*, vol. 66, pp. 671–684, 1944.

TABLE A-10 Laminar friction constants
fN_R for rectangular and triangular ducts

Rectangular		Isosceles triangle	
b/a	fN_{RD_h}	θ, deg	fN_{RD_h}
0.0	96.00	0	48.0
0.05	89.91	10	51.6
0.1	84.68	20	52.9
0.125	82.34	30	53.3
0.167	78.81	40	52.9
0.25	72.93	50	52.0
0.4	65.47	60	51.1
0.5	62.19	70	49.5
0.75	57.89	80	48.3
1.0	56.91	90	48.0

TABLE A-11 Resistance coefficients $K = \dfrac{h_m}{v^2/2g}$ for open valves, elbows, and trees

Nominal diameter, in	Screwed				Flanged				
	$\frac{1}{2}$	1	2	4	1	2	4	8	20
Valves (fully open):									
Globe	14	8.2	6.9	5.7	13	8.5	6.0	5.8	5.5
Gate	0.30	0.24	0.16	0.11	0.80	0.35	0.16	0.07	0.03
Swing check	5.1	2.9	2.1	2.0	2.0	2.0	2.0	2.0	2.0
Angle	9.0	4.7	2.0	1.0	4.5	2.4	2.0	2.0	2.0
Elbows:									
45° regular	0.39	0.32	0.30	0.29					
45° long radius					0.21	0.20	0.19	0.16	0.14
90° regular	2.0	1.5	0.95	0.64	0.50	0.39	0.30	0.26	0.21
90° long radius	1.0	0.72	0.41	0.23	0.40	0.30	0.19	0.15	0.10
180° regular	2.0	1.5	0.95	0.64	0.41	0.35	0.30	0.25	0.20
180° long radius					0.40	0.30	0.21	0.15	0.10
Tees:									
Line flow	0.90	0.90	0.90	0.90	0.24	0.19	0.14	0.10	0.07
Branch flow	2.4	1.8	1.4	1.1	1.0	0.80	0.64	0.58	0.41

TABLE A-12 Increased losses of partially open valves

	Ratio K/K(open condition)	
Condition	Gate value	Globe value
Open	1.0	1.0
Closed, 25%	3.0–5.0	1.5–2.0
50%	12–22	2.0–3.0
75%	70–120	6.0–8.0

TABLE A-13 Values of n in Manning's formula
Prepared by R. E. Horton and others

	n	
Nature of surface	Min	Max
Neat cement surface	0.010	0.013
Wood-stave pipe	0.010	0.013
Plank flumes, planed	0.010	0.014
Vitrified sewer pipe	0.010	0.017
Metal flumes, smooth	0.011	0.015
Concrete, precast	0.011	0.013
Cement mortar surfaces	0.011	0.015
Plank flumes, unplaned	0.011	0.015
Common-clay drainage tile	0.011	0.017
Concrete, monolithic	0.012	0.016
Brick with cement mortar	0.012	0.017
Cast iron—new	0.013	0.017
Cement rubble surfaces	0.017	0.030
Riveted steel	0.017	0.020
Corrugated metal pipe	0.021	0.025
Canals and ditches, smooth earth	0.017	0.025
Metal flumes, corrugated	0.022	0.030
Canals:		
Dredged in earth, smooth	0.025	0.033
In rock cuts smooth	0.025	0.035
Rough beds and weeds on sides	0.025	0.040
Rock cuts, jagged and irregular	0.035	0.045
Natural streams:		
Smoothest	0.025	0.033
Roughest	0.045	0.060
Very weedy	0.075	0.150

TABLE A-14 Typical values of the Hazen-Williams Coefficient, C

Extremely smooth and straight pipes	140
New steel or cast iron	130
Wood; concrete	120
New riveted steel; vitrified	110
Old cast iron	100
Very old and corroded cast iron	80

TABLE A-15 Values of C from the Kutter formula

Slope S	n	\multicolumn{15}{c}{Hydraulic Radius R in Feet}
		0.2	0.3	0.4	0.6	0.8	1.0	1.5	2.0	2.5	3.0	4.0	6.0	8.0	10.0	15.0
.00005	.010	87	98	109	123	133	140	154	164	172	177	187	199	207	213	220
	.012	68	78	88	98	107	113	126	135	142	148	157	168	176	182	189
	.015	52	58	66	76	83	89	99	107	113	118	126	138	145	150	159
	.017	43	50	57	65	72	77	86	93	98	103	112	122	129	134	142
	.020	35	41	45	53	59	64	72	80	84	88	95	105	111	116	125
	.025	26	30	35	41	45	49	57	62	66	70	78	85	92	96	104
	.030	22	25	28	33	37	40	47	51	55	58	65	74	78	84	90
.0001	.010	98	108	118	131	140	147	158	167	173	178	186	196	202	206	212
	.012	76	86	95	105	113	119	130	138	144	148	155	165	170	174	180
	.015	57	64	72	81	88	92	103	109	114	118	125	134	140	143	150
	.017	48	55	62	70	75	80	88	95	99	104	111	118	125	128	135
	.020	38	45	50	57	63	67	75	81	85	88	95	102	107	111	118
	.025	28	34	38	43	48	51	59	64	67	70	77	84	89	93	98
	.030	23	27	30	35	39	42	48	52	55	59	64	72	75	80	85
.0002	.010	105	115	125	137	145	150	162	169	174	178	185	193	198	202	206
	.012	83	92	100	110	117	123	133	139	144	148	154	162	167	170	175
	.015	61	69	76	84	91	96	105	110	114	118	124	132	137	140	145
	.017	52	59	65	73	78	83	90	97	100	104	110	117	122	125	130
	.020	42	48	53	60	65	68	76	82	85	88	94	100	105	108	113
	.025	30	35	40	45	50	54	60	65	68	70	76	83	86	90	95
	.030	25	28	32	37	40	43	49	53	56	59	63	69	74	77	82
.0004	.010	110	121	128	140	148	153	164	171	174	178	184	192	197	193	203
	.012	87	95	103	113	120	125	134	141	145	149	153	161	165	168	172
	.015	64	73	78	87	93	98	106	112	115	118	123	130	134	137	142
	.017	54	62	68	75	80	84	92	98	101	104	110	116	120	123	128
	.020	43	50	55	61	67	70	77	83	86	88	94	99	104	106	110
	.025	32	37	42	47	51	55	60	65	68	70	75	82	85	88	92
	.030	26	30	33	38	41	44	50	54	57	59	63	68	73	75	80
.001	.010	113	124	132	143	150	155	165	172	175	178	184	190	195	197	201
	.012	88	97	105	115	121	127	135	142	145	149	154	160	164	167	171
	.015	66	75	80	88	94	98	107	112	116	119	123	130	133	135	141
	.017	55	63	68	76	81	85	92	98	102	105	110	115	119	122	127
	.020	45	51	56	62	68	71	78	84	87	89	93	98	103	105	109
	.025	33	38	43	48	52	55	61	65	68	70	75	81	84	87	91
	.030	27	30	34	38	42	45	50	54	57	59	63	68	72	74	78
.01	.010	114	125	133	143	151	156	165	172	175	178	184	190	194	196	200
	.012	89	99	106	116	122	128	136	142	145	149	154	159	163	166	170
	.015	67	76	81	89	95	99	107	113	116	119	123	129	133	135	140
	.017	56	64	69	77	82	86	93	99	103	105	109	115	118	121	126
	.020	46	52	57	63	68	72	78	84	87	89	93	98	102	105	108
	.025	34	39	44	49	52	56	62	65	68	70	75	80	83	86	90
	.030	27	31	35	39	43	45	51	55	58	59	63	67	71	73	77

TABLE A-16 One-dimensional isentropic relations†

N_M	A/A^*	p/p_0	ρ/ρ_0	T/T_0	N_M	A/A^*	p/p_0	ρ/ρ_0	T/T_0
0.00	0........	1.000	1.000	1.000	1.10	1.01	0.468	0.582	0.805
0.01	57.87	0.9999	0.9999	0.9999	1.12	1.01	0.457	0.571	0.799
0.02	28.94	0.9997	0.9999	0.9999	1.14	1.02	0.445	0.561	0.794
0.04	14.48	0.999	0.999	0.9996	1.16	1.02	0.434	0.551	0.788
0.06	9.67	0.997	0.998	0.999	1.18	1.02	0.423	0.541	0.782
0.08	7.26	0.996	0.997	0.999	1.20	1.03	0.412	0.531	0.776
0.10	5.82	0.993	0.995	0.998	1.22	1.04	0.402	0.521	0.771
0.12	4.86	0.990	0.993	0.997	1.24	1.04	0.391	0.512	0.765
0.14	4.18	0.986	0.990	0.996	1.26	1.05	0.381	0.502	0.759
0.16	3.67	0.982	0.987	0.995	1.28	1.06	0.371	0.492	0.753
0.18	3.28	0.978	0.984	0.994	1.30	1.07	0.361	0.483	0.747
0.20	2.96	0.973	0.980	0.992	1.32	1.08	0.351	0.474	0.742
0.22	2.71	0.967	0.976	0.990	1.34	1.08	0.342	0.464	0.736
0.24	2.50	0.961	0.972	0.989	1.36	1.09	0.332	0.455	0.730
0.26	2.32	0.954	0.967	0.987	1.38	1.10	0.323	0.446	0.724
0.28	2.17	0.947	0.962	0.985	1.40	1.11	0.314	0.437	0.718
0.30	2.04	0.939	0.956	0.982	1.42	1.13	0.305	0.429	0.713
0.32	1.92	0.932	0.951	0.980	1.44	1.14	0.297	0.420	0.707
0.34	1.82	0.923	0.944	0.977	1.46	1.15	0.289	0.412	0.701
0.36	1.74	0.914	0.938	0.975	1.48	1.16	0.280	0.403	0.695
0.38	1.66	0.905	0.931	0.972	1.50	1.18	0.272	0.395	0.690
0.40	1.59	0.896	0.924	0.969	1.52	1.19	0.265	0.387	0.684
0.42	1.53	0.886	0.917	0.966	1.54	1.20	0.257	0.379	0.678
0.44	1.47	0.876	0.909	0.963	1.56	1.22	0.250	0.371	0.672
0.46	1.42	0.865	0.902	0.959	1.58	1.23	0.242	0.363	0.667
0.48	1.38	0.854	0.893	0.956	1.60	1.25	0.235	0.356	0.661
0.50	1.34	0.843	0.885	0.952	1.62	1.27	0.228	0.348	0.656
0.52	1.30	0.832	0.877	0.949	1.64	1.28	0.222	0.341	0.650
0.54	1.27	0.820	0.868	0.945	1.66	1.30	0.215	0.334	0.645
0.56	1.24	0.808	0.859	0.941	1.68	1.32	0.209	0.327	0.639
0.58	1.21	0.796	0.850	0.937	1.70	1.34	0.203	0.320	0.634
0.60	1.19	0.784	0.840	0.933	1.72	1.36	0.197	0.313	0.628
0.62	1.17	0.772	0.831	0.929	1.74	1.38	0.191	0.306	0.623
0.64	1.16	0.759	0.821	0.924	1.76	1.40	0.185	0.300	0.617
0.66	1.13	0.747	0.812	0.920	1.78	1.42	0.179	0.293	0.612
0.68	1.12	0.734	0.802	0.915	1.80	1.44	0.174	0.287	0.607
0.70	1.09	0.721	0.792	0.911	1.82	1.46	0.169	0.281	0.602
0.72	1.08	0.708	0.781	0.906	1.84	1.48	0.164	0.275	0.596
0.74	1.07	0.695	0.771	0.901	1.86	1.51	0.159	0.269	0.591
0.76	1.06	0.682	0.761	0.896	1.88	1.53	0.154	0.263	0.586
0.78	1.05	0.669	0.750	0.891	1.90	1.56	0.149	0.257	0.581
0.80	1.04	0.656	0.740	0.886	1.92	1.58	0.145	0.251	0.576
0.82	1.03	0.643	0.729	0.881	1.94	1.61	0.140	0.246	0.571
0.84	1.02	0.630	0.719	0.876	1.96	1.63	0.136	0.240	0.566
0.86	1.02	0.617	0.708	0.871	1.98	1.66	0.132	0.235	0.561
0.88	1.01	0.604	0.698	0.865	2.00	1.69	0.128	0.230	0.556
0.90	1.01	0.591	0.687	0.860	2.02	1.72	0.124	0.225	0.551
0.92	1.01	0.578	0.676	0.855	2.04	1.75	0.120	0.220	0.546
0.94	1.00	0.566	0.666	0.850	2.06	1.78	0.116	0.215	0.541
0.96	1.00	0.553	0.655	0.844	2.08	1.81	0.113	0.210	0.536
0.98	1.00	0.541	0.645	0.839	2.10	1.84	0.109	0.206	0.531
1.00	1.00	0.528	0.632	0.833	2.12	1.87	0.106	0.201	0.526
1.02	1.00	0.516	0.623	0.828	2.14	1.90	0.103	0.197	0.522
1.04	1.00	0.504	0.613	0.822	2.16	1.94	0.100	0.192	0.517
1.06	1.00	0.492	0.602	0.817	2.18	1.97	0.097	0.188	0.513
1.08	1.01	0.480	0.592	0.810	2.20	2.01	0.094	0.184	0.508

TABLE A-16 One-dimensional isentropic relations (continued)

N_M	A/A^*	p/p_0	ρ/ρ_0	T/T_0	N_M	A/A^*	p/p_0	ρ/ρ_0	T/T_0
2.22	2.04	0.091	0.180	0.504	2.74	3.31	0.040	0.101	0.400
2.24	2.08	0.088	0.176	0.499	2.76	3.37	0.039	0.099	0.396
2.26	2.12	0.085	0.172	0.495	2.78	3.43	0.038	0.097	0.393
2.28	2.15	0.083	0.168	0.490	2.80	3.50	0.037	0.095	0.389
2.30	2.19	0.080	0.165	0.486	2.82	3.57	0.036	0.093	0.386
2.32	2.23	0.078	0.161	0.482	2.84	3.64	0.035	0.091	0.383
2.34	2.27	0.075	0.157	0.477	2.86	3.71	0.034	0.089	0.379
2.36	2.32	0.073	0.154	0.473	2.88	3.78	0.033	0.087	0.376
2.38	2.36	0.071	0.150	0.469	2.90	3.85	0.032	0.085	0.373
2.40	2.40	0.068	0.147	0.465	2.92	3.92	0.031	0.083	0.370
2.42	2.45	0.066	0.144	0.461	2.94	4.00	0.030	0.081	0.366
2.44	2.49	0.064	0.141	0.456	2.96	4.08	0.029	0.080	0.363
2.46	2.54	0.062	0.138	0.452	2.98	4.15	0.028	0.078	0.360
2.48	2.59	0.060	0.135	0.448	3.00	4.23	0.027	0.076	0.357
2.50	2.64	0.059	0.132	0.444					
2.52	2.69	0.057	0.129	0.441	3.10	4.66	0.023	0.0685	0.342
					3.20	5.12	0.020	0.062	0.328
2.54	2.74	0.055	0.126	0.437	3.3	5.63	0.0175	0.0555	0.315
2.56	2.79	0.053	0.123	0.433	3.4	6.18	0.015	0.050	0.302
2.58	2.84	0.052	0.121	0.429					
2.60	2.90	0.050	0.118	0.425	3.5	6.79	0.013	0.045	0.290
					3.6	7.45	0.0114	0.041	0.278
2.62	2.95	0.049	0.115	0.421	3.7	8.17	0.0099	0.037	0.2675
2.64	3.01	0.047	0.113	0.418	3.8	8.95	0.0086	0.0335	0.257
2.66	3.06	0.046	0.110	0.414					
2.68	3.12	0.044	0.108	0.410	3.9	9.80	0.0075	0.030	0.247
2.70	3.18	0.043	0.106	0.407	4.0	10.72	0.0066	0.028	0.238
2.72	3.24	0.042	0.103	0.403					

†For a perfect gas with constant specific heat, $k = 1.4$

TABLE A-17 One-dimensional normal-shock relations†

N_{M1}	N_{M2}	$\dfrac{p_2}{p_1}$	$\dfrac{T_2}{T_1}$	$\dfrac{(p_0)_2}{(p_0)_1}$	N_{M1}	N_{M2}	$\dfrac{p_2}{p_1}$	$\dfrac{T_2}{T_1}$	$\dfrac{(p_0)_2}{(p_0)_1}$
1.00	1.000	1.000	1.000	1.000	2.04	0.571	4.689	1.720	0.702
1.02	0.980	1.047	1.013	1.000	2.06	0.567	4.784	1.737	0.693
1.04	0.962	1.095	1.026	1.000	2.08	0.564	4.881	1.754	0.683
1.06	0.944	1.144	1.039	1.000	2.10	0.561	4.978	1.770	0.674
1.08	0.928	1.194	1.052	0.999	2.12	0.558	5.077	1.787	0.665
1.10	0.912	1.245	1.065	0.999	2.14	0.555	5.176	1.805	0.656
1.12	0.896	1.297	1.078	0.998	2.16	0.553	5.277	1.822	0.646
1.14	0.882	1.350	1.090	0.997	2.18	0.550	5.378	1.839	0.637
1.16	0.868	1.403	1.103	0.996	2.20	0.547	5.480	1.857	0.628
1.18	0.855	1.458	1.115	0.995	2.22	0.544	5.583	1.875	0.619
1.20	0.842	1.513	1.128	0.993	2.24	0.542	5.687	1.892	0.610
1.22	0.830	1.570	1.140	0.991	2.26	0.539	5.792	1.910	0.601
1.24	0.818	1.627	1.153	0.988	2.28	0.537	5.898	1.929	0.592
1.26	0.807	1.686	1.166	0.986	2.30	0.534	6.005	1.947	0.583
1.28	0.796	1.745	1.178	0.983	2.32	0.532	6.113	1.965	0.575
1.30	0.786	1.805	1.191	0.979	2.34	0.530	6.222	1.984	0.566
1.32	0.776	1.866	1.204	0.976	2.36	0.527	6.331	2.003	0.557
1.34	0.766	1.928	1.216	0.972	2.38	0.525	6.442	2.021	0.549
1.36	0.757	1.991	1.229	0.968	2.40	0.523	6.553	2.040	0.540
1.38	0.748	2.055	1.242	0.963	2.42	0.521	6.666	2.060	0.532
1.40	0.740	2.120	1.255	0.958	2.44	0.519	6.779	2.079	0.523
1.42	0.731	2.186	1.268	0.953	2.46	0.517	6.894	2.098	0.515
1.44	0.723	2.253	1.281	0.948	2.48	0.515	7.009	2.118	0.507
1.46	0.716	2.320	1.294	0.942	2.50	0.513	7.125	2.138	0.499
1.48	0.708	2.389	1.307	0.936	2.52	0.511	7.242	2.157	0.491
1.50	0.701	2.458	1.320	0.930	2.54	0.509	7.360	2.177	0.483
1.52	0.694	2.529	1.334	0.923	2.56	0.507	7.479	2.198	0.475
1.54	0.687	2.600	1.347	0.917	2.58	0.506	7.599	2.218	0.468
1.56	0.681	2.673	1.361	0.910	2.60	0.504	7.720	2.238	0.460
1.58	0.675	2.746	1.374	0.903	2.62	0.502	7.842	2.260	0.453
1.60	0.668	2.820	1.388	0.895	2.64	0.500	7.965	2.280	0.445
1.62	0.663	2.895	1.402	0.888	2.66	0.499	8.088	2.301	0.438
1.64	0.657	2.971	1.416	0.880	2.68	0.497	8.213	2.322	0.431
1.66	0.651	3.048	1.430	0.872	2.70	0.496	8.338	2.343	0.424
1.68	0.646	3.126	1.444	0.864	2.72	0.494	8.465	2.364	0.417
1.70	0.641	3.205	1.458	0.856	2.74	0.493	8.592	2.396	0.410
1.72	0.635	3.285	1.473	0.847	2.76	0.491	8.721	2.407	0.403
1.74	0.631	3.366	1.487	0.839	2.78	0.490	8.850	2.429	0.396
1.76	0.626	3.447	1.502	0.830	2.80	0.488	8.980	2.451	0.389
1.78	0.621	3.530	1.517	0.821	2.82	0.487	9.111	2.473	0.383
1.80	0.617	3.613	1.532	0.813	2.84	0.485	9.243	2.496	0.376
1.82	0.612	3.698	1.547	0.804	2.86	0.484	9.376	2.518	0.370
1.84	0.608	3.783	1.562	0.795	2.88	0.483	9.510	2.541	0.364
1.86	0.604	3.869	1.577	0.786	2.90	0.481	9.645	2.563	0.358
1.88	0.600	3.957	1.592	0.777	2.92	0.480	9.781	2.586	0.352
1.90	0.596	4.045	1.608	0.767	2.94	0.479	9.918	2.609	0.346
1.92	0.592	4.134	1.624	0.758	2.96	0.478	10.055	2.632	0.340
1.94	0.588	4.224	1.639	0.749	2.98	0.476	10.194	2.656	0.334
1.96	0.584	4.315	1.655	0.740	3.00	0.475	10.333	2.679	0.328
1.98	0.581	4.407	1.671	0.730					
2.00	0.577	4.500	1.688	0.721					
2.02	0.574	4.594	1.704	0.711					

†For a perfect gas with $k = 1.4$.

TABLE A-18 Some expansion factors Y for compressible flow through flow-nozzles and venturi meters

p_2/p_1	k	Ratio of Diameters (d_2/d_1)				
		0.30	0.40	0.50	0.60	0.70
0.95	1.40	0.973	0.972	0.971	0.968	0.962
	1.30	.970	.970	.968	.965	.959
	1.20	.968	.967	.966	.963	.956
0.90	1.40	0.944	0.943	0.941	0.935	0.925
	1.30	.940	.939	.936	.931	.918
	1.20	.935	.933	.931	.925	.912
0.85	1.40	0.915	0.914	0.910	0.902	0.887
	1.30	.910	.907	.904	.896	.880
	1.20	.902	.900	.896	.887	.870
0.80	1.40	0.886	0.884	0.880	0.868	0.850
	1.30	.876	.873	.869	.857	.839
	1.20	.866	.864	.859	.848	.829
0.75	1.40	0.856	0.853	0.846	0.836	0.814
	1.30	.844	.841	.836	.823	.802
	1.20	.820	.818	.812	.798	.776
0.70	1.40	0.824	0.820	0.815	0.800	0.778
	1.30	.812	.808	.802	.788	.763
	1.20	.794	.791	.784	.770	.745

For $p_2/p_1 = 1.00$, $Y = 1.00$.

TABLE A-19 Discharge coefficients for vertical sharp-edged circular orifices
For Water at 60 °F Discharging into Air at Same Temperature

Head in Feet	Orifice Diameter in Inches					
	0.25	0.50	0.75	1.00	2.00	4.00
0.8	0.647	0.627	0.616	0.609	0.603	0.601
1.4	.635	.619	.610	.605	.601	.600
2.0	.629	.615	.607	.603	.600	.599
4.0	.621	.609	.603	.600	.598	.597
6.0	.617	.607	.601	.599	.597	.596
8.0	.614	.605	.600	.598	.596	.595
10.0	.613	.604	.600	.597	.596	.595
12.0	.612	.603	.599	.597	.595	.595
14.0	.611	.603	.598	.596	.595	.594
16.0	.610	.602	.598	.596	.595	.594
20.0	.609	.602	.598	.596	.595	.594
25.0	.608	.601	.597	.596	.594	.594
30.0	.607	.600	.597	.595	.594	.594
40.0	.606	.600	.596	.595	.594	.593
50.0	.605	.599	.596	.595	.594	.593
60.0	.605	.599	.596	.594	.593	.593

Source: F. W. Medaugh and G. D. Johnson, Civil Engr., July 1940, p. 424.

TABLE A-20 Drag of two-dimensional bodies at $N_R \geq 10^4$

Shape	C_D based on frontal area	Shape	C_D based on frontal area
Plate:	2.0	Half-cylinder:	1.2
Square cylinder:	2.1		1.7
	1.6	Equilateral triangle:	1.6
Half tube:	1.2		2.0
	2.3		

Elliptical cylinder:		Laminar	Turbulent
1:1		1.2	0.3
2:1		0.6	0.2
4:1		0.35	0.15
8:1		0.25	0.1

TABLE A-21 Drag of three-dimensional bodies at $N_R \geq 10^4$

Body	Ratio	C_D based on frontal area

Cube:

1.07

0.81

60° cone:

0.5

Disk:

1.17

Cup:

1.4

0.4

Parachute (low porosity):

1.2

Rectangular plate:

b/h	
1	1.18
5	1.2
10	1.3
20	1.5
∞	2.0

Flat-faced cylinder:

L/d	
0.5	1.15
1	0.90
2	0.85
4	0.87
8	0.99

Ellipsoid:

L/d	Laminar	Turbulent
0.75	0.5	0.2
1	0.47	0.2
2	0.27	0.13
4	0.25	0.1
8	0.2	0.08

TABLE A-22 Drag coefficients for three-dimensional bodies

Shape		C_D	Shape		C_D Laminar flow	C_D Turbulent flow
Disc	→ |	1.17	Sphere: → ○		0.47	0.27
60° cone	→ ◁	0.49	Ellipsoidal body of revolution:			
Cube	→ □	1.05				
	→ ◇	0.80	2:1 → ◦		0.27	0.06
Hollow cup	→ (	0.38	4:1 → ⬭		0.20	0.06
	→)	1.42	8:1 → ⬭		0.25	0.13
Solid hemisphere	→ ◗	0.38				
	→ ◗	1.17				

TABLE A-23 Typical drag coefficients for various cylinders in two-dimensional flow

Body shape			C_D	Reynolds number
Circular cylinder	→	○	1.2	10^4 to 1.5×10^5
Elliptical cylinder	→	⬭ 2:1	0.6	4×10^4
			0.46	10^5
	→	⬭ 4:1	0.32	2.5×10^4 to 10^5
	→	⬭ 8:1	0.29	2.5×10^4
			0.20	2×10^5
Square cylinder	→	□	2.0	3.5×10^4
	→	◇	1.6	10^4 to 10^5
Triangular cylinders	→	▷120°	2.0	10^4
	→	◁120°	1.72	10^4
	→	▷90°	2.15	10^4
	→	◁90°	1.60	10^4
	→	▷60°	2.20	10^4
	→	◁60°	1.39	10^4
	→	▷30	1.8	10^5
	→	◁30°	1.0	10^5
Semitubular	→	)	2.3	4×10^4
	→	(	1.12	4×10^4

From W. F. Lindsey, *NACA Tech. Rep. 619*, 1938.